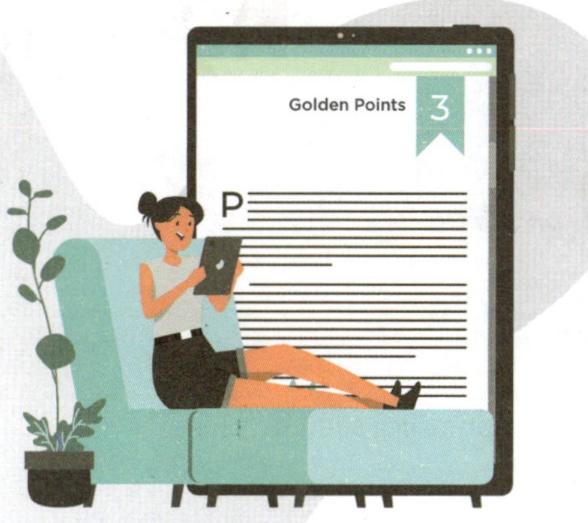

Reading is not enough, Now ***Listen, Learn & Practice*** Every Chapter

i. Golden Points

Get Chapter-wise One Liners in **PODCAST form** (audio) for quick revision of chapters

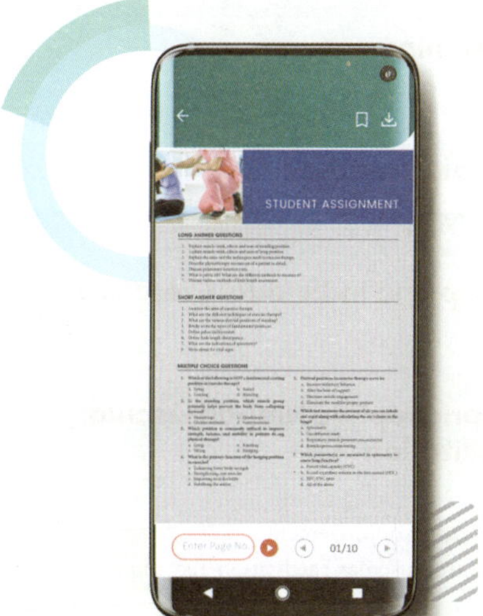

ii. Solved Exercises

Subjective and Objective exercises of the book have been given with their solution (in PDF) to evaluate and assess the complete chapter knowledge

- **120+** Long Answer Questions
- **120+** Short Answer Questions
- **250+** Multiple Choice Questions

iii. MCQs

200+ Chapter-wise Multiple Choice Questions in Practice and Review mode to provide in-depth concept clarity

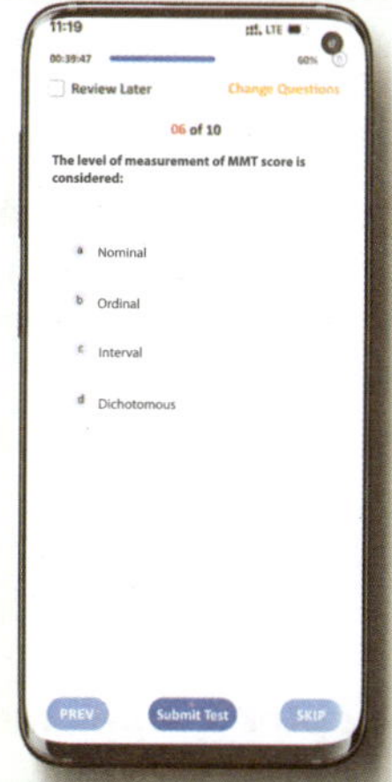

from Google Play store to access the content

CBS Physiobrid Books >
Textbook of Exercise Therapy Fundamentals & Practices
The Hybrid Edition

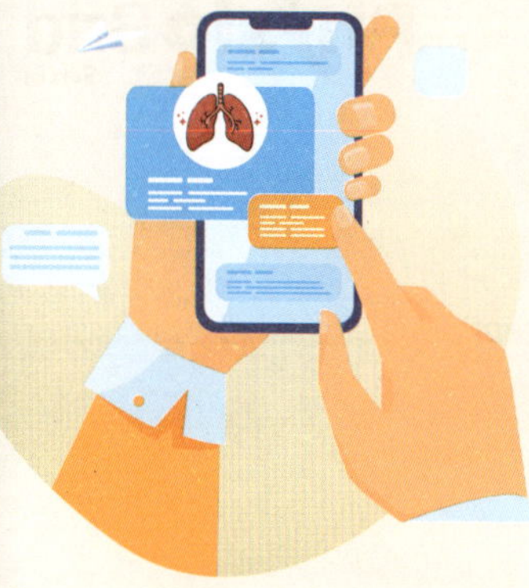

High Yield Topics

Revise on the Go

Get 50+ Topic-wise Selective Images & Tables with their descriptions for LMR and Quick reference, based on the topics of University examination

List of High Yield Topics

and many more...

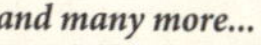

CBS Physiobrid Books >
Textbook of Exercise Therapy Fundamentals & Practices
The Hybrid Edition

Add Ons

Dil Mange More Content

Don't settle for less, go beyond with
Recent Updates, E-books & much more...

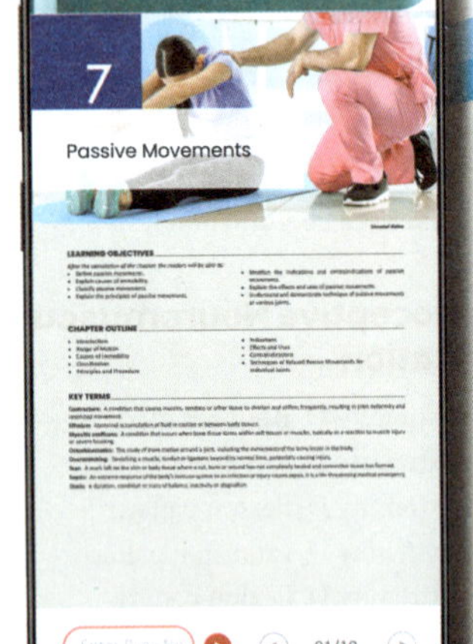

i. Recent Update

Regular updates related to **Recent advancement** and **Book Errata**

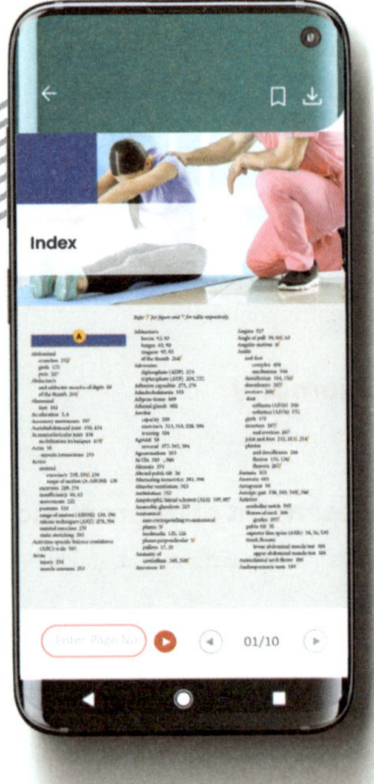

ii. E-book

Get PDFs of important chapters/sections **(Annexures/ Appendices)** of book

(This feature is for Institutions and pro-users)

Textbook of

Exercise Therapy

Fundamentals & Practices

As per Physiotherapy Curriculum of All Universities of India and NCAHP, Ministry of Health & Family Welfare

Sheetal Kalra
BPT, MPT (Sports), PhD
Associate Professor and HOD
School of Physiotherapy
Delhi Pharmaceutical Sciences and Research University
New Delhi

Foreword
Narkeesh Arumugam

CBS Publishers & Distributors Pvt Ltd
• New Delhi • Bengaluru • Chennai • Kochi • Kolkata • Lucknow • Mumbai
• Hyderabad • Jharkhand • Nagpur • Patna • Pune • Uttarakhand

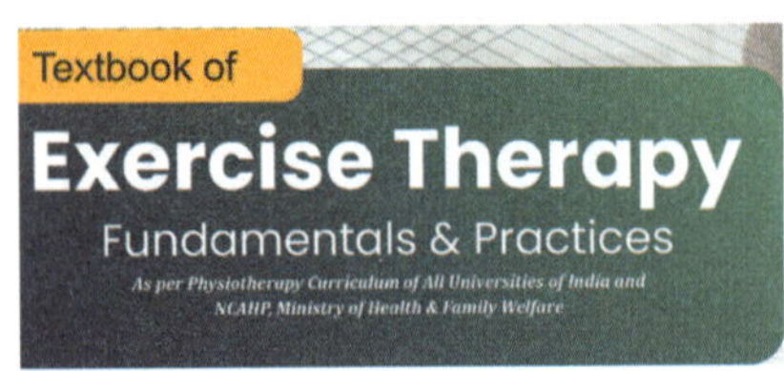

ISBN: 978-81-977500-6-9

First Edition: 2025

Published by **Satish Kumar Jain** and produced by **Varun Jain** for

CBS Publishers & Distributors Pvt Ltd

4819/XI Prahlad Street, 24 Ansari Road, Daryaganj, New Delhi 110 002, India.

Ph: +91-11-23289259, 23266861, 23266867 Website: www.cbspd.com
Fax: 011-23243014

e-mail: delhi@cbspd.com; cbspubs@airtelmail.in.

Corporate Office: 204 FIE, Industrial Area, Patparganj, Delhi 110 092

Ph: +91-11-4934 4934 Fax: 4934 4935

e-mail: feedback@cbspd.com; bhupesharora@cbspd.com

Branches

- **Bengaluru:** Seema House 2975, 17th Cross, K.R. Road, Banasankari 2nd Stage, Bengaluru-560 070, Karnataka
  Ph: +91-80-26771678/79 Fax: +91-80-26771680 e-mail: bangalore@cbspd.com
- **Chennai:** 7, Subbaraya Street, Shenoy Nagar, Chennai-600 030, Tamil Nadu
  Ph: +91-44-26680620, 26681266 Fax: +91-44-42032115 e-mail: chennai@cbspd.com
- **Kochi:** 68/1534, 35, 36-Power House Road, Opp. KSEB, Cochin-682018, Kochi, Kerala
  Ph: +91-484-4059061-65 Fax: +91-484-4059065 e-mail: kochi@cbspd.com
- **Kolkata:** Hind Ceramics Compound, 1st Floor, 147, Nilganj Road, Belghoria, Kolkata-700056, West Bengal
  Ph: +033-2563-3055/56 e-mail: kolkata@cbspd.com
- **Lucknow:** Basement, Khushnuma Complex, 7-Meerabai Marg (Behind Jawahar Bhawan), Lucknow-226001, Uttar Pradesh
  Ph: +0522-4000032 e-mail: tiwari.lucknow@cbspd.com
- **Mumbai:** PWD Shed, Gala No. 25/26, Ramchandra Bhatt Marg, Next to J.J. Hospital Gate No. 2, Opp. Union Bank of India, Noor Baug, Mumbai-400009, Maharashtra
  Ph: +91-22-66661880/89 Fax: +91-22-24902342 e-mail: mumbai@cbspd.com

Representatives

- **Hyderabad** +91-9885175004
- **Jharkhand** +91-9811541605
- **Nagpur** +91-9421945513
- **Patna** +91-9334159340
- **Pune** +91-9623451994
- **Uttarakhand** +91-9716462459

Printed at : Goyal Offset Works Pvt. Ltd. Haryana

Dedicated to

the Almighty God and my beloved family

Foreword

Exercise therapy plays the role of foundation stone in the field of Physiotherapy bridging the gap between injury, rehabilitation, and optimal physical function. For many individuals, the journey to recovery goes beyond clinical treatments and involves a dedicated effort to restore the body through well-structured, evidence-based exercises. While the human body is inherently resilient, it often requires specialized support to recover from injuries or manage the complexities of chronic conditions. Exercise therapy is crucial in restoring balance and functional capacity. In the context of today's multidisciplinary rehabilitation approach, exercise therapy ensures a holistic and effective path to recovery.

With great excitement, I recommend *Textbook of Exercise Therapy—Fundamentals and Practices* that is a thorough resource tailored for physiotherapy students, educators, and practitioners. Exercise therapy is a dynamic and ever-evolving field, enriched by new techniques, insights, and evidence that continuously enhance the understanding of movement and recovery. This book bridges foundational concepts with advanced therapeutic methods, ensuring readers are equipped with the essential knowledge and practical skills needed to provide exceptional care in the field of physiotherapy.

The primary aim of this book is to offer a comprehensive information regarding fundamentals of exercise therapy. It provides readers with a deep understanding of the key principles behind human movement, muscle function, and therapeutic interventions. By covering all techniques of practice, *Textbook of Exercise Therapy—Fundamentals and Practices* ensures that physiotherapy professionals gain a well-rounded perspective of the core components of exercise therapy.

The book also includes essential practical skills—such as Manual Muscle Testing (MMT), goniometry, and neuromuscular assessments—empowering practitioners to enhance their clinical proficiency. The structure of the book has been carefully crafted to balance theoretical knowledge with real-world clinical application.

- The first section, *Essentials of Exercise Therapy*, provides a comprehensive introduction to the core principles that govern human movement.
- *Examination/Testing Techniques* takes a deeper dive into the diagnostic tools and assessment methods that are crucial for evaluating neuromuscular function and joint mobility.
- The *Application of Principles of Exercise Therapy* section focuses on the practical application of exercise therapy techniques, aimed at enhancing key aspects of physical health, including mobility, strength, and flexibility.
- The *Therapeutic Techniques* section explores advanced interventions, including proprioceptive neuromuscular facilitation (PNF) and hydrotherapy—evidence-based approaches with clinical relevance.
- The *Movement and Alignment* section discusses posture, gait, balance and yoga.

Textbook of Exercise Therapy—Fundamentals and Practices is designed to be a trusted, evidence-backed resource that supports the development of physiotherapists at all stages of their career. This book provides clear, practical guidance, modern techniques, and invaluable insights that will elevate the practice and ongoing professional growth of physiotherapy professionals and students.

As you engage with the content of this book, I encourage you to view exercise therapy not just as a set of techniques, but as a transformative tool for improving patient outcomes. By blending theoretical knowledge with hands-on application, this book will empower to enhance both the well-being and functionality of the patients, advancing the clinical skills and impact on the field of physiotherapy.

Narkeesh Arumugam
BPT, MPT (Neurology) PhD, DO (Spain),
DOMTP (OCO Canada),
PG Diploma in Osteopathy (UK),
PG Diploma in Chiropractor (Sweden)
Professor and Former Head
Department of Physiotherapy
Former Dean
Faculty of Medicine
Punjabi University, Patiala, Punjab

The Advisory Board

The Advisory Board

About the Author

Sheetal Kalra, *BPT, MPT (Sports), PhD,* is working as Associate Professor and HOD, School of Physiotherapy, Delhi Pharmaceutical Sciences and Research University, New Delhi. She is a Graduate from DAV College, Yamunanagar, Haryana and a Postgraduate from SBSPGI, Dehradun, Uttarakhand. She has got over 20 years of experience in academia, industry and research. She has around 60 publications till date in peer reviewed National and International journals and has got 6 books in her name. Recently, she has also been granted a project from Indian Council of Social Science and Research (ICSSR), New Delhi. Dr Sheetal Kalra is the recipient of many awards like Academic Excellence Award, Medanta and Young Achievers Award, INCPT AIIMS. She has been invited as speaker in many national and international conferences which includes International Conference at AIIMS (2015, 2016, 2017, 2018, 2023) and International Conference at Singapore, Jakarta and many more.

Preface

When the publishers initially approached me with the idea of authoring *Textbook of Exercise Therapy—Fundamentals and Practices*, I was both honored and skeptical. The prospect of creating a comprehensive text on a subject so vital yet ever-evolving felt daunting. However, as the project slowly took shape, the journey turned into a rewarding collaborative endeavor.

This book is the culmination of inputs from experts in the field, including seasoned academicians and professionals from the physiotherapy fraternity. Together, we meticulously developed chapter outlines, aiming to balance foundational concepts with advanced topics. The result is a blend of theoretical and research-based evidence, enriched with practical insights, which I believe will be immensely useful for students, educators, and practitioners alike.

The book is organized into five sections. The first section, *Essentials of Exercise Therapy*, lays the groundwork with chapters on mechanics, muscle action, and the basics of therapeutic exercises. These chapters aim to provide readers with a strong foundation of the principles of exercise therapy.

The second section, *Examination/Testing Techniques*, delves into essential diagnostic tools such as manual muscle testing, goniometry, and assessment of neuromuscular efficiency. These chapters emphasize the practical application of assessment methods, equipping readers with the skills necessary for clinical practice.

In the third section, *Application of Principles of Exercise Therapy*, we explore passive and active movements, resisted exercises, stretching, and aerobic exercises. This section also focuses on functional re-education, coordination exercises, and the structuring of both group and individual exercise programs. The chapters here are designed to bridge the gap between theoretical knowledge and its real-world application.

The fourth section, *Therapeutic Techniques*, highlights advanced modalities like proprioceptive neuromuscular facilitation, manual therapy and peripheral joint mobilization, suspension therapy and hydrotherapy.

The fifth section, *Movement and Alignment*, includes topics such as posture, gait, and balance and yoga, providing a holistic perspective on rehabilitation and therapeutic interventions.

Sheetal Kalra

Acknowledgments

First of all, I would like to thank the Almighty God. The *Textbook of Exercise Therapy—Fundamentals and Practices* would not have been possible without His blessings.

I am thankful for the invaluable contributions of my colleagues and fellow physiotherapists, many of whom shared their specialized knowledge and experiences to enrich individual chapters. My sincere gratitude goes to the Institute, School of Physiotherapy, Delhi Pharmaceutical Sciences and Research University, New Delhi and the physiotherapy fraternity from adjacent colleges, whose collaboration and expertise were instrumental in shaping this work.

I extend my heartfelt thanks to the publishers for their patience and unwavering support throughout this endeavor. I hope that *Textbook of Exercise Therapy—Fundamentals and Practices* will serve as a reliable resource for its readers, inspiring both learning and application in the dynamic field of physiotherapy.

I extend my special thanks to Mr Satish Kumar Jain (Chairman) and Mr Varun Jain (Managing Director), M/s CBS Publishers and Distributors Pvt Ltd for their wholehearted support in publication of this book. I have no words to describe the role, efforts, inputs and initiatives undertaken by Mr Bhupesh Aarora (Sr. Vice President – Publishing & Marketing (Health Sciences Division)] for helping and motivating me.

My special thanks are due to Dr Divya Gupta, PT (Project Manager & Editorial [Scientific] Head – Physiotherapy) and Dr Apurva Chatterjee, PT (Content Strategist – Physiotherapy) for their valuable support, suggestions and advice that have helped me in refining the text and making it more comprehensive.

I sincerely thank the entire CBS team for bringing out the book with utmost care and attractive presentation. I would like to thank Ms Nitasha Arora (Assistant General Manager Publishing – Medical and Nursing), Ms Daljeet Kaur (Assistant Publishing Manager) and Dr Anju Dhir (Product Manager and Medical Development Editor) for their publishing support. I would also extend my thanks to Ms Surbhi Gupta (Sr. English Editor), Mr Ashutosh Pathak (Sr. Proofreader cum Team Coordinator) and all the production team members for devoting laborious hours in designing and typesetting the book.

Last but not least, I am thankful to my colleagues, peers, family and friends without whose support this book would not have been possible.

Contributors and Reviewers

CONTRIBUTORS

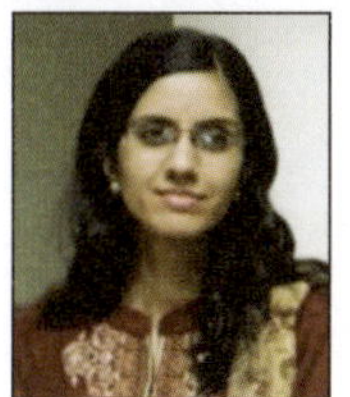

Archana Khanna
BPT, MPT (Sports), PhD
Associate Professor
Department of Physiotherapy
Sharda School of Allied Health Sciences
Sharda University
Greater Noida, Uttar Pradesh

Deepak Raghav
D Pharma, BPT, MPT (Orthopedics), PhD
Professor and Principal
Santosh College of Physiotherapy
Santosh Medical College and Hospital
Ghaziabad, Uttar Pradesh

Jitender Munjal
BPT, MPT (Sports)
Assistant Professor
School of Physiotherapy
Delhi Pharmaceutical Sciences & Research University
New Delhi

Kalpana Zutshi
BPT, MPT (Sports)
Associate Professor
Department of Physiotherapy
Jamia Hamdard University
New Delhi

Kopal Pajnee
BPT, MPT (Sports), PhD
Visiting Faculty
School of Physiotherapy
Delhi Pharmaceutical Sciences & Research University
New Delhi

Founder and Owner
Rehab N Recovery Physiotherapy Clinic
New Delhi

Paridhi Sharma
BPT, MPT (Neurology), PhD Pursuing
Scholar
School of Physiotherapy
Delhi Pharmaceutical Sciences & Research University
New Delhi

The names of the contributors and reviewers are arranged in alphabetical order.

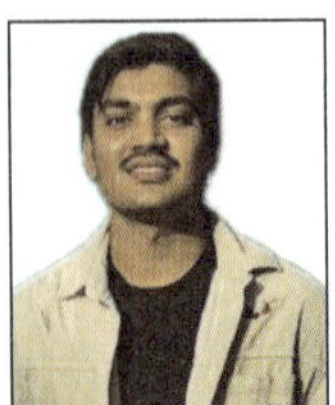

Pradeep Kumar
BPT (Pursuing)
Student
School of Physiotherapy
Delhi Pharmaceutical Sciences & Research University
New Delhi

Sheetal Kalra
BPT, MPT (Sports), PhD
Associate Professor and Head
School of Physiotherapy
Delhi Pharmaceutical Sciences & Research University
New Delhi

Puneeta Ajmera
BPTh, MBA (Hospital and Healthcare Management), PhD
Associate Professor
School of Allied Health Sciences
Department of Public Health
Delhi Pharmaceutical Sciences & Research University
New Delhi

Sheetal Yadav
BPTh, MBA (Hospital and Healthcare Management), PhD
Assistant Professor
School of Allied Health Sciences
Department of Public Health
Delhi Pharmaceutical Sciences & Research University
New Delhi

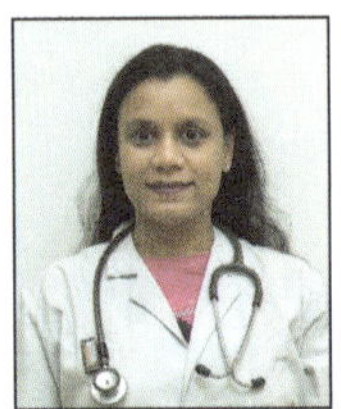

Richa Rai
BPT, MPTh (Cardiopulmonary), MBA (Finance), CFA, PhD, PGDHE
Professor
School of Physiotherapy
Delhi Pharmaceutical Sciences & Research University
New Delhi

Sonia Pawaria
BPT, MPT (Cardiopulmonary), PhD
Assistant Professor
Department of Physiotherapy
Gurugram University
Gurugram, Haryana

Sajjan Pal
BPT, MPT (Sports), PhD
Associate Professor
Faculty of Physiotherapy
SGT University
Gurugram, Haryana

Varsha Chorasiya
BPT, MPT (Neurology), PhD, ICMR-Fellow, PHRF, PGDM (Applied Epidemiology)
Researcher
New Delhi

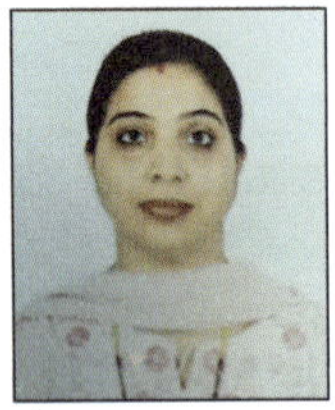

Savita Tamaria
BPT, MPT (Musculoskeletal)
Assistant Professor
School of Physiotherapy
Delhi Pharmaceutical Sciences & Research University
New Delhi

REVIEWERS

Aditi Singh
BPT, MPT (Neurology), PhD
Associate Professor
Department of Physiotherapy
Jagannath University
Jaipur, Rajasthan

Ahmad Merajul Hasan Inam
BPT, MPT (Musculoskeletal)
Associate Professor
Integral University
Lucknow, Uttar Pradesh

The names of the contributors and reviewers are arranged in alphabetical order.

Akshita Juneja
BPT, MPT (Cardiopulmonary)
Head of Department
Faculty of Paramedical and Allied Health Sciences
Mewar University
Chittorgarh, Rajasthan

Anjali Agarwal
BPT, MPT (Neurology), PhD Scholar
Demonstrator
Department of Physiotherapy, Faculty of Paramedical Sciences
Uttar Pradesh University of Medical Sciences (UPUMS), Saifai
Etawah, Uttar Pradesh

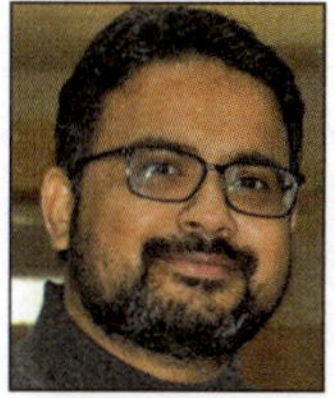

Ankit Bhargava
BPT, MPT (Musculoskeletal), PhD (Physiotherapy), PhD (Health Sciences)
Dean and Professor
Faculty of Physiotherapy and Diagnostics
Jayoti Vidyapeeth Women's University
Jaipur, Rajasthan

Anushree Rai
BPT, MPT (Orthopedics)
Assistant Professor
Jamia Millia Islamia
New Delhi

Apurva Chatterjee
BPT, MPT (Neurology), PhD Scholar
Content Strategist (Physiotherapy)
Nursing Next Exam Prep Pvt Ltd.
Noida, Uttar Pradesh

Avinash Singh
BPT, MPT (Neurology), PhD Scholar
Professor cum HOD
Department of Physiotherapy
Rabindranath Tagore University
Bhopal, Madhya Pradesh

Bhawna Sharma
BPT, MPT (Neurology)
Assistant Professor
Department of Physiotherapy
ITS Institute of Health and Allied Sciences
Ghaziabad, Uttar Pradesh

Charu Sharma
BPT, MPT (Neurology)
Head of Department
College of Healthcare Professions
DIT University
Dehradun, Uttarakhand

Deptee Warikoo
BPT, MPT (Musculoskeletal), PhD Scholar
HOD cum Associate Professor
Department of Physiotherapy
Dolphin PG Institute of Biomedical and Natural Sciences
Dehradun, Uttarakhand

Divya Gupta
BPT, MPT (Pediatrics), MPH, PG Dip (Yoga)
Project Manager and Editorial (Scientific)
Head – Physiotherapy
CBS Publishers & Distributors Pvt. Ltd
New Delhi

Farukh Mohammad Pinjara
BPT, MPT (Sports)
Associate Professor
Pacific College of Physiotherapy
Pacific Medical University
Udaipur, Rajasthan

Grishma Agrawal
BPT, MPT (Neurology)
Professor
Rajeev Gandhi College
Bhopal, Madhya Pradesh

Hemang Kumar Sudhakarbhai Jani
BPT, MPT (Cardiovascular and Pulmonary), PhD
Professor and Principal Incharge
Institute of Physiotherapy
Ganpat University
Mehsana, Gujarat

Jafar Khan
BPT, MPT (Orthopedic), MSc, PhD (Med. Micro.)
Dean and HOD
Pacific College of Physiotherapy
Pacific Medical University
Udaipur, Rajasthan

The names of the contributors and reviewers are arranged in alphabetical order.

Kamini Pathak
BPT, MPT (Musculoskeletal), PhD Scholar
Director
MS Hospital and Research Centre
Lucknow, Uttar Pradesh

Khushboo Arora
BPT, MPT (Orthopedics), PhD Scholar
Senior Physiotherapist
Department of Physiotherapy
University of Technology
Udaipur, Rajasthan

M Vijayakumar
BPT, MPT (Sports), PhD
Professor and Head
Dr. D. Y. Patil College of Physiotherapy
Dr. D. Y. Patil Vidyapeeth (DPU)
Pune, Maharashtra

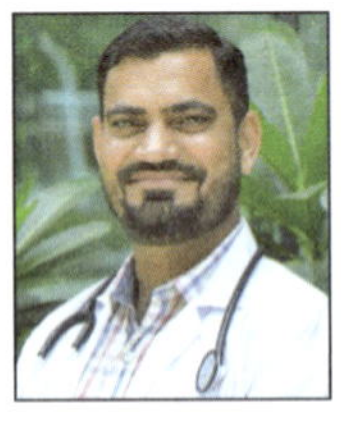

Mahesh Kumar Shou
BPT, D. Acu, MPT (Rehabilitation), D Pharma, PhD
Sr. Professor cum HOD
Swasthya Kalyan College and Rajasthan Hospital
Jaipur, Rajasthan

Manoj K Deshmukh
BPT, MPT (Neurology), PhD
Faculty
Government Physiotherapy College
Pt. D.D.U.M.H.S and Ayush University
Raipur, Chhattisgarh

Namrata Suri
BPT, MPT (Musculoskeletal), PhD Scholar
Assistant Professor
Integral Institute of Allied Health Sciences and Research
Integral University
Lucknow, Uttar Pradesh

Nitin Dhar
BPT, MPT (Cardiorespiratory)
Former Assistant Professor
KR Mangalam University
Gurugram, Haryana
Founder and Consultant Physiotherapist
Ergowork Healthcare
Gurugram, Haryana

Prajakta Sahasrabudhe
BPT, MPT (Cardiorespiratory)
Professor and Head of Department
Sancheti Institute for Orthopedics and Rehabilitation College of Physiotherapy
Pune, Maharashtra

Priyadarshini Mishra
BPT, MPT (Orthopedics), PhD, FRCT
Associate Professor
Abhinav Bindra Sports Medicine and Research Institute (ABSMARI)
Bhubaneswar, Odisha

Purusotham Chippala
BPT, MPT (Neurology), PhD
Professor
Nitte Institute of Physiotherapy
Mangaluru, Karnataka

Raghumahanti Raghuveer
BPT, MPT (Neurology), PhD
Professor and Head
Ravi Nair Physiotherapy College
Datta Meghe Institute of Higher Education and Research (Deemed to be University)
Wardha, Maharashtra

Richa Kashyap
BPT, MPT (Orthopedics), PhD
Principal cum Professor
ITS College of Health and Wellness Sciences
Greater Noida, Uttar Pradesh

Roopa Rao
BPT, MPT (Musculoskeletal)
Assistant Professor
BKL Walawalkar College of Physiotherapy
Chiplun, Maharashtra

Ruchi Desai
BPT, MPT (CBR), PhD Scholar
Assistant Professor
LJ Institute of Physiotherapy
LJ University
Ahmedabad, Gujarat

The names of the contributors and reviewers are arranged in alphabetical order.

Sampada S Jahagirdar
BPT, MPT (Neurology), PhD
Assistant Professor
Amar Jyoti Institute of Physiotherapy
University of Delhi
Delhi

Seemi Aniket Retharekar
BPT, MPTh (Cardiorespiratory)
PhD Scholar
Principal and Professor
College of Physiotherapy
Suryadatta Institute of Health Sciences
Pune, Maharashtra

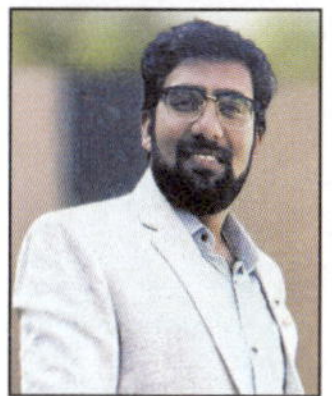

Shashank Apte
BPT, MPT (Neurology)
Assistant Professor
College of Life Science Center Hospital and Research Institute
Jiwaji University
Gwalior, Madhya Pradesh

Shweta Kumar
BPT, MPT (Neurology)
Assistant Professor
School of Physiotherapy and Rehabilitation Sciences
KR Mangalam University
Gurugram, Haryana

Sonam Verma
BPT, MPT (Neurology), PhD Scholar
Associate Professor
Department of Physiotherapy
Suresh Gyan Vihar University
Jaipur, Rajasthan

Surbhi Bhadauria
BPT, MPT (Cardiology)
Assistant Professor and In charge of Research Coordinator
S.I.M.S. College of Physiotherapy
Guntur, Andhra Pradesh

Thiagarajan Subramanian
BPT, MPT (Cardiorespiratory), PhD
Professor
Lovely Professional University
Phagwara, Punjab

Urvashi Jadeja
BPT, MPT (Pediatrics), PhD Scholar
Principal (I/C) cum Associate Professor
Shree Swaminarayan Physiotherapy College
Swaminarayan University
Kalol, Gujarat

Vaibhav Agarwal
BPT, MPT (Neurology)
Lecturer
Department of Physiotherapy
Swami Rama Himalayan University
Dehradun, Uttarakhand

Vipra Manthan Shah
BPT, MPT (Community Health and Rehabilitation), PhD Scholar
Principal
Apollo Institute of Physiotherapy
Ahmedabad, Gujarat

Vivek Gupta
BPT, MPT (Musculoskeletal)
Senior Teacher
Faculty of Paramedical Sciences
King George Medical University
Lucknow, Uttar Pradesh

The names of the contributors and reviewers are arranged in alphabetical order.

From the Publisher's Desk

Dear Readers,

Physiotherapy education has evolved significantly over the years, transitioning from traditional classroom-based learning to a more technology-integrated approach. Initially, students relied on lectures, reference books, and clinical training for knowledge acquisition. However, with the advancement of digital learning tools, physiotherapy education now incorporates innovative methods that enhance the learning experience and prepare students for clinical practice.

At CBS Publishers & Distributors, we have always recognized the vast potential in the Medical and Nursing fields. Our commitment to innovation in Health Sciences has enabled us to make significant and impactful contributions. With a vision to enhance learning in the Physiotherapy domain, we have introduced a range of books under our Physiobrid Series, carefully designed to align with student needs and course curricula. Several titles in this series, including well-known works like *B D Chaurasia's Human Anatomy (Physiotherapy edition), LPR's Fundamentals of Medical Physiology by L Prakasam Reddy, The Principles of Exercise Therapy by M Dena Gardner, Electrotherapy—Principles and Practice by R L Meena, Decode Physiotherapy through MCQs by Swati A Bhise and Rajashree Lad, CBS Physiotherapy Dictionary* and many more have already gained widespread recognition.

As a dedicated publisher in the field of *Medical Science, Physiotherapy, and Allied Health Sciences*, we have consistently contributed to academic excellence by supporting Indian authors and curating high-quality content. Through our Physiobrid series, we aim to provide students and professionals with comprehensive, up-to-date resources that not only strengthen their theoretical foundation but also enhance their practical skills. Our commitment to advancing education is reflected in:

- Developing well-researched books authored by experienced physiotherapy educators and practitioners
- Developing content as per Physiotherapy Curriculum of All Universities of India and NCAHP, Ministry of Health & Family Welfare
- Offering a diverse range of titles covering undergraduate, postgraduate, and specialized physiotherapy topics
- Introducing Digital and Hybrid Books to facilitate seamless learning
- Presenting text with various pedagogical features as per the relevance

We continue to refine and expand our physiotherapy book collection to meet the evolving needs of students, educators, and professionals.

With the growing demand for skilled physiotherapists in a rapidly evolving healthcare landscape, we remain committed to publishing high-quality academic resources that empower the next generation of physiotherapists.

Together, let's build a stronger foundation for physiotherapy learning and practice, ensuring that future professionals are well-equipped to excel in patient care and rehabilitation.

Mr Bhupesh Aarora
(Sr Vice President – Publishing & Marketing)
bhupeshaarora@cbspd.com| +91 95553 53330

Special Features of the Book

Learning Objectives in the beginning of every Chapter help readers understand the purpose of the chapter.

> **LEARNING OBJECTIVES**
>
> *After the completion of the chapter, the readers will be able to:*
> - Define force, speed, velocity, work, energy, power, acceleration, momentum, friction and inertia.
> - Explain the components of forces and determine its magnitude by parallelogram of forces.
> - Define gravity, center of gravity, line of gravity, and correlate its relevance with human body mechanics.

> **CHAPTER OUTLINE**
>
> - Introduction
> - Kinematics
> - Kinetics
> - Force and its Composition
> - Levers
> - Anatomical Pulleys

Chapter Outline gives a glimpse of the content covered in the chapter.

Key Terms are added in each chapter to help understand difficult scientific terms in easy language.

> **KEY TERMS**
>
> **Acceleration:** The rate at which velocity changes in relation to time is known as acceleration in mechanics. Its SI unit is m/s^2.
>
> **Base of Support (BoS):** The term "base of support" refers to the space that is covered beneath an object or a person, i.e., the space between each point of contact the object or person makes with the supporting surface.

TABLE 22.2: Causes of Trendelenburg gait

Causes	Specific conditions
Failure of the fulcrum	Osteonecrosis of the hip
	Legg-Calvé-Perthes disease
	Developmental dysplasia of the hip
	Chronically dislocated hips secondary to trauma
	Chronically dislocated hips secondary to infections (e.g., tuberculosis of the hip)

Numerous **Tables** have been used in the chapters to facilitate learning in a quick way.

The book is well illustrated with practices/techniques of exercise therapy for better understanding of the theoretical concepts.

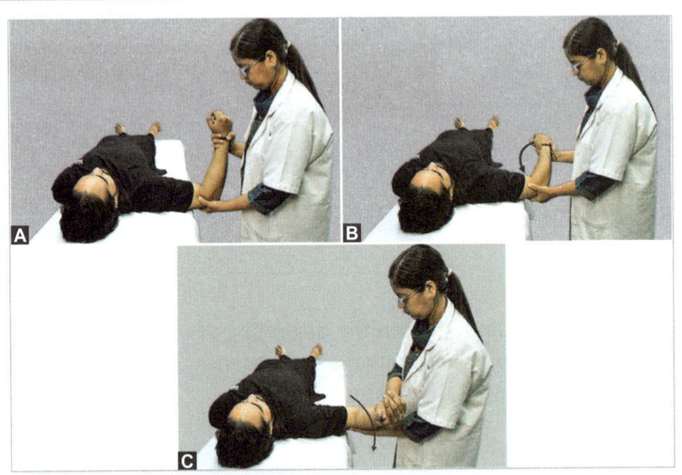

Figs 7.4A to C: Shoulder joint: **A.** Rotation starting position; **B.** Internal rotation; **C.** External rotation

Did You Know?

Suspension therapy also engages the proprioceptive system, which involves sensory receptors in the muscles and joints that provide feedback to the brain about body position and movement. This engagement helps improve proprioception, which can enhance coordination, balance, and overall body awareness, contributing to better movement patterns and injury prevention.

Did You Know? boxes give an overview of important facts and terms of the concerned topic.

Evolving conceptual details for application in clinical situations are depicted in **Clinical Correlation** boxes.

Clinical Correlation

Suspension therapy can also be adapted for postural correction and rehabilitation. By suspending the body in a controlled manner, therapists can manipulate the positioning of the body to promote proper alignment and muscle activation. This can be especially helpful for individuals with postural imbalances or conditions such as scoliosis, where targeted exercises and positioning can help alleviate pain and improve overall posture and spinal alignment.

SUMMARY

- Suspension therapy is a therapeutic exercise that involves suspending a person's body in the air while performing various movements and exercises.
  - This technique can be administered while sitting or lying down, and the suspension devices can be ropes or slings. It is often used by physical therapists, chiropractors, and other medical practitioners to increase joint mobility, flexibility, and strength.
  - The theory behind suspension therapy is that by suspending the person in midair, the force of gravity is diminished, which can help decompress joints and relieve strain on painful joints.

Important takeaway points of respective chapters have been highlighted under **Summary** boxes.

FURTHER READINGS

- El-Meniawy GH, Kamal HM, Elshemy SA. Role of treadmill training versus suspension therapy on balance in children with Down syndrome. Egyptian Journal of Medical Human Genetics. 2012;13(1):37-43.
- Gao, B., Rong, X., Liang, D. and Li, L. (2008) The Effect of Sling Exercise Therapy on Low Back Pain Caused by Exercises Training. Chinese Journal of Rehabilitation Medicine, 23, 1095-1097.

To give extra edge to the study, **Further Readings** have been included at the end of every chapter.

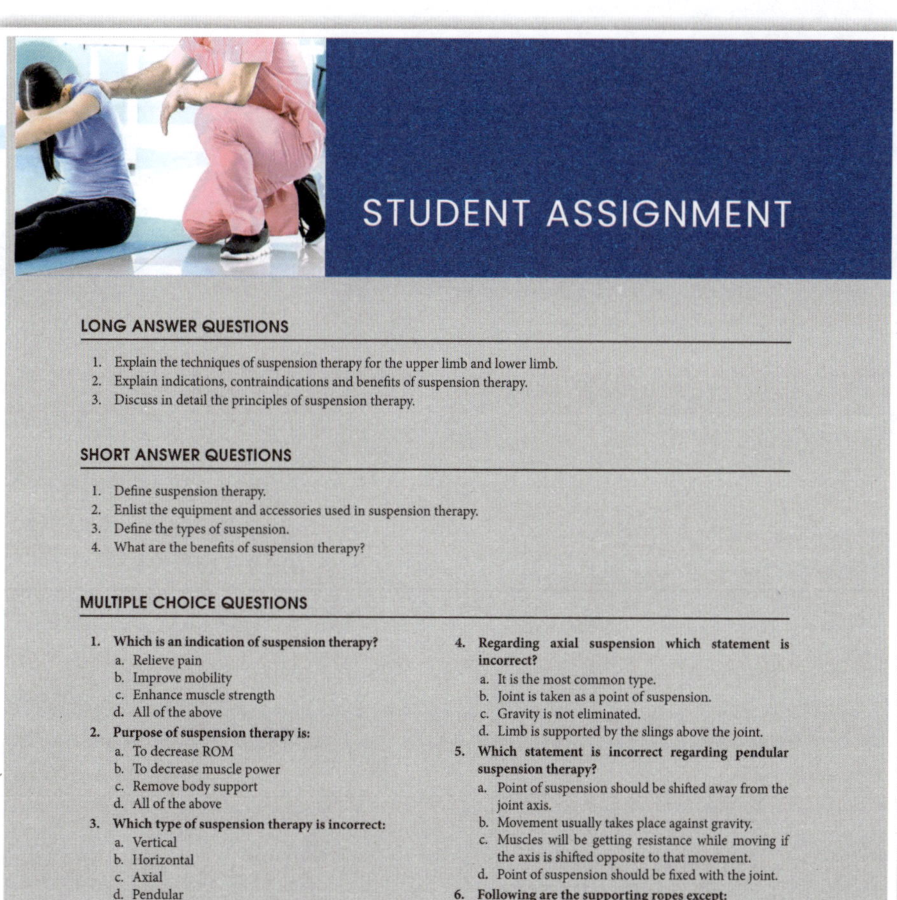

STUDENT ASSIGNMENT

LONG ANSWER QUESTIONS

1. Explain the techniques of suspension therapy for the upper limb and lower limb.
2. Explain indications, contraindications and benefits of suspension therapy.
3. Discuss in detail the principles of suspension therapy.

SHORT ANSWER QUESTIONS

1. Define suspension therapy.
2. Enlist the equipment and accessories used in suspension therapy.
3. Define the types of suspension.
4. What are the benefits of suspension therapy?

MULTIPLE CHOICE QUESTIONS

1. **Which is an indication of suspension therapy?**
   a. Relieve pain
   b. Improve mobility
   c. Enhance muscle strength
   d. All of the above
2. **Purpose of suspension therapy is:**
   a. To decrease ROM
   b. To decrease muscle power
   c. Remove body support
   d. All of the above
3. **Which type of suspension therapy is incorrect:**
   a. Vertical
   b. Horizontal
   c. Axial
   d. Pendular
4. **Regarding axial suspension which statement is incorrect?**
   a. It is the most common type.
   b. Joint is taken as a point of suspension.
   c. Gravity is not eliminated.
   d. Limb is supported by the slings above the joint.
5. **Which statement is incorrect regarding pendular suspension therapy?**
   a. Point of suspension should be shifted away from the joint axis.
   b. Movement usually takes place against gravity.
   c. Muscles will be getting resistance while moving if the axis is shifted opposite to that movement.
   d. Point of suspension should be fixed with the joint.
6. **Following are the supporting ropes except:**
   a. Single rope
   b. Double rope
   c. Pulley rope
   d. Triple rope

At the end of chapters, **Student Assignment** section is given which contains frequently asked questions in exams and multiple choice questions to help students attain mastery over the subject.

EXERCISE THERAPY

THEORY

Basics of Exercise Therapy

1. **Mechanics**

   Define the following terms and describe the principles involved with suitable examples.
   - **Force:** Composition of force, parallelogram of forces.
   - **Equilibrium:** Stable, unstable, neutral.
   - **Gravity:** Center of gravity, line of gravity.
   - **Levers:** 1st, 2nd and 3rd order. Their examples in the human body and their practical application in physiotherapy, forces applied to the body levers.
   - **Pulleys:** Fixed, movable.
   - **Springs:** Series, parallel.
   - Tension.
   - **Elasticity:** Hooks law.
   - **Axis:** Sagittal, frontal, and horizontal.
   - **Planes:** Sagittal, frontal, and horizontal.
   - **Definitions:** Speed, velocity, work, energy, power, acceleration, momentum, friction and inertia.

2. **Introduction to Exercise Therapy**
   - The aims of exercise therapy, the techniques of exercise therapy, approach to patient's problems, assessment of patient's condition—measurements of vital parameters.
   - Starting positions—fundamental positions—muscle work, effect and uses.
   - Derived positions, planning of treatment.

3. **Muscle Action**
   - **Muscle work:** Isotonic (concentric, eccentric), isometric (static).
   - **Group actions:** Agonists (prime movers), antagonists, synergists, fixators.
   - Angle of muscle pull, mechanical efficiency of the muscles.
   - Active and Passive insufficiency.

4. **Passive Movements**
   - Causes of immobility, Classification of passive movements, and specific definitions related to passive movements, principles of giving passive movements, indications, contraindications, effects and uses, techniques of giving passive movements.

5. **Active Movements**
   - Definition of strength, power and work, endurance, muscle actions.
   - **Physiology of muscle performance:** Structure of skeletal muscle, chemical and mechanical events during contraction and relaxation, muscle fiber types, motor unit, force gradation.
   - Causes of decreased muscle performance.
   - **Physiologic adaptation to training:** Strength and power, endurance.
   - Types of active movements.
   - **Free exercise:** Classification, principles, techniques, indications, contraindications, effects and uses.

- **Active assisted exercise:** Principles, techniques, indications, contraindications, effects and uses.
- **Assisted-resisted exercise:** Principles, techniques, indications, contraindications, effects and uses.
- **Resisted exercise:** Definition, principles, indications, contraindications, precautions and techniques, effects and uses.
- **Types of resisted exercises:** Manual and mechanical resistance exercise, isometric exercise.
- **Dynamic exercise:** Concentric and eccentric, constant versus variable resistance, isokinetic exercise, open-chain and closed-chain exercise.
- Specific exercise regimens.
- **Isotonic:** de Lormes, Oxford, Macqueen, circuit weight training.
- **Isometric:** Brief resisted isometric exercise (BRIME), multiple angle isometrics, isokinetic regimens.
- **Progressive resisted exercises:** Describe the following exercises, their advantages and disadvantages and demonstrate the techniques of the following types of PREs: Fractional system, Macqueen's set system, Mcqueen's power system.
- **Demonstrate practically each system using:** Delorme boot, dumbells, sand bags in pulleys, powder board and suspension therapy.

6. **Functional Re-education/Mat Activities**
   - **Lying to sitting:** Activities on the mat/bed, movement and stability at floor level; sitting activities and gait; lower limb and upper limb activities.
   - Demonstrate common mat activities.
   - Rolling, prone on elbows, prone on hands, Hook lying, bridging, quadruped position, long sitting, short sitting, kneeling, half kneeling, standing, walking.
   - Describe the term re-education of muscles and the techniques, spatial summation and temporal summation.
   - Demonstrate the various re-education techniques and facilitating methods for various groups of muscles.
   - Demonstrate the progressive exercises in strengthening by using various applications from (according to their muscle power) Grade-I to Grade-IV.
7. **Aerobic Exercise**
   Definition and key terms; physiological response to aerobic exercise, examination and evaluation of aerobic capacity—exercise testing, determinants of an exercise program, the exercise program, normal and abnormal response to acute aerobic exercise, physiological changes that occur with training, application of principles of an aerobic conditioning program for patients—types and phases of aerobic training.
8. **Stretching**
   - Definition of terms related to stretching; tissue response toward immobilization and elongation, determinants of stretching exercise, effects of stretching, inhibition and relaxation procedures, precautions and contraindications of stretching, techniques of stretching.
   - Special emphasis on stretching of pectoral major, biceps brachii, triceps brachii, and long flexors of fingers, rectus femoris, iliotibial band, gastrocnemius-soleus, hamstrings, hip abductors, iliopsoas, sternocleidomastoid.
9. **Balance**
   - Definition.
   - **Physiology of balance:** Contributions of sensory systems, processing sensory information, generating motor output.
   - Components of balance (sensory, musculoskeletal, biomechanical).
   - Causes of impaired balance, examination and evaluation of impaired balance, activities for treating impaired balance with mode posture, movement, precautions and contraindications, types.
   - Balance retraining.
10. **Coordination Exercise**
    - Anatomy and physiology of cerebellum with its pathways.
    - Definitions: Coordination, incoordination.
    - Explain the mechanism of neuromuscular coordination.
    - Describe the incoordination due to lower motor neuron lesions (flaccidity) upper motor neuron lesions (spasticity) cerebellar lesions, loss of kinesthetic sense (tabes dorsalis, syringomyelia, leprosy), imbalance due to muscular disease.
    - Causes for incoordination.
    - Tests for coordination: Equilibrium test, nonequilibrium test, principles of coordination exercise.
    - **Frenkel's exercise:** Uses of Frenkel's exercise, technique of Frenkel's exercise, progression, home exercise.

Examination/Testing Techniques

11. **Manual Muscle Testing**
    - **Introduction to MMT, principles and aims, various methods, indications and limitations, techniques of MMT for group and individual muscles:** Techniques of MMT for upper limb/techniques of MMT for lower limb/techniques of MMT for spine.

- Describe the types of muscle grading, key to muscle grading, demonstrate the skill to grade the individual and group muscles of upper and lower limb, neck and trunk muscles.

12. **Goniometry**
    - **Measurement of joint range:** ROM-definition, normal ROM for all peripheral joints and spine, goniometer-parts, types, principles, uses, limitations of goniometry, techniques for measurement of ROM for all peripheral joints
    - Describe the normal range of various joints. Describe goniometer, range of measuring systems for trunk and head, techniques of goniometer.
    - Demonstrate measuring of individual joint range using goniometer.
    - Demonstrate measurement of limb girth (using measuring tape for arm, forearm, thigh, calf.

13. **Tests for Neuromuscular Efficiency**
    - Electrical tests.
    - **Anthropometric measurements:** Muscle girth—biceps, triceps, forearm, quadriceps, calf.
    - Static power test.
    - Dynamic power test.
    - Endurance test.
    - Speed test
      - Tests for coordination.
      - Tests for sensation.
      - Pulmonary function tests.
      - **Measurement of limb length:** True limb length, apparent limb length, segmental limb length.
      - Measurement of the angle of pelvic inclination.
    - Describe—pelvic tilts, alterations from normal-anterior tilt (forward), posterior tilt (backward) lateral tilt. Muscles responsible for alterations and pelvic rotation. Identification of normal pelvic tilts, pelvic rotation and altered tilts and their correction.

Therapeutic Techniques

14. **Proprioceptive Neuromuscular Facilitation**
    - Definitions and goals.
    - **Basic neurophysiologic principles of PNF:** Muscular activity, diagonals patterns of movement: upper limb, lower limb.
    - **Procedure:** Components of PNF.
    - Techniques of facilitation.
    - **Mobility:** Contract relax, hold relax, rhythmic initiation.
    - **Strengthening:** Slow reversals, repeated contractions, timing for emphasis, rhythmic stabilization stability: alternating isometric, rhythmic stabilization.
    - **Skill:** Timing for emphasis, resisted progression endurance: slow reversals, agonist reversal.
    - Patterns of movement in detail.

15. **Suspension Therapy**
    - Definition, principles, equipment and accessories, Indications and contraindications, benefits of suspension therapy. Types of suspension therapy—axial, vertical, pendula, eccentric fixation (anterior, posterior, medial and lateral). Explain the indications and technique for each type of suspension. Demonstrate axial and eccentric fixation for mobilizing, strengthening and re-education of various muscles and joints.
    - Techniques of suspension therapy for upper limb, techniques of suspension therapy for lower limb.

16. **Manual Therapy and Peripheral Joint Mobilization**
    - **Basics in manual therapy and applications with clinical reasoning:**
      - Examination of joint integrity
      - Contractile tissues
      - Noncontractile tissues
      - Mobility—assessment of accessory movement and end feel.
      - Assessment of articular and extra-articular soft tissue status.
      - Myofascial assessment.
      - Acute and chronic muscle hold.
      - Tightness.
      - Pain-original and referred.
      - Basic principles, indications and contraindications of mobilization skills for joints and soft tissues.
      - Maitland
      - Mulligan
      - McKenzie
      - Muscle energy technique
      - Myofascial stretching
      - Cyriax
      - Neuro dynamic testing
    - Schools of manual therapy, principles, grades, indications and contraindications, effects and uses—Maitland, Kaltenborn, Mulligan
    - Biomechanical basis for mobilization, effects of joint mobilization, indications and contraindications, grades of mobilization, principles of mobilization, techniques of mobilization for upper limb, lower limb, precautions.

17. **Massage**
    - History and classification of massage technique principles, indications and contraindications technique of massage manipulations.

- Physiological and therapeutic uses of specific manipulations, history of massage.
- Mechanical points to be considered while giving massage; techniques, indications and contraindications.
- Physiological effects of massage on various systems of body. Effects on: Excretory system, Circulatory system, Muscular system, Nervous system and Metabolic system.
- Define and describe the various manipulation techniques, effects, uses and contraindications used in massage.
  - **Stroking manipulation:** Effleurage, stroking.
  - **Pressure manipulations:** Kneading, squeezing, stationary, circular ironing (reinforced kneading), finger kneading, petrissage (picking up, wringing, rolling), frictions.
  - **Percussion manipulation:** Tapotement, hacking, clapping, beating and pounding.
  - **Shaking manipulations:** Vibration, shaking.
- **Demonstrate the following techniques on patients/ models:**
  - **Massage for upper limb:**
    - Scapular region
    - Shoulder joint
    - Upper arm
    - Elbow joint
    - Forearm
    - Wrist joint
    - Hand.
  - **Massage for lower limb:**
    - Thigh
    - Knee joint
    - Leg
    - Foot (including ankle joints and toes)
  - **Massage for back:**
    - Neck and upper back
    - Middle and lower back
    - Gluteal region, arm and leg.
  - **Massage for the face**

18. Hydrostatics and Hydrodynamics

- History
- Properties of water, specific gravity, hydrostatic pressure
- Archimedes principle, buoyancy-law of floatation
- Effect of buoyancy on movements performed in water
- Equilibrium of a floating body, Bernoulli's theorem
- Physiological effects of exercise in water

19. Hydrotherapy

- Definitions, goals and indications, precautions and contraindications, properties of water, use of special equipment, techniques, effects and uses, merits and demerits.
- Describe the dress of patients and the therapist and necessary hydrotherapy equipment.
- **Types of hydrotherapy:** Sterile pool contrast bath, whirlpool bath, Hubbard tank.
- **Construction of hydrotherapy tank:** Design of construction, safety features, cleaning the pool, water heating systems, hygiene of patient and pool.

Posture, Gait, Miscellaneous

20. Posture

- Definition, active and inactive postures, postural mechanism, patterns of posture, principles of re-education, corrective methods and techniques, patient education.
- **Describe posture:** Posture (static and dynamic), definition of good posture, muscles responsible for good posture. Postural mechanisms, definition of abnormal posture (kyphosis, scoliosis, lordosis, kyphoscoliosis, kypholordosis), assessment of posture (inspection, scoliosis, lordosis, kyphoscoliosis, kypholordosis), assessment of posture (inspection, measurement—length of legs, width of pelvis, plumb line—ROM of trunk in flexion, extension, side flexion and rotation).
- Describe and demonstrate postural correction by strengthening of muscles, mobilization of trunk, relaxation. Active correction of the deformities, passive correction (traction), postural awareness, abdominals and back extensors.
- Outline principles in bracing of the trunk and surgical correction.
- **Demonstrate practically:** Identification of abnormal posture, and postural corrective measures.

21. Gait

- Define gait and center of gravity of the human body.
- Describe muscles responsible for normal gait, six determinants of gait (pelvic rotation, pelvic tilt, hip flexion, lateral displacement of pelvis, knee flexion, in stance phase, normal foot pattern during walking).
- **Describe the gait cycle:** Stance (heel strike, foot flat, mid stance, and foot off), Swing (acceleration, mid swing and deacceleration).
- **Describe the following pathological gaits:** Gluteus medius gait, gluteus maximus gait, hip flexor weakness gait, quardriceps weakness gait, foot drop gait, hemiplegic gait, ataxic waddling gait, equinus gait, calcaneus gait, equinovarus gait.
- Demonstrate skill in identifying pathological gait and proper gait training.

22. Gait Training and Walking Aids
- Definition, different methods of gait training, gait training in parallel bars.
- **Walking aids:** Types: Crutches, canes, frames; principles and training with walking aids.
- Describe components of a crutch, types of classification of crutches, characters of good crutch, preparing a patient for crutch walking, crutch walking muscles, measurement of crutches (axillary piece, hand piece), crutch stance, crutch palsy, types of crutch walking (4 point, 3 point, 3 point (nonweight bearing and partial weight bearing), modified 3 point (paraplegic and shuffling gait, swing to and swing through).
- Demonstrate crutch measurement (sitting, standing and lying positions) and various types of crutch walking (even ground, stairs and ramps).

23. Individual and Group Exercises

Advantages and disadvantages, organization of group exercises, recreational activities and sports.

24. Complications of Bed Rest
- Describe the complications of patients on prolonged bed rest.
- Buerger exercises.
- Demonstrate maintenance exercises for patients on prolonged bed rest.

25. Relaxation
- **Definitions:** Muscle tone, postural tone, voluntary movement, degrees of relaxation, pathological tension in muscle, stress mechanics, types of stresses, effects of stress on the body mechanism, description of fatigue and spasm, general causes, signs and symptoms of fatigue, indications of relaxation, rationale of relaxation techniques.
- Methods and techniques of relaxation—principles and uses for general, local, Jacobson's, Mitchel's, additional methods.

26. Introduction to Yoga
- Physiology and therapeutic principles of yoga.
- Yoga Sana for physical culture, relaxation and medication.
- Application of yoga Sana in physical fitness, flexibility.
- Therapeutic application of yoga. Yoga a holistic approach.

PRACTICAL

The students of exercise therapy are to be trained in practical laboratory work for all the topics discussed in theory. The student must be able to evaluate and apply judiciously the different methods of exercise therapy techniques on the patients. They must be able to:
- Demonstrate the technique of measuring using goniometry.
- Demonstrate muscle strength using the principles and technique of MMT.
- Demonstrate the techniques for muscle strengthening based on MMT grading.
- Demonstrate the PNF techniques.
- Demonstrate exercises for training coordination—Frenkel exercise.
- Demonstrate the techniques of massage manipulations.
- Demonstrate techniques for functional re-education.
- Assess and train for using walking aids, crutch gaits, gait training using various walking aids.
- Demonstrate mobilization of individual joint regions.
- Demonstrate to use the technique of suspension therapy for mobilizing and strengthening joints and muscles.
- Demonstrate the techniques for muscle stretching.
- Assess and evaluate posture and gait.
- Demonstrate to apply the technique of passive movements.
- Demonstrate various techniques of active movements.
- Demonstrate techniques of strengthening muscles using resisted exercises.
- Demonstrate techniques for measuring limb length and body circumference.
- Various types breathing exercises, chest mobilization exercises, postural drainage.

Contents

SECTION II EXAMINATION/TESTING TECHNIQUES

SECTION III APPLICATION OF PRINCIPLES OF EXERCISE THERAPY

SECTION V MOVEMENT AND ALIGNMENT

Section I

Essentials of Exercise Therapy

SECTION OUTLINE

1 Mechanics

Sheetal Kalra, Paridhi Sharma

LEARNING OBJECTIVES

After the completion of the chapter, the readers will be able to:

- Define force, speed, velocity, work, energy, power, acceleration, momentum, friction and inertia.
- Explain the components of forces and determine its magnitude by parallelogram of forces.
- Define gravity, center of gravity, line of gravity, and correlate its relevance with human body mechanics.
- Describe levers and correlate their practical application in physiotherapy.
- Explain pulleys and its types, with biomechanical relevance.
- Describe springs and their types.
- Explain tension and elasticity.

CHAPTER OUTLINE

- Introduction
- Kinematics
- Kinetics
- Force and its Composition
- Levers
- Anatomical Pulleys
- Spring

KEY TERMS

Acceleration: The rate at which velocity changes in relation to time is known as acceleration in mechanics. Its SI unit is m/s^2.

Base of Support (BoS): The term "base of support" refers to the space that is covered beneath an object or a person, i.e., the space between each point of contact the object or person makes with the supporting surface.

Displacement: A vector quantity, displacement is the shortest path between starting and ending positions of an object in motion. Its SI unit is meter (m).

Free-body diagram (FBD): A representation of a segment of a whole system that has been separated to ascertain the forces acting on that rigid body is called a free-body diagram.

Inertia: Until an outside force acts upon an object, it maintains its state of rest or uniform motion in a straight line due to the property of matter known as inertia. Newton's first law of motion describes it.

INTRODUCTION

- **Mechanics:** Mechanics basically is the branch of science which studies about various forces acting upon an object and the effects thereby produced by these forces. The study of forces acting upon or produced by human body is described as biomechanics. As we all know, humans can produce various gestures and postures, which require the generation and response to systems of forces acting upon the body. In order to produce movements of body segments and to support body loads, human musculoskeletal system is responsible.
- **Kinesiology:** Kinesiology, which is the biomechanics' parent discipline, is the science that aids in movement investigation. It can be classified into mechanical and anatomical aspects of human movement. The mechanical aspect can further be divided into statics and dynamics.
- **Statics:** Statics is the mechanics branch that investigates masses, bodies, and forces in equilibrium or at rest.
- **Dynamics:** Dynamics investigates bodies and forces in motion. Biomechanical investigation is broadly dissected into two subgroups: Kinetics and Kinematics. Kinematics provides a detailed description of movements without any consideration to the forces responsible to produce that movement. Kinematic analysis of motion involves measurement of velocity, acceleration, and position of either one or more than one body parts. Linear and angular measurements of specific parts or performing their combination can be used to quantify the spatial and temporal properties of movements of multiple body parts simultaneously. However, kinetics investigates the action of forces in producing or changing the motion of masses.

KINEMATICS

Originating from the Greek term "kinesis", which signifies motion, comes the word "kinematics". Body system movements are studied in the subject of kinematics, which does not directly take into account any potential fields or forces that may affect motion.

Definition: Kinematics is the study of the exchange of energy and momentum between interacting bodies. A body segment's motion and displacement can be explained by the following five kinematic variables:

1. Type of motion or displacement.
2. Location of moving segment in space.
3. Direction of displacement of the segment.
4. Magnitude of displacement.
5. Rate of displacement or rate of change of displacement.

Types of Motion or Displacement

Motion can be described as change in the position of a body (or a segment) with respect to time. Linear and angular motions are particularly two basic types of motion which are considered "pure" motion. In general, the complex movements performed by the human body segments in daily activities are achieved by combination of these two types of movements. **Linear motion**, also known as translatory motion or translation is the one in which all the segments are moving in same direction and at same speed (Fig. 1.1). In other words, motion along a line is translatory or linear motion. If the movement occurs along a straight line, it is termed as **rectilinear** motion and if the path is curved, it is termed as **curvilinear**. Rotation around a fixed imaginary axis known as the axis of rotation, which is perpendicular to the plane in which rotation happens, is known as **angular motion** (Fig. 1.2). In a true angular motion, each point of

Fig. 1.1: Linear motion

Fig. 1.2: Angular motion

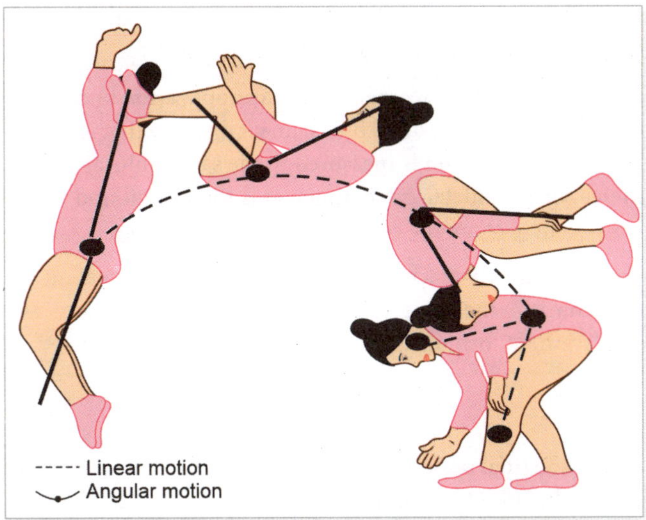

Fig. 1.3: General motion: Combination of linear and angular motion

segment is moved through same time, at same angle and at constant distance from the center of rotation. The human body is made up of several segments that move together to provide a **general motion** that includes simultaneous angular and linear motions (Fig. 1.3). For example, while walking the trunk moves linearly, whereas the segments of the specific extremity are in rotational motion.

Clinical Correlation

The principles of linear and angular motion help in creating prosthetic limbs, braces, and orthotic devices that provide optimal support and mobility for individuals with limb loss or musculoskeletal disorders.

Location of Moving Segment in Space

The position of the moving segment in space can be determined by three-dimensional Cartesian coordinate system, which includes three mutually perpendicular lines known as axes which are in turn perpendicular to three mutually perpendicular planes. A plane can be described as the imaginary surface on which the movement occurs, whereas axis is an imaginary line along which the movement occurs. Planes are positioned perpendicularly through the body, and intersect at center of mass of body. There are three imaginary planes (Fig. 1.4) namely:

1. The **sagittal plane**, dividing the body vertically into left and right halves and is also known as anteroposterior plane.
2. The **frontal plane** splits the body vertically into front and back halves and is also known as coronal plane.
3. The **transverse plane** is the plane parallel to the ground and divides the body into top and bottom halves and is also known as horizontal plane as shown in Figure 1.4.

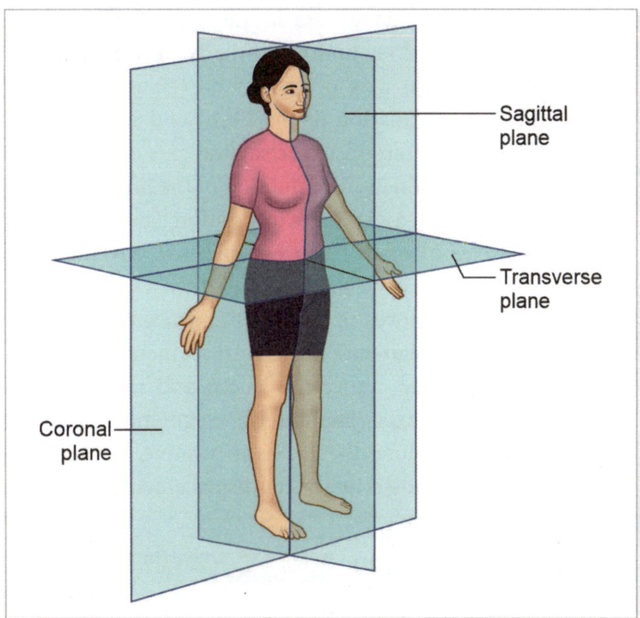

Fig. 1.4: Three anatomical planes perpendicular to each other

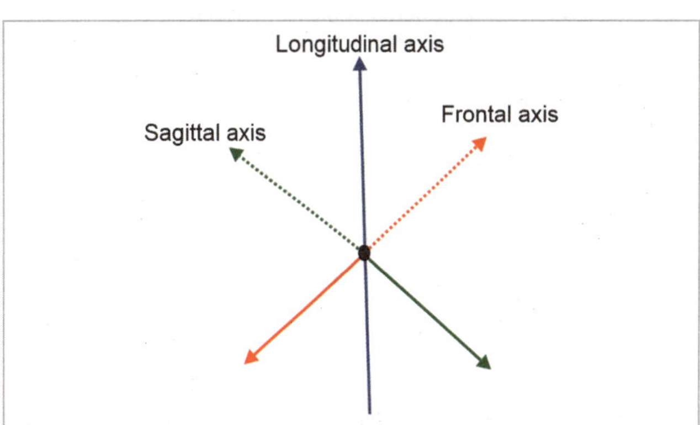

Fig. 1.5: Three anatomical axes corresponding to anatomical planes

When a human segment moves, it rotates around an imaginary axis of rotation that passes through a joint to which it is attached. To describe human motion, there are three reference axes which are positioned perpendicular to the above-mentioned three planes.

1. The **frontal axis** (mediolateral) is perpendicular to sagittal plane.
2. The **sagittal axis** (anteroposterior) is oriented perpendicular to frontal plane.
3. **Longitudinal axis** is perpendicular to transverse plane as shown in Figure 1.5.

Direction of Displacement of the Segment

Rotation of a body segment is described as occurring parallel to one of the three cardinal planes, and around one of the three possible axes. Rotational movement of a segment about an axis in opposite directions in one plane (uniaxial or single plane) refers to rotation of a segment about the frontal/coronal axis, that occurs in the sagittal plane (e.g., flexion, extension of a limb). Similarly, rotation about a vertical/longitudinal axis occurs in the transverse plane (e.g., medial, lateral rotation of the limb) and rotation of a segment about the anteroposterior/sagittal axis occurs in the frontal plane (e.g., abduction, adduction of the limb).

Similar to rotation, a segment's translation can occur along one of three axes in one of two directions. Conventionally a segment's linear displacement along the X-axis is seen as positive when it moves to the right and negative when it moves to the left. A segment's linear displacement is regarded as positive, if it occurs above the Y-axis and negative, if it occurs below the Y-axis. The anterior segment's (anterior) linear displacement along the Z-axis is positive, whereas the posterior segment's (posterior) linear displacement is negative.

Clinical Correlation

By recognizing how body segments move in relation to the three cardinal planes and axes, physiotherapists can better diagnose issues and prescribe targeted exercises to improve range of motion, strength, and functional movement patterns.

Magnitude of Displacement

The International System of Units [SI] units of radians or degrees can be used to describe the magnitude of segment rotatory motion. A complete circle is made up of 360° or 6.28 radians, by an object. Which means:

$$2\pi = 360°$$

and,

$$\pi = 180°$$

Hence, from the above equation, we can say, 180° is equal to π radian.

Radian is always represented in terms of Pi, where the value of Pi is equal to 22/7 or 3.14.

Example: Convert 90° to radians.

Solution: Given, 90° is the angle. Angle in radian = Angle in degree × (π/180)

$$= 90 \times \left(\frac{\pi}{180}\right)$$

$$= \frac{\pi}{2}$$

Hence, 90° is equal to π/2 in radian.

In a nutshell:

Degrees × (π/180) = Radians
Radians × (180/π) = Degrees
360° = 2π Radians
180° = π Radians

Range or motion (ROM): The magnitude of the rotatory motion that body segment moves through or can move through is known as its ROM. The most widely used standardized clinical method of measuring available joint ROM is goniometry, with units given in degrees.

Displacement: A segment's translation, also known as displacement, is measured by the linear distance that it moves. Translational motion is measured in the same units as length. Meter (or millimeter or centimeter) is the SI unit and in the US system, feet are the comparable unit (or inches).

To calculate displacement: We find the displacement by determining the object's initial and final position. The displacement formula is given by:

$$\Delta x = x_f - x_i$$

where,

x_f is the final position
x_i is the initial position
Δx is the change in the position of the object.

Linear displacements of the entire body are often measured clinically.

Example: 6 minutes walk test, which assesses functional status in individuals with cardiorespiratory problems, measures the distance (in feet or meters) that a person walks in 6 minutes. Segment displacements can also be measured by using three-dimensional motion analysis systems.

Rate of Change of Displacement

The rate of change in the segment's position (the displacement per unit of time) is just as significant as the displacement's magnitude.

In terms of linear motion: Since velocity is a vector quantity and displacement is a vector quantity, **velocity** is defined as the displacement per unit time in a certain direction. In contrast, speed a scalar quantity and it is, the distance per unit time, independent of direction. That means:

$$v = \frac{ds}{dt}$$

where,

s = displacement
v = velocity
t = time

Linear velocity (velocity of a translating segment) is expressed as meter per second (m/sec) in SI units or feet per second (ft/sec) in US units.

Now, if the velocity changes over time, then it is important to know about rate of change of velocity which is termed as **acceleration**. That means:

If v_i is the initial velocity of the body and v_f being the final velocity of the body covered within total time t (second), then as per the definition given above:

$$a = \frac{(v_f - v_i)}{t}$$

In mathematical terms:

$$a = \frac{dv}{dt}$$

where,

a = acceleration
v = velocity
t = time

And dv/dt is the rate of change of velocity.

Linear acceleration (acceleration of a translating segment) is expressed as meters per second square (m/sec^2) in SI units or feet per second (ft/sec^2) in US units.

It is important to know that how these variables defined above play an important role in kinematics, hence Newton gave the kinematic equation of motions which are as follows:

$$s = ut + \tfrac{1}{2} at^2$$
$$v = u + at$$
$$v^2 = u^2 + 2\,as$$

where,

s = displacement
v = final velocity
u = initial velocity
a = acceleration
t = time

In a nutshell: Kinematics provides us with the description of motion but it does not explain the reason behind occurrence of that motion. Whether the body segment remains at rest or at motion, totally depends on the force acting on that segment. The consideration of this force, which is responsible for the occurrence of motion of segment comes under the field of Kinetics.

KINETICS

Force

Force is an external agent which changes the state of body either into rest or in motion. In other words, we can describe force as an external push or pull applied to an object having some mass that results in its change of velocity, i.e.,

$$F = \frac{mdv}{dt}$$

that means $F = ma$

(where 'm' is the mass of the object and 'a' is the acceleration);

$$a = \frac{dv}{dt}$$

The unit for a force in the SI system is Newton (N), which is the amount of force required to accelerate 1 kg of mass at 1 m/s^2:

$$1\text{ N} = (1\text{ kg})\ (1\text{ m/s}^2).$$

The unit in the US system is pound (lbf). A pound of force is the amount of force necessary to accelerate a mass of 1 slug at 1 ft/s^2, and 1 lb is equal to 4.45 N:

$$1\text{ lbf} = (1\text{ slug})\ (1\text{ ft/s}^2).$$

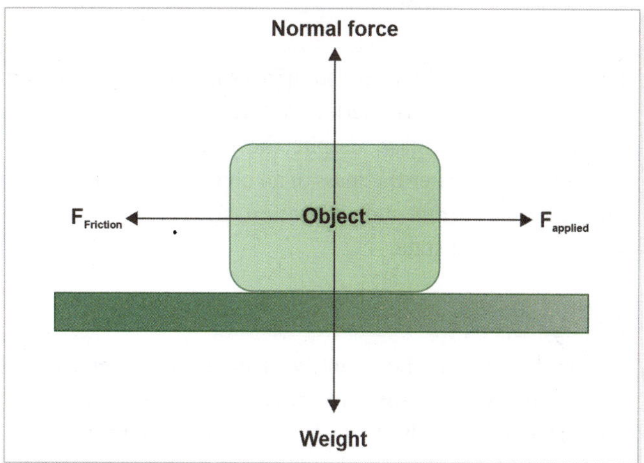

Fig. 1.6: Free-body diagram showing system of forces acting on an object

When examining the impacts of forces on a body or system of interest, creating a free body diagram is usually the initial step because many forces work simultaneously in most circumstances. Any object, person or body portion that is the subject of analysis is referred to as a free body. Force vector (represented by arrows in Fig. 1.6) is characterized by magnitude, direction and point of application.

In a nutshell: Force vectors are represented by arrows that (1) are referenced to the object on which they act (the point of application), and (2) have a shaft and arrowhead that indicate the direction of the applied force, and (3) has a length that represents the amount (magnitude) of the force applied.

It is crucial to realize that the net force or the vector sum of all operating forces, determines the overall impact of several forces acting on a system or free body, as one force rarely operates alone. The object stays in its initial state of motion—either at rest or moving at a steady speed—when all acting forces are balanced or cancel each other. This is known as the net force. An object experiences acceleration proportional to the size of the net force and moves in the direction of the net force when there is one. A useful framework for comprehending the consequences of force is provided by a grasp of the notions of inertia, mass, weight, and torque.

Clinical Correlation

Clinicians use force analysis techniques to evaluate gait abnormalities, joint stability, and biomechanical imbalances, guiding treatment options for conditions such as osteoarthritis and sports injuries. These techniques involve tools such as force plates, pressure mapping systems, and dynamometers to measure forces and pressure during movement. By providing insights into joint mechanics, muscle strength, and movement patterns, they help detect abnormalities, plan effective treatments, and enhance performance in both clinical and athletic settings.

Inertia

The body's propensity to resist changes in its rest or moving condition is known as inertia. A body's inertia is directly proportional to its mass, despite the fact that it cannot be measured. The greater the mass of an object, the more it tends to maintain its current state of motion, and the more difficult it is to disrupt that state.

Mass

Mass is defined as "the quantity of matter contained in an object". The common unit of mass in the metric system is kilogram (kg), with the English unit of mass being the slug, which is much larger than 1 kg.

Center of Mass/Center of Gravity

The point at which a body's weight is evenly balanced regardless of its position is known as the **center of mass or center of gravity**. The body's response to outside forces during movement is determined by the location of the center of mass.

The term "the amount of gravity acting on a body" refers to weight. This definition of force is an algebraic version of the standard definition. Mass (m) times gravitational acceleration (g) equals weight in this case.

Weight = Mass (m) × gravitational acceleration (g)

Units of weight are equivalent to units of force (N or lb) since weight is force. The weight increases in proportion to the weight of the body. Gravity-related acceleration is -9.81 m/s^2, which is the coefficient of proportionality. If there is a negative sign, it means that the Earth's center is being approached or dropped by the gravitational acceleration.

Since weight is a force, its characteristics change with its size, direction, and point of action. Weight always moves in the direction of the Earth's center. The center of gravity of an object is the location on the free body diagram where the weight vector seems to act, since this is the point at which weight is assumed to act on the object. Figures 1.7A and B shows how center of gravity acts on an object and human body, respectively.

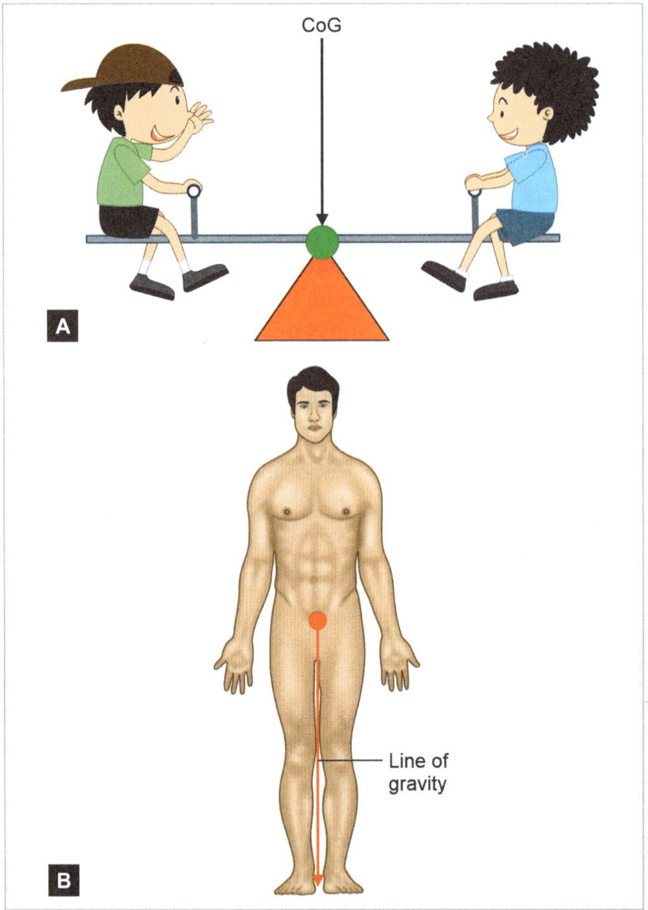

Figs 1.7A and B: A. Center of gravity (CoG), a point around which the body's weight is equally balanced; **B.** Center of gravity in human body lies at 2nd sacral vertebra

Clinical Correlation

Therapists should pay attention to the alignment of the body during exercises to optimize muscle recruitment and joint mechanics. Proper alignment helps distribute forces evenly across the joints and muscles, reducing the risk of overuse injuries or joint strain.

Torque

Torque (τ), also referred to as moment of force, is the rotating effect caused by an eccentric force. The angle equivalent of a linear force is called torque, and it can be thought of as a rotational force. Torque can be expressed algebraically as the product of the force (F) and the vertical distance (d) between the force's line of action and the axis of rotation.

$$\tau = d \times F$$

More the torque applied to a rotating shaft, the greater its tendency to rotate. Both metric and imperial torque units follow algebraic definitions. This is the unit, where force is multiplied by the unit of distance:

Newton meter (Nm) or foot pound (ft-lb)

The notion is significant because it implies that a muscle's moment of force is essential for its ability to sustain weight-bearing. To sustain sufficient knee extension for weight bearing, the patella, for instance, forms an effective moment with the quadriceps near the center of the knee's rotation.

Clinical Correlation

Understanding the concept of torque is crucial for prescribing exercises that target specific muscle groups while minimizing joint stress. For example, in knee rehabilitation, therapists may choose exercises that produce minimal torque at the knee joint while still effectively engaging the quadriceps and hamstrings to promote stability and strength.

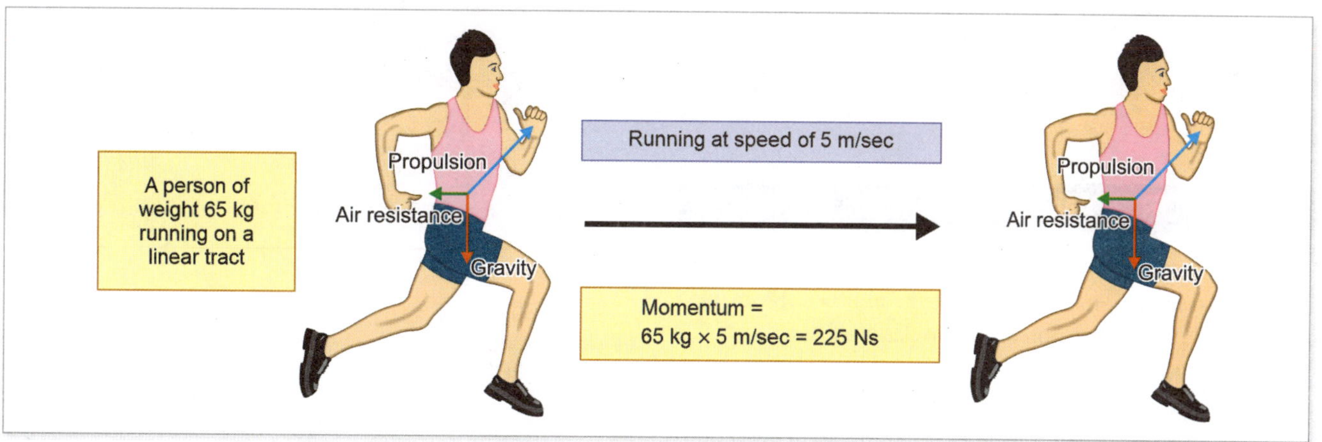

Fig. 1.8: Diagram representing calculation of momentum

Momentum

Momentum is the quantity which is being used in description of the moving body having non-zero mass. Therefore, momentum can be applied to all moving objects. If m is the mass of an object and v is the velocity with which this body travels, then momentum can be expressed as:

$$p = mv$$

where p is the denotation for momentum.

Since, velocity is considered to be a vector quantity, momentum is considered vector, too. This means, momentum has a significant magnitude and direction. For example, when a ball with a given mass is traveling at a particular speed, it possesses momentum. The moment the ball hits wall, it comes to rest and therefore, transfers its momentum to the wall. Therefore, we can conclude from the above scenario that momentum is always conserved. Figure 1.8 shows an example of how momentum is being calculated.

Work

To do a work, in a scientific sense, force must be applied, and displacement must be produced in direction of force. With this said, we can say that **work** is "multiplication of component of force in direction of displacement and magnitude of this displacement." Mathematically, the above statement is expressed as follows:

$$W = (F \cos \theta)d = Fd \text{ (product of F and d)}$$

where, W is the work done by the force;

F is the force;

d is the displacement caused by force; and

θ is the angle between the force vector and the displacement vector.

SI unit of work is Joules (J).

Example: A weight lifter lifts a barbell weighing 25 kg and displaces it from the ground by 2 m. Here, the work done upon the barbell is against gravity. The work done upon the weight against gravity can be calculated as follows:

Work done = (Mass × acceleration due to gravity) × Displacement
= (25 × 9.8) × 2 J = 490 J.

Energy

Energy is "the caliber to perform work". This could be said that more a man works, greater the energy he possesses which also implies that reverse is true, as well. Energy is quantitative in nature that must be converted to another form to make the work easier.

Example: A generator converts mechanical energy into electrical energy. Energy exists in different forms in nature viz. Potential, kinetic, electrical, nuclear, thermal, atomic, etc. Like work, energy is also a scalar quantity. This means that energy has no direction but magnitude.

Power

Energy and power are not the same. This is so because **power** is "the rate of performing work", but energy is the capacity to execute the greatest amount of work. Here, it is evident that power is the ability of an individual to accomplish any activity in a certain amount of time; on the other hand, energy is the maximum amount of work that an individual can accomplish without taking time into account or to put it another way, time has no bearing on energy. Power is defined as the quantity of energy converted or transferred per unit of time in physics. Watt, the SI unit of power, is equivalent to one Joule per second (J/s). Power is the rate of work relative to time; it is the work's time derivative:

$$P = \frac{dW}{dt}$$

where,

P = power
W = work
t = time

FORCE AND ITS COMPOSITION

Force, in biomechanics, is a way to represent load and is defined as an action of one object on another. From biomechanical perspective, forces can be classified as:

- **External forces:** Gravity, wind, objects, other people.
- **Internal forces:** Muscles, connective tissue (elastic), bone.
- **Reactionary forces:** Ground reaction, joint reaction, gliding/shear/friction.

Primary Rules of Forces

Levangie and Norkin reiterated that there are three primary rules of forces:

1. A force that acts on a segment must come from something.
2. Anything that contacts a segment must create a force on that segment.
3. Gravity is considered to have a force effect on all objects.

In nature, two or more forces are found to act simultaneously at a single point. But the effect of single force is observed. That single force is resultant of all the other forces and the other forces are its components. Process to determine the net force is known as **composition of forces**.

If there is a net unbalanced force (or forces) in a segment, it can be found using the force synthesis procedure. The reason for this is that it determines whether the section is in motion or not. Furthermore, segment motion type and direction are determined by the direction, location, and strength of the net imbalance force. The way forces interact—that is, whether they are in a parallel, simultaneous, or linear system—determines how the composition process works.

Forces that act by forming an angle from coordinate axes could get resolved into *components* which are mutually perpendicular forces. The component of force parallel to X-axis is called x-component, parallel to Y-axis is called y-component, and so on.

Components of a Force in XY Plane

As given in Figure 1.9, F_y represents the y-component of force F and F_x represent the x-component of force F, both the components being mutually perpendicular to each other. However, the resultant of both the components is F, which is the net directional force that provides both the magnitude and direction of the combination of these forces.

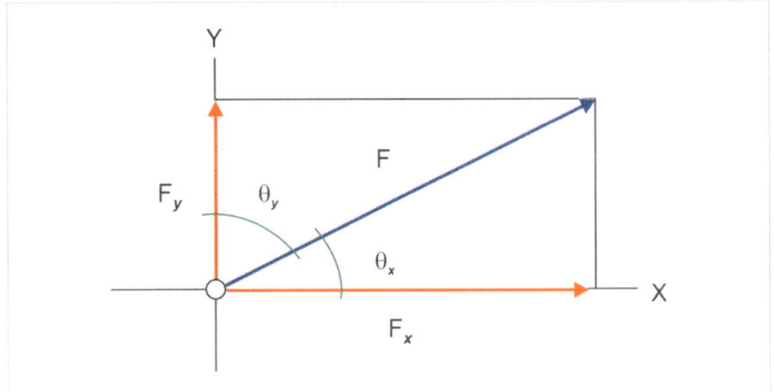

Fig. 1.9: Resolution of force F in its *x* and *y* components

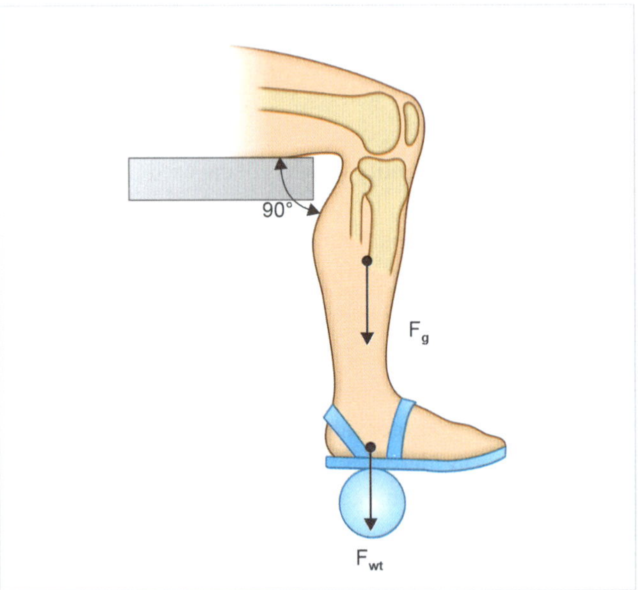

Fig. 1.10: The forces of gravity F_g and weight F_{wt} are in the same linear force system when the leg-foot segment is at 90° of knee flexion

Linear Force System

When all forces occur along same action line, it is called **linear force system** (Fig. 1.10). Forces may act in same direction or in opposite direction. As a result, it may produce tension or compression effects. For example, in cervical traction, two forces act in opposite direction. Trapezius muscles on both sides act along the same action line, but in opposite directions. Weakness on one side causes the resultant force to be bigger on the other side, resulting in lateral deviation of the spine.

Generally, if $\vec{F_1}, \vec{F_2}, \vec{F_3}$ are the forces acting on a body, their resultant force $\vec{F}$ is given by:

$$\vec{F} = \vec{F_1} + \vec{F_2} + \vec{F_3} + \cdots$$

Depending on the number forces acting, the resultant can be obtained geometrically by applying triangle law or parallelogram law of vector addition. It is important to note that the magnitudes of the forces cannot be added or subtracted unless they are collinear. Let F_1, F_2 be the forces with magnitudes 50 *N*, 10 *N*, acting in right direction and F_3 with magnitude 70 *N*, acting in opposite (left) direction. Then,

$$F_1 = 50\ N, F_2 = 10\ N, \text{ and } F_3 = -70\ N$$

The resultant force:

$$\begin{aligned} F &= F_1 + F_2 + F_3 \\ &= 50 + 10 - 70 \\ &= -10\ N \end{aligned}$$

Where, $F = -10\ N$ means, the resultant force is of magnitude 10 *N*, acting toward the left.

In a nut-shell: Forces applied up (Y-axis), forward or anterior (Z-axis), or to the right (X-axis) will be assigned positive signs, whereas forces applied down, back, or posterior, or to the left will be assigned negative signs. The magnitudes of vectors in opposite directions should always be assigned opposite signs.

Parallel Force System

When all the forces are coplanar (acting at the same plane), at two different points, and parallel to each other, but do not share the same action line, are called parallel force system. These forces produce rotatory effects (Fig. 1.11). For example, in hamstring muscles components, i.e., medial (semitendinosus and semimembranosus) and lateral (biceps femoris), the medial and lateral forces act in the same direction to produce knee flexion. If the forces are equal to each other the resultant is located in the middle producing knee flexion.

Concurrent Forces System

When all forces meet at same point of application but do not lie along same line of action, by forming an angle with each other, are called concurrent force system. For example, rectus femoris and vastus muscles of quadriceps. As shown in Figure 1.12, anterior cruciate ligament (ACL), posterior cruciate ligament (PCL) and gravity together acting on a same point of application but in different directions forms a concurrent force system.

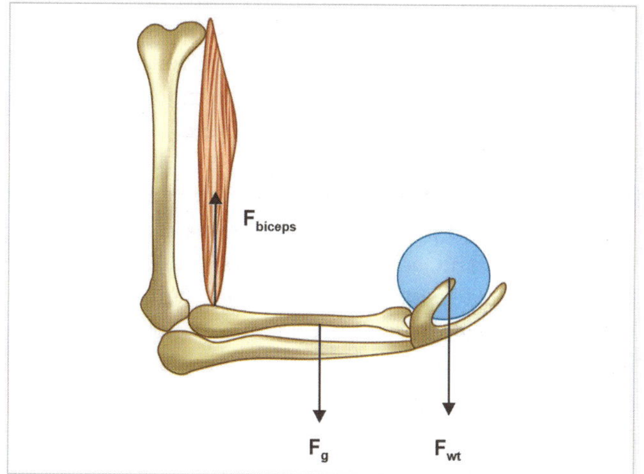

Fig. 1.11: The forces of biceps muscle F_{biceps}, weight F_{wt} and gravity F_g together forming the parallel force system on the forearm

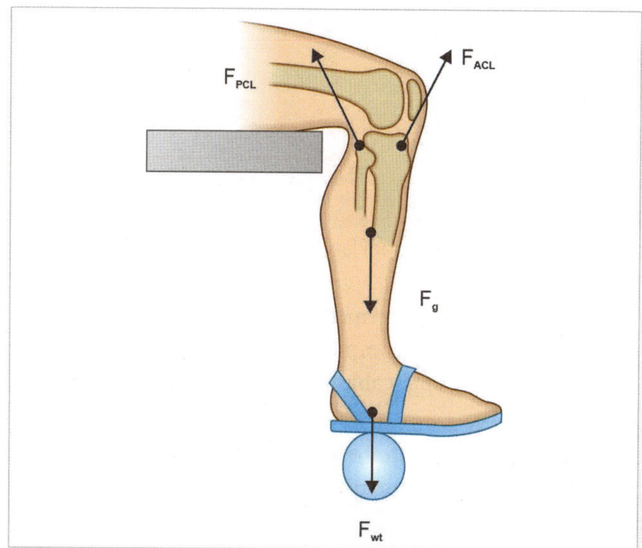

Fig. 1.12: Concurrent forces of anterior cruciate ligament, posterior cruciate ligament and gravity acting on knee joint

Abbreviations: F_{ACL}, force of anterior cruciate ligament; F_g, force of gravity; F_{PCL}, force of posterior cruciate ligament; F_{wt}, force of weight

Determining Resultant of Forces

To determine resultant of forces described above, there are two laws of vector addition:

1. Triangle law of vector addition (Fig. 1.13)
2. Parallelogram law of vector addition (Fig. 1.14)

Triangle Law of Vector Addition

One of the main vector addition laws is triangle law for adding vectors. We can define vector addition as geometrical sum of two or more vectors which do not obey regular algebraic laws.

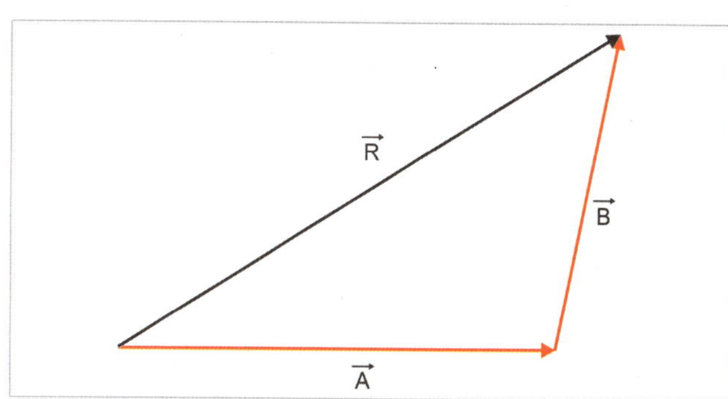

Fig. 1.13: Triangle law of vector addition $\vec{R} = \vec{A} + \vec{B}$

SECTION I Essentials of Exercise Therapy

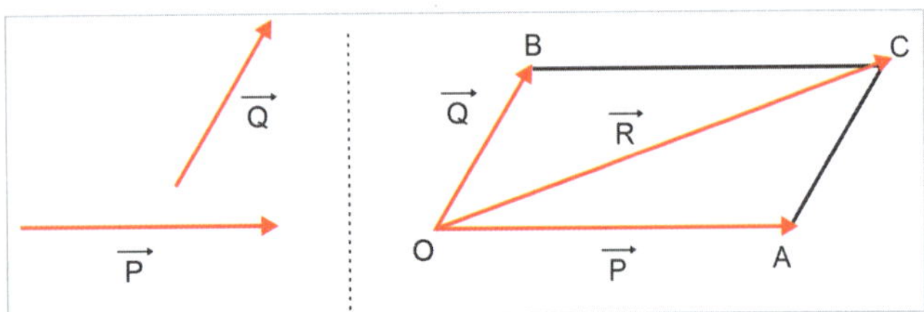

Fig. 1.14: Parallelogram law of vector addition $\vec{R} = \vec{P} + \vec{Q}$

The resultant vector is termed as composition of a vector as explained above in the **composition of forces**. There are a few conditions that are applicable for any vector addition, they are:

- Scalars and vectors can never be added.
- For any two vectors to be added, they must be of the same nature. Example, velocity can be added to velocity and not with force.

Triangle law of vector addition states that when two vectors are represented as two sides of the triangle with the order of magnitude and direction, then the third side of the triangle represents the magnitude and direction of the resultant vector (Fig. 1.13).

Parallelogram Law of Vector Addition

The parallelogram law states that the sum of the squares of the length of the four sides of a parallelogram is equal to the sum of the squares of the length of the two diagonals. In Euclidean geometry, it is necessary that the parallelogram should have equal opposite sides. If ABCD is a parallelogram, then AB = DC and AD = BC. Then as per the definition given above, it is stated as:

$$2(AB)^2 + 2\,(BC)^2 = (AC)^2 + (BD)^2.$$

In case the parallelogram is a rectangle, then the law is stated as:

$$2(AB)^2 + 2\,(BC)^2 = 2(AC)^2$$

This is because in a rectangle, two diagonals are of equal lengths, i.e., (AC = BD) (Fig. 1.15)

If two vectors are acting simultaneously at a point, then it can be represented both in magnitude and direction by the adjacent sides drawn from a point. Therefore, the resultant vector is completely represented both in direction and magnitude by the diagonal of the parallelogram passing through the point.

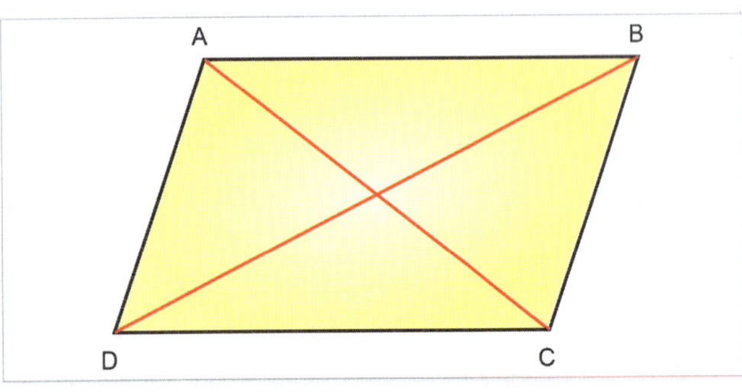

Fig. 1.15: Parallelogram ABCD

Consider the Figure 1.14, the vector P and vector Q represents the sides, OA and OB, respectively. According to the parallelogram law as explained above, the side OC of the parallelogram represents the resultant vector R.

$$OA + OB = OC$$

or

$$\vec{P} + \vec{Q} = \vec{R}$$

> **Did You Know?**
>
> The steps for the parallelogram law of addition of vectors are:
> 1. Draw a vector using a suitable scale in the direction of the vector.
> 2. Draw the second vector using the same scale from the tail of the first vector.
> 3. Treat these vectors as the adjacent sides and complete the parallelogram.
> 4. Now, the diagonal represents the resultant vector in both magnitude and direction.

To have a broad understanding of biomechanics of movement, one should have a thorough knowledge of Newton's Laws of Motion.

Newton's Laws of Motion

Law of Inertia

According to Newton's First Law of Inertia, objects usually resist changes in motion. An object in motion will continue to move, while an object at rest will continue to remain still unless moved upon by a force. For instance, because of the action of an outside force, an ice skater will continue to glide over the ice in the same direction and at the same speed. However, in actuality, the two factors that tend to slow skaters and other moving bodies are friction and air resistance. Static analysis is another field where this law is applicable. An engineering technique for analyzing the forces and moments created when objects interact is called static analysis. This concept is applied in biomechanics for estimation of unknown forces of muscle and joint reaction in musculoskeletal system.

Law of Acceleration

A force's ability to produce motion is explained by Newton's Second Law. The acceleration an object experiences is inversely related to its mass (F_{object}= ma) and proportionate to the amount of the force acting on it. Acceleration is the tendency of a body to change shape or speed. A ball, for instance, tends to move in the direction of the applied force's line of action when it is thrown, kicked or struck. The ball will move faster the more force is given to it. Athletes will be more adept at using their legs to propel their body forward during exercise while keeping their body mass constant if they focus on strengthening their legs. This will increase their speed and agility. The concept helps athletes use their technique more effectively by allowing them to exert more force for longer periods of time, like in shot put.

Recalling the definition of momentum studied earlier, $p = mv$.

Differentiating with respect to time:

$$\frac{dp}{dt} = \frac{d}{dt}(mv)$$

$$\frac{dp}{dt} = \frac{m.dv}{dt} + \frac{dm}{dt.v}\frac{dp}{dt} = m.a + 0.v$$

$$\frac{dp}{dt} = ma = F$$

Hence, Newton's second law of motion also suggests that the rate of a change of a particle's momentum p is given by the force acting on the particle, i.e., $F = dp/dt$.

Law of Reaction

According to the Third Law, there is an equal and opposite reaction force to every action (force). This indicates that forces work in equal and opposite pairs between interacting bodies rather than operating alone. Because the Earth is significantly more massive than the player, the player accelerates and moves quickly while the Earth does not actually accelerate or move at all. For instance, the force produced by a player's legs "pushing" against the ground results in ground reaction forces, which cause the ground to "push back" and allow the runner to cross the court. This action-reaction also happens when a racket and ball make contact. A force equal to and opposite from the force applied to the ball is applied to the racket.

Clinical Correlation

Newton's Third Law helps therapists understand how forces are distributed throughout the body during movement and exercise. By considering action-reaction pairs, therapists can identify potential areas of stress or imbalance and design interventions to optimize movement patterns and prevent injury.

Gravitational, Friction and Ground Reaction Forces

Gravitational Force

The force of attraction that pulls two objects in space together is termed as **gravitational force**. It is directly proportional to mass of objects and inversely proportional to distance between them. According to Universal Law of Gravitation stated by Newton, the gravitational force is directly proportional to product of mass 'M' of two objects and inversely proportional to square of distance 'r' between them that is:

$$F \propto M_1 M_1/r^2$$

Gravity

Gravity is a natural phenomenon that acts between the Earth's surface and objects in space. The force that exists between the Earth and nearby objects is known as **gravity**. Gravity operates upon every mass that constitutes an item, in contrast to other forces which are limited to acting on a particular point or region of contact. To keep things simple, we will assume that the center of mass (CoM) of an object or segment is where the gravitational force operating on it applies. CoM is a place that is determined with respect to the object, system of objects, or center of gravity (CoG). In a uniform gravitational field, mass and gravity have the same center. The geometric center of an item is where the CoM is found in symmetric objects (Fig. 1.16A). Since the CoM will be at the heavier end of an asymmetric object, the mass should be dispersed equally around it (Fig. 1.16B).

The gravitational force exerted on an item is always directed vertically downward toward the Earth's core, even though the direction and orientation of most forces depend on their source. Gravity vectors are commonly referred to as lines of gravity (LoG). A rope with a weight at the end (vertically) that is fastened to the object's CoM is the best way to conceptualize the LoG. While an object's size cannot be determined,

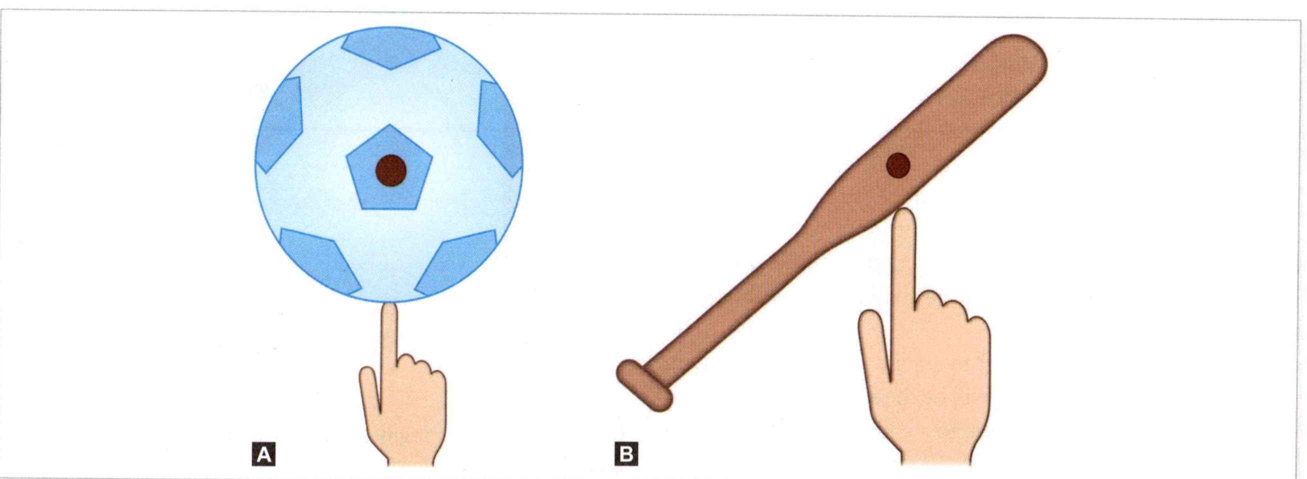

Figs 1.16A and B: Center of mass: **A.** Symmetrical object; **B.** Asymmetrical object

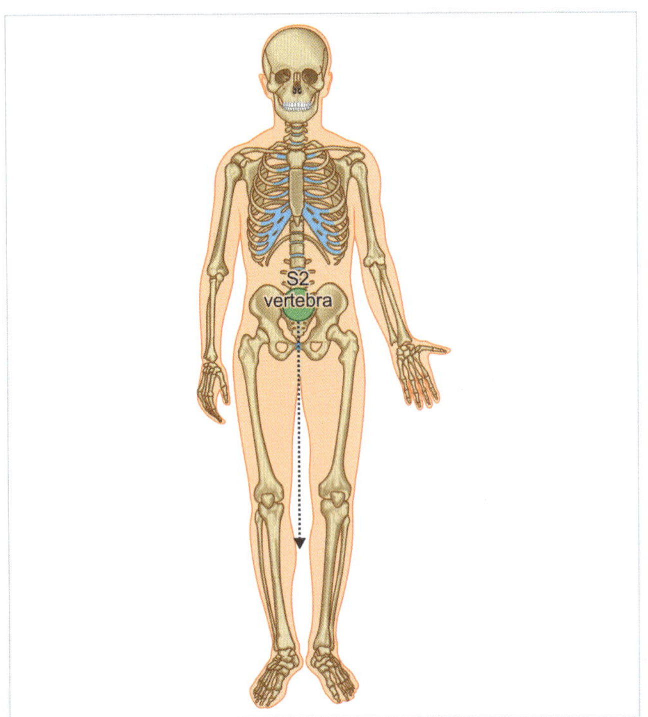

Fig. 1.17: Center of gravity in human body (green point)

Fig. 1.18: Center of mass (CoM) of the man's left leg and the right leg combine to form the new CoM for the lower limbs (denoted by blue point). The new CoM combines with the upper trunk CoM (blue point marked on trunk) to produce the CoM for the entire body (green point). The LoG from the combined CoM falls well outside the man's base of support (BoS). Thus, he is unstable and cannot maintain this position

the vertical lines displayed on its CoM provide a precise representation of the object's point of application, direction, and gravity. One can think of each body segment as having its own CoM and LoG, and can calculate the results using the same method that is explained for force configuration.

Center of Mass of Human Body

The center of mass (CoM) of the body is located roughly anterior to the second sacral vertebra when all body segments are united and regarded as a single rigid object in anatomic position (Fig. 1.17). The CoM's exact location for an individual in anatomic position is determined by their weight distribution and proportions. The LoG also varies within the body, changing relative to the base of support (BoS), which is the person's feet, and the trunk and limbs when the individual is standing in anatomic position. The LoG protruding from the CoM of the body lies perpendicular to the trunk and limbs while the individual is lying down (still in anatomic posture), as opposed to parallel when they are upright.

Biomechanically, factors like base of support (BoS), center of gravity (CoG), body mass and friction between the body and the surface in contact affect the stability of that body. In nutshell:

- BoS is directly proportional to stability, that means larger the base of support, greater will be the stability of that body.
- CoG is inversely proportional to stability that means lower the center of gravity to the BoS, more stable or balanced the body will be (Fig. 1.18).
- An object cannot be stable unless its line of gravity is located within BoS (Fig. 1.19).

Fig. 1.19: The man's segments are arranged identically to those in Figure 1.18, but by fixing his foot against the wall, he has expanded his base of support and is now stable

Clinical Correlation

It has been noted that obese people are more likely to fall because of dynamic rather than static factors. One could clarify by pointing out that, in a static scenario, individuals with obesity have no real cause to experience a balancing disadvantage as long as their center of gravity stays within the base of support. However, because of the increased body weight that must be moved and the obese person's comparatively lower mass specific lower limb power, gathering one's balance becomes even more difficult than it is for people of normal weight once their center of gravity is affected by obesity and falls farther than their base of support.

Chronic low back pain: Kim DH published a study in the Journal of Back and Musculoskeletal Rehabilitation, where it was observed that patients having chronic low back pain tend to have their center of gravities that are located excessively toward the back. In this study, patients were diagnosed with decreased low back curve, and decreased spinal extensors strength. The research group also came to the conclusion, that in order to regain postural control, those with persistent low back pain whose center of gravity is too far back, may need to overcome difficulties with strength and balance.

Friction

Friction can be explained as a resistance or drag offered by surface on body that happens to be in contact with it while moving. Friction aids in traction that is required to walk without being slipped. Friction is helpful in most cases. However, it also offers a great measure of opposition to the motion. As we have discussed in free body diagram in Figure 1.6, the friction is also being acted on object. It is important to note while creating of an object that Frictional force always acts opposite to the direction of applied force. Friction occurs because of irregularities among both surfaces happen to be in contact. So, when one body moves over another, these irregularities get entangled, which give rise to friction. More the irregularities, greater is the roughness that leads to significant friction. There are four types of friction, namely:

1. Static friction
2. Sliding friction
3. Rolling friction
4. Fluid friction

There are two forms of magnitude of frictional force: One when body is at rest (static friction), the other when in motion (kinetic friction). When the bodies are at rest, the magnitude of static friction fs is $[fs \le \mu s N]$, where μs is coefficient of static friction and N is the magnitude of the normal force (the force perpendicular to the surface).

Once body is moving, magnitude of kinetic friction fk is given by $[fk = \mu_k N]$, where μ_k is the coefficient of kinetic friction. A system where $fk = \mu_k N$ is depicted as a system where friction behaves normally.

Ground Reaction Force (GRF)

Force experienced by body (which is in contact with ground), exerted by ground, and is equivalent to body weight. It is the force that opposes the force of gravity and allows one to stand upright. The ground reaction force acts in the opposite direction of the body's motion. It is this force that propels the body forward when walking. The faster one walks, the greater the ground reaction force will be. The GRF also helps to keep the body upright. When leaning forward, the ground reaction force pushes back against the body, keeping the person from falling over.

Clinical Correlation

Having a thorough knowledge of all these forces discussed above, is useful to know how these forces together form a machine to work (generating movement) efficiently. The musculoskeletal system is a set of simple machines that work together to support loads and generate movement. In human musculoskeletal system, lever and pulley serve as simple machine, and have three main functions, which are: Changing the direction of the applied force, amplification of force and motion. However, in the musculoskeletal system, most of these machines are designed to amplify motion rather than force.

LEVERS

Lever is basically a rigid segment that rotates around an axis (fulcrum). In this system two forces are applied on a segment to produce opposing torques. In humans, motion is inherently complex, involving forces applied on one or more segments (with segments interchangeably used as bony levers). In a lever system, the force that is producing the resultant torque (the force acting in the direction of rotation) is called the **effort force** (EF) and other force must be creating an opposing torque, it is known as the **resistance force** (RF), are acting on a segment with a fulcrum, which is a point of support, or axis, about which a lever rotates. The moment arm (MA) of the RF is named as **resistance arm** (RA) and similarly, MA of EF is termed as **effort arm** (EA).

First Class Lever

A first-class lever is a lever system in which the axis lies between the point of application of the effort force and the point of application of the resistance force, without regard to the size of EA or RA. As long as fulcrum lies between points of application of EF and RF, EA can be equal to RA (Fig. 1.20A), greater than RA (Fig. 1.20B), or lesser than RA (Fig. 1.20C).

Example:

- An example of first-class lever in human body is head and neck during neck flexion and extension. The atlanto-occipital joint serves as fulcrum, which lies in between the

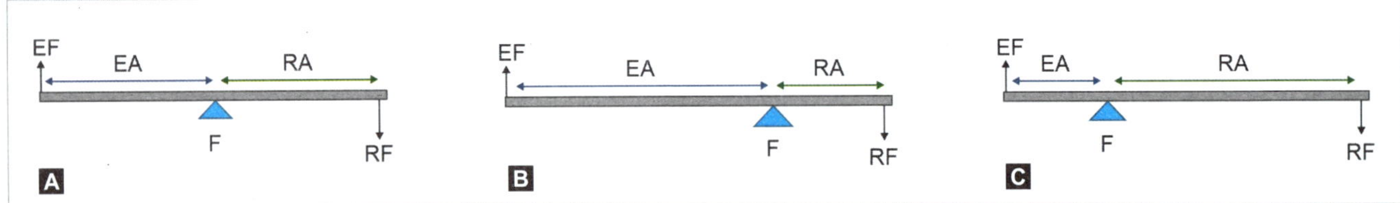

Figs 1.20A to C: First-order levers where **A.** EA = RA; **B.** EA > RA; **C.** RA > EA

front of the skull (where gravity serves as resistance) and neck extensor muscles, which provide the effective force (EF). The extensor muscles are attached to the posterior part of the skull, allowing larger effort arm. This allows the skull to nod forward, backward, and side to side, acting as first-class lever.

- A further illustration of a muscle acting on a first-class lever is the supraspinatus's pull on the humerus. Just above the shoulder joint, on the other side of the composite axis of rotation for the glenohumeral joint from the CoM of upper extremity, is where the supraspinatus attaches to the larger tubercle of the humerus. Whether the supraspinatus is contracting eccentrically (as the RF) or concentrically (as the EF), this lever still functions as a first-class lever because both the muscle and the gravitational force are located on either side of the joint axis.

Second Class Lever

In a second-class lever system, EA is always greater than RA because the resistance force's application point is situated between the axis and the effort force's application point.

Example: Jumping off the ground or pushing the block in a spring start, are the examples of second-class lever. While standing on toes, the ball of the foot serves as fulcrum, body weight acts as load and contraction of the gastrocnemius muscle provides the effective force.

Third Class Lever

When an effort force application point is consistently located between the axis/fulcrum and the resistance force application point, the mechanism is referred to as a third-class lever. Consequently, the RA consistently exceeds the EA.

Examples:

- An example of third-class lever in human body is during a biceps curl, where elbow joint serves as fulcrum, contraction of biceps provides effective force and the weight of forearm or the person may be holding, provides resistance.
- The action of quadriceps muscle on the leg-foot segment against the resistance of gravity and the weight on foot serves as a typical example.

Mechanical Advantage of Levers

The relative efficacy of the effort force relative to the resistance force is known as **mechanical advantage** or MAd, which quantifies the mechanical efficiency of the lever. The ratio of an effort arm's (MA of the effort force) to resistance arm's (MA of the resistance force) mechanical advantage is represented by this relationship, i.e.,

$$MAd = EA/RA$$

when EA is larger than RA, the MAd will be >1, hence it is said to have high mechanical advantage. In levers with high mechanical advantage, even a small amount of effort can lift large loads.

Did You Know?

Second-class levers always have high mechanical advantage since in second-order lever, EA is always greater than RA. If the fulcrum is close to load in first-class lever, large mechanical advantage can be achieved. However, in third-class levers, the MAd will always be <1 because EA is always smaller than RA. Thus, lever is at a "disadvantage" or is "mechanically inefficient" since the magnitude of effort force must always be greater than the magnitude of the resistance force for the torque of the EF to exceed the torque of the RF.

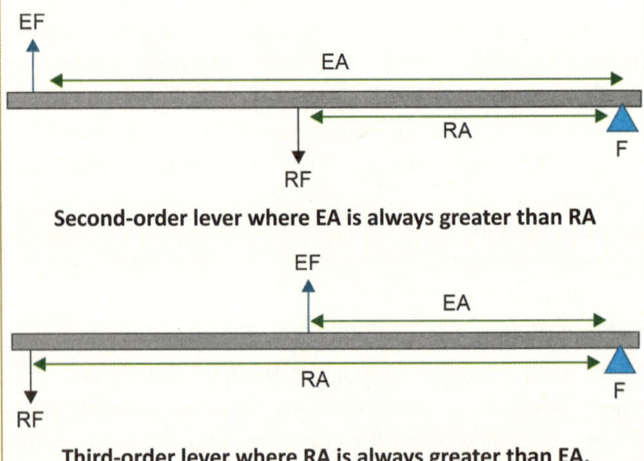

Second-order lever where EA is always greater than RA

Third-order lever where RA is always greater than EA.

Clinical Correlation

Incorporating functional training exercises that mimic real-life movements and challenges faced by the patient can help improve movement efficiency, coordination, and stability, reducing the risk of injury during daily activities or sports.

ANATOMICAL PULLEYS

A system of a wheel which carries a flexible cord or rope on its rim is known as **pulley**. The pulley's main function is to make a task easier by redirecting the force. A pulley allows the direction of lifting force to be reversed by pulling down a rope which is looped over a wheel which makes it easier to lift weight. With a two-wheel pulley, one can minimize the force applied to lift same weight. Hence, the weight can be lifted at half strength. This is known as the mechanical advantage (MAd). The bigger the mechanical advantage, the less force is required to lift a weight. There are mainly two types of pulleys: (1) Fixed pulley and (2) Movable pulley.

Clinical Correlation

The concept of anatomical pulleys enhances the mechanical advantage of muscles in the human body and has implications beyond biomechanics. Understanding the role of anatomical pulleys can also aid medical professionals in diagnosing and treating various musculoskeletal conditions.

Fixed Pulley

A fixed pulley is the one in which axle and wheel are fixed (Fig. 1.21). It changes the direction of force applied on rope which moves along its wheel. For example, while pulling out a bucket full of water from a well, one can pull sideways with a firm grip to pull it even more comfortably than pulling it vertically in usual manner. The water bucket still feels like the same weight, but it's more convenient to lift.

Movable Pulley

A movable pulley is a pulley free to move up and down which consists of an axle in movable block being supported by

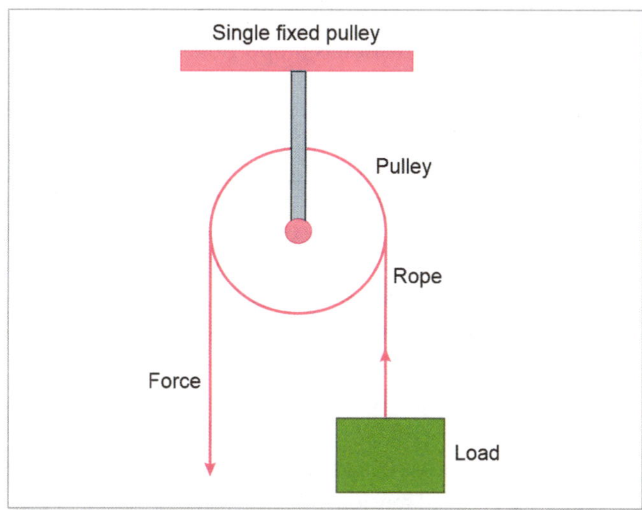

Fig. 1.21: Fixed pulley

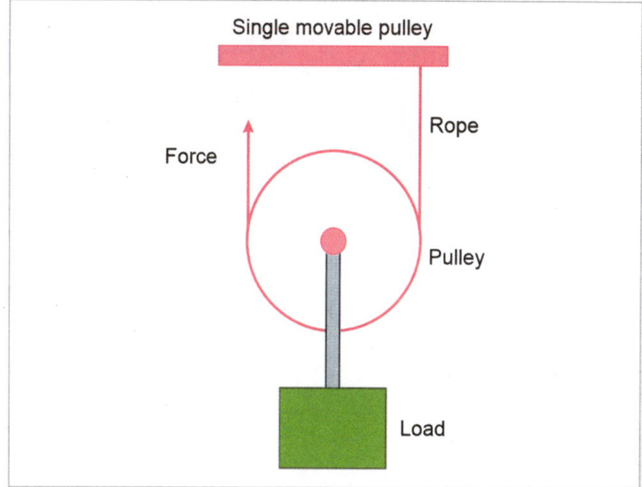

Fig. 1.22: Movable pulley

two lengths of same rope (Fig. 1.22) having a mechanical advantage of 2. The direction of applied force remains unchanged; however, the load will "feel" lighter than it was earlier.

An upgraded form of axle and wheel is **anatomical pulley**. Rotating a bodily part is the "task" associated with human movement. Anatomical pulleys facilitate the task of increasing the mechanical advantage of the mechanical force by diverting the muscle's action line from the joint axis. Often, a bony prominence will deflect or wrap around a bone in the fibers of a muscle or tendon. The bone or bony prominence causes the deflection of the moment arm of the muscle to form an anatomic pulley.

Few examples of anatomical pulleys are:

- To prevent bowstring, the flexor retinaculum of hand act as pulley for finger flexors. Without this, these muscles won't be effective enough to move these joints. Here, ligament serves as pulley.
- The angle of insertion of biceps muscle is increased because of its size increment. For the muscle which is passing over it, the muscle underneath serves as a pulley. For example, brachialis provides better angle of insertion to the biceps by raising it.
- In human eye, trochlea allows superior oblique muscle when it passes over it to rotate the eye obliquely. The muscle would have little effect on eye due to absence of this pulley.
- In the knee complex, presence of patella improves efficiency of quadriceps muscle as pulley will increase angle of insertion of the patellar ligament into the tibial tuberosity. Pulley alters the muscle action at knee joint (Fig. 1.23).
- Lateral malleolus acts as pulley for peroneus longus muscle. If it were not there, this muscle would have produced ankle dorsiflexion and eversion rather than plantar flexion and inversion. This is because, the presence of lateral malleolus changes the path of the muscle as the muscle now passes

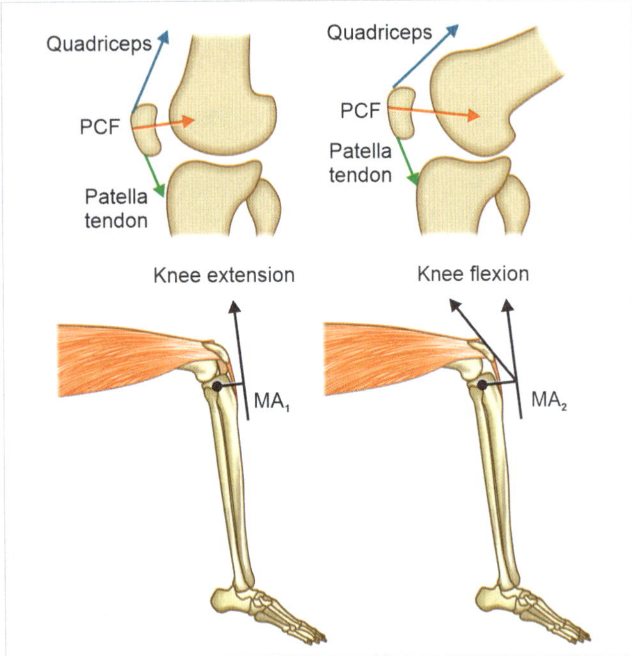

Fig. 1.23: With the patella's pulley effect, the line of pull of the muscle is deflected, increasing the MA of muscle

Abbreviations: MA, mechanical advantage; PCF, patellofemoral compressive force

behind the malleolus and inserted in the base of first metatarsal, which otherwise would have passed in front of ankle joint. So, here the pulley is this bone.

Clinical Correlation

Educating patients about the role of anatomical pulleys in maintaining musculoskeletal health can empower them to adopt preventive measures and modify activities to reduce the risk of injury or overuse.

SPRING

Spring is an elastic element that gets compressed under the application of load and regains its original form as soon as the load is being removed. In other words, it is made of a material possessing very high yield strength for restoring elasticity. It mainly helps in absorbing shocks, also resist transferring shocks and vibrations to nearby segments or structures. In a system of springs, if connected end to end, then the combination is known to be in series and if these are connected side by side, the system is known to be in parallel. In both cases, combination acts as a single spring (Figs 1.24A and B).

In humans, skeletal muscles are organized in multinucleated myofibers, whose function is to generate length and velocity dependent forces for movement or stability. As described by Bahler AS, the skeletal muscles could be organized in different components based on their function and architecture namely:

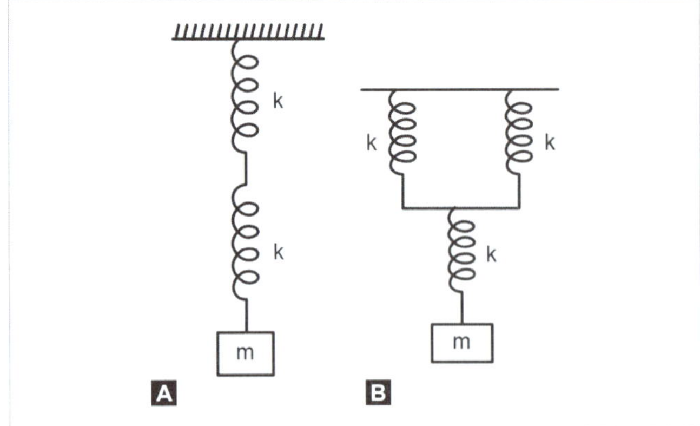

Figs 1.24A and B: Springs: **A.** In series; **B.** In parallel

- The Series Elastic Component (SEC)
- The Parallel Elastic Component (PEC)

Parallel elastic component is considered to consist of membranes covering contractile components that comprise the sarcoplasmic retinaculum, sarcolemma, the perimysium and the epimysium, while the series elastic components reside in the tendons and aponeuroses.

Deformation

Moving from consideration of forces that affect motion of object (such as friction and gravity) to those that affect an object's shape, if a hammer is struck on a metal sheet, the sheet will not move rather there will be a noticeable change in its shape. This change in shape because of applied force is called **deformation**. Even a small amount of force is sufficient to produce some deformation. For minute deformations, two important characteristics are to be observed. Firstly, the body regains its original shape after the force is removed-that means, deformation is elastic for minute deformations. Secondly, size of deformation is proportional to force.

Tension

Tension is a tensile force acting on a body that has two components namely tensile stress and tensile strain. It can also be argued that the body experiences a pull to stretch itself. When this force is applied to the body, a tension corresponding to this applied force is created, which leads to an increase in the length of the body, but a decrease in the cross-sectional area.

When there are loads (forces) or material; applied to a structure, internal resistance and internal deformation occur in the structure or material. This internal reaction to the applied force is called tension.

Stress

Stress is defined as the force per unit cross-sectional area of the material and can be expressed mathematically by the following formula,

$$S = \frac{F}{A}$$

where,

S = stress

F = applied force

A = area

Stress cannot be measured directly, but is calculated from the measured applied to the material of forces. Stress is expressed in Pascal units (N/m^2).

Strain

The relative deformation (change in shape, length or width) of a structure or material that follows stress is called **strain**. Elongation is the amount of deformation that occurs relative to the original length of the material. We cannot measure elongation directly because it is a numerical index, but we can calculate it mathematically by finding the ratio of the change in configuration to the original configuration. Mathematically:

$$\text{Strain} = \frac{(L_2 - L_1)}{L_1}$$

where,

L_1 = initial length and

L_2 = final length

Strain is a relative measure expressed as a percentage, so it has no units.

When two externally applied forces are equal and act on opposite sides of a structure in the same direction, they form a compressive load and a compressive stress, and as a result, for example, compressive stress occurs in the structure. When two external forces are equal, parallel and directed in opposite directions, but not in line with each other, they constitute a shear load. When two parallel and opposite forces are applied perpendicular to the longitudinal axis of the structure, they constitute a torsional load.

Clinical Correlation

Changes in the mechanical properties of tissues can indicate disease progression or treatment response. For example, in cardiology, alterations in the load-strain curve of myocardial tissue may indicate the onset or progression of heart failure, guiding therapeutic interventions and monitoring disease management.

Stress/Load-Strain Curve

The load-strain curve (Fig. 1.25) provides information about the elasticity, ductility, tensile strength, and stiffness of a material, as well as the amount of energy the material can store before failure. The elastic region is the curved area between points A and B. In this area, the material deformation is transient, and the structure instantly reverts to its initial dimensions upon the removal of the load. The elastic zone ends at Point B, also known as the yield point. After this, the material may recover over time but won't instantly revert to its initial state when the load is released. The plastic area is the following section of the curve that goes from B to C. When the load is removed in this area, the material deforms consistently yet the structure holds together. It is assumed that the synthesis and rearrangement of new tissue components is required for the restoration of the original structure following the removal of the stress. The material deforms until it reaches the eventual failure point C as loading moves through the plastic range. The load applied, when this point is reached, is the failure load.

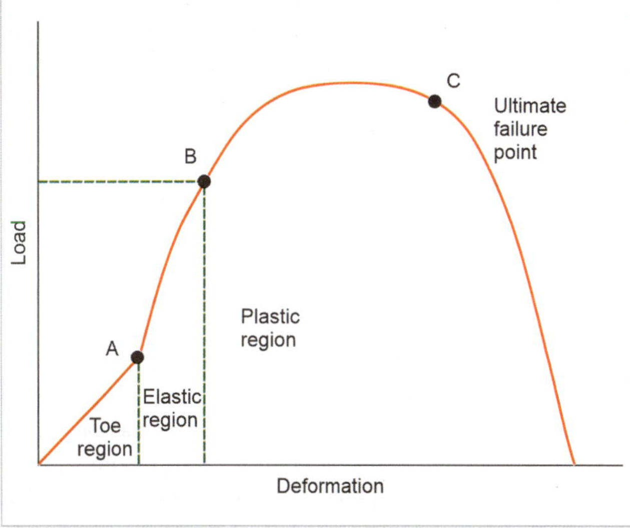

Fig. 1.25: Load deformation curve for connective tissue

Clinical Correlation

The plastic region of the curve, where deformation occurs without immediate recovery, is relevant in wound healing and tissue repair. Clinicians monitor how tissues respond to mechanical stress during rehabilitation processes. For instance, in orthopedic rehabilitation, understanding the plastic behavior of bone tissue helps in designing effective exercises to promote healing and restore function.

According to Hooke's law, a material will deform in proportion to the applied stress while yet remaining within its elastic bounds. Thus, strain and stress can be used to express the Hooke's law equation.

Excessive stress or Stress/Strain = Constant (E)

Here, E is the elastic modulus also called Young's modulus.

By studying the stress-strain relationship under various loads, the elastic properties of materials (e.g., ligaments and tendons). Stress-strain curves provide insight into the stress-strain relationship in tissues can be. Stress and corresponding strain values are displayed on a stress-strain curve.

Different regions of the stress-strain diagram:

Toe Region

The region of the stress-strain curve that follows Hooke's law. In this case, the relationship between stress and strain gives the proportionality constant "Young's modulus". In the graph, "OA" represents the toe area. In this region, large deformations can be produced with minimal force. The toe area may be the same area where the evaluator clinically tests the integrity of the ligament by applying tension or slack to the tendon that the muscle must accommodate before the tendon begins to move the bone.

Elastic Limit

The "AB" in the graph, represents the elastic region. This is the critical area where the material can regain its original shape. Figure 1.26 is example of stress-strain curve. The results are independent of tissue dimensions and therefore reflect the material from which the tissue is made.

A-B – elastic region
B-C – plastic region

Failure typically occurs at approximately 8–10% strain. The load acting on it is completely eliminated. Beyond this limit, the body cannot regain its shape and a state of plastic deformation begins within it. In this region of the curve, collagen fibrils stretch to resist the applied force.

Yield Strength

Yield strength is defined as the point at which plastic deformation begins. After the yield point, the body becomes permanently deformed and plastic. The two yield strengths that can be achieved are: (1) Upper yield strength (2) Lower yield strength.

Ultimate Stress Point

This is the point at which the body can withstand maximum stress. Loads exceeding this point will cause failure. In the plastic range (B-C), progressive destruction of collagen fibers begins and the ligament or tendon can no longer return to its original length. Plasticity can be considered a sign of microcracks. Recovery from this level of stress will take a significant amount of time because it involves, among other things, healing aspects such as the synthesis of new tissue and the cross-linking of collagen molecules. Past this point failure occurs.

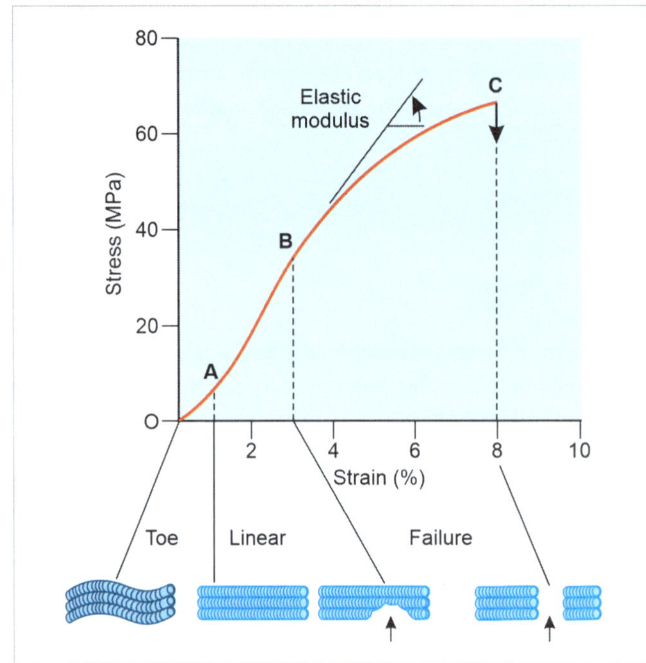

Fig. 1.26: An example of a stress-strain curve

Failure or Point of Failure

The point at which breakage/destruction of an object is achieved is called the point of failure. Collagen fibrils, under the influence of continuous forces, increase stress and continuously break down, resulting in macro-destruction of the tissue. In the case of ligaments or tendons, destruction of connective tissue due to a tear that occurs in the middle part of the structure is called a rupture. During this failure, the bone attachment of the ligament or tendon is broken resulting in an avulsion. When bone tissue breaks, it is called a fracture. Low loading rates result in delamination or fracture, while high loading rates result in intermediate material rupture.

SUMMARY

- Mechanics studies forces and their effects on objects; biomechanics focuses on forces on the human body to enable movement. Motion involves linear (translatory) and angular (rotational) types; human body movements combine both in daily activities.
- Kinesiology delves into movement, divided into mechanical (statics and dynamics) and anatomical aspects, while biomechanics explores kinetics (forces' action) and kinematics (movement description).
- Linear motion moves along a line (straight or curved), while angular motion involves rotation around a fixed axis, seen in various body segments during activities like walking.
- A three-dimensional Cartesian system uses mutually perpendicular axes and planes to locate positions in space.
- Imaginary planes like sagittal, frontal, and transverse planes divide the body into halves and intersect at its center of mass, aiding in spatial orientation.
- Three reference axes—frontal, sagittal, and longitudinal—are perpendicular to the body planes, aiding in describing human motion by their orientations.
- Body segment rotations occur around specific axes and planes (e.g., flexion, extension) while linear displacements follow Cartesian axes (x, y, z) with conventions for positive and negative directions.
- Rotatory motions quantify angular displacement (degrees or radians) with ROM measured in degrees; translatory motions measure linear displacement in meters/feet and rate of change is expressed as velocity (m/sec or ft/sec) and acceleration (m/sec^2 or ft/sec^2).
- Force induces changes in an object's state, described as a push or pull affecting mass and acceleration (F = ma), measured in Newton (N) in the SI system and pounds (lb) in the US system.
- Net force, the sum of acting forces, determines the body's motion; inertia (resistance to change), mass, weight (gravitational force on mass), torque (rotary force), momentum (mass in motion), work (force causing displacement), energy (capacity for work), and power (rate of work) all contribute to understanding physical dynamics.
- Forces act simultaneously, requiring free body diagrams to analyze effects; concepts like inertia, mass, weight, and torque impact motion and force application, while momentum is conserved, work involves force and displacement, and energy manifests in various forms with power representing work rate.
- Forces in biomechanics are categorized into external (gravity, wind, objects) and internal (muscles, connective tissue), with reactionary forces (ground reaction, joint reaction) observed.
- Composition of forces involves resolving forces into components (x, y) to determine net unbalanced forces, with linear, parallel, and concurrent force systems impacting the body's motion.
- Vector addition laws (triangle law, parallelogram law) aid in determining resultant forces from multiple forces' magnitudes and directions, crucial in understanding biomechanical movements and Newton's Laws of Motion.
- Newton's First Law (Law of Inertia) asserts objects maintain their state of motion unless acted upon by an external force, pivotal in static analysis in biomechanics and evident in scenarios like an ice skater's constant glide.
- Newton's Second Law (Law of Acceleration) links force, mass, and acceleration (F = ma), influencing sports techniques by highlighting how force impacts speed and agility, showcased in athletes enhancing leg strength for improved acceleration.
- Newton's Third Law (Law of Reaction) demonstrates that every action has an equal and opposite reaction, crucial in understanding interactions like ground reaction forces during running or impact forces in sports like tennis.
- Gravitational force is proportional to mass and inversely proportional to distance, acting downwards toward Earth's center, while the line of gravity influences stability based on center of mass (CoM) and base of support (BoS).
- Friction opposes motion due to irregularities between surfaces, exhibiting static and kinetic forms, vital for traction, with ground reaction force (GRF) opposing gravity, supporting upright posture and aiding movement.
- Levers employ effort and resistance forces around a fulcrum; first-class levers have various EA and RA scenarios, while second-class levers always position EA > RA, and third-class levers have RA > EA, influencing mechanical advantage and efficiency.
- Pulleys redirect forces; fixed pulleys change force direction, while movable pulleys decrease felt load, with anatomical pulleys enhancing mechanical advantage by altering muscle action paths around bony structures, seen in body systems like the hand, eye, knee, and ankle.
- Anatomical pulleys enhance muscle efficiency by altering angle of insertion or deflection paths, seen in structures like the flexor retinaculum in the hand, biceps muscle, trochlea in the eye, patella in the knee, and lateral malleolus in the ankle.
- Skeletal muscles encompass Series Elastic Components (SEC) in tendons and aponeuroses and Parallel Elastic Components (PEC) including membranes around contractile elements.
- Deformation involves a material's change in shape due to applied force; for small deformations, it's elastic and proportional to the force applied. Tension causes elongation, exhibiting stress (force per area) and strain (relative deformation). Load-deformation curves showcase material properties: The elastic region (reversible deformation), yield point (permanent deformation initiation), ultimate stress point (max stress endurance), and fracture point (failure). Hooke's law relates stress and strain within elastic limit, and stress-strain curves illustrate behavior under varying loads.

FURTHER READINGS

- Bahler A S. Series elastic component of mammalian skeletal muscle. American Journal of Physiology-Legacy Content. 1967 Dec 1;213(6):1560–4.
- Godfrey, Alan and Conway, Richard and Meagher, David and ÓLaighin, Gearóid. (2009). Direct measurement of human movement by accelerometry. Medical engineering and physics.
- Hall S J (2011). Basic Biomechanics (6th ed.). McGraw-Hill Education.
- Hussein E M A: Collision Kinematics. In: Hussein E M A, editor. Radiation Mechanics: Elsevier; 2007, p. 67–151.
- Hussein, E M A (2007). Collision Kinematics. In E M A. Hussein (Ed.), Radiation Mechanics (p. 67–151). Elsevier.
- Kim D H, Park J K, Jeong M K. Influences of posterior-located center of gravity on lumbar extension strength, balance, and lumbar lordosis in chronic low back pain. J Back Musculoskelet Rehabil. 2014;27(2):231–7. doi:10.3233/BMR-130442
- Levangie P K, Norkin C C. Joint Structure and function: Acomprehensiveanalysis. 4th. Philadelphia: FA. Davis Company. 2005.
- Malik S S, Malik S S. Orthopaedic Biomechanics Made Easy. Cambridge University Press; 2015 May 28.
- Pataky Z, Armand S, Müller-Pinget S, Golay A, Allet L. Effects of obesity on functional capacity. Obesity. 2014 Jan;22(1):56–62.
- Rode C, Siebert T, Herzog W, Blickhan R. The effects of parallel and series elastic components on the active cat soleus force-length relationship. Journal of Mechanics in Medicine and Biology. 2009 Mar; 9(01):105–22.
- Staurt B P. Tidy's Physiotherapy. Saunders Elsevier 2013.
- Watkins J. Fundamental biomechanics of sport and exercise. Routledge; 2014 Mar 26.

STUDENT ASSIGNMENT

LONG ANSWER QUESTIONS

1. Explain levers in detail with its practical implication in physiotherapy.
2. Define moment arm. How does it affect the ability of a force to rotate a segment?
3. Explain force and its composition. How can we determine the net force of a segment when a system of forces are acting on it?
4. A man carries a briefcase weighing 5 kg in his right hand. How does this external weight affect his balance with respect to CoG and LoG? Explain in detail.
5. Explain the stress-strain relationship in detail, including the Hooke's law.
6. Explain the Newton's Laws of Motion.
7. Give the detailed description of axes and planes.

SHORT ANSWER QUESTIONS

1. What is the mechanical advantage of a lever? How does it affect the leverage of a joint in human?
2. Define pulley. Give at least two examples of anatomical pulley.
3. What is an avulsion fracture?
4. Differentiate between stress and strain.
5. Define friction. What are its types?

MULTIPLE CHOICE QUESTIONS

1. Friction is the resistive force offered by the surface, when one surface moves over the other, which is:
   a. Directly proportional to the area of the surface in contact
   b. Nature of the surface
   c. Weight of the moving object
   d. All of the above

2. Second-order lever is the lever of_______.
   a. Stability b. Instability
   c. Speed d. Efficiency

3. Standing on toes is an example of _______ order lever.
   a. 1st b. 2nd
   c. 3rd d. 4th

4. In our body more numbers of _______ order levers are present.
   a. 1st b. 2nd
   c. 3rd d. 4th

5. It is the branch of mechanics which deals with the study of bodies in motion:
   a. Statics b. Dynamics
   c. Kinetics d. None of these

ANSWER KEY

1. d **2.** a **3.** b **4.** c **5.** b

2 Basics of Exercise Therapy

Sheetal Kalra

LEARNING OBJECTIVES

After the completion of the chapter, the readers will be able to:

- Understand the aims of exercise therapy.
- Explain the techniques used in exercise therapy.
- Understand the approach to addressing a patient's problems.
- Demonstrate assessment of patient's condition and measurements of vital parameters.
- Describe pulmonary function tests.
- Explain tests for measuring limb length.
- Describe measurement of the angle of pelvic inclination.
- Describe normal pelvic tilts, deviations of pelvic tilt such as anterior tilt (forward), posterior tilt (backward) and lateral tilt.
- Explain various starting positions, including fundamental positions, their muscle work, effect and uses.
- Explain various derived positions.

CHAPTER OUTLINE

- Introduction
- Aims
- Techniques Used in Exercise Therapy
- Patient-Centric Approach
- Patient Assessment—Physiotherapy Perspective
- Vital Signs
- Pulmonary Function Tests
- Leg Length Measurement
- Pelvic Tilt
- Fundamental Positions
- Derived Positions

KEY TERMS

Flexibility: It is the ability of the muscles, joints and soft tissues to move unrestricted through a pain-free range of motion. It allows for smooth and efficient movement.

Proprioception: Proprioception refers to awareness of the position and movement of the body. It involves conscious or unconscious awareness of joint position, movement or kinesthesia and actions of parts of the body.

Quality of life (QoL): WHO defines (QoL) as an individual's perception of his/her position in life in the context of the culture and value systems in which he/she lives and in relation to his/her goals, expectations, standards and concerns.

INTRODUCTION

Exercise therapy, is a type of treatment that uses movements and physical exercises to improve and restore a person's physical function, mobility, strength, flexibility, and overall well-being. Exercise therapy is commonly recommended by healthcare experts, such as physical therapists, to address a wide range of musculoskeletal, neuromuscular, cardiovascular, and pulmonary disorders. The main goal of exercise therapy is to help individuals recover from injuries, manage chronic conditions, prevent future injuries, and enhance their physical performance. It can be customized to meet the specific needs and capabilities of each individual, making it a highly personalized approach to rehabilitation and wellness.

Exercise therapy typically involves a variety of exercises and techniques, which may include stretching, strengthening, endurance training, balance and coordination, functional exercises, etc. Exercise therapy can benefit in a number of conditions, like orthopedic injuries, neurological disorders, cardiovascular diseases, respiratory issues, chronic pain, and sports-related injuries. The specific exercises and intensity will vary depending on the individual's condition, age, fitness level, and treatment goals.

Did You Know?

Exercise therapy is not just about physical rehabilitation. It can also play a crucial role in mental health. Engaging in regular physical activity through exercise therapy has been shown to reduce symptoms of depression, anxiety, and stress, while also improving mood and overall well-being. This holistic approach highlights the interconnectedness of physical and mental health, making exercise therapy a valuable tool for promoting comprehensive wellness.

AIMS

The purpose of exercise therapy is to achieve various physical, functional, and psychological goals to improve an individual's overall health and well-being. The specific aims can vary depending on the person's condition and goals. The following are some common aims of exercise therapy:

- **Pain relief:** Exercise therapy can assist in reducing pain brought on by musculoskeletal injuries, ongoing illnesses, and other physical disorders. Cardiovascular exercises, stretching, and strengthening routines might encourage the release of endorphins, which have analgesic properties.
- **Increases range of motion and improved flexibility:** Exercise therapy increases the range of motion and improved flexibility of muscles and joints. These are stretching exercises benefits the muscles and joints, which also help to reduce stiffness and increase mobility.
- **Enhance muscular fitness:** By focusing on particular muscle groups, it helps to develop the muscular fitness like strength and endurance. Injuries can be avoided in the future and overall physical performance can be improved.
- **Enhance cardiovascular fitness:** Activities that improve heart and lung function, such as walking, jogging, cycling or swimming, can enhance cardiovascular fitness.
- **Enhance balance and coordination:** Certain exercises can help individuals become more coordinated and balanced, which lowers their chance of falling and increases stability while doing everyday tasks.
- **Encourage functional independence:** Functional exercises are designed to assist people regain or increase their independence in doing daily chores. They do this by simulating real-life activities.
- **Accelerate recovery from injuries:** Exercise therapy can expedite the healing process after injuries by promoting circulation, reducing inflammation, and maintaining joint and muscle integrity.
- **Manage chronic conditions:** Regular exercise can be beneficial for individuals with chronic conditions like arthritis, diabetes, or heart disease. It can help manage symptoms, improve overall health, and enhance quality of life.
- **Prevent future injuries:** Strengthening and conditioning the body through exercise therapy can help prevent injuries, particularly in athletes and individuals involved in physically demanding activities.
- **Enhance mental well-being:** Exercise has been shown to have beneficial impact on mental health, reducing stress, anxiety, and depression while promoting a sense of well-being and increased self-esteem.
- **Support weight management:** Regular physical activity can assist in weight management by burning calories and improving metabolism.
- **Improve overall quality of life:** By achieving the above aims, exercise therapy can lead to an improved overall quality of life, allowing individuals to engage in activities they enjoy and participate more fully in daily life.

TECHNIQUES USED IN EXERCISE THERAPY

Exercise therapy encompasses a wide range of techniques and approaches tailored to individual needs and conditions. The following are some common techniques used in exercise therapy:

- **Stretching exercises:** Various stretching techniques are employed to improve flexibility and increase the range of motion in muscles and joints, functional movements and sports performance.
- **Muscle strengthening exercises:** These exercises target specific muscle groups to improve strength and endurance. Resistance training, weightlifting, bodyweight exercises, and the use of resistance bands or weights are examples of strengthening techniques.

- **Cardiovascular exercises:** Activities that elevate the heart rate and improve cardiovascular fitness, such as walking, running, cycling, swimming, and aerobics, are often prescribed.
- **Balance and coordination improving exercises:** Techniques that focus on improving balance and coordination, often through exercises that challenge stability and proprioception.
- **Core stabilization exercises:** Exercises designed to strengthen the muscles of the core, including the abdomen, lower back, and pelvic floor, to improve posture and support the spine.
- **Functional exercises:** These exercises mimic real-life activities to help individuals regain or enhance their ability to perform daily tasks and functional movements effectively.
- **Endurance training:** It is characterized by the muscle contractions to lift or lower a light weight for multiple repetitions or by sustaining a muscle contraction for an extended period.
- **Isometric exercises:** These exercises involve muscle contractions without joint movement, which can help improve strength and stability in specific positions.
- **Plyometric exercises:** Explosive and rapid movements that enhance power, agility, and neuromuscular coordination.
- **Hydrotherapy:** Therapeutic exercises performed in water to reduce joint impact and provide resistance, commonly used for rehabilitation purposes.
- **Postural re-education:** Techniques to improve posture and body mechanics, especially in cases of poor posture or imbalances.
- **Neuromuscular re-education:** Exercises that retrain the brain and muscles to work together effectively, commonly used in cases of neurological injuries or conditions.
- **Flexibility and mobility training:** A combination of stretching and mobility exercises to improve joint flexibility and function.
- **Relaxation and breathing techniques:** These techniques are employed to reduce stress and promote relaxation, which can aid in pain management and overall well-being.
- **Balance training:** Specific exercises that challenge the balance system to improve stability and reduce the risk of falls.
- **Gait training:** Techniques to improve walking patterns and restore normal gait mechanics.
- **Manual therapy:** Hands-on techniques performed by a physical therapist to improve joint mobility, reduce muscle tension, and promote healing.
- **Proprioceptive training:** Exercises to enhance proprioception (the body's sense of position and movement) and improve body awareness.

PATIENT-CENTRIC APPROACH

The approach to address patients' problems in exercise therapy should be comprehensive, individualized, and evidence-based. Healthcare professionals, such as physical therapists, typically follow a systematic approach to address patients' needs effectively. Here are the key elements of a patient-centered approach in exercise therapy:

1. **Assessment and evaluation:** Start by conducting a thorough assessment of the patient's condition, medical history, functional restrictions, and therapeutic goals. Perform physical examinations, tests, and screenings as necessary to understand the underlying issues.
2. **Individualization:** Recognize that each patient is unique and may have different needs, capabilities, and preferences. Tailor the exercise therapy program to suit the patient's specific requirements and take into account any contraindications or limitations.
3. **Goal setting:** Collaborate with the patient to establish realistic and achievable short- and long-term goals. These goals should be specific, measurable, attainable, relevant, and time-bound (SMART).
4. **Evidence-based practice:** Base the exercise therapy program on current researches, best practices, and clinical guidelines. Utilize evidence-based interventions that are proven to be effective for similar conditions.
5. **Gradual progression:** Implement a progressive exercise plan, gradually increasing the intensity, duration, and complexity of exercises as the patient improves. This approach helps prevent overexertion and reduces the risk of injury.
6. **Holistic approach:** Consider the patient's overall health and lifestyle factors that may impact his/her exercise therapy outcomes. Address any psychosocial or environmental barriers that could make it difficult for the patient to follow the regimen.
7. **Safety first:** Prioritize the safety of the patient during exercise therapy. Supervise the exercises, provide appropriate instructions, and be prepared to modify the program, if any issues or adverse effects arise.
8. **Communication:** Maintain open and clear communication with the patient to address any concerns, answer their queries, and provide guidance. Effective communication enhances the patient's understanding and compliance with the exercise program.
9. **Education:** Educate the patients about their condition, the rationale behind the prescribed exercises, and the expected benefits of the therapy. This empowers the patients to actively participate in their rehabilitation and take ownership of their health.
10. **Motivation and support:** Encourage and support the patients throughout their exercise therapy journey.

Positive reinforcement, praise, and acknowledging progress can help keep the patients motivated and engaged.

11. **Monitoring and adjustments:** Regularly assess the patient's progress and reassess his/her condition. Adjust the exercise therapy program as needed to accommodate changes in the patient's condition or goals.
12. **Continuity of care:** It is important to ensure smooth transition of care, if necessary, between different healthcare providers or settings. Collaboration with other healthcare professionals involved in the patient's care is essential for a holistic and coordinated approach.

PATIENT ASSESSMENT—PHYSIOTHERAPY PERSPECTIVE

The assessment process conducted by a physiotherapist is a systematic and comprehensive approach to understanding the patient's condition, identifying physical impairments, functional limitations, and treatment needs. Here is a step-by-step process of how a physiotherapist assesses a patient.

Subjective Evaluation

Patient Interview and History Taking

- Begin by introducing yourself and establishing rapport with the patient.
- Gather information about the patient's chief complaint, medical history, past injuries, surgeries, and any relevant pre-existing medical conditions.
- Ask about the onset of symptoms, factors that aggravate or alleviate the problem, and any previous treatments or interventions.
- Further explore the patient's current condition and functional limitations through open-ended questions. Inquire about the impact of the problem on their daily activities, work, and recreational pursuits.
- Assess the patient's goals and expectations from physiotherapy.

Objective Examination

- Conduct a physical examination to assess the patient's posture, joint mobility, muscle strength, flexibility, balance, coordination, and overall functional abilities.
- Use standardized assessment tools and tests, as well as specific measurements, to quantify the patient's impairments and functional limitations.

Vital Parameters

Vital signs include vital physiological functions that are monitored to detect or track health issues in medical settings or at home. Blood pressure, respiration rate, pulse rate, and body temperature are usually among them.

Diagnostic Tests and Investigations

If needed, perform specific special tests or use diagnostic tools, such as imaging (e.g., X-rays, MRI) or laboratory tests, to aid the diagnosis and understanding of the patient's condition.

Assessment

Clinical Reasoning

- Analyze the information gathered from the subjective and objective examination to develop a clinical hypothesis and potential causes of the patient's problem.
- Identify any red flags or signs that may require immediate medical attention or further investigation.

Diagnosis

- Based on the clinical reasoning and examination findings, provide a clear and accurate diagnosis of the patient's condition.
- The diagnosis should be specific, yet comprehensive, and it should guide the subsequent treatment plan.

Plan

Goal Setting

Set realistic and achievable treatment goals. These goals should be based on the diagnosis and the patient's functional limitations and expectations.

Treatment Planning

- Develop an individualized treatment plan that includes specific interventions, exercises, modalities, and techniques to address the patient's impairments and achieve the desired goals.
- Consider the patient's preferences, lifestyle, and any other relevant factors in formulating the treatment plan.

Patient Education

- Educate the patient about his condition, treatment plan, and the importance of active participation in his rehabilitation.
- Provide instructions on proper body mechanics, home exercises, and self-management strategies to enhance the patient's recovery.

Treatment Implementation

- Initiate the treatment plan, which may include therapeutic exercises, manual therapy, electrotherapy, hydrotherapy, and other modalities as deemed appropriate.

- As the treatment progresses, evaluate how patient is responding to the treatment and make any necessary adjustments to the plan.

Reassessment and Progress Tracking

- Regularly reassess the patient's condition and functional abilities to track progress toward the treatment goals.
- Modify the treatment plan accordingly, taking into account changes in the patient's condition or goals.

Documentation

- Maintain accurate and thorough documentation of the assessment findings, treatment plan, interventions provided, and the patient's progress.
- Documentation is essential for communication with other healthcare professionals and for insurance purposes.

VITAL SIGNS

Rapidly identifying changes in vital signs generally coincides with quicker recognition of changes in a patient's cardio-pulmonary state, allowing for prompt escalation of therapy, if necessary. Vital sign readings are heavily used in physiotherapy examinations, serving as critical checkpoints to detect possible problems and steer treatment. These metrics are referred to as "vital" because of their foundational importance in initiating and guiding clinical examinations. They serve as crucial indicators of an individual's overall health state, confirming the proper functioning of the circulatory, respiratory, neurological, and endocrine systems when within normal ranges.

Blood Pressure

Blood pressure or the force exerted by circulating blood on artery walls, is measured in two ways—systolic (during peak heart contraction) and diastolic (between heartbeats). Due to the impracticality of direct intra-arterial measurement, it is typically recorded as systolic over diastolic values. Over the last two decades, oscillometry-based automated equipment have replaced traditional auscultation with a stethoscope and sphygmomanometer. Blood pressure (BP) is often measured at the brachial artery and remains an important element of health monitoring.

Body Temperature

Body temperature is important for therapeutic decisions, which in healthy adults, hovers around 98.6°F or 37.0°C, fluctuating between 97.7°F and 99.5°F (36.5°C and 37.5°C). Each method has various advantages and accuracy levels when measured at sites such as the oral, rectal, axillary, and tympanic utilizing electronic or infrared equipment. Oral readings from the sublingual pocket are simple and reliable, whereas rectal measurements, while inconvenient, are the gold standard for accuracy. Temperature changes are substantially influenced by the time of day, circadian rhythms, and individual characteristics such as age and fitness. Understanding these subtleties is critical for making accurate clinical assessments and interpretations.

Pulse Rate

The radial, ulnar, brachial, posterior tibialis, dorsalis pedis, femoral, and carotid pulses are all common pulse locations, with the radial pulse being the most commonly measured. Rate (60–100 beats/min. in adults), rhythm regularity, loudness, and symmetry are critical assessment parameters. Irregularities such as sinus arrhythmia, which is characterized by rate variations during breathing, or unusually irregular patterns may indicate specific disorders such as atrial fibrillation. Checking radial and femoral pulses at the same time detects abnormalities such as aortic coarctation, while measuring pulse volume indirectly predicts systolic blood pressure. Symmetry tests aid in the detection of problems such as aortic dissection.

Respiration

The respiratory rate/breathing rate is defined as one breath for each passage of air in and out of the lungs. Adults' respiratory rates range from 12 to 20 breaths/min. Abnormal rates—tachypnea (>20 breaths/min.) or bradypnea (12 breaths/min.)—can be caused by physiological variables such as exercise or pathological disorders such as pneumonia, asthma, or anxiety states. Hyperpnea (increased depth), hyperventilation (both rate and depth), and hypoventilation (decreased rate and depth) are all examples of depth fluctuations seen in a variety of circumstances ranging from exercise to metabolic derangements. Changes in breathing patterns, such as Cheyne-Stokes, or Kussmaul breathing, indicate a variety of diseases ranging from high intracranial pressure to diabetic ketoacidosis. Congestive heart failure is indicated by orthopnea (difficulty lying flat), but paradoxical ventilation occurs with diaphragmatic paralysis or chest trauma.

PULMONARY FUNCTION TESTS

When there are risk factors for lung disease, occupational exposures, or pulmonary toxicity, assessment of patients' respiratory function using pulmonary function tests (PFTs) becomes necessary.

The PFTs also enable doctors to gauge how serious a patient's pulmonary condition is, monitor it over time, and evaluate how well it is responding to treatment.

Various lung function tests:

- **Spirometry:** The amount of air one can inhale and exhale is measured by spirometry. Additionally, it calculates how much air is present in the lungs.
- **Body plethysmography or lung volumes:** The varied amounts of air in the lungs at various moments after inhaling and exhaling, are measured by lung volumes or body plethysmography.
- **Gas diffusion study:** How much oxygen and other gases go from the lungs to the blood is determined by a gas diffusion study.
- **Respiratory muscle pressure measurement**
- **Bronchoprovocation testing**
- **Cardiopulmonary exercise testing (CPET):** A CPET evaluates the efficiency of heart, lungs, and muscles during exercise.

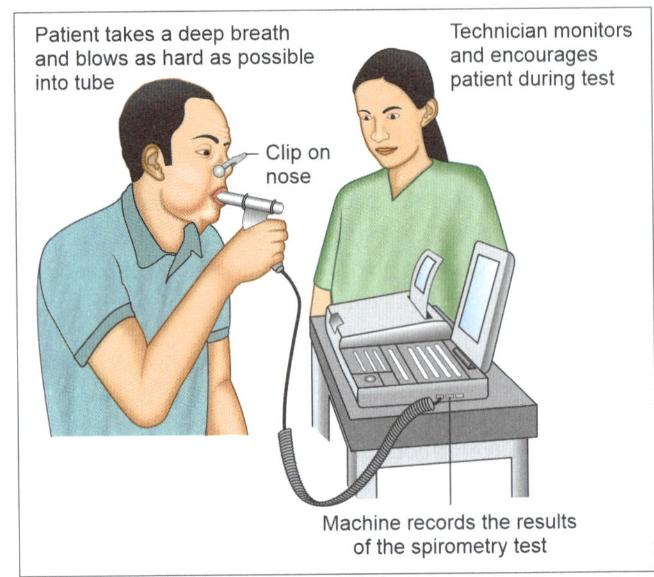

Fig. 2.1: Spirometry procedure

Spirometry

- **Principle:** Spirometry is a test that checks how well someone can breathe in and out over time. Asthma and chronic obstructive pulmonary disease (COPD) are two common conditions for which spirometry is a diagnostic test. Additionally, it helps track how different respiratory disorders are developing. Spirometry produces three primary outcomes forced vital capacity (FVC), forced expiratory volume in the first second (FEV_1), and the ratio of FEV_1 to FVC. These measurements provide important information about lung function and respiratory health.
- **Procedure:** Spirometry is a three-phase process maximal inspiration, a "blast" of expiration, and continued complete exhalation to the end of the test.

The spirometry procedure is often carried out while seated normally (Fig. 2.1). However, in some neuromuscular disorders, supine spirometry measurement may be necessary.

Clinical Correlation

Lying down can result in lower FVC and FEV_1. In patients with spinal cord damage at T6 and above, indicating an increased risk of breathing issues during sleep.

Lung Volumes

- When calculating lung volumes, important factors such as functional reserve capacity (FRC), vital capacity (VC), slow vital capacity (SVC), expiratory reserve volume (ERV), and residual volume (RV) are taken into account. When variations in FVC are seen in spirometry, these measurements are critical.
- **Functional reserve capacity (FRC):**
  - FRC is the amount of gas remained in the lungs following a normal exhale during normal breathing.
  - Additionally, ERV and RV add up to FRC. All other volumes can be calculated after the FRC has been determined.
- **ER and RV:**
  - ERV is the maximum amount of gas exhaled following end-inspiratory tidal breathing. After a maximum exhalation, the gas volume in the airways is known as RV. The volume of gas released between full inspiration and residual volume is known as VC. The FVC is comparable, but the patient exhales quickly and forcefully.
  - RV and ERV are supplemented by tidal volume (TV) and inspiratory reserve volume (IRV). The maximum inhaled volume during normal breathing is represented by IRV, whereas the usual amount of air breathed or expelled every breath is represented by TV.
- **SVC:**
  - Slow vital capacity (SVC) is the amount of air exhaled following a 15-second calm expiration from complete inhalation.
  - It is useful when FVC is low and there is airway blockage.
- **TLC:**
  - Total lung capacity (TLC) denotes the amount of air in the lungs at maximum inhalation.
  - TLC, which is calculated by adding RV and VC or FRC and IC, is the gold standard for detecting restrictive lung disorders.
  - A TLC <80% of the anticipated value implies a restrictive ventilatory dysfunction.
- **Procedure:**
  - Lung volumes are estimated using both **body plethysmography and gas dilution techniques** (nitrogen washout or inert gas dilution).

- Gas dilution employs nitrogen or helium, inert gases that are poorly soluble in lung tissues, allowing the FRC to be calculated after the equilibrium inhaled gas is evacuated.
- In body plethysmography, subjects breathe against a shutter valve in a body box, and FRC is calculated using Boyle's Law based on gas volume and pressure.

Clinical Correlation

Individuals with obstructive lung disease frequently have higher FRCs when measured using body plethysmography rather than gas dilution methods.

Diffusion Capacity

- Gases that diffuse over the alveolar-capillary membranes are the subject of diffusion research. Carbon monoxide (CO) is used in its measurement to determine the pulmonary diffusion capacity and it is called Diffusing Capacity of the Lung for Carbon Monoxide (DLCO). The classic single-breath diffusion technique is the most widely used. Milliliters per minute per mm Hg (mL/min./mm Hg) is the unit of measurement (Fig. 2.2).
- Spirometry and lung volumes are used in conjunction with the DLCO to interpret results. Obesity, intrapulmonary hemorrhage, and high DLCO are all related. When pulmonary vascular disorders, early ILD, or emphysema are present, normal spirometry and lung volumes with low DLCO may also be present. Low DLCO and an obstructive ventilatory defect point to lymphangiomyomatosis or emphysema.

Respiratory Muscle Pressures

- The respiratory muscle strength is measured using the maximal inspiratory pressure (MIP) and maximal expiratory pressure (MEP) (Fig. 2.3). The diaphragm and other inspiratory muscles' strength is revealed by the MIP, whereas the abdominal and other expiratory muscles' strength is revealed by the MEP. Three measurements of MIP and MEP are taken, and the maximum value is then reported. Normal values:
  - For individuals aged 18–65, the Maximum Inspiratory Pressure (MIP) should be under –90 cm H_2O in men and –70 cm H_2O in women.
  - For those over 65, it is recommended to be <–65 cm H_2O in men and –45 cm H_2O in women.
  - Meanwhile, the Maximum Expiratory Pressure (MEP) in men should exceed 140 cm H_2O and in women, surpass 90 cm H_2O.
  - MEP below 60 cm H_2O may indicate challenges in coughing and clearing secretions.

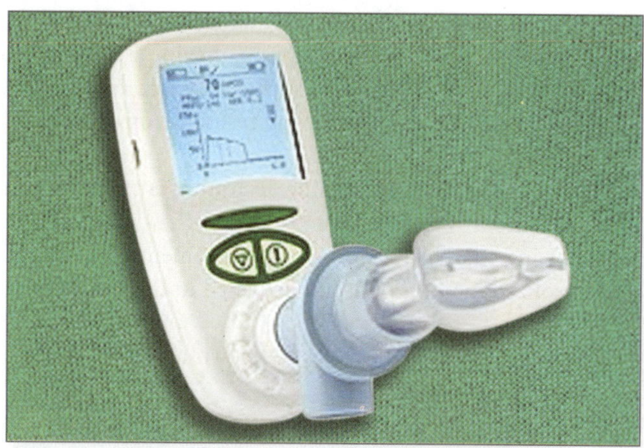

Fig. 2.3: Respiratory muscle pressures meter

Bronchoprovocation Testing

- Bronchoprovocation testing is used to assess airway responsiveness and diagnose asthma, using techniques such as pharmacological or exercise challenges (Fig. 2.4).

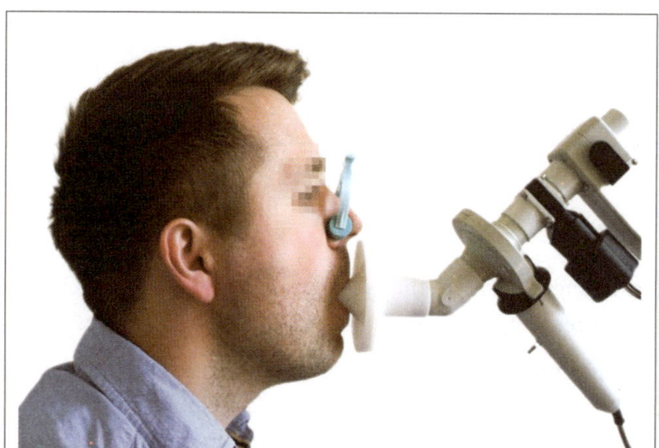

Fig. 2.2: Measurement of diffusion capacity of the lungs

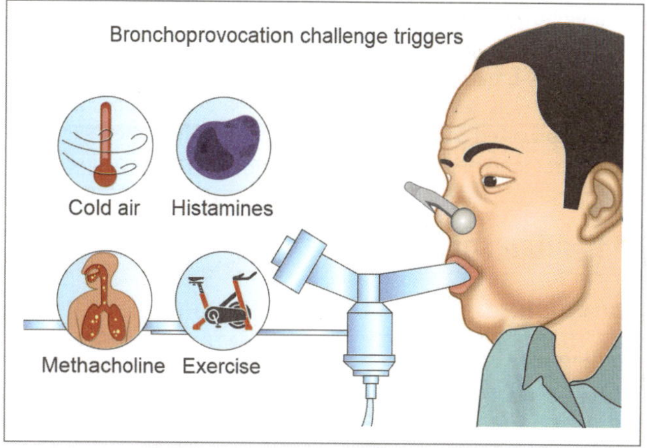

Fig. 2.4: Bronchoprovocation testing

- Precautions must be taken prior to doing this test, including recognizing severe bronchospasm, having rapid-acting bronchodilators on hand, ensuring FEV has returned to baseline after the test, and having resuscitation equipment and qualified people on hand.
- Methacholine, an acetylcholine derivative, is often utilized in bronchoprovocation tests.
- Following baseline spirometry, incremental doses of aerosolized methacholine are delivered until FEV_1 falls by >20%.
- The dose that causes this drop, known as PD20, denotes airway hyperresponsiveness.
- A PD20 of 8 mg/mL or less is regarded positive for airway hyperresponsiveness, whereas a PD20 >16 mg/mL is deemed negative.

Cardiopulmonary Exercise Testing

- Involves applying an external workload to the cardiovascular and musculoskeletal systems using a treadmill or cycle ergometer.
- Basic exercise tests involve patients exerting effort while noninvasive measurements are taken, such as measuring heart rate and rhythm with an electrocardiogram (ECG).
- Other simple, noninvasive measurements are blood pressure and respiratory rate monitoring.
- Analysis of exhaled gas is noninvasive, but the patient has to breathe through a mouthpiece or mask.
- Ventilation and tidal volume (VT) can be estimated by collecting the exhaled air.
- Analysis of expired gases permits oxygen consumption and CO_2 production to be measured.
- When invasive measures (blood gas analysis, arterial catheters, pulmonary artery catheters) are used, the entire range of physiologic variables that affect exercise can be monitored.

Clinical Correlation

Computerized exhaled gas analysis allows sophisticated measurements to be made rapidly while the patient continues to exercise (breath-by-breath gas analysis). Simple exercise tests, such as the 6-minute walk test (6 MWT), has become popular because the distance walked correlates well with more sophisticated exercise measurements and with clinical outcomes in a variety of diseases.

Indications of PFT

Spirometry

To diagnose lung disease:

- Based on pulmonary symptoms like dyspnea, wheezing, cough or chest discomfort
- Identification of physical signs or abnormalities.
- Support from abnormal laboratory findings or imaging studies.
- Prior to engaging in vigorous physical activities.

To measure lung function in existing diseases:

- Chronic obstructive pulmonary disease (COPD), asthma, cystic fibrosis, interstitial diseases.
- Assessment in cardiac or neuromuscular diseases.

Assess effects of exposure: For smokers or those in hazardous work environments.

Evaluate therapy impact:

- Monitor effects of medications like bronchodilators, steroids or cardiac drugs.
- Assess outcomes post-lung surgery or rehabilitation.

Assess risks for surgical procedures:

- Prior to lung or thoracic surgeries.
- For upper abdominal procedures.

Evaluate disability or impairment:

- For social security, compensation or legal purposes.
- As part of cardiopulmonary rehabilitation assessment.

In epidemiologic or clinical research: Participation in studies related to lung health or disease.

Lung Volumes

Diagnosing or assessing restrictive lung diseases (reduced TLC): Distinguishing between obstructive and restrictive disease patterns.

Assessing responses to therapies such as:

- Medications like bronchodilators, steroids.
- Interventions like lung transplantation, resection, radiation or chemotherapy.
- Preoperative assessment for patients with compromised lung function.
- Evaluating hyperinflation extent.
- Assessing gas trapping by comparing different lung volume measures.
- Standardizing other lung function measures (e.g., specific conductance).

DLCO Measurements

Diffusing capacity of the lungs for carbon monoxide (DLCO) measurements are recommended for various purposes.

Evaluating or monitoring progress of parenchymal lung diseases: Due to exposure to dusts (asbestos, silica, metals), organic agents or drugs.

Assessing pulmonary involvement in systemic diseases:
- Rheumatoid arthritis, sarcoidosis, systemic lupus erythematosus (SLE), systemic sclerosis, mixed connective tissue disease

Evaluating obstructive lung diseases:
- Monitoring disease progression in emphysema, cystic fibrosis
- Distinguishing types of obstruction (emphysema, chronic bronchitis, asthma)
- Predicting arterial desaturation during exercise in COPD

Assessing cardiovascular diseases:
- Primary pulmonary hypertension, pulmonary thromboembolism, pulmonary edema, congestive heart failure
- Quantifying disability associated with interstitial lung diseases
- Evaluating conditions like pulmonary hemorrhage, polycythemia, or left-to-right shunts (increased DLCO)

Exercise Testing

Documenting or diagnosing exercise limitations due to fatigue, dyspnea or pain:
- Assessing exercise intolerance or fitness level
  - Cardiovascular diseases:
    - Myocardial ischemia, dyskinesis
    - Cardiomyopathy, congestive heart failure
    - Peripheral vascular disease
    - Evaluating candidates for heart transplantation
  - Pulmonary diseases:
    - Airway obstruction (including cystic fibrosis), hyperreactivity
    - Interstitial lung disease, pulmonary vascular disease
  - Conditions with mixed cardiovascular and pulmonary causes
  - Unexplained dyspnea

Conducting exercise assessments for cardiac or pulmonary rehabilitation:
- Monitoring exercise-related desaturation/hypoxemia
- Determining oxygen requirements
- Assessing preoperative risks, especially for lung resection or reduction
- Evaluating disability, particularly related to occupational lung diseases
- Evaluating outcomes post-therapeutic interventions like heart or lung transplantation

LEG LENGTH MEASUREMENT

Need

Leg length inequality is frequently linked to compensatory gait errors and may cause lumbar spine and lower extremities degenerative arthritis. Leg length discrepancy (LLD) patients frequently demonstrate not only soft tissue contractures in the same or opposite limb, but also angular and torsional deformities that affect the functional length of their legs. Flexion contractures at the knee and hip, for example, can visually shorten the leg, but abduction contractures near the hip and equinus deformities near the ankle tend to increase the functional length of the affected limb. In addition to clinical assessment, a number of imaging methods have been described to measure LLD. To effectively treat a patient, it is crucial to apply the proper clinical techniques and imaging modalities for assessing the LLD.

Leg Length Discrepancy

There are two different discrepancies in leg length.

1. **True leg length discrepancy or true shortening:** It is brought by a structural or anatomical change in lower leg as a result of trauma (such as a fracture) or congenital maldevelopment (such as adolescent coxa vara, congenital hip dysplasia, or bony abnormalities). The outcome is an anatomical short leg, which frequently affects the spine and pelvis, causing lateral pelvic tilt and scoliosis (Fig. 2.5A).
2. **Functional or apparent leg length discrepancy:** It is also known as functional shortening, is the second form of leg length discrepancy. It is the outcome of compensating for a change that may have happened due to posture rather than structure. When measuring leg length from the umbilicus (Fig. 2.5B) for instance, a functional leg length discrepancy may be present due to unilateral foot pronation, spinal scoliosis, or pelvic deformities.

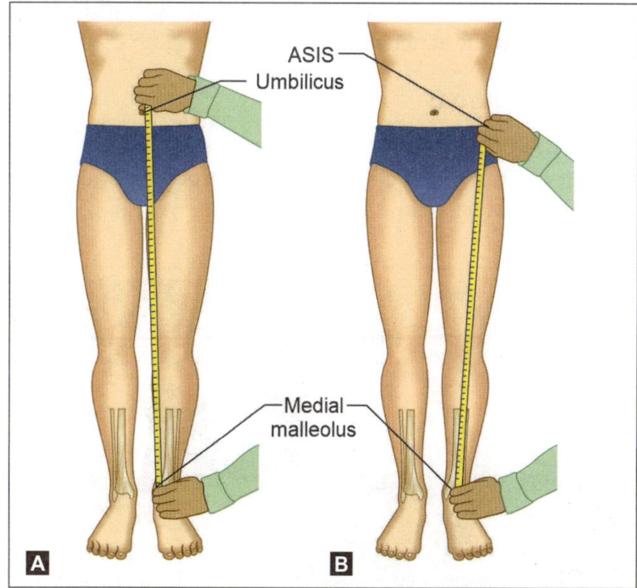

Figs 2.5A and B: Direct tape measurement to assess leg length. **A.** Apparent leg length; **B.** True leg length
Abbreviation: ASIS, anterior superior iliac spine

Measurement

Tape Method

- **Procedure:** The patient is positioned supine on the examination table with the pelvis level [with the anterior superior iliac spine (ASIS) in a straight line and the lower extremities parallel to the table] in order to estimate the genuine leg length inequality (Fig. 2.6).
- The symmetrical placement of the legs should be parallel to one another and spaced roughly 10–20 cm apart.
- On either side, measurements can be taken from the ASIS to the medial malleolus. Most patients often present with an acceptable 1–2 cm difference in leg length. If there is a difference in leg length from the:
  - The ASIS to the greater trochanter
  - The greater trochanter to the knee joint
  - The knee joint to the medial malleolus.

Comparing these measurements with the contralateral side to determine the location of the discrepancy.

Measurements can be made from a fixed location in the body's center, such as the umbilicus or xiphoid process, to analyze apparent leg length disparities.

Indirect Method

- **Procedure:** Standing on blocks, placing on known height beneath the heel of the short leg to level the pelvis allows "indirect" measurement of leg length discrepancy.
- It is another way to test LLD. This method uses a variety of block heights to determine the extra length required for the patient to obtain a sense of balance, taking into account the difference in foot height between the two limbs.
- It also helps to determine the functional LLD, which may differ from the actual LLD.

Imaging Methods

The orthoroentgenogram, scanogram, and teleoroentgenogram are three basic radiographic procedures used to detect LLD.

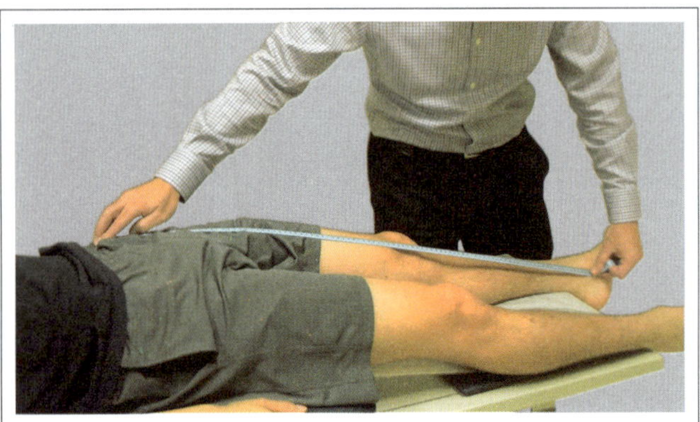

Fig. 2.6: Measurement of leg length in supine position

- **Orthoroentgenogram:** To reduce magnification error, three radiographic exposures are used at the hip, knee, and ankle joints. While lying down, a single big cassette is placed beneath the patient between each exposure.
- **Scanogram:** Three radiographic exposures are also used to limit magnification error at the hip, knee, and ankle joints. In contrast to the orthoroentgenogram, a standard-length cassette is shifted for three exposures while the patient is supine near a calibrated ruler.
- **Teleoroentgenogram:** A full-length AP radiograph of the lower limbs is taken while standing. It captures both lower extremities in a single exposure, with the subject standing erect and both patellae pointing forward. To level the pelvis, a lift is inserted beneath the shorter leg; measuring the lift's height under the shorter limb helps determine LLD if both iliac crests are equal, indicating LLD.
- **Computed radiography:** Use of computed radiography (CR) to measure leg length discrepancy is becoming more common. Three 35 × 43 cm CR storage phosphor cassettes are inserted vertically in a cassette holder on specialized long-length imaging equipment. At the CR reader console, these three images are later combined using specialized software. Picture archiving and communication systems (PACS) and other automated systems can further edit the digitally transferred composite image to make a film radiograph. By altering image characteristics on the computer, final image can be improved.
- **Microdose digital radiography:** When compared to traditional radiographic procedures, microdose digital radiography significantly reduces patient radiation exposure. The patient stands in front of the X-ray assembly within a vertical gantry with this method, remains stationary for a brief 20-second scan. The patient is exposed to very less radiations during the entire scan due to the detector's excellent effectiveness in capturing and processing X-ray photons from this single source. Because of the low exposure, this technique is especially useful for serial radiograph evaluations needed for conditions like progressive leg length inequalities.
- **Ultrasound:** Ultrasound was used by European researchers to assess LLD. An ultrasound transducer is used to locate bone landmarks at the hip, knee, and ankle joints.
- **CT scanogram:** LLD is measured using digitalized CT scan pictures, which provide exact assessments.
- **MRI scan:** MRI has gained appeal for evaluating bone disorders, despite its traditional application for soft tissue imaging. T1-weighted spin echo sequences generate MRI images that are used to standardize femur length evaluations by recognizing bone landmarks in chosen coronal images such as the femoral head and medial femoral condyle.

PELVIC TILT

Types of pelvic tilt are as follows:
1. Anterior pelvic tilt (Fig. 2.7B)
2. Posterior pelvic tilt (Fig. 2.7C)
3. Lateral pelvic tilt (Fig. 2.8)

Causes

Anterior Pelvic Tilt

- Tight hip flexors can pull the pelvis forward causing anterior pelvic tilt.
- **Weakness in the gluteal muscles:** These muscles are responsible for stabilization of pelvis and maintenance of proper posture. Their weakness can result into tilting of the pelvis forward, i.e., in an anterior direction.
- **Weakness in the abdominal muscles:** Weakness of the abdominals can contribute to pelvic tilt in anterior direction by failing to properly support the pelvis.

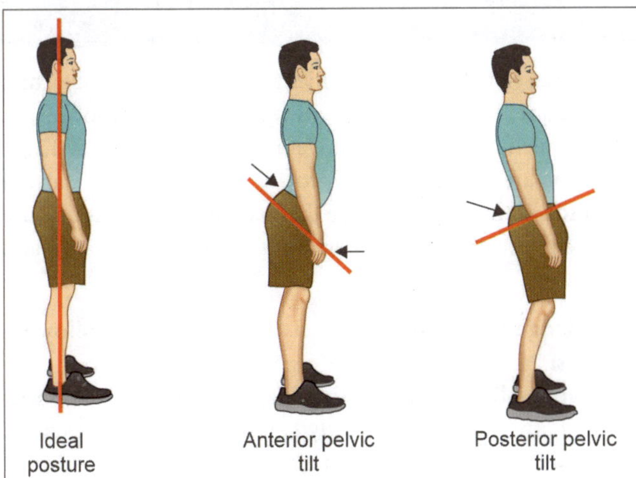

Fig. 2.7: Types of pelvic tilt

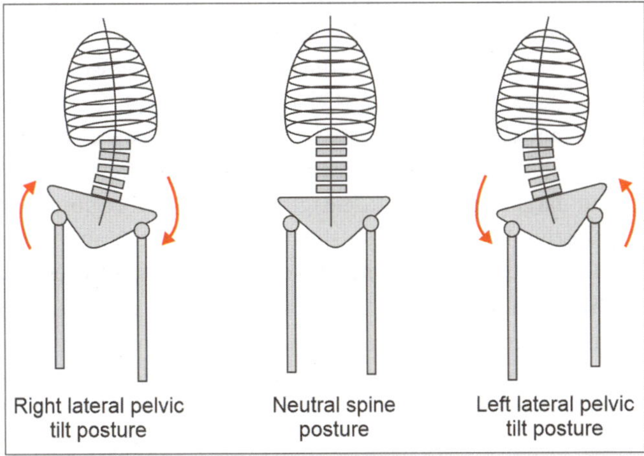

Fig. 2.8: Lateral pelvic tilts

- **Pregnancy:** The growing weight of the uterus in pregnant female can cause shifting of the center of gravity (COG) in a forward direction causing anterior pelvic tilt.
- **Trauma to pelvis or lower back** can result into muscular imbalances causing anterior pelvic tilt.

Posterior Pelvic Tilt

- Lengthened or weak hip flexor can cause pelvis to tilt posteriorly.
- Tight abdominals can lead to post pelvic tilt.
- Tight hamstrings can lead the posteriorly tilted pelvic.
- **Trauma to the pelvis or back** can result into imbalances in the muscles resulting into posterior pelvic tilt.
- **Poor posture:** Such as slouching or standing with an exaggerated arch can cause posterior pelvic tilt.
- **Scoliosis:** In this condition the spine bends to one side which can cause the pelvis also to tilt backward on the opposite side.

Lateral Pelvic Tilt

- **Leg length discrepancy:** In case one leg is shorter than the other, there can be a lateral pelvic tilt.
- **Scoliosis:** In this, the spine curves to one side. This can cause pelvis to shift laterally.
- **Muscle imbalances:** Muscles attached to pelvis can become tight or weak, leading to muscular imbalances and lateral tilt of pelvis. For example, a tight muscle on one side of the hip can pull the pelvis in an upward direction.
- **Poor postural habits** like standing with uneven weight distribution can cause lateral pelvic tilt.
- **Trauma:** Injuries to pelvis or lower back can cause muscular imbalances, resulting into lateral tilt of pelvis.

Identification

Normal Pelvic Tilt and Rotation

Identification of normal pelvic tilt and rotation requires a physical examination to be done by the physiotherapist. Following should be checked:

- **Pelvic tilt:** From a side view, a natural pelvic tilt is slightly anterior or 'S' shaped, i.e., the front of the pelvis is slightly lower than the back. Exaggerated anterior tilt is present when the front of pelvis is significantly lower than the back of pelvis and posterior tilt can be seen when the back of the pelvis is significantly lower than the front.
- **Pelvic rotation:** When the pelvis is seen from antero-posterior view the two iliac crests should be at the same level indicating that the pelvis is not rotated. In case, one iliac crest is higher than the other, it indicates rotation of the pelvis.

Altered Pelvic Tilt

There are different methods that a physiotherapist can use for the identification of altered pelvic tilt:

Posture Analysis

In this method the patient's posture is assessed in standing, sitting and lying positions to look for pelvic rotation or tilt.

Palpation Method

The physiotherapist can use their hands to feel for muscular imbalances or tenderness in the region of lower back, hips and pelvis that may be leading to an altered pelvic tilt.

Functional Movement Assessment

In this method therapist assess the ability of the patient to perform functional movements such as stepping or lunging to identify any limitations or compensation related to an altered pelvic tilt.

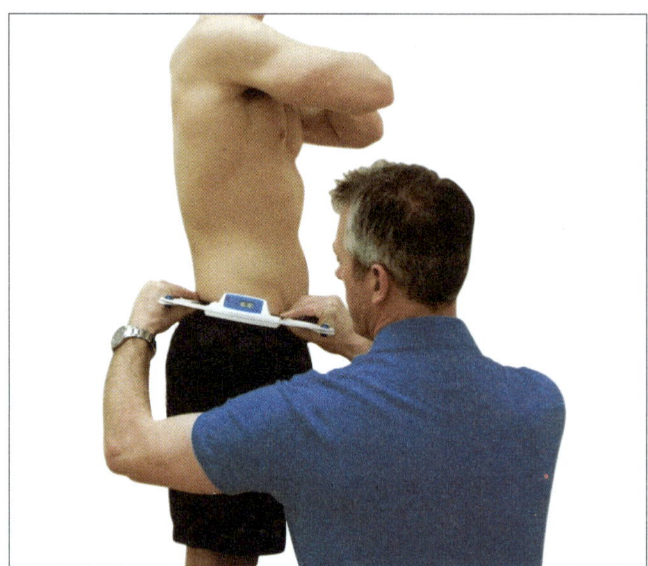

Fig. 2.9: Pelvic inclination measurement

Measurements

Pelvic Tilt Angle

"Pelvic tilt angle" is used to measure anterior pelvic tilt. This measurement is typically calculated as the angle between a line connecting the anterior superior iliac spines (ASIS) and a vertical line. A goniometer could be used to measure the angle, with the patient lying supine on a table and the examiner palpating the ASIS to locate the landmarks for measurement.

Pelvic Inclinometer

A pelvic inclinometer is a specialist equipment used to evaluate pelvic tilt or inclination in physical therapy and orthopedics. It is intended to measure the pelvic angle in reference to a horizontal plane. It has a digital display that allows practitioners to reliably measure the anterior or posterior tilt, lateral tilt, or rotation of the pelvis. Anteroposterior inclination of the pelvis may be measured by means of a pelvic inclinometer (Fig. 2.9).

- **Procedure:** The inclinometer is positioned on particular bony landmarks of the pelvis, such as the anterior superior iliac spine (ASIS) or posterior superior iliac spine (PSIS), to quantify pelvic tilt. The patient is often positioned in a standing position to effectively isolate and assess the pelvic tilt. The measurements from the inclinometer indicate the degree of tilt or inclination.

Recent Updates

Artificial intelligence (AI): Schwarz et al. investigated the use of AI in evaluating pelvic radiographs in a recent article published in the journal of International Orthopedics (2023). The researchers examined over 500 pelvic radiographs using a sophisticated deep learning technique. The findings demonstrated the AI software's exceptional accuracy in detecting fractures and other anomalies in images, which exceeded 90%.

FUNDAMENTAL POSITIONS

In exercise therapy, fundamental positions or starting positions refer to the initial body postures; or positions from which specific exercises or movements are performed. These positions are crucial as they provide a standardized starting point for exercises and ensure that patients or individuals are in the correct alignment to perform the movements safely and effectively.

Knowledge of starting positions is essential for following reasons:

- **Proper alignment:** Starting positions help ensure that the body is aligned correctly before initiating the exercise or any special test. Correct alignment reduces the risk of injury and optimizes the effectiveness of the exercise.
- **Consistency:** By using standardized starting positions, physiotherapist can consistently measure and track the progress of patients over time. It allows them to compare the patient's performance during subsequent sessions and monitor improvements or changes in strength, flexibility, or range of motion, etc.
- **Targeted muscle engagement:** Different starting positions can target specific muscle groups or body regions. By selecting the appropriate starting position, therapists can focus on the intended muscles or areas during the exercise.
- **Progression:** Starting positions can be modified or progressed to accommodate different levels of ability or to make exercises more challenging as the patient's strength and abilities improve.
- **Safety:** Ensuring proper starting positions reduces the risk of injury during exercise therapy. Correct alignment and posture help to prevent strain on muscles and joints.

Did You Know?

Proper alignment is not just about physical safety, but it also enhances the effectiveness of the exercise. When the body is aligned correctly, muscles can engage more efficiently, leading to better outcomes in terms of strength, flexibility, and overall function. So, paying attention to alignment isn't just about avoiding injury; it is also about maximizing the benefits of each exercise session.

Common starting positions used in exercise therapy include:

- Movement can originate from a variety of starting or static positions. Starting positions are the initial postures, which can be either passive or active. To sustain static contraction and joint stability, they depend on stabilizing muscles and outside assistance. Positions that differ from these basic positions are called derived positions since they are adaptations or modifications of them.

  There are five fundamental starting positions:

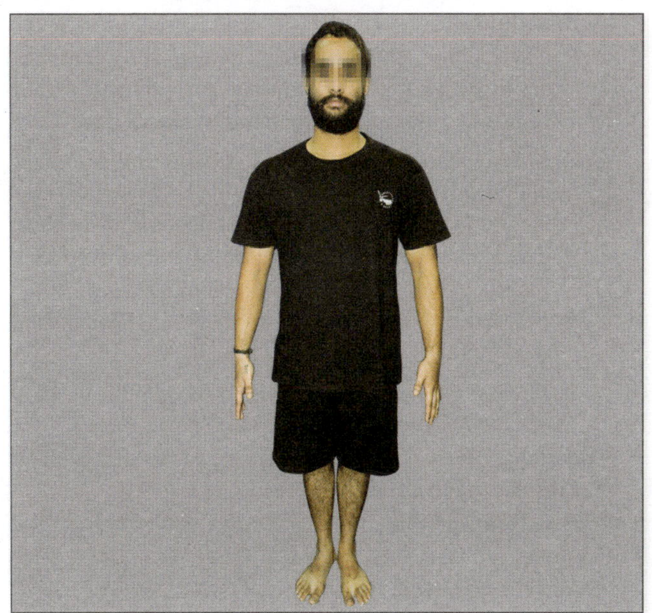

Fig. 2.10: Standing position

1. Standing

In standing, the whole body must be balanced and stabilized in correct alignment on the feet with a small base of support by coordinated work of many muscle groups (Fig. 2.10).

While in the standing position, keep the toes slightly apart and place the heels in a straight line. The hips are extended and slightly externally rotated, with knees kept straight and together. As the spine stretches to its maximum length, it creates a lengthened posture by supporting the pelvis on the femoral heads. In order to maintain level ears and a forward focus, the head is raised simultaneously. In order to improve general alignment and stability in this posture, the shoulders are purposefully positioned downward and back. Arms are kept by the side of the body with palms facing inwards.

Muscle Work

In a standing position, the muscles of the body work together to maintain stability and support the body against the force of gravity.

The joints involved in the standing position include the ankles, knees, hips, and spine. Here is a breakdown of how muscles work joint-wise in the standing position:

- **Feet:**
  - **Muscles involved:** Intrinsic muscles of the foot
  - **Function:** Contracts to stabilize foot and prevents toe curling.
- **Ankles:**
  - **Muscles involved:** Gastrocnemius, soleus, tibialis anterior, peroneals
  - **Function:**
    - The calf muscles (Gastrocnemius and Soleus) contract to maintain ankle stability and prevent the body from falling forward.
    - The tibialis anterior and peroneal muscles work to maintain balance and control the foot movement.
- **Knees:**
  - **Muscles involved:** Quadriceps, hamstrings
  - **Function:**
    - The quadriceps muscles on the front of the thigh work to straighten the knee joint.
    - The hamstrings work to stabilize and control the bending of the knee.
- **Hips:**
  - **Muscles involved:** Gluteus maximus, gluteus medius, hip adductors, hip abductors.
  - **Functions:**
    - The gluteus maximus is the main hip extensor and helps to maintain the upright posture by keeping the body from collapsing forward.
    - The gluteus medius and hip abductors stabilize the pelvis and prevent excessive side-to-side tilting.
    - The hip adductors help maintain stability and balance by preventing the legs from moving too far apart.
- **Spine:**
  - **Muscles involved:** Erector spinae, Abdominal muscles (rectus abdominis, obliques), Transverse abdominis, prevertebral neck muscles, flexors and extensors of atlanto-occipital joint.

- **Functions:**
  - The erector spinae muscles, which run along the spine, work to maintain an upright posture by preventing the spine from bending forward.
  - The abdominal muscles, particularly the rectus abdominis and obliques, contract to support the spine and prevent excessive arching of the lower back.
  - The transverse abdominis, known as the deep core muscles, provide stability to the lumbar region.
  - Prevertebral muscles of neck straighten the neck and prevent overactivity of neck extensors.
  - Flexors and extensors muscles of atlanto-occiptal joint work to balance the head.
- **Mandible:**
  - **Muscles involved:** Elevators muscles
  - **Function:** Close the mouth
- **Scapula:**
  - **Muscles involved:** Retractors
  - **Function:** Pull the scapula backwards

Effects and Uses

The standing position has various effects and uses, both physiological and practical, that play important roles in our daily lives. Here are some of the effects and uses of the standing position:

- **Postural support:** Standing provides structural support to the body against the force of gravity. The alignment of the skeletal system, along with the tension in various muscles, helps maintain an upright posture, preventing the body from collapsing.
- **Muscle activation:** Standing engages several muscles, including those in the legs, core, and back. This helps strengthen these muscles and improve overall stability and balance.
- **Circulation:** When standing, the muscles in the legs and lower body work against gravity to assist in venous return, which improves blood circulation back to the heart. This can reduce the risk of blood pooling in the legs and decrease the likelihood of developing certain circulatory issues.
- **Caloric expenditure:** Standing burns more calories compared to sitting or lying down, although the difference may not be significant. Nonetheless, prolonged standing can contribute to a slight increase in energy expenditure over time.
- **Physical rehabilitation:** In physical therapy and rehabilitation, the standing position is often used to improve strength, balance, and mobility in patients recovering from injuries or surgeries.
- **Reduced sedentary behavior:** Alternating between sitting and standing throughout the day can help reduce sedentary behavior, which has been associated with various health risks.
- **Exercise:** Many physical activities, such as weightlifting, yoga, and stretching, involve the standing position as part of the movements or poses.

Clinical Correlation

However, it is important to note that prolonged standing without proper breaks can also have negative effects, such as fatigue, foot discomfort, and an increased risk of developing musculoskeletal issues like lower back pain or varicose veins. It is essential to strike a balance between sitting and standing and incorporate movement and breaks throughout the day for optimal health and well-being.

2. Kneeling

The kneeling position is a posture in which a person supports his body weight on his knees. The legs are resting on the floor or couch with feet either in plantar flexed position or in mid position at the edge of the couch (Fig. 2.11).

Muscle Work

- **Knees:**
  - **Muscles involved:** Quadriceps, hamstrings
  - **Functions:**
    - The quadriceps muscles, work to extend the knee joint and keep the leg straight while kneeling.
    - The hamstrings, on the back of the thigh, act as stabilizers to support the knee joint during this position.
- A tension in the rectus femoris which gets stretched across knee and hip joints might pull the lumbar spine into hyperextension. Gluteus maximus and the abdominal muscles, particularly the rectus abdominis and obliques, contract more strongly to support the spine and prevent excessive arching of the lower back. The remaining muscle work is identical to standing.

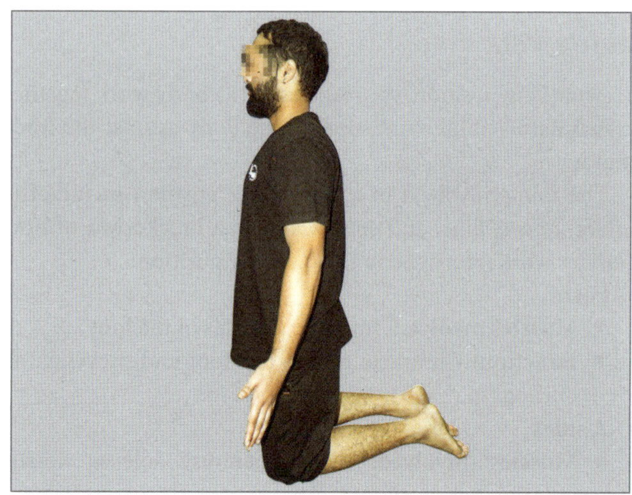

Fig. 2.11: Kneeling position

Effects and Uses

- The position is more stable than standing. The COG in kneeling is slightly lower than standing position.
- It is used for backwards movements in sagittal plane. Also can be used for training balance and coordination.

3. Sitting

With the feet flat on the floor and the hips and knees bent to a straight angle, this position is assumed while sitting on a chair or stool (Fig. 2.12).

Muscle Work

- **Hip:**
  - **Muscles involved:** Hip flexors, gluteus maximus
  - **Functions:**
    - The hip flexors, including muscles like the iliopsoas and rectus femoris, work to flex the hips and bring the thighs forward, allowing you to sit down.
    - The gluteus maximus, the large muscle of the buttocks, helps stabilize the hips and supports the upper body while seated.
- Remaining muscle work is identical to standing.

Effects and Uses

- **Exercise:** Several exercises can be performed in sitting position like spine, and leg exercises, breathing exercises and chair aerobics.
- **Rest and relaxation:** Sitting allows individuals to rest and take a break from standing or physical activity, promoting relaxation and conserving energy.
- **Support and stability:** Sitting provides a stable base of support for various activities. It is a comfortable position that people with limited strength and coordination can use to attain a more difficult starting position.

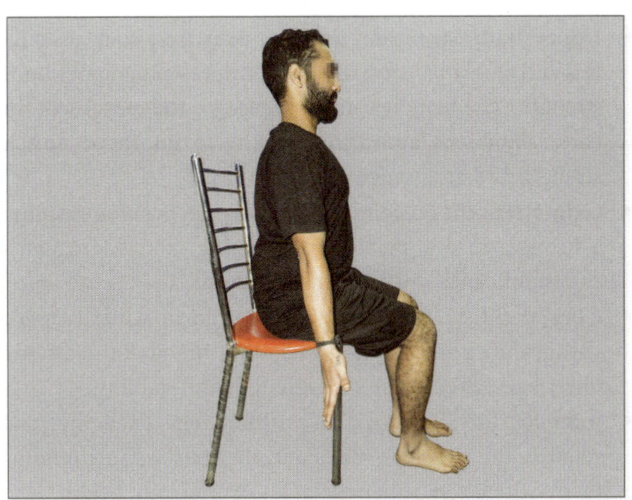

Fig. 2.12: Sitting position

- **Joint relief:** For people with certain health conditions or injuries, sitting can relieve pressure on the knees, hips, and spine compared to standing or weight-bearing positions.
- **Meditation and mindfulness:** Sitting in a relaxed position is commonly practiced for meditation and mindfulness exercises, promoting mental clarity and relaxation.

4. Lying

The lying position, also known as the supine position, is a posture in which a person lies flat on their back with their face and torso facing upward. The arms and legs can be placed in various positions, such as by the sides, crossed over the chest or extended outward (Fig. 2.13). This is a comfortable position, does not require much muscle work. When assumed on a firm surface following muscle groups are active:

Muscle Work

- **Hips:**
  - **Muscles involved:** Hip extensors, abductors and, adductors
  - **Functions:**
    - The hip extensors, majorly the gluteus maximus, work to extend the hips and prevent hollowing of the back.
    - The hip abductors and adductors, located on the outer and inner thighs, respectively, assist in stabilizing the hips during lying positions.
- **Neck:**
  - **Muscles involved:** Neck rotators on both sides.
  - **Function:** Contracts to maintain the head in a neutral position.

Effects and Uses

- **Spinal alignment:** Lying on the back allows the spine to rest in a more neutral position, which can help relieve pressure on the vertebrae and reduce strain on the back.
- **Muscle relaxation:** The lying position promotes relaxation of muscles throughout the body, especially the neck, shoulders, and back.
- **Improved circulation:** Lying on the back with legs elevated can aid venous return, helping blood flow back to the heart and reducing swelling in the lower extremities.

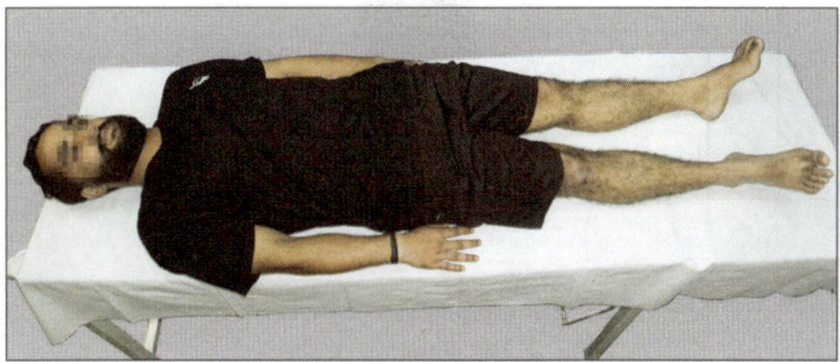

Fig. 2.13: Lying position

- **Reduced pressure points:** The lying position distributes body weight evenly across the surface, reducing pressure on specific areas, such as the hips and shoulders.
- **Relaxation and meditation:** Lying down in a quiet and comfortable environment can promote relaxation and mindfulness practices.
- **Physical therapy and rehabilitation:** The lying position is often used in various physical therapy exercises and rehabilitation programs to target specific muscle groups and promote recovery after injuries or surgeries.
- **Pregnant women:** Lying on the back with the legs elevated (supine with legs elevated) can be a recommended position for pregnant women to alleviate pressure on the lower back and improve circulation.
- **Breathing exercises:** The lying position may be used during certain breathing exercises to help individuals focus on their breath and promote relaxation.

5. Hanging

The hanging position refers to the act of suspending the body from an overhead object, such as a bar or rings, with the arms supporting the body weight. The arms should be straight and shoulder width apart. The scapulae are pulled down and retracted. The trunk and legs are straight and ankles are plantar flexed (Fig. 2.14). It is commonly used in various forms of exercises and fitness training, such as hanging exercises, gymnastics, and calisthenics. The hanging position primarily involves the muscles and joints of the upper body.

Muscle Work

Here is an explanation of the muscle work joint-wise in the hanging position:

- **Shoulders:**
  - **Muscles involved:** Latissimus dorsi, teres major, posterior deltoids, rotator cuff muscles.
  - **Function:** The latissimus dorsi and teres major muscles are the primary movers responsible for pulling the body up during hanging exercises. The rotator cuff muscles stabilize the shoulder joint during hanging movements.

Fig. 2.14: Hanging position

- **Elbows:**
  - **Muscles involved:** Biceps brachii, brachialis, brachio-radialis.
  - **Function:** The biceps brachii, brachialis, and brachiora-dialis muscles are the primary elbow flexors, enabling the bending of the arms while hanging. These muscles are engaged when initiating the pull-up motion in hanging exercises.
- **Wrists and grip:**
  - **Muscles involved:** Forearm flexors and extensors, hand intrinsic muscles.
  - **Function:** The muscles of the forearms, including the flexors and extensors, play a crucial role in maintaining grip strength while hanging. The hand intrinsic muscles help to control finger movements and contribute to overall grip stability.
- **Core and abdominals:**
  - **Muscles involved:** Rectus abdominis, transverse abdominis, obliques
  - **Function:** The core muscles, including the transverse abdominis, rectus abdominis and the obliques, work to control the trunk and prevent excessive swinging or arching of the back during hanging exercises.
- **Hips:**
  - **Muscles involved:** Hip extensors
  - **Function:** Prevents the arching of back and hip adductors to keep the legs together.
- **Scapula:**
  - **Muscles involved:** Depressors and medial rotators
  - **Function:** Keep the scapula braced to upper back.

Effects and Uses

- **Upper body strength:** Hanging exercises, such as pull-ups and chin-ups, are excellent for developing upper body strength. The hanging position engages the muscles of the back, shoulders, and arms, leading to improved muscle strength and endurance in these areas.
- **Grip strength:** Sustaining the hanging position requires a strong grip, and performing hanging exercises regularly can significantly enhance grip strength.
- **Core stability:** Maintaining proper form while hanging engages the core muscles, including the abdominals and lower back, promoting core strength and stability.
- **Joint stability:** The hanging position helps strengthen and stabilize the shoulder joints and the muscles surrounding them, which can be beneficial for overall joint health.

- **Scapular mobility:** During hanging exercises, scapular mobility is engaged as the shoulder blades move and rotate. This improves shoulder function and reduces the risk of shoulder injuries.
- **Body awareness:** Hanging exercises require body awareness and control, as individuals need to maintain proper form and balance while suspended in the air.
- **Strength training:** Hanging exercises chin-ups, pull ups and hanging leg raises, are staples in strength training routines for developing upper body and core strength.
- **Bodyweight training:** Bodyweight exercises that involve hanging, such as inverted rows or hanging knee tucks, are popular in bodyweight training programs.
- **Calisthenics:** The hanging position is a fundamental component of calisthenics, a type of exercise that uses body-weight movements for building strength and flexibility.
- **Gymnastics:** In gymnastics training, hanging exercises are essential for building upper body strength and performing various advanced skills on bars and rings.
- **Rehabilitation:** In physical therapy and rehabilitation settings, hanging exercises can be used to improve shoulder stability, mobility, and overall upper body strength.

DERIVED POSITIONS

The term "derived positions" describes a variety of stances or postures that are altered or derived from a fundamental or initial position (Table 2.1). These positions require modifying the body's posture or limb placements while keeping some characteristics of the initial position. Derived positions are frequently employed in a variety of contexts, including as daily chores, sports, physical activity, and particular disciplines like yoga or dance. Their benefits can include, enhancing flexibility and aiding functional motions in addition to offering stability and balance.

Did You Know?

Purpose of derived positions is to:

- Alter the base of support
- Alter position of COG
- Recruit certain muscle groups
- Increase or decrease the leverage
- Achieve local or general body relaxation
- Provide a starting position for performing a particular activity or exercise.

TABLE 2.1: Positions derived from fundamental positions

Positions derived from standing (by alteration of arms)		
Derived position	**Position and muscle work**	**Effects and uses**
1. Wing standing	• Both the hands are resting on the iliac crest with fingers and thumb positioned anterior and posterior to ASIS respectively. • Bilateral adductor muscles of the shoulder and extensor muscles of the elbows work to press the hands against the trunk.	• Various trunk exercises are performed in this position. • This is used as a posture for testing balance.
2. Bend standing 	• Standing with feet together and weight of body on balls of the feet. • Muscles that work in this position are: ▪ The shoulder adductors and lateral rotators work together. ▪ The scapular depressors and retractors ▪ The forearm supinators and elbow flexors ▪ Flexors of the fingers and wrist.	This position can be utilized for arm stretching and strengthening exercises

Contd...

Derived position	Position and muscle work	Effects and uses
3. Reach standing 	• Standing upright while reaching out with arms. The shoulder flexor muscles maintain the position against gravity. • The transverse muscles of the back control the forward movement of the scapulae. • The extensors of the elbows, radial flexors of the wrist and extensors of the fingers contract to keep the arm straight.	The position can be used for arm and trunk exercises and balance exercises.
4. Half-stretch standing 	• Standing while extending muscles or body components to their full length. • The abductors, extensors and lateral rotator muscles of the shoulder work strongly. • The lateral rotators of the scapulae, work to stabilize the arm in position. • The extensors of the elbows keep them straight	This position is used for stretching of the lateral trunk muscles and general fitness exercise.
5. Yard standing 	• Standing with both arms abducted at 90°. • The abductors, extensors and lateral rotator muscles of the shoulder stabilize the arms. • The extensors of the elbows, wrists and fingers work to stabilize the upper limbs. • Depressors of the scapula control the elevation of the shoulder girdle.	The position is used for arm, trunk and balance exercises.

Contd...

Positions derived from standing (by alteration of legs)		
Derived position	**Position and muscle work**	**Effects and uses**
1. Toe standing 	• Toe standing is standing on tiptoes, which requires balance and engages the calf muscles. • The plantar flexor muscles of the ankle joint work strongly against gravity to keep the heels elevated. All the leg muscles work more strongly than in the fundamental position.	• This position is used in the treatment of postural flat feet, strengthening of calf muscles, balance training. • This position is used for manual muscle testing of calf muscles.
2. Stride standing 	• Standing with legs apart, is stride standing. • The adductor muscles of the hips contract to prevent the legs from sliding further apart.	• This is an easy and stable position to perform exercises, especially those in frontal plane. • It can be used as the starting position for performing arm and trunk exercises.
3. Walk standing 	• To stand with a ready or poised posture, as if one is about to begin walking. • The extensor muscles of the hip and knee of the posterior leg work strongly to maintain the position.	• The base is much wider in the antero-posterior direction stabilizing the body for exercises in the sagittal plane. • The position can be used for performing balance exercises.

Contd...

Derived position	Position and muscle work	Effects and uses
4. Half standing 	• A position in which the person is standing on one leg and the other is bent at hip and knee. • The standing leg's abductors of the hip assist in keeping the center of gravity over the base. • The opposing side's lumbar lateral flexors assist in aligning the trunk. • For the purpose of sustaining the extra weight and preserving balance, all muscles in the supporting leg contracts more forcefully than when standing.	• After various abdominal operations, the leg that is liberated from body weight can be utilized to rest in a variety of ways, such as on a stool with its hips and knees bent. • This helps to relieve stress on the side of the abdominal wall. It can also be practiced for balance training.
5. Step standing 	• Step standing is the position of elevating one foot to stand on a platform or step. • The leg positioned forward (lead leg) primarily targets the quadriceps, hamstrings, and glutes, as they bear the body's weight. • The rear leg helps stabilize the body and activates the calf muscles and hip stabilizers	• It is used to increase ankle strength and mobility, test balance, activate leg muscles, and enhance stability. • Exercises which are designed to improve lower limb proprioception, balance, and strength, frequently use this position.

Positions derived from standing (by alteration of trunk)

Derived position	Position and muscle work	Effects and uses
1. Stoop standing 	• Standing with waist slightly bent forward is known as stoop standing. The muscles of the feet work strongly to hold the position. • The extensors of the knees work to counteract the tension of the hamstrings. • The back muscles, extensors of elbows and shoulder flexors maintain the position against the pull of gravity. • The posterior neck muscles, controlled by the anterior neck muscles, support the head.	By promoting alignment and strengthening the core muscles. It is used in specific exercises and activities to activate the lower back muscles, increase flexibility, and improve posture.

Contd...

Derived position	Position and muscle work	Effects and uses
2. Lax stoop standing	• The muscles of the spine contract to hold this position. • The dorsiflexors stabilize the position of the joint, while the intrinsic foot muscles grip the floor.	The pose can help with expiration and be used to train upper body local relaxation.

Positions derived from kneeling		
Derived position	**Position and muscle work**	**Effects and uses**
1. Half-kneeling 	• With one knee on the floor and the other leg at a 90° angle with the foot flat on the ground, one can assume the half-kneeling position. • The abductor muscles of the hip joint of the supporting leg and the lumbar side flexors of the opposite side work to hold the position. • The extensor muscles of the hip and knee of the forward leg work to maintain balance.	It is used in therapy to strengthen lower body muscles, increase hip mobility, improve balance, and exercises for leg, hip or lower back issues.
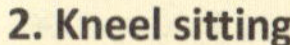**2. Kneel sitting** 	• Sitting on one's knees with the buttocks resting on the heels or halfway between the heels and thighs is known as the kneel sitting position. • The muscles around the shoulder and hip joints work to stabilize the supporting limbs at right angles to the trunk. • The spine muscles work to maintain erect posture. • The extensors of the neck and head, are balanced by the anterior neck muscles keep the head in alignment.	• It is frequently used in meditation, prayer or as a therapeutic posture for spine alignment and back relief. • It also helps to engage core muscles and create better posture.

Contd...

Derived position	Position and muscle work	Effects and uses
3. Side sitting 	• Sitting on the floor with the legs bent to one side to form a right angle at the hips and knees is known as the side-sitting position. • This pose works the obliques, as well as the hips, thighs, and core muscles.	• It helps with hip mobility, stretches the muscles in the inner thighs, and enhances lateral stability. • Yoga and pilates practitioners frequently utilize the side-sitting position for lateral stretches, spinal mobility exercises, and meditation techniques. • It strengthens the core and increases flexibility.
4. Prone kneeling 	In order to maintain stability, one must kneel on the ground with his torso facing downward and contract the gluteal muscles, thighs, lower back, and core muscles.	• By strengthening the muscles that keep the spine neutral, this position helps to improve spinal alignment and posture. • To increase core strength, stability, and balance, the prone kneeling posture is frequently utilized in physical therapy and exercise regimens. • It is beneficial for functional training programs and rehabilitation, as it also helps to improve proprioception and body awareness.
Positions derived from sitting		
Derived position	**Position and muscle work**	**Effects and uses**
1. Ride sitting 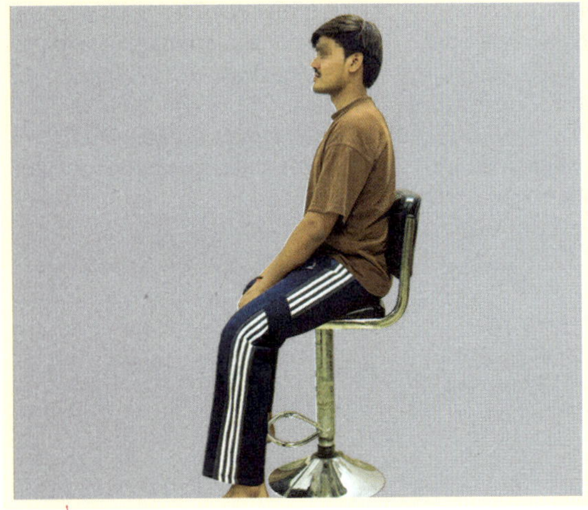	• A seated position that resembles riding posture, typically with an erect torso and engaged core muscles. • The adductor muscles of the hips can be used to hold the appropriate equipment such as a gymnastic ball, while the patient sits on it. • If the pose is performed on a tall plinth, there may be no leg muscle activity if the thighs are strapped to the plinth for further fixation.	This position can be used to exercise head, arms, and trunk.

Contd...

Derived position	Position and muscle work	Effects and uses
2. Crook sitting 	• Sitting in a hunched or bent position, usually with a curved or rounded back. The hip flexors exert significant force in supporting the thighs and preventing excessive bending of the lumbar region. • To enable further leg fixation, the plantar flexors of the ankles and the flexors of the knees may also be used. • Strong work is done by the transverse and longitudinal back muscles to keep the trunk in an upright position.	• This position can be used for strengthening of the core muscles. • Strong work for the extensors of the thoracic spine to hold the position is of value in training their efficiency. • It can be used as a position for testing strength and endurance of abdominal muscles.
3. High sitting 	• The fundamental sitting position is taken on a high plinth chair or table but the feet remain unsupported. • The elevated seating posture of the high sitting position is achieved by flexing the knees and hips. • Muscles of back, abdomen, and hips work to support this posture.	• It helps with posture corrections, breathing, and relieving pressure on the lower back during specific therapeutic exercises or activities. • This position can be used for performing lower limb exercises.
4. Cross sitting 	• In the cross sitting position, the legs are crossed at the knees or ankles. • Hip adductors, lower back muscles, and core stabilizers work to maintain the posture and balance.	• It is frequently included in relaxation exercises, meditation or as a casual sitting position. • It helps to increase flexibility, support hip mobility, and improve posture.

Contd...

Derived position	Position and muscle work	Effects and uses
5. Long sitting 	• Sitting up straight with legs out in front is the long sitting position. • Hip flexors, lower back, core stabilizers, and hamstring muscles work to maintain this posture.	It is used in physical therapy exercises aimed with improving posture, hamstring flexibility, core strength, and lower limb muscles, joints or spinal rehabilitation.

Positions derived from lying

Derived position	Position and muscle work	Effects and uses
1. Crook lying 	• Lying on the back with knees bent and feet flat on the ground is known as the crook lying position. • Adductors and medial rotators of hip contract to hold the position.	It facilitates pelvic tilt exercises, lessens lower back pain, and improves core muscles engagement during specific therapeutic motions or exercises.
2. Crook lying with pelvis lifted 	• In crook lying position pelvis is lifted up. • The extensors of the hips contract to hold the position.	The position is useful for pelvic floor muscle reeducation, and as exercise for low back pain management.
3. Half lying 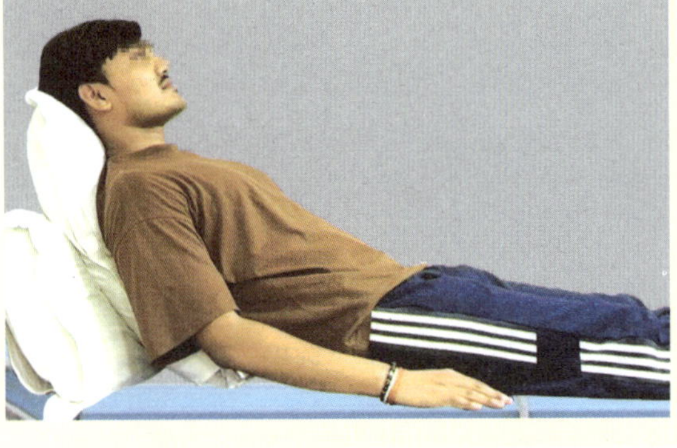	• It involves lying down with the upper body partially elevated, frequently supported by cushions or a reclining platform. • Muscles used for support and stability often include the core, including the abdominal and lower back muscles, as well as those in the legs and hips.	In addition to reducing back pain and offering a supportive resting posture for recuperation and relaxation, half-lying helps with breathing and digestion.

Contd...

Derived position	Position and muscle work	Effects and uses
4. Prone lying 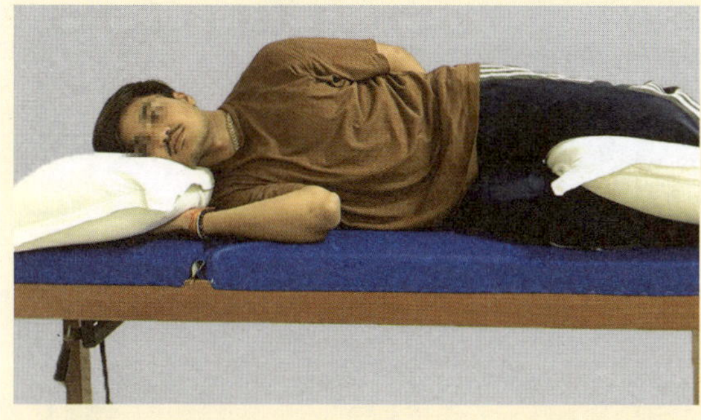	• Prone lying involves lying face down, engaging the muscles in the back, shoulders, and posterior chain for support and stability. • The retractors and depressors of the scapulae work to brace the upper back.	The position is used for the management of low back pain by straightening back muscles, spinal alignment, and for respiratory assistance.
5. Side lying 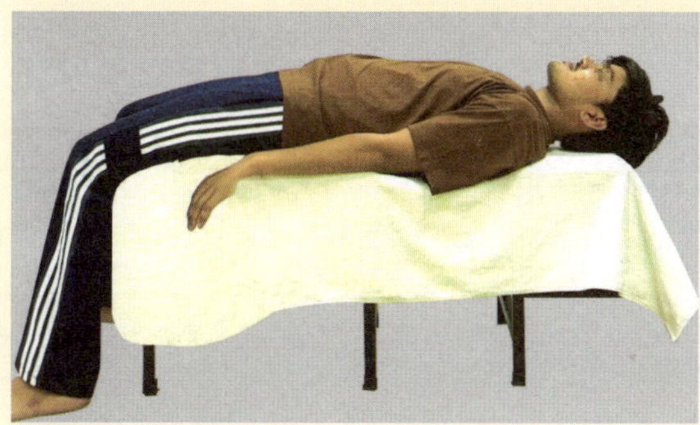	• Lying on one's side provides comfort, supports proper alignment of the body, and facilitates therapeutic exercises or rest. • To maintain alignment, the obliques, hip abductors, and shoulder stabilizers are utilized.	The side lying posture helps with stability, pressure alleviation, and the rehabilitation of a variety of diseases. This is also used for relaxation.
6. Sit lying 	• The patient is in a supine position, with the lower leg dangling over the end of the plinth vertically and the knees bent. • The core and hip muscles work to hold the position.	This is used for strengthening chest and upper limb muscles.
7. Leg prone lying	• The extensors of the hips, the longitudinal and transverse back muscles, and the prevertebral and posterior neck muscles all exert force to keep the trunk in position against gravity. • The elbows and shoulder extensors keep the arm at the sides. • The lumbar region, which has a tendency to hollow out, is controlled by the flexors of the lumbar spine.	This position can be used for muscle strength testing, mobility and strengthening of various muscles.

Contd...

Positions derived from hanging		
Derived position	**Position and muscle work**	**Effects and uses**
1. Half hanging	• The "half hanging" position is most likely one of partial suspension or support, with upper body muscles engaged for stabilization and strength. The muscles of the wrist, elbow, and shoulder help to release tension from these joints as the flexors of the fingers grasp the bar. • The trunk is pulled upward between the arms by the powerful retractors of the scapulae. • The cervical spines and the atlanto-occipital joints flexors stop the head from falling backward. • The trunk is supported by the transversal and longitudinal back muscles. • The plantar flexors press the feet to the floor, and the hip extensors maintain the alignment of the trunk. Strong back muscles, particularly the scapulae retractors, which defy gravity and the body's weight, are necessary for this position.	This position is used for strengthening of the trunk muscles.

SUMMARY

- Exercise therapy encompasses a range of exercises tailored to improve physical function, mobility, and overall well-being, often recommended by healthcare professionals like physical therapists. It aims to address various conditions, from musculoskeletal injuries to chronic diseases, offering pain relief, enhancing strength, flexibility, and cardiovascular fitness.
  - Customized to individual needs, it includes stretching, strengthening, and functional exercises, promoting recovery, preventing future injuries, and boosting mental well-being, ultimately improving quality of life.
  - A patient-centered approach involves assessment, individualized programs, gradual progression, education, and continuous monitoring to achieve specific goals while prioritizing safety and holistic care.
- Physiotherapy assessments involve patient interviews, subjective and objective examinations, aiming to understand medical history and current physical limitations.
  - Vital signs like temperature, pulse rate, respiration, and blood pressure play crucial roles, indicating overall health and guiding treatment.
  - The assessment process guides diagnosis, goal-setting, and individualized treatment plans, emphasizing continual reassessment and patient-centered care for optimal rehabilitation outcomes.
- PFTs evaluate lung health by assessing air intake, lung capacity, gas exchange, and respiratory muscle strength.
  - Spirometry evaluates inhalation and exhalation abilities and tracks conditions such as COPD, asthma, and disease progression. Lung volumes assess the amount of air in the lungs, suggesting the degree of lung disease and airway obstruction.
  - The gas diffusion study assesses oxygen transfer from the lungs to the blood. Respiratory muscle pressures (RMPs) determine strength, which is essential for coughing and secretion clearance. Bronchoprovocation testing detects asthma by assessing airway response.
  - Cardiopulmonary exercise testing (CPET) measures the efficiency of the heart, lungs, and muscles during activity, which aids in disease diagnosis and monitoring. These tests are critical in the diagnosis, management, and monitoring of pulmonary diseases.
- LLDs can be real or functional, with true discrepancies occurring from structural changes and functional differences deriving from compensatory processes.
  - Tape measurements from anatomical landmarks like the ASIS to the medial malleolus, evaluating apparent length variations by standing on blocks, or utilizing imaging techniques like radiography, CT, MRI, or ultrasound are all ways to estimate leg length.
  - Pelvic tilt, like anterior, posterior, or lateral tilt, is frequently caused by muscular imbalances, poor posture, trauma, or structural problems. Normal pelvic tilt is identified by viewing the pelvis from various angles and levels.
  - Visual inspection, palpation, posture analysis, and functional movement tests are all used to detect abnormal pelvic tilt.
  - Pelvic tilt angles can be measured using tools such as goniometers and pelvic inclinometers. Furthermore, AI has remarkable potential in analyzing pelvic radiographs, offering precise insights into pelvic anomalies.
- Starting positions in exercise therapy lay the groundwork for safe, effective movements. They provide proper alignment, allowing for targeted muscle engagement, progression, and continuous progress tracking.
  - The basic positions—standing, kneeling, sitting, lying, and hanging—all have different functions.
  - Standing uses multiple muscle groups to maintain the body against gravity and improves circulation, whereas kneeling is more stable and helps with posterior motions.
  - Sitting promotes relaxation and improved spinal alignment, whereas lying promotes relaxation and increased joint stability.
  - Each position engages different muscle groups and has different consequences, ranging from muscular engagement to joint relief and rehabilitation.
- Derived postures are derived from fundamental postures and serve to adjust posture or limb placement in daily work, sports, and other activities.
  - For flexibility, stability, and functional movement, they change support, center of gravity, and muscle recruitment.
  - These postures are designed to maximize leverage, relaxation, and to provide particular starting points for various tasks or exercises.

FURTHER READINGS

- Gardiner Dena M. Principles of Exercise Therapy. CBS Publishers and Distributors Pvt Ltd; 4th ed., 2023.
- Hollis M, Fletcher-Cook P. Practical Exercise Therapy. Blackwell Science; 2018 Jul 17.

STUDENT ASSIGNMENT

LONG ANSWER QUESTIONS

1. Explain muscle work, effects and uses of standing position.
2. Explain muscle work, effects and uses of lying position.
3. Explain the aims and the techniques used in exercise therapy.
4. Describe physiotherapy assessment of a patient in detail.
5. Discuss pulmonary function tests.
6. What is pelvic tilt? What are the different methods to measure it?
7. Discuss various methods of limb length assessment.

SHORT ANSWER QUESTIONS

1. Mention the aims of exercise therapy.
2. What are the different techniques of exercise therapy?
3. What are the various derived positions of standing?
4. Briefly write the types of fundamental positions.
5. Define pelvic inclinometer.
6. Define limb length discrepancy.
7. What are the indications of spirometry?
8. Write about the vital signs.

MULTIPLE CHOICE QUESTIONS

1. **Which of the following is NOT a fundamental starting position in exercise therapy?**
   a. Lying
   b. Seated
   c. Leaning
   d. Standing
2. **In the standing position, which muscle group primarily helps prevent the body from collapsing forward?**
   a. Hamstrings
   b. Quadriceps
   c. Gluteus maximus
   d. Gastrocnemius
3. **Which position is commonly utilized to improve strength, balance, and mobility in patients during physical therapy?**
   a. Lying
   b. Kneeling
   c. Sitting
   d. Hanging
4. **What is the primary function of the hanging position in exercise?**
   a. Enhancing lower body strength
   b. Strengthening core muscles
   c. Improving neck flexibility
   d. Stabilizing the ankles
5. **Derived positions in exercise therapy serve to:**
   a. Increase sedentary behavior
   b. Alter the base of support
   c. Decrease muscle engagement
   d. Eliminate the need for proper posture
6. **Which test measures the amount of air you can inhale and expel along with calculating the air volume in the lungs?**
   a. Spirometry
   b. Gas diffusion study
   c. Respiratory muscle pressures measurement
   d. Bronchoprovocation testing
7. **Which parameter(s) are measured in spirometry to assess lung function?**
   a. Forced vital capacity (FVC)
   b. Forced expiratory volume in the first second (FEV_1)
   c. FEV_1/FVC ratio
   d. All of the above

8. **Which lung volume denotes the amount of gas remained in the lungs following a normal exhale during normal breathing?**
   a. Residual volume (RV)
   b. Tidal volume (TV)
   c. Functional reserve capacity (FRC)
   d. Inspiratory reserve volume (IRV)
9. **What is the primary function of the diffusion capacity test?**
   a. Measures the strength of respiratory muscles
   b. Measures the amount of oxygen and other gases transferred from lungs to blood
   c. Measures lung volumes during calm exhalation
   d. Measures the airway responsiveness in asthma
10. **Which test is utilized to evaluate the efficiency of the heart, lungs, and muscles during exercise?**
    a. Gas diffusion study
    b. Bronchoprovocation testing
    c. Cardiopulmonary exercise testing (CPET)
    d. Respiratory muscle pressures measurement

ANSWER KEY

1. c **2.** c **3.** b **4.** b **5.** b **6.** a **7.** d **8.** c **9.** b **10.** c

3 Muscle Action

Sheetal Kalra

LEARNING OBJECTIVES

After the completion of the chapter, the readers will be able to:

- Explain types of muscles, muscle work and its types.
- Explain group action of muscles.
- Understand angle of muscle pull and mechanical efficiency of muscles.
- Explain and demonstrate active and passive insufficiency of muscles.

CHAPTER OUTLINE

- Introduction
- Types of Muscles
- Muscle Contraction
- Group Action of Muscles
- Angle of Pull and Mechanical Efficiency of Muscles
- Active and Passive Insufficiency

KEY TERMS

Actin: Actin is a highly abundant intracellular protein present in all eukaryotic cells and has a pivotal role in muscle contraction, as well as, in cell movements. A sarcomere is composed of two main protein filaments which are the active structures responsible for muscular contraction. Actin forms the thin filament.

Myosin: A protein found in muscle tissue as a thick filament and is made up of an aggregate of similar proteins. Myosin and actin form the contractile units (sarcomeres) of skeletal muscle.

Sarcomere: It is the functional unit of striated muscles defined by the area between two Z-lines. It is the basic contractile unit of a myocyte or a muscle fiber.

INTRODUCTION

Muscles are specialized tissues essential for movement, posture, and bodily function. They contract and relax in response to neural stimuli, enabling various physiological processes. The three types of muscle—skeletal, cardiac, and smooth—each have distinct roles, with skeletal muscles controlling voluntary movement and cardiac and smooth muscles regulating involuntary functions like heartbeat and organ function. Efficient muscle function is integral to maintaining mobility, stability, and overall health. Basic structure of a muscle is described in chapter 8.

TYPES OF MUSCLES

Muscle cells are designed to produce force and movement. Skeletal muscles attach to bones and move them relative to one another (Fig. 3.1). The heart's muscle is responsible for pumping blood through the vasculature. Skeletal and cardiac muscles are classified as striated because, their actin and myosin filaments form repeating arrays known as sarcomeres, giving them a striated look under the microscope. Smooth muscle does not have sarcomeres, but instead contracts its actin and myosin filaments to move blood vessels and hollow organs in the body.

MUSCLE CONTRACTION

Muscle contraction is a physiological process in which muscles are capable of generating tension when stimulated by the nervous system or by means of electrical impulses. The length of the muscle changes during a muscle contraction, which defines the type of contraction.

When we perform an activity, the muscles tighten, shorten, or lengthen as a result of the contraction. When we stretch, lift weights or hold something in our hands, or pick something up, the muscle contracts. Muscle relaxation is when contracted muscles return to their precontracted state, and is frequently followed by muscle contraction.

Skeletal muscle contraction begins with a nerve impulse that triggers the release of calcium ions from the sarcoplasmic reticulum. These calcium ions bind to troponin, causing a conformational change that moves tropomyosin away from actin's binding sites, allowing myosin heads to attach to actin and perform a power stroke. This interaction, powered by ATP, shortens the muscle fiber, resulting in contraction. The general mechanism of skeletal muscle contraction explained by Gash et al. is depicted in Figure 3.2.

Muscle contraction occurs:

- To offer stability to joints and connective tissues.
- To create heat to keep the body at a constant temperature. About 40% of the body's heat is converted into muscle work. When shivering, the skeletal muscles contract to warm the body as a reaction to the cold.
- To maintain the posture by using the muscles to hold positions like sitting or standing.

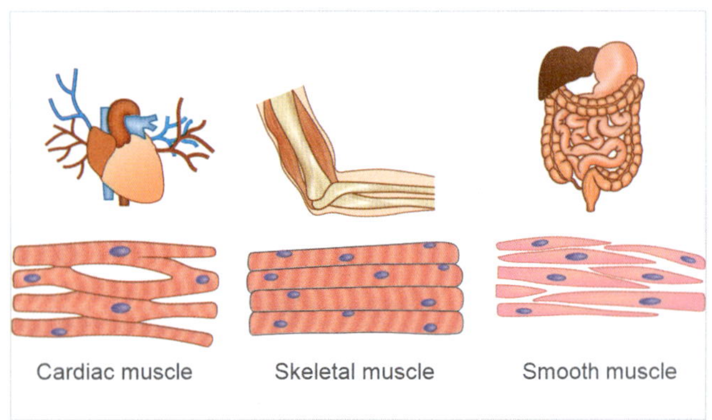

Fig. 3.1: Types of muscles

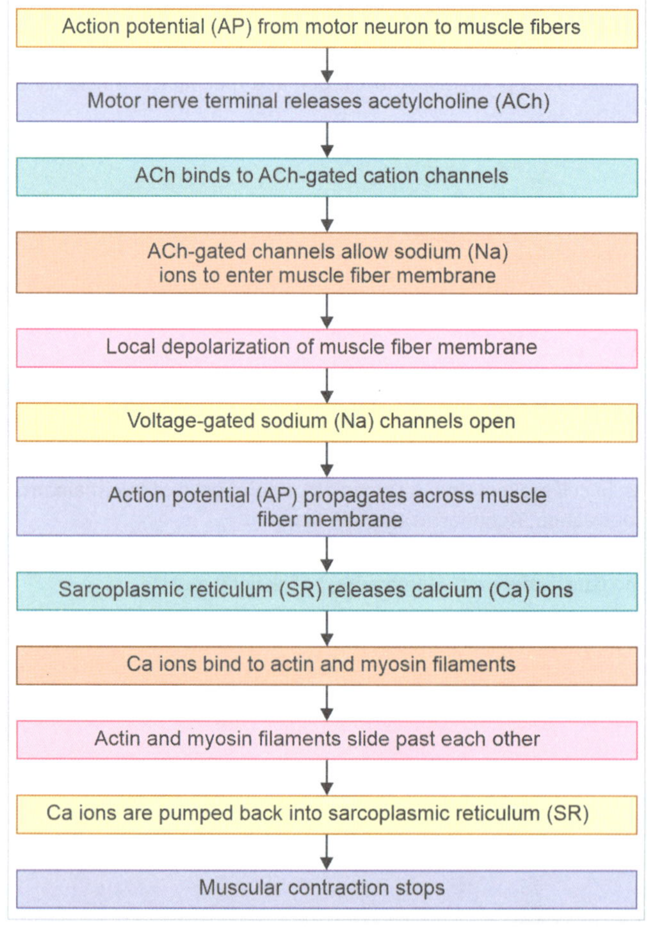

Fig. 3.2: General mechanism of skeletal muscle contraction

Types of Muscle Contraction

Striated muscles can undergo various types of contractions which are distinguished by changes in muscle length and tension during the contraction (Fig. 3.3). They are discussed as follows.

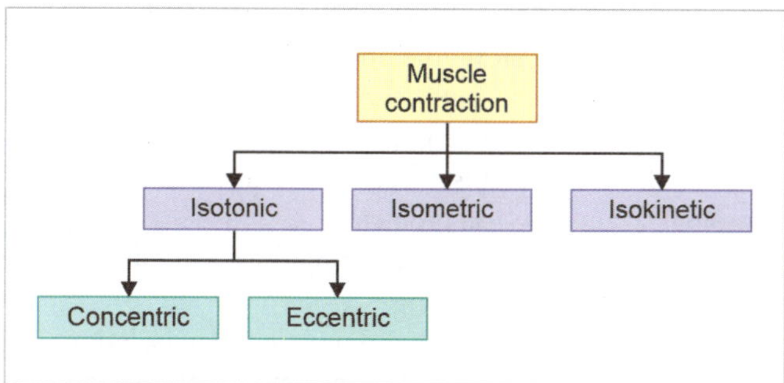

Fig. 3.3: Types of muscle contraction

Isometric Muscle Contractions

A process in which muscle contracts without a change in the length of the muscle is called an isometric contraction (Fig. 3.4A). An isometric muscular contraction would be necessary to grip a heavy object without moving it before raising. Isometric contractions occur while pushing against an immovable object or attempting to raise a weight that is excessively heavy. To preserve posture, isometric contractions are routinely utilized. Work done in isometric contractions is zero.

Isotonic Muscle Contractions

When a muscle contracts isotonically, the tension stays the same while the muscle length changes. This sort of contraction happens when the contraction force equals the entire load on the muscle. Walking, running, and crouching involve

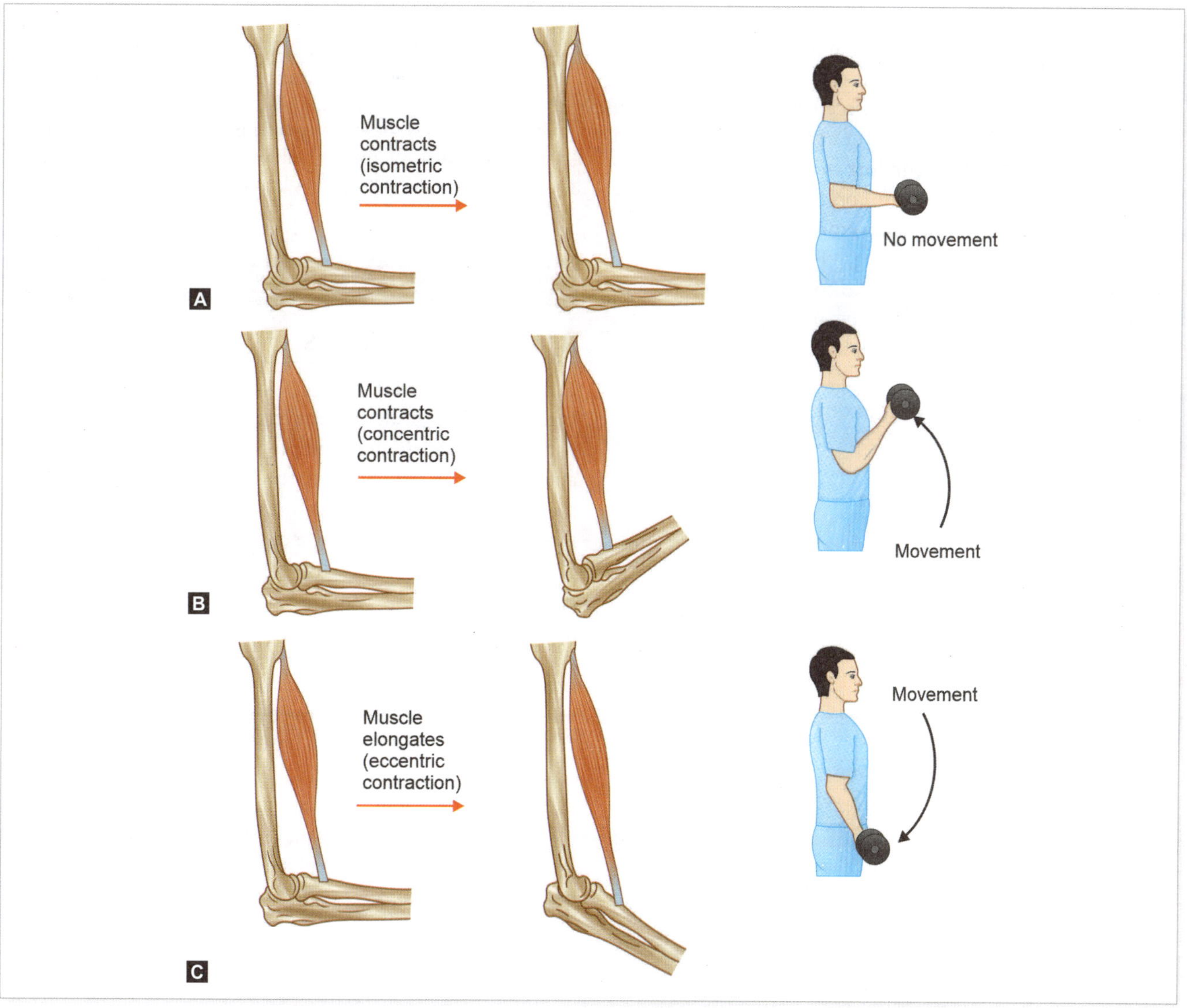

Figs 3.4A to C: **A.** Isometric contraction; **B.** Concentric contraction; **C.** Eccentric contraction

isotonic contractions. Work done in such contraction is either positive or negative.

Concentric or eccentric muscular contractions are both possible with isotonic forces, which were once referred as dynamic contractions.

Concentric Contraction

It is a contraction in which muscle shortens and the muscle attachments move toward each other (Fig. 3.4B). Concentric muscle contraction occurs when muscle tension is adequate to overcome the load, causing the muscle to contract and shorten. In this, positive work is done by the muscle. Sarcomere, muscle fiber, and muscle are all compressed by cross-bridge cycling. A concentric contraction of the biceps, for instance, would force the arm to bend at the elbow when lifting a heavy object, moving the object toward the shoulder. Quadriceps muscle contracts concentrically while ascending stairs.

Eccentric Contraction

An eccentric contraction causes a muscle to lengthen while it is still producing force; as a result, resistance is greater than the force produced (Fig. 3.4C). Eccentric contractions can be forced or voluntary. This type of contraction can happen involuntarily (e.g., while attempting to move a weight that is too heavy for the muscle to raise) or consciously (e.g., when the muscle is 'smoothing out' a movement or opposing gravity, such as during downhill walking). Eccentric contractions serve as a braking force in contrast to concentric contractions, protecting joints from harm. Negative muscle work is done in this type of contraction. Quadriceps muscle work eccentrically while descending stairs.

Difference between three types of muscle contractions is shown in Table 3.1.

Isokinetic Muscle Contraction

Another type of muscle contraction is isokinetic contraction. When the muscle length varies but the contraction velocity stays constant, this is known as an isokinetic muscular contraction. The muscle's force is variable and depends on the subject's engagement level and the joint's position within its range of motion.

TABLE 3.1: Work done by different types of muscle contractions

Types of contractions	Distance change	Function	Work
Concentric	Shortening (+*D*)	Acceleration	Positive $W = F \times (+D)$
Isometric	No change (0 *D*)	Fixation	Zero
Eccentric	Lengthening (–*D*)	Deceleration	Negative $W = F \times (-D)$

It is possible to load isokinetic muscles eccentrically or concentrically. The muscle contracts in an isokinetic concentric contraction while being loaded. In reality, the speed at which it occurs distinguishes it from other kinds of muscular contraction, and it can only be produced by a specific piece of equipment called isokinetic dynamometer. Isokinetic contractions are generally performed in fitness centers or in physical therapy settings. Despite being uncommon, isokinetic contractions are thought to accelerate the development of strength, endurance, and muscular growth. Physical therapists also utilize them to treat some neurological and physical problems.

The advantages of "accommodating resistance" include:

- Complete muscular loading.
- Effective joint movements and muscle group isolation.
- Presenting resistance equal to the amount of force applied by the subject.

Did You Know?

Isokinetic muscle work provides a unique advantage by offering accommodating resistance. This means that the resistance provided matches the force applied by the subject throughout the entire range of motion, allowing for complete muscular loading and effective isolation of muscle groups.

GROUP ACTION OF MUSCLES

Agonist/Prime Movers

The muscles which are responsible for initiation and maintenance of the desired movement are called prime movers or agonists. Even though multiple muscles may be active during an activity, the primary muscle is referred to as the prime mover or agonist. The main muscle responsible for forearm flexion, such as when raising a cup, is the biceps brachii.

Antagonist

An antagonist is a muscle that works in the opposite direction from the prime mover. Muscles must perform two crucial tasks in order to perform properly:

1. Maintaining body or limb position, such as holding an arm out or standing straight
2. Controlling rapid movement, e.g., triceps muscle that extends elbow joint is antagonist to brachialis, which is the prime mover for flexion of elbow joint.

The quadriceps femoris, contract when the knee is extended and is called the agonists of leg extension at the knee. To stop or slow down the movement, a group of antagonists called the hamstrings in the posterior compartment of the thigh is engaged.

Synergist

Synergist muscles are those that work together to produce or facilitate movement. They support the prime mover (agonist) muscles by either directly supplying force or stabilizing joints to allow for efficient movement. The concept of synergist muscles is well established in biomechanics and functional anatomy.

Synergist muscles are grouped into two primary types:

1. *Prime Synergists (Synergistic Agonist)*

These muscles aid the primary mover in achieving the desired movement. They frequently serve a comparable or overlapping function to the primary mover. In elbow flexion, the biceps brachii is the primary mover, while the brachialis serves as the primary synergist.

2. *Stabilizer Synergists*

These muscles help to support joints, allowing the prime mover to work more efficiently. They don't necessary add much force to the movement, but they do provide support and control.

Clinical Correlation

- During a squat, back muscles such as the erector spinae maintain the spine, allowing the quadriceps to act as prime movers.
- By targeting specific muscle groups through exercise therapy, practitioners can help restore balance, improve movement patterns, and prevent injuries.
- Additionally, recognizing the synergistic relationships between muscles can aid in designing more effective training programs and treatment protocols.

Fixator

The muscles that are responsible for stabilizing the proximal attachment of a muscle so that it can act efficiently at the distal joint are known as fixators, e.g., muscles that attach shoulder girdle to trunk are fixators, as they allow deltoid to abduct the shoulder joint.

Stabilizers

Stabilizer muscles are those that co-contract with other muscles to cause joint stiffness and activate quickly in response to perturbations using either a feed-forward or feedback regulatory mechanism.

For example, the shoulder stabilizer muscles (such as the deltoids, rotator cuff muscles, and latissimus dorsi muscles) aid in keeping the shoulder in place while allowing the elbow to move freely during biceps curls.

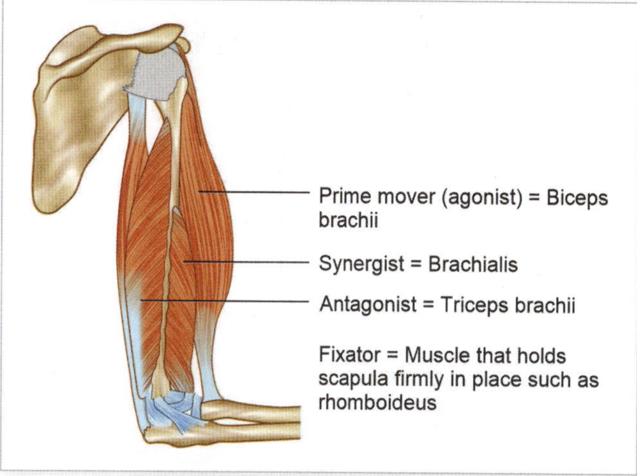

Fig. 3.5: Muscle actions during a biceps curl

The group action of muscles while performing a 'biceps curl' can be seen in Figure 3.5.

ANGLE OF PULL AND MECHANICAL EFFICIENCY OF MUSCLES

The angle of pull is the angle formed by the muscle's pull line and the bone it inserts into. The direction from the insertion point to the origin point must always be used when drawing the line of pull. When the direction of movement changes, the angle of pull also changes. Additionally, the force required to move something depends on the angle of pull. For example, the case of elbow flexion is explained here in detail.

Every time a muscle inserts into a bone, there is a vertical component (F_v) that is parallel to the bone. This part is referred as the translatory component. The component that acts perpendicular to the bone is referred as the rotatory or rotary component (Fig. 3.6A).

Figure 3.6B shows angle of pull of biceps brachii muscle with elbow flexion of 30°.

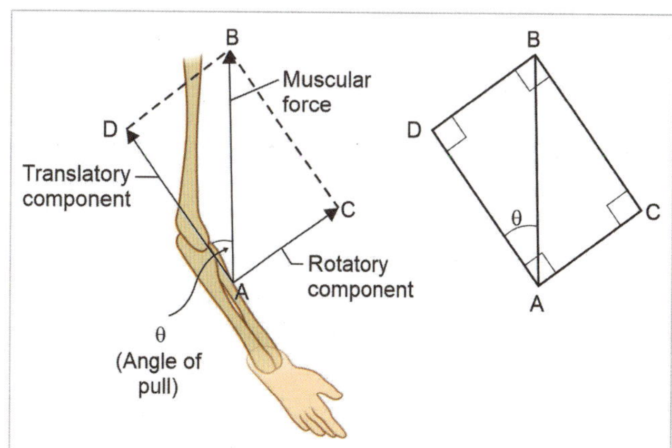

Fig. 3.6A: Components of muscle force

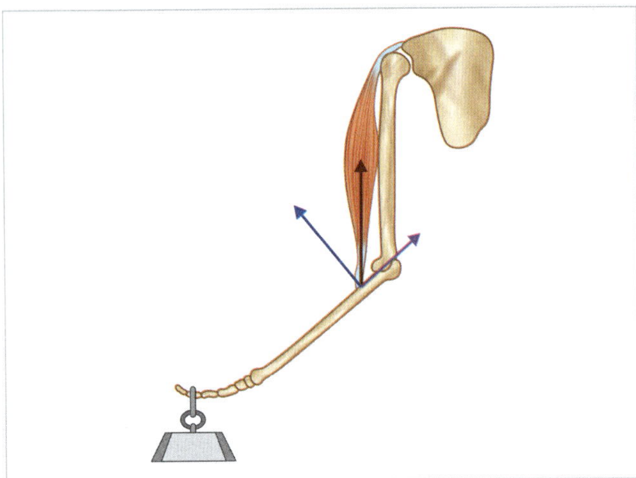

Fig. 3.6B: Angle of pull <90°

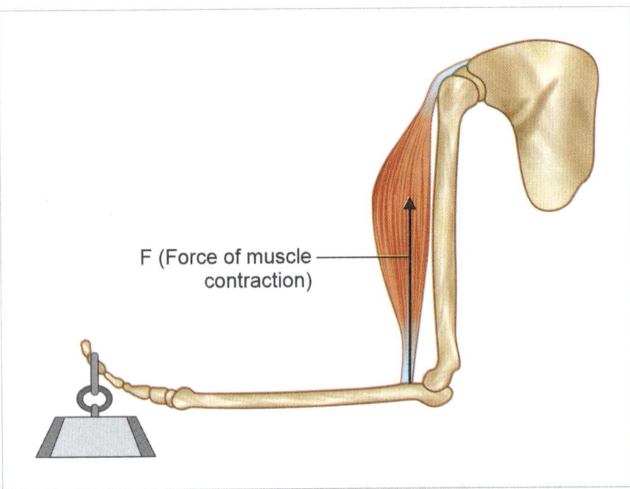

Fig. 3.6C: Angle of pull at 90°

The line of pull of the biceps brachii equals the rotational component when the elbow flexes exactly 90°. Therefore, the movement is being produced by the rotary component which is 100% of the force, i.e., greatest mechanical efficiency. The angle of pull in Figure 3.6B is <90°. There is another component of force acting on it in addition to the rotating component (blue arrow) which is represented by the purple arrow. Angles of pull that are <90° generate stabilizing forces since translatory component of the muscular force is directed at joints (Fig. 3.6B).

The angle of pull in Figure 3.6D is >90°. There is additional force acting on it, in addition to the rotating component (blue arrow), which is represented by the green arrow. When the angle of the pull exceeds 90°, the translatory component of muscular force generated is directed away from the joint, and it becomes a dislocating force.

- When the moment arm is at its longest and the muscle's angle of pull is 90°, the maximum amount of torque is generated. All of the muscle's effort is focused on producing rotation in this position (Fig. 3.6C).
- The amount of force that contributes to rotational motion (rotational force) decreases and the amount of force that does not contribute to rotation (nonrotational force) increases as the angle of pull of the muscle changes from 90°. As a result, the amount of force that can produce rotational motion (or torque) decreases.
- The angle of pull and the length of the moment arm at that angle determine how much of a muscle's force is used as rotational force (vector) and how much is used as nonrotational force (vector).
- Depending on the angle of pull, the muscle's nonrotational force tends to either stabilize the joint by providing compression or destabilize it by producing a distraction force (Fig. 3.6E). The muscle's force stabilizes or destabilizes the joint more when the angle of pull deviates from 90°, and rotates the joint less as a result.

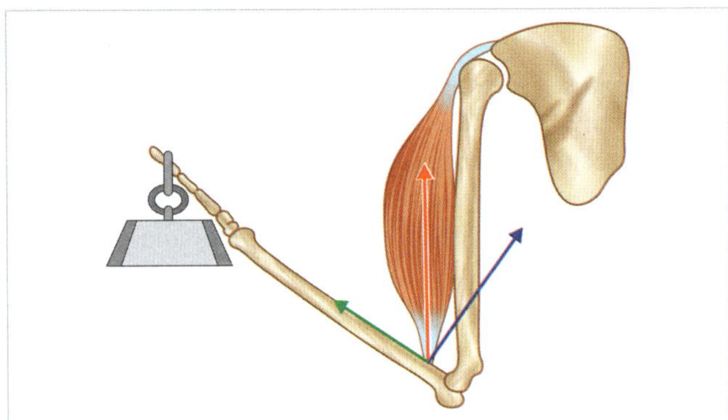

Fig. 3.6D: Angle of pull >90°

ACTIVE AND PASSIVE INSUFFICIENCY

Only muscles with several joints can experience the functional states of active and passive insufficiency. A multijoint muscle experiences active insufficiency when it contracts over both joints at once, creating so much slack that the muscular tension is virtually dissipated.

When a multijoint muscle is stretched to its maximum length at both joints, while also restricting the complete range of motion (ROM) of each joint it crosses, this condition is known as passive insufficiency.

Active Insufficiency

When a muscle that crosses two or more joints moves simultaneously in all of those joints and shortens to the point where it is unable to create effective tension, that muscle is said to be in active insufficiency. Active insufficiency is the state that occurs when an agonist (prime mover) is shortened to

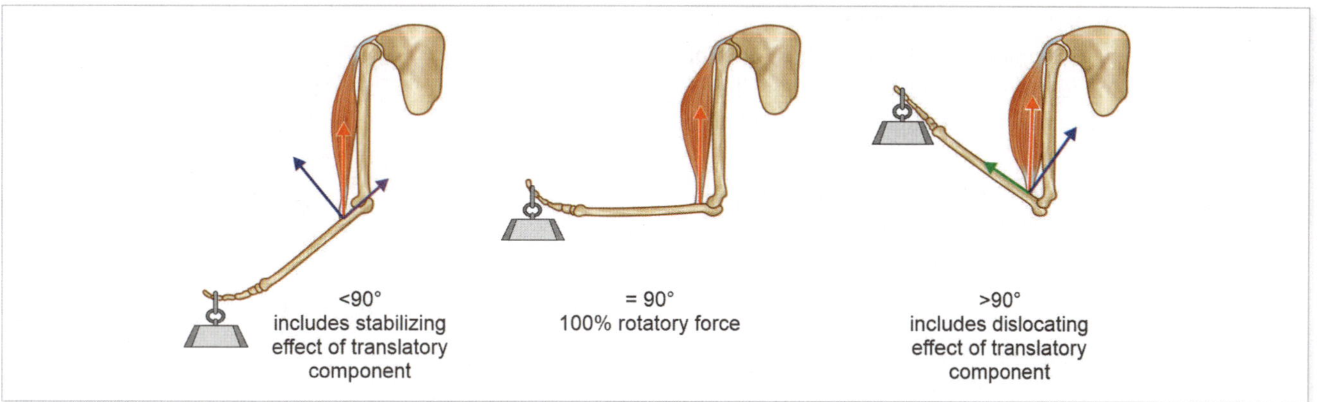

Fig. 3.6E: Effect of translatory component due to angle of pull

the point that it is unable to produce or sustain active tension. This limitation arises because a muscle is most effective at producing force while it is at an intermediate length, and its capacity to do so declines as it shortens further.

Key points concerning active insufficiency are:

- **Maximum shortening:** Muscles have an ideal length at which they can exert the most force. When a muscle is shortened beyond its ideal length, its ability to create force is reduced.
- **Multijoint muscles:** Active insufficiency is more noticeable in muscles that span several joints. When a muscle crosses two joints and shortens at one, it may be unable to generate adequate force at the other joint owing to active insufficiency.
- **Functional implications:** Active insufficiency can have functional consequences for movements that involve several joints. For example, a muscle that is maximally shortened at one joint may have limited ability to generate force at another joint during the same exercise.

For example, active insufficiency can be seen in the hamstring muscles. When the hip is extended and knee is flexed simultaneously, the hamstrings are shortened at both joints. In this configuration, they may generate less force, restricting their capacity to resist further knee flexion effectively (Fig. 3.7).

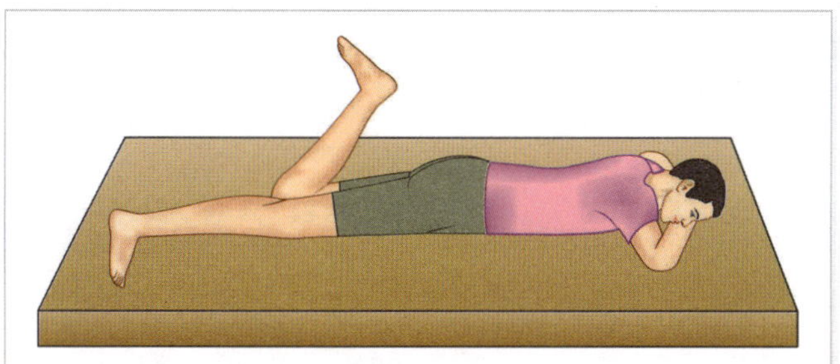

Fig. 3.7: Simultaneous hip and knee flexion creates active insufficiency of hamstrings muscles

Clinical Correlation

Acute insufficiency of biceps brachii

When shoulder and elbow are flexed simultaneously due to the shortening of the biceps brachii, maximal elbow flexion and maximal shoulder flexion cannot be accomplished at the same time (Fig. 3.8).

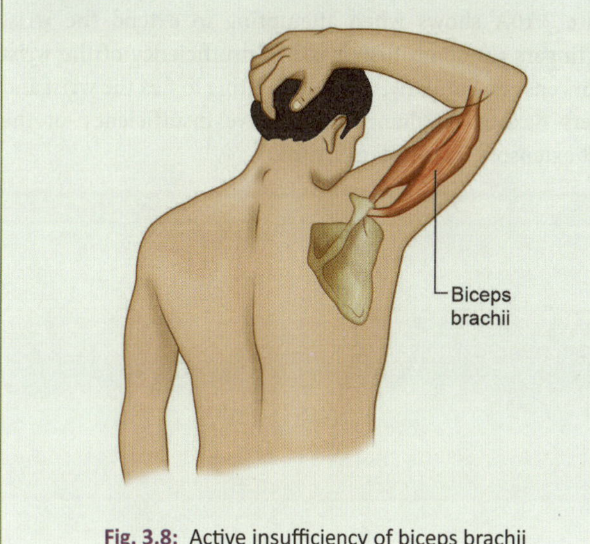

Fig. 3.8: Active insufficiency of biceps brachii

Passive Insufficiency

When a muscle that crosses multiple joints is stretched to its maximum length at all joints restricting the whole range of motion at each joint it crosses, this condition is known as passive insufficiency.

Did You Know?

Passive insufficiency can affect movement efficiency and performance in activities that require simultaneous stretching of a muscle across multiple joints.

Key points concerning passive insufficiency are:

Passive insufficiency is common in muscles that span multiple joints. When a muscle traverses numerous joints, it may become passively deficient when attempting to extend all of them at the same time.

- **Length-tension relationship:** Muscles have an ideal length at which they can exert the most force. Passive insufficiency occurs when a muscle is overstretched or unduly elongated, reducing its ability to generate force effectively.
- **Functional implications:** Passive insufficiency can impair joint range of motion and functional actions that need muscles to stretch across multiple joints. It is particularly important in instances where a muscle is stretched over both its origin and insertion simultaneously.
- **Example:** The hamstrings can also serve as an example of passive insufficiency. When the knee is extended and the hip is flexed, the hamstrings are passively insufficient because it cannot lengthen together across both the joints. Attempting to further extend the knee in this position might be limited due to the elongated state of the hamstrings (Fig. 3.9).

Figure 3.10A shows when attempting to extend the wrist and fingers simultaneously passive insufficiency of the wrist flexors ensues. Similarly, while attempting to flex the wrist and fingers flexors simultaneously passive insufficiency of the wrist extensors is seen (Fig. 3.10B).

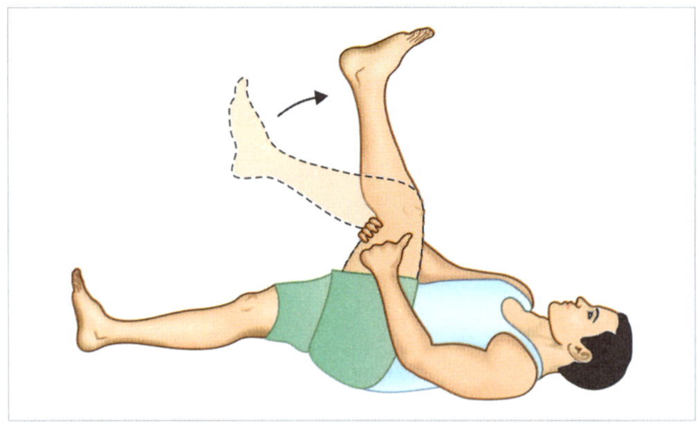

Fig. 3.9: Passive insufficiency of hamstrings

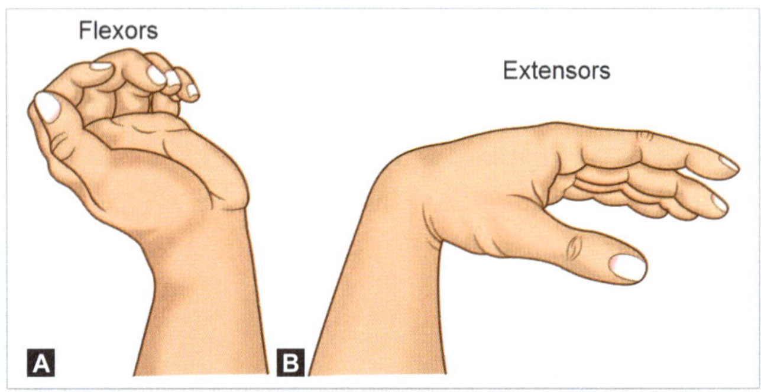

Figs 3.10A and B: Passive insufficiency of flexors and extensors at wrist and fingers

SUMMARY

- There are three types of muscles that contract: (1) Skeletal, (2) Cardiac, and (3) Smooth. Skeletal and cardiac muscles are striated, whereas, smooth muscles lack sarcomeres. Muscle contraction provides a variety of purposes, including joint stability, heat generating, and posture maintenance.
- Muscle contractions are classified into two types: Isometric (static contraction) and dynamic contraction.
- Dynamic contractions include isotonic (concentric and eccentric) and isokinetic contractions, each with a unique functional purpose.
- Isokinetic contractions, which require specific equipment, maintain a consistent contraction velocity, providing benefits in strength growth and rehabilitation.
- During movement, muscles work in groups: Agonists begin, antagonists oppose, synergists assist, fixators stabilize proximal attachments, and stabilizers work together to maintain joint stability.
- The angle of pull has an impact on mechanical efficiency; a 90° angle maximizes force for movement, angles <90° give stability, and angles >90° may diminish efficiency.
- Understanding these muscle movements and angles is essential for improving biomechanics and movement efficiency in a variety of tasks.
- Active insufficiency occurs when a muscle shortens, decreasing its ability to generate effective tension. This is most obvious in muscles that span many joints.
- Passive insufficiency: Caused by straining an antagonist muscle beyond its limit, which prevents further movement; prevalent in muscles that span many joints.
- Both include limits in muscle function, which affect joint motions and functional tasks.
- Hamstrings may exhibit active insufficiency during hip and knee flexion, as well as passive insufficiency during knee extension and hip flexion.
- Understanding these ideas is critical for optimizing biomechanics and avoiding limits in muscle performance.

FURTHER READINGS

- Feher J J. Quantitative Human Physiology: An Introduction. Academic Press; 2017 Jan 2.
- Gardiner Dena M. Principles of Exercise Therapy. CBS Publishers and Distributors Pvt Ltd; 4th ed., 2023.
- Gash M C, Kandle PF, Murray IV, Varacallo M. Physiology, muscle contraction. InStatPearls [Internet] 2022 May 2. StatPearls Publishing.
- Kisner C and Kolby L A. Therapeutic Exercise. Foundation and Technique Jaypee Publishers, New Delhi 1996.
- Sangwan S, Green R A, Taylor NF. Characteristics of Stabilizer Muscles: A Systematic Review. Physiother Can. 2014 Fall;66(4):348-58. doi: 10.3138/ptc.2013-51. PMID: 25922556; PMCID: PMC4403366.
- Sweeney H L, Hammers D W. Muscle Contraction. Cold Spring Harbor Perspectives in Biology. 2018 Feb 1;10(2):a023200.

STUDENT ASSIGNMENT

LONG ANSWER QUESTIONS

1. Define muscle contraction and its types with examples.
2. Describe active and passive insufficiency with examples.
3. Discuss angle of pull and mechanical efficiency with example of different joint angles.

SHORT ANSWER QUESTIONS

1. How do skeletal muscles contract?
2. Write about agonists, antagonists and synergists with relevant examples.
3. What are the advantages of 'accommodating resistance.'

MULTIPLE CHOICE QUESTIONS

1. **What is active insufficiency?**
   a. Shortening of a muscle beyond its optimal length
   b. Contraction of a muscle over both joints simultaneously
   c. Stretching of an antagonist muscle past its limit
   d. Lengthening of a muscle during movement
2. **When does passive insufficiency occur in a multi-joint muscle?**
   a. During contraction over both joints
   b. When the muscle is stretched to maximum length at one joint
   c. When the muscle is shortened at one joint
   d. While maintaining a stable position without movement
3. **Which statement is true about active insufficiency?**
   a. It occurs in single-joint muscles only
   b. It leads to increased muscular tension during contraction
   c. It is unrelated to joint movement limitations
   d. It involves stretching an antagonist muscle.
4. **What happens during active insufficiency in a multi-joint muscle?**
   a. The muscle contracts with optimal tension
   b. Tension is dissipated due to simultaneous contraction over both joints
   c. The muscle remains unaffected by joint movement.
   d. It results in excessive muscle lengthening.
5. **Passive insufficiency occurs when a multijoint muscle is stretched to:**
   a. Its optimal length at both joints
   b. Maximum length at both joints
   c. Maximum length at one joint only
   d. Minimum length at both joints
6. **Which type of muscle contraction involves a constant muscle length with no visible movement?**
   a. Concentric b. Isometric
   c. Eccentric d. Isotonic
7. **During which type of contraction does the muscle lengthen while still producing force?**
   a. Concentric b. Isometric
   c. Eccentric d. Isokinetic
8. **What characterizes isotonic muscle contraction?**
   a. Constant muscle length with no change in tension
   b. Muscle shortening with increased tension
   c. Muscle lengthening with increased tension
   d. Variable muscle length with constant tension
9. **In concentric contraction, muscles:**
   a. Lengthen while producing force
   b. Shorten while producing force
   c. Remain at a constant length with no tension change
   d. Lengthen with no tension change
10. **Which type of contraction involves a constant contraction velocity with variable force?**
    a. Concentric b. Eccentric
    c. Isometric d. Isokinetic

ANSWER KEY

1. b **2.** b **3.** c **4.** b **5.** b **6.** b **7.** c **8.** b **9.** b **10.** d

Section II

Examination/Testing Techniques

SECTION OUTLINE

4 Manual Muscle Testing

Sheetal Kalra

LEARNING OBJECTIVES

After the completion of the chapter, the readers will be able to:

- Define MMT, its principles, aims, and various methods.
- Explain the indications and limitations of MMT.
- Explain the techniques of MMT for group and individual muscles.
- Demonstrate the techniques of MMT for upper limb, lower limb, spine and face muscles.
- Explain the techniques for respiratory muscle strength and pelvic floor muscle strength assessment.

CHAPTER OUTLINE

- Introduction
- Principles of Manual Muscle Testing
- Aims of Manual Muscle Testing
- Techniques
- Principles of Application of Resistance
- Indications of Manual Muscle Testing
- Limitations
- Guidelines
- Grades
- Techniques for Upper Limb Muscles
- Techniques for Lower Limb Muscles
- Muscles of Spine
- Techniques for Spinal Muscles
- Techniques for Facial Muscles
- Respiratory Muscle Strength Assessment
- Assessment of Pelvic Floor Muscle Strength

KEY TERMS

Break testing: Resistance is applied to the body component near the conclusion of its range of motion during manual muscle testing. The reason it is termed the "break test" is that the patient's goal is to prevent the physiotherapist from "breaking" their muscle contraction when they apply resistance.

Make testing: In a "make test", the patient must push back against a fixed resistance that the examiner provides at their maximum voluntary effort.

Manual muscle testing (MMT): A technique which is used in physical therapy and rehabilitation to evaluate the strength and functionality of specific muscles or muscle groups.

MRC scale: A commonly used tool for evaluating muscular strength is the Medical Research Council (MRC) Scale, which ranges from Grade 5 (normal) to Grade 0 (no apparent contraction).

INTRODUCTION

Manual Muscle Testing (MMT) is a method used to assess the strength and function of specific muscles or muscle groups. It is essential for diagnosing muscle weakness, imbalances, or neuromuscular conditions and plays a crucial role in planning effective rehabilitation strategies. MMT helps track patient progress and ensures targeted interventions for optimal recovery.

In the MMT procedure, the patient's motions and muscle activity are carefully observed, and the muscle is palpated to assess its size, tone, and texture. To isolate the muscle being tested, the patient is typically positioned in a precise way. To provide a correct assessment of muscular strength, the resistance is typically applied in a specified direction and at a specific angle.

The MMT can be applied in a range of clinical contexts, including neurology, sports medicine, occupational therapy, and physical therapy. In order to diagnose and treat muscle and movement disorders, monitor patient progress, and create individualized treatment programs, it is frequently used in conjunction with other evaluation instruments.

Did You Know?

In addition to examining muscular strength, manual muscle testing (MMT) is an essential tool for evaluating neurological function. Healthcare providers can learn a great deal about the health of the neuromuscular system by evaluating a patient's capacity to produce and regulate muscular movements in the face of opposition. Because of this reason, MMT is an effective diagnostic technique for a variety of illnesses, including neurological disorders like stroke and spinal cord injury as well as musculoskeletal problems.

The MMT can be defined as a technique which is used in physical therapy and rehabilitation to evaluate the strength and functionality of specific muscles or muscle groups. To assess a muscle's capacity to produce force, specific resistance is applied against the muscle's motion. It can be helpful in determining the difference between muscle weakness and imbalance or inadequate muscle endurance.

PRINCIPLES OF MANUAL MUSCLE TESTING

- **Muscle grading:** MMT measures muscle strength using a scale that ranges from 0 (no muscle contraction) to 5 (normal strength against full resistance). The physiotherapist provides a grade based on the amount of resistance the patient is able to overcome during a particular movement. Each grade corresponds to a particular level of muscular function.
- **Muscle isolation:** The physiotherapist uses MMT to isolate particular muscles or muscle groups in order to assess their strength and functionality. This involves placing the patient in a targeted position that emphasizes the particular muscle or muscle group being assessed and making sure that other muscles aren't covering up for the targeted muscle.
- **Testing position:** Accurate and reliable results from MMT depend on the testing position employed. The posture should give the patient full range of motion, be comfortable for them, and give the physiotherapist appropriate leverage for exerting resistance.
- **Stabilization:** The physiotherapist may employ a variety of stabilization strategies, such as holding the limb being tested or utilizing straps or other supports, to make sure the patient can maintain a stable position during MMT.
- **Palpation:** A crucial part of MMT is palpation, which is the act of touching and feeling the muscle being evaluated. This enables the physiotherapist to evaluate the muscle's tone, texture, and soreness, which can reveal more details regarding the function of the muscle.
  Examiner should start with the side that is not dominant or damaged and keep applying pressure across the entire range. As holding a breath during the test can give a forced result, make sure the patient is breathing normally.
- **Gravity and resistance:** Physiotherapists should always assess in an antigravity position initially. As long as the muscles are really weak, one can test opposite gravity by looking at a horizontal plan.

This resistance may be supplied directly by the physiotherapist or against gravity (for example, by lifting a limb against gravity). To prevent injury, the resistance should be given gradually and at a level adequate for the grade of muscle being tested. Regardless of the muscle being tested, resistance should be applied in the direction that is exactly counter to the "line of pull" of that muscle.

AIMS OF MANUAL MUSCLE TESTING

- Evaluation of strength and function of a particular muscle or muscle groups.
- **Identify muscle weakness:** Physiotherapists can use MMT to detect muscle weakness or dysfunction. The physiotherapist can detect regions of weakness and modify treatment as necessary by rating the strength of particular muscles or muscle groups.
- **Track progress:** MMT can be used to track a patient's advancement over time, for instance, following an injury or throughout therapy. The physiotherapist can monitor improvements in muscle strength and function and modify treatment as necessary by performing the muscle tests often.
- **Guide in planning of treatment:** Using MMT results, the physiotherapist can create a treatment plan that is customized to the patient's individual requirements. This could involve strengthening exercises for weak muscles, or other measures to deal with any diagnosed muscle dysfunction.

- **Determine preparedness for return to activities:** MMT can also be used to determine whether a patient is ready to resume a certain activity, such as employment or sports. The physiotherapist can establish if the patient is prepared to return to that activity safely by assessing muscle strength and function in relation to the demands of the activity.

Clinical Correlation

Starting with a small amount of resistance and progressively increasing it as needed is crucial when applying resistance during MMT. This reduces the patient's risk of damage or unnecessary discomfort while enabling the physician to precisely estimate the patient's strength.

Clinicians should compare the strength of matching muscles on both sides of the body when doing a bilateral muscle assessment. Notable variations in strength could point to underlying problems that need more analysis.

TECHNIQUES

- **Break testing:** The most popular MMT technique asks the patient to resist the examiner applying pressure to a muscle or group of muscles. The "line of pull" of the involved muscle or muscles should always be followed while applying manual resistance. The patient is instructed to hold the part at that position and not to allow the physiotherapist to "break" the hold using physical resistance. This can be done near the end of the range of motion, or at a point in the range where the muscle is most taxed.
- **Make testing:** In this technique, the patient is asked to contract a muscle or set of muscles while the examiner applies resistance. While the patient tries to contract the muscle, the examiner exerts tension on the area to stop it from moving. In contrast to Break Testing, where the examiner applies a force to overcome the patient's resistance, in "Make Testing", the examiner matches the patient's generated force without exceeding it. When a patient cannot actively move a joint because of an injury or paralysis, this technique is used.
- **Gravity eliminated testing:** This technique involves carefully placing the patient's limb or body to remove gravity's influence from the muscle under test. When a patient has extreme weakness or paralysis, this can be helpful in determining their muscular strength.

PRINCIPLES OF APPLICATION OF RESISTANCE

In order to ensure consistency of method, external force (resistance) is applied to one-joint muscles at the end of their range during manual muscle testing. Usually, two-joint muscles are examined in the mid-range, where length-tension is more advantageous.

The area near the distal end of the segment to which the muscle attaches is where the physiotherapist should apply resistance to an extremity or portion. The scapular muscles and the hip abductors are two common exceptions to this rule.

The preferred point of resistance for assessing the vertebro-scapular muscles, such as the rhomboids, is the arm as opposed to the scapula where these muscles insert. The longer lever more accurately reflects the functional requirements that take into account the arm's weight. A painful condition that needs to be avoided or a healing wound in a location where resistance might otherwise be applied are two more exceptions to the general rule of applying distal resistance.

To reach the highest tolerated force intensity, the physiotherapist should apply resistance while the patient is fully conscious, fairly slowly and gradually, somewhat exceeding the muscle's force as it develops over 2–3 seconds.

When resistance is applied in the direction opposite to the muscular force or torque, it is possible to measure the strength of the muscles. Grades 2, 1, and 0 involve testing weak muscles in a plane parallel to the direction of gravity while supporting the body part on a smooth, flat surface with the least amount of friction possible. In order to reduce friction, a powder board could be utilized. Resistance is applied perpendicular to the line of gravity (grades 4 and 5) for stronger muscles that can perform a full range of motion in a direction against the pull of gravity (grade 3).

INDICATIONS OF MANUAL MUSCLE TESTING

- **Musculoskeletal injuries:** After a musculoskeletal injury, MMT can be performed to assess muscle strength and function. The physiotherapist can create a treatment plan to enhance muscle function and lower the risk of further injury by identifying areas of weakness.
- **Neurological diseases:** MMT can also be used to assess muscular function and strength in people with neurological conditions including Parkinson's disease, multiple sclerosis, or stroke. The physiotherapist can create a treatment plan to enhance function and quality of life by identifying areas of muscular dysfunction or weakness.
- **Postsurgical rehabilitation:** MMT is frequently used to assess muscular function and strength as part of post-surgical rehabilitation and to direct treatment strategies. A muscle evaluation at the surgical site or in other potential surgically impacted parts of the body may fall under this category.
- **Sports performance:** MMT can be used to assess muscle function and strength in athletes, both to pinpoint areas of injury risk and to direct training and conditioning regimens to enhance performance.

- **Occupational therapy:** MMT is also utilized in occupational therapy to assess muscular function and strength in individuals who have sustained job-related injuries or disabilities, as well as to create interventions to enhance function and assist a return to work.

LIMITATIONS

- **Subjectivity:** The findings of MMT can vary depending on how the examiner interprets the strength and makes judgments about it. The outcomes of the test can be affected by a number of variables, including the patient's position, the examiner's force, and the patient's effort.
- **Exclusivity:** MMT's limited scope excludes information on other elements of muscle function, such as coordination, endurance, or power, and only assesses the strength of specific muscles or muscle groups. It should, therefore, be used in conjunction with other examinations and assessments to offer a thorough assessment of muscle function.
- **Sensitivity:** MMT may not be sensitive enough to pick up on minute changes in muscular strength, especially in patients who already have modest or early-stage muscle weakening. In certain circumstances, other testing techniques like electromyography (EMG) or dynamometry can be more suitable.
- **Reliability:** A number of variables, including examiner expertise, patient effort, and modifications in testing procedures, can have an impact on MMT's reliability. As a result, it is crucial to guarantee that the examiner has training and experience in MMT techniques and that the testing processes are uniform and fair.
- **Safety issues:** Using stress on muscles during MMT can be uncomfortable or painful for certain individuals, especially those who have recently sustained injuries or have diseases like fibromyalgia. During testing, care should be taken to prevent aggravating discomfort or reinjuring an existing injury.

GUIDELINES

1. **Preparation of the patient:** The patient should be positioned so that they are comfortable, with the muscle group physiotherapist wants to test in a relaxed state. For instance, the patient should have their arm extended and should be at ease when testing the strength of the biceps muscle. The patient should be wearing loose clothing so that the limb can be moved in the available range.
2. **Describe the process:** Inform the patient about what will be done and what will be examined. Let them know that as they flex their muscles, resistance will be applied.
3. **Measure the resistance to gravity:** Check the resistance to gravity first. Asking the patient to move a limb against gravity is part of this. Ask the patient to straighten their leg while lying on their back, for instance, to gauge the strength of the quadriceps muscle. If they can perform the movement without difficulty, move on to step 4.
4. **Apply resistance:** The physiotherapist should ask the patient to contract their muscle against the physiotherapist's hand by placing the hand on the end of the limb that is farthest from the body. This will apply resistance to the muscle group being evaluated. To determine whether the muscle is powerful enough to overcome it, provide enough resistance.
5. **Grade the strength:** Use a grading scale to determine the muscle's strength. The scale for grading runs from 0–5, with 0 denoting complete loss of muscle strength and 5 denoting typical strength.
6. **Record the outcomes:** Note the tested muscle group, the strength rating, and any further comments regarding the patient's capacity to do the test.
7. **Repeat the test:** To ensure correctness, repeat the test numerous times. Before doing another test if the patient is tired, let them rest.

GRADES

Daniels and Worthingham's Muscle Grading Scale

The Daniels and Worthingham's Muscle Grading Scale is a widely used system to assess and document muscle strength during MMT. The scale ranges from 0 to 5, with each grade corresponding to specific levels of muscle function and resistance:

Grade	Meaning	Interpretation
Grade 5	Normal	The muscle can move the joint through its full range of motion (ROM) against gravity and holds against maximum resistance applied by the examiner
Grade 4	Good	The muscle can move the joint through its full ROM against gravity and holds against moderate resistance, but not maximal
Grade 3	Fair	The muscle can move the joint through its full ROM against gravity but cannot hold against any additional resistance
Grade 2	Poor	The muscle can move the joint through its full ROM, but only when gravity is eliminated or minimized
Grade 1	Trace	There is no movement of the joint, but a slight muscle contraction can be palpated or observed
Grade 0	Zero	No muscle contraction is observed or palpated

Oxford Scale

The Oxford scale, also known as the medical research council (MRC) Manual Muscle Testing scale, is the most widely used technique for assessing muscle strength. This technique involves measuring the strength of the patient's major muscles in the upper and lower extremities against the examiner's resistance and evaluating it from 0 to 5.

Grade	Interpretation
0–No contraction	No visible or palpable movement in the muscle
1–Flicker	Evidence of muscle contraction with no joint movement
2–Movement with gravity eliminated	Muscle can move if the joint is positioned appropriately to counteract gravity
3–Movement against gravity	Muscle can move against gravity, but not against resistance
4–Movement against some resistance	Muscle can move against moderate resistance applied by the examiner
5–Normal movement against full resistance	Muscle can move against strong resistance without evident fatigue

Kendall Muscle Grading Scale

The Kendall Muscle Grading Scale is another system for assessing muscle strength, similar to the Oxford and Daniels & Worthingham scales but with slight variations. It evaluates the ability of a muscle or muscle group to perform a specific motion against gravity and resistance, typically using grades ranging from 0 to 10. Here is an overview of its structure:

Kendall Muscle Grading Scale

Feature	Functions of the muscle	Muscle grades			Kendall scale
No movement	No contractions felt or seen in the muscle	0	0	Zero	No visible or palpable contraction
	Tendon becomes prominent or feeble contraction felt in the muscle, but no visible movement of the part	T	1	Trace	Visible or palpable contraction (No ROM)
Test movement	**Movement in horizontal plane**				
	Moves through partial range of motion	1	2–	Poor–	Partial ROM, gravity eliminated
	Moves through complete range of motion	2	2	Poor	Full ROM, gravity eliminated
	Antigravity position				
	Moves through partial range of motion	3	2+	Poor+	Gravity eliminated/slight resistance or <1/2 range against gravity
Test position	Gradual release from test position	4	3–	Fair–	>1/2 but <Full ROM, against gravity
	Holds test position (no added pressure)	5	3	Fair	Full ROM against gravity
	Holds test position against slight pressure	6	3+	Fair+	Full ROM against gravity, slight resistance
	Holds test position against slight to moderate pressure	7	4–	Good–	Full ROM against gravity, mild resistance
	Holds test position against moderate pressure	8	4	Good	Full ROM against gravity, moderate resistance
	Holds test position against moderate to strong pressure	9	4+	Good+	Full ROM against gravity, almost full resistance
	Holds test position against strong pressure	10	5	Normal	Normal, maximal resistance

Modified from 1993 Florence P. Kendall. Author grants permission to reproduce this chart

Clinical Correlation

Clinicians should use standardized procedures and consistent criteria throughout evaluations when employing grading scales like the Oxford Scale for Manual Muscle Testing (MMT) or the Medical Research Council (MRC). Accurate tracking of changes in muscle strength and function is made possible by this uniformity, which guarantees consistent and dependable results throughout time. Moreover, when evaluating MMT results, doctors should take into account the unique needs and capabilities of each patient, accounting for variables including age, comorbidities, and individual differences in muscle morphology and physiology respectively. Healthcare professionals can improve the therapeutic usefulness of MMT in patient care and preserve inter-rater reliability by receiving regular training and calibration.

A quick checklist before applying MMT

- Physiotherapist should explain to the patient all the test's components as well as the outcomes.
- Apply pressure consistently and start with the side that is not affected.
- Remind the patient to breathe normally throughout the test because holding breath can cause a forced result to be more noticeable.
- Ensure the patient has complete range of motion and is wearing comfortable attire.
- To enable them to fully focus their effort on the body area being tested, place the patient in a well-supported position.
- Always perform the initial test while antigravity. The muscles are next put to the test in the horizontal plane to see if they are too frail to work against gravity.
- It is necessary to apply resistance immediately opposite the "line of pull" of the muscles being tested.
- To gain the most realistic image of strength and/or impairment, always test both sides and compare the strength or muscle grade of both limbs.

TECHNIQUES FOR UPPER LIMB MUSCLES

Muscles of Shoulder Flexion

Muscles	Origin	Insertion	Action	Nerves supply
Deltoid	Lateral one-third of the clavicle, acromion, the lower lip of the crest of the spine of the scapula	Deltoid tuberosity of the humerus	Abducts arm; anterior fibers flex and medially rotate the arm; posterior fibers extend and laterally rotate the arm	Axillary nerve (C5, 6) from the posterior cord of the brachial plexus
Coracobrachialis	Coracoid process of the scapula	Medial side of the humerus at mid-shaft	Flexes and adducts the arm	Musculocutaneous nerve (C5, 6)

MMT Techniques for Shoulder Flexors

Muscles	Patient's position	Physiotherapist's position	Techniques	Demonstrations
• Anterior deltoid • Coracobrachialis	• **For grades 3–5:** ▪ **Group muscle testing:** The patient sits or stand with arm at the side. The elbow is kept straight. ▪ **Coracobrachialis:** Elbow completely flexed, forearm in supination.	Physiotherapist to **stand on the side** of the patient	The patient is asked to **actively flex** the shoulder. • **For grades 4–5:** ▪ Resistance is applied over distal humerus just above the elbow in the direction opposite to shoulder flexion.	Resistance Movement **Shoulder flexors grades 3–5**

Contd...

Muscles	Patient's position	Physiotherapist's position	Techniques	Demonstrations
	• **For grades 0–2:** The patient is in side lying position, the elbow flexed to 90° and supported by the physiotherapist		▪ The patient must be able to move through the full range of motion (active resistance testing) or maintain an end point range (break testing) against the highest level of resistance in order to have normal muscle strength, grade 5. • **For grades 2 and 1:** Movement is assessed with reduced resistance, focusing on partial range of motion for grade 2 or detecting a slight contraction without visible movement for grade 1.	**Coracobrachialis grades 4–5** **Shoulder flexors grades 0–2**

Muscles of Shoulder Extension

Muscles	Origin	Insertion	Action	Nerves supply
Latissimus dorsi (LD)	Vertebral spines from T7 to the sacrum, posterior third of the iliac crest, lower 3 or 4 ribs, sometimes from the inferior angle of the scapula	Floor of the intertubercular groove	Extends the arm and rotates the arm medially	Thoracodorsal nerve (C7, 8) from the posterior cord of the brachial plexus
Deltoid	Lateral one-third of the clavicle, acromion, the lower lip of the crest of the spine of the scapula	Deltoid tuberosity of the humerus	Abducts arm; anterior fibers flex and medially rotate the arm; posterior fibers extend and laterally rotate the arm	Axillary nerve (C5, 6) from the posterior cord of the brachial plexus
Teres major	Dorsal surface of the inferior angle of the scapula	Crest of the lesser tubercle of the humerus	Adducts the arm, medially rotates the arm, assists in arm extension	Lower subscapular nerve (C5, 6) from the posterior cord of the brachial plexus

MMT Techniques for Shoulder Extensors

Muscles	Patient's position	Physiotherapist's position	Techniques	Demonstrations
• **Latissimus dorsi (LD)** • **Deltoid** • **Teres major (TM)**	• **For grades 3–5:** ▪ **Group muscle testing:** The patient sits or stands with arm at the side. The elbow is kept straight. ▪ **LD and TM:** The patient is in prone position with elbow extended, arm free to move, palm directed up in direction of internal rotation.	The physiotherapist **stands on the test side** of the patient	The patient is asked to extend the arm through available range • **For grades 4 and 5:** ▪ The resistance is applied over posterior surface of distal humerus in a direction opposite to extension and adduction for LD and TM.	**Shoulder extensors grades 3–5**

Contd...

Muscles	Patient's position	Physiotherapist's position	Techniques	Demonstrations
	• **For grades 0–2:** Patient is in side lying position with upper extremity supported by the physiotherapist		▪ The patient must be able to move through the full range of motion (active resistance testing) or maintain an end point range (break testing) against the highest level of resistance in order to have normal muscle strength, grade 5. • **For grades 2 and 1:** Movement is assessed with reduced resistance, focusing on partial range of motion for grade 2 or detecting a slight contraction without visible movement for grade 1.	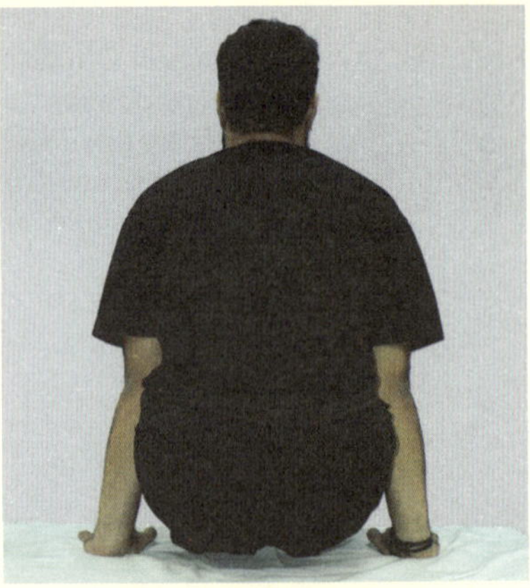 **Latissimus dorsi (bilateral) alternate position** 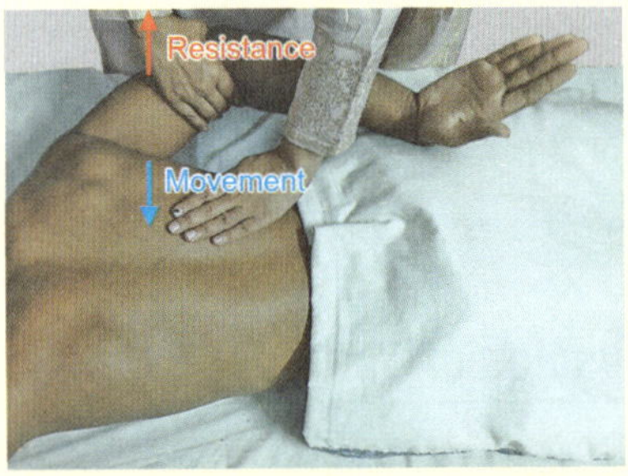 **Latissimus dorsi and teres major grades 3–5 (Physiotherapist is palpating LD)** 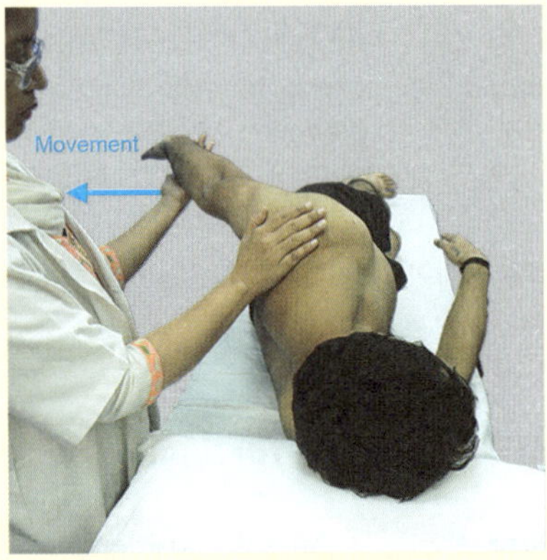 **Shoulder extensors grades 0–2**

Muscles of Shoulder Abduction

Muscles	Origin	Insertion	Action	Nerves supply
Deltoid	Lateral one-third of the clavicle, acromion, the lower lip of the crest of the spine of the scapula	Deltoid tuberosity of the humerus	Abducts arm; anterior fibers flex and medially rotate the arm; posterior fibers extend and laterally rotate the arm	Axillary nerve (C5, 6) from the posterior cord of the brachial plexus
Supraspinatus	Supraspinatus fossa	Greater tubercle of the humerus (highest facet)	Abducts the arm (initiates abduction)	Suprascapular nerve (C5, 6) from the superior trunk of the brachial plexus

MMT Techniques for Shoulder Abductors

Muscles	Patient's position	Physiotherapist's position	Techniques	Demonstrations
• Deltoid • Supraspinatus	• **For grades 3–5:** ▪ **Deltoid:** The patient is in sitting position, elbow flexed at right angle and shoulder abducted to 90° (middle deltoid), shoulder abducted in slight flexion and external rotation (anterior deltoid), shoulder abducted in slight extension and medial rotation (posterior deltoid) ▪ **Supraspinatus:** The patient is in a standing position. The arm is at the side of the body. • **For grades 0–2:** The patient is in supine position. The test arm is at the side in neutral rotation with the elbow extended.	The physiotherapist **stands on the test side of the patient**. Scapula stabilization is provided by the physiotherapist.	The patient is asked to **abduct the shoulder** through available range • **For grades 4–5:** ▪ The physiotherapist provides resistance over distal humerus in the direction opposite to shoulder abduction in the scapular plane. ▪ The patient must be able to move through the full range of motion (active resistance testing) or maintain an end point range (break testing) against the highest level of resistance in order to have normal muscle strength, grade 5. • **For grades 2 and 1:** Movement is assessed with reduced resistance, focusing on partial range of motion for grade 2 or detecting a slight contraction without visible movement for grade 1.	**Supraspinatus grades 3–5** **Anterior fibers of deltoid MMT grades 3–5** 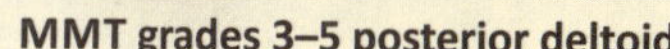 **MMT grades 3–5 posterior deltoid** 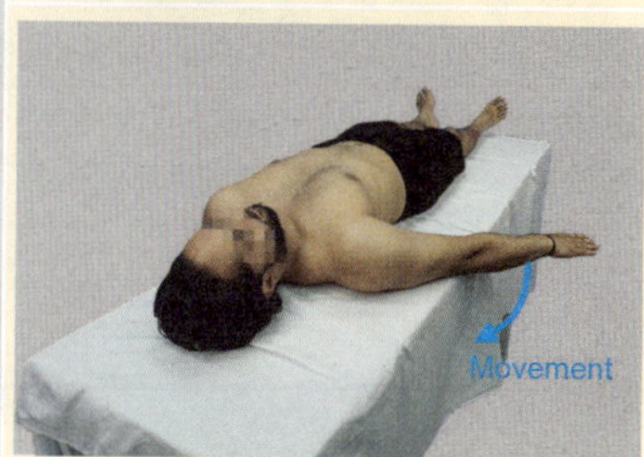 **Shoulder abductors grades 0–2**

Muscles of Shoulder Horizontal Adduction

Muscle	Origin	Insertion	Action	Nerve supply
Pectoralis major	Medial 1/2 of the clavicle, manubrium and body of sternum, costal cartilages of ribs 2–6, sometimes from the rectus sheath of the upper abdominal wall	Crest of the greater tubercle of the humerus	Flexes and adducts the arm, medially rotates the arm	Medial and lateral pectoral nerves (C5–T1)

MMT Techniques for Shoulder Horizontal Adductors

Muscle	Patient's position	Physiotherapist's position	Techniques	Demonstrations
Pectoralis major	• **For grades 3–5:** ▪ The patient lies in a supine position with the shoulder abducted to the desired angle based on the fibers being tested. ▪ **Clavicular fibers:** Shoulder abducted to 60°. ▪ **Complete muscle testing:** Shoulder abducted to 90°. ▪ **Sternal fibers:** Shoulder abducted to 120°. • **For grades 0–2:** Patient is in sitting position with shoulder in flexed position.	The physiotherapist **stands on the test side** of the patient	• The patient is asked to **horizontally adduct** the shoulder (across the midline of the chest) through full available range. • **For complete muscle testing at 90° abduction:** The patient horizontally moves the arm across chest • **For clavicular head testing at 60° abduction:** The patient moves the arm up and across the chest. • **For Sternal head testing at 120° abduction:** The patient moves the arm down and across the chest • **For grades 4 and 5:** Resistance is applied to forearm, just proximal to wrist in a direction opposite to horizontal adduction ▪ The patient must be able to move through the full range of motion (active resistance testing) or maintain an end point range (break testing) against the highest level of resistance in order to have normal muscle strength, grade 5. • **For grades 2 and 1:** Movement is assessed with reduced resistance, focusing on partial range of motion for grade 2 or detecting a slight contraction without visible movement for grade 1.	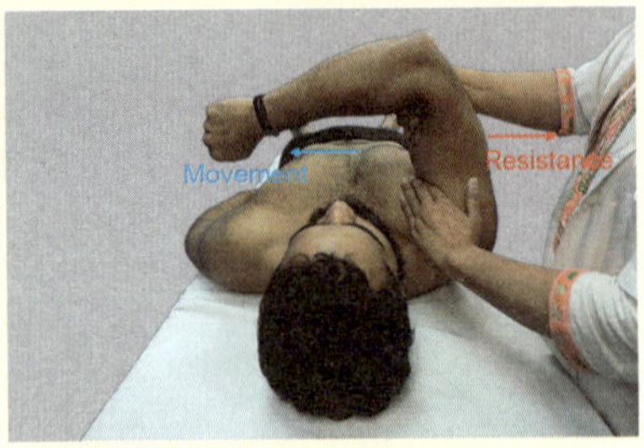 **Shoulder horizontal adductors (complete muscle testing) grades 3–5** 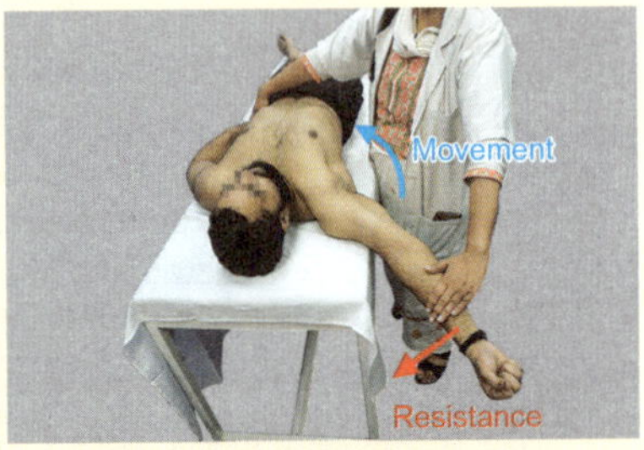 **Shoulder horizontal adductors (sternal fibers) grades 3–5** 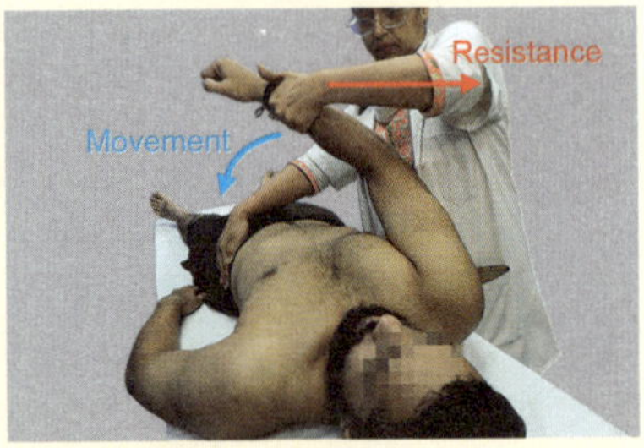 **Shoulder horizontal adductors (sternal fibers) grades 3–5** 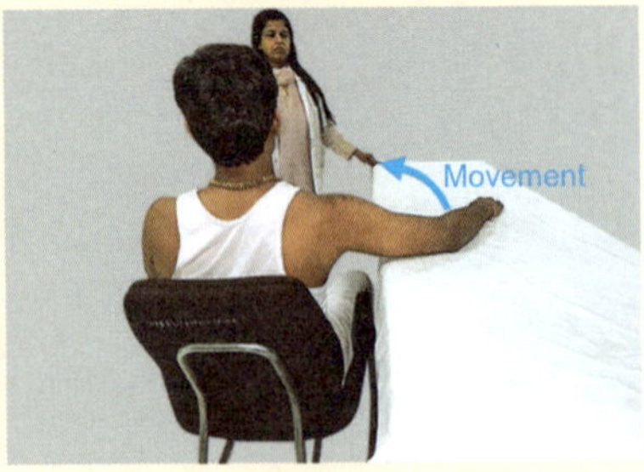 **Shoulder horizontal adductors grades 0–2**

Muscles of Shoulder Medial Rotation

Muscle	Origin	Insertion	Action	Nerve supply
Subscapularis	Subscapular fossa of scapula	Lesser tubercle of humerus	Shoulder internal rotation	Upper and lower subscapular nerves (C5–C6)

MMT Techniques for Shoulder Medial Rotators

Muscle	Patient's position	Physiotherapist's position	Techniques	Demonstrations
• **Subscapularis** • **Latissimus dorsi** • **Pectoralis major** • **Teres major**	• **For grades 3–5:** The patient is in prone position with arm supported on the table and elbow flexed at right angle • **For grades 0–2:** Patient is in sitting position with arm by the side of body, elbow flexed and physiotherapist stabilizing the arm	The physiotherapist **stands on the test side** of the patient. Stabilizes distal humerus	Patient is asked to actively **medially rotate** the shoulder. • **For grades 4 and 5:** ▪ The physiotherapist applies the resistance in the direction of laterally rotating the shoulder ▪ The patient must be able to move through the full range of motion (active resistance testing) or maintain an end point range (break testing) against the highest level of resistance in order to have normal muscle strength, grade 5. • **For grades 2 and 1:** Movement is assessed with reduced resistance, focusing on partial range of motion for grade 2 or detecting a slight contraction without visible movement for grade 1.	Movement Resistance **Shoulder medial rotators grades 3–5** Movement **Shoulder medial rotators grades 0–2**

Muscles of Shoulder Lateral Rotation

Muscles	Origin	Insertion	Action	Nerves supply	Blood supply
Infraspinatus	Infraspinous fossa of scapula	Greater tubercle of humerus	Shoulder joint: Arm external rotation	Suprascapular nerve	Thoracodorsal branch of subscapular artery and posterior circumflex humeral artery
Teres minor	Inferior angle and lower part of the lateral border of the scapula	Intertubercular sulcus (medial lip) of the humerus	Extension and internal rotation of the humerus (arm)	Lower subscapular nerve	

MMT Techniques for Shoulder Lateral Rotators

Muscles	Patient's position	Physiotherapist's position	Techniques	Demonstrations
• **Infraspinatus** • **Teres minor**	• **For grades 3–5:** The patient is in prone position with arm supported on the table and elbow flexed at right angle • **For grades 0–2:** Patient is in sitting position with arm by the side of body, elbow flexed and physiotherapist stabilizing the arm	The physiotherapist **stands on the test side** of the patient. Stabilizes distal humerus	Patient is asked to actively **laterally rotate** the shoulder. • **For grades 4 and 5:** ▪ The physiotherapist applies the resistance in the direction of medially rotating the shoulder ▪ The patient must be able to move through the full range of motion (active resistance testing) or maintain an end point range (break testing) against the highest level of resistance in order to have normal muscle strength, grade 5. • **For grades 2 and 1:** Movement is assessed with reduced resistance, focusing on partial range of motion for grade 2 or detecting a slight contraction without visible movement for grade 1.	Resistance Movement **Shoulder lateral rotators grades 3–5** Movement **Shoulder lateral rotators grades 0–2**

Muscles of Scapula Elevation

Muscles	Origin	Insertion	Action	Nerves supply
Trapezius	Medial third of the superior nuchal line, external occipital protuberance, ligamentum nuchae, spinous processes of vertebrae C7–T12	Lateral third of the clavicle, medial side of the acromion and the upper crest of the scapular spine, tubercle of the scapular spine	Elevates and depresses the scapula (depending on which part of the muscle contracts) rotates the scapula superiorly; retracts scapula	Motor: Spinal accessory (XI), proprioception: C3–C4
Levator scapulae	Transverse processes of C1–C4 vertebrae	Medial border of the scapula from the superior angle to the spine	Elevates the scapula	Dorsal scapular nerve (C5); the upper part of the muscle receives branches of C3 and C4

MMT Techniques for Scapula Elevators

Muscles	Patient's position	Physiotherapist's position	Techniques	Demonstrations
• **Upper trapezius** • **Levator scapula**	• **For grades 3–5:** ▪ **Combined testing:** The patient is in a sitting position with arms relaxed in lap. ▪ **Levator scapula:** The patient is in prone position, arm is adducted, elbow flexed and forearm pronated. • **For grades 0–2:** The patient is in prone position with arm supported on the table.	The physiotherapist **stands behind the patient** for combined testing and on test side for levator scapula testing.	Patient is asked to **actively elevate** the scapula • **For grades 4 and 5:** ▪ Physiotherapist gives resistance downward in the direction opposite to elevation. ▪ The patient must be able to move through the full range of motion (active resistance testing) or maintain an end point range (break testing) against the highest level of resistance in order to have normal muscle strength, grade 5. • **For grades 2 and 1:** Movement is assessed with reduced resistance, focusing on partial range of motion for grade 2 or detecting a slight contraction without visible movement for grade 1.	**Scapula elevators grades 4–5** **Scapula elevators grade 3** **Levator scapula grades 3–5** **Scapula elevators grades 0–2**

Muscles of Scapula Adduction/Retraction

Muscles	Origin	Insertion	Action	Nerves supply
Trapezius	Medial third of the superior nuchal line, external occipital protuberance, ligamentum nuchae, spinous processes of vertebrae C7–T12	Lateral third of the clavicle, medial side of the acromion and the upper crest of the scapular spine, tubercle of the scapular spine	Elevates and depresses the scapula (depending on which part of the muscle contracts) rotates the scapula superiorly; retracts scapula	Motor: Spinal accessory (XI), proprioception: C3–C4
Rhomboid major	Spines of vertebrae T2–T5	Medial border of the scapula inferior to the spine of the scapula	Retracts, elevates and rotates the scapula inferiorly	Dorsal scapular nerve (C5)

MMT Techniques for Scapula Adductors/Retractors

Muscles	Patient's position	Physiotherapist's position	Techniques	Demonstrations
• **Middle trapezius** • **Lower trapezius** • **Rhomboid major**	• **For grades 3–5:** ▪ The patient is in prone position. The test side arm shoulder is in 90° abduction, shoulder lateral rotation with thumb facing up for **middle trapezius**. ▪ The test side arm shoulder is in 130 abduction, shoulder lateral rotation with thumb facing up for **lower trapezius**. ▪ Shoulder is internally rotated with thumb facing down for **rhomboids**. ▪ **Alternate position for rhomboids:** Patient is prone lying, arm by the side of the body, elbow is flexed and forearm pronated. • **For grades 0–2:** ▪ The patient is in prone position. The physiotherapist supports the arm. ▪ Alternatively, the patient is in sitting position with arm supported on a flat surface such as a plinth ('gravity minimal' position)	The physiotherapist **stands at test side**.	The patient is asked to perform **scapula adduction "lift tip of elbow toward the ceiling"** • **For grades 4 and 5:** ▪ The physio-therapist provides resistance over distal forearm or humerus in the direction opposite to scapular adduction. ▪ The patient must be able to move through the full range of motion (active resistance testing) or maintain an end point range (break testing) against the highest level of resistance in order to have normal muscle strength, grade 5. • **For grades 2 and 1:** Movement is assessed with reduced resistance, focusing on partial range of motion for grade 2 or detecting a slight contraction without visible movement for grade 1.	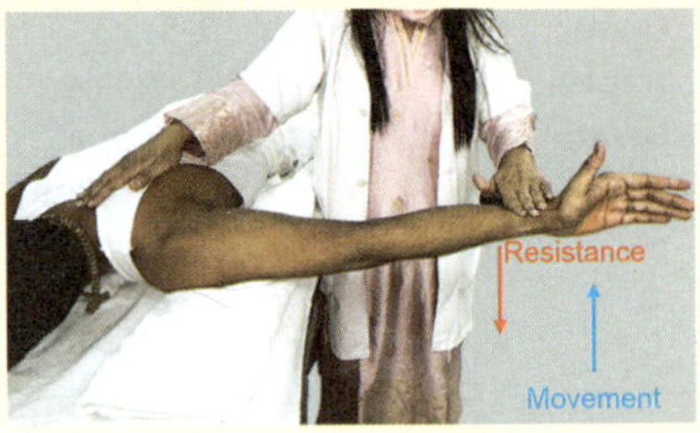 Resistance Movement **Middle trapezius grades 3–5** 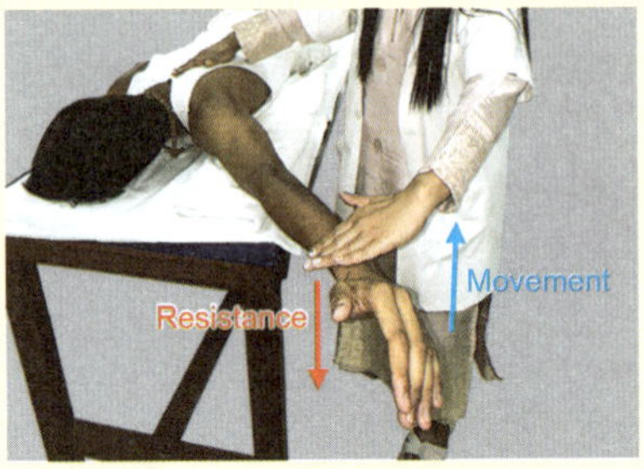 Resistance Movement **Lower trapezius grades 3–5** 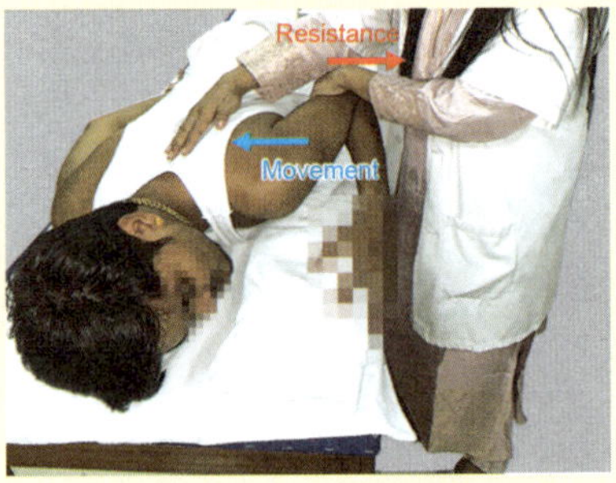 Resistance Movement **Rhomboids grades 3–5** 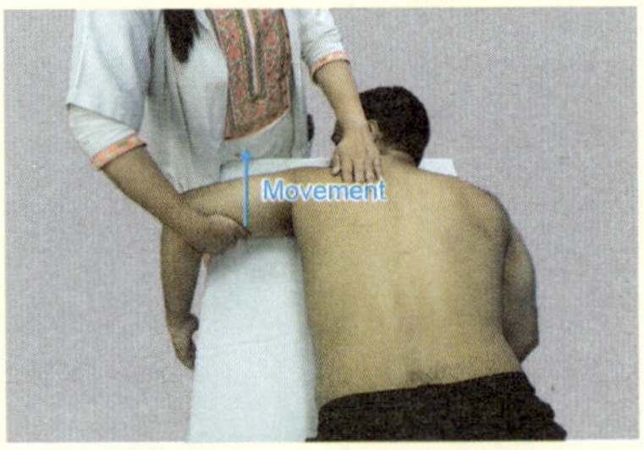 Movement **Scapula adductors/retractors grades 0–2** 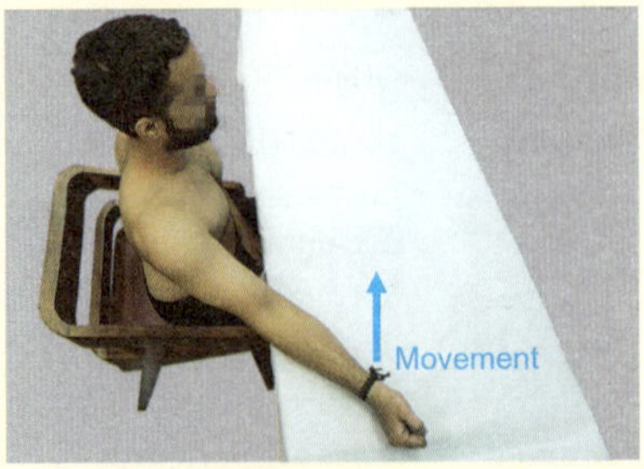 Movement **Scapula adductors/retractors grades 0–2 (alternative position)**

Muscles of Scapular Abduction

Muscle	Origin	Insertion	Action	Nerve supply
Serratus anterior	• **Superior part:** Ribs 1–2, Intercostal fascia • **Middle part:** Ribs-3–6 • **Inferior part:** Ribs 7–8/9 [variably extends to rib 10 (+ external oblique muscle)]	• **Superior part:** Anterior surface of superior angle • **Middle part:** Anterior surface of medial border • **Inferior part:** Anterior surface of inferior angle and medial border	Rotates scapula (draws inferiorly angle laterally)	Long thoracic nerve

MMT Techniques for Scapular Abductors

Muscle	Patient's position	Physiotherapist's position	Techniques	Demonstrations
Serratus anterior	• **For grades 3–5:** The patient is in supine position. The shoulder is flexed to 90° with slight horizontal adduction and the elbow is extended • **For grade 2:** The patient is in sitting position. The shoulder is flexed to 120° with slight horizontal adduction, and the elbow is extended. The physiotherapist supports the weight of the upper extremity	The physiotherapist stands at test side of the patient	The patient is asked to protract the scapula through full, ROM • **For grades 4 and 5:** ▪ The physiotherapist provides resistance on the distal forearm or hand near wrist in the direction of scapular adduction. ▪ The patient must be able to move through the full range of motion (active resistance testing) or maintain an end point range (break testing) against the highest level of resistance in order to have normal muscle strength, grade 5. • **For grades 2 and 1:** Movement is assessed with reduced resistance, focusing on partial range of motion for grade 2 or detecting a slight contraction without visible movement for grade 1.	 **Scapular abductors grades 3–5** **Scapula abductors grades 0–2**

Muscles of Elbow Flexion

Muscles	Origin	Insertion	Action	Nerves supply
Biceps brachii	Short head: Tip of the coracoid process of the scapula; long head: Supraglenoid tubercle of the scapula	Tuberosity of the radius	Flexes the forearm, flexes arm (long head), supinates	Musculocutaneous nerve (C5, 6)
Brachialis	Anterior surface of the lower one-half of the humerus and the associated intermuscular septa	Coronoid process of the ulna	Flexes the forearm	Musculocutaneous nerve (C5, 6)
Brachioradialis	Upper two-thirds of the lateral supracondylar ridge of the humerus	Lateral side of the base of the styloid process of the radius	Flexes the elbow, assists in pronation and supination	Radial nerve

MMT Techniques for Elbow Flexors

Muscles	Patient's position	Physiotherapist's position	Techniques	Demonstrations
• **Biceps brachii** • **Brachialis** • **Brachioradialis**	• **For grades 3–5:** The patient is in sitting position, shoulder adducted with different positions of forearm. • **For grades 0–2:** ▪ The patient is in sitting position with the arm supported on the couch. The shoulder is abducted to 90°, the elbow is extended. • **Forearm positioning:** ▪ **Biceps:** Supinated ▪ **Brachialis:** Pronated ▪ **Brachioradialis:** Midposition	The physiotherapist **stands at the test side**	The patient is asked to flex the elbow • **For grades 4 and 5:** ▪ The resistance is applied over flexor surface at the distal forearm with force in the direction opposite to flexion. ▪ The patient must be able to move through the full range of motion (active resistance testing) or maintain an end point range (break testing) against the highest level of resistance in order to have normal muscle strength, grade 5. • **For grades 2 and 1:** Movement is assessed with reduced resistance, focusing on partial range of motion for grade 2 or detecting a slight contraction without visible movement for grade 1.	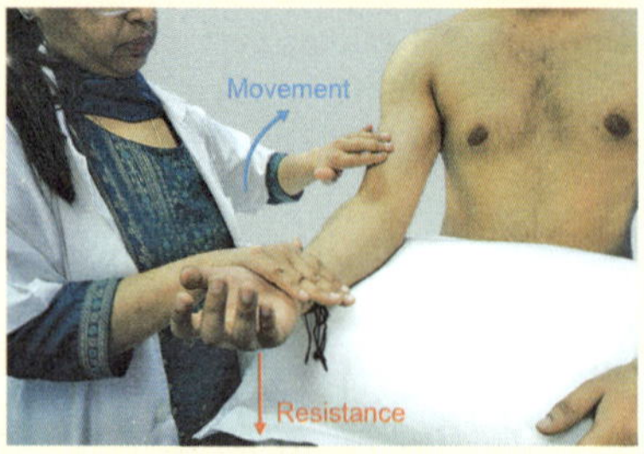 **Elbow flexors (biceps brachii) grades 3–5** 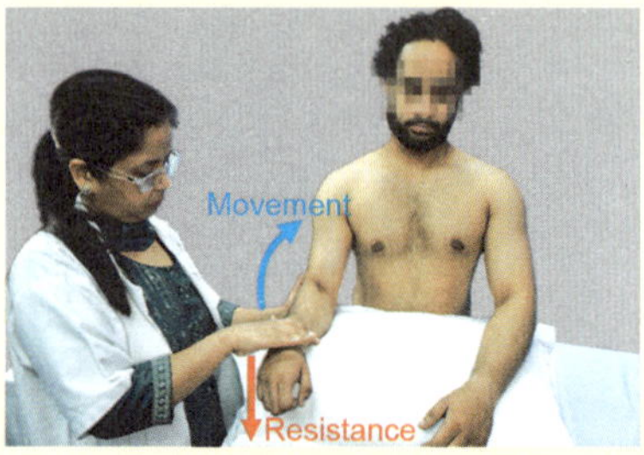 **Elbow flexors (brachialis) grades 3–5** 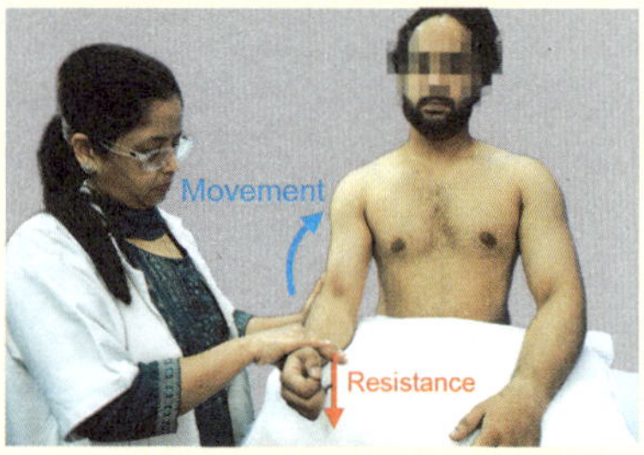 **Elbow flexors (brachioradialis) grades 3–5** 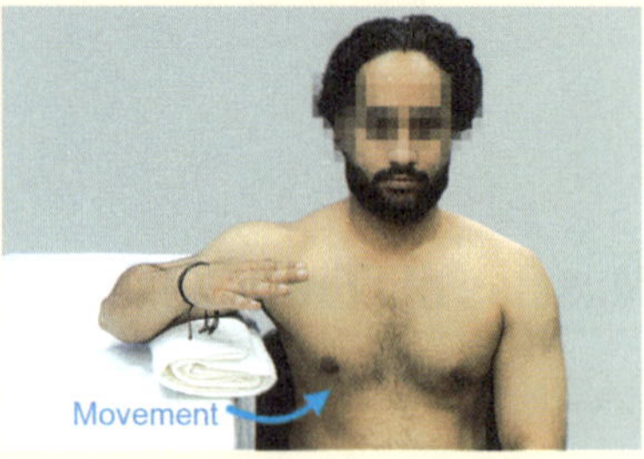 **Elbow flexors grades 0–2**

Muscles of Elbow Extension

Muscles	Origin	Insertion	Action	Nerves supply
Triceps brachii	Long head: Infraglenoid tubercle of the scapula; lateral head: Posterolateral humerus and lateral intermuscular septum; medial head: Posteromedial surface of the inferior 1/2 of the humerus	Olecranon process of the ulna	Extends the forearm; the long head extends and adducts arm	Radial nerve Nerve to anconeus, from the radial nerve
Anconeus	Lateral epicondyle of the humerus	Lateral side of the olecranon and the upper one-fourth of the ulna	Extends the forearm	

MMT Techniques for Elbow Extensors

Muscles	Patient's position	Physiotherapist's position	Techniques	Demonstrations
• **Triceps** • **Anconeus**	• **For grades 3–5:** ▪ The patient is in prone position, test side shoulder in 90° abduction position. ▪ Patient's test side forearm is hanging over side of plinth; upper arm supported on plinth to elbow. • **For grades 0–2:** The patient is sitting. The shoulder is abducted to 90°, the elbow is flexed, and the forearm is in mid position.	• The physiotherapist **stands at the test side**. • Stabilization is provided just above elbow	• Patient is asked to actively **extend elbow to full available range**. • **For grades 4 and 5:** ▪ The physiotherapist applies resistance over the dorsal forearm in the direction opposite to extension. ▪ The patient must be able to move through the full range of motion (active resistance testing) or maintain an end point range (break testing) against the highest level of resistance in order to have normal muscle strength, grade 5. • **For grades 2 and 1:** Movement is assessed with reduced resistance, focusing on partial range of motion for grade 2 or detecting a slight contraction without visible movement for grade 1.	**Elbow extensors grades 3–5** **Elbow extensors grades 0–2**

Muscles of Pronation of the Forearm

Muscles	Origin	Insertion	Action	Nerves supply
Pronator teres	• Humeral head, medial supracondylar ridge of humerus • Ulnar head: Coronoid process of ulna	Lateral surface of radius (distal to supinator)	Pronation of forearm at the proximal radioulnar joint, flexion of the forearm at the elbow joint	Median nerve
Pronator quadratus	Distal anterior surface of ulna	Distal anterior surface of radius	Proximal radioulnar joint: Forearm pronation	Median nerve

MMT Techniques for Pronators of the Forearm

Muscles	Patient's position	Physiotherapist's position	Techniques	Demonstrations
• **Pronator teres** • **Pronator quadratus**	• **For grades 0–5:** ▪ The patient is in supine position. The arm is at the side, the elbow is flexed, and the forearm is supinated ▪ Pronator Teres and Pronator Quadratus: Elbow partially flexed. ▪ Pronator Quadratus: Elbow completely flexed. • **For grades 0–2:** The patient is sitting, the shoulder and the elbow are flexed to 90°, and the forearm is supinated.	• The physiotherapist **stands at the test side** • Stabilization is provided just above elbow	Patient is asked to actively **pronate the forearm** in the full available range • **For grades 4 and 5:** ▪ The physiotherapist applies resistance above wrist in the direction of supination ▪ The patient must be able to move through the full range of motion (active resistance testing) or maintain an end point range (break testing) against the highest level of resistance in order to have normal muscle strength, grade 5. • **For grades 2 and 1:** Movement is assessed with reduced resistance, focusing on partial range of motion for grade 2 or detecting a slight contraction without visible movement for grade 1.	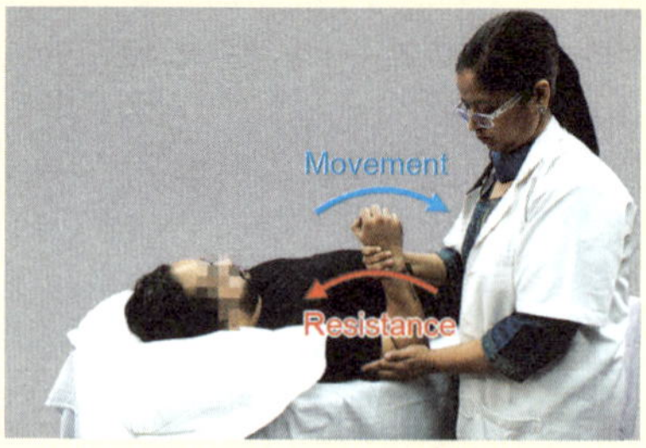 **Pronator quadratus grades 4–5** 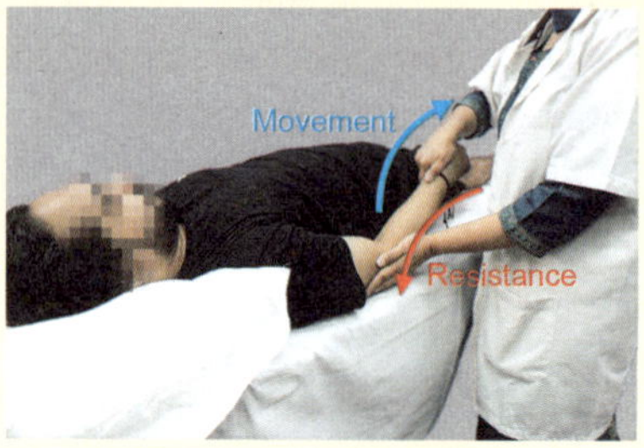 **Pronators teres and quadratus (combined testing) grades 4–5** 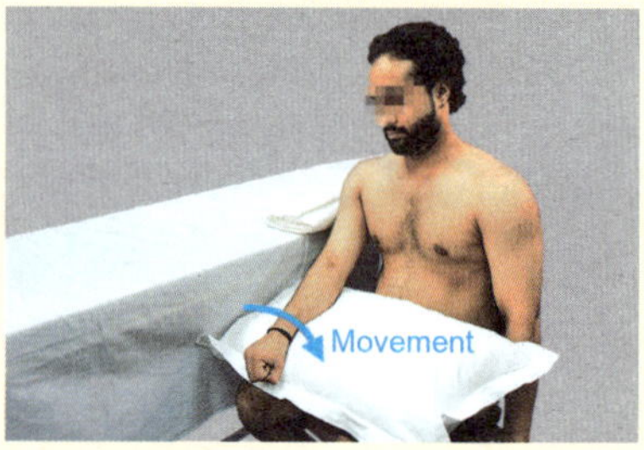 **Forearm pronators (end position) grades 0–2**

Muscles of Supination of Forearm

Muscle	Origin	Insertion	Action	Nerve supply
Supinator	Lateral epicondyle of humerus	Lateral, posterior, and anterior surfaces of proximal third of radius	Forearm supination	Posterior interosseous nerve

MMT Techniques for Supinator of Forearm

Muscle	Patient's position	Physiotherapist's position	Techniques	Demonstrations
Supinator	• **For grades 3–5:** The patient is in supine lying/ sitting position. The arm is at the side, the elbow may be kept slightly flexed the forearm is pronated.	• The physiotherapist **stands toward the test side**. • Stabilization is provided just above the elbow.	Patient is asked to **actively supinate the forearm** in the full available range	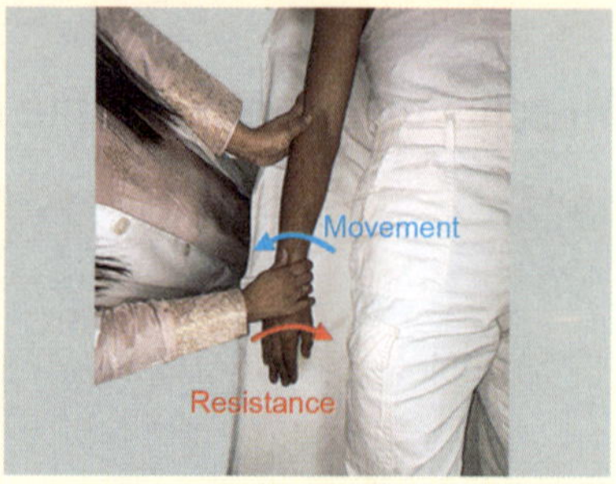 **Supinator grades 3–5**

Contd...

Muscle	Patient's position	Physiotherapist's position	Techniques	Demonstrations
	• **For grades 0–2:** The forearm is kept in mid-position with thumb facing upwards.		• **For grades 4 and 5:** ▪ The physiotherapist applies resistance above wrist in the direction of supination. ▪ The patient must be able to move through the full range of motion (active resistance testing) or maintain an end point range (break testing) against the highest level of resistance in order to have normal muscle strength, grade 5. • **For grades 2 and 1:** Movement is assessed with reduced resistance, focusing on partial range of motion for grade 2 or detecting a slight contraction without visible movement for grade 1.	 **Supinators grades 0–2**

Muscles of Flexion of MCP and Extension of IP Joints

Muscle	Origin	Insertion	Action	Nerve supply
Lumbricals	Tendons of flexor digitorum profundus muscle	Extensor expansion of hand	• Metacarpophalangeal joints 2–5: Finger flexion • Interphalangeal joints 2–5: Finger extension	• Lumbricals 1–2: Median nerve (C8–T1) • Lumbricals 3–4: Ulnar nerve (C8–T1)

MMT Techniques for Flexors of MCP and Extensors of IP Joints

Muscle	Patient's position	Physiotherapist's position	Techniques	Demonstrations
Lumbricals	• **For grades 3–5:** The patient can be in sitting or supine position with elbow extended and wrist in neutral position. • **For grades 0–2:** The patient is in sitting or supine position. The forearm is in midposition.	The physiotherapist **stands at the test side.**	Patient is asked to actively flex MCP joints and extend the interphalangeal (IP) joints. • **For grades 4 and 5:** ▪ The physiotherapist applies resistance first against the dorsum of middle and distal phalanx in the direction of flexion and then on the proximal phalanx in the direction of extension ▪ The patient must be able to move through the full range of motion (active resistance testing) or maintain an end point range (break testing) against the highest level of resistance in order to have normal muscle strength, grade 5.	 **Lumbrical MCP flexion grades 3–5** **Lumbricals IP extension grades 3–5**

Contd...

Muscle	Patient's position	Physiotherapist's position	Techniques	Demonstrations
			• **For grades 2 and 1:** Movement is assessed with reduced resistance, focusing on partial range of motion for grade 2 or detecting a slight contraction without visible movement for grade 1.	**Lumbricals MCP flexion grades 0–2** **Lumbricals IP extension grades 0–2**

Muscles of Wrist Flexion

Muscles	Origin	Insertion	Action	Nerves supply
Flexor carpi radialis	Common flexor tendon from the medial epicondyle of the humerus	Base of the second and third metacarpals	Flexes the wrist, abducts the hand	Median nerve
Flexor carpi ulnaris	Common flexor tendon and (ulnar head) from medial border of olecranon and upper 2/3 of the posterior border of the ulna	Pisiform, hook of hamate, and base of fifth metacarpal	Flexes wrist, adducts hand	Ulnar nerve

MMT Techniques for Wrist Flexors

Muscles	Patient's position	Physiotherapist's position	Techniques	Demonstrations
• **Flexor carpi radialis (FCR)** • **Flexor carpi ulnaris (FCU)**	• **For grades 3–5:** The patient is in sitting position, dorsal surface of forearm supported in supination. • **For grades 0–2:** The patient is in sitting or supine position. The forearm is in slight pronation for FCR and slight supination for FCU.	The physiotherapist is in **sitting position** to access test-side.	Patient is asked to actively **flex the wrist through range toward radial side for FCR and toward ulnar side for FCU**	**Wrist flexors grades 3–5**

Contd...

Muscles	Patient's position	Physiotherapist's position	Techniques	Demonstrations
			• **For grades 4 and 5:** ▪ The physiotherapist applies resistance through palm in a direction of extension toward ulnar side for FCR and in the direction of extension toward radial side for FCU. ▪ The patient must be able to move through the full range of motion (active resistance testing) or maintain an end point range (break testing) against the highest level of resistance in order to have normal muscle strength, grade 5. • **For grades 2 and 1:** Movement is assessed with reduced resistance, focusing on partial range of motion for grade 2 or detecting a slight contraction without visible movement for grade 1.	Movement **Wrist flexors grades 0–2**

Muscles of Wrist Extension

Muscles	Origin	Insertion	Action	Nerves supply
Extensor carpi radialis longus	Lower one-third of the lateral supracondylar ridge of the humerus	Dorsum of the second metacarpal bone (base)	Extends the wrist; abducts the hand	Radial nerve
Extensor carpi radialis brevis	Common extensor tendon (lateral epicondyle of humerus)	Dorsum of the third metacarpal bone (base)	Extends the wrist; abducts the hand	Deep radial nerve
Extensor carpi ulnaris	Common extensor tendon and the middle one-half of the posterior border of the ulna	Medial side of the base of the 5th metacarpal	Extends the wrist; adducts the hand	Deep radial nerve

MMT Techniques for Wrist Extensors

Muscles	Patient's position	Physiotherapist's position	Techniques	Demonstrations

• **Extensor carpi radialis longus (ECRL)**
• **Extensor carpi radialis brevis (ECRB)**
• **Extensor carpi ulnaris (ECU)**
	• **For grades 3–5:** The patient is in sitting position with palmar surface of forearm supported in pronation, elbow slightly flexed, wrist is neutral.
	• **For grades 0–2:**
	  ▪ The patient is in sitting or supine position.
	  ▪ The forearm is in slight supination and supported on a surface
		The physiotherapist is in **sitting position** to access test-side	The patient is asked to **actively extend and abduct the wrist through range for ECRL and ECRB** and extend and adduct the wrist for ECU
			• **For grades 4 and 5:**
			  ▪ The physiotherapist applies resistance through dorsal aspect of hand in a direction of flexion toward ulnar side for ECRL and ECRB and in the direction of flexion toward radial side for ECU
			  ▪ The patient must be able to move through the full range of motion (active resistance testing) or maintain an end point range (break testing) against the highest level of resistance in order to have normal muscle strength, grade 5.
			• **For grades 2 and 1:** Movement is assessed with reduced resistance, focusing on partial range of motion for grade 2 or detecting a slight contraction without visible movement for grade 1.
				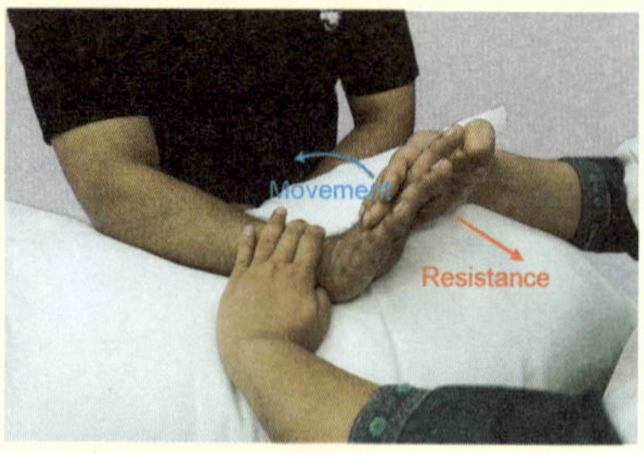 Movement
				Resistance
				Wrist extensors (ECRL and ECRB) grades 3–5 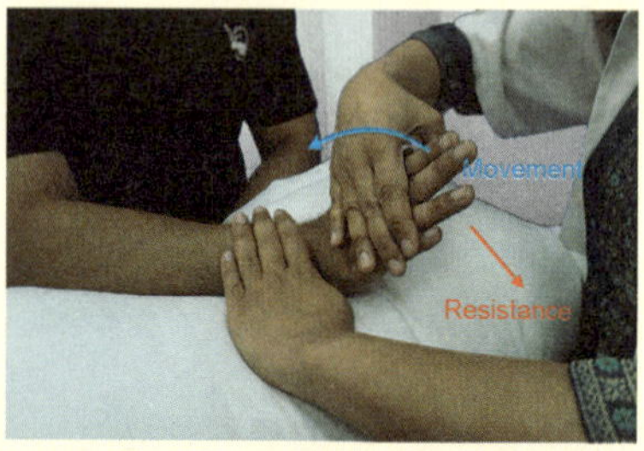
				Movement
				Resistance
				Wrist extensors (ECU) grades 3–5 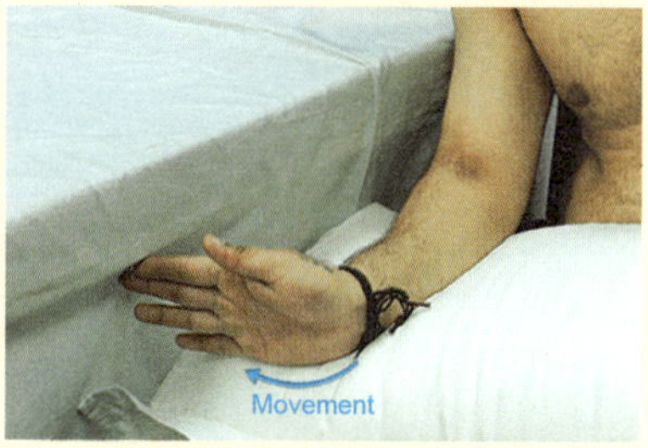
				Movement
				Wrist extensors grades 0–2

Abductor and Adductor Muscles of Digits

Muscles	Origin	Insertion	Action	Nerves supply
Dorsal interossei	Adjacent sides of metacarpal bones 1–5	• 1 and 2: Radial bases of proximal phalanges/extensor expansions of digits 2 and 3 • 3 and 4: Ulnar bases of proximal phalanges/extensor expansions of digits 3 and 4	• Metacarpophalangeal joints 2–4: Finger abduction, finger flexion • Interphalangeal joints 2–4: Finger extension	Deep branch of ulnar nerve (C8–Th1)
Palmar interossei	Ulnar side of metacarpal bone 2, Radial side of metacarpal bones 4 and 5	• 1: Ulnar base of proximal phalanx/extensor expansion of digit 2 • 2 and 3: Radial base of proximal phalanges/extensor expansions of digits 4 and 5	• Metacarpophalangeal joints 2, 4 and 5: Finger adduction, Finger flexion • Interphalangeal joints 2, 4 and 5: Finger extension	Deep branch of ulnar nerve

MMT Techniques for Abductor and Adductor Muscles of Digits

Muscles	Patient's position	Physiotherapist's position	Techniques	Demonstrations
• **Dorsal interossei** • **Palmar interossei**	• **For grades 0–5:** The patient is in sitting or supine position. Elbow is flexed. • **For grades 0–2:** The patient is in sitting position with shoulder abducted, elbow flexed on a supporting surface	• The physiotherapist is in **sitting position** to access test-side. • Stabilization of adjacent digit is required to give fixation of digit toward which finger is to be moved.	**For Dorsal interossei:** The patient is asked to **abduct the index finger toward the thumb, the middle finger** toward the index finger and the **ring finger toward the little finger**. **For palmar interossei**, the patient is asked to **adduct the index, ring, and little finger toward the middle finger**. • **For grades 3–5:** ▪ The physiotherapist applies resistance in the direction of adduction for dorsal interossei. The resistance is applied in the direction of abduction for palmar interossei ▪ The patient must be able to move through the full range of motion (active resistance testing) or maintain an end point range (break testing) against the highest level of resistance in order to have normal muscle strength, grade 5. • **For grades 2 and 1:** Movement is assessed with reduced resistance, focusing on partial range of motion for grade 2 or detecting a slight contraction without visible movement for grade 1.	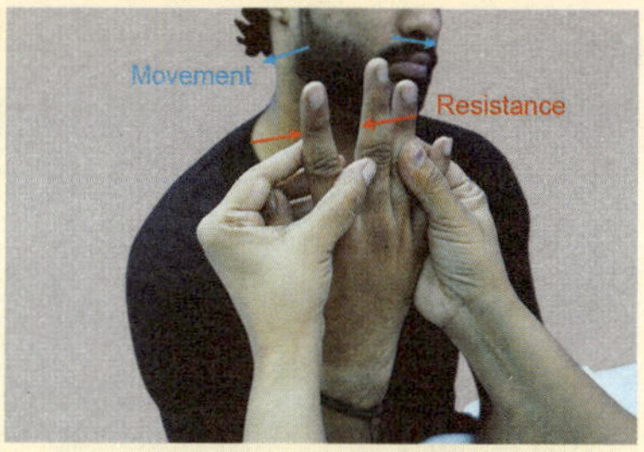 Movement, Resistance **Dorsal interossei grade 4–5** 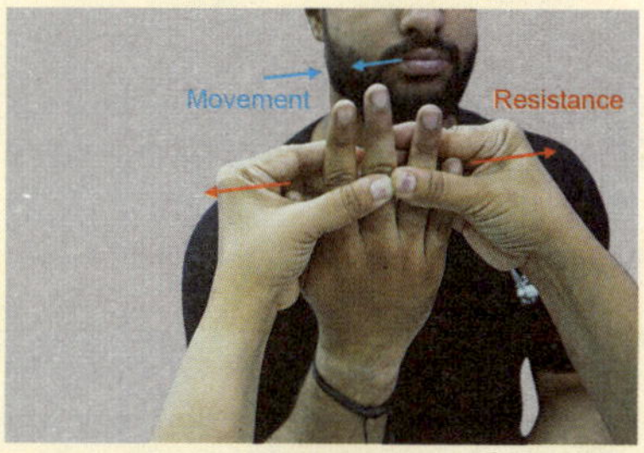 Movement, Resistance **Palmar interossei grade 4–5** 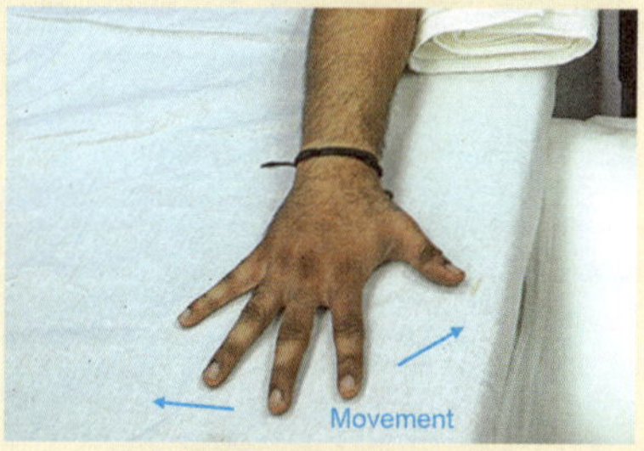 Movement **Dorsal interossei grade 0–2** 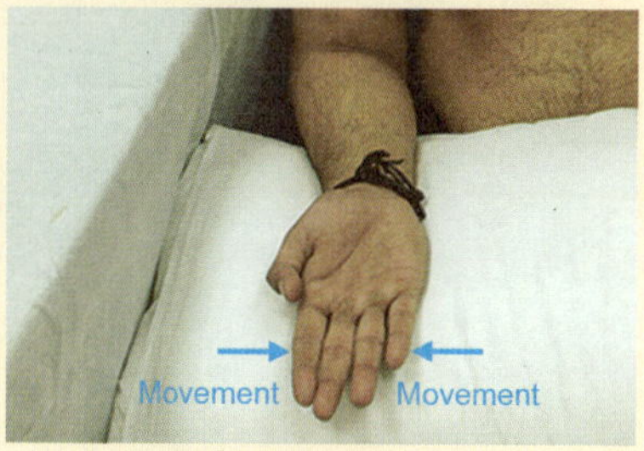 Movement, Movement **Palmar interossei grade 0–2**

TECHNIQUES FOR LOWER LIMB MUSCLES

Muscles of Hip Flexion

Muscles	Origin	Insertion	Action	Nerves supply
Iliacus	Iliac fossa and iliac crest; ala of sacrum	Lesser trochanter of the femur	Flexes the thigh; if the thigh is fixed it flexes the pelvis on the thigh	Femoral nerve
Psoas major	Bodies and transverse processes of lumbar vertebrae	Lesser trochanter of femur (with iliacus) *via* iliopsoas tendon	Flexes the thigh; flexes and laterally bends the lumbar vertebral column	Branches of the ventral primary rami of spinal nerves L2–L4
Rectus femoris	Straight head: Anterior inferior iliac spine; reflected head: Above the superior rim of the acetabulum	Patella and tibial tuberosity (*via* the patellar ligament)	Extends the leg, flexes the thigh	Femoral nerve

MMT Techniques for Hip Flexors

Muscles	Patient's position	Physiotherapist's position	Techniques	Demonstrations
• **Iliacus** • **Psoas major** • **Rectus femoris**	• **For grades 3–5:** ▪ **Iliopsoas:** The patient is in short sitting position with thighs supported. ▪ **Group muscle testing:** The patient is in supine position, knee extended. • **For grades 0–2:** The patient is in side lying position.	The physiotherapist **stands toward the test side**	The patient is asked to actively **flex the hip joint** • **For grades 4 and 5:** ▪ The physiotherapist applies resistance over the distal femur for Iliopsoas and on distal leg for group testing in a direction opposite to flexion. ▪ The patient must be able to move through the full range of motion (active resistance testing) or maintain an end point range (break testing) against the highest level of resistance in order to have normal muscle strength, grade 5. • **For grades 2 and 1:** Movement is assessed with reduced resistance, focusing on partial range of motion for grade 2 or detecting a slight contraction without visible movement for grade 1.	 **Hip flexors group muscle testing grades 4–5** **Hip flexors (Iliopsoas) grades 4–5** **Hip flexors grade 2**

Muscles of Hip Extension

Muscles	Origin	Insertion	Action	Nerves supply
Gluteus maximus	Posterior gluteal line, posterior surface of sacrum and coccyx, sacrotuberous ligament	Upper fibers: Iliotibial tract; lower fibers: Gluteal tuberosity of the femur	Extends the thigh; laterally rotates the femur	Inferior gluteal nerve
Semitendinosus	Lower, medial surface of ischial tuberosity (common tendon with biceps femoris m.)	Medial surface of tibia (*via* pes anserinus)	Extends the thigh, flexes the leg	Tibial nerve
Semimembranosus	Upper, outer surface of the ischial tuberosity	Medial condyle of the tibia	Extends the thigh, flexes the leg	Tibial nerve
Biceps femoris	Long head: Ischial tuberosity; short head: Lateral lip of the linea aspera	Head of fibula and lateral condyle of the tibia	Extends the thigh, flexes the leg	Long head: Tibial nerve; short head: Common fibular (peroneal) nerve

MMT Techniques for Hip Extensors

Muscles	Patient's position	Physiotherapist's position	Techniques	Demonstrations
• Gluteus maximus • Semitendinosus • Semimembranosus • Biceps femoris	• **For grades 3–5:** ▪ **Group muscle testing:** The patient is in prone position. Knee is extended ▪ **For gluteus maximus testing**, knee should be in flexed position • **For grades 0–2:** The patient is in side lying position.	• The physiotherapist **stands at the side of limb** to be tested at level of pelvis. • The hand used to provide resistance is contoured over the posterior thigh just above the knee.	Patient is asked to **actively extend the hip** through entire available range of motion. • **For grades 4 and 5:** ▪ The physiotherapist applies resistance through distal thigh in the direction opposite to hip extension. ▪ The patient must be able to move through the full range of motion (active resistance testing) or maintain an end point range (break testing) against the highest level of resistance in order to have normal muscle strength, grade 5. • **For grades 2 and 1:** Movement is assessed with reduced resistance, focusing on partial range of motion for grade 2 or detecting a slight contraction without visible movement for grade 1.	**Group muscle testing grade 3–5** **Hip extensors (glutues maximus) grades 4–5** **Hip extensors grade 2**

Muscles of Hip Abduction

Muscles	Origin	Insertion	Action	Nerves supply
Gluteus medius	External surface of the ilium between the posterior and anterior gluteal lines	Greater trochanter of the femur	Abducts the femur; medially rotates the thigh	Superior gluteal nerve
Gluteus minimus	External surface of the ilium between the anterior and inferior gluteal lines	Greater trochanter of the femur	Abducts the femur; medially rotates the thigh	Superior gluteal nerve

MMT Techniques for Hip Abductors

Muscles	Patient's position	Physiotherapist's position	Techniques	Demonstrations
• **Gluteus medius** • **Gluteus minimus**	• **For grades 3–5:** The patient is in side lying position with the test side up and pelvis in neutral position for Gluteus minimus and slight forward rotation for gluteus medius • **For grades 0–2:** The patient is in side lying position.	• The physio-therapist **stands behind the patient** in side lying or on test side in supine • Provide stabilization at the pelvis in side lying	Patient is asked to actively **abduct the hip** through available range of motion • **For grades 4 and 5:** ▪ The physiotherapist provides resistance in the direction opposite to abduction ▪ The patient must be able to move through the full range of motion (active resistance testing) or maintain an end point range (break testing) against the highest level of resistance in order to have normal muscle strength, grade 5. • **For grades 2 and 1:** Movement is assessed with reduced resistance, focusing on partial range of motion for grade 2 or detecting a slight contraction without visible movement for grade 1.	Movement Resistance **Hip abductors grades 4–5** Movement **Hip abductors grade 2**

Muscles of Hip Adduction

Muscles	Origin	Insertion	Action	Nerves supply
Adductor magnus	Ischiopubic ramus and ischial tuberosity	Linea aspera of the femur; the ischiocondylar part inserts on the adductor tubercle of the femur	Adducts, flexes, and medially rotates the femur; extends the femur (ischiocondylar part)	Posterior division of the obturator nerve; tibial nerve (ischiocondylar part)
Adductor brevis	Inferior pubic ramus	Pectineal line and linea aspera (deep to the pectineus and adductor longus muscles)	Adducts, flexes, and medially rotates the femur	Anterior division of the obturator nerve
Adductor longus	Medial portion of the superior pubic ramus	Linea aspera of the femur	Adducts, flexes, and medially rotates the femur	Anterior division of the obturator nerve
Pectineus	Pecten of the pubis	Pectineal line of the femur	Adducts, flexes, and medially rotates the thigh	Femoral nerve and possibly the anterior division of the obturator nerve

MMT Techniques for Hip Adductors

Muscles	Patient's position	Physiotherapist's position	Techniques	Demonstrations
• **Adductor magnus** • **Adductor brevis** • **Adductor longus** • **Pectineus**	• **For grades 3–5:** The patient is in side lying with top leg (non-test side) in abducted position up to 25°. Test side is the bottom leg in side lying. • **For grades 0–2:** The patient is in supine position, the non-test side is abducted 25°.	The physiotherapist **stands behind the patient** in side lying or on test side in supine	The patient is asked to actively **adduct the hip** through available range of motion • **For grades 4 and 5:** ▪ The physiotherapist provides resistance over the medial femur in the direction opposite to adduction ▪ The patient must be able to move through the full range of motion (active resistance testing) or maintain an end point range (break testing) against the highest level of resistance in order to have normal muscle strength, grade 5. • **For grades 2 and 1:** Movement is assessed with reduced resistance, focusing on partial range of motion for grade 2 or detecting a slight contraction without visible movement for grade 1.	Movement Resistance **Hip adductors grades 4–5 (right side)** Movement **Hip adductors grade 2**

Muscles of Hip External Rotation

Muscles	Origin	Insertion	Action	Nerves supply
Gluteus maximus	Posterior gluteal line, posterior surface of sacrum and coccyx, sacrotuberous ligament	Upper fibers: Iliotibial tract; lowermost fibers: Gluteal tuberosity of the femur	Extends the thigh; laterally rotates the femur	Inferior gluteal nerve
Piriformis	Anterior surface of sacrum	Upper border of greater trochanter of femur	Laterally rotates and abducts thigh	Ventral rami of S1–S2
Quadratus femoris	Lateral border of the ischial tuberosity	Quadrate line of the femur below the intertrochanteric crest	Laterally rotates the thigh	Nerve to the quadratus femoris
Obturator externus	The external surface of the obturator membrane and the superior and inferior pubic rami	Trochanteric fossa of the femur	Laterally rotates the thigh	Obturator nerve
Obturator internus	The internal surface of the obturator membrane and margin of the obturator foramen	Greater trochanter on its medial surface above the trochanteric fossa	Laterally rotates and abducts the thigh	Obturator nerve
Gemellus superior	Ischial spine	Obturator internus tendon	Laterally rotates the femur	Nerve to the obturator internus muscle
Gemellus inferior	Ischial tuberosity	Obturator internus tendon	Laterally rotates the femur	Nerve to the quadratus internus muscle

MMT Techniques for Hip External Rotators

Muscles	Patient's position	Physiotherapist's position	Techniques	Demonstrations
• **Gluteus maximus** • **Piriformis** • **Quadratus femoris** • **Obturator externus** • **Obturator internus** • **Gemellus superior** • **Gemellus inferior**	• **For grades 3–5:** The patient is in short sitting on plinth with test-side knee flexed to 90° • **For grades 0–2:** The patient is in prone position with knee flexed to 90°	The physiotherapist **stands or sits towards the test side.**	The patient is asked to actively **externally rotate the hip** through available range of motion. • **For grades 4 and 5:** ▪ The physiotherapist applies resistance through the medial ankle through range in a direction opposite to external rotation ▪ The patient must be able to move through the full range of motion (active resistance testing) or maintain an end point range (break testing) against the highest level of resistance in order to have normal muscle strength, grade 5. • **For grades 2 and 1:** Movement is assessed with reduced resistance, focusing on partial range of motion for grade 2 or detecting a slight contraction without visible movement for grade 1.	Movement Resistance **Hip external rotators grades 4–5** Movement **Hip external rotators grade 2**

Muscles of Hip Internal Rotation

Muscles	Origin	Insertion	Action	Nerves supply
Gluteus minimus	External surface of the ilium between the anterior and inferior gluteal lines	Greater trochanter of the femur	Abducts the femur; medially rotates the thigh	Superior gluteal nerve
Gluteus medius	External surface of the ilium between the posterior and anterior gluteal lines	Greater trochanter of the femur	Abducts the femur; medially rotates the thigh	Superior gluteal nerve
Tensor fascia lata	Anterior part of the iliac crest, anterior superior iliac spine	Iliotibial tract	Flexes, abducts, and medially rotates the thigh	Superior gluteal nerve

MMT Techniques for Hip Internal Rotators

Muscles	Patient's position	Physiotherapist's position	Techniques	Demonstrations
• **Gluteus minimus** • **Gluteus medius** • **Tensor fascia lata**	• **For grades 3–5:** The patient is in short sitting position on plinth with test-side knee flexed to 90°. • **For grades 0–2:** The patient is in prone position.	The physiotherapist **stands or sits** toward the test side.	The patient is asked to actively **internally rotate the hip** through available range • **For grades 4 and 5:** ▪ The physiotherapist applies resistance through the lateral ankle through range in a direction opposite to internal rotation ▪ The patient must be able to move through the full range of motion (active resistance testing) or maintain an end point range (break testing) against the highest level of resistance in order to have normal muscle strength, grade 5. • **For grades 2 and 1:** Movement is assessed with reduced resistance, focusing on partial range of motion for grade 2 or detecting a slight contraction without visible movement for grade 1.	 **Hip internal rotators grades 4–5** **Hip internal rotators grade 2**

Muscles of Knee Flexion

Muscles	Origin	Insertion	Action	Nerves supply
Semitendinosus	Lower, medial surface of ischial tuberosity (common tendon with biceps femoris muscle)	Medial surface of tibia (*via* pes anserinus)	Extends the thigh, flexes the leg	Tibial nerve
Semimembranosus	Upper, outer surface of the ischial tuberosity	Medial condyle of the tibia	Extends the thigh, flexes the leg	Tibial nerve
Biceps femoris	Long head: Ischial tuberosity; short head: Lateral lip of the linea aspera	Head of fibula and lateral condyle of the tibia	Extends the thigh, flexes the leg	Long head: Tibial nerve; short head: Common fibular (peroneal) nerve

MMT Techniques for Knee Flexors

Muscles	Patient's position	Physiotherapist's position	Techniques	Demonstration
Medial hamstring • Semitendinosus • Semimembranosus **Lateral hamstring** • Biceps femoris	• **For grades 3–5:** The patient is in prone position with the knee in 50°–70° of flexion. ▪ **For medial hamstring**, the patient holds the leg in medial rotation and ▪ **For lateral hamstring** in lateral rotation. • **For grades 0–2:** The patient is in side lying position, knee extended.	The physiotherapist **stands at test side in prone** If patient is **in side lying, the physiotherapist stands behind the patient** supporting the lower leg	The patient is asked to **actively flex the knee** through range. • **For grades 4 and 5:** ▪ The physiotherapist applies resistance through the distal tibia and fibula in a direction opposite to flexion ▪ The patient must be able to move through the full range of motion (active resistance testing) or maintain an end point range (break testing) against the highest level of resistance in order to have normal muscle strength, grade 5. • **For grades 2 and 1:** Movement is assessed with reduced resistance, focusing on partial range of motion for grade 2 or detecting a slight contraction without visible movement for grade 1.	Resistance Movement **Knee flexors grades 4–5** Movement **Knee flexors grades 0–2**

Muscles of Knee Extension

Quadriceps which includes the following muscles:

Muscles	Origin	Insertion	Action	Nerves supply
Vastus medialis	Medial intermuscular septum, medial lip of the linea aspera	Patella and medial patellar retinaculum	Extends leg	Femoral nerve
Vastus lateralis	Lateral intermuscular septum, lateral lip of the linea aspera and the gluteal tuberosity	Patella and medial patellar retinaculum	Extends leg	Femoral nerve
Vastus intermedius	Anterior and lateral surface of the femur	Patella	Extends leg	Femoral nerve
Rectus femoris	Straight head: Anterior inferior iliac spine; reflected head: above the superior rim of the acetabulum	Patella and tibial tuberosity (*via* the patellar ligament)	Extends the leg, flexes the thigh	Femoral nerve

MMT Techniques for Knee Extensors

Muscles	Patient's position	Physiotherapist's position	Techniques	Demonstrations
Quadriceps: • **Vastus medialis** • **Vastus lateralis** • **Vastus inter-medius** • **Rectus femoris**	• **For grades 3–5:** The patient is in short sitting position. • **For grades 0–2:** The patient is in side lying position.	The physio-therapist **stands or sits towards the test side or behind the patient for grade 2**.	The patient is asked to actively **extend the knee** through available range of motion. • **For grades 4 and 5:** ▪ The physiotherapist applies resistance through the distal tibia and fibula in a direction opposite to extension ▪ The patient must be able to move through the full range of motion (active resistance testing) or maintain an end point range (break testing) against the highest level of resistance in order to have normal muscle strength, grade 5. • **For grades 2 and 1:** Movement is assessed with reduced resistance, focusing on partial range of motion for grade 2 or detecting a slight contraction without visible movement for grade 1.	Movement Resistance **Knee extensors grades 4–5** Movement **Knee extensors grades 0–2**

Muscles of Ankle Dorsiflexion

Muscles	Origin	Insertion	Action	Nerves supply
Tibialis anterior	Lateral tibial condyle and the upper lateral surface of the tibia	Medial surface of the medial cuneiform and the 1st metatarsal	Dorsiflexes and inverts the foot	Deep fibular (peroneal) nerve
Extensor digitorum longus	Lateral condyle of the tibia, anterior surface of the fibula, lateral portion of the interosseous membrane	Dorsum of the lateral 4 toes *via* extensor expansions (central slip inserts on base of middle phalanx, lateral slips on base of distal phalanx)	Extends the metatarsophalangeal, proximal interphalangeal and distal interphalangeal joints of the lateral 4 toes	Deep fibular (peroneal) nerve
Extensor hallucis longus	Middle half of the anterior surface of the fibula and the interosseous membrane	Base of the distal phalanx of the great toe)	Extends the metatarsophalangeal interphalangeal joints of the great toe	Deep fibular (peroneal) nerve
Peroneus tertius	Distal part of the anterior surface of the fibula	Dorsum of the shaft of the 5th metatarsal bone	Everts the foot	Deep fibular (peroneal) nerve

MMT Techniques for Ankle Dorsiflexors

Muscles	Patient's position	Physiotherapist's position	Techniques	Demonstrations
• **Tibialis anterior** • **Peroneus tertius** • **Extensor hallucis longus**	• **For grades 3–5:** The patient is in supine position, ankle hanging off the plinth • **For grades 0–2:** The patient is in side lying position.	The physiotherapist is **standing toward the test side** of the patient	The patient is asked to actively **dorsiflex the foot** through the range. • **For grades 4 and 5:** ▪ The physiotherapist gives resistance to dorsiflexion and slight inversion movement ▪ The patient must be able to move through the full range of motion (active resistance testing) or maintain an end point range (break testing) against the highest level of resistance in order to have normal muscle strength, grade 5. • **For grades 2 and 1:** Movement is assessed with reduced resistance, focusing on partial range of motion for grade 2 or detecting a slight contraction without visible movement for grade 1.	Movement Resistance **Ankle dorsiflexor grades 4–5** Movement **Ankle dorsiflexor grade 2**

Muscles of Ankle Plantar Flexion

Muscles	Origin	Insertion	Action	Nerves supply
Gastrocnemius	Femur; medial head: Above the medial femoral condyle; lateral head: Above the lateral femoral condyle	Dorsum of the calcaneus *via* the calcaneal (Achilles') tendon	Flexes leg; plantar flexes foot	Tibial nerve
Soleus	Posterior surface of head and upper shaft of the fibula, soleal line of the tibia	Dorsum of the calcaneus *via* the calcaneal (Achilles') tendon	Plantar flexes the foot	Tibial nerve

MMT Techniques for Ankle Plantar Flexors

Muscles	Patient's position	Physiotherapist's position	Techniques	Demonstrations
• **Gastrocnemius** • **Soleus**	**Combined testing of all ankle plantar flexors:** • **For grades 3–5:** ▪ The patient is in prone position with knee extended and test foot over the end of the table ▪ **For testing soleus only:**	The physiotherapist **stands toward the test side** of the patient.	The patient is asked to **actively plantar flex the ankle joint** in available range.	Movement Resistance **Ankle plantar flexor grades 4–5**

Contd...

Muscles	Patient's position	Physiotherapist's position	Techniques	Demonstrations
	• **For grades 3–5:** ▪ The patient is in prone position with knee flexed at 90°. ▪ **For grades 0–2:** The patients is in side-lying position, knee extended with knee. ▪ **For grades 0–2:** The patient is in side-lying position.		• **For grades 4 and 5:** ▪ The physiotherapist applies resistance against the forefoot in the direction of dorsiflexion ▪ The patient must be able to move through the full range of motion (active resistance testing) or maintain an end point range (break testing) against the highest level of resistance in order to have normal muscle strength, grade 5. • **For grades 2 and 1:** Movement is assessed with reduced resistance, focusing on partial range of motion for grade 2 or detecting a slight contraction without visible movement for grade 1.	 **Soleus grades 3–5** **Ankle plantar flexor grade 2**

Muscles of Eversion

Muscles	Origin	Insertion	Action	Nerves supply
• **Peroneus longus** • **Peroneus brevis**	• Upper 2/3 of the lateral surface of the fibula • Lower 1/3 of the lateral surface of the fibula	• After crossing the plantar surface of the foot deep to the intrinsic muscles, it inserts on the medial cuneiform and the base of the 1st metatarsal bone • Tuberosity of the base of the 5th metatarsal	• Extends (plantar flexes) and everts the foot • Extends (plantar flexes) and everts the foot	• Superficial fibular (peroneal) nerve • Superficial fibular (peroneal) nerve

MMT Techniques for Evertors

Muscles	Patient's position	Physiotherapist's position	Techniques	Demonstration
• **Peroneus longus** • **Peroneus brevis**	• **For grades 3–5:** The patient is supine or side lying position, test-side ankle off the edge of the plinth. • **For grades 0–2:** Patient can be supine or short sitting.	The physiotherapist is in **standing position**.	Patient is asked to **actively evert the foot** through the range • **For grades 4 and 5:** ▪ The physiotherapist gives resistance toward inversion and slight dorsiflexion	 **Evertors grades 3–5**

Contd...

Muscles	Patient's position	Physiotherapist's position	Techniques	Demonstration
			▪ The patient must be able to move through the full range of motion (active resistance testing) or maintain an end point range (break testing) against the highest level of resistance in order to have normal muscle strength, grade 5. • **For grades 2 and 1:** Movement is assessed with reduced resistance, focusing on partial range of motion for grade 2 or detecting a slight contraction without visible movement for grade 1.	

Muscle of Inversion

Muscle	Origin	Insertion	Action	Nerve supply
Tibialis posterior	Interosseous membrane, posteromedial surface of the fibula, posterolateral surface of the tibia	Tuberosity of the navicular and medial cuneiform, metatarsals 2–4	Plantar flexes the foot; inverts the foot	Tibial nerve

MMT Techniques for Invertors

Muscle	Patient's position	Physiotherapist's position	Techniques	Demonstration
Tibialis posterior	• **For grades 3–5:** The physiotherapist is supine or side lying position with ankle off the edge of the plinth. • **For grades 0–2:** Patient can be in supine or short sitting.	The physiotherapist is in **standing position** towards the test side	The patient is asked to actively **invert the foot** through the range. • **For grades 4 and 5:** ▪ The physiotherapist provides resistance toward eversion and slight dorsiflexion ▪ The patient must be able to move through the full range of motion (active resistance testing) or maintain an end point range (break testing) against the highest level of resistance in order to have normal muscle strength, grade 5. • **For grades 2 and 1:** Movement is assessed with reduced resistance, focusing on partial range of motion for grade 2 or detecting a slight contraction without visible movement for grade 1.	Movement Resistance **Invertors grades 3–5**

Flexor Muscles of the Toes

Muscles	Origin	Insertion	Action	Nerves supply
Flexor digitorum brevis	Tuberosity of the calcaneus, plantar aponeurosis, intermuscular septae	Base of the middle phalanx of digits 2–5 after splitting to allow passage of the flexor digitorum longus tendons	Flexes the metatarsophalangeal and proximal interphalangeal joints of digits 2–5	Medial plantar nerve
Flexor digitorum longus	Middle half of the posterior surface of the tibia	Bases of the distal phalanges of digits 2–5	Flexes the metatarsophalangeal, proximal interphalangeal and distal interphalangeal joints of digits 2–5; plantar flexes the foot	Tibial nerve
Flexor hallucis longus	Lower 2/3 of the posterior surface of the fibula	Base of the distal phalanx of the great toe	Flexes the metatarsophalangeal and proximal interphalangeal joints of the great toe; plantar flexes the foot	Tibial nerve

MMT Techniques for Flexor Muscles of the Toes

Muscles	Patient's position	Physiotherapist's position	Techniques	Demonstrations
• **Flexor digitorum brevis (FDB)** • **Flexor digitorum longus (FDL)** • **Flexor hallucis longus (FHL)**	• **For grades 3–5:** The patient will be in supine lying position • **For grades 0–2:** The patient will be in side lying position	The physiotherapist is in **standing** position toward the test side of patient	• Patient is asked to **flex the PIP joints of 2nd to 5th digits for FDB, flex the IP joint of great toe for FHL** • **Flex the DIP of 2nd to 5th digits for FDL** • The patient must be able to move through the full range of motion (active resistance testing) or maintain an end point range (break testing) against the highest level of resistance in order to have normal muscle strength, grade 5.	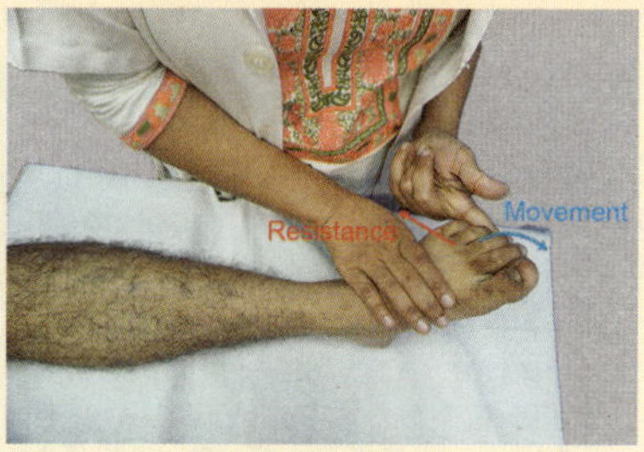 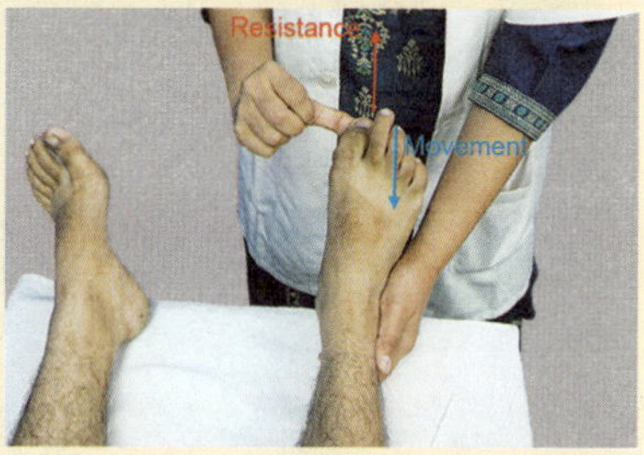 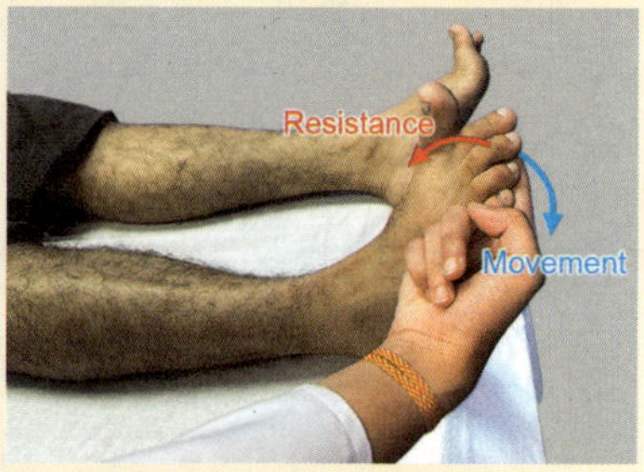 Resistance / Movement **Toe flexors** Resistance / Movement **FHL** Resistance / Movement **FDL (2nd to 5th digits flexion of DIP) grades 3–5**

Extensor Muscles of the Toes

Muscles	Origin	Insertion	Action	Nerves supply
Extensor digitorum longus	Lateral condyle of the tibia, anterior surface of the fibula, lateral portion of the interosseous membrane	Dorsum of the lateral 4 toes *via* extensor expansions (central slip inserts on base of middle phalanx, lateral slips on base of distal phalanx)	Extends the metatarsophalangeal, proximal interphalangeal and distal interphalangeal joints of the lateral 4 toes	Deep fibular (peroneal) nerve
Extensor hallucis longus	Middle half of the anterior surface of the fibula and the interosseous membrane	Base of the distal phalanx of the great toe	Extends the metatarsophalangeal interphalangeal joints of the great toe	Deep fibular (peroneal) nerve
Extensor hallucis brevis	Superolateral surface of the calcaneus	Dorsum of base of proximal phalanx of the great toe	Extends the great toe	Deep fibular (peroneal) nerve

MMT Techniques for Extensor Muscles of the Toes

Muscles	Patient's position	Physiotherapist's position	Techniques	Demonstrations
• **Extensor digitorum longus** • **Extensor hallucis longus** • **Extensor hallucis brevis**	• **For grades 3–5:** The patient will be in supine lying or sitting position • **For grades 0–2:** The patient will be in side lying position • **For grades 3–5:** The patient will be in supine lying position • **For grades 0–2:** The patient will be in side lying position	The physiotherapist is in standing position toward the test side.	• The patient is asked to **extend all joints of 2nd to 5th toe for EDL, all joints of great toe for EHL and MTP joint of 5th toe for EHB** • The physiotherapist applies resistance on the dorsal surface of 2nd to 5th toe in the direction of flexion for EDL Resistance is applied on the distal phalanx of great toe in the direction of flexion for EHL • The patient must be able to move through the full range of motion (active resistance testing) or maintain an end point range (break testing) against the highest level of resistance in order to have normal muscle strength, grade 5. • The patient must be able to move through the full range of motion (active resistance testing) or maintain an end point range (break testing) against the highest level of resistance in order to have normal muscle strength, grade 5.	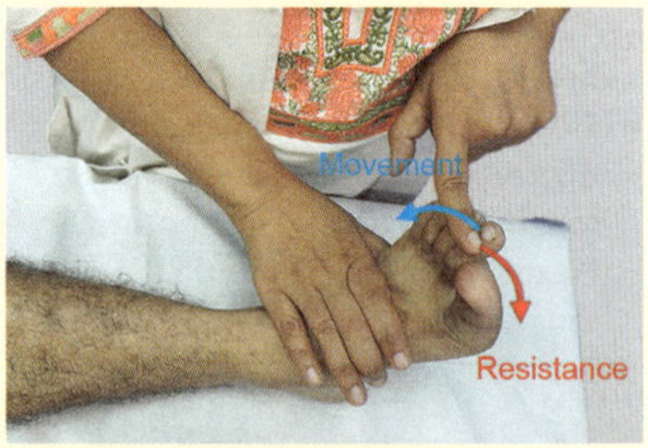 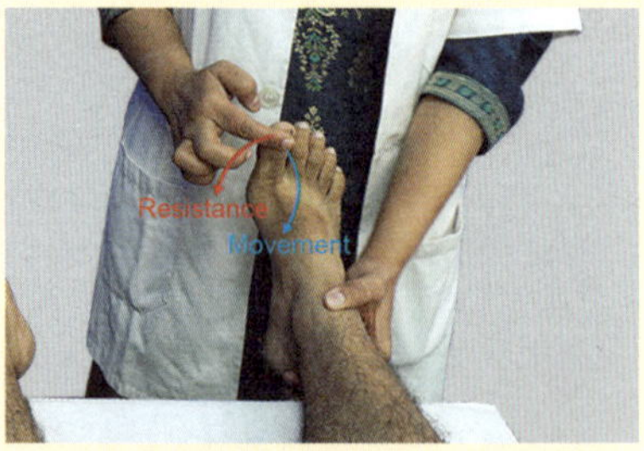 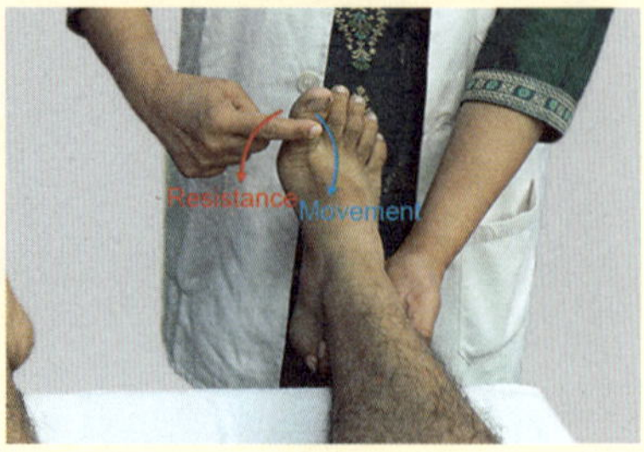 Movement Resistance Toe extensors Resistance Movement EHL Resistance Movement EHB

MUSCLES OF SPINE

Muscle	Origin	Insertion	Action	Innervation
Erector spinae	Iliac crest, sacrum, transverse and spinous processes of vertebrae and supraspinal ligament	Angles of the ribs, transverse and spinous processes of vertebrae, posterior aspect of the skull	Extends and laterally bends the trunk, neck and head	Segmentally innervated by dorsal primary rami of spinal nerves C1–S5
Iliocostalis	Iliac crest and sacrum	Angles of the ribs	Extends and laterally bends the trunk and neck	Dorsal primary rami of spinal nerves C4–S5
Interspinales	Upper border of spinous process	Lower border of spinous process above	Extends trunk and neck	Dorsal primary rami of spinal nerves C1–L5
Intertransversarii	Upper border of transverse process	Lower border of transverse process above	Laterally bends trunk and neck	Dorsal primary rami of spinal nerves C1–L5
Longissimus	Transverse process at inferior vertebral levels	Transverse process at superior vertebral levels and mastoid process	Extends and laterally bends the trunk, neck and head	Dorsal primary rami of spinal nerves C1–S1
Multifidus	Sacrum, transverse processes of C3–L5	Spinous processes 2–4 vertebral levels superior to their origin	Extends and laterally bends trunk and neck, rotates to opposite side	Dorsal primary rami of spinal nerves C1–L5
Obliquus capitis inferior	Spinous process of the axis	Transverse process of atlas	Rotates the head to the same side	Suboccipital nerve (DPR of C1)
Obliquus capitis superior	Transverse process of atlas	Occipital bone above inferior nuchal line	Extends the head, rotates the head to the same side	Suboccipital nerve (DPR of C1)
Rectus capitis posterior major	Spinous process of axis	Inferior nuchal line	Extends the head, rotate to same side	Suboccipital nerve (DPR of C1)
Rectus capitis posterior minor	Posterior tubercle of atlas	Inferior nuchal line medially	Extends the head	Suboccipital nerve (DPR of C1)
Rotatores	Transverse processes	Long rotatores: Spines 2 vertebrae above origin; Short rotatores: Spines 1 vertebrae above origin	Rotates the vertebral column to the opposite side	Dorsal primary rami of spinal nerves C1–L5
Semispinalis	Transverse processes of C7–T12	Capitis: Back of skull between nuchal lines; Cervicis and thoracis: Spines 4–6 vertebrae above origin	Extends the trunk and laterally bends the trunk, rotates the trunk to the opposite side	Dorsal primary rami of spinal nerves C1–T12
Spinalis	Spinous processes at inferior vertebral levels	Spinous processes at superior vertebral levels and base of the skull	Extends and laterally bends trunk and neck	Dorsal primary rami of spinal nerves C2–L3
Splenius	Ligamentum nuchae and spines C7–T6	Capitis: Mastoid process and superior nuchal line laterally; cervicis: Posterior tubercles of C1–C3 vertebrae	Extends and laterally bends neck and head; rotates head to same side	Dorsal primary rami of spinal nerves C2–C6
Splenius capitis	Ligamentum nuchae and spines of C7–T6 vertebrae	Mastoid process and lateral end of the superior nuchal line	Extends and laterally bends the neck and head, rotates head to the same side	Dorsal primary rami of spinal nerves C2–C6
Splenius cervicis	Ligamentum nuchae and spines of C7–T6 vertebrae	Posterior tubercles of the transverse processes of C1–C3 vertebrae	Extends and laterally bends neck and head, rotates head to the same side	Dorsal primary rami of spinal nerves C2–C6

TECHNIQUES FOR SPINAL MUSCLES

Anterior Trunk Flexors: Upper Abdominal Muscle Test

- **Muscles:** Rectus abdominis, iliopsoas, rectus femoris, internal abdominal oblique, and external abdominal oblique.
- **Patient position:** Supine lying
- **Physiotherapist position:** By the side of the patient
- **Movement:** The patient is asked to flex the spine and keep it flexed while entering the hip-flexion phase and coming to a sitting position
- **Resistance:** The body weight offers resistance.
- **Grades:**
  - **Grade 5:** Patient is capable of successfully completing a test movement while holding hands behind the head (Fig. 4.1A).
  - **Grade 4:** Patient is capable of completing test movement successfully while crossing arms at the chest (Fig. 4.1B).
  - **Grade 3:** Patient is capable of flexing the vertebral column and performing a posterior pelvic tilt with the arms extended, but unable to maintain trunk flexion when attempting to enter the hip-flexion phase of the test movement (Fig. 4.1C).
  - **Grade 2:** Patient is capable of maintaining a posterior pelvic tilt but unable to hold it as the head is elevated from the table.
  - **Grade 1:** The anterior abdominal muscles contract when the patient tries to lower the chest or tilt the pelvis posteriorly, but the pelvis and thorax do not approximate.
  - **Grade 0:** No palpable muscle contraction.

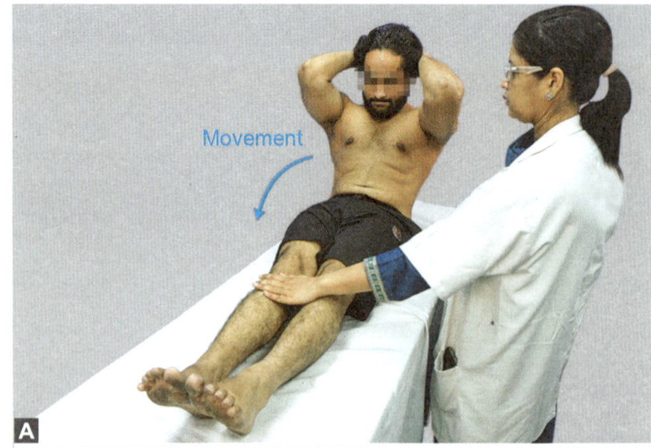

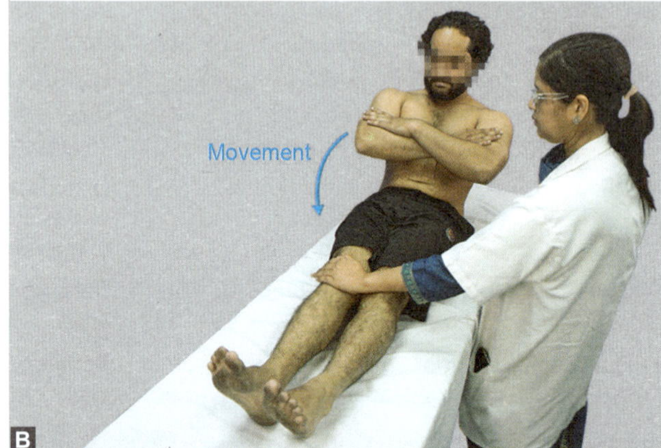

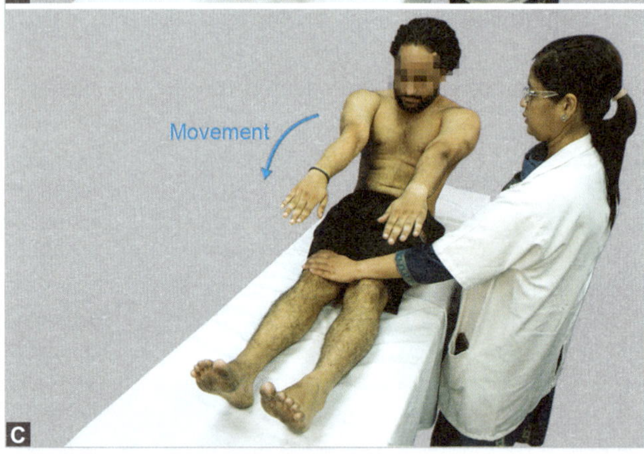

Figs 4.1A to C: A. Upper abdominals grade 5; **B.** Upper abdominals grade 4; **C.** Upper abdominals grade 3

Anterior Trunk Flexors: Lower Abdominal Muscle Test

- **Muscles:** Rectus abdominis, external oblique, internal oblique (Figs 4.2A to C)
- **Patient position:** Supine lying with hips flexed at 90° with both legs raised straight.
- **Physiotherapist position:** By the side of the patient
- **Movement:** The patient is asked to hold low back flat during leg lowering.
- **Resistance:** The body weight offers resistance
- **Grades:**
  - **Grade 5:** Patient is capable of lowering the legs to the level of the table while performing a posterior pelvic tilt and holding the low back flat on the table (Fig. 4.2C).
  - **Grade 4:** Patient is capable of performing a posterior pelvic tilt and holding the low back flat against the table while lowering the knees at a 30° angle (Fig. 4.2B).
  - **Grade 3:** Patient is capable of lowering the legs to a 75° angle with the table while performing a posterior pelvic tilt and holding the low back flat on the table (Fig. 4.2A).
- **Grade 2:** Patient can initiate posterior pelvic tilt but cannot maintain it while lowering the legs.
- **Grade 1:** Palpable contraction of abdominal muscles without movement of the pelvis (Fig. 4.3).
- **Grade 0:** No palpable contraction of abdominal muscles.

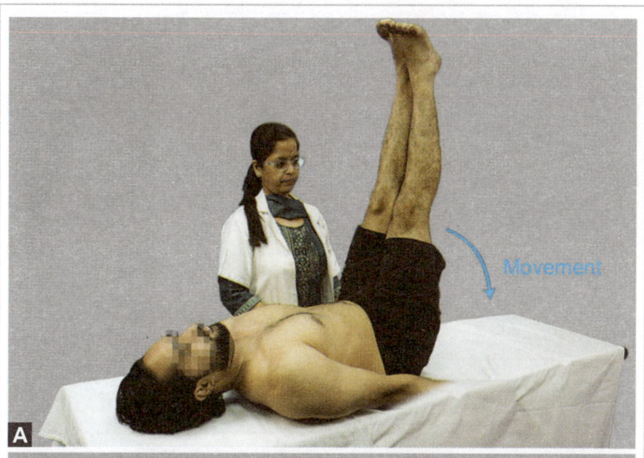

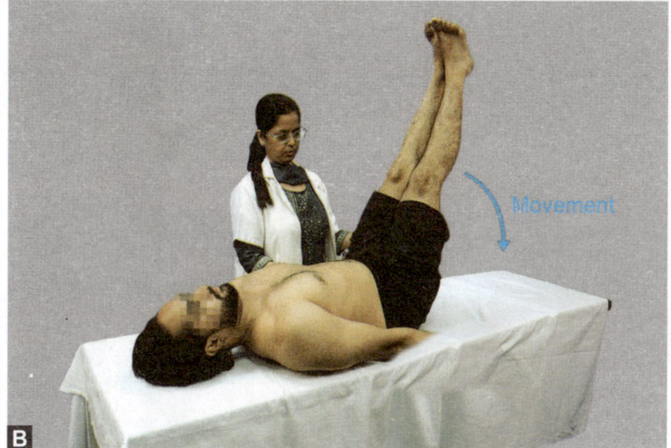

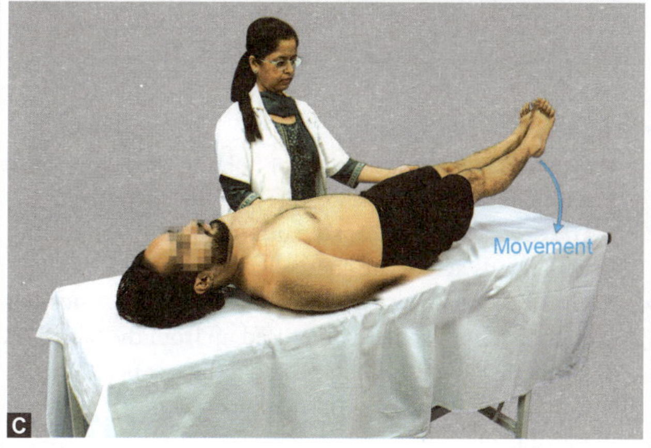

Figs 4.2A to C: **A.** Lower abdominals starting position; **B.** Lower abdominals grade 3; **C.** Lower abdominals grade 5;

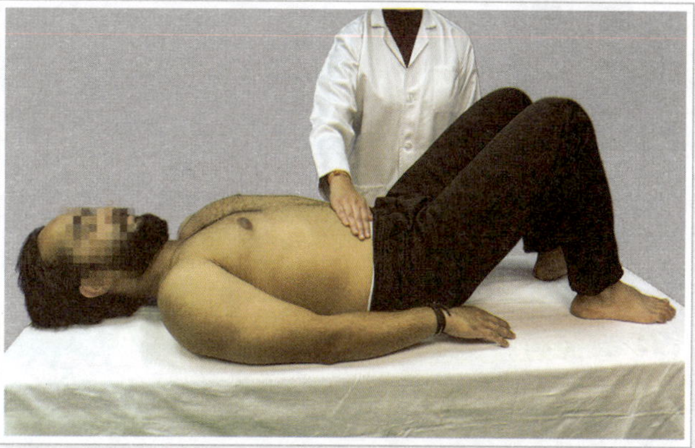

Fig. 4.3: Anterior trunk flexors lower abdominal muscle test grade 1

- **Grades:**
  - **Grade 5:** Patient is capable of completing test movement while holding hands behind the head (Fig. 4.4A).
  - **Grade 4:** Patient is capable of performing test movement while holding hands behind the back (Fig. 4.4B).
  - **Grade 3:** Patient is capable of partially completing test movement with hands clasped behind the back and xiphoid process lifted slightly.
  - **Grade 2:** The patient lifts the head and upper sternum off the plinth with the arms at the sides.
  - **Grade 1:** Palpable contraction of muscle without movement.
  - **Grade 0:** No visible contraction.

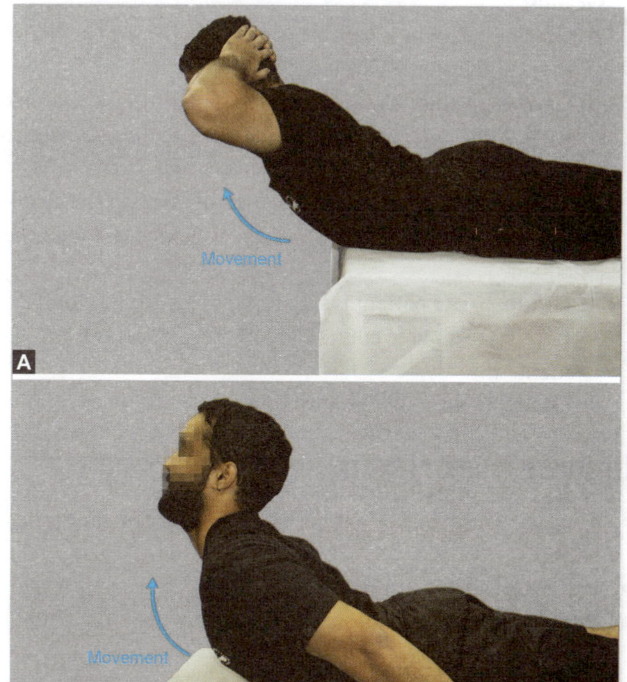

Figs 4.4A and B: **A.** Back extensors grade 5; **B.** Back extensors grade 4

Back Extensors

- **Muscles:** External abdominal oblique, internal abdominal oblique, rectus abdominis
- **Patient position:** Prone lying position
- **Physiotherapist position:** Standing toward the buttocks of patient stabilizing both lower limbs
- **Movement:** The patient is asked to extend the trunk to the full range of motion
- **Resistance:** The body weight offers resistance

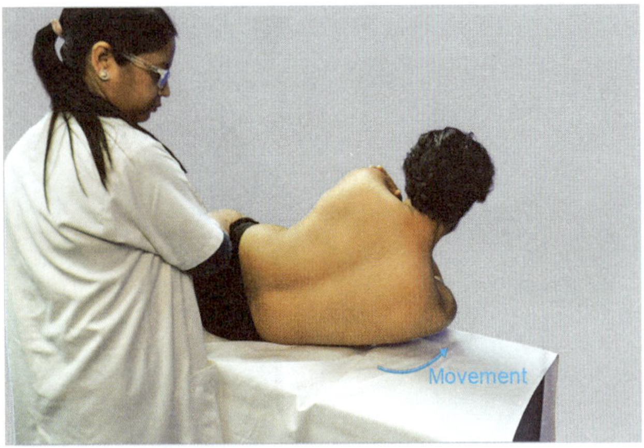

Fig. 4.5: Lateral trunk flexors grade 3

Lateral Trunk Flexors

- **Muscles:** The quadratus lumborum, oblique abdominals, and erector spinae
- **Patient position:** Side lying with head, trunk, pelvis and lower limbs in straight line. The under arm is held across the chest
- **Physiotherapist position:** By the side of the patient stabilizing the pelvis.
- **Movement:** The patient is asked to raise the trunk sideways without rotating it
- **Resistance:** The body weight offers resistance
- **Grades:**
  - **Grade 5:** Patient is capable of laterally flexing the trunk to its maximum ROM (Fig. 4.5).
  - **Grade 4:** Patient is capable of laterally flexing the trunk so that the underneath shoulder is 4 inches above the couch.
  - **Grade 3:** Patient is capable of laterally flexing the trunk so that the underneath shoulder is 2 inches above the couch.
  - **Grade 2:** The patient is able to lift head off the floor.
  - **Grade 1:** Palpable contraction of muscle without movement.
  - **Grade 0:** No visible contraction.

Oblique Trunk Flexors

- **Muscles:** External abdominal oblique, internal abdominal oblique
- **Patient position:** Supine lying
- **Physiotherapist position:** Towards the feet of patient stabilizing the legs.
- **Movement:** The patient is asked to maintain a position of trunk flexion and rotation.
- **Resistance:** The body weight offers resistance.

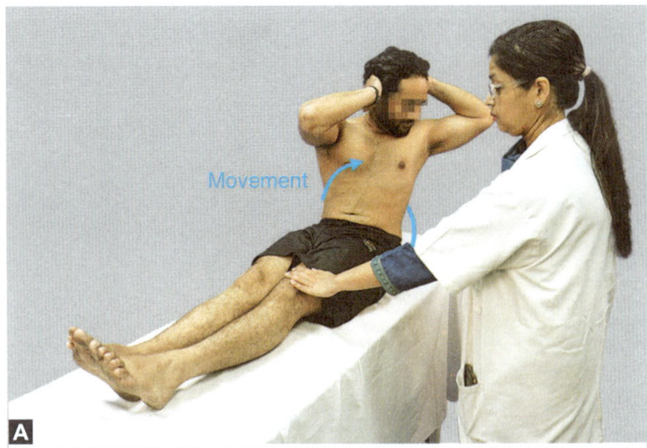

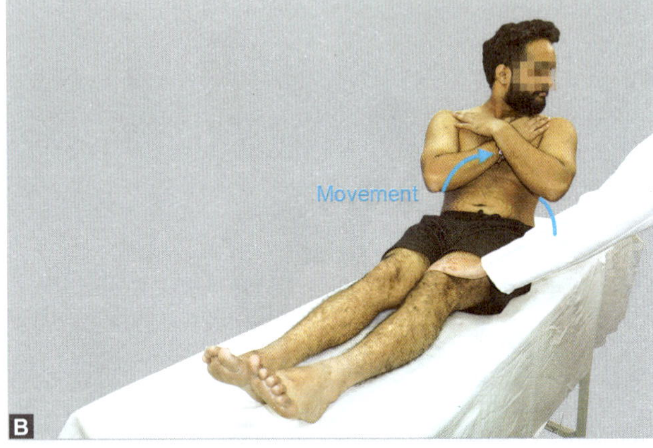

Figs 4.6A and B: A. Oblique trunk flexors grade 5; B. Oblique trunk flexors grade 4

- **Grades:**
  - **Grade 5:** Patient is capable of holding the test position with hands behind the head (Fig. 4.6A).
  - **Grade 4:** Patient is capable of holding the test position with hands across the chest (Fig. 4.6B).
  - **Grade 3:** Patient is capable of holding the test position such that both scapula are raised up from the table.
  - **Grade 2:** Patient is unable to hold the test position.
  - **Grade 1:** Palpable contraction of muscle without movement.
  - **Grade 0:** No visible contraction.

Anterior Flexors of Neck

- **Muscles:** Rectus capitis anterior, longus capitis, longus colli, scalenus anterior, sternomastoid
- **Patient position:** Supine lying position with both arms resting on the plinth
- **Physiotherapist position:** Standing toward head end of patient
- **Movement:** The patient is instructed to flex the cervical spine by lifting head from table with chin toward sternum

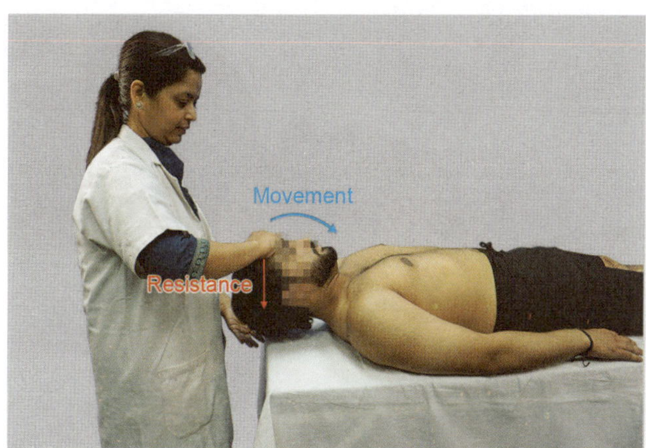

Fig. 4.7: Anterior flexors of neck grades 4–5

- **Resistance:** The physiotherapist applies resistance against the forehead in the direction of extension
- **Grades:** For normal muscle strength, i.e., grade 5, the patient must have the ability to move through complete range of motion (active resistance testing) or maintain an end point range (break testing) against maximum resistance.
  - **Grade 5:** Patient is capable of holding the test position opposite to the maximum resistance force (Fig. 4.7).
  - **Grade 4:** Patient is capable of holding the test position opposite to the moderate resistance force.
  - **Grade 3:** Patient is able to tuck chin without raising the head from the examination table.
  - **Grade 2:** Patient finishes the partial motion in the available range.
  - **Grade 1:** Palpable contraction of muscle without movement.
  - **Grade 0:** No visible contraction.

Neck Extensors

- **Muscles:** Semispinalis capitis, rectus capitis posterior (major and minor), obliquus capitis (inferior and superior), splenius capitis, semispinalis cervicis, longissimus capitis and cervicis, splenius cervicis, spinalis capitis and cervicis, and iliocostalis cervicis
- **Patient position:** Prone lying position with head off the table at the edge and arms by the side
- **Physiotherapist position:** Standing toward head end of patient
- **Movement:** The patient is instructed to extend the cervical spine by lifting head from table
- **Resistance:** The physiotherapist applies resistance against the occiput in the direction of flexion
- **Grades:** For normal muscle strength, i.e., grade 5, the patient must have the ability to move through complete range of motion (active resistance testing) or maintain an end point range (break testing) against maximum resistance.

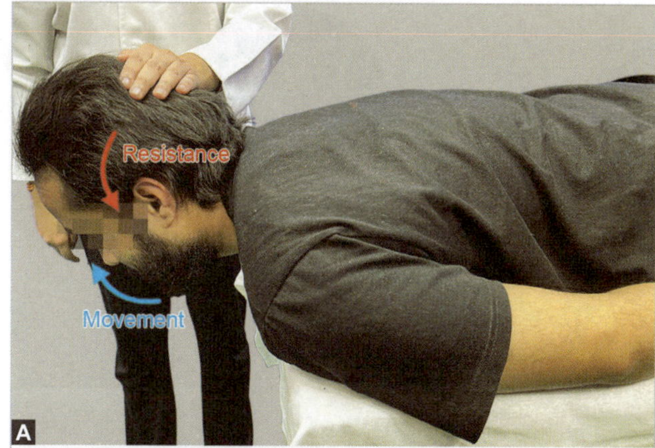

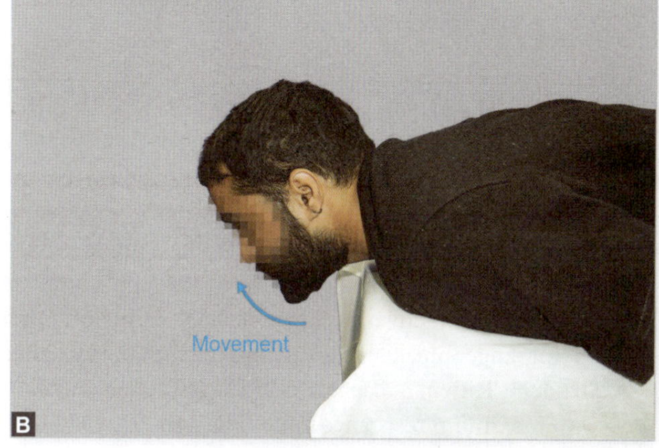

Figs 4.8A and B: **A.** Neck extensors grades 4–5; **B.** Neck extensors grade 3

  - **Grade 5:** Patient is capable of maintaining test position opposite to the strong resistance force (Fig. 4.8A).
  - **Grade 4:** Patient is capable of holding the test position opposite to the moderate resistance force.
  - **Grade 3:** Patient does the neck extension without looking up or tilting their chin (Fig. 4.8B).
  - **Grade 2:** Patient moves through partial range of neck extension.
  - **Grade 1:** Palpable contraction of muscle without movement.
  - **Grade 0:** No visible contraction.

Cervical Rotators

- **Muscles:** Semispinalis cervicis, multifidus scalene anterior, splenius cervicis and capitis, sternocleidomastoid, inferior oblique (head only), rectus capitis posterior major (head only).
- **Patient's position:** For grades 5–3, the patient should be in supine lying position. Head is resting on a plinth with the face rotated to one side as far as possible.
- **Physiotherapist's position:** Standing at the head of the couch.

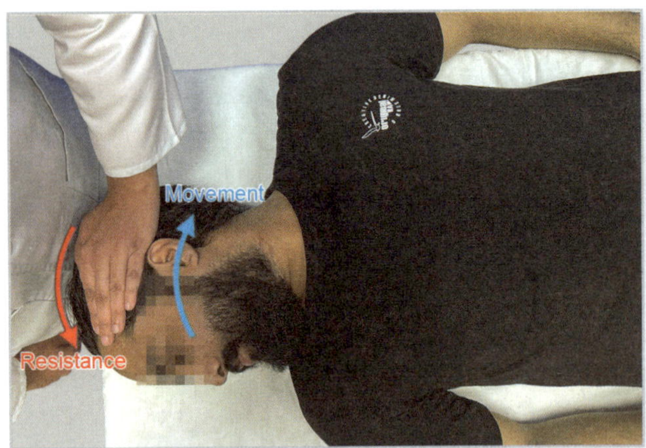

Fig. 4.9: Cervical rotators grades 4–5

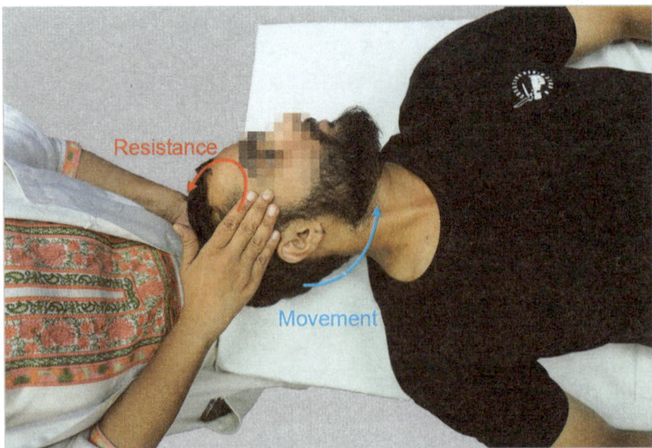

Fig. 4.10: Sternocleidomastoid grades 4–5

- **Movement:** The patient is asked to rotate the head and face the ceiling
- **Resistance:** The physiotherapist applies resistance on the side of patients head over the ear.
- **Grades:** For normal muscle strength, i.e., grade 5, the patient must have the ability to move through complete range of motion (active resistance testing) or maintain an end point range (break testing) against maximum resistance
  - **Grade 5:** Patient is capable of holding the test position with maximum resistance force (Fig. 4.9).
  - **Grade 4:** Patient is capable rotating the head with moderate resistance force.
  - **Grade 3:** Patient should turn head throughout the full present range of motion to both sides right and left without any resistance force.
  - **Grade 2:** Patient is unable to complete the range of motion.
  - **Grade 1:** Slight palpable contraction of muscle with no movement.
  - **Grade 0:** No visible contraction.

Anterolateral Neck Flexor (Sternocleidomastoid)

- **Patient's position:** For testing right sternocleidomastoid, patient is in supine with head supported on table and turned to the left.
- **Physiotherapist's position:** Standing toward head end of patient.
- **Movement:** The patient is instructed to flex the head while keeping it turned.
- **Resistance:** The physiotherapist applies resistance against the against the temporal area just above the ear.
- **Grades:** For normal muscle strength, i.e., grade 5, the patient must have the ability to move through complete range of motion (active resistance testing) or maintain an end point range (break testing) against maximum resistance.
  - **Grade 5:** Patient is capable of holding the test position against strong resistance (Fig. 4.10).
  - **Grade 4:** Patient is capable of holding the test position against moderate resistance.
  - **Grade 3:** Patient completes the available range of motion without resistance.
  - **Grade 2:** The patient completes partial ROM.
  - **Grade 1:** Palpable contraction of muscle without movement.
  - **Grade 0:** No visible contraction.

TECHNIQUES FOR FACIAL MUSCLES

The facial muscles (Fig. 4.11), also called craniofacial muscles, are a group of about 20 flat skeletal muscles lying underneath the skin of the face and scalp. Most of them originate from the bones or fibrous structures of the skull and radiate to insert on the skin.

Clinical Correlation

Loss of strength in the face muscles can occur due to various conditions, including:

- **Bell's palsy:** Bell's palsy is a condition that causes the facial muscles on one side of the face to suddenly weaken or paralyse. It frequently results in facial muscle weakness or paralysis due to inflammation or compression of the face nerve (cranial nerve VII).
- **Stroke:** Usually affecting one side of the face, a stroke can cause facial muscle paralysis or weakening. It happens when there is a disruption in the blood flow to the brain, which damages the area in charge of controlling the muscles of the face.
- **Guillain-Barré syndrome:** This immune-mediated condition may cause muscular wasting or paralysis, including of the facial muscles.
- **Myasthenia gravis:** Myasthenia gravis is a chronic autoimmune disease that affects the neuromuscular junction, causing muscle weakness. Facial muscles can be involved in this condition.

Contd...

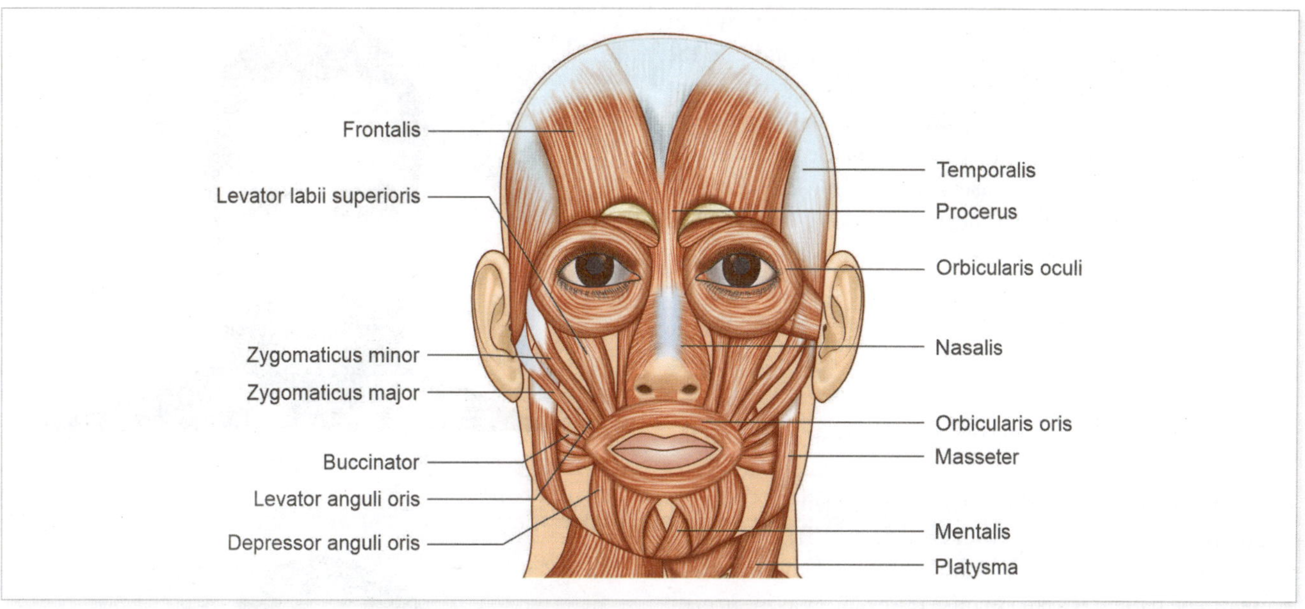

Fig. 4.11: The facial muscles

- **Amyotrophic lateral sclerosis (ALS):** ALS is a progressive neurological disorder that affects the nerve cells responsible for controlling voluntary muscles. It can lead to muscle weakness and eventually paralysis, including the face muscles.
- **Facial muscular disorders:** Some particular muscular conditions might affect the strength of the facial muscles. For instance: Facial dystonia, often called craniofacial dystonia or Meige syndrome, is a movement disease characterized by uncontrollable facial muscle spasms, which cause aberrant movements and facial muscle weakness.
- **Facial myopathies:** Uncommon hereditary muscular diseases that mostly affect the muscles of the face and can cause paralysis or weakening.

House-Brackmann Facial Nerve Grading System

The House-Brackmann Facial Nerve Grading System is a system for grading facial muscle strength. It is frequently used to evaluate the function and strength of the facial muscles in those who have facial nerve paralysis or weakening. The House-Brackmann scale assigns a letter grade to facial muscle function, with I denoting normal function and VI denoting complete paralysis. The grading scheme is tabulated in Table 4.1.

TABLE 4.1: House-Brackmann Facial Nerve Grading System

Grade	Meaning	Interpretation
Grade I	Normal function	No facial weakness or asymmetry at rest or with movement.
Grade II	Mild dysfunction	• Slight weakness noticeable only on close inspection. • Forehead wrinkles are equal or nearly equal on both sides. • Eye closure is complete with minimal effort. • Asymmetry may be present with maximal effort or during prolonged eye closure.
Grade III	Moderate dysfunction	• Obvious weakness or asymmetry noticeable at rest. • Forehead wrinkles are less distinct on the affected side. • Incomplete eye closure with effort. • Complete eye closure is possible. • Mouth symmetry is maintained at rest but there is asymmetry with maximal effort (smiling).
Grade IV	Moderately severe dysfunction	• Obvious weakness or asymmetry at rest. • Incomplete eye closure with effort. • Incomplete eye closure is possible. • Forehead wrinkles are absent on the affected side. • Asymmetry of the mouth at rest and with smiling.
Grade V	Severe dysfunction	• Only slight voluntary movement present. • Incomplete eye closure is possible. • No forehead movement. • Asymmetry of the mouth at rest and with smiling.
Grade VI	Total paralysis	• No movement of facial muscles. • No eye closure or movement of the forehead or mouth.

Contd...

The House-Brackmann Facial Nerve Grading System provides a standardized way to assess and communicate the degree of facial muscle weakness or paralysis. It is important to note that this scale focuses primarily on the function of the facial nerve and may not capture muscle weakness or impairment due to other causes, such as muscle or nerve damage unrelated to facial nerve function. It is always recommended to consult with a healthcare professional, such as a neurologist or otolaryngologist, for a comprehensive evaluation and accurate grading of facial muscle strength.

Frontalis Muscle

- **Origin:** Galea aponeurotica (a tendonous structure on the top of the skull).
- **Insertion:** Skin of the eyebrows and forehead.
- **Nerve supply:** Facial nerve (cranial nerve VII).
- **Action:** Raises the eyebrows and wrinkles the forehead.
- **Test:** The patient is asked to raise the eyebrow as in surprise (Fig. 4.12).

Corrugator Supercilii

- **Origin:** Medial end of superciliary arches, Fibers of orbicularis oculi muscle
- **Insertion:** Skin above middle of supraorbital margin
- **Nerve supply:** Temporal branches of facial nerve
- **Action:** Creates vertical wrinkles over glabella
- **Test:** The patient is asked to draw the eyebrows together as in frowning (Fig. 4.13).

Orbicularis Oculi Muscle

- **Origin:** Medial wall of the orbit (bones around the eye).
- **Insertion:** Tissues surrounding the eyelids.
- **Nerve supply:** Facial nerve (cranial nerve VII).

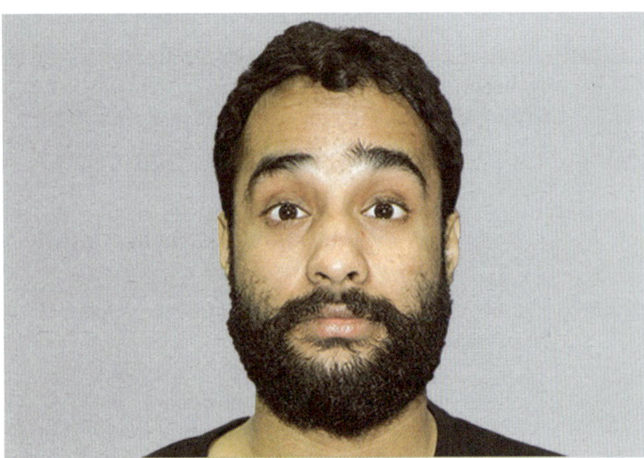

Fig. 4.12: MMT for frontalis muscles

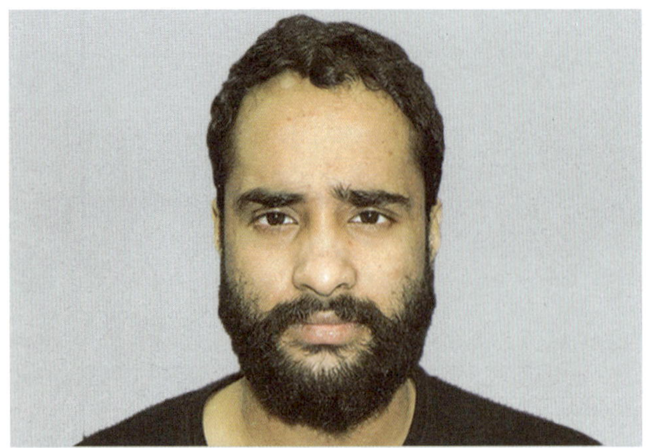

Fig. 4.13: MMT for corrugator supercilii muscles

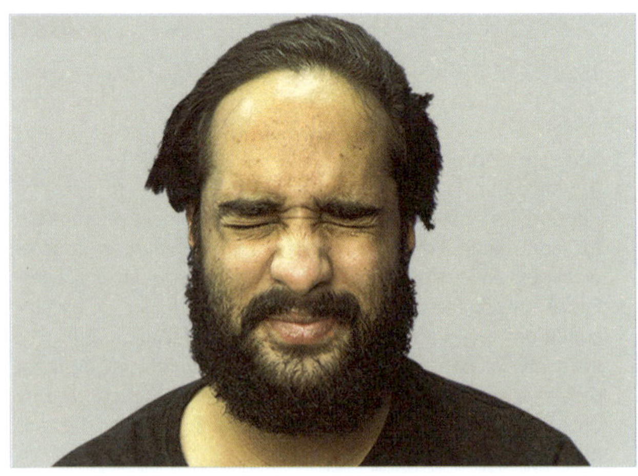

Fig. 4.14: MMT for orbicularis oculi muscles

- **Action:** Closes the eyelids, produces blinking, squinting, and contributes to the formation of crow's feet wrinkles.
- **Test:** The patient is asked to close the eyes as tightly as possible (Fig. 4.14).

Buccinator Muscle

- **Origin:** Alveolar processes of the maxilla and mandible (upper and lower jawbones) and pterygomandibular raphe (connective tissue in the mouth).
- **Insertion:** Orbicularis oris muscle (around the mouth).
- **Nerve supply:** Facial nerve (cranial nerve VII).
- **Action:** Compresses the cheeks inward, assists in sucking and blowing, and helps with mastication (chewing).
- **Test:** The patient is asked to compress the cheek firmly against the side teeth pulling back the angle of mouth (Fig. 4.15).

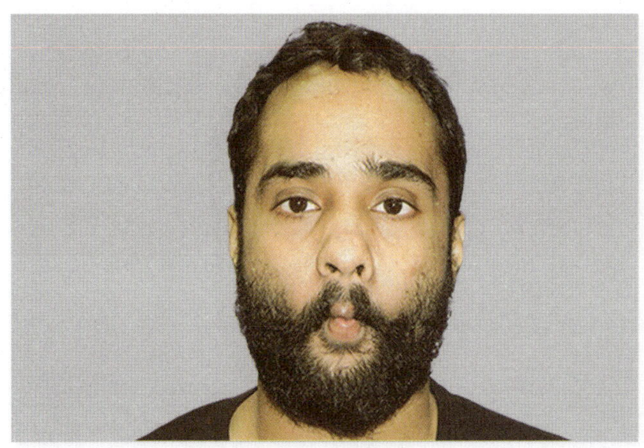

Fig. 4.15: MMT for buccinator muscles

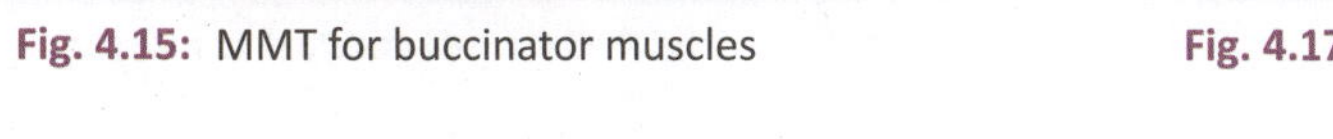

Orbicularis Oris Muscle

- **Origin:** Maxilla (upper jawbone) and mandible (lower jawbone).
- **Insertion:** Lips and surrounding tissues.
- **Nerve supply:** Facial nerve (cranial nerve VII).
- **Action:** Closes and purses the lips, helps with facial expressions and speaking.
- **Test:** The patient is asked to close the lips as in whistling (Fig. 4.16).

Zygomaticus Major and Minor Muscles

- **Origin:** Zygomatic bone (cheekbone).
- **Insertion:** Skin and muscle at the corner of the mouth.
- **Nerve supply:** Facial nerve (cranial nerve VII).
- **Action:** Pulls the corner of the mouth upward and laterally, producing smiling and facial expressions of happiness.
- **Test:** The patient is asked to draw the angle of mouth upward and outward as in smiling (Fig. 4.17).

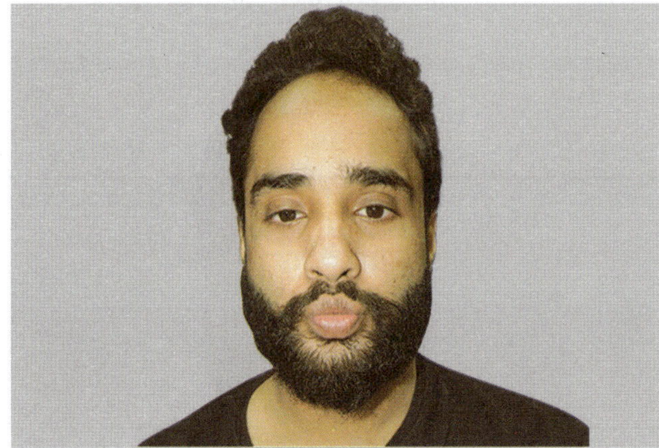

Fig. 4.16: MMT for orbicularis oris muscles

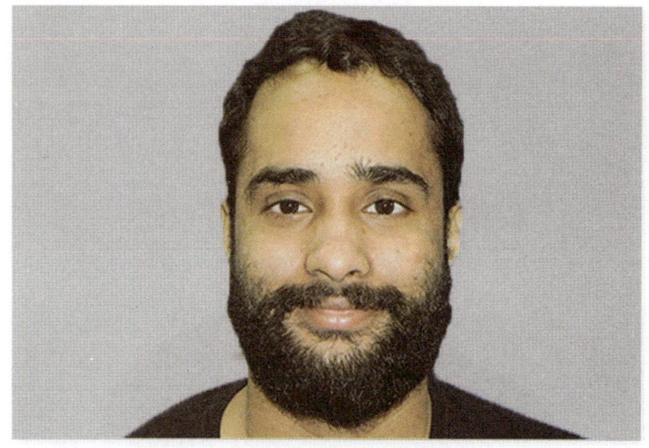

Fig. 4.17: MMT for zygomaticus major and minor muscles

Masseter Muscle

- **Origin:** Zygomatic arch (bony arch at the side of the skull).
- **Insertion:** Lateral surface of the mandibular ramus (lower jawbone).
- **Nerve supply:** Masseteric branch of the mandibular division of the trigeminal nerve (cranial nerve V3).
- **Action:** Elevates the mandible (closing the jaw) during chewing.
- **Test:** The patient is asked to bite firmly as in clenching the teeth (Fig. 4.18).

Platysma Muscle

- **Origin:** Fascia of the upper chest and shoulder.
- **Insertion:** Mandible (lower jawbone) and skin of the lower face.
- **Nerve supply:** Facial nerve (cranial nerve VII).
- **Action:** Depresses the mandible, tenses the skin of the neck, and assists in facial expressions like frowning.

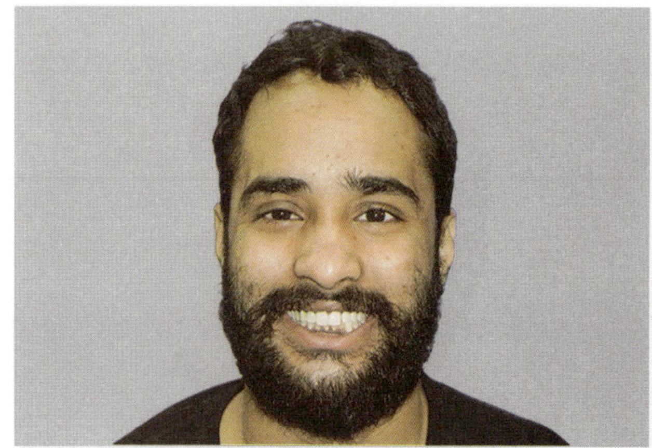

Fig. 4.18: MMT for masseter muscles

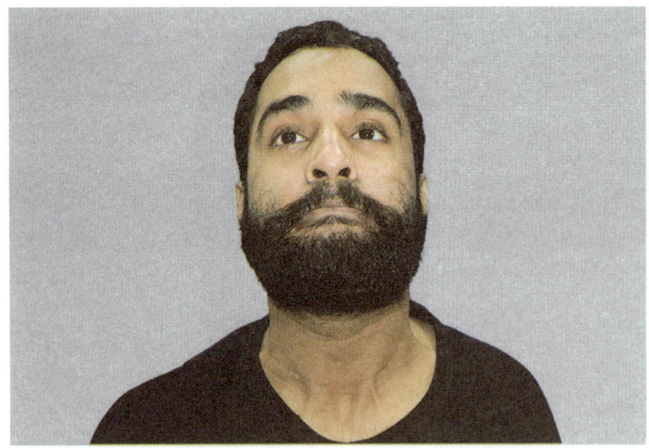

Fig. 4.19: MMT for platysma muscles

- **Test:** The patient is asked to the angle of mouth and lower lip downward and outward as if trying to tense the skin over the neck (Fig. 4.19).

Temporalis Muscle

- **Origin:** Temporal fossa (depression on the side of the skull).
- **Insertion:** Coronoid process of the mandible (lower jawbone).
- **Nerve supply:** Anterior deep temporal branches of the mandibular division of the trigeminal nerve (cranial nerve V3).
- **Action:** Closes the jaw and aids in retracting (pulling back) the mandible.
- **Test:** The patient is asked to bite firmly as in clenching the teeth (Fig. 4.20).

Nasalis

- **Origin alar part:** Frontal process of maxilla (superior to lateral incisor)

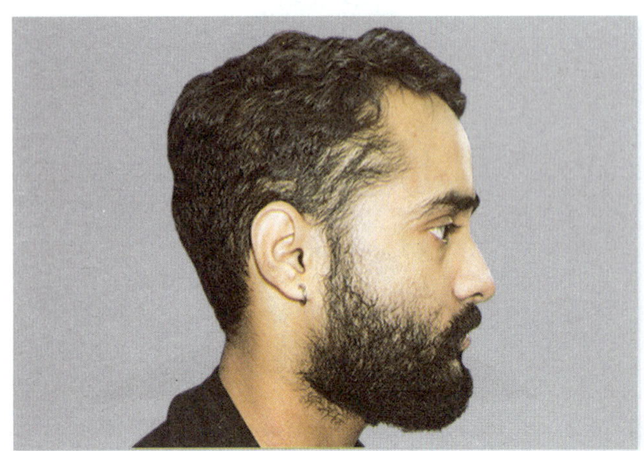

Fig. 4.20: MMT for temporalis muscles

Fig. 4.21: MMT for nasalis muscles

- **Transverse part:** Maxilla (superolateral to incisive fossa)
- **Insertion:** Alar part: Skin of ala;
- **Transverse part:** Merges with counterpart at dorsum of nose
- **Nerve supply:** Buccal branch of facial nerve
- **Action:**
  - **Alar part:** Depresses ala laterally, dilates nostrils
  - **Transverse part:** Wrinkles skin of dorsum of nose
- **Test:** The patient is asked to widen the apertures of the nostrils imitating forced breathing (Fig. 4.21).

Procerus

- **Origin:** Nasal bone, (superior part of) lateral nasal cartilage
- **Insertion:** Skin of glabella, fibers of frontal belly of occipitofrontalis muscle
- **Action:** Depresses medial end of eyebrow, wrinkles skin of glabella
- **Nerve supply:** Temporal, lower zygomatic or buccal branches of facial nerve
- **Test:** The patient is asked to pull the nose upwards as if trying to make transverse wrinkles on nose (Fig. 4.22).

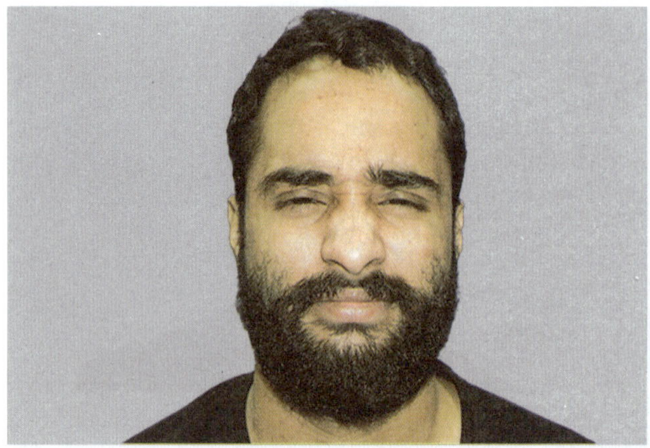

Fig. 4.22: MMT for procerus muscles

Levator Anguli Oris

- **Origin:** Canine fossa of maxilla
- **Insertion:** Modiolus
- **Nerve supply:** Zygomatic and buccal branches of facial nerve
- **Action:** Elevates angle of mouth
- **Test:** The patient is asked to draw the angle of mouth upwards (Fig. 4.23).

Risorius

- **Origin:** Parotid fascia, Buccal skin, zygomatic bone (variable)
- **Insertion:** Modiolus
- **Nerve supply:** Buccal branch of facial nerve (CN VII)
- **Action:** Extends angle of mouth laterally
- **Test:** The patient is asked to draw the angle of mouth backwards (Fig. 4.24).

Levator Labi Superioris

- **Origin:** Zygomatic process of maxilla, maxillary process of zygomatic bone
- **Insertion:** Blends with muscles of upper lip
- **Action:** Elevates upper lip, exposes maxillary teeth
- **Nerve supply:** Zygomatic and buccal branches of facial nerve
- **Test:** The patient is asked to lift and protrude the upper lip as if trying to show upper gums (Fig. 4.25).

Depressor Anguli Oris

- **Origin:** Mental tubercle and oblique line of mandible (continuous with platysma muscle)
- **Insertion:** Modiolus
- **Action:** Depresses angle of mouth
- **Nerve supply:** Buccal and mandibular branches of facial nerve
- **Test:** The patient is asked to draw angle of mouth downwards (Fig. 4.26).

Fig. 4.23: MMT for levator anguli oris muscles

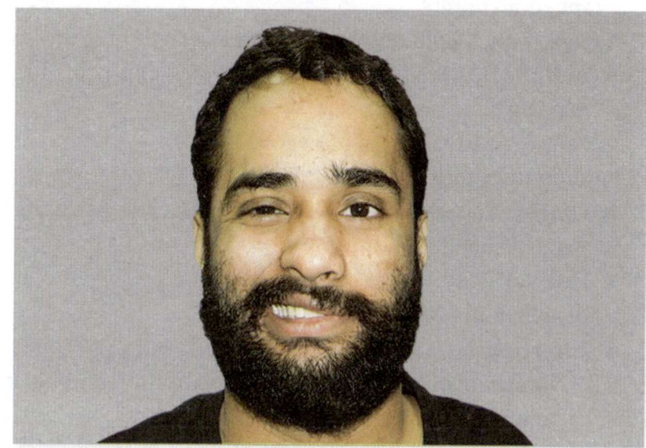

Fig. 4.25: MMT for levator labi superioris muscles

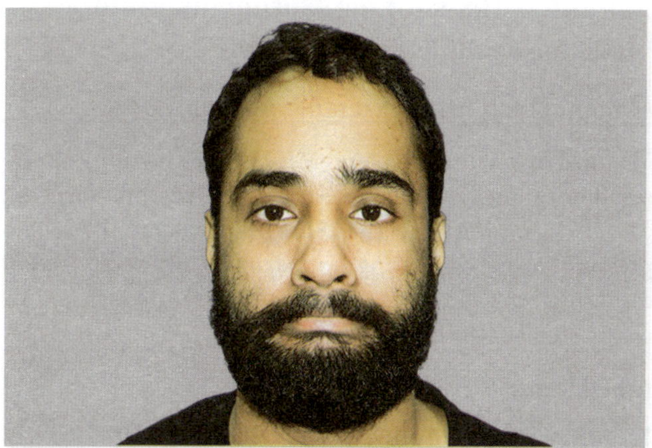

Fig. 4.24: MMT for risorius muscles

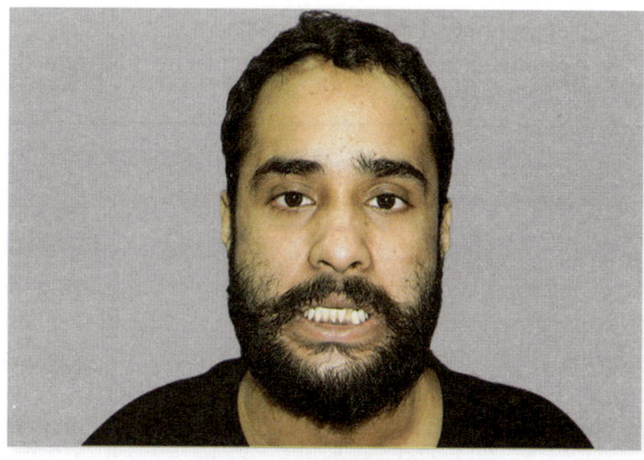

Fig. 4.26: MMT for depressor anguli oris muscles

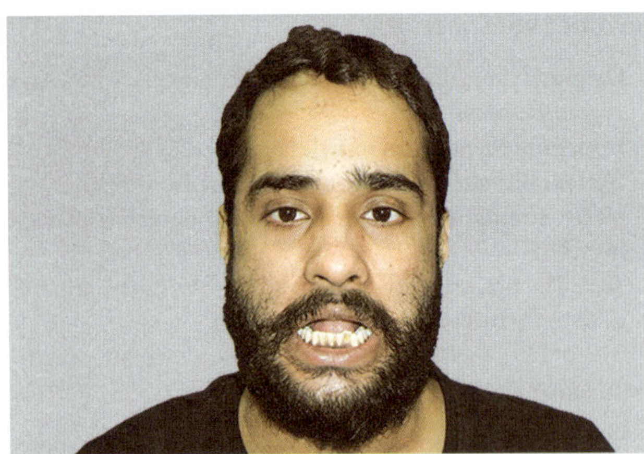

Fig. 4.27: MMT for medial and lateral pterygoid muscles

Medial and Lateral Pterygoid

- **Medial pterygoid:**
  - **Origin:** Superficial part: Tuberosity of maxilla, Pyramidal process of palatine bone;
  - **Deep part:** Medial surface of lateral pterygoid plate of sphenoid bone
  - **Insertion:** Medial surface of ramus and angle of mandible
  - **Action:** Bilateral contraction—elevates and protrudes mandible
  - **Nerve supply:** Medial pterygoid nerve
- **Lateral pterygoid:**
  - **Origin:** Superior head: Infratemporal crest of greater wing of sphenoid bone
  - **Inferior head:** Lateral surface of lateral pterygoid plate of sphenoid bone
  - **Insertion:**
    - **Superior head:** Joint capsule of temporomandibular joint
    - **Inferior head:** Pterygoid fovea on neck of condyloid process of mandible
  - **Action:** Bilateral contraction-Protrudes and depresses mandible
  - **Nerve supply:** Lateral pterygoid nerve
  - **Combined testing:** Ask the patient to protrude the lower jaw (Fig. 4.27).

RESPIRATORY MUSCLE STRENGTH ASSESSMENT

Assessing respiratory muscle strength involves measuring the pressure generated during inspiration or expiration. These measurements, usually expressed in kilopascals (kPa) or centimeters of water (cmH_2O), reflect the overall force exerted by respiratory muscles but are not specific to individual muscles. Accurate interpretation requires understanding various influencing factors.

Key Determinants

- **Measurement basis:**
  - Respiratory muscle strength is indirectly measured through pressure changes against atmospheric pressure.
  - The pressures reflect the combined output of all involved muscles, not individual muscle strength.
- **Potential causes of reduced pressure:**
  - **Reduced pressures may stem from issues at any level of the respiratory pathway:** Cerebral, spinal cord, peripheral nerves, neuromuscular junctions, or muscle fibers.
  - Not all reductions in pressure are due to respiratory muscle dysfunction.
- **Geometric considerations:**
  - **Diaphragm geometry:** The diaphragm's pressure generation is influenced by its shape, length, and attachment to the rib cage. For example, a flattened diaphragm in emphysema patients may not generate normal pressures despite good muscle condition.
  - **Laplace law:** The pressure generated by a muscle is inversely related to its radius.
- **Length-tension relationship:**
  - Respiratory muscles have an optimal length (L0) for generating maximum tension.
  - If muscles are at suboptimal lengths (e.g., due to hyperinflation), they produce less tension, affecting pressure measurements.
- **Lung volume effects:**
  - **Functional residual capacity (FRC):** At FRC, the elastic recoil pressures of the lungs and chest wall balance each other, so any additional pressure reflects muscle activation.
  - **Total lung capacity (TLC) and residual volume (RV):** Pressures at TLC include both expiratory muscle effort and elastic recoil, while pressures at RV reflect inspiratory muscle effort and chest wall expansion.

Volitional Tests for Measuring Inspiratory Muscle Strength

Maximal Inspiratory Pressure Measurement

Maximal inspiratory pressure (MIP) is a key test for assessing respiratory muscle strength, especially in patients suspected of muscle weakness (Fig. 4.28). MIP measures the highest pressure achieved during a maximal inspiratory effort, capturing both the force of the inspiratory muscles and the elastic recoil of the lungs and chest wall.

Advantages

- **Cost-effective and portable:** Uses inexpensive, portable equipment.
- **Simple and quick:** The test is straightforward and fast.

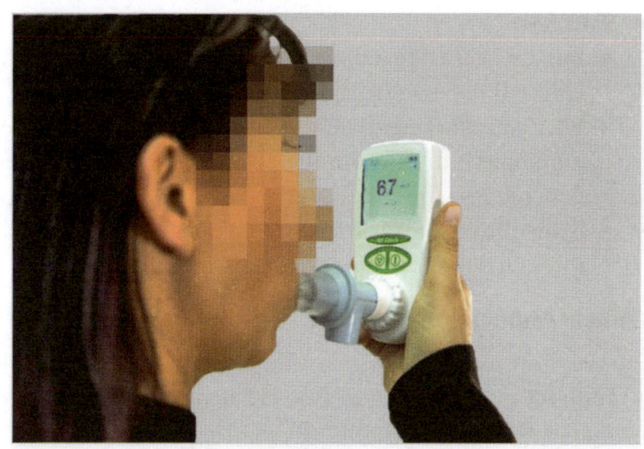

Fig. 4.28: MIP measurement

- **Noninvasive:** No need for invasive procedures.
- **Well-established norms:** Reference values are well-documented (e.g., 60 cmH_2O for females and 80 cmH_2O for males).
- **Early detection:** Can identify inspiratory muscle weakness before lung volume changes become evident.

Disadvantages

- **Requires cooperation:** Patient cooperation is crucial; low values may result from poor effort rather than true muscle weakness.
- **Variability:** High intraindividual and interindividual variation (10–13%).
- **Limited predictive accuracy:** Less reliable for predicting extubation success in ventilated patients.

Measurement Procedure

- **Preparation:** Measure from either RV or FRC. MIP from RV is higher (about 30% more) but FRC measurements are more reproducible.
- **Equipment:** Prefer digital manometers over analog ones for accuracy. Use a mouthpiece with a 2-mm wide opening to prevent overestimation due to glottal closure or mouth muscle pressure.
- **Techniques:**
  1. Have the patient exhale to RV.
  2. Perform a maximal inspiratory effort for 1–2 seconds.
  3. Repeat the maneuver 3–8 times, recording the highest value.
- **In intubated patients:**
  - Use a one-way valve to measure MIP, which allows for accurate assessment despite patient uncooperativeness.

Clinical Applications

- Diagnosing inspiratory muscle weakness in various diseases (pulmonary, cardiac, neuromuscular).

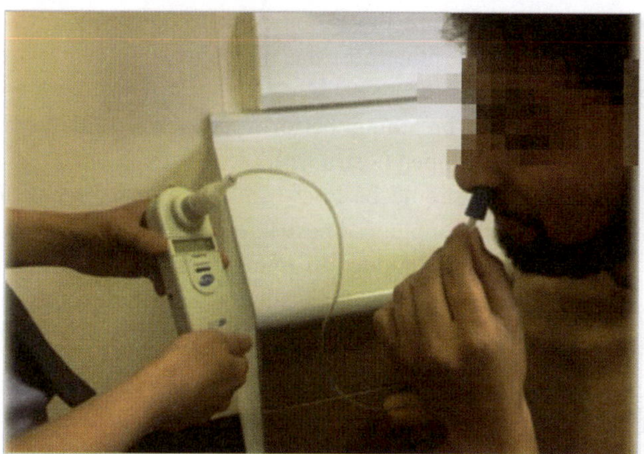

Fig. 4.29: Measurement of SNIP

- Differentiating dyspnea causes and obstructive lung diseases.
- Monitoring response to physiotherapy and respiratory muscle training.
- Assessing readiness for weaning from mechanical ventilation.

Sniff Nasal Inspiratory Pressure

Sniff nasal inspiratory pressure (SNIP) is a noninvasive method for measuring inspiratory muscle strength, specifically evaluating the joint activity of the diaphragm and other inspiratory muscles. It reflects esophageal pressure (Pes) and offers a simpler alternative to methods involving esophageal balloons (Fig. 4.29).

Advantages

- **Noninvasive and simple:** Uses basic pressure manometers and requires only a nasal sniff, making it easy and reproducible.
- **Well-established norms:** Normal values are established (e.g., 60 cmH_2O for females and 70 cmH_2O for males).
- **Complementary to MIP:** Provides additional insights into inspiratory muscle strength, reducing false-positive results when used alongside MIP.

Disadvantages

- **Patient cooperation required:** Dependent on patient's ability to perform the sniff maneuver correctly.
- **Not suitable for ventilated patients:** Cannot be used in mechanically ventilated patients.
- **Underestimation in airway obstruction:** May provide lower values in patients with significant airway obstruction or nasal blockage.

Measurement Procedure

- **Positioning SNIP:** Can be performed in any body position, typically sitting.

- **Nasal preparation:** Block one nostril completely with a plug; ensure the other nostril is fully open.
- **Maneuver:**
  - After quiet breathing, perform a fast, deep inhalation through the open nostril with the mouth closed.
  - Each sniff should be brief (≤500 ms) and forceful to collapse the unplugged nostril.
  - **Repetitions:** Conduct 10 maneuvers, with additional repetitions (up to 20) if values vary significantly or if inspiratory muscle weakness is suspected due to neuromuscular disease.

Clinical Applications

SNIP is highly specific for assessing inspiratory muscle strength and is valuable in diagnosing and monitoring muscle weakness in various conditions, including neuromuscular and pulmonary diseases. It is especially useful when combined with MIP to provide a comprehensive evaluation.

Inspiratory Mouth Pressure

Inspiratory mouth pressure (Pm) measures pressure at the mouth during inspiration, using a pressure sensor attached to a mouthpiece or tracheal tube. It is primarily used in three clinical scenarios: as an indirect measure of esophageal pressure (Pes) during a sniff, to verify esophageal catheter placement, and to measure P0.1, the pressure generated in the first 100 milliseconds of an inspiratory effort against a closed airway.

Advantages

- **Noninvasive:** Uses simple, noninvasive equipment.
- **Versatile equipment:** Utilizes the same instruments as MIP and SNIP.
- **Alternative to esophageal catheters:** Useful for patients where esophageal catheter use is contraindicated or impractical due to anatomical or clinical reasons.

Disadvantages

- **Patient difficulty:** Mouth measurements are generally more challenging for patients than nasal measurements.
- **Limited insight:** Like MIP, Pm does not specify which respiratory muscles are affected.
- **Pressure transmission issues:** In cases of severe expiratory flow limitation or lung disease, pressure transmission may be compromised, affecting accuracy.

Measurement Procedure

- **Equipment:** Use a mouthpiece with a wide cross-sectional area to avoid errors from the Bernoulli effect and support the cheeks with hands to prevent distortion.
- **Sniff test:**
  - Pm can be used to indirectly measure Pes during a sniff, especially when esophageal catheters are not available.
  - **P0.1 measurement:** Close the distal end of the mouthpiece to measure the pressure generated in the first 100 ms of inspiratory effort.
  - **Nonvolitional measurement:** Pm can also be measured through phrenic nerve stimulation if needed.

Clinical Application

Pm is mainly used for:

- **Indirect measurement of Pes:** Helps confirm inspiratory muscle weakness during a sniff.
- **Esophageal balloon verification:** Assists in confirming the correct positioning of the esophageal balloon using the Baydur test.

Transdiaphragmatic Pressure

Transdiaphragmatic pressure (Pdi) measures the force generated by the diaphragm, calculated as the difference between gastric pressure (Pga) and esophageal pressure (Pes) (Pdi = Pga – Pes). It specifically assesses diaphragm strength, a critical component of inspiratory muscle function.

Advantages

- **Targeted measurement:** Provides a direct assessment of diaphragm strength, which is responsible for 60–70% of tidal volume during normal breathing.
- **Established norms:** Reference values are well-documented for different populations (e.g., 70 cmH_2O for females and 80 cmH_2O for males during a sniff).

Disadvantages

- **Invasive:** Requires placement of catheters through the nose into the esophagus and stomach, which is invasive and requires specialized equipment.
- **Expertise required:** Accurate catheter placement depends on skilled practitioners.
- **Limited availability:** Equipment and materials are not widely available in all hospitals.

Measurement Procedure

- **Catheter placement:**
  - **Balloon catheters:** Insert one catheter into the esophagus and another into the stomach, or use a single catheter with dual balloons.
  - **Microtransducer catheters:** Utilize a single catheter for both Pes and Pga, offering better patient tolerance and quicker response for accurate measurements.

- **Verification:**
  - Ensure correct catheter placement by checking that Pes and Pga show opposing pressure changes during inspiration and expiration.
  - Use the Baydur test to compare Pes with mouth pressure (Pm) to confirm accurate esophageal placement. Proper positioning should reflect at least 80% of Pm variation.
  - **Measurement context:** Pdi can be assessed during normal breathing, maximal inspiratory maneuvers, or phrenic nerve stimulation.

Clinical Applications

Pdi is primarily used to evaluate diaphragm strength, particularly in patients with airway obstruction where other methods like Pm or SNIP might be less accurate. Its invasive nature and complexity limit its use to specialized settings.

Nonvolitional Methods for Assessing Inspiratory Muscle Strength

Phrenic Nerve Stimulation

This technique involves stimulating the phrenic nerve to induce diaphragmatic contractions. The resulting pressures are measured to assess diaphragm function.

Advantages

Provides direct measurement of diaphragm strength without patient effort; useful in cases where voluntary testing is not feasible.

Clinical Application

Often used in research and specialized clinical settings to evaluate diaphragm function in neuromuscular disorders.

Magnetic Stimulation

Magnetic stimulation of the phrenic nerve or diaphragm muscle uses transcranial magnetic stimulation (TMS) or magnetic pulses to evoke diaphragmatic contractions.

Advantages

It is noninvasive and can assess the central nervous system's control over the diaphragm; provides insights into both neural and muscular function.

Application

Utilized in research to study respiratory control mechanisms and in clinical settings to evaluate diaphragmatic function.

Electrical Stimulation

Electrical stimulation involves applying electrical impulses to the diaphragm or phrenic nerve to induce muscle contractions.

Advantages

Allows for direct measurement of muscle strength and response to stimulation; useful for assessing neuromuscular function.

Application

Common in research and clinical trials to evaluate diaphragm and respiratory muscle function.

Respiratory Muscle Ultrasound

Uses ultrasound imaging to assess diaphragm thickness and motion, indirectly evaluating muscle strength.

Advantages

It is noninvasive and provides real-time visualization of muscle function and structural changes.

Clinical Application

Increasingly used in clinical practice to monitor diaphragm function in various respiratory conditions.

Impulse Oscillometry (IOS)

Measures airway resistance and reactance using sound waves to assess the mechanical properties of the respiratory system, including muscle function.

Advantages

It is noninvasive and can provide information about respiratory mechanics and muscle function indirectly.

Application

Used in research and clinical settings to evaluate respiratory mechanics and response to interventions.

Volitional Tests to Measure Maximum Expiratory Pressure

Maximum Expiratory Pressure

Maximum expiratory pressure (MEP) measures the highest pressure a person can generate during a maximal expiratory effort, reflecting the strength of the expiratory muscles.

Procedure

The patient exhales forcefully into a pressure transducer connected to a mouthpiece or tracheostomy tube. The highest recorded pressure is noted.

Advantages

- **Noninvasive and simple:** Easy to perform with portable, low-cost equipment.
- **Useful for assessing expiratory muscle strength:** Valuable in diagnosing conditions affecting respiratory muscles.

Disadvantages

- **Depends on patient effort:** Inconsistent results if the patient cannot perform optimally.
- **High variability:** Results can vary based on patient effort, body position, and lung volume.

Clinical Application

MEP helps diagnose and monitor conditions like COPD and neuromuscular disorders, and is used in assessing respiratory muscle function and guiding rehabilitation.

Cough Gastric Pressure

Cough gastric pressure is an indirect measure of the abdominal pressure generated during a cough. It reflects the force produced by the abdominal muscles and can provide insights into the strength of the cough reflex and respiratory muscle function.

Procedure

A pressure sensor is placed in the stomach (usually *via* an intragastric balloon catheter) to measure the pressure generated during a cough. This measurement is used to assess the effectiveness of the cough and the strength of the abdominal muscles involved.

Advantages

Direct assessment of abdominal muscle function: Provides a clear picture of the pressure exerted by the abdominal muscles during a cough, which is crucial for evaluating cough effectiveness and respiratory muscle strength.

Disadvantages

Invasive procedure: Requires the placement of a catheter, which may be uncomfortable for the patient and not suitable for all clinical settings.

Gastric Pressure After Magnetic Stimulation of the Abdominal Wall Muscles

This method measures expiratory muscle strength by evaluating gastric pressure (Pga) following magnetic stimulation of the abdominal wall muscles. A circular coil is placed over the dorsal spine, between the eighth and tenth thoracic vertebrae (T8–T10), to induce muscle contractions.

Advantages

Independent of patient cooperation: The technique does not rely on patient effort or cooperation, making it useful for uncooperative or critically ill patients.

Disadvantages

- **Invasive:** Requires the insertion of a catheter into the stomach, which can be uncomfortable.
- **Limited reference values:** Only one study has reported reference values, making interpretation less standardized.

Procedure

1. **Catheter placement:** A balloon or microtransducer catheter is inserted into the stomach. Correct positioning is verified by observing pressure curves or manual compression of the epigastrium.
2. **Magnetic stimulation:** Magnetic stimulation is applied over the T8–T10 region of the spine. Pressure variations in the stomach are recorded.
3. **Measurement:** Perform about 5 measurements at 30-second intervals to prevent muscle potentiation.

Clinical Application

This method is particularly useful for assessing expiratory muscle strength in patients who cannot perform volitional tests, such as those with neuromuscular diseases or severe illness.

Electromyography

Electromyography (EMG) assesses muscle activity by measuring electrical signals generated during muscle contractions. It can be performed using surface electrodes, needle electrodes, or esophageal catheter electrodes.

Advantages

- **Surface EMG:** Noninvasive and useful for continuous monitoring. Highly sensitive for detecting muscle activity.
- **Needle EMG:** Minimally invasive, providing more specific data than surface electrodes.

Disadvantages

- **Surface EMG:** Subject to interference from other muscle groups and lacks standardized signal analysis.
- **Needle EMG:** Invasive and can be painful, with limited standardization for population reference values.
- **Esophageal EMG:** Highly specific but invasive, with limited availability and no established population reference values.

Procedure

- **Surface EMG:** Electrodes are placed on the skin over the muscle after cleaning to improve signal transmission.
- **Needle EMG:** Fine needles are inserted into the muscle, often guided by ultrasound to minimize complications.
- **Esophageal EMG:** Electrodes are placed 1–3 cm above the esophageal-gastric junction to measure diaphragm activity.

Clinical Application

EMG is valuable for diagnosing muscle weakness and monitoring respiratory muscle function, with various methods suited to different patient needs and clinical scenarios.

ASSESSMENT OF PELVIC FLOOR MUSCLE STRENGTH

Direct Methods

Internal Vaginal Digital Palpation

Internal vaginal digital palpation is a widely used method for assessing pelvic floor muscle (PFM) strength. This technique involves the insertion of one or two fingers into the vaginal cavity or rectum to directly feel the muscle contractions as the patient performs various maneuvers (Fig. 4.30).

Grading Systems

- **Modified Oxford scale:** The modified Oxford scale is a commonly used grading system for PFM strength assessment. It rates muscle contractions on a scale from 0 to 5:

0: No contraction
1: Flicker of contraction
2: Weak contraction (muscle contraction felt but no lift)
3: Moderate contraction (muscle contracts and lifts slightly)
4: Strong contraction (muscle contracts and lifts well, but may not hold fully)
5: Very strong contraction (muscle contracts and lifts maximally, holding with maximal force)

Perfect Scheme—Laycock

Developed by Laycock, the PERFECT Scheme provides a detailed assessment of PFM strength and function:

- **Power:** Rated using the Modified Oxford Scale (0–5).
- **Endurance:** Assessed by how long the patient can hold a maximal contraction, typically up to 10 seconds.
- **Repetitions:** Measures the number of maximal contractions the patient can perform with rest between them, up to 10 repetitions.
- **Fast:** Counts the number of 1-second maximal contractions performed consecutively, up to 10.
- **Every contraction timed:** Ensures all contractions are timed to standardize measurement.

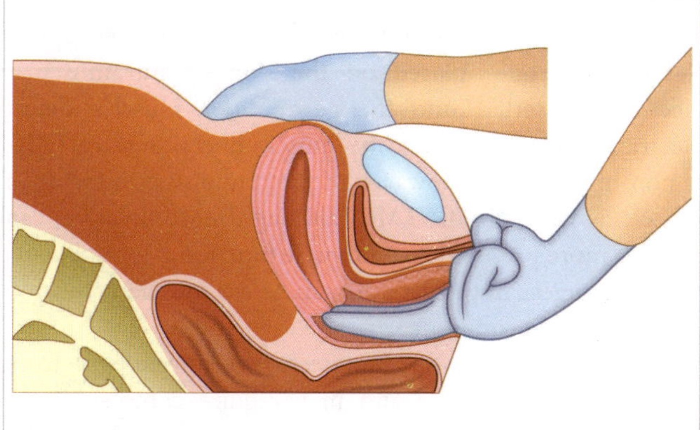

Fig. 4.30: Internal vaginal digital palpation method

Advantages and Challenges

- **Advantages:** Digital palpation is valued for its flexibility, cost-effectiveness, and ability to assess various aspects of muscle function, including resting tone, contraction force, endurance, and coordination.
- **Challenges:** The method is subjective and relies on the examiner's skill, which can lead to variability in scoring. The Modified Oxford Scale is often used, but discrepancies in the descriptive terms for each score can affect consistency. Intra-rater reliability is good, meaning the same examiner tends to get consistent results, while inter-rater reliability is only fair, due to differences between examiners.

Dynamometry

Dynamometry provides an objective measure of PFM strength by quantifying the force exerted by the muscles. This method involves using a dynamometer, which measures both active and passive forces exerted by the pelvic floor muscles (Fig. 4.31).

- **Advantages:** Dynamometry is particularly useful for evaluating the viscoelastic properties of the muscles and for assessing how muscle strength relates to the length-tension principle.
- **Challenges:** Although it offers precise measurements, the technique has limitations, such as being restricted to supine positioning and not distinguishing between right and left muscle functions. Additionally, the measurement can be influenced by intra-abdominal pressure and accessory muscle use, which need to be controlled or accounted for analysis.

Vaginal Cones

Vaginal cones assess PFM strength by requiring the patient to retain weighted cones in the vagina while performing dynamic movements, such as walking for one minute. This method primarily measures muscular endurance rather than maximal force, as it relies on the ability of the PFM to hold the

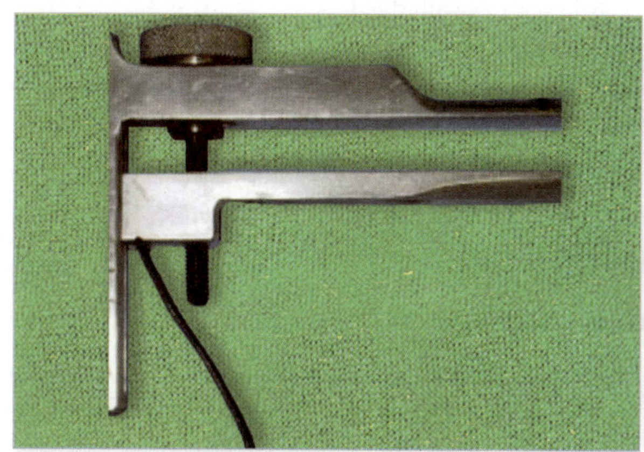

Fig. 4.31: Dynamometer

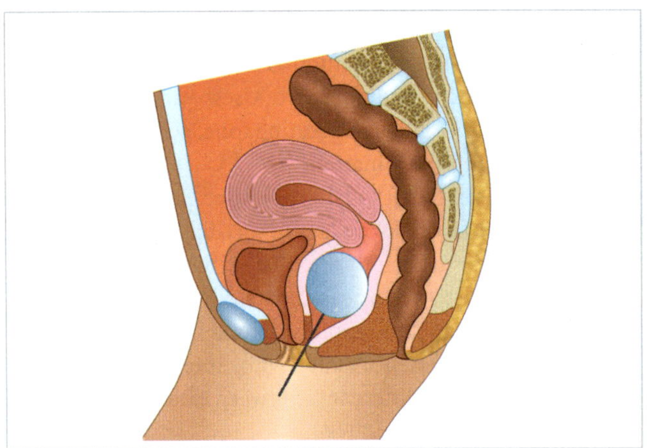

Fig. 4.32: Colpexin pull test

cone against gravity for an extended period. The majority of pelvic floor muscle fibers are slow-twitch, which aligns with the endurance aspect of this test.

Challenges: Vaginal cones have limitations related to anatomical variations that can affect insertion and retention. Studies have also noted issues with the cones' rotation and retention without increasing contraction intensity, questioning the method's reliability.

Colpexin Pull Test

The **Colpexin pull test** offers a novel, objective method for assessing pelvic floor muscle (PFM) strength and contractility (Fig. 4.32). Developed to address the limitations of subjective assessments, this test involves inserting a sphere attached to a force gauge into the vagina. During the test, the patient contracts the pelvic floor muscles while the sphere is pulled out at a steady rate. The maximum force required to remove the sphere is recorded and provides a direct measure of PFM strength.

- **Advantages:** This method has shown promise, with early studies indicating significant improvements in muscle strength over 16 weeks in patients performing regular Kegel exercises with the device.
- **Challenges:** Despite its advantages, the Colpexin pull test has limitations, such as potential issues with sphere size accommodating anatomical variations and its inability to detect muscle asymmetry or differentiate between true PFM strength and accessory muscle use.

Indirect Methods

Ultrasound Imaging

- **Real-time ultrasound**, whether performed transabdominally or transperineally, is an effective noninvasive method for assessing pelvic floor muscle contractions.
- **Transabdominal ultrasound** involves placing the probe suprapubically to capture images of the pelvic floor muscles. It is particularly useful for evaluating the **squeeze and lift** aspects of muscle contractions and for assessing **endurance**. This method is valued for its non-invasive nature, making it suitable for patients who may be uncomfortable with internal examinations, such as children. Additionally, it provides immediate visual feedback to patients, aiding in the correct execution of pelvic floor contractions. However, while it can assess muscle coordination and endurance, it does not accurately measure the strength of the contractions.
- **Transperineal ultrasound** is performed by placing the probe externally on the perineum. This method shares many of the advantages of transabdominal ultrasound, including its noninvasive nature and ability to provide visual feedback. It can be particularly useful for evaluating pelvic floor function in various patient populations. Nevertheless, interpreting ultrasound images for muscle strength and function requires specialized expertise and equipment, and the complexity of muscle strength assessment based on the imaging can be a limitation (Figs 4.33A and B).

Magnetic Resonance Imaging

Magnetic resonance imaging (MRI) offers detailed anatomical images of the pelvic floor muscles and can be used to assess muscle morphology, positioning, and function. Functional MRI can evaluate muscle activity by observing changes in muscle volume and contraction dynamics during specific tasks.

- **Advantages:** MRI provides high-resolution images that can help diagnose structural abnormalities and guide treatment planning.
- **Challenges:** The primary limitations are its high cost, limited availability, and the need for patient cooperation, as the procedure requires the patient to remain still.

Electromyography

Electromyography (EMG) measures the electrical activity of the pelvic floor muscles during contraction and relaxation. Surface electrodes placed on the skin or intramuscular electrodes inserted into the pelvic floor muscles detect electrical signals generated by muscle fibers. Surface EMG is non-invasive and allows for the continuous monitoring of muscle activity. Needle EMG, although more invasive, can provide more precise data on muscle function.

- **Advantages:** Both methods can assess muscle activation patterns and endurance.
- **Challenges:** EMG can be affected by interference from other muscle groups and may require complex signal interpretation.

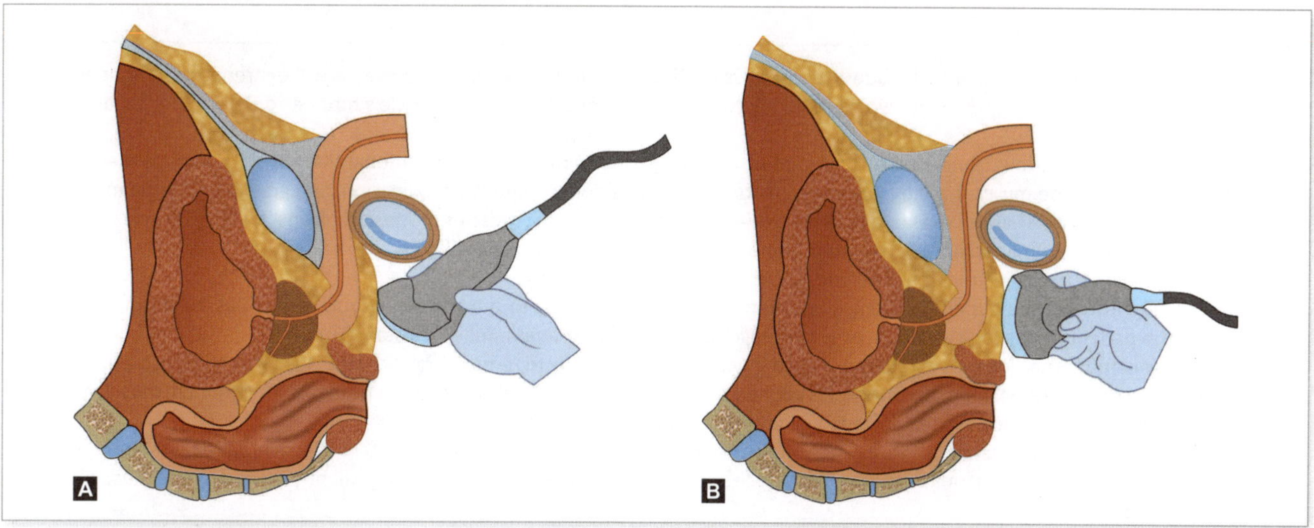

Figs 4.33A and B: Transperineal ultrasound

Pressure Measurement Devices

Pressure measurement devices, such as intravaginal or intra-anal pressure sensors, assess the pressure generated by pelvic floor muscles during contractions (Fig. 4.34). These devices measure the internal pressure exerted by the pelvic floor muscles against a fixed resistance or during various maneuvers.

- **Advantages:** Pressure sensors provide indirect measures of muscle strength and can be used to evaluate changes over time or in response to treatment.
- **Challenges:** While effective in assessing pressure generation, these devices can be invasive and may not differentiate between muscle strength and the effect of other factors like intra-abdominal pressure.

Functional Testing

Functional testing involves assessing the pelvic floor muscles' ability to perform specific tasks or maneuvers, such as the ability to stop urine flow or perform a cough. Tests like the "Knack" maneuver, where the patient is asked to contract the pelvic floor muscles before coughing or sneezing, provide indirect measures of muscle strength and coordination.

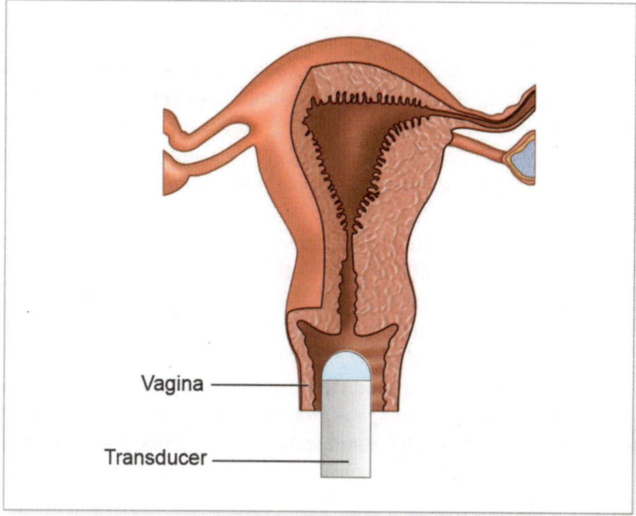

Fig. 4.34: Pressure measurement using transducer

Challenges: Functional tests are simple and can be performed in a clinical setting, but they may not provide a comprehensive measure of muscle strength or endurance and can be influenced by patient technique and cooperation.

SUMMARY

- Manual muscle testing (MMT) is a technique used in physical therapy and rehabilitation to evaluate the strength and functionality of specific muscles or muscle groups. It involves applying specific resistance against the muscle's motion to determine its capacity to produce force.
- MMT can be applied in various clinical contexts, including neurology, sports medicine, occupational therapy, and physical therapy. The principles of MMT include muscle grading, muscle isolation, testing position, stabilization, palpation, and gravity and resistance.
- The main objectives of MMT are to evaluate the strength and function of specific muscles or muscle groups, identify muscle weakness, track progress, guide treatment planning, and determine preparedness for return to activities.
  - Techniques include break testing, make testing, and gravity eliminated testing.
  - Break testing involves the patient resisting the examiner applying pressure to a muscle or group of muscles, while make testing involves the patient contracting a muscle or set of muscles while the examiner applies resistance.
  - Gravity eliminated testing involves carefully placing the patient's limb or body to remove gravity's influence from the muscle under test.
- To ensure consistency, external force (resistance) is applied to one-joint muscles at the end of their range during manual muscle testing. Two-joint muscles are examined in the mid-range, where length-tension is more advantageous.
  - The preferred point of resistance for assessing vertebro-scapular muscles is the arm, as it better reflects functional requirements.
  - To reach the highest tolerated force intensity, resistance should be applied slowly and gradually, slightly exceeding the muscle's force over 2–3 seconds.
- Musculoskeletal injuries, neurological diseases, post-surgical rehabilitation, sports performance, and occupational therapy are all areas where MMT can be used to assess muscle strength and function.
  - MMT can help physiotherapists create treatment plans to enhance muscle function and reduce the risk of further injury. It is also used in postsurgical rehabilitation to direct treatment strategies.
  - MMT is also used in sports performance to identify injury risk and to direct training and conditioning regimens to enhance performance.
  - In occupational therapy, it is used to assess muscle function and strength in individuals who have sustained job-related injuries or disabilities. However, MMT has limitations such as subjectivity, limited scope, insensitivity, reliability, and safety issues.
- The steps of MMT include patient preparation, description of the process, measurement of resistance to gravity, application of resistance, grading of strength, recording outcomes, and repeating the test.
  - It is crucial to ensure the examiner has training and experience in MMT techniques and that the testing processes are uniform and fair.
  - Safety issues should be considered during testing to prevent aggravating discomfort or reinjuring an existing injury.
- Before applying MMT, the physiotherapist should explain the test's components and outcomes to the patient. Consistent pressure should be applied, starting with the side not dominant or damaged.
  - The patient should breathe normally throughout the exam, have complete range of motion, be in a well-supported position, and perform the initial test while antigravity.
  - Apply resistance immediately opposite the "line of pull" of the muscles being tested.
  - To gain a realistic image of strength and impairment, test both sides and compare the strength or muscle grade of both limbs.
- The Medical Research Council scale (MRC) is a popular approach for rating muscle strength, with a numerical grade from 0 to 5.
  - The Oxford Scale measures the strength of the patient's major muscles in the upper and lower extremities against the examiner's resistance and evaluates it from 0 to 5.
  - The MRC scale is used to assess patients' muscular function and strength.
- Respiratory muscle strength is assessed *via* pressure measurements against atmospheric pressure, reflecting combined muscle output.
  - Factors affecting this include diaphragm geometry, length-tension relationships, and lung volumes.
  - Maximal Inspiratory Pressure (MIP) measures peak inspiratory effort and is cost-effective but requires patient cooperation.
  - Sniff Nasal Inspiratory Pressure (SNIP) is noninvasive and provides complementary insights but isn't suitable for ventilated patients.
  - Inspiratory Mouth Pressure (Pm) and Transdiaphragmatic Pressure (Pdi) offer additional methods for assessing muscle strength, with Pdi specifically measuring diaphragm strength but being invasive.
- Nonvolitional methods for assessing inspiratory muscle strength include phrenic nerve stimulation, which directly measures diaphragm contractions without patient effort; magnetic stimulation, which uses magnetic pulses to evaluate diaphragm and neural function; electrical stimulation, inducing muscle contractions to assess strength and response; and respiratory muscle ultrasound, which provides real-time imaging of diaphragm thickness and motion.
  - These methods are useful in specialized clinical settings or research, offering valuable insights when voluntary testing is impractical or unreliable.
- Pelvic floor muscle strength can be assessed using various methods. Direct methods include internal vaginal digital palpation, which uses manual examination and grading systems like the Modified Oxford Scale and the PERFECT Scheme, and dynamometry, which quantifies muscle force.
  - Other direct techniques are vaginal cones for endurance and the Colpexin pull test for objective strength measurement.
  - Indirect methods involve ultrasound imaging, which visualizes muscle contractions and endurance, MRI for detailed anatomical and functional assessment, electromyography (EMG) for electrical activity measurement, and pressure measurement devices that gauge internal pressure exerted by muscles.
  - Functional testing assesses muscle ability during specific tasks but may not fully capture strength or endurance.

FURTHER READINGS

- Caruso, Pedro et al. Diagnostic methods to assess inspiratory and expiratory muscle strength. Jornal brasileiro de pneumologia: publicacao oficial da Sociedade Brasileira de Pneumologia e Tisilogia vol. 41,2 (2015): 110–23. doi:10.1590/S1806-37132015000004474
- Deegan EG, Stothers L, Kavanagh A, Macnab AJ. Quantification of pelvic floor muscle strength in female urinary incontinence: A systematic review and comparison of contemporary methodologies. Neurourology and Urodynamics. 2018 Jan;37(1):33–45.
- Hislop, Helen J. and Jacqueline Montgomery. Daniels and Worthingham's Muscle Testing: Techniques of Manual Examination. 11th ed./ St. Louis, Mo., Saunders/Elsevier, 2025.
- House, J. W., and Brackmann, DE (1985). Facial nerve grading system. Otolaryngology-Head and Neck Surgery, 93(2), 146–147.
- Kendall, Florence Peterson, et al. Muscles, Testing and Function. 3rd ed. Baltimore, Williams & Wilkins, 1983.
- Lowry F. New Device Described as Objective Test of Pelvic Floor Musculature. Women's Health. 2006 May 15.
- Parezanović-Ilić K, Jevtić M, Jeremić B, Arsenijević S. Muscle strength measurement of pelvic floor in women by vaginal dynamometer. Srpski arhiv za celokupno lekarstvo. 2009;137(9–10):511–7.
- Troosters T, Gosselink R, Decramer M. Respiratory muscle assessment. European respiratory monograph. 2005 Apr 1;31:57.

STUDENT ASSIGNMENT

LONG ANSWER QUESTIONS

1. Discuss the principles, aims, indications and limitations of MMT.
2. Explain the technique of MMT for shoulder muscles.
3. Elaborate the technique of MMT for elbow muscles.
4. Explain the technique of MMT for hip muscles.
5. Discuss the technique of MMT for knee muscles.
6. What are the techniques of MMT for lumbar and thoracic spine muscles?
7. Explain the technique of MMT for ankle muscles.
8. Discuss the technique of MMT for neck muscles.
9. Explain the technique of MMT for facial muscles.

SHORT ANSWER QUESTIONS

Write notes on:

1. The different grading systems of MMT.
2. The volitional and non-volitional tests for measuring inspiratory muscle strength.
3. The volitional and non-volitional tests for measuring expiratory muscle strength.
4. The direct methods to assess the pelvic floor muscle strength.
5. The indirect methods to assess the pelvic floor muscle strength.

MULTIPLE CHOICE QUESTIONS

1. **Regarding MMT, which statement is correct?**
   a. Break test is less reliable than Make test.
   b. Stabilization is only required for weak muscles.
   c. Break test requires the tester to exert greater force than the patient.
   d. MMT could be used in severe spastic upper motor neuron lesion.
2. **The gravity minimized position in MMT is the position that allows the movement to be:**
   a. Vertical with gravity
   b. Vertical against gravity
   c. Horizontal parallel to the ground
   d. Diagonal from vertical to horizontal
3. **MMT is intended to assess muscle function as:**
   a. An agonist
   b. An antagonist
   c. A stabilizer
   d. A conjoint muscle
4. **Grade 5 break muscle testing is awarded if the following criteria is fulfilled:**
   a. Yielding against maximal resistance
   b. Holding against maximal resistance
   c. Full ROM against moderate resistance
   d. Holding against maximal resistance
5. **MMT reliability:**
   a. Interrater reliability is less than intrarater reliability
   b. Intrarater reliability is less than interrater reliability
   c. Intrarater reliability equals interrater reliability
   d. MMT is not reliable at all
6. **The level of measurement of MMT score is considered:**
   a. Nominal
   b. Ordinal
   c. Interval
   d. Dichotomous

ANSWER KEY

1. c **2.** c **3.** a **4.** b **5.** a **6.** b

5 Goniometry

Sheetal Kalra, Sajjan Pal

LEARNING OBJECTIVES

After the completion of the chapter, the readers will be able to:

- Define goniometry, the principles, aims and various methods.
- Explain indications and limitations of goniometry.
- Explain the techniques of using a goniometer for all joints.
- Demonstrate the techniques of measuring joint ROM for the joints of upper limb, lower limb and spine.

CHAPTER OUTLINE

- Introduction
- Principles
- Aims
- Types of Goniometers
- Psychometric Properties of Goniometers
- Methods of Joint Measurement
- Indications
- Procedure

KEY TERMS

Anatomical landmarks: These are the points on an organism that have biological significance.

End feel: It is a term used to describe the quality or sensation felt as a joint reaches the limit of its range of motion.

Goniometry: A technique which is used in physical therapy and rehabilitation for measurement of joint angles in the human body.

Inclinometer: It functions similarly to a goniometer and is frequently used in physiotherapy to measure the range of motion (ROM) of joints.

Range of motion: A limit to which a body part can move around a joint or fixed point.

INTRODUCTION

Goniometry can be defined as a technique used in physical therapy and rehabilitation to measure joint angles in the human body, with the help of an equipment called **Goniometer**. A goniometer normally has two arms, one of which is stationary and the other is mobile, with a scale resembling a protractor between them. The therapist or medical professional uses goniometry by aligning the movable arm with the segment being measured, such as the limb or body part, while the stationary arm is aligned with a reference point on the body, such as the joint axis. Therapist then use the scale on the goniometer to observe and calculate the angle created by the two arms. Typically, the measurement is expressed in degrees.

The range of motion of different joints in the body, such as the shoulder, elbow, wrist, hip, knee, and ankle, can be evaluated using goniometry. It is frequently used to assess joint flexibility, spot any limits or mobility limitations, monitor recovery progress, and assess the efficacy of therapeutic approaches. It is important to remember that goniometry is only one element of a thorough musculoskeletal examination. To get a more complete picture of the patient's condition and create a suitable treatment plan, healthcare providers may also take into account additional characteristics like pain, strength, functional restrictions, and patient-reported results.

Did You Know?

Goniometry was first introduced in the late 19th century by French physician Charles Féré, who developed the first goniometer for measuring joint angles. Since then, goniometry has evolved significantly, with modern advancements in technology leading to the development of electronic goniometers and computerized motion analysis systems, further enhancing its applications in healthcare and research.

PRINCIPLES

The principles of using a goniometer effectively and accurately include the following:

Standardization

Ensure that the patient's body position and the placement of the goniometer are consistent and standardized for reliable and reproducible measurements. This helps in comparing measurements over time and between different evaluators.

Anatomical Landmarks

Identify and mark the relevant anatomical landmarks on the patient's body. These landmarks are used as reference points for aligning the goniometer accurately, such as joint centers or bony prominences.

Joint Alignment

Position the stationary arm of the goniometer along the axis of the joint being measured. Align the movable arm with the segment of the body that is moving. This ensures that the goniometer accurately measures the joint angle.

Stabilization

Stabilize proximal and distal joints or segments that are not involved in the movement being measured. This helps isolate the specific joint and prevents compensatory movements that could affect the accuracy of the goniometer reading.

Range of Motion

Observe the patient's movement and assist them, if needed, to achieve their maximum range of motion (ROM) without causing discomfort or pain. This allows for a comprehensive assessment of the joint's full range.

Patient Cooperation

Instruct the patient to actively participate in the movement being measured and provide feedback on any pain, discomfort, or limitations they may experience. Cooperation and effort from the patient are crucial for obtaining accurate measurements.

Reliability and Consistency

Perform multiple measurements and average the results to ensure reliability and minimize measurement errors. Consistency in technique and documentation is essential when monitoring progress or comparing measurements over time.

Documentation

Record the goniometer measurements accurately, including the joint being measured, the patient's position, the range of motion, and any other relevant observations or limitations. Accurate documentation facilitates tracking progress, communicating findings, and informing treatment planning.

Did You Know?

Goniometry offers measurable information that individuals may comprehend and utilize to monitor their own development. Patients are empowered to participate actively in their rehabilitation and make knowledgeable decisions about their care when healthcare personnel involve them in the measurement process and explain the relevance of goniometric data.

AIMS

The aims of using a goniometer in healthcare and rehabilitation settings include the following:

Evaluation of Joint Function

Goniometers are generally used to gauge and categorize the mobility at different joints throughout the body. Goniometry, which measures joint angles, gives precise information about the amount of movement, both that which is actively performed by the patient and that which is passively attained with assistance. With the aid of this data, one may assess joint mobility, pinpoint constraints or limitations, and monitor advancement over time.

Aid in Diagnosis

The diagnosis and evaluation of musculoskeletal diseases depend heavily on goniometry. Healthcare experts can spot joint abnormalities, muscle imbalances, contractures, or other dysfunctions by comparing the recorded range of motion with normal values. Goniometry aids in evaluating the condition's severity, tracking its development, and evaluating the success of interventions or treatments.

Set Treatment Goals

Goniometer readings are used to create treatment plans and establish precise objectives for therapy or rehabilitation. Healthcare providers can create tailored therapies to enhance joint function, promote flexibility, restore regular movement patterns, or address particular deficits by determining the patient's baseline range of motion and any limits. Goniometry offers a quantitative foundation for monitoring development and making necessary adjustments to treatment plans.

Monitoring Progress and Outcomes

The progress of patients receiving therapy or rehabilitation can be monitored with the help of goniometric measurements. The range of motion variations over time can be reliably evaluated by therapists by frequently measuring and recording joint angles. This aids in assessing the success of interventions, identifying the need for treatment strategy adjustments, and informing patients of their progress.

Research and Evidence-Based Practice

The use of goniometry is crucial in research investigations and helps the development of evidence-based healthcare practices. Researchers are able to measure and compare joint movements, evaluate the effectiveness of interventions or therapies, and explore the connection between range of motion and different outcomes, including functional capacity, pain thresholds, or quality of life.

Clinical Correlation

Goniometry is not only used in clinical settings but also finds applications in various other fields. For instance, it is utilized in ergonomics to assess and optimize workstation setups to prevent musculoskeletal disorders. Additionally, goniometry is employed in sports science and biomechanics to analyze athletes' movements and enhance performance or prevent injuries. Its versatility highlights its importance beyond healthcare, making it a valuable tool in multiple disciplines focused on human movement and function.

TYPES OF GONIOMETERS

There are several types of goniometers available, each designed for specific purposes and joints. Some common types of goniometers are discussed ahead.

Universal Goniometer

This kind of goniometer, which has two arms and resembles a protractor, is the most popular one. In order to measure range of motion, it normally has a fixed arm and a moving arm that can be aligned with joint axes. Universal goniometers are flexible and useful for a range of joints (Fig. 5.1).

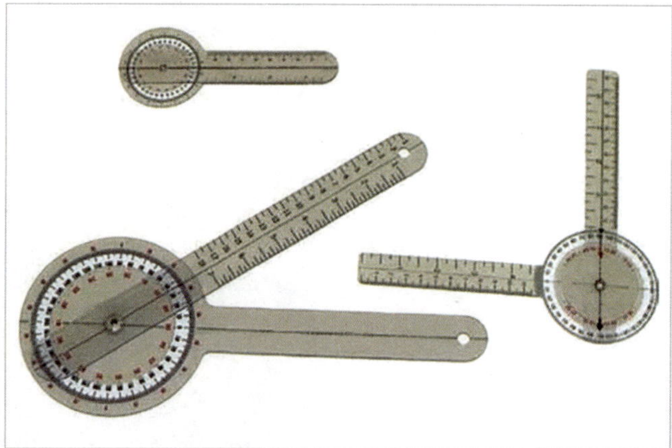

Fig. 5.1: Universal goniometers

Gravity Inclinometer

This kind of goniometer measures joint angles using a weighted pendulum or an inclinometer. It is frequently used to quantify the spine's range of motion, such as the cervical or lumbar spine. Gravity inclinometers offer more accurate measurement by calculating the angle using the force of gravity (Fig. 5.2).

Digital Goniometer

Digital goniometers measure joint angles using electronic sensors. A digital display that gives a quick reading of the joint angle is frequently present. Digital goniometers provide accurate readings and are very helpful in research, healthcare settings, or other situations where accuracy is required (Fig. 5.3).

Electrogoniometer

Electrogoniometers detect joint angles using electrical signals and sensors. They are made up of flexible or rigid sensors that can be sewn into wearable technology or applied directly to the skin. Electrogoniometers are frequently employed in gait analysis, biomechanics research, and other situations that call for ongoing joint motion monitoring (Fig. 5.4).

Finger Goniometer

For the purpose of determining the range of motion in the finger joints, finger goniometers were created. They are more compact and smaller than universal goniometers, which enable accurate assessment of finger flexion, extension, and other movements (Fig. 5.5).

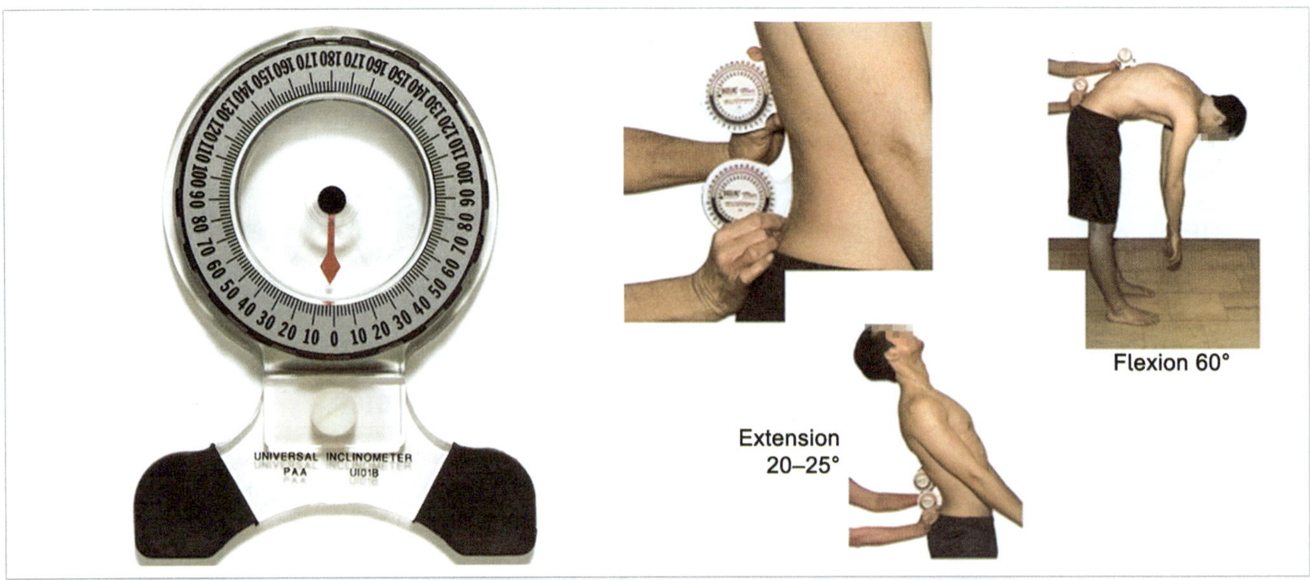

Fig. 5.2: Gravity inclinometer

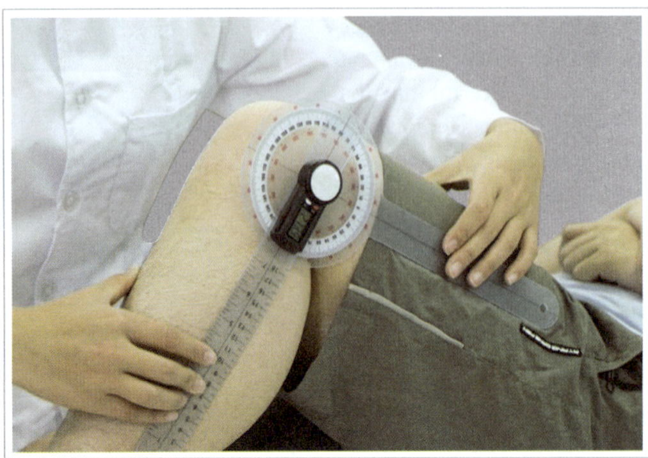

Fig. 5.3: Digital goniometer

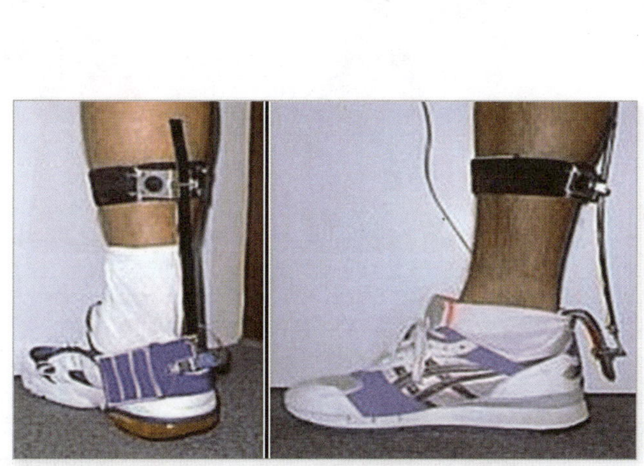

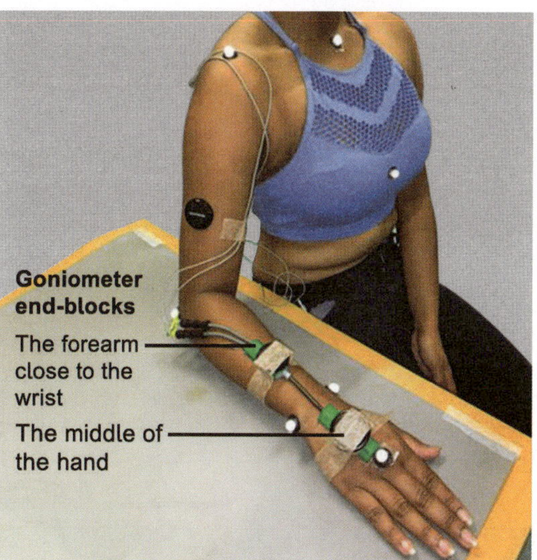

Fig. 5.4: Electrogoniometer

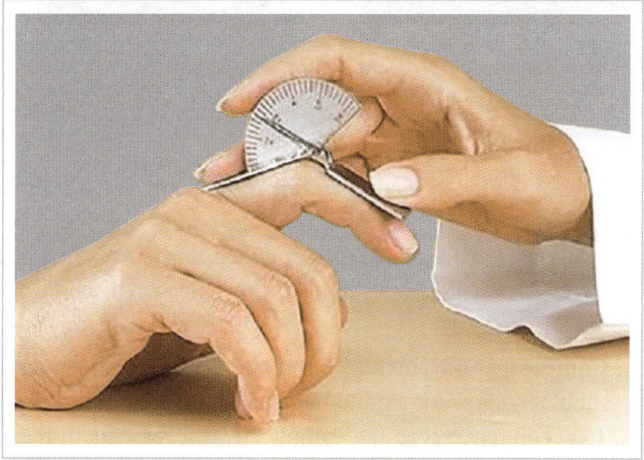

Fig. 5.5: Finger goniometer

Pronation-Supination Goniometer

The forearm's range of motion, specifically the rotation of the radius around the ulna, is measured with pronation-supination goniometers. They are made up of a spherical or semispherical disc with a rotating arm that is parallel to the forearm's rotational axis (Fig. 5.6A).

Cervical Range of Motion Devices

Specialized cervical goniometers called cervical range of motion (CROM) devices are used to evaluate the neck's range of motion. In most cases, they are designed to measure various motion of the neck and have distinct inclinometers to measure cervical flexion, extension, lateral bending, and rotation (Fig. 5.6B).

PSYCHOMETRIC PROPERTIES OF GONIOMETERS

Validity

Validity is the extent to which a goniometer measures what it is intended to measure. It is the precision with which a goniometer measures joint angles and accurately depicts the full range of motion in the context of goniometry.

Reliability

The consistency and accuracy of measurements made using a goniometer are referred to as reliability. It refers to how well results from repeated measurements using the same goniometer hold up over time.

Reliability of goniometry can be evaluated in several ways:

- **Intra-rater reliability:** This evaluates the accuracy of readings made by the same evaluator on various occasions. By comparing measurements made by the same examiner at several intervals and estimating the level of agreement between them, intra-rater reliability is ascertained.
- **Inter-rater reliability:** When measuring joint angles using the same goniometer, this gauges the level of agreement amongst various evaluators. The degree of agreement between measures made by various evaluators on the same subjects is compared to determine inter-rater reliability.
- **Test-retest reliability:** This tests the consistency of measurements made on the same person during various sessions by the same examiner using the same goniometer. By comparing the measurements made at several time points and estimating the level of agreement between them, test-retest reliability is assessed.

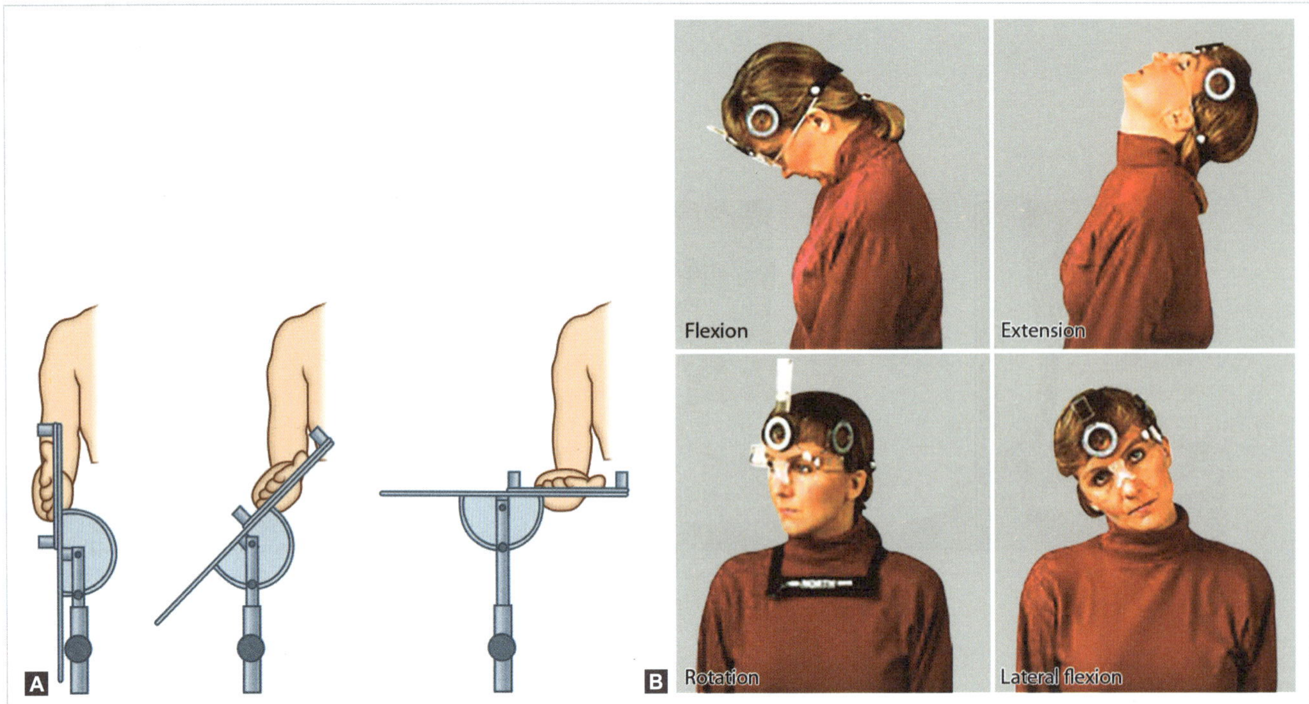

Figs 5.6A and B: A. Pronation-supination goniometer; **B.** Portable cervical range of motion instrument

METHODS OF JOINT MEASUREMENT

To perform goniometry, a variety of procedures or methods are employed. Depending on the precise joint being assessed and the patient's position, these techniques may change. Here are a few typical ways to use a goniometer:

- **Active range of motion (AROM):** The patient's capacity to freely move a joint through its range of motion without help is assessed using this technique. While the goniometer is set up to measure the joint angle, the patient actively executes the movement.
- **Passive range of motion (PROM):** With this technique, the patient is not required to actively participate when the examiner rotates the joint through its range of motion. As the examiner moves the joint, the goniometer is utilized to measure the joint angle.
- **Active assisted range of motion (A-AROM):** Both active and passive movements are combined in A-AROM. With some help from the examiner or a tool, such as a pulley system or a therapeutic band, the patient completes the exercise. The joint angle is measured using a goniometer as the patient and the help are working together.
- **Gravity-eliminated range of motion:** This method is employed when it is necessary to reduce the gravitational force in order to facilitate joint movement. For instance, to lessen the effect of gravity when seated, the arm may be supported on a table or suspended in a sling. The joint angle during the movement is then measured using the goniometer.
- **Composite movements:** Some joints, like the hip and shoulder, allow for simultaneous motion in several planes. To accurately record the complicated movements in these situations, goniometry may call for measuring angles in different planes or employing a mix of goniometers.

INDICATIONS

Evaluation of Joint Flexibility

Goniometry is a technique used to measure the range of motion of joints and spot any constraints or limitations in the motion. It aids in determining the range of motion a joint can achieve, including flexion, extension, abduction, adduction, rotation, and more. The diagnosis of conditions that impair joint mobility, such as arthritis, contractures, or ligamentous injuries, depends on the evaluation of joint flexibility.

Did You Know?

Joint flexibility: The ability of a joint to move through its full range of motion freely and without discomfort.

- **Hypomobility:**
  - Hypomobility refers to reduced or restricted joint mobility, typically characterized by a decreased ability of a joint to move through its normal range of motion. This limitation in joint movement can result from various factors, including injury, inflammation, muscle imbalances, or structural abnormalities within the joint.
  - Hypomobility is often associated with conditions such as osteoarthritis, joint contractures, or the consequences of immobilization due to surgery or injury. Research on hypomobility may focus on understanding its causes, consequences, and potential treatments or interventions to improve joint function.
- **Hypermobility:**
  - Hypermobility, refers to an excessive or abnormal range of motion in a joint. Individuals with hypermobility can move their joints beyond the typical range, often due to laxity in ligaments, joint capsules, or connective tissues. It is sometimes referred to as "double-jointedness."
  - Hypermobility can be a result of genetic factors or conditions like Ehlers-Danlos syndrome or benign joint hypermobility syndrome. Research in hypermobility may explore the effects of joint instability, potential joint-related complications, and interventions to manage and stabilize hypermobile joints.

Monitoring Rehabilitation

In therapeutic and rehabilitation settings, goniometry is essential. It aids in monitoring changes in joint range of motion over time and tracking the progress of patients receiving rehabilitation therapies. Therapists can assess the efficacy of therapies and modify treatment plans by measuring joint angles at various phases of treatment. In orthopedic rehabilitation, sports medicine, and postoperative care, goniometry is especially helpful.

Prevention of Deformities

Goniometry is used to monitor the progression of musculoskeletal conditions, such as joint degeneration, muscular imbalances, or postural abnormalities. Regular measurements help healthcare professionals track changes in joint range of motion and detect any worsening or improvement of the condition. This information aids in treatment planning, setting realistic goals, and adjusting interventions as needed.

Developing Treatment Plans

Goniometry is commonly performed before and after surgical interventions involving joints or musculoskeletal structures. Preoperatively, goniometry provides a baseline measurement of joint mobility, which helps in surgical planning and goal setting. Postoperatively, it is used to assess the immediate and long-term effects of surgery on joint range of motion and guide rehabilitation protocols.

Physical Fitness and Sports Performance Evaluation

Goniometry is utilized in assessing joint mobility and flexibility in athletes and individuals involved in physical fitness activities. By measuring joint angles, healthcare professionals can identify any asymmetries, restrictions, or limitations that may impact performance, predispose to injuries, or affect movement efficiency. Goniometry assists in designing personalized exercise programs and stretching routines to optimize athletic performance and prevent injuries.

Research and Clinical Studies

Goniometry is widely employed in research studies investigating joint range of motion, musculoskeletal conditions, and interventions. It provides quantitative data for evaluating the efficacy of treatment methods, comparing different interventions, and establishing normative values for specific populations. Goniometric measurements contribute to evidence-based practice and support the development of clinical guidelines and protocols.

PROCEDURE

Preparation

Before applying a goniometer, it is helpful to go through a quick checklist to ensure a smooth and accurate measurement process. Here are some key items to consider:

- **Verify patient readiness:** Verify that the patient is ready and at ease for the goniometry evaluation. Get their approval for the process and answer any queries or concerns they may have.
- **Review the patient's medical history:** Review the patient's medical history, paying particular attention to any pertinent details on previous illnesses, operations, or injuries that might have an impact on joint range of motion or the goniometry evaluation.
- **Gather necessary equipment:** Make sure the goniometer you're using is the right one for the joint you are measuring. Verify that the goniometer is accessible, in good condition, and clean. Have accessible any additional goods required for stabilization or patient's comfort.
- **Ensure proper lighting and visibility:** To verify that the joint and goniometer are clearly visible, check the illumination in the evaluation area. Proper alignment

and correct reading of the goniometer scale are aided by adequate lighting.

- **Communicate and instruct the patient:** Give the patient a thorough explanation of the goniometry process, including the intended movement and their part in the evaluation. Explain the movement to them and what to expect from the measurement procedure. Encourage them to participate fully and, as necessary, reassure and support them.
- **Maintain professionalism and respect:** Throughout the assessment, foster a respectful and professional environment. Maintain proper drapery, respect patient privacy, and use clear, compassionate communication. Take into account the patient's comfort level and any cultural or individual aspects that may have an impact on the assessment process.

Steps

The following steps outline a general procedure for using a goniometer to measure joint range of motion:

1. **Prepare the patient:** Ensure that the patient is positioned comfortably. Position of the patient will depend on the joint being evaluated and the condition of the patient.
2. **Identify and mark anatomical landmarks:** Find the necessary anatomical landmarks connected to the joint under evaluation. Bony prominences, joint centers, or particular muscle attachments are examples of common landmarks. To use as a guide when placing the goniometer, palpate and note these landmarks using a pen or marker.
3. **Select the appropriate goniometer:** Select a goniometer that is appropriate for measuring the joint. To accommodate diverse joints and measurement requirements, goniometers are available in a variety of sizes and designs. Make sure that the goniometer is in good working order and is clean.
4. **Align the goniometer:** Align the goniometer's stationary arm with the joint's axis before measuring it. The segment of the body that is moving during the joint motion should be aligned with the movable arm. To get an accurate measurement, make sure the goniometer is perpendicular to the joint axis.
5. **Perform the joint movement:** Tell the patient to move the joint in the desired way. The patient may be requested to move the joint on their own or with the help from the examiner, depending on the joint and the measuring method (active, passive, etc.). If required, assist the patient in making the movement by offering support or resistance.
6. **Read and record the measurement:** To find the joint angle when the joint movement is finished, read the goniometer scale. Make a note of the measurement by either telling an assistant about it or by writing it down immediately on a form or electronic device.
7. **Repeat measurements, if needed:** It is frequently advised to take the measurement twice or three times, average the results, and repeat for precision. Repeat the process, being careful to line the goniometer correctly, stabilize the patient, and give the patient precise directions. Note any additional measurements that are required.
8. **Document the findings:** In the patient's medical or therapeutic file, note the goniometer measurements. Include pertinent details about the joint being assessed, the patient's position, any restrictions or observations, and the degree of range of motion attained. An appropriate progress monitoring system and communication between healthcare providers are both supported by proper documentation.

Clinical Correlation

In clinical assessments, understanding and documenting the active range of motion, passive range of motion, and the end feel of a joint are essential for diagnosing musculoskeletal conditions, tracking progress during rehabilitation, and planning appropriate interventions or treatments.

- **End feel:** End feel is a term used to describe the quality or sensation experienced at the end of a joint's range of motion when it reaches its limit. It helps assess the integrity of the joint and its surrounding structures. There are different types of end feel, which include:
  - **Hard end feel:** This occurs when the joint's movement is limited by the contact of bone against bone, such as in the elbow's extension.
  - **Soft end feel:** This occurs when the joint's movement is limited by the compression of soft tissues, such as in the knee's flexion, where the thigh muscle comes into contact with the calf.
  - **Firm end feel:** This occurs when the joint's movement is limited by the tension in the joint's ligaments, such as the resistance felt when extending the fingers.
  - **Empty end feel:** This occurs when pain prevents the completion of the joint's range of motion, and the examiner cannot determine the true end feel due to the patient's discomfort.

Upper Limb Goniometry

Shoulder Flexion

Plane : Sagittal
Axis : Frontal
ROM : 0° to 165°–180°
End feel : Firm

To measure shoulder flexion using a goniometer, follow these steps (Figs 5.7A and B):

1. **Prepare the patient:** Ensure that the patient is in a comfortable position, such as supine. Shoulder is placed by the side of the body in neutral rotation. Forearm is placed in mid position with palm facing the body. Weight of the trunk will stabilize the thorax.
2. **Identify anatomical landmarks:** Locate the relevant anatomical landmarks associated with the shoulder joint.

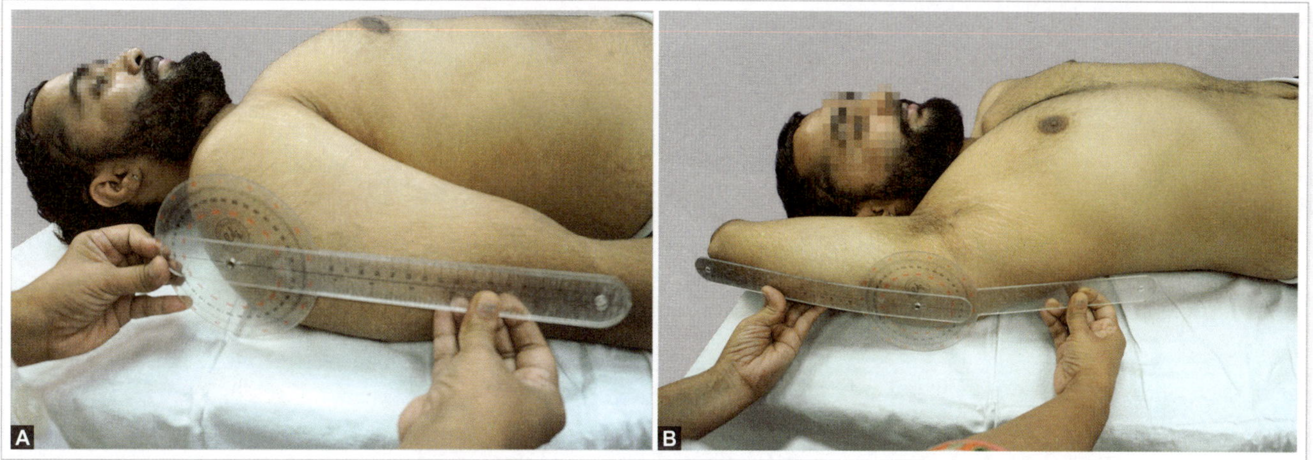

Figs 5.7A and B: Shoulder flexion: **A.** Starting position; **B.** Final position

The commonly used landmarks for shoulder flexion measurement are the acromion process and the lateral condyle of the humerus. Palpate and mark these landmarks with a pen or marker for reference.

3. **Select the appropriate goniometer:** Choose a goniometer suitable for measuring shoulder flexion. A universal goniometer with sufficient size and range of motion is typically used. Ensure that the goniometer is clean, in good working condition, and within reach.
4. **Position the goniometer:** The fulcrum of the goniometer is placed at the lateral aspect of greater tuberosity of humerus. Align the stationary arm of the goniometer with the lateral midline of the trunk, parallel to the spine. Position the movable arm along the lateral midline of the humerus, parallel to the upper arm.
5. **Instruct the patient:** Explain the procedure to the patient and demonstrate the desired movement. Instruct the patient to elevate his arm forward, flexing at the shoulder joint. Encourage the patient to move their shoulder to the maximum extent without pain or discomfort.
6. **Read and record the measurement:** As the patient performs the shoulder flexion movement, observe the range of motion. Read the goniometer scale at the point where the stationary and movable arms intersect. This indicates the angle of shoulder flexion. Record the measurement in degrees.
7. **Repeat if necessary:** For accuracy and reliability, it is recommended to repeat the measurement two or three times and calculate an average. Ensure proper goniometer alignment and patient stabilization for each repetition. Record all measurements and calculate the average for documentation.
8. **Document the findings:** Record the goniometer measurements, including the patient's position, any limitations or observations, and the range of motion achieved. Document the results in the patient's medical or therapy record for reference and future comparison.

Shoulder Extension

Plane : Frontal
Axis : Sagittal
ROM : 0° to 50°–60°
End feel : Firm

To measure shoulder extension using a goniometer, follow these steps (Figs 5.8A and B):

1. **Prepare the patient:** Ensure that the patient is in a comfortable position, such as prone with head in neutral position. Shoulder is placed by the side of the body in neutral rotation, elbow slightly flexed, forearm in neutral rotation and palm facing the body. Weight of the trunk stabilizes the thorax. The therapist can also stabilize the trunk to prevent rotation at spine.
2. **Identify anatomical landmarks:** Locate the relevant anatomical landmarks associated with the shoulder joint. The commonly used landmarks for shoulder extension measurement are the acromion process and the lateral condyle of the humerus. Palpate and mark these landmarks with a pen or marker for reference.
3. **Select the appropriate goniometer:** Choose a goniometer suitable for measuring shoulder extension. A universal goniometer with sufficient size and range of motion is typically used. Ensure that the goniometer is clean, in good working condition, and within reach.
4. **Position the goniometer:** The fulcrum of the goniometer is placed over the lateral aspect of greater tuberosity of humerus. Align the stationary arm of the goniometer with the lateral midline of the trunk, parallel to the spine.

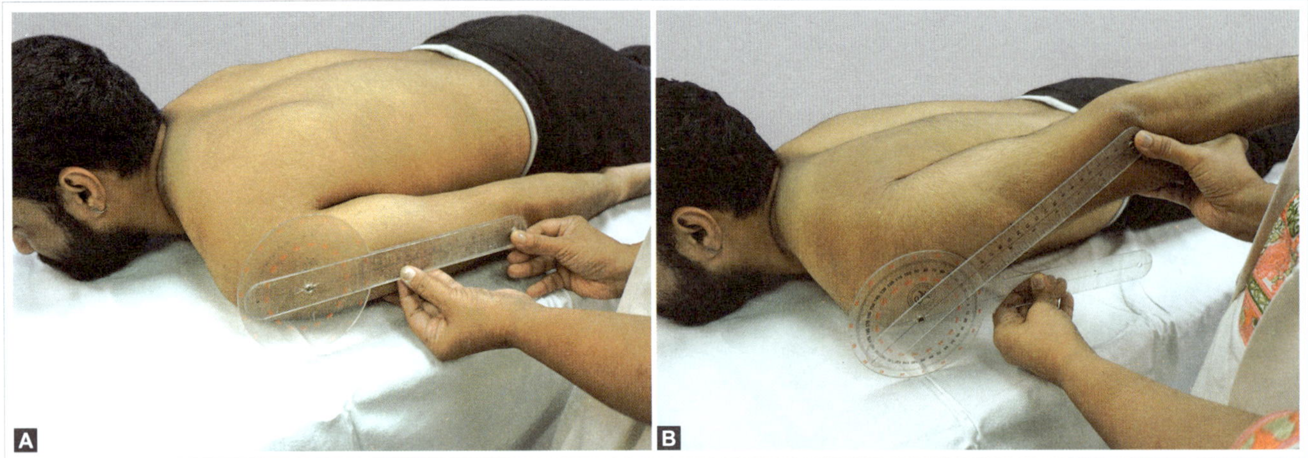

Figs 5.8A and B: Shoulder extension: **A.** Starting position; **B.** Final position

Position the movable arm along the lateral midline of the humerus, parallel to the upper arm.

5. **Instruct the patient:** Explain the procedure to the patient and demonstrate the desired movement. Instruct the patient to extend his arm backward, extending the shoulder joint. Encourage the patient to move his shoulder to the maximum extent without pain or discomfort.
6. **Read and record the measurement:** As the patient performs the shoulder extension movement, observe the range of motion. Read the goniometer scale at the point where the stationary and movable arms intersect. This indicates the angle of shoulder extension. Record the measurement in degrees.
7. **Repeat if necessary:** For accuracy and reliability, it is recommended to repeat the measurement two or three times and calculate an average. Ensure proper goniometer alignment and patient stabilization for each repetition. Record all measurements and calculate the average for documentation.
8. **Document the findings:** Record the goniometer measurements, including the patient's position, any limitations or observations, and the range of motion achieved. Document the results in the patient's medical or therapy record for reference and future comparison.

Shoulder Abduction and Adduction

Plane : Frontal
Axis : Sagittal
ROM : 0° to 170°–180°
End feel : Firm

Since adduction in the frontal plane is the return to the zero starting position following complete abduction, it is typically not measured or documented.

To measure shoulder abduction using a goniometer, follow these steps (Figs 5.9A and B):

1. **Prepare the patient:** Ensure that the patient is in a comfortable position, such as supine position with shoulder by the side of the body, laterally rotated and palm

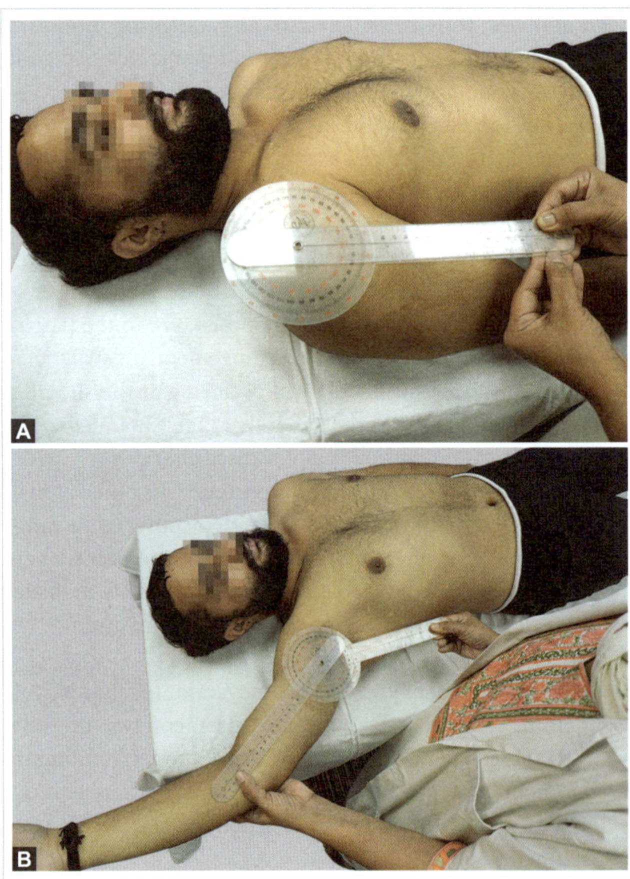

Figs 5.9A and B: Shoulder abduction: **A.** Starting position; **B.** Final position

facing anteriorly. The elbow should be extended. The therapist stabilizes scapula to prevent rotation and elevation.

2. **Identify anatomical landmarks:** Locate the relevant anatomical landmarks associated with the shoulder joint. The commonly used landmarks for shoulder abduction measurement are the acromion process and the lateral condyle of the humerus. Palpate and mark these landmarks with a pen or marker for reference.
3. **Select the appropriate goniometer:** Choose a goniometer suitable for measuring shoulder abduction. A universal goniometer with sufficient size and range of motion is typically used. Ensure that the goniometer is clean, in good working condition, and within reach.
4. **Position the goniometer:** The anterior aspect of the acromion process is where the goniometer's fulcrum is located. Place the goniometer's stationary arm parallel to the spine and in line with the trunk's midline. Place the movable arm parallel to the upper arm along the humerus's midline.
5. **Instruct the patient:** While demonstrating the intended movement, explain the technique to the patient. Give the patient instructions to abduct their shoulder joint by raising his arm out to the side. Encourage the patient to move his shoulder to the maximum extent without pain or discomfort.
6. **Read and record the measurement:** As the patient performs the shoulder abduction movement, observe the range of motion. Read the goniometer scale at the point where the stationary and movable arms intersect. This indicates the angle of shoulder abduction. Record the measurement in degrees.
7. **Repeat if necessary:** For accuracy and reliability, it is recommended to repeat the measurement two or three times and calculate an average. Ensure proper goniometer alignment and patient stabilization for each repetition. Record all measurements and calculate the average for documentation.
8. **Document the findings:** Record the goniometer measurements, including the patient's position, any limitations or observations, and the range of motion achieved. Document the results in the patient's medical or therapy record for reference and future comparison.

Shoulder Medial Rotation

Plane : Transverse
Axis : Vertical
ROM : 0° to 70°–90°
End feel : Firm

To measure shoulder medial rotation using a goniometer, follow these steps (Figs 5.10A and B):

1. **Prepare the patient:** Ensure that the patient is in a comfortable position, such as supine with arm in 90° abducted position. The forearm is in mid prone position, at 90° to the supporting surface with palm facing the feet.
2. **Identify anatomical landmarks:** Locate the relevant anatomical landmarks associated with the shoulder joint. The commonly used landmarks for shoulder medial rotation measurement are the olecranon process and the ulnar styloid process. Palpate and mark these landmarks with a pen or marker for reference.
3. **Select the appropriate goniometer:** Choose a goniometer suitable for measuring shoulder medial rotation. A universal goniometer with sufficient size and range of motion is typically used. Ensure that the goniometer is clean, in good working condition, and within reach.
4. **Position the goniometer:** The fulcrum of the goniometer is placed over the olecranon process of the elbow. Align the stationary arm of the goniometer with the ulnar styloid process, parallel to the forearm. Position the movable arm along the ulnar midline of the forearm, parallel to the ulna bone.
5. **Instruct the patient:** Explain the procedure to the patient and demonstrate the desired movement. Instruct the patient to internally rotate his shoulder, bringing the palm

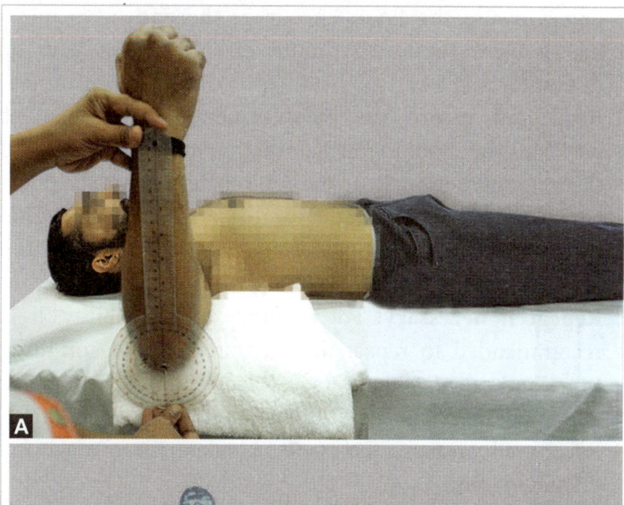

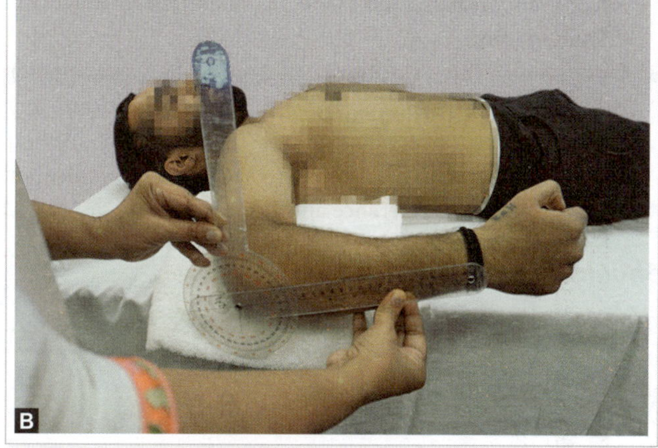

Figs 5.10A and B: Shoulder medial rotation: **A.** Starting position; **B.** Final position

down toward the floor. Encourage the patient to move his shoulder to the maximum extent without pain or discomfort.

6. **Read and record the measurement:** As the patient performs the shoulder medial rotation movement, observe the range of motion. Read the goniometer scale at the point where the stationary and movable arms intersect. This indicates the angle of shoulder medial rotation. Record the measurement in degrees.
7. **Repeat if necessary:** For accuracy and reliability, it is recommended to repeat the measurement two or three times and calculate an average. Ensure proper goniometer alignment and patient stabilization for each repetition. Record all measurements and calculate the average for documentation.
8. **Document the findings:** Record the goniometer measurements, including the patient's position, any limitations or observations, and the range of motion achieved. Document the results in the patient's medical or therapy record for reference and future comparison.

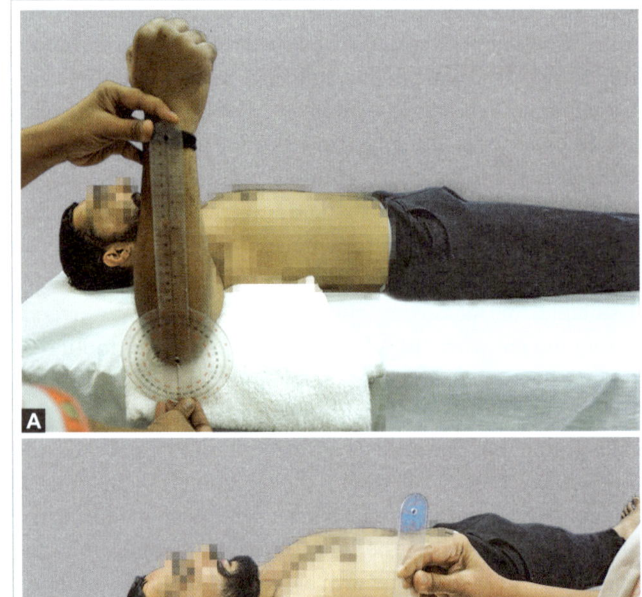

Figs 5.11A and B: Shoulder lateral rotation: **A.** Starting position; **B.** Final position

Shoulder Lateral Rotation

Plane : Transverse
Axis : Vertical
ROM : 0° to 80°–90°
End feel : Firm

To measure shoulder lateral rotation using a goniometer, follow these steps (Figs 5.11A and B):

1. **Prepare the patient:** Ensure that the patient is in a comfortable position, such as supine with arm in 90° abducted position. The forearm is in mid prone position, at 90° to the supporting surface with palm facing the feet.
2. **Identify anatomical landmarks:** Locate the relevant anatomical landmarks associated with the shoulder joint. The commonly used landmarks for shoulder lateral rotation measurement are the olecranon process and the ulnar styloid process. Palpate and mark these landmarks with a pen or marker for reference.
3. **Select the appropriate goniometer:** Choose a goniometer suitable for measuring shoulder lateral rotation. A universal goniometer with sufficient size and range of motion is typically used. Ensure that the goniometer is clean, in good working condition, and within reach.
4. **Position the goniometer:** The fulcrum of the goniometer is placed over the olecranon process of the elbow. Align the stationary arm of the goniometer with the ulnar styloid process, parallel to the forearm. Position the movable arm along the ulnar midline of the forearm, parallel to the ulna bone.
5. **Instruct the patient:** Explain the procedure to the patient and demonstrate the desired movement. Instruct the patient to externally rotate his shoulder, moving the dorsum of the palm toward the floor. Encourage the patient to move his shoulder to the maximum extent without pain or discomfort.
6. **Read and record the measurement:** As the patient performs the shoulder lateral rotation movement, observe the range of motion. Read the goniometer scale at the point where the stationary and movable arms intersect. This indicates the angle of shoulder lateral rotation. Record the measurement in degrees.
7. **Repeat if necessary:** For accuracy and reliability, it is recommended to repeat the measurement two or three times and calculate an average. Ensure proper goniometer alignment and patient stabilization for each repetition. Record all measurements and calculate the average for documentation.
8. **Document the findings:** Record the goniometer measurements, including the patient's position, any limitations or observations, and the range of motion achieved. Document the results in the patient's medical or therapy record for reference and future comparison.

Elbow Flexion and Extension

For elbow flexion:

Plane : Sagittal
Axis : Frontal
ROM : 0° to 140°–150°
End feel : Soft (compression of muscle bulk of arm)/firm (tension in the posterior joint capsule, the lateral and medial heads of the triceps muscle, and the anconeus muscle).

For elbow extension:

Plane : Sagittal
Axis : Frontal
ROM : Full elbow flexion to 0°
End feel : Hard

To measure elbow flexion using a goniometer, follow these steps (Figs 5.12A and B):

1. **Prepare the patient:** Ensure that the patient is in a comfortable position, such as supine. The shoulder is placed by the side of the body with neutral rotation. The forearm is in supinated position with palm facing the ceiling. A towel roll can be placed under the distal arm for stabilization.
2. **Identify anatomical landmarks:** Locate the relevant anatomical landmarks associated with the elbow joint. The commonly used landmarks for elbow flexion measurement are the lateral condyle of the humerus and the styloid process of the radius. Palpate and mark these landmarks with a pen or marker for reference.
3. **Select the appropriate goniometer:** Choose a goniometer suitable for measuring elbow flexion. A universal goniometer with sufficient size and range of motion is typically used. Ensure that the goniometer is clean, in good working condition, and within reach.
4. **Position the goniometer:** The fulcrum of the goniometer is placed over the lateral condyle of the humerus. Align the stationary arm of the goniometer with the lateral midline of the humerus, parallel to the upper arm. Position the movable arm along the lateral midline of the forearm, parallel to the radius bone.
5. **Instruct the patient:** Explain the procedure to the patient and demonstrate the desired movement. Instruct the patient to flex his elbow, bringing his hand toward his shoulder. Encourage the patient to move their elbow to the maximum extent without pain or discomfort. For elbow extension the patient is asked to straighten the elbow from flexed position and the angle is recorded.
6. **Read and record the measurement:** As the patient performs the elbow flexion movement, observe the range of motion. Read the goniometer scale at the point where the stationary and movable arms intersect. This indicates the angle of elbow flexion. Record the measurement in degrees.
7. **Repeat if necessary:** For accuracy and reliability, it is recommended to repeat the measurement two or three times and calculate an average. Ensure proper goniometer alignment and patient stabilization for each repetition. Record all measurements and calculate the average for documentation.
8. **Document the findings:** Record the goniometer measurements, including the patient's position, any limitations or observations, and the range of motion achieved. Document the results in the patient's medical or therapy record for reference and future comparison.

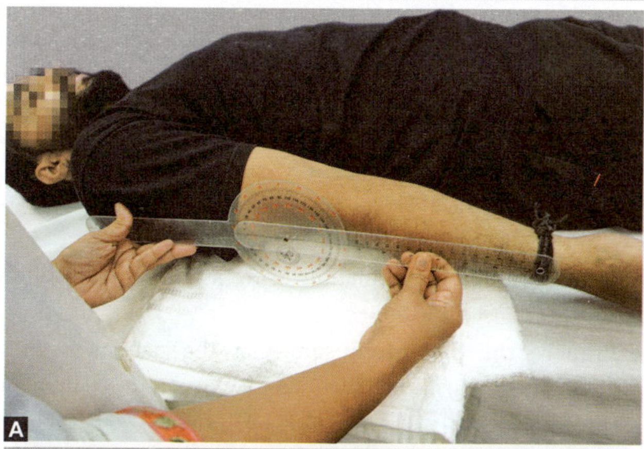

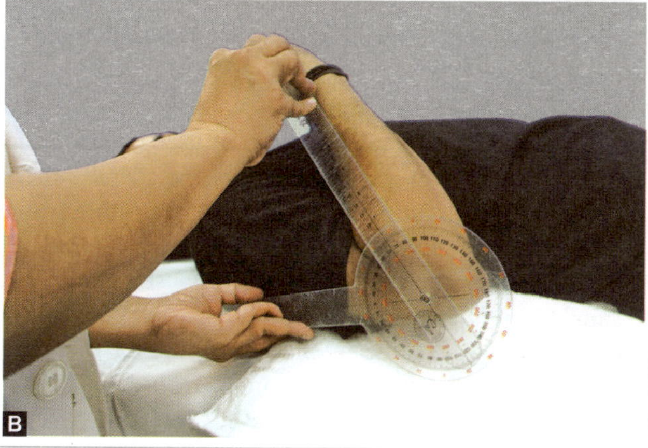

Figs 5.12A and B: Elbow flexion: **A.** Starting position; **B.** Final position

Forearm Pronation

Plane : Frontal
Axis : Anteroposterior
ROM : 0° to 70°–80°
End feel : Hard (contact between ulna and radius)/Firm (tension in dorsal radioulnar ligament, supinator muscle).

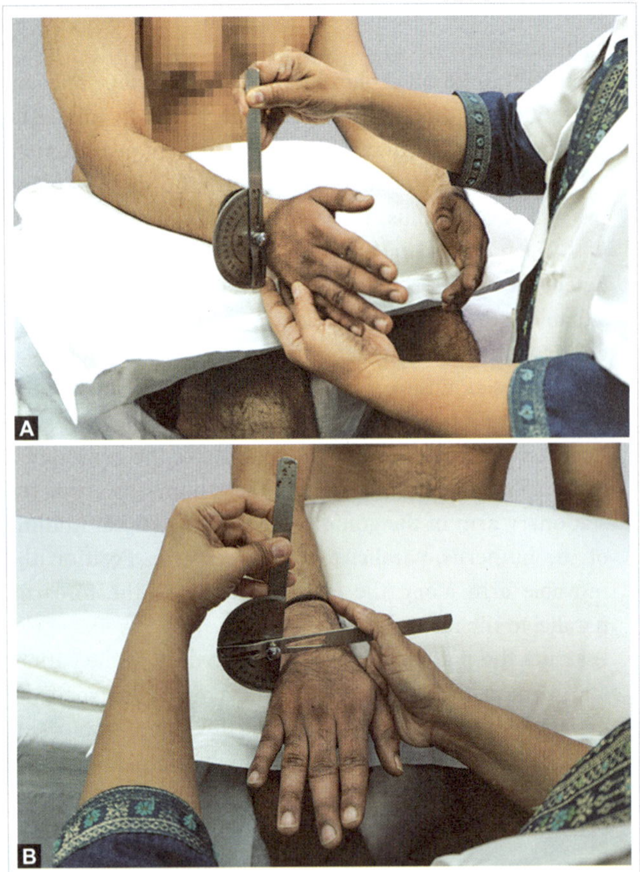

Figs 5.13A and B: Forearm pronation: **A.** Starting position; **B.** Final position

To measure forearm pronation using a goniometer, follow these steps (Figs 5.13A and B):

1. **Prepare the patient:** Ensure that the patient is in a comfortable position, such as sitting on a chair. The shoulder is held close to the body with neutral rotation. The elbow is flexed at 90° and forearm between pronation and supination. The therapist stabilizes the distal end of the humerus.
2. **Identify anatomical landmarks:** Locate the relevant anatomical landmarks. The commonly used landmarks for forearm pronation measurement are the styloid process of radius. Palpate and mark these landmarks with a pen or marker for reference.
3. **Select the appropriate goniometer:** Choose a goniometer suitable for measuring forearm pronation. A universal goniometer with sufficient size and range of motion is typically used. Ensure that the goniometer is clean, in good working condition, and within reach.
4. **Position the goniometer:** The goniometer's fulcrum is positioned laterally, close to the ulnar styloid process. Place the goniometer's stationary arm parallel to the humerus' anterior midline. Place the movable arm in a position parallel to the styloid process of the ulna and radius over the dorsal surface of the forearm.
5. **Instruct the patient:** Explain the procedure to the patient and demonstrate the desired movement. Instruct the patient to pronate the forearm by moving the distal radius so that the palm faces the floor. Encourage the patient to move his forearm to the maximum extent without pain or discomfort.
6. **Read and record the measurement:** As the patient performs the forearm pronation movement, observe the range of motion. Read the goniometer scale at the point where the stationary and movable arms intersect. This indicates the angle of forearm pronation. Record the measurement in degrees.
7. **Repeat if necessary:** For accuracy and reliability, it is recommended to repeat the measurement two or three times and calculate an average. Ensure proper goniometer alignment and patient stabilization for each repetition. Record all measurements and calculate the average for documentation.
8. **Document the findings:** Record the goniometer measurements, including the patient's position, any limitations or observations, and the range of motion achieved. Document the results in the patient's medical or therapy record for reference and future comparison.

Forearm Supination

Plane : Frontal
Axis : Antero-posterior
ROM : 0° to 80°–90°
End feel : Firm

To measure forearm supination using a goniometer, follow these steps (Figs 5.14A and B):

1. **Prepare the patient:** Ensure that the patient is in a comfortable position, such as sitting on a chair. The shoulder is held close to the body with neutral rotation. The elbow is flexed at 90° and forearm between pronation and supination. The therapist stabilizes the distal end of the humerus.
2. **Identify anatomical landmarks:** Locate the relevant anatomical landmarks. The commonly used landmarks for forearm supination measurement are the styloid process of ulna. Palpate and mark these landmarks with a pen or marker for reference.
3. **Select the appropriate goniometer:** Choose a goniometer suitable for measuring forearm supination. A universal goniometer with sufficient size and range of motion is typically used. Ensure that the goniometer is clean, in good working condition, and within reach.
4. **Position the goniometer:** The goniometer's fulcrum is positioned medially, close to the ulnar styloid process. Place the goniometer's stationary arm parallel to the humerus

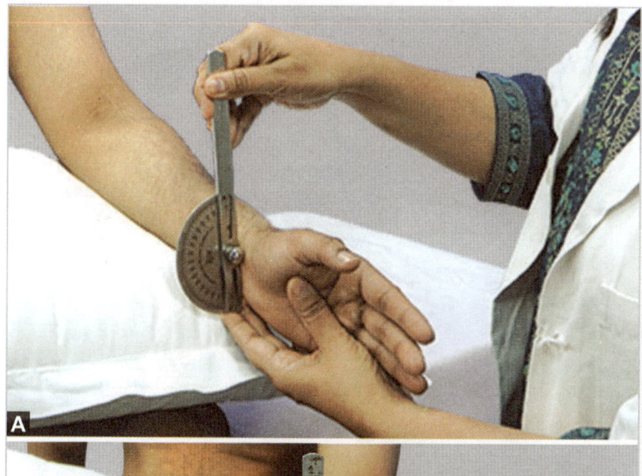

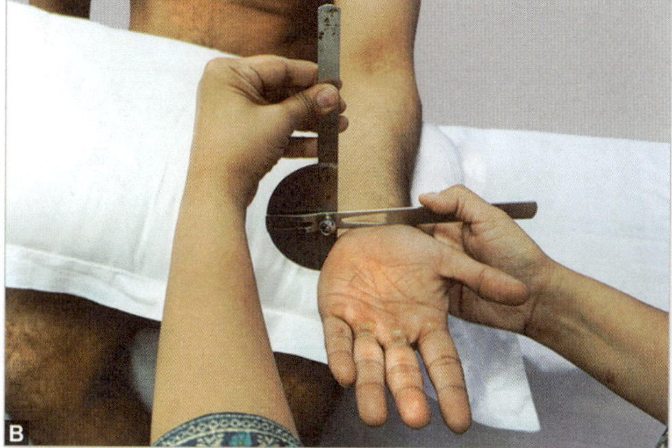

Figs 5.14A and B: Forearm supination: **A.** Starting position; **B.** Final position

anterior midline. Place the movable arm in a position parallel to the styloid process of the ulna and radius over the ventral surface of the forearm.

5. **Instruct the patient:** While demonstrating the intended movement, go over the technique with the patient. Direct the patient to move their distal radius such that their dorsum is facing the floor in order to supinate his forearm. Encourage the patient to move his forearm to the maximum extent without pain or discomfort.
6. **Read and record the measurement:** As the patient performs the forearm supination movement, observe the range of motion. Read the goniometer scale at the point where the stationary and movable arms intersect. This indicates the angle of forearm supination. Record the measurement in degrees.
7. **Document the findings:** Record the goniometer measurements, including the patient's position, any limitations or observations, and the range of motion achieved. Document the results in the patient's medical or therapy record for reference and future comparison.

Wrist Flexion

Plane : Sagittal
Axis : Frontal
ROM : 0° to 60°–80°
End feel : Firm

To measure wrist flexion using a goniometer, follow these steps (Figs 5.15A and B):

1. **Prepare the patient:** Ensure that the patient is in a comfortable position, such as sitting on a chair. Forearm is comfortably placed on a supporting surface, hand is free to move. The elbow is placed in 90° flexed position and wrist in neutral position. The therapist stabilizes the distal forearm to prevent pronation and supination.
2. **Identify anatomical landmarks:** Locate the relevant anatomical landmarks. The commonly used landmarks for wrist flexion measurement are triquetrum and ulnar styloid process. Palpate and mark these landmarks with a pen or marker for reference.

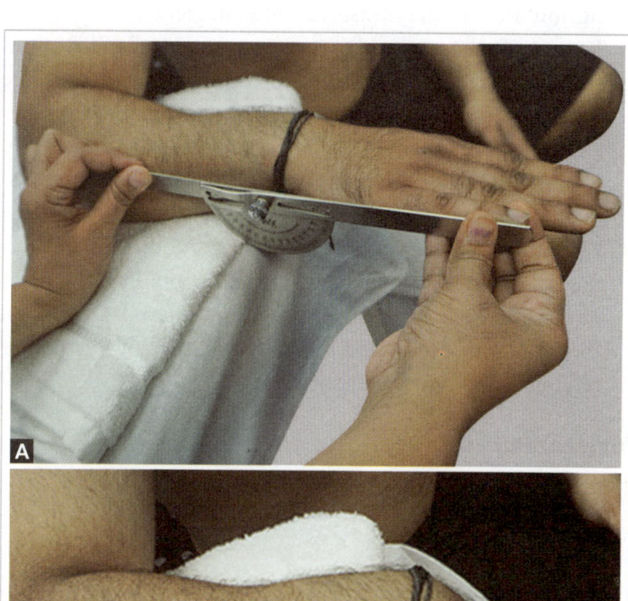

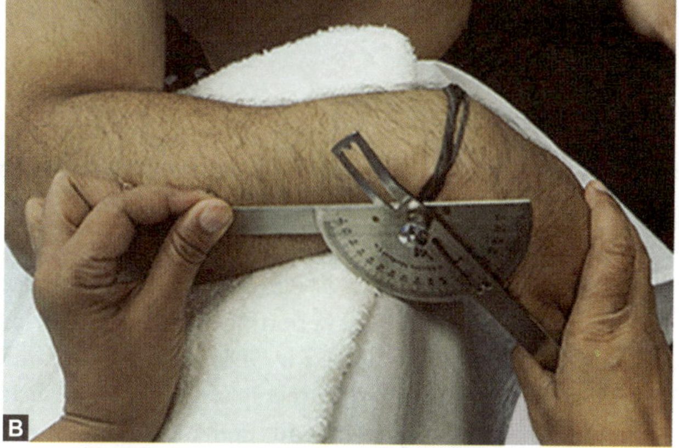

Figs 5.15A and B: Wrist flexion: **A.** Starting position; **B.** Final position

3. **Select the appropriate goniometer:** Choose a goniometer suitable for measuring forearm supination. A universal goniometer with sufficient size and range of motion is typically used. Ensure that the goniometer is clean, in good working condition, and within reach.
4. **Position the goniometer:** The goniometer's fulcrum is positioned over the triquetrum, lateral to the wrist. Using the ulnar styloid process as a point of reference, align the goniometer's stationary arm parallel to the lateral midline of the ulna. Place the movable arm in line with the fifth metacarpal's lateral midline.
5. **Instruct the patient:** Explain the procedure to the patient and demonstrate the desired movement. Instruct the patient to flex the wrist by moving the palm toward the floor. Encourage the patient to move his wrist to the maximum extent without pain or discomfort.
6. **Read and record the measurement:** As the patient performs the movement, observe the range of motion. Read the goniometer scale at the point where the stationary and movable arms intersect. This indicates the angle of wrist flexion. Record the measurement in degrees.
7. **Repeat if necessary:** For accuracy and reliability, it is recommended to repeat the measurement two or three times and calculate an average. Ensure proper goniometer alignment and patient stabilization for each repetition. Record all measurements and calculate the average for documentation.
8. **Document the findings:** Record the goniometer measurements, including the patient's position, any limitations or observations, and the range of motion achieved. Document the results in the patient's medical or therapy record for reference and future comparison.

Wrist Extension

Plane : Sagittal
Axis : Frontal
ROM : 0° to 60°–75°
End feel : Firm

To measure wrist extension using a goniometer, follow these steps (Figs 5.16A and B):

1. **Prepare the patient:** Ensure that the patient is in a comfortable position, such as sitting on a chair. Forearm is comfortably placed on a supporting surface, hand is free to move. The elbow is placed in 90° flexed position and wrist in neutral position. The therapist stabilizes the distal foreram to prevent pronation and supination.
2. **Identify anatomical landmarks:** Locate the relevant anatomical landmarks. The commonly used landmarks for wrist extension measurement are triquetrum and ulnar styloid process. Palpate and mark these landmarks with a pen or marker for reference.

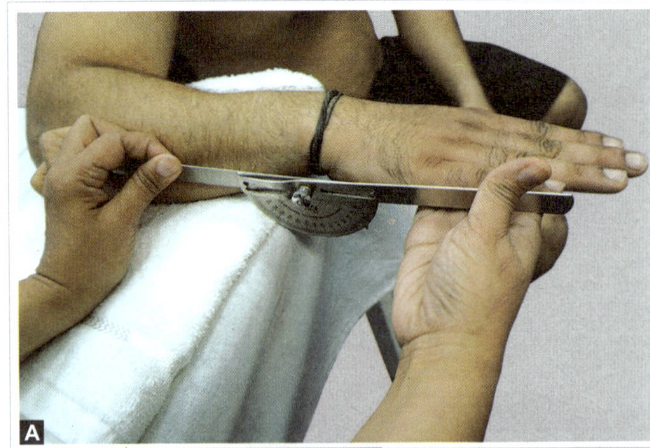

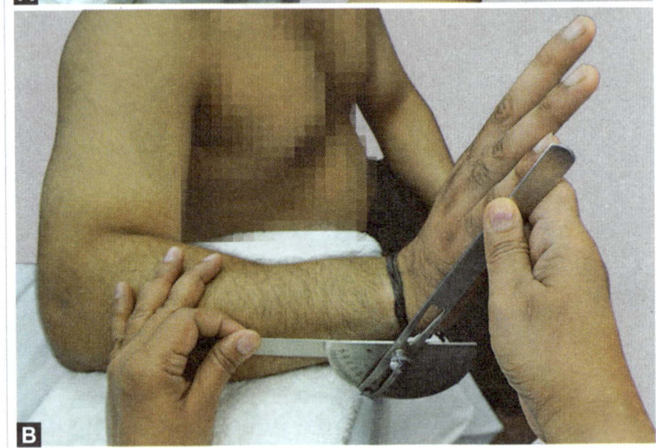

Figs 5.16A and B: Wrist extension: **A.** Starting position; **B.** Final position

3. **Select the appropriate goniometer:** Choose a goniometer suitable for measuring forearm supination. A universal goniometer with sufficient size and range of motion is typically used. Ensure that the goniometer is clean, in good working condition, and within reach.
4. **Position the goniometer:** The goniometer's fulcrum is positioned over the triquetrum, lateral to the wrist. Using the ulnar styloid process as a point of reference, align the goniometer's stationary arm parallel to the lateral midline of the ulna. Place the movable arm in line with the fifth metacarpal's lateral midline.
5. **Instruct the patient:** Explain the procedure to the patient and demonstrate the desired movement. Instruct the patient to extend the wrist by moving the dorsum of hand upward toward the ceiling. Encourage the patient to move their wrist to the maximum extent without pain or discomfort.
6. **Read and record the measurement:** As the patient performs the wrist extension movement, observe the range of motion. Read the goniometer scale at the point where the stationary and movable arms intersect. This indicates the angle of wrist extension. Record the measurement in degrees.

7. **Repeat if necessary:** For accuracy and reliability, it is recommended to repeat the measurement two or three times and calculate an average. Ensure proper goniometer alignment and patient stabilization for each repetition. Record all measurements and calculate the average for documentation.
8. **Document the findings:** Record the goniometer measurements, including the patient's position, any limitations or observations, and the range of motion achieved. Document the results in the patient's medical or therapy record for reference and future comparison.

Wrist Radial Deviation

Plane : Frontal
Axis : Sagittal
ROM : 0° to 20°–25°
End feel : Firm

To measure wrist radial deviation using a goniometer, follow these steps (Figs 5.17A and B):

1. **Prepare the patient:** Ensure that the patient is in a comfortable position, such as sitting on a chair. Forearm is comfortably placed on a supporting surface, hand is free to move. The therapist stabilizes the distal forearm to prevent pronation and supination.
2. **Identify anatomical landmarks:** Locate the relevant anatomical landmarks. The commonly used landmarks for wrist radial deviation measurement are capitate and third metacarpal. Palpate and mark these landmarks with a pen or marker for reference.
3. **Select the appropriate goniometer:** Choose a goniometer suitable for measuring wrist ulnar deviation. A universal goniometer with sufficient size and range of motion is typically used. Ensure that the goniometer is clean, in good working condition, and within reach.
4. **Position the goniometer:** The goniometer's fulcrum is positioned above the capitate on the dorsum of the wrist. Place the goniometer's stationary arm parallel to the forearm's dorsal midline. Place the movable arm in line with the third metacarpal's dorsal midline.
5. **Instruct the patient:** Explain the procedure to the patient and demonstrate the desired movement. Instruct the patient to move the wrist radially toward the midline of the body. Encourage the patient to move their wrist to the maximum extent without pain or discomfort.
6. **Read and record the measurement:** As the patient performs the wrist radial deviation movement, observe the range of motion. Read the goniometer scale at the point where the stationary and movable arms intersect. This indicates the angle of wrist radial deviation. Record the measurement in degrees.
7. **Repeat if necessary:** For accuracy and reliability, it is recommended to repeat the measurement two or three times and calculate an average. Ensure proper goniometer alignment and patient stabilization for each repetition. Record all measurements and calculate the average for documentation.
8. **Document the findings:** Record the goniometer measurements, including the patient's position, any limitations or observations, and the range of motion achieved. Document the results in the patient's medical or therapy record for reference and future comparison.

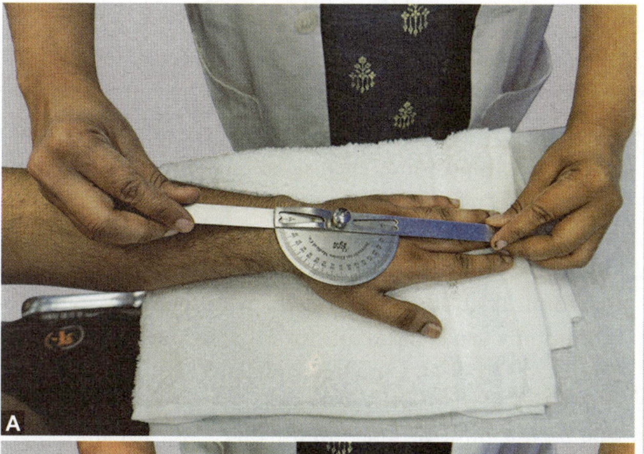

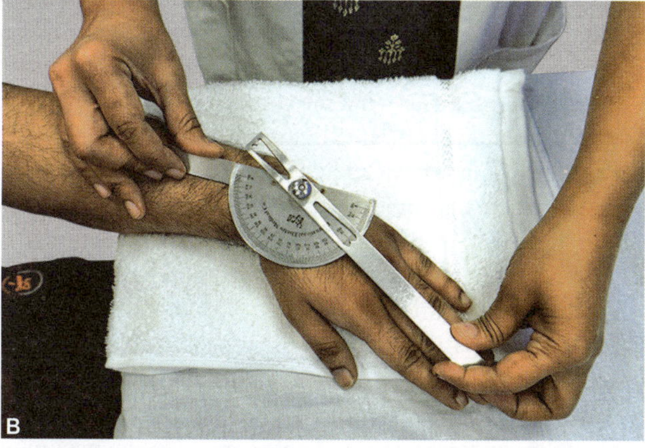

Figs 5.17A and B: Wrist radial deviation: **A.** Starting position; **B.** Final position

Wrist Ulnar Deviation

Plane : Frontal
Axis : Sagittal
ROM : 0° to 30°–40°
End feel : Firm

To measure wrist ulnar deviation using a goniometer, follow these steps (Figs 5.18A and B):

1. **Prepare the patient:** Ensure that the patient is in a comfortable position, such as sitting on a chair. Forearm is comfortably placed on a supporting surface; hand is free to

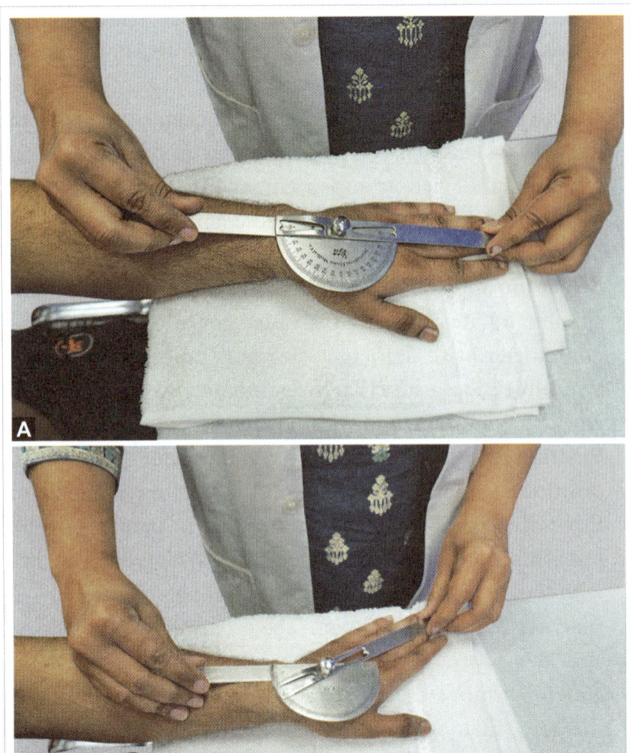

Figs 5.18A and B: Wrist ulnar deviation: **A.** Starting position; **B.** Final position

move. The therapist stabilizes the distal forearm to prevent pronation and supination.

2. **Identify anatomical landmarks:** Locate the relevant anatomical landmarks. The commonly used landmarks for wrist ulnar deviation measurement are capitate and third metacarpal. Palpate and mark these landmarks with a pen or marker for reference.
3. **Select the appropriate goniometer:** Choose a goniometer suitable for measuring wrist radial deviation. A universal goniometer with sufficient size and range of motion is typically used. Ensure that the goniometer is clean, in good working condition, and within reach.
4. **Position the goniometer:** The goniometer's fulcrum is positioned above the capitate on the dorsum of the wrist. Place the goniometer's stationary arm parallel to the forearm's dorsal midline. Place the movable arm in line with the third metacarpal's dorsal midline.
5. **Instruct the patient:** Explain the procedure to the patient and demonstrate the desired movement. Instruct the patient to move the wrist in ulnar direction away from the midline of the body. Encourage the patient to move their wrist to the maximum extent without pain or discomfort.
6. **Read and record the measurement:** As the patient performs the wrist ulnar deviation movement, observe the range of motion. Read the goniometer scale at the point where the stationary and movable arms intersect. This indicates the angle of wrist ulnar deviation. Record the measurement in degrees.
7. **Repeat if necessary:** For accuracy and reliability, it is recommended to repeat the measurement two or three times and calculate an average. Ensure proper goniometer alignment and patient stabilization for each repetition. Record all measurements and calculate the average for documentation.
8. **Document the findings:** Record the goniometer measurements, including the patient's position, any limitations or observations, and the range of motion achieved. Document the results in the patient's medical or therapy record for reference and future comparison.

Metacarpophalangeal (MCP) Joint Flexion

Plane : Sagittal
Axis : Frontal
ROM : 0° to 90°–100°
End feel : Hard (contact between the palmar aspect of the proximal phalanx)/Firm (tension in dorsal joint capsule and the collateral ligaments)

To measure MCP flexion using a goniometer, follow these steps (Figs 5.19A and B):

1. **Prepare the patient:** Ensure that the patient is in a comfortable position, such as sitting. Forearm is resting on supporting surface. Forearm is pronated and wrist in neutral position. The therapist stabilizes the metacarpal.
2. **Identify anatomical landmarks:** Locate the relevant anatomical landmarks associated with the MCP joint. The commonly used landmarks for MCP flexion measurement are metacarpal and proximal phalanx. Palpate and mark these landmarks with a pen or marker for reference.
3. **Select the appropriate goniometer:** Choose a goniometer suitable for measuring MCP flexion. A universal goniometer with sufficient size and range of motion is typically used. Ensure that the goniometer is clean, in good working condition, and within reach.
4. **Position the goniometer:** Over the dorsum of the MCP joint is where the goniometer's fulcrum is positioned. Place the stationary arm in line with the metacarpal's dorsal midline. Place the movable arm in relation to the proximal phalanx's dorsal midline.
5. **Instruct the patient:** Explain the procedure to the patient and demonstrate the desired movement. Instruct the patient to flex the MCP joint by moving the finger toward the palm. Encourage the patient to move to the maximum extent without pain or discomfort.

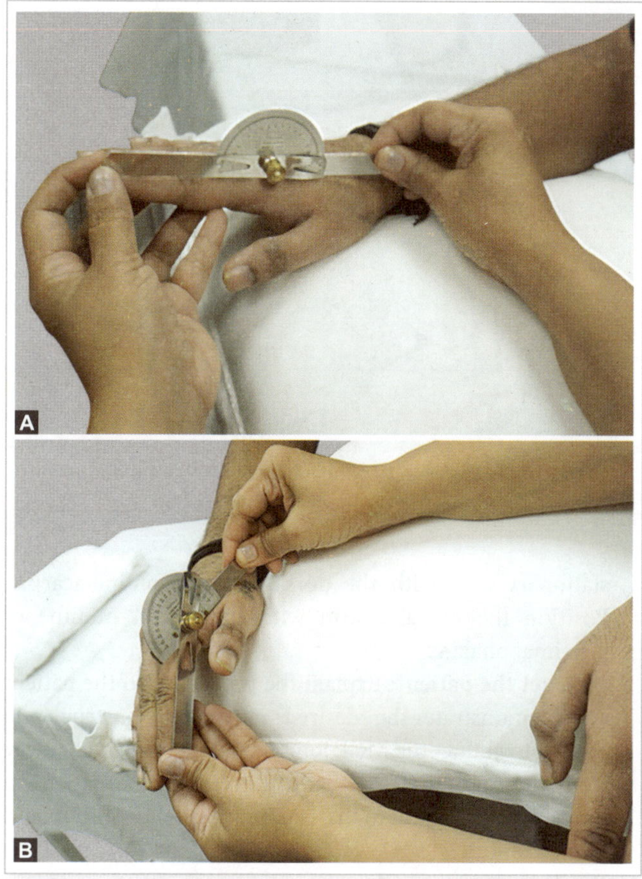

Figs 5.19A and B: Metacarpophalangeal (MCP) joints flexion: A. Starting position; B. Final position

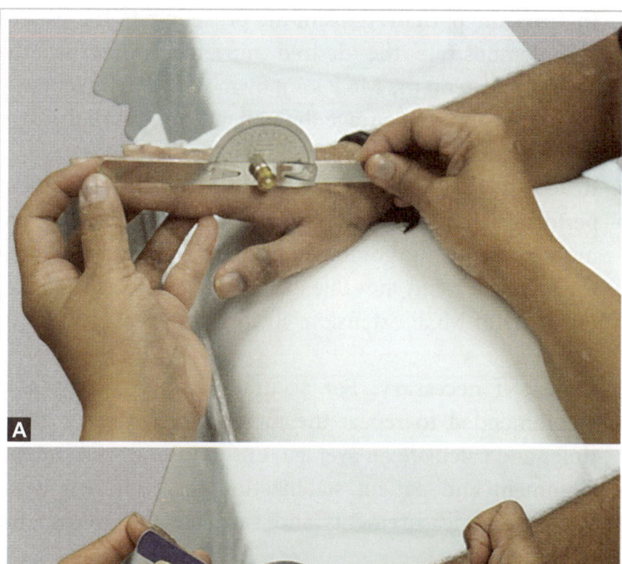

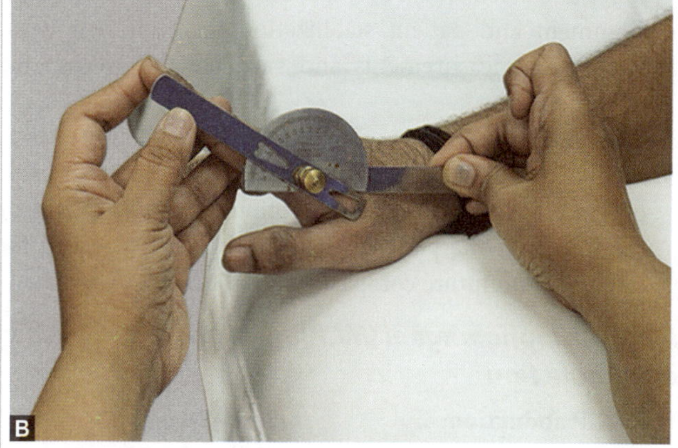

Figs 5.20A and B: Metacarpophalangeal (MCP) joints extension: A. Starting position; B. Final position

6. **Read and record the measurement:** As the patient performs the MCP flexion movement, observe the range of motion. Read the goniometer scale at the point where the stationary and movable arms intersect. This indicates the angle of MCP flexion. Record the measurement in degrees.
7. **Repeat if necessary:** For accuracy and reliability, it is recommended to repeat the measurement two or three times and calculate an average. Ensure proper goniometer alignment and patient stabilization for each repetition. Record all measurements and calculate the average for documentation.
8. **Document the findings:** Record the goniometer measurements, including the patient's position, any limitations or observations, and the range of motion achieved. Document the results in the patient's medical or therapy record for reference and future comparison.

Metacarpophalangeal (MCP) Joint Extension

Plane : Sagittal
Axis : Frontal
ROM : 0° to 20°–45°
End feel : Firm

To measure MCP extension using a goniometer, follow these steps (Figs 5.20A and B):

1. **Prepare the patient:** Ensure that the patient is in a comfortable position, such as sitting. Forearm is resting on supporting surface. Forearm is pronated and the wrist is in neutral position. The therapist stabilizes the metacarpal to avoid wrist movement.
2. **Identify anatomical landmarks:** Locate the relevant anatomical landmarks associated with the MCP joint. The commonly used landmarks for MCP extension measurement are metacarpal and proximal phalanx. Palpate and mark these landmarks with a pen or marker for reference.
3. **Select the appropriate goniometer:** Choose a goniometer suitable for measuring MCP extension. A universal goniometer with sufficient size and range of motion is typically used. Ensure that the goniometer is clean, in good working condition, and within reach.
4. **Position the goniometer:** The dorsum of the MCP joint is where the goniometer's fulcrum is positioned. Align the stationary arm with the metacarpal's dorsal midline. Align the movable arm with the proximal phalanx's dorsal midline.

5. **Instruct the patient:** Explain the procedure to the patient and demonstrate the desired movement. Instruct the patient to extend the MCP joint by moving the finger away from the palm. Encourage the patient to move their MCP to the maximum extent without pain or discomfort.
6. **Read and record the measurement:** As the patient performs the MCP movement, observe the range of motion. Read the goniometer scale at the point where the stationary and movable arms intersect. This indicates the angle of MCP extension. Record the measurement in degrees.
7. **Repeat if necessary:** For accuracy and reliability, it is recommended to repeat the measurement two or three times and calculate an average. Ensure proper goniometer alignment and patient stabilization for each repetition. Record all measurements and calculate the average for documentation.
8. **Document the findings:** Record the goniometer measurements, including the patient's position, any limitations or observations, and the range of motion achieved. Document the results in the patient's medical or therapy record for reference and future comparison.

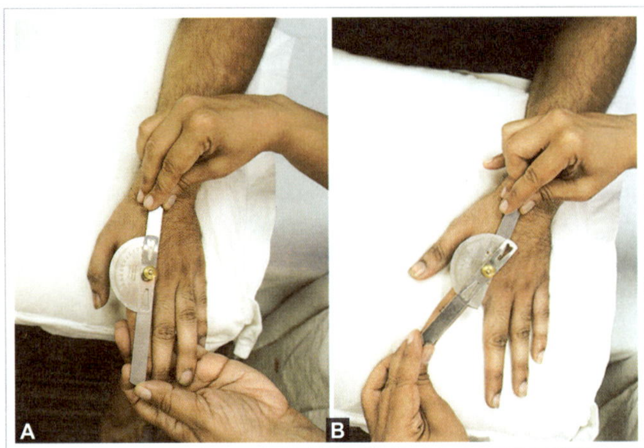

Figs 5.21A and B: Metacarpophalangeal (MCP) joint abduction: **A.** Starting position; **B.** Final position

Metacarpophalangeal (MCP) Joint Abduction and Adduction

For MCP abduction:

Plane : Frontal
Axis : Sagittal
ROM : 0° to 20°–45°
End feel : Firm

Since metacarpophalangeal adduction is the return from full abduction to the 0° starting position, it is typically not measured and recorded separately from MCP abduction.

To measure MCP abduction using a goniometer, follow these steps (Figs 5.21A and B):

1. **Prepare the patient:** Ensure that the patient is in a comfortable position, such as sitting. Forearm is resting on supporting surface. Forearm is in pronated position and wrist and MCP in neutral position. The therapist stabilizes the metacarpal to avoid wrist movement
2. **Identify anatomical landmarks:** Locate the relevant anatomical landmarks associated with the MCP joint. The commonly used landmarks for MCP abduction measurement are metacarpal and proximal phalanx. Palpate and mark these landmarks with a pen or marker for reference.
3. **Select the appropriate goniometer:** Choose a goniometer suitable for measuring MCP abduction. A universal goniometer with sufficient size and range of motion is typically used. Ensure that the goniometer is clean, in good working condition, and within reach.
4. **Position the goniometer:** The fulcrum of the goniometer is placed over the dorsum of the MCP joint. Align the stationary arm with the dorsal midline of metacarpal. Position the movable arm with the dorsal midline of proximal phalanx.
5. **Instruct the patient:** Explain the procedure to the patient and demonstrate the desired movement. Instruct the patient to abduct the MCP joint by moving the finger away from the midline of the hand. Encourage the patient to move the metacarpal to the maximum extent without pain or discomfort.
6. **Read and record the measurement:** As the patient performs the MCP abduction movement, observe the range of motion. Read the goniometer scale at the point where the stationary and movable arms intersect. This indicates the angle of MCP abduction. Record the measurement in degrees.
7. **Repeat if necessary:** For accuracy and reliability, it is recommended to repeat the measurement two or three times and calculate an average. Ensure proper goniometer alignment and patient stabilization for each repetition. Record all measurements and calculate the average for documentation.
8. **Document the findings:** Record the goniometer measurements, including the patient's position, any limitations or observations, and the range of motion achieved. Document the results in the patient's medical or therapy record for reference and future comparison.

Fingers Proximal Interphalangeal (PIP) Joint Flexion and Extension

For PIP flexion:

Plane : Sagittal
Axis : Frontal
ROM : 0° to 100°–110°

Since proximal interphalangeal extension is the return from full flexion to the 0° starting position, it is typically not measured and recorded separately from PIP flexion.

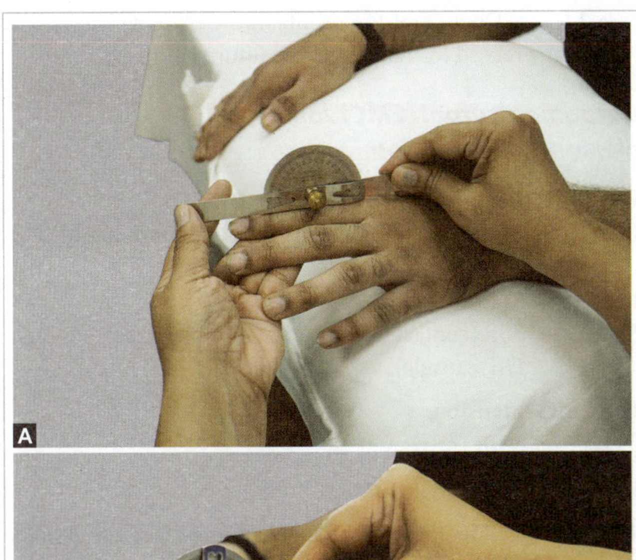

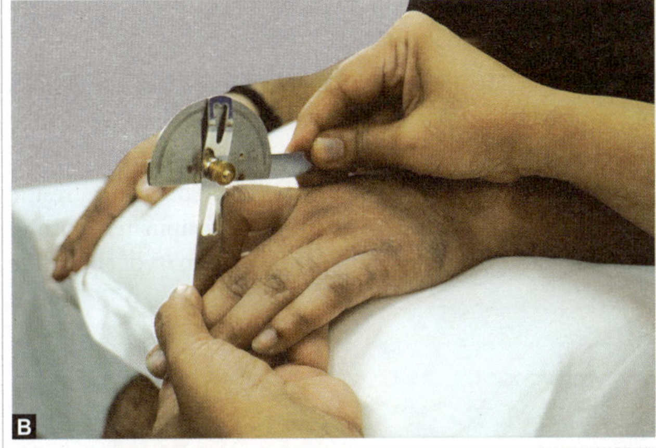

Figs 5.22A and B: Proximal interphalangeal (PIP) joint flexion: **A.** Starting position; **B.** Final position

To measure PIP flexion, follow these steps (Figs 5.22A and B):

1. **Prepare the patient:** Ensure that the patient is in a comfortable position, such as sitting. Forearm is resting on supporting surface. Forearm is in pronated and wrist and MCP in neutral position. The therapist stabilizes the proximal phalanx to prevent movement of metacarpal.
2. **Identify anatomical landmarks:** Locate the relevant anatomical landmarks associated with the PIP joint. The commonly used landmarks for PIP ROM measurement are the proximal phalanx and the middle phalanx. Palpate and mark these landmarks with a pen or marker for reference.
3. **Select the appropriate goniometer:** Choose a goniometer suitable for measuring PIP flexion. A finger goniometer with sufficient size and range of motion is typically used. Ensure that the goniometer is clean, in good working condition, and within reach.
4. **Position the goniometer:** The fulcrum of the goniometer is placed over the dorsal aspect of the PIP joint. Align the stationary arm with the dorsal midline of the proximal phalanx. Position the movable arm with the dorsal midline of middle phalanx.
5. **Instruct the patient:** Explain the procedure to the patient and demonstrate the desired movement. Instruct the patient to flex the PIP joint by moving the finger toward the palm. For PIP extension, the finger is moved away from the palm.
6. **Read and record the measurement:** As the patient performs the PIP flexion movement, observe the range of motion. Read the goniometer scale at the point where the stationary and movable arms intersect. This indicates the angle of finger PIP flexion. Record the measurement in degrees.
7. **Repeat if necessary:** For accuracy and reliability, it is recommended to repeat the measurement two or three times and calculate an average. Ensure proper goniometer alignment and patient stabilization for each repetition. Record all measurements and calculate the average for documentation.
8. **Document the findings:** Record the goniometer measurements, including the patient's position, any limitations or observations, and the range of motion achieved. Document the results in the patient's medical or therapy record for reference and future comparison.

Fingers Distal Interphalangeal (DIP) Joint Flexion and Extension

For DIP flexion:

Plane : Sagittal
Axis : Frontal
ROM : 0° to 70°–90°

Since distal interphalangeal extension is the return from full flexion to the 0° starting position, it is typically not measured and recorded separately from DIP flexion.

To measure DIP flexion, follow these steps (Figs 5.23A and B):

1. **Prepare the patient:** Ensure that the patient is in a comfortable position, such as sitting. Forearm is resting on supporting surface. Forearm is in pronated position and wrist and MCP in neutral position. The therapist stabilizes the middle and the proximal phalanx.
2. **Identify anatomical landmarks:** Locate the relevant anatomical landmarks associated with the DIP joint. The commonly used landmarks for DIP ROM measurement are metacarpal and middle phalanx. Palpate and mark these landmarks with a pen or marker for reference.
3. **Select the appropriate goniometer:** Choose a goniometer suitable for measuring finger DIP flexion. A finger goniometer with sufficient size and range of motion is typically used. Ensure that the goniometer is clean, in good working condition, and within reach.
4. **Position the goniometer:** The fulcrum of the goniometer is placed over the dorsal aspect of the PIP joint. Align the

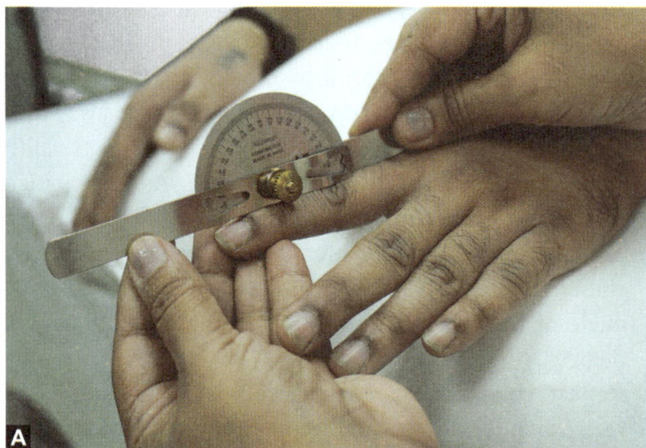

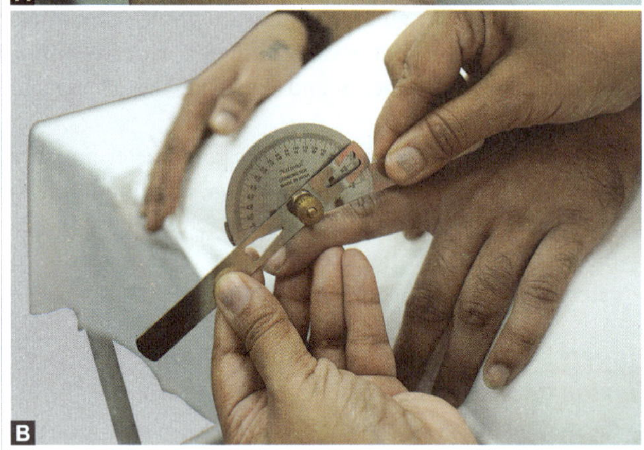

Figs 5.23A and B: Distal interphalangeal (DIP) joint flexion: **A.** Starting position; **B.** Final position

stationary arm with the dorsal midline of middle phalanx. Position the movable arm with the dorsal midline of distal phalanx.

5. **Instruct the patient:** Explain the procedure to the patient and demonstrate the desired movement. Instruct the patient to flex the DIP joint by moving the finger toward the palm. For DIP extension, the patient moves the finger away from the palm.
6. **Read and record the measurement:** As the patient performs the DIP movement, observe the range of motion. Read the goniometer scale at the point where the stationary and movable arms intersect. This indicates the angle of finger DIP flexion. Record the measurement in degrees.
7. **Repeat if necessary:** For accuracy and reliability, it is recommended to repeat the measurement two or three times and calculate an average. Ensure proper goniometer alignment and patient stabilization for each repetition. Record all measurements and calculate the average for documentation.
8. **Document the findings:** Record the goniometer measurements, including the patient's position, any limitations or observations, and the range of motion achieved. Document the results in the patient's medical or therapy record for reference and future comparison.

Carpometacarpal (CMC) Joint Flexion and Extension

For CMC flexion:

Plane : Frontal (Motion occurs in the plane of hand)
Axis : Sagittal
ROM : 0° to 15°–25°
End feel : Soft (contact between muscle bulk)/firm (tension in DIP capsule).

For CMC extension:

ROM : 0° to 15°–35°
End feel : Firm

To measure CMC flexion and extension using a goniometer, follow these steps (Figs 5.24A and B):

1. **Prepare the patient:** Ensure that the patient is in a comfortable position, such as sitting. Forearm is resting on supporting surface. Forearm is in supinated position and wrist in neutral position and CMC in 0° of abduction.

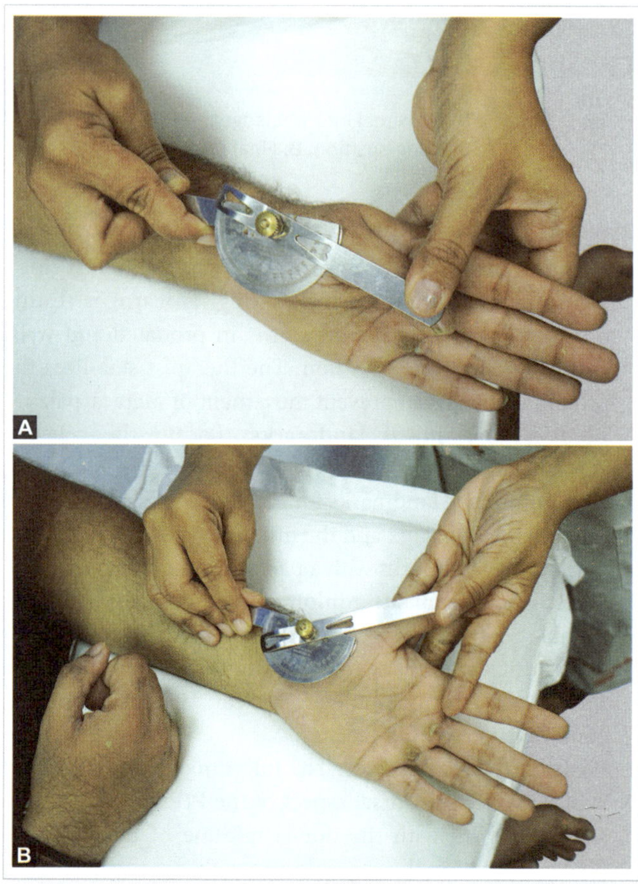

Figs 5.24A and B: Carpometacarpal (CMC) joint: **A.** Flexion; **B.** Extension

The therapist stabilizes the wrist of the patient to avoid its motion.

2. **Identify anatomical landmarks:** Locate the relevant anatomical landmarks associated with the CMC joint. The commonly used landmarks for CMC, ROM measurement are 1st metacarpal and the styloid process of the radius. Palpate and mark these landmarks with a pen or marker for reference.
3. **Select the appropriate goniometer:** Choose a goniometer suitable for measuring CMC flexion/extension. A finger goniometer with sufficient size and range of motion is typically used. Ensure that the goniometer is clean, in good working condition, and within reach.
4. **Position the goniometer:** The fulcrum of the goniometer is placed over the palmar aspect of the CMC joint. Align the stationary arm with the ventral midline of radius with radial styloid as the reference. Position the movable arm with the ventral midline of the first metacarpal.
5. **Instruct the patient:** Explain the procedure to the patient and demonstrate the desired movement. Instruct the patient to flex the CMC joint of the thumb by moving it toward the medial aspect of palm. For CMC extension, the patient is asked to move the thumb away from the medial aspect of the palm.
6. **Read and record the measurement:** As the patient performs the CMC flexion movement, observe the range of motion. Read the goniometer scale at the point where the stationary and movable arms intersect. This indicates the angle of CMC flexion/extension. Record the measurement in degrees.
7. **Repeat if necessary:** For accuracy and reliability, it is recommended to repeat the measurement two or three times and calculate an average. Ensure proper goniometer alignment and patient stabilization for each repetition. Record all measurements and calculate the average for documentation.
8. **Document the findings:** Record the goniometer measurements, including the patient's position, any limitations or observations, and the range of motion achieved. Document the results in the patient's medical or therapy record for reference and future comparison.

Carpometacarpal (CMC) Abduction and Adduction

Plane : Sagittal (Motion occurs in the plane of hand)
Axis : Frontal
ROM : 0° to 40°–50°
End feel : Firm

Since adduction of the thumb's CMC joint is the return to the 0° starting position following full abduction, it is not frequently measured and recorded separately.

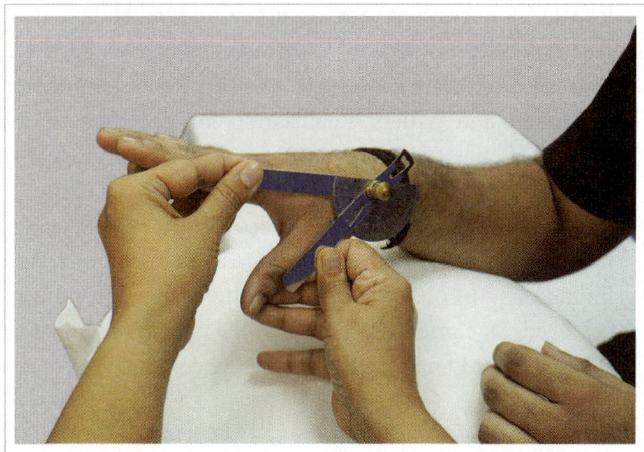

Fig. 5.25: Carpometacarpal abduction final position

To measure CMC abduction using a goniometer, follow these steps (Fig. 5.25):

1. **Prepare the patient:** Ensure that the patient is in a comfortable position, such as sitting. Forearm is resting on supporting surface. Forearm is in pronated position, wrist in neutral position and CMC, MCP and IP joints in neutral position. The therapist stabilizes the wrist of the patient to avoid its motion.
2. **Identify anatomical landmarks:** Locate the relevant anatomical landmarks associated with the CMC joint. The commonly used landmarks for CMC abduction measurement are metacarpal and MCP joint. Palpate and mark these landmarks with a pen or marker for reference.
3. **Select the appropriate goniometer:** Choose a goniometer suitable for measuring CMC abduction. A finger goniometer with sufficient size and range of motion is typically used. Ensure that the goniometer is clean, in good working condition, and within reach.
4. **Position the goniometer:** The fulcrum of the goniometer is placed over the lateral aspect of scaphoid. Align the stationary arm with the lateral midline of 2nd metacarpal with 2nd MCP as the reference. Position the movable arm with the dorsal midline of thumb with 1st MCP joint as reference.
5. **Instruct the patient:** Explain the procedure to the patient and demonstrate the desired movement. Instruct the patient to abduct the CMC joint of the thumb by moving it away from the palm.
6. **Read and record the measurement:** As the patient performs the CMC abduction movement, observe the range of motion. Read the goniometer scale at the point where the stationary and movable arms intersect. This indicates the angle of CMC abduction. Record the measurement in degrees.

7. **Repeat if necessary:** For accuracy and reliability, it is recommended to repeat the measurement two or three times and calculate an average. Ensure proper goniometer alignment and patient stabilization for each repetition. Record all measurements and calculate the average for documentation.
8. **Document the findings:** Record the goniometer measurements, including the patient's position, any limitations or observations, and the range of motion achieved. Document the results in the patient's medical or therapy record for reference and future comparison.

Thumb Opposition

(Combination of abduction, flexion, medial rotation and adduction at the CMC joints of the thumb).
End feel : Firm

The normal range of motion for thumb opposition refers to the ability of the thumb to move across the palm of the hand and make contact with the tips of the fingers. This movement is essential for activities like grasping objects and holding them securely.

To measure thumb opposition, following method is used:

Ruler Method (Fig. 5.26)

1. Ensure that the patient is in a comfortable position, such as sitting. Forearm is resting on supporting surface. Forearm is fully supinated, and wrist is in neutral position.
2. The therapist may stabilize the medial aspect of palmar surface of hand to avoid motion at wrist and 5th MCP.
3. Explain the procedure to the patient and demonstrate the desired movement. Explain the patient how to move the first metacarpal away from the palm (abduction), then from the ulnar (flexion and adduction) toward the base of the little finger, allowing the first metacarpal to medially rotate.
4. The shortest distance between the tip of the thumb and the center of the little finger's proximal digital crease at the conclusion of opposition is frequently measured using a linear ruler. An impaired distance is one of <8 centimeters.
5. Repeat if necessary. For accuracy and reliability, it is recommended to repeat the measurement two or three times and calculate an average.
6. Ensure proper alignment and patient stabilization for each repetition.
7. Record all measurements and calculate the average for documentation.
8. Document the findings.

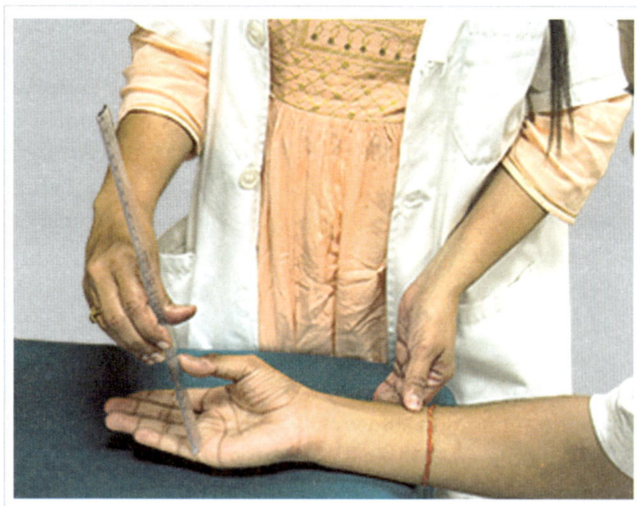

Fig. 5.26: Carpometacarpal opposition ruler method

Lower Limb Goniometry

Hip Flexion

Plane : Sagittal
Axis : Frontal
ROM : 0° to 120°–140°
End feel : Soft

To measure hip flexion using a goniometer, follow these steps (Figs 5.27A to C):

1. **Prepare the patient:** Ensure that the patient is in a comfortable position, such as supine. The hip and pelvis are in neutral rotation. The knee is in extended position. The therapist stabilizes the pelvis to prevent rotation.
2. **Identify anatomical landmarks:** Locate the relevant anatomical landmarks associated with the hip flexion. The commonly used landmarks for hip flexion measurement are the lateral condyle of the femur and the greater trochanter of hip. Palpate and mark these landmarks with a pen or marker for reference.
3. **Select the appropriate goniometer:** Choose a goniometer suitable for measuring hip flexion. A universal goniometer with sufficient size and range of motion is typically used. Ensure that the goniometer is clean, in good working condition, and within reach.
4. **Position the goniometer:** The fulcrum of the goniometer is placed over the lateral aspect of hip joint over greater trochanter. Align the stationary arm with the lateral midline of the pelvis. Position the movable arm with the lateral midline of femur with lateral condyle as reference.
5. **Instruct the patient:** Explain the procedure to the patient and demonstrate the desired movement. Instruct the

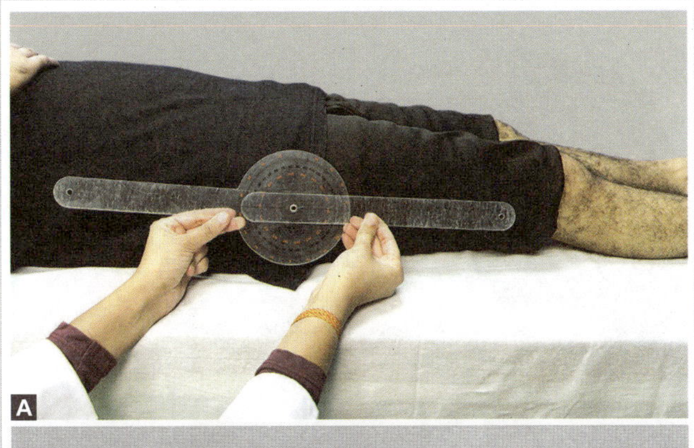

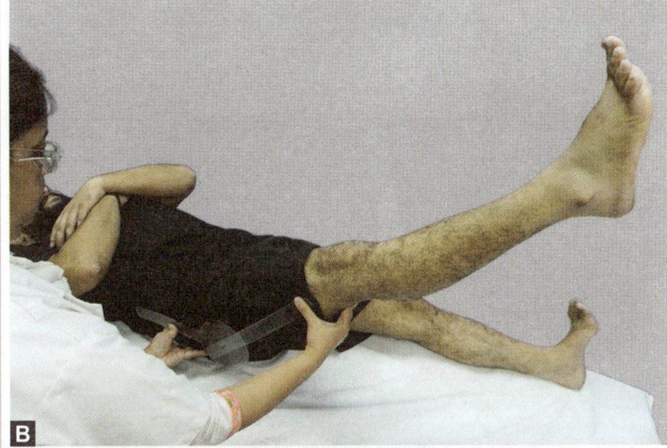

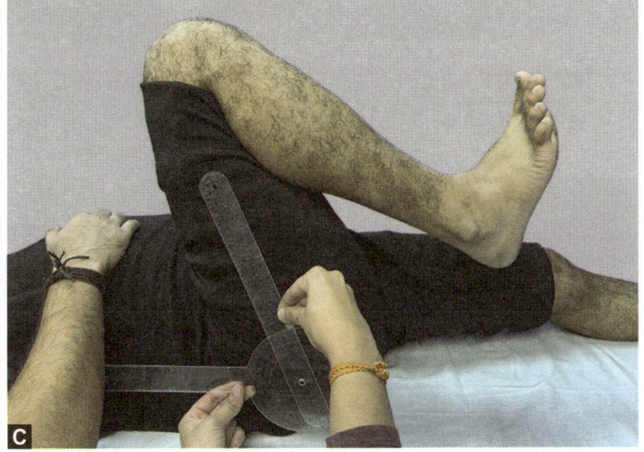

Figs 5.27A to C: Hip flexion: **A.** Starting position; **B.** With knee extension (final position); **C.** With knee flexion (final position)

patient to flex the thigh by lifting it off the table. Encourage the patient to move their hip to the maximum extent without pain or discomfort. Alternatively, it can be done in knee flexed position as well.

6. **Read and record the measurement:** As the patient performs the hip flexion movement, observe the range of motion. Read the goniometer scale at the point where the stationary and movable arms intersect. This indicates the angle of hip flexion. Record the measurement in degrees.
7. **Repeat if necessary:** For accuracy and reliability, it is recommended to repeat the measurement two or three times and calculate an average. Ensure proper goniometer alignment and patient stabilization for each repetition. Record all measurements and calculate the average for documentation.
8. **Document the findings:** Record the goniometer measurements, including the patient's position, any limitations or observations, and the range of motion achieved. Document the results in the patient's medical or therapy record for reference and future comparison.

Hip Extension

Plane : Sagittal
Axis : Frontal
ROM : 0° to 10°–30°
End feel : Firm

To measure hip extension using a goniometer, follow these steps (Figs 5.28A and B):

1. **Prepare the patient:** Ensure that the patient is in a comfortable position, such as prone. The hip and pelvis is in neutral rotation. The knee is in extended position. The therapist stabilizes the pelvis to prevent rotation.

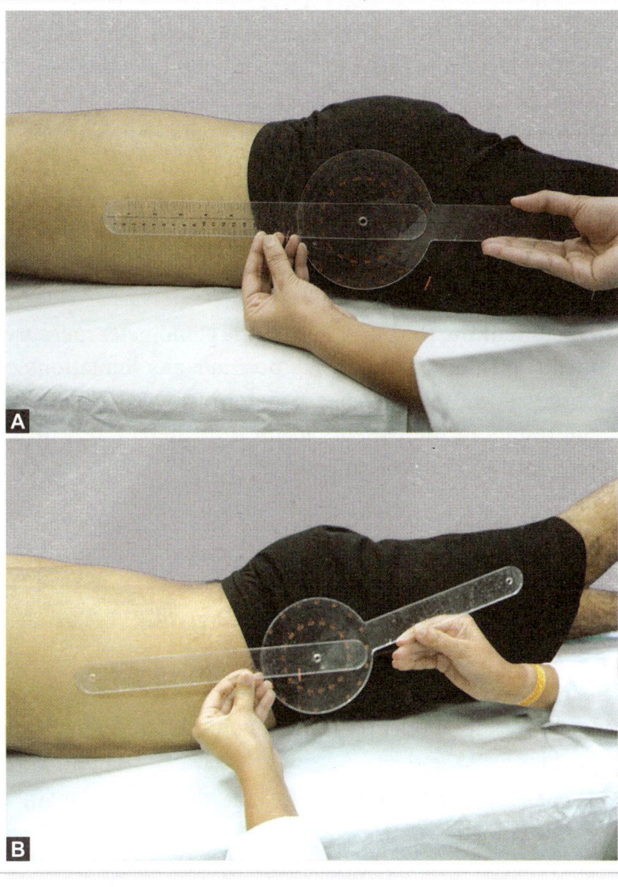

Figs 5.28A and B: Hip extension: **A.** Starting position; **B.** Final position

2. **Identify anatomical landmarks:** Locate the relevant anatomical landmarks associated with the hip joint. The commonly used landmarks for hip extension measurement are the greater trochanter and lateral condyle of femur. Palpate and mark these landmarks with a pen or marker for reference.
3. **Select the appropriate goniometer:** Choose a goniometer suitable for measuring hip extension. A universal goniometer with sufficient size and range of motion is typically used. Ensure that the goniometer is clean, in good working condition, and within reach.
4. **Position the goniometer:** The fulcrum of the goniometer is placed over the lateral aspect of hip joint over greater trochanter. Align the stationary arm with the lateral midline of the pelvis. Position the movable arm with the lateral midline of femur with lateral condyle as reference.
5. **Instruct the patient:** Explain the procedure to the patient and demonstrate the desired movement. Instruct the patient to extend the lower limb by lifting it off the table. Encourage the patient to move their hip to the maximum extent without pain or discomfort.
6. **Read and record the measurement:** As the patient performs the hip extension movement, observe the range of motion. Read the goniometer scale at the point where the stationary and movable arms intersect. This indicates the angle of hip extension. Record the measurement in degrees.
7. **Repeat if necessary:** For accuracy and reliability, it is recommended to repeat the measurement two or three times and calculate an average. Ensure proper goniometer alignment and patient stabilization for each repetition. Record all measurements and calculate the average for documentation.
8. **Document the findings:** Record the goniometer measurements, including the patient's position, any limitations or observations, and the range of motion achieved. Document the results in the patient's medical or therapy record for reference and future comparison.

Hip Abduction

Plane : Frontal
Axis : Sagittal
ROM : 0° to 40°–55°
End feel : Firm

To measure hip abduction using a goniometer, follow these steps (Figs 5.29A and B):

1. **Prepare the patient:** Ensure that the patient is in a comfortable position, such as supine. The hip and pelvis are in neutral rotation. The knee is in extended position. The therapist stabilizes the pelvis to prevent rotation.

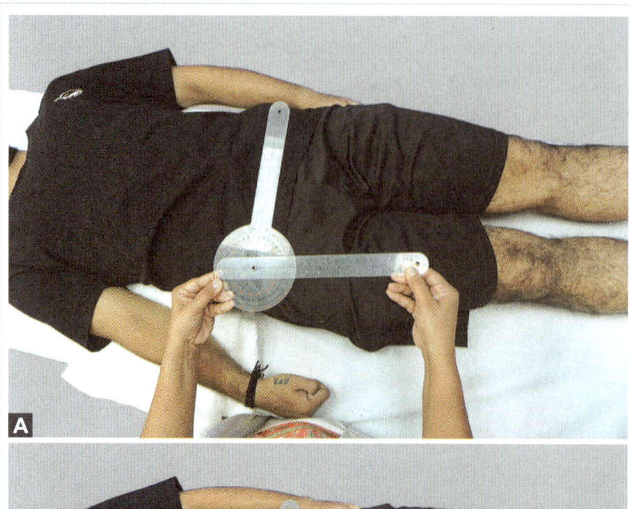

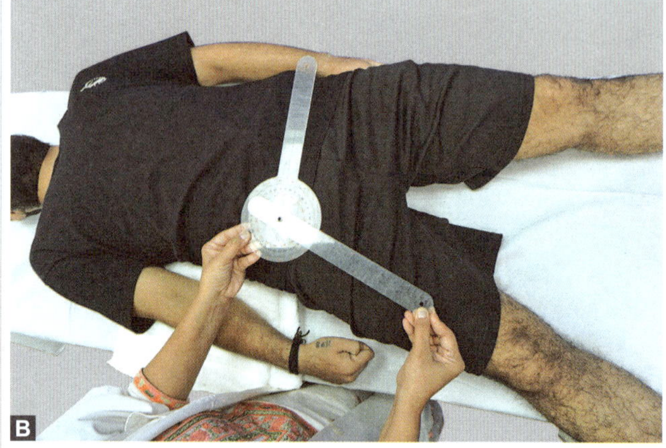

Figs 5.29A and B: Hip abduction: **A.** Starting position; **B.** Final position

2. **Identify anatomical landmarks:** Locate the relevant anatomical landmarks associated with the hip joint. The commonly used landmarks for hip abduction measurement are ASIS and patella. Palpate and mark these landmarks with a pen or marker for reference.
3. **Select the appropriate goniometer:** Choose a goniometer suitable for measuring hip abduction. A universal goniometer with sufficient size and range of motion is typically used. Ensure that the goniometer is clean, in good working condition, and within reach.
4. **Position the goniometer:** The fulcrum of the goniometer is placed over ASIS of the test side. Align the stationary arm with an imaginary horizontal line pointing toward ASIS of the other side. Position the movable arm with the anterior midline of femur with midline of the patella as reference.
5. **Instruct the patient:** Explain the procedure to the patient and demonstrate the desired movement. Instruct the patient to abduct the lower limb by moving it laterally. Encourage the patient to move their hip to the maximum extent without pain or discomfort.

6. **Read and record the measurement:** As the patient performs the hip abduction movement, observe the range of motion. Read the goniometer scale at the point where the stationary and movable arms intersect. This indicates the angle of hip abduction. Record the measurement in degrees.
7. **Repeat if necessary:** For accuracy and reliability, it is recommended to repeat the measurement two or three times and calculate an average. Ensure proper goniometer alignment and patient stabilization for each repetition. Record all measurements and calculate the average for documentation.
8. **Document the findings:** Record the goniometer measurements, including the patient's position, any limitations or observations, and the range of motion achieved. Document the results in the patient's medical or therapy record for reference and future comparison.

Hip Adduction

Plane : Frontal
Axis : Sagittal
ROM : 0° to 20°–30°
End feel : Firm

To measure hip adduction (Figs 5.30A and B) using a goniometer, follow these steps:

1. **Prepare the patient:** Position the patient in supine on a flat surface. The opposite leg should be abducted to avoid interference. Ensure the pelvis is stabilized, and the hip is in neutral rotation.
2. **Identify anatomical landmarks:** Palpate and mark the anterior superior iliac spine (ASIS) and the midline of the femur toward the center of the patella. These landmarks will guide goniometer placement.
3. **Select the appropriate goniometer:** Use a standard universal goniometer, ensuring it is clean and functional.
4. **Position the goniometer:** The fulcrum of the goniometer is positioned over the ASIS of the leg being measured. Align the stationary arm with an imaginary line connecting both ASISs. Position the movable arm along the midline of the femur, using the patella as the reference point.
5. **Instruct the patient:** Explain the procedure and demonstrate the desired movement. Instruct the patient to bring the leg toward the midline of the body (adduction) as far as possible without pain or compensatory movements.
6. **Read and record the measurement:** Observe the movement and note the angle on the goniometer at the end of the motion. This represents the ROM for hip adduction.
7. **Repeat if necessary:** For accuracy, repeat the measurement 2–3 times and calculate the average. Ensure consistent alignment and patient stabilization during each attempt.

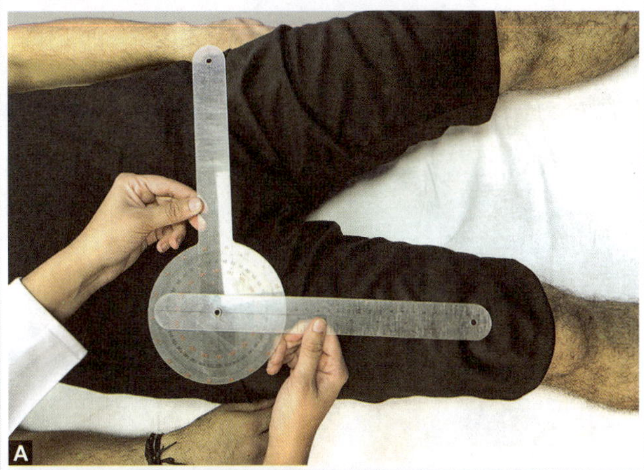

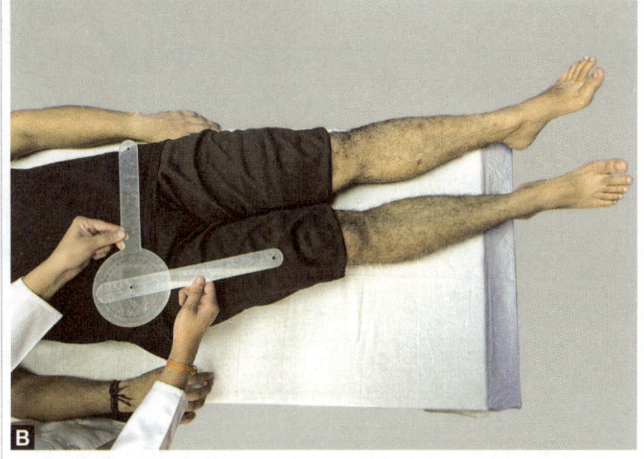

Figs 5.30A and B: Hip adduction (on right side): **A.** Starting position; **B.** Final position

8. **Document the findings:** Record the measurement, patient positioning, and any limitations or observations. Include these details in the patient's medical record for reference and future comparison.

Hip Medial Rotation

Plane : Transverse
Axis : Vertical
ROM : 0° to 30°–45°
End feel : Firm

To measure hip medial rotation using a goniometer, follow these steps (Figs 5.31 and 5.32):

1. **Prepare the patient:** Ensure that the patient is in a comfortable position, such as sitting. The hip and knee are kept in 90° flexed position. The other lower extremity is kept in abducted position to allow space for full hip adduction. A towel roll can be kept under the distal part of thigh. The therapist stabilizes the distal femur.

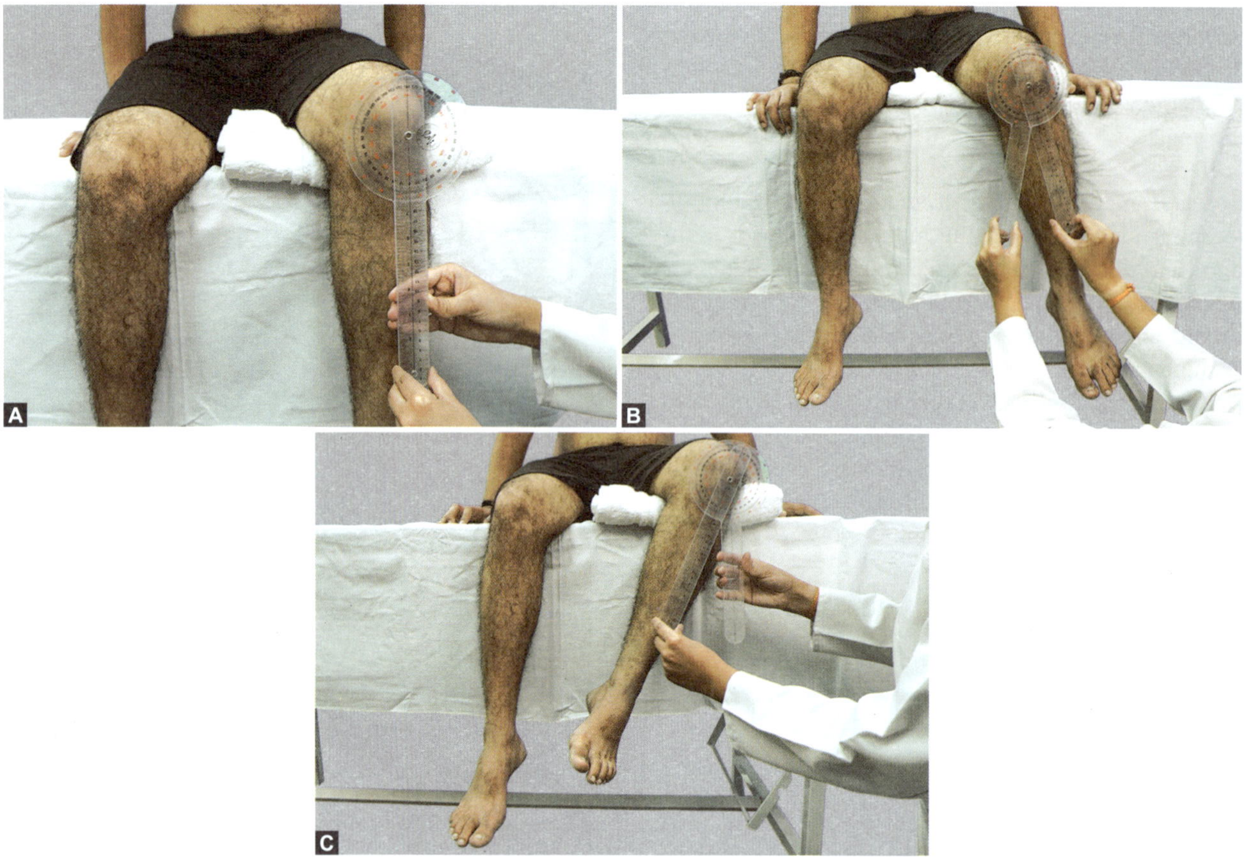

Figs 5.31A to C: Hip rotation in high sitting: **A.** Starting position; **B.** Medial rotation (final position); **C.** Lateral rotation (final position)

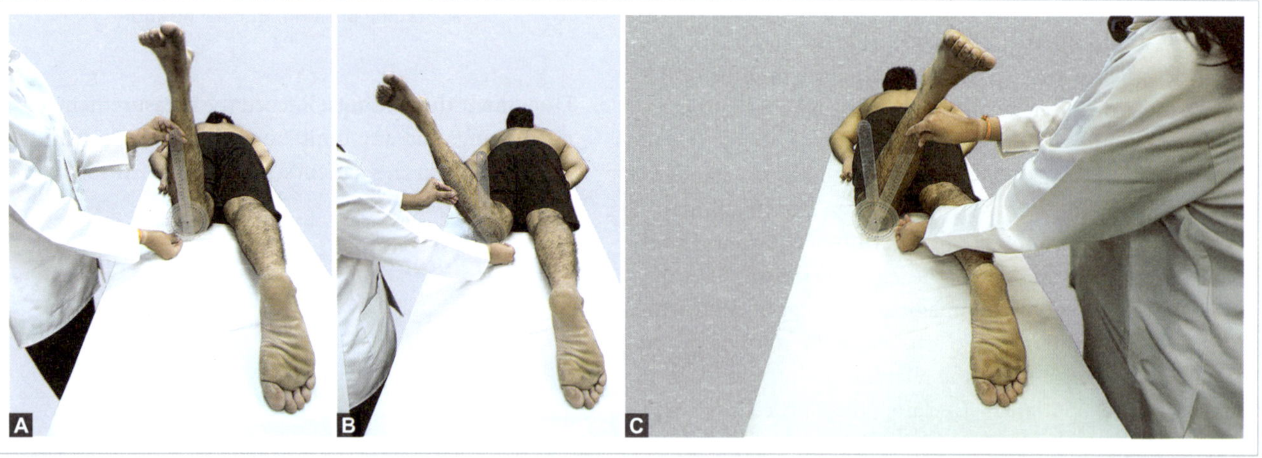

Figs 5.32A to C: Hip rotation in prone lying: **A.** Starting position; **B.** Medial rotation (final position); **C.** Lateral rotation (final position)

2. **Identify anatomical landmarks:** Locate the relevant anatomical landmarks associated with the hip joint. The commonly used landmarks for hip medial rotation measurement are patella and medial and lateral malleolus. Palpate and mark these landmarks with a pen or marker for reference.
3. **Select the appropriate goniometer:** Choose a goniometer suitable for measuring hip medial rotation. A universal goniometer with sufficient size and range of motion is typically used. Ensure that the goniometer is clean, in good working condition, and within reach.
4. **Position the goniometer:** The fulcrum of the goniometer is placed over anterior patella. Align the stationary arm

perpendicular to the floor. Position the movable arm with the anterior midline of the lower leg with midpoint between two malleolus as reference.

5. **Instruct the patient:** Explain the procedure to the patient and demonstrate the desired movement. Instruct the patient to move the lower leg laterally. Encourage the patient to move their leg to the maximum extent without pain or discomfort.
6. **Read and record the measurement:** As the patient performs the hip medial rotation movement, observe the range of motion. Read the goniometer scale at the point where the stationary and movable arms intersect. This indicates the angle of hip medial rotation. Record the measurement in degrees.
7. **Repeat if necessary:** For accuracy and reliability, it is recommended to repeat the measurement two or three times and calculate an average. Ensure proper goniometer alignment and patient stabilization for each repetition. Record all measurements and calculate the average for documentation.
8. **Document the findings:** Record the goniometer measurements, including the patient's position, any limitations or observations, and the range of motion achieved. Document the results in the patient's medical or therapy record for reference and future comparison.

Hip Lateral Rotation

Plane : Transverse
Axis : Vertical
ROM : 0° to 30°–50°
End feel : Firm

To measure hip lateral rotation (Fig. 5.31C and 5.32C) using a goniometer, follow these steps:

1. **Prepare the patient:** Ensure that the patient is in a comfortable position, such as sitting. The hip and knee are kept in 90° flexed position. The other lower extremity is kept in abducted position to allow space for full hip adduction. A towel roll can be kept under the distal part of thigh. The therapist stabilizes the distal femur.
2. **Identify anatomical landmarks:** Locate the relevant anatomical landmarks associated with the hip joint. The commonly used landmarks are patella and medial and lateral malleolus. Palpate and mark these landmarks with a pen or marker for reference.
3. **Select the appropriate goniometer:** Choose a goniometer suitable for measuring hip lateral rotation. A universal goniometer with sufficient size and range of motion is typically used. Ensure that the goniometer is clean, in good working condition, and within reach.
4. **Position the goniometer:** The fulcrum of the goniometer is placed over anterior patella. Align the stationary arm perpendicular to the floor. Position the movable arm with the anterior midline of the lower leg with midpoint between 2 malleolus as reference.
5. **Instruct the patient:** Explain the procedure to the patient and demonstrate the desired movement. Instruct the patient to move the lower leg medially. Encourage the patient to move their leg to the maximum extent without pain or discomfort.
6. **Read and record the measurement:** As the patient performs the hip lateral rotation movement, observe the range of motion. Read the goniometer scale at the point where the stationary and movable arms intersect. This indicates the angle of hip lateral rotation. Record the measurement in degrees.
7. **Repeat if necessary:** For accuracy and reliability, it is recommended to repeat the measurement two or three times and calculate an average. Ensure proper goniometer alignment and patient stabilization for each repetition. Record all measurements and calculate the average for documentation.
8. **Document the findings:** Record the goniometer measurements, including the patient's position, any limitations or observations, and the range of motion achieved. Document the results in the patient's medical or therapy record for reference and future comparison.

Note: Alternatively, hip rotations can be measured in prone lying position with knee flexed to 90° as shown in Fig. 5.32.

Knee Flexion and Extension

For knee flexion:

Plane : Sagittal
Axis : Frontal
ROM : 0° to 130°–140°
End feel : Soft

For knee extension:

Plane : Sagittal
Axis : Frontal
ROM : 0°
End feel : Firm

To measure knee flexion using a goniometer, follow these steps (Figs 5.33A to D):

1. **Prepare the patient:** Ensure that the patient is in a comfortable position, such as supine. The hip and knee are kept in extended position. Ensure neutral rotation at hip. The therapist stabilizes the femur to prevent rotation.
2. **Identify anatomical landmarks:** Locate the relevant anatomical landmarks associated with the knee joint. The commonly used landmarks for knee flexion measurement are the lateral condyle of femur, lateral malleolus and greater trochanter. Palpate and mark these landmarks with a pen or marker for reference.
3. **Select the appropriate goniometer:** Choose a goniometer suitable for measuring knee flexion. A universal goniometer

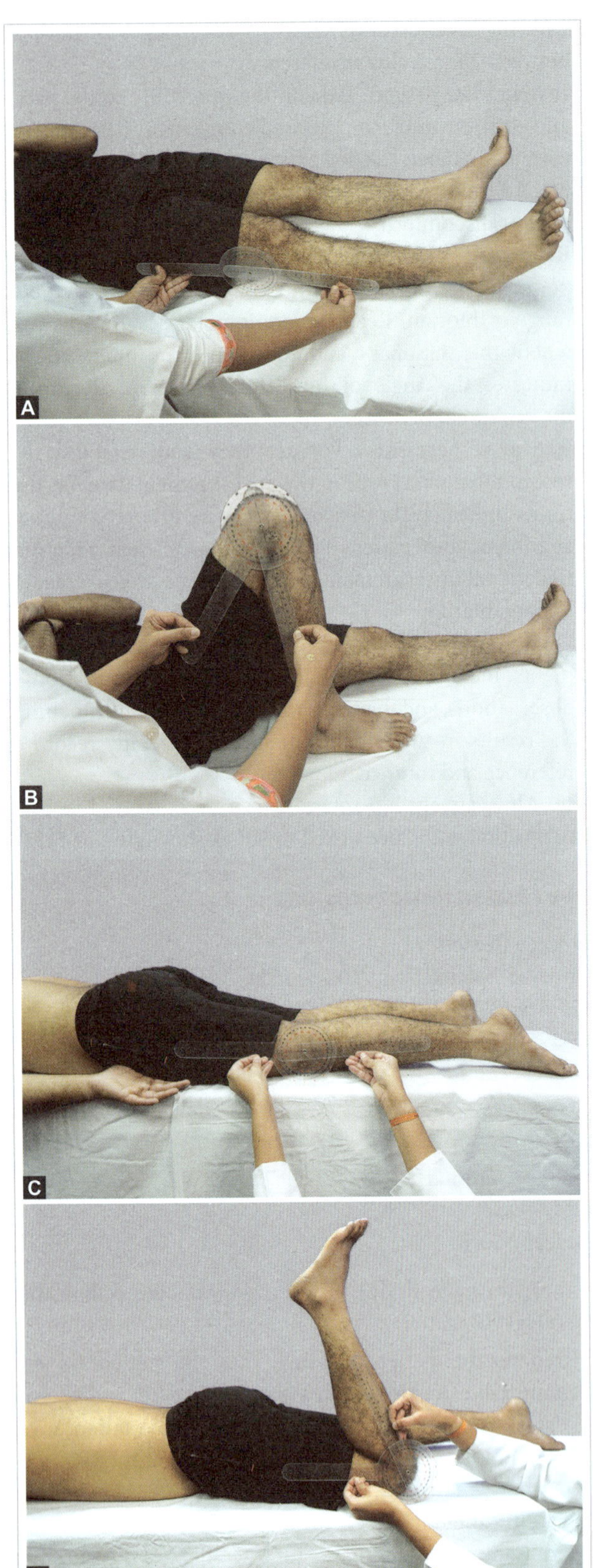

Figs 5.33A to D: **A.** Knee flexion with hip flexion starting position—supine; **B.** Knee flexion with hip flexion final position—supine; **C.** Knee flexion with hip extension starting position—prone; **D.** Knee flexion with hip extension final position—prone

with sufficient size and range of motion is typically used. Ensure that the goniometer is clean, in good working condition, and within reach.

4. **Position the goniometer:** The fulcrum of the goniometer is placed over lateral condyle of femur. Align the stationary arm with lateral midline of femur with greater trochanter as reference. Position the movable arm with the lateral midline of the fibula with lateral malleolus as reference.
5. **Instruct the patient:** Explain the procedure to the patient and demonstrate the desired movement. Instruct the patient to move the hip and knee into flexion. Encourage the patient to move their knee to the maximum extent without pain or discomfort.
6. **Read and record the measurement:** As the patient performs the knee flexion movement, observe the range of motion. Read the goniometer scale at the point where the stationary and movable arms intersect. This indicates the angle of knee flexion. Record the measurement in degrees.
7. **Repeat if necessary:** For accuracy and reliability, it is recommended to repeat the measurement two or three times and calculate an average. Ensure proper goniometer alignment and patient stabilization for each repetition. Record all measurements and calculate the average for documentation.
8. **Document the findings:** Record the goniometer measurements, including the patient's position, any limitations or observations, and the range of motion achieved. Document the results in the patient's medical or therapy record for reference and future comparison.

Note: Alternatively, knee flexion can be measured in prone lying position as well, as shown in Figs 5.33C and D.

Ankle Dorsiflexion

Plane : Sagittal
Axis : Frontal
ROM : 0° to 15°–20°
End feel : Firm

To measure ankle dorsiflexion using a goniometer, follow these steps (Figs 5.34A to D):

1. **Prepare the patient:** Ensure that the patient is in a comfortable position, such as sitting. The knee is kept in 90° flexion position. Ensure neutral rotation at ankle. The therapist stabilizes the lower leg to prevent movement at knee.
2. **Identify anatomical landmarks:** Locate the relevant anatomical landmarks associated with the ankle joint. The commonly used landmarks for ankle dorsiflexion measurement are the lateral malleolus, 5th metatarsal, head of fibula. Palpate and mark these landmarks with a pen or marker for reference.
3. **Select the appropriate goniometer:** Choose a goniometer suitable for measuring ankle dorsiflexion. A universal

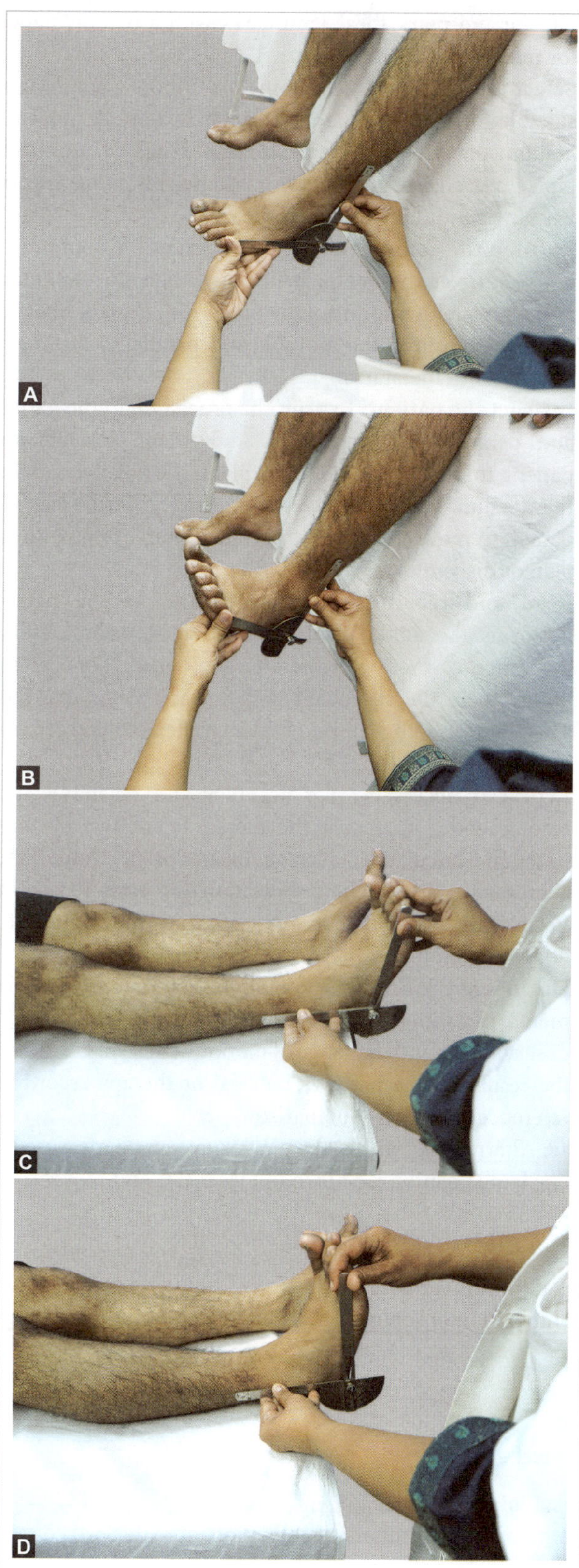

Figs 5.34A to D: Ankle dorsiflexion: A. Starting position (in sitting); B. Final position (in sitting); C. Starting position (in supine); D. Final position (in supine)

goniometer with sufficient size and range of motion is typically used. Ensure that the goniometer is clean, in good working condition, and within reach.

4. **Position the goniometer:** The lateral malleolus is situated above the goniometer's fulcrum. Align the stationary arm with the fibula's lateral midline, using the head of the fibula as a point of reference. Place the movable arm across the fifth metatarsal's lateral aspect.
5. **Instruct the patient:** Explain the procedure to the patient and demonstrate the desired movement. Instruct the patient to move the foot into dorsiflexion by moving the dorsum of the foot toward the leg. Encourage the patient to move his foot to the maximum extent without pain or discomfort.
6. **Read and record the measurement:** As the patient performs the ankle dorsiflexion movement, observe the range of motion. Read the goniometer scale at the point where the stationary and movable arms intersect. This indicates the angle of ankle dorsiflexion. Record the measurement in degrees.
7. **Repeat if necessary:** For accuracy and reliability, it is recommended to repeat the measurement two or three times and calculate an average. Ensure proper goniometer alignment and patient stabilization for each repetition. Record all measurements and calculate the average for documentation.
8. **Document the findings:** Record the goniometer measurements, including the patient's position, any limitations or observations, and the range of motion achieved. Document the results in the patient's medical or therapy record for reference and future comparison.

Note: Alternatively, ankle dorsiflexion can be measured in supine lying position as shown in Figs 5.34C and D.

Ankle Plantar Flexion

Plane : Sagittal
Axis : Frontal
ROM : 0° to 45°–55°
End feel : Firm

To measure ankle plantar flexion using a goniometer, follow these steps (Figs 5.35A to D):

1. **Prepare the patient:** Ensure that the patient is in a comfortable position, such as sitting. The knee is kept in 90° flexion position. Ensure neutral rotation at ankle. The therapist stabilizes the lower leg to prevent movement at knee.
2. **Identify anatomical landmarks:** Locate the relevant anatomical landmarks associated with the ankle plantar flexion. Identify anatomical landmarks. The commonly used landmarks for ankle plantar flexion measurement are the lateral malleolus, 5th metatarsal, head of fibula.

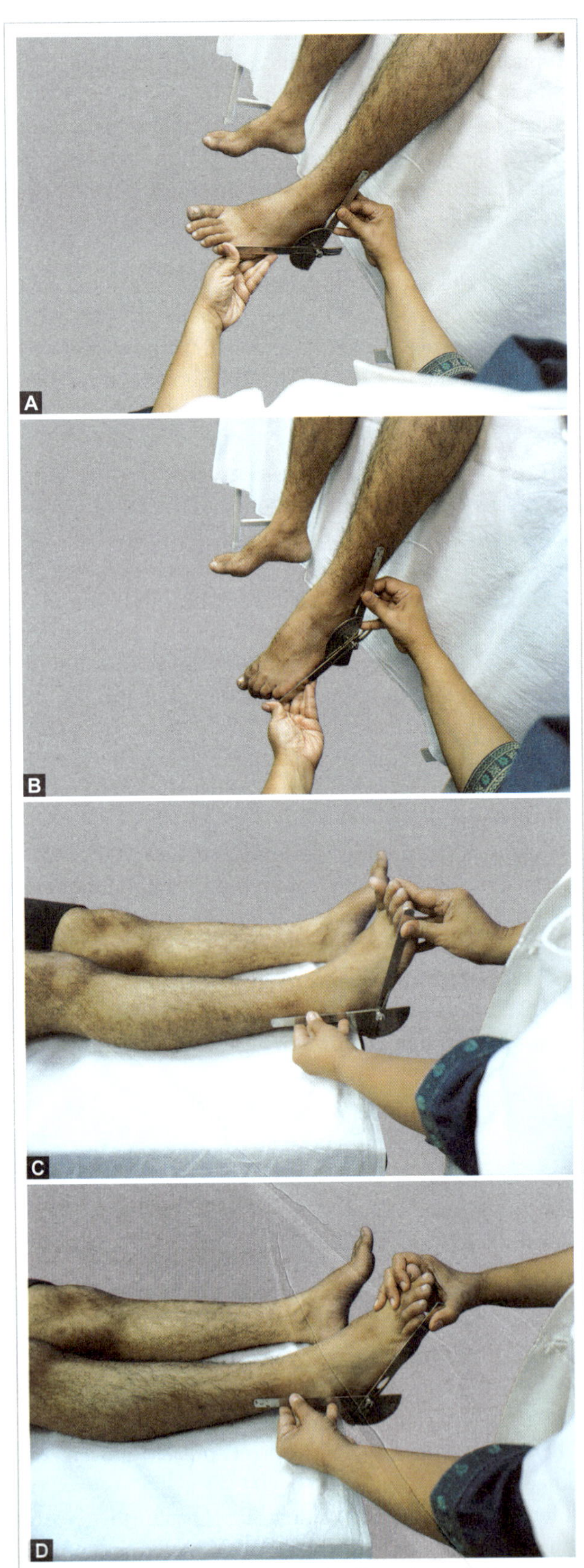

Figs 5.35A to D: Ankle plantar flexion: **A.** Starting position (in sitting); **B.** Final position (in sitting); **C.** Starting position (in supine); **D.** Final position (in supine)

Palpate and mark these landmarks with a pen or marker for reference.

3. **Select the appropriate goniometer:** Choose a goniometer suitable for measuring ankle plantar flexion. A universal goniometer with sufficient size and range of motion is typically used. Ensure that the goniometer is clean, in good working condition, and within reach.
4. **Position the goniometer:** The fulcrum of the goniometer is placed over the lateral malleolus. Align the stationary arm with lateral midline of the fibula with head of fibula as reference. Position the movable arm parallel to the lateral aspect of 5th metatarsal.
5. **Instruct the patient:** Explain the procedure to the patient and demonstrate the desired movement. Instruct the patient to move the foot into plantarflexion by moving the dorsum of the foot away from the leg. Encourage the patient to move their foot to the maximum extent without pain or discomfort.
6. **Read and record the measurement:** As the patient performs the ankle plantar flexion movement, observe the range of motion. Read the goniometer scale at the point where the stationary and movable arms intersect. This indicates the angle of plantar flexion. Record the measurement in degrees.
7. **Repeat if necessary:** For accuracy and reliability, it is recommended to repeat the measurement two or three times and calculate an average. Ensure proper goniometer alignment and patient stabilization for each repetition. Record all measurements and calculate the average for documentation.
8. **Document the findings:** Record the goniometer measurements, including the patient's position, any limitations or observations, and the range of motion achieved. Document the results in the patient's medical or therapy record for reference and future comparison.

Note: Alternatively, ankle plantar flexion can be measured in supine lying position as shown in Figs 5.35C and D.

Tarsal Joint Inversion and Eversion

Tarsal joint inversion:

(Combination of supination, adduction and plantar flexion)

Plane : Frontal
Axis : Sagittal
ROM : 0° to 30°–35°
End feel : Firm

Tarsal joint eversion:

(Combination of pronation, abduction, dorsiflexion)

Plane : Frontal
Axis : Sagittal
ROM : 0° to 15°–20°
End feel : Hard

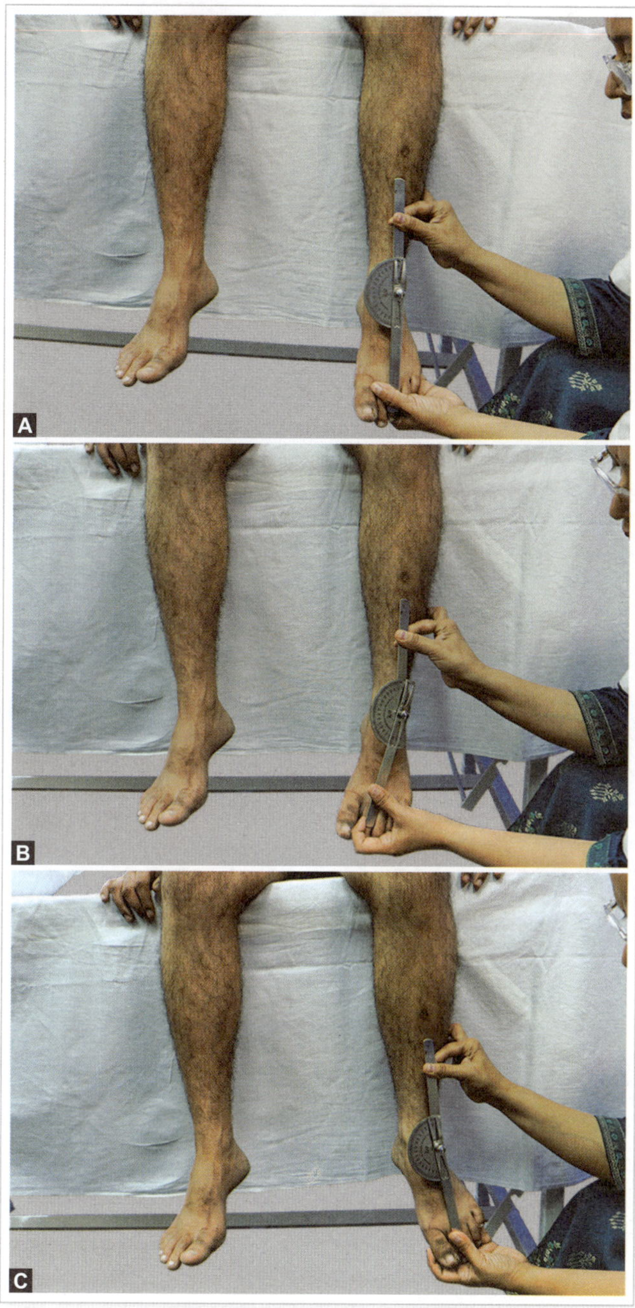

Figs 5.36A to C: Tarsal joint: **A.** Starting position; **B.** Inversion (final position); **C.** Eversion (final position)

To measure tarsal joint inversion or eversion using a goniometer, follow these steps (Figs 5.36A to C):

1. **Prepare the patient:** Ensure that the patient is in a comfortable position, such as sitting or supine. The knee is kept in 90° flexion position. Ensure neutral rotation at ankle. The therapist stabilizes the lower leg to prevent movement at knee.
2. **Identify anatomical landmarks:** Locate the relevant anatomical landmarks associated with the tarsal joint. The commonly used landmarks for tarsal joint inversion measurement are the medial and lateral malleolus and tibial tuberosity. Palpate and mark these landmarks with a pen or marker for reference.
3. **Select the appropriate goniometer:** Choose a goniometer suitable for measuring tarsal joint inversion or eversion. A universal goniometer with sufficient size and range of motion is typically used. Ensure that the goniometer is clean, in good working condition, and within reach.
4. **Position the goniometer:** The fulcrum of the goniometer is placed over the anterior aspect of ankle midway between 2 malleolus. Align the stationary arm with anterior midline of lower leg with tibial tuberosity as reference. Position the movable arm parallel to the anterior midline of 2nd metatarsal.
5. **Instruct the patient:** Explain the procedure to the patient and demonstrate the desired movement. For inversion, instruct the patient to move the forefoot medially and downwards into adduction and plantar flexion. The sole of the foot is turned medially into supination. For eversion, instruct the patient to move the forefoot laterally and upwards into abduction and dorsiflexion. The sole of the foot is turned laterally into pronation. Encourage the patient to move their foot to the maximum extent without pain or discomfort.
6. **Read and record the measurement:** As the patient performs the tarsal joint inversion or eversion movement, observe the range of motion. Read the goniometer scale at the point where the stationary and movable arms intersect. This indicates the angle of tarsal joint inversion or eversion, according to the movement performed. Record the measurement in degrees.
7. **Repeat if necessary:** For accuracy and reliability, it is recommended to repeat the measurement two or three times and calculate an average. Ensure proper goniometer alignment and patient stabilization for each repetition. Record all measurements and calculate the average for documentation.
8. **Document the findings:** Record the goniometer measurements, including the patient's position, any limitations or observations, and the range of motion achieved. Document the results in the patient's medical or therapy record for reference and future comparison.

Subtalar Joint Inversion and Eversion

Subtalar joint inversion:

(Combination of supination, adduction and plantar flexion)

Plane : Frontal
Axis : Sagittal
ROM : 0°–5°
End feel : Hard (contact between calcaneum and sinus tarsi) or firm (tension in deltoid ligament and tibialis posterior muscle)

Subtalar joint eversion:
(Combination of pronation, abduction, and dorsiflexion)

Plane : Frontal
Axis : Sagittal
ROM : 0° to 5°–12°
End feel : Firm (due to tension in the lateral ligaments and peroneus longus and brevis muscles) or hard (due to contact between the calcaneus and the floor of the sinus tarsi)

To measure subtalar inversion or eversion using a goniometer, follow these steps (Figs 5.37A to C):

1. **Prepare the patient:** Ensure that the patient is in a comfortable position, such as prone with foot at the edge of the surface. The hip and knee are extended and in neutral position. The therapist stabilizes the lower leg to prevent movement at knee.
2. **Identify anatomical landmarks:** Locate the relevant anatomical landmarks associated with the subtalar joint. The commonly used landmarks for subtalar joint measurement are the medial and lateral malleolus. Palpate and mark these landmarks with a pen or marker for reference.
3. **Select the appropriate goniometer:** Choose a goniometer suitable for measuring subtalar inversion or eversion. A universal goniometer with sufficient size and range of motion is typically used. Ensure that the goniometer is clean, in good working condition, and within reach.
4. **Position the goniometer:** The fulcrum of the goniometer is placed over the posterior aspect of ankle midway between two malleolus. Align the stationary arm with posterior midline of lower leg. Position the movable arm parallel to posterior midline of calcaneus.
5. **Movement:** The therapist pulls the calcaneum into adduction and supination thereby producing subtalar inversion. For producing subtalar eversion, the therapist pulls the calcaneum into abduction and pronation.
6. **Read and record the measurement:** As the subtalar joint inversion or eversion is produced, observe the range of motion. Read the goniometer scale at the point where the stationary and movable arms intersect. This indicates the angle of subtalar inversion or eversion, according to the movement performed. Record the measurement in degrees.
7. **Repeat if necessary:** For accuracy and reliability, it is recommended to repeat the measurement two or three times and calculate an average. Ensure proper goniometer alignment and patient stabilization for each repetition. Record all measurements and calculate the average for documentation.
8. **Document the findings:** Record the goniometer measurements, including the patient's position, any limitations or observations, and the range of motion achieved. Document the results in the patient's medical or therapy record for reference and future comparison.

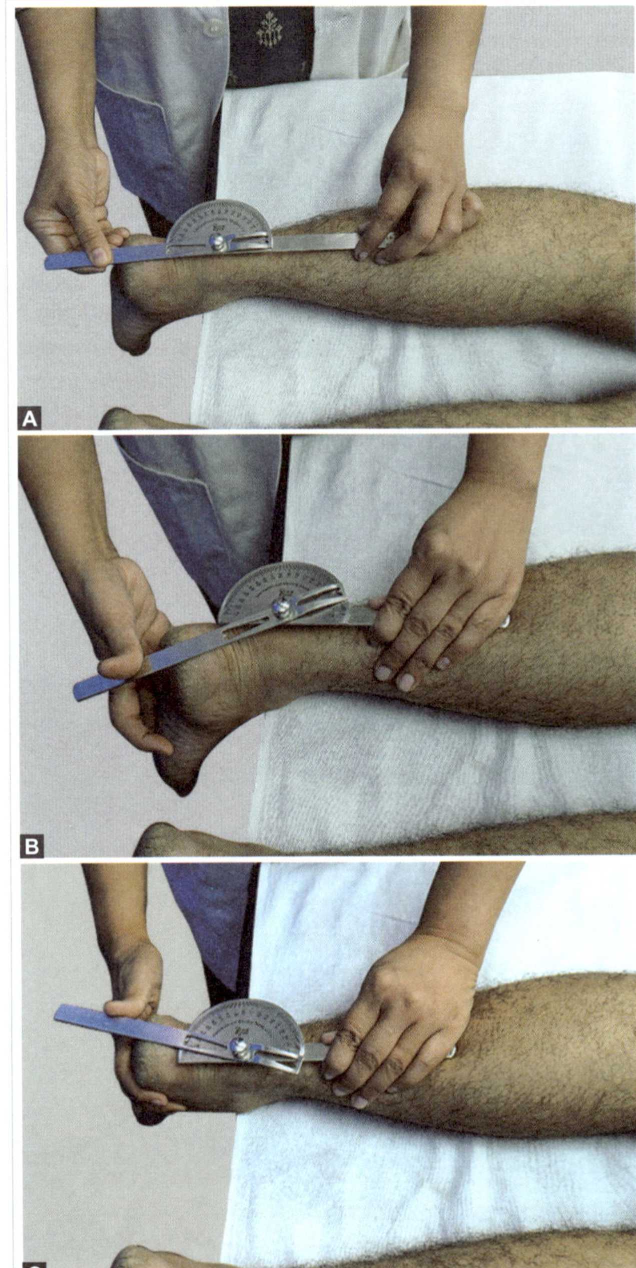

Figs 5.37A to C: Subtalar joint: **A.** Starting position; **B.** Inversion (final position); **C.** Eversion (final position)

Metatarsophalangeal Joint Flexion and Extension

For metatarsophalangeal joint flexion:

Plane : Sagittal
Axis : Frontal
ROM : 0° to 30°–45°
End feel : Firm

For metatarsophalangeal joint extension:

Plane : Sagittal
Axis : Frontal
ROM : 0° to 70°–80°
End feel : Firm

To measure MTP flexion or extension using a goniometer, follow these steps (Figs 5.38A to C):

1. **Prepare the patient:** Ensure that the patient is in a comfortable position, such as sitting or supine. The ankle and MTP joint should be in neutral position. The therapist stabilizes the metatarsals to prevent plantar flexion at ankle and foot.
2. **Identify anatomical landmarks:** Locate the relevant anatomical landmarks associated with the MTP joint. The commonly used landmarks for MTP flexion or extension measurement are metatarsal and proximal phalanx. Palpate and mark these landmarks with a pen or marker for reference.
3. **Select the appropriate goniometer:** Choose a goniometer suitable for measuring MTP flexion or extension. A universal goniometer with sufficient size and range of motion is typically used. Ensure that the goniometer is clean, in good working condition, and within reach.
4. **Position the goniometer:** The fulcrum of the goniometer is placed over the lateral aspect of MTP joint. Align the stationary arm with dorsal midline of the metatarsal. Position the movable arm parallel to the dorsal midline of proximal phalanx.
5. **Instruct the patient:** Explain the procedure to the patient and demonstrate the desired movement. Instruct the patient to move the plantar surface of the great toe downward into flexion. For extension, the great toe is moved toward the dorsum of the foot. Encourage the patient to move their foot to the maximum extent without pain or discomfort.
6. **Read and record the measurement:** As the patient performs the MTP flexion or extension as per the movement, observe the range of motion. Read the goniometer scale at the point where the stationary and movable arms intersect. This indicates the angle of MTP flexion or extension, as per the movement. Record the measurement in degrees.
7. **Repeat if necessary:** For accuracy and reliability, it is recommended to repeat the measurement two or three times and calculate an average. Ensure proper goniometer alignment and patient stabilization for each repetition. Record all measurements and calculate the average for documentation.
8. **Document the findings:** Record the goniometer measurements, including the patient's position, any limitations or observations, and the range of motion achieved. Document the results in the patient medical or therapy record for reference and future comparison.

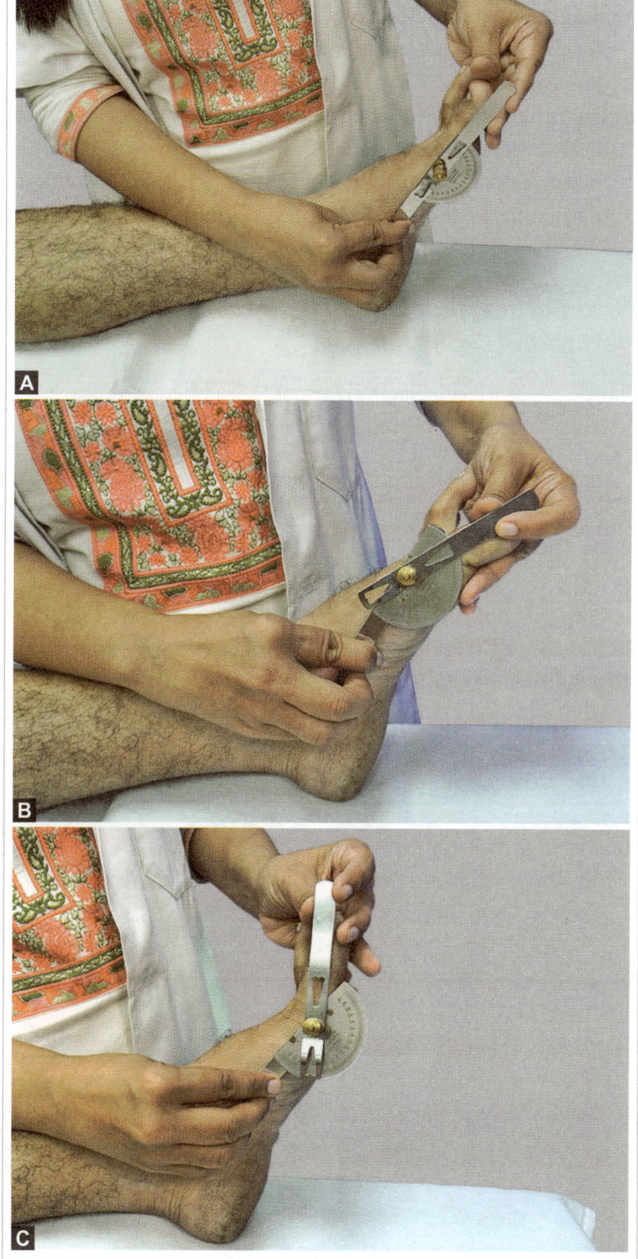

Figs 5.38A to C: Metatarsophalangeal joint: **A.** Starting position; **B.** Flexion (final position); **C.** Extension (final position)

Metatarsophalangeal (MTP) Joint Abduction and Adduction

Plane : Transverse
Axis : Vertical
ROM : 0° to 10°–15°
End feel : Firm

Adduction, which is typically unmeasured, is the return from abduction to the 0° starting position.

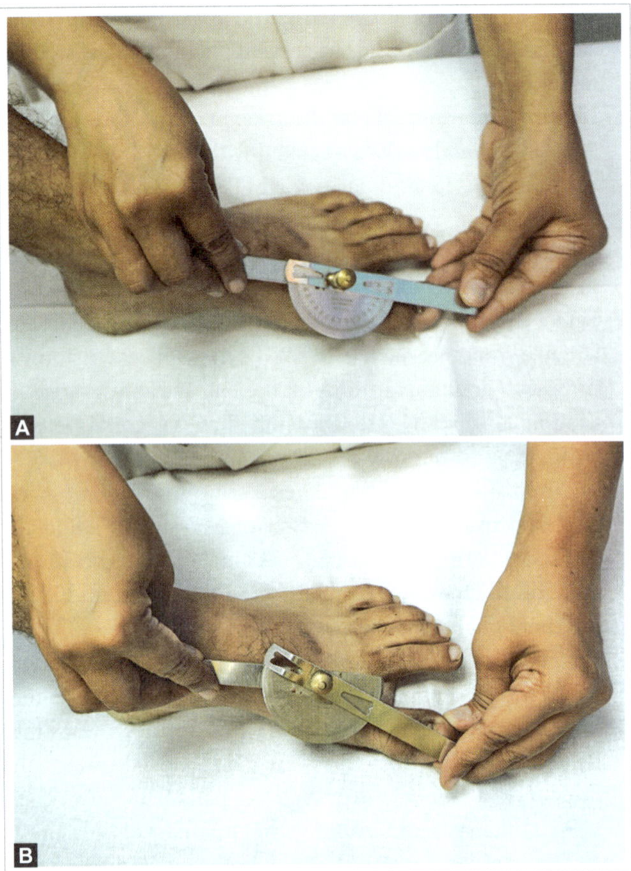

Figs 5.39A and B: Metatarsophalangeal joint abduction: A. Starting position; B. Final position

To measure MTP abduction using a goniometer, follow these steps (Figs 5.39A and B):

1. **Prepare the patient:** Ensure that the patient is in a comfortable position, such as sitting or supine. The ankle and MTP joint should be in neutral position. The therapist stabilizes the metatarsals to prevent foot movement.
2. **Identify anatomical landmarks:** Locate the relevant anatomical landmarks associated with the MTP joint. The commonly used landmarks for MTP abduction measurement are metatarsal and proximal phalanx. Palpate and mark these landmarks with a pen or marker for reference.
3. **Select the appropriate goniometer:** Choose a goniometer suitable for measuring MTP abduction. A universal goniometer with sufficient size and range of motion is typically used. Ensure that the goniometer is clean, in good working condition, and within reach.
4. **Position the goniometer:** The fulcrum of the goniometer is placed over the dorsum of MTP joint. Align the stationary arm with dorsal midline of the metatarsal. Position the movable arm parallel to the dorsal midline of proximal phalanx.
5. **Instruct the patient:** The therapist pulls the proximal phalanx of toe away from midline of the body into abduction.
6. **Read and record the measurement:** As the proximal phalanx is moved into abduction, observe the range of motion. Read the goniometer scale at the point where the stationary and movable arms intersect. This indicates the angle of MTP abduction. Record the measurement in degrees.
7. **Repeat if necessary:** For accuracy and reliability, it is recommended to repeat the measurement two or three times and calculate an average. Ensure proper goniometer alignment and patient stabilization for each repetition. Record all measurements and calculate the average for documentation.
8. **Document the findings:** Record the goniometer measurements, including the patient's position, any limitations or observations, and the range of motion achieved. Document the results in the patient's medical or therapy record for reference and future comparison.

Proximal Interphalangeal (PIP) Joints of the Toes Flexion and Extension

Plane : Sagittal
Axis : Frontal
ROM of 1st toe : 0–90°
ROM lesser 4 toes : 0–35°
End feel : Soft

Since the extension motion involves a return from flexion to the initial position of 0°, it is typically not measured.

To measure PIP flexion using a goniometer, follow these steps (Fig. 5.40):

1. **Prepare the patient:** Ensure that the patient is in a comfortable position, such as sitting or supine. The ankle,

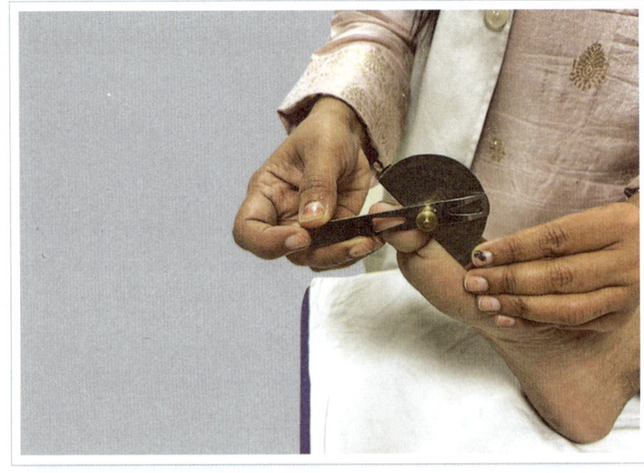

Fig. 5.40: Measuring flexion at proximal interphalangeal joints of the toes

foot and MTP joint should be in neutral position. The therapist stabilizes the metatarsals and the proximal phalanx to prevent foot and ankle movement.

2. **Identify anatomical landmarks:** Locate the relevant anatomical landmarks associated with the PIP joint. The commonly used landmarks are PIP joint and proximal phalanx. Palpate and mark these landmarks with a pen or marker for reference.
3. **Select the appropriate goniometer:** Choose a goniometer suitable for measuring PIP flexion. A universal goniometer with sufficient size and range of motion is typically used. Ensure that the goniometer is clean, in good working condition, and within reach.
4. **Position the goniometer:** The fulcrum of the goniometer is placed over the dorsum of IP joint of respective toe. Align the stationary arm with dorsal midline of the proximal phalanx. Position the movable arm parallel to the dorsal midline of phalanx distal to the joint to be tested.
5. **Instruct the patient:** Explain the procedure to the patient and demonstrate the desired movement. Instruct the patient to move the distal phalanx of 1st toe or middle phalanx of lesser 4 toes downward into flexion. Encourage the patient to move their foot to the maximum extent without pain or discomfort.
6. **Read and record the measurement:** As the patient performs the PIP flexion movement, observe the range of motion. Read the goniometer scale at the point where the stationary and movable arms intersect. This indicates the angle of PIP flexion. Record the measurement in degrees.
7. **Repeat if necessary:** For accuracy and reliability, it is recommended to repeat the measurement two or three times and calculate an average. Ensure proper goniometer alignment and patient stabilization for each repetition. Record all measurements and calculate the average for documentation.
8. **Document the findings:** Record the goniometer measurements, including the patient's position, any limitations or observations, and the range of motion achieved. Document the results in the patient's medical or therapy record for reference and future comparison.

Distal Interphalangeal (DIP) Joints of the Toes Flexion and Extension

Plane : Sagittal
Axis : Frontal
ROM : 0° to 30°–60°
End feel : Firm

Since this extension motion involves a return from flexion to the initial position of 0, it is typically not measured.

To measure extension using a goniometer, follow these steps:

1. **Prepare the patient:** Ensure that the patient is in a comfortable position, such as sitting or supine. The ankle and foot should be in neutral position. The therapist stabilizes the metatarsals and proximal and middle phalanx to prevent foot and toe movement.
2. **Identify anatomical landmarks:** Locate the relevant anatomical landmarks associated with the DIP joint. The commonly used landmarks are DIP joint and proximal phalanx. Palpate and mark these landmarks with a pen or marker for reference.
3. **Select the appropriate goniometer:** Choose a goniometer suitable for measuring DIP flexion. A universal goniometer with sufficient size and range of motion is typically used. Ensure that the goniometer is clean, in good working condition, and within reach.
4. **Position the goniometer:** The fulcrum of the goniometer is placed over the dorsum of DIP joint. Align the stationary arm with dorsal midline of the middle phalanx. Position the movable arm parallel to the dorsal midline of distal phalanx.
5. **Instruct the patient:** Explain the procedure to the patient and demonstrate the desired movement. Instruct the patient to move the distal phalanx of toes downward toward the plantar aspect of foot. Encourage the patient to move their foot to the maximum extent without pain or discomfort.
6. **Read and record the measurement:** As the patient performs the DIP flexion movement, observe the range of motion. Read the goniometer scale at the point where the stationary and movable arms intersect. This indicates the angle of DIP flexion. Record the measurement in degrees.
7. **Repeat if necessary:** For accuracy and reliability, it is recommended to repeat the measurement two or three times and calculate an average. Ensure proper goniometer alignment and patient stabilization for each repetition. Record all measurements and calculate the average for documentation.
8. **Document the findings:** Record the goniometer measurements, including the patient's position, any limitations or observations, and the range of motion achieved. Document the results in the patient's medical or therapy record for reference and future comparison.

Goniometry of Spine

Cervical Spine Flexion

Plane : Sagittal
Axis : Frontal
ROM : 0–40°
End feel : Firm

Goniometer Method

1. **Patient positioning and stabilization:**
   - Ask the patient to sit upright with his feet at shoulder width, maintaining a neutral spine and providing proper thoracolumbar spine support.
   - Stabilize the patient's shoulder girdle and chest to prevent any movement in the thoracic and lumbar spine.
   - Place the fulcrum of the goniometer over the external auditory meatus (Fig. 5.41A).
   - Align the fixed arm parallel to the trunk.
   - Position the distal arm parallel to the patient's nose.
2. **Instruct the patient:** Instruct the patient to flex his neck by moving the chin downward toward the chest.
3. **Recording and documentation:**
   - Read the goniometer scale at the point where the stationary and movable arms intersect, indicating the angle of cervical flexion (Fig. 5.41B).
   - Record the measurement in degrees.
   - For enhanced accuracy and reliability, it is advisable to repeat the measurement two or three times and calculate an average.
   - Ensure that proper goniometer alignment and patient stabilization are maintained for each repetition.
   - Document all measurements and calculate the average, which should be recorded in the patient's medical or therapy record for reference and future comparisons.

Tape Method

1. Ask the person to tuck his chin toward his chest and bend his head forward until they feel resistance to moving any further.
2. **Procedure:**
   - Mark the lower edge of the sternal notch and the middle of the tip of the chin with a skin marking pencil while the subject is in the starting testing posture (Fig. 5.41C).
   - Measure the distance between the marks at the end of the joint range of motion from the tip of the chin and the mark at the lower border of the sternum (Fig. 5.41D).

Single Inclinometer Method

1. Make sure the inclinometer's dial is set to 0° before placing it on the person's head (Fig. 5.41E). Ask the subject to bend his head forward.
2. Throughout the motion, the examiner must keep solid contact between the inclinometer and the subject's head.
3. Read and record the degrees on the inclinometer's dial at the conclusion of the motion (Fig. 5.41F).

Double Inclinometer Method

1. Assure that one inclinometer is set to 0° before placing it directly on the spinous process of the T1 vertebra. Ask the subject to bend his head forward.

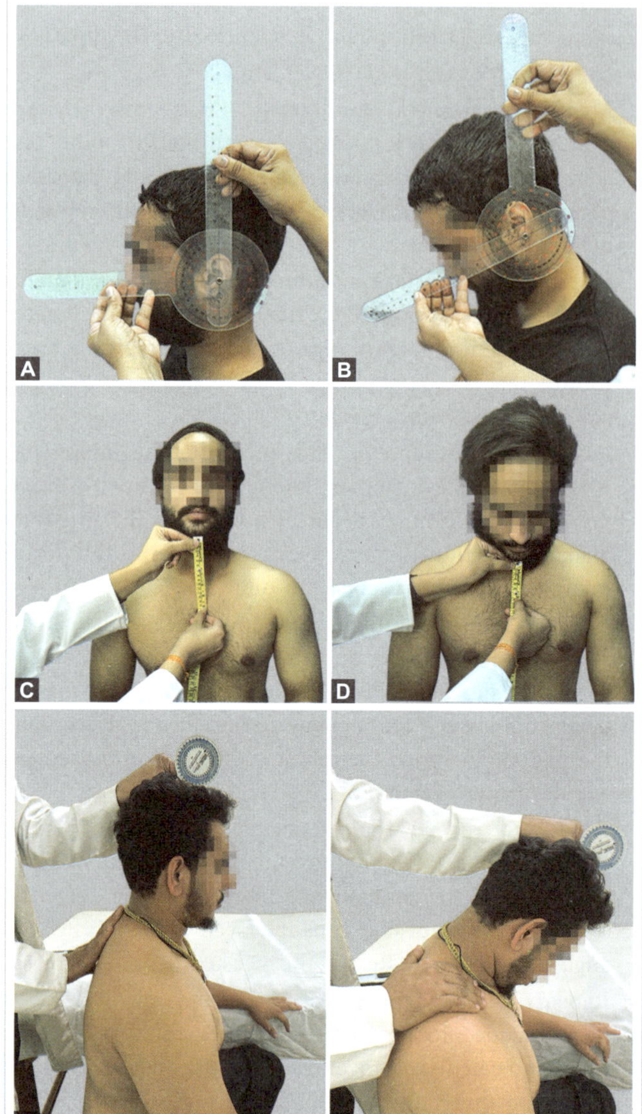

Figs 5.41A to F: Cervical spine flexion: **A.** Goniometer method (starting position); **B.** Goniometer method (final position); **C.** Tape method (starting position); **D.** Tape method (final position); **E.** Single Inclinometer method (starting position); **F.** Single inclinometer method (final position)

2. Throughout the motion, the examiner must keep a tight grip on the vertebra and inclinometer.
3. The therapist firmly positions the second inclinometer on top of the head, making sure that it is set to 0°. Throughout the motion, the examiner must maintain a firm connection between the inclinometer and the head.
4. Check the dials of each inclinometer to see the degrees at the conclusion of the move.

The ROM is the difference between the readings from the two inclinometers.

Cervical Spine Extension

Plane : Sagittal
Axis : Frontal
ROM : 0°–50°
End feel : Firm

Measurement Using Goniometer (Figs 5.42A and B)

1. **Patient positioning and stabilization:**
   - Make the patient sit in an upright position with feet at shoulder width, maintaining a neutral spine position, and providing proper support for the thoracolumbar spine.
   - Stabilize the patient's shoulder girdle and chest to prevent any movement in the thoracic and lumbar spine.
   - Place the fulcrum of the goniometer over the external auditory meatus.
   - Align the fixed arm parallel to the trunk.
   - Position the distal arm parallel to the patient's nose.
2. **Instruct the patient:** Instruct the patient to extend his neck by lifting his chin upward and tilting the head backward as far as possible.
3. **Recording and documentation:**
   - Read the goniometer scale at the point where the stationary and movable arms intersect, indicating the angle of cervical extension.
   - Record the measurement in degrees.
   - For increased accuracy and reliability, it is advisable to repeat the measurement two or three times and calculate an average.
   - Ensure proper goniometer alignment and patient stabilization for each repetition.
   - Document all measurements and calculate the average, which should be recorded in the patient's medical or therapy record for reference and future comparisons.

Tape Method (Figs 5.42C and D)

1. **Instruct the patient:** Begin in the neutral head position and instruct the individual to tuck in the chin before looking up and back as much as possible without moving the trunk.
2. **Procedure:**
   - Mark the lower edge of the sternal notch and the middle of the tip of the chin with a skin marking pencil while in the starting testing posture.
   - At the conclusion of the cervical extension, measure the distance between the mark at the lower edge of the sternal notch and the mark at the tip of the chin.
   - Make sure the person's mouth stays shut throughout the motion.

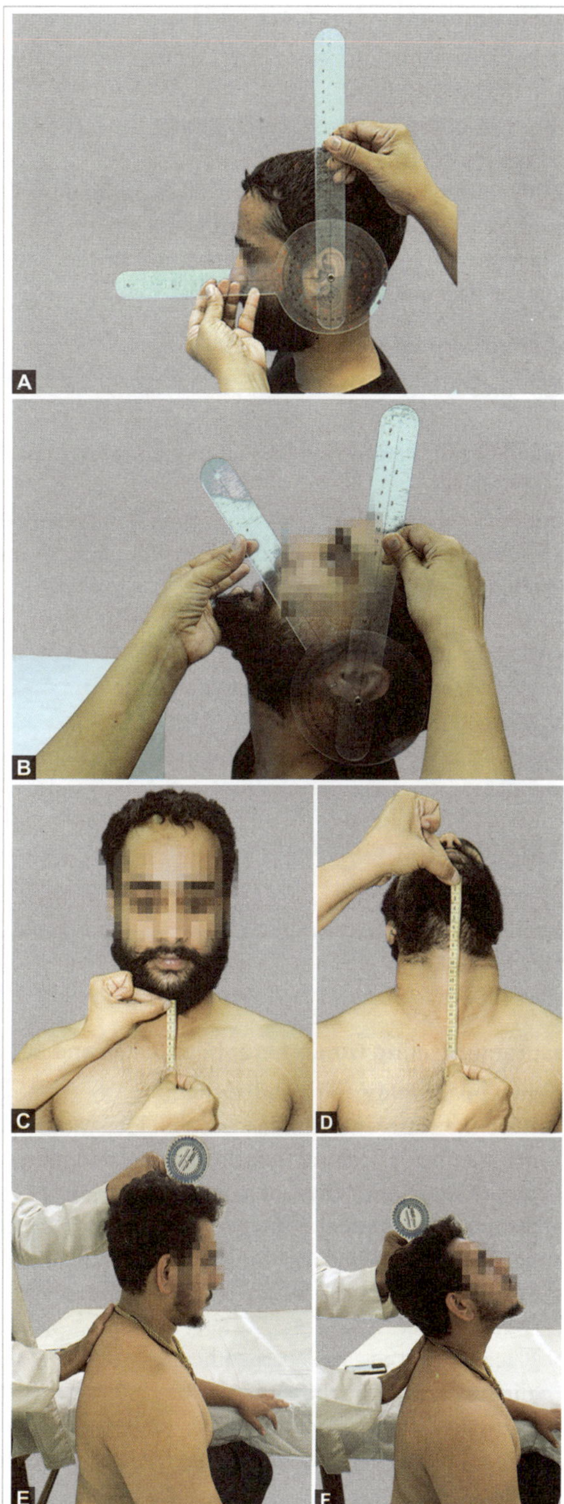

Figs 5.42A to D: Cervical spine extension: **A.** Goniometer method (starting position); **B.** Goniometer method (final position); **C.** Tape method (starting position); **D.** Tape method (final position); **E.** Single Inclinometer method (starting position); **F.** Single Inclinometer method (final position)

Single Inclinometer (Figs 5.42E and F)

1. Make sure the inclinometer's dial is set to 0° before placing it on the person's head.
2. Throughout the motion, securely secure the inclinometer to the subject's head.
3. Tell the person to elevate his chin before pushing his head as far back as he can without moving his trunk.
4. When the person reaches the conclusion of the motion, note the data on the inclinometer's dial.

Double Inclinometer Method

1. One inclinometer should be placed lateral to the T1 vertebra or over the scapular spine. The inclinometer's dial should be adjusted to read 0°. Ask the subject to move the head backward.
2. The back of the head may come into touch with the inclinometer if it is positioned over the first thoracic vertebra.
3. Set the additional inclinometer firmly on the top of the head and that it displays 0°.
4. After the motion, read and write down the data on each inclinometer's dial.
5. The ROM is the difference between the two inclinometers' readings.

Cervical Spine Lateral Flexion

Plane : Frontal
Axis : Sagittal
ROM : 0°–50°
End feel : Firm

Measurement Using Goniometer (Figs 5.43A and B)

1. **Patient positioning and stabilization:**
   - Have the patient sit with his feet shoulder-width apart, maintaining a neutral spine position, and ensuring proper support for the thoracolumbar spine.
   - Stabilize the patient's shoulder girdle and chest to prevent any movement at the thoracic and lumbar spine.
   - Position the fulcrum of the goniometer over the C7 spinous process.
   - Place the fixed arm perpendicular to the ground, aligning it with the thoracic vertebra.
   - Align the distal arm with the dorsal midline of the head, using the external occipital protuberance as a reference point.
2. **Instruct the patient:** Instruct the patient to perform side flexion of the head by bringing his ear close to his shoulder while maintaining the seated position.
3. **Recording and documentation:**
   - Read the goniometer scale at the point where the stationary and movable arms intersect. This reading indicates the angle of cervical side flexion.

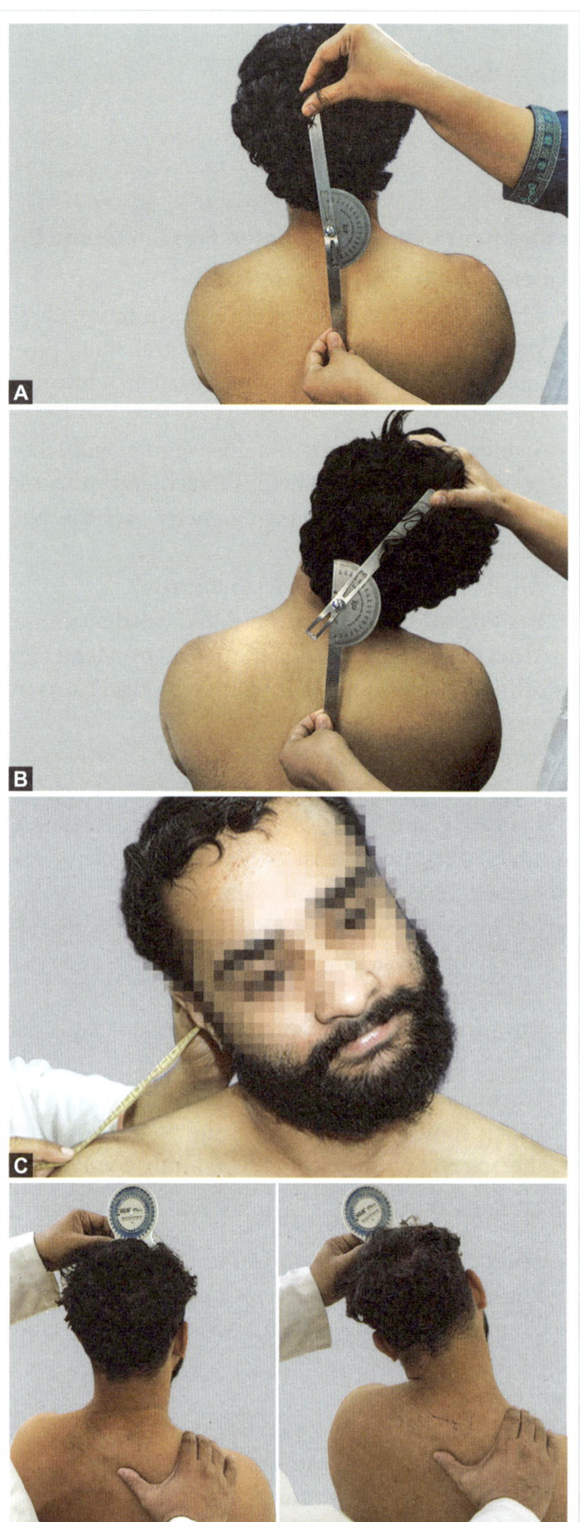

Figs 5.43A to E: Cervical spine lateral flexion: **A.** Goniometer method (starting position); **B.** Goniometer method (final position on the right side); **C.** Tape method (final position on the right side); **D.** Single Inclinometer method (starting position); **E.** Single Inclinometer method (final position on the left side)

- Record the measurement in degrees.
- For enhanced accuracy and reliability, it is advisable to repeat the measurement two or three times and calculate an average.
- Ensure that proper goniometer alignment and patient stabilization are maintained for each repetition.
- Document all measurements and calculate the average, which should be recorded in the patient's medical or therapy record for reference and future comparisons.

Tape Method

1. **Patient positioning:** Hold the individual's head up and to the side (opposite to the motion's direction). Turn the head toward the shoulder. During the move, do not allow the head to twist, forward flex, or stretch (Fig. 5.43C).
2. **Procedure:**
   - Mark the individual's lateral tip of the acromion process and the mastoid process with a skin-marking pencil.
   - At the conclusion of cervical lateral flexion, measure the distance between the two markings.

Single Inclinometer Method (Figs 5.43D and E)

1. Position the inclinometer at the crown of the person's head and adjust the dial so that the bubble is centered at zero.
2. Maintain a firm grip on the inclinometer throughout the movement. Ask the subject to bend the neck sideways.
3. Record the number of degrees indicated on the inclinometer's dial once the motion is complete.

Double Inclinometer Method

1. For this method, place one inclinometer directly over the T1 vertebra's spinous process and adjust its dial to zero the bubble.
2. Simultaneously, position the second inclinometer on the person's head and zero its dial as well. Ask the subject to bend the neck sideways.
3. After completing the movement, note and record the readings from both inclinometers.
4. The range of motion is determined by calculating the difference between the two inclinometer readings.

Cervical Spine Rotation

Plane : Transverse
Axis : Sagittal
ROM : 0°–50°
End feel : Firm.

Measurement Using Goniometer

1. **Patient positioning and stabilization:**
   - Make the patient sit in an upright position with his feet shoulder-width apart and the spine in a neutral alignment.
   - Ensure proper support for the thoracolumbar spine.
   - The therapist should stabilize the shoulder girdle and chest to prevent movement at the thoracic and lumbar spine.
   - Place the fulcrum of the goniometer over the center of the cranial aspect of the head (Fig. 5.44A).
   - Position the fixed arm of the goniometer parallel to an imaginary line drawn between the right and left acromion (shoulder) points.
   - Align the distal arm with the tip of the patient's nose.
2. **Instruct the patient:** Instruct the patient to gently rotate their head by moving his chin toward his shoulder while maintaining an upright posture.
3. **Recording and documentation:**
   - Read the goniometer scale at the point where the stationary and movable arms intersect. This reading indicates the angle of cervical rotation (Fig. 5.44B).
   - Record the measurement in degrees.
   - For accuracy and reliability, it is advisable to repeat the measurement two or three times and calculate an average.
   - Ensure that proper goniometer alignment and patient stabilization are maintained for each repetition.
   - Document all measurements and calculate the average, which should be recorded in the patient's medical or therapy record for reference and future comparisons.

Tape Method

1. **Patient positioning and stabilization:** Ask the subject to look straight ahead and then turn the individual's head to the right as far as feasible without rotating the trunk (Fig. 5.44C).
2. **Procedure:**
   - Begin by marking two points on the person's face with a skin-marking pencil, one at the acromial process and the other at the tip of the chin.
   - The individual should initially look straight ahead and then turn his head as far to the right as possible without rotating their upper body.
   - After the motion is completed, measure the distance between the two marked points.

Single Inclinometer Method

1. Have the person lie in a supine position with his head in a neutral position (Fig. 5.44D).
2. Zero the inclinometer and place it at the center of the subject's forehead.
3. Securely hold the inclinometer while instructing the individual to smoothly rotate his head through its range of motion without tilting it to the left, right, or backward (Figs 5.44E1 and 5.44E2).
4. The inclinometer will provide the range of motion reading once the motion is complete.

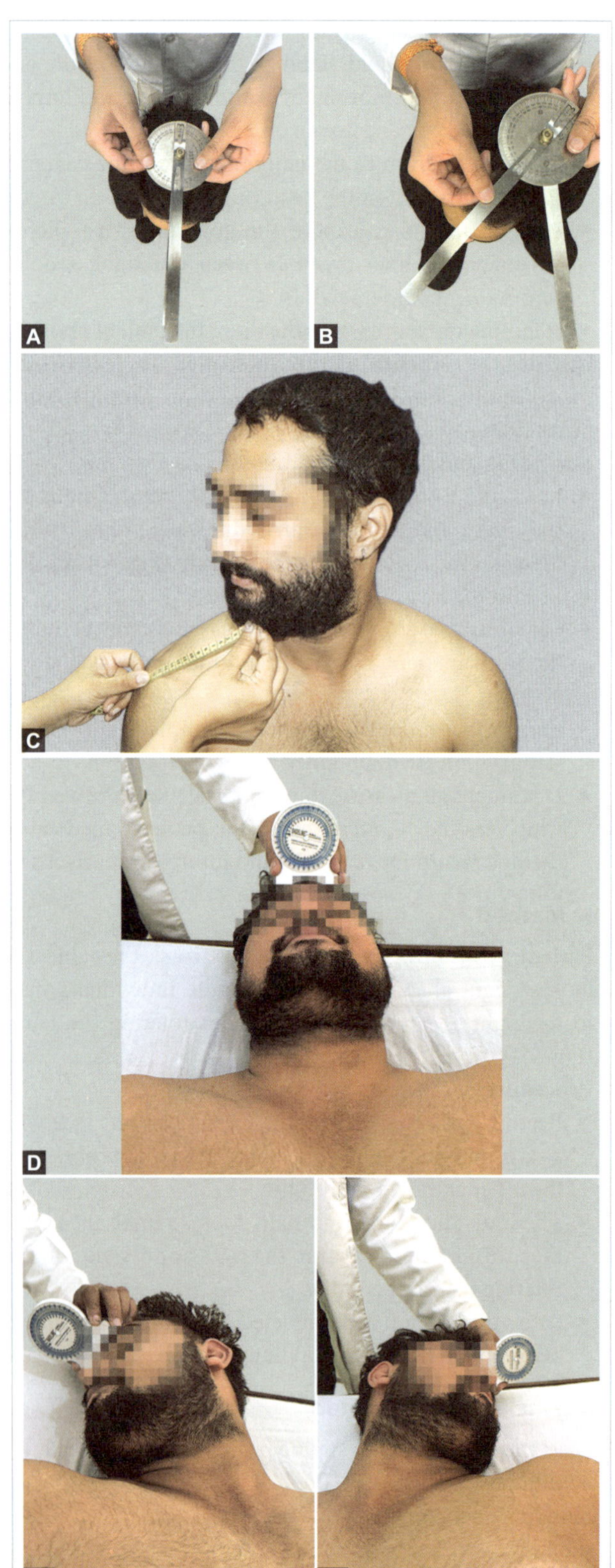

Figs 5.44A to E: Cervical spine rotation: **A.** Goniometer method (starting position); **B.** Goniometer method (final position); **C.** Tape method (final position on the right); **D.** Single Inclinometer method (starting position); **E1 and E2.** Single Inclinometer method (final position-E1 on the right and E2 on the left)

Thoracolumbar Spine Flexion

Plane : Frontal
Axis : Sagittal
ROM : 0° to 50°–60°
End feel : Firm

Average distance between T1 and S2: 10 cm.

Tape Method (Figs 5.45A and B)

1. **Patient positioning and stabilization:**
   - The patient stands with feet shoulder-width apart and maintains a neutral spine position. The therapist ensures pelvic stability to prevent any rotation.
2. **Procedure:**
   - While the person is standing, mark the locations of the T1 and S2 vertebrae's spinous processes using a skin-marking pencil.
   - Place a tape measure between the marked spinous processes (T1 and S2) and record the initial measurement at the beginning of the range of motion.
   - As the person performs the flexion range of motion, keep the tape measure in position and note the measurement at the end of the ROM.
   - The difference between the first and second measurements represents the thoracolumbar flexion range of motion.

Fingertip-to-Floor Method

1. **Patient positioning and stabilization:** In this method, the patient is instructed to bend forward and attempt to touch the ground with his fingertips. The physical therapist uses a standard measuring tape to determine the distance between the floor and the patient's right long finger (Fig. 5.45C1).
2. **Procedure:**
   - Instruct the patient to bend forward and attempt to touch the floor with his fingertips (Fig. 5.45C2).
   - Use a standard measuring tape to measure the distance between the floor and the patient's right long finger.

Double Inclinometer Method

1. The individual should be positioned in a standing posture.
2. Using a skin-marking pencil, mark the locations of the spinous processes of the T1 vertebra and the S2 vertebra (Fig. 5.45D).
3. Then, set up two inclinometers, one positioned above the spinous process of the T1 and the other aligned with the S2 vertebra on the sacrum.
4. After positioning them, reset both inclinometers to zero.

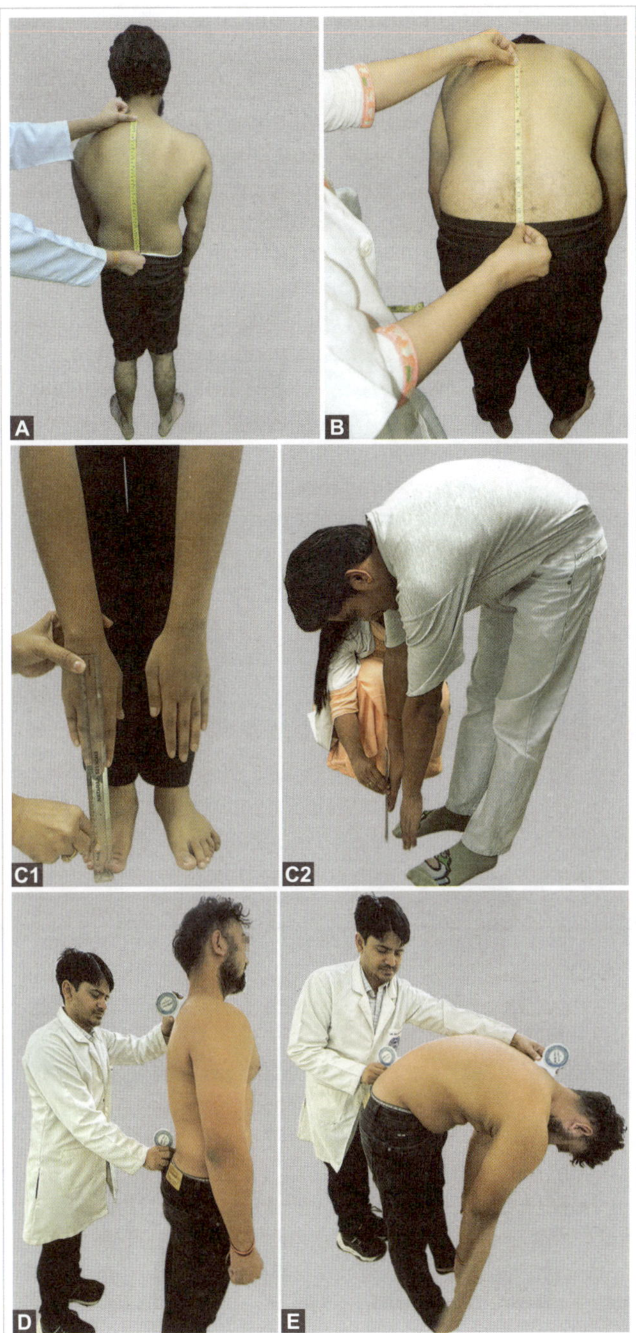

Figs 5.45A to E: Thoracolumbar spine flexion: **A.** Tape method (starting position); **B.** Tape method (final position); **C1 and C2.** Fingertip-to-floor method; **D.** Double Inclinometer method (starting position); **E.** Double Inclinometer method (final position)

5. Ask the subject to bend forward. During the motion, read and record the values displayed on both inclinometers at the end of the movement (Fig. 5.45E).
6. The range of motion for thoracolumbar flexion is determined by calculating the difference between the two inclinometer readings.

Thoracolumbar Spine Extension

Plane : Frontal
Axis : Sagittal
ROM : 0°–25°
End feel : Firm.

Tape Method

1. **Patient positioning and stabilization:** The patient should stand with his feet at shoulder-width apart, maintaining a neutral spine position. The therapist's role is to provide pelvic stabilization to prevent any rotation.
2. **Procedure:**
   - Ask the individual to stand with his feet shoulder-width apart and knees fully extended.
   - Use a skin-marking pencil to mark the locations of the T1 and S2 vertebrae's spinous processes.
   - Measure the distance between these two marked spinous processes using a tape measure and make a note of this measurement (Fig. 5.46A).
   - Instruct the person to gently lean backward, ensuring that the tape measure remains properly aligned throughout the movement.
   - After the range of motion is completed, record the measurement once again (Fig. 5.46B).
   - The amount of thoracolumbar extension is calculated by finding the difference between the initial and final measurements taken during the motion.

Double Inclinometer Method

1. The individual should stand in an upright position.
2. Using a skin-marking pencil, mark the locations of the T1 and S2 vertebrae's spinous processes (Fig. 5.46C).
3. Set up two inclinometers, one positioned above the spinous process of T1 and the other aligned with the S2 level on the sacrum.
4. After positioning him, reset both inclinometers to zero.
5. Ask the subject to bend backward. Upon completing the motion, read and record the values displayed on both inclinometers (Fig. 5.46D).
6. The degree of thoracolumbar extension is determined by calculating the difference between the readings on the two inclinometers at the conclusion of the movement.

Thoracolumbar Spine Lateral Flexion

Plane : Frontal
Axis : Sagittal
ROM : 0°–25°
End feel : Firm

Measurement Using Goniometer (Figs 5.47A and B)

1. Instruct the individual to stand upright. Mark the locations of the T1 and S2 vertebrae's spinous processes with a skin-marking pencil.

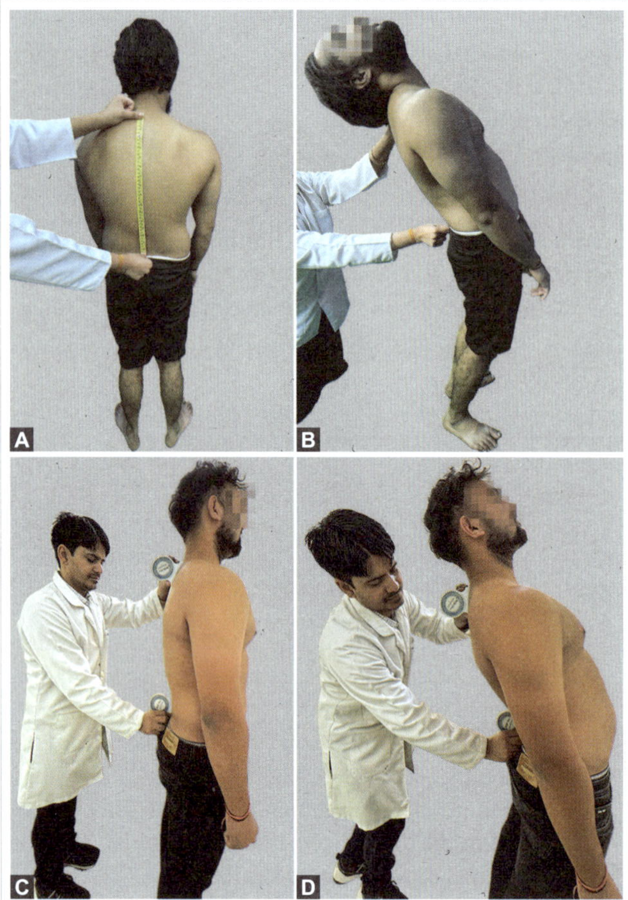

Figs 5.46A to D: Thoracolumbar spine extension: **A.** Tape method (starting position); **B.** Tape method (final position); **C.** Double Inclinometer method (starting position); **D.** Double inclinometer method (final position)

2. Position the fulcrum of the goniometer over the posterior part of S2's spinous process.
3. Align the proximal arm of the goniometer vertically to the ground.
4. Adjust the distal arm to parallel the back of T1's spinous process. Ask the subject to bend to one side as much as possible.
5. Measure the angle, read the goniometer, and record the result.

Tape Method (Figs 5.47C and D)

1. Ask the person to stand with his back against a wall, feet shoulder-width apart, and arms at their sides.
2. Instruct him to bend as far to the side as they comfortably can while maintaining contact between his back and shoulders against the wall, keeping his feet flat on the ground, and his knees extended.
3. At the conclusion of the range of motion, mark the level of the leg using the tip of the middle finger.
4. Then, utilize a ruler or tape to measure the distance between that mark and the floor.

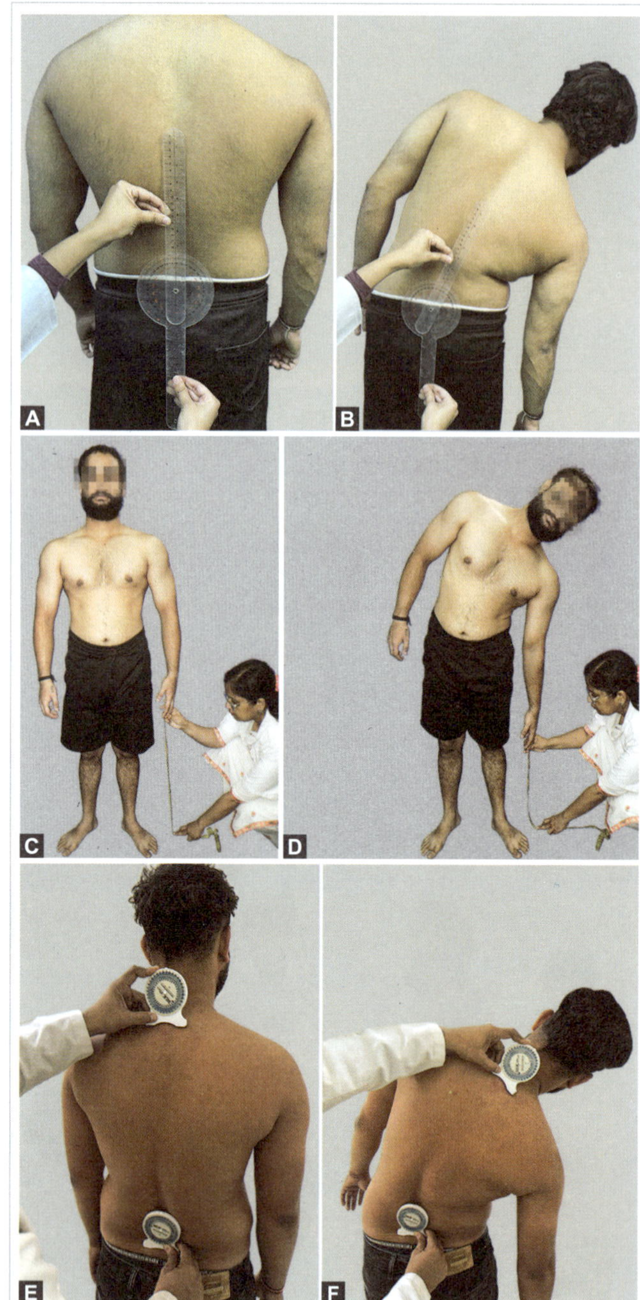

Figs 5.47A to F: Thoracolumbar spine lateral flexion: **A.** Goniometer method (starting position); **B.** Goniometer method (final position); **C.** Tape method (starting position); **D.** Tape method (final position); **E.** Double Inclinometer method (starting position); **F.** Double Inclinometer method (final position)

Fingertip-to-Thigh Method

1. Ask the person to stand with his back against a wall, feet shoulder-width apart, and arms hanging naturally by his sides.
2. Make a mark on the person's thigh at the location where his third fingertip rests.

3. Instruct the person to bend as far to the side as they can while maintaining a straight back, keeping his shoulders against the wall, and his feet flat on the ground with knees fully extended.
4. Use the tip of the middle finger to place a second mark on the leg at the point reached at the conclusion of the range of motion.
5. Measure the distance between the two marks on the leg using a ruler or tape measure.

Double Inclinometer Method (Figs 5.47E and F)

1. With the person standing, use a skin-marking pencil to mark the positions of the T1 and S2 vertebrae's spinous processes.
2. Set up two inclinometers, placing one over the T1 spinous process and the other at the level of the S2 sacrum, and reset both inclinometers to zero (Fig. 5.47E).
3. Ask the person to bend as far to the side as they can while keeping both knees straight and both feet securely planted.
4. Read and record the values displayed on both inclinometers at the conclusion of the range of motion (Fig. 5.47F).
5. To calculate the lateral flexion range of motion, subtract the degrees on the sacral inclinometer from the degrees on the thoracic inclinometer.
6. Repeat the same procedure to assess the range of motion in lateral flexion on the opposite side.

Thoracolumbar Spine Rotation

Plane : Transverse
Axis : Vertical
ROM : 0°–45°
End feel : Firm

Goniometer Method

1. **Patient positioning and stabilization:** Make the patient sit in an unsupported position with feet shoulder-width apart, maintaining a neutral spine.
2. **Procedure:**
   - Use the center of the cranial aspect of the subject's head as the fulcrum for the goniometer.
   - Position the proximal arm parallel to a line drawn between the two tubercles on the iliac crest.
   - Align the distal arm with an imaginary line connecting the two acromion processes.
   - Ask the person to rotate his body as far as possible to one side while keeping an upright posture and both feet flat on the ground.
   - The examiner should be aware of the moment when the pelvis begins to rotate, indicating the end of the motion.

Tape Method (Figs 5.48A and B)

1. **Patient positioning and stabilization:** Ask the subject to sit upright with his arms crossed over his chest and maintain a stable pelvis. Ensure the individual does not lean or bend while performing the movement.
2. **Procedure:** Begin by marking two points with a skin-marking pencil:
   - One point at the lateral acromion process of the shoulder on the side of rotation.
   - The second point at the posterior superior iliac spine (PSIS) of the pelvis on the opposite side.
   - Measure the initial distance between the two points while the subject is in a neutral seated position.
   - Instruct the subject to rotate his torso to the right as far as possible without moving his pelvis.
   - After completing the motion, measure the new distance between the two points.

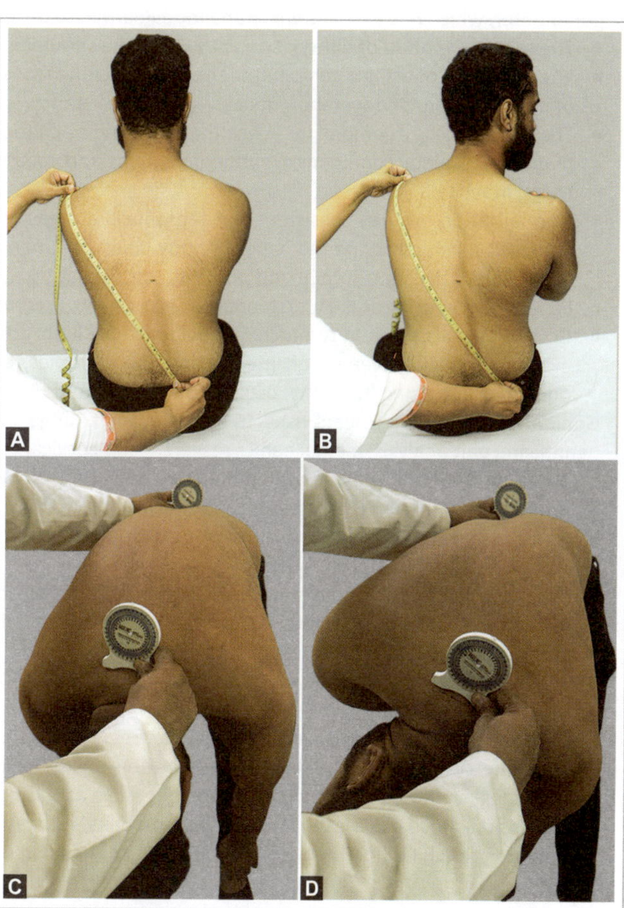

Figs 5.48A to D: Thoracolumbar spine rotation: **A.** Tape method (starting position); **B.** Tape method (final position); **C.** Double Inclinometer method (starting position); **D.** Double inclinometer method (final position)

- Subtract the rotated measurement from the baseline measurement to determine the range of motion.
- Repeat the procedure for the left side.

Double Inclinometer Method

1. Mark the positions of the T1 and S2 vertebrae's spinous processes using a skin-marking pencil.
2. Stand the person up with his feet slightly apart and their back parallel to the ground.
3. Set up two inclinometers, placing one over the sacrum at the level of S2 and the other over the spinous process of T1. Reset both inclinometers to zero (Fig. 5.48C).
4. Ask the person to rotate his trunk as far as he can without extending his arms.
5. The examiner should maintain a firm contact between the inclinometers and the subject's back throughout the motion.
6. After the motion is completed, note the degrees displayed on the inclinometers (Fig. 5.48D).
7. The rotation range of motion is determined by calculating the difference between the readings on the inclinometers.

Clinical Correlation

By adhering to the steps outlined for using a goniometer, healthcare providers can optimize the clinical utility of goniometric assessments and enhance the quality of care delivered to patients undergoing rehabilitation or treatment for musculoskeletal conditions.

SUMMARY

- Goniometry is a technique used in physical therapy and rehabilitation to measure joint angles in the human body. It involves using a goniometer, which has two stationary and mobile arms, to observe and calculate the angle created by the two arms.
  - The range of motion of different joints, such as the shoulder, elbow, wrist, hip, knee, and ankle, can be evaluated using goniometry.
  - The principles of using a goniometer effectively include standardization, anatomical landmarks, joint alignment, stabilization, range of motion, patient cooperation, reliability and consistency, communication, and documentation.
  - The aims of using a goniometer in healthcare and rehabilitation settings include range of motion evaluation, diagnosis and evaluation of musculoskeletal conditions, treatment planning and goal setting, monitoring progress and outcomes, and research and evidence-based practice.
  - Normalization ensures consistent and reproducible measurements, while anatomical landmarks help identify and mark relevant anatomical landmarks on the patient's body. Joint alignment ensures accurate measurement of the joint angle, while stabilization helps isolate the specific joint and prevents compensatory movements.
  - Patient cooperation and understanding are crucial for accurate measurements.
  - Goniometry also aids in research and evidence-based practice, allowing researchers to measure and compare joint movements, evaluate the effectiveness of interventions or therapies, and explore the connection between range of motion and different outcomes, including functional capacity, pain thresholds, and quality of life.
- Goniometers are instruments used to measure joint range of motion, which is crucial for various applications such as research, healthcare, gait analysis, biomechanics research, and cervical goniometry.
  - There are several types of goniometers, including universal, gravity inclinometer, digital, electrogoniometer, finger goniometer, pronation-supination goniometer, and cervical goniometer.
  - Goniometers have validity, which refers to the precision with which they measure joint angles and accurately depict the full range of motion.
  - Reliability refers to the consistency and accuracy of measurements made using a goniometer over time. It can be evaluated through intra-rater reliability, inter-rater reliability, and test-retest reliability.
  - Various methods of goniometry are employed, including active range of motion (AROM), passive range of motion (PROM), active-assisted range of motion (A-AROM), resisted range of motion (RROM), gravity-eliminated range of motion, and composite movements.
  - These methods assess the patient's ability to freely move a joint through its range of motion without help.
  - Goniometry has numerous applications in various fields, including joint flexibility evaluation, rehabilitation and therapy, monitoring musculoskeletal conditions, pre- and postoperative assessment, and physical fitness and sports performance evaluation.
  - It helps diagnose diseases that impair joint mobility, monitor changes in joint range of motion over time, track progress of patients receiving rehabilitation therapies, monitor the progression of musculoskeletal conditions, and guide treatment plans.

FURTHER READINGS

- Johnson, Peter & Jonsson, Per & Hagberg, Mats. (2002). Comparison of measurement accuracy between two wrist goniometer systems during pronation and supination. Journal of electromyography and kinesiology: Official journal of the International Society of Electrophysiological Kinesiology. 12. 413-20. 10.1016/S1050-6411(02)00031-7.
- Tousignant, Michel & Smeesters, Cécil & Breton, Anne-Marie & Breton, Emilie & Corriveau, Hélène. (2006). Criterion Validity Study of the Cervical Range of Motion (CROM) Device for Rotational Range of Motion on Healthy Adults. The Journal of orthopaedic and sports physical therapy. 36. 242-8. 10.2519/jospt.2006.36.4.242.

STUDENT ASSIGNMENT

LONG ANSWER QUESTIONS

1. Discuss the principles, aims, indications, and limitations of goniometry.
2. Explain the technique of goniometry for the shoulder joint.
3. Write about the technique of goniometry for the elbow joint.
4. Discuss the technique of goniometry for the joint.
5. Explain the technique of goniometry for the joint.
6. Write about explain the technique of goniometry for the joint.
7. Explain the technique of goniometry for the lumbar and thoracic spine joint.
8. Discuss explain the technique of goniometry for the ankle joint.
9. Explain the technique of goniometry for the neck joint.

SHORT ANSWER QUESTIONS

1. Write a note on goniometer.
2. What are the indications of using goniometer?
3. Write the principles of goniometry.
4. What are the aims of goniometry?
5. Write a note on inclinometer.

MULTIPLE CHOICE QUESTIONS

1. **Which of the following joints is commonly assessed using goniometry?**
   a. Axial joint
   b. Synovial joint
   c. Fibrous joint
   d. Cartilaginous joint
2. **During goniometric measurement of joint ROM, what is the starting position typically referred to?**
   a. Neutral position
   b. Resting position
   c. End range
   d. Maximum resistance
3. **Which of the following techniques can help ensure accurate goniometric measurements?**
   a. Stabilizing the proximal joint segment
   b. Using consistent anatomical landmarks
   c. Avoiding measurement over clothing
   d. All of the above
4. **In goniometric measurement, which term refers to the quality of sensation when joint reaches its limit?**
   a. End feel
   b. Capsular pattern
   c. Resting position
   d. Starting angle

ANSWER KEY

1. b **2.** a **3.** d **4.** a

6 Assessment of Neuromuscular Efficiency

Sheetal Kalra, Archana Khanna

LEARNING OBJECTIVES

After the completion of the chapter, the readers will be able to:

- Define neuromuscular efficiency.
- Explain electrical tests for neuromuscular efficiency.
- Explain anthropometric measurements like limb circumferences, body composition and adiposity.
- Explain static and dynamic muscle power assessment.
- Describe the tests for coordination.
- Describe the sensory examination.

CHAPTER OUTLINE

- Introduction
- Tests of Anthropometry
- Muscle Strength/Power Assessment
- Sensory Assessment
- Tests for Incoordination
- Electrical Tests

KEY TERMS

Anthropometry: Analysis of body length, width, circumference, and skinfold thickness which provides a simple, reliable and accurate way to calculate body size and proportions.

Muscle power: It is the result of force multiplied by velocity, or speed of contraction of a muscle.

Neuromuscular incoordination: Neuromuscular incoordination, which is frequently observed in diseases, like cerebral palsy, multiple sclerosis, or muscular dystrophy, etc. is characterized by poor synchronization between the neurological system and muscles, impairing movement control.

INTRODUCTION

Neuromuscular efficiency (NME) denotes the effective collaboration between the nervous system and muscles to produce precise, coordinated, and powerful movements. This process involves communication among the brain, nerves, and muscles, ensuring the correct recruitment, sequence, and force of muscle contractions.

A strength and conditioning principle, known as neuromuscular efficiency, describes the capacity to recruit muscles to produce power or carry out a task. The ability of the central nervous system (CNS) to activate muscles more or less effectively determines whether a person has a greater or lower NME. According to the National Academy of Sports Medicine, neuromuscular efficiency characterizes how effectively the neural system coordinates muscle recruitment to produce, control, and stabilize the body across many planes of motion. It involves the exact activation of muscles in three-dimensional movements to produce, manage, and stabilize forces.

A person with a high NME can efficiently activate his CNS and exert a lot of force during movement. A high NME athlete such as a sprinter, is an excellent example.

Neuromuscular efficiency tests may be performed by electrical, manual or mechanical methods.

This chapter provides a detailed exploration of neuromuscular efficiency, covering its critical components to offer an in-depth understanding of this important concept. It begins with an overview of electromyography, which assesses muscle activation, including motor unit recruitment and the timing of muscle coordination. Strength production is examined through tools like the strength-duration curve, while electrical efficiency is evaluated using advanced methods such as the H-reflex and evoked potentials, including motor and somatosensory evoked potentials. The chapter also highlights the role of anthropometry in neuromuscular assessment, providing data on body dimensions and composition that influence performance. Measurements like limb circumferences and segment lengths are linked to muscle strength, joint leverage, and movement efficiency, while body composition metrics indicate fat distribution and its effect on muscle function. Muscle strength/power assessment is included to evaluate functional capabilities, and neuromuscular coordination is analyzed through reflex responses and skill-based tests. Finally, the chapter examines joint stability and proprioception, emphasizing their importance in maintaining balance, posture, and movement.

Did You Know?

Neuromuscular efficiency plays a crucial role not only in athletic performance but also in everyday activities. Whether lifting groceries, climbing stairs, or simply maintaining good posture, the neuromuscular system is constantly at work to ensure efficient muscle activation and movement control. It is like the behind-the-scenes manager of the body's movements, optimizing coordination and stability across various planes of motion.

TESTS OF ANTHROPOMETRY

Measurements of body size, structure, and composition are part of anthropometric tests. The main components of anthropometry include height, weight, waist, hip, and limb circumferences to measure adiposity, as well as skinfold thickness.

Circumference or Girth Measurement

Girths are measurements of the circumference at typical anatomical places on the body. It is useful for determining body composition and size, as well as for tracking changes in these parameters.

An inextensible tape measure is used to determine the circumference of a limb at a specific place. For girth measurement, mark the measurement site first. Generally, it is the right side of the body where the girth measurement is done but comparison should be done on both sides.

- The limb should be relaxed and free from clothing that might interfere with the measurement.
- Identify the area by using anatomical landmarks to determine the measurement site.
- Common sites include thigh, calf, and upper arm.
- Maintain appropriate tape tension—not too tight or too loose—to ensure it lays flat on the skin and remains horizontally positioned for accurate measurements.

Various Areas for Girth Measurement

- **Head girth:** It is the greatest measurement obtained around the head with a tape measure at the back of the head and above the brows (Fig. 6.1A).
- **Neck girth:** Tape is placed above the larynx, perpendicular to the neck's long axis (Fig. 6.1B).
- **Upper limb girth**
  - **Mid-arm girth:** This is the circumference of the upper arm, also known as the relaxed arm, or biceps circumference (Fig. 6.1C).

- **Forearm girth:**
  - Measure the maximal circumference of the forearm in the upper third.
  - Straighten the arm and slightly flex the elbow, measure the circumference around the fullest part of the forearm.
  - Repeat to ensure accuracy (Fig. 6.1D).
- **Wrist girth:** Measure the circumference of the wrist from the styloid processes of the ulna and radius distally (Fig. 6.1E).
- **Palm dimensions:**
  - Hand length is measured from the palm crease to the tip of the longest finger (Fig. 6.1F).
  - The width across the widest point, where the fingers and palm meet.
  - Circumference of the hand is measured around the dominant hand's palm (excluding the thumb) just below the knuckles (Fig. 6.1G).

- **Chest girth:** Measure around the chest, ensuring the individual sits or stands straight and no clothing is blocking the measurement, passing over the xiphoid process and under the axilla (Fig. 6.1H).
- **Waist girth:** The narrowest point of the body, halfway between the iliac crest and the 12th rib, is where the measurement is taken. The person must not be contracting any muscles and should stand comfortably upright with hands by sides.
- **Abdominal girth:** Measures the circumference of the torso at the umbilicus level. Contracting the abdominal muscles is not advised (Fig. 6.1I).
- **Lower limb girth**
  - **Gluteal girth:** Subject stands with their feet together, wearing only the barest minimum of clothing. The soft tissue is not compressed by the placement of tape on top of the garment. It is the area around the gluteals' larger posterior protuberance. It is sometimes referred as the hip or buttock circumference (Fig. 6.1J).
  - **Proximal/upper thigh girth:** Measure horizontally along the femur's long axis about 1 cm below the gluteal fold (Fig. 6.1K).
  - **Mid-thigh girth:** Distance between the inguinal crease and the upper margin of the patella.
  - **Distal thigh girth:** Measure from the femoral condyles immediately above the knee joint.
  - **Knee girth:** For joint swelling, measure at the patella level, (Fig. 6.1L) and 5 cm above and below for muscle atrophy.
  - **Calf girth:** Measure around the thickest part of the calf (Fig. 6.1M).
  - **Ankle girth:** The smallest circumference on the lower leg above the malleoli (Fig. 6.1N).
  - **Foot girth:** The length of the foot from the top of the foot, around and over the highest point of the foot arch, and back to the top of the foot (Fig. 6.1O).

Various sites for girth measurement are shown in Figures 6.1A to O.

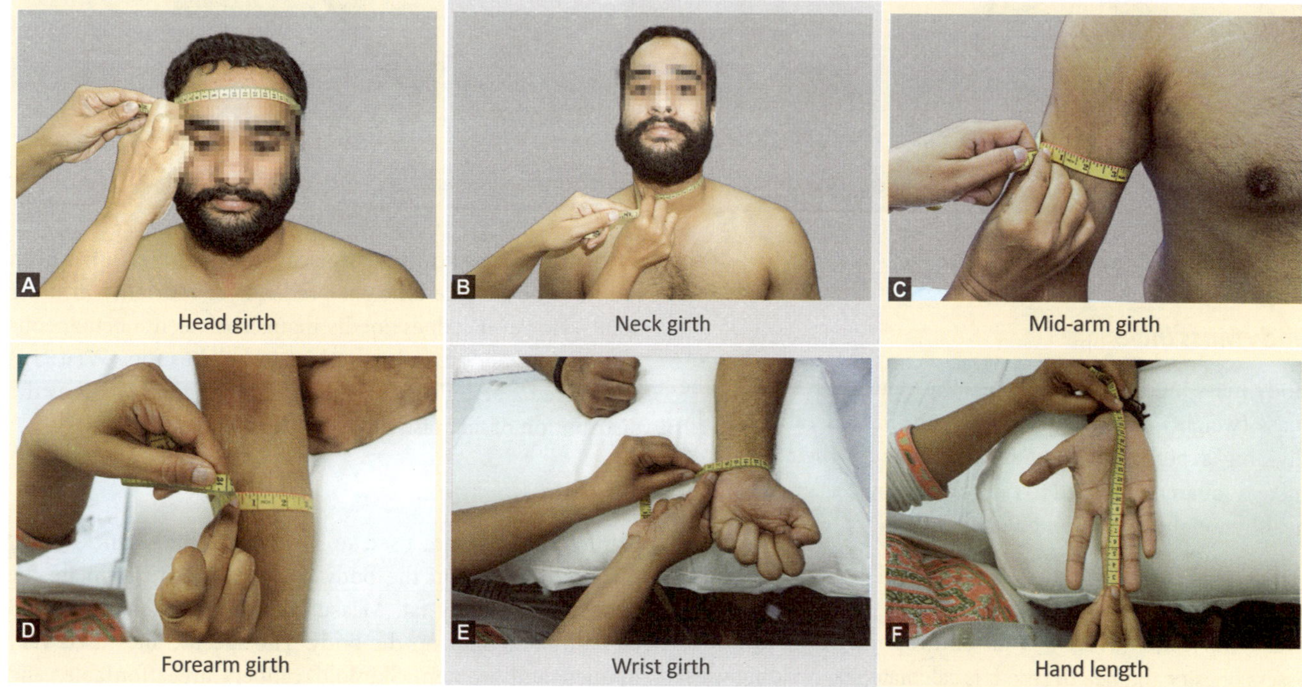

A Head girth; B Neck girth; C Mid-arm girth; D Forearm girth; E Wrist girth; F Hand length

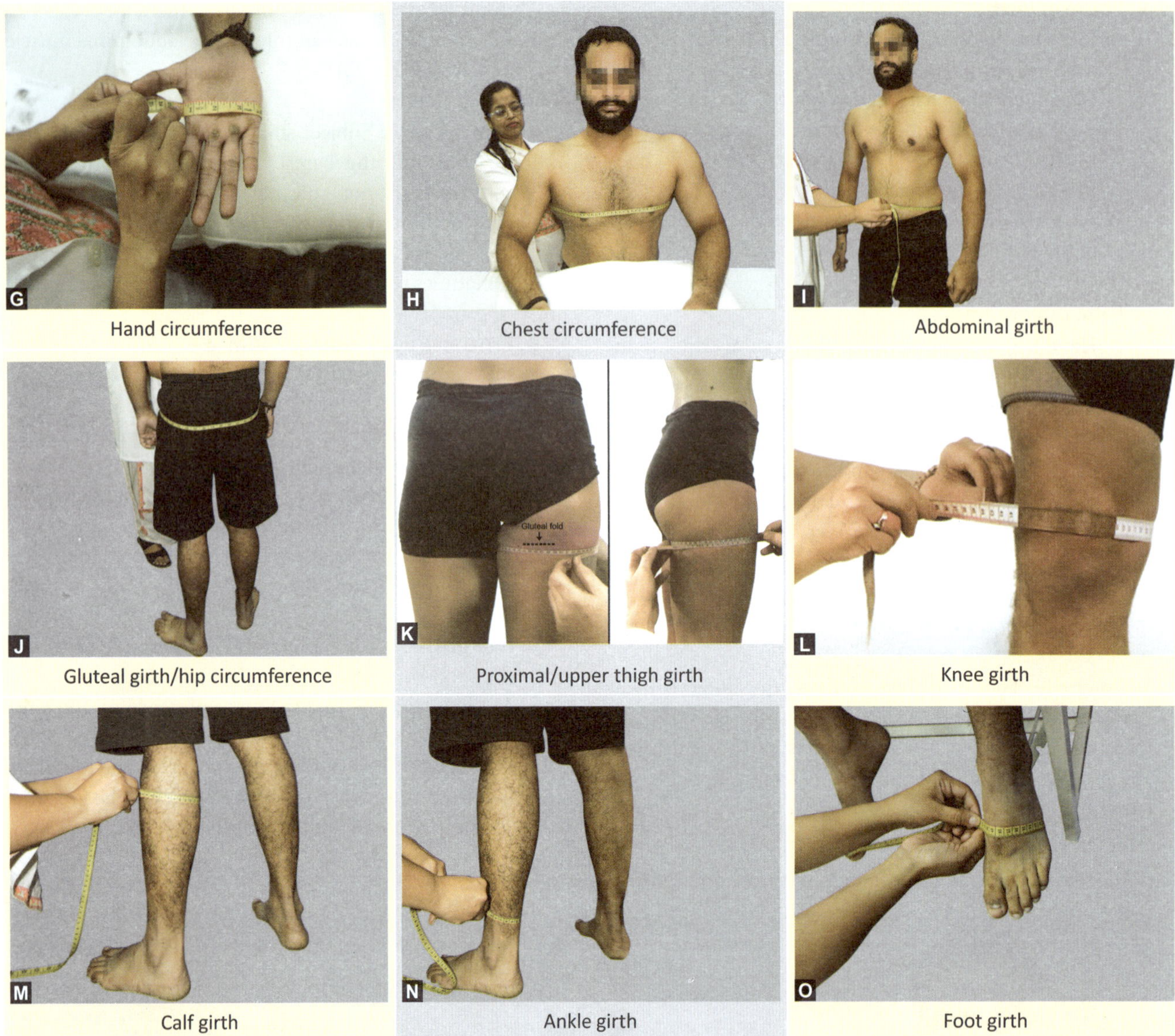

Figs 6.1A to O: Various sites for girth measurements

Other Measuresment Methods

Body Mass Index

Body mass index (BMI) is a numeric value derived from the mass (weight) and height of a person. BMI is calculated by dividing a person's weight in kilograms by their height in meters squared. It is used to categorize people as underweight, healthy weight, overweight and obese.

Waist Hip Ratio

Waist hip ratio (WHR) is an anthropometric measure to assess obesity and health risk. It is calculated by dividing waist circumference by hip circumference measured in the same units. WHR is a more accurate predictive indicator that BMI for the risk of cardiovascular disease and diabetes. It has been found to be a more efficient predictor of mortality in older adults. However, it does not distinguish between subcutaneous and visceral abdominal fat. Waist to height ratio (WHtR) is considered a better predictor of health risk than WHR as the distribution of fat mass varies between individuals.

Skinfold Thickness

Skinfold calipers are used to measure skinfold thickness at certain locations on the body and are used to estimate the percentage of body fat. Measurements are taken at either 3 or 7 different sites on the body. The specific sites used vary in men and women. Skinfold calipers are affordable, and measurements can be taken quickly. They can be used at home and are also portable.

Body Fat Percentage

There are many ways to assess body fat besides BMI, circumference measurements, WHR and skin fold thickness. Assessing body fat can be done by using dual-energy X-ray absorptiometry, hydrostatic weighing and bioelectrical impedance methods.

- **Dual-Energy X-ray Absorptiometry (DEXA):** Provides accurate and detailed information, including a breakdown of different body regions and bone density readings. However, it is often unavailable to the general population, as it is fairly expensive and not feasible for regular testing.
- **Hydrostatic weighing:** Also known as underwater weighing or hydrodensitometry, estimates body composition based on its density. It is accurate and relatively quick.
- **Bioelectrical Impedance Analysis (BIA):** The device works by sending small electrical currents through the body to see how easily they travel through the tissues. The electrical currents move through muscle easier than fat due to the higher water content present in the muscle.

MUSCLE STRENGTH/POWER ASSESSMENT

Tests for Static Muscle Strength/Power

Static tests focus on muscle strength or endurance in a stationary position, without significant movement, such as during isometric contractions.

Isometric Squat Hold

Assesses static lower body muscle endurance by having the participant hold a squat position for as long as possible at a set knee angle (e.g., 90°).

Isometric Leg Press

The participant presses against a leg press machine in a fixed position, and the maximum force exerted is recorded, evaluating leg muscle strength without movement.

Plank Hold

Measures core strength and endurance by having the participant hold a plank position for as long as possible, focusing on the static stabilization of the body.

Group Strength Test Using Handheld Dynamometer

A handheld dynamometer is extensively used nowadays for quantifying the strength of the muscle. The dynamometer measures the force that the muscle produces during maximal contraction. It can be used for different muscles of upper limb, lower limb, and spine (Figs 6.2A and B).

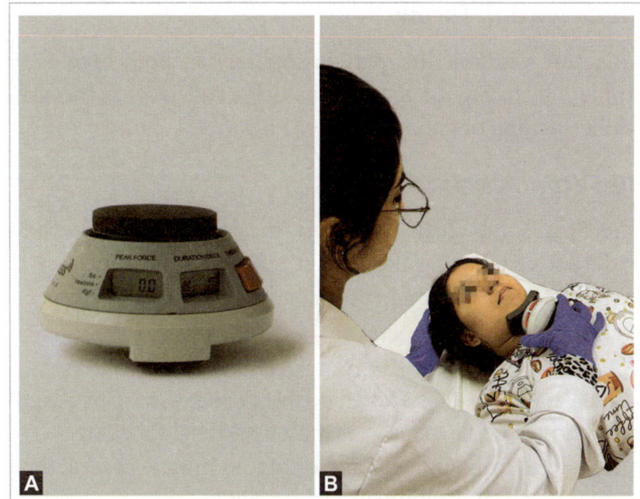

Figs 6.2A and B: Handheld digital (myometer) dynamometer and its use

Pinch Strength Testing

This method uses a small handheld pinch strength tester. The instrument is held between the subject's fingers and thumb, and they pinch it with maximum effort.

Tests for Dynamic Muscle Strength/Power

One Repetition Maximum Testing

This is the method for assessing dynamic muscle strength. The technique involves the subject performing a maximum effort of a specific weight for one repetition. The weight is progressively increased until the subject can no longer lift it with proper form.

Vertical Jump Testing

This is one of the famous tests used for the assessment of dynamic muscle power of the lower limb muscles. This technique involves subjects performing a jump with maximum efforts and touching the wall to the maximum height. The height of the jump is then measured using a measuring tape.

Standing Long Jump Test

In this method, the subject jumps horizontally using the maximum power of their lower limb muscles. The distance of the jump is measured using a measuring tape.

Wingate Anaerobic Test

This 30-second all-out cycling test evaluates muscle endurance and power output during high-intensity cycling in order to determine peak anaerobic power.

Medicine Ball Throw

This test assesses the participant's upper body explosive strength by having them toss a medicine ball as far as they can while standing or sitting.

30-Second Chair Stand Test

The 30-second chair stand test measures leg power and endurance by having the subject stand up and sit down as many times as they can in 30 seconds.

Manual Muscle Testing

Manual muscle testing is a method that evaluates muscle strength on a six-point scale to measure musculoskeletal function.

- It ranges from Grade 0, indicating no muscle contraction, to Grade 5, indicating normal strength.
- Each grade represents varying degrees of muscle response, from a single flicker (Grade 1) to movement against gravity (Grade 3), and finally strong movements against resistance (Grade 4).
- This classification system provides therapists with a standardized method to measure and document muscular strength, assess levels of disability, and monitor progress in rehabilitation or treatment plans.
- Accurate assessments rely on the examiner applying resistance to specific muscles or muscle groups while the patient contracts them, with proper technique and adherence to protocols ensuring precision and reliability (Fig. 6.3).

Clinical Correlation

The six-point muscle strength scale provides therapists with a standardized method for assessing and documenting muscular function. This scale, ranging from Grade 0 to Grade 5, allows for precise evaluation of muscle contraction levels, aiding in the determination of severity and disability of the monitoring of rehabilitation progress.

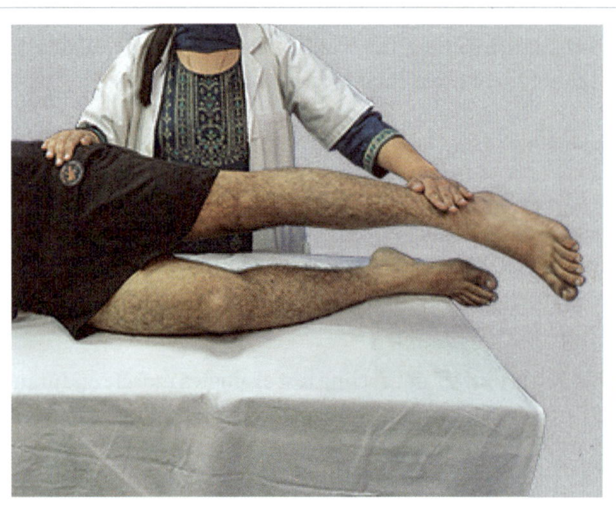

Fig. 6.3: Manual muscle testing of the hip abductors

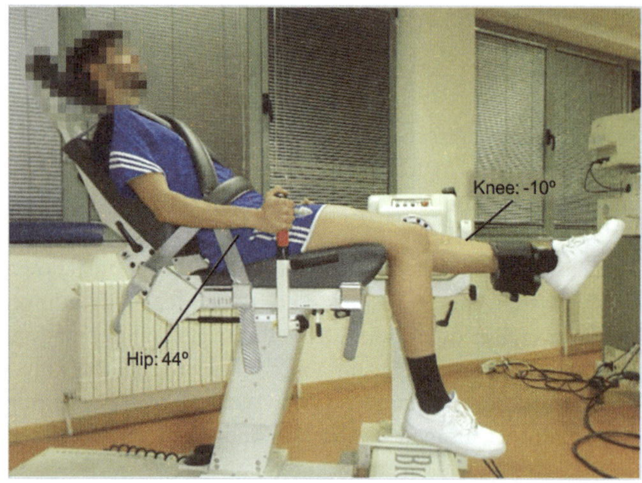

Fig. 6.4: Isokinetic testing

Isokinetic Testing

Isokinetic testing is a technique that involves measurement of muscle strength during a concentric muscle contraction at a constant speed (Fig. 6.4).

SENSORY ASSESSMENT

Different Types of Sensations

Superficial Sensations

The surface-level sensations come from exteroceptors. They are stimulated by the environment outside through the skin and subcutaneous tissue. The experience of pain, warmth, light touch, and pressure is mediated by exteroceptors.

Deep Sensations

Proprioceptors are important for receiving input from muscles, tendons, and joints, providing critical information for position sense, joint awareness, and movement perception (kinesthesia).

Combined Cortical Sensations

These are a synthesis of superficial and deep sensory systems, requiring input from exteroceptive and proprioceptive receptors, as well as intact cortical sensory association areas. Stereognosis, two-point discrimination, barognosis, graphesthesia, tactile localization, texture identification, and double simultaneous stimulation all require coordinated inputs for perception, integrating information from various sensory pathways and cortical areas for accurate interpretation and recognition.

Examples of different types of sensations have been depicted in Table 6.1.

TABLE 6.1: Examples of different types of sensations

Superficial sensation	Deep sensation	Combined cortical sensation
• Pain perception • Temperature awareness • Touch awareness • Pressure perception	• Proprioception • Kinesthesia Awareness • Vibration Perception	• Stereognosis • Tactile localization • Two-Point Discrimination • Double simultaneous Stimulation • Graphesthesia • Recognition of texture • Barognosis

Superficial Sensation Testing

Pain

Pain evaluates the ability to differentiate sharp from dull sensations, reflecting spinothalamic tract function.

Patient Instructions

- "We will be examining your sensitivity to pain using a sharp object"
- "I will lightly press a sharp tool against various parts of your body. Please let me know if you feel the pain stimulus" (Fig. 6.5).

Steps for Testing

1. Apply the sharp point to the skin's surface gently, without applying too much pressure. To avoid penetrating the skin, use a safety pin or a splintered wooden tongue depressor with sharp edges.
2. Examine various locations from proximal to distal regions, including the shoulders, arms, and legs, on both sides of the body.
3. Instruct the patient to signal if they feel the pain during the assessment.
4. To avoid accidental damage or harm, dispose of the instrument carefully after use.
5. Ensure a complete evaluation of pain sensitivity throughout the body regions.
6. Additional testing may be needed in any area where the patient reports sensory loss or hypersensitivity. Patients with reduced superficial pain perception may report a dull or nonexistent response to the stimuli.

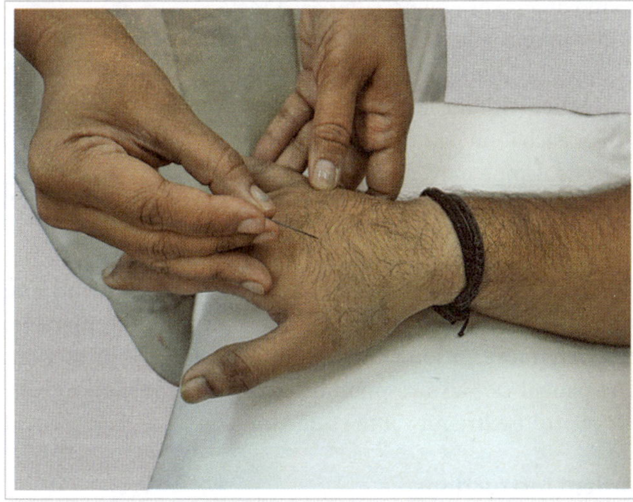

Fig. 6.5: Testing for pain

Temperature

Temperature measures the ability to sense warm and cold stimuli, assessing the spinothalamic pathway.

Patient Instructions

- "We will be testing your ability to perceive warmth and coolness on your skin"
- "I will use test tubes containing warm and cold water to touch different areas of your skin. Please let me know if you feel warm or cold" (Fig. 6.6).

Steps for Testing

1. Prepare two test tubes, one with warm water and one with ice water, with temperatures ranging from 41°F (5°C) to 50°F (10°C) for cold and 104°F (40°C) to 113°F (45°C) for warmth.
2. To elicit appropriate reactions, make sure the sides of the test tubes, not only the distal ends, make contact with the skin.
3. Apply the test tubes to various skin surfaces at random to ensure even coverage.
4. After each application, ask the patient whether they feel warmth or coolness.
5. Take precautions to prevent temperatures that might induce pain, as this could compromise the accuracy of the test.

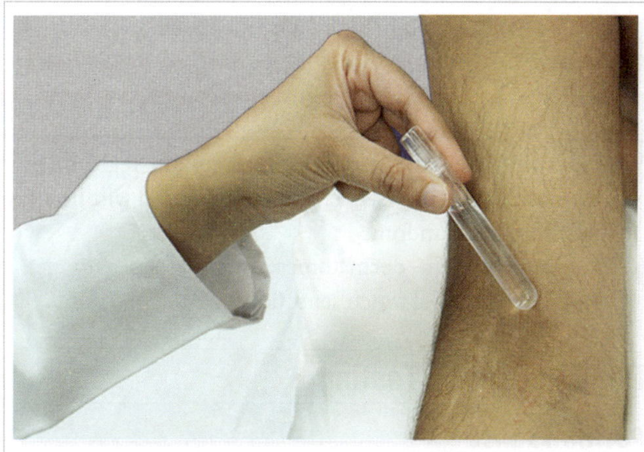

Fig. 6.6: Testing for temperature

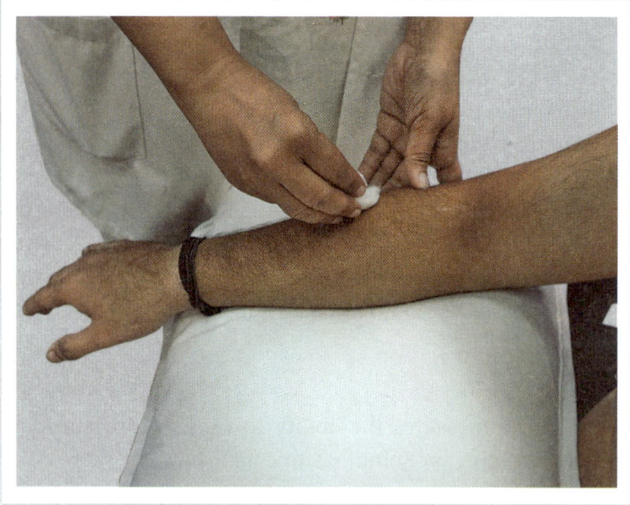

Fig. 6.7: Testing for light touch

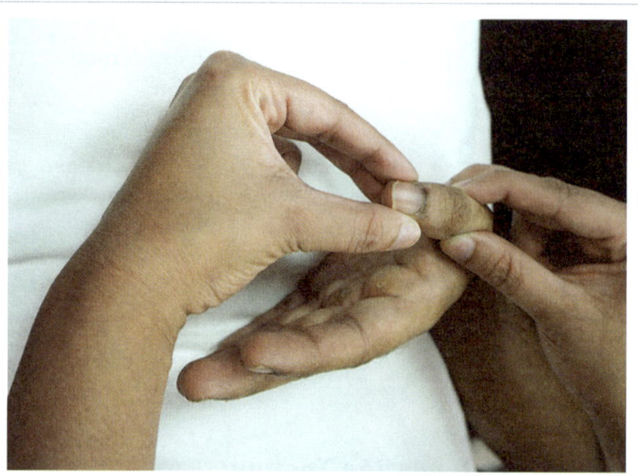

Fig. 6.8: Position sense testing in upper limb

Light Touch

Light touch assesses the ability to perceive gentle tactile stimuli on the skin, testing the integrity of fine touch pathways.

Patient Instructions

- "We will be conducting a touch sensation test using a soft cotton wisp."
- "I will apply a gentle touch to different areas of your body. Please let me know if you feel anything, and if it feels similar or different from a reference point such as the sternum or forehead" (Fig. 6.7).

Steps for Testing

1. Show the cotton wisp to the patient to acquaint them with the testing object.
2. Apply the cotton swab to several dermatomal locations and ask the patient if they feel the stimulus and if it matches the reference sensation on either the sternum or forehead.
3. To generate accurate sensory responses, avoid stroking or tickling, and instead focus on a gentle touch.
4. Caudocranial testing should be performed by comparing the left and right sides while ensuring consistency between the locations evaluated.
5. To confirm that the patient's reactions are real and not automatic affirmations, alternate applying and withholding the stimulus at random.
6. Examine different dermatomes on both sides to get a complete picture of touch sensation.

Deep Sensation Testing

Position Sense

Test the patient's ability to recognize the movement direction as the therapist moves the part up or down while keeping the patient's eyes closed. It is crucial to hold the digit on both sides rather than the top and bottom when moving it. This is because, even if proprioception is compromised, pressure sensation on the top or bottom of the digit could aid the patient in determining position awareness. Repeat the test multiple times on both sides, using the thumb for the upper extremities. If an aberrant response is obtained, examine a more proximal joint (such as the wrist or ankle) (Fig. 6.8).

Patient Instructions

- "While you are aware of your joint movements, let us check if you can perceive the specific direction the joint is moving."
- "I will be slowly moving your thumb or big toe. I want you to pay attention to whether it is moving up or down-up being flexion and down being extension."

Steps for Testing

1. Slowly flex and extend the thumb or toe, asking "is that moving up or down?" after each move.
2. Move one side as the patient copies the position with the opposite limb (e.g., move the right toe while the patient mimics with the left) (Fig. 6.9).
3. For each joint pair (ankle/wrist, knee/elbow, hip/shoulder), evaluate one movement plane.

Kinesthesia

Kinesthesia is the term for the awareness of movement. The patient is asked to describe the movement as the therapist passively moves a joint through a modest range of motion. The patient can also respond by replicating the motion with the opposite extremity at the same time.

Patient Instructions

- "I am going to see if you can detect movement in your joint without looking at it."

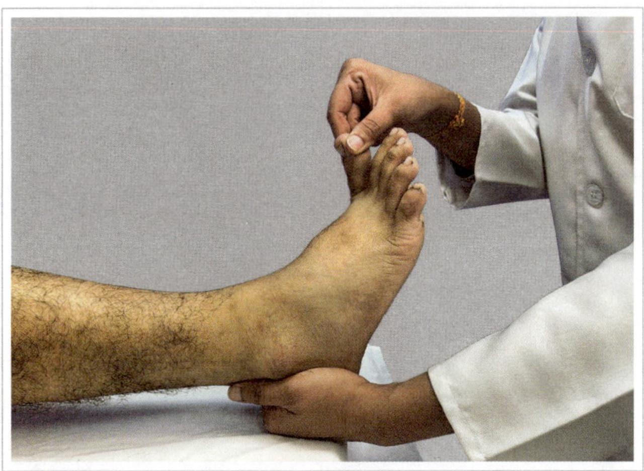

Fig. 6.9: Position sense testing for the lower limb

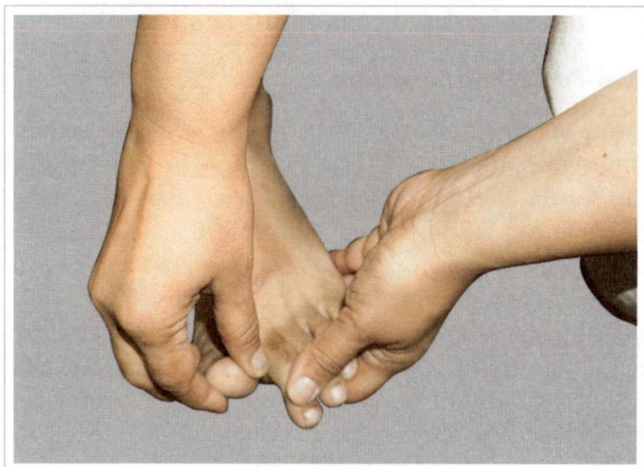

Fig. 6.10: Testing for kinesthesia

- "I will move your big toe-thumb, and I want you to let me know when it moves" (Fig. 6.10).
- Show the big toe's (flexion/extension) movement on the metatarsophalangeal joint (MTPJ) or the thumb's movement on carpometacarpal (CMC) joint.
- "Please close your eyes."

Steps for Testing

1. Slowly flex and extend the toe or thumb at arbitrary intervals.
2. Check the patients ability to correctly identify the movement without visual input.
3. Only one plane of movement needs to be evaluated for each joint when testing the ankle/wrist, knee/elbow, and hip/shoulder.

Did You Know?

To avoid providing the patient with additional sensory clues, apply a gentle grip to the medial and lateral portions of the proximal phalanx.

Vibration

Vibration tests deep sensation using a tuning fork to detect vibration on bony prominences, involving the dorsal column pathway.

Patient Instructions

- "During this test, I will apply a vibrating tuning fork to specific areas on your body" (Fig. 6.11).
- "I want you to tell me if and when you feel the vibration and whether it is strong or faint."

Steps for Testing

1. Apply the vibrating tuning fork to certain regions, asking the patient to signal when they feel the vibration and explain how strong it is.
2. The tuning fork can be used to assess various places on the body, such as bony prominences or nerve pathways.
3. Take note of the patient's reactions to various vibration intensities and places.
4. Repeat the process for a thorough review, ensuring that all anatomical regions are assessed.
5. Check the patient ability to feel the vibration.
6. The vibratory feeling in the feet and ankles may deteriorate or disappear with age. When compared to pain perception, a decrease in vibratory sensation and proprioception often indicates dorsal column illness.

   Clinical examinations typically reveals sensory abnormalities in conditions such as tabes dorsalis, vitamin B_{12} deficiency, multiple sclerosis, or demyelinating neuropathies.

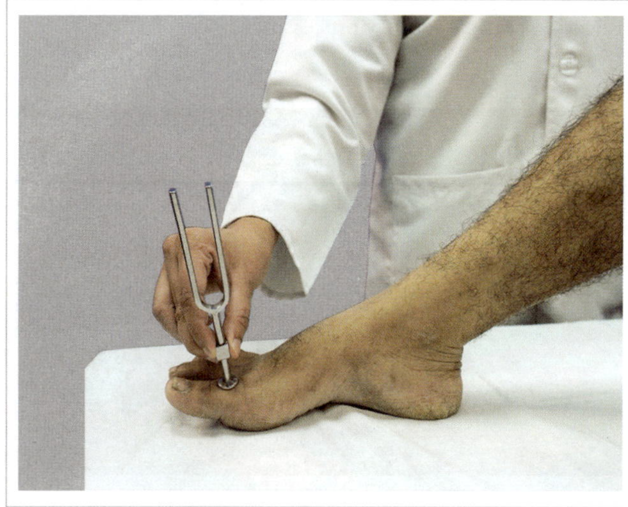

Fig. 6.11: Testing for vibration

Combined Cortical Sensation Testing

Two-Point Discrimination

Two-point discrimination establishes the ability to perceive two spots on the skin at the same time. The tools used for this test include an aesthesiometer or a circular two-point discriminator. The two-point discrimination test measures the capability to perceive two stimuli applied to the skin simultaneously, determining the minimum distance at which the stimuli are discerned as separate. This test's sensitivity reflects the density of skin innervation and the representation within the somatosensory cortex, with sensory information transmitted *via* the posterior column-medial lemniscus pathway to the central nervous system.

Patient Instructions

- "We will assess your ability to discern between two distinct points of touch on your skin."
- "Using an aesthesiometer or a circular two-point discriminator, two tips will make simultaneous contact with your skin. Please let me know whether you feel one or two points" (Fig. 6.12).

Steps for Testing

1. To apply two points on the skin at the same time, use an aesthesiometer or a circular two-point discriminator.
2. After each application, gradually bring the two points closer together until the patient perceives them as a single point.
3. Calculate the shortest distance between the points that are still recognized as two separate points which will indicate the patient's tactile discriminating ability.

Stereognosis

Stereognosis tests the ability to recognize objects by touch, indicating cortical sensory processing.

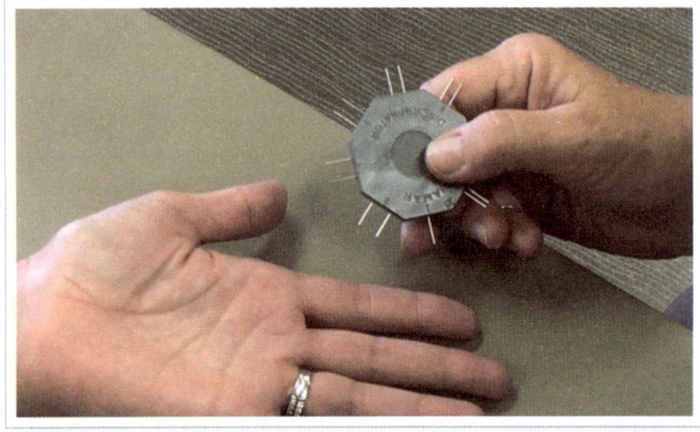

Fig. 6.12: Two point discrimination using aesthesiometer

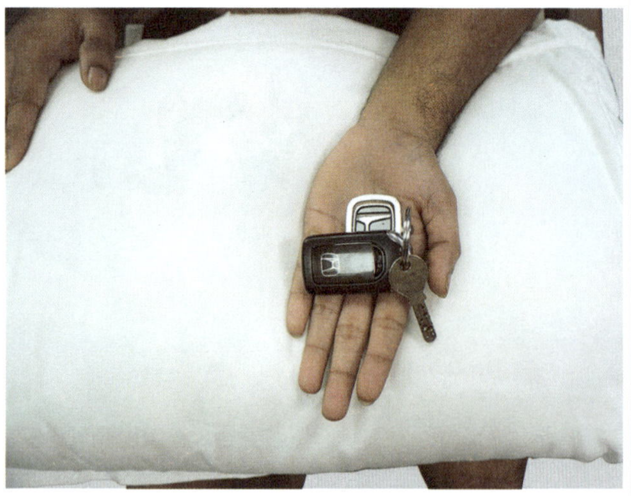

Fig. 6.13: Testing for stereognosis

Patient Instructions

- "We will assess your ability to recognize objects by touch."
- "I will give you various objects such as keys, coins, or combs."
- "Without looking, your duty is to feel and describe what you are holding."

Steps for Testing

1. Provide a variety of everyday objects in varied sizes and forms (keys, money, combs, etc.).
2. Instruct the patient to investigate each object by touch and orally describe it without the use of a visual aid (Fig. 6.13).

Tactile Localization

The capacity to localize touch sensation on the skin is tested during the test. This test is conducted concurrently with either pressure perception or touch awareness, not independently.

Patient Instructions

- "We will assess your ability to pinpoint where you feel touch on your skin."
- "While experiencing touch or pressure, your task is to identify and indicate the exact location where you feel it."

Steps for Testing

1. Prompt the patient to identify and specify the particular spot on his skin where he senses the touch sensation while doing pressure perception or touch awareness evaluations.
2. Ascertain that the patient can appropriately localize and enunciate the location of the touch or pressure stimulation (Figs 6.14A and B).

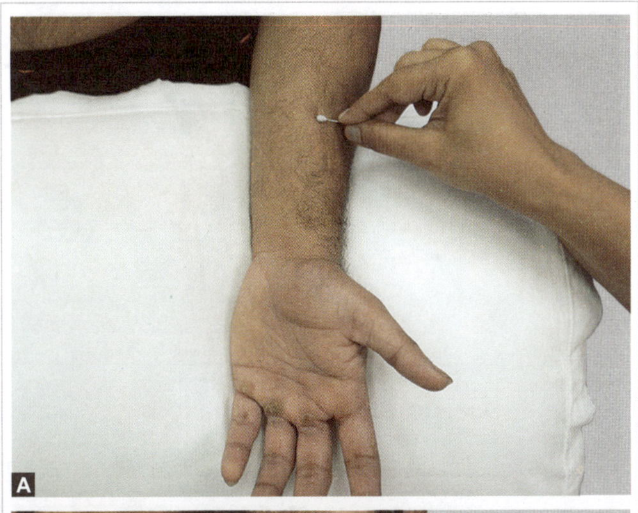

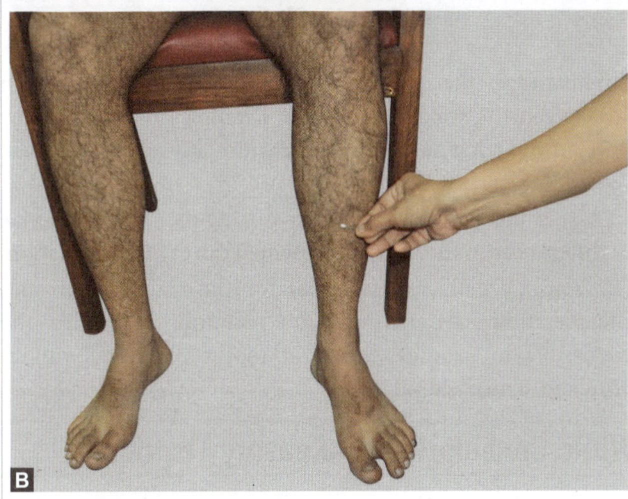

Figs 6.14A and B: Testing for tactile localization

Double Simultaneous Stimulation

Double Simultaneous Stimulation (DSS) tests the capacity to perceive a touch stimuli simultaneously on both sides of the body, on one extremity proximally and distally or on one side of the body.

Patient Instructions

- "We will evaluate your ability to perceive touch sensations simultaneously on different parts of your body or extremities."
- "You may experience touch sensations on both sides or in different areas at the same time." Please respond when you experience these touches."

This test evaluates simultaneous touch perception across many body areas.

Steps for Testing

1. Apply tactile stimuli to both sides of the body at the same time or in separate regions, such as proximally and distally on one extremity or on one side of the body.

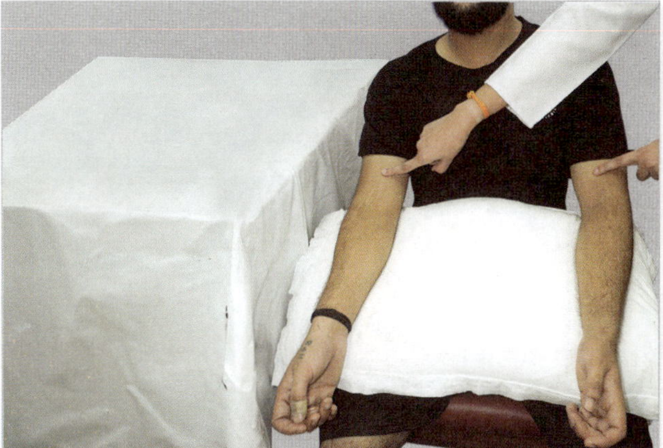

Fig. 6.15: Testing for double simultaneous stimulation

2. Prompt the patient to identify and confirm when they detect simultaneous touches on the appropriate regions.
3. Examine the patient's capacity to sense and recognize simultaneous touches on different regions of their body or extremities (Fig. 6.15).

Graphesthesia (Traced Figure Identification)

The use of a fingertip or the pencil rubber end to recognize letters, numbers, or drawings sketched on the skin is tested. The subject is vocally questioned regarding the drawings on their skin.

Patient Instructions

- "We will assess your ability to recognize letters, numbers, or shapes drawn on your skin."
- "I will trace shapes or symbols on your skin with my fingertip or the rubber end of a pencil. Identify and describe the sensation.

When asked about the patterns or symbols made on the skin, the patient expresses what they felt.

Steps for Testing

1. Draw letters, numbers, or shapes on the patient's skin with the fingertip or the rubber end of a pencil.
2. Prompt the patient to identify and vocally describe the traced figures. Examples of this include: "What letter or shape do you feel?" or "what shape do you identify?" or "Can you recognize this number?"
3. Examine the patient's ability to correctly identify and describe the tracings on their skin (Fig. 6.16).

Recognition of Texture

The capacity to distinguish between different textures, such as cotton, wool, or silk, is tested on the test.

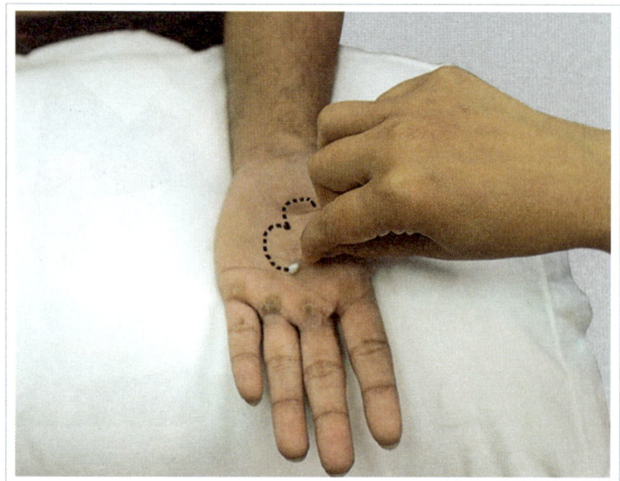

Fig. 6.16: Testing for graphesthesia

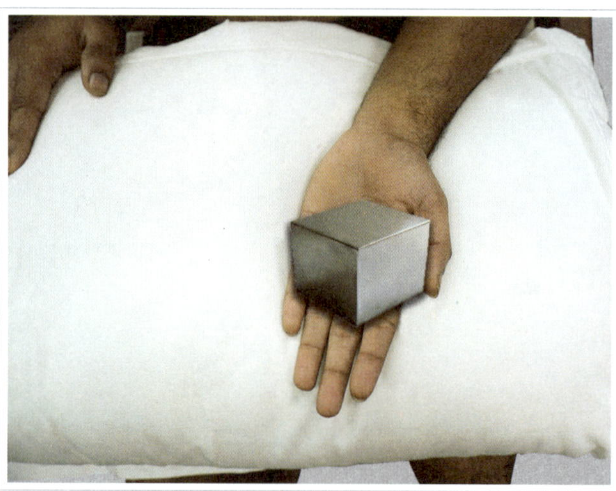

Fig. 6.17: Testing for barognosis

Patient Instructions

"We will test your ability to tell the difference between different textures, such as cotton, wool, and silk. I will place various fabrics or materials on your skin, and you'll identify and describe the textures you feel. Concentrate on identifying and describing texture differences."

Steps for Testing

1. Apply different fabrics or materials with different textures to the patient's skin, such as cotton, wool, silk, or other textiles.
2. Instruct the patient to identify and explain the textures they are experiencing. Examples of this include: "Can you tell what material this is?" or even "How would you describe the feeling of this fabric?"
3. Assess the patient's ability to appropriately detect and describe the texture distinctions presented during the evaluation.

Barognosis

Different weights are used for the test to determine barognosis (weight recognition). The therapist may decide to simultaneously place a separate weight in each hand or apply a sequence of different weights, one at a time, to the same hand.

Patient Instructions

- "We will assess your ability to recognize different weights."
- "You will feel various weights either simultaneously in both hands or one at a time in the same hand."

Concentrate on recognizing and indicating weight discrepancies.

Steps for Testing

1. Use varied weights, either by placing separate weights in each hand at the same time or by applying a succession of varying weights to the same hand one after the other (Fig. 6.17).
2. Encourage the patient to recognize and distinguish between the weights they feel. Examples of this include: "Can you tell if one side feels heavier?" as well as "Are these two weights different?"
3. Assess the patient's ability to distinguish and recognize differences in the weights presented during the evaluation.

The sensory pathway is crucial for the transmission and interpretation of environmental stimuli. It would be impossible to produce the right motor reactions without sufficient sensory input.

While Considering for Sensory Testing

- Ensure to test the unaffected side first before testing the affected side, for all sensory assessment methods. This keeps testing from being compromised by the last stimulus's preserved memory.
- Comparing proximal to distal regions as well as symmetrical parts on both sides of the body.

TESTS FOR INCOORDINATION

Fundamental principles of coordination and coordination exercises are described in Chapter 13.

In all the tests, accuracy, smoothness and coordination of the movement are checked

Tests for upper limb incoordination:

- Finger-to-nose test
- Finger-to-finger test
- Rapid alternating movement

Tests for lower limb incoordination:

- Finger-to-toe test
- Heel-to-shin test
- Romberg test
- The foot tapping test
- Lower extremity motor coordination test

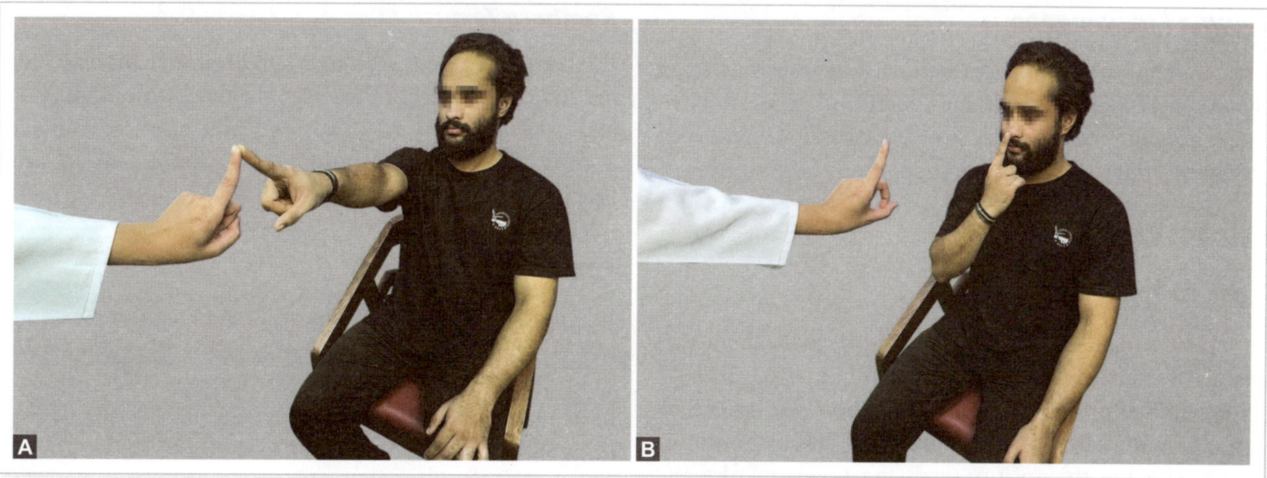

Figs 6.18A and B: Finger-to-nose test

Tests for Upper Limb

Finger-to-Nose Test

The subject is instructed to touch their nose with the index finger of one hand and then the tip of the index finger of the therapist, alternating between the two (Figs 6.18A and B).

Patients with cerebellar pathologies have trouble touching their noses, often showing wavy or oscillating motions. When the patient has posterior column disease, they can accurately touch their nose when their eyes are open but not when the eyes are closed.

Finger-to-Finger Test

The subject performs the finger-to-finger test by raising both hands toward the midline, extending both elbows and abducting their shoulders. Then they are instructed to touch the index finger of one hand to the index finger of the other (Fig. 6.19).

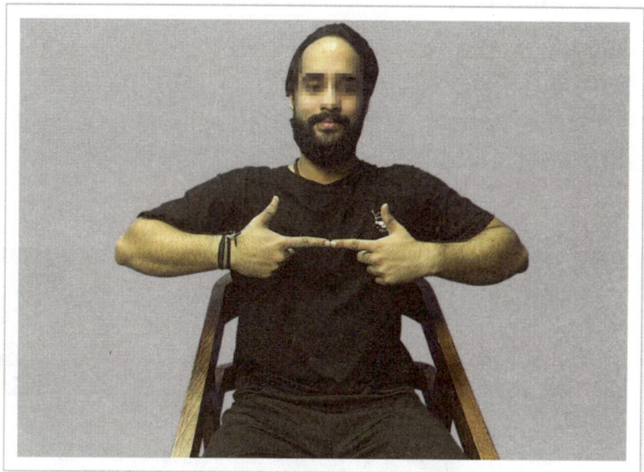

Fig. 6.19: Finger-to-finger test

Rapidly Alternating Movement

For this test, the patient is required to tap their hands and alternately pronate and supinate the forearms (Figs 6.20A and B).

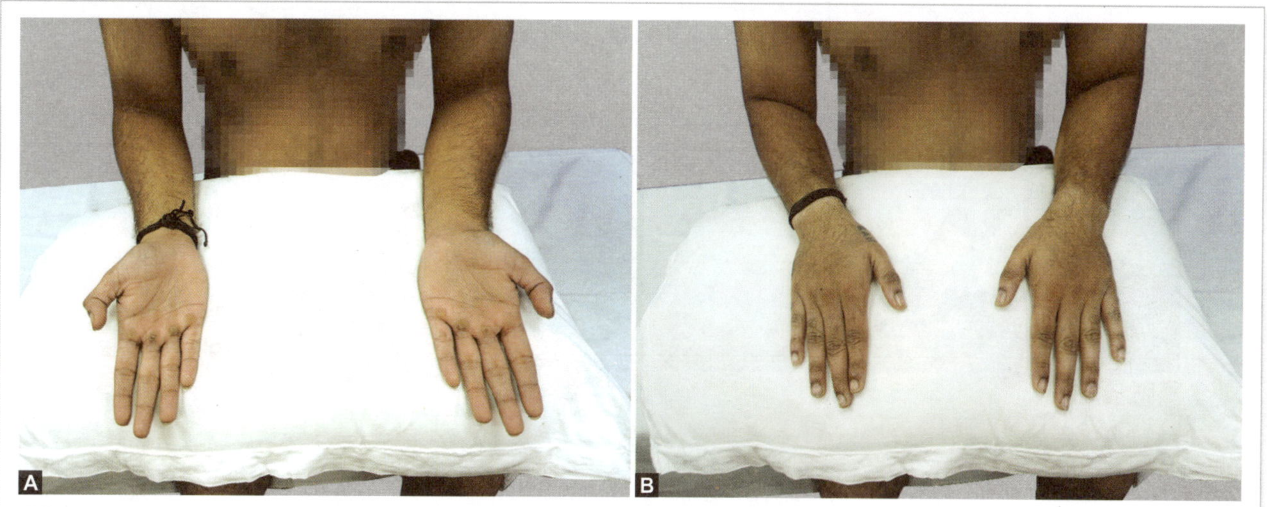

Figs 6.20A and B: Rapidly alternating movement

Rebound Phenomenon

The rebound phenomenon occurs when, after resisting a force, a sudden relaxation of resistance results in an instinctive, exaggerated movement in the opposite direction, frequently causing the limb to move or jerk suddenly.

Tests for Lower Limb

Finger-to-Toe Test

The patient is requested to touch their big toe to the therapist's finger as the therapist points their finger up to two feet above the patient's big toe, testing coordination and precision (Figs 6.21A and B). Coordination is tested by repeating the test with the therapist moving the finger in different directions on separate occasions.

Heel-to-Shin Test

The subject is instructed to touch their great toe with the heel of the opposite leg while sliding down the shin. The subject is instructed to do the same test without rubbing their skin (Figs 6.22A and B).

Romberg Test

Romberg test assesses balance and proprioception by having the subject stand with their feet together, arms at their sides, and eyes closed. If they sways or loses balance, it indicates a difficulty with proprioception or inner ear disorders impacting balance (Figs 6.23A and B).

Foot Tapping Test

In this test, the subject sits on a stool or chair extending the hips and knees to a roughly 90° angle, and taps the floor firmly with the toes for ten seconds while keeping the heels planted. The taps are counted separately for each side (Figs 6.24A and B).

Lower Extremity Motor Coordination Test

To assess lower limb coordination, participants must sit on a chair with height adjustment, touch proximal and distal targets spaced 30 cm apart alternatively with the big toe for 20 seconds, and record the number of targets touched (Figs 6.25A and B).

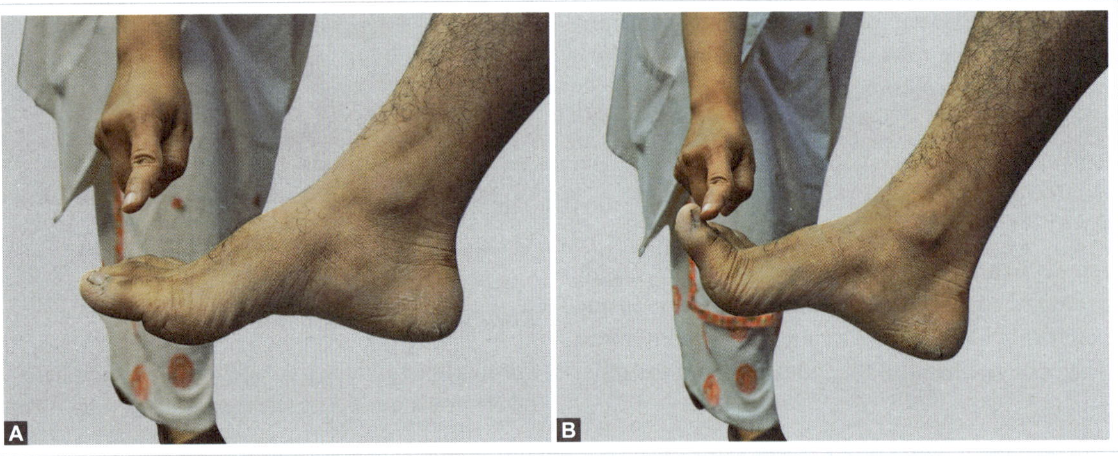

Figs 6.21A and B: Finger-to-toe test

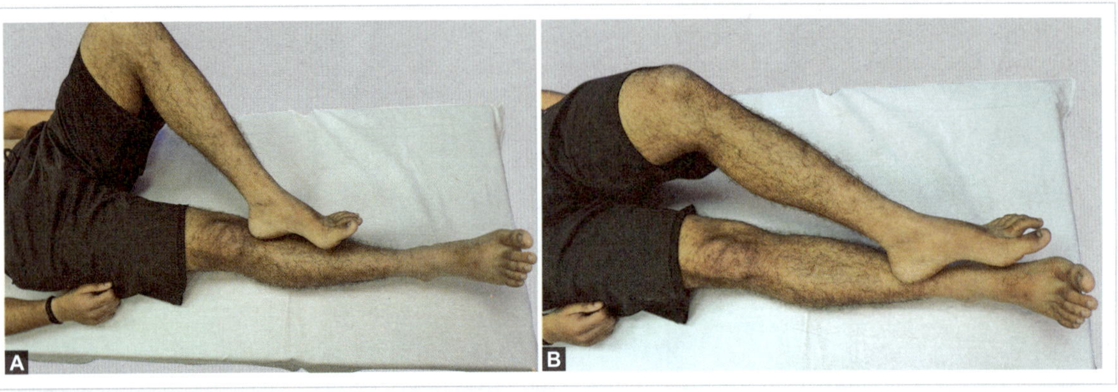

Figs 6.22A and B: Heel-to-shin test

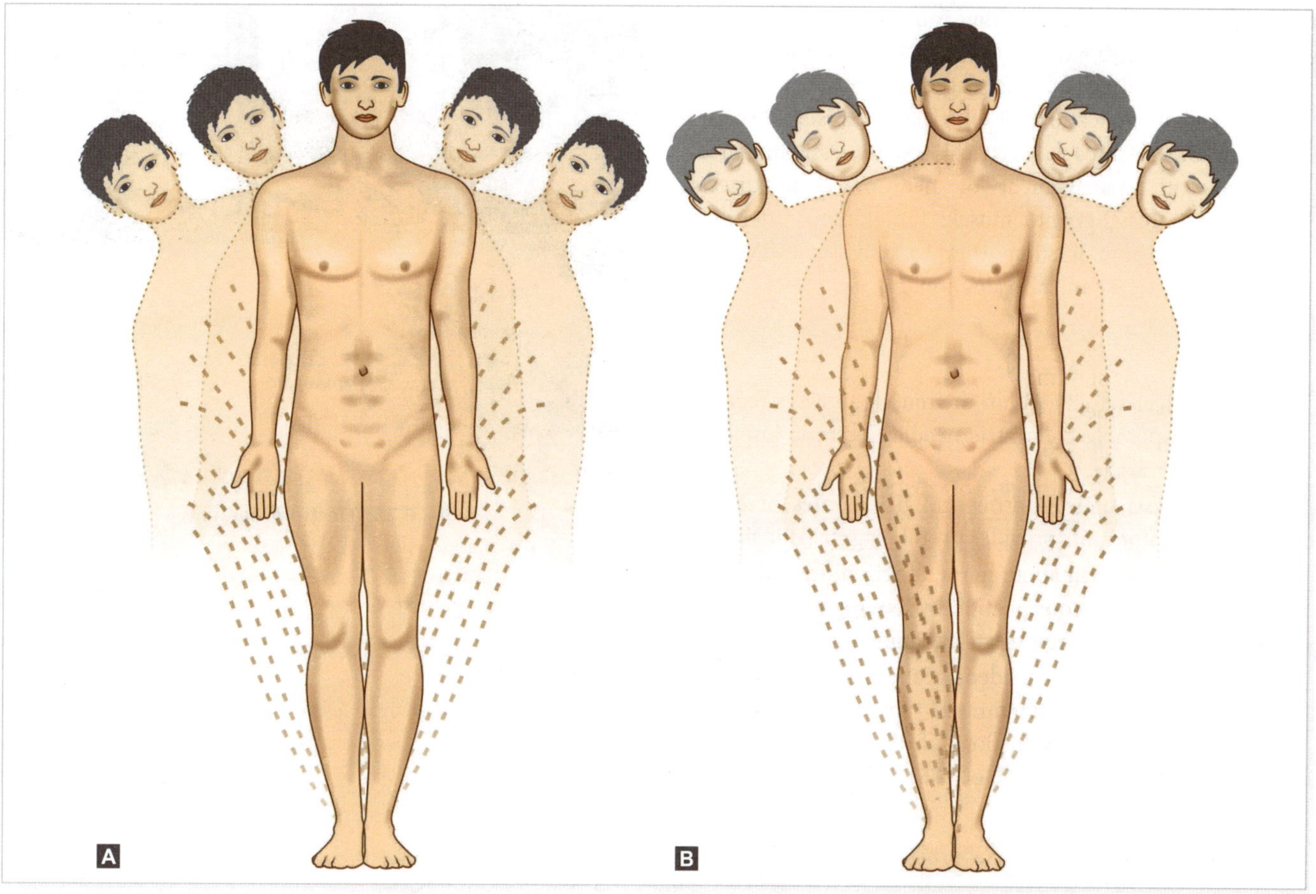

Figs 6.23A and B: Romberg test with eyes open and closed

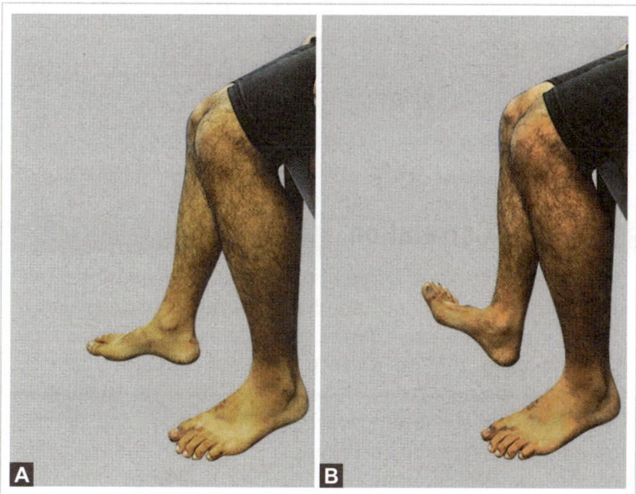

Figs 6.24A and B: Foot tapping test

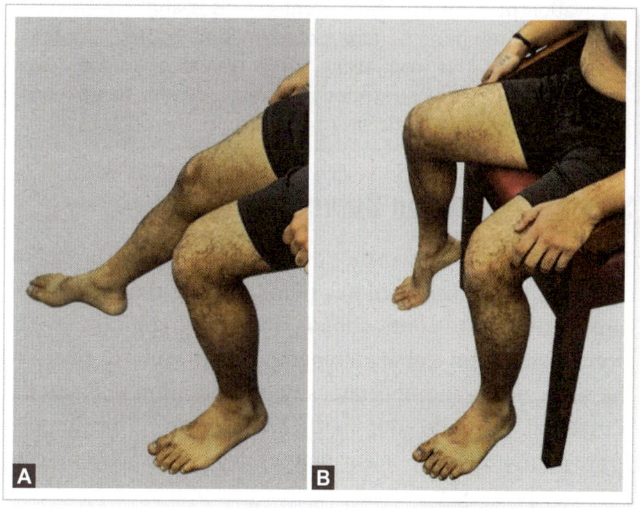

Figs 6.25A and B: Lower extremity motor coordination test

ELECTRICAL TESTS

Electrical test can be performed by using the strength-duration curve or the electromyography. They are especially useful for diagnosing problems.

Electromyography

Electromyography (EMG) it is a technique for monitoring and documenting the electrical activity generated by skeletal muscles in order to establish their status. The nervous system is always in charge of controlling muscular movements like relaxation or contraction. The precise interpretation of EMG data is critical in assessing neuromuscular diseases and associated applications. It enables specialists to evaluate and comprehend muscle electrical activity, assisting in the diagnosis and management of neuromuscular diseases by providing insights into muscle function and nerve-to-muscle transmission. It is important to think about EMG as a continuation of the clinical evaluation.

Procedure

- The electrical activity within the muscle tissue is recorded using a disposable concentric needle electrode implanted within the muscle.
- EMG analysis of waveforms and firing rates of one or multiple motor units provides useful diagnostic insights.
- Clinicians can learn about nerve function, muscle health, and nerve-muscle coordination by looking at the patterns and frequencies of these electrical impulses.
- This analysis aids in the identification of neuromuscular conditions, the understanding of motor unit recruitment patterns, and the evaluation of the nervous system's connection with muscles (Fig. 6.26).

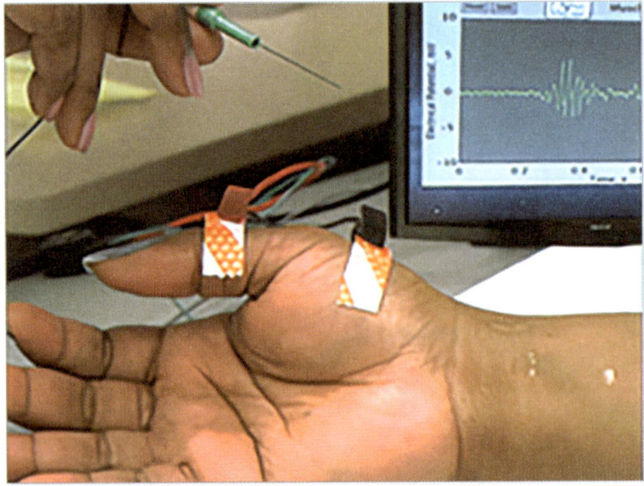

Fig. 6.26: Electromyography

Did You Know?

Electromyography is not only used in diagnosing neuromuscular disorders but also has applications in sports science and biomechanics. Researchers utilize EMG to analyze muscle activation patterns during different exercises and movements, helping to optimize training programs and improve athletic performance. This technology offers valuable insights into muscle recruitment strategies, allowing athletes to enhance their neuromuscular efficiency and maximize their potential.

Strength-Duration Curve

The relationship between the power of an electrical stimulation delivered to a muscle's motor point and the time needed to cause a minimum contraction is depicted graphically by the strength-duration curve. Strength in this context refers to the vertical axis stimulus intensity, and duration refers to the horizontal axis pulse duration. The integrity of the muscle-nerve complex is graphically represented by the S-D curve, to put it simply (Fig. 6.27). This could be helpful as an additional technique to evaluate peripheral neuropathy, lower motor neuron lesion, motor nerve dysfunction.

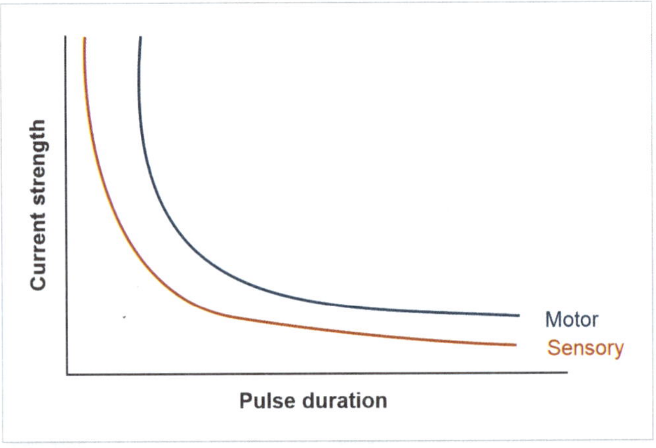

Fig. 6.27: Strength-duration (SD) curve

Clinical Correlation

The strength-duration curve not only aids in diagnosing nerve and muscle disorders but also has applications in determining the optimal parameters for electrical stimulation therapy. By understanding the relationship between stimulus intensity and pulse duration, clinicians can tailor electrical stimulation treatments to effectively target specific muscles and nerves, optimizing therapeutic outcomes. This highlights the versatility of the strength-duration curve beyond diagnostic purposes, demonstrating its importance in rehabilitation and physical therapy practices.

Other Electrical Methods

- **H-Reflex** testing is a kind of electromyographic testing used to help diagnose damage to a nerve. It measures reflex response of muscle to a nerve stimulus.
- **Evoked potentials** are the electrical signals generated in response to sensory stimulus like sound or light. They are helpful in diagnosing nerve dysfunction.
- **Somatosensory evoked potentials** are helpful in diagnosing nerve damage. They are generated in response to sensory stimulation of skin, muscle tissue, etc.
- **Motor evoked potentials** help in diagnosing motor dysfunction. They are generated in response to electrical stimulation of motor cortex in the brain.

Did You Know?

The evoked potentials, including somatosensory and motor evoked potentials, are valuable tools for diagnosing nerve dysfunction. These electrical signals, generated in response to sensory stimuli or motor cortex stimulation, provide insights into nerve and motor function, assisting clinicians in the diagnosis and management of neurological conditions.

SUMMARY

- Neuromuscular efficiency (NME) refers to the coordination between the nervous system and muscles to produce precise and powerful movements, with the central nervous system (CNS) playing a key role in activating muscles.
  - NME is assessed through various methods such as electromyography (EMG), which records muscle electrical activity to evaluate motor unit recruitment and coordination, aiding in diagnosing neuromuscular conditions.
  - Advanced tests like the strength-duration curve, H-reflex, and evoked potentials assess electrical efficiency and nerve function.
  - Additionally, anthropometric data and muscle strength/power assessments contribute to understanding muscle performance, while reflex responses and proprioception are important for evaluating neuromuscular coordination, joint stability, and overall movement efficiency.
- Anthropometric tests involve measuring body size, structure, and composition, including height, weight, waist, hip, and limb circumferences, as well as skinfold thickness, to assess adiposity and body composition.
  - Girth measurements, taken with an inextensible tape, help evaluate body size and composition at specific anatomical sites such as the head, neck, arms, chest, waist, hips, thighs, and calves.
  - Other methods include Body Mass Index (BMI), which categorizes weight status, and Waist-Hip Ratio (WHR), a better predictor of cardiovascular and diabetes risks. Skinfold thickness, measured with calipers at various body sites, estimates body fat percentage.
  - Advanced techniques like Dual-Energy X-ray Absorptiometry (DEXA), hydrostatic weighing, and Bioelectrical Impedance Analysis (BIA) provide more detailed assessments of body fat and composition.
- Static muscle strength tests, like the Isometric Squat Hold and Leg Press, assess endurance and strength in fixed positions, while dynamic tests, such as One Repetition Maximum and Vertical Jump, measure maximal effort and power. Handheld dynamometers, pinch strength tests, and the 30-Second Chair Stand Test evaluate muscle strength and endurance. Isokinetic testing measures muscle strength at constant speeds, and Manual Muscle Testing uses a six-point scale to assess muscle function for rehabilitation.
- Coordination tests assess the precision and control of movements, with exercises for both upper and lower limbs.
  - Upper limb tests include the Finger-to-Nose and Finger-to-Finger tests, which assess coordination and accuracy, and the Rapidly Alternating Movement test, which evaluates fine motor control. The Rebound Phenomenon observes exaggerated movements after resistance.
  - Lower limb tests such as the Finger-to-Toe and Heel-to-Shin tests evaluate coordination and precision, while the Romberg Test assesses balance and proprioception. The Foot Tapping Test and Lower Extremity Motor Coordination Test measure coordination and motor function in the lower limbs. These tests highlight the importance of sensory input for proper motor responses.
- Sensory testing involves assessing different types of sensations, including superficial, deep, and combined cortical sensations.
  - Superficial sensations, like pain, temperature, light touch, and pressure, are tested through various methods to evaluate the integrity of exteroceptive pathways.
  - Deep sensations, such as proprioception, kinesthesia, and vibration, assess the body's awareness of position and movement.
  - While combined cortical sensations involve more complex tests, such as stereognosis, two-point discrimination, tactile localization, and graphesthesia, which require integration of multiple sensory pathways.
  - Proper testing involves following specific steps to ensure accuracy and comparing both sides of the body to identify abnormalities in sensation.

FURTHER READINGS

- Mills KR. The basics of electromyography. Journal of Neurology, Neurosurgery & Psychiatry 2005;76:ii32-ii35.
- O'Sullivan SB, Schmitz TJ, Fulk G. Physical Rehabilitation. FA Davis; 2019 Jan 25.
- Friedli WG, Meyer M. Strength-duration curve: A measure for assessing sensory deficit in peripheral neuropathy. Journal of Neurology, Neurosurgery & Psychiatry. 1984 Feb 1;47(2):184-9.
- Nelson RM, Hunt GC. Strength-duration curve: Intrarater and interrater reliability. Physical therapy.
- Kendall FP, McCreary EK, Provance PG, Rodgers MM, Romani WA. Muscles: Testing and Function, with Posture and Pain. Lippincott Williams and Wilkins, Pennsylvania. determination of footedness. J. Phys. Med. Rehabil. 1983;2:835–41.
- Klein GR, Sharkey PF. Evaluation of hip pain in the young adult. InSeminars in Arthroplasty 2005 Mar 1 (Vol. 16, No. 1, p. 2–9). WB Saunders.
- In Surgical Treatment of Hip Arthritis 2009 Jan 1 (p. 3–8). WB Saunders.
- David J. Magee PhD, BPT, CM, in Orthopedic Physical Assessment, 2021.
- Wang J, Thornton JC, Kolesnik S, Pierson Jr RN. Anthropometry in body composition: An overview. Annals of the New York Academy of Sciences. 2000 May;904(1):317–26.

STUDENT ASSIGNMENT

LONG ANSWER QUESTIONS

1. Describe various tests for assessing coordination.
2. Describe sensory examination.

SHORT ANSWER QUESTIONS

1. Define neuromuscular efficiency.
2. Write about the electrical tests for neuromuscular efficiency.
3. What are the anthropometric measurements for the mid-arm, forearm, knee, and calf?
4. Enumerate the static and dynamic power tests.
5. Write briefly about various electrical tests for assessing neuromuscular efficiency.

MULTIPLE CHOICE QUESTIONS

1. **Which test assesses balance and proprioception by having the subject stand with his feet together, arms at its sides, and eyes closed?**
   a. Finger-to-nose test
   b. Romberg's test
   c. Rapidly alternating movement test
   d. Heel-to-shin test
2. **What does the finger-to-finger test primarily assess?**
   a. Pain perception
   b. Superficial sensations
   c. Coordination abnormalities
   d. Proprioception
3. **Which sensory test involves evaluating the ability to recognize different textures, such as cotton, wool, or silk?**
   a. Stereognosis
   b. Graphesthesia
   c. Recognition of texture
   d. Two-point discrimination
4. **What is assessed in the test involving the ability to perceive two separate points on the skin?**
   a. Tactile localization
   b. Stereognosis
   c. Two-point discrimination
   d. Graphesthesia
5. **Which incoordination test is commonly affected in individuals with cerebellar ataxia?**
   a. Heel-to-shin test
   b. Finger-to-nose test
   c. Finger-to-finger test
   d. Romberg's test
6. **What type of sensory evaluation involves testing pain, temperature, and touch awareness?**
   a. Superficial sensation testing
   b. Deep sensation testing
   c. Combined cortical sensation testing
   d. Discriminative sense testing
7. **Which test evaluates the ability to perceive simultaneous touch sensations on different parts of the body or extremities?**
   a. Two-point discrimination
   b. Double simultaneous stimulation
   c. Tactile localization
   d. Stereognosis
8. **Which of the following is true regarding Electromyography (EMG)?**
   a. Measures the electrical activity of muscles at rest
   b. Evaluates nerve conduction in peripheral nerves
   c. Assesses muscle strength through manual testing
   d. Monitors and records the electrical activity of skeletal muscles

9. **What does Nerve Conduction Velocity (NCV) primarily assess?**
   a. Muscle strength
   b. Muscle coordination
   c. Sensory perception
   d. Functioning of peripheral nerves

10. **Which condition might EMG testing help diagnose?**
   a. Asthma
   b. Osteoarthritis
   c. Neuromuscular diseases
   d. Allergic reactions

11. **What does an abnormal NCV result indicate?**
   a. Slowed or impaired nerve function
   b. Increased muscle coordination
   c. Enhanced sensory perception
   d. Improved motor neuron activity

12. **Which measurement is used to determine waist circumference?**
   a. Measure at the widest part of the hips
   b. Measure between the iliac crest and the 12th rib
   c. Measure around the chest over the xiphoid process
   d. Measure at the midpoint between the olecranon and acromion processes

13. **What is the purpose of circumference or girth measurement in anthropometry?**
   a. To evaluate skin thickness
   b. To assess joint flexibility
   c. To determine body composition and size
   d. To measure bone density

14. **Which area is typically measured during anthropometric assessment for upper limb girth?**
   a. Forehead
   b. Elbow
   c. Knee
   d. Ankle

ANSWER KEY

1. b **2.** c **3.** c **4.** c **5.** b **6.** a **7.** b **8.** d **9.** d **10.** c
11. c **12.** b **13.** c **14.** b

Section III

Application of Principles of Exercise Therapy

SECTION OUTLINE

7 Passive Movements

Sheetal Kalra

LEARNING OBJECTIVES

After the completion of the chapter, the readers will be able to:

- Define passive movements.
- Explain causes of immobility.
- Classify passive movements.
- Explain the principles of passive movements.
- Mention the indications and contraindications of passive movements.
- Explain the effects and uses of passive movements.
- Understand and demonstrate technique of passive movements at various joints.

CHAPTER OUTLINE

- Introduction
- Range of Motion
- Causes of Immobility
- Classification
- Principles and Procedure
- Indications
- Effects and Uses
- Contraindications
- Techniques of Relaxed Passive Movements for Individual Joints

KEY TERMS

Contracture: A condition that causes muscles, tendons or other tissue to shorten and stiffen; frequently, resulting in joint deformity and restricted movement.

Effusion: Abnormal accumulation of fluid in cavities or between body tissues.

Myositis ossificans: A condition that occurs when bone tissue forms within soft tissues or muscles, typically as a reaction to muscle injury or severe bruising.

Osteokinematics: The study of bone motion around a joint, including the movements of the bony levers in the body.

Overstretching: Stretching a muscle, tendon or ligament beyond its normal limit, potentially causing injury.

Scar: A mark left on the skin or body tissue where a cut, burn or wound has not completely healed and connective tissue has formed.

Sepsis: An extreme response of the body's immune system to an infection or injury causes sepsis. It is a life-threatening medical emergency.

Stasis: A duration, condition or state of balance, inactivity or stagnation.

INTRODUCTION

Passive movement (PM) is an intervention in which an outside force, usually by a therapist or a mechanical device, repeatedly moves a person's joints through their range of motion (ROM). These are performed during periods of muscular inactivity or when range of motion is restricted due to various reasons. Physiology and medicine have long recognized the benefits of passive exercise and movement. Early clinical uses of passive exercise/movement stimulated the vasculature and elicited changes in blood flow to prevent complications caused by stasis and vascular disease.

The basic objective of PMs is to influence the extensibility of soft tissues surrounding joints in order to preserve or improve joint mobility.

They are also used to lessen side effects brought on by cartilage deterioration. People who are unable to move certain joints on their own due to paralysis, severe pain, or limited consciousness typically receive PMs for a short period of time. Passive movements can take 20–30 minutes to administer. In some patients with neurological disabilities, these are administered every day throughout a person's life.

RANGE OF MOTION

The definition of ROM varies among published sources; Kapandji and colleagues described ROM as "the extent of osteokinematic motion available for movement activities, functional or otherwise, with or without assistance".

In its most basic form, movement—which is required to carry out functional tasks—can be understood as the result of muscles or outside forces moving bones in different ways or ranges of motion. The central nervous system is responsible for the complex regulation of muscle activity that drives or regulates motion when a person moves. At the joints, bones move relative to one another. The amount of mobility that can take place between any two bones depends on the structure of the joints as well as the flexibility of the soft tissues that cover the joints. The segments must periodically move across their ranges, to maintain appropriate ROM. It is acknowledged that there are so many variables, like systemic, joint, neurological, or muscle illnesses; traumatic or surgical; or physical inactivity or immobilization for any reason, that can lead to decreased ROM. Therapeutically, ROM activities are administered to maintain joint and soft tissue mobility to minimize loss of tissue flexibility and contracture formation.

The ROM can be reduced by inactivity or immobilization for any reason. Therapeutic ROM exercises are used to keep soft tissues and joints mobile in order to prevent contracture formation and tissue stiffening.

Types of ROM Exercises

- **Passive ROM:** Physiology and medicine have a long and rich history with passive exercise/movement, which is simply defined as the movement of the body or a limb (such as the leg or arm) without deliberate effort or muscle activity. The external force could be from gravity, a machine or by someone else.
- **Active ROM:** The movement of a segment inside the unrestricted ROM is known as active range of motion (AROM), and it is caused by the active contraction of the muscles that span that joint.
- **Active assisted ROM:** Because the prime mover muscles require assistance, active-assistive range of motion (A-AROM) is a kind of AROM in which assistance is given manually or mechanically by an outside source while the muscle also contracts to complete the movement.

CAUSES OF IMMOBILITY

The inability to move or difficulty in moving can be caused by a number of factors. Listed below are a few typical reasons:

- **Musculoskeletal injuries:** Immobility may result from sprains, fractures, or strains. Casts or braces may be necessary for immobilization in cases of severe injuries.
- **Neurological conditions:** Immobility can be a consequence of neurological disorders such as multiple sclerosis, Parkinson's disease, spinal cord injuries, stroke, or other nervous system disorders.
- **Joint issues:** Disorders such as arthritis or inflammation of the joints can impair mobility and cause pain.
- **Chronic pain:** Resulting from neurological problems or musculoskeletal conditions, chronic pain can make a person immobile and make it difficult for them to move around comfortably and go about their everyday lives.
- **Surgery:** Following some surgeries, patients may become immobile for a while as they heal.
- **Obesity:** Carrying too much weight around can strain muscles and joints, making movement difficult.
- **Circulatory issues:** Disorders that impair blood flow, like peripheral artery disease or deep vein thrombosis, can make a person immobile.
- **Breathing problems:** Illnesses such as severe chronic obstructive pulmonary disease (COPD) or specific respiratory infections can cause weakness and fatigue making it difficult to move.
- **Infections:** Severe infections can cause immobility, particularly if they affect the joints or bones.
- **Age-related problems:** As people age, their muscles, bones, and joints might change, which can lead to decreased mobility in older persons.

- **Trauma:** Severe trauma, like that from a fall or an automobile accident, can cause injuries that restrict movement.
- **Psychological factors:** Disorders such as anxiety or depression can cause a person to lose desire and energy, which in turn might cause them to engage in less physical exercise.

Clinical Correlation

Integrating mental health support into treatment plans is crucial to address underlying mood disorders and motivational barriers, promote adherence to rehabilitation programs, and improve overall well-being and quality of life.

CLASSIFICATION

Relaxed Passive Movements

These are the movements which are carried out in the range of motion that is available, with precision, rhythm, and fluidity. The movements are carried out within the same direction and plane as active movements. The joint is moved through its current range of motion and just as far as it causes pain.

Accessory Movements

Any normal joint has accessory movements, but in abnormal circumstances they could be restricted or nonexistent. They consist of gliding or rotating movements that can be separated by the physiotherapist but cannot be done independently as voluntary movements.

Passive Manual Mobilization Techniques

Mobilization of Joints

These are often small, repeated, rhythmical, oscillatory, localized accessory, or functional motions carried out by the physiotherapist in a variety of amplitudes. The motions can be controlled by the patient and are within the available range of motion.

Manipulation of Joints

These are precisely localized, single, swift, and decisive movements that are finished at high speeds before the patient can stop them. These movements enhance range of motion in a stiff joint by breaking or stretching the restricting structures such as adhesions. A sudden forceful movement is given at the limit of the motion. Accessory movements can be restored at the joints which otherwise cannot be localized actively.

Precautions:

- Patient must be encouraged to relax and should be provided with care.
- Manipulation should be performed only by trained doctors and therapists.
- Anesthesia is used during these maneuvers by a surgeon or doctor to eliminate pain and protective spasm so as to increase range of motion.

Controlled Sustained Stretching of Passive Structures

Movement can be achieved by lengthening the muscle by inhibiting the tendon protective reflex or by stretching adhesions in these tissues. The collagen fibers which are strong, inextensible and inelastic will elongate when subjected to prolonged stress. The resistance of tight structures can be overcome by this method.

Continuous Passive Motion

A device that promotes continuous passive motion keeps a joint moving (Figs 7.1A and B). Constant motion reduces stiffness and pain. It can be carried out mechanically or manually. Continuous passive motion (CPM) is usually the name for device assisted passive motion carried out by a

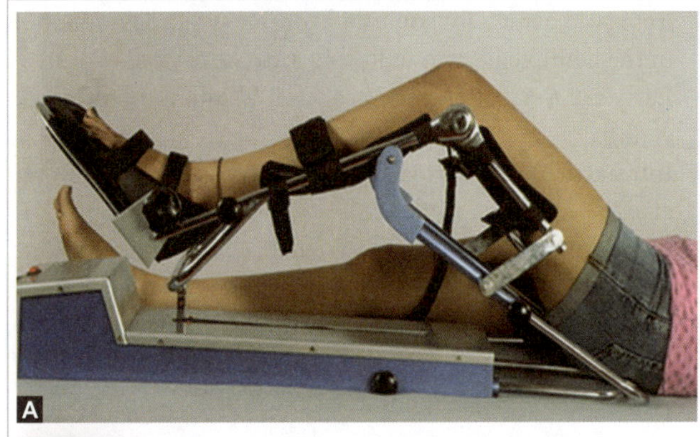

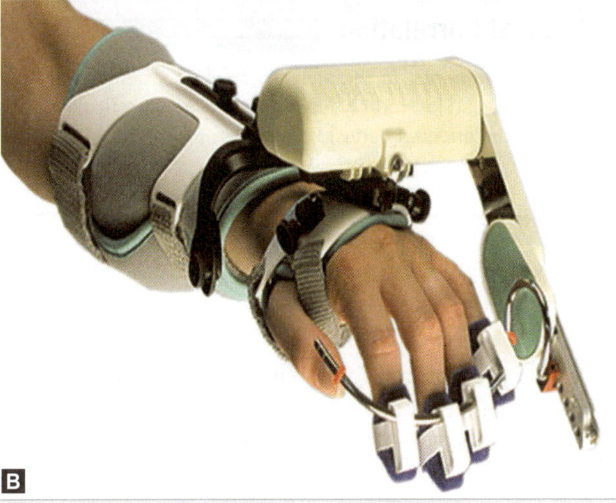

Figs 7.1A and B: **A.** Lower limb continuous passive motion machine; **B.** Wrist and hand continuous passive motion machine

mechanical device that gently and continuously moves a joint through a controlled range of motion. It has good healing effects on diseased or injured joint structures and soft tissues in animal and clinical tests. Mechanical devices are now available for almost every joint in the body.

Continuous passive motion must be used with an appropriate range of motion without causing distress. Based on the patient's comfort level and the required number of repetitions in a certain amount of time, the speed of movement through the range of motion should be determined. It should be one cycle every 45 seconds.

Clinical Correlation

The CPM therapy is particularly beneficial in the early stages following surgery or injury, helping to minimize joint hemarthrosis, periarticular edema, and aiding in the removal of blood from the joint and surrounding tissues.

Indications

- Treating early scar contracture after hand burns (Zhao et al.).
- Flexion contractures (Lindenhovious et al.).
- Continuous passive range of motion may provide an improved benefit for some unfavorable symptom reduction in the hemiplegic arm following a stroke (Lynch et al.).
- Early recovery after periosteal transplantation (Alfredson et al.).
- Intra-articular anterior cruciate reconstructions (Rosen et al.).
- Intra-articular fractures of the knee, elbow, and ankle, etc. (Hill et al.).
- Following a total knee and hip arthroplasty (Harvey et al.).
- Post cartilage repair (Howard et al.).
- After repair of rotator cuff (Lastayo et al.).

Clinical Correlation

Passive movements in hemiplegic stroke patients activate patterns crucial for motor function recovery, particularly by activating the affected hemisphere.

Precautions

- **Correct alignment:** In order to avoid putting more strain on joints or soft tissues, make sure the patient's limb is properly positioned inside the CPM device.
- **Patient positioning:** To prevent undue strain on the joints and surrounding tissues, maintain the patient's optimal positioning during CPM therapy.
- **Contraindications:** Recognize which medical conditions, such as particular fractures or surgical operations, make CPM contraindicated. Make sure that CPM is only applied when medical professionals suggest it is suitable.
- **Monitoring comfort levels:** Throughout CPM sessions, periodically gauge the patient's level of comfort. As appropriate, modify the range of motion or stop the CPM if there is severe pain or discomfort.
- **Individualized settings:** Adjust the CPM settings to the patient's unique requirements, accounting for their pain threshold, range of motion, and kind of surgery or injury.
- **Limit range of motion:** To avoid overstretching or stressing the joint, gradually increase the range of motion within the boundaries set by the healthcare professional.
- **Patient education:** Inform the patient about the goals and procedures of CPM, stressing the need of adhering to prescription guidelines and reporting any uncomfortable or odd sensations.
- **Regular supervision:** In order to ensure appropriate usage and make any required adjustments, healthcare practitioners should initially keep a careful eye on CPM sessions. As soon as the patient gets comfortable using the gadget, less supervision is required.
- **Skin integrity:** Check for irritation, redness, or pressure sores on a regular basis on the skin surrounding the joint. It could be required to reposition or pad in order to preserve skin integrity.
- **Joint stability:** Make sure the joint doesn't move excessively or is unstable during CPM sessions. If it does, it should be swiftly adjusted.

Contraindications

- **Uncontrolled pain:** It might not be appropriate if the patient has uncontrollably high pain levels during CPM sessions. It is best to treat pain properly before thinking about CPM.
- **Fracture with displacement:** CPM is usually not recommended in fracture instances involving considerable bone fragment displacement.
- **Unstable fractures:** When a fracture is unstable or when the stability of the fracture site may be jeopardized by the use of CPM, CPM usage is not recommended.
- **Compromised soft tissue:** If the soft tissues surrounding the joint are damaged or compromised in any way, such as by severe contusions, open wounds, or infections, CPM may not be appropriate.
- **Joint infections:** There may be contraindications due to active joint infections or the possibility of infection spreading to the joint through CPM.
- **Deep vein thrombosis (DVT):** Because of the possibility of displacing blood clots, the use of CPM may not be appropriate in cases of suspected or confirmed DVT.
- **Peripheral vascular disease (PVD):** Because CPM can worsen circulation problems, it might not be appropriate for people with peripheral vascular disease.
- **Compromised vascular supply:** When the blood supply to the afflicted joint or limb is impaired, CPM may not be appropriate.

- **Neurological impairments:** The use of CPM may be contraindicated in certain cases of neurological disorders or impairments, particularly if the patient is unable to participate actively or communicate properly.
- **Lack of cooperation:** Individuals unable to comprehend or assist with the CPM procedure, including CPM as part of the treatment must be avoided.

Effects and Uses

The CPM and passive exercises are successful therapies to stimulate movement in various orthopedic, neurological and sports injuries like hemiplegic stroke, post fracture stiffness, burns, sprains, strains, etc.

- CPM exercises have a local circulatory effect by raising blood flow to the skeletal muscles.
- Edema, spasticity, rigidity, and motor function have all been found to be considerably improved by CPM therapies (Brenner et al., 2018).
- Moving a joint through CPM leads to a different sensory input, which could reduce pain.
- CPM lessens the perception of pain in addition to reducing pain, avoiding the negative effects of immobilization, and enabling the "healing power of motion" to occur (Mccarthy et al., 1992).
- Animal studies have found use of CPM promotes development of hyaline cartilage in rabbits with full thickness defects of articular cartilage, accelerates the clearance of blood, decreases erythrocyte trapping in the synovium in induced effusion and hemarthrosis, enhances ROM and transport of metabolites out of joint.
- To minimize joint hemarthrosis and periarticular edema and aid in the removal of blood from the joint and surrounding tissues, CPM treatment is most helpful in the early stages following an accident or surgery.
- CPM causes a sinusoidal oscillation in intra-articular pressure, facilitating in the effective clearance of hemarthrosis. The displacement from the maximal volume position reduces the elevated intra-articular pressure brought on by joint effusion.
- CPM is effective in increasing blood clearance from the joint and periarticular tissues, hence limiting further accumulation of edema. This is confirmed by radiolabeled erythrocyte tracking.
- CPM is most helpful during the first several days and hours following surgery (the first and second stages of stiffness). In the third and fourth stages, it becomes less helpful (Driscoll et al.).
- Applying CPM promotes the metabolism of chondrocyte Proteoglycan 4 (PRG4). One way that CPM may help cartilage and joint health during postoperative rehabilitation is by stimulating PRG4 synthesis (Nugent et al.).

PRINCIPLES AND PROCEDURE

Planning for Examination, Assessment and Treatment

1. Examine and evaluate patient's general condition, level of impairment and function before administering passive movements.
2. Determine the motion that can be safely applied to the patients.
3. Documentation, re-evaluation and modification of intervention may be required based on patient's response.

Patient Preparation

1. Describes the strategy and technique used to achieve the goals to the patient.
2. Remove all constrictive garments, splints, and dressings from the area. Cover the patient as needed.
3. Place the patient in a relaxed position that allows moving the segment through its range of motion while maintaining good body alignment and stability.

Technique of Relaxed Passive Movements

1. **Relaxation:** It can be achieved by a brief explanation to the patient of what is going to happen. A suitable starting position is selected for comfort and support of patient before starting passive movements. It is important to gain patient's confidence in maintaining relaxation till the passive movement is completed.
2. **Fixation:** The bone proximal to the joint where movement is to be performed should be fixed. This helps in localizing movement to the joint being moved and avoiding compensatory movements.
3. **Support:** To gain patient's confidence and relaxation throughout the range of motion the part being moved should be comfortably supported. Firm but comfortable grip of physiotherapist is required to provide support. Alternatively slings or axial suspension can also be used for supporting heavy limbs.
4. **Stance:** A firm and comfortable stance of physiotherapist is required to ensure movement is performed properly. Both feet should be apart and in line with the direction of movement.
5. **Traction:** Traction should be given along the long axis of the joint to facilitate movement. The fixation of the bone proximal to the joint provides an opposing force to the sustained pull on the distal bone.
6. **Movement:** The physiotherapist performs the movement similar to that of natural movement taking care that range is as full as possible without provoking pain in adjacent areas. To ensure full range slight overpressure can be applied in the end of the range.

7. **Speed and duration:** The speed must be gradual and rhythmic with appropriate repetitions of the action since it is crucial that relaxation be maintained throughout the movement.
8. **Repetition:** Perform passive movement at least 5–10 times smoothly and rhythmically. However the number of repetitions will depend upon condition of the patient.

INDICATIONS

- Passive motion is advantageous in the area where there is acute, inflamed tissue and when aggressive motion would be harmful to the healing process. (After an injury or operation, inflammation often lasts 2–6 days.)
- Patients with paralysis like hemiplegia, paraplegia.
- Postoperative cases like TKR, THR, etc.
- Reconstructive surgeries like ligament repair, injuries around joints, dislocation, fractures, etc.
- Comatose or bed ridden patients.
- Critically ill patients like patients in intensive care. To prevent deterioration, improve condition, and shorten intensive care unit (ICU) and inpatient stays, early mobilization and rehabilitation are advised (Boulain et al., 2002).
- Treatment and prevention of contractures (Prabhu et al.).

EFFECTS AND USES

- Preserve joint and soft tissue flexibility.
- Passive motions stop adhesions from forming around and inside the soft tissues of joints. This could be achieved by blocking collagen's production of cross-bridges (Prabhu et al., 2013).
- Help with neuromuscular reeducation
- Improve synovial circulation
- Prevent shortening and contracture while maintaining the natural characteristics of the muscle (extensibility, elasticity, etc.). Prevent myofascial adhesions.
- Help by activating the kinesthetic receptors, which preserve and maintain memory of the movement patterns (Paillard et al., 1968).
- The venous and lymphatic return is improved by the mechanical pressure created by the stretching of the thin walled veins that pass over the moving joints (improving circulation).
- To train for relaxation because the soothing effects of the rhythmic, continuous passive movements can help relax more and fall asleep while also enhancing sense of position and movement (Paillard et al., 1968).
- Increases in metabolic and hemodynamic demands, such as increased venous return and stroke volume, as well as a potential impact on the inflammatory response are all potential consequences of PMs.
- In a small study with healthy people, passive leg activity was found to increase minute ventilation, and it may also enhance oxygen consumption (Clini et al., 2005).
- Improve collagen re-organization after injury, maintain joint range of motion, and minimize contracture formation (Prabhu et al., 2013).
- Repeated, externally imposed, sequential flexion-extension movements of the elbow decreased the elbow flexor stretch reflex in spastic hypertonia patients (Schmit et al., 2020).
- In the paretic lower limb muscles of chronic stroke patients, passive leg movements have the power to stimulate muscular activity and improve oxygen metabolism. This kind of exercise may be a helpful and effective way to stop the lower limb paretic muscles of chronic stroke patients from degenerating metabolically (Jigjid et al., 2008).
- Deeply sedated patients who get passive exercise have improved muscle fiber function and reduced muscle atrophy, suggesting a possible positive impact for immobile ICU patients (Llano et al., 2012).
- In critical patients on mechanical ventilation in ICUs, early rehabilitation with a passive cycle ergometer can preserve the morphology of the lower limb and respiratory muscles (diaphragm) and hence prevent neuromuscular complications (dos Santos et al., 2015).
- Before clinical recovery, passive movements in hemiplegic stroke patients evoke activation patterns that may be crucial for the recovery of motor function. Particularly, it appears that persistent and early activation of the afflicted hemisphere, as shown with a transcranial doppler, prefigures the improvement of a motor deficiency (Matteis et al., 2003).
- Passive Range of Motion (PROM) is utilized to assess mobility restrictions, joint stability, muscle flexibility, and other soft tissue elasticity when a therapist examines inert structures.
- To show the desired motion when a therapist is teaching an active exercise workout program, PROM is employed.
- PROM is frequently used before passive stretching treatments when a therapist is getting a patient ready for stretching.

Clinical Correlation

Early rehabilitation with passive cycle ergometer in critically ill patients on mechanical ventilation preserves lower limb and respiratory muscle morphology, preventing neuromuscular complications.

CONTRAINDICATIONS

- Immediately after surgery or acute injury like muscle/ligament tear.
- Conditions like myositis ossificans.
- Unhealed or recent fractures.

Did You Know?

Common effects and uses of passive movement

Besides the major effects and uses of passive movements, some more common effects and uses are as follows:

- Prevents the formation of adhesions and contractures, which lead to joint stiffness.
- Stimulates the healing of tendons and ligaments.
- Promotes the healing of incisions over moving joints.
- Increases synovial fluid lubrication of the joint, which speeds up the rate of intra-articular cartilage healing and regeneration.
- Prevents the debilitating effects of immobilization.
- Provide a quicker return of ROM.

TECHNIQUES OF RELAXED PASSIVE MOVEMENTS FOR INDIVIDUAL JOINTS

Techniques of performing relaxed passive movements by the physiotherapist at individual joints for specific movements are described as follows.

Shoulder Joint

Flexion and Extension

- **Patient position:** The patient should be comfortably positioned in supine lying for flexion, and supine or prone lying for extension.
- **Hand placement:** The therapist typically stands or sits beside the patient on the side on which passive movement is to be performed. Hold the patient's arm beneath the elbow with the upper hand. With the other hand, reach over and grip the patient's wrist and palm.
- **Procedure:**
  1. Elevate the arm through its available range of flexion (Fig. 7.2A) and bring it back down through extension (Fig. 7.2B).
  2. Execute the movements smoothly and rhythmically, respecting the natural range of motion of each joint.
  3. Hold the end position briefly, allowing the muscles and joints to adapt to the stretch.

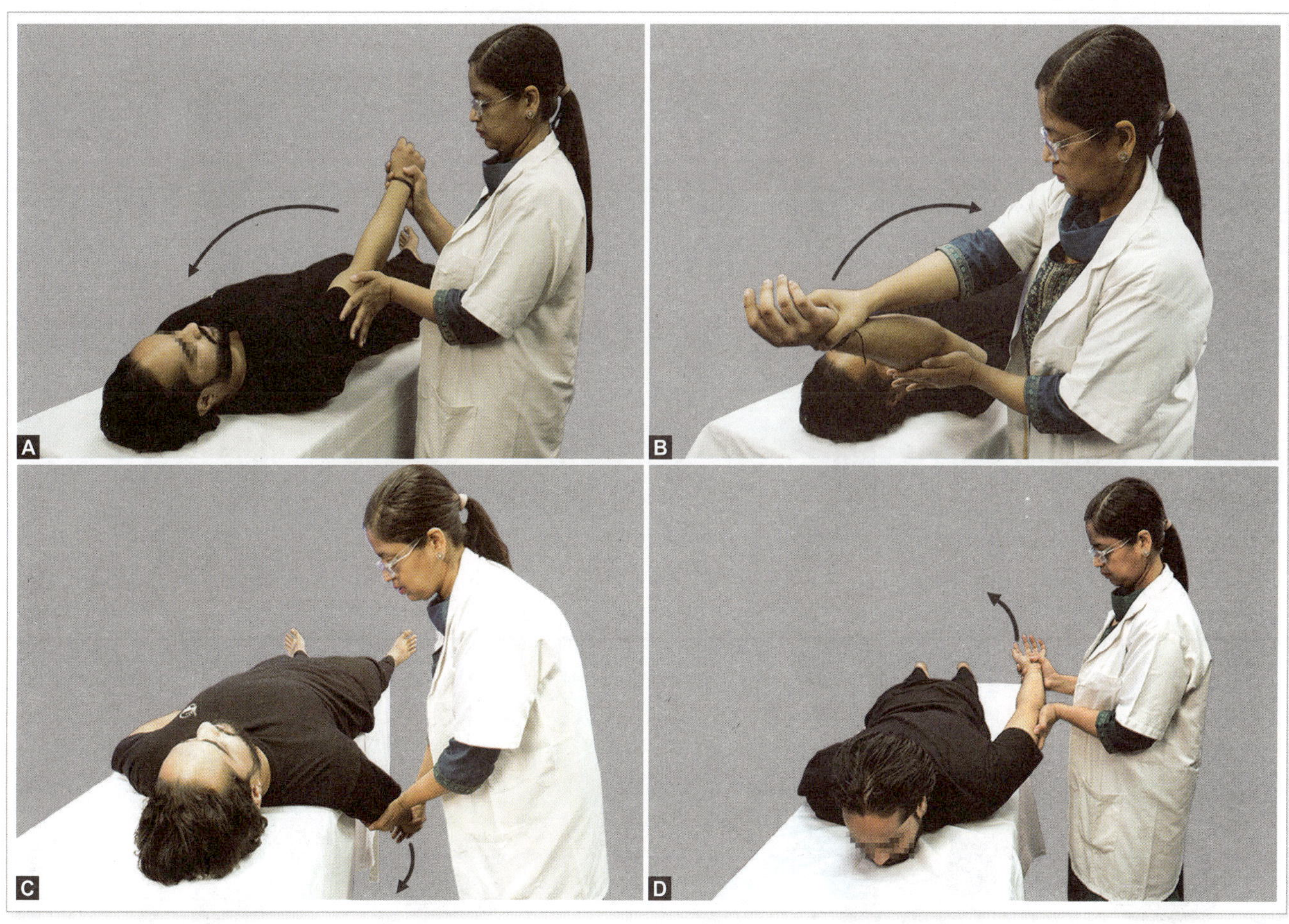

Figs 7.2A to D: Shoulder joint: **A.** Flexion; **B.** Extension; **C.** Hyperextension in supine; **D.** Hyperextension in prone

4. Return the limb to the starting position in a controlled manner.
5. Repeat the movement within a comfortable range, following the prescribed number of repetitions.

Hyperextension

Extension beyond zero can be achieved when the patient's shoulder is at the bed's edge in the supine position (Fig. 7.2C) or if the patient is in a side-lying, prone (Fig. 7.2D), or seated posture.

Abduction and Adduction

- **Patient position:** For shoulder abduction and adduction, the patient should be comfortably seated or lying down.
- **Hand placement:** Utilize the same hand placement as in flexion.
- **Procedure:**
  1. Move the patient's arm sideways (Fig. 7.3A). The elbow can be flexed.
  2. To achieve full abduction range, external rotation of the humerus and upward rotation of the scapula are essential.
  3. Execute the movements smoothly and rhythmically, respecting the natural range of motion of each joint. Hold the end position briefly, allowing the muscles and joints to adapt to the stretch.
  4. Return the limb to the starting position in a controlled manner (Fig. 7.3B).
  5. Repeat the movement within a comfortable range, following the prescribed number of repetitions.

Horizontal Abduction and Adduction

- **Patient's position:** For shoulder horizontal abduction and adduction, the patient should be comfortably seated or lying down. For complete horizontal abduction, the patient's shoulder should be at the table's edge. Commence with the arm flexed or abducted at 90° (Fig. 7.3C).
- **Hand placement:** Hand placement mirrors that of flexion.
- **Procedure:**
  1. Move the patient's arm outward and then across the body for horizontal adduction (Fig. 7.3C).
  2. Execute the movements smoothly and rhythmically, respecting the natural range of motion of each joint.

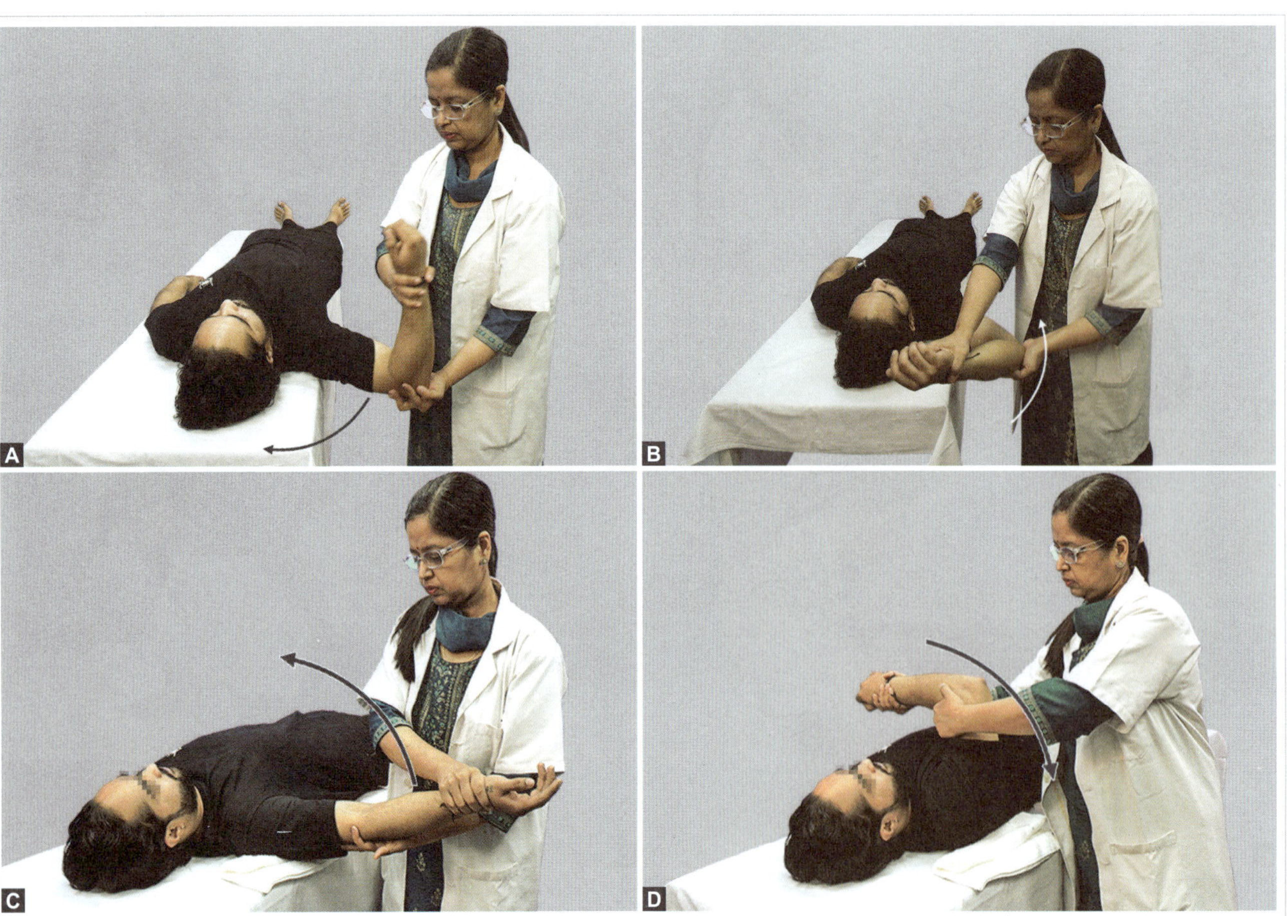

Figs 7.3A to D: Shoulder joint: **A.** Abduction; **B.** Adduction (with external rotation); **C.** Horizontal abduction; **D.** Horizontal adduction

3. Hold the end position briefly, allowing the muscles and joints to adapt to the stretch.
4. Return the limb to the starting position in a controlled manner for horizontal abduction (Fig. 7.3D).
5. Repeat the movement within a comfortable range, following the prescribed number of repetitions.

Internal (Medial) and External (Lateral) Rotation

- **Patient's position:** For shoulder medial or lateral rotation, the patient should be comfortably seated or lying down. Ideally, the arm is abducted to 90°, the elbow is flexed at 90°, and the forearm is held neutrally (Fig. 7.4A). Internal rotation may also be achieved with the arm at the thorax's side, but it's limited in this position.
- **Hand placement:** The therapist typically stands or sits beside the patient. Grasp the hand and wrist with one hand. Secure the elbow with the other hand.
- **Procedure:**
  1. Rotate the humerus by moving the forearm inward for medial rotation (Fig. 7.4B) and outward for lateral rotation (Fig. 7.4C) in a spoke-like fashion.
  2. Execute the movements smoothly and rhythmically, respecting the natural range of motion of each joint.
  3. Hold the end position briefly, allowing the muscles and joints to adapt to the stretch.
  4. Return the limb to the starting position in a controlled manner.
  5. Repeat the movement within a comfortable range, following the prescribed number of repetitions.

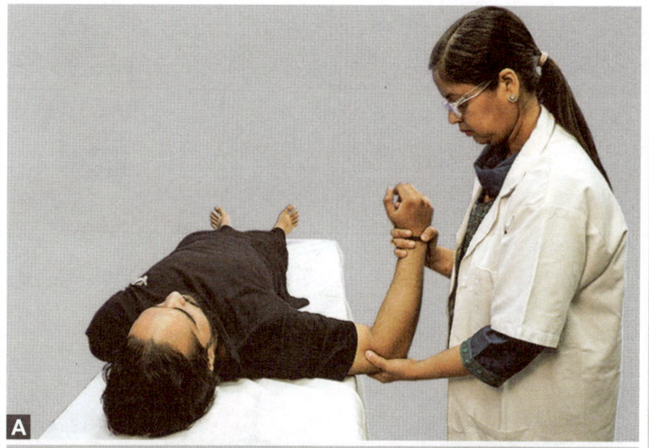

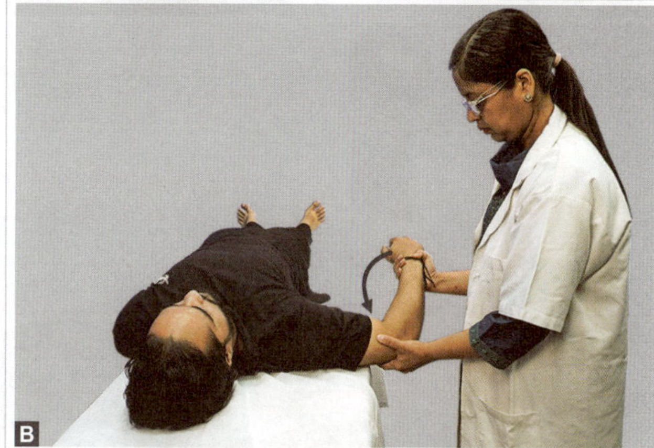

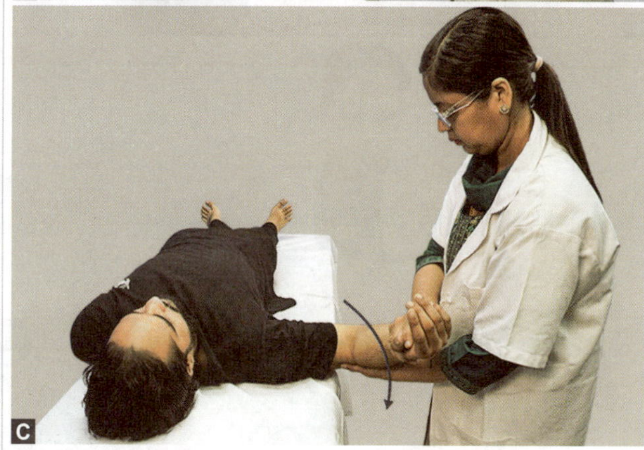

Figs 7.4A to C: Shoulder joint: **A.** Rotation starting position; **B.** Internal rotation; **C.** External rotation

Scapulothoracic Joint

Elevation and Depression

- **Patient's position:** Position the patient either prone lying or side lying with patient facing the therapist.
- **Hand placement:** The therapist typically stands or sits beside the patient when performing scapula elevation and depression allowing for better control and guidance of the movements. Cup one hand over the acromion process and the other around the inferior angle of the scapula. Alternatively, hands can be clasped across the scapula.
- **Procedure:**
  1. For elevation and depression of scapula the therapist moves the hand upward and downward respectively (Figs 7.5A and B).
  2. Execute the movements smoothly and rhythmically, respecting the natural range of motion of each joint.
  3. Hold the end position briefly, allowing the muscles and joints to adapt to the stretch.
  4. Return the limb to the starting position in a controlled manner.
  5. Repeat the movement within a comfortable range, following the prescribed number of repetitions.

Protraction and Retraction

- **Patient position:** Position the patient either prone lying or side lying with patient facing the therapist.
- **Hand placement:** The therapist typically stands or sits beside the patient when performing hand placements for scapular upward and downward rotation, allowing

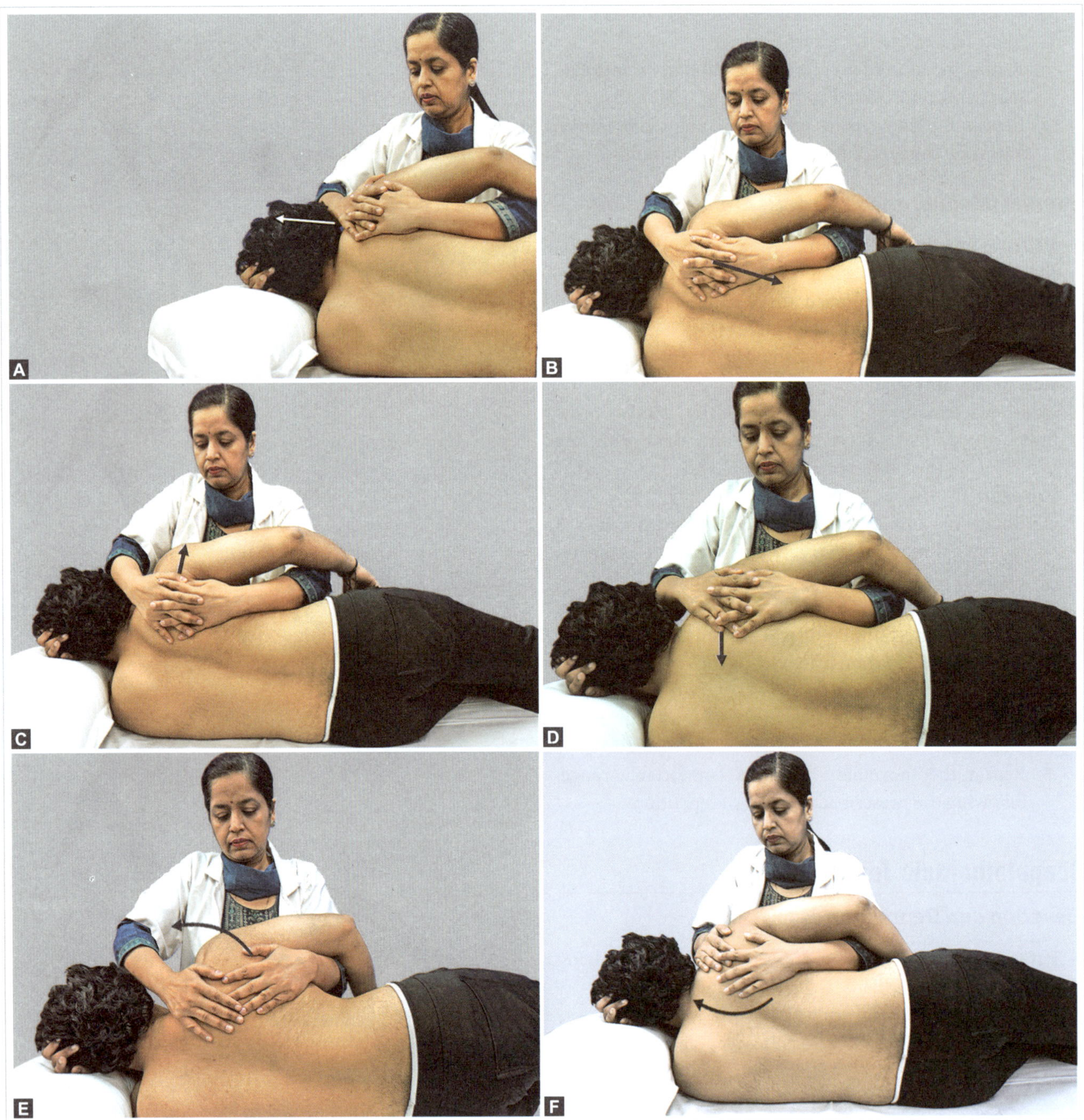

Figs 7.5A to F: Scapulothoracic joint: **A.** Elevation; **B.** Depression; **C.** Protraction; **D.** Retraction; **E.** Upward rotation; **F.** Downward rotation

for better control and guidance of the movements. The therapist positions hands at the medial border of the scapula.

- **Procedure:**
  1. To initiate scapular protraction, place hands on the medial borders of the scapulae and guide them away from the spine (Fig. 7.5C).
  2. For retraction, position hands on the lateral borders and guide the scapulae toward the spine (Fig. 7.5D).
  3. Execute the movements smoothly and rhythmically, respecting the natural range of motion of each joint.
  4. Hold the end position briefly, allowing the muscles and joints to adapt to the stretch.
  5. Return the limb to the starting position in a controlled manner.
  6. Repeat the movement within a comfortable range, following the prescribed number of repetitions.

Upward and Downward Rotation

- **Patient's position:** Position the patient either prone lying or side lying with patient facing the therapist.
- **Hand placement:** The therapist typically stands or sits beside the patient when performing hand placements for scapular upward and downward rotation, allowing for better control and guidance of the movements. The therapist positions hands at inferior angle of scapula and at acromion process (Fig. 7.5E).
- **Procedure:**
  1. To rotate, direct scapular motions at the inferior angle of the scapula while pressing the acromion in the opposite direction, creating a force pair turning effect.
  2. Inferior angle of scapula moves laterally for upward rotation (Fig. 7.5E) and medially for downward rotation (Fig. 7.5F).

Elbow and Forearm

Flexion and Extension

- **Patient position:** Patient should be comfortably seated or lying down with the arm supported. Ensure a relaxed position with a slightly flexed elbow for optimal movement.
- **Hand placement:** Therapist stands or sits facing the patient. While supporting the patient's forearm, grasp the hand.
- **Procedure:**
  1. Lift and lower the forearm gently through its available range for elbow flexion (Fig. 7.6A).
  2. Execute the movements smoothly and rhythmically, respecting the natural range of motion of each joint.
  3. Hold the end position briefly, allowing the muscles and joints to adapt to the stretch.
  4. Return the limb to the starting position in a controlled manner for elbow extension (Fig. 7.6B).
  5. Repeat the movement within a comfortable range, following the prescribed number of repetitions.

Pronation and Supination

- **Patient position:** The patient should be comfortably seated or lying down. The elbow can be flexed or extended based on the desired range of motion.
- **Hand placement:** Therapist stands or sits facing the patient. Secure the patient's wrist, placing the thumb on the dorsal aspect and fingers on the volar side (Figs 7.6C and D). The elbow is stabilized with the other hand. Alternatively, handshake grasp may also be used (Figs 7.6E to G)

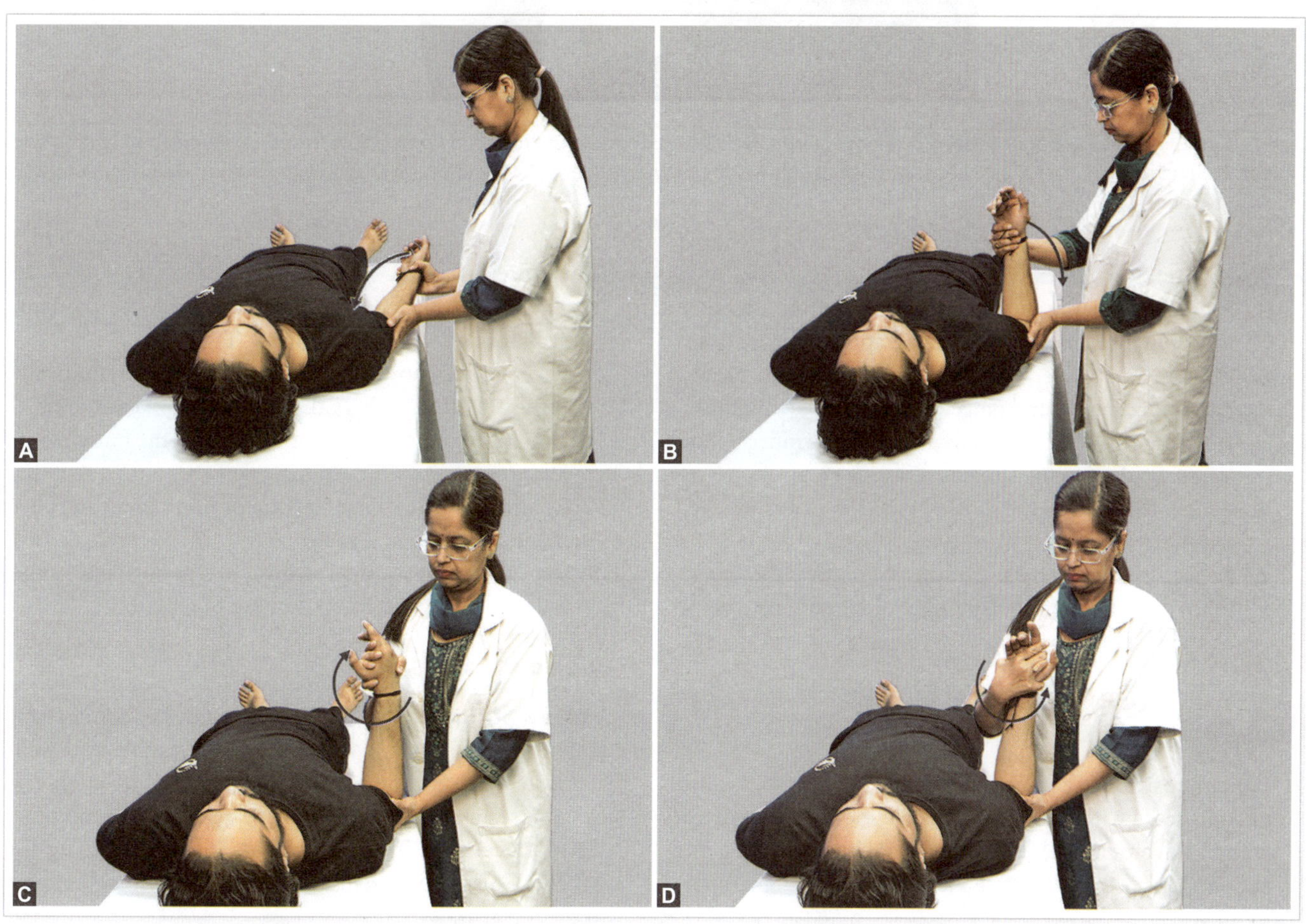

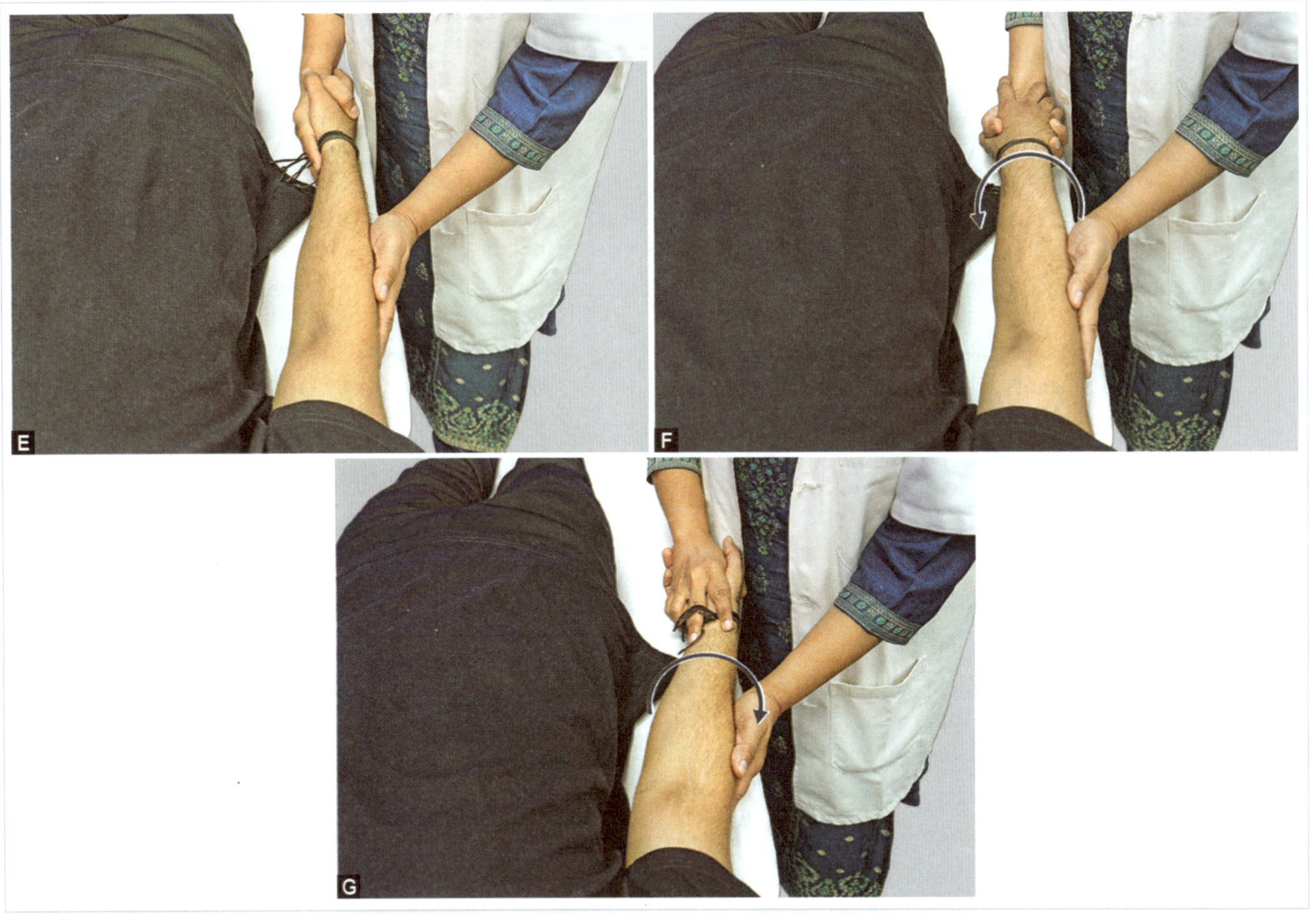

Figs 7.6A to G: Elbow joint and forearm: **A.** Elbow flexion; **B.** Elbow extension; **C.** Pronation (with elbow flexed); **D.** Supination (with elbow flexed); **E.** Starting position for pronation and supination (with elbow extended); **F.** Pronation (with elbow extended); **G.** Supination (with elbow extended)

- **Procedure:**
  1. Rotate the volar aspect of forearm away from therapist for supination and toward the therapist for pronation.
  2. Execute the movements smoothly and rhythmically, respecting the natural range of motion of each joint.
  3. Hold the end position briefly, allowing the muscles and joints to adapt to the stretch.
  4. Return the limb to the starting position in a controlled manner.
  5. Repeat the movement within a comfortable range, following the prescribed number of repetitions.

Wrist Joint

Radial and Ulnar Deviation

- **Patient position:** The patient should be comfortably seated or lying down with the hand accessible and well-supported.
- **Hand placement:** Therapist stands or sits facing the patient. Grasp the patient's hand while providing support to the forearm.
- **Procedure:**
  1. Therapist applies gentle pressure to deviate the hand radially (Fig. 7.7A) and ulnarly (Fig. 7.7B).
  2. Execute the movements smoothly and rhythmically, respecting the natural range of motion of each joint.
  3. Hold the end position briefly, allowing the muscles and joints to adapt to the stretch.
  4. Return the limb to the starting position in a controlled manner.
  5. Repeat the movement within a comfortable range, following the prescribed number of repetitions.

Flexion and Extension

- **Patient position:** The patient should be comfortably seated or lying down with the hand accessible and well-supported. Ensure the elbow is in a relaxed position and forearm supinated, allowing for the desired wrist motion.
- **Hand placement:** The therapist stands or sits facing the patient. Grasp the patient's hand while providing support to the forearm.

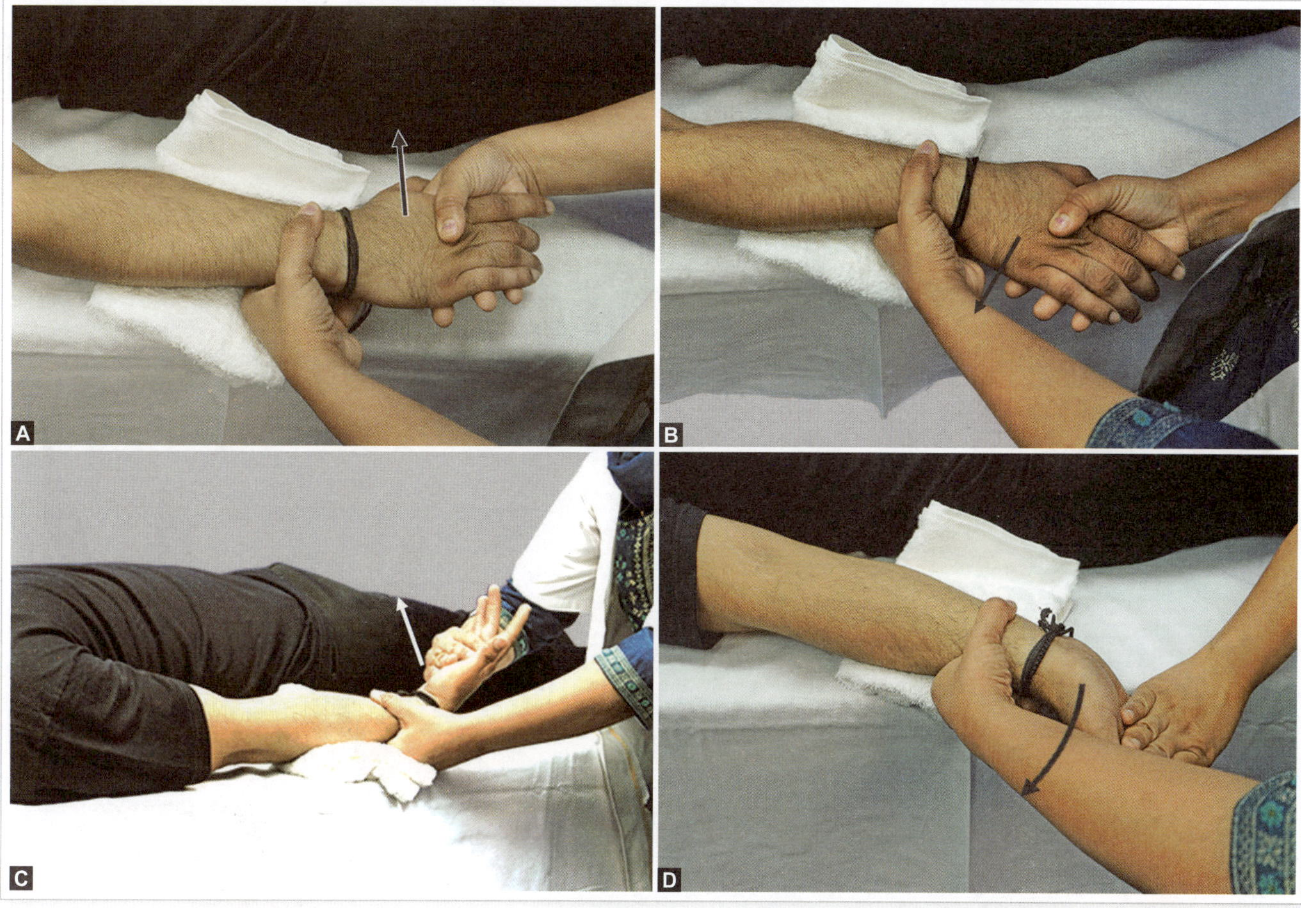

Figs 7.7A to D: Wrist joint: A. Radial deviation; B. Ulnar deviation; C. Flexion; D. Extension

- **Procedure:**
  1. Move the wrist and hand upwards for flexion (Fig. 7.7C) and downwards for extension (Fig. 7.7D) movements of the wrist.
  2. Execute the movements smoothly and rhythmically, respecting the natural range of motion of each joint.
  3. Hold the end position briefly, allowing the muscles and joints to adapt to the stretch.
  4. Return the limb to the starting position in a controlled manner.
  5. Repeat the movement within a comfortable range, following the prescribed number of repetitions.

Carpometacarpal (CMC) and Intermetacarpal Joints

Cupping and Flattening the Arch of the Hand

- **Patient position:** The patient should be comfortably seated or lying down with the hand accessible and well-supported.
- **Hand placement:** The therapist sits or stand facing the patient's hand, positioning the fingers of both the hands into the patient's palm, with thenar eminences on the posterior aspect.
- **Procedure:**
  1. Roll the metacarpals palmar-ward to enhance the arch (Fig. 7.8A), and dorsal-ward to flatten it (Fig. 7.8B).
  2. Execute the movements smoothly and rhythmically, respecting the natural range of motion of each joint.
  3. Hold the end position briefly, allowing the muscles and joints to adapt to the stretch.
  4. Return the limb to the starting position in a controlled manner.
  5. Repeat the movement within a comfortable range, following the prescribed number of repetitions.

Flexion, Extension, Abduction and Adduction at CMC Joint

- **Patient position:** The patient should be comfortably seated or lying down with the forearm supported on a table or armrest, hand relaxed.
- **Hand placement:** The therapist sits or stand facing the patient's hand. One hand stabilizes the base of the first metacarpal (at the wrist), and the other hand holds the thumb.

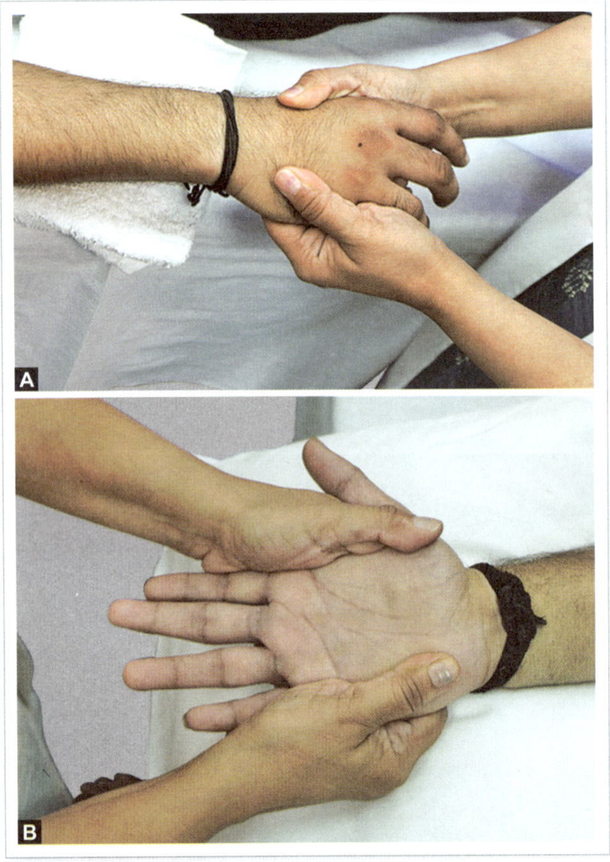

Figs 7.8A and B: Carpometacarpal and intermetacarpal joints: **A.** Hand cupping; **B.** Hand flattening

- **Procedure:**
  1. To perform passive CMC joint movements, guide the thumb toward the palm for flexion (Fig. 7.9A), away from the palm for extension (Fig. 7.9B), outward from the palm for abduction (Fig. 7.9C), and back toward the palm for adduction (Fig. 7.9D), ensuring smooth and comfortable motion throughout.
  2. Execute the movements smoothly and rhythmically.
  3. Hold the end position briefly and then return the limb to its starting position in a controlled manner.
  4. Repeat the movement within a comfortable range, following the prescribed number of repetitions.

Metacarpophalangeal (MCP) and Interphalangeal (IP) Joints

Flexion and Extension

- **Patient position:** The patient should be comfortably seated or lying down with the hand accessible and well-supported.
- **Hand placement:** The therapist stands or sits in front of the patient. Place one hand on the dorsum (back) of the patient's hand, just proximal to the MCP joint, with fingers

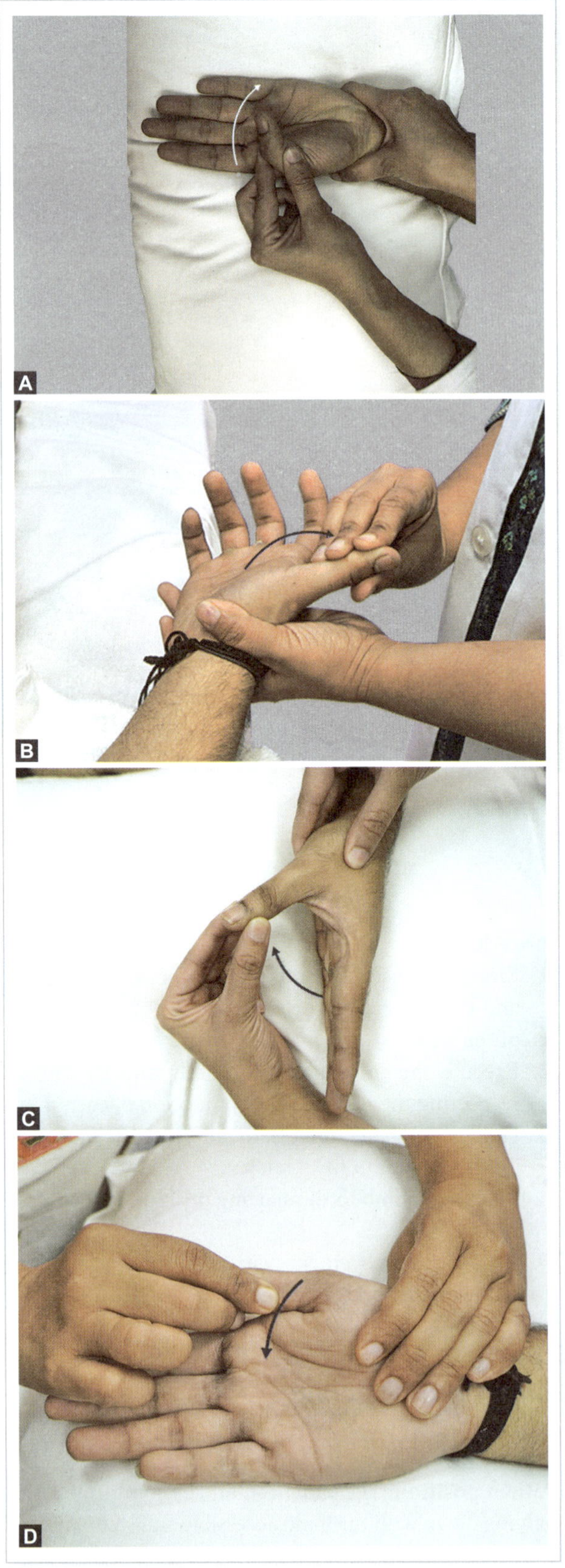

Figs 7.9A to D: Passive movements at continuous passive motion joint: **A.** Flexion; **B.** Extension at continuous passive motion joint; **C.** Abduction; **D.** Adduction

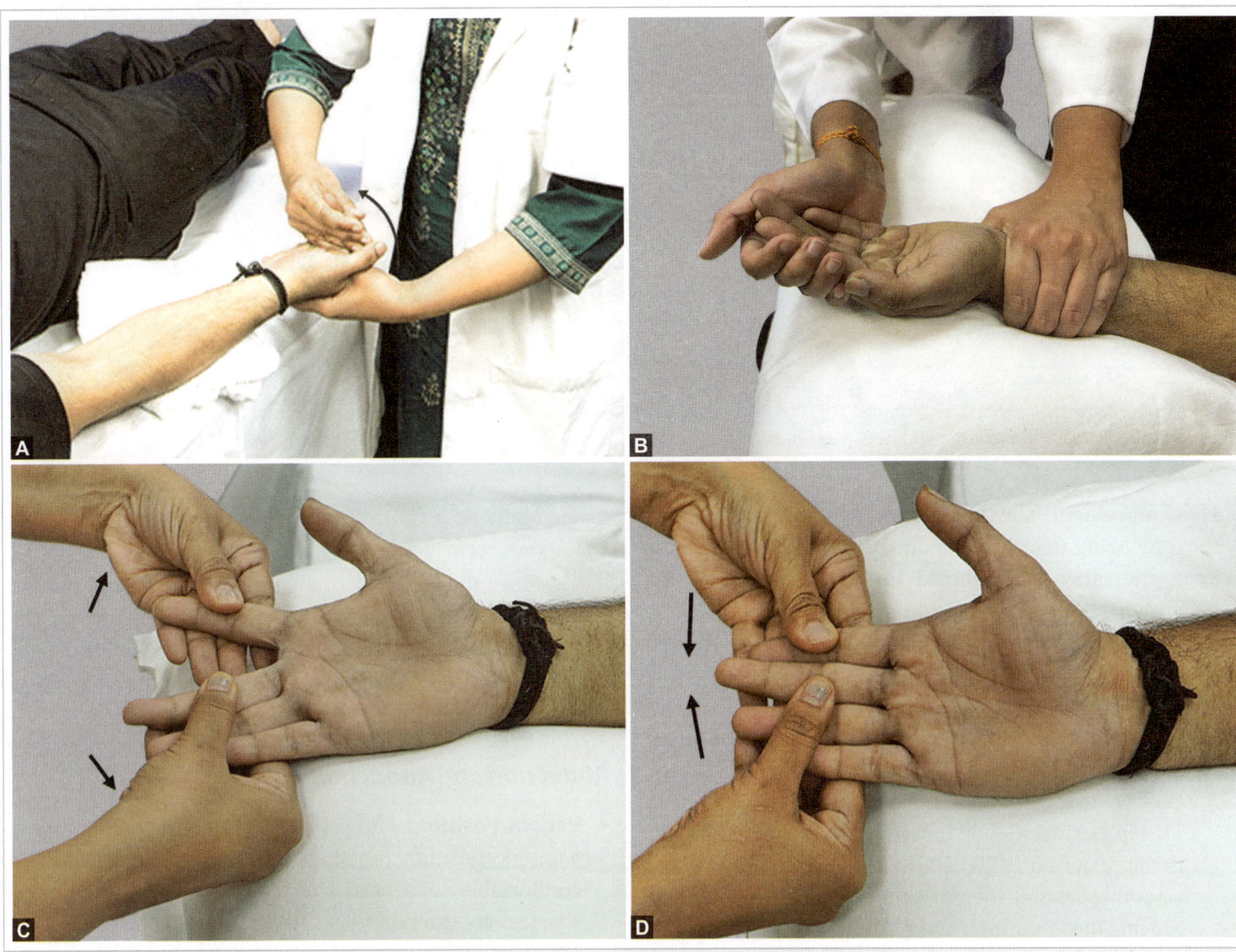

Figs 7.10A to D: Metacarpophalangeal and interphalangeal joints: **A.** Finger flexion; **B.** Finger extension; **C.** Metacarpophalangeal abduction; **D.** Metacarpophalangeal adduction

resting on the metacarpal bones. Use the other hand to support the proximal phalanges (finger bones) at the MCP joint.

- **Procedure:**
  1. For MCP joint flexion, gently guide the fingers toward the palm using controlled motion (Fig. 7.10A). Simultaneously, for MCP joint extension, guide the fingers back to their extended position (Fig. 7.10B).
  2. To perform IP joint flexion, carefully bend the fingers at the proximal or distal interphalangeal joint.
  3. For IP joint extension, guide the fingers to straighten at the interphalangeal joints.
  4. Repeat the movement within a comfortable range, following the prescribed number of repetitions.

Abduction and Adduction

- **Patient position:** Ensure the patient is comfortably seated or lying down, with the hand accessible and well-supported.
- **Hand placement:** The therapist stands or sits in front of the patient. Begin by securely holding the patient's hand, maintaining a stable grip.
- **Procedure:**
  1. For MCP joint abduction, gently guide the fingers away from the midline (Fig. 7.10C).
  2. Simultaneously, for MCP joint adduction, guide the fingers back toward the midline (Fig. 7.10D).
  3. Repeat the movement within a comfortable range, following the prescribed number of repetitions.

Hip Joint

Flexion and Extension

- **Patient position:** The patient should be positioned comfortably on a treatment table or bed, lying in a supine (on their back) position.

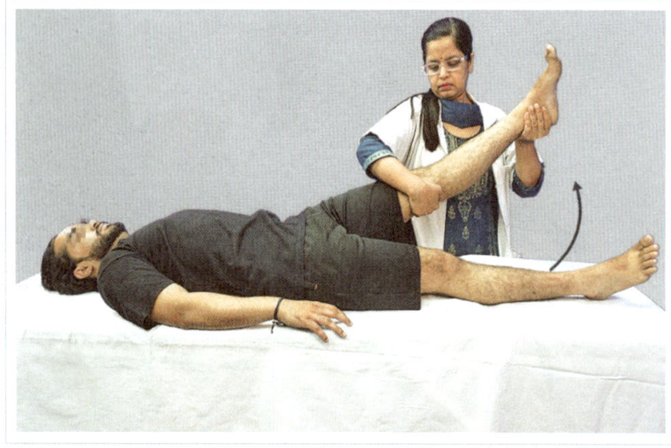

Fig. 7.11: Hip flexion with knee in extension

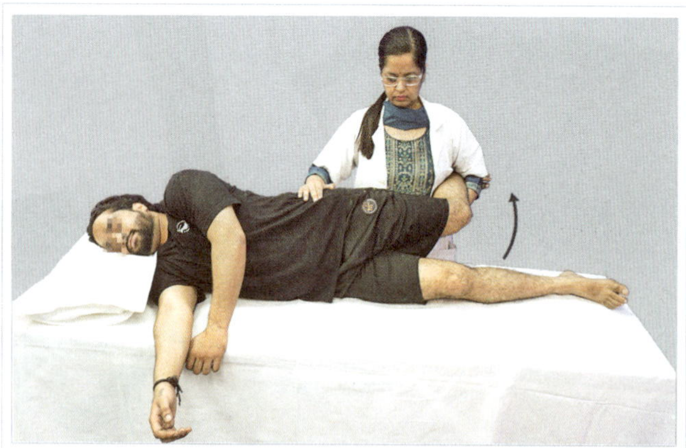

Fig. 7.12: Hip extension in side lying

- **Hand placement:** The therapist should stand on the side of the limb being treated. Place one hand under the patient's knee, the other hand should be positioned under the patient's heel, the hands should be in a position to allow controlled and smooth movement through the desired range of motion.
- **Procedure:**
  1. With one hand supporting the ankle and the other supporting the knee, gently lift the lower limb off the bed or treatment table keeping the knee extended for hip flexion. Gently move the limb down and back or hip extension (Fig. 7.11). Alternatively, slowly and smoothly flex both the hip and knee joints simultaneously, moving the lower limb toward the patient's chest for hip flexion and down and back for hip extension.
  2. Move the limb through a comfortable range of motion, taking care not to force the joint beyond its natural limits.
  3. Repeat the movement within a comfortable range, following the prescribed number of repetitions.

Hyperextension

- **Patient position:** The patient lies on the side on a treatment table, with the hip and knee of the lower leg slightly flexed for comfort. The upper leg is kept straight, and a small pillow may be placed between the knees for support.
- **Hand placement:** The therapist stands at the side of the treatment table, at the back of the patient. Place one hand over the patient's hip for support, near the joint. The other hand is positioned under the patient's thigh. Do not fully flex the knee to avoid the two-joint rectus femoris to restrict the movement. Hands should be arranged to allow controlled and smooth movement during the passive hip extension.
- **Procedure:**
  1. Gradually extend the hip joint, moving the leg backward (Fig. 7.12).
  2. Maintain a controlled movement, avoiding any abrupt or forceful actions.
  3. Move the leg through a comfortable range of motion, considering the individual's flexibility and any existing limitations.
  4. Repeat the movement within a comfortable range, following the prescribed number of repetitions.

Abduction and Adduction

- **Patient position:** The patient lies on a treatment table in a supine position (on their back). Ensure the patient is comfortable, with head and trunk adequately supported. The legs are extended.
- **Hand placement:** The therapist stands on the side of the limb being treated. For hip abduction and adduction place one hand under the patient's knee or thigh. The other hand supports the ankle or heel.
- **Procedure:**
  1. With one hand supporting the knee and the other supporting the ankle or heel, gently lift the lower limb off the bed.
  2. Gradually move the leg away from the midline, performing hip abduction (Fig. 7.13A) and toward the midline, performing hip adduction (Fig. 7.13B).
  3. Move the limb through a comfortable range of motion, avoiding any forceful actions.
  4. Hold the end position briefly to allow the muscles and joints to adapt to the stretch.
  5. Repeat the movement within a comfortable range, following the prescribed number of repetitions.

Internal and External Rotation

- **Patient position:** The patient lies on a treatment table in a supine position (on their back). Ensure the patient

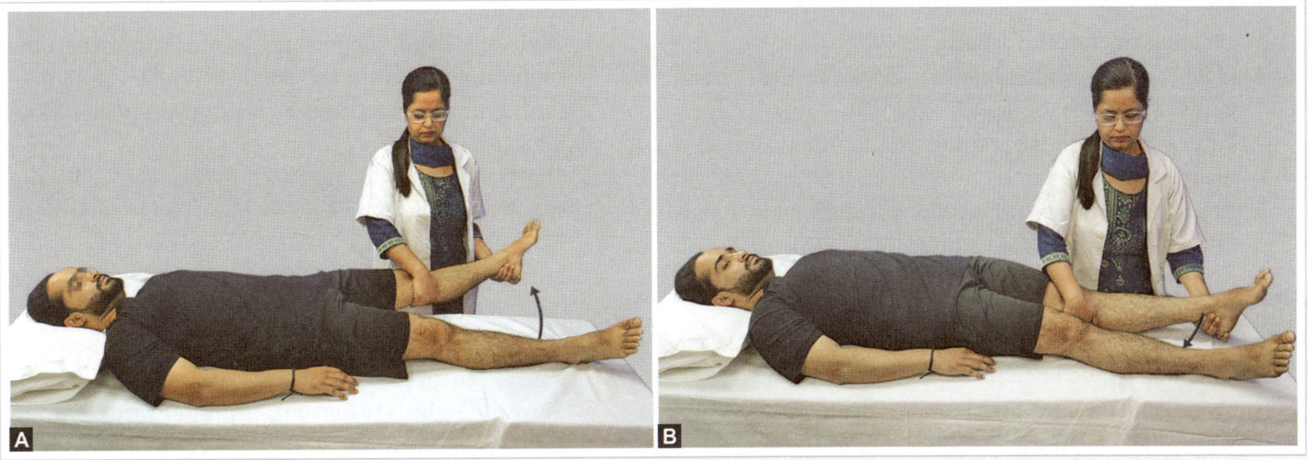

Figs 7.13A and B: Hip joint: **A.** Abduction; **B.** Adduction

is comfortable, with the head and trunk adequately supported. The legs are extended.

- **Hand placement:** The therapist stands on the side of the limb being treated. For hip external and internal rotation place one hand on distal part of thigh and other hand supporting lower part of leg.
- **Procedure:**
  1. With one hand supporting the knee and the other supporting the outer ankle or heel, gently rotate the lower limb inward for medial rotation (Fig. 7.14A) and outward for lateral rotation (Fig. 7.14B).

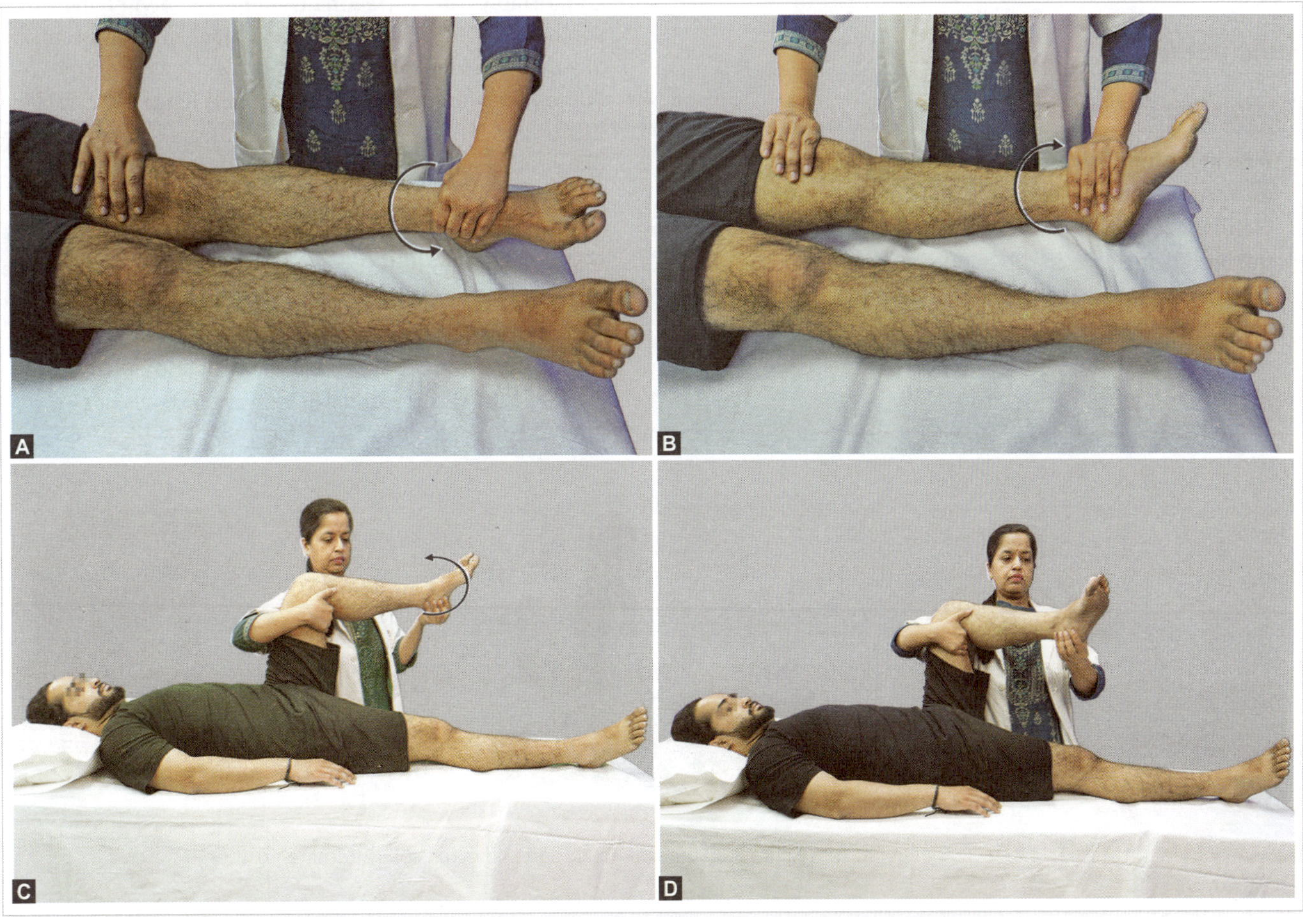

Figs 7.14A to D: Hip joint: **A.** Internal rotation; **B.** External rotation; **C.** Alternative position for hip internal rotation; **D.** Alternative position for hip external rotation

2. Move the limb through a comfortable range of motion, avoiding any forceful actions.
3. Hold the end position briefly to allow the muscles and joints to adapt to the stretch.
4. Return the leg to the starting position in a controlled manner.
5. Repeat the movement within a comfortable range, following the prescribed number of repetitions.

Alternate Method

1. In the procedure for rotation with the hip and knee flexed, the patient's hip and knee are flexed to approximately 90°.
2. The top hand is positioned under the patient's knee, offering initial support and stabilization to the knee joint and the other hand grasps the ankle or the heel. If there is instability in the knee, the bottom hand cradles the thigh and provides additional support to the proximal calf and knee.
3. The femur is then rotated by moving the leg in a pendulum-like motion (Figs 7.14C and D).
4. While this hand placement offers support to the knee, caution should be exercised, particularly in cases of knee instability.

Knee Joint

Flexion and Extension

- **Patient position:** The patient should be positioned comfortably on a treatment table or bed, lying in a prone position.
- **Hand placement:** The therapist should stand on the side of the limb being treated. Place one hand under the patient's knee or at the distal thigh, the other hand grasps the patient's ankle, the hands should be in a position to allow controlled and smooth movement through the desired range of motion.
- **Procedure:**
  1. With one hand supporting the thigh and the other supporting the lower leg, gently lift the lower limb off the bed or treatment table by bending the patient's knee joint (Fig. 7.15A).
  2. Slowly and smoothly flex both the hip and knee joints simultaneously, moving the lower limb toward the patient's chest.
  3. Move the knee and leg through a comfortable range of motion, avoiding forceful actions.
  4. Hold the end position briefly to allow the muscles and joints to adapt to the stretch.
  5. Move the leg back to the starting position in a controlled manner for knee extension (Fig. 7.15B).
  6. Alternatively, the patient can be in a supine lying position (Fig. 7.15C).
  7. Repeat the movement within a comfortable range, following the prescribed number of repetitions.

Ankle Joint and Foot

Dorsiflexion and Plantar Flexion

- **Patient position:** The patient should be positioned comfortably on a treatment table or bed, lying in a supine position. Ensure that the patient can communicate any discomfort during the procedure.
- **Hand placement:** The therapist stands toward the test side of the patient. For ankle dorsiflexion and plantarflexion place one hand on the heel so that the therapist's forearm supports the plantar surface of the foot (Figs 7.16A and B). The other hand is used to stabilize the lower leg around the malleoli. Alternatively, the hands can be clasped across the dorsum of the foot (Figs 7.16C and D).
- **Procedure:**
  1. Gently lift the foot toward the shin and downward for ankle dorsi and plantar flexion respectively.

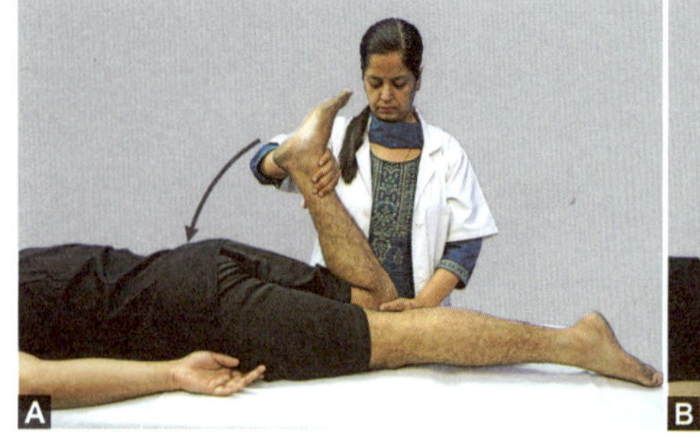

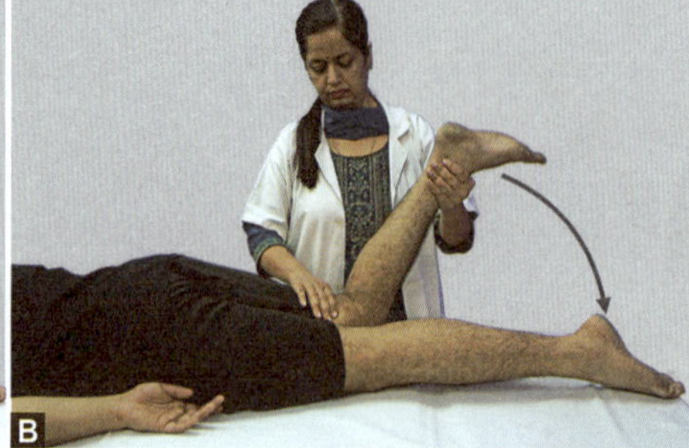

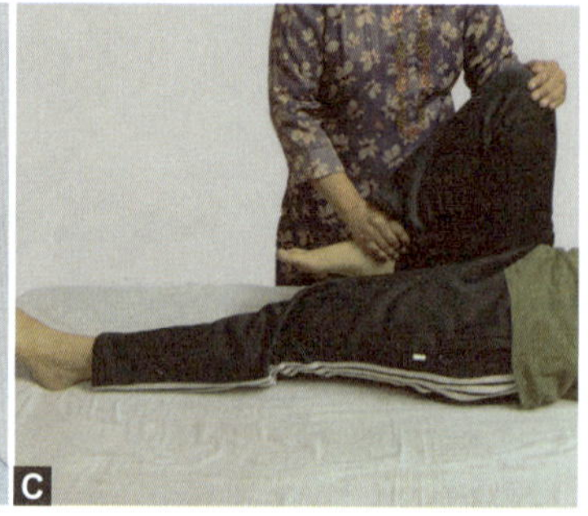

Figs 7.15A to C: Knee joint: **A.** Flexion; **B.** Extension; **C.** Flexion and extension (Alternate position)

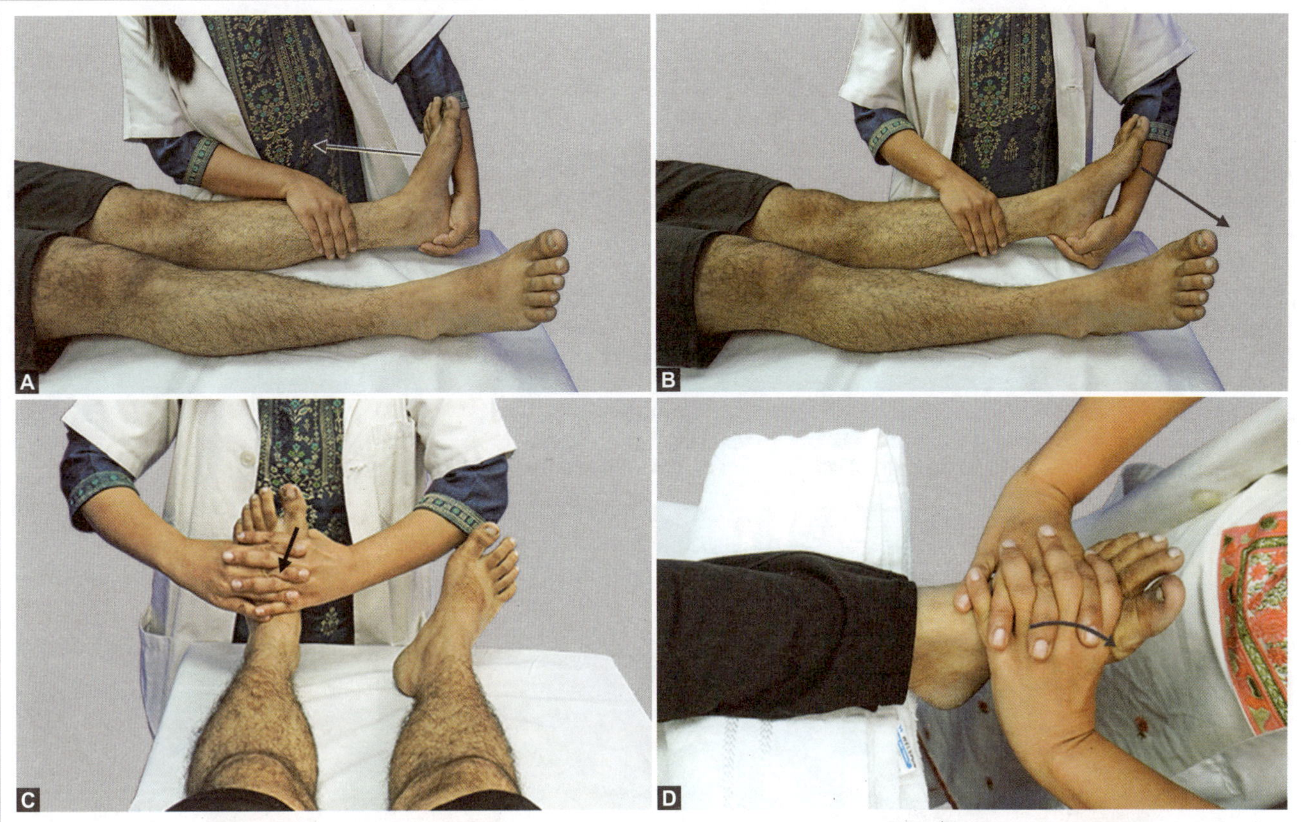

Figs 7.16A to D: Ankle joint and foot: **A.** Dorsiflexion; **B.** Plantarflexion; **C.** Dorsiflexion (Alternate grasp); **D.** Plantarflexion (Alternate grasp)

2. Move the foot through a comfortable range of motion, avoiding forceful actions.
3. Hold the end position briefly to allow the muscles and joints to adapt to the stretch.
4. Move the foot back to the starting position in a controlled manner.
5. Repeat the movement within a comfortable range, following the prescribed number of repetitions or therapist's guidance.

Inversion and Eversion

- **Patient position:** The patient is typically seated on a treatment table or lying in a supine position. Ensure the patient is comfortable, and their lower leg and foot are adequately supported.
- **Hand placement:** The therapist stands beside the patient. For subtalar joint inversion and eversion place thumb of one hand on the inside of the heel and fingers on the outer side, supporting the lateral aspect of the foot. The other hand is used to stabilize the lower leg if needed.
- **Procedure:**
  1. With one hand supporting the heel and the lateral aspect of the foot, gently tilt the heel inward for inversion (Fig. 7.17A) and outward for eversion (Fig. 7.17B).
  2. Move the foot through a comfortable range of motion, avoiding forceful actions.
  3. Hold the end position briefly to allow the subtalar joint to adapt to the stretch.
  4. Return the foot to the starting position in a controlled manner.
  5. Repeat the movement within a comfortable range, following the prescribed number of repetitions.

Metatarsophalangeal (MTP) and Interphalangeal (IP) Joints Flexion and Extension

- **Patient position:** The patient should be comfortably seated or lying down with the leg extended and supported, foot relaxed.
- **Hand placement:** The therapist sits or stands toward the affected side. One hand stabilizes the metatarsal or proximal phalanx, while the other hand holds the toe (Figs 7.18A to D).
- **Procedure:**
  1. To perform passive MTP joint movements, guide the toe toward the sole for flexion and away from the sole for extension.
  2. For IP joint movements, guide the toe toward the sole for flexion and away from the sole for extension.

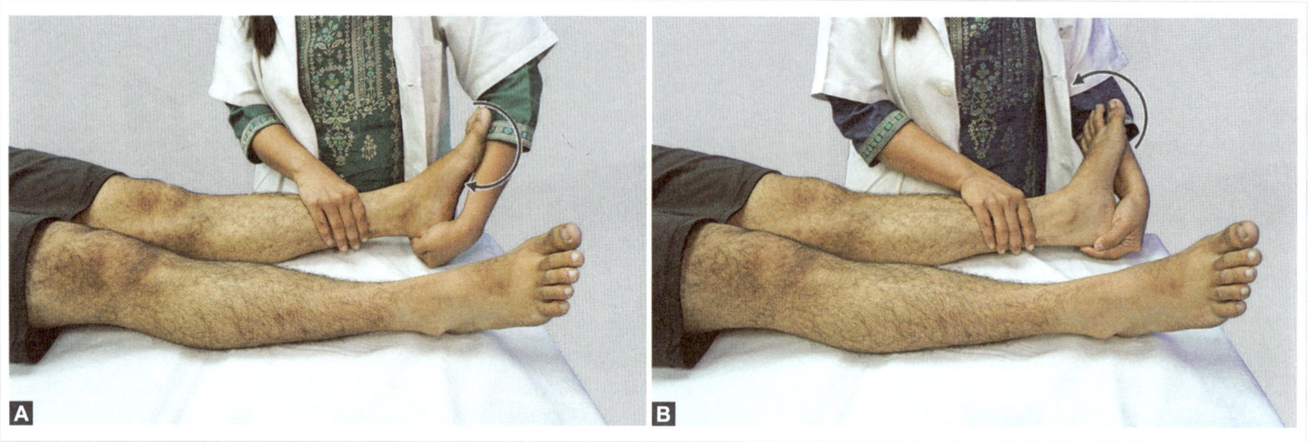

Figs 7.17A and B: Ankle joint and foot: **A.** Subtalar inversion; **B.** Subtalar eversion

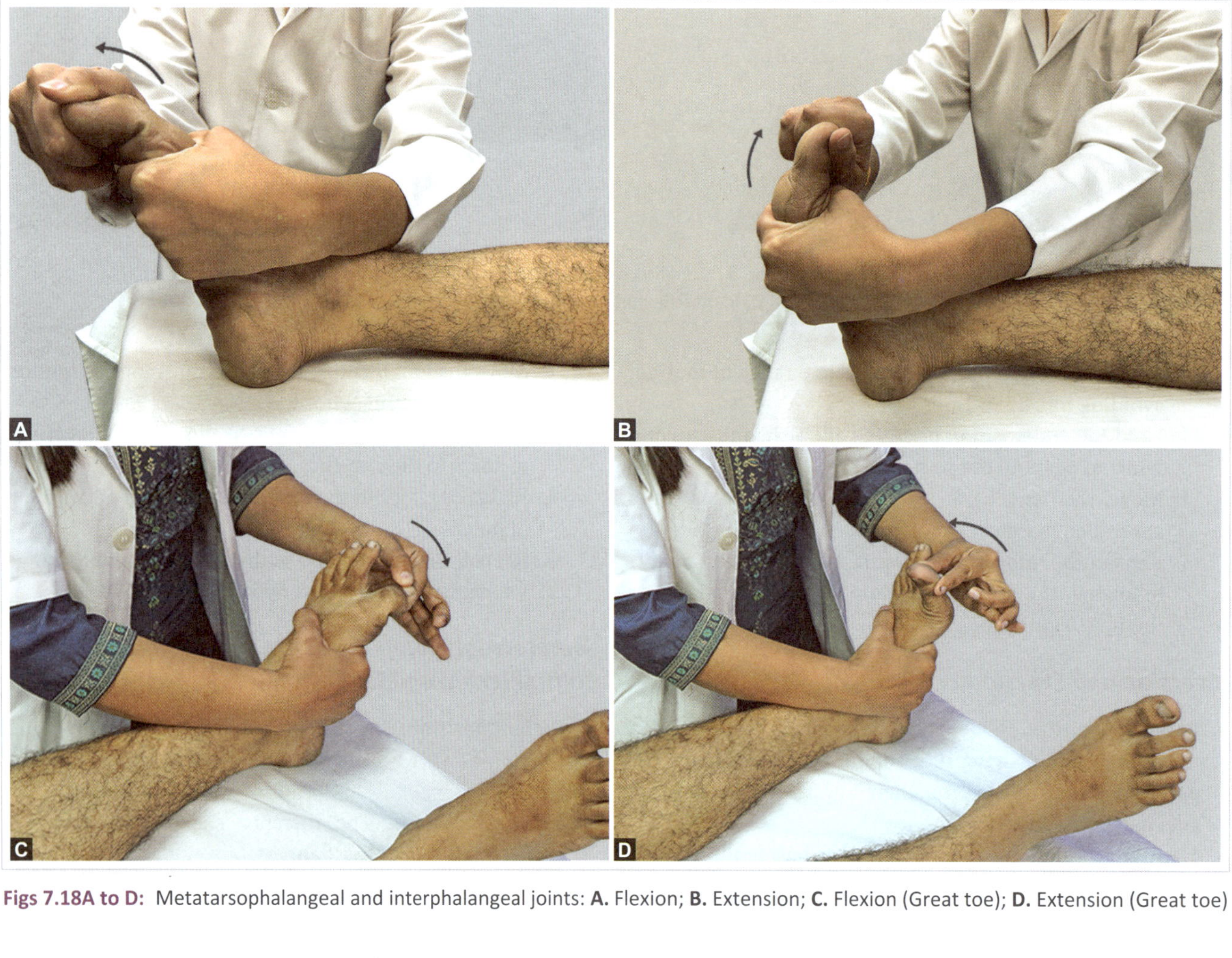

Figs 7.18A to D: Metatarsophalangeal and interphalangeal joints: **A.** Flexion; **B.** Extension; **C.** Flexion (Great toe); **D.** Extension (Great toe)

3. Execute the movements smoothly and rhythmically, holding the end position briefly before returning the toe to its starting position in a controlled manner.
4. Repeat the movement within a comfortable range, following the prescribed number of repetitions.

Cervical Spine

Flexion and Extension

- **Patient position:** The patient lies on their back in a supine position on a treatment table with the head out of the edge of the table.
- **Hand placement:** The therapist stands at the head of the table. Place both hands on the patient's occiput (the back of the head), with the fingers supporting the base of the skull.
- **Procedure:**
    1. Gently guide the patient's head forward, flexing the cervical spine (Fig. 7.19A).
    2. Move the chin toward the chest in a controlled manner.
    3. Move the head through a comfortable range, avoiding any sudden or forceful actions.
    4. Ensure the patient feels a gentle stretch without discomfort.
    5. Hold the flexed position briefly to allow the cervical spine to adapt to the stretch.
    6. Gradually bring the patient's head back to the starting position in a controlled manner.
    7. Ensure smooth and controlled movement throughout.
    8. Repeat the movement within a comfortable range, considering the patient's tolerance and any specific guidelines provided by the therapist.
    9. For cervical spine extension move the head backward (Fig. 7.19B).
    10. Patient position and hand placement remains same.

Lateral Flexion and Rotation

- **Patient position:** The patient lies on their back in a supine position on a treatment table.

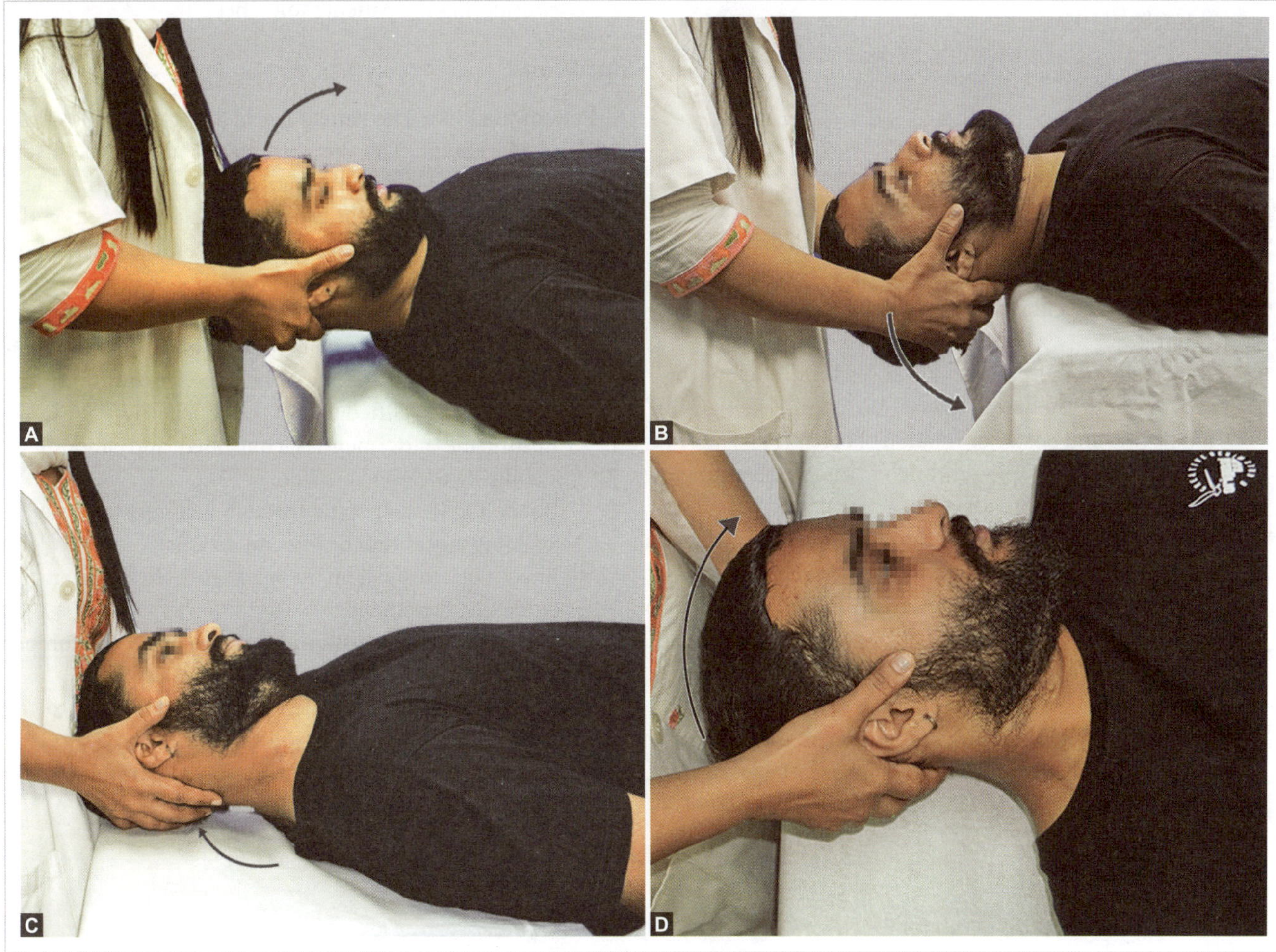

Figs 7.19A to D: Cervical spine: **A.** Flexion; **B.** Extension; **C.** Lateral flexion (left side); **D.** Rotation (left side)

- **Hand placement:** The therapist stands at the head of the table. Place both hands on the patient's occiput (the back of the head), with the fingers supporting the base of the skull and the thumbs resting on the upper neck or around the jaw.
- **Procedure:**
  1. Gently move the patient's head and neck side to side for laterally flexing the cervical spine (Fig. 7.19C).
  2. Move in a controlled manner through a comfortable range, avoiding any sudden or forceful actions.
  3. Ensure the patient feels a gentle stretch without discomfort.
  4. Hold the laterally flexed position briefly to allow the cervical spine to adapt to the stretch.
  5. Gradually bring the patient's head back to the starting position in a controlled manner.
  6. Ensure smooth and controlled movement throughout.
  7. Repeat the movement within a comfortable range, considering the patient's tolerance and any specific guidelines provided by the therapist.
  8. For cervical spine rotation, rotate the head from side to side (Fig. 7.19D).

Lumbar Spine

Flexion

- **Patient position:** The patient lies on their back in a supine position on a treatment table with legs extended.
- **Hand placement:** The therapist stands at the side of the table. The therapist grasps the legs and knees.
- **Procedure:**
  1. Gently guide the patient's lower back into flexion by moving both the knees toward the chest causing full hip flexion and posterior rotation of pelvis (Fig. 7.20A).
  2. Move the lumbar spine through a comfortable range, avoiding any abrupt or forceful actions.
  3. Ensure the movement is smooth and controlled, focusing on the flexibility of the lumbar region.
  4. Hold the flexed position briefly to allow the lumbar spine to adapt to the stretch.
  5. Gradually bring the patient's lower back to the neutral position in a controlled manner.
  6. Ensure smooth and controlled movement throughout.
  7. Repeat the movement within a comfortable range, considering the patient's tolerance.

Extension

- **Patient position:** The patient lies on their abdomen in a prone position on a treatment table.
- **Hand placement:** The therapist stands at the side of the table. Place one hand at the patient's lumbar spine, providing gentle support, typically around the lower back. The other hand is placed under the thighs.
- **Procedure:**
  1. Gently guide the patient's lower back into extension, by lifting the thighs upward causing the pelvis to rotate anteriorly, encouraging a slight arch in the lumbar spine (Fig. 7.20B).
  2. Move the lumbar spine through a comfortable range, avoiding any abrupt or forceful actions.
  3. Ensure the movement is smooth and controlled, focusing on the extension of the lumbar region.
  4. Hold the extended position briefly to allow the lumbar spine to adapt to the stretch.
  5. Gradually bring the patient's lower back to the neutral position in a controlled manner.
  6. Repeat the movement within a comfortable range, considering the patient's tolerance.

Rotation

- **Patient position:** The patient lies in a hook lying position.
- **Hand placement:** The therapist stands at the side of the table. Place one hand on the patient's knees to stabilize and guide the movement. The other hand is positioned on the opposite shoulder or chest, depending on the direction of rotation.
- **Procedure:**
  1. Gently guide the patient's lower body into rotation by pushing both the knees in one direction until the opposite side is lifted up, off the couch (Fig. 7.20C).
  2. Ensure the movement is controlled and smooth by stabilizing the opposite side.
  3. Rotate the lumbar spine through a comfortable range, avoiding any abrupt or forceful actions.
  4. Encourage the patient to relax and let the therapist guide the movement.
  5. Hold the rotated position briefly to allow the lumbar spine to adapt to the stretch.
  6. Gradually bring the patient's lower body back to the neutral position in a controlled manner.
  7. Repeat the movement within a comfortable range, considering the patient's tolerance.

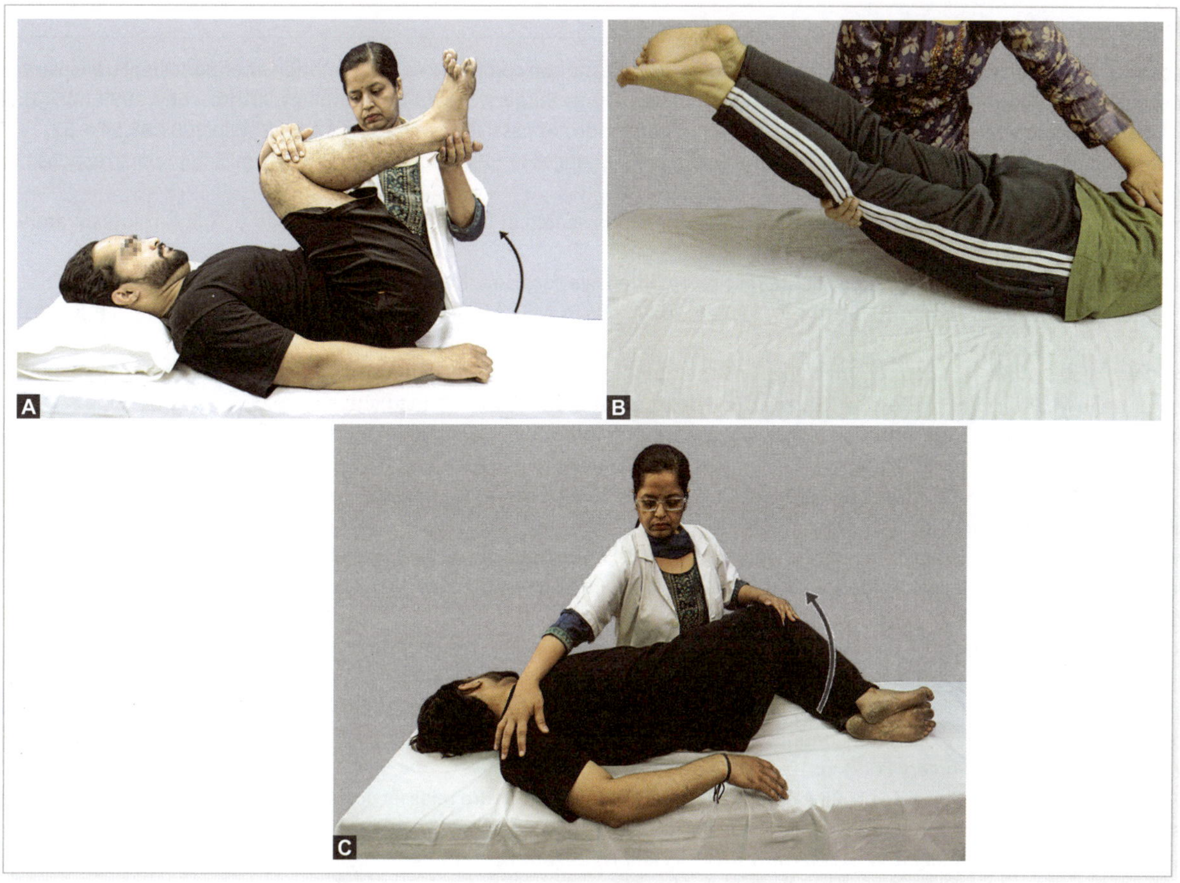

Figs 7.20A to C: Lumbar spine: A. Flexion; B. Extension; C. Rotation (left side)

SUMMARY

- Range of motion (ROM) is the amount of osteokinematic motion accessible for movement activities, which is governed by factors such as joint shape and soft tissue elasticity.
- Immobility can be caused by musculoskeletal injuries, neurological conditions, joint issues, chronic pain, surgery, obesity, circulatory or respiratory problems, infections, age-related changes, trauma, or psychological factors, all of which necessitate therapeutic range of motion exercises to prevent contracture formation and maintain mobility.
- Passive motions (PMs) are when external forces, such as therapists or mechanical devices, move a person's joints to maintain or enhance joint mobility.
- PMs, which include mobilization, manipulation, and controlled stretching, aim to impact soft tissue extensibility and are especially effective for people who have limited motion owing to conditions such as paralysis, pain etc.
- Continuous passive motion (CPM) devices, which are a subset of PMs, provide therapeutic benefits by encouraging joint movement, reducing stiffness, improving circulation, and preventing problems like adhesions, but they must be carefully monitored and contraindicated.
- Principles and procedures for applying passive movements involve thorough planning, patient assessment, and documentation.
- Patient preparation includes explaining techniques, removing constrictive garments, and positioning for optimal movement.
- The technique for relaxed passive movements emphasizes patient relaxation, joint fixation, proper support, stance, traction, and controlled movement with gradual speed and appropriate repetitions.
- Indications for passive movements include acute inflammation, paralysis, postoperative cases, reconstructive surgeries, comatose or bedridden patients, and critical care situations.
- The effects of passive movements encompass preserving joint flexibility, preventing adhesions, aiding neuromuscular reeducation, improving synovial motion, and various physiological benefits.
- However, contraindications include immediate postsurgery or acute injury, conditions like myositis ossificans, unhealed fractures, and life-threatening situations.
- Overall, passive movements play a crucial role in rehabilitation and healthcare across diverse medical conditions.

FURTHER READINGS

- Alfredson H, Lorentzon R. Superior results with continuous passive motion compared to active motion after periosteal transplantation: a retrospective study of human patella cartilage defect treatment. Knee Surgery, Sports Traumatology, Arthroscopy. 1999 Jul;7(4):232–8.
- Bell H, Ramsaroop DM, Duffin J. The respiratory effects of two modes of passive exercise. Eur J Appl Physiol2003;88:544–52.
- Boulain T, Achard JM, Teboul JL, Richard C, Perrotin D, Ginies G. Changes in BP induced by passive leg raising predict response to fluid loading in critically ill patients. Chest 2002;121:1245–52.
- Chartered Society of Physiotherapy. Priorities for physiotherapy research in the UK: Topics prioritised by the cardiorespiratory panel. London: Chartered Society of Physiotherapy; 2002.
- Clini E, Ambrosino N. Early physiotherapy in the respiratory intensive care unit. Respir Med 2005;99:1096–104.
- Dos Santos, L.J., de Aguiar Lemos, F., Bianchi, T., Sachetti, A., Acqua, A.M.D., da Silva Naue, W., Dias, A.S. and Vieira, S.R.R., 2015. Early rehabilitation using a passive cycle ergo.
- Frank CAW, Woo S, Amiel D, Coutts R. Physiology and therapeutic value of passive joint motion. Clin OrthopRelat Res 1984;185:113–25
- Gardiner Dena M. Principles of Exercise Therapy. CBS Publishers and Distributors Pvt Ltd; 4th ed., 2023.
- Harvey LA, Brosseau L, Herbert RD. Continuous passive motion following total knee arthroplasty in people with arthritis. Cochrane Database of Systematic Reviews. 2014(2).
- Hill AD, Palmer MJ, Tanner SL, Snider RG, Broderick JS, Jeray KJ. Use of continuous passive motion in the postoperative treatment of intra-articular knee fractures. JBJS. 2014 Jul 16;96(14):e118.
- Howard JS, Mattacola CG, Romine SE, Lattermann C. Continuous Passive Motion, Early Weight Bearing, and Active Motion following Knee Articular Cartilage Repair: Evidence for Clinical Practice. Cartilage. 2010 Oct;1(4):276–86. doi: 10.1177/1947603510368055. PMID: 26069559; PMCID: PMC4297055.
- Jigjid E, Kawashima N, Ogata H, et al. Effects of passive leg movement on the oxygenation level of lower limb muscle in chronic stroke patients. Neurorehabil Neural Repair. 2008;22(1):40-49. doi:10.1177/1545968307302927
- Kisner, Carolyn., and Lynn Allen Colby. Therapeutic Exercise: Foundations and Techniques. 5th ed. Philadelphia: F.A. Davis, 2007.
- Lastayo PC, Wright T, Jaffe R, Hartzel J, Continuous Passive Motion after Repair of the Rotator Cuff. A Prospective Outcome Study. The Journal of Bone and Joint Surgery 80(7): p. 1002–11, July 1998.
- Lindenhovius AL, van de Luijtgaarden K, Ring D, Jupiter J. Open elbow contracture release: postoperative management with and without continuous passive motion. The Journal of hand surgery. 2009 May 1;34(5):858–65.
- Llano-Diez M, Renaud G, Andersson M, Marrero HG, Cacciani N, Engquist H, et al. Mechanisms underlying ICU muscle wasting and effects of passive mechanical loading. Crit Care 2012;16(5):R209.
- Lynch D, Ferraro M, Krol J, Trudell CM, Christos P, Volpe BT. Continuous passive motion improves shoulder joint integrity following stroke. Clinical Rehabilitation. 2005;19(6):594–599.
- McCarthy MR, O'Donoghue PC, Yates CK, Yates-McCarthy JL. The clinical use of continuous passive motion in physical therapy. Journal of Orthopaedic and Sports Physical Therapy. 1992 Mar;15(3):132–40.
- Matteis M, Vernieri F, Troisi E, Pasqualetti P, Tibuzzi F, Caltagirone C, Silvestrini M. Early cerebral hemodynamic changes during passive movements and motor recovery after stroke. Journal of neurology. 2003 Jul;250(7):810–7.
- Nugent-Derfus GE, Takara T, O'neill JK, Cahill SB, Görtz S, Pong T, Inoue H, Aneloski NM, Wang WW, Vega KI, Klein TJ. Continuous passive motion applied to whole joints stimulates chondrocyte biosynthesis of PRG4. Osteoarthritis and cartilage. 2007 May 1;15(5):566–74.
- O Driscoll SW, Giori NJ. Continuous passive motion (CPM): Theory and principles of clinical application. Journal of rehabilitation research and development. 2000 Apr;37(2):179–88.
- Paillard J, Brouchon M. Active and passive movements in the calibration of position sense. The neuropsychology of spatially oriented behavior. 1968;11:37–55.
- Prabhu RK, Swaminathan N, Harvey LA. Passive movements for the treatment and prevention of contractures. Cochrane Database of Systematic Reviews. 2013(12).
- Rosen MA, Jackson DW, Atwell EA. The efficacy of continuous passive motion in the rehabilitation of anterior cruciate ligament reconstructions. The American journal of sports medicine. 1992 Mar;20(2):122–7.
- Salter RB, Clements ND, Ogilvie-Harris D, Bogoch ER, Wong DA, Bell RS, Minster R: The healing of articular tissues through continuous passive motion. Essence of the first ten years of experimental investigations. I Bone Surg 64B(6):640, 1982
- Schmit, B.D., Dewald, J.P. and Rymer, W.Z., 2000. Stretch reflex adaptation in elbow flexors during repeated passive movements in unilateral brain-injured patients. Archives of physical medicine and rehabilitation, 81(3), p.269–278.
- Stockley RC, Hughes J, Morrison J, Rooney J. An investigation of the use of passive movements in intensive care by UK physiotherapists. Physiotherapy. 2010 Sep 1;96(3):228–33.
- Winkelman C, Higgins PA, Chen YJ, Levine AD. Cytokines in chronically critically ill patients after activity and rest. Bio Res Nurs2007;8: 261–71.
- Zhao HY, Han JT, Liu JQ, Wang HT, Zhou Q, Zhu C, Lu Y, Hu DH. Effects of hand continuous passive motion system combined with functional training and pressure gloves in treating early scar contracture after burn on the back of the hand. Zhonghua shao shang za zhi= Zhonghua shaoshang zazhi= Chinese journal of burns. 2021 Apr 1;37(4):319–26.

STUDENT ASSIGNMENT

LONG ANSWER QUESTIONS

1. Explain and classify passive movements.
2. Explain the indications, contraindications, effects and uses of passive movements.
3. Discuss technique of relaxed passive movement with suitable example.

SHORT ANSWER QUESTIONS

1. Write short note on continuous passive motion.
2. Write about effects and uses of relaxed passive motion.
3. Classify range of motion exercises.

MULTIPLE CHOICE QUESTIONS

1. **____________ are small, repeated, rhythmical, oscillatory, localized accessory, or functional motions carried out by the physiotherapist in a variety of amplitudes while the patient is in control.**
   a. Manipulation b. Mobilization
   c. Stretching d. All of these
2. **Passive motion cannot be given in:**
   a. Postoperative cases like TKR, THR etc
   b. Reconstructive surgeries like ligament repair
   c. Comatosed or bed ridden patients
   d. Immediately after surgery
3. **_________ are precisely localized, single, swift, and decisive movements that are finished at high speeds before the patient can stop them.**
   a. Manipulation b. Mobilization
   c. Stretching d. None of these
4. **What does CPM stand for in the context of rehabilitation?**
   a. Continuous passive motion
   b. Controlled physical manipulation
   c. Comfortable passive movement
   d. Continuous progressive mobilization
5. **In the context of passive movements, what is the primary objective of fixation during the technique?**
   a. Enhancing patient comfort
   b. Localizing movement to the joint
   c. Preventing muscle contractions
   d. Facilitating rapid movements
6. **What is the recommended speed for performing passive movements to ensure patient relaxation?**
   a. Slow and abrupt b. Fast and erratic
   c. Gradual and rhythmic d. Sudden and forceful
7. **Indications for passive movements include:**
   a. Chronic inflammation
   b. Stable fractures
   c. Conscious and active patients
   d. Critical care situations
8. **Which physiological effect is NOT associated with passive movements?**
   a. Improved circulation
   b. Prevention of myofascial adhesions
   c. Reduced joint flexibility
   d. Activation of kinesthetic receptors
9. **What is a contraindication for applying passive movements?**
   a. Chronic inflammation
   b. Unhealed fractures
   c. Critical care situations
   d. Controlled muscle contractions
10. **In the context of continuous passive motion, what is the purpose of adjusting the settings based on the patient's unique requirements?**
    a. To maintain joint stability
    b. To increase discomfort
    c. To maximize overstretching
    d. To prevent appropriate usage

11. **When is passive motion considered most helpful in postsurgery or postinjury situations?**
    a. During the acute inflammation phase
    b. After 2–6 days of inflammation
    c. In the presence of stable fractures
    d. In conscious and active patients
12. **What is the main objective of passive movements in critically ill patients in intensive care?**
    a. To induce pain for alertness
    b. To prolong ICU stays
    c. To prevent neuromuscular complications
    d. To exacerbate inflammation
13. **Which term refers to the technique that involves externally moving a patient's joints without their deliberate muscle activity?**
    a. Controlled physical manipulation
    b. Passive exercise
    c. Controlled progressive mobilization
    d. Active assisted range of motion

ANSWER KEY

1. b	**2.** d	**3.** a	**4.** a	**5.** b	**6.** c	**7.** d	**8.** c	**9.** b	**10.** a
11. a	**12.** c	**13.** b							

8 Active Movements

Sheetal Kalra

LEARNING OBJECTIVES

After the completion of the chapter, the readers will be able to:

- Understand structure of skeletal muscle, chemical and mechanical events during contraction and relaxation, muscle fiber type, motor unit, force gradation.
- Explain causes of decreased muscle performance.
- Define strength, power, work and endurance.
- Understand physiologic adaptations to training: Strength and power, endurance.
- Explain active movements and it is types.
- Describe free exercise, its principles, techniques, indications, contraindications, effects and uses.
- Explain active assisted exercise, its principles, techniques, indications, contraindications, effects and uses.
- Explain assisted-resisted exercise, its principles, techniques, indications, contraindications, effects and uses.

CHAPTER OUTLINE

- Introduction
- Basic Structure of a Muscle
- Muscle Performance
- Physiological Adaptations
- Active Exercises

KEY TERMS

Athetosis: Involuntary, slow-moving writhing movements, usually involving the hands or fingers. It is observed in conditions such as cerebral palsy.

Chorea: Quick, jerky, uncontrollable movements that seem to happen at random times and affect different body parts. The condition of chorea is linked to Huntington's disease.

Contraction: Movement of a muscle that makes it tense and sometimes painful.

Dystonia: Repetitive or twisted movements caused by involuntary contractions of the muscles. Anomalies in posture may result from this. One example is spasmodic torticollis, often known as cervical dystonia.

Endurance: The ability to maintain long-term muscle work or activity.

Myoclonus: Quick, unexpected, shock-like spasms in the muscles. One common symptom of myoclonus is tics.

Organelles: Like an organ in the body, an organelle is a subcellular structure that has one or more specific functions in a cell.

Tremors: Uncontrollably occurring rhythmic oscillations in a body part. Examples include Parkinsonian tremor and essential tremor.

INTRODUCTION

Active movements are movements performed by a person using their own muscles, without any help from outside forces. These movements are an important part of daily activities and are essential for keeping joints flexible, muscles strong, and movements coordinated. Active movements also help improve fluid dynamics, blood circulation and body awareness.

In exercise therapy, active movements are used to help people recover from injuries, prevent further problems, and improve overall fitness. They can be adjusted to suit different levels of ability, from simple exercises for those with limited movement to more advanced ones for athletes. Understanding how to use active movements is important for creating effective exercise plans that match each person's needs and goals.

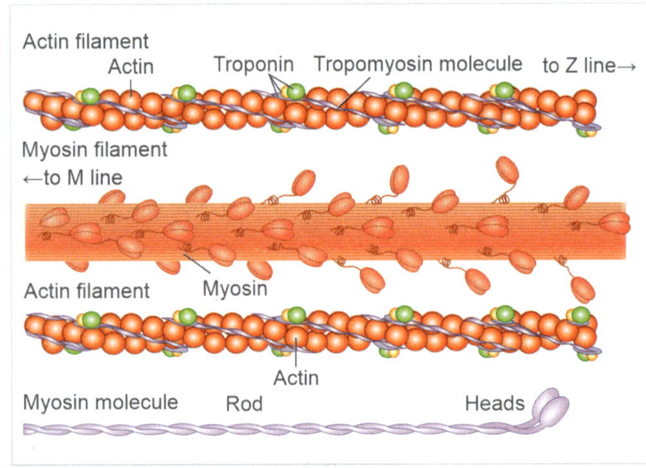

Fig. 8.1: Muscle protein

BASIC STRUCTURE OF A MUSCLE

Muscle fibers, also known as myofibers or muscle cells, and related connective tissue are arranged in skeletal muscle in a very specific and well-described manner. The total size of a muscle is mostly determined by the quantity and dimensions of its individual muscle fibers that are postmitotic and multinucleated. Nucleus of each muscle fiber typically controls the kind of protein produced in that specific cell.

The epimysium is a layer of connective tissue that envelops each individual muscle. Fiber clusters of muscles are grouped in bundles and encircled by the perimysium, an additional layer of connective tissue. A single muscle fiber has a sarcolemma or cell membrane encircling it. A complex of several proteins, such as the myosin and actin, specifically the actin protein found in the thin filament, is physically linked to the internal myofilament structure and is associated with the sarcolemma. Sarcolemma damage, muscular weakening, and atrophy can occur when one of these proteins is absent (wholly or partially) or functions improperly. As an example, the dystrophin protein, which is either completely absent or partially present in certain neuromuscular diseases such as Duchenne and Becker muscular dystrophies, is found in this complex.

Muscle Proteins

Without accounting for its water content, the majority of composition of a single muscle fiber (80%) is made up of contractile, regulatory, and cytoskeletal proteins, and 8% is sarcoplasm. Estimates suggest that each muscle fiber has thousands of myofibrils and billions of myofilaments. The formation of fundamental contractile units of skeletal muscles, i.e., sarcomeres results from the arrangement of myofilaments in a highly ordered and characteristic pattern. The two most common myofilaments (proteins) in a single fiber are actin and myosin, which together account for about 70–80% of the overall protein composition (Fig. 8.1).

The sarcomere and sarcoplasm include a large number of extra proteins that aid in the formation of the cytoskeleton, couple the processes of excitation and contraction, release energy, and generate force and power. Particularly relevant are regulation proteins that are attached to the actin filament and play crucial roles in the activation process leading to myofilament sliding and force production. Examples of these proteins are tropomyosin and the calcium-dependent troponin complex, which consists of troponins C, I, and T. Two more proteins that help maintain the physiological and mechanical properties of muscle are called titin and nebulin. Large and elastic, titin binds myosin to the sarcomeric Z-disc, helping to align and stabilize the thick filament. On the other hand, nebulin forms the actin-containing thin filaments when it combines with other proteins. These proteins sustain sarcomere integrity, influence the passive tension and stiffness characteristics of individual cells, and may be crucial for myofibril assembly and cell signaling. One of the proteins in the sarcomeric Z-disc is actinin, which also serves as an attachment point for the actin myofilament.

Clinical Correlation

Understanding the structure and organization of muscle fibers, as well as the role of key proteins like dystrophin, is important for diagnosing and managing neuromuscular disorders such as Duchenne and Becker muscular dystrophies. Loss or dysfunction of these proteins can lead to muscle weakness, atrophy, and other symptoms characteristic of these conditions.

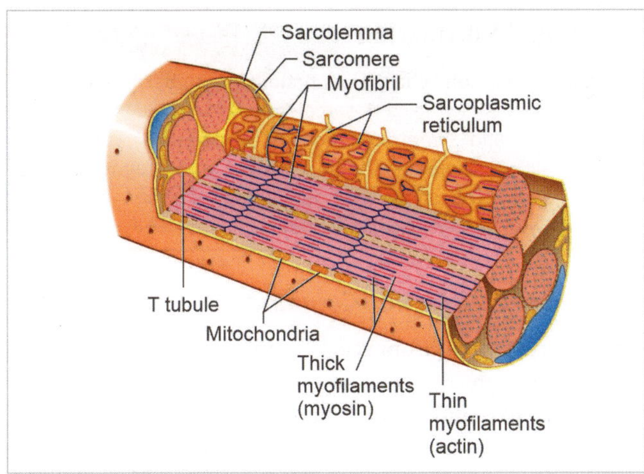

Fig. 8.2: Muscle fiber organelles

Muscle Fiber Organelles

The sarcoplasm of muscle fibers also contains a mitochondrial network, the sarcoplasmic reticulum, and the transverse tubular system (T tubule) (Fig. 8.2). One essential function of the T tubule system, an invagination of the sarcolemma, is the transmission of the nerve action potential to the interior of the cell. Since this tubular network is in contact with the outside of fiber, it guarantees that excitation can propagate uniformly throughout the fiber. The T tubule membrane contains a necessary protein known as dysferlin, which is responsive to changes in the intracellular calcium concentration. The sarcoplasmic reticulum stores, releases, and reabsorbs calcium following activation. Calcium is stored in the terminal cisternae (sarcoplasmic reticulum ends), which are located near the transverse tubule system. The two cisternae on either side of the T tubule form the triad, a structure consisting of two cisternae plus one tubule. Two types of proteins identified in the sarcoplasmic reticulum that aid in maintaining calcium homeostasis include calsequestrin and sarco/endoplasmic reticulum Ca^2 -ATPase (SERCA). The former is responsible for calcium absorption into the cisterna after muscular activation, while the latter loosely binds calcium in the sarcoplasmic reticulum.

When oxygen is available to muscle fibers, mitochondria form a three-dimensional network throughout the cell to generate the energy needed for muscle activity, rather than existing as isolated organelles (Frontera et al.).

Skeletal Muscle Fibers

Skeletal muscle fibers are the cells that make up skeletal muscles. They are the smallest contractile units of muscles.

- **Structure (Fig. 8.3):**
  - Skeletal muscle fibers are long, cylindrical cells that are striated and multinucleated.

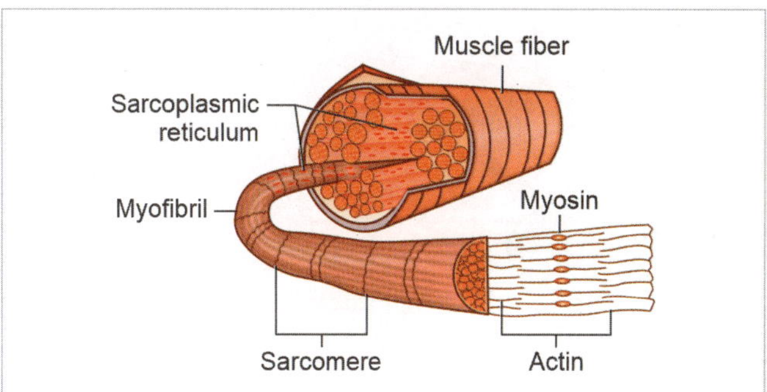

Fig. 8.3: Structure of skeletal muscle

  - They are made up of myofibrils, which are the basic units of the muscle fiber.
  - Myofibrils contain contractile proteins called thick and thin filaments, which are arranged into units called sarcomeres.
- **Size:** Skeletal muscle fibers range from less than half an inch to just over three inches in diameter and can be many centimeters long.
- **Organization:**
  - Skeletal muscles are made up of thousands of muscle fibers wrapped together by connective tissue sheaths called fasciculi. The connective tissue sheaths that over them are called perimysium.
  - The outermost connective tissue sheath that surrounds the entire muscle is called the epimysium.

Types of Skeletal Muscle Fibers

- **Type 1:** Slow oxidative (SO) fibers produce ATP through aerobic respiration (oxygen and glucose) and contract relatively slowly. They are slow to get tired and produce low power contractions for extended periods of time.
- **Type 2 A:** Fast oxidative (FO) fibers have rapid contractions and primarily engage in aerobic respiration. Since they can transition to anaerobic respiration (glycolysis), they are susceptible to fatigue more quickly than SO fibers.
- **Type 2 B:** Anaerobic glycolysis is largely used by fast glycolytic (FG) fibers, which contract quickly. FG fibers deteriorate quicker than other fibers

Electrical Events in Muscle Contraction

Cells in muscle fibers can be excited. The cell membrane (sarcolemma) contains the voltage-gated ion channels necessary for the generation of an action potential in addition to the ion channels and pumps needed to maintain a very low resting membrane potential (Fig. 8.4). As with all excitable cells, the membrane potential of muscle cells results from the net electrochemical gradients of ions that the membrane is

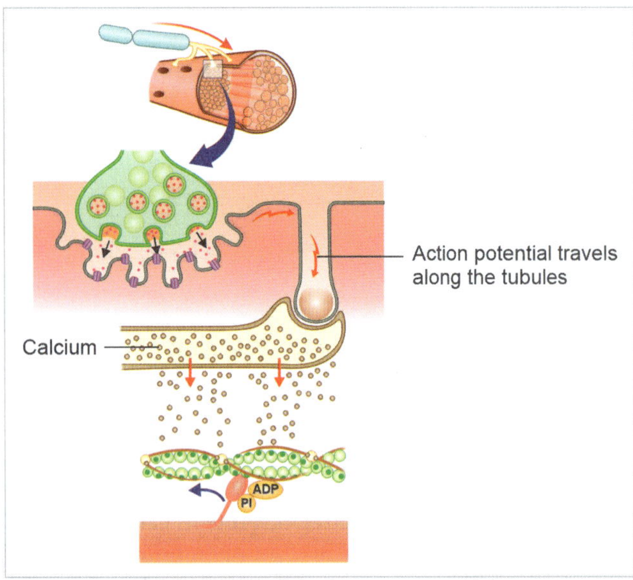

Fig. 8.4: Electrical events in muscle contraction

permeable to at any given time. The immediate permeability of the sarcolemma to ions that are distributed unevenly across it is mostly determined by the open ion-selective membrane channels. When the concentration and electrochemical gradients of a permeable ion are identical, the equilibrium potential is the potential difference between the intracellular and extracellular compartments.

Resting membrane potential of skeletal muscle cells, which ranges from 70 mV to 90 mV, is comparable to that of neurons. It is significantly influenced by Cl conductance, in contrast to nerve cells, where K permeability dominates the resting membrane potential. K^+ is lost from the cell when a muscle contracts.

The gradient of K concentration across the sarcolemma runs down with repeated exercise. The muscle would not repolarize sufficiently to renew the active state of the channels responsible for generating subsequent action potentials without the Cl current to sustain resting membrane potential.

Action potential of a skeletal muscle is generated when the motor endplate potential is high enough to elevate the surrounding sarcolemmal potential above the threshold for activation of the voltage-gated Na channels, which are extensively dispersed throughout the sarcolemma. The membrane is quickly depolarized toward the Nernst potential for Na when these channels are active. As soon as an action potential is created, it propagates over the sarcolemma as a wave. Transverse- or t-tubules, which extend deep within the body of the cell and run perpendicular to the surface of cell, give skeletal muscle its sarcolemma appearance. The action potential is transported to the structures in charge of converting an electrical signal into a chemical signal that activates the contractile components by traveling down the t-tubular membrane (Hopkins et al.).

Clinical Correlation

Disruption of ion channel function or imbalance in ion concentrations across the sarcolemma can lead to muscle dysfunction and symptoms such as weakness or contractile abnormalities. Clinicians should consider the integrity of ion channels and the stability of ion gradients when assessing patients with neuromuscular disorders.

Excitation-Contraction Coupling

The process that connects stimulation of the plasma membrane with the generation of cross-bridge forces is known as excitation-contraction coupling. At the neuromuscular junction, the muscle synaptically receives a brain signal and transforms it into mechanical force. The release of acetylcholine (ACh) is the chemical signal. Calcium signals are started by muscle action potentials created by the excitation-contraction coupling. This triggers muscle action potentials, which lead to the release of calcium ions from the sarcoplasmic reticulum. The released calcium ions then enable muscle contraction. The activation of the cross-bridge cycle is referred to as contraction. By binding to troponin, Ca^{2+} activates the attractive forces between the actin and myosin filaments (Fig. 8.5). The high-energy bonds of adenosine triphosphate (ATP), which are converted to adenosine diphosphate (ADP), are the source of the energy needed for contraction to continue.

The heads of myosin molecules create cross-bridges with active sites on the fine actin filaments in the presence of Ca^{2+} and ATP. The resulting energy causes a conformational change in the myosin head region, which in turn causes the actin filament to experience a directional force. Actin filaments are then dragged toward the center of sarcomere where they overlap the myosin filament. Sarcomere shortening or muscular contraction is the end outcome. Each cycle of the cross-bridge requires one ATP.

Excitation-contraction coupling relaxation follows cross-bridge creation during the contraction phase. Muscle relaxation results from the pumping of calcium back into the Sarcoplasmic reticulum (SR) by Ca-ATPase, which also lowers calcium levels. To enable the release of Ca^{2+} from troponin, the concentration of calcium must decrease. The elastic connective tissues within the muscle and titin help the filaments slide back to their original place.

Did You Know?

The excitation-contraction coupling is a highly regulated process that ensures precise control over muscle contraction. It involves the coordinated release of calcium ions, binding of calcium to troponin, and subsequent activation of the cross-bridge cycle. This intricate mechanism allows for the finely-tuned control of muscle force production, enabling us to perform tasks ranging from delicate movements to powerful actions like lifting heavy objects.

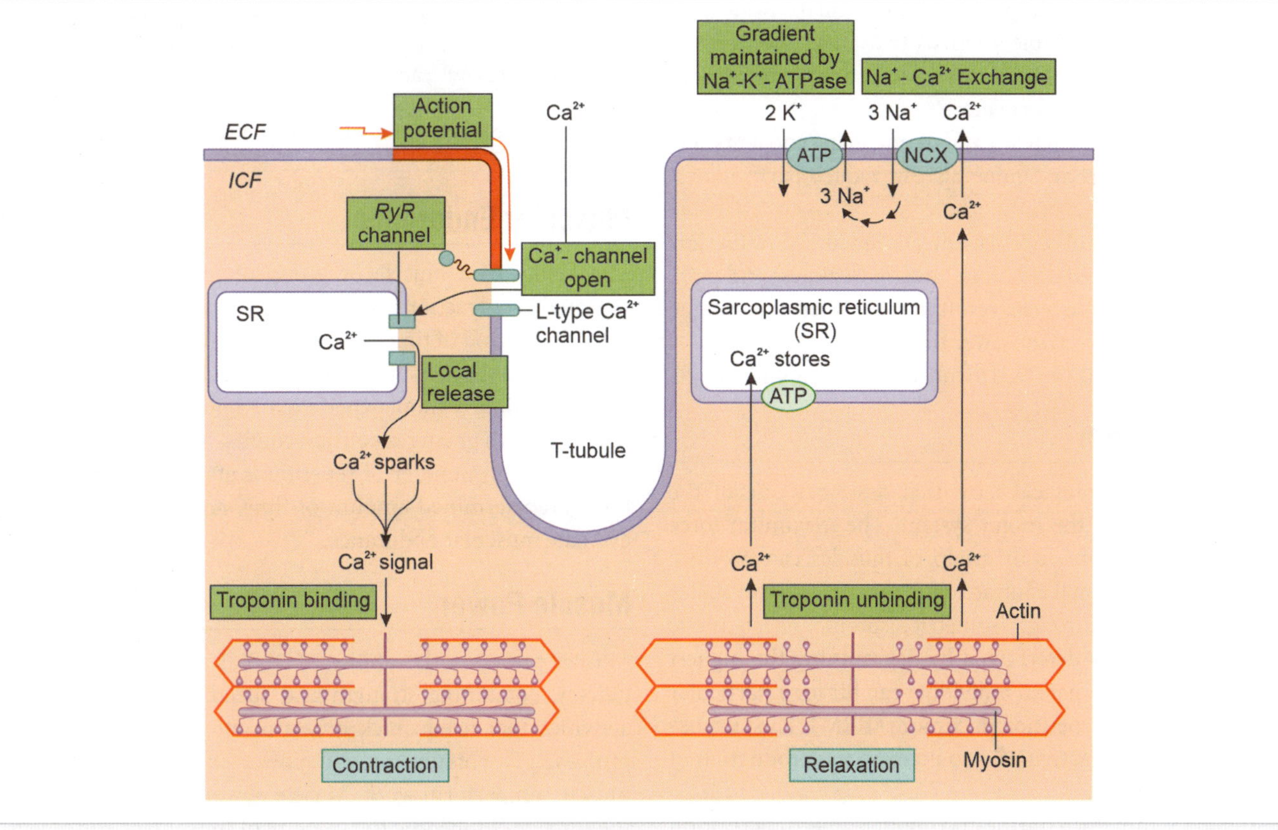

Fig. 8.5: Excitation-contraction coupling mechanism

MUSCLE PERFORMANCE

Muscle performance is the capacity of the muscles to do work. The key elements of muscle performance are strength, power and endurance.

Physiology of Muscle Performance

One of the human body's most flexible and active tissues is skeletal muscle, which contains 50–75% of the body's total protein content, and makes up around 40% of the total body weight. Water makes up the majority of muscle tissue (75%) followed by protein (20%) and then additional elements such as inorganic salts, minerals, fat, and carbohydrates (5%). Protein synthesis and breakdown must be balanced for muscles to grow, and both processes are affected by a variety of things, including dietary intake, hormonal balance, physical activity, and health conditions.

The main function of skeletal muscle according to biomechanics, is to convert chemical energy into mechanical energy in order to generate force and power, maintain posture, and produce movement that influences activity, permits engagement in social and professional contexts, maintains or enhances health, and aids in an individual's ability to function independently. In addition to contributing to basic energy metabolism and serving as a store for necessary substrates like amino acids and carbohydrates, skeletal muscle also produces heat to regulate body temperature and uses up the majority of the oxygen and fuel used during exercise and physical activity. Moreover, the release of amino acids from muscles during hunger aids in maintaining blood glucose levels. A decrease in muscle mass affects the body's capacity to react to stress, which is relevant to the prevention of disease and the preservation of health.

Causes of Decreased Muscle Performance

- **Age:** With age, muscle mass and strength decline due to various factors like hormonal changes and decline in physical activity.
- **Physical inactivity:** Reduction in physical activity levels can result into muscle atrophy which is loss of muscle mass and strength.
- **Injuries:** Musculoskeletal injuries such as that of muscles or joints can result in decreased performance of the muscles which cannot contract properly or are inhibited because of pain.

- **Diseases:** Certain diseases like muscular dystrophy, Parkinson's and multiple sclerosis that affect the nervous or the musculoskeletal system of the body can lead to decreased muscle performance.
- **Nutritional deficiencies:** Lack of adequate nutrients like vitamins, proteins and minerals can reduce performance of the muscle.
- **Medications:** Use of certain medications like chemotherapy drugs for a long time can cause muscle weakness and pain.
- **Stress:** Chronic stress can result in muscle tension, fatigue and reduced performance over time.
- **Dehydration:** This can lead to muscle cramping and fatigue.

Muscle Strength

Strength is a physiological term that describes one of the output capacities of the motor system. The maximum force or tension that a muscle or group of muscles can produce during a single maximal contraction is referred to as muscular strength measured through isometric, isotonic, or isokinetic tests. This is usually determined by determining the highest resistance or weight that an individual can lift for a particular exercise, like a one-repetition maximum (1 RM). The purpose of strength training activities is to increase the production of muscle force.

Strength is thought to be the maximum isometric activation of the motor system. Strength measurements are typically limited to the activity around one joint at a time for the sake of simplicity. Despite being one of the most straightforward measures of human performance that can be established (i.e., isometric and single joint), strength is still the result of a complex interaction between all neuromuscular components which are mechanical, muscular and neurological components. Neuronal factors associated with motor unit activity include recruitment and discharge frequency modulation. The muscular variables are the cross-sectional area and length of the muscle(s) at the time of measurement. As the force that a person applies to a load depends on the torques acting on the system, particularly muscular torque, the mechanical factors also include the moment arms connected to the various forces (Rantanen et al.).

Did You Know?

- As we get older, our strength and ability to carry out daily chores wane. Functional measures like walking speed are substantially correlated with muscle strength.
- Increased muscle strength has been associated with better potentiation effects and a decrease in injury rates, as well as better force-time properties (such as rate of force production and external mechanical power) and general athletic skill performance (such as jumping, sprinting, and direction changing).

Contd...

- Muscle quality is a valuable indicator of physiological qualities of skeletal muscle tissues which impacts strength and function. Muscle quality can be defined as all muscle-associated physiological factors that influence the force-generating capability of skeletal muscle tissue (Kuschel et al.).

Muscular Endurance

The ability of a muscle or group of muscles to withstand repeated contractions or to withstand fatigue over an extended period of time is known as muscular endurance. It is the ability to sustain submaximal contractions for a prolonged amount of time without experiencing a noticeable loss of force. Exercises with greater repetition counts, including completing the maximum number of repetitions at a given resistance or for a predetermined amount of time, are frequently used to evaluate muscular endurance.

Muscle Power

Muscle power is the ability of a muscle to produce force quickly, combining strength and speed. It is important for activities that need quick bursts of effort, like sprinting or jumping. Essentially, muscle power measures how fast a muscle can perform work. Muscle power can be explained as the force multiplied by the velocity of movement. Since work is defined as force times distance traveled, divided by time required, power can alternatively be expressed as work done per unit of time (Newton et al.).

$$W = F \times D$$

$$V = \frac{D}{T}$$

Therefore,

$$P = F \times \frac{D}{T}$$

$$= \frac{W}{T}$$

Here,

W = Work done
F = Force
D = Distance
T = Time taken
P = Power

Muscle Work

Work is a result of force acting on a body and that body being moved in the same direction as the force. The work performed by a constant force acting parallel to the displacement of a body is defined as:

$$W = Fd$$

where W is work measured in Joules (J), F is the magnitude of force measured in Newton (N), and d is displacement of the body measured in meters (m).

When a muscle works, its energy cost rises, whereas when a muscle works on itself, its energy cost falls.

Did You Know?

To describe the work that is done when a force is applied to a body, three things must be understood:
1. The average force applied to a body.
2. The direction of force.
3. Body displacement in the direction of the acting force while the force acts on the body.

When the body moves in the same direction as the force, positive work is produced. A few instances include the weight lifter raises the barbell, the javelin thrower moves the javelin in the direction of the throw, and the ski jumper's lift off has a beneficial effect on the ski jump surface.

PHYSIOLOGICAL ADAPTATIONS

Strength training is a long-term fitness program focused on resistance exercises designed to shape and develop muscles. To be effective, routines should target all major muscle groups and be regularly updated, as muscles require continuous stimulation to reach their full potential. Strength and power training present a potent stimulus to the musculoskeletal system. This type of stress elicits a wide variety of physiological responses and subsequent adaptations instrumental to the increase in muscular strength, power, hypertrophy and local muscular endurance observed during resistance training.

Changes in Muscular Strength

Changes can be categorized under the following headings:
- Changes in muscle size
- Muscle fiber size
- Fiber type transition
- Changes in muscle architecture
- Connective tissue adaptations
- Metabolic adaptations

The characteristic features of muscular strength and power changes are summarized in Table 8.1

Changes in Muscle Power

Muscle power training is a type of exercise that focuses on the development of explosive power in muscles which can be defined as the ability to generate force with respect to time.

TABLE 8.1: Physiological adaptations to strength training

Aspects	Changes/adaptations
Changes in muscle size	• Skeletal muscle hypertrophy, or enlargement. • Linear evolution of muscle hypertrophy in response to regular RT.
Muscle protein and fiber size	• Resistance training (RT) results in increased muscle protein content. • RT produces hypertrophy by accumulating new contractile myofilaments and biosynthesizing them. • RT increases cross-sectional area (CSA) and myofibril proliferation, particularly in fast-twitch type II fibers.
Fiber type transition	• RT encourages the changeover of more oxidative (type 2A) fibers from glycolytic (type 2X) fibers. • The percentage of type I fiber is essentially unchanged.
Changes in muscle architecture	• Increased pennation angles allow for greater deposition of contractile material and force generation in hypertrophied muscles. • Excessive hypertrophy and increase in pennation angle may have a deleterious impact on force output.
Connective tissue adaptations	• Resistance training strengthens muscles and also improves the connective tissues (tendons, ligaments, and intramuscular tissue) that support them. • Bone mass, density and architecture adjust to mechanical strain during RT; increased tendon stiffness may improve elastic energy consumption.
Metabolic adaptations	• Depletion of ATP, PCr, and glycogen stores, increase in blood lactate after heavy RT. • Increased activity of anaerobic enzymes and intramuscular PCr/glycogen levels with chronic RT. • Improvement in oxidative capacity of skeletal muscle with chronic RT. • Enhanced insulin sensitivity with persistent RT, especially in trained muscles. • Maintenance or slight decrease in mitochondrial content after RT. • Possible improvement or maintenance of blood pressure and blood lipid profiles after chronic RT. • Resistance-trained athletes may experience concentric hypertrophy of the left ventricular wall. • Improved body composition with gains in fat-free mass and losses in fat mass after long-term RT.

TABLE 8.2: Physiological adaptations to power training

Benefits	Description
Improved neuromuscular coordination	Muscle power training enhances coordination between the nervous system and muscles, optimizing muscle recruitment and activation.
Increased muscle fiber recruitment	Power training boosts force production and power output by recruiting more muscle fibers during explosive movements.
Improved quality of muscle fibers	Muscle power training induces changes in muscle fiber structure and type, enhancing the ability to generate muscle power and speed.
Improved energy transfer	Power training facilitates efficient transfer of energy between muscle groups, enabling coordinated and effective movement patterns
Increased metabolic functions	Muscle power training enhances anaerobic capacity and lactate threshold, improving high-intensity exercise performance and speed.

Exercises like plyometric, jump training and Olympic lifts, etc. help in the development of muscle power. Physiological adaptations to power training are given as follows in Table 8.2.

Changes in Muscle Endurance

Endurance training leads to significant physiological adaptations that enhance the body's ability to sustain prolonged activity. These are enlisted in Table 8.3.

TABLE 8.3: Physiological adaptations to endurance training

Benefits	Description
Improved cardiovascular function	Endurance training increases stroke volume, capillary density, and oxygen uptake (VO_2 max), while lowering resting heart rate.
Enhanced respiratory efficiency	Training improves lung efficiency and strengthens respiratory muscles, making breathing more effective during prolonged activity.
Increased mitochondrial density	Endurance training boosts mitochondrial density in muscles, enhancing the body's ability to produce energy aerobically.
Improved oxidative capacity	Training enhances oxidative enzyme activity and myoglobin content in muscles, improving the ability to utilize oxygen for energy.
Enhanced fat utilization	The body becomes more efficient at using fat as an energy source, helping to conserve glycogen during extended activities.
Increased glycogen storage	Endurance training improves the muscles' ability to store and use glycogen, providing a more sustained energy supply during prolonged exercise.
Reduced lactate production	Endurance training decreases lactate accumulation, helping to delay fatigue and improve performance during high-intensity efforts.
Improved thermoregulation	The body adapts to better dissipate heat, increasing tolerance to heat and preventing overheating during long-duration activities.
Enhanced metabolic efficiency	The body's overall energy efficiency improves, allowing for better performance and endurance during extended physical activity.
Hormonal adaptations	Increased endorphin levels and improved insulin sensitivity support better recovery, mood, and overall metabolic function.

Contd...

ACTIVE EXERCISES

Exercises or movements that involve active bodily motions and individual efforts to generate the action are referred as **active exercises**. These are the exercises or motions that the muscles execute on their own accord. They are produced when a person's muscles contract. The importance of active movement and exercise for physical health and well-being has been studied extensively. The risk factors for chronic diseases including diabetes, heart disease, etc. have been found to be reduced by regular performance of active exercise, especially vigorous motions. They are typically advised for the rehabilitation of those whose physical function has been negatively impacted by an illness or injury.

Indications

- **Musculoskeletal injuries:** Active exercises are typically advised for people healing from musculoskeletal ailments such as fractures, sprains, and strains. These exercises aid in regaining range of motion, strength, endurance, and flexibility in the muscles as well as encourage tissue repair and pain relief.
- **Neurological conditions:** Patients with neurological conditions including Parkinson's disease and multiple sclerosis benefit from active exercises. The workouts enhance physical function, balance, and coordination.
- **Cardiovascular conditions:** Active exercises can help to control cardiovascular risk factors like high blood pressure, diabetes, and obesity, among others.

- **Chronic pain:** Patients with chronic pain conditions like fibromyalgia and arthritis are advised to engage in active workouts.
- **Postsurgical rehabilitation:** Active exercises are prescribed as part of postsurgical rehabilitation program to help individuals regain muscle strength, range of motion and function following surgery.
- **General fitness:** Active exercises also help individuals improve their physical fitness (strength, endurance, flexibility, etc.) and overall physical function.

Types

1. **Voluntary:** These are movements initiated and controlled consciously by the individual. Examples are lifting an arm, walking, or writing. Different types of active exercises include:
   - Free exercises
   - Active assisted exercises
   - Assisted resisted exercises
   - Active resisted exercises
2. **Involuntary:** Motions or acts that take place without conscious thought or intention are referred to as involuntary movements. Usually controlled by the autonomic nervous system, these movements are not directly influenced by the volition.
   - Reflex
   - Pathological

Free Exercises

Free exercises are performed without the use of any external resistance and instead rely on the body's own weight or the gravitational pull of the object being moved or stabilized. These are the workouts that can be done anywhere and are a practical method to maintain joint mobility or increase strength without needing expensive equipment.

Benefits

Free exercises can promote the following:
- Relaxation due to the rhythmical or pendular character of the exercise
- Maintenance of muscle tone
- Coordination by way of natural pattern
- Confidence to perform and control movement.

Types

1. **Localized free exercises:** Are those exercises that are designed to produce local and specific effects. For example, mobilization of a particular joint, strengthening of a particular muscle, e.g., pendular exercises of the shoulder (Fig. 8.6).

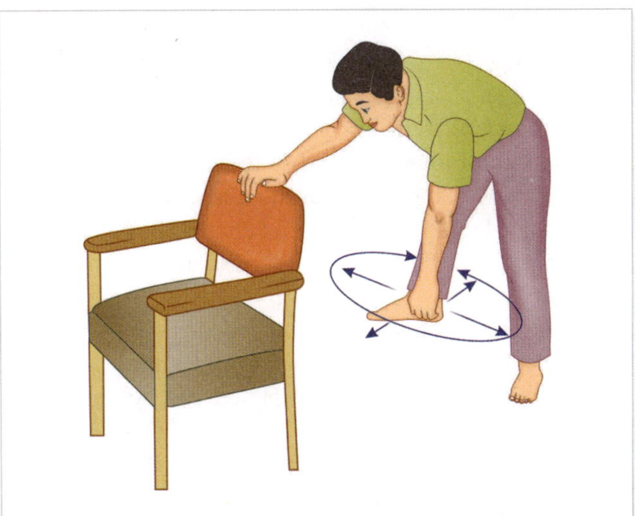

Fig. 8.6: Pendular exercise of the shoulder

2. **General free exercises:** Are those exercises that involve multiple joints and muscles across the body. Their effect is widespread, e.g., walking, running, etc., General exercises are further classified as subjective and objective.
   - **Subjective exercises** consist of anatomical movements performed in full available range with patient focusing on pattern of exercise and its accuracy.
   - **Objective exercises** are performed with patient's attention focused on achievement of a particular goal, for example, patient in sitting position stretches their arm forward to lift a cup of tea kept in front of them on a table.

Principles and Techniques

- Choosing an appropriate starting position based on the patient's condition and the desired level of muscle contraction.
- Before beginning any workout, warm up a little. Stretching activities can be used as a warm-up. The muscles and joints can benefit from these workouts to get ready for the exercise.
- It is crucial to employ the correct form and technique.
- Control must be the main focus. Many free exercises call for controlled, slow motion rather than quick, explosive motion. This ensures that the targeted muscles are correctly engaged and that there are no injuries.
- Gradually increase the intensity. It is crucial to start out cautiously and build up intensity over time. Increasing the number of repetitions, exercise length, or intensity may be necessary.
- Exercise progression should depend on the patient's capacity to complete a challenging assignment.

Effects and Uses

- **Relaxation:** By executing motions that are rhythmical and pendular in character, one can relax the muscles and the structures around them.
- **Strength and tone:** In response to tension that builds up in the muscles, their strength and tone can be increased or maintained. Muscle strength and endurance are increased by contracting them against the resistance of the body's weight.
- **Enhancement of ROM:** Limitation of range of motion can be extended by conducting rhythmical swinging movements after removal of plaster cast following any surgical procedure.
- **Neuromuscular (NM) coordination:** By facilitating the conduction of impulses across the neuromuscular pathways, repetition of the actions aids in the improvement of coordination.
- Improves strength and endurance.
- Improves flexibility and mobility.
- Muscle re-education.

Contraindications

- Acute pain and inflammation.
- **Lack of supervision or instruction:** Free exercises should not be performed without proper instructions of supervision as they can lead to injury.
- **Cardiovascular or respiratory issues:** It is important to seek medical advice before starting an exercise program by an individual with cardiorespiratory disease.
- **Pregnancy:** There are certain exercises which may not be suitable for pregnant women.
- **Recent surgical procedure:** Passive or of active assisted exercises may be initiated first.

Indications

- Postoperative rehabilitation (e.g., joint replacement, tendon repair)
- Muscle strains and sprains
- Frozen shoulder (adhesive capsulitis)
- Chronic pain conditions (e.g., fibromyalgia)
- Post-cast or immobilization recovery
- Muscle atrophy due to inactivity
- Stroke recovery (neurological rehabilitation)

Active Assisted Exercises

Active assisted (AA) exercises are the exercises in which patients can execute active exercises on their own or with external aid from therapists, pulleys, cords, or weight and pulley circuits. When the movement cannot be produced by the voluntary muscular activity alone, external force must be introduced. In addition to mechanical aids like slings, pulleys, finger ladder, skate boards and wands, patients can self-assist themselves with the use of their sound limb or receive manual assistance from a therapist. The direction of muscle action must be followed when applying an external force. The mechanical advantage of helping force must be sufficient to supplement muscular activity but not to completely replace it.

Benefits

- These are typically utilized in rehab settings to assist patients in regaining strength and mobility following an injury.
- These exercises help to complete a full range-of-motion; reduce resistance to the limb, by supporting or helping to move the limb when the patient becomes fatigued and can no longer safely control the limb.
- When strength loss is typically at its worst, active-assisted exercises are typically started early in the rehabilitation process.
- Examples of active assisted exercises include:
  - The wall slides, in which one slides their arms on the wall down to increase shoulder mobility (Fig. 8.7).
  - Therapist assists the child to perform pelvic bridging (Fig. 8.8).
  - Shoulder elevation being performed by the patient with the assistance of unaffected side in spinal lying (Fig. 8.9).
  - Use of wand to increase shoulder internal and external rotation (Fig. 8.10).

Principles and Techniques

- Depending on the patient's condition and the desired level of muscle activation, an appropriate starting posture is chosen.
- To ensure that the assistant can effectively assist with ROM exercises, make sure patient is positioned comfortably. In case, a device is chosen for assistance, ensure that it is positioned comfortably.

Fig. 8.7: Wall slide exercise

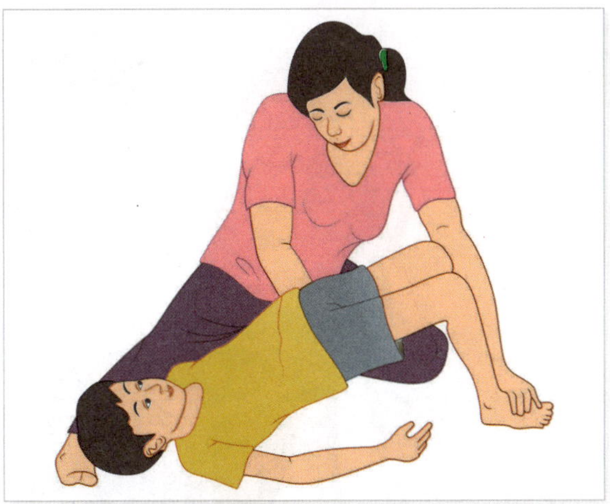

Fig. 8.8: Therapist assisting child to perform pelvic bridging

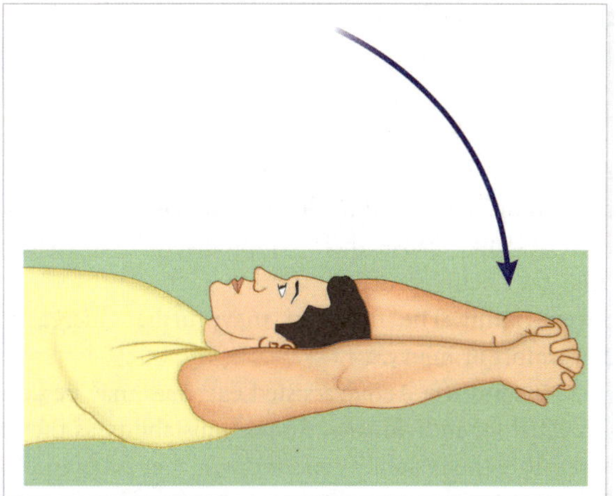

Fig. 8.9: Active assisted exercise (shoulder elevation) being performed by the patient using healthy/unaffected upper limb

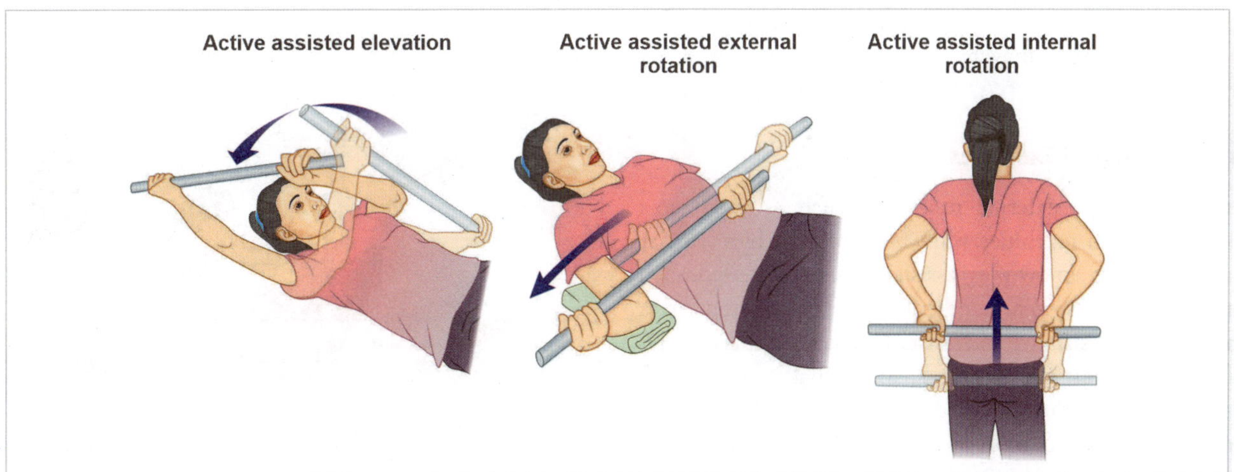

Fig. 8.10: Active assisted shoulder exercises using wand

- All ROM exercises should be done gently, evenly, and slowly. To avoid possibly injuring muscles or joints, refrain from jerking, pushing, or overstretching muscles.
- The force supplied to help muscles contraction must be exerted in the direction of muscular contraction.
- During range of motion exercises, avoid overstretching the muscles or joints as this can cause injury instead of therapeutic benefits.
- Move the joint to the point of resistance, once this is achieved, stop. By doing this, excess tension on the joint and its supporting tissues is reduced.
- Repetition of the movements should be performed depending on the patient's requirement and condition or as prescribed by the therapist.

Effects and Uses

- Enhanced neuromuscular coordination
- Increased strength and endurance of the muscles.
- Enhanced coordination of the movements
- Increased range of motion.
- Confidence in the performance of movement is established.

Indications

- **Rehabilitation after injury or surgery:** Active exercises can be used to improve strength and range of motion in injured or surgically repair joints.
- **Improving range of motion:** Active assisted exercises can be used to improve flexibility and range of motion in stiff joints.
- **Coordination of movement:** By frequent repetitions of the movement, the memory of coordinated pattern of movement is stimulated. Repeated movements help in facilitation of conduction of impulses in the neuromuscular pathways.
- **Strengthening of the muscles:** Active assisted exercises help strengthen muscles by maintaining their natural properties and promoting muscle growth.
- Preparing for more advanced exercises.

- Improving overall fitness.
- Pain management.

Contraindications

- **Acute injuries:** Active assisted exercises may be contraindicated for individuals with acute injuries as the exercises may cause their symptoms to exacerbate and delay the healing process.
- **Severe pain:** The exercises may further increase the discomfort or injury.
- **Joint instability:** Active assisted exercises may be contraindicated for individuals with joint instability as they may exacerbate the instability and increase the risk of injury.
- **Recent surgery:** In patients who have undergone recent surgery, active exercises may be contraindicated, where focus is on healing and minimizing movements.
- **Uncontrolled medical conditions:** In many individuals these exercises may increase heart rate and blood pressure. Medical clearance may be necessary before starting an exercise program.

Assisted Resisted Exercises

Exercises that use external resistance, such as weights, a resistance band or machines, while being assisted by a partner or a trainer are known as assisted resisted exercises. In simpler words, these can be described as exercises that are a combination of assistance and resistance in a single movement.

The additional resistance or force that the external resistance offers to the muscles during exercise can help them become stronger and more resilient. For example, assisted resisted exercises using resistance bands.

Example of shoulder exercises with partner assistance:

During the workout, the therapist applies extra resistance by pushing or pulling on the resistance band (Figs 8.11 to 8.13).

The technique of application will remain same as active assisted exercise except that the during the performance of the exercise, there will be a combination of assistance and resistance throughout the range.

Indications

- Muscle weakness and imbalance
- Post-surgical rehabilitation
- Enhancing neuromuscular control and coordination
- Restricted strength or mobility
- Individuals recovering from injury or illness
- Athletes aiming to develop their sport-specific strength.

Contraindications

- People who suffer from conditions like arthritis, osteoporosis or joint pain.
- Musculoskeletal injuries such as rotator cuff tears.

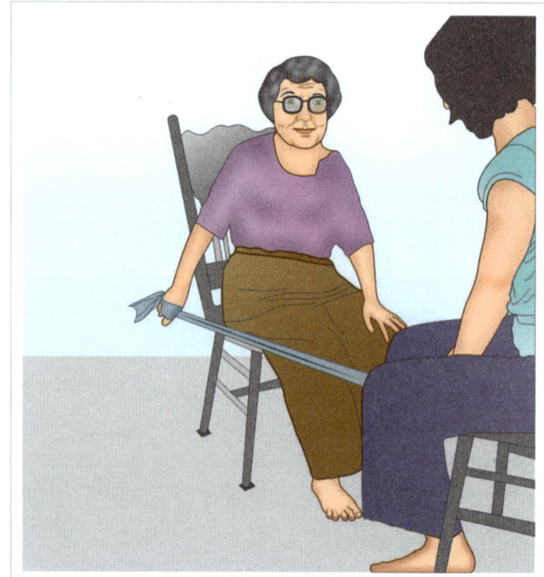

Fig. 8.11: Shoulder exercises with partner assistance

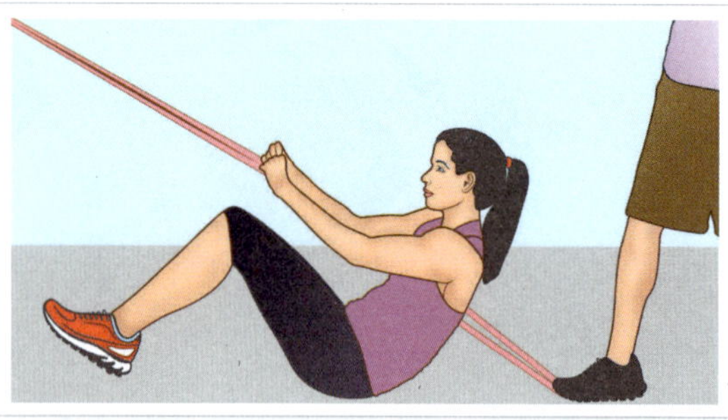

Fig. 8.12: Abdominal crunches with partner assistance

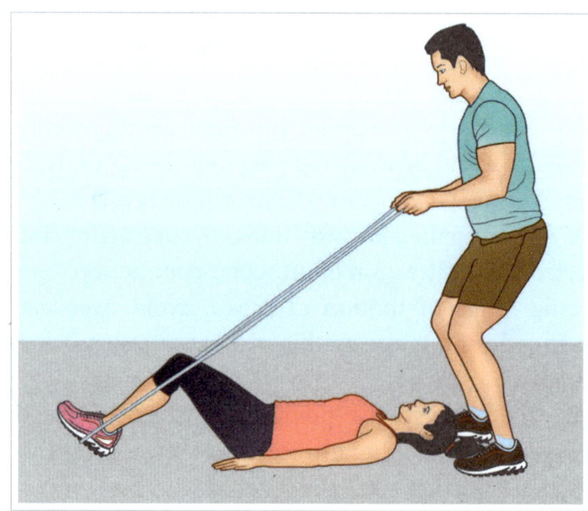

Fig. 8.13: Core strengthening with partner assistance

Effects and Uses

- Strength and muscle building
- Endurance training
- Sport-specific training

- Fat loss and weight management
- Improving overall fitness

Active Resisted Exercises

Exercises that involve active muscle contraction while resistance is provided physically or by equipment (such as weights, machines, or resistance bands) are known as active resisted exercises. By allowing the user to work against the resistance during the movement and combining voluntary muscle contraction with external resistance, these workouts aid in muscular strengthening. These exercises are essential parts of resistance training regimens and are meant to cause changes in the musculoskeletal system that will enhance functional ability and general fitness. Resisted exercises can focus on different muscular groups and can include isometric, concentric, or other muscle motions based on the particular training objective and activity. Resisted exercises are discussed in detail in chapter 9 Resisted Exercises.

Reflex Movements

- **Stretch reflex:** This kind of reflex happens when a muscle is stretched. A muscle contracts quickly when it is stretched because the activation of the muscle spindles occurs. The knee-jerk reflex or patellar reflex is a typical illustration.
- **Withdrawal reflex:** In reaction to a painful stimulation, a body part is quickly withdrawn as a protective mechanism. For instance, rapidly removing the hand from contact with a hot object.

Pathological Movements

- **Tremors:** Uncontrollably occurring rhythmic oscillations in a body component. Examples include Parkinsonian tremor and essential tremor.
- **Dystonia:** Repetitive or twisted movements caused by involuntary contractions of the muscles. Anomalies in posture may result from this. One example is spasmodic torticollis, often known as cervical dystonia.
- **Chorea:** Quick, jerky, uncontrollable motions that seem to happen at random times and affect different body parts. The condition chorea is linked to Huntington's disease.
- **Athetosis:** Involuntary, slow-moving writhing movements, usually involving the hands or fingers. It is observed in ailments such as cerebral palsy.
- **Myoclonus:** Quick, unexpected, shock-like spasms in the muscles. One common symptom of myoclonus is tics.

SUMMARY

- Making up a sizable amount of the body, skeletal muscle is a dynamic tissue essential to movement, posture maintenance, and force production.
  - Muscle, which is mostly composed of water (20%) and protein (20%), has an impact on energy metabolism, substrate storage, heat production, and oxygen use during exercise.
  - Muscle growth and function are determined by the processes of protein synthesis and degradation, which are impacted by nutrition, hormones, and physical activity.
- The layers of connective tissue that envelop multinucleated fibers in muscles are called the perimysium, sarcolemma, and epimysium.
  - Nuclear domains of fibers control the synthesis of proteins. Muscle health depends on proteins like dystrophin, which are essential for the integrity of the sarcolemma.
  - The main types of proteins found in muscle fibers include cytoskeletal, regulatory, and contractile proteins.
  - Myofilaments, which are made up of actin and myosin, make up 70–80% of all fiber protein.
  - The contraction, stability, and integrity of muscles are facilitated by sarcomere proteins such as nebulin, titin, troponin, and tropomyosin.
- Important roles are played by cellular constituents such as mitochondria, sarcoplasmic reticulum, and T tubules. To ensure equal stimulation, nerve action potential is transmitted by T tubules.
  - Calcium is stored in the sarcoplasmic reticulum and released when contraction occurs. When oxygen is present, mitochondrial networks produce energy for muscular activity.
  - In the sarcolemma, voltage-gated channels, pumps, and ion channels are used to excite muscle fibers. Ion-selective channels affect the resting membrane potential, which is necessary for the production of action potentials.
  - When the motor endplate potential in skeletal muscles exceeds the threshold, voltage-gated Na channels are activated, triggering the action potential.
  - Through t-tubules, the depolarization wave spreads throughout the sarcolemma.
  - The creation of cross-bridge force is connected to plasma membrane stimulation through excitation-contraction coupling. Acetylcholine release at the neuromuscular junction triggers the cross-bridge cycle and calcium signals *via* signaling muscle action potentials.
  - Actin-myosin interactions are triggered by calcium binding to troponin, which results in ATP-fueled muscle contraction.
  - When muscles relax, titin and elastic connective tissues cause filaments to slide back and absorb calcium into the sarcoplasmic reticulum.

- In the musculoskeletal system, strength and power training cause a variety of physiological reactions and adaptations.
  - Muscle hypertrophy is one of the most noticeable improvements; after many months of resistance training (RT).
  - The production of contractile myofilaments, especially in fast-twitch type II fibers, results in a rise in the size of muscle fibers.
  - Program design, with an emphasis on heavy repetitions, short rest intervals, and several sets, has an impact on RT-induced muscle growth.
- Muscle strength, determined by maximal contractions such as the one-repetition maximum (1 RM), is the ability to apply force against external resistance.
  - It is the outcome of intricate interactions between neuromuscular components that include neurological, muscular, and mechanical elements including control of discharge frequency and recruitment of motor units. Contrarily, muscular endurance is the capacity to perform repeated contractions or resist fatigue over a prolonged length of time.
  - Muscle power is measured as work done per unit of time and is the result of multiplying force with movement velocity.
  - Work is the outcome of a force acting on a body that is moving in the same direction as the force, as determined by the body's displacement, average force, and direction.
- Muscle power training focuses on developing explosive strength through exercises like Olympic lifts and plyometrics.
  - It induces physiological changes, including enhanced neuromuscular coordination, improved muscle fiber recruitment and quality, efficient energy transfer between muscle groups, and heightened metabolic functions.
  - These adaptations collectively contribute to increased power, speed, and endurance during high-intensity activities.
- Active exercises involve voluntary bodily motions, relying on individual effort to generate actions. These exercises, including isometric and isotonic movements, are vital for physical health, reducing the risk of chronic diseases, and aiding rehabilitation after injuries or illnesses.
  - Indications for active movements encompass various conditions, from musculoskeletal injuries to neurological issues and cardiovascular concerns.
  - The types of active movements, both voluntary and involuntary, cater to specific needs, including free exercises that utilize body weight for relaxation, muscle tone maintenance, and improved coordination.
- Active assisted exercises, which are frequently used in rehabilitation, involves using external forces to assist joint motions. This may include apparatus, pulleys, or therapy support.
  - These improve range of motion, muscle strength, and neuromuscular coordination.
  - While they are beneficial for strengthening muscles, increasing flexibility, and facilitating rehabilitation, they should not be used in cases of acute injuries, excruciating pain, unstable joints, recent surgery, or cardiovascular problems.
- Exercises that combine external resistance—such as weights or resistance bands—with support from a partner or trainer are known as assisted resisted exercises.
  - Exercises for the shoulders and core that involve a partner using a resistance band to create extra resistance are some examples. This method is similar to active assisted workouts, except that resistance and support happen at the same time across the entire range of motion.
  - Beginners, people with limited strength or mobility, people healing from injuries, and athletes honing skills particular to their activity can benefit from these exercises.
  - Strength and muscular gain, endurance training, conditioning specialized to a sport, fat loss, and an increase in general fitness are the main results.

FURTHER READINGS

- Rantanen T. Muscle Strength, Disability and Mortality. Scandinavian Journal of Medicine and Science in Sports. 2003 Feb; 13(1):3–8.
- Newton, Robert U. MHMS, CSCS; Kraemer, William J. PhD, CSCS. Developing Explosive Muscular Power: Implications for a Mixed Methods Training Strategy. Strength and Conditioning 16(5): p. 20–31, October 1994.
- Evans WJ. Exercise Strategies Should be Designed to Increase Muscle Power. The Journals of Gerontology Series A: Biological Sciences and Medical Sciences. 2000 Jun 1;55(6):M309–10.
- Frontera WR, Ochala J. Skeletal Muscle: A brief review of structure and function. Calcified tissue international. 2015 Mar; 96:183–95.
- Philip M Hopkins, Skeletal Muscle Physiology, Continuing Education in Anaesthesia Critical Care and Pain, Volume 6, Issue 1, February 2006, p. 1–6, https://doi.org/10.1093/bjaceaccp/mki062
- Kuschel LB, Sonnenburg D, Engel T. Factors of Muscle Quality and Determinants of Muscle Strength: A systematic literature review. InHealthcare 2022 Oct 3 (Vol. 10, No. 10, p. 1937). MDPI.
- Ahtiainen JP. Physiological and molecular adaptations to strength training. Concurrent Aerobic and Strength Training: Scientific Basics and Practical Applications. 2019:51–73.
- Gardiner Dena M. Principles of Exercise Therapy. CBS Publishers and Distributors Pvt Ltd; 4th ed., 2023.

STUDENT ASSIGNMENT

LONG ANSWER QUESTIONS

1. Explain the structure of skeletal muscle and mechanical events during contraction and relaxation of a muscle.
2. Explain physiological adaptation to training for muscle strength, power and endurance.
3. Describe free exercises under the following headings: Principles, techniques, indications, contraindications, effects and uses.
4. Explain active assisted exercises under the following headings: Principles, techniques, indications, contraindications, effects and uses.
5. Explain assisted-resisted exercises in detail.

SHORT ANSWER QUESTIONS

1. Define muscle strength, power, work and endurance.
2. Define muscle actions.
3. What are the causes of decreased muscle performance?
4. Write briefly about types of active movements.
5. Write in brief about excitation-contraction coupling mechanism of muscle contraction.

MULTIPLE CHOICE QUESTIONS

1. **What is muscular strength?**
   a. Capacity to endure fatigue
   b. Ability to resist external force
   c. Capacity to exert force against external resistance
   d. Ability to endure repeated contractions
2. **How is muscular strength commonly measured?**
   a. Work done
   b. One-Repetition Maximum (1 RM)
   c. Muscle power
   d. Endurance capacity
3. **What is the primary factor in calculating muscle power?**
   a. Force b. Displacement
   c. Time d. Distance
4. **What is the definition of muscular endurance?**
   a. Maximum force generation in a single contraction
   b. Ability to withstand repeated contractions without fatigue
   c. Force multiplied by movement velocity
   d. Capacity to exert force against external resistance
5. **What is the result of force acting on a body moving in the same direction as the force?**
   a. Distance b. Time
   c. Power d. Work
6. **What is the primary function of the T tubule system in muscle fibers?**
   a. Energy production
   b. Nerve action potential transmission
   c. Calcium storage
   d. Protein synthesis
7. **Which protein is crucial for sarcolemma integrity and is partially or entirely lacking in some neuromuscular illnesses?**
   a. Actinin b. Titin
   c. Dystrophin d. Nebulin
8. **What is the primary role of sarcoplasmic reticulum in muscle physiology?**
   a. Protein synthesis b. Oxygen generation
   c. Calcium homeostasis d. Substrate storage

9. **What are the main components of skeletal muscle?**
   a. Water and carbohydrates
   b. Water and lipids
   c. Water and protein
   d. Water and vitamins
10. **Which protein is crucial for the integrity of the sarcolemma?**
   a. Actin b. Myosin
   c. Dystrophin d. Troponin
11. **What is the resting potential range of skeletal muscle cells?**
   a. 50–60 mV b. 60–70 mV
   c. 70–90 mV d. 90–100 mV
12. **Which fiber type transition occurs in response to resistance training (RT)?**
   a. Type 1 to Type 2X b. Type 2X to Type 1
   c. Type 2X to Type 2A d. Type 2A to Type 2X
13. **What is the primary stimulus for the transition of muscle fibers during RT?**
   a. Oxygen availability
   b. Glycogen levels
   c. Force generation
   d. ATP synthesis
14. **Which metabolic adaptation occurs in response to resistance training?**
   a. Increase in ATP levels
   b. Reduction in anaerobic enzymes
   c. Decrease in PCr and glycogen
   d. Enhanced insulin resistance
15. **What is the primary factor responsible for muscle hypertrophy in resistance training?**
   a. Long rest intervals
   b. Low repetitions
   c. Heavy resistance
   d. Limited sets

ANSWER KEY

1. c **2.** b **3.** c **4.** b **5.** d **6.** b **7.** c **8.** c **9.** c **10.** c
11. c **12.** c **13.** c **14.** b **15.** c

9

Resisted Exercises

Sajjan Pal, Sheetal Kalra

LEARNING OBJECTIVES

After the completion of the chapter, the readers will be able to:

- Define resisted exercises.
- Classify resisted exercises.
- Explain the principles of resisted exercises.
- Understand various equipment used for strength training.
- Explain physiological adaptation with resisted exercises.
- Explain the precautions of resisted exercises.
- Mention the indications and contraindications of resisted exercises.
- Understand and demonstrate techniques of resisted exercises at various joints.

CHAPTER OUTLINE

- Introduction
- Resisted Training
- Classification
- Various Equipment for Resistance Training
- Determinants
- Principles
- Physiological Adaptations
- Precautions
- Contraindications
- Exercise Regimens
- Technique of Application of Resisted Exercises

KEY TERMS

Cycle ergometer: It is a stationary device utilized in rehabilitation settings to facilitate passive, active or resisted exercises.

DeLorme principle: In accordance with the DeLorme principle, individuals engage in a program of progressive resistance exercise (PRE) based on a 10-repetition maximum (10 RM). Subjects perform the first set of 10 repetitions at 50% of their 10 RM, the second at 75% of their 10 RM, and the third (final set) at 100% of their 10 RM.

Eccentric: In an eccentric training session, the weight is controlled as it descends by using target muscles to control the weight as it moves down.

Isokinetic: Isokinetic training is a type of strength training where the speed of movement remains the same but the resistance varies.

Isometric: An isometric exercise involves the static contraction of the muscle without visible movement at the angle of the joint.

Plyometric: Plyometric exercises are intense and are a series of explosive bodyweight resistance workout exercises that enhance physical performance by utilizing muscle fiber. The stretch-shortening cycle (SSC) of muscle fibers are used to increase physical performance.

INTRODUCTION

Fitness, sports performance, and metabolic health are all significantly impacted by skeletal muscles. Both in healthy persons and those with medical issues, low muscle mass is linked to increased mortality rates. Internal and external factors, including biological processes occurring within the muscle, have a role in the regulation of muscle mass. External influences include resistance exercise training (RET). For building muscle, RET is acknowledged as the most successful nondrug approach. At the molecular level, RET-induced muscle hypertrophy involves the accumulation of various cellular proteins within existing muscle fibers. Over the past two decades, resistance training has gained popularity due to its diverse benefits for sports performance and overall health. Originally favored by athletes and bodybuilders, it is now widely acknowledged for enhancing muscular strength, power, speed, hypertrophy, local muscle endurance, motor performance, balance, and coordination.

Leading health organizations like the American College of Sports Medicine and the American Heart Association now recommend resistance training for various populations, including adolescents, healthy adults, the elderly, and individuals with cardiovascular or neuromuscular diseases.

RESISTED TRAINING

By definition, "resistance training (RT)" is a specialized approach to physical conditioning that includes several training modalities, such as medicine balls, weight machines, free weights (barbells, dumbbells, etc.), elastic bands, and plyometric exercises, along with the progressive application of a wide range of resistive loads and movement velocities.

The RT is a crucial component of rehabilitation programs for those with reduced function, promoting health and well-being. These exercises involve resisting static or dynamic muscular contractions with an external force, which can be applied manually or mechanically. Mechanical resistance exercises involve applying load using workout equipment. Common resistance training regimens include progressive resistance exercise (PRE), circuit weight training, plyometric training, and isokinetic training. These exercise regimens develop muscle strength, power, and endurance, which are crucial for health and athletic fitness.

Effects

- **Boosts muscular strength:** Strength, which is inversely proportional to the degree of tension in the contracting muscle, is defined as the force that the muscle produces.

  A muscle must be loaded or subjected to resistance in order for an increase in tension to form as a result of hypertrophy and the activation of new muscle fibers.
- **Increases physical endurance:** The capacity of a muscle to continue working at a low level for a long time is known as endurance.

  Exercises that are performed repeatedly while facing light resistance help to increase muscular endurance.
- **Increases the power:** Resistance workouts help to increase muscle power. Maximizing muscle power is critical for success in explosive athletic exercises such as throwing and jumping. To do this, athletes use a variety of training approaches designed to improve their dynamic performance. Among these strategies, manipulating training loads is critical in determining the type of training stimuli and subsequent adaptations.

According to research, different training loads cause specific adaptations in muscle power, emphasizing the need of choosing optimal loads for maximum power production. Incorporating periodization and mixed training methodologies into resistance-training regimens can also promote muscle power development and improve overall athletic performance.

Benefits

- Minimize musculoskeletal alterations associated with aging, contributing to the health and well-being of older individuals (Ciolac et al.).
- Prevent and control the development of musculoskeletal chronic diseases (Jorge et al.).
- Rehabilitation of musculoskeletal disorders (Jorge et al.).
- Resistance exercise (RE) is an excellent therapy method for osteosarcopenia because it stimulates bone mass accretion by subjecting bone tissues to mechanical pressures that exceed normal activities, hence retaining both bone and muscle mass efficiently (Hong et al.).
- Improve muscular fitness in adolescents (Dorgo et al.).
- Enhance cardiorespiratory fitness and muscle strength (Paoli et al.).
- Improve fitness and sports performance (Hrysomallis et al.).
- Enhance physical fitness, function, and independence in older adults with musculoskeletal disorders, leading to improved quality of life.
- Improve pain and grip strength in patients with conditions like lateral epicondylitis (Raman et al.).
- Improve neuromuscular functioning in patients with dementia.
- Improve muscle volume, muscle force, and functional status in persons with Parkinson's disease as a result of high-force eccentric resistance training.
- Improve muscle strength, cardiovascular function, and gross motor skill performance in children with cerebral palsy.
- Improve disability, physical performance, and pain in older disabled individuals with knee osteoarthritis (Ettinger et al.).

- Develop muscular strength, endurance, and mass, as well as maintain basal metabolic rate and prevent falls in the elderly (Pollock et al.).
- Improve function in frail, and elderly patients through both upper- and lower-body exercises (Pollock et al.).
- Enhance basic motor abilities, functional muscle strength, and walking efficiency in children with conditions like spastic diplegia (Liao et al.).
- Improve self-efficacy in people with human immunodeficiency virus through supervised resistance exercise programs (Fillipas et al.).
- Boost self-esteem, physical fitness, body composition, and chemotherapy completion rates in breast cancer survivors (Courneya et al.).
- Act as an adjunct to standard care to improve glycemic control and address metabolic syndrome abnormalities in high-risk older adults with type 2 diabetes (Castaneda et al.).

CLASSIFICATION

Based on Types of Muscle Work/Muscle Action

Strengthening Exercises

Strengthening exercises can be divided into two main categories:

1. **Static exercises** (also known as **isometric exercises**) involve maintaining a constant muscle length. An example is "quadriceps setting" or the wall push isometric exercise which involves pressing palms against a wall with elbows bent, and holding the position to engage and strengthen the chest, shoulders, and arms.
2. **Dynamic exercises**, on the other hand, dynamic exercises include movements that involve joint motion.

   Types of dynamic muscle actions:
   - **Isotonic actions:** In these actions, the muscle maintains a consistent force throughout the movement. For example, squatting, where knees are bent and hips lowered, then returning to standing to engage the quadriceps, gluteals, and hamstrings.
   - **Isokinetic actions:** These involve moving at a constant speed and usually require special exercise machines. They are often used in rehabilitation settings to ensure controlled movements.
   - **Concentric actions:** This is when the muscle shortens while producing force. For example, when one lifts a dumbbell during a bicep curl, the biceps are doing a concentric action.
   - **Eccentric actions:** In contrast, eccentric actions occur when the muscle lengthens while still generating force. Lowering the dumbbell back down is an example of this.

Importance of eccentric muscle actions: Eccentric muscle actions are particularly interesting because:

- **Efficiency:** They can generate more force compared to concentric actions with the same effort.
- **Muscle damage and recovery:** After a heavy bout of eccentric exercise, healthy individuals might experience some muscle soreness. This soreness is linked to a temporary increase in creatine kinase (CK), a protein that indicates muscle damage.
- **Muscle repair and hypertrophy:** For healthy individuals, this muscle damage can be beneficial. It acts as a signal for the body to repair and strengthen muscles, leading to hypertrophy (muscle growth). However, it's important to allow time for recovery after such workouts.
- **Concerns for patients with muscle diseases:** In individuals with certain muscle diseases—like dystrophinopathies and sarcoglycanopathies—eccentric exercises can pose a risk. These conditions affect the muscle cell membranes, making it harder for the muscles to repair themselves after damage.
- **Postexercise discomfort:** Eccentric training may result in higher post-exercise discomfort, peaking at 48 hours and decreasing with successive sessions. It is not the dominant mode in most programs, but it provides variety.

Isometric Resistance Exercises

Isometric exercise involve muscle contractions without changing muscle length. It is often done against immovable objects or self-generated tension. Maximal contractions are more effective, with gains from both short or longer duration exercises. Training at multiple joint angles improves strength and dynamic power. Isometrics are used in injury rehab and for strength improvements in sports. They are often part of a larger resistance training program.

Benefits

Isometric strength training (IST) has various advantages over dynamic strength training, including:

- Less fatigue and higher joint angle-specific strength.
- It has also been demonstrated to improve dynamic sports performance, such as running, jumping, and cycling.
- Integrating IST into athletes' training routines can prevent undue fatigue while also encouraging neuromuscular changes.
- It is notably useful for increasing strength in biomechanically disadvantaged joint positions and improving motions that require largely isometric contractions, particularly for athletes who have limited mobility owing to injuries.

Technique

It is recommended that IST be performed at 70–75% of maximum voluntary contraction (MVC) for the purpose of muscular growth. For pain-free holds, the recommended duration is 5–10 seconds. For hypertrophy, the sustained contraction duration should be of 80–150 seconds per session, while for maximum strength, it should be 30–90 seconds per session. Hypertrophy requires >70% intensity of maximum voluntary contraction (MVC), and for maximum strength, the intensity should be 80–100% MVC.

Various Isometric Workouts

- **Muscular-setting exercises:** Involve low-intensity contractions to relieve muscular soreness and increase circulation after an injury, which helps with early mobility maintenance.
- **Stabilization exercises:** Use sustained co-contractions, improve postural or joint stability and often include body weight or manual resistance.
- **Multiple-angle isometrics:** Aims to increase strength across the range of motion by providing resistance at various joint locations, which is especially effective when dynamic exercises are difficult.
- **Brief resisted isometric maximal exercise (BRIME):** It is a type of exercise in which brief, powerful contractions are performed against resistance in an isometric (static) fashion. In BRIME, the muscles are maximally stimulated and kept in a static position for a brief period of time, often ranging from a few seconds to about 10 seconds, depending on the protocol.

Examples

- **Isometric strengthening for neck extensors:** Head pushes back the clasped hand behind the head. There is no head or neck movement, but the tone of the neck extensors gets stronger (Fig. 9.1).

Did You Know?

Isometric exercises have been used not only in sports training but also in various fields, including rehabilitation and even space travel. Astronauts, for instance, use isometric exercises to maintain muscle strength and prevent muscle atrophy while in space where traditional resistance training methods are limited. This highlights the versatility and effectiveness of isometric training across different contexts.

- **Plank:** This exercise works the core muscles, as well as the shoulders, back, and legs (Fig. 9.2).

Clinical Correlation

Variable resistance machines offer a valuable tool in rehabilitation settings and can be particularly beneficial for patients with joint impairments or muscle imbalances, allowing for targeted strengthening of specific muscle groups while minimizing stress on vulnerable areas. Machines that change resistance across the range of motion in an effort to match gains and reductions in strength (strength curve) throughout the range of motion are used for strength training. Theoretically, this leads the muscle to contract maximally across the range of motion, resulting in greater strength improvements.

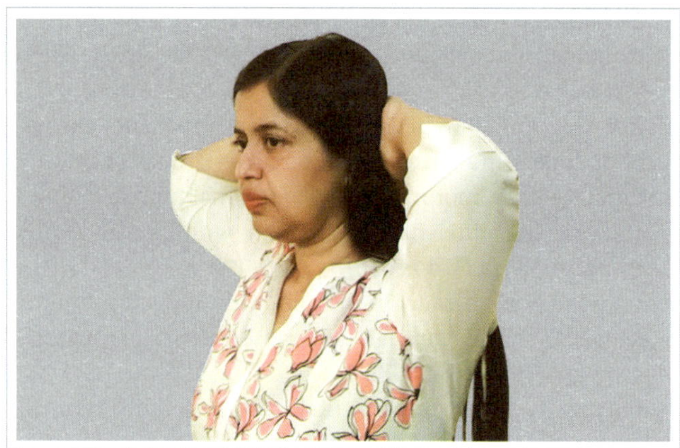

Fig. 9.1: Isometric strengthening for neck extensors

Fig. 9.2: Plank

Isokinetic Training

Isokinetic exercise consists of dynamic motions in which the velocity of muscle shortening or lengthening remains constant, as controlled by an isokinetic dynamometer. Isokinetic training differs from standard resistance training in that the velocity of limb movement is adjusted rather than the weight itself. This technique, also known as accommodating resistance exercise, allows for variable but maximal force output across the whole range of motion.

Isokinetic training provides a wide range of exercise velocities, from slow to fast, to meet various functional demands. Despite its ability to improve muscular performance characteristics, training impacts on functional skills are limited. Mismatch between isokinetic training velocities and varied speeds seen in regular activity creates challenges. Furthermore, isokinetic exercise often isolates single muscles or muscle groups, which may not adequately address the complexities of multijoint, multi-planar movements in functional tasks.

While isokinetic training has been shown to improve speed, there is little evidence of considerable overflow or mode specificity. Despite isolating specific muscle areas, the strategy has advantages, particularly in rehabilitating strength deficits.

Isokinetic Regimens

Velocity spectrum rehabilitation (VSR) is a strategy for overcoming the limitation of training effects transferring from one velocity to another. Typically, medium (60°–90°–180°/sec) and rapid (180°–360°/sec) angular velocities are used. It should be noted that, while isokinetic devices allow for testing and training at speeds >360/sec, the highest speeds are frequently avoided due to minimal muscular resistance across the range of motion.

A typical VSR protocol may include one or two sets of 8–10 or up to 20 repetitions at various velocities for the agonist and antagonist muscle groups. Rest intervals of 15–20 seconds between sets and 60 seconds between exercise velocities are recommended, with a maximum of three sessions per week.

Warm-up exercises on the dynamometer use submaximal effort, whereas strength or power workouts are performed at maximum intensity. Slow-speed exercise is tapered out as the program develops, with a focus on maximum-effort, high-velocity training during the advanced rehabilitation stages. However, because of the high shear stresses across joint surfaces, maximum-effort, slow-velocity training is not recommended.

Eccentric isokinetic exercise, albeit less prevalent, can be used in the later stages of rehabilitation to address isolated strength and power deficits. Medium training velocities are chosen for safety reasons, as sudden, rapid movements of the dynamometer's torque arm on a limb might result in harm.

Clinical Correlation

Proper warm-up exercises using submaximal effort are essential to prepare muscles and joints for the intensity of isokinetic training. Similarly, incorporating cool-down exercises can help prevent muscle soreness and promote recovery post-training.

For details on isotonic exercises and its types, readers can refer to Chapter 3.

Based on Types of Resistance

Manual Resistance Exercises

It is when a therapist applies resistance force to a dynamic or isometric muscle contraction. Manual resistance training (MRT), integrates accommodating resistance techniques, which may result in equivalent strength increases to standard weight resistance training.

The MRT requires working in couples, with one person supplying resistance to the other throughout the exercise. While typical equipment is not required, certain items such as benches and straps may be used. The research on MRT is limited, although it reveals possible benefits. For example, a research that compared MRT to calisthenics found that MRT produced greater strength improvements, in healthy college students.

Advantages

- Despite its limits, MRT has several advantages, including minimum equipment requirements and changing resistance, which may increase muscular tension over the entire range of motion. This strategy may be extremely beneficial for programs with budget constraints or limited access to conventional weight facilities.
- Another advantage of this method is its versatility, which allows for resistance modifications to accommodate varied strength levels.
- It provides a simple and economical method of muscle building muscle strength and endurance.
- It allows for targeted strengthening of specific muscle groups.

Limitations

However, its success is dependent on the therapist's abilities and effort, as resistance cannot be exactly quantified. This method also requires a major time and effort commitment because therapists may only treat one patient at a time.

Clinical Correlation

For patients with limited access to equipment or in settings with budget constraints, MRT offers a cost-effective solution with minimal equipment requirements. However, therapists must ensure proper technique and exertion levels, as the success of MRT depends on their abilities and effort.

Mechanical Resistance Exercises

They involve using equipment to provide resistance against isotonic or isometric muscle contractions. A variety of mechanical devices, ranging from simple to advanced and affordable to expensive, can be employed depending on the individual's needs, exercise objectives, and equipment accessibility.

Common options include:

- Free weights, such as dumbbells and barbells, which allow for a wide range of movements and muscle engagement.
- Weight machines provide a controlled environment for targeting specific muscle groups, while resistance bands offer versatile and portable training solutions.
- Cable machines are also popular for their ability to provide continuous tension throughout exercises.
- Additionally, medicine balls and kettle bells can enhance functional strength and stability.
- Bodyweight exercises such as push ups, squats, lunges, etc. use gravity as resistance and are effective for building strength without equipment.
- Suspension training systems, like TRX, utilize body weight and leverage to challenge muscles in dynamic ways.

- Hydraulic resistance involves resistance being applied using hydraulic cylinders or pistons, which are commonly found in hydraulic workout machines.

Advantages

The benefits of mechanical resistance exercises are numerous:

- Firstly, they allow for the objective measurement of patient progress, providing clear feedback on strength gains and improvements over time.
- These exercises are not dependent on the therapist's strength levels, allowing patients to work at their own pace and intensity. This independence can lead to greater engagement and motivation during the workout sessions.
- Mechanical resistance exercises make efficient use of the therapist's time and effort, as they can oversee multiple patients simultaneously.

Limitations

Mechanical resistance exercises do have some limitations:

- One significant drawback is the inability to adjust resistance levels throughout the entire range of motion.
- The resistance remains fixed, which may not fully accommodate the varying strength capabilities of individuals or provide optimal resistance for different phases of the exercise.
- Implementing mechanical resistance exercises may require adequate space for equipment setup and may entail costs associated with acquiring and maintaining the necessary equipment.

Based on Functions

- **Strength training:** Aims to increase muscle strength and power through exercises such as heavy lifting and low-repetition sets.
- **Hypertrophy training:** Aims to increase muscle size with moderate to high repetitions and volume.
- **Endurance training:** Increases muscle endurance and stamina through high-repetition sets with lighter weights or bodyweight movements (Table 9.1).

TABLE 9.1: Loading recommendations for muscle strength along the repetition continuum

Repetition range	Load (% of 1 RM)	Primary adaptation
1–5 repetitions	80–100%	Optimizes strength increases
8–12 repetitions	60–80%	Optimizes hypertrophic gains
15+ repetitions	Below 60%	Optimizes local muscular endurance improvements

Based on Muscle Groups/Number of Joints Targeted

1. **Upper body exercises:** Exercises that focus on muscles in the upper body, such as the chest, back, shoulders, and arms.
2. **Lower body exercises:** Exercises that focus on muscles in the lower body, such as the legs, glutes, and calves.
3. **Core exercises:** Exercises that target the core muscles, such as the abs, obliques, and lower back.
4. **Single-joint (SJ) exercises:** Exercises target only one joint and typically involve the movement of one major muscle group. Examples include dumbbell flies for the chest and knee flexion exercises for the quadriceps.
5. **Multijoint (MJ) exercises:** Exercises involve the movement of multiple joints and engage multiple major muscle groups simultaneously. Examples include bench presses for the chest, shoulders, and triceps, and deadlifts for the lower back, glutes, hamstrings, and quadriceps. In a study, participants in the MJ group performed exercises like bench presses and deadlifts.
6. **Open chain exercises:** In open chain exercises, the component that is farthest from the body—such as the hand or foot—moves freely and is not attached to anything. One of the main advantages of open chain exercises is their enhanced capacity to isolate particular muscles. This function is helpful for training for sports that require open-chain exercises and targeted muscle restoration (Fig. 9.3A).
7. **Closed chain exercises:** Exercises or movements involving the distal aspect of the extremities fixed to a stationary object are known as closed kinetic chain (CKC) exercises. Resistance training is provided concurrently to the proximal and distal parts because, with the distal component immobile, movement at any one joint in the kinetic chain necessitates action at the other joints in the kinetic chain (Fig. 9.3B).

Did You Know?

Incorporating these diverse resistance training methods into clinical practice allows healthcare professionals to design individualized rehabilitation programs tailored to patients' unique needs, promoting optimal recovery and long-term musculoskeletal health.

VARIOUS EQUIPMENT FOR RESISTANCE TRAINING

Over the past 160 years, equipment has been developed to apply external resistance to skeletal muscles for resistance exercise. Each equipment has its benefits and drawbacks. Some disadvantages may be mechanical, such as how resistance

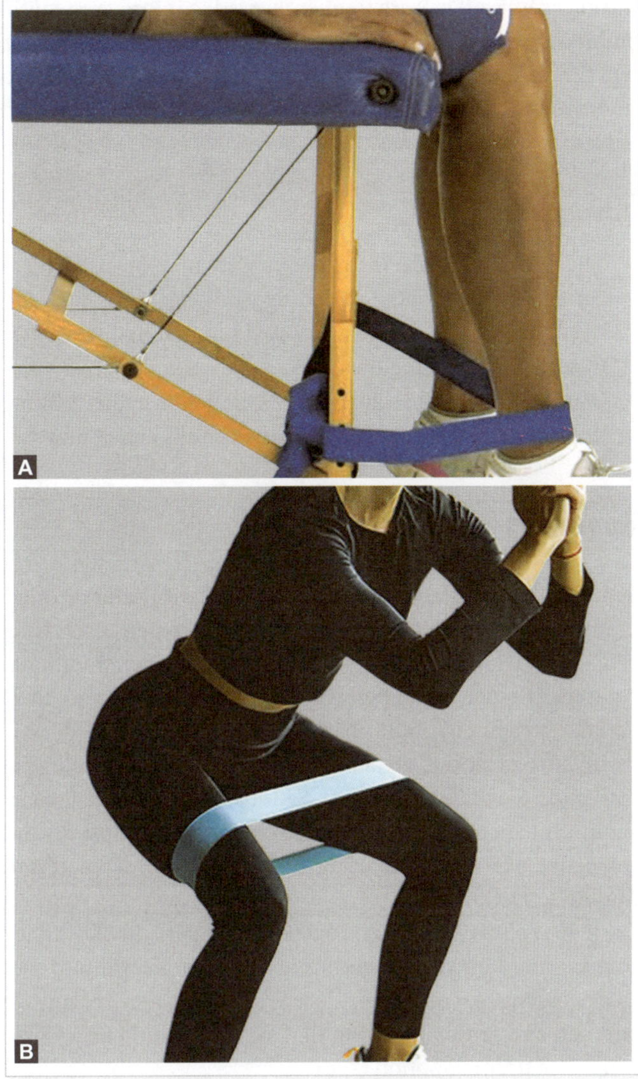

Figs 9.3A and B: A. Open chain resisted exercise; B. Closed chain resisted exercise

is communicated to the body. Some disadvantages may be practical, such as machine size, weight, and cost. Various equipment used for performing resisted exercises are shown in Figures 9.4A to D.

Free Weights

Free weights use isotonic resistance, which offers consistent resistance across the range of motion. This allows for balanced movement in many planes. Examples of free weights include dumbbells, barbells, kettlebells, weight plates, medicine balls, and weighted vests.

Advantages

- Versatility
- Suitability for ballistic workouts

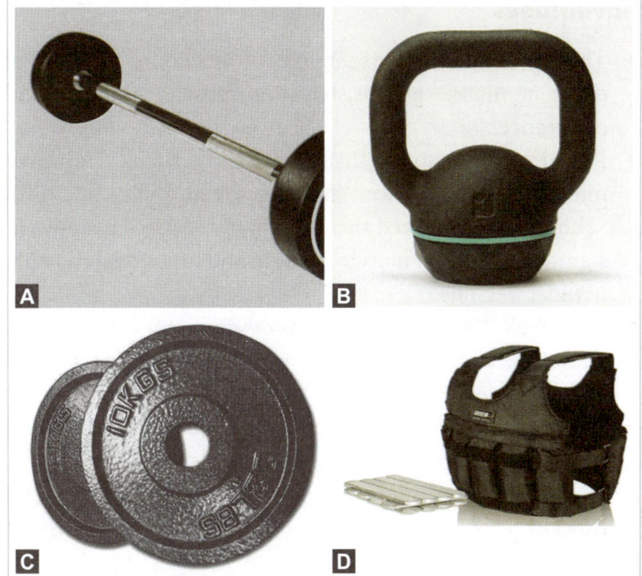

Figs 9.4A to D: A. Barbell; B. Kettle bell; C. Weight plates; D. Weighted vests

- The ability to replicate real-world movements
- The involvement of numerous muscle groups
- A reduced cost.

Disadvantages

- Intimidation for some users, limited resistance direction, time-consuming adjustments, the danger of injury if used incorrectly, and the requirement for proper workout technique.
- Some routines are difficult to accomplish with free weights and may demand more room than single-station machines.

Dumbbells

Dumbbells are portable free weights made out of a handle with weighted ends that are often used for strength and resistance exercises (Fig. 9.5).

Fig. 9.5: Dumbbells

Advantages

- They provide a diverse range of workouts that target different muscle groups, boosting muscle, strength, and endurance.
- Research has shown that dumbbell exercise improves muscle strength, power, and functional ability. In 2016, Schott et al., observed that dumbbell exercises resulted in significant gains in muscle strength and functional capacity in older persons.
- Dumbbells are adaptable, providing for progressive resistance and routines tailored to specific fitness levels and goals.
- They are commonly used in gyms, rehabilitation clinics, and at home training programs, providing a convenient and effective technique to increase strength and general fitness.

Medicine Ball

A medicine ball is a multifunctional exercise item made up of a weighted sphere, usually packed with sand or another dense material, and weighing anywhere from a few pounds to >20 (Fig. 9.6). It is used in a variety of exercises designed to increase strength, power, and coordination. Throws, smashes, and twists work many muscle groups, particularly those in the core, upper body, and lower body.

Advantages

This flexible piece of equipment improves functional fitness, athletic performance, and overall physical conditioning by focusing on different muscle groups and encouraging balance, coordination, and flexibility.

Weight Cuffs

Weight cuffs are adjustable bands filled with weights that wrap over limbs, usually ankles or wrists, to provide resistance during exercise. They are frequently used to improve strength, endurance, and muscle tone in rehabilitation, physical therapy, and fitness settings. Weight cuffs have been found in studies to improve muscle strength, power, and functional ability when used in workout routines.

Advantages

- They are adaptable instruments that may be used for a variety of workouts, including walking, running, and strength training, while also providing tailored resistance to certain muscle regions.
- Weight cuffs are handy, portable, and adjustable, making them suitable for people of all fitness levels and abilities.

Sandbags

Sandbags for resistance training are filled with sand or other granular materials that offer resistance when exercising (Fig. 9.7). They provide distinct advantages over traditional equipment, such as kettlebells or dumbbells, due to their variable weight distribution.

In 2016, Hackett et al., investigated acute physiological responses to sandbag training. Sandbag exercises resulted in significant increase in heart rate, oxygen consumption, and subjective exertion, indicating high metabolic demand and muscle activation.

In 2017, Dawes et al., conducted a comparison of sandbag training and standard weightlifting activities. They discovered that sandbag exercises produced similar gains in muscle strength and power, implying that sandbags could be a viable resistance training option.

Advantages

- Sandbag training works many muscular groups concurrently, improving strength, stability, and functional fitness.

Fig. 9.6: Medicine ball

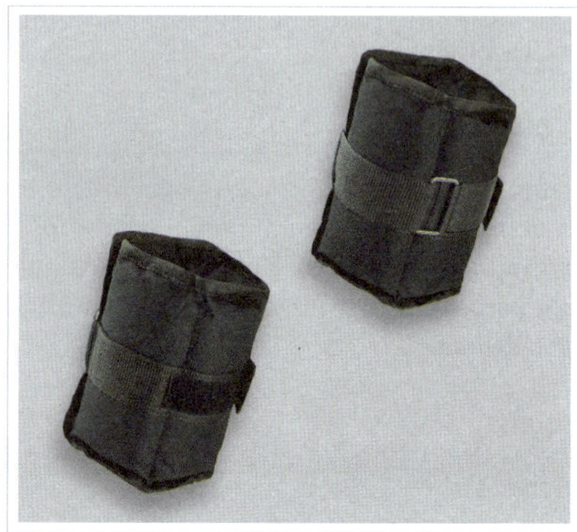

Fig. 9.7: Sandbags

- They come in a variety of sizes and forms, providing for a range of exercises that target different muscle groups and movement patterns.
- They are portable, inexpensive, and adaptable to many locations, making them popular among fitness enthusiasts, military people, and athletes seeking variation and challenge in their workouts.

DeLorme's Boot

The DeLorme weight boot, developed by American Physician Thomas L DeLorme, is a simple but efficient instrument for strength training and rehabilitation. It firmly attaches weights to the foot and ankle, which promotes muscular growth and strength improvements. Developed in the 1940s, the boot was motivated by DeLorme's own experience with muscle atrophy caused by rheumatic sickness. It has a simple design, consisting of an aluminum foot plate, collar, straps, and a rod for attaching weights. The boot targets quadriceps, ankle dorsiflexors, inverters, evertors, and hamstrings. DeLorme's progressive resistance training routine, which included the boot, became an integral part of physiotherapy rehabilitation treatments. The regimen normally consists of three sets of 10 repetitions, allowing for gradual weight increases. This method was effective in rehabilitating injured warriors returning from World War II.

Suspension

Suspension therapy is a therapeutic strategy in which body parts are suspended in the air. This type of therapy can be done standing, sitting, or lying down, with ropes or slings as suspension devices. Physical therapists, and other medical practitioners frequently use suspension therapy to improve joint mobility, flexibility, and strength. The topic is discussed separately in Chapter 19.

Pulley Systems

Pulley systems provide versatile resistance training options with constant load throughout the range of motion (Fig. 9.8).

Advantage

They offer adjustable resistance and allow for varied exercise positions, targeting specific muscle groups.

Disadvantage

Controlled movements are essential for safety and effectiveness.

Fig. 9.8: Pulley system

Weight Cable System

Variable-resistance weight cable machines incorporate a cam mechanism designed to adjust the load applied to muscles throughout the range of motion (ROM), even with a consistent weight selection (Fig. 9.9).

Advantages

- Cable machines remain vital in gyms due to their versatility in engaging multiple muscle groups through various planes of motion.
- They enable functional exercises targeting key muscle groups such as the chest, back, shoulders, and arms while enhancing overall stability and balance.
- With features like multiple weight stacks and adjustable angles, users can tailor their workouts for optimal results, including exercises like pulldowns, chest presses, and rows.
- These machines support functional and core exercises like squats and planks, contributing to overall fitness progression for individuals of all levels.

Disadvantage

While the cam aims to mimic the natural torque curve of muscles, its effectiveness in providing truly accommodating resistance across the full ROM is subject to debate.

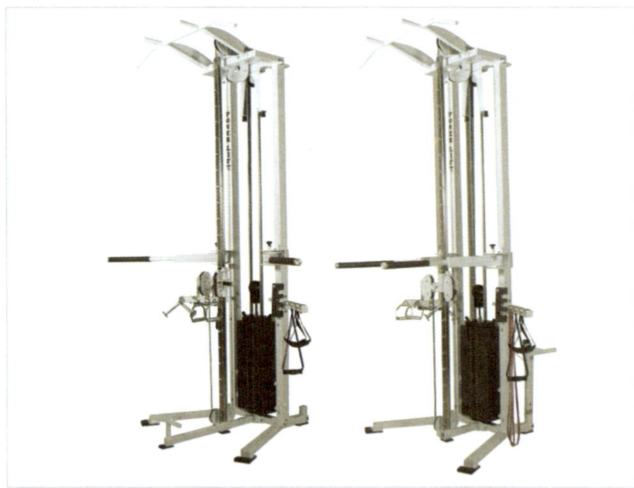

Fig. 9.9: Weight cable system

Elastic Bands and Tubes

Elastic resistance training (ERT) offers a versatile and practical alternative for strength training, utilizing bands or tubes made of elastic materials. Popular since the 1980s, ERT has gained attention due to its benefits, including improved functional capacity, increased strength and endurance, enhanced muscle activation, and better body composition and quality of life.

Compared to no intervention, ERT are beneficial in improving functional performance and muscle strength in healthy individuals as per research by de Oliviera et al.

Advantages

- They can be utilized at home and in smaller outpatient clinics as a workable substitute for weight training.
- They are reasonably priced, space-efficient, adaptable, and portable.
- These devices offer easy intensity modulation, making them cost-effective and low-maintenance.

Disadvantages

- Injuries may occur if elastic resistance is not used properly. It complicates finding the right level of resistance because users must consult a table of figures for quantitative information about each color-coded grade of material.
- Lack of stabilization and control of extraneous movements is a worry, similar to free weights, necessitating muscle stabilization to maintain proper movement patterns.
- Although material fatigue is often low during clinical use, elastic bands and tubing should be replaced on a regular basis to ensure patient safety and effectiveness.
- Some elastic items include latex, potentially limited use for those with latex allergy. However, latex-free alternatives are available.

Precautions

- Maintain appropriate posture and alignment.
- Use the specified band or tubing.
- Include a warm-up and cool-down.
- Exercises should be performed gently and with control.
- Avoid joint hyperextension and locking.
- Breathe smoothly and exhale throughout exercise.
- One key aspect is to increase the number of repetitions, starting with a range of 8–12 reps and gradually moving to 12–15 as strength improves. Additionally, progression can be achieved by selecting thicker bands, typically indicated by color; for instance, transitioning from lighter bands (often yellow) to medium bands (red) and then to heavier bands (green or blue).

Clinical Correlation

- Elastic bands and tubes are widely used in physiotherapy and fitness for resistance training. The resistance levels of elastics bands and tubes are typically color-coded. While the exact resistance levels may vary slightly depending on the manufacturer, the general resistance chart is as follows:
  - **Yellow and red:** Early-stage rehabilitation, low-intensity strength training, and warm-ups.
  - **Green and blue:** General strengthening exercises and rehabilitation for intermediate stages.
  - **Black, silver, and gold:** For athletes, advanced users, or those focusing on strength and power.

Technique

1. Before using the elastic band or tubing, make sure it is properly linked to the hand or foot to avoid slippage and harm. Always check for good attachment before exercising.
2. Use the "double wrapping" approach to increase security. This involves winding the band around the hand or foot more than once to increase stability.

Grip Wrap

1. With the end facing the little finger, place the band flat in the palm (Fig. 9.10A).
2. Encircle the back of the hand with the band's extended end (Fig. 9.10B).
3. To ensure a snug fit, wrap the band as many times as required (Fig. 9.10C).
4. Firmly grip the band to maintain control throughout the workout.

Palm Wrap

1. The hands should be facing up as one starts, and the band should end between the thumb and palm (Fig. 9.11A).

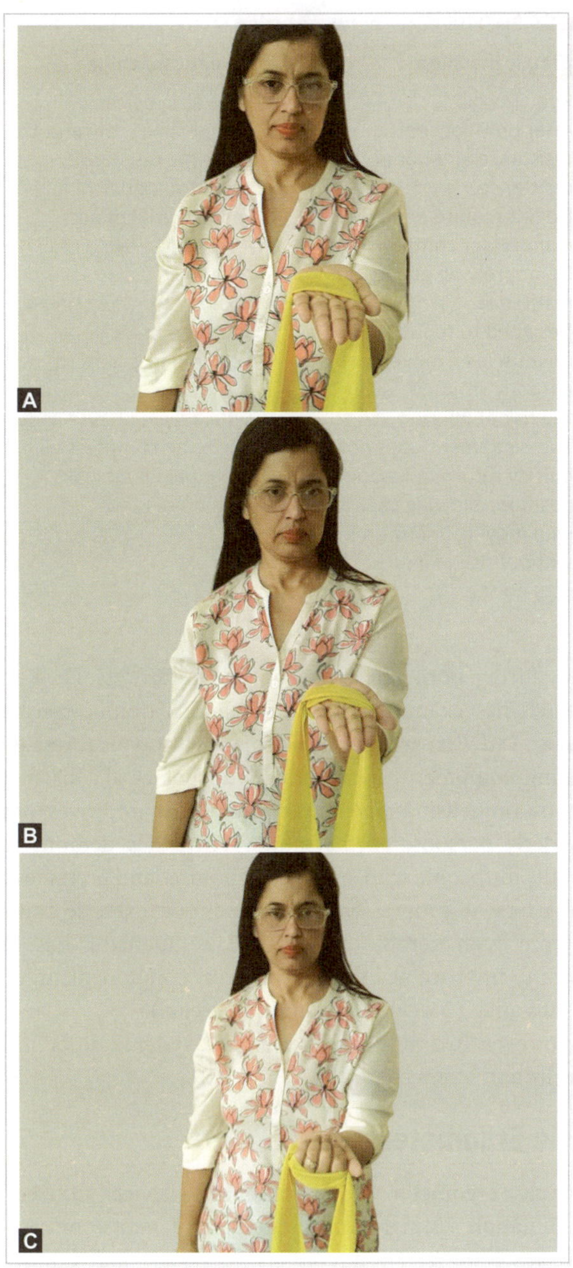

Figs 9.10A to C: Grip wrap

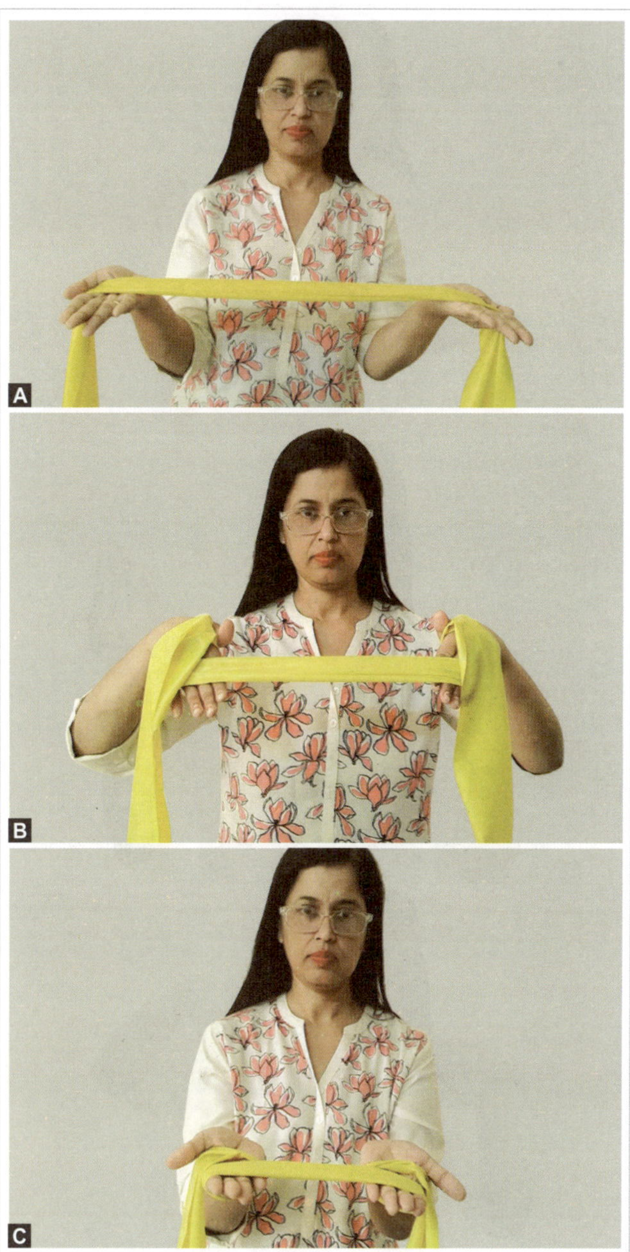

Figs 9.11A to C: Palm wrap

2. Wrap the band over the backs of the hands while rotating the palms inward (Fig. 9.11B).
3. Re-wrap as needed (Fig. 9.11C).
4. Clasp the band securely to guarantee a firm grasp throughout exercises.

Euro Wrap

1. Start with the palm facing forward and the band finishes between the thumb and palm (Fig. 9.12A).
2. Turn the arm inwards, then encircle the hand's back with the band (Fig. 9.12B).
3. Let the band around the back of the hand as one turns the hand downward (Fig. 9.12C).
4. Place the band between the thumb and fingers as one brings the palm back to the front (Fig. 9.12D).
5. To ensure stability and control when exercising, hold onto the band firmly.

Hydraulic/Pneumatic Resistance

There are currently several hydraulic and pneumatic workout devices available for strengthening musculature that provide unilateral or bilateral action as well as different types of resistive

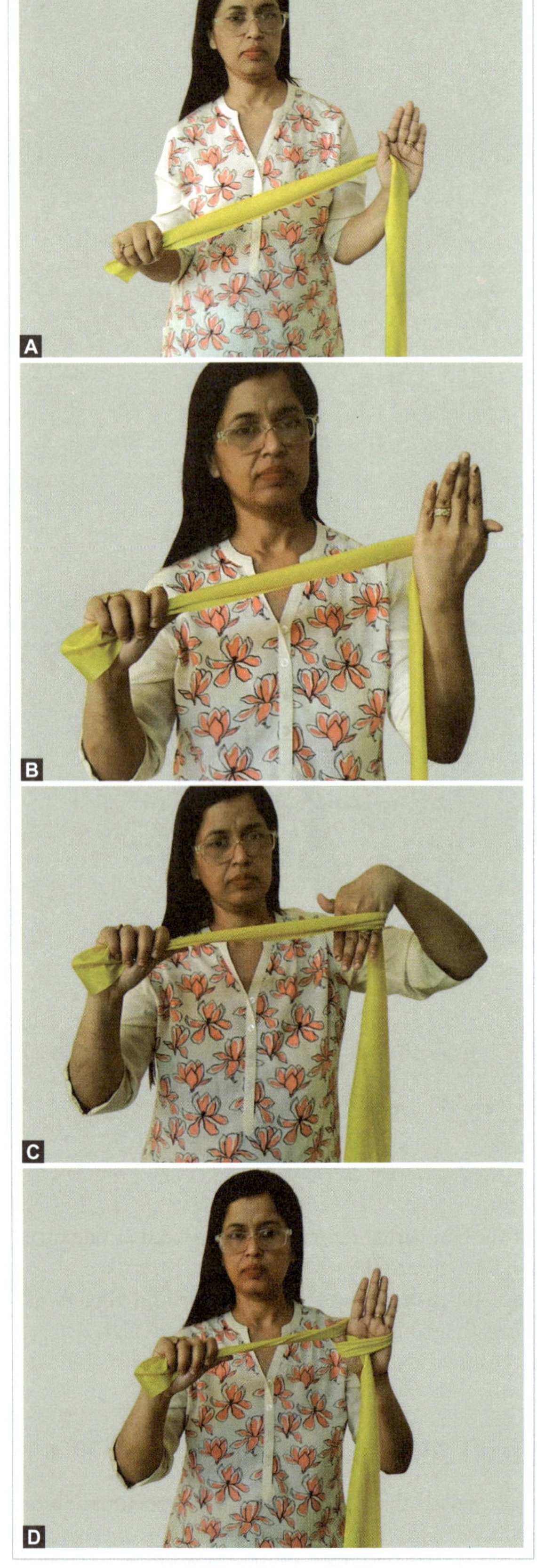

Figs 9.12A to D: Euro wrap

TABLE 9.2: Differences between hydraulic and pneumatic machines

Hydraulic machine	Pneumatic machine
• Requires oil. • After only concentric resistance and not eccentric resistance. • Do not require energy or compressor components. • Do not rely on gravity for resistance. • Designed to reduce stress on muscles if movement is abruptly stopped, lowering the risk of injury. • The equipment does not quickly return to its original position, allowing user to stop moving at any time without risk of injury. • Less noise.	• Uses air pressure. • Provide both concentric and eccentric resistance, e.g., weight slack machine. • Lacks inertial effect. • It relies on gravity of resistance. • Poor control; consistent speed and precise positioning is hard to maintain. • Difficultly in precise control due to air compressibility, potential for air leaks. • High noise level.

force during both the extension and return stroke. Recent research has examined pneumatic and hydraulic-resistance exercise (HDRE) systems as alternatives to typical resistance training equipment. For example, de Vos et al., discovered that training using pneumatic equipment improves power in elderly persons. Furthermore, HDRE has demonstrated benefits for people with spinal cord injuries and prepubescent youngsters. A progressive adaptive circuit exercise training program with aerobic and HDRE components has been shown to improve health-related fitness components in older persons after 12 weeks of training (Lee et al).

However, hydraulic and pneumatic systems differ mechanically and biomechanically (Table 9.2).

Cycle Ergometer

A cycle ergometer is a stationary device utilized in rehabilitation settings to facilitate passive, active, or resisted exercises (Fig. 9.13). It consists of pedals and adjustable

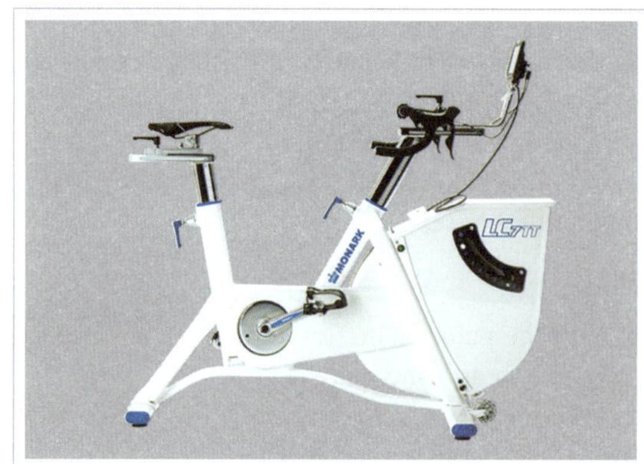

Fig. 9.13: Cycle ergometer

resistance mechanisms, allowing for a range of therapeutic interventions aimed at improving physical function and promoting recovery.

Evidence-Based Uses

- **Stroke rehabilitation:** Studies by Diserens et al., in 2007 and Santos et al., in 2015 have shown that cycle ergometer training (CET) can enhance muscular strength and range of motion in chronic poststroke patients. It is considered an important tool in the early mobilization and rehabilitation of individuals in the acute post-stroke phase (dos Santos et al., 2015).
- **Intensive care unit (ICU) rehabilitation:** CET has been integrated into standard care protocols in ICUs to improve outcomes for patients with intensive care unit-acquired weakness. It has been associated with enhancements in walking ability, lower limb muscle strength, cardiovascular endurance, and overall quality of life during inpatient rehabilitation (Veldema et al.).
- **Elderly rehabilitation:** Cycle ergometer training has demonstrated positive effects on the functional status, muscle strength, cardiovascular health, and cognitive performance of older adults aged over 70. This intervention enables older individuals to maintain or regain independence in daily activities (Bouaziz W et al.).
- **Chronic obstructive pulmonary disease (COPD) management:** Home-based CET has shown physiological training effects in patients with moderate to severe COPD. These effects include increased peak oxygen uptake and decreased heart rate and minute ventilation during exercise, indicating its potential as an adjunctive therapy for managing COPD and improving respiratory function (Larson et al).

Clinical Correlation

By incorporating cycle ergometer training into rehabilitation programs, clinicians can leverage its benefits to enhance physical function, promote recovery, and improve quality of life across diverse patient populations.

DETERMINANTS

Kraemer et al., gave the determinants of resisted exercises. Resistance training can enhance various aspects of physical performance, including strength, power, and local muscular endurance, depending on the program design. These improvements stem from physiological adaptations that occur with consistent resistance training. The effectiveness of resistance training programs is optimized when tailored to individual training goals.

The degree of these benefits hinges on factors such as initial performance and health status, as well as specific program design variables like type of exercise, order, loading, rest interval, number of sets and frequency.

Types of Exercises

In a resistance exercise program, the mode of exercise encompasses various factors such as the type of muscle contraction, form of exercise, and method of resistance application. This can include dynamic or static movements, weight-bearing or nonweight-bearing positions, and manual or mechanical resistance application. The selection of exercise modes is influenced by several factors, with further details provided in subsequent sections.

The types of muscle contractions in resistance exercise programs include isometric (static) and dynamic contractions, with the latter further categorized into concentric (shortening) and eccentric (lengthening) contractions. Isokinetic contractions, where limb movement velocity is controlled, are also considered a form of dynamic contraction. Additionally, resistance training exercises can be classified as single-joint or multiple-joint movements, each offering unique benefits and considerations. Multiple-joint exercises engage multiple muscles and joints, making them effective for strength and power development, while movements involving large muscle groups impact metabolic and hormonal responses, influencing muscle endurance and body composition goals.

Order of Exercises

Exercises should be performed in the following order: Larger muscle groups usually come first, followed by smaller muscle groups within a body part. This prioritizing is based on the idea that stimulating larger muscle groups first results in a stronger stimulation.

Some sequencing strategies for strength and power training include:

Training all major muscle groups in a single session:

1. Large muscle group exercises should be performed first, followed by small muscle group workouts.
2. Prioritize multi-joint exercises above single-joint activities.
3. Prioritize total-body workouts for power training, working the way down from the most complicated to the least complex.
4. Alternate between upper and lower body exercises, or opposing muscle group exercises.

When dividing upper and lower body exercises:

1. Before moving on to smaller muscle groups, start with major muscle group workouts.
2. Prioritize multiple-joint exercises above single-joint exercises.
3. Alternate between exercises that target opposing muscle groups.

When working on certain muscle groups:

1. Prioritize multi-joint exercises above single-joint activities.
2. Before moving on to lower-intensity activities, begin with higher-intensity workouts (closer to the one-repetition maximum).

Loading

Delorme's method of measuring exercise intensity is called repetition maximum (RM).

The RM is the maximum weight that a muscle can pass through its range of motion (ROM) a given number of times.

A muscle's 1 RM is the most weight (resistance) it can lift at once to complete its full range of motion. The muscle will not be able to perform the repetition for a second time. The greatest weight (resistance) that a muscle can contract over its complete ROM ten times is known as its 10 RM. The muscle won't be able to complete the eleventh repetition. The RM concept is used to assess the baseline dynamic strength of the muscle before starting a training program and for determining the exercise load to be used during the training. Percentage of 1 RM is generally used in a training program.

The amount of weight lifted during resistance training significantly impacts metabolic, hormonal, neural, and cardiovascular responses. Lighter loads (45–50% of 1 RM or less) are effective for beginners, improving motor learning and coordination. As training progresses, heavier loads (>80–85% of 1 RM) are needed to induce further neural adaptations and recruit high-threshold motor units crucial for maximal strength development. Loading varies based on individual goals and exercise selection.

Repetitions and Intensity

There is an inverse relationship between weight lifted and repetitions performed. Heavier loads (>80–85% of 1 RM) are optimal for increasing maximal dynamic strength, while moderate loads (60–80% of 1 RM) are effective for hypertrophy. Lighter loads (<60% of 1 RM) improve local muscular endurance. Mixing loading zones is crucial to prevent training plateaus and overtraining. When training is at a certain RM load, it is recommended that 2–10% increase in load be provided when the individual can accomplish the present workload for one to two repetitions more than the desired number American College of Sports Medicine (ACSM).

Power Training

To maximize power, two loading strategies are used: Heavy loads recruit fast-twitch motor units, which are helpful in generating heavy force while light-to-moderate loads performed explosively improve velocity. For power training, the recommended exercise intensity typically ranges from 70% to 90% of one-repetition maximum (1 RM) with fewer repetitions.

Rest Interval

Rest intervals between sets and exercises are important when designing a resistance training program. Shorter rest periods may increase reliance on glycolytic energy sources, resulting in elevated lactate concentrations. Rest periods must be carefully managed to reduce undue stress and maximize performance, particularly in sport-specific training programs.

Short rest intervals (1 minute or less) are good for hypertrophy and local muscular endurance, but they may reduce force and power generation. Longer rest intervals (3–5 minutes) are indicated for absolute strength or power training, whilst moderate rest intervals (2–3 minutes) are beneficial for strength improvements. Shorter rest intervals (1–2 minutes) promote anabolic hormone release and metabolite formation, which increases muscle protein synthesis. Rest interval length should be varied based on the training goal and exercise exhaustion level for best performance and adaptation.

Repetition Velocity

The rate at which repetitions are completed influences neuronal, hypertrophic, and metabolic responses to resistance training. Isokinetic training is velocity-specific, with moderate velocities (180°–240°/sec) resulting in the highest strength improvements. Dynamic continuous external resistance training involves both purposeful and inadvertent slow velocities, which affect force production and neural activation. Slow velocities and submaximal loads may initially help local muscle endurance and hypertrophy training, but they may not maximize strength increases in resistance-trained individuals. Moderate and fast velocities provide better muscle performance and strength improvements. Power training necessitates explosive exercises to optimize power production. For local muscle endurance, rapid velocities are recommended for isokinetic training while both slow and moderate-to-fast velocities are effective during dynamic constant external resistance training.

Number of Sets

Training outcomes are directly influenced by the number of sets used, with several sets often yielding greater strength improvements than single sets. Beginners can begin with one or two sets, while more advanced athletes benefit from three to six sets with suitable rest intervals.

Frequency of Resistance Training

Resistance training frequency varies based on individual recovery and training experience. Traditionally, three workouts per week have been recommended, however some people may benefit from higher frequency. To promote maximum recovery and adaptability, program design should take into account elements such as soreness and fatigue.

Duration of Exercise

Exercise duration refers to the length of time, usually in weeks or months, that a resistance training program is implemented. While some people may only need a month or two to regain muscle function or activity levels, others may require ongoing care to maintain optimal function. While neural adaptations that result in initial strength increase can occur within 2–3 weeks, significant muscular changes, such as hypertrophy or increased vascularization, typically require at least 6–12 weeks of constant resistance training.

Variation in Training

To ensure long-term effectiveness, training stimuli should be varied over time through periodization. Periodization models vary, with the goal of adjusting volume and intensity to enhance strength and power growth while avoiding over training.

PRINCIPLES

The principles of resisted exercises guide the effective and safe implementation of strength training programs. These principles are rooted in exercise science and are essential for achieving optimal results, whether in rehabilitation, fitness, or athletic training. Here are the key principles:

Overload

The overload principle is essential in training, aiming to provide enough challenge to stimulate physical, physiological, or performance improvements. To achieve this, exercises must exceed typical levels of physical activity; otherwise, the body won't adapt effectively.

- The specificities of overload—like strength (intensity), frequency, and duration—can vary. For example, intensity refers to how hard you work (often measured as a percentage of maximum effort), while frequency is how often you train, and duration is how long each session lasts.
- Training volume, which measures the total amount of work done and energy used, is crucial for understanding overload. Intensity relates to the speed and effort required to complete the exercises.
- Research supports this principle, showing that progressively increasing the load leads to better strength and performance gains. For instance, studies have found that incremental increases in weight or intensity can significantly enhance muscle hypertrophy and overall athletic performance. By applying the overload principle consistently, individuals can maximize their training results and achieve their fitness goals more effectively.

Variation

- Variation refers to the right adjustment of exercise selection, volume, speed of movement, and training intensity.
- Appropriate variation is essential for maintaining consistent progress in adaptations over time.
- Periodized sequencing of exercise volume, intensity, and selection, including exercises focused on building speed and strength, can also lead to a higher degree of progress in a variety of performance skills.
- Although resistance offered by machines vary in intensity, they can only provide a limited range of movement patterns, application, sequencing, and variety for speed-strength and speed-oriented exercises.
- These limitations are due to the inherent restrictions in the movement patterns and features of machine-based exercises.

Specificity

- The specific adaptation to imposed demands (SAID) principle is a fundamental concept in exercise physiology that emphasizes the significance of designing exercise regimens to achieve specific goals. It holds that the body adapts to the exact pressures applied on it, resulting in targeted gains in strength, endurance, or other desirable characteristics. This theory assists therapists in selecting exercises that closely mimic the demands of functional activities, ensuring that training benefits are tailored to individual needs.
- Training specificity includes elements such as mode, velocity, and movement patterns, in addition to exercise selection. For example, if the goal is to increase the ability to climb stairs, exercises should mimic the muscle motions and speed required. By aligning these variables with functional requirements, therapists can optimize the transfer of training effects to real-world activities. While the SAID principle promotes specificity, there has been some evidence of training transfer. This phenomenon, known as cross-training or overflow, implies that progress in one activity or task may transfer to another.

Joint Angle Specificity

- Improvements from isometric strength training mainly happen at the specific joint angle targeted during training. As assessments are made at angles farther from the training angle, the increase in isometric maximum strength becomes smaller.
- Also, strength increases with variable-resistance devices peak at the joint angle where the most resistance is applied and may diminish at other angles—a problem that does not exist with free-swinging or freely moving devices. The success of

variable resistance devices in matching resistance to human strength curves is not well supported by data.

Reversibility Principle

- The principle of reversibility, sometimes known as the "use it or lose it" theory, states that physiological adaptations achieved by exercise would eventually decline if activity is halted. In essence, removing the training stimulus causes the body to revert to its pretraining state. This idea emphasizes the significance of exercising consistently in order to preserve the benefits of training.
- Strength training is a good example of the reversibility concept. When a person follows a strength training program, their muscles adapt by growing in size and strength. However, if the individual discontinues strength training, they will eventually lose the muscular growth and strength improvements they have made. This regression might develop rather quickly when compared to the time.

PHYSIOLOGICAL ADAPTATIONS

The following physiological adaptations due to resistive training have been described by Deschenes et al.:

Neural

- Early-phase adaptation in neuromuscular system occurs due to a rise in neuronal factors, and a change in protein quality (greater potential to recruit fast twitch motor).
- Adaptation in the late stages increase hypertrophy of the muscles.
- Neural adaptations after normal training are caused by alterations in the neuromuscular junction (NMJ), reduced Golgi tendon organ (GTO) sensitivity, and decreased CNS inhibition, increased recruitment of motor units and increased firing rate.

Neural adaptations continue to contribute to strength improvements much after the initial period of training, probably between the sixth and twelfth months.

Other neurological modifications include decreased antagonist muscle co-contraction, neuromuscular junction enlargement, and greater synchrony in motor unit discharge.

Muscular

- Increased myofibril density per muscle fiber.
- Fiber splitting results in an increase in the cross sectional area of the muscle.
- Expand the number of muscle fibers.
- Expand the amount of protein in each muscle fiber.
- Increased levels of adenosine triphosphate (ATP) and creatine phosphate as well as the energy source needed to power muscular contraction.
- **Hyperplasia:** The longitudinal splitting of a tiny percentage of muscle fibers may result in an increase in number.
- **Fiber type specific adaptations:** Resistance training causes fiber type conversions in the trained muscle.
  - Typically, type IIA fibers increase while type IIB fibers decrease, with no effect on type I fibers.
  - Hypertrophy occurs in all major fiber types (I, IIA, and IIB), with type IIA fibers exhibiting the most growth, followed by IIB and lastly type I.
  - Heavy resistance training completes the change to type IIA fibers, making type IIB fibers rare.
  - Resistance training alters the myosin adenosine triphosphatase fiber type profile and myosin heavy chain composition, particularly in type II subtypes. There are no notable changes between type I and type II muscle fibers.
- **Vascular and metabolic adjustments:**
  - Reduction in the capillary bed density
  - Decrease of mitochondrial density
  - Storage of ATP and CP rises
  - Increase in the storage of myoglobin, a protein that helps store and transport oxygen in muscles.
  - Additionally, the activity of myokinase (MK), which regulates ATP production in muscles, and creatine phosphokinase (CPK), which aids in ATP production from creatine phosphate, both increase.

Endocrinal

Resistance exercise can cause several changes in hormone levels in the body, which play important roles in muscle growth and recovery:

- **Testosterone:** After a workout, resistance training can lead to short-term increases in testosterone levels, a hormone that helps build muscle.
- **Growth hormone (GH):** The response of GH to resistance exercise varies. Bodybuilding workouts, which focus on higher repetitions and moderate weights, tend to produce greater increases in GH compared to powerlifting workouts, which use heavier weights with fewer repetitions.
- **Insulin-like growth factor-1 (IGF-1):** IGF-1 is another hormone that helps muscles grow. It mainly works locally in the muscle tissue through autocrine (acting on the same cell) and paracrine (acting on nearby cells) mechanisms. However, the overall levels of IGF-1 in the blood may not consistently increase after resistance exercise, and long-term training might not change the baseline levels.

- **Cortisol:** Cortisol is a hormone that can break down muscle tissue (catabolic). It often rises after resistance training but may decrease with consistent, long-term training. Lower cortisol levels over time can actually help promote muscle growth (hypertrophy).

Bioenergetic

Resistance exercise affects muscle chemistry and metabolism in specific ways:

- **Phosphagen levels:** Resistance training has little impact on the levels of phosphagen, which includes ATP and phosphocreatine—important energy sources for short bursts of activity.
- **Glycogen levels:** Glycogen, the stored form of carbohydrates in muscles, may not consistently increase with long-term resistance training.
- **Glycolysis enzymes:** Key enzymes involved in breaking down glucose, like phosphorylase and phosphofructokinase, do not show much change with resistance training.
- **Lipid depots:** Fat stores used for energy in muscles might decrease during resistance training.
- **Oxidative metabolism enzymes:** The enzymes that help with oxidative metabolism (using oxygen to produce energy) typically do not change much in muscles trained for resistance, and some may even decrease.
- **Mitochondrial density:** Muscles that are built up through resistance training might have less mitochondrial density (the structures that produce energy) because of increased muscle size, which can limit their ability to use oxygen effectively.
- **Myoglobin levels:** Myoglobin, a protein that helps store oxygen in muscles, might actually decrease rather than increase in resistance-trained muscles.
- **Capillary density:** Long-term resistance training may reduce the number of capillaries (small blood vessels) in muscles due to hypertrophy (muscle growth). However, it could improve the connection between capillaries and muscle fibers.
- **Maximal oxygen uptake:** Overall, resistance training does not significantly enhance the muscles' ability to use oxygen, leading to minimal improvements in maximal oxygen uptake.

Clinical Correlation

While resistance training may not have a substantial impact on certain metabolic and physiological markers traditionally associated with aerobic exercise, such as muscle phosphagen levels, glycogen levels, and oxidative metabolism enzymes, it has shown to decrease cardiovascular strain during daily tasks, especially among older individuals, even without significant increases in maximal oxygen uptake. This reduction in cardiovascular stress can enhance quality of life and independence in activities of daily living.

PRECAUTIONS

- The use of valsalva maneuver should be avoided during resistive training especially by a patient with cardiopulmonary disease, recent abdominal, intervertebral disc or eye surgery.
- **Acute muscle soreness:** Arises during or after intense exercise due to factors like inadequate blood flow and the buildup of metabolites in muscles. It causes temporary burning or aching sensations but typically subsides quickly with rest and active recovery.
- **DOMS:** Resistance training especially eccentric may result in delayed onset muscle soreness (DOMS). During resistive training especially with eccentric contraction and increase exercise dose, minor lesions of the muscle and inflammation occur resulting in delayed onset muscle soreness. Soreness or injury can affect athletes' performance and preparation, particularly after strenuous resisted exercises.
  - DOMS causes mechanical disturbance to sarcomeres and stimulate an inflammatory response. This response involves prostaglandin and leukotriene production, which causes discomfort and swelling. Fast-twitch fibers may be more vulnerable to injury, and exercising with longer muscle lengths can exacerbate symptoms.
  - DOMS symptoms include strength loss, soreness, tenderness, stiffness, and edema, with varied peaks and durations.
  - Treatments focus on minimizing the inflammatory response postexercise, and frequently use pharmacological, therapeutic, or nutritional interventions.
- **Growth plate injuries:** Youth resistance training may pose a risk of injury to the physis or growth plate. The growth plate is typically three to five times weaker than surrounding connective tissue and may be less resistant to shear and tension forces. Injuries to this part can hinder training, cause discomfort, and disrupt growth.
  - Injuries to the growth cartilage in children are mostly caused by faulty lifting practices, inappropriate training loads, or inadequate adult supervision.
- **Elevated blood pressure:** Strength training, particularly with big weights, causes health dangers due to forceful breathing against constricted airways, which results in elevated chest pressures of 100–200 mm Hg.
  - While it stabilizes the spine and improves muscular attachment, it can also induce internal vein congestion and raise blood pressure.
  - While it is generally safe for healthy people, those with pre-existing cardiovascular diseases face hazards such as cardiac rhythm issues or stroke.
  - To lower blood pressure and avoid back discomfort, older persons should engage in controlled and moderate strength training, focusing on big muscle groups at

around 50% of maximum force, with a mandatory sports medical consultation for those at risk.

CONTRAINDICATIONS

- **Acute injury:** Engaging in resisted exercises while recovering from recent injuries can exacerbate damage.
- **Cardiovascular issues:** Those with unstable heart conditions or uncontrolled hypertension should avoid heavy resistance exercises.
- **Severe joint pain:** Individuals experiencing acute or chronic joint pain may need to avoid resistance exercises to prevent further discomfort or injury.
- **Neuromuscular disorders:** Conditions like muscular dystrophy or certain neurological disorders may contraindicate heavy resistance training.
- **Inflammation:** Inflammatory neuromuscular pathology or acute muscle inflammation
- **Uncontrolled hypertension:** High-intensity resistance training can lead to a significant increase in blood pressure during exercise.
- **Severe osteoporosis:** Individuals with severe osteoporosis may be at risk of fractures due to the stress placed on the bones during resistance exercises.
- **Joint pain or injury:** Resistance exercises may exacerbate joint pain or existing injuries, particularly in conditions like osteoarthritis or rheumatoid arthritis.
- **Recent surgical procedures:** Individuals who have undergone recent surgical procedures should avoid resistance training until they have fully recovered and received clearance from their healthcare provider.
- **Uncontrolled diabetes:** Resistance training can affect blood glucose levels, and individuals with uncontrolled diabetes may experience fluctuations during exercise.
- **Pregnancy:** While resistance training can be safe during pregnancy, precautions should be taken to avoid excessive strain.
- **Chronic respiratory conditions:** Individuals with severe chronic respiratory conditions may experience difficulty with breathing during resistance exercises.

EXERCISE REGIMENS

Progressive Resisted Exercise

Exercise is considered progressive and resisted when the amount of load applied is increased over time as the body adapts to the demand that it is placed under on it.

Almost sixty years after DeLorme and Watkins first described progressive resistance exercise (PRE) to increase muscle force output, the basic ideas behind the practice have not altered much. It's a dynamic training method where external resistance is progressively increased to improve muscle strength:

- Work out in modest increments until one becomes fatigued
- Give enough time to recover between exercises
- Raise the resistance as the force-generating capacity rises.

It is advised to lift loads in one to three sets, training two or three days a week, in order to achieve an 8–12 repetition maximum (RM). The amount of weight that can be lifted through the available range of motion eight to twelve times before needing to rest is known as an 8 RM to 12 RM load.

Young adults in good health have traditionally taken PRE to enhance their athletic performance. Recent assessments, however, have highlighted the possible health advantages of incorporating PRE into community-wide efforts to promote physical activity. Combining PRE into a general exercise program may have positive effects on health by lowering risk factors for osteoporosis and conditions like diabetes and cardiovascular disease.

The PRE may be a helpful intervention in physical therapy due to its health benefits. Physical therapists frequently treat patients who have diminished muscle force production as a result of trauma, disease, or inactivity. Physical therapists should utilize the PRE principles when creating treatment plans if a patient's incapacity to perform daily tasks is caused by an impairment involving the inability of muscles to generate force.

Dynamic Training Method

In dynamic training method, external resistance is progressively increased to improve muscle strength. It comprises three techniques:

1. Delorme and Watkins
2. Macqueen
3. Oxford

The resistance in PRE is determined and progressed based on the RM, which is the maximum weight a muscle can move through a specific number of repetitions. For example, the one repetition maximum (1 RM) is the heaviest weight that can be lifted once with correct form, while the 10 repetition maximum (10 RM) is the weight that can be lifted 10 times with proper form. These parameters are often used in athletes to enhance strength and endurance and are occasionally tested in healthy individuals.

DeLorme Principle

- Developed by Thomas DeLorme in the 1940s.
- Based on the concept of progressive overload.
- Begins with a warm-up set at 50% of the 10-repetition maximum (10 RM), followed by sets at 75% and then 100% of 10 RM.
- Aims to overload the muscle gradually and promotes strength gains over time.
- Warm-up sets are crucial for preventing muscle soreness and teaching proper exercise technique.

Oxford Principle

- Introduced as a regressive loading technique.
- First set performed at 100% of 10 RM, followed by subsequent sets at 75% and then 50% of 10 RM.
- Each set aims to exercise the muscle to its maximum capacity while preserving the overload principle.
- Both approaches have their merits and can be effective in strengthening muscles. The DeLorme Principle focuses on gradually increasing resistance, while the Oxford Principle emphasizes the replication of muscle fatigue. The choice between these methods may depend on individual preferences and specific training goals.

Daily Adjusted Progressive Resistive Exercise Program

The daily adjusted progressive resistive exercise (DAPRE) program provides an organized approach to strength training that is appropriate for athletic rehabilitation and strength and conditioning environments. Kenneth Knight developed DAPRE in 1979, which focuses on using the 6-repetition maximum (6-RM) as a foundation for resistance increases.

The DAPRE program is discussed as follows:

Versatility and Objectivity

- The DAPRE offers a diverse and objective approach to raising resistance in strength training.
- It ensures the recruitment of both fast- and slow-twitch muscle fibers, which is necessary for strength and power improvements.

Lifting Bouts and Progress

- Each exercise is covered in four lifting bouts throughout each session.
- Resistance increases are proportional to the amount of repetitions accomplished in Set 3.
- **Establishing the working weight 6-RM:**
  - The athlete's working weight 6-RM is calculated using the two-failure criterion for 1-RM testing.
  - This method takes into account neuromuscular activation as the athlete approaches their 1-RM, which results in a more accurate assessment.
  - Once the 1-RM has been determined, it is transformed to a 6-RM using a conversion ratio of 1.20.

Adaptation to Rehabilitation

- To minimize undue strain, athletes in rehabilitation may benefit from determining their 10-RM.
- The 10-RM can be converted to a 6-RM with a conversion factor of 1.36.

Warm-Up and Preparation

While the DAPRE program includes a warm-up period that begins at 50–75% of the athlete's 6-RM, a low-intensity, high-repetition bout (about 15 reps) is recommended for additional preparation.

Delorme	Oxford	DAPRE
Starting with a 10 RM of 50 lbs	Starting with a 10 RM of 50 lbs	Starting with a 6 RM of 60 lbs
Performing the first set of 10 reps at 50% of 10 RM (25 lbs)	Performing the first set of 10 reps at 100% of 10 RM (50 lbs)	Performing four sets of each exercise per session.
The second set at 75% of 10 RM (37.5 lbs)	The second set at 75% of 10 RM (37.5 lbs)	Subsequent increases are determined based on the number of repetitions performed in the third set.
The third set at 100% of 10 RM (50 lbs).	The third set at 50% of 10 RM (25 lbs).	For example, if 8 reps are completed in the third set, the weight is increased for the next session.

Resistance Circuit-Based Training

Resistance circuit-based training is a popular concurrent training technique that combines muscular endurance, and strength adaptations all in one workout session using mechanical resistance in different exercises. This method involves performing single or many sets of various exercises in succession with little break between them.

Circuit training (CT) combines aerobic conditioning, muscular endurance, and strength adaptations in a single workout. It typically involves performing multiple exercises in succession with minimal rest between them. An exercise "circuit" is one completion of all set exercises in the program. When one circuit is completed, one begins the first exercise again for the next circuit. Traditionally, the time between exercises in circuit training is short and often with rapid movement to the next exercise.

Exercises can be done at varying intensities: Low <60% of 1-RM, moderate (60–80% of 1-RM), or high (>80% of 1-RM). The number of repetitions can vary, generally ranging from 12–15 for higher reps to <12 for lower reps, or can be structured around a set duration (e.g., 30 seconds) with very short rest periods (around 30 seconds) between exercises.

Circuit training is especially recommended for untrained individuals or those with lower fitness levels, as it can effectively improve both upper body strength and cardiovascular capacity (VO_2 max). Additionally, it is commonly used as an introductory training method for athletes to enhance their cardiovascular conditioning and increase their 1-RM during periodization programs. Researchers have found that circuit-based resistance training is highly effective for enhancing various aspects of fitness, including maximum oxygen consumption, pulmonary ventilation, functional capacity, and strength, while also improving body composition. This method is particularly appealing because it allows individuals to achieve significant health and fitness improvements in a time-efficient manner.

Key aspects

- **Exercise variety and sequence:**
  - The CT is executing a sequence of exercises that target different muscle groups sequentially.
  - Exercises can be done at different loads (low, moderate, or high) based on the percentage of 1-RM.
- **Repetition range and timing:**
  - Repetition ranges vary from high (12–15 repetitions) to low (<12 repeats) to time-based (e.g., 30 seconds).
  - The CT frequently includes short rest periods between exercises (e.g., 30 seconds).
  - Exercise order in circuit training should alternate between upper and lower body movements. Prioritize larger muscle groups and multi-joint movements to minimize fatigue and reduce the risk of injury.

Plyometric Training

Plyometric training is a dynamic workout strategy that aims to improve neuromuscular efficiency, motor learning, and force generation. It consists of quick and strong motions that take use of the body's stretch-shortening cycle to generate maximum force in the shortest amount of time. Plyometric, which includes motions like jumps, bounds, and hops, trains the body to efficiently transition from eccentric (muscle lengthening) to concentric (muscle shortening) muscular activities.

Phases

The three phases of plyometric exercise—eccentric, amortization, and concentric—are critical in maximizing muscle function.

1. During the eccentric phase, muscles pre-stretch, storing potential energy in the muscle's elastic components.
2. This energy is subsequently released during the concentric phase, which results in increased force generation.
3. The amortization phase, also known as the transition phase, is crucial for the effective transfer of energy from eccentric to concentric muscle actions. A rapid transition in this phase leads to more powerful muscular responses.

Mechanism

One of the most important physiological aspects driving plyometric training is the use of muscles' elastic and proprioceptive qualities. Mechanoreceptors, which include muscle spindles and Golgi tendon organs (GTOs), are essential for giving sensory feedback to the central nervous system. This feedback changes muscle tone, motor execution, and kinesthetic awareness, resulting in improved neuromuscular efficiency and functional strength.

Effects

- **Improved power and strength:** By rapidly lengthening and contracting muscles, plyometrics boost explosive strength, essential for activities requiring quick bursts of energy.
- **Neuromuscular adaptations:** Plyometric exercises prompt neural adaptations, refining coordination, motor unit recruitment, and activation of muscle fibers, leading to more efficient movement patterns.
- **Bone health and density:** The impact nature of plyometric movements stimulates bone density enhancements, which are crucial for reducing the risk of osteoporosis and maintaining skeletal health.
- **Improved explosiveness:** Plyometric exercises boost muscle power, enabling athletes to rapidly generate force and propel themselves off the ground with greater speed and force. This explosiveness is invaluable for activities like sprinting, jumping, and executing rapid direction changes, enhancing performance in dynamic sports scenarios.
- **Faster reaction times:** Plyometric training hones athletes' ability to swiftly respond to external cues or stimuli, a critical skill in sports that demand quick decision-making and reflexive actions. By sharpening reaction times,

athletes can anticipate and react to opponents' movements more effectively, gaining a competitive edge.

- **Enhanced athletic performance:** Athletes integrating plyometric into their training regimen often witness significant improvements in overall performance within their respective sports. Whether it's achieving greater speed, agility or power, the dynamic nature of plyometric exercises translates directly to enhanced athletic capabilities across various domains.
- **Injury prevention:** Plyometric workouts not only build muscular strength but also fortify tendons and ligaments, providing essential support and stability during high-impact activities. Strengthening these structures reduces the likelihood of injuries such as sprains or strains, promoting longevity and resilience in athletic pursuits.

TECHNIQUE OF APPLICATION OF RESISTED EXERCISES

Assessment:

1. Perform a comprehensive assessment, including health history, systems review, and strength measurements.
2. Determine qualitative and quantitative baselines of strength, endurance, and range of motion.
3. Interpret findings to assess appropriateness of resistance exercise and identify functional goals.

Selection of the exercise:

1. Select appropriate forms of resistance exercise based on individual needs and expected outcomes.
2. If using mechanical resistance, ensure necessary equipment is available.
3. Review goals and procedures with the patient, ensuring understanding and consent.

Application of resistance:

1. Warm-up with light, dynamic movements specific to the exercise.
2. Apply resistance to the distal end of the muscle segment being targeted.
3. Adjust resistance placement if discomfort occurs.
4. Apply resistance opposite to desired motion during concentric exercise and in the same direction during eccentric exercise.
5. Gradually increase intensity and resistance as tolerated by the patient.
6. Use 8–12 repetitions per set against moderate resistance, adjusting as needed.
7. Provide clear verbal or written instructions using simple language.
8. Monitor patient responses throughout the exercise session.
9. Cool down with rhythmic movements and gentle stretching after completing resistance exercises.

Strength Training Guidelines

For Children and Adolescents

Medical evaluation before starting strength training:

- Identify risk factors for injury and discuss previous injuries, medical conditions, and training goals.
- Discuss motives for starting the program with both the child and parents.

Start of strength training:

- Wait until balance and postural control skills mature, typically by 7–8 years of age.
- Ensure children have reached a certain level of skill proficiency in their sport before beginning a formal strength-training program.

Exercise selection and equipment:

- Utilize resistance training modalities suitable for youth, such as machines, free weights, medicine balls, and elastic bands.
- Modify equipment designed for adults to fit the child's size.

Exercise technique:

- Emphasize proper technique and strict supervision to reduce the risk of injury.
- Begin with low-resistance exercises until proper technique is perfected.

Progression of training:

- Start with one or two sets of 8–15 repetitions with light to moderate loads.
- Increase weight in 10% increments when 8–15 repetitions can be performed.
- Increase repetitions of lighter resistance to improve endurance.

Training variables:

- Warm-up and cool-down: Include a 10–15 minutes warm-up and cool-down with appropriate stretching techniques.
- Choice and order of exercise: Select exercises that target all muscle groups and progress over time.
- Training intensity and volume: Begin with light to moderate loads and adjust volume and intensity as needed.
- Repetition tempo: Emphasize controlled movement speeds to maintain proper technique.
- Rest interval: Adjust rest intervals based on intensity, volume, and individual fitness capabilities.
- Training frequency: Recommend training 2–3 times per week on nonconsecutive days for beginners.

Plyometric exercises:

- Incorporate plyometric exercises to improve strength and power safely.

Long-term benefits:

- Combine strength training with aerobic training for long-term health benefits.
- Focus on skill development in sports for performance improvement.

For Older Adults

Starting point:
- Begin with light weights and good posture.
- Avoid holding breath during lifting.

Progressive overload:
- Increase weight gradually as strength improves.
- Aim for 8–15 repetitions at 60–80% of 1-RM.

Training intensity:
- Higher intensities (85–100% of 1-RM) can lead to greater gains but may increase injury risk in older adults.

Set and repetition scheme:
- Prescribe 2–3 sets of 8–15 repetitions.
- Slowly perform each repetition through full range of motion.

Frequency:
- Train 2–3 days/week for older adults.
- Focus on large muscle groups for functional strength.

Equipment preference:
- Use variable resistance machines with weight stacks.

Upper Limb Muscles

Patient Position

- **For shoulder exercises:** Seated, standing or lying with arms relaxed by the sides or slightly abducted.
- **For elbow, forearm, and wrist exercises:** Seated or standing, with the forearm supported on a stable surface.

Therapist Position

Facing the patient, standing or sitting as appropriate, positioned to provide stabilization and resistance.

Stabilization

Stabilization may involve supporting the patient's limb or body to ensure proper alignment and isolation of the targeted muscle group.

Application of Resistance

Shoulder Flexors (Figs 9.14A to D)

- **Therapist hand position:** Placing one hand on the patient's forearm.
- **Command to patient:** "Lift your arms forward, keeping them straight. Push against my hands as you lift."

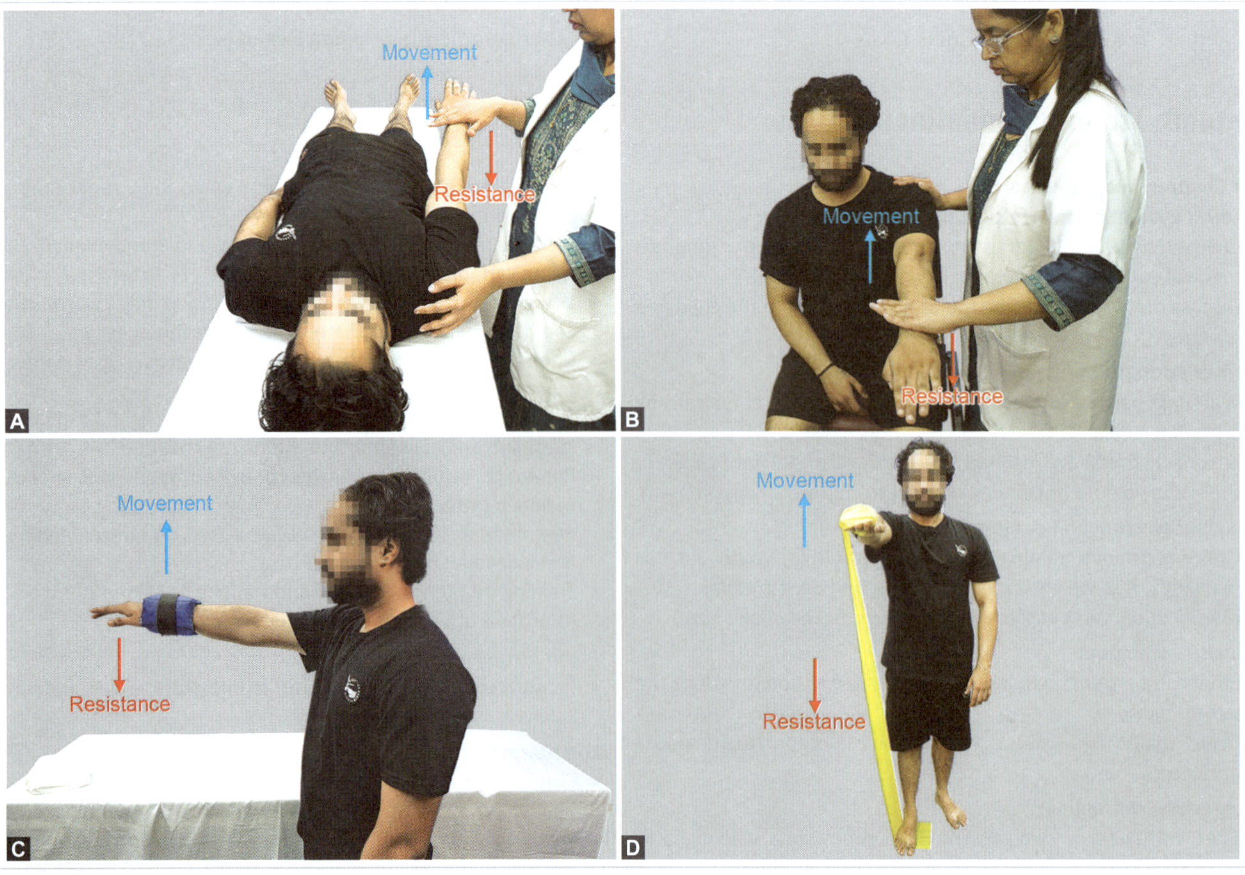

Figs 9.14A to D: Resisted exercises for shoulder flexors

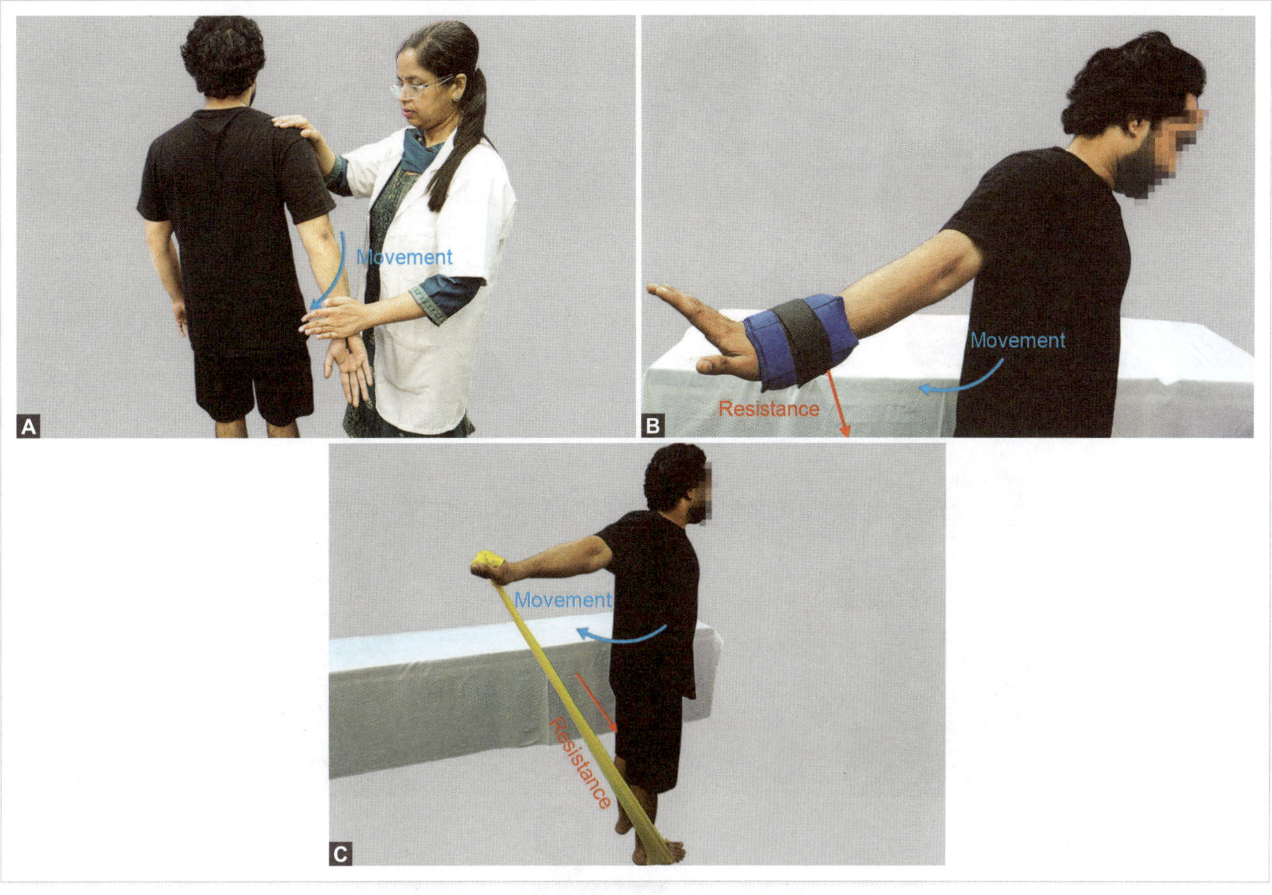

Figs 9.15A to C: Resisted exercises for shoulder extensors

Shoulder Extensors (Figs 9.15A to C)

- **Therapist hand position:** Placing one hand on the patient's forearms or upper arms.
- **Command to patient:** "Pull your arms backward. Push against my hands as you pull."

Shoulder Abductors (Figs 9.16A to E)

- **Therapist hand position:** Applying pressure inward on the patient's forearms or upper arms.
- **Command to patient:** "Raise your arms out to the sides, keeping them straight. Resist against my pressure as you lift."

Shoulder Adductors (Fig. 9.17)

- **Therapist hand position:** Applying pressure outward on the patient's forearms or upper arms.
- **Command to patient:** "Bring your arms inward across your body. Push against my hands as you bring them together."

Internal and External Rotators (Figs 9.18 and 9.19)

- **Therapist hand position:** Providing resistance against the patient's forearm during rotation.
- **Command to patient (internal rotation):** "Rotate your arms inward, like you are trying to turn doorknobs toward your body. Push against my hands."
- **Command to patient (external rotation):** "Rotate your arms outward, like you are trying to turn doorknobs away from your body. Push against my resistance."

Elbow Flexors and Extensors (Figs 9.20 and 9.21)

- **Therapist hand position:** Providing resistance against the patient's hands or wrists.
- **Command to patient (flexors):** "Bend your elbows, bringing your hands toward your shoulders. Push against my hands as you bend."
- **Command to patient (extensors):** "Straighten your elbows, pushing your hands downward. Resist against my pressure as you extend."

Forearm Pronators and Supinators (Figs 9.22 and 9.23)

- **Therapist hand position:** Applying resistance to the patient's hands or wrists during pronation and supination.
- **Command to patient (pronator):** "Turn your palm outward, as if moving toward my hands."

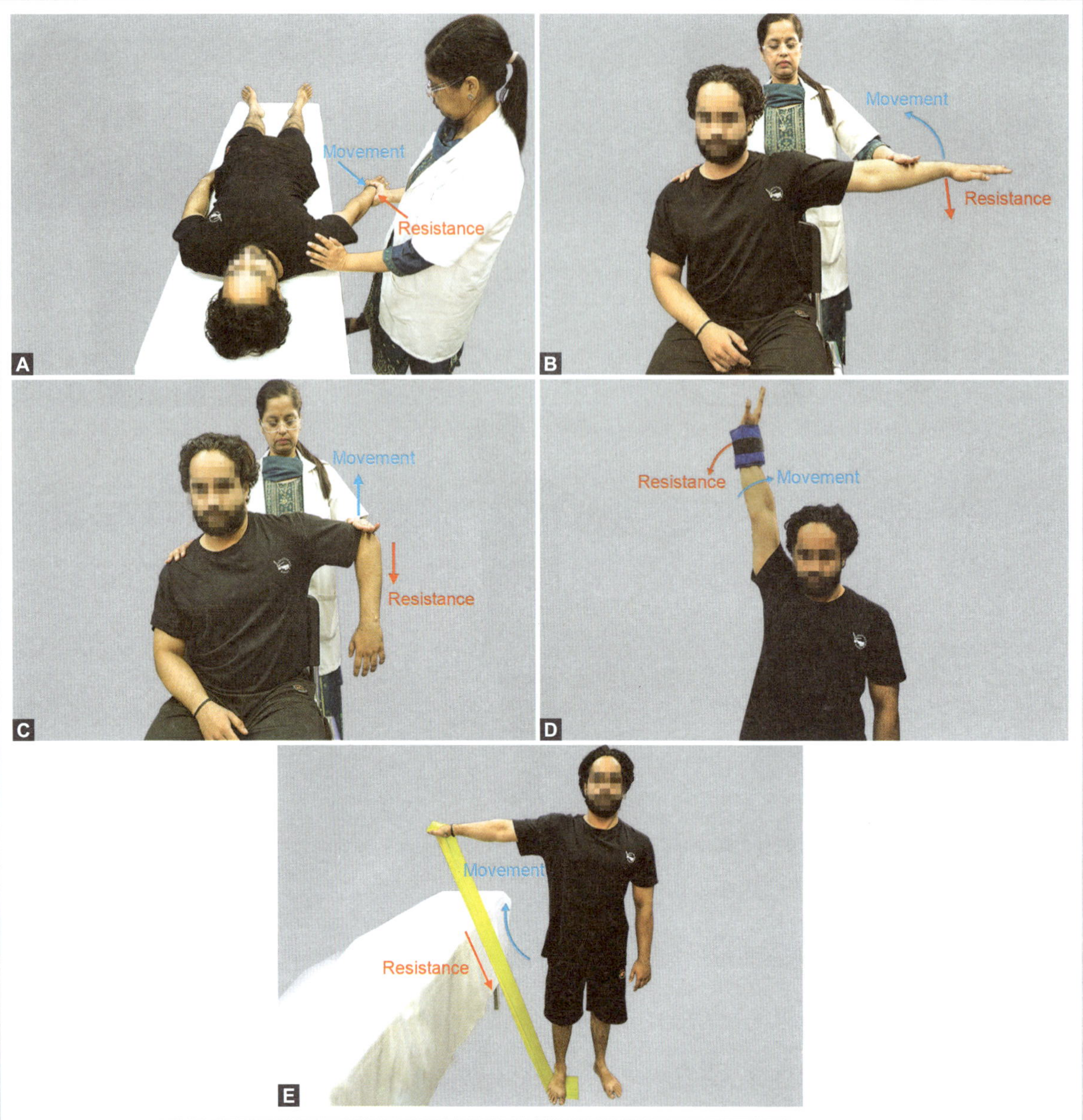

Figs 9.16A to E: Resisted exercises for shoulder abductors

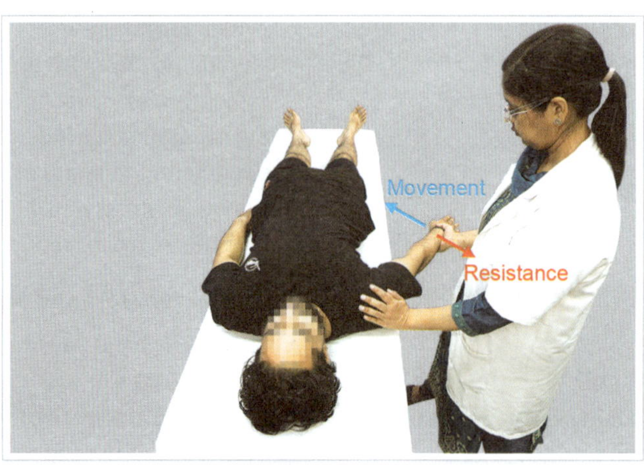

Fig. 9.17: Resisted exercises for shoulder adductors

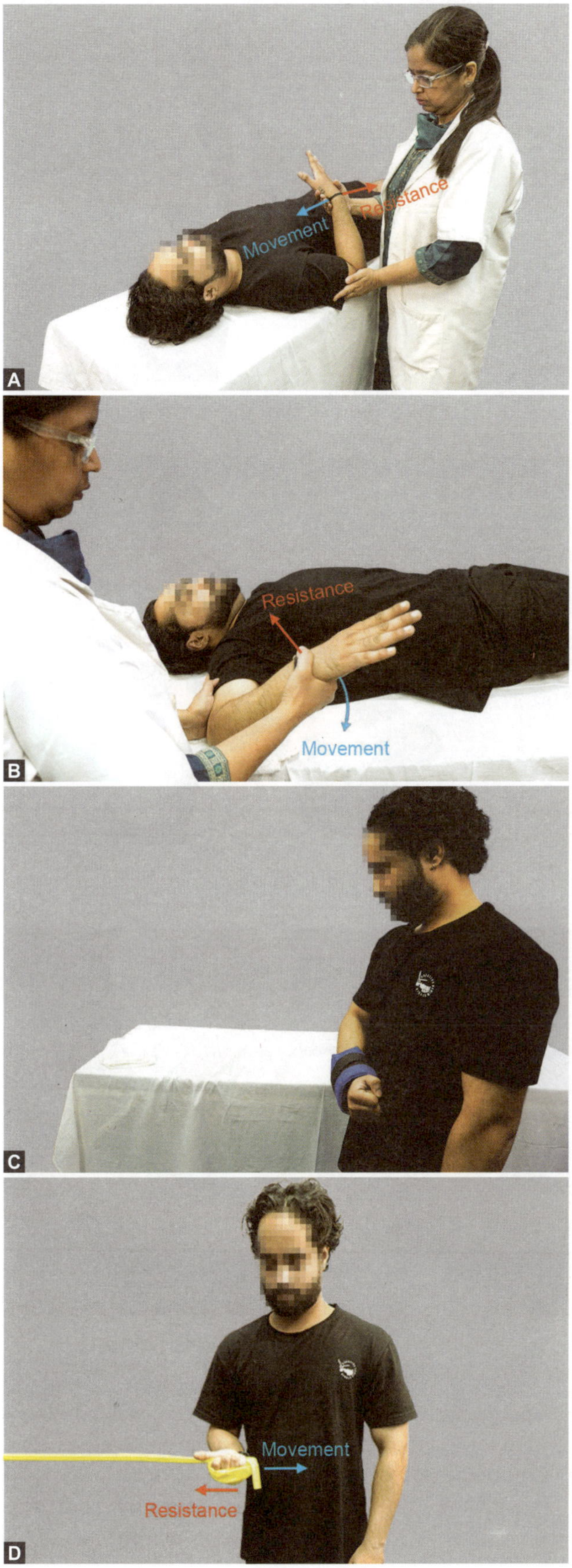

Figs 9.18A to D: Resisted exercises for shoulder internal rotators

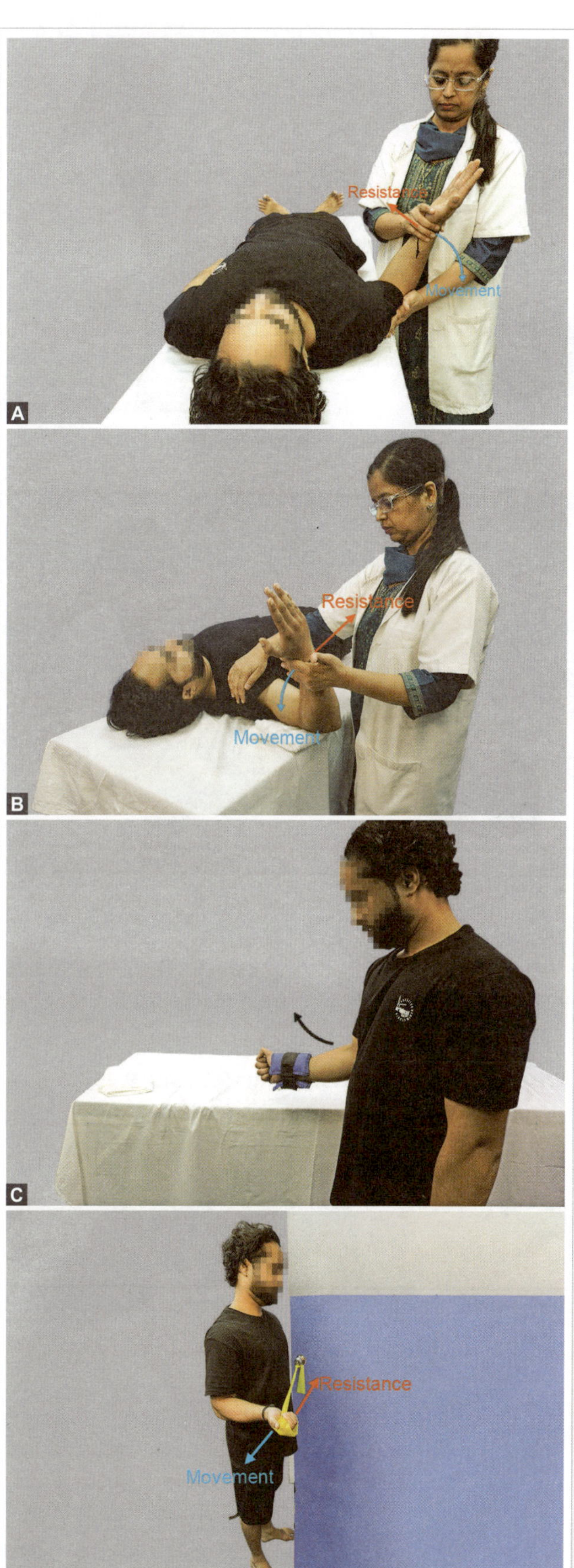

Figs 9.19A to D: Resisted exercises for shoulder external rotators

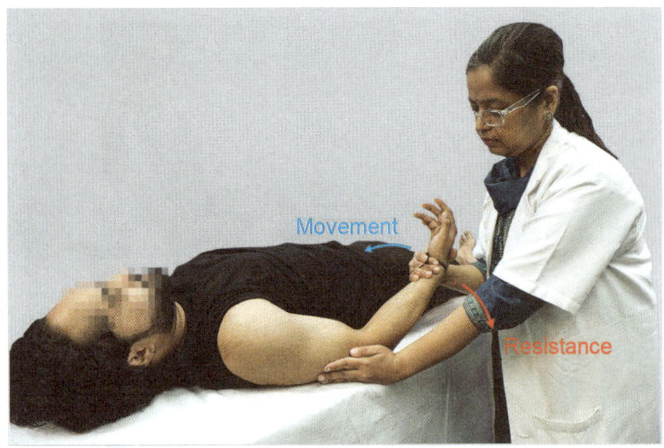

Fig. 9.20: Resisted exercises for elbow flexors

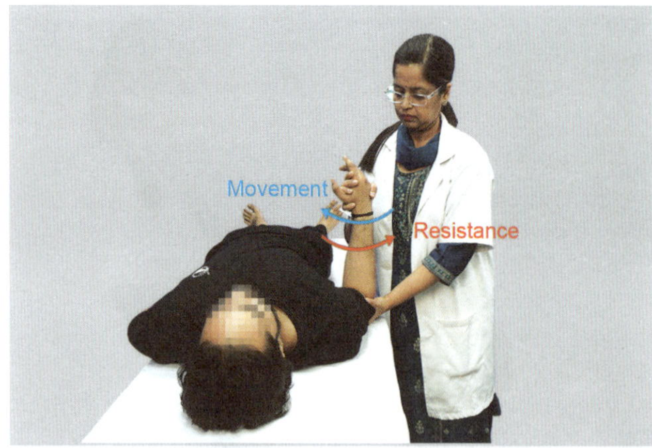

Fig. 9.22: Resisted exercises for pronators

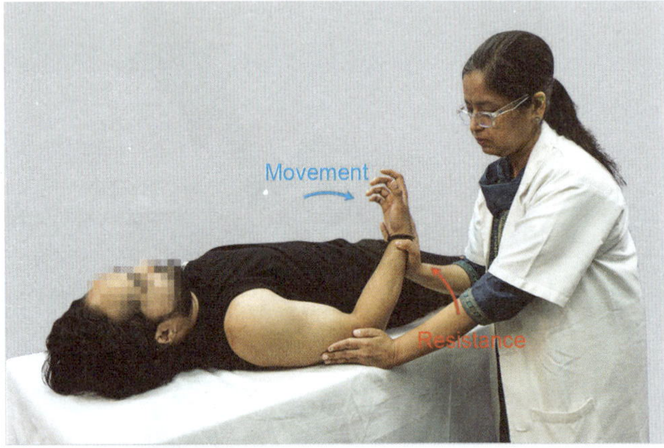

Fig. 9.21: Resisted exercises for elbow extensors

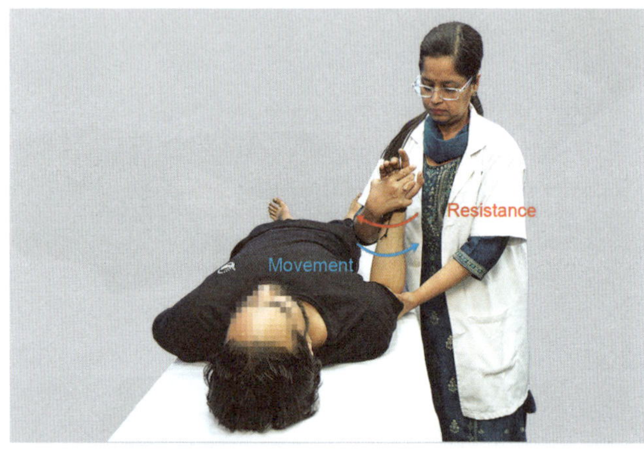

Fig. 9.23: Resisted exercises for supinators

- **Command to patient (supinator):** "Turn your palm inwards as if moving away from my hands. Resist against my resistance as you rotate."

Wrist Flexors and Extensors (Figs 9.24 and 9.25)

- **Therapist hand position:** Providing resistance against the patient's hands or wrists.

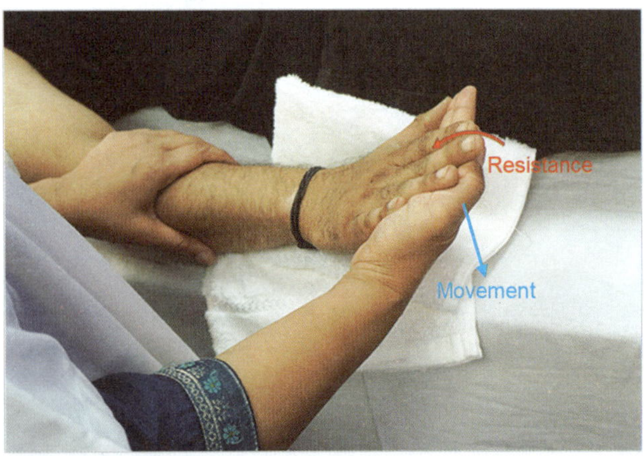

Fig. 9.24: Wrist flexors (manually resisted)

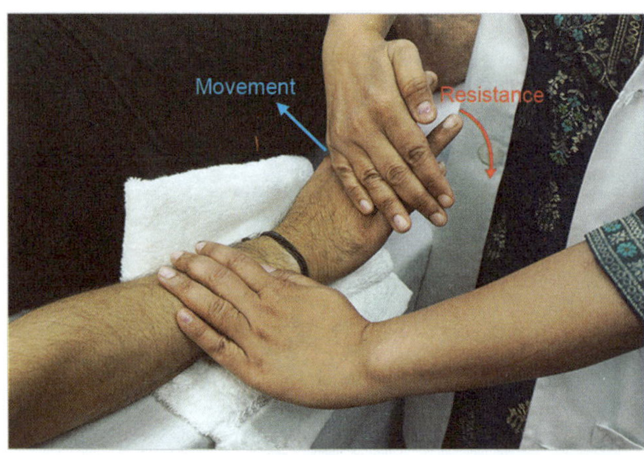

Fig. 9.25: Wrist extensors (manually resisted)

- **Command to patient (flexors):** "Curl your hands downward toward the floor. Push against my hands as you flex."
- **Command to patient (extensors):** "Extend your hands upward toward your body. Resist against my pressure as you extend."

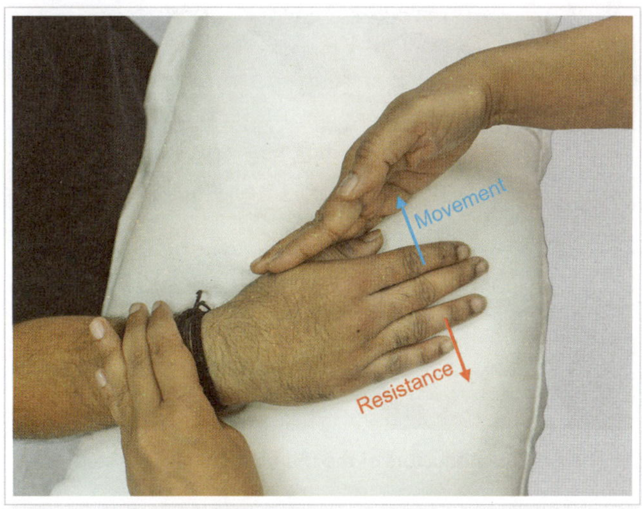

Fig. 9.26: Radial deviators (manually resisted)

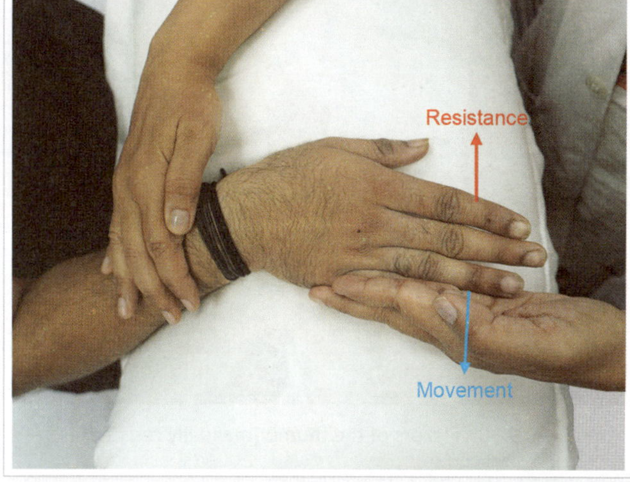

Fig. 9.27: Ulnar deviators (manually resisted)

Radial and Ulnar Deviators (Figs 9.26 and 9.27)

- **Therapist hand position:** Applying pressure against the patient's hands or wrists during radial and ulnar deviation.
- **Command to patient (radial deviation):** "Move your hands toward the thumb side of your forearm. Push against my resistance."
- **Command to patient (ulnar deviation):** "Move your hands toward the pinky side of your forearm. Resist against my pressure as you deviate."

Finger Flexors (Fig. 9.28)

- **Therapist hand position:** Provide resistance against the volar (front) surface of the patient's fingers, near the proximal phalanx, or as needed based on the level of resistance.
- **Command to patient:** "Curl your fingers down toward your palm. Push against my resistance as you flex".

Finger Extensors (Fig. 9.29)

- **Therapist hand position:** Provide resistance against the dorsal (back) surface of the patient's fingers, near the distal phalanx, or as needed based on the level of resistance.
- **Command to patient:** "Straighten your fingers out, away from your palm. Resist against my pressure as you extend".

Thumb Flexors and Extensors (Figs 9.30 and 9.31)

- **Therapist hand position:** Providing resistance against the patient's thumb.
- **Command to patient (flexors):** "Bend your thumb across your palm. Push against my resistance as you flex."
- **Command to patient (extensors):** "Straighten your thumb away from your palm. Resist against my pressure as you extend."

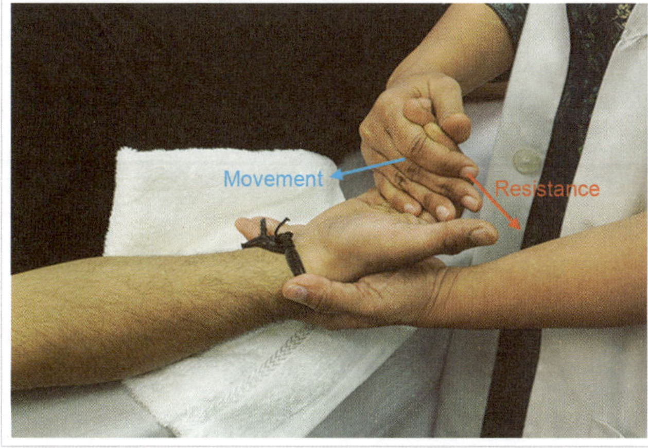

Fig. 9.28: Finger flexors (manually resisted)

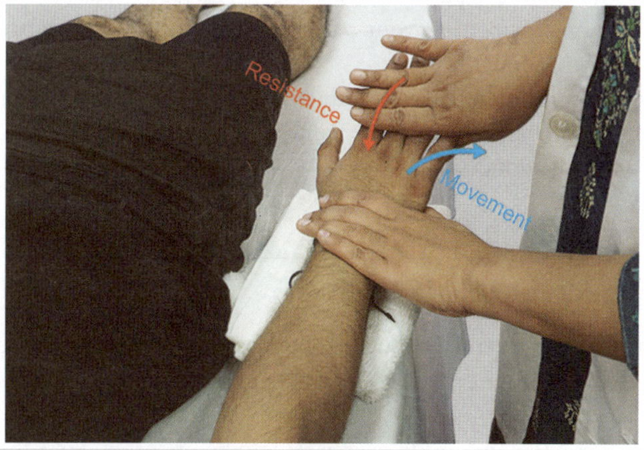

Fig. 9.29: Finger extensors (manually resisted)

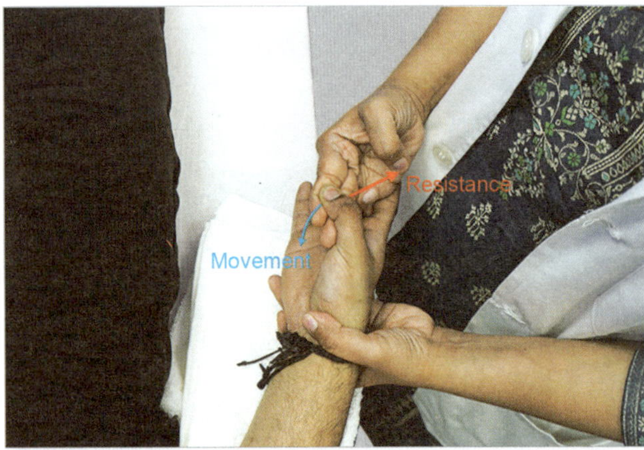

Fig. 9.30: Flexors of the thumb (manually resisted)

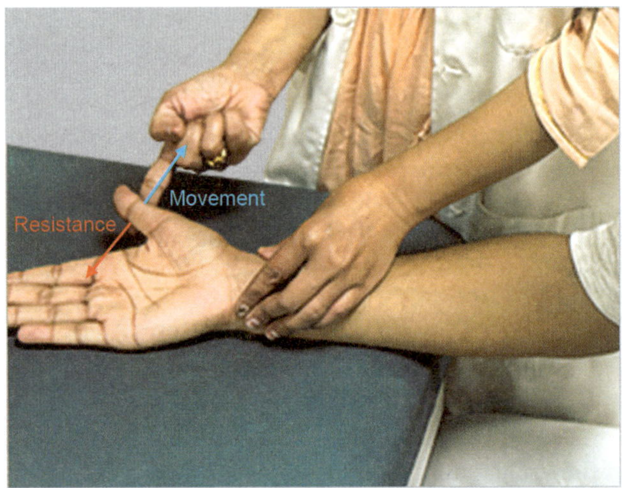

Fig. 9.31: Extensors of the thumb (manually resisted)

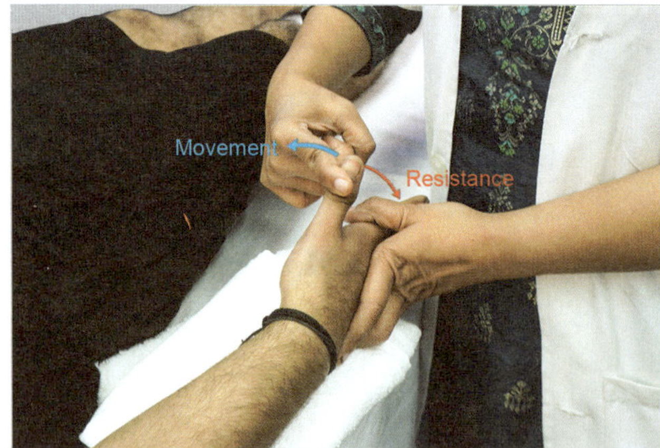

Fig. 9.32: Abductors of the thumb (manually resisted)

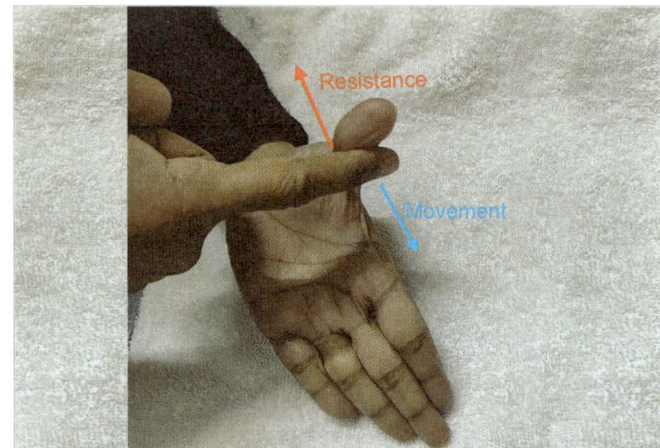

Fig. 9.33: Adductors of the thumb (manually resisted)

Thumb Abductors and Adductors

- **Therapist hand position:** Providing resistance against the lateral and medial sides of the patient's thumb.
- **Command to patient (abductors) (Fig. 9.32):** "Move your thumb away from your palm. Push against my resistance as you abduct."
- **Command to patient (adductors) (Fig. 9.33):** "Bring your thumb back toward your palm. Resist against my pressure as you adduct."

Lower Limb Muscles

Patient Position

- **For hip exercises:** Supine (lying on the back) or prone (lying on the stomach), with legs extended or bent as needed.
- **For knee and ankle exercises:** Supine or seated, with the knees and ankles positioned comfortably.

Therapist Position

Positioned beside or in front of the patient, ready to provide stabilization and resistance as required.

Stabilization

Stabilization may involve supporting the patient's pelvis, thighs, or lower legs to maintain proper alignment and isolate the targeted muscle groups.

Application of Resistance

Hip Flexors (Fig. 9.34)

- **Therapist hand position:** Applying pressure against the patient's thighs or shins during hip flexion.
- **Command to patient:** "Lift your leg upward toward your ceiling. Push against my hands as you lift."

Hip Extensors (Figs 9.35A and B)

- **Therapist hand position:** Providing resistance against the patient's lower legs or ankles during hip extension.

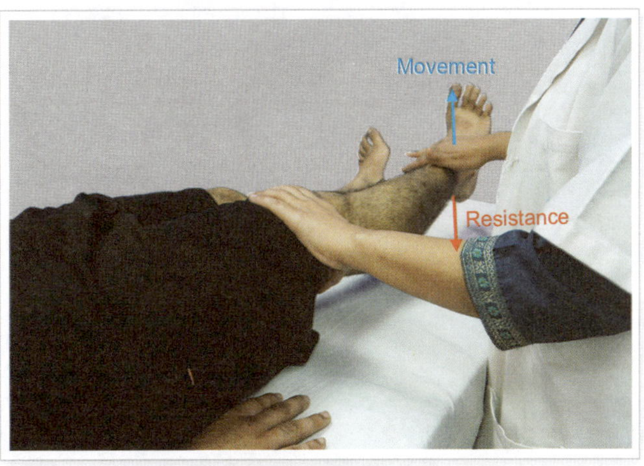

Fig. 9.34: Hip flexors (manually resisted)

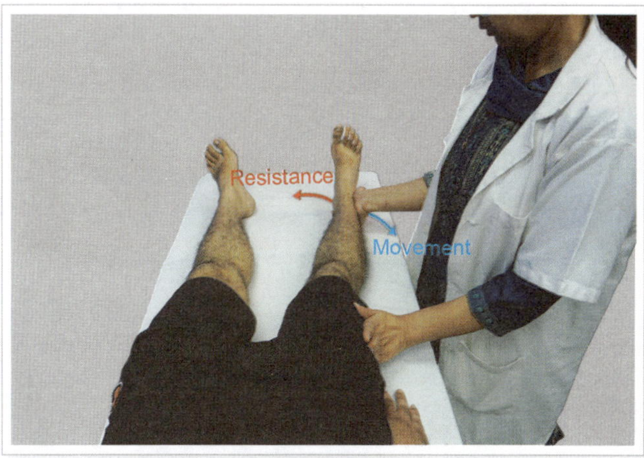

Fig. 9.36: Hip abductors (manually resisted)

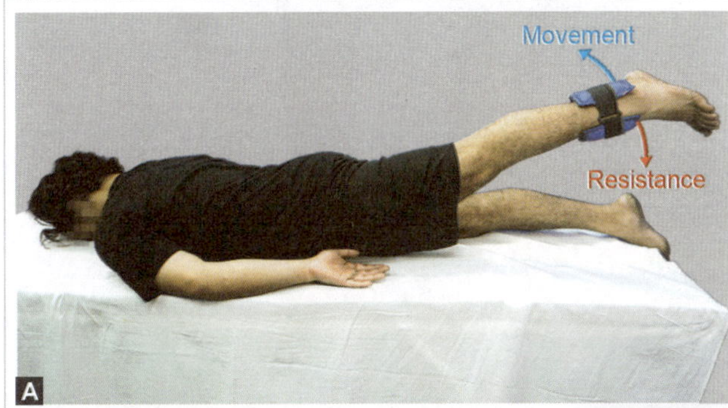

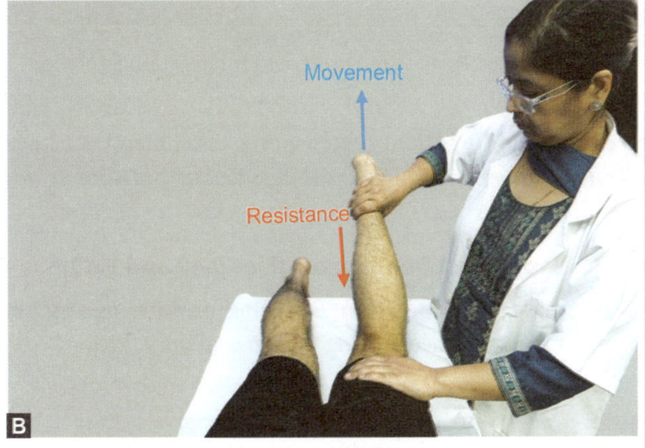

Figs 9.35A and B: Resisted exercises for hip extensors

- **Command to patient:** "Push your leg backward, straightening it behind you. Resist against my pressure as you extend."

Hip Abductors (Fig. 9.36)

- **Therapist hand position:** Applying resistance against the patient's outer thigh or leg during hip abduction.
- **Command to patient:** "Lift your leg out to the side, away from your body. Push against my hands as you lift."

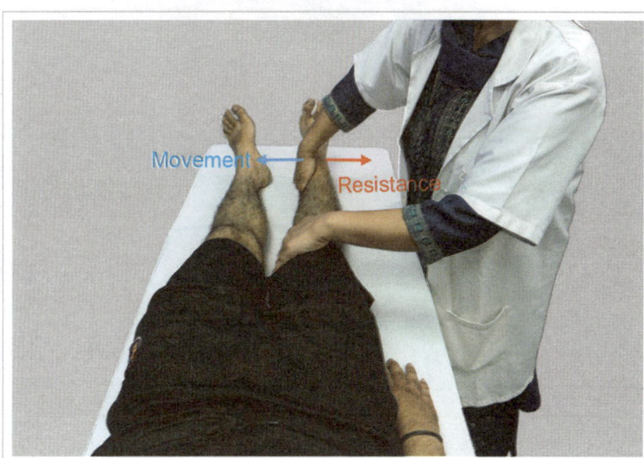

Fig. 9.37: Hip adductors (manually resisted)

Hip Adductors (Fig. 9.37)

- **Therapist hand position:** Providing resistance against the patient's inner thigh or leg during hip adduction.
- **Command to patient:** "Bring your leg inward toward the midline of your body. Resist against my pressure as you adduct."

Internal and External Rotators (Figs 9.38 and 9.39)

- **Therapist hand position:** Applying resistance to the patient's lower leg or ankle during internal and external rotation.
- **Command to patient (external rotation):** "Rotate your leg inward, like you're trying to point your toes toward the opposite leg. Push against my hands."
- **Command to patient (internal rotation):** "Rotate your leg outward, like you're trying to point your toes away from the opposite leg. Resist against my resistance."

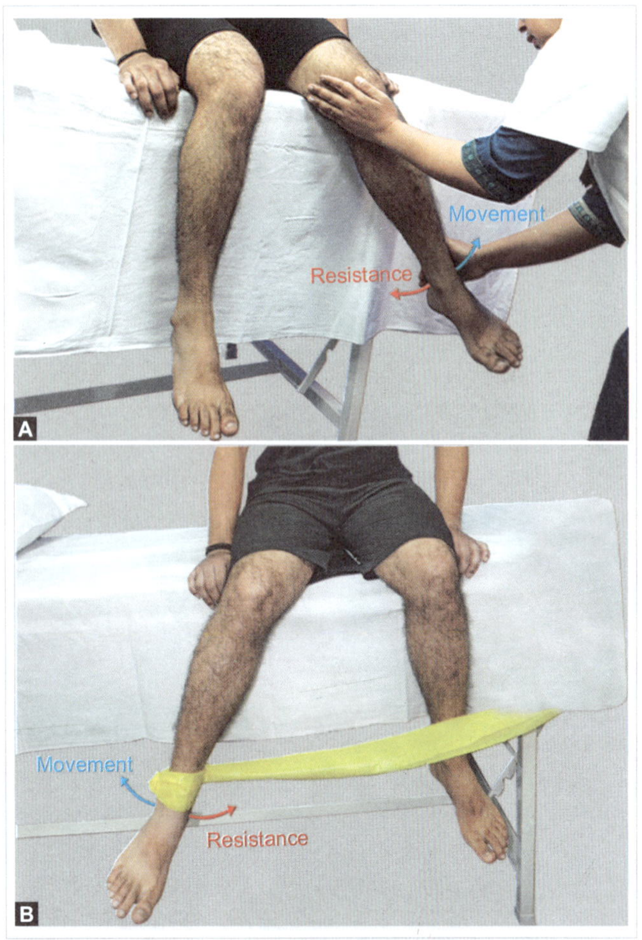

Figs 9.38A and B: Resisted exercises for hip internal rotators

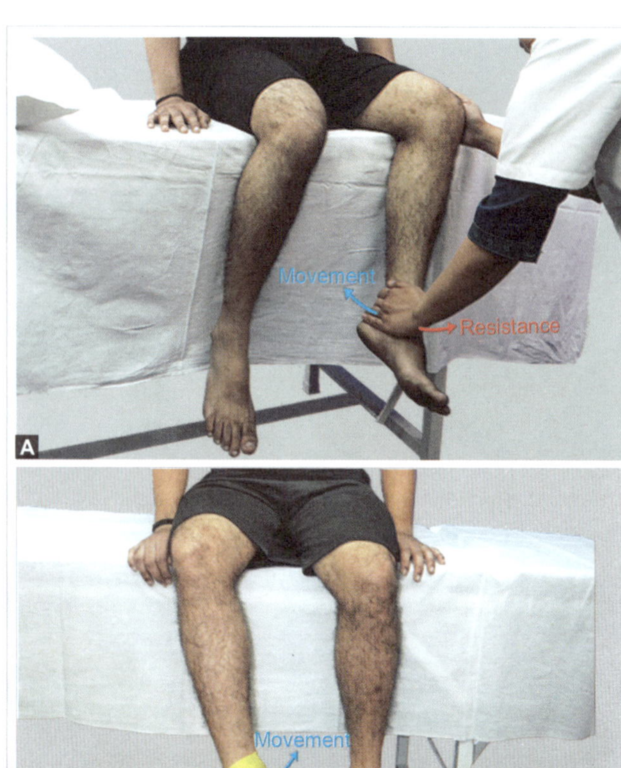

Figs 9.39A and B: Resisted exercises for hip external rotation

Knee Flexors and Extensors (Figs 9.40 and 9.41)

- **Therapist hand position:** Providing resistance against the patient's lower leg or ankle.
- **Command to patient (flexors):** "Bend your knee, bringing your heel toward your buttocks. Push against my hands as you flex."
- **Command to patient (extensors):** "Straighten your knee, up toward the ceiling. Resist against my pressure as you extend."

Ankle Plantar and Dorsiflexors (Figs 9.42 and 9.43)

- **Therapist hand position:** Applying pressure against the patient's foot or heel during plantar and dorsiflexion.

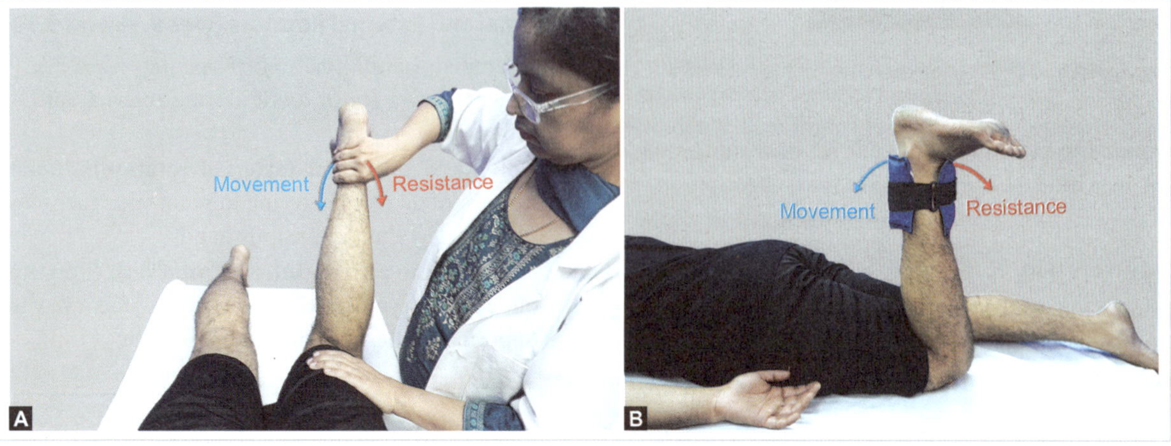

Figs 9.40A and B: Resisted exercises for knee flexors

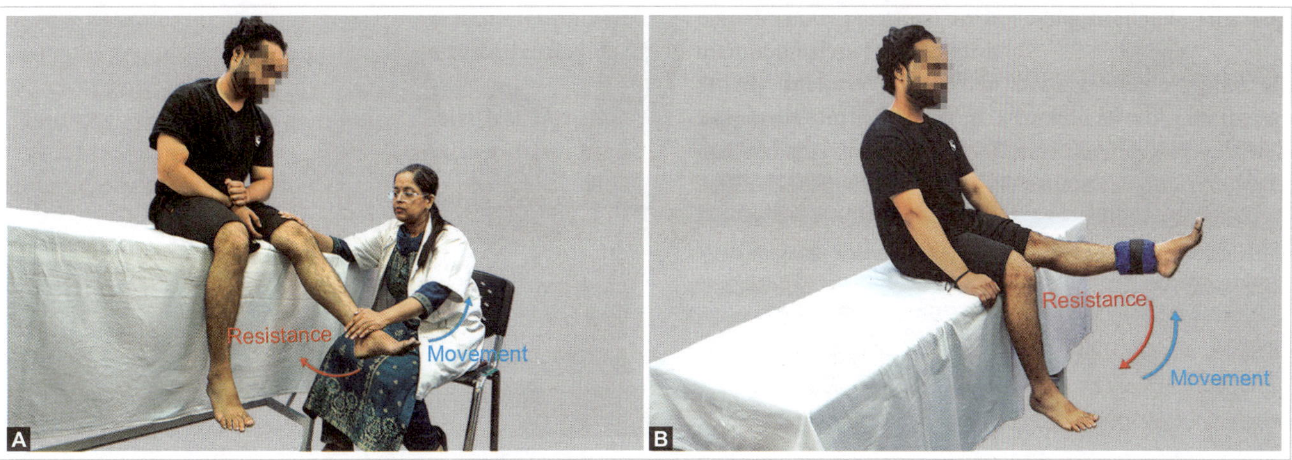

Figs 9.41A and B: Resisted exercises for knee extensors

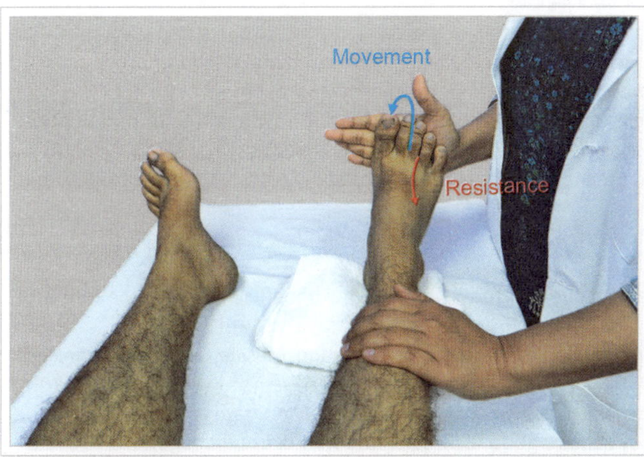

Fig. 9.42: Ankle plantar flexors (manually resisted)

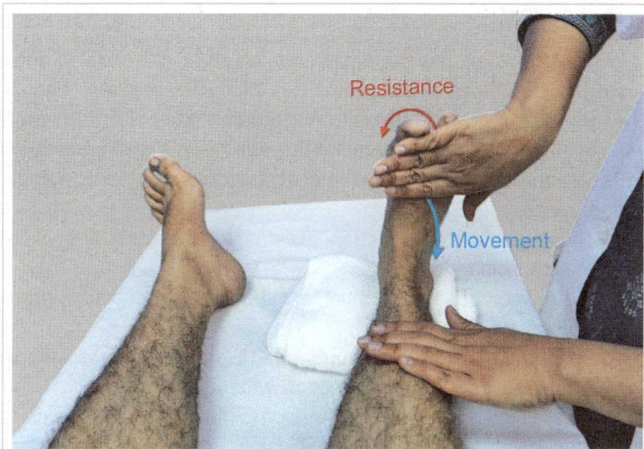

Fig. 9.43: Ankle dorsiflexors (manually resisted)

- **Command to patient (plantar flexors):** "Point your toes downward, like you are pressing down on a pedal. Push against my hands."
- **Command to patient (dorsiflexors):** "Lift your toes upward toward your shin. Resist against my pressure as you dorsiflex."

Ankle Invertors and Evertors (Figs 9.44 and 9.45)

- **Therapist hand position:** Providing resistance against the patient's foot or ankle during inversion and eversion.
- **Command to patient (invertors):** "Turn your foot inward, like you are trying to tilt the sole toward the opposite foot. Push against my hands."
- **Command to patient (evertors):** "Turn your foot outward, like you are trying to tilt the sole away from the opposite foot. Resist against my resistance."

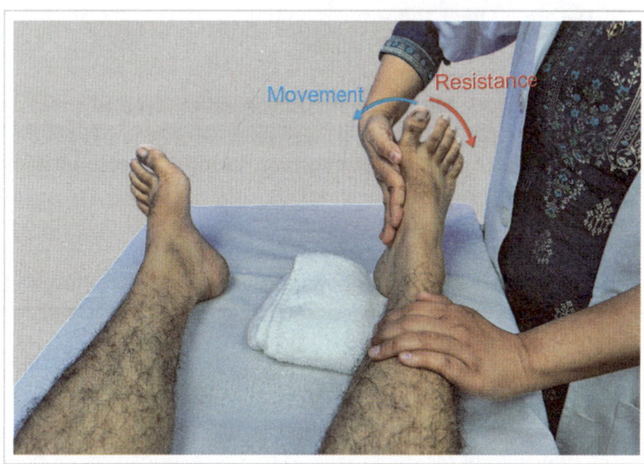

Fig. 9.44: Ankle invertors (manually resisted)

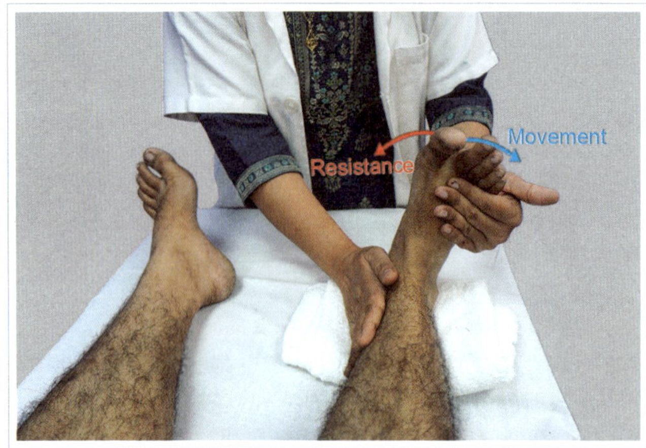

Fig. 9.45: Ankle evertors (manually resisted)

SUMMARY

- Resistance training is the systematic use of resistive loads, different movement velocities, and training modalities to improve muscle strength, endurance, and power. It is critical for recovery, health promotion, and sports performance improvement.
  - Benefits include improved musculoskeletal health, increased fitness, and a higher quality of life for a wide range of populations, including teenagers, older adults, and people with chronic conditions.
- Resisted workouts include isometric, isotonic, isokinetic, manual, and mechanical approaches for building strength, hypertrophy, and endurance.
  - They differ by muscle action, resistance type, and functional classification, addressing upper/lower body, core, and single/multi-joint movements.
  - These activities are essential for rehabilitation, sports performance, and overall fitness, with several benefits across populations and training objectives.
  - Resistance training promotes health and functional capability by developing strength, power, and muscle adaptation through isometric holds and dynamic movements.
- There are several types of resistance training equipment available, each with its own set of advantages and disadvantages.
  1. Free weights, such as dumbbells and barbells, are versatile, but they require good technique and may cause damage.
  2. Weight cuffs offer adjustable resistance and are ideal for targeted muscle training.
  3. Sandbags provide varying weight distribution, which improves stability and functional fitness.
  4. Medicine balls are versatile tools for strengthening and coordination workouts.
  5. DeLorme's boot and suspension therapy targets specific muscle groups and aids in recovery.
- Kraemer et al. define determinants of resisted workouts as a set of criteria crucial to enhancing training efficacy and obtaining targeted goals.
  - These include frequency, intensity, duration, type, alignment and stability, volume, periodization, rest interval, and sequencing.
  - The interconnectedness of these parameters emphasizes the significance of a comprehensive and individualized training routine.
  - The type of exercise is important, taking into account factors such as muscle contraction type, workout form, and resistance application method.
  - Loading, as measured by characteristics such as repetition maximum (RM) and percentage of 1 RM, has a substantial impact on training adaptations, with varied loads targeting distinct physiological responses.
  - Repetition velocity and rest periods influence neuromuscular, hypertrophic, and metabolic responses.
  - Furthermore, the number of sets, training frequency, and duration are critical elements for program design that adapt to individual needs and goals.
  - Periodization incorporates variance to ensure continuous improvement and prevent plateauing.
  - Overall, understanding and managing these factors is critical to creating effective resistance training regimens that improve strength, power, and muscle endurance.
- The principles of resistant exercises include overload, variation, specificity, repetition maximum, and the reversibility principle.
  - Overload is the process of exceeding average levels of physical performance in order to produce adaptation, with elements such as intensity, frequency, and duration serving as stimuli.
  - Variation requires modifying exercise selection, volume, speed of movement, and intensity throughout training sessions in order to maximize adaptation.
  - Specificity focuses on designing workout programs to mirror the demands of desired activities, resulting in focused increases in strength and endurance.

- The RM measures exercise intensity, but the reversibility principle emphasizes the significance of regular training for maintaining physiological adaptations.
- These principles govern the design and implementation of effective resistance training programs, promoting progressive improvement while reducing the possibility of detraining effects.

- Resistance training causes physiological modifications in the nervous system, muscular, fiber type-specific, vascular/metabolic, endocrine, and bioenergetic systems.
  - Neural adaptations occur early in training, resulting in enhanced motor unit recruitment and firing rate, which contribute to initial strength improvements.
  - Muscle responses include increased myofibril density, fiber splitting, and hypertrophy, with type IIA fibers showing the greatest growth.
  - Vascular and metabolic adaptations include alterations in capillary density, mitochondrial density, and ATP/CP storage.
  - Endocrine responses include immediate increases in testosterone and growth hormone levels during exercise, although bioenergetic adjustments have little impact on muscle phosphagen levels and glycolysis-related enzymes.
  - Resistance exercise may not considerably increase maximal oxygen uptake, but it can lower cardiovascular strain during normal activities.
  - These adaptations improve muscular strength, endurance, and general functional ability, emphasizing resistance training's numerous physiological benefits.
- Precautions should be taken when engaging in resisted exercise to mitigate potential risks and ensure safety, particularly for individuals with specific health conditions or circumstances.
  - Avoiding the Valsalva maneuver, especially for those with cardiopulmonary issues or recent surgeries, is crucial to prevent adverse effects.
  - Acute muscle soreness and delayed onset muscle soreness (DOMS) may occur after intense exercise, necessitating proper rest and recovery strategies.
  - Growth plate injuries are a concern in youth resistance training due to the vulnerability of the growth plate to excessive forces. Individuals with cardiovascular disorders should also exercise caution due to the potential increase in blood pressure and risk of cardiac events.
  - Those with inflammation, uncontrolled hypertension, severe osteoporosis, joint discomfort or injury, recent surgeries, uncontrolled diabetes, pregnancy, or chronic respiratory diseases should exercise caution when engaging in resistance training.
  - Adhering to these precautions can reduce the risk of injury and promote safe participation in resistance exercise programs.
- Progressive resisted exercise (PRE) is a dynamic training strategy that improves muscle strength by gradually increasing external resistance over time.
  - The PRE principles, first suggested by Thomas L Delorme roughly 60 years ago, include doing a small number of repetitions until fatigue, giving enough rest between exercises for recuperation, and increasing resistance as force generation improves.
  - This approach has typically been utilized by young, healthy adults to improve athletic performance, but it is now being recognized for its potential health advantages in lowering risk factors for osteoporosis, cardiovascular disease, and diabetes.
- The PRE uses a variety of strategies, including Delorme and Oxford's concepts, which emphasize progressive stress and muscular fatigue.
  - Delorme's concept begins with a warm-up set at 50% of the 10-RM and gradually increases resistance, whereas Oxford's principle begins with the first set at 100% of 10-RM and decreases resistance in following sets to simulate muscle fatigue.
- Kenneth Knight devised the daily adjusted progressive resistance exercise (DAPRE) program in 1979.
  - The DAPRE uses the 6-repetition maximum (6-RM) as a basis for resistance increases and offers an organized approach to strength training that is appropriate for sports rehabilitation and strength and conditioning settings.
  - The program consists of four lifting bouts every session, with resistance increasing dependent on the number of repetitions completed.
- Resistance circuit-based training is a dynamic workout method that combines muscular endurance and strength adaptations in one session.
  - It involves performing sets of various exercises in succession with minimal rest between them, completing one "circuit" before starting again.
  - Key aspects include exercise variety, alternating between upper and lower body movements, and prioritizing larger muscle groups.
- Plyometric training is a strategy aimed at improving neuromuscular efficiency and force generation through quick and strong movements, such as jumps and bounds.
  - It utilizes the stretch-shortening cycle to transition efficiently between eccentric and concentric muscle actions.
  - The three phases—eccentric, amortization, and concentric—are crucial for maximizing muscle function and power.
  - Plyometrics enhance explosive strength, neuromuscular adaptations, bone health, and athletic performance while also aiding in injury prevention.

FURTHER READINGS

- American College of Sports Medicine. American College of Sports Medicine position stand. Progression models in resistance training for healthy adults. Med Sci Sports Exerc. 2009 Mar;41(3):687–708. doi: 10.1249/MSS.0b013e3181915670. PMID: 19204579.
- Bouaziz W, Schmitt E, Kaltenbach G, Geny B, Vogel T. Health benefits of cycle ergometer training for older adults over 70: A review. European Review of Aging and Physical Activity. 2015 Dec;12:1–3
- Castaneda C, Layne JE, Munoz-Orians L, Gordon PL, Walsmith J, Foldvari M, Roubenoff R, Tucker KL, Nelson ME. A randomized controlled trial of resistance exercise training to improve glycemic control in older adults with type 2 diabetes. Diabetes care. 2002 Dec 1;25(12):2335–41.
- Ciolac EG, Rodrigues-da-Silva JM. Resistance training as a tool for preventing and treating musculoskeletal disorders. Sports Medicine. 2016 Sep;46:1239–48.
- Colado, J. C., and García-Massó, X. (2009). Technique and safety aspects of resistance exercises: A systematic review of the literature. The Physician and sportsmedicine, 37(2), 104–111.
- Coudeyre E, Jegu AG, Giustanini M, Marrel JP, Edouard P, Pereira B. Isokinetic muscle strengthening for knee osteoarthritis: A systematic review of randomized controlled trials with meta-analysis. Annals of physical and rehabilitation medicine. 2016 Jun 1;59(3):207–15.
- Courneya KS, Segal RJ, Mackey JR, Gelmon K, Reid RD, Friedenreich CM, Ladha AB, Proulx C, Vallance JK, Lane K, Yasui Y. Effects of aerobic and resistance exercise in breast cancer patients receiving adjuvant chemotherapy: A multicenter randomized controlled trial. J Clin Oncol. 2007 Oct 1;25(28):4396–404.
- DA BV. Isokinetic Dynamometry Applications and Limitations. Sports Medicine. 1989;8(2):101–16.
- Dawes, J., Lentz, D., and Bishop, E. (2017). Comparison of sandbag and traditional weightlifting techniques to traditional weightlifting exercises: A cross-sectional study. International Journal of Sports Science and Coaching, 12(2), 225–234.
- de Oliveira PA, Blasczyk JC, Junior GS, Lagoa KF, Soares M, de Oliveira RJ, Gutierres Filho PJ, Carregaro RL, Martins WR. Effects of elastic resistance exercise on muscle strength and functional performance in healthy adults: A systematic review and meta-analysis. Journal of physical activity and health. 2017 Apr 1;14(4):317–27.
- Deschenes MR, Kraemer WJ. Performance and physiologic adaptations to resistance training. American Journal of Physical Medicine and Rehabilitation. 2002 Nov 1;81(11):S3–16.
- Dibble LE, Hale TF, Marcus RL, Droge J, Gerber JP, LaStayo PC. High-intensity resistance training amplifies muscle hypertrophy and functional gains in persons with Parkinson's disease. Movement disorders: Official journal of the Movement Disorder Society. 2006 Sep;21(9):1444–52.
- Dorgo S, King GA, Candelaria NG, Bader JO, Brickey GD, Adams CE. Effects of manual resistance training on fitness in adolescents. The Journal of Strength and Conditioning Research. 2009 Nov 1;23(8):2287–94.
- Dorgo S, King GA, Rice CA. The effects of manual resistance training on improving muscular strength and endurance. The Journal of Strength and Conditioning Research. 2009 Jan 1;23(1):293–303.
- Ettinger WH, Burns R, Messier SP, Applegate W, Rejeski WJ, Morgan T, Shumaker S, Berry MJ, O'Toole M, Monu J, Craven T. A randomized trial comparing aerobic exercise and resistance exercise with a health education program in older adults with knee osteoarthritis: The Fitness Arthritis and Seniors Trial (FAST). Jama. 1997 Jan 1;277(1):25–31.
- Fillipas S, Oldmeadow LB, Bailey MJ, Cherry CL. A six-month, supervised, aerobic and resistance exercise program improves self-efficacy in people with human immunodeficiency virus: A randomised controlled trial. Australian Journal of Physiotherapy. 2006 Jan 1;52(3):185–90.
- Fish DE, Krabak BJ, Johnson-Greene D, DeLateur BJ. Optimal resistance training: Comparison of DeLorme with Oxford techniques. American journal of physical medicine and rehabilitation. 2003 Dec 1;82(12):903–9.
- Foust TL. The DAPRE technique. International Journal of Athletic Therapy and Training. 1997 Nov 1;2(6):13–4.
- Hackett, D. A., Johnson, N. A., and Chow, C. M. (2016). Training practices and ergogenic aids used by male bodybuilders. Journal of Strength and Conditioning Research, 30(2), 322–327.
- Haff G.G., Triplett N.T. Essentials of Strength and Conditioning. Human Kinetics; Champaign, IL, USA: 2015.
- Haff GG. Roundtable discussion: Machines versus free weights. Strength and Conditioning Journal. 2000 Dec 1;22(6):18.
- Hong AR, Kim SW. Effects of resistance exercise on bone health. Endocrinology and Metabolism. 2018 Dec;33(4):435.
- Hrysomallis C. The effectiveness of resisted movement training on sprinting and jumping performance. The Journal of Strength and Conditioning Research. 2012 Jan 1;26(1):299–306.
- Jorge RT, Souza MC, Jones A, Lombardi Júnior I, Jennings F, Natour J. Progressive resistance training in chronic musculoskeletal disorders. Revista Brasileira de Reumatologia. 2009;49:726–34.
- Kelly M, Darrah J. Aquatic exercise for children with cerebral palsy. Developmental Medicine and Child Neurology. 2005 Dec;47(12):838–42.
- Kraemer WJ, Fleck SJ, Deschenes M. Exercise Physiology Corner: A Review: Factors in exercise prescription of resistance training. Strength and Conditioning Journal. 1988 Oct 1;10(5):36–42.
- Kraemer WJ, Ratamess NA. Fundamentals of resistance training: Progression and exercise prescription. Medicine and science in sports and exercise. 2004 Apr 1;36(4):674–88.
- Larson JL, Covey MK, WIrtz SE, Berry JK, Alex CG, Langbein WE, Edwards L. Cycle ergometer and inspiratory muscle training in chronic obstructive pulmonary disease. American Journal of Respiratory and Critical Care Medicine. 1999 Aug 1;160(2):500–7.

- Lee S, Islam MM, Rogers ME, Kusunoki M, Okada A, Takeshima N. Effects of hydraulic-resistance exercise on strength and power in untrained healthy older adults. The Journal of Strength and Conditioning Research. 2011 Apr 1;25(4):1089–97.
- Liao HF, Liu YC, Liu WY, Lin YT. Effectiveness of loaded sit-to-stand resistance exercise for children with mild spastic diplegia: A randomized clinical trial. Archives of physical medicine and rehabilitation. 2007 Jan 1;88(1):25–31.
- Lim C, Nunes EA, Currier BS, Mcleod JC, Thomas AC, Phillips SM. An Evidence-Based Narrative Review of Mechanisms of Resistance Exercise–Induced Human Skeletal Muscle Hypertrophy. Medicine and science in sports and exercise. 2022 Sep;54(9):1546.
- MA Williams, et al., Resistance exercise in individuals with and without cardiovascular disease: 2007 update: A scientific statement from the American Heart Association Council on Clinical Cardiology and Council on Nutrition, Physical Activity, and Metabolism, Circulation, vol. 116, no. 5, p. 572–584, 2007.
- Marshall EM, Parks JC, Tai YL, Kingsley JD. The effects of machine-weight and free-weight resistance exercise on hemodynamics and vascular function. International Journal of Exercise Science. 2020;13(2):526.
- Minshull, C., and Gleeson, N. (2017). Considerations of the principles of resistance training in exercise studies for the management of knee osteoarthritis: A systematic review. Archives of Physical Medicine and Rehabilitation, 98(9), 1842–1851.
- Naunton J, Street G, Littlewood C, Haines T, Malliaras P. Effectiveness of progressive and resisted and nonprogressive or non-resisted exercise in rotator cuff related shoulder pain: A systematic review and meta-analysis of randomized controlled trials. Clinical Rehabilitation. 2020 Sep;34(9):1198–216.
- Nicholas F Taylor, Karen J Dodd, Diane L Damiano, Progressive Resistance Exercise in Physical Therapy: A Summary of Systematic Reviews, Physical Therapy, Volume 85, Issue 11, 1 November 2005, p. 1208–1223, https://doi.org/10.1093/ptj/85.11.1208
- Paoli A, Gentil P, Moro T, Marcolin G, Bianco A. Resistance training with single vs. multi-joint exercises at equal total load volume: Effects on body composition, cardiorespiratory fitness, and muscle strength. Frontiers in physiology. 2017 Dec 22;8:322368.
- Pollock ML, Franklin BA, Balady GJ, Chaitman BL, Fleg JL, Fletcher B, Limacher M, Piña IL, Stein RA, Williams M. and Bazzarre T, 2000. Resistance exercise in individuals with and without cardiovascular disease: Benefits, rationale, safety, and prescription an advisory from the committee on exercise, rehabilitation, and prevention, council on clinical cardiology, American Heart Association. Circulation, 101(7), p. 828–833.
- Raman J, MacDermid JC, Grewal R. Effectiveness of different methods of resistance exercises in lateral epicondylosis—a systematic review. Journal of Hand Therapy. 2012 Jan 1;25(1):5–26.
- Ramos-Campo DJ, Andreu Caravaca L, Martinez-Rodriguez A, Rubio-Arias JÁ. Effects of resistance circuit-based training on body composition, strength and cardiorespiratory fitness: A systematic review and meta-analysis. Biology. 2021 Apr 28;10(5):377.
- Stone MH, Collins D, Plisk S, Haff G, and Stone ME (2000). Training principles: Evaluation of modes and methods of resistance training. Strength and Conditioning Journal, 22(3), 65.
- Thomas VS, Hageman PA. Can neuromuscular strength and function in people with dementia be rehabilitated using resistance-exercise training? Results from a preliminary intervention study. The Journals of Gerontology Series A: Biological Sciences and Medical Sciences. 2003 Aug 1;58(8):M746–51.
- Veldema J, Bösl K, Kugler P, Ponfick M, Gdynia HJ, Nowak DA. Cycle ergometer training vs resistance training in ICU-acquired weakness. Acta Neurologica Scandinavica. 2019 Jul;140(1):62–71.
- Wilke J, Stricker V, Usedly S. Free-weight resistance exercise is more effective in enhancing inhibitory control than machine-based training: A randomized, controlled trial. Brain sciences. 2020 Oct 3;10(10):702.

STUDENT ASSIGNMENT

LONG ANSWER QUESTIONS

1. Explain physiological adaptations to resisted exercises.
2. Explain in detail principles of resisted exercises.

SHORT ANSWER QUESTIONS

1. Define resisted exercises and their determinants.
2. Write about types of resisted exercises with example.
3. What are the effects and benefits of resisted exercises?

MULTIPLE CHOICE QUESTIONS

1. What is the primary goal of resistance training?
   a. Enhancing flexibility
   b. Improving cardiovascular endurance
   c. Increasing muscle strength, endurance, and power
   d. Promoting relaxation

2. Which of the following is NOT a type of resisted exercise?
   a. Isotonic b. Isometric
   c. Aerobic d. Isokinetic

3. Which equipment offers adjustable resistance and is ideal for targeted muscle training?
   a. Free weights b. Medicine balls
   c. Weight cuffs d. Sandbags

4. What determines the effectiveness of resistance training programs according to Kraemer et al.?
   a. Technique only
   b. Frequency, intensity, duration, and other factors
   c. Exercise type only
   d. Rest period duration only

5. Which principle of resistance training emphasizes modifying exercise selection and volume to maximize adaptation?
   a. Overload b. Variation
   c. Specificity d. Reversibility

6. Which phase of plyometric exercise involves the rapid transition from eccentric to concentric muscle actions?
   a. Eccentric b. Amortization
   c. Concentric d. Proprioception

7. What is the primary aim of the DeLorme method in progressive resistance exercise?
   a. Gradually decreasing resistance
   b. Maximizing resistance in the first set
   c. Reducing the number of sets
   d. Gradually increasing resistance

8. What is a key aspect of resistance circuit-based training?
   a. Long rest periods between exercises
   b. Focusing solely on upper body movements
   c. Performing sets of various exercises with minimal rest
   d. Prioritizing isolation exercises

9. Which phase of plyometric exercise involves pre-stretching muscles to store potential energy?
   a. Eccentric b. Amortization
   c. Concentric d. Proprioception

10. What is the primary determinant of resisted workouts according to Kraemer et al.?
   a. Equipment type
   b. Volume only
   c. Alignment and stability only
   d. Frequency, intensity, duration, and other factors

11. Which principle of resistance training focuses on exceeding average levels of physical performance to produce adaptation?
   a. Overload b. Variation
   c. Specificity d. Reversibility

12. What type of adaptations occur early in resistance training, contributing to initial strength improvements?

a. Vascular and metabolic adaptations
b. Bioenergetic adaptations
c. Neural adaptations
d. Muscle fiber-specific adaptations

13. Which type of equipment is versatile but requires good technique and may cause damage?

a. Medicine balls
b. Weight cuffs
c. Sandbags
d. Free weights

14. What is the primary aim of plyometric training?

a. Enhancing flexibility
b. Improving aerobic endurance
c. Improving neuromuscular efficiency and force generation
d. Promoting relaxation

15. Which principle of resistance training emphasizes designing workout programs to mirror the demands of desired activities?

a. Overload
b. Variation
c. Specificity
d. Reversibility

ANSWER KEY

1. c	**2.** c	**3.** c	**4.** b	**5.** b	**6.** b	**7.** d	**8.** c	**9.** a	**10.** d
11. a	**12.** c	**13.** d	**14.** c	**15.** c					

10 Stretching

Sheetal Kalra, Kopal Pajnee (Stretching)
Jitender Munjal, Savita Tamaria, Kalpana Zutshi, Varsha Chorasiya (Myofascial Release)

LEARNING OBJECTIVES

After the completion of the chapter, the readers will be able to:

- Understand the concept of mobility, flexibility, stretching and related terms.
- Understand tissue response toward immobilization and the effects of stretching.
- Explain the determinants of stretching.
- Explain the indications, contraindications, precautions and benefits of stretching.
- Demonstrate the techniques of stretching of different muscles.
- Demonstrate the procedure of self-stretching.
- Understand benefits and biomechanics of stretching.
- Understand the principle and mechanism of action of myofascial release.
- Demonstrate the techniques of application of myofascial release.

CHAPTER OUTLINE

- Introduction
- Joint Mobility
- Interventions to Improve Soft Tissue Mobility
- Stretching Exercises
- Determinants
- Classification of Stretching
- Procedure
- Techniques
- Self-Stretching
- Myofascial Release

KEY TERMS

Adhesive capsulitis: This condition involves inflammation and thickening of the shoulder joint capsule, causing pain and severe restriction of shoulder movement.

Autoimmune disorders: These occur when the body's immune system unintentionally targets and destroys healthy body tissue.

Contractile tissues: Within the musculoskeletal system, contractile tissue is a type of soft tissue with the ability to contract and relax.

Contractures: Prolonged immobility due to conditions like stroke, paralysis or being bedridden can lead to contractures, where muscles, tendons, and other soft tissues become shortened, reducing joint range of motion (ROM).

Creep: It refers to the gradual deformation of a tissue that occurs when it is subjected to a constant mechanical load over an extended period.

Degenerative changes: Age-related wear and tear on joints and tissues can lead to reduced joint mobility.

Ergonomics: The study of individuals in their work environments.

Joint mobility: It refers to the ROM which a specific joint can achieve without causing pain or discomfort.

Muscle imbalance: Weak or tight muscles around a joint can cause an imbalance in forces, leading to limited movement.

Noncontractile tissues: Ligaments, the joint capsule, and fascia—all made of collagen and elastin fibers—are the noncontractile components of muscle. They are unable to relax and contract.

INTRODUCTION

Healthy joint mobility ensures that one can maintain an active lifestyle, engage in physical exercise, and partake in recreational activities. It also plays a significant role in preventing injuries, as flexible joints can absorb impact and stress better. Moreover, joint mobility supports good posture, which has a positive impact on one's musculoskeletal health and overall appearance. As one ages, maintaining joint mobility becomes increasingly important to ensure independence and quality of life. It allows one to perform self-care tasks, navigate our environment, and engage socially without limitations.

Clinical Correlation

Maintaining optimal joint flexibility and mobility improves mental wellness, as well. Frequent movement and physical exercise can help elevate mood, lower stress levels, and improve cognitive performance. Thus, maintaining muscle suppleness and joint mobility not only benefits physical health but also general mental and emotional well-being.

JOINT MOBILITY

Joint mobility refers to the range of motion (ROM) a specific joint can achieve without causing pain or discomfort. Joint mobility is crucial in daily life for several reasons. It enables one to perform basic activities with ease and comfort, such as walking, reaching, bending, and lifting.

Contractile and Noncontractile Tissue of Muscle

Muscle tissue is composed of two primary types contractile and noncontractile tissue.

1. Contractile tissue, primarily made up of muscle fibers, is responsible for generating force and movement through contraction. These muscle fibers contain the proteins actin and myosin, which interact to create the sliding filament mechanism—the basis of muscle contraction. When stimulated by nerve signals, these fibers contract, generating tension and enabling movements such as lifting, walking, and even internal processes like digestion.
2. On the other hand, noncontractile tissue includes components like blood vessels, nerves, connective tissue, and other structural elements that support and facilitate the function of the muscle. Blood vessels provide oxygen and nutrients to muscle cells, while nerves carry signals that initiate muscle contractions (Figs 10.1A to D). Connective tissue, such as fascia and tendons, help anchor muscles to bones and transmit forces generated during contraction. Additionally, noncontractile tissue contributes to the overall stability, nourishment, and protection of muscle tissue. When contractures develop, adhesions in and between collagen fibers resist and restrict movement.

Relationship Between Joint Mobility and Musculoskeletal Flexibility

Joint mobility and flexibility of musculoskeletal structures are closely related and interdependent aspects of musculoskeletal health. It is influenced by factors like the shape of the bones at the joint, the flexibility of surrounding muscles and tendons, and the condition of ligaments and connective tissues. Good joint mobility ensures that the joint can move freely in various directions, allowing for functional movements like bending, rotating, and extending. Flexibility refers to the ability of muscles, tendons, and other soft tissues to lengthen and stretch. Adequate flexibility allows these tissues to extend to their full length, enabling a broader ROM around a joint.

The relationship between joint mobility and flexibility is intricate. Optimal joint mobility relies on the flexibility of the surrounding muscles and tissues. When muscles are too tight, they can limit the joint's ROM. Conversely, if the joint is stiff due to injury or a medical condition, it can impact the surrounding muscles' flexibility. Therefore, both joint mobility and flexibility need to be addressed for overall musculoskeletal health. Stretching exercises and regular movement help maintain and improve flexibility. This is particularly important for preventing muscle imbalances, reducing the risk of strain and injury, and ensuring proper alignment of the body during movements.

Hypomobility

Hypomobility refers to a condition that limits movement in certain areas of the body, particularly the joints. This may manifest as stiffness, difficulty in movement, and a restricted ROM.

Causes

Hypomobility or restricted motion at a joint can be caused by various factors, including:

- **Joint injury or trauma:** Fractures, dislocations, ligament tears, and other injuries can result in scar tissue formation and joint instability, limiting the joint's ROM.
- **Muscle imbalance:** Weak or tight muscles around a joint can cause an imbalance in forces, leading to limited movement. This often occurs when one set of muscles becomes overactive while its opposing muscles weaken or lengthen.
- **Adhesive capsulitis (Frozen shoulder):** This condition involves inflammation of the capsule of the shoulder joint causing pain and severe restriction of shoulder movement.

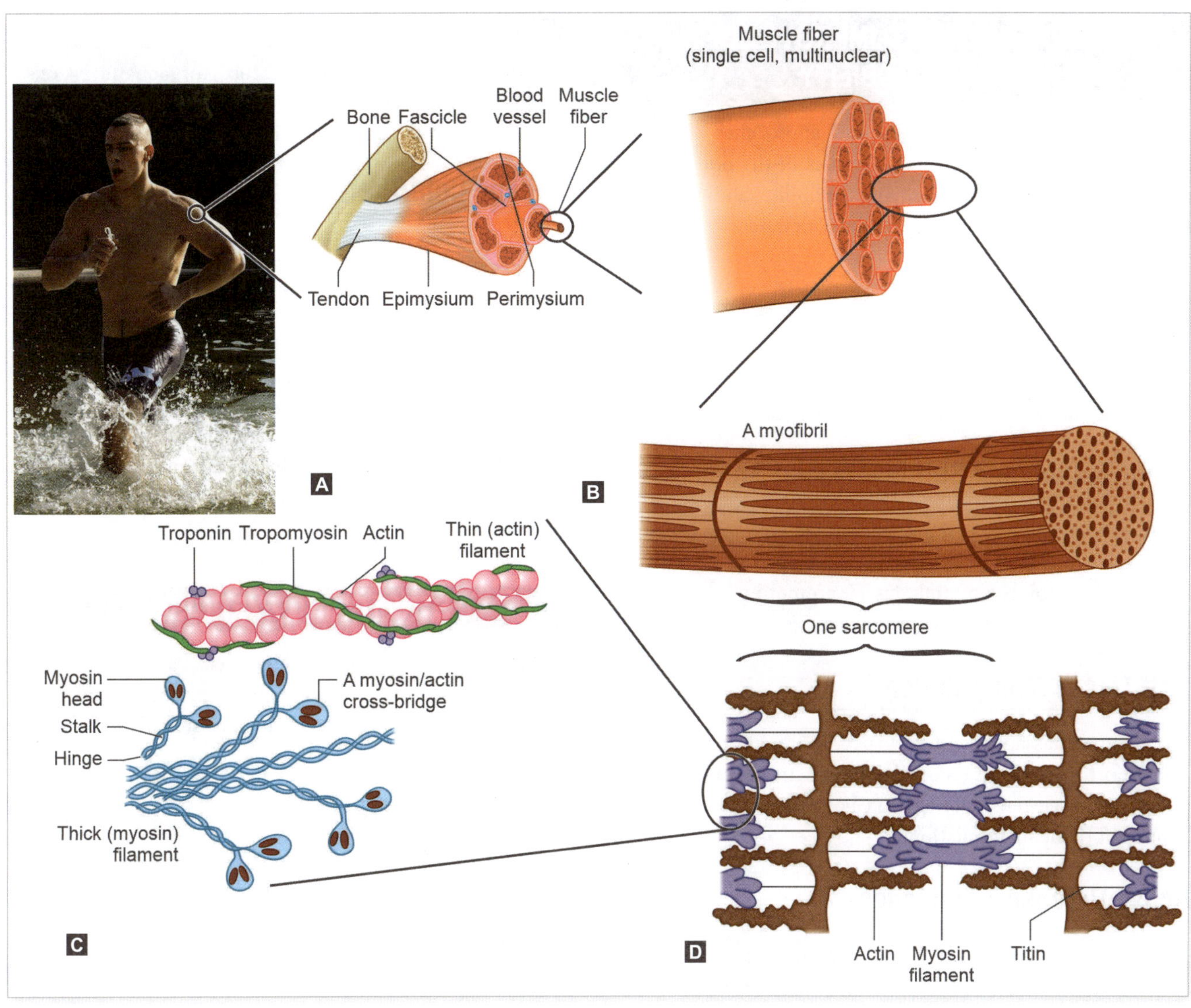

Figs 10.1A to D: Contractile and noncontractile tissues of muscle: **A.** Muscle; **B.** Muscle fiber; **C.** Sarcomere; **D.** Thick and thin filament

- **Contractures:** Prolonged immobility due to conditions like stroke, paralysis, or being bedridden can lead to contractures, where muscles, tendons, and other soft tissues become shortened, reducing joint ROM.
- **Degenerative changes:** Age-related wear and tear on joints and tissues can lead to reduced joint mobility. Conditions like degenerative disc disease in the spine can contribute to limited movement.
- **Neurological disorders:** Conditions affecting the nervous system, such as cerebral palsy or Parkinson's disease, can result in muscle stiffness and restricted joint mobility.
- **Postsurgical scarring:** Surgery can lead to scar tissue formation, which may restrict joint movement if not managed properly during rehabilitation.
- **Autoimmune disorders:** Conditions like lupus or ankylosing spondylitis can cause inflammation and stiffness in multiple joints.
- **Psychological factors:** Stress, anxiety, and emotional tension can contribute to muscle tightness and reduced joint mobility.

It is important to note that early intervention and appropriate treatment can help manage and improve joint hypomobility. Physical therapy, medication, and in some cases, surgical interventions can address the underlying causes and enhance joint function.

Clinical Correlation

In addition to managing existing hypomobility, healthcare professionals should also focus on preventive strategies to reduce the risk of future mobility issues. This may involve educating patients about proper body mechanics, injury prevention techniques, and lifestyle modifications to promote musculoskeletal health.

INTERVENTIONS TO IMPROVE SOFT TISSUE MOBILITY

There are several interventions and approaches that can help increase the mobility of soft tissues in the body. These interventions focus on improving flexibility and reducing muscle tension. Here are some effective methods:

Regular Stretching

Regular stretching routines, including static, dynamic, and proprioceptive neuromuscular facilitation (PNF) stretching, can help elongate muscles and improve soft tissue flexibility.

Foam Rolling and Self-Myofascial Release

Foam rolling involves using a foam roller to apply pressure to muscles and fascia. This technique helps release muscle knots, trigger points, and tight fascia, promoting better tissue mobility. The technique is discussed later in this chapter.

Soft Tissue Manipulation

Professional massages can target specific areas of tightness or discomfort, promoting relaxation, blood circulation, and the release of muscle tension. Refer to Chapter 17 for more details about soft tissue manipulation.

Active Release Techniques

Active release techniques (ART) is a manual therapy approach that combines movement and pressure to address soft tissue restrictions, scar tissue, and adhesions.

Joint Mobilization and Manipulation

Manual techniques performed by trained professionals, such as chiropractors or physical therapists, can help improve joint mobility by addressing restrictions in the joint capsule and surrounding tissues. Refer to Chapter 16 for more details about peripheral joint mobilization.

Complementary Therapies (Yoga and Pilates)

Mindfulness, meditation, and relaxation techniques can help reduce stress and tension, indirectly promoting soft tissue mobility. These practices emphasize controlled movements, stretches, and poses that enhance flexibility, balance, and posture, thus improving soft tissue mobility. Refer to Chapter 25 for more details about yoga.

Hydrotherapy

Immersion in warm water, hot tubs, or aquatic exercises can help relax muscles and improve joint mobility. Refer to Chapter 20 for more details about hydrotherapy.

Heat and Cold Therapy

Heat applications (such as warm showers or heating pads) can relax muscles, while cold applications (like ice packs) can reduce inflammation and pain, potentially improving tissue mobility.

Postural Awareness and Ergonomics

Maintaining proper posture and ergonomics throughout daily activities can prevent muscle imbalances and reduce the strain on soft tissues. Refer to Chapter 21 for more details about posture.

STRETCHING EXERCISES

Stretching is a physical exercise that can help improve flexibility ROM, and performance in physical activities. It helps to prevent injury, reduce muscle soreness and improve posture.

Biomechanical Effects

Tissue lengthening is influenced by the mechanical features of both contractile and noncontractile soft tissues as well as the neurophysiological qualities of contractile tissue. The foundation for choosing and implementing the safest, most efficient stretching techniques in a therapeutic exercise program for patients with limited mobility is an awareness of these tissues' characteristics and how they react to immobilization and stretching.

Contractile Unit of Muscle

Biomechanics play a crucial role in understanding the effects of stretching on the contractile unit of muscle. When a muscle is stretched, the biomechanical principles of tension and elasticity come into play within its contractile unit. This unit comprises the interaction between the actin and myosin filaments within muscle fibers. As a muscle is elongated during stretching, the sarcomeres—the functional units of muscle fibers—lengthen. Force transduction happens both longitudinally as well as laterally during passive stretching. Tension increases substantially when the series of elastic (connective tissue) component first lengthens. The sarcomeres abruptly lengthen after a certain point due to mechanical disruption of the cross-bridges caused by changes in the nervous system and in the body's chemistry. This phenomenon is known as "sarcomere give". The individual sarcomeres revert to their resting length upon relaxation of the stretch force.

However, it is essential to note that while gentle stretching can enhance the flexibility of muscles and improve the ROM, excessive or aggressive stretching can potentially disrupt the balance between tension and elasticity, leading to muscle strain or injury.

Biomechanics of Stretching on Muscles

Immobilized in Shortened Position

Muscle's reaction to immobilization depends on its composition; in tonic (slow-twitch) postural muscle fibers, atrophy occurs more rapidly and extensively than in phasic (fast-twitch) fibers. The degree of atrophy and loss of strength and power is also influenced by the length of time and position of immobilization. Muscle atrophy and loss of functional strength increase with the period of immobilization.

When a muscle becomes immobile in a restricted position, several biomechanical effects occur within the contractile unit of the muscle. This situation can arise due to factors such as prolonged immobilization after injury or surgery. The biomechanical consequences of muscle shortening include changes in muscle architecture, altered force production, and potential adaptations at the cellular level.

- **Adaptive shortening:** Immobilization in a shortened position can lead to adaptive changes in the muscle fibers' resting length and sarcomere alignment. Over time, the muscle fibers may physically shorten to match the imposed shortened position. This adaptive shortening can result in decreased flexibility and limited ROM.
- **Loss of sarcomeres:** Immobilization in a shortened position may cause a reduction in the number of sarcomeres within muscle fibers. Sarcomeres are the contractile units responsible for muscle contraction. With prolonged immobilization, the body might reabsorb some sarcomeres, making the muscle overall less extensible.
- **Decreased force production:** Muscles immobilized in a shortened position often experience reduced force-generating capacity. The muscle fibers lose their optimal length-tension relationship, which is crucial for generating maximal force. As a result, the muscle may become weaker and less capable of producing forceful contractions.
- **Connective tissue changes:** Immobilization can lead to changes in the connective tissue within and around the muscle. This can include an increase in the stiffness of fascia, tendons, and other connective structures. These changes can further limit the muscle's ability to stretch and contract efficiently.
- **Altered neural control:** Prolonged immobilization can affect the neural control of muscles. The neuromuscular connections between the nervous system and muscle fibers may become less efficient, potentially leading to decreased coordination and motor control.
- **Impaired blood flow:** Immobilization can lead to reduced blood flow to the muscle tissue, affecting the delivery of oxygen and nutrients. This can lead to muscle atrophy (loss of muscle mass) and a decrease in overall muscle health.
- **Increased risk of injury:** Muscles that are immobilized in a shortened position are more prone to injury when suddenly subjected to stretching or lengthening movements. The altered muscle properties and reduced flexibility can make the muscle vulnerable to strains and tears.

Biomechanics of Stretching on Muscle Tissues

Immobilized in a Lengthened Position

When a muscle is immobilized in a lengthened position, several biomechanical effects can occur within the contractile unit of the muscle.

- **Loss of sarcomeres:** Immobilization in a lengthened position can lead to a loss of sarcomeres within the muscle fibers. When a muscle is lengthened for an extended period, some sarcomeres may be reabsorbed. This can lead to a decrease in muscle contractile efficiency and a reduction in force-generating capacity.
- **Decreased force production:** Muscles immobilized in a lengthened position may experience a reduction in force production capacity. The length-tension relationship, which is critical for optimal muscle contraction and force generation, can be disrupted. As a result, the muscle may struggle to generate forceful contractions.
- **Altered muscle fiber arrangement:** Immobilization in a lengthened position can lead to changes in the arrangement of muscle fibers. The fibers may align in a way that is less conducive to effective force transmission, further reducing the muscle's ability to generate force.
- **Reduced stiffness:** Muscles immobilized in a lengthened position may experience a reduction in muscle stiffness. This can affect the muscle's ability to stabilize joints and provide support during movement.
- **Increased risk of atrophy:** Prolonged immobilization in a lengthened position can lead to muscle atrophy, which is the loss of muscle mass. Muscle fibers may lose their cross-sectional area and become weaker due to lack of use and reduced mechanical loading.
- **Altered neural control:** Immobilization can affect the neural control of muscles. The connection between the nervous system and muscle fibers may become less efficient, leading to decreased motor control and coordination.
- **Changes in muscle fiber type:** Extended immobilization in a lengthened position can potentially lead to shifts in muscle fiber types. Fast-twitch muscle fibers, responsible for powerful contractions, may be more affected by atrophy, which can further compromise muscle function.

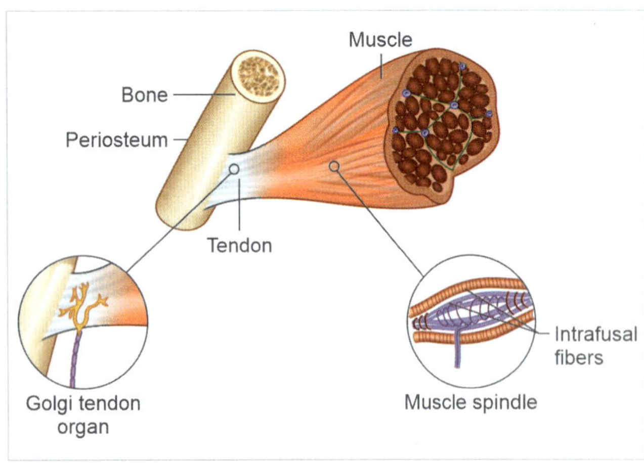

Fig. 10.2: Muscle spindle and Golgi tendon

- **Joint stiffness:** Immobilization in a lengthened position can also impact the surrounding joints, potentially leading to decreased ROM and altered joint stiffness.

Biomechanics of Stretching on Muscle Spindles and Golgi Tendon Organs

Muscle spindles and Golgi tendon organs are specialized sensory receptors within muscles that play crucial roles in proprioception—the body's ability to sense its position and movements in space. These receptors are essential components of the neuromuscular system, providing feedback to the central nervous system to regulate muscle tone, coordination, and protective reflexes (Fig. 10.2).

Muscle spindles are stretch-sensitive structures located within the muscle belly, parallel to the muscle fibers. Each muscle spindle consists of several intrafusal muscle fibers encapsulated by a connective tissue sheath. These fibers are innervated by both sensory and motor neurons. When a muscle is stretched, the muscle spindle's sensory neurons are activated, detecting the degree and rate of stretch. This information is rapidly relayed to the central nervous system, where it contributes to the control of muscle length and the initiation of stretch reflexes. The stretch reflex is an automatic response that involves the muscle contracting in response to a sudden stretch, helping maintain muscle tone and stability during movement. If the stretch is rapid or exceeds a certain threshold, the muscle spindle triggers the stretch reflex. In this reflex, the muscle contracts in response to the stretch to prevent excessive elongation and protect the muscle from potential injury. This reflexive contraction is an automatic response that helps maintain muscle tone and stability, particularly in situations where a sudden stretch might occur. In the context of stretching, this reflex can sometimes limit the extent of the stretch if it's too rapid or forceful.

Golgi tendon organs, on the other hand, are located near the muscle-tendon junction. They consist of sensory nerve endings intertwined within the collagen fibers of a tendon. Golgi tendon organs are highly sensitive to changes in muscle tension and force. When muscle contraction generates tension in the tendon, the Golgi tendon organ's sensory neurons are activated. This information is sent to the central nervous system, where it plays a crucial role in regulating muscle force output. The Golgi tendon reflex is a protective mechanism that prevents excessive force generation by causing muscle relaxation when tension becomes too high. This reflex helps prevent muscle and tendon damage during activities that involve intense force generation.

As a muscle is stretched during a stretching exercise, the Golgi tendon organs come into play. These receptors are located at the muscle-tendon junction and are sensitive to changes in muscle tension. When the muscle is stretched, tension is transmitted through the tendon and activates the sensory neurons of the Golgi tendon organs. This information is sent to the central nervous system, and in response, the Golgi tendon reflex can be triggered. In this reflex, the muscle relaxes to decrease tension and protect the muscle and tendon from excessive force. This reflex serves as a safety mechanism to prevent damage caused by too much force during stretching (Fig. 10.3).

Biomechanics of Stretching on Connective Tissues

Stretching has significant biomechanical effects on connective tissues throughout the body. Connective tissues include structures like tendons, ligaments, and fascia, which play essential roles in providing structural support, transmitting forces, and maintaining the integrity of joints and muscles. When subjected to stretching, these tissues undergo various changes that can impact their mechanical properties and overall function.

Tendons

Tendons are dense connective tissues that attach muscles to bones. Stretching can influence the mechanical behavior of tendons by affecting their stiffness, elasticity, and overall flexibility. Gradual stretching can lead to increased tendon length, allowing for greater joint ROM. However, excessive or sudden stretching can potentially damage tendons and compromise their structural integrity. Tendons are sensitive to the rate and magnitude of stretch; controlled and gradual stretching can promote positive adaptations in tendon tissues, while sudden or forceful stretching might lead to microtears or inflammation.

Ligaments

Ligaments are tough bands of connective tissue that connect bones to each other, stabilizing joints. Stretching can impact the flexibility and strength of ligaments. Controlled stretching can help maintain or improve ligament flexibility, which is

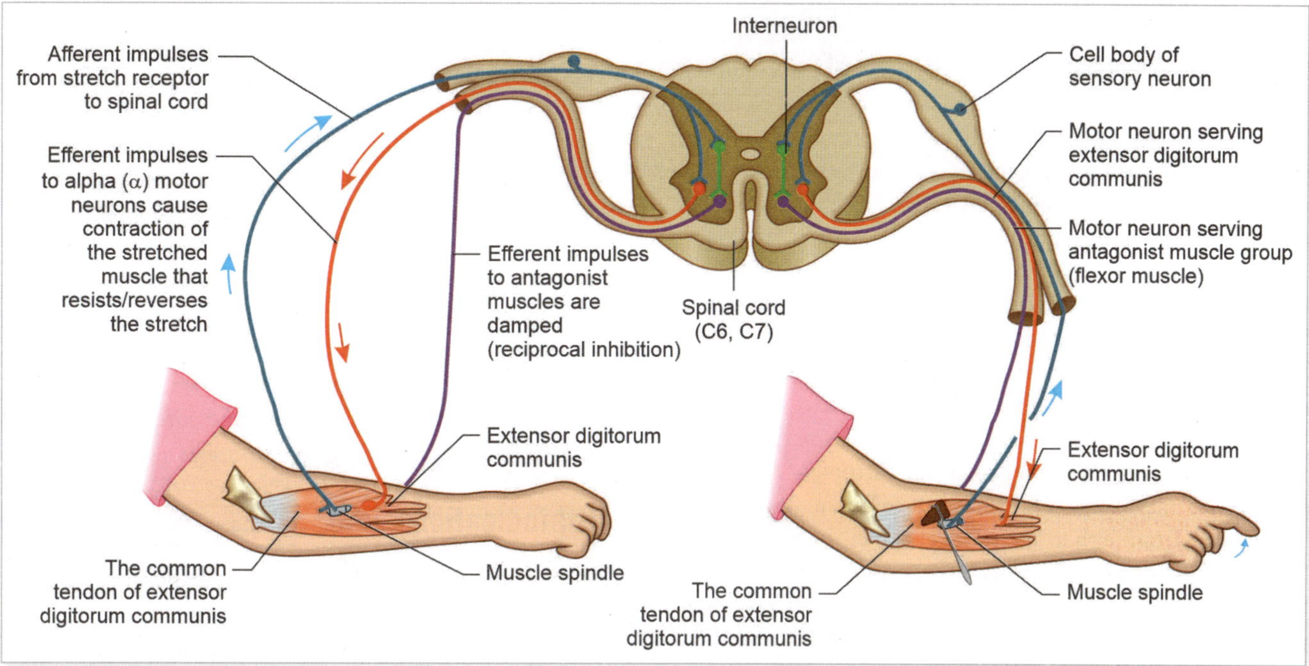

Fig. 10.3: Stretch reflex

essential for joint mobility. However, excessive stretching can strain ligaments, potentially leading to laxity or instability in the joint. Balancing stretching to promote flexibility without compromising joint stability is crucial.

Fascia

Fascia is a complex network of connective tissue that envelops muscles, organs, and other structures. Stretching affects fascia by promoting improved tissue mobility and reducing restrictions that might impede movement. Myofascial stretching techniques, such as foam rolling or manual therapies, target fascial tissue to release tension and enhance tissue gliding. Stretching can also impact the viscoelastic properties of fascia, potentially leading to increased tissue extensibility.

Adhesions and Scar Tissue

Connective tissues can sometimes develop adhesions or scar tissue due to injury or lack of movement. Stretching can help break down these adhesions and increase tissue mobility. However, care must be taken to avoid overly aggressive stretching that might exacerbate scar tissue or cause further damage.

Connective tissue deformation (stretch) occurs to different degrees at different intensities of force and at different rates of application. It requires breaking of collagen bonds and realignment of the fibers for there to be permanent elongation or increased flexibility. Failure of tissue begins as micro failure of fibrils and fibers before complete failure of the tissue occurs. Complete tissue failure can occur as a single maximal event (acute tear from a traumatic injury or manipulation that exceeds the failure point) or from repetitive submaximal stress (fatigue or stress failure from cyclic loading). Micro failure (needed for permanent lengthening) also occurs with creep, stress-relaxation, and controlled cyclic loading.

Clinical Correlation

At varying force strengths and application rates, there are varying degrees of connective tissue deformation, or stretch. For there to be a lasting elongation or improved flexibility, the collagen bonds must be broken and the fibers must realign. Prior to the tissue failing completely, there is a micro failure of the fibers and fibrils. Complete tissue failure can result from repeated submaximal stress (fatigue or stress failure from cyclic loading) or from a single maximal event (acute tearing from a traumatic injury or manipulation that surpasses the failure point). Micro failure, which is required for permanent lengthening, can also be caused by controlled cyclic loading, creep, and stress-relaxation.

Did You Know?

The biomechanical properties of muscle tissue can be influenced not only by stretching but also by factors such as hydration, temperature, and even time of the day. Staying well-hydrated, warming up properly before stretching, and considering the time of day when performing stretching exercises can all impact the effectiveness of the stretching routine.

Impact of Stretching on Joint Extensibility

The quantity of ROM in joints is necessary for human mobility. Joints and muscles are the two anatomical elements that can generally limit ROM. Joint congruency and geometry, as well as the capsuloligamentous structures encircling the joint, are instances of joint constraints. Both passive and active tension are produced by muscles; passive tension is based on the structural characteristics of the muscle and the surrounding fascia, whereas active tension is produced by dynamic muscular contraction. The neuroreflexive characteristics of muscle, notably peripheral motor neuron innervation (alpha motor neuron) and reflexive activation (gamma motor neuron), give rise to active tension. Although there are undoubtedly other causes and factors for decreased joint ROM, muscle tightness is just one of them. An increase in tension from either active or passive sources causes "tightness" in the muscles. Muscles may shorten passively as a result of scarring or postural adaptation, or actively as a result of spasm or contraction. Whatever the reason, tightness restricts ROM and can lead to an imbalance in the muscles.

Taking into account the underlying cause of muscle tension, physiotherapists must choose the appropriate intervention. Stretching is generally used to enhance flexibility and decrease stiffness by elongating the musculotendinous unit, which is simply the space between a muscle's place of origin and insertion. Stretching plays a pivotal role in enhancing joint mobility and flexibility by targeting the muscles and soft tissues that surround a joint. When muscles are tight or shortened they can restrict joint movement and lead to reduced flexibility. Regular stretching routines help elongate and relax these muscles, allowing them to reach their optimal length. This results in improved joint mobility, as the muscles no longer impede the joint's natural ROM. Additionally, stretching encourages better blood flow to the stretched muscles, promoting tissue health and suppleness. It also stimulates the production of synovial fluid, which lubricates the joint and facilitates smoother movement. By preventing muscle imbalances and reducing tension, stretching contributes to a balanced musculoskeletal system, resulting in better posture and less strain on joints.

Clinical Correlation

Stretching not only improves joint mobility and flexibility but also helps alleviate muscle soreness and stiffness. Stretching after exercise or physical activity can aid in the recovery process by promoting blood flow to the muscles, reducing lactic acid build-up, and enhancing tissue repair. So, incorporating stretching into the post-workout routine can contribute to faster recovery and better overall muscle health.

Indications

- To improve flexibility.
- To increase ROM.
- Reduce injuries, strains and damage: Stretching exercises before exercise are thought to help the body get ready for exercise and athletic events by increasing joint ROM, which enhances performance and lowers the risk of injury.
- Before and after exercise to lessen muscle soreness.
- To improve muscle performance.
- To promote better circulation.
- To reduce the chance of muscular injuries.
- To help prevent muscular stiffness and soreness.
- To help reduce tension and to encourage relaxation.

Contraindications

- Limited movement at the joint due to the presence of a bony block.
- Directly after fracture, when it has not completely healed.
- Acute inflammation or, infection.
- Sharp pain during joint movement.
- Hematoma or other soft tissue trauma.
- Hypermobility.

Guidelines

A successful stretch program should:

- Involve all the major groups from head to toe: Neck, shoulders, chest, trunk, lower back, hips, legs and ankles.
- Be designed with a goal to:
  - Stretch 2–3 days/week
  - Complete 2–4 repetitions per stretch.
- Have enough time allotted to:
  - Hold each stretch for a minimum of 15–30 seconds.
- Stretch in a slow relaxed manner.
- Stretch until there is a slight discomfort in the muscle.
- Never stretch to the point of feeling pain.
- Do breathe slowly, rhythmically and under control.
- Avoid bouncing or jerking.
- Do not continue to stretch if it causes sudden, sharp or intense pain.
- Do not begin a stretch program in acute joint or back pain.

Benefits

- **Increases flexibility:** Stretching on a regular basis can improve flexibility, which is important for general health. Increased flexibility can not only make it easy to carry out daily tasks, but it can also help delay the potential loss of mobility that comes with aging.

- **Increases range of motion (ROM):** More ROM is available when a joint can move through its whole range. Regular stretches can aid in improving ROM.
- **Improves performance in physical activities:** Dynamic stretches, also known as moving stretches, have been found to be beneficial before engaging in physical activity. It might also help to perform better during an exercise or athletic event.
- **Increases blood flow to muscles:** Regular stretching exercises might help with circulation. Enhanced blood flow to the muscles due to improved circulation helps lessen muscular soreness also known as delayed onset muscle soreness (DOMS) and speed up the recovery.
- **Improves posture:** Poor posture can result from muscle imbalances, which are frequent. Reducing musculoskeletal pain and promoting appropriate alignment can be achieved through the combination of stretching and strengthening targeted muscle groups. The posture may then get better as a result of it.
- **Relaxation:** Engaging in a consistent stretching regimen can improve flexibility and promote mental calmness.

DETERMINANTS

The effectiveness and safety of stretching are influenced by several determinants that play significant roles in how tissues respond to stretching forces. Understanding these determinants is crucial for designing stretching exercise program that promotes flexibility, enhances performance, and minimizes the risk of injury. Here are the key determinants of stretching:

Proper Stabilization and Positioning

Proper stabilization and positioning of a patient are essential aspects of ensuring safe and effective stretching routines, especially in clinical or therapeutic settings. When performing stretches on patients, attention to stabilization and positioning is crucial to prevent injury and promote positive outcomes. Stabilization involves securing and supporting the patient's body in a way that minimizes the risk of uncontrolled movements or undue stress on joints and tissues. Here's how proper stabilization and positioning are achieved:

Stabilization

Proper stabilization ensures that the patient maintains a stable and controlled position during stretching, preventing accidental falls or movements that could lead to injury. Techniques for stabilization include.

- **Assistance from therapist:** A therapist can provide manual stabilization by gently holding or supporting the patient's body parts that are not being stretched. This helps maintain balance and control.
- **Sturdy surface:** Performing stretches on a stable and supportive surface, such as a treatment table or therapy mat, enhances patient safety. Ensure that the surface is well-padded and appropriate for the patient's size and mobility.
- **Use of straps or belts:** Using straps or belts can help stabilize the patient's limbs during stretches. These tools can prevent excessive movement and allow the therapist to control the intensity of the stretch.

Positioning

Positioning involves arranging the patient's body to target specific muscles or muscle groups accurately. Proper positioning maximizes the effectiveness of the stretch while minimizing discomfort. Consider these factors.

- **Anatomical alignment:** Position the patient's body in anatomically neutral positions whenever possible to reduce stress on joints. This ensures that the stretch primarily targets the intended muscles.
- **Joint positioning:** Carefully position joints to maintain proper alignment and prevent overstretching. For example, when stretching a leg, the knee joint should be straight or slightly bent as appropriate for the specific stretch.
- **Supportive pillows or cushions:** Use pillows or cushions to support body parts that are not actively involved in the stretch. This enhances comfort and relaxation during the stretching session.

Duration

The duration for which a stretch is held affects the tissue's response. Static stretches are typically held for around 15–60 seconds. Longer durations allow for greater tissue elongation and adaptation. However, excessively long stretches can potentially lead to tissue damage or excessive flexibility. Balancing the duration is essential to achieve the desired effects without causing harm. In a treatment session when stretches are repeated more than once (stretch cycle), the accumulative duration of all such cycles will be considered total duration of stretch defined total elongation time.

Frequency

How often stretching is performed impacts the rate of flexibility improvement. Regular stretching sessions are necessary to maintain and enhance flexibility over time. Performing stretches several times a week allows tissues to adapt progressively without subjecting them to unnecessary strain. The underlying reason of decreased mobility, the quality and extent of tissue repair, the chronicity and severity of a condition, the patient's age, usage of corticosteroids, and the patient's previous reaction to stretching all are factors of the frequency of stretching that is advised. The number of sessions each week varies between two and five, with rest periods in between to promote tissue healing and lessen pain after exercise.

Intensity

The intensity of stretching refers to the degree of force applied during a stretch. Gentle, controlled stretching is more effective and safer than aggressive or forceful stretching. Overstretching can lead to tissue damage or injury, so it's important to respect the tissue's limits and avoid pushing too hard. The best ROM improvement comes from gentle, sustained stretching that protects immobilized tissues from overstress and possible damage. It has also been demonstrated that low-intensity stretching is superior to high-intensity stretching in terms of elongating thick connective tissue, which is a major cause of chronic contractures, while causing less soft tissue damage and postexercise soreness than a high-intensity stretch.

Warm-Up

Engaging in a proper warm-up before stretching increases tissue temperature and blood flow, enhancing tissue extensibility and reducing the risk of injury. Warm muscles and connective tissues are more responsive to stretching and less likely to experience strain.

Speed

The speed at which stretching is performed, often referred to as the stretching velocity or speed of elongation, is an important factor that can influence the effectiveness and safety of stretching exercises. The evidence on stretching speed suggests that different speeds can have varying effects on muscle and connective tissue responses.

Many studies and experts recommend slow and controlled stretching movements. Slow stretching allows the muscles and connective tissues to adapt gradually to the elongation, minimizing the risk of injury or overstretching. This approach is particularly beneficial for static stretching, where the muscle is held in a stretched position for a specific duration. Slow stretching can lead to improved flexibility and reduced muscle tension while promoting relaxation. Rapid or ballistic stretching involves using momentum to perform stretching movements. Research indicates that this type of stretching, characterized by bouncing or jerking movements, can lead to muscle strains, microtears, and injury. Ballistic stretching may activate the stretch reflex, causing muscles to contract in response to the sudden stretch. Therefore, this approach is generally discouraged due to the increased risk of harm. Dynamic stretching involves controlled, fluid movements that take joints and muscles through a full ROM. Evidence supports the use of dynamic stretching as part of a warm-up routine before physical activity. Dynamic stretching can enhance blood flow, increase body temperature, and improve neuromuscular coordination. The speed of dynamic stretching should be moderate and controlled to avoid overloading the tissues.

Type

Different types of stretching techniques, such as, dynamic, proprioceptive neuromuscular facilitation (PNF), which combines muscle contraction and stretching can be used for unique advantages and may be customized to meet various needs and fitness objectives. These include self-stretching, manual and mechanical stretching, and classifications such as passive, aided, or active stretching. Types of stretching are explained later in this chapter.

Individual Variability

Each individual's anatomical structure, genetics, and fitness level can influence how they respond to stretching. Some people naturally have greater flexibility, while others may need more time and effort to achieve similar results. Tailoring stretching routines to individual needs is essential for safe and effective outcomes.

Age and Gender

Age and gender can affect tissue elasticity. Generally, younger individuals tend to have more flexible tissues, while older individuals may require more focused and consistent stretching to maintain or improve flexibility.

Injury History and Health Conditions

Past injuries and underlying health conditions can impact the choice and application of stretching exercises. Individuals with certain injuries or medical conditions may need modified stretches or specific precautions to avoid exacerbating their condition.

Did You Know?

Recommendations

For a general fitness program, the American College of Sports Medicine recommends static stretching for most individuals that is preceded by an active warm-up, at least 2–3 days/week. Each stretch should be held 15–30 seconds and repeated 2–4 times.

CLASSIFICATION OF STRETCHING

1. **Static stretching:**
   - **Definition:** Holding a muscle or group of muscles in a stretched position for a period of time.
   - **Duration:** Typically 15–60 seconds.
   - **Types:**
     - **Active static stretching:** Using one's own muscles to hold the stretch.
     - **Passive static stretching:** Using assistance from a partner or prop.
2. **Static progressive stretching:**
   - **Definition:** A controlled method that gradually increases the stretch over time.
   - **Duration:** Typically 30–60 seconds, with sustained low-level force applied incrementally.
   - **Applications:** Effective in rehabilitation and flexibility training.
3. **Cyclic (intermittent) stretching:**
   - **Definition:** Alternating periods of stretching with periods of rest.
   - **Duration:** Stretch phase usually held between 5 and 10 seconds.
   - **Applications:** Improves flexibility while allowing recovery between stretches.
4. **Manual stretching (assisted stretching):**
   - **Definition:** Involves external force applied by a partner or therapist.
   - **Benefits:** Immediate gains in flexibility; effective for both active and passive applications.
5. **Self-stretching:**
   - **Definition:** Performed independently without assistance.
   - **Duration:** Typically 30–60 seconds per repetition.
   - **Focus:** Maintaining correct body alignment for effectiveness.
6. **Mechanical stretching:**
   - **Definition:** Use of external mechanical forces to induce elongation in tissues.
   - **Duration:** Can range from 15 minutes to several hours or continuous throughout the day.
   - **Methods:** Includes the use of devices or serial casts.
7. **Ballistic stretching:**
   - **Definition:** Involves bouncing or jerking movements to force a muscle beyond its normal ROM.
   - **Characteristics:** High-speed movements aiming for maximum motor unit recruitment.
   - **Applications:** Typically used in sports for performance enhancement.
8. **Dynamic stretching:**
   - **Definition:** Active movements that take a joint or muscle through a full ROM.
   - **Characteristics:** Controlled and mimics specific sport movements.
   - **Benefits:** Increases muscle temperature, blood flow, and prepares the body for physical activity.
9. **Proprioceptive neuromuscular facilitation (PNF) stretching:**
   - **Definition:** A technique that lengthens muscles by causing them to relax through contraction.
   - **Techniques:**
     - **Contract-relax:** Stretching a muscle, then isometrically contracting it.
     - **Hold-relax:** Similar to Contract-Relax, focusing on muscle relaxation after contraction.
     - **Hold-relax agonist:** Involves contracting the opposing muscle to promote relaxation of the target muscle.

Static Stretching

In static stretching, a particular muscle or group of muscles is stretched by holding that stretch for a period of time. It involves slowly stretching a muscle to its farthest point and holding that position for a period of time, typically around 15–60 seconds. Static stretching has been extensively recommended before engaging in physical exercise since the early 1980s as a way to reduce the risk of injury and enhance performance during exercise. Static stretching is most beneficial for athletes requiring flexibility for their sports (e.g., gymnastics, dance, etc.). Static stretching can be done actively (using one's own muscles to hold the stretch) or passively (with a partner or prop), both of which help to increase flexibility. Static stretching lasting from several hours to several days or weeks, if a mechanical device is used to create the static stretch, has also been described.

Effects

- Improves ROM in the joint
- Reduces stiffness and pain in muscles
- Reduces the risk of muscle strains and other injuries
- Improves postural awareness and body posture
- Increases circulation which in return decreases recovery period after exercises.

Static Progressive Stretching

Static progressive stretching is an evidence-based technique that involves a gradual and controlled approach to improve flexibility and joint ROM. This method employs the principles of static stretching but adds an element of progressive adaptation over time. Through this approach, the muscle or muscle group is stretched to a point of gentle tension and then held in that position for a specified duration, usually around 30–60 seconds. However, what sets static progressive stretching apart is the subsequent application of low-level, sustained force to incrementally increase the stretch. This force is usually applied using external devices like straps, weights, or specialized equipment.

Effects

- Research supports the effectiveness of static progressive stretching in various scenarios, such as postinjury rehabilitation, reducing muscle tightness, and managing musculoskeletal conditions.
- It allows for targeted, controlled adaptation of tissues, gradually elongating them while minimizing the risk of injury.
- The evidence suggests that this method can lead to significant improvements in joint mobility and muscle flexibility over time, making it a valuable addition to rehabilitation protocols and flexibility training regimens.

Cyclic or Intermittent Stretching

Cyclic or intermittent stretching refers to a stretching technique that involves alternating periods of stretching with periods of rest or relaxation. As per research each cycle of stretch is generally held between 5 and 10 seconds. This approach is used to gradually increase tissue length and flexibility while allowing tissues to recover between stretching cycles.

This alternating pattern is repeated multiple times during a session. The specific duration of the stretch and rest phases, as well as the total number of cycles, can vary based on individual goals and the targeted muscle group.

Effects

- Research into cyclic stretching has shown promising results in terms of improving muscle flexibility and joint ROM.
- By allowing tissues to recover during the rest phases, cyclic stretching aims to reduce the risk of tissue damage or overstretching that can occur with prolonged static stretching.
- The alternating pattern may help prevent the activation of the stretch reflex and promote relaxation within the stretched muscle.
- Additionally, cyclic stretching has been explored as a potential method for improving muscle performance and reducing muscle stiffness.
- It has been studied in the context of conditions such as contractures (limited joint mobility), muscle strains, and post surgery rehabilitation.

Manual Stretching

Manual stretching, also known as assisted stretching, involves the application of external force by a partner or therapist to elongate muscles and increase joint flexibility. Research shows its effectiveness in improving flexibility and ROM, especially when compared to self-stretching or no stretching. The therapist manually controls the site. Manual stretching can be performed actively by the patient, passively, or with assistance from the patient, or even independently by the patient. The act of manually stretching the body can be done passively, with the patient's help, or even on their own.

Effects

- Studies have demonstrated that manual stretching can lead to immediate gains in joint flexibility, with sustained application resulting in lasting improvements.
- The controlled force applied during manual stretching promotes muscle relaxation and may reduce discomfort.

Indications

- **Limited flexibility:** Individuals with limited flexibility or reduced joint ROM can benefit from manual stretching. Assisted stretching can help target specific muscles or muscle groups that are resistant to self-stretching techniques.
- **Rehabilitation:** Manual stretching can be an effective component of rehabilitation programs for individuals recovering from injuries, surgeries, or musculoskeletal conditions. It can aid in restoring functional movement and preventing muscle or joint stiffness during recovery.
- **Sports performance:** Athletes often use manual stretching to enhance their flexibility and mobility, which can improve performance and reduce the risk of injuries. Specific stretches can be tailored to the demands of the athlete's sport.
- **Muscle imbalances:** Manual stretching can address muscle imbalances that result from overuse or improper training techniques. Assisted stretching can help lengthen tight muscles and restore balance between opposing muscle groups.
- **Postexercise recovery:** Incorporating manual stretching after intense workouts can help prevent muscle tightness, alleviate soreness, and promote relaxation. It is particularly effective when applied by a skilled practitioner who understands the muscle groups involved.
- **Chronic conditions:** Individuals with chronic conditions that affect mobility, such as arthritis, may find manual stretching beneficial in maintaining joint function and reducing discomfort.
- **Preventive maintenance:** Regular manual stretching can serve as a preventive measure to maintain flexibility and joint ROM, potentially reducing the risk of future injuries.

- **Enhancing relaxation:** Manual stretching can promote relaxation and stress reduction by targeting tense muscles and helping the individual unwind.
- **Neurological conditions:** Some neurological conditions can lead to muscle stiffness or spasticity. Manual stretching, when performed by trained professionals, can help manage these symptoms and improve comfort.
- **Preventing muscle contractures:** Individuals at risk of developing muscle contractures due to immobilization or prolonged bed rest can benefit from manual stretching to prevent loss of muscle length.

Self-Stretching

Self-stretching is a type of stretching, where individuals perform stretches on their own without external assistance. Its beneficial effects have been seen on flexibility, muscle function, and overall health. For self-stretching to be effective, one must maintain the correct alignment of the body. Static stretching is thought to be the safest method for self-stretching regimens, with each repetition lasting 30–60 seconds.

- **Flexibility improvement:** Self-stretching has been shown to effectively improve flexibility and joint ROM. Regular and consistent self-stretching routines can lead to gradual increases in muscle length and improved overall flexibility.
- **Static versus dynamic self-stretching:** Both static and dynamic self-stretching have demonstrated benefits. Both approaches have shown effectiveness in improving flexibility, but their application can depend on the individual's goals and the context in which they are performed.
    - Static self-stretching involves holding a stretch at the end of the ROM, while dynamic self-stretching involves controlled movement through a full ROM.
    - **Warm-up and performance:** Dynamic self-stretching is often used as part of a warm-up routine before physical activity. Research suggests that dynamic stretching can enhance muscle performance, increase blood flow, and improve neuromuscular coordination when performed correctly.

Effects

- **Muscle soreness and recovery:** Self-stretching, especially after intense exercise, may help reduce muscle soreness and promote recovery. Gentle stretches that target the muscles used during the workout can alleviate post-exercise discomfort.
- **Injury prevention:** Incorporating self-stretching into a regular fitness routine can contribute to injury prevention by maintaining optimal muscle length and joint mobility. It may help prevent muscle imbalances and improve overall functional movement.
- **Effect on muscle performance:** Some studies suggest that static self-stretching before certain activities, particularly activities requiring explosive strength, might temporarily reduce muscle strength and power. However, dynamic self-stretching or post-exercise stretching does not seem to have a significant negative impact on muscle performance.
- **Relaxation and flexibility maintenance:** The timing and duration of self-stretching sessions can impact its effectiveness. Pre-exercise dynamic stretching is recommended before physical activities that require dynamic movements, while static self-stretching is often performed post-exercise for relaxation and flexibility maintenance.
- **Managing chronic conditions:** Self-stretching can play a role in managing chronic conditions that affect flexibility and joint health, such as arthritis. It can also contribute to maintaining musculoskeletal health as individual's age.

Mechanical Stretching

Mechanical stretching refers to the application of external mechanical forces to tissues, such as muscles, tendons, ligaments, and joints, with the goal of inducing elongation, improving flexibility, promoting tissue adaptation, and influencing physiological responses. This type of stretching involves using mechanical devices, equipment, or resistance to create controlled tension or compression in the targeted tissues. Mechanical stretching can be performed manually by a practitioner, such as a physical therapist, or through the use of specialized devices designed to apply specific forces.

Studies on the effectiveness of the two categories of mechanical loading devices base their effectiveness on the quick-acting soft tissue qualities of creep or stress- relaxation as well.

A patient may be able to regulate and alter the load (stretch force) during a stretching session with the help of certain devices, such as the Joint Active Systems™ adjustable orthosis. With other devices, the load is set before the splint is applied, and it stays that way while the splint is in place. Compared to manual stretching or self-stretching activities, mechanical stretching requires a significantly longer overall stretch duration. With the exception of time spent away from the device for exercise and personal hygiene, mechanical stretches have been reported to last anywhere from 15–30 minutes to as long as 8–10 hours at a time. Serial casts are put on and taken off repeatedly over the course of days or weeks.

Effects

There are various methods and modalities of mechanical stretching, each with its own approach, effect and purpose:

- **Stretching devices:** Some devices are designed to mechanically stretch specific tissues. For example, devices for spinal traction apply controlled forces to the spine to alleviate pressure on discs and nerves.
- **Orthopedic devices:** Medical braces, splints, and orthoses can provide controlled mechanical stretching to aid in tissue healing, prevent contractures, and maintain joint mobility during rehabilitation.
- **Biomechanical models:** Researchers use computer simulations and biomechanical models to study the effects of mechanical stretching on tissues virtually, providing insights into how forces are distributed within the body.

Mechanical stretching takes advantage of the body's ability to adapt to mechanical stimuli. The application of mechanical stretching can vary widely based on individual needs, goals, and conditions. Seeking guidance from healthcare professionals or qualified practitioners is recommended to ensure safe and effective use of mechanical stretching techniques.

When tissues experience controlled mechanical loading, they respond by remodeling, increasing protein synthesis, and adapting to the applied forces. This can lead to improved tissue integrity, increased flexibility, and enhanced overall function. However, it's important to apply mechanical stretching with care and under appropriate guidance to prevent overstretching, tissue damage, or other adverse effects.

Ballistic Stretching

Ballistic stretches involves using bouncing or jerking movements to force a muscle beyond its normal ROM. The goal to move at maximum speed and the acceleration of a mass during the course of a movement are characteristics of ballistic exercise (BE). While they can be carried out without bulky, heavy equipment, these activities nonetheless aim to accomplish maximum motor unit recruitment.

Effects

The BE shows promise in improving performance in high-intensity activities like sprints and jumps.

Risks

Since ballistic stretching can activate the stretch reflex, many people have postulated that ballistic stretching has a greater potential to cause muscle or tendon damage, especially in the tightest muscles.

Dynamic Stretching

Dynamic stretching is a more useful form of stretching when the limbs are stretched beyond their normal ROM through the use of sport-specific movements. In general, swinging, jumping or excessive motions that cause the limbs to go to or beyond their normal ROM and trigger a proprioceptive reflex reaction are characteristics of dynamic stretching. The nerves that triggered the muscle cells may be facilitated by the appropriate activation of the proprioceptors. The muscle can contract more quickly and forcefully as a result of this facilitation, which speeds up nerve firing. Dynamic stretching is demonstrated to be beneficial for enhancing sports performance because it raises muscle temperature and proprioceptive activation. Ballistic stretching is not to be confused with dynamic stretching.

Even though they both require repetitive motions, ballistic motions are quick, bouncy motions with limited ranges of motion toward the conclusion of the ROM. For athletes requiring running or jumping performance during their sport such as basketball players or sprinters, dynamic structure are extensively used in training program. Dynamic stretching involves active movements that take a joint or muscle through a full ROM. These stretches are typically performed in a controlled manner and mimic the movements of a specific activity or sport.

Effects

- Restore flexibility and bodily functionality
- Increase neuromuscular control by doing repeating movements, which will improve muscle compliance, motor control, and the speed at which nerve messages are transmitted.
- Raises the body's core temperature
- Quickens the production of energy
- Enhances performance metrics like strength and speed.

Proprioceptive Neuromuscular Facilitation Stretching

Proprioceptive neuromuscular facilitation (PNF) stretching, is a popular flexibility training technique that lengthens muscles by causing them to relax. Originally designed to treat neurological dysfunctions, this technique was created in the 1940s with the goal of increasing muscular length and strength through regulated contractions. Strength, joint stability, and neuromuscular control are among its many advantages; these are especially helpful for disorders requiring knee, shoulder, hip, and ankle rehabilitation. PNF stretching is well known for its potential to improve general functional capacity.

Types of PNF Stretching

1. **Contract relax PNF:** To avoid fatigue or damage, the contract relax PNF stretching technique involves stretching the tight muscle and then contracting it isometrically for

at least 3 seconds at submaximal effort. To get a longer stretch, the patient releases the muscle after contracting it and engages the opposing muscle.

2. **Hold relax technique:** Similar to contract relax, but applied in situations where the agonist muscle is insufficiently strong, the hold relax technique depends on the Golgi tendon organ to allow the muscle to relax after a prolonged contraction.
3. **Hold-relax agonist:** Another PNF technique called hold-relax agonist lengthens tense muscles by contracting the opposing muscle, which promotes relaxation and elongation by reciprocal inhibition. This method is based on cues such as "Hold" and its goal is to increase passive ROM.

Clinical Correlation

Factors affecting stretching

1. **Viscoelastic properties:** Muscles and tendons exhibit viscoelastic properties, meaning their behavior combines elements of both elasticity and viscosity. Elasticity refers to the ability of a tissue to return to its original shape after being stretched, while viscosity refers to its resistance to deformation. Stretching helps improve the extensibility of these tissues, allowing them to withstand greater loads and elongation without causing damage.
2. **Plastic deformation:** If a stretch is taken beyond the tissue's elastic limit, plastic deformation may occur. This means that the tissue undergoes permanent structural changes and may result in injury. It is essential to stretch within a safe and comfortable range, respecting the body's limitations.
3. **Temperature and warm-up:** Temperature plays a role in stretching biomechanics. Warm muscles and soft tissues are more pliable and less prone to injury. This is why it is recommended to perform a warm-up before stretching to increase tissue temperature and prepare the body for more extensive ranges of motion.

Effects

The PNF stretching improves muscle flexibility and relaxation by using physiological responses. Autogenic inhibition is achieved by using isometric and concentric contractions prior to a passive stretch. This causes the same muscle part to relax. The target muscle is further relaxed via reciprocal inhibition, which is attained by contractions in opposing muscle groups. Initially, these methods involve maintaining a passive stretch for around ten seconds. The mechanism of PNF stretching is explained as follows:

1. **Activation of muscle spindles:** Muscles contain specialized sensory receptors called muscle spindles. These receptors detect changes in muscle length and the rate of change. When a muscle is stretched, the muscle spindles are activated and send signals to the spinal cord.
2. **Stretch reflex:** The signals from the muscle spindles trigger a reflex called the stretch reflex. The spinal cord receives the signals and immediately sends back a response to the stretched muscle, causing it to contract. This reflex contraction is an automatic protective mechanism designed to prevent overstretching and potential injury.
3. **Reciprocal inhibition:** Along with the stretch reflex, another phenomenon called reciprocal inhibition occurs. When a muscle is actively contracted, the antagonistic muscle (the one opposing the movement) relaxes. During stretching, the antagonist muscle relaxes through reciprocal inhibition, allowing the target muscle to elongate further.
4. **Tendon stretch:** In addition to the muscles, stretching also affects tendons, which connect muscles to bones. Tendons are less elastic than muscles and offer resistance to stretching. As tension is applied, the tendons gradually elongate and contribute to the overall stretching effect.

PROCEDURE

1. Select the muscle to be stretched and the most appropriate stretching technique based on the goals and desired outcomes.
2. Explain the purpose and procedure to the patient to ensure relaxation and confidence.
3. Application of warm-up to increase the extensibility of the tissues to be stretched and to avoid injury.
4. Identify the best position for stretching to be performed. Ensure patient is wearing a loose comfortable clothing.
5. Gently move the body part through full range till the point of restriction.
6. Start the stretch gently and gradually increase the intensity. Avoid bouncing or jerking, as this can lead to injury. As a therapist, hold the stretch, a mild tension or pull may be experienced, but it should not be painful. The direction of the stretching movement is directly opposite the line of pull of the range-limiting muscle. Hold the stretch for minimum 30 seconds or longer.
7. Stabilize the proximal body part and apply gentle stretch at the distal segment. For a multi-joint muscle, stretch the muscle over one joint at a time and then over all joints simultaneously until the optimal length of soft tissues is achieved. During the movement, ensure a firm yet comfortable grasp.
8. When tension reduces, move the part a little farther to progressively lengthen the hypomobile tissues.
9. Gradually release the stretch force and allow the patient and therapist to rest momentarily while maintaining the range-limiting tissues in a comfortably elongated position. Then repeat the sequence several times.

10. Maintaining the tissues in a comfortably extended posture, gradually reduce the stretching effort.

Clinical Correlation

Clinicians should recognize that individuals may respond differently to stretching based on factors such as age, tissue elasticity, and previous injury history. Adjusting stretching parameters and techniques based on individual characteristics can improve outcomes and minimize the risk of adverse effects.

TECHNIQUES

Upper Limb Muscles

Shoulder Extensors

- **Goal:** To increase shoulder flexion range.
- **Position of the patient:** Supine lying.
- **Method:**
  - The therapist grasps the back of the lower arm just above the elbow.
  - For the latissimus dorsi, support the lateral aspect of the thorax and upper pelvis part of pelvis.
  - For the teres major, stabilize the axillary border of the scapula.
  - To lengthen the shoulder extensors, fully flex the patient's arm upward (Fig. 10.4).

Shoulder Flexors

- **Goal:** To increase shoulder extension range
- **Position of the patient:** Prone lying
- **Method:**
  - The therapist holds the forearm with one hand.
  - Scapula may be stabilized with the other hand.
  - To lengthen the shoulder flexors, extend the patient's arm in upward direction.

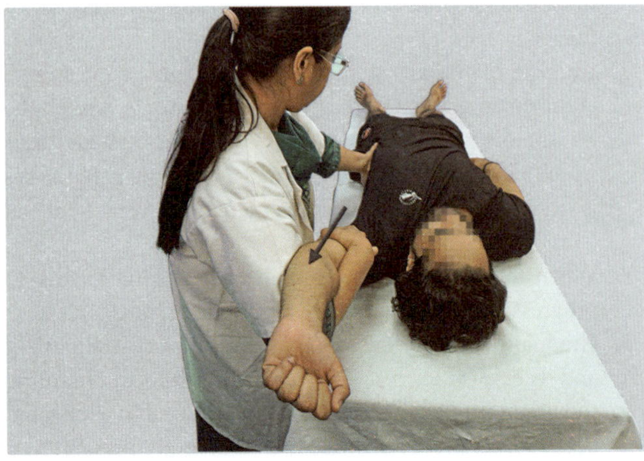

Fig. 10.4: Shoulder extensors stretching—supine position

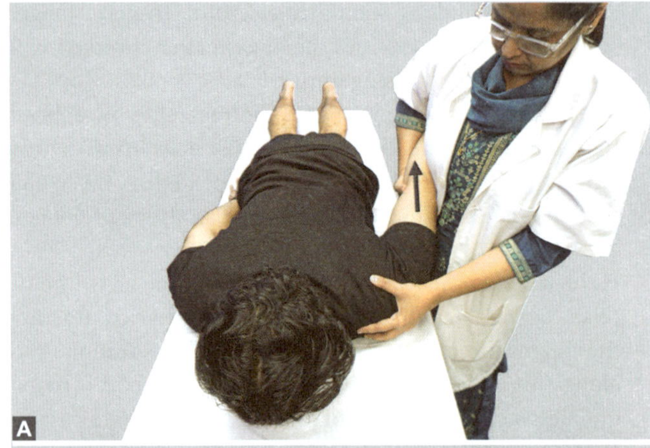

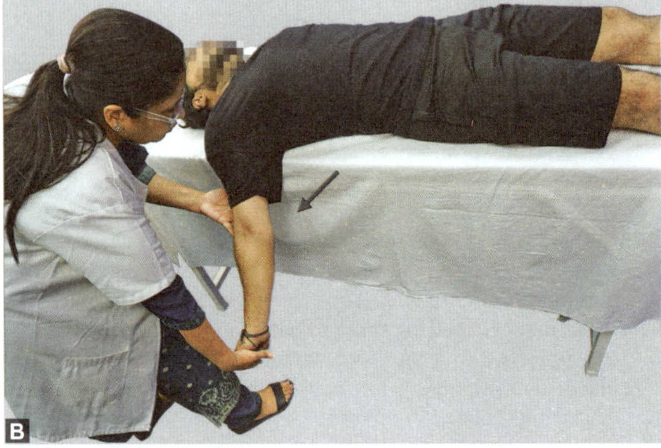

Figs 10.5A and B: Shoulder flexors stretching: **A.** Prone lying; **B.** Supine position

- **Alternative position:** The patient is in supine lying position.
- **Method:**
  - The therapist holds the patient's distal forearm with one hand and supports back of the lower arm with other hand.
  - To fully stretch the shoulder flexors, gently extend the patient's arm in a downward direction while keeping the elbow straight, ensuring that the shoulder is fully extended (Figs 10.5A and B).

Shoulder Adductors

- **Goal:** To increase shoulder abduction range
- **Position of the patient:** Supine lying
- **Method:**
  - Bend the patients elbow to a 90° angle.
  - The therapist holds the back of the lower arm with one hand.
  - The other hand provides stability at the axillary border of the scapula.
  - To lengthen the shoulder adductors, raise the patient's arms into full abduction (Figs 10.6A and B).

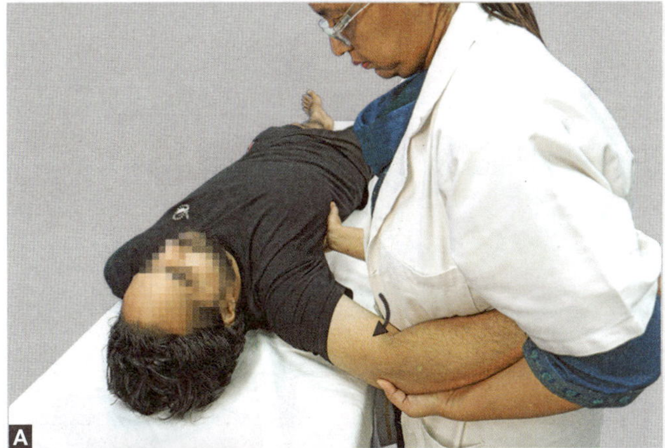

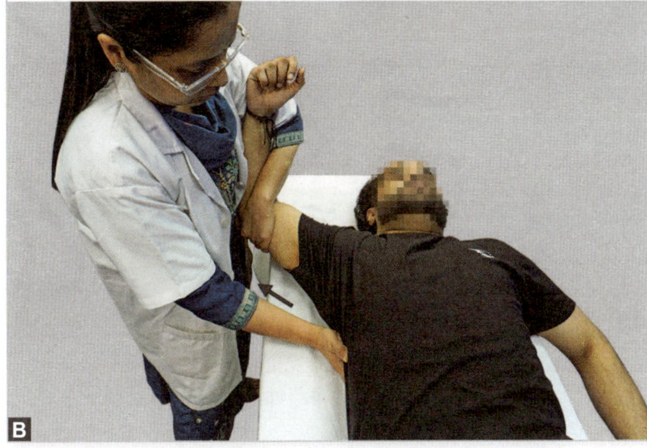

Figs 10.6A and B: Shoulder adductors stretching—supine position

Shoulder Abductors

- **Goal:** To increase shoulder adduction range
- **Position of the patient:** Supine lying
- **Method:**
  - The therapist supports the patient's arm at the elbow and holds the back of the lower arm.
  - To stabilize the scapula, the therapist uses the other hand.
  - To stretch the shoulder abductors, gently move the patient's arm across the body toward the opposite side, keeping the elbow straight.
  - Ensure that the shoulder is not elevated during the stretch.

Shoulder Internal Rotators

- **Goal:** To increase shoulder external rotation range
- **Position of the patient:** Supine lying
- **Method:**
  - Place the arm by the patient's side or elevate the shoulder to a comfortable angle (30° or 45°, increasing to 90° in the event that the GH joint is secure). Bend the elbow to a 90° angle.

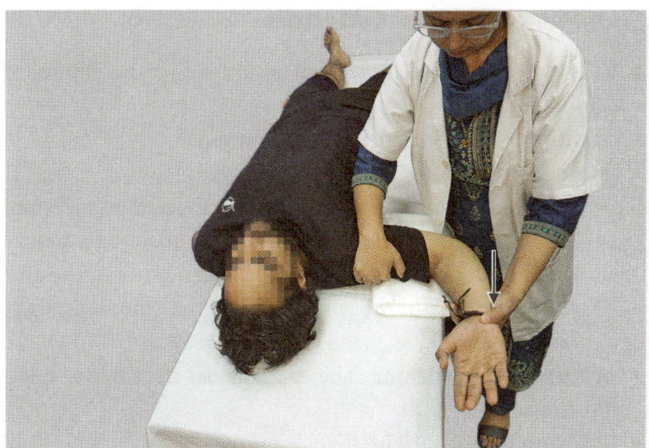

Fig. 10.7: Shoulder internal rotators stretching—supine position

  - With one hand, the therapist grasps the distal forearm. The table offers stability to the scapula.
  - To lengthen the internal rotators, externally rotate the patient's forearm bringing it closer to the table (Fig. 10.7).

Shoulder External Rotators

- **Goal:** To increase shoulder internal rotation range
- **Position of the patient:** Supine lying
- **Method:**
  - Place the arm by the patient's side or elevate the shoulder to a comfortable angle (30° or 45°, increasing to 90° in the event that the GH joint is secure).
  - Bend the elbow to a 90° angle.
  - With one hand the therapist grasps the distal forearm. The table offers stability to the scapula.
  - To lengthen the external rotators, internally rotate the patient's forearm bringing it closer to the table (Fig. 10.8).

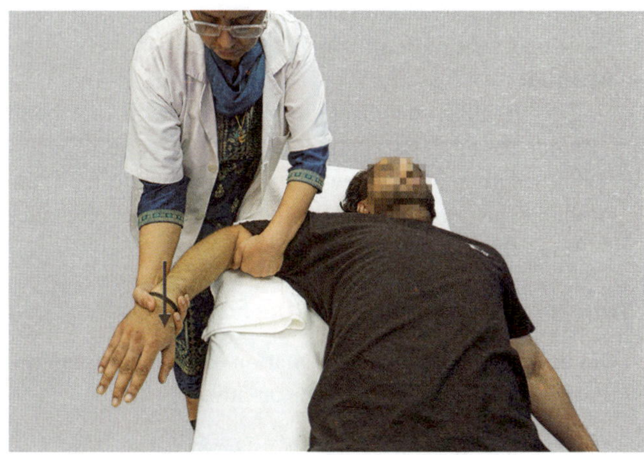

Fig. 10.8: Shoulder external rotators stretching—supine position

Shoulder Horizontal Adductors

- **Goal:** To increase the horizontal abduction range of shoulder
- **Position of the patient:** Sitting or supine
- **Method:**
  - With both arms, the therapist holds the patient's elbows.
  - To stretch the horizontal adductors bilaterally, gently move both arms out to the sides and slightly backward, away from the body, while keeping the elbows straight (Fig. 10.9A).
- **Alternative position:** The patient is in supine lying position.
- **Position of the patient:**
  - Place the patient's shoulder at the edge of the table with 60°–90° abduction.
  - Flex the elbow optionally to attain full horizontal abduction.
- **Method:**
  - The therapist grasps the front of the lower arm.
  - With the other hand, the therapist stabilizes the shoulder.
  - To lengthen horizontal adductors, move the patient's arm fully horizontally abducted below the edge of the table (Fig. 10.9B).

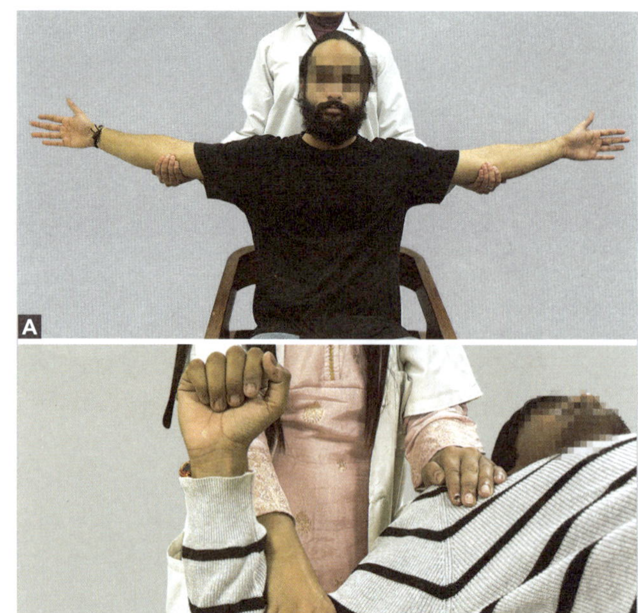

Figs 10.9A and B: Shoulder horizontal adductors stretching: **A.** Sitting position; **B.** Supine position

Elbow Extensors

- **Goal:** To increase elbow flexion range
- **Position of the patient:** Supine lying or sitting
- **Method:**
  - The therapist stabilizes the shoulder in maximum flexion with one hand and grasps the distal forearm or hand with the other hand.
  - To lengthen the extensors, flex the elbow and hold onto the forearm while keeping the shoulder in flexion (Fig. 10.10).

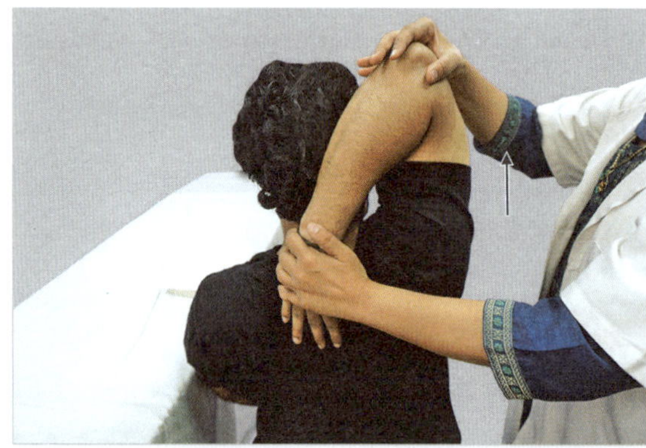

Fig. 10.10: Elbow extensors stretching—sitting position

Elbow Flexors

- **Goal:** To increase elbow extension range
- **Position of the patient:** Supine
- **Method:**
  - The therapist stabilizes the shoulder with one hand while the other hand grasps the distal forearm.
  - To lengthen the elbow flexors, straighten/extend the elbow just a bit past the point of tissue resistance (Fig. 10.11).

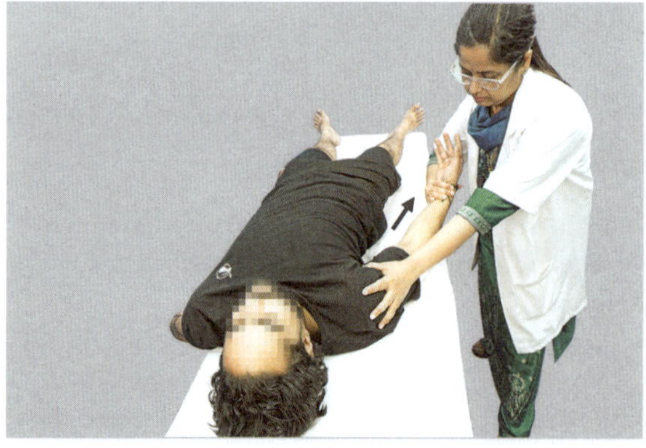

Fig. 10.11: Elbow flexors stretching—supine position

Pronators and Supinators in the Forearm

- **Goal:** To increase forearm pronation or supination range
- **Position of the patient:** Supine or sitting
- **Method:**
  - With the elbow flexed to 90° and the humerus resting on the supporting surface, the therapist grasps the distal forearm.

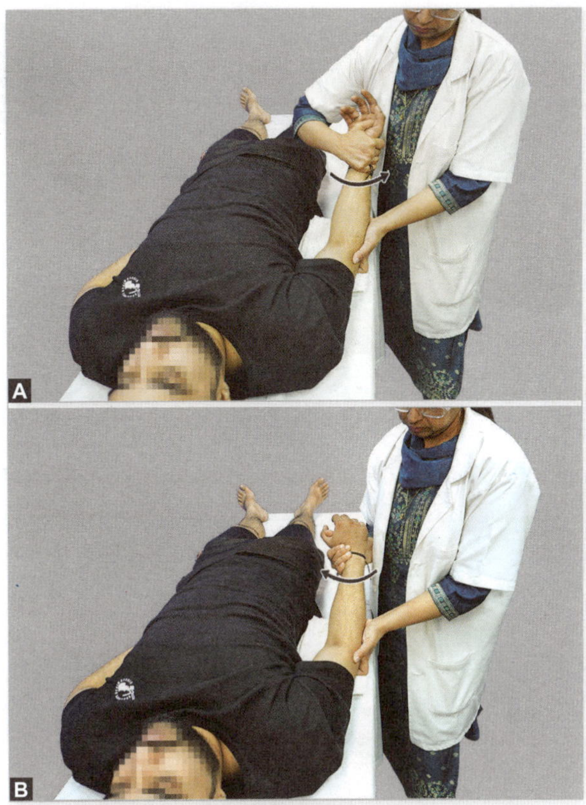

Figs 10.12A and B: A. Pronators stretching—supine position; B. Supinators stretching—supine position

- The other hand may be used to stabilize the elbow into flexion.
- To lengthen either muscles, rotate the volar aspect of the forearm away from the therapist for pronators stretching (increasing supination range) and toward the therapist for supinators stretching (increasing pronation range) (Figs 10.12A and B).

Wrist Extensors

- **Goal:** To increase wrist flexion.
- **Position of the patient:** Supine or sitting
- **Method:**
  - Place the forearm in supinated position.
  - The therapist grasps the patient's hand from the dorsal aspect with one hand while other hand stabilizes the distal forearm.
  - To lengthen wrist extensors, flex the patient's wrist past the tissue resistance (Fig. 10.13).

Wrist Flexors

- **Goal:** To increase wrist extension
- **Position of the patient:** Supine or sitting
- **Method:**
  - Place the forearm in pronated position.

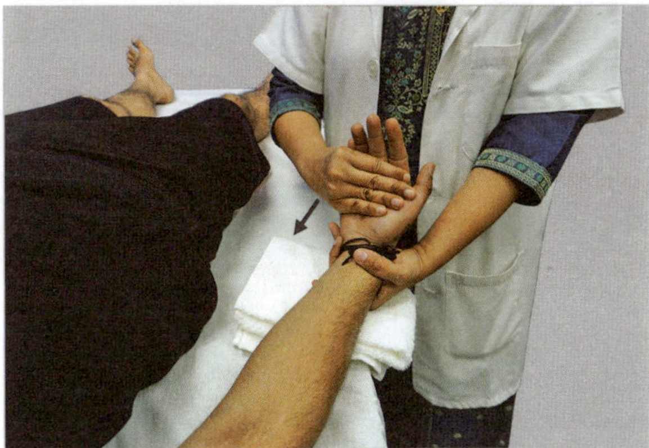

Fig. 10.13: Wrist extensors stretching—supine position

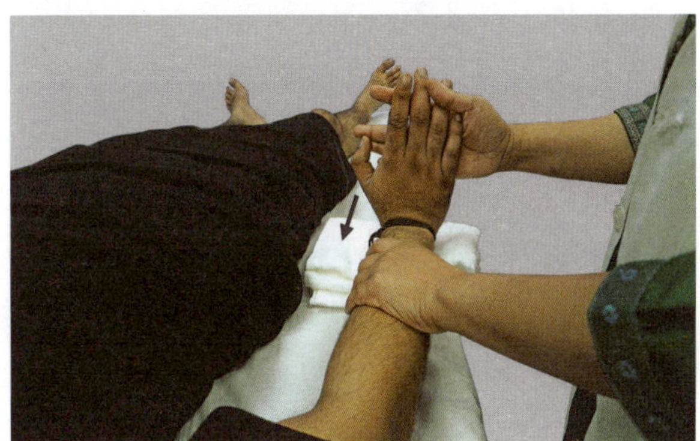

Fig. 10.14: Wrist flexors stretching—supine position

  - The therapist grasps the patient's hand from the volar aspect with one hand while other hand stabilizes the distal forearm.
  - To lengthen wrist flexors extend the patient's wrist past the tissue resistance (Fig. 10.14).

Ulnar Deviators

- **Goal:** To increase wrist radial deviation range
- **Position of the patient:** Supine or sitting
- **Method:**
  - With one hand, the therapist stabilizes the distal forearm while the other hand grasps the radial aspect of the hand.
  - To lengthen the ulnar deviators, move the wrist radially while maintaining forearm stability (Fig. 10.15).

Radial Deviators

- **Goal:** To increase ulnar deviation range
- **Position of the patient:** Supine or sitting
- **Method:**
  - With one hand, the therapist stabilizes the distal forearm while the other hand grasps the radial aspect of the hand.
  - To lengthen the radial deviators, ulnarly deviate the wrist while maintaining forearm stability (Fig. 10.16).

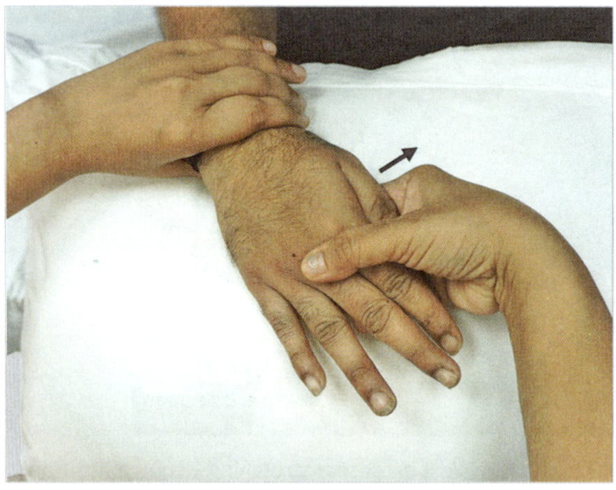

Fig. 10.15: Ulnar deviators stretching—sitting position

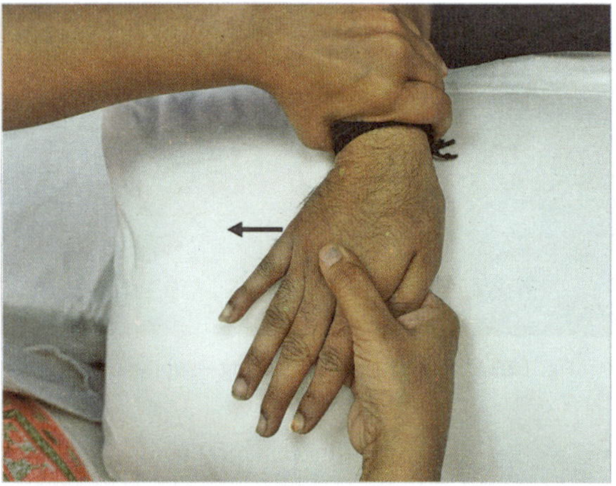

Fig. 10.16: Radial deviators stretching—sitting position

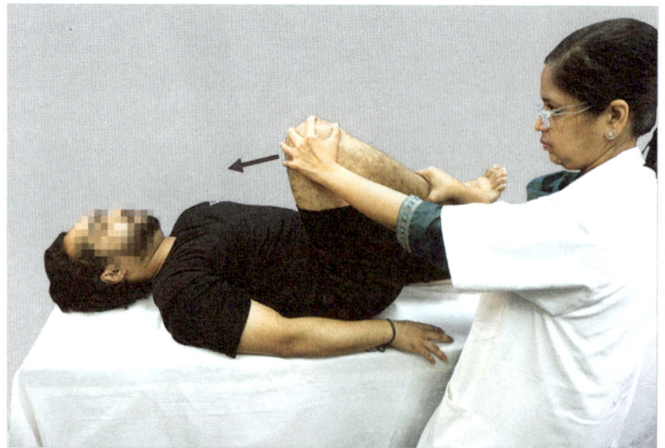

Fig. 10.17: Hip extensor (gluteus maximus) stretching—supine position

Fig. 10.18: Hip extensor stretching (hamstring muscles)—supine position

Lower Limb Muscles

Hip Extensor (Gluteus Maximus)

- **Goal:** To increase hip flexion range with knee flexed
- **Position of the patient:** Supine lying
- **Method:**
  - With one hand, the therapist grasps the distal part of the leg and with other supports the distal thigh.
  - To lengthen hip extensors, flex the knee and hip at the same time (Fig. 10.17).

Hip Extensors (Hamstring Muscles)

- **Goal:** To increase hip flexion range with knee extended
- **Position of the patient:** Supine lying
- **Method:**
  - With one hand, the therapists-stabilizes the knee in extension while the other hand holds the distal part of the leg
  - To lengthen hip extensors, flex the hip as far as possible with knee kept extended (Fig. 10.18).

Hip Flexors (Iliopsoas)

- **Goal:** To increase hip extension
- **Patient position:** Supine lying
- **Method:**
  - With one hand, the therapist stabilizes the opposite extremity in the direction of the chest by holding it from the distal thigh.
  - To lengthen the hip flexors, the therapist presses downward on the distal thigh with the other hand (Fig. 10.19).

Hip Flexors (Rectus Femoris)

- **Goal:** To increase hip extension
- **Patient position:** Prone lying with knee in flexion

Fig. 10.19: Hip flexors (Iliopsoas) stretching—supine position

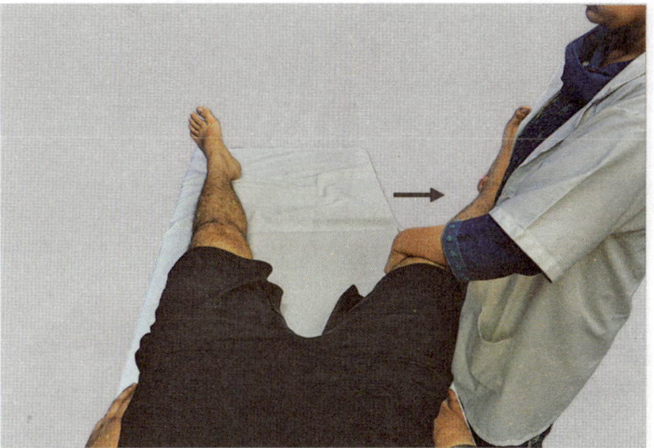

Fig. 10.21: Hip adductors stretching—supine position

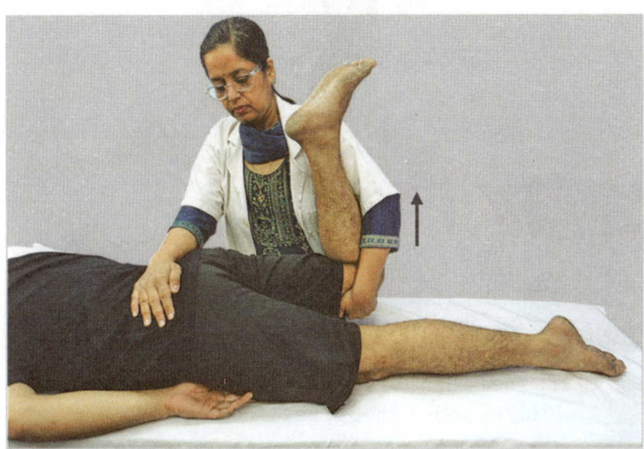

Fig. 10.20: Hip flexors (Rectus femoris) stretching—prone position

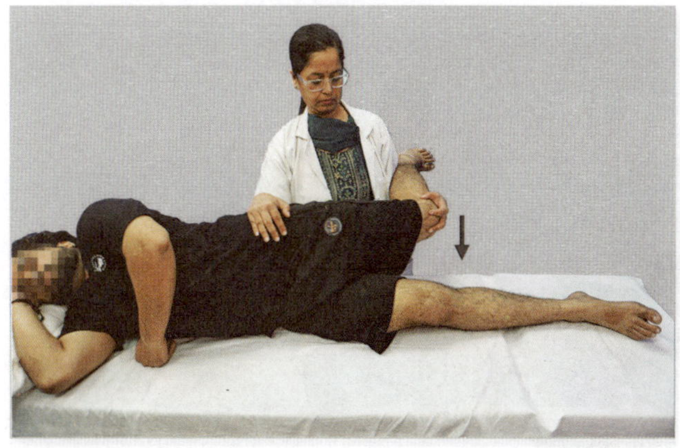

Fig. 10.22: Hip abductor (TFL) stretching—side lying position

- **Method:**
  - With one hand, the therapist stabilizes the pelvis while the other hand holds the distal part of the thigh.
  - To lengthen the hip flexors, the therapist extends the hip as far as possible (Fig. 10.20).

Hip Adductors

- **Goal:** To increase hip abduction range
- **Patient position:** Supine lying
- **Method:**
  - With one hand, the therapist stabilizes the knee in extension while the other hand holds the distal part of the leg.
  - To lengthen the hip adductors, the therapist abducts the extremity by moving it outwards (Fig. 10.21).

Hip Abductors Tensor Fascia Latae (TFL)

- **Goal:** To increase hip adduction range
- **Position of the patient:** The patient should lie on their side with their hips extended, knee and bottom hip flexed.
- **Methods:**
  - With one hand, the therapist stabilizes the pelvis while the other hand holds the distal femur.
  - To lengthen TFL, extend and adduct the hip while applying pressure on distal femur in downward direction (Fig. 10.22).

Hip Internal Rotators

- **Goal:** To increase hip external rotation range
- **Position of the patient:** Prone, hips extended, knees flexed 90°.
- **Method:**
  - With one hand, the therapist stabilizes the pelvis while the other hand graps the distal part of leg.
  - To lengthen internal rotators, apply pressure from the outside to rotate the hip externally (Fig. 10.23).

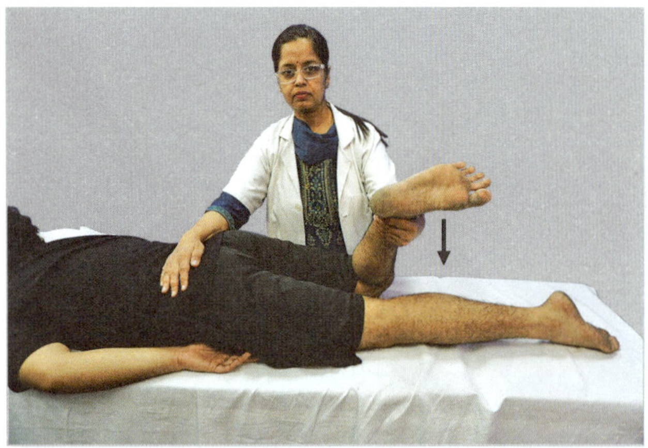

Fig. 10.23: Hip internal rotators stretching—prone position

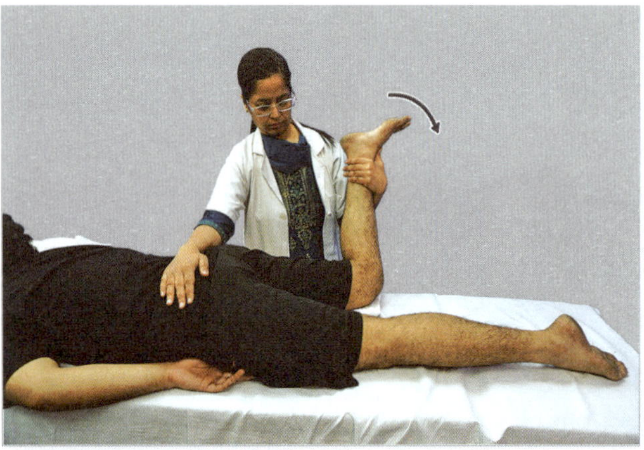

Fig. 10.24: Hip external rotators stretching—prone position

Hip External Rotators

- **Goal:** To increase hip internal rotation range
- **Position of the patient:** Prone, hips extended, knees flexed to 90°.
- **Method:**
  - With one hand, the therapist stabilizes the pelvis while the other hand grasps the distal part of leg.
  - To lengthen external rotators, apply pressure from the inside to rotate the hip internally (Fig. 10.24).

Hip External Rotators (Piriformis)

- **Goal:** To increase hip internal rotation range
- **Position of the patient:** Supine lying
- **Method:**
  - The therapist grasps the patient's distal leg with one hand, while the other hand holds the distal thigh.
  - To lengthen piriformis, gently guide the patient's knee across the body toward the opposite shoulder (Fig. 10.25).

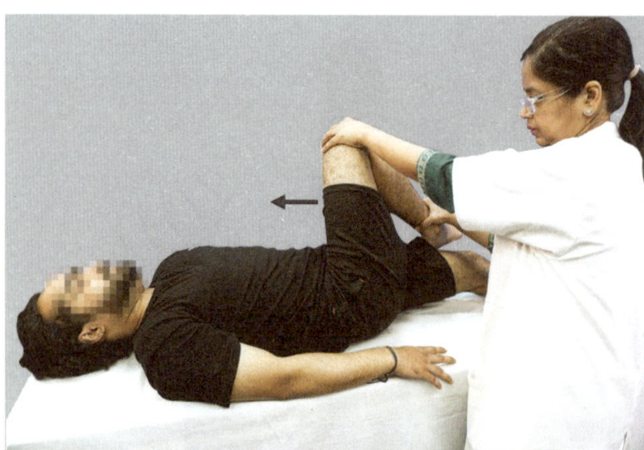

Fig. 10.25: Hip external rotator (Piriformis) stretching—supine position

Fig. 10.26: Knee extensors stretching (Quadriceps)—prone lying, hip extended, knee to be flexed beyond 90°

Knee Extensors (Quadriceps)

- **Goal:** To increase knee flexion range.
- **Position of the patient:** Prone, hips extended, knees flexed beyond 90°.
- **Method:**
  - Place a towel roll beneath the leg above the knee.
  - With one hand, the therapist stabilizes the pelvis while the other hand grasps the distal part of leg.
  - To lengthen knee extensors, flex the knee while holding onto the anterior portion of the distal tibia (Fig. 10.26).

Knee Flexors

- **Goal:** To increase knee extension range
- **Position of the patient:** Prone lying, hips extended, knees flexed 90°.
- **Method:**
  - Place a towel roll beneath the leg above the knee.
  - With one hand, the therapist stabilizes the pelvis while the other hand grasps the distal part of leg.

- To lengthen knee flexors, extend the knee while holding onto the anterior portion of the distal tibia (Fig. 10.27).
- **Alternative position:** The patient is in supine lying position.
- **Method:**
  - Place a towel roll beneath the leg above the knee.
  - With one hand the therapist stabilizes the distal thigh while the other hand grasps the distal part of leg.
  - To lengthen knee flexors extend the knee while holding onto the distal portion of the tibia (Fig. 10.28).

Plantar Flexors (Gastrocnemius Muscle)

- **Goal:** To increase ankle dorsiflexion while extending the knee.
- **Position of the patient:** Supine lying
- **Method:**
  - With one hand, the therapist grasps the patient's heel while maintaining the subtalar joint in a neutral position.
  - The forearm supports the plantar aspect of the foot.
  - Using the other hand, stabilize the tibia's front aspect.

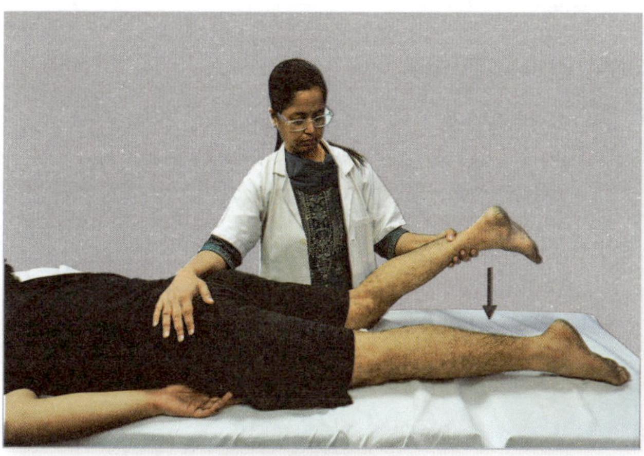

Fig. 10.27: Knee flexors stretching—prone position

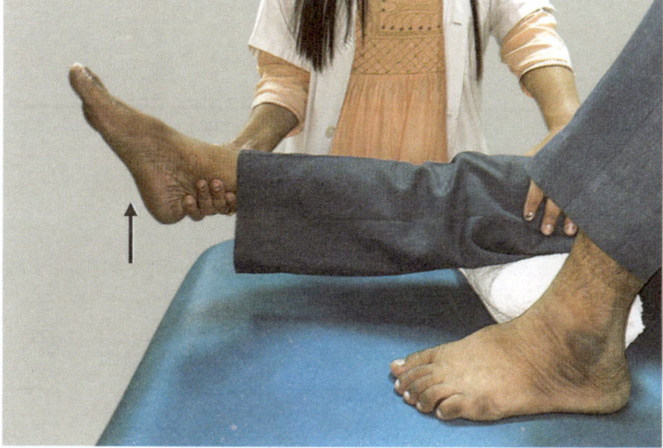

Fig. 10.28: Knee flexors end range stretching—supine position

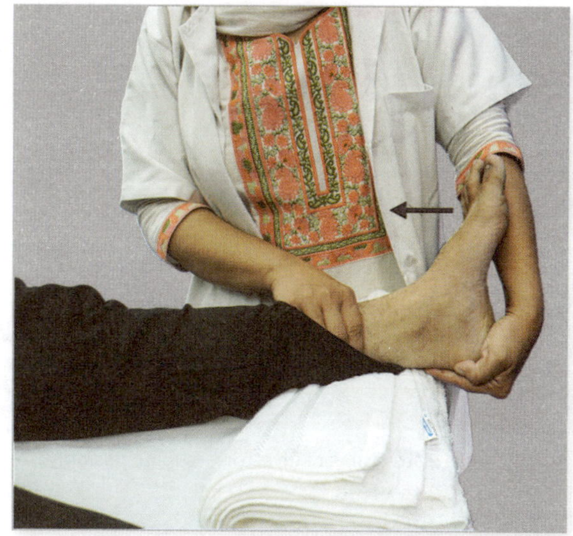

Fig. 10.29: Plantar flexor stretching—supine position

  - To lengthen plantar flexors, the therapist pushes the calcaneal in downward direction and pushes the foot upward in dorsiflexion using forearm (Fig. 10.29).
  - For stretching soleus, keep the knee in flexed position. Rest of the procedure will remain same.

Dorsiflexors

- **Goal:** To increase ankle plantarflexion.
- **Position of the patient:** Supine lying
- **Method:**
  - The therapist uses one hand to support the posterior aspect of the distal tibia.
  - With the other hand, the therapist grasps the foot between the metatarsal and tarsal regions.
  - To stretch the dorsiflexors and increase ankle plantarflexion, apply a gentle stretch force to the anterior portion of the foot while simultaneously plantar flexing the foot as much as possible (Fig. 10.30).

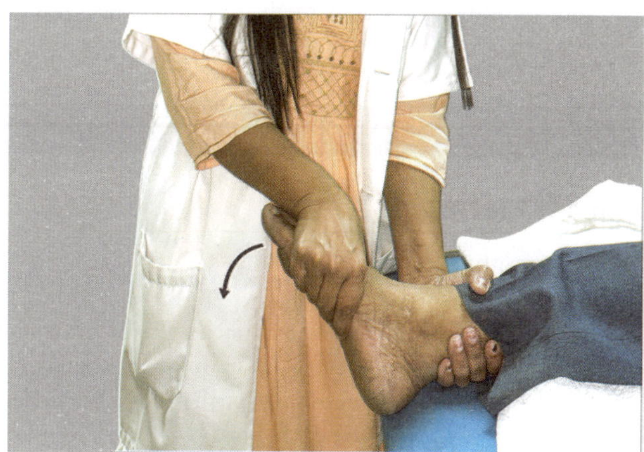

Fig. 10.30: Stretching of dorsiflexors—supine position

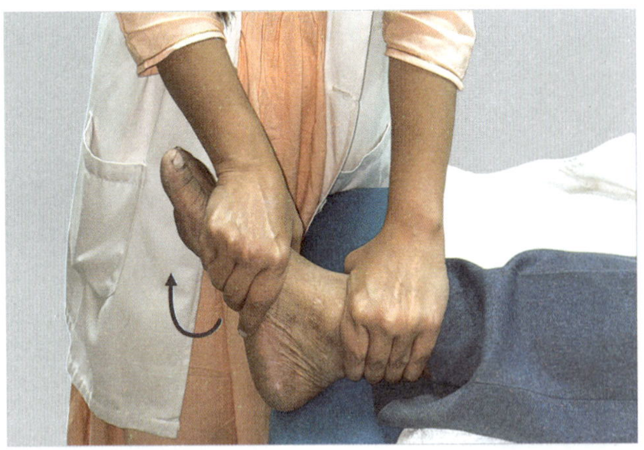

Fig. 10.31: Stretching of invertors—supine position

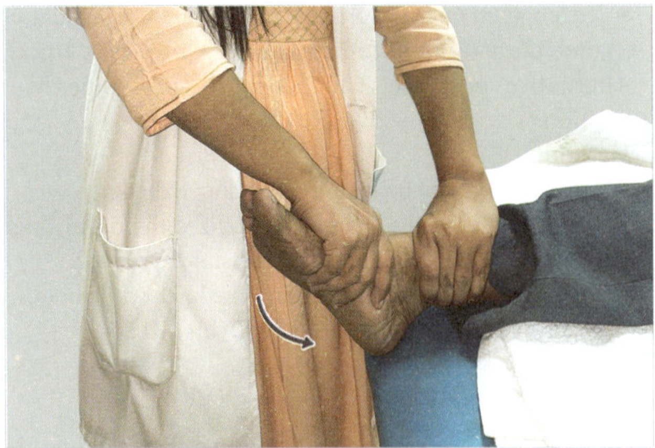

Fig. 10.32: Stretching of evertors

Invertors

- **Goal:** To increase ankle eversion range.
- **Position of the patient:** Supine lying.
- **Method:**
  - The therapist supports the distal tibia with one hand, stabilizing the ankle.
  - With the other hand, the therapist grasps the foot at the metatarsal region.
  - To stretch the invertors, gently apply a force to evert the foot by moving the foot laterally away from the midline (Fig. 10.31).

Evertors

- **Goal:** To increase ankle inversion range.
- **Position of the patient:** Supine lying.
- **Method:**
  - The therapist supports the distal tibia with one hand, stabilizing the ankle.
  - With the other hand, the therapist grasps the foot at the metatarsal region.
  - To stretch the evertors, gently apply a force to invert the foot by moving the foot medially toward the midline (Fig. 10.32).

SELF-STRETCHING

Upper Trapezius

- **Position:** Sitting
- **Method:**
  - Sit up straight on a chair or on couch with the head and neck in a neutral position, ears in line with shoulders. Hold the edge of the chair/couch with the upper extremity of the side to be stretched.
  - Keep the contralateral hand on the parietal area. Now side flex the neck to the contralateral side and rotate the neck to the same side.
  - Apply downward pressure from contralateral hand until stretch is felt.
  - Repeat it for 3–5 times or as instructed by the therapist (Fig. 10.33).

Levator Scapulae

- **Position:** Sitting
- **Method:**
  - Sit up straight on a chair or on couch with the head and neck in a neutral position, ears in line with shoulders. Hold the edge of the chair/couch with the upper extremity of the side to be stretched.
  - Keep the contralateral hand on the parietal area.

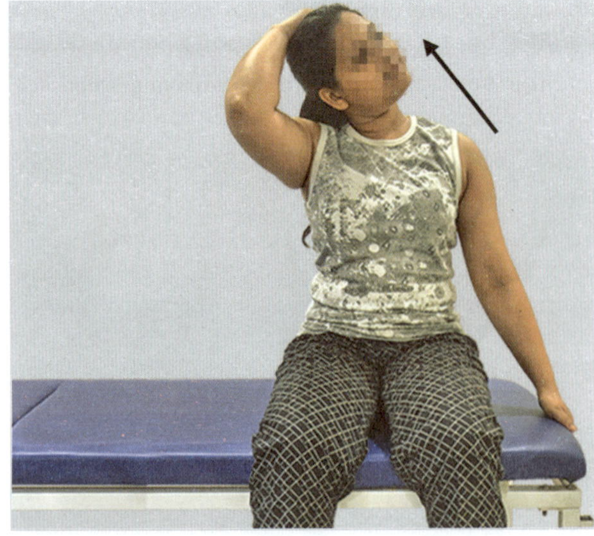

Fig. 10.33: Upper trapezius—self-stretching

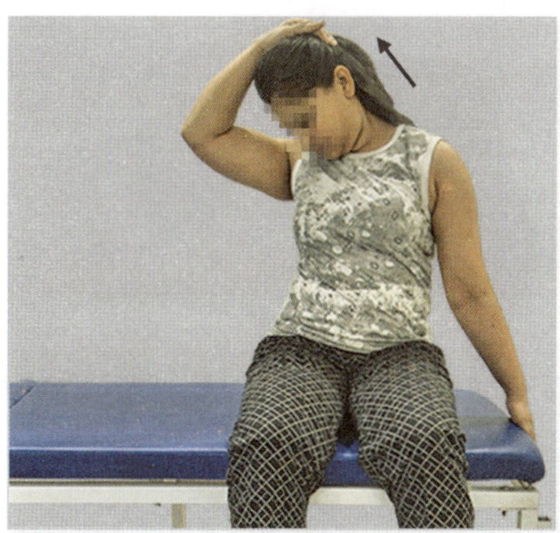

Fig. 10.34: Levator scapulae—self-stretching

- Now flex the neck and rotate the neck to opposite side as if looking toward opposite armpit.
- Apply downward pressure from contralateral hand until stretch is felt.
- Repeat it for 3–5 times or as instructed by the therapist (Fig. 10.34).

Scalenii

- **Position:** Sitting
- **Method:**
  - Sit up straight on a chair or on couch with the head and neck in a neutral position, ears in line with shoulders.
  - Hold the edge of the chair/couch with the upper extremity of the side to be stretched.
  - Keep the contralateral hand on the parietal area.
  - Now side flex the neck to the contralateral side until stretch is felt.
  - Repeat it for 3–5 times or as instructed by the therapist (Fig. 10.35).

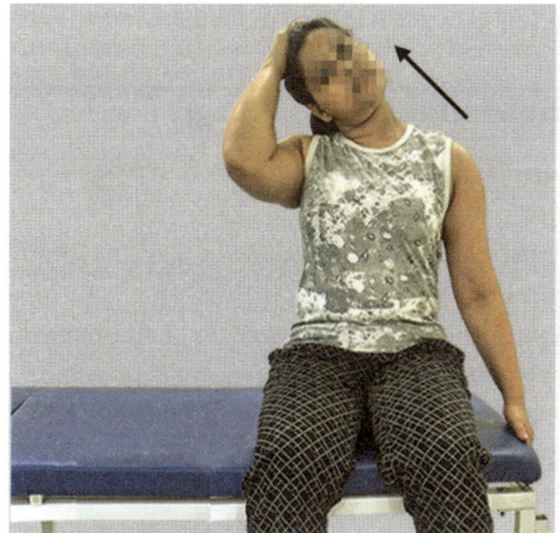

Fig. 10.35: Scalenii—self-stretching

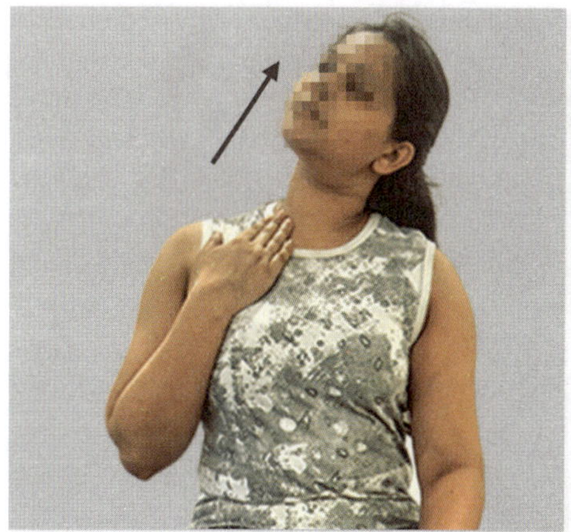

Fig. 10.36: Sternocleidomastoid—self-stretching

Sternocleidomastoid

- **Position:** Sitting
- **Method:**
  - Sit up straight on a chair or on couch with the head and neck in a neutral position, ears in line with shoulders.
  - Keep the contralateral hand over the origin of sternocleidomastoid muscle and side flex the neck to the opposite side and rotate to the ipsilateral side until stretch is felt.
  - Repeat it for 3–5 times or as instructed by the therapist (Fig. 10.36).

Pectoralis Major

- **Position:** Stride standing
- **Method:**
  - The patient is in stride standing position with arm up to the side, abducted at 90° with palms facing forward and resting on the door frame.
  - The patient shifts the weight on the forward leg until the stretch is felt in the chest area (Fig. 10.37).

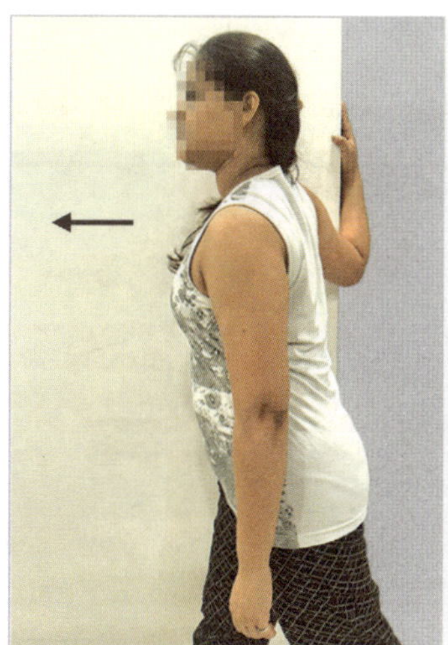

Fig. 10.37: Pectoralis major—self-stretching

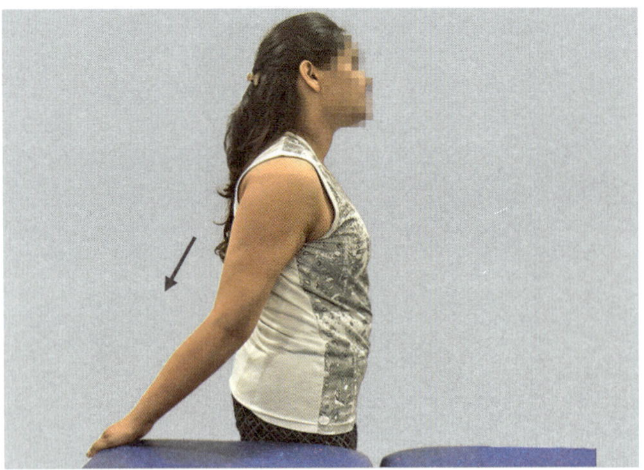

Fig. 10.38: Biceps brachii—self-stretching

Biceps Brachii

- **Position:** Stride standing
- **Method:**
  - The patient is in stride standing position facing away from the edge of the couch and holds the edge with shoulder in full extension.
  - The patient is instructed to shift the weight forward until stretch is felt in the muscle.
  - Repeat it for 3–5 times or as instructed by the therapist (Fig. 10.38).

Triceps Muscle

- **Position:** Sitting or standing
- **Method:**
  - Patient is in sitting or standing position. The patient then abducts the shoulder bending at elbow and keeps the hand at the middle of upper back.
  - The patient places the contralateral hand on the elbow and applies a downward force.
  - Repeat it for 3–5 times or as instructed by the therapist (Fig. 10.39).

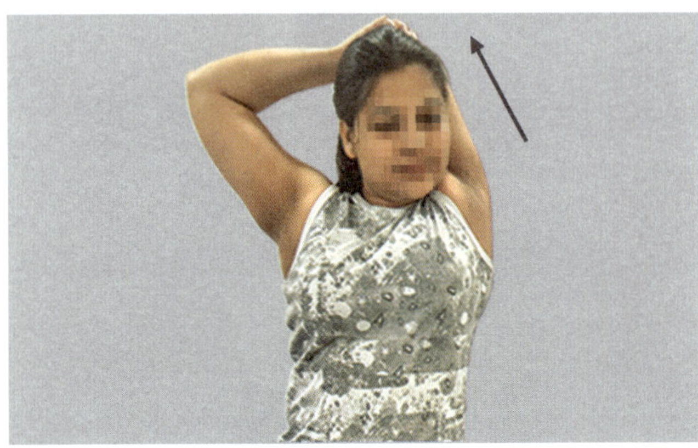

Fig. 10.39: Triceps muscle—self-stretching

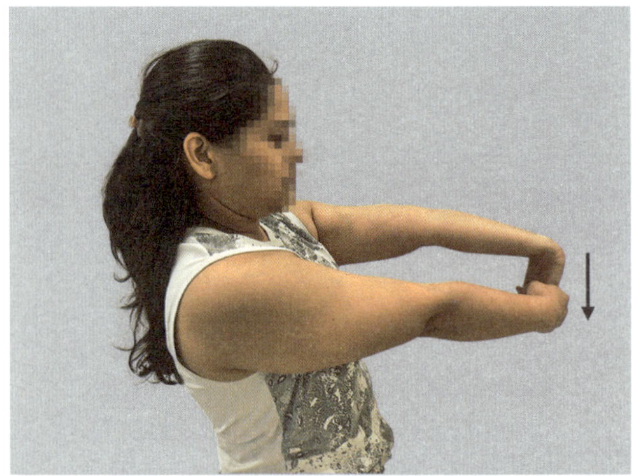

Fig. 10.40: Wrist flexors—self-stretching

Wrist Flexors

- **Position:** Sitting or standing
- **Method:**
  - Patient is in sitting or standing position and the shoulder is flexed at 90°, the elbow is extended, forearm supinated and the wrist extended.
  - The patient holds the palm with the contralateral hand and applies a force in the backward direction till a stretch is felt.
  - Repeat it for 3–5 times or as instructed by the therapist (Fig. 10.40).

Wrist Extensors

- **Position:** Sitting or standing
- **Method:**
  - Patient is in sitting or standing position and the shoulder is flexed at 90°, the elbow is extended, forearm pronated and the wrist flexed.
  - The patient holds the palm with the contralateral hand and applies a force in the backward direction till a stretch is felt.
  - Repeat it for 3–5 times or as instructed by the therapist (Fig. 10.41).

Quadratus Lumborum

- **Position:** Lying on side to be stretched
- **Method:**
  - The patient lies on the side to be stretched and stabilizes upper body on the forearm.
  - Keeping the pelvis stabilized to the couch, the patient tries to push the torso as up right as possible.
  - The patient feels a stretch on the side close to the couch (Fig. 10.42).

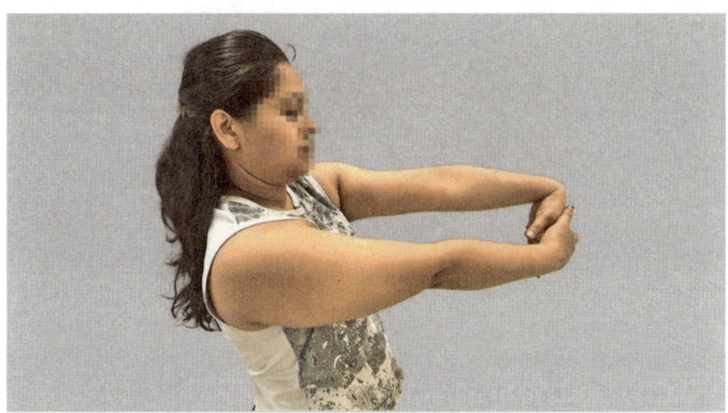

Fig. 10.41: Wrist extensors—self-stretching

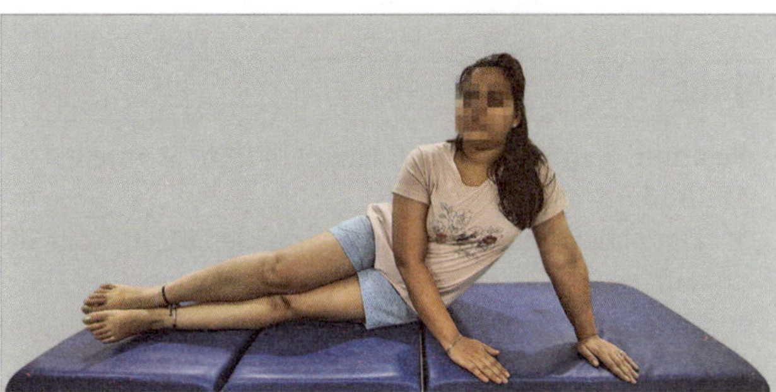

Fig. 10.42: Quadratus lumborum—self-stretching

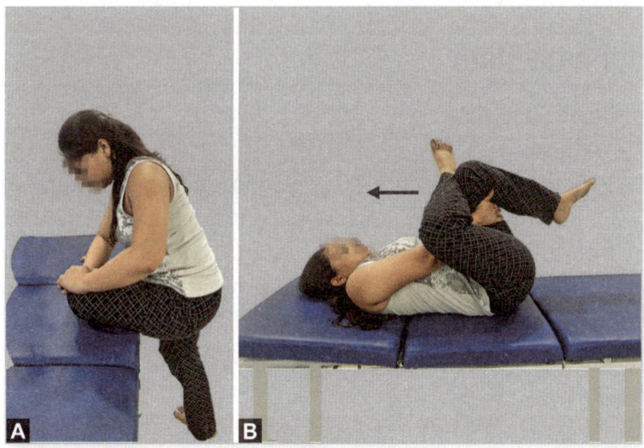

Figs 10.43A and B: Gluteus maximus—self-stretching

Gluteus Maximus

- **Position:** Standing
- **Method:**
  - The patient is in standing position with the leg to be stretched kept on chair.
  - The patient bends forward until a stretch is felt in posterior hip.
  - Repeat it for 3–5 times or as instructed by the therapist (Figs 10.43A and B).

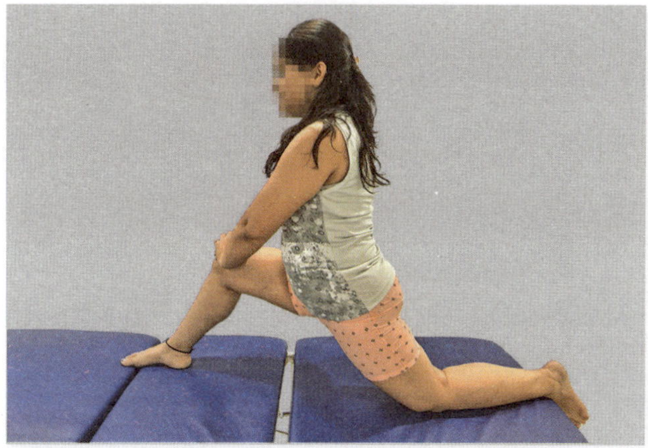

Fig. 10.44: Iliopsoas—self-stretching

Iliopsoas

- **Position:** Half kneeling
- **Method:**
  - The patient is in half kneeling position with the leg to be stretched kept behind. Both the hands are kept on the knees of the front leg.
  - The patient then shifts the weight on the front leg so that stretch is felt on the back leg.
  - Repeat it for 3–5 times or as instructed by the therapist (Fig. 10.44).

Rectus Femoris

- **Position:** Standing
- **Method:**
  - The patient is in standing position at the edge of the couch.
  - The patient flexes the knee and grabs the leg by holding at the ankle.
  - The stretch force is applied by adding hip extension until the stretch is felt.
  - Repeat it for 3–5 times or as instructed by the therapist (Fig. 10.45).

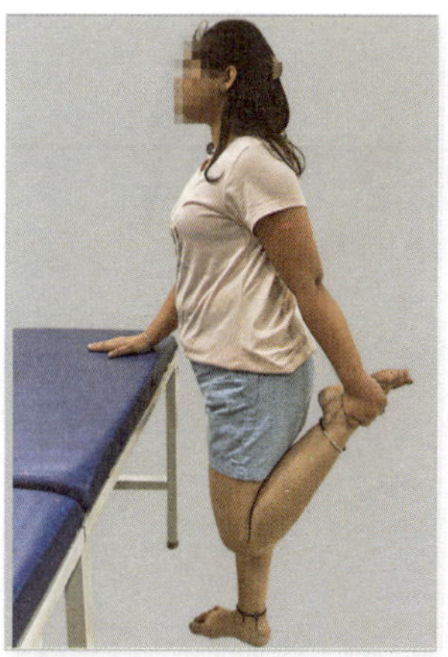

Fig. 10.45: Rectus femoris—self-stretching

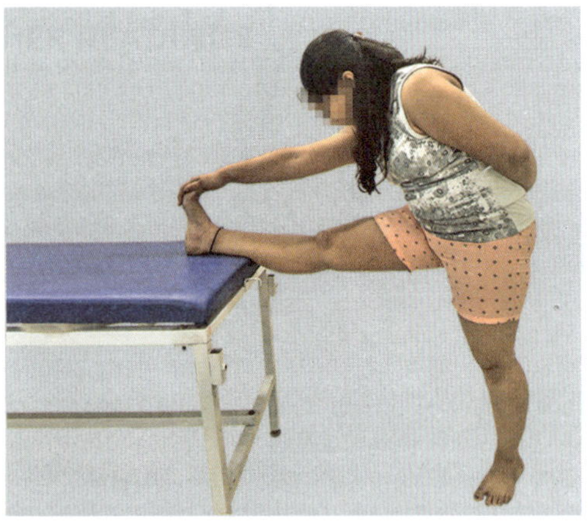

Fig. 10.46: Hamstrings—self-stretching

Hamstrings

- **Position:** Standing
- **Method:**
  - The patient is in standing position and keeps the leg to be stretched on a chair or a higher surface.
  - The patient then bends forwards and tries to touch the toes.
  - A stretch is felt at the back of thigh.
  - Repeat it for 3–5 times or as instructed by the therapist (Fig. 10.46).

Gastrocnemius

- **Position:** Standing
- **Method:**
  - The patient stands about an arm's length in front of a wall.
  - The leg to be stretched is kept behind. Reach both arms to the wall.
  - Shift the body's weight on to the front leg, while keeping the back heel pressed into the floor until a stretch is felt in posterior leg (Fig. 10.47).

Fig. 10.47: Gastrocnemius—self-stretching

Fig. 10.48: Soleus—self-stretching

Soleus

- **Position:** Stands at an arm's length in front as a well
- **Method:**
  - The patient stands about an arm's length in front of a wall.
  - The leg to be stretched is kept behind. Reach both arms to the wall.
  - Shift the body's weight on to the front leg, while keeping the back heel pressed into the floor and the back knee slightly bent until a stretch is felt in posterior leg (Fig. 10.48).

Hip Adductors

- **Position:** Sitting
- **Method:**
  - The patient is in sitting position and pulls the heels toward the groin, as close as comfortable.
  - Put the hands on the knees, and gently push them to perform abduction and lateral rotation of the thigh at the hip joint (Fig. 10.49).

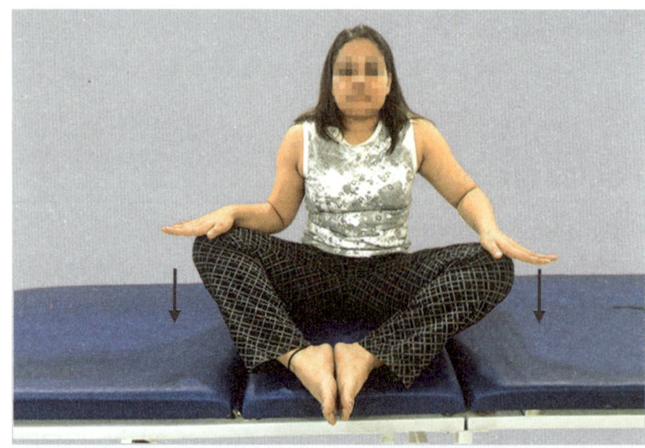

Fig. 10.49: Hip adductors—self-stretching

MYOFASCIAL RELEASE

Myofascial release (MFR) is a manual therapy technique designed to alleviate pain and restore mobility by targeting the fascia—the connective tissue surrounding muscles, bones, and organs. This therapeutic approach is grounded in the understanding that restrictions and tightness within the fascia can lead to myofascial pain syndrome (MPS), resulting in significant discomfort and functional impairment. Myofascial release is notable for its wide application in treating various conditions such as chronic pain, sports injuries, and postural imbalances, with practitioners employing techniques to identify and manipulate specific trigger points (tight knots in the muscle that can cause localized pain or referred pain to other areas of the body) to promote healing and improve overall physical function.

The mechanism of action in myofascial release involves applying gentle, sustained pressure to the affected fascia and muscle tissue, which helps to relax contracted muscles, reduce abnormal tension, and enhance blood circulation. Clinical studies suggest that MFR can be effective in addressing chronic pain syndromes, back pain, and conditions related to stress and anxiety, thereby providing both physical and psychological benefits to patients. Furthermore, self-myofascial release techniques, such as using foam rollers, have gained popularity among athletes and fitness enthusiasts as a means to enhance recovery and prevent injuries.

Despite its advantages, myofascial release is not suitable for everyone; specific contraindications include acute injuries, certain skin conditions, and severe vascular disorders, necessitating a thorough evaluation by healthcare professionals before initiating treatment. Additionally, individual responses to therapy may vary, highlighting the importance of tailored treatment plans, and when appropriate, integrating MFR with other therapeutic modalities for optimal outcomes.

Mechanism of Action

The underlying mechanism of myofascial release is based on reducing abnormal pressure, tone, and spasms within the fascia and muscles (Fig. 10.50). By applying gentle, sustained pressure on the trigger points and surrounding tissues, myofascial release aims to relax the contracted muscles and restore normal blood flow, thereby facilitating the movement of the affected structures. This approach can also help correct postural imbalances and improve overall physical function by promoting efficient movement patterns.

Benefits

Clinically, myofascial release is beneficial for various conditions, including chronic pain, sports injuries, and postural issues. It has been utilized to treat back pain, neck pain, shoulder pain, hip pain, knee pain, and conditions like plantar fasciitis. The technique is typically well-tolerated by patients and is considered safe when performed by a trained professional. Furthermore, self-myofascial release techniques, such as using foam rollers, can also be incorporated to enhance recovery and reduce muscle soreness after physical activities.

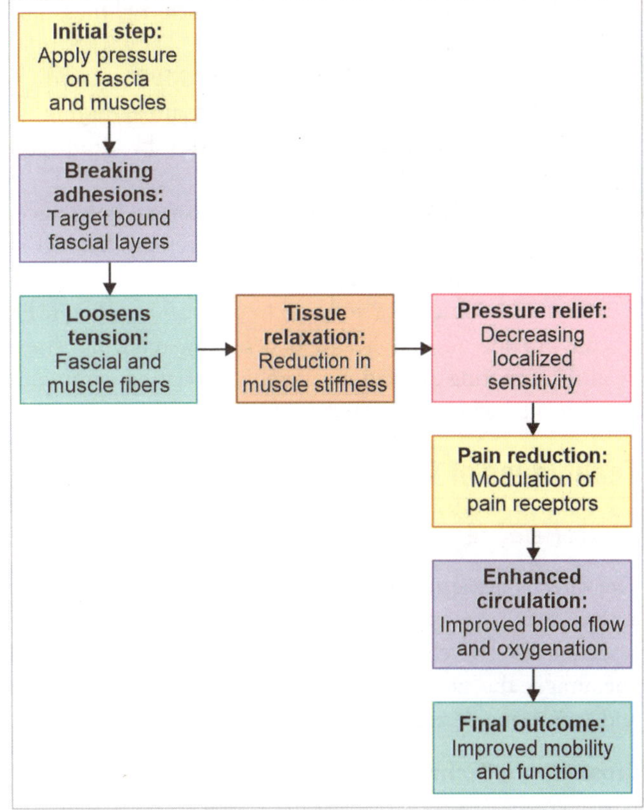

Fig. 10.50: Principle and mechanism of action of myofascial release

Indications

The MFR therapy is indicated for various conditions associated with MPS and other musculoskeletal disorders. The therapy aims to alleviate pain, improve range of motion, and enhance overall function through the application of gentle, sustained pressure on affected fascia and trigger points.

The MFR has shown efficacy in treating a range of conditions, including but not limited to:

- **Chronic pain conditions:** MFR is beneficial for patients suffering from chronic pain syndromes, including MPS, fibromyalgia, and tension-type headaches. These conditions often involve the presence of trigger points that can be addressed through myofascial techniques.
- **Back and joint pain:** Individuals experiencing back pain, particularly lower back pain, and joint pain, can find

relief through MFR, which targets muscle tightness and restrictions in the fascia.
- **Sports injuries:** Athletes may utilize MFR to address injuries resulting from overuse or trauma, aiding in recovery and preventing future injuries by improving flexibility and muscle function.
- **Postural issues:** MFR can help correct postural imbalances that lead to chronic discomfort by addressing tension in the fascia and promoting proper alignment.
- **Stress and anxiety:** In addition to physical benefits, MFR can provide mental health benefits by reducing stress and promoting relaxation, thereby alleviating symptoms associated with anxiety and depression.

Techniques of Application

Direct Release

The direct release method involves the therapist identifying a tight area and applying sustained pressure in a specific direction to stretch and release the fascia. This method encourages the tissue to elongate over time, easing tension and promoting relaxation.

Cross-Hands Technique

In the cross-hands technique, the therapist places their hands on either side of the restricted area and gently pulls in opposite directions. This creates a sustained stretching effect, guiding the tissue without force, which helps facilitate the release (Fig. 10.51).

Deep Tissue Massage

Deep tissue massage (Fig. 10.52) is often used in conjunction with myofascial release to address deeper layers of muscle and fascia. By applying firm pressure, practitioners aim to break down adhesions and restore the tissue's natural pliability.

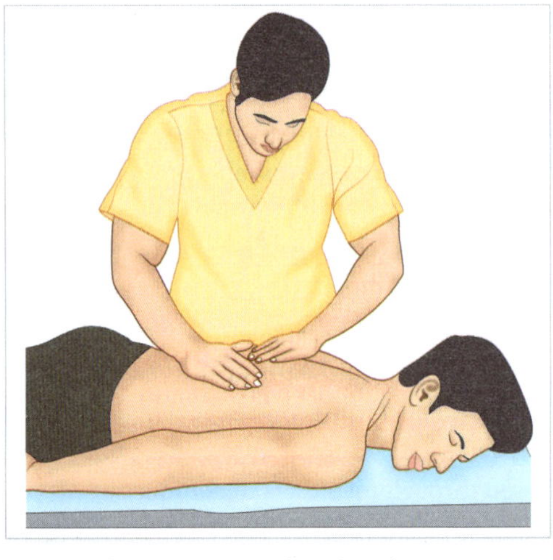

Fig. 10.51: Cross-hands technique

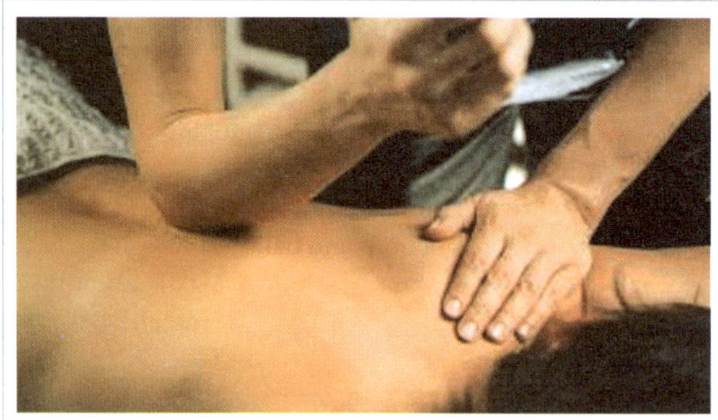

Fig. 10.52: Deep tissue massage technique

Stretching

Stretching techniques involve gentle stretching of the fascia and underlying muscles which help improve flexibility and range of motion.

Indirect Release

Indirect myofascial techniques (IMTs) are used to treat restrictions in the connective tissue, or fascia, by working with the body's inherent movements rather than directly opposing the tension in the tissue. These techniques are particularly suited for patients with acute pain, hypersensitivity, or fragile tissues, as they are gentle and nonaggressive. IMTs aim to relieve pain, improve range of motion, and restore the body's natural balance and alignment.

Positional Release Technique

Positional release technique (PRT) is a manual therapy approach that aims to relieve pain and muscle tension by placing the body in a position of comfort. This technique focuses on restoring normal muscle function and reducing discomfort. Table 10.1 enlists the differences between direct and indirect myofascial release techniques (D & I MRI).

Active Release Techniques

A specific type of MFR combines manual therapy with active patient movement. Active release techniques (ART) are a soft tissue method that focuses on relieving tissue tension *via* the removal of fibrosis/adhesions which can develop in tissues as a result of overload due to repetitive use. These disorders may lead to muscular weakness, numbness, aching, tingling and burning sensations.

Tool-Assisted Techniques

Instrument Assisted Soft Tissue Mobilization

Instrument assisted soft tissue mobilization (IASTM) is a skilled myofascial intervention used for soft-tissue treatment

TABLE 10.1: Differences between direct and indirect myofascial release techniques

Aspect	Direct myofascial release techniques	Indirect myofascial release techniques
Definition	A technique that applies sustained force directly against the restriction to stretch and release of fascial tightness	A gentle technique that follows the path of least resistance, working with the body's natural movement to release tension
Principle	Targets and works against the tissue restriction	Moves with direction of ease or the tissues natural motion
Force applied	Requires firm, sustained pressure to stretch or release fascia	Uses minimal force and light touch to guide the fascia
Goal	Breaks adhesions and remodel fascial tissue directly	Allows the fascia to relax and self-correct
Approach	Aggressive or intense, focusing on stretching and restructuring	Gentle and noninvasive, respecting the tissue inherent dynamics
Techniques included	Myofascial release (direct), deep tissue massage, Graston technique	Myofascial unwinding, positional release therapy, balanced ligamentous tension
Patient suitability	Effective for chronic restrictions, dense adhesion, or thickened tissue	Ideal for acute pain, hypersensitive areas, fragile tissue or high inflammation
Sensory feedback	May cause temporary discomfort or "therapeutic pain" during application	Typically painless, patients often feel a sense of relaxation
Tissue interaction	Physically stretches and elongates fascia to remove restrictions	Facilitates the fascial intrinsic release without forcing
Speed of release	Often more immediate but may cause soreness	Gradual, allowing the body to adapt over time
Nervous system effect	Stimulates more active response, potentially activating protective reflexes	Engages the parasympathetic nervous system, promotes relaxation
Clinical application	• Chronic fascial restriction • Scar tissue breakdown • Postural imbalance	• Acute injuries • Pain hypersensitivity • Patients with low pain tolerance or inflammatory conditions
Practitioners role	Applies direct pressure and manipulates tissue to achieve release	Guides and supports the tissue as it releases itself

(Fig. 10.53). It is based on the principles by James Cyriax of cross friction massage. The purpose of friction massage is to maintain the mobility within the soft tissue structures of ligament, tendon, and muscle and prevent adherent scars from forming. The massage is deep and must be applied transversely to the specific tissue involved unlike the superficial massage given in the longitudinal direction parallel to the vessels which enhances circulation and return of fluid.

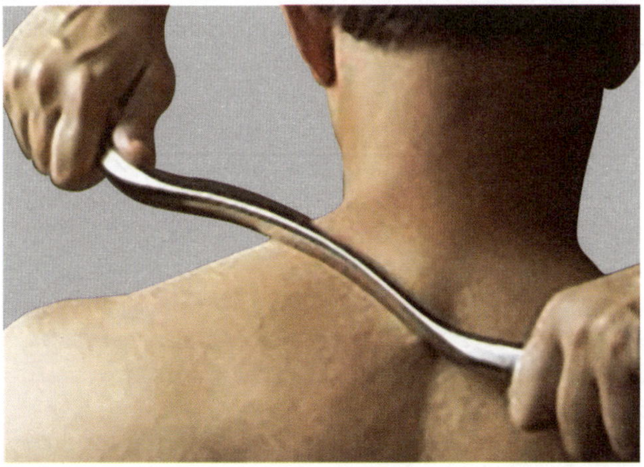

Fig. 10.53: Instrument assisted soft tissue mobilization (IASTM): Graston technique

Self-Myofascial Release

Tools like foam rollers or massage balls are utilized to allow individuals to perform MFR on themselves. This method can be integrated into daily routines to maintain fascial health.

Foam Rollers

Foam rollers (Fig. 10.54) are cylindrical devices that users roll under various parts of the body, applying their body weight to manipulate the fascia. They are particularly effective for larger muscle groups, like the back, thighs, and calves, improving blood circulation and enhancing muscle recovery.

Massage Balls

Massage balls (Fig. 10.55) provide a more targeted approach to myofascial release, ideal for addressing specific trigger points and smaller muscle groups. They can reach areas that foam rollers might not, such as around the shoulder blades or the arches of the feet.

Types of myofascial release techniques are given in Figure 10.56.

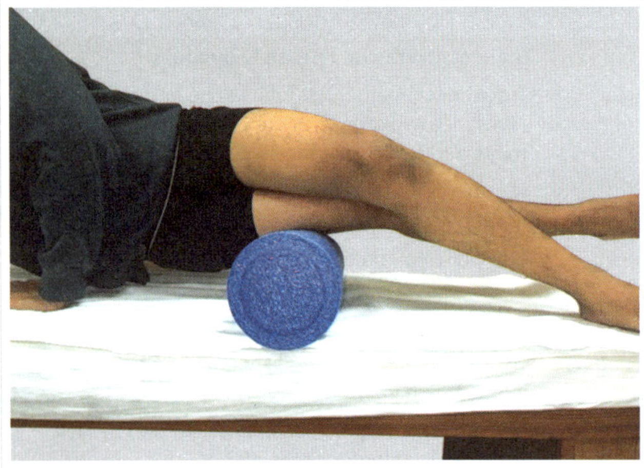
Fig. 10.54: Self-release technique using foam roller

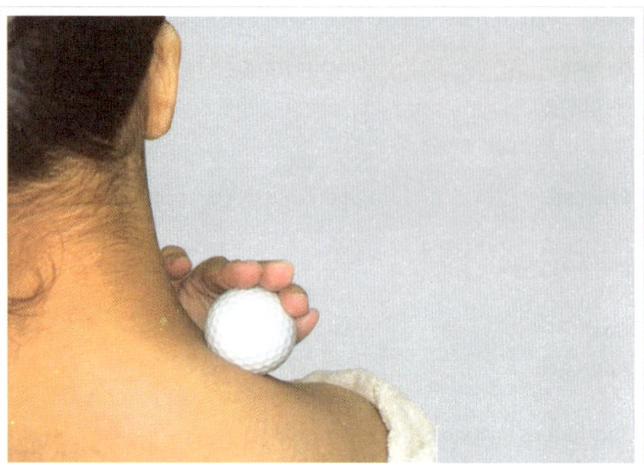
Fig. 10.55: Self-release techniques using massage ball

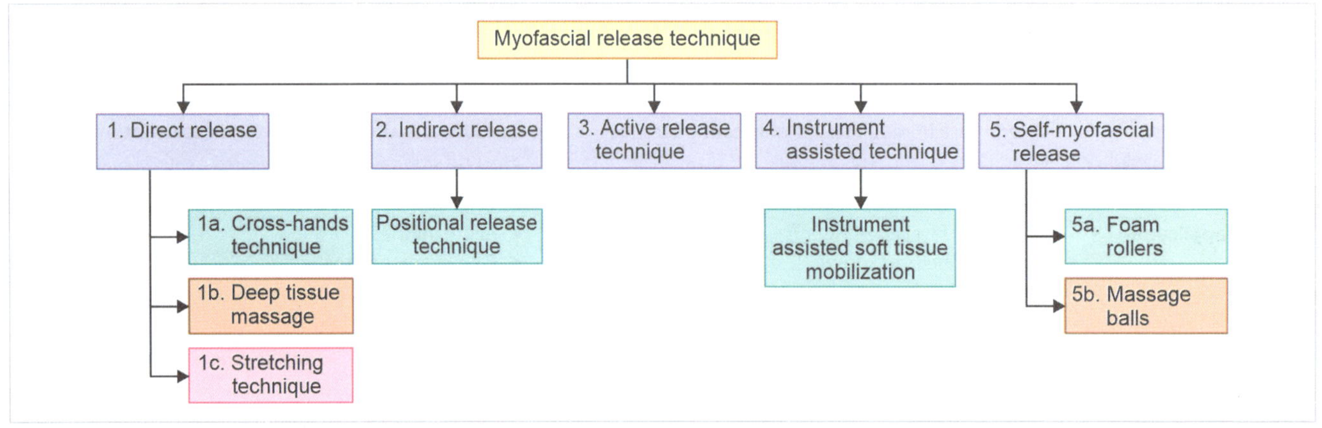

Fig. 10.56: Types of myofascial release techniques

Procedure

The goal of MFR is to decrease fascial restrictions, improve movement, and reduce pain. Proper application requires attention to both the therapist's technique and patient preparation. Below is a detailed explanation of the process, from patient preparation to the application of MFR.

Patient Preparation

1. **Patient assessment:**
   - Perform a thorough assessment to identify fascial restrictions, pain areas, and limited mobility.
   - Use palpation to assess the texture, tightness, and elasticity of the fascia.
   - Record the patient's pain level and range of motion to track progress.
2. **Educate the patient:**
   - Explain the purpose, process, and expected outcomes of MFR.
   - Address concerns about potential discomfort during the technique (if using direct MFR).
   - Highlight that MFR is a gradual process and may require multiple sessions for optimal results.
3. **Positioning the patient:**
   - Ensure the patient is in a comfortable position, either lying down, sitting, or standing, depending on the area to be treated.
   - Use bolsters, pillows, or other supports to avoid unnecessary strain on other body parts.
   - Expose the part to be treated with the consent of the patient and maintain privacy.
4. **Environment:**
   - Create a calm and quiet setting to help the patient relax.
   - Maintain a comfortable room temperature, as fascia responds better when the body is warm.
5. **Informed consent:**
   - Obtain the patient's consent before starting the therapy.
   - Discuss any contraindications, such as open wounds, recent fractures, or infections.

Application

1. **Contact the tissue:**
   - Use the palms, fingers, or knuckles to make contact with the targeted area.
   - Apply light to moderate pressure to sink into the superficial layers of the fascia without sliding on the skin.
   - Ensure that the touch is firm but not painful, creating a connection with the tissue.
2. **Assess fascial restriction:**
   - Move the tissue gently in different directions (e.g., up, down, side-to-side or rotational) to identify the direction of restriction.
   - Evaluate the tissue's resistance to movement and note the areas of tension or "stiffness".
3. **Apply the technique:** Depending on the chosen method (direct or indirect), follow these steps:
   - **For direct MFR:** Apply sustained pressure into the restricted fascia. Gradually increase the stretch by moving the fascia toward the restriction. Hold the stretch for 90 seconds to 3 minutes or until a release is felt (e.g., a melting sensation, softening of the tissue).
   - **For indirect MFR:** Gently guide the tissue in the direction of least resistance or ease. Hold the position for 90 seconds to 3 minutes, allowing the fascia to release naturally.
4. **Monitor tissue response:**
   - Observe changes in the tissue texture, such as softening or increased pliability.
   - Pay attention to the patient's feedback; they may report sensations, like warmth, tingling, or relief as the fascia releases.
5. **Maintain communication:**
   - Regularly check in with the patient about their comfort level and sensations during the treatment.
   - Adjust pressure or technique based on their responses.
6. **Reassess:**
   - After releasing a specific restriction, reassess the tissue to evaluate changes in mobility and tension.
   - Compare the treated side with the untreated side to gauge the effectiveness of the technique.

Postsession Care

Patient Education

- Advise the patient on post-session care, such as gentle stretching or hydration to facilitate recovery.
- Inform them that some soreness may occur after the session, which is normal and temporary.

Re-evaluation

- Check the patient's pain level and range of motion after the session.
- Document changes to monitor progress over subsequent treatments.

Follow-up Recommendations

- Provide a treatment plan outlining the frequency and duration of MFR sessions.
- Suggest complementary therapies (e.g., strengthening exercises, posture correction), if needed.

Limitations and Considerations

While MFR can offer significant benefits, it is essential to note that individual responses to therapy can vary based on several factors, including the severity of symptoms and the patient's overall health condition. Therefore, a healthcare professional should always be consulted prior to commencing MFR or any new therapeutic regimen. Furthermore, it is recommended that MFR be integrated with other therapeutic modalities, such as exercise and manipulation, for optimal outcomes.

Contraindications

The MFR is a therapeutic technique used to alleviate pain and restore function by targeting the fascia, the connective tissue surrounding muscles. However, there are specific contraindications to consider before applying this treatment.

General Contraindications

The MFR should be approached with caution or avoided in patients with certain conditions.

- **Acute injuries:** Individuals with recent injuries, particularly those involving inflammation or swelling, should refrain from MFR until the acute phase has passed, as the technique may exacerbate pain and discomfort.
- **Fractures and bone pathologies:** Patients with fractures, osteoporosis, or other bone-related issues may be at risk of further injury if subjected to myofascial techniques that involve deep pressure or manipulation.
- **Skin conditions:** Active skin infections, rashes, or other dermatological issues present a risk of exacerbation or spreading infection during treatment, thus contraindicating the use of MFR in these scenarios.
- **Severe vascular conditions:** Individuals with significant vascular disorders, such as deep vein thrombosis or severe varicosities, should avoid MFR due to the potential risk of disrupting blood flow or causing complications.

Specific Health Conditions

Certain health conditions require careful assessment prior to receiving MFR:

- **Cancer:** Patients with a history of cancer should consult their healthcare provider before undergoing MFR, as manipulation of tissue can potentially promote the spread of malignancies in certain cases.

- **Neurological disorders:** Conditions such as multiple sclerosis, spinal cord injuries, or neuropathies may complicate the response to MFR, making it crucial to obtain a thorough medical evaluation beforehand.
- **Pregnancy:** While many pregnant individuals can benefit from gentle therapies, MFR should be used with caution and ideally avoided during the first trimester due to the sensitivity of the developing fetus.
- **Cognitive impairment:** Patients with cognitive impairments may not be able to communicate discomfort or adverse reactions effectively, which poses a risk during myofascial techniques that require patient feedback.

SUMMARY

- Joint mobility refers to the ROM a joint can achieve without pain, playing a vital role in daily activities like walking, reaching, and lifting. Healthy joint mobility is essential for maintaining an active lifestyle, preventing injuries, and supporting good posture, which impacts overall musculoskeletal health.
  - As we age, preserving this mobility becomes increasingly important for independence and quality of life, enabling individuals to perform self-care and engage socially without limitations.
- The relationship between joint mobility and flexibility is intricate; good mobility relies on the flexibility of surrounding muscles and connective tissues.
  - When muscles are tight, they can restrict a joint's ROM, while stiff joints can lead to decreased muscle flexibility. Addressing both aspects through regular stretching and movement is crucial for preventing muscle imbalances and ensuring proper body alignment, which contributes to overall musculoskeletal health.
- Various interventions can enhance soft tissue mobility, including stretching exercises, foam rolling, massage therapy, and yoga.
  - Stretching plays a key role in improving flexibility by elongating muscles and soft tissues, thereby enhancing joint mobility.
- Regular stretching not only increases blood flow to the muscles but also stimulates synovial fluid production, facilitating smoother movement.
- The biomechanics of stretching involve the interactions between contractile and noncontractile tissues, as well as the neurophysiological responses of muscle fibers.
  - Muscle tissue consists of two primary types: Contractile tissue, composed mainly of muscle fibers that generate force through contraction, and noncontractile tissue, which includes connective tissues, blood vessels, and nerves that support and stabilize muscle function.
  - Understanding the mechanical properties of these tissues is essential for selecting effective stretching techniques, especially for patients with limited mobility.
- When muscles are immobilized in either a shortened or lengthened position, various biomechanical changes occur. Immobilization in a shortened position can lead to adaptive shortening of muscle fibers, a loss of sarcomeres, decreased force production, and altered neural control.
  - These changes can result in reduced flexibility and an increased risk of injury upon sudden stretching.
  - Conversely, immobilization in a lengthened position can also lead to loss of sarcomeres, decreased force generation, altered muscle fiber arrangement, and joint stiffness.
  - Both scenarios can contribute to muscle atrophy and impaired coordination due to reduced mechanical loading and altered neural control.
- During stretching, biomechanical principles like tension and elasticity are activated within the contractile unit of muscle.
  - Stretching elongates the sarcomeres, facilitating changes in muscle length and force production.
  - While gentle stretching can enhance flexibility and ROM, excessive stretching may disrupt the delicate balance of tension and elasticity, potentially resulting in muscle strain or injury.
- Muscle spindles and Golgi tendon organs are crucial sensory receptors in the neuromuscular system, aiding proprioception—the body's awareness of its position and movement.
  - Muscle spindles, located within the muscle belly, detect stretch and its rate, triggering protective reflexes like the stretch reflex to prevent injury.
  - In contrast, Golgi tendon organs, found at the muscle-tendon junction, respond to tension; they facilitate muscle relaxation when excessive force is detected, helping to protect the muscle during intense activities.
- During stretching, muscle spindles can limit how far a muscle can be stretched by initiating a protective contraction if the stretch is too rapid. Meanwhile, Golgi tendon organs promote relaxation, allowing for safe elongation.
  - Together, they help maintain muscle tone and stability while enabling increased flexibility.
- Stretching also affects connective tissues, such as tendons and ligaments, through processes like creep and stress relaxation. Creep allows tissues to gradually elongate under sustained load, while stress relaxation reduces the force needed to maintain that length as the tissue adapts.

- The effectiveness and safety of stretching hinge on several key determinants that influence how tissues respond to stretching forces.
  - Proper stabilization and positioning are crucial, especially in therapeutic contexts, as they help prevent injury and enhance stretch efficacy.
  - Techniques like therapist assistance and the use of supportive surfaces ensure that patients are securely positioned.
  - The duration of stretches significantly impacts tissue adaptation; typically, holding static stretches for 15–60 seconds promotes elongation, while frequency—ideally 2–5 times a week—allows for gradual improvement in flexibility.
  - Stretching intensity is another critical factor; gentle, controlled stretches are more effective and safer than aggressive ones, reducing the risk of injury.
  - Additionally, engaging in a proper warm-up increases tissue temperature and blood flow, enhancing extensibility and minimizing strain.
  - Lastly, the speed of stretching also matters; slower, sustained stretches are generally more beneficial for muscle and connective tissue, ensuring a safer stretching routine.
- Stretching can be classified into several types, each with distinct techniques and applications.
  - **Static stretching** involves holding a muscle in a stretched position for 15–60 seconds, and can be active (using one's own strength) or passive (with assistance).
  - **Static progressive stretching** gradually increases the stretch over time, while **Cyclic stretching** alternates between stretching and resting periods.
  - **Manual stretching** employs external force from a partner or therapist, and **Self-stretching** is performed independently, focusing on body alignment.
  - **Mechanical stretching** uses devices for prolonged tissue elongation, whereas **Ballistic stretching** incorporates bouncing movements to extend range of motion.
  - **Dynamic stretching** consists of active movements that mimic sports activities, enhancing blood flow and readiness.
  - **Proprioceptive neuromuscular facilitation (PNF)** combines muscle contraction and relaxation techniques to improve flexibility through methods like Contract-Relax and Hold-Relax.
- Select the muscle to be stretched and the appropriate technique based on goals, explaining the procedure to the patient for relaxation and confidence. Apply a warm-up, gently move the body part through its range, stabilize the proximal segment, and gradually increase the stretch intensity, holding the stretch for at least 30 seconds while ensuring comfort and avoiding injury.
- Myofascial release (MFR) is a manual therapy technique designed to alleviate pain and restore mobility by targeting the fascia.
  - By applying gentle, sustained pressure on the trigger points and surrounding tissues, myofascial release aims to relax the contracted muscles and restore normal blood flow, thereby facilitating the movement of the affected structures.
  - Clinically, myofascial release is beneficial for various conditions, including chronic pain, sports injuries, and postural issues.
  - Techniques of application include: Direct release, indirect release, active release, tool-assisted and self myofascial release.

FURTHER READINGS

- A comprehensive guide to myofascial release | Stretch therapy. (n.d.). Stretch Therapy. https://stretchtherapy.net/myofascial-release/
- Bentz M. (2024, February 2). Unlock your Wellness: A deep dive into myofascial release therapy | Everest Therapeutics. Everest Therapeutics Massage Therapy. https://www.everesttherapeutics.com/massage-techniques/myofascial-release-therapy
- Cht, LGO. (n.d.). Myofascial release therapy. Spine-health. https://www.spine-health.com/treatment/physical-therapy/myofascial-release-therapy
- Conditions Treated with Myofascial Release | MFR Therapists / The Best of Myofascial Release. (n.d.). https://mfrtherapists.com/conditions-treated-with-myofascial-release/
- Desforges S. (2024b, November 18). How massage therapy can alleviate chronic Pain - ACMA Association. ACMA Association. https://www.acma-association.com/how-massage-therapy-can-alleviate-chronic-pain/
- Dpt, MB (n.d.). MFR Conditions Library | MFR Therapists / The Best of Myofascial Release. https://mfrtherapists.com/mfr-conditions-library/
- Gleason M. (2020, September 23). A Beginner's guide to myofascial release | OhioHealth. OhioHealth. https://blog.ohiohealth.com/a-beginners-guide-to-myofascial-release/
- He P, Fu W, Shao H, Zhang M, Xie Z, Xiao J, Li, L, Liu, Y, Cheng Y, & Wang Q. (2023). The effect of therapeutic physical modalities on pain, function, and quality of life in patients with myofascial pain syndrome: a systematic review. BMC Musculoskeletal Disorders, 24(1). https://doi.org/10.1186/s12891-023-06418-6
- Holland, K. (2024, September 26). What is myofascial release, and how does it work? Healthline. https://www.healthline.com/health/chronic-pain/myofascial-release
- Kisner C and Colby LA. Therapeutic Exercises: Foundations and Techniques. 7th Edition, 2019.
- Myofascial release therapy: Benefits, techniques, and uses. (2024, November 7). Peak sPerformance Sports and Spine Center. https://peakperformancesd.com/blog/myofascial-release-therapy-benefits-techniques-and-uses
- Victoria GD, Carmen EV, Alexandru S, Antoanela O, Florin C, Daniel D. The PNF (Proprioceptive Neuromuscular Facilitation) Stretching Technique-A Brief Review. Ovidius University Annals, Series Physical Education and Sport/Science, Movement and Health. 2013 Jul 2;13.
- Wellness, PW. (2023, November 13). What is myofascial release therapy and how does it work - PEACEFUL WARRIORS WELLNESS CENTER. Peaceful Warriors Wellness Center. https://peacefulwarriorswellness.com/what-is-myofascial-release-therapy/

STUDENT ASSIGNMENT

LONG ANSWER QUESTIONS

1. Explain the determinants of stretching.
2. What are the key determinants of an individual's flexibility, and how do factors like age and genetics influence it?
3. Explain the concept of the stretch reflex and its role in determining an individual's maximum ROM.
4. Discuss indications, contraindications, and precautions of stretching.
5. What are some contraindications to stretching, and why should they be considered when designing a stretching routine?
6. List some precautions that should be taken into account when performing advanced stretching exercises.
7. Explain benefits of stretching and types.

SHORT ANSWER QUESTIONS

1. Define mobility and flexibility. How do they differ from each other?
2. What is the role of joint mobility in overall physical health and performance?
3. Write briefly about tissue response to immobilization and stretching.
4. Which physiological changes occur in muscles and connective tissues during immobilization?
5. What are the effects of regular stretching on muscle and tendon tissues?
6. State indications and contraindications for MFR.

MULTIPLE CHOICE QUESTIONS

1. **What is a primary cause of hypomobility in joints?**
   a. Excessive flexibility
   b. Joint inflammation
   c. Improved muscle strength
   d. Increased joint ROM
2. **Which of the following interventions can help improve soft tissue mobility?**
   a. Immobilization
   b. Dynamic movement training
   c. Prolonged rest
   d. Isolation of affected muscles
3. **What type of stretching involves holding a muscle in a stretched position for a period of time?**
   a. Dynamic stretching b. Ballistic stretching
   c. Static stretching d. PNF stretching
4. **What biomechanical effect occurs when a muscle is immobilized in a shortened position?**
   a. Increased force production
   b. Adaptive shortening
   c. Enhanced blood flow
   d. Decreased muscle stiffness
5. **Which of the following is a contraindication for stretching?**
   a. Improved flexibility
   b. Acute inflammation
   c. Increased ROM
   d. Stress relief
6. **What is the primary reason that gentle stretching is recommended over aggressive stretching for improving flexibility?**
   a. It is quicker to perform
   b. It leads to greater tissue damage
   c. It is safer and reduces the risk of injury
   d. It does not require a warm-up.
7. **Muscle spindles primarily detect changes in which of the following?**
   a. Muscle tension
   b. Muscle length and rate of stretch
   c. Joint angle
   d. Muscle contraction strength

8. **What is the main goal of static stretching?**
   a. To increase muscle strength
   b. To improve flexibility by holding a stretch for a period of time
   c. To enhance cardiovascular endurance
   d. To prepare for explosive movements
9. **Which type of stretching involves bouncing movements to push a muscle beyond its normal range?**
   a. Static stretching
   b. Dynamic stretching
   c. Ballistic stretching
   d. PNF stretching
10. **What is the primary purpose of proprioceptive neuromuscular facilitation (PNF) stretching?**
    a. To enhance muscle hypertrophy
    b. To improve muscular endurance
    c. To increase flexibility through muscle relaxation techniques
    d. To promote joint stability
11. **Which of the following stretches would be appropriate for increasing the ROM in the hip flexors?**
    a. Stretching the gluteus maximus
    b. Stretching the quadriceps
    c. Stretching the iliopsoas
    d. Stretching the hamstrings

ANSWER KEY

1. b **2.** b **3.** c **4.** b **5.** b **6.** c **7.** b **8.** b **9.** c **10.** c
11. c

11

Aerobic Exercises

Sonia Pawaria, Richa Rai, Sheetal Kalra

LEARNING OBJECTIVES

After the completion of the chapter, the readers will be able to:

- Understand the concept of aerobic exercises.
- Understand the physiological response to aerobic exercise.
- Demonstrate examination and evaluation of aerobic exercises.
- Explain exercise testing.
- Explain determinants of an exercise program.
- Understand abnormal response to acute aerobic exercises.

CHAPTER OUTLINE

- Introduction
- Types of Aerobic Exercises
- Physiological Responses to Aerobic Exercises
- Exercise Testing
- Exercise Prescription
- Aerobic Training
- Overtraining Syndrome
- Abnormal Responses to Acute Aerobic Exercise

KEY TERMS

Aerobic exercise: Exercise that is repetitive and structured necessitating the body's metabolic process to consume oxygen in order to produce energy is referred to as aerobic exercise.

Deconditioning: The reversible changes that occur in the body as a result of physical inactivity and neglect. The most significant and consistent aspect of deconditioning is the reduction in muscle strength.

Oxygen debt: Refers to the elevated oxygen consumption that occurs after strenuous physical activity. Oxygen debt represents the oxygen required to metabolize the metabolic byproducts and restore the body to its pre-exercise state. It contributes to postexercise energy expenditure and is associated with various physiological processes, including replenishing ATP stores, restoring muscle oxygen levels, and clearing metabolic byproducts. This phenomenon underscores the importance of postexercise recovery in optimizing performance and adaptation to exercise stress.

VO_2 max: Maximum amount of oxygen that an individual can utilize during intense exercise, measured in milliliters of oxygen per kilogram of body weight per minute (mL/kg/min). This measurement is a key indicator of cardiovascular fitness and aerobic endurance.

INTRODUCTION

According to the American College of Sports Medicine (ACSM 2022), aerobic exercise is defined as any physical activity that engages large muscle groups, can be performed continuously, is rhythmic in nature, and relies on the body's metabolic system to utilize oxygen for energy production. This type of exercise involves major muscle groups, such as those in the legs, arms, and core, ensuring the body can perform the activity efficiently. The rhythmic, repetitive motions in these activities, like walking, running, cycling, and swimming, help maintain a steady pace and enhance movement coordination. Central to aerobic exercise is the use of the aerobic metabolic pathway, which utilizes oxygen to convert nutrients into energy, supporting long-duration activities.

TYPES OF AEROBIC EXERCISES

Figure 11.1 shows different types of aerobic exercises.

Clinical Correlation

Regular engagement in aerobic exercise improves oxygen utilization, metabolic efficiency, and overall cardiovascular, respiratory health and physical fitness of the individual.

Fig. 11.1: Types of common aerobic exercises

PHYSIOLOGICAL RESPONSES TO AEROBIC EXERCISES

Immediate Effects

- Preparatory effect of exercise reduces vagal tone and causes an increase in heart rate.
- Increase in sympathetic tone and skeletal muscle activity increases the heart rate to 180 beats/min during moderate exercise and up to 260 beats/min during severe exercise.
- There is a proportionate increase in cardiac output with moderate to severe exercise which is from 20 L/min to 35 L/min due to increase in venous return caused by skeletal muscle pumping, respiratory pump and splanchnic vasoconstriction.
- Blood flow to the skeletal muscle increases from 80 mL during moderate to 120 mL during severe exercises.
- Blood becomes acidic due to increase in carbon dioxide content and the blood count increases due to hypoxia near glomerulus apparatus and release of erythropoietin.
- Systolic blood pressure increases with exercise; however, diastolic pressure reduces due to vasodilatation caused by accumulation of metabolites.
- Blood volume reduces due to sweating; and severe exercise may lead to dehydration.

Long-Term Effects

Engaging in regular aerobic exercise yields a large number of long-term beneficial physiological effects.

Cardiovascular

Aerobic exercise significantly impacts the cardiovascular system, leading to various beneficial adaptations.

- **Improved cardiac output:** During aerobic exercise, the heart rate increases to pump more blood per minute (cardiac output). Stroke volume, the amount of blood ejected per beat, also increases due to enhanced myocardial contractility, and increased venous return (According to Frank-Starling Law). These changes result in greater oxygen delivery to muscles and tissues, improving overall cardiovascular efficiency. Over time, the heart becomes more efficient, often evidenced by a lower resting heart rate and increased stroke volume. Regular aerobic exercise increases cardiac output, which is crucial for enhanced physical performance and cardiovascular health.
- **Lower resting heart rate:** With consistent aerobic training, the heart adapts by becoming more efficient at pumping blood, leading to a lower resting heart rate. This adaptation occurs due to increased stroke volume and improved autonomic regulation, with a relative increase in

parasympathetic (vagal) tone. A lower resting heart rate is associated with reduced cardiac workload and improved heart function, which can decrease the risk of cardiovascular diseases. Aerobic exercise contributes to a significant reduction in resting heart rate, indicating improved cardiac efficiency and health.

- **Enhanced vascular function:** Aerobic exercise stimulates the production of nitric oxide, a vasodilator that improves endothelial function and arterial flexibility. It also increases capillary density in muscle tissues, enhancing blood flow and nutrient delivery. These adaptations help lower blood pressure, reduce arterial stiffness, and improve overall vascular health, reducing the risk of hypertension and atherosclerosis. Regular aerobic exercise enhances endothelial function and increases arterial compliance, contributing to better vascular health.
- **Reduced blood pressure:** The improved efficiency of the cardiovascular system and enhanced endothelial function result in lower systolic and diastolic blood pressure. Aerobic exercise also reduces sympathetic nervous system activity and increases parasympathetic activity, contributing to lower blood pressure. Lower blood pressure reduces the risk of developing hypertension and associated cardiovascular complications, such as stroke and heart attack.

Did You Know?

Studies have shown that regular aerobic exercise significantly reduces blood pressure, contributing to the prevention and management of hypertension.

- **Increased high-density lipoprotein cholesterol and decreased low-density lipoprotein cholesterol:** Aerobic exercise affects lipid metabolism by increasing the activity of enzymes involved in lipid transport and breakdown, such as lipoprotein lipase. This results in increased levels of high-density lipoprotein (HDL) cholesterol and decreased levels of low-density lipoprotein (LDL) cholesterol and triglycerides. Improved lipid profiles are associated with a lower risk of atherosclerosis and coronary artery disease, contributing to better long-term cardiovascular health. Regular aerobic exercise positively impacts lipid profiles, increasing HDL cholesterol and decreasing LDL cholesterol and triglycerides.
- **Enhanced myocardial oxygen supply:** Aerobic exercise induces angiogenesis, the formation of new blood vessels, in the myocardium (heart muscle). It also improves coronary artery dilation and blood flow, ensuring better oxygen supply to the heart muscle. Improved myocardial oxygen supply enhances heart function, particularly during physical exertion, and reduces the risk of ischemic heart disease. Aerobic training increases myocardial perfusion and oxygenation, which is critical for preventing ischemic conditions and improving heart health.

Respiratory

Aerobic exercise has significant effects on the respiratory system, enhancing both its function and efficiency.

- **Increased lung capacity:** Aerobic exercise strengthens the respiratory muscles, including the diaphragm and intercostal muscles. This increased muscular strength enhances the ability of the lungs to expand, thus increasing lung capacity or total lung volume. An increased lung capacity allows for greater oxygen intake during each breath, improving overall oxygen availability to the body, which is particularly beneficial during intense physical activities. Regular aerobic exercise leads to increased lung capacity, which improves the body's ability to take in and utilize oxygen efficiently.
- **Enhanced alveolar ventilation:** Aerobic exercise increases the efficiency of gas exchange in the alveoli (tiny air sacs in the lungs where oxygen and carbon dioxide are exchanged). This is achieved by increasing the surface area of the alveoli and enhancing capillary networks within the lungs. Enhanced alveolar ventilation means more oxygen is transferred into the blood and more carbon dioxide is expelled from the body, improving overall respiratory efficiency. Aerobic exercise significantly enhances alveolar ventilation, resulting in improved oxygen uptake and carbon dioxide removal.

Did You Know?

Aerobic exercise not only enhances gas exchange efficiency in the alveoli but also stimulates the production of surfactant, a substance that reduces surface tension in the alveoli, preventing their collapse and promoting better gas exchange. This additional factor contributes to the overall improvement in respiratory function with regular aerobic exercise.

- **Increased respiratory rate and tidal volume:** During aerobic exercise, the body's demand for oxygen increases, leading to an increased respiratory rate (number of breaths per minute) and tidal volume (amount of air inhaled and exhaled per breath). These changes enable the body to meet the heightened oxygen demands of muscles during exercise and facilitate the removal of carbon dioxide produced by active tissues. Aerobic exercise increases both respiratory rate and tidal volume, which are crucial for meeting the metabolic demands during physical activity.
- **Improved respiratory muscle endurance:** Regular aerobic exercise enhances the endurance of respiratory muscles, reducing the effort required for breathing. This is achieved through improved mitochondrial density and capillary supply in the respiratory muscles. Improved respiratory muscle endurance reduces the perception of breathlessness during exercise and everyday activities, enhancing overall respiratory efficiency. Aerobic training improves the

endurance of respiratory muscles, leading to more efficient breathing and reduced fatigue.

- **Increased oxygen diffusion capacity:** Aerobic exercise increases the surface area and reduces the thickness of the alveolar-capillary membrane, enhancing the diffusion capacity of the lungs. This allows for more efficient transfer of oxygen from the alveoli to the blood. Increased oxygen diffusion capacity ensures that more oxygen is available to be transported to working muscles, improving aerobic capacity and performance. Regular aerobic exercise enhances the oxygen diffusion capacity of lungs, which is essential for improved aerobic performance.

Musculoskeletal

Aerobic exercise brings about several positive adaptations in the musculoskeletal system, enhancing muscle function, endurance, and overall skeletal health.

- **Improved muscle endurance:** Aerobic exercise increases mitochondrial density within muscle fibers. Mitochondria are the powerhouses of cells, responsible for producing energy through oxidative phosphorylation. Aerobic training also enhances the activity of oxidative enzymes. Increased mitochondrial density and oxidative enzyme activity improve the ability of muscles to produce energy aerobically, which delays the onset of fatigue and increases endurance. Regular aerobic exercise significantly improves muscle endurance by enhancing mitochondrial density and oxidative capacity.
- **Enhanced capillary density:** Aerobic exercise stimulates angiogenesis, the formation of new capillaries, within muscle tissues. This increases the capillary-to-muscle fiber ratio. Greater capillary density improves blood flow to muscles, ensuring a more efficient supply of oxygen and nutrients while facilitating the removal of metabolic waste products. This supports sustained muscular activity and recovery. Aerobic training enhances capillary density in skeletal muscles, leading to improved oxygen and nutrient delivery.
- **Increased muscle glycogen storage:** Aerobic exercise enhances the ability of muscles to store glycogen, the stored form of glucose. This occurs through increased insulin sensitivity and improved glucose uptake by muscle cells. Greater glycogen stores provide a readily available energy source during prolonged exercise, enhancing endurance and performance.

  Regular aerobic exercise increases muscle glycogen storage capacity, contributing to improved endurance performance.
- **Muscle fiber type transformation:** Aerobic exercise induces a shift in muscle fiber composition, favoring type I (slow-twitch) muscle fibers, which are more efficient at using oxygen to generate ATP through aerobic metabolism. There may also be adaptations in type II (fast-twitch) fibers to enhance their oxidative capacity. Type I muscle fibers are less prone to fatigue and better suited for sustained, long-duration activities, leading to improved endurance. Aerobic training promotes a shift toward type I muscle fibers, enhancing muscular endurance and efficiency.
- **Increased bone density and strength:** Weight-bearing aerobic exercises, such as running and brisk walking, apply mechanical stress to bones, stimulating osteoblast activity (cells responsible for bone formation). Increased bone formation and remodeling result in greater bone density and strength, reducing the risk of osteoporosis and fractures. Weight-bearing aerobic exercise increases bone density and strength, contributing to better skeletal health.
- **Improved joint function and flexibility:** Regular aerobic exercise maintains joint mobility by promoting the production and circulation of synovial fluid, which lubricates joints. It also strengthens the muscles surrounding the joints, providing better support and stability. Enhanced joint function and flexibility reduce the risk of joint pain and stiffness, improving overall mobility and quality of life. Aerobic exercise helps maintain joint function and flexibility, reducing the risk of joint-related issues.

Nutritional

Aerobic exercise has significant beneficial effects on the lipid profile, which includes various blood lipids such as cholesterol and triglycerides. These changes contribute to improved cardiovascular health and reduced risk of metabolic diseases.

- **Increased high-density lipoprotein cholesterol:** Aerobic exercise stimulates the activity of enzymes involved in lipid metabolism, such as lipoprotein lipase, which enhances the breakdown of triglycerides in lipoproteins. It also increases the transfer of cholesterol to HDL particles. Higher levels of HDL cholesterol are beneficial because HDL helps transport cholesterol from the arteries to the liver for excretion, reducing the risk of atherosclerosis and cardiovascular disease. Regular aerobic exercise increases HDL cholesterol levels, which is associated with a reduced risk of cardiovascular diseases.
- **Decreased low-density lipoprotein cholesterol:** Aerobic exercise reduces LDL cholesterol by enhancing the ability of liver to clear LDL from the bloodstream and by decreasing the production of LDL particles. Lower LDL cholesterol levels are crucial because high levels of LDL can lead to the buildup of cholesterol in arteries, increasing the risk of atherosclerosis and heart disease. Aerobic exercise is effective in reducing LDL cholesterol levels, contributing to better cardiovascular health.

- **Reduced triglycerides:** Aerobic exercise increases the activity of lipoprotein lipase, which breaks down triglycerides in the blood. It also enhances fatty acid oxidation in muscles, reducing the circulating triglyceride levels. Lower triglyceride levels reduce the risk of cardiovascular disease and pancreatitis, as high levels of triglycerides are associated with these conditions. Regular aerobic exercise significantly reduces triglyceride levels, which is beneficial for overall cardiovascular health.
- **Improved lipoprotein particle size:** Aerobic exercise affects the size and density of lipoprotein particles. It tends to increase the size of LDL particles, making them less atherogenic, and increases the size of HDL particles, enhancing their protective effects. Larger, less dense LDL particles are less likely to penetrate the arterial wall and form plaques, while larger HDL particles are more efficient at reverse cholesterol transport. Exercise modifies lipoprotein particle size, making LDL less atherogenic and HDL more effective in cholesterol transport.
- **Enhanced postprandial lipid metabolism:** Aerobic exercise enhances the body's ability to clear lipids from the bloodstream after meals. This is achieved by increasing the activity of enzymes involved in lipid metabolism and improving insulin sensitivity. Better postprandial lipid clearance reduces the risk of lipid accumulation in blood vessels, which can lead to atherosclerosis. Regular aerobic exercise improves postprandial lipid metabolism, thereby reducing the risk of cardiovascular disease.

Neurophysiological

Aerobic exercise has profound neurophysiological effects that enhance brain function, improve mood, and protect against cognitive decline. These effects are mediated through various mechanisms that influence brain structure and function, neurochemical balance, and cognitive performance.

- **Enhanced brain function and cognitive performance:** Aerobic exercise increases cerebral blood flow, enhancing oxygen and nutrient delivery to the brain. It also promotes neurogenesis (the creation of new neurons) and synaptogenesis (the formation of new synapses) in the hippocampus, a brain region critical for learning and memory. These changes result in improved cognitive functions such as memory, attention, and executive function. Enhanced brain plasticity supports better learning and adaptability. Aerobic exercise has been shown to improve cognitive performance and increase hippocampal volume, which is associated with better memory and learning.
- **Mood enhancement and reduced symptoms of depression and anxiety:** Aerobic exercise stimulates the release of endorphins, which are natural mood lifters. It also increases levels of neurotransmitters such as serotonin and norepinephrine, which play key roles in regulating mood. Additionally, exercise reduces the levels of cortisol, a stress hormone. These neurochemical changes contribute to improved mood, reduced symptoms of depression and anxiety, and a general sense of well-being. Regular aerobic exercise has been demonstrated to significantly reduce symptoms of depression and anxiety, largely through its effects on neurochemical balance.
- **Increased brain-derived neurotrophic factor levels:** Aerobic exercise boosts the production of brain-derived neurotrophic factor (BDNF), a protein that supports the survival, growth, and maintenance of neurons. BDNF is crucial for long-term potentiation (LTP), a mechanism underlying learning and memory. Higher BDNF levels enhance synaptic plasticity, cognitive function, and resilience against neurodegenerative diseases. Exercise-induced increase in BDNF levels is associated with improved cognitive function and neuroprotection.
- **Enhanced executive function:** Aerobic exercise improves the connectivity and efficiency of neural networks involved in executive functions, such as the prefrontal cortex. This region of the brain is responsible for processes like planning, decision-making, and impulse control. Improved executive function helps in better decision-making, problem-solving, and multitasking abilities. Aerobic exercise has been shown to enhance executive functions, likely due to improved prefrontal cortex activity and connectivity.
- **Reduction in age-related cognitive decline:** Aerobic exercise mitigates age-related brain atrophy, particularly in the prefrontal cortex and hippocampus. It also reduces oxidative stress and inflammation, which are linked to cognitive decline. These protective effects help maintain cognitive functions in older adults and reduce the risk of developing neurodegenerative diseases such as Alzheimer's. Regular aerobic exercise is associated with slower cognitive decline and reduced risk of dementia in older adults.

Endocrinal/Hormonal

Aerobic exercise has a profound impact on the regulation of stress hormones, contributing to improved mood, reduced anxiety, and enhanced overall well-being.

- **Reduction in cortisol levels:** Cortisol, known as the "stress hormone", is released by the adrenal glands in response to stress. Aerobic exercise helps to modulate cortisol levels by improving the efficiency of the hypothalamic-pituitary-adrenal (HPA) axis, which controls stress responses. Regular aerobic exercise leads to lower baseline cortisol levels and a more balanced response to acute stress, reducing the negative effects of chronic stress on the body. Aerobic exercise has been shown to reduce cortisol levels, which helps mitigate the harmful effects of chronic stress.

- **Increase in endorphin production:** Aerobic exercise stimulates the release of endorphins, which are endogenous opioids produced in the brain and nervous system. These neurochemicals act on the opiate receptors in the brain to reduce pain and induce feelings of pleasure or euphoria. Increased endorphin production during and after aerobic exercise contributes to the "Runner's High", a state of wellbeing and reduced perception of stress and pain. Regular aerobic exercise increases endorphin levels, promoting a sense of well-being and reducing stress perception.
- **Increased production of serotonin:** Aerobic exercise increases the availability of tryptophan, a precursor to serotonin, and enhances the activity of serotonin pathways in the brain. Serotonin is a neurotransmitter that plays a key role in mood regulation and anxiety reduction. Higher levels of serotonin are associated with improved mood and reduced stress and anxiety, contributing to better mental health. Aerobic exercise increases serotonin levels, which helps improve mood and reduce stress and anxiety.
- **Improved regulation of the hypothalamic pituitary adrenal axis:** The HPA axis controls the body's response to stress. Aerobic exercise enhances the regulation of the HPA axis by promoting a more balanced and efficient hormonal response to stress. Improved HPA axis regulation leads to a more adaptive stress response, with quicker recovery from stress and reduced overall stress hormone levels.
- **Enhanced insulin sensitivity:** Aerobic exercise increases the translocation of glucose transporter type 4 (GLUT4) to the cell membrane, facilitating glucose uptake by muscle cells. This improves glycemic control and reduces the risk of type 2 diabetes.

Immunological

Moderate aerobic exercise mobilizes immune cells such as neutrophils and natural killer cells, enhancing their activity and circulation. This boosts the body's ability to fight infections and reduces the incidence of chronic inflammation-related diseases.

EXERCISE TESTING

Exercise testing is essential to understand different methods of measuring aerobic exercise capacity and conducting exercise testing. These methods are crucial for assessing an individual's cardiovascular fitness, designing appropriate exercise programs, and monitoring progress. The methods for testing aerobic exercise capacity are discussed as follows:

VO_2 Max

The most common way of assessment of aerobic capacity of an individual is determination of VO_2 max. Though the other peripheral and central adaptations to exercise also determine the capacity.

Factors Affecting VO_2 Max

Following are the factors that affect the maximal oxygen uptake.

Nonmodifiable factors affecting VO_2 max include:
- Gender
- Age
- Heredity

Modifiable factors affecting VO_2 max include:
- Body size and composition
- Mode and intensity of exercise
- State of training of an individual

There are many methods of testing for VO_2 max. Few of them are discussed below.

Maximal Oxygen Uptake (VO_2 Max) Testing

VO_2 max is the maximum amount of oxygen the body can utilize during intense exercise. It is considered the gold standard for measuring aerobic capacity.

Procedure

- The individual performs graded exercise on a treadmill, cycle ergometer, or other ergometers.
- The intensity of exercise gradually increases until volitional exhaustion occurs.
- During the test, oxygen consumption and carbon dioxide production are measured using a metabolic cart.

Interpretation

Higher VO_2 max values indicate greater aerobic capacity and cardiovascular fitness.

Submaximal Exercise Testing

Submaximal exercise testing involves assessing an individual's aerobic capacity without the requirement for exercise to the point of exhaustion. These tests estimate the maximal oxygen uptake (VO_2 max) based on the heart rate response to submaximal workloads. Submaximal tests are safer, less strenuous, and more practical for a wider range of populations, including those with medical conditions, older adults, and sedentary individuals. There are number of tests used to perform submaximal exercise testing.

Astrand-rhyming Cycle Ergometer Test

It is a single-stage test performed on a cycle ergometer to estimate VO_2 max from heart rate responses.

Procedure:

- The individual cycles at a steady workload for 6 minutes.
- Workload is adjusted based on the individual's fitness level and gender.
- Heart rate is measured during the last minute of the exercise.
- VO_2 max is estimated using a nomogram or predictive equations.

Interpretation: The lower the heart rate at a given workload, the higher the estimated VO_2 max, indicating better cardiovascular fitness.

Step Tests

"The step test is a practical, low-cost exercise test used to evaluate cardiovascular fitness, particularly aerobic capacity and endurance. It is commonly used in clinical settings, rehabilitation programs, and fitness assessments."

Procedure

Step tests involve stepping up and down on a standardized platform at a set cadence.

1. **Harvard step test:** Step up and down on a 20-inch (50.8 cm) bench at 30 steps/min for 5 minutes or until exhaustion. Then measure heart rate at 1-, 2-, and 3-minutes postexercise.

   Calculate the fitness index:

   Fitness Index = (duration of exercise in seconds × 100)/(sum of heart rates at 1, 2, and 3 minutes).

2. **Queens college step test:**
   - Step up and down on a 16.25-inch (41.3 cm) bench for 3 minutes at a rate of 22 steps/min for women and 24 steps/min for men. Measure heart rate for 15 seconds immediately after the test.
   - Estimate VO_2 max using gender-specific equations.

Interpretation

The heart rate response post-exercise is used to estimate aerobic capacity.

Rockport Walking Test

Rockport walking test involves walking one mile as quickly as possible on a flat surface.

Procedure

- The individual walks one mile at a steady, brisk pace.
- Record the time taken to complete the walk and the heart rate at the end of the walk.
- **Use the following formula to estimate VO_2 max:**
  VO_2 max (mL/kg/min) = 132.853 – (0.0769 × weight in pounds) – (0.3877 × age) + (6.315 × gender, where 1 for male and 0 for female) – (3.2649 × walk time in minutes) – (0.1565 × heart rate).

Interpretation

Faster walking time and lower heart rate indicate better aerobic capacity.

VO_2 Max scores, rated by healthy standards for each age group are as follows:

Highest death risk (from any cause)		49% lower	64% lower	76% lower
Age	Poor	Fair	Good	Excellent
*18–19	<37.9	38 → 45.4	45.5 → 48.9	>49
*20–29	<36.3	36.4 → 41.9	42 → 47.9	>48
30–39	<35.2	35.3 → 39.1	39.2 → 45.4	>45.5
40–49	<34.6	34.7 → 38.4	38.5 → 43.7	>43.8
50–59	<28.9	29 → 34.9	35 → 39.8	>39.9
60–69	<24.7	24.8 → 29.7	29.8 → 34.9	>35
70–79	<21.3	21.4 → 24.4	24.5 → 29.7	>29.8
80+	<18.1	18.2 → 22.0	22.1 → 25.5	>25.6
10-year survival rate (from middle age/50)	77%	91%	93.5%	96%

*If you are 22-year-old or younger, multiply your score by 0.85.

Otherwise your score comes out too high. The test was originally designed for the age group 30–69.

Clinical Correlation

Exercise testing can also serve as a tool for risk stratification, aiding in the identification of individuals who may be at higher risk for cardiovascular events such as heart attacks or strokes.

Field Tests for Aerobic Capacity

Field tests are practical, easy-to-administer methods for assessing aerobic capacity. They require minimal equipment and are suitable for large groups, making them ideal for use in various settings, including schools, fitness centers, and clinical environment.

Cooper 12-Minute Run Test

Developed by Dr Kenneth Cooper in 1968, this test measures the distance an individual can cover in 12 minutes of running or walking. It is widely used to estimate aerobic capacity (VO_2 max).

Procedure

- The individual warms up for 5–10 minutes.
- The test is conducted on a standard 400-meter track or a flat, measured course.
- The individual runs or walks as far as possible within 12 minutes.
- Distance covered is recorded in meters.

Interpretation

VO_2 max can be estimated using the formula:
- VO_2 max (mL/kg/min) = (distance in meters – 504.9)/44.73.
- Performance is compared to normative data to assess cardiovascular fitness levels.

1.5 Mile Run Test

Mile run test measures how quickly an individual can run 1.5 miles. It is commonly used by the military and other organizations to assess aerobic fitness.

Procedure

- The individual warms up for 5–10 minutes.
- The test is conducted on a standard 400-meter track or a flat, measured course.
- The individual runs 1.5 miles as quickly as possible.
- Time taken to complete the distance is recorded in minutes and seconds.

Interpretation

VO_2 max can be estimated using the formula: VO_2 max (mL/kg/min) = 3.5 + 483/(completion time in minutes). Results are compared to normative data for different age and gender groups to evaluate fitness levels.

6-Minute Walk Test

The 6-Minute Walk Test (6 MWT) measures the distance an individual can walk in 6 minutes. It is often used for clinical populations, including those with chronic respiratory or cardiovascular conditions.

Procedure

- The test is conducted on a flat, straight, indoor course with a length of 30 m.
- The individual walks back and forth along the course for six minutes.
- Total distance walked is recorded in meters.
- The individual is allowed to self-pace and rest, if necessary, but the timer continues to run.

Interpretation

The distance walked is compared to reference values based on age, sex, height, and weight. It is used to assess functional exercise capacity and monitor disease progression or response to treatment.

20-Meter Shuttle Run Test (Beep Test)

Shuttle run tests involve running back and forth between two lines set 20 meters apart at increasing speeds until exhaustion.

Procedure

- The individual runs back and forth between 2 lines 20 meters apart, keeping pace with audio beeps that gradually increase in frequency until exhaustion.
- The test ends when the individual can no longer keep pace with the beeps.

Interpretation

The level and number of shuttles completed are used to estimate VO_2 max using established formulas.

Heart Rate Recovery Tests

Heart rate recovery (HRR) tests measure the rate at which heart rate declines after exercise. A faster decline indicates better cardiovascular fitness.

Procedure

- The individual performs a bout of exercise (e.g., treadmill, cycle ergometer).
- Heart rate is recorded immediately at the cessation of exercise and at specific intervals during recovery (e.g., 1 minute, 2 minutes).

Interpretation

A faster decrease in heart rate post-exercise is associated with higher aerobic capacity and better autonomic function.

Lactate Threshold Testing

Lactate threshold testing measures the exercise intensity at which lactate begins to accumulate in the blood. It is an important marker of endurance performance.

Procedure

- The individual performs graded exercise while blood samples are taken at intervals to measure lactate concentration.
- Exercise intensity is increased progressively until a marked rise in blood lactate is observed.

Interpretation

The lactate threshold provides information on the maximal sustainable intensity of exercise and helps in tailoring training programs.

Metabolic Equivalent of Task Testing

Metabolic equivalent of task (MET) testing involves calculating the energy expenditure of various activities relative to resting metabolic rate (Ainsworth et al., 2000). Figures 11.2 and 11.3 show energy required (in terms of METs) for light, moderate, and vigorous exercise.

Procedure

- Activities are classified based on their intensity, and energy expenditure is expressed in METs.
- Exercise capacity can be estimated by determining the maximal MET level an individual can sustain.

Interpretation

Higher maximal MET levels indicate better aerobic capacity.

> **Clinical Correlation**
>
> Exercise testing can also serve as a tool for risk stratification, aiding in the identification of individuals who may be at higher risk for cardiovascular events such as heart attacks or strokes.

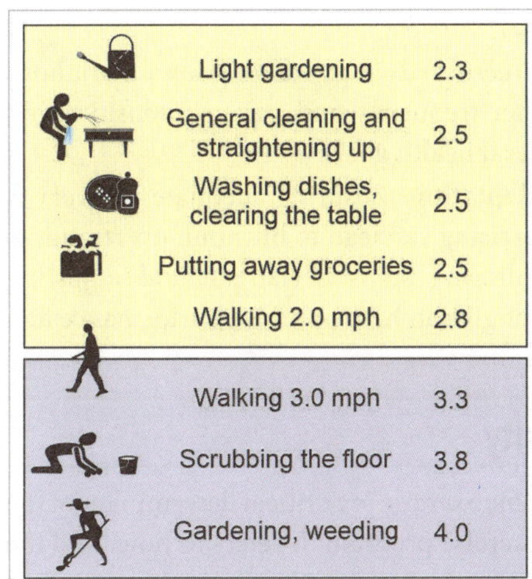

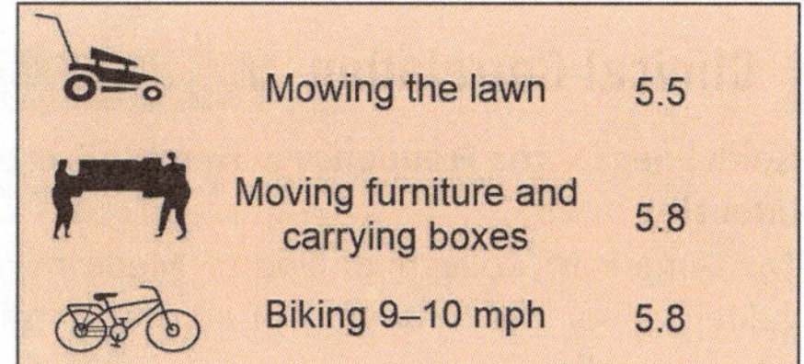

**Energy ratings are based on METs (metabolic equivalent).*
Light exercise is <3.0 METs.
Moderate exercise is 3.0–5.9 METs.
Vigorous exercise is 6.0 METs and above.

Intensity levels in sports and activities vary from person to person.

Fig. 11.2: Energy required (in terms of METs) for light to moderate exercises from the compendium of physical activities

Activity	METs
Hiking (light pack)	6.0
Weightlifting (vigorous)	6.0
Ballet or modern dance (vigorous)	6.8
Stationary bike 100 watts	6.8
Backpacking	7.8
Boxing, sparring	7.8
Climbing stairs	8.0
Biking 13 mph	*8.0
Competitive beach volleyball	8.0
Basketball	8.0

Activity	METs
Tennis, singles	8.0
Rock climbing	8.0
Circuit training/body weight exercises	8.0
Swimming 50 yd/min	8.3
Rowing 150 or 8.6 mph	8.5
Running 6 mph or 10:00/mile	9.8
Rollerblading 11 mph	9.8
Biking 15 mph	10.0
Competitive soccer	10.0
Swimming 75 yd/min	10.0

Activity	METs
Breaststroke	10.3
Martial arts	10.3
Jumping rope	11.0
Rowing 200w or 9.3 mph	12.0
Biking 17–18 mph	12.0
Running 8.6 mph or 7 min/mile	12.3
Rollerblading 13+ mph	12.3

Intensity levels in sports vary from person to person.

Fig. 11.3: Energy required (in terms of METs) for vigorous and intense exercise from the compendium of physical activities

EXERCISE PRESCRIPTION

The prescription of aerobic exercise involves several key components, which can be systematically described using the FITT principle.

FITT stands for:

F : Frequency/exercise frequency
I : Intensity/exercise intensity
T : Time/exercise time
T : Type/exercise type

Exercise Frequency

The frequency of aerobic exercise refers to how often should exercise sessions occur within a given timeframe, typically measured in days per week. Determining the appropriate frequency is essential for achieving the desired health and fitness outcomes while also allowing adequate recovery time.

Clinical Correlation

Guidelines for the Frequency of Aerobic Exercises based on the Intensity

The American College of Sports Medicine (ACSM) provides guidelines for the frequency of aerobic exercise based on the intensity of the activity:

- **Mild intensity aerobic exercises:**
  - Recommended frequency 5–7 days/week
  - Activities that can be easily integrated into daily routines (e.g., walking during breaks)
- **Moderate-intensity aerobic exercise:**
  - Recommended frequency is at least 5 days/week.
  - Examples of moderate-intensity activities include brisk walking, recreational swimming, and gardening.
- **Vigorous-intensity aerobic exercise:**
  - Recommended frequency is at least 3 days/week.
  - Examples of vigorous-intensity activities include running, aerobic dancing, and high-intensity interval training (HIIT).
- **Combination of moderate- and vigorous-intensity aerobic exercise:**
  - Recommended frequency is 3–5 days/week.

The goal is to accumulate at least 150 minutes of moderate-intensity exercise, 75 minutes of vigorous-intensity exercise, or an equivalent combination of both per week.

Frequency for Different Populations

Beginners

For individuals new to exercise or returning after a long period of inactivity, starting with lower frequency (e.g., 3 days/week) and gradually increasing can help prevent injury and ensure adherence.

Older Adults

Older adults may benefit from more frequent, shorter sessions. For example, engaging in moderate-intensity activity for 30 minutes on 5 days/week can be more manageable and effective in maintaining functional capacity and reducing fall risk.

Individuals with Chronic Conditions

For those with chronic conditions such as cardiovascular disease or diabetes, more frequent, moderate-intensity sessions depending on patients' functional capacity may be beneficial. For example, daily moderate-intensity exercise can help manage blood glucose levels in individuals with diabetes.

Adjusting Frequency Based on Individual Needs

The frequency of aerobic exercise should be personalized based on several factors:

- **Fitness level:** More fit individuals may tolerate higher frequencies of vigorous-intensity exercise. Less fit individuals may need to start with lower frequencies and gradually increase.
- **Goals:** Specific fitness goals (e.g., training for a marathon) may require higher frequency and varied intensity levels compared to general health goals.
- **Recovery and adaptation:** Ensuring adequate recovery is crucial. Over-exercising can lead to burnout, overtraining, and injury. Monitoring signs of fatigue and adjusting frequency accordingly can help optimize performance and health outcomes.

Exercise Intensity

The intensity of aerobic exercise is a critical determinant of the effectiveness of an exercise program. It refers to how hard the body is working during physical activity. Properly prescribed exercise intensity ensures that the individual is training at a level that is safe, effective, and aligned with their fitness goals.

Methods of Measuring Intensity

There are several ways to measure and monitor exercise intensity, each with its own advantages and applications:

Heart Rate

- **Maximum Heart Rate (HRmax):**
  A common method to determine exercise intensity is by using a percentage of maximum heart rate, which can be estimated using the formula:

  $$\text{HRmax} = 220 - \text{age}$$

- **Heart rate reserve (HRR):** This method considers both resting heart rate (HRrest) and maximum heart rate, providing a more individualized target:

  $$\text{Target HR} = (\text{HRmax} - \text{HRrest}) \times \text{Intensity} + \text{HRrest}$$

  - **Target heart rate zones:** Exercise intensity is often categorized into zones:

- Moderate intensity: 50–70% of HRmax or 40–60% of HRR
- Vigorous intensity: 70–85% of HRmax or 60–80% of HRR

- **Rate of perceived exertion (RPE):** The borg RPE scale ranges from 6 to 20, correlating to heart rates. Moderate-intensity exercise typically corresponds to an RPE of 12–14, while vigorous intensity corresponds to an RPE of 15–17.
- **Metabolic equivalents (METs):** METs measure the energy cost of physical activities. One MET is the energy expended at rest. Moderate-intensity activities range from 3 to 6 METs, and vigorous-intensity activities are above 6 METs.
- **Talk test:** This is a practical method where an individual should be able to carry on a conversation during moderate-intensity exercise but will struggle to do so at vigorous intensities.
- **VO_2 max:** Exercise intensity can be prescribed as a percentage of VO_2 max. This method is particularly useful for athletes and individuals who undergo VO_2 max testing.

Categorization of Intensity of Exercise Based on Percentage of VO_2 Max

Low intensity:

- <40% of VO_2 max.
- Suitable for warm-ups, cool-downs, and initial conditioning for very deconditioned individuals.
- Activities include light walking or gentle cycling.

Moderate intensity:

- 40–59% of VO_2 max.
- Suitable for most healthy adults aiming to improve general fitness.
- Activities include brisk walking, recreational swimming, or easy cycling.

Vigorous intensity:

- **60–89% of VO_2 max:** Suitable for well-conditioned individuals aiming to improve cardiovascular endurance and athletic performance.
- Activities include running, competitive swimming, or high-intensity interval training (HIIT).

High intensity:

- ≥90% of VO_2 max.
- Used in high-intensity interval training and for athletes during peak performance training.
- Activities include sprinting and maximal effort intervals

Determining Exercise Intensity

Low-intensity exercise:

- Typically <50% of HRmax or <40% of HRR.
- Activities might include leisurely walking, light gardening, or gentle yoga.
- Benefits include improvements in health for sedentary individuals and initial conditioning for deconditioned people.

Moderate-intensity exercise:

- 50–70% of HRmax or 40–60% of HRR.
- Examples include brisk walking, recreational swimming, or dancing.
- Associated with substantial health benefits, including cardiovascular fitness, weight management, and reduced risk of chronic diseases.

Vigorous-intensity exercise:

- 70–85% of HRmax or 60–80% of HRR.
- Examples include running, cycling at a fast pace, or aerobic dancing.
- Provides greater improvements in cardiovascular fitness and higher caloric expenditure, but also requires more recovery and may not be suitable for all individuals, especially those with certain health conditions.

High-intensity interval training (HIIT):

- Involves alternating short periods of very high-intensity exercise (>85% of HRmax) with periods of low-intensity recovery or rest.
- HIIT has been shown to improve cardiovascular fitness, glucose metabolism, and fat loss more efficiently than continuous moderate-intensity exercise in some studies.

Adjusting Intensity Based on Individual Needs

- **Fitness level:** Beginners should start at a lower intensity and gradually increase as their fitness improves. More conditioned individuals can train at higher intensities to continue improving fitness.
- **Health status:** Individuals with cardiovascular disease, hypertension, diabetes, or other chronic conditions may need to exercise at a lower intensity to avoid adverse events. Regular monitoring and gradual progression are key.
- **Goals:** Weight loss may benefit from higher-intensity exercise to increase caloric burn. Improved endurance may require sustained moderate to vigorous intensity.
- **Preferences and enjoyment:** Adherence to exercise improves when individuals enjoy the activity. Selecting enjoyable exercises at a preferred intensity can enhance long-term adherence.

Exercise Time/Duration

There is no threshold duration which induces the training response; however, it is dependent on the total accumulated work by a combination of the intensity and frequency depending on the initial fitness level of an individual. However, a continuous bout of 20–30 minutes of exercise has shown to demonstrate better results. For deconditioned patients, a 3–5 minutes duration of exercise progressing to an accumulated 10 minutes interval training daily exercise has also shown good results.

Long-term continuation of aerobic exercise, spanning weeks, months, and years, is essential for achieving and maintaining health benefits. By following a structured progression through the initial, adaptation, improvement, and maintenance phases, individuals can enhance their cardiovascular fitness, achieve specific goals, and integrate physical activity into their lifestyle for lifelong health and well-being.

Exercise Type/Mode

The type or mode of aerobic exercise refers to the specific activities performed to improve cardiovascular fitness and overall health. Choosing the appropriate exercise type is crucial for ensuring the exercise program is effective, enjoyable, and sustainable.

AEROBIC TRAINING

Aerobic exercises are characterized by their ability to be sustained over extended periods, from 20 minutes to several hours, promoting cardiovascular endurance as the heart and lungs work steadily.

Principles

- **Specificity:** The training program should be specific to the individual's goals, fitness level, and health status. For instance, someone aiming to improve running performance would have a different program than someone aiming to improve heart health.
- **Progression:** Gradual increase in intensity, duration, and frequency is crucial to prevent overuse injuries and to ensure continuous improvement. Progression may occur by manipulating variables like speed, duration, and frequency of exercise.
- **Overload:** To see improvement in aerobic capacity, the body needs to be challenged beyond its current capacity. This involves exercising at intensities that require the heart and lungs to work harder than usual.
- **Reversibility:** The principle of reversibility states that gains made from exercise are lost when training ceases. Thus, consistency is key to maintaining aerobic fitness levels.
- **Individuality:** Everyone responds differently to exercise due to factors like genetics, age, and health status. Therefore, programs should be individualized to meet the unique needs and abilities of each person.

Types

1. **Continuous training:** Involves sustained, steady-state exercise performed at a constant intensity.
   Examples: Walking, jogging, cycling, swimming.
2. **Interval training:** Alternates between periods of high-intensity exercise and low-intensity recovery or rest.
   Examples: High-intensity interval training (HIIT), fartlek training.
3. **Circuit training:** Combines aerobic exercises with resistance training in a circuit format, moving from one exercise to the next with minimal rest.
   Examples: Circuit classes, boot camp workouts.

Phases

1. **Initial phase (base building):** It focuses on building an aerobic base by gradually increasing duration and intensity. The emphasis is on improving cardiovascular endurance and building a foundation for more intense training.
2. **Improvement phase:** It introduces more structured workouts with specific goals such as increasing speed, endurance, or performance. This phase may include interval training, tempo runs or hill repeats to challenge the cardiovascular system.
3. **Maintenance phase:** Once fitness goals are achieved, the focus shifts to maintaining aerobic capacity through consistent training. This phase involves regular workouts to sustain cardiovascular fitness levels and prevent detraining effects.
4. **Tapering phase:** Prior to competitions or events, athletes undergo a tapering phase where training volume decreases while intensity remains moderate. This allows the body to recover fully and peak performance on the day of the event.

By adhering to the principles and implementing phases of aerobic training, individuals can optimize their cardiovascular health, improve endurance, and achieve their fitness goals effectively and safely.

Examples

- **Walking:** It is low-impact and accessible for most fitness levels. It is a simple and effective way to get started with aerobic exercise. It can be performed indoors on a treadmill or outdoors. It is suitable for beginners and those with joint issues.
- **Running/jogging:** It has higher impact and intensity compared to walking. It is suitable for improving cardiovascular fitness and endurance. It can be varied with different terrains and speeds.
- **Cycling:** It is also low-impact and suitable for all fitness levels. It can be performed on stationary bikes indoors or

on roads/trails outdoors. It is good for building lower body strength and endurance.

- **Swimming:** It is nonweight-bearing and low-impact, making it ideal for individuals with joint issues. It engages multiple muscle groups and improves overall strength and cardiovascular fitness. It is also suitable for all fitness levels.
- **Rowing:** It is low-impact and engages both upper and lower body which can be performed on rowing machines indoors or on water outdoors. It is effective for building cardiovascular endurance and muscle strength.
- **Dancing:** It is a fun and engaging way to improve cardiovascular fitness. It includes various styles such as Zumba, aerobic dance classes, or recreational dancing. It is suitable for all fitness levels.
- **Elliptical training:** It is also low-impact, simulating walking or running without joint stress which is performed on an elliptical machine, often with adjustable resistance and incline. It is suitable for all fitness levels.
- **Group exercise classes:** It includes structured aerobic classes like step aerobics, spinning, or cardio kickboxing. This exercise provides social interaction and motivation. It is suitable for various fitness levels, with modifications available for different abilities.

The type or mode of aerobic exercise is a crucial component of an effective exercise program. By considering individual preferences, fitness levels, health status, and accessibility, individuals can select appropriate aerobic activities that enhance enjoyment, adherence, and overall health benefits. Incorporating a variety of aerobic exercise types can help maintain motivation and ensure a well-rounded fitness routine.

Aerobic conditioning programs are designed to improve cardiovascular health, endurance, and overall fitness levels by rhythmic, continuous, and moderate to high-intensity exercises. Physiotherapist should focus on tailoring programs that suit individual needs while adhering to established principles.

Components

- **Engagement of large muscle groups:** Aerobic exercise involves the use of major muscle groups, such as those in the legs, arms, and core. These muscle groups are essential for maintaining sustained movement and ensuring that the body can perform the exercise efficiently. For instance:
  - **Leg muscles:** Used extensively in activities like walking, running, cycling, and swimming.
  - **Arm muscles:** Engaged during swimming, rowing, and certain aerobic fitness classes.
  - **Core muscles:** Stabilize the body during all forms of aerobic exercise, ensuring balance and proper form.
- **Continuous performance:** The hallmark of aerobic exercise is its ability to be maintained continuously over an extended period. This contrasts with anaerobic exercises, which are typically short bursts of high-intensity activity. Continuous performance means:
  - **Sustained activity:** Aerobic exercises can last from a minimum of 20 minutes to several hours, depending on the intensity and the individual's fitness level.
  - **Endurance:** This aspect helps build cardiovascular endurance, as the heart and lungs are required to work steadily over time.
- **Rhythmic nature:** Aerobic exercises are characterized by their rhythmic, repetitive motions, which help in maintaining a steady pace. The rhythmic nature includes:
  - **Consistent movements:** Activities like walking, running, cycling, and swimming involve repetitive motions that can be performed smoothly and regularly.
  - **Flow and coordination:** This rhythm helps in coordinating the movements of different muscle groups, enhancing overall efficiency, and reducing the risk of injury.
- **Utilization of the body's metabolic system to use oxygen for energy production:** Aerobic exercise relies on the aerobic metabolic pathway, which uses oxygen to convert nutrients (primarily carbohydrates and fats) into energy. This process includes:
  - **Oxygen consumption:** During aerobic exercise, the body increases its intake of oxygen to meet the heightened energy demands.
  - **Energy production:** The aerobic system is efficient for sustained energy production, making it ideal for long-duration activities.
  - **Metabolic efficiency:** Over time, regular aerobic exercise enhances the body's ability to utilize oxygen more efficiently, improving overall metabolic health.

OVERTRAINING SYNDROME

Overtraining syndrome (OTS), in the context of aerobic training, is a critical issue for endurance athletes and individuals engaging in high-volume, high-intensity aerobic exercise without sufficient recovery. Understanding the causes, symptoms, prevention, and treatment of OTS in aerobic training is essential for physical therapists to ensure the health and performance of athletes and active individuals.

Causes

- **Excessive training volume and intensity:** Prolonged periods of high-intensity aerobic exercise, such as long-distance running, cycling, or swimming, can lead to excessive stress on the cardiovascular and musculoskeletal systems. Inadequate recovery between intense training sessions can prevent the body from repairing and adapting, leading to cumulative fatigue and overtraining.

- **Inadequate recovery:** Insufficient sleep, poor nutrition, and lack of rest days contribute to inadequate recovery, which is critical for adaptation and performance improvement. Continuous training without adequate rest disrupts the balance between training load and recovery, leading to OTS.
- **Psychological stress:** Psychological stress from personal life, work, or competition can exacerbate the physical stress of aerobic training, increasing the risk of OTS. Mental fatigue and emotional strain can impair recovery and reduce the body's ability to handle training stress.

Symptoms

Performance-Related

- **Decreased performance:** A notable decline in endurance, speed, and overall athletic performance despite continued or increased training efforts.
- **Increased perceived effort:** Activities that were previously manageable feel significantly more challenging.

Physiological

- **Chronic fatigue:** Persistent tiredness that does not improve with rest.
- **Sleep disturbances:** Difficulty falling asleep or staying asleep, leading to poor sleep quality.
- **Frequent illnesses:** Increased frequency of infections, particularly upper respiratory tract infections, due to suppressed immune function.
- **Musculoskeletal pain:** Persistent muscle soreness, joint pain, and increased incidence of overuse injuries.
- **Altered heart rate:** Elevated resting heart rate and abnormal heart rate responses to exercise, such as a delayed return to resting heart rate after exertion.

Psychological

- **Mood disturbances:** Increased irritability, anxiety, depression, and a sense of burnout.
- **Loss of motivation:** Reduced enthusiasm for training and competition.
- **Concentration difficulties:** Impaired cognitive function and difficulty focusing during training and daily activities.

Diagnosis

Diagnosing OTS in aerobic athletes involves a combination of subjective and objective measures:

- **Medical history and physical examination:** Evaluating symptoms, training history, and overall health.
- **Performance tests:** Monitoring changes in endurance performance metrics over time.
- **Laboratory tests:** Assessing biomarkers such as cortisol levels, immune function markers, and other indicators of systemic stress and inflammation.

Prevention

- **Periodization of training:** Implement structured training programs that include cycles of varying intensity and volume to allow for adequate recovery. It includes phases of reduced training load or active recovery to prevent chronic overreaching.
- **Adequate rest and recovery:** Ensure sufficient sleep, rest days, and active recovery sessions to allow the body to repair and adapt. Incorporate relaxation techniques and stress management strategies to support psychological recovery.
- **Balanced nutrition:** Provide adequate caloric intake, macronutrients, and micronutrients to support training demands and recovery. Emphasize the importance of hydration and nutrient timing to optimize performance and recovery.
- **Monitoring training load:** Use tools such as training logs, heart rate monitors, and perceived exertion scales to track and adjust training intensity. Monitor early signs of overtraining, such as elevated resting heart rate, decreased performance, and mood changes.
- **Psychological support:** Address mental health and stress management through counseling and relaxation techniques. Foster a supportive training environment that encourages open communication about training stress and recovery needs.

Did You Know?

Prolonged and excessive aerobic training without adequate rest can lead to alterations in hormones such as cortisol and testosterone, which play essential roles in regulating metabolism, muscle growth, and recovery.

ABNORMAL RESPONSES TO ACUTE AEROBIC EXERCISE

Abnormal responses can indicate underlying health issues or inappropriate exercise intensity and necessitate immediate attention to prevent serious health consequences.

Cardiovascular

- **Abnormal heart rate responses:**
  - **Bradycardia:** An abnormally slow heart rate (below 60 beats/min) during exercise can indicate conduction problems, such as sick sinus syndrome or an atrioventricular block.

  - **Tachycardia:** An excessively high heart rate (above 100 beats/min at rest or beyond the expected range during exercise) may be due to arrhythmias, dehydration, or overexertion.
  - **Heart rate irregularities:** Irregular heartbeats or palpitations can suggest arrhythmias, such as atrial fibrillation or ventricular tachycardia, which requires medical evaluation.
- **Blood pressure abnormalities:**
  - **Hypertensive response:** Excessively high blood pressure (e.g., systolic >250 mm Hg or diastolic >115 mm Hg) during exercise can increase the risk of cardiovascular events like stroke or myocardial infarction.
  - **Hypotensive response:** A drop in systolic blood pressure below pre-exercise levels or a failure of blood pressure to rise appropriately can indicate cardiac issues, such as ischemia or heart failure.
- **Chest pain (angina):** Chest pain or discomfort during exercise may indicate myocardial ischemia, particularly if it is relieved by rest or nitroglycerin. Immediate cessation of exercise and medical evaluation are required.
- **Dyspnea:** Abnormal shortness of breath, especially if disproportionate to the exercise intensity, can signal underlying cardiovascular or pulmonary conditions, such as heart failure or chronic obstructive pulmonary disease (COPD).

Metabolic

Hypoglycemia: Symptoms such as dizziness, confusion, sweating, and weakness can indicate low blood sugar, particularly in individuals with diabetes or those who have not eaten adequately before exercise. Monitoring blood glucose levels and ensuring proper nutritional intake are crucial.

Musculoskeletal

- **Muscle cramps and spasms:** Painful muscle contractions during or after exercise may be due to dehydration, electrolyte imbalances, or overexertion. Ensuring proper hydration and electrolyte balance, as well as appropriate warm-up and cool-down routines, can help prevent these issues.
- **Musculoskeletal pain:** Sharp or severe pain in muscles, joints, or connective tissues can indicate overuse injuries, sprains, or strains. Persistent pain requires medical evaluation and possible modifications to the exercise regimen.

Respiratory

- **Exercise-induced bronchoconstriction:** Symptoms include wheezing, coughing, and shortness of breath during or after exercise. Individuals with exercise-induced bronchoconstriction (EIB) or asthma should use pre-exercise medications and avoid triggers.
- **Hyperventilation:** Rapid or shallow breathing may be due to anxiety, overexertion, or underlying respiratory issues. Ensuring a controlled breathing pattern and appropriate exercise intensity can help manage this response.

Neurological

- **Dizziness or syncope:** Lightheadedness, dizziness, or fainting during exercise can indicate issues such as orthostatic hypotension, dehydration, or cardiac conditions. Immediate cessation of exercise and medical evaluation are necessary.
- **Visual disturbances:** Blurred vision, seeing spots, or temporary blindness can signal inadequate blood flow to the brain or eyes, possibly due to hypotension or other cardiovascular issues. Immediate medical attention is required.

Gastrointestinal

- **Nausea and vomiting:** These symptoms during or after exercise can result from overexertion, dehydration, or inadequate nutrition. Ensuring proper hydration and avoiding heavy meals before exercise can help prevent these issues.
- **Abdominal pain:** Cramping or pain in the abdomen can be related to gastrointestinal distress, often due to improper eating or hydration before exercise. If pain is severe or persistent, medical evaluation is necessary.

Management and Prevention

- **Pre-exercise screening:** It is recommended to conduct thorough health screenings and risk assessments, particularly for individuals with known health conditions or risk factors (Table 11.1).
- **Gradual progression:** Ensure a gradual increase in exercise intensity and duration, especially for beginners or those returning to exercise after a hiatus.
- **Proper warm-up and cool-down:** It includes adequate warm-up and cool-down periods to prepare the cardiovascular and musculoskeletal systems for exercise and facilitate recovery.
- **Monitoring:** It is recommended to use heart rate monitors, blood pressure cuffs, and other tools to monitor physiological responses during exercise, particularly in high-risk individuals.
- **Hydration and nutrition:** Ensure adequate hydration and appropriate nutritional intake before, during, and after exercise to maintain energy levels and prevent metabolic disturbances.
- **Individualized exercise programs:** There should be tailored exercise prescriptions to individual needs, considering health status, fitness levels, and personal goals.

TABLE 11.1: Example of questions asked during pre-exercise screening

Sl. no	Question	Response	
1.	Has your doctor ever told you that you have a heart condition or have you ever suffered a stroke?	Yes	No
2.	Do you ever experience unexplained pain in your chest at rest or during physical activity/exercise?	Yes	No
3.	Do you ever feel faint or have spells of dizziness during physical activity/exercise that causes you to lose balance?	Yes	No
4.	Have you had an asthma attack requiring immediate medical attention at any time over the last 12 months?	Yes	No
5.	If you have diabetes (type I or type II) have you had trouble controlling your blood glucose in the last 3 months?	Yes	No
6.	Do you have any diagnosed muscle, bone or joint problems that you have been told could be made worse by participating in physical activity/exercise?	Yes	No
7.	Do you have any other medical condition(s) that may make it dangerous for you to participate in physical activity/exercise?	Yes	No

SUMMARY

- Aerobic exercise is a type of physical activity that engages large muscle groups, can be performed continuously, and relies on the body's metabolic system for energy production.
  - It involves major muscle groups like legs, arms, and core, ensuring efficient performance.
  - The aerobic metabolic pathway is central to aerobic exercise, which uses oxygen to convert nutrients into energy, supporting long-duration activities.
- Regular aerobic exercise yields numerous long-term beneficial physiological effects, including improved cardiac output, lower resting heart rate, enhanced vascular function, reduced blood pressure, increased HDL cholesterol and decreased LDL cholesterol, and enhanced myocardial oxygen supply.
  - It also affects lipid metabolism by increasing the activity of enzymes involved in lipid transport and breakdown, resulting in increased HDL cholesterol and decreased LDL cholesterol.
- In terms of the respiratory system, aerobic exercise increases lung capacity, enhances alveolar ventilation, increases respiratory rate and tidal volume, improves respiratory muscle endurance, and increases oxygen diffusion capacity. It also enhances the musculoskeletal system by increasing mitochondrial density, capillary density, muscle glycogen storage, muscle fiber type transformation, bone density and strength, joint function and flexibility, and lipid profile.
- Aerobic exercise has significant neurophysiological effects that enhance brain function, improve mood, and protect against cognitive decline.
  - These effects are mediated through various mechanisms that influence brain structure and function, neurochemical balance, and cognitive performance.
- Aerobic capacity is an essential factor in determining an individual's cardiovascular fitness, with factors such as gender, age, heredity, body size and composition, mode and intensity of exercise, and state of training.
  - There are several methods used to measure aerobic exercise capacity, including Maximal Oxygen Uptake (VO_2 Max) Testing, Submaximal Exercise Testing, Step Tests, Field Tests, Shuttle Run Tests, and Heart Rate Recovery (HRR) Tests.
- The prescription of aerobic exercise involves several key components, which can be systematically described using the FITT principle.
- Exercise intensity is crucial for determining the right type of exercise for an individual's fitness level and health status.
  - Aerobic exercise, which involves a combination of intensity and frequency, can improve cardiovascular fitness and overall health.
  - Common types include continuous training, interval training, and circuit training.
  - Choosing the appropriate exercise type is essential for an effective exercise program.
- Abnormal responses to acute aerobic exercise can indicate underlying health issues or inappropriate exercise intensity.
  - To manage these, it is recommended to conduct thorough health screenings, gradually increase exercise intensity and duration, and ensure proper warm-up and cool-down periods.
  - Proper nutrition and hydration can also help prevent overuse injuries.
- Overtraining syndrome (OTS) is a significant issue in aerobic training, particularly for endurance athletes and those engaging in high-volume, high-intensity exercise without sufficient recovery.
  - Prevention involves periodization of training, adequate rest and recovery, balanced nutrition, monitoring training load, and psychological support.

FURTHER READINGS

- Ahlskog, J. E., Geda, Y. E., Graff-Radford, N. R., and Petersen, R. C. (2011). Physical exercise as a preventive or disease-modifying treatment of dementia and brain aging. Mayo Clinic proceedings, 86(9), 876–884. https://doi.org/10.4065/mcp.2011.0252
- Ainsworth, B. E., Haskell, W. L., Whitt, M. C., Irwin, M. L., Swartz, A. M., Strath, S. J., ... and Leon, A. S. (2000). Compendium of physical activities: An update of activity codes and MET intensities. Medicine and Science in Sports and Exercise, 32 (9, Suppl 1), S498–S504.
- American College of Sports Medicine. (2022). ACSM's guidelines for exercise testing and prescription. Wolters Kluwer.
- ATS Committee on Proficiency Standards for Clinical Pulmonary Function Laboratories. (2002). ATS statement: Guidelines for the six-minute walk test. American Journal of Respiratory and Critical Care Medicine, 166 (1), 111-117.
- Benedetti, M. G., Furlini, G., Zati, A., and Letizia Mauro, G. (2018). The Effectiveness of Physical Exercise on Bone Density in Osteoporotic Patients. BioMed Research International, 2018, 4840531. https://doi.org/10.1155/2018/4840531
- Best J. R. (2010). Effects of Physical Activity on Children's Executive Function: Contributions of Experimental Research on Aerobic Exercise. Developmental review: DR, 30(4), 331–551. https://doi.org/10.1016/j.dr.2010.08.001
- Bird, S. R., and Hawley, J. A. (2017). Update on the effects of physical activity on insulin sensitivity in humans. BMJ open sport and exercise medicine, 2(1), e000143. https://doi.org/10.1136/bmjsem-2016-000143
- Borg, G. (1998). Borg's perceived exertion and pain scales. Human Kinetics.
- Brooks, G. A., Fahey, T. D., and Baldwin, K. M. (2005). Exercise physiology: Human bioenergetics and its applications (4th ed.). McGraw-Hill.
- Chodzko-Zajko, W. J., Proctor, D. N., Fiatarone Singh, M. A., Minson, C. T., Nigg, C. R., Salem, G. J., and Skinner, J. S. (2009). Exercise and physical activity for older adults. Medicine and Science in Sports and Exercise, 41 (7), 1510–1530.
- Coggan, A. R., and Coyle, E. F. (1991). Carbohydrate ingestion during prolonged exercise: Effects on metabolism and performance. Exercise and sport sciences reviews, 19, 1–40.
- Cooper K. H. (1968). A means of assessing maximal oxygen intake. Correlation between field and treadmill testing. JAMA, 203(3), 201–204.
- Cornelissen, V. A., and Smart, N. A. (2013). Exercise training for blood pressure: A systematic review and meta-analysis. Journal of the American Heart Association, 2(1), e004473. https://doi.org/10.1161/JAHA.112.004473
- Erickson, K. I., Voss, M. W., Prakash, R. S., Basak, C., Szabo, A., Chaddock, L., Kim, J. S., Heo, S., Alves, H., White, S. M., Wojcicki, T. R., Mailey, E., Vieira, V. J., Martin, S. A., Pence, B. D., Woods, J. A., McAuley, E., and Kramer, A. F. (2011). Exercise training increases size of hippocampus and improves memory. Proceedings of the National Academy of Sciences of the United States of America, 108(7), 3017–3022. https://doi.org/10.1073/pnas.1015950108
- Fagard R. H. (2001). Exercise characteristics and the blood pressure response to dynamic physical training. Medicine and science in sports and exercise, 33(6 Suppl), S484–S494. https://doi.org/10.1097/00005768-200106001-00018
- Garber, C. E., Blissmer, B., Deschenes, M. R., Franklin, B. A., Lamonte, M. J., Lee, I. M., ... and Swain, D. P. (2011). Quantity and quality of exercise for developing and maintaining cardiorespiratory, musculoskeletal, and neuromotor fitness in apparently healthy adults: Guidance for prescribing exercise. Medicine and Science in Sports and Exercise, 43 (7), 1334–1359.
- Gibala, M. J., Little, J. P., Macdonald, M. J., and Hawley, J. A. (2012). Physiological adaptations to low-volume, high-intensity interval training in health and disease. Journal of Physiology, 590*(5), 1077–1084.
- Gill, J. M., and Hardman, A. E. (2003). Exercise and postprandial lipid metabolism: An update on potential mechanisms and interactions with high-carbohydrate diets. The Journal of Nutritional Biochemistry, 14(3), 122-132.
- Green, D. J., Maiorana, A., O'Driscoll, G., and Taylor, R. (2004). Effect of exercise training on endothelium-derived nitric oxide function in humans. The Journal of Physiology, 561(Pt 1), 1–25. https://doi.org/10.1113/jphysiol.2004.0681974.
- Haskell, W. L., Lee, I. M., Pate, R. R., Powell, K. E., Blair, S. N., Franklin, B. A., Macera, C. A., Heath, G. W., Thompson, P. D., and Bauman, A. (2007). Physical activity and public health: Updated recommendation for adults from the American College of Sports Medicine and the American Heart Association. Medicine and science in sports and exercise, 39(8), 1423–1434. https://doi.org/10.1249/mss.0b013e3180616b27
- Holloszy J. O. (2008). Regulation by exercise of skeletal muscle content of mitochondria and GLUT4. Journal of Physiology and Pharmacology: An official journal of the Polish Physiological Society, 59 Suppl 7, 5–18
- Howley, E. T., and Thompson, D. L. (2020). Fitness Professional's Handbook* (7th ed.). Human Kinetics.
- https://sites.google.com/site/compendiumofphysicalactivities/home
- Imai, K., Sato, H., Hori, M., Kusuoka, H., Ozaki, H., Yokoyama, H., ... and Takeda, H. (1994). Vagally mediated heart rate recovery after exercise is accelerated in athletes but blunted in patients with chronic heart failure. Journal of the American College of Cardiology, 24 (6), 1529–1535.
- Jette, M., Sidney, K., and Blümchen, G. (1990). Metabolic equivalents (METs) in exercise testing, exercise prescription, and evaluation of functional capacity. Clinical Cardiology, 13(8), 555-565.
- Kelley, G. A., and Kelley, K. S. (2009). Impact of progressive resistance training on lipids and lipoproteins in adults: A meta-analysis of randomized controlled trials. Preventive medicine, 48(1), 9–19. https://doi.org/10.1016/j.ypmed.2008.10.010
- Kellmann, M. (2010). Preventing overtraining in athletes in high-intensity sports and stress/recovery monitoring. Scandinavian Journal of Medicine and Science in Sports, 20 (Suppl 2), 95-102. https://doi.org/10.1111/j.1600-0838.2010.01192.x

- Kodama, S., Tanaka, S., Saito, K., Shu, M., Sone, Y., Onitake, F., Suzuki, E., Shimano, H., Yamamoto, S., Kondo, K., Ohashi, Y., Yamada, N., and Sone, H. (2007). Effect of aerobic exercise training on serum levels of high-density lipoprotein cholesterol: A meta-analysis. Archives of internal medicine, 167(10), 999–1008. https://doi.org/10.1001/archinte.167.10.999
- Kraus, W. E., Houmard, J. A., Duscha, B. D., Knetzger, K. J., Wharton, M. B., McCartney, J. S., Bales, C. W., Henes, S., Samsa, G. P., Otvos, J. D., Kulkarni, K. R., and Slentz, C. A. (2002). Effects of the amount and intensity of exercise on plasma lipoproteins. The New England Journal of Medicine, 347(19), 1483–1492. https://doi.org/10.1056/NEJMoa020194
- Kreher, J. B., and Schwartz, J. B. (2012). Overtraining syndrome: A practical guide. Sports Health, 4 (2), 128–138. https://doi.org/10.1177/1941738111434406
- Laughlin, M. H., and McAllister, R. M. (1992). Exercise training-induced coronary vascular adaptation. Journal of Applied Physiology (Bethesda, Md. : 1985), 73(6), 2209–2225. https://doi.org/10.1152/jappl.1992.73.6.2209.
- McArdle, W. D., Katch, F. I., and Katch, V. L. (2015). Exercise physiology: Nutrition, energy, and human performance (8th ed.). Lippincott Williams and Wilkins.
- McKenzie D. C. (2012). Respiratory physiology: Adaptations to high-level exercise. British Journal of Sports Medicine, 46(6), 381–384. https://doi.org/10.1136/bjsports-2011-090824
- Meeusen, R., Duclos, M., Foster, C., Fry, A., Gleeson, M., Nieman, D., Raglin, J., Rietjens, G., Steinacker, J., Urhausen, A.; European College of Sport Science; American College of Sports Medicine. (2013). Prevention, diagnosis, and treatment of the overtraining syndrome: Joint consensus statement of the European College of Sport Science and the American College of Sports Medicine. Medicine and Science in Sports and Exercise, 45 (1), 186–205. https://doi.org/10.1249/MSS.0b013e318279a10a.
- Powers, S. K., and Howley, E. T. (2012). Exercise physiology: Theory and application to fitness and performance (8th ed.). McGraw-Hill.
- Powers, S. K., and Howley, E. T. (2018). Exercise physiology: Theory and application to fitness and performance (10th ed.). McGraw-Hill Education.
- Schuch, F. B., Vancampfort, D., Richards, J., Rosenbaum, S., Ward, P. B., and Stubbs, B. (2016). Exercise as a treatment for depression: A meta-analysis adjusting for publication bias. Journal of Psychiatric Research, 77, 42–51. https://doi.org/10.1016/j.jpsychires.2016.02.023
- Sheel A. W. (2002). Respiratory muscle training in healthy individuals: Physiological rationale and implications for exercise performance. Sports medicine (Auckland, N.Z.), 32(9), 567–581. https://doi.org/10.2165/00007256-200232090-00003
- Smith L. L. (2004). Tissue trauma: The underlying cause of overtraining syndrome?. Journal of Strength and Conditioning Research, 18(1), 185–193. https://doi.org/10.1519/1533–4287(2004)018<0185:tttuco>2.0.co;2
- Tsatsoulis, A., and Fountoulakis, S. (2006). The protective role of exercise on stress system dysregulation and comorbidities. Annals of the New York Academy of Sciences, 1083, 196–213. https://doi.org/10.1196/annals.1367.020
- Uthman O A, van der Windt D A, Jordan J L, Dziedzic K S, Healey E L, Peat G M et al. Exercise for lower limb osteoarthritis: Systematic review incorporating trial sequential analysis and network meta-analysis BMJ 2013; 347:f 5555 doi:10.1136/bmj.f5555
- Voss, M. W., Vivar, C., Kramer, A. F., and van Praag, H. (2013). Bridging animal and human models of exercise-induced brain plasticity. Trends in cognitive sciences, 17(10), 525–544. https://doi.org/10.1016/j.tics.2013.08.001

STUDENT ASSIGNMENT

LONG ANSWER QUESTIONS

1. Describe the physiological response to aerobic exercise.
2. Demonstrate examination and evaluation of aerobic exercises.
3. Explain aerobic conditioning program under the following heading—principles, types and phases of aerobic training.

SHORT ANSWER QUESTIONS

1. Define aerobic exercises.
2. Write a short note on exercise testing.
3. Write briefly about overtraining syndrome.

MULTIPLE CHOICE QUESTIONS

1. **Which of the following statements accurately describes the primary metabolic system utilized during aerobic exercise?**
    a. The anaerobic metabolic system relies on oxygen to convert nutrients into energy.
    b. Aerobic exercise primarily utilizes the anaerobic metabolic system.
    c. Aerobic exercise relies on the body's aerobic metabolic system, which uses oxygen to convert nutrients into energy.
    d. Anaerobic exercise relies on the body's aerobic metabolic system for energy production.
2. **Which of the following is NOT a key component of aerobic exercise?**
    a. Engagement of major muscle groups
    b. Continuous and rhythmic movement
    c. High-intensity bursts of activity
    d. Promotion of cardiovascular endurance
3. **Which term refers to the maximum amount of oxygen one can consume during intense activity, indicating cardiovascular fitness?**
    a. Anaerobic threshold
    b. Oxygen debt
    c. VO_2 max
    d. Aerobic capacity
4. **What immediate effect does aerobic exercise have on blood flow to active muscles?**
    a. Decreases blood flow
    b. No change in blood flow
    c. Slightly increases blood flow
    d. Surges blood flow
5. **Which of the following adaptations occurs in the cardiovascular system as a result of regular aerobic exercise?**
    a. Decreased cardiac output
    b. Increased resting heart rate
    c. Reduced blood pressure
    d. Decreased vascular function
6. **Which testing method involves graded exercise on a treadmill or cycle ergometer until exhaustion to measure maximum oxygen uptake?**
    a. Submaximal exercise testing
    b. Field tests
    c. VO_2 max testing
    d. Heart rate recovery tests
7. **What is the recommended frequency of vigorous-intensity aerobic activities per week for optimal health?**
    a. 1–2 days
    b. 3–4 days
    c. 5–7 days
    d. None of these

8. **Which principle of exercise programming involves gradually increasing exercise intensity, duration, and frequency to prevent overuse injuries?**
   a. Overload
   b. Reversibility
   c. Specificity
   d. Progression
9. **Which type of aerobic exercise involves alternating periods of high-intensity exercise with periods of lower intensity or rest?**
   a. Continuous training
   b. Interval training
   c. Fartlek training
   d. Cross-training
10. **What is the primary goal of tapering phase in aerobic training?**
    a. Building cardiovascular endurance
    b. Maximizing muscle strength
    c. Allowing the body to recover fully before competition
    d. Increasing exercise intensity

ANSWER KEY

1. c **2.** c **3.** c **4.** d **5.** c **6.** c **7.** b **8.** d **9.** b **10.** c

12 Functional Re-Education

Paridhi Sharma, Pradeep Kumar

LEARNING OBJECTIVES

After the completion of the chapter, the readers will be able to:

- Understand the concept of functional re-education.
- Demonstrate lying to sitting: Activities on the mat/bed, movement and stability at floor level; sitting activities and gait; lower limb and upper limb activities.
- Demonstrate common mat activities.

CHAPTER OUTLINE

- Introduction
- Benefits of Functional Re-Education Program
- Components
- MAT/BED Activities
- Mobility and Stability Activities at the Floor

KEY TERMS

Bridging: In supine lying position, the patient is instructed to lift the trunk off the ground or from the couch, with both knees flexed and the feet resting on the couch. The trunk, hips and knees should be aligned in a straight line.

Functional re-education: It involves teaching people to perform familiar actions that are affected by ailments.

Hiking: Hiking in exercise therapy usually refers to a form of cardiovascular exercise that involves walking over natural terrain such as trails, hills or mountains. It is a low-impact activity that provides both aerobic conditioning and muscular engagement, particularly in the lower body. Hiking can be adapted to various fitness levels and can be a great way to improve cardiovascular health, endurance, and strength.

Hitching: It typically refers to a movement pattern involving a slight pause or interruption in the motion. This could involve pausing momentarily at certain points during an exercise movement or intentionally adding brief pauses to create variability and challenge in the exercise. Hitching can help improve stability, control, and muscle engagement by disrupting the typical rhythm of movement.

Quadruped position: Other names for it include the four-footed position and prone kneeling. In contrast to prone lying on one's hands, this position has a lower base of support and a higher center of gravity.

INTRODUCTION

Re-education, as the name suggests, implies educating something which has already been known by a person. Hence, functional re-education means training an individual for a particular action or function which they are already aware of. Here, the individual is familiar with the activity or function that has to be performed but is unable to perform properly due to some ailment or pathology. The main motto of this program is to make a person independent.

In this training, the sequence of progression of positions (for example, lying to sitting to walking) is taught to the patient, just like the developmental milestones of a child from lying to walking. As soon as the child is born, it assumes supine position and once its stability is established, the side lying position can be achieved by the child. This side lying position is further advanced to prone. Once in prone position, the child tries to lift up its head and trunk with the help of the upper limb, achieving prone on elbows position. Soon after this, child further tries and gains prone on hand position. In this position, child pulls his legs upward and adopts the quadruped position, from which it begins first movement, i.e., crawling. From this quadruped position, the child further progresses to kneeling and later, kneel-walking with some support say wall, furniture, etc. After attaining enough stability, child progresses to the standing position with the help of support in the kneeling position.

BENEFITS OF FUNCTIONAL RE-EDUCATION PROGRAM

A functional re-education program can be created for a patient based on their level of independence and any accompanying conditions. The sequence of activities in this program can be varied from patient to patient. Multiple postures can be overlapped during the program. A functional re-education training program helps in:

- Improving the strength and endurance of the muscles.
- Gaining static and dynamic stability.
- Enhancing pelvic stability.
- Improving balance and coordination.
- Improving postural stability.
- Enhancing proprioceptive function.
- Improving the ambulatory skills of the patient.

COMPONENTS

Functional re-education consists of bed/mat activities (pre-ambulatory training) and ambulatory training. This includes:

- Rolling
  - From supine to side lying
  - From side lying to prone lying
  - From prone lying to side lying
  - From side lying to supine
- Prone-on-elbows lying
- Prone-on-hands lying
- Side lying with elbow support
- Quadruped position
- Side sitting
- Sitting
- Kneeling
- Kneel sitting
- Half-kneeling
- Standing
- Walking

In each of the aforementioned postures, many exercises can be practiced for progression in order to improve the stability and mobility. To the maximal extent possible, all of the abilities of the patient are employed to minimize the effects of inactivity and assist in helping themselves. This chapter will discuss some recommendations for basic functional activity training. When the patient can complete the tasks independently, they can practice alone or with others to speed up performance and enhance endurance.

Clinical Correlation

Functional re-education draws inspiration from the developmental milestones of infants, mirroring the natural progression from lying to standing that occurs during early childhood development. By retraining patients in a similar sequence of activities, therapists aim to restore functional abilities and promote independence, much like a child learns to navigate their environment through successive stages of mobility. This approach not only addresses physical impairments but also taps into the innate motor learning mechanisms of the human body.

MAT/BED ACTIVITIES

Rolling from Supine Lying to Side Lying

Rolling from supine lying to side lying (Figs 12.1A and B) requires complete body flexion-with-rotation, started and guided by the head and neck. When it is possible, strong limb action is used to help, such as when rolling forward and to the left.

- A firm pull on the left hand, which is holding onto a fixture at the side of the bed, helps the trunk rotate. The left shoulder retracting makes the right shoulder protract easier.
- It is possible to drag, push, or swing the right arm across to the opposing hip.
- The left leg can be looped over the side of the bed and used to help rotate the lower trunk and pelvis, by flexing and adducting.

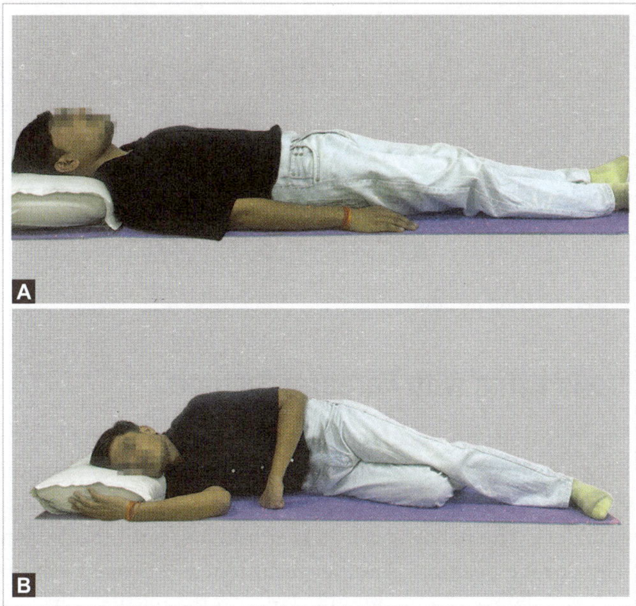

Figs 12.1A and B: Rolling from supine to side lying

- Rolling motion of the lower trunk can be finished by bending the knee so that pressure on the right foot can lift and push the right side of the pelvis higher and over.

The body can be placed back in supine lying position when the movements are reversed. The return is started by pushing off with the arm or retracting the right shoulder, but to maintain control, the head should stay forward until the final position is attained.

Uses

- Effective at avoiding pressure sores.
- Helpful in assisting nursing activities including bed making, assessing back, and postural drainage, among others.
- Aids the patients in gaining some kind of independence so that they can turn as necessary.
- Makes it simple for the patients to proceed to the next posture.

Rolling from Side Lying to Prone Lying

- Left elbow is extended along with the adduction of the left shoulder and is placed underneath the torso.
- Rotating the upper trunk with the right hand while holding the head end next to bars or the bed end.
- The entire left upper extremity is extended.
- Rotate the lower trunk by rotating the right knee, which is flexed, as the foot pushes the mat as shown in Figure 12.2.

Uses

- The most crucial aspect of the prone position is that it aids to counteract and combat the serious impacts of prolonged bed rest.

Fig. 12.2: Prone lying

- According to a study by Meftahi et al., both massage and prone lying alone can significantly lower systolic blood pressure (SBP).

Clinical Correlation

It has been suggested that adopting the prone position for conscious COVID-19 patients needing basic respiratory assistance may help them by enhancing oxygenation, lowering the requirement for invasive ventilation, and possibly even reducing death (Bamford et al.).

Rolling Prone Lying to Side Lying

- The upper trunk and head can be elevated by applying pressure to the mat with the right hand held sideways.
- Right knee bent; lower trunk rotated by applying pressure to mat.
- The body is propelled toward the left side by the left hand, which is holding the bedside rails or the right-side bed end.
- Here, the entire body rotates 90°. Right hand now ascends, while left hand descends.

Bridging

In supine lying position, the patient is instructed to lift their trunk off the ground or from the couch, with both knees flexed and the feet resting on the couch (crook lying position). The trunk, hips and knees should be aligned in a straight line (Figs 12.3A and B).

Uses

- Bridging makes bedpan routines easier for everyone involved with a patient who is confined to bed.
- Sensitive pressure points are relieved from body weight by raising the lower back off the mattress.
- When rotational and side-flexional movements are incorporated into the lifting motion, the weight may shift to one buttock or the other as it is lowered to the bed (preliminary training for transfers and ambulation).

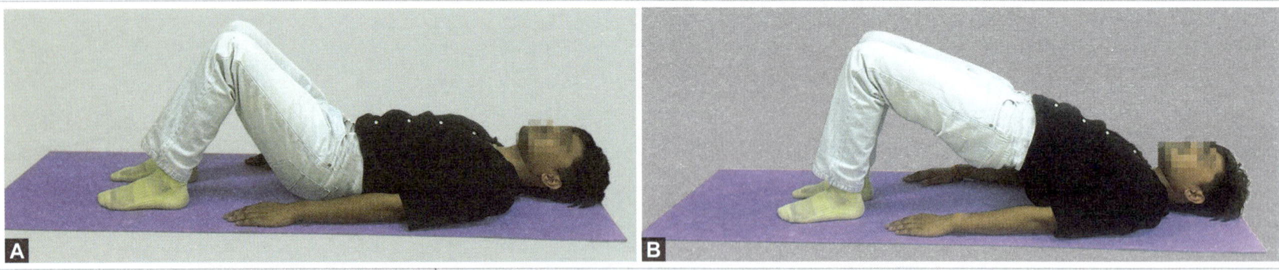

Figs 12.3A and B: Bridging from crook lying position

- In order to counteract the long-term effects of flexor circumstances and maintain the effectiveness of the extensor muscles, extensibility of the hip and lumbar regions is combined.
- When dressing in bed, several tasks, including pulling on pants, are made simpler by the ability to bridge.
- Patient can feel pressure on the bottoms of their feet, which improves proprioception.

Forearm Supported Side Lying

The typical method for getting into this posture is to roll about one side and then push up with the elbow to support the upper trunk with the entire forearm. When both shoulders are in the same plane and the supporting arm is vertical from elbow to shoulder, the upper trunk position is at its most stable. By bending one leg, the pelvis is kept stable (Fig. 12.4).

Uses

- This position is useful when transitioning to sitting from the lying position.
- In this position, one can reach over to a bedside table without getting out of bed.
- It is a rather comfortable position that is great for reading and taking in the scenery while sitting on the floor or couch.
- This position can be helpful in the treatment of some shoulder disorders because applying pressure across the shoulder joint (approximation) increases activity throughout the shoulder area.

Prone Lying with Forearm Support

This position can be attained from forearm supported side-lying position, and moving the free elbow to a position shoulder width apart from the other elbow to support both shoulders. The pelvis and legs keep moving until the person is in a prone position. Another technique is a prone push-up with the elbows. To maintain balance in the posture with the least amount of effort, the upper arms must be vertical, and the weight should be distributed evenly down the entire length of the forearm and hand. The forearms stay in the pronation and the wrist and hands are extended with the palms supported flat on the surface.

Once enough stability is achieved, patient can further progress to prone on hand position by lifting the elbows off the couch. This is also called the prone on elbows position (Fig. 12.5).

Uses

- To keep the head aligned with the upper trunk and prevent stooping posture when there is only a slight loss of normal extensibility, there must be sufficient activation of the extensor muscles of the head and through to the upper trunk.
- All of the muscles of the shoulder region are stimulated when both arms are supported, helping to sustain the position. Certain shoulder issues can benefit from the rocking motion that transfers weight from one arm to the other.

Fig. 12.4: Side lying supported on forearm

Fig. 12.5: Prone on elbows

- Strong coordinated movement, similar to that utilized for other kinds of ambulation, is provided by creeping movements that propel the body along the floor utilizing the arms, pelvic movements, and legs, if possible.
- This position aids in maintaining the extensibility of the lumbar spine and hip joints.
- This position gives the patient a chance to hold and read a magazine, newspaper, etc.

Sitting on the Bedside

The body is propelled upright from the forearm support side lying posture by extending the elbow while the legs are raised and swung over the side of the bed. The body pivots on one buttock during the motion until it reaches the sitting position, at which point the weight is evenly distributed over both buttocks. When sitting, stability is achieved by pressing the thighs firmly against the mat or bed and, if possible, placing the feet on the floor (Figs 12.6A to C).

Uses

- Standing is preceded by this stance.
- It gives many people a welcome chance to "lift the weight off their feet".
- Provides the patient a level of independence by setting the upper limb free to perform various tasks.

Hitching and Hiking

Transfers for wheelchair patients, for instance to move from a bed to a chair, require the capacity to bear weight on the arms to lift and move the pelvis. After the arms provide support, exercise moves the pelvis forward, backward, and sideways while rotating it to get ready to lift the buttocks. The hands and buttocks can be moved alternately, or the latter can be moved one after the other, to achieve advancement in any direction (like walking on the buttocks) (Figs 12.7A and B).

Quadruped Position

Other names for quadruped position include the four-footed position and prone kneeling (Fig. 12.8). In contrast to prone lying on one's hands, this position has a lower base of support and a higher center of gravity. In the re-education program, it is the initial posture where weight bearing *via* the hip joint occurs. It can be achieved from side-sitting or reclining prone on one's hands.

- **Quadruped position from prone on hands:** From the hand-prone position, the body is brought up with or

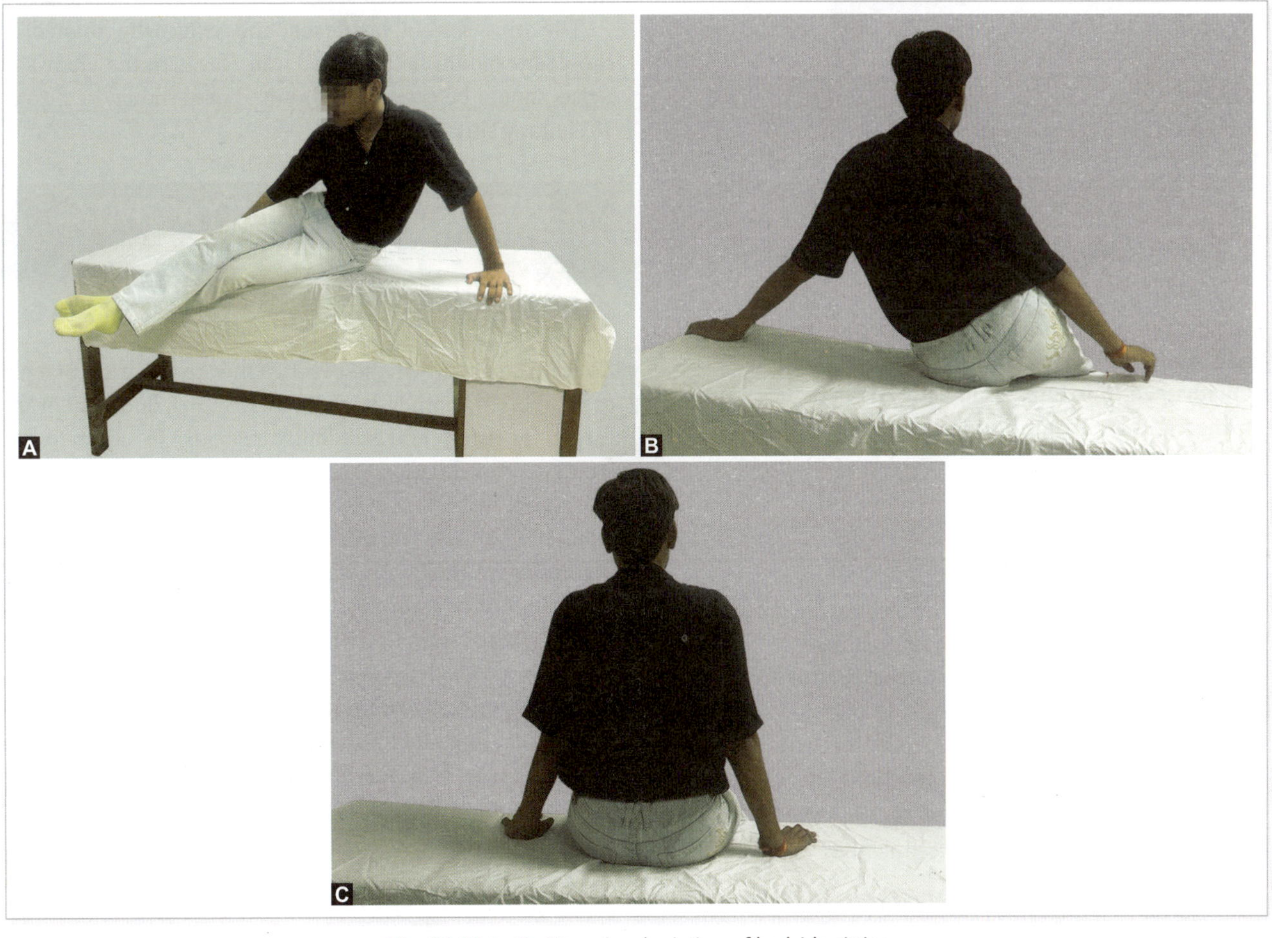

Figs 12.6A to C: Stepwise depiction of bedside sitting

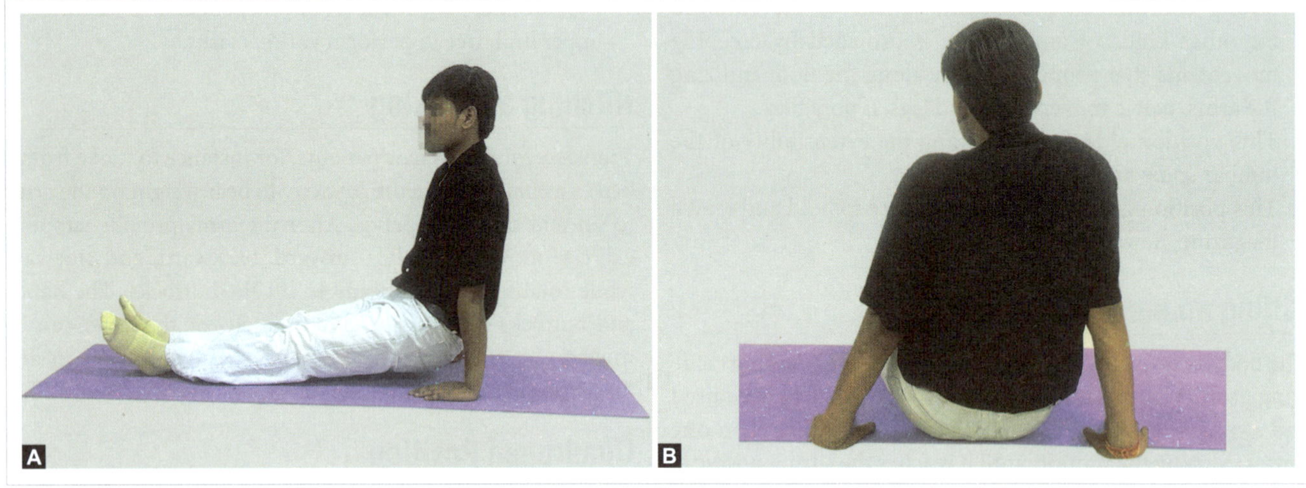

Figs 12.7A and B: Hitching and Hiking

Fig. 12.8: Quadruped position

without the use of an external support by flexing the hips and knees and raising the pelvis to knee level.

- **Quadruped position from side sitting:** The trunk is turned and raised up during side sitting. The hands and knees can bear weight since both upper limbs are placed in front.

Uses

- It can be used for activities performed on the ground level, such as playing with children or gardening.
- It enables individuals who are unable to walk to ambulate within and outside of the house.
- This is where the patient begins to "crawl", which allows them to move across the floor in any direction they choose. Patients with vertigo and others who are temporarily unable to bear weight on their feet may find it to be of great benefit.

Long Sitting

Long sitting is a highly secure position to hold for an extended amount of time. This can be executed from sitting on the side (side sitting). Both upper limbs are supported by keeping them on each side of the trunk, which is held in erect posture by the trunk muscles, which should have adequate power and strength to prevent back sagging.

The hips should be flexed and externally rotated, the shoulders should be abducted and elevated, the elbows and wrists should be extended, and the knees should be flexed to 90° in long sitting.

MOBILITY AND STABILITY ACTIVITIES AT THE FLOOR

Getting Down to the Floor

To learn to get down and practice movement from one place or position to another, it is best to do it on a floor mat or thick carpet for safety and suitability reasons. In order to build strength, coordination, endurance, and confidence without having to worry excessively about balance or the fear of falling, the muscle work is vast and varied. Both patients and therapists are frequently discouraged by the initial challenge of getting to the floor, but, like so many challenges, this can typically be surmounted with aid and proper planning. Having strong and knowledgeable helpers on hand is necessary for passive lifting and maximum support, but many patients are capable of doing a lot for themselves with the right instruction. Following ideas may be helpful in determining the best course of action for a certain patient.

- For the patient who needs assistance to stand, move to face the chair or bed, bending forward so that both hands can

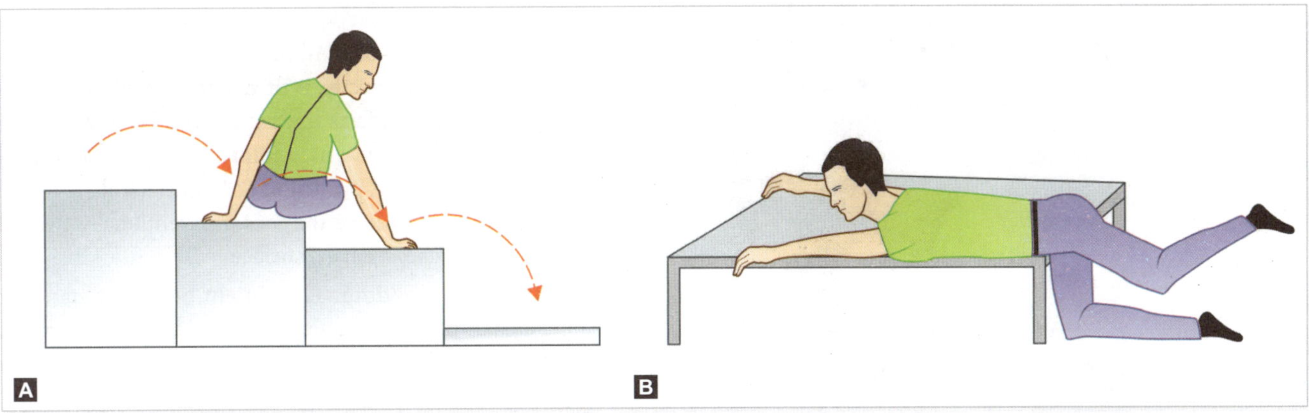

Figs 12.9A and B: Transferring down to the floor

support them, kneel with one hand on the floor far to the side, and then sit down next to it.

- Change positions by moving from a chair to a low stool, then to the floor. The therapist can assemble any number of chairs or wooden crates to create shallow steps. Patients with bi-lateral amputees should particularly consider this approach (Fig. 12.9A).
- On a high mat or wide bed, roll prone. Rotate so that the feet hang over the edge. Then, slide backward, bending the knees until they are resting on the floor. Kneel down and take a seat on the ground (Fig. 12.9B).

Side Sitting

As the trunk is pivoted to sit upright, the push up to side sitting on the floor differs from the push up to sitting on the side of the bed in that there is little to no rotation. The legs are still bent and resting on the ground. Unless substantial lateral trunk and abductor muscular action on one side is available to liberate the hands for other uses, the arms may be used as support to sit up straight (Fig. 12.10).

Fig. 12.10: Side sitting

Uses

- Retaining the position without using the arms highlights how the lateral trunk and abductor muscles are working on one side of the body. This could aid in resolving any imbalances brought on by the inefficiency of these muscles.
- From a side sitting position, an ambulatory movement is started that enables the patient to move from one region of a room to another. This might be helpful for those who fall and need to reach a chair or other piece of furniture for support in getting back up. It can also be helpful when weight-bearing on the legs is not permitted for whatever reason.

Crawling

Crawling can be achieved as a progression from prone kneeling position. Patients can start practicing lifting a hand or a knee off the floor to balance "on three legs" if balance and stability in the quadruped position have been established. Since the weight-free limb can be lifted and moved into a new position before resuming the weight, the crawling motion can now be performed in any desired direction. If feasible, it is best to let the sequence of limb movement and support develop naturally because giving patients specific directions about which and where to move the limbs frequently confuses them (Fig. 12.11).

Uses

- Crawling improve bodily coordination throughout, including the arms and legs moving in unison, as is necessary for walking. The distribution and intensity of the neuromuscular activity used in the crawl are determined by the direction of the crawl, such as forward, backward, or sideways.

Fig. 12.11: Crawling

- This activity is helpful for mobilization and/or for learning control of excessive mobility because the spine is weight-free in the horizontal position, increasing the possible range of movement in these joints.
- Crawling backward is used to retrain the function of the limbs, for example, to regain knee flexion and arm elevation. Weight bearing on the arms and legs increases activity in the region of these joints by approximating joint surfaces.
- For people with very poor balance in an upright posture, crawling provides a safe and efficient way to move from one region of a space to another. People who are not too proud to use the floor not only accomplish their goal of moving, but they also do so on their own.

Kneeling

One can either go down to kneel or squat down. It is advisable to check two things before the patient tries to kneel. The surface on which to kneel must be comfortable enough for the patient to accept pressure on the knees; and there must be enough range of knee flexion, at least 100°. All furniture that will be used as support is positioned correctly, is firm, and is immovable. Any other impairment that affects the patient has been taken into account, such as restricted ankle joint movement or painful toes. A chair or bed is best pushed back against a wall. It can be really uncomfortable to have the toes unintentionally twisted under or pinched against a hard surface.

- **Kneeling from standing:** The patient begins by standing facing a chair or another suitable support, leaning forward to place some weight on one or both hands as the legs bend to lower first one knee, then the other. Alternately, both legs can be bent and both knees can be rocked to the ground at the same time by stepping further away from the support. The rest of the body is brought into alignment in the erect position after the knees are in place.
- **Kneeling down from sitting:** The patient leans to one side while sitting, shifting weight between two hands. The twisting motion continues as the patient's knees descend to the floor and the trunk stays prone and supported by the bed. Making sure the knees are on the floor next to the bed (and not underneath it). Hip extension and the assistance of the arms are used to align the body in the erect position once the knees are firmly planted on the ground (Figs 12.12A and B).
- **Kneeling up from prone kneeling:** The kneeling position is attained by leaning back on the heels and stretching the hips and knees. The lower legs serve as a bracket, distributing the body weight equally, and the remainder of the body is held upright. If necessary, the arms can be employed to regulate or aid the movement.
- **Kneeling up from side sitting:** The turn and raise of the pelvis that are needed to get into the prone kneeling position can be combined with extension in a continuous movement that ends in the kneeling posture.

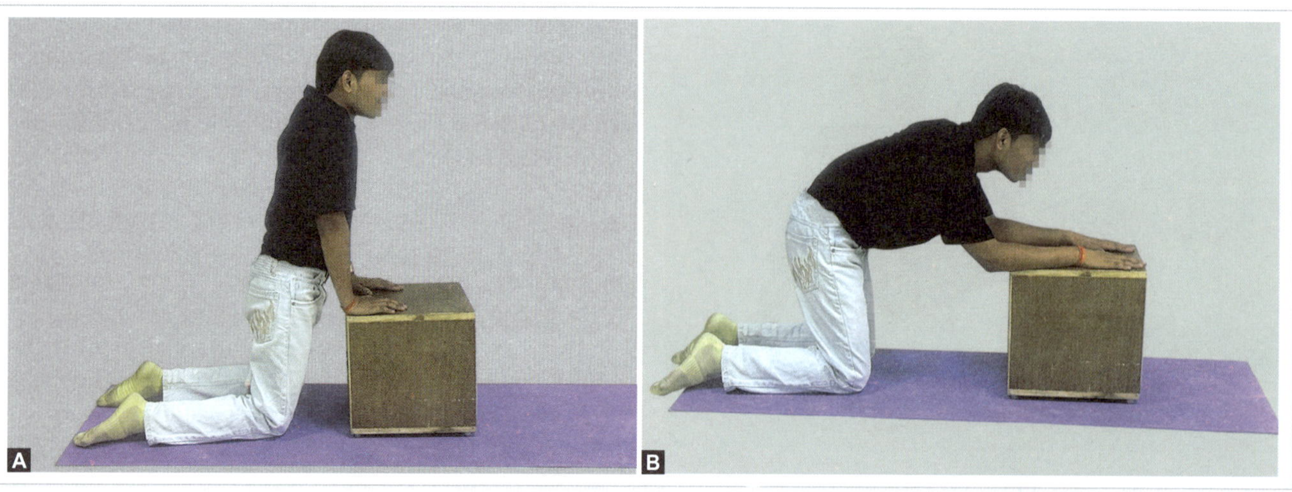

Figs 12.12A and B: Kneeling

Uses

- Kneeling serves as a connecting point in the chain of events that moves the body from a horizontal to a vertical position.
- It is appropriate for re-educating or correcting hip and lumbar control, for instance, when hip and lumbar flexion persists from habitual situations as a result of prolonged sitting.
- Establishing control while bearing weight on one leg (by the knee) can be seen as a precursor to walking, which requires bearing weight on one leg for a sufficient amount of time to allow for the other to be replaced in a different position. This also holds true when standing up after bending one leg.
- Without changing the body's position, weight can be rhythmically transferred from one knee to the other. One can also practice standing on one leg while using their arms as support if necessary. Steps taken in a sideways motion can establish progression.

Half-Kneeling

Kneeling into this pose requires supporting the body weight on one knee while lifting and bringing the other leg forward until the foot is on the ground (Fig. 12.13). Patients frequently find this action challenging, even when the arms are being utilized for support, since the supporting hip cannot maintain extension and/or the joints of the moving leg do not have the range of flexion to allow the leg to be folded as it is moved forward. From prone kneeling, some people find that this is simpler because the supporting limb is steady and balance is not a concern. Once the legs are in place, the body may be elevated. It serves as an intermediate posture between standing and kneeling.

Fig. 12.13: Half-kneeling

Getting Up from the Floor

Encouragement should be given to each patient to choose the best and most efficient way to get up. Furthermore, if needed, exercise caution. Following strategies can serve as a starting point for activity planning:

Getting Up from Sitting Position

- When seated sideways next to a big, sturdy chair, the patient leans on the seat and, with its assistance, bends down to kneel. Next, one leg is raised and brought forward into a half-kneeling position before both legs help the body rise up. The arms continue to support the body as the buttocks are turned to sit on the chair.
- For people who have strong arms, a series of transfers utilizing the hands to pull the body upward, backward, or sideways to a low stool, then to another slightly higher, and lastly to chair level is a beneficial strategy.
- Patients with bilateral amputees find this treatment to be highly effective.
- A push onto the bed with one or both feet from a kneeling position, facing a bed of a reasonable height, and with the body completely supported, may be enough to force the pelvis onto the bed. The patient is then able to roll over in bed and sit up all at once.

Getting Up from Lying Position

One can get into the sitting position by standing up from a lying position, such as from a bed, or by getting up from the floor. The fundamental position necessitates active head and trunk balance and stability in an upright position. Prior to walking, every patient must have this skill because trunk stability provides the necessary foundation for efficient movement of the extremities.

Standing

Sitting to Standing

Standing up straight is a huge morale booster and is seen as a sign of success for many patients. But from the perspective of independence, being able to stand or walk serves little benefit unless one can get to a standing posture alone.

Although each patient has unique difficulties, there are a few things to keep in mind even before trying to stand. These elements consist of:

- **Proper footwear:** They ought to be supportive, snug, and firm. For the elderly in particular, walking on a cold, hard floor with uncomfortable slippers or bare feet is more of a hindrance than an aid.
- **Proper clothing:** Slacks or a track suit that is simple to put on seem to be more practical than long, silky nightgowns, dressing gowns, or ill-fitting pyjamas.

- **Joint ROM:** The range of motion in the ankles, knees, hips, and lumbar spine should be examined to ensure that it is sufficient to allow the body segments to be brought into alignment for balance in the erect posture. Restricted range of any of these joints will necessitate postural corrections in order to maintain balance. For instance, a rapid cork "heel-raise" can compensate for limited ankle dorsiflexion and allow for a heel-strike without the need for excessive knee extension and hip flexion. In this instance, there is an added benefit because the pressure on the heel stimulates dorsiflexion.
- **Sustained support:** Any equipment utilized as support (for example hand-bars, bed, chair, and table) needs to have its stability examined and tested.

This pose is an intermediate between walking and half-kneeling. Long sitting and half-kneeling can be used to achieve it. To begin walking, one must first be standing.

Moving to Standing Position and Returning to Sit

With the patient's weight, feet, and hands supported by a chair or bar, total body extension begins when the patient presses downward with both hands and feet. This process is repeated until the patient's legs are vertically aligned and the patient stands. The therapist is there to provide assistance or support as needed, keeping in mind that too much assistance can impede the development of independence, while too little assistance can undermine confidence. It is safe for the patient to practice alone only if the therapist is confident that they can stand, balance while standing, and return to sitting. In order to evaluate the presence and stability of chair before sitting down once more, the patient places one hand on the chair's arm before flexing entire body forward and then backward to sit. Before the back is straightened, the buttocks can then be "walked" or raised to the rear of the chair seat (Figs 12.14A to C).

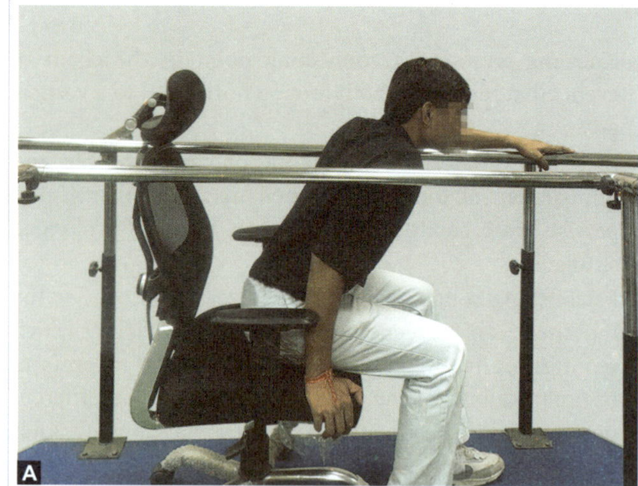

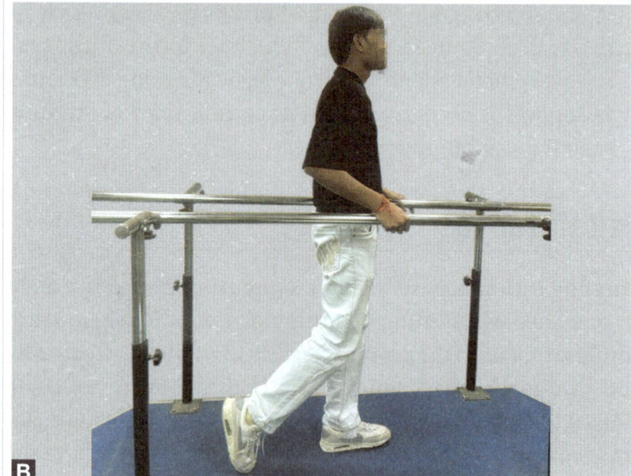

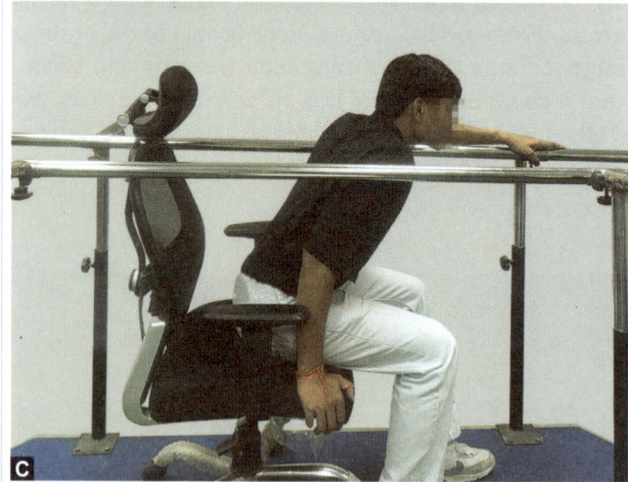

Figs 12.14A to C: Stepwise depiction of sitting to standing and returning to sitting

Parallel Bar Walking

Parallel bar exercises can be introduced as soon as standing posture motor control is attained. The therapist must provide the patient the appropriate directions throughout the parallel bar training. The therapist might offer support on the side limb that is weaker to promote stability. The patient can perform the following progressive activities to become better:

- Initially, the patient is made to stand with the assistance of the parallel bar without the therapist's assistance.
- Without changing the location of the hands in the parallel bar, shifting the weight to the lateral, anterior, and posterior sides.
- The patient is then advised to stand with one leg supported and put the whole on that leg.
- Patient is made to stand with one hand support after removing one hand taken from the parallel bar.
- The patient is then instructed to do a hip hiking motion, which aids in lifting the foot off the ground and propelling the body ahead while walking.

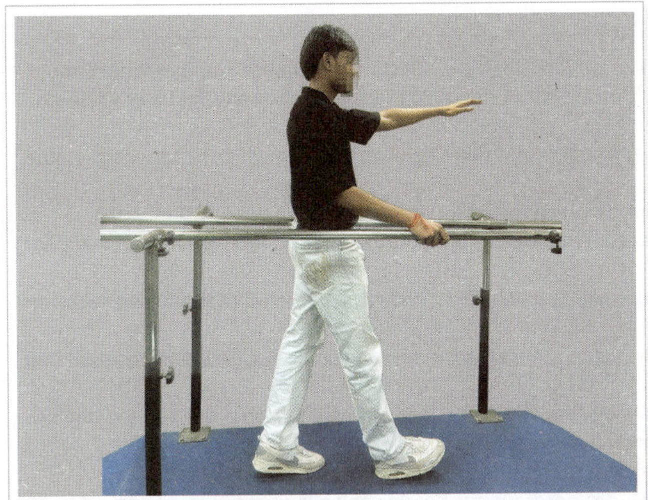

Fig. 12.15: Parallel bar walking

- The patient then practices moving one leg forward and returning to the neutral position, as well as taking one step backward and returning to the neutral position by remaining in the same position (Fig. 12.15).

Walking

Cooperation and coordination of the entire body is required to facilitate the intricate motion of walking. While teaching a patient to walk once more, some general guidelines are as follows:

- The patient needs to pick up the right walking pattern early on.
- Enough assistance and/or support must be given to make it possible to walk in the right pattern on a daily basis.
- The patient must be able to demonstrate that they can walk properly with less support or without assistance before the amount of support or help is reduced or changed.

Mechanism of Walking

While walking, the use of each leg alternates; one leg bears the weight of the body, while the other swings forward to replace the foot in a new position, ready to hold the body weight when it is transferred. Leg function is reversed when this happens. In contrast to "swing", which is when the limb is weight-free, the term "stance" denotes that the limb is bearing weight (Figs 12.16A to C).

The body always tries its hardest to fulfill the goals of an activity. Since this new pattern quickly becomes habitual, it frequently lingers long after the necessity for it has passed away. If it is prohibited or hindered from adopting the habitual pattern, it substitutes another (so-called "trick movement"), such as a limping gait. The automatic redesign of a walking pattern is frequently caused by pain, neuromuscular inefficiency, or a restriction in joint range; unless these issues are addressed or compensated for, the improper pattern will continue. The key to a successful therapy session is the therapist's ability to identify the primary cause(s) of the altered pattern and then take precise action to address it.

Climbing Stairs or Ramps

The patient's freedom frequently depends on climbing stairs or ramps. The initial stage is to attempt a shallow step or slope, and practice is then given to balance on one leg or getting the step ready for weight transference. It is necessary to cover both up- and down-moving as well as controlling crutches and sticks. If there is a hand rail, it is advised to use it; in this case, the free hand is used to hold crutches or walking sticks. The activity must be repeated frequently in order to develop endurance, speed, and confidence.

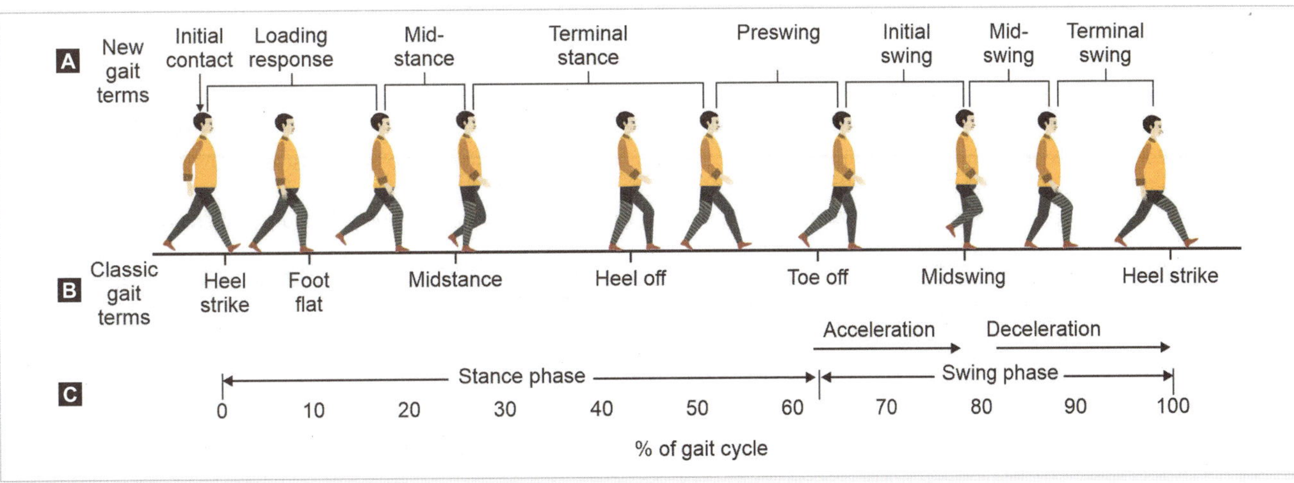

Figs 12.16A to C: Mechanism of walking

SUMMARY

- Functional re-education involves teaching people to perform familiar actions that are affected by ailments, progressing through different postures such as rolling, sitting, kneeling, and standing, with the goal of improving independence and mobility while also addressing muscle strength, stability, balance, and coordination.
- Rolling and other mat/bed exercises make it easier to transition between supine and side sleeping positions, which helps with pressure sore prevention and nursing care.
- Prone lying alleviates the consequences of prolonged bed rest and may improve respiratory function.
- Bridging or elevating the trunk while reclining facilitates bedpan procedures, reduces pressure points, and increases hip and lumbar flexibility for tasks such as dressing and transfers.
- Forearm-supported side lying makes it easier to shift to sitting, reach for nightstand objects, and read comfortably. It can also help with shoulder rehabilitation by boosting muscular activation.
- Prone lying with forearm support helps to keep the head and trunk aligned, engages the shoulder muscles, facilitates coordinated movement, and keeps the hip and lumbar flexible.
- Sitting on the bedside requires extending the elbow and swinging the legs over, with the thighs pressed against the bed for stability. This position precedes standing, provides relaxation from standing, and allows for upper-limb flexibility.
- Hitching and hiking help with wheelchair transfers by transferring weight and rotating the pelvis.
- The quadruped position, obtained from prone on hands or side sitting, promotes weight-bearing and allows activities at ground level, ambulation, and crawling, aiding persons with vertigo or temporary weight-bearing problems.
- Long sitting involves sitting with both upper limbs supported on either side of the trunk while maintaining an upright posture and enough trunk muscle power. Transitioning to the floor involves a variety of approaches, including kneeling with support or taking tiny steps.
- Side sitting improves lateral trunk and abductor muscle function and makes ambulatory motions easier for people with restricted weight-bearing capacity.
- Crawling, which progresses from prone kneeling, improves coordination and joint mobility while allowing safe movement for people with weak balance. It helps with mobility and limb function rehabilitation by including different limb movements and weight distribution. Kneeling, whether standing, sitting, or prone, serves as a transition between horizontal and vertical positions, promoting hip and lumbar control, weight-bearing control, and mobility advancement.
- Half-kneeling is an intermediate posture between standing and kneeling in which one knee supports the body weight, while the other leg is brought forward to a kneeling position. Getting up from the floor can be done in a variety of ways, including utilizing a chair as support or shifting to a low stool. Proper footwear, clothes, and joint range of motion are all important considerations before attempting to stand.
- Standing requires continuous support and alignment of body segments for balance. To stand, press downward with the hands and feet until one reaches vertical alignment, and to sit, bend the body forward and backward with support.
- Parallel bar walking can be introduced after standing posture control has been established, advancing from standing with support to walking with one hand or leg support. Hip hiking motion and practicing forward and backward leg movements aid in gait rehabilitation.
- Walking consists of alternating leg function between bearing weight (stance) and swinging forward (swing), with the goal of achieving normal walking patterns with little support or aid to resolve changed patterns caused by variables such as pain or neuromuscular inefficiency.
- Climbing stairs or ramps is critical to patient freedom. Begin with modest stairs or slopes, then balance on one leg and gradually transfer weight. Use handrails as support while carrying crutches or sticks in the free hand. Repeat regularly to increase endurance, speed, and confidence.

FURTHER READINGS

- Bamford P, Bentley A, Dean J, Whitmore D, Wilson-Baig N. ICS Guidance for Prone Positioning of the Conscious COVID-19 Patient 2020. Intensive Care Society. 2020 Apr 12.
- Meftahi N, Bervis S, Taghizadeh S, Ghafarinejad F. The Effect of Lying in Prone Position on Blood Pressure and Heart Rate with and without Massage.
- Meftahi N, Bervis S, Taghizadeh S, Ghafarinejad F. Effects of Massage in Prone Position on Blood Pressure and Heart Rate in Healthy Women. Journal of Rehabilitation Sciences and Research. 2014 Jun 1;1(2):40–3.

STUDENT ASSIGNMENT

LONG ANSWER QUESTIONS

1. Define re-education and explain how to re-educate walking in a hemiplegic patient.
2. Describe in detail various mat activities.
3. How to re-educate a wheelchair bound patient to stand? Explain with diagrams.
4. What is the purpose of mat activities? Explain techniques for rolling in detail.

SHORT ANSWER QUESTIONS

1. What is quadruped position? How is it achieved from prone lying?
2. Differentiate between prone kneeling and half-kneeling.
3. Write in briefly how to re-educate a patient to sit on the bedside.
4. Write the technique for parallel bar walking.

MULTIPLE CHOICE QUESTIONS

1. **What is the primary goal of functional re-education training?**
   a. To introduce new activities to patients
   b. To make patients dependent on caregivers
   c. To train patients in activities they are already familiar with
   d. To restrict patients' movement
2. **Which position is typically used to transition to sitting from the lying position?**
   a. Prone lying with forearm support
   b. Side lying supported on forearm
   c. Bridging from crook lying position
   d. Quadruped position
3. **What is the purpose of rolling from supine to side lying position?**
   a. To relieve pressure on sensitive pressure points
   b. To facilitate the bedpan routine
   c. To assist with nursing activities
   d. To avoid pressure sores
4. **Which position helps in maintaining the extensibility of the lumbar spine and hip joints?**
   a. Prone lying with forearm support
   b. Bridging from crook lying position
   c. Side lying supported on forearm
   d. Prone lying on elbows
5. **What is the next progression after achieving prone lying on hands position?**
   a. Side sitting
   b. Sitting
   c. Quadruped position
   d. Standing
6. **What is the primary purpose of the quadruped position in functional re-education?**
   a. To practice sitting on the bedside
   b. To facilitate crawling movements
   c. To improve long sitting posture
   d. To assist in rolling from supine to side lying position
7. **Which position helps in improving lateral trunk and abductor muscle efficiency?**
   a. Bridging
   b. Side sitting
   c. Prone lying with forearm support
   d. Long sitting
8. **How is sitting on the bedside typically achieved from the side lying posture?**
   a. By extending the elbow and raising the legs
   b. By bending the knees and rotating the trunk
   c. By flexing the hips and knees and raising the pelvis
   d. By pivoting on one buttock while extending the arms

9. **What is the primary benefit of crawling in functional re-education?**
   a. Improving static and dynamic stability
   b. Enhancing proprioceptive function
   c. Facilitating mobilization and controlling excessive mobility
   d. Improving balance and coordination
10. **Which movement is crucial for transferring from a chair to the floor?**
   a. Rolling prone
   b. Pushing up to side sitting
   c. Crawling backward
   d. Hitching and hiking
11. **What is a crucial consideration before a patient attempts to kneel?**
   a. The number of repetitions required
   b. The availability of handrails
   c. The comfort of the kneeling surface and the range of knee flexion
   d. The height of the surrounding furniture
12. **How is kneeling down from sitting typically accomplished?**
   a. By leaning forward and placing hands on the floor for support
   b. By bending both knees and rocking them to the ground simultaneously
   c. By extending the elbows while raising the legs
   d. By twisting the body while sitting and shifting weight between the hands
13. **What is the primary purpose of half-kneeling in functional re-education?**
   a. To facilitate crawling movements
   b. To serve as an intermediate posture between standing and kneeling
   c. To improve static and dynamic stability
   d. To assist in climbing stairs or ramps
14. **What is an essential consideration before attempting to stand from a sitting position?**
   a. The availability of a walking aid
   b. The range of motion in the ankles, knees, hips, and lumbar spine
   c. The presence of a therapist for support
   d. The time of day
15. **What is emphasized as crucial for a successful therapy session regarding walking?**
   a. The number of repetitions
   b. The speed of progression
   c. The identification and precise action to address the primary cause(s) of altered patterns
   d. The use of assistive devices

ANSWER KEY

1. c	**2.** b	**3.** d	**4.** a	**5.** d	**6.** b	**7.** b	**8.** a	**9.** c	**10.** d
11. c	**12.** d	**13.** b	**14.** b	**15.** c					

13 Coordination Exercises

Sheetal Kalra, Puneeta Ajmera, Sheetal Yadav

LEARNING OBJECTIVES

After the completion of the chapter, the readers will be able to:

- Understand anatomy and physiology, functions of cerebellum with its pathways.
- Explain physiology/mechanism of neuromuscular coordination.
- Define coordination and incoordination.
- Explain causes for incoordination.
- Describe the incoordination due to lower motor neuron lesions (flaccidity), upper motor neuron lesions (spasticity), cerebellar lesions, loss of kinesthetic sense (tabes dorsalis, syringomyelia, leprosy).
- Define and explain test for coordination: equilibrium test, nonequilibrium test.
- Explain principles of coordination exercise.
- Describe the uses and technique of Frenkel's exercises.

CHAPTER OUTLINE

- Introduction
- Anatomy of Cerebellum
- Physiology of Neuromuscular Coordination
- Incoordination
- Coordination Tests
- Coordination Exercises
- Frenkel's Exercises

KEY TERMS

Coordination: Coordination, or the ability to choose the right muscle at the right moment with the right intensity to perform the right action, can also be defined as the capability to carry out accurate, regulated, and smooth motor responses—the optimal interplay of muscle activity.

Cytoarchitecture: The examination of the cellular makeup of the tissues that make up the central nervous system under a microscope is known as cytoarchitecture.

Lower motor neuron lesion: A lesion that affects nerve fibers that go from the spinal cords anterior horn to the corresponding muscle or muscles is known as a lower motor neuron lesion.

Tandem walking: Tandem gait is a way of walking in which the toes of the rear or supporting foot are barely touched or immediately in front of the heel of the forward-stepping foot with each tiny stride in a straight line.

Upper motor neuron lesion: A lesion of the neural pathway above the anterior horn of the spinal cord or the motor nuclei of the cranial nerves is referred to as an upper motor neuron lesion.

INTRODUCTION

Coordination in the human body is the seamless interaction of the nervous, muscular, and skeletal systems, enabling precise and efficient movement. This process relies on sensory input, motor activation, and central integration by the brain and spinal cord to execute functional tasks. Coordination is essential for daily activities, sports, and injury prevention, and its impairment can hinder independence and quality of life.

Coordination is the capacity to perform smooth, accurate, and controlled motions that require the precise activation of several muscles and joints at the right time and with the appropriate amount of force. It is based on somatosensory, visual, and vestibular input, as well as a fully functional neuromuscular system extending from the brain to the spinal cord. Coordinated movements exhibit proper speed, distance, direction, timing, and muscle tension.

Coordination exercises are a fundamental aspect of physiotherapy, enhancing neuromuscular control, balance, and proprioception while aiding recovery, preventing injury, and optimizing performance. This chapter introduces the principles and applications of these exercises for effective rehabilitation.

Fig. 13.1: Anatomy of cerebellum

ANATOMY OF CEREBELLUM

In all vertebrates, the cerebellum occupies a sizable portion of the brain. In humans, it makes up around ten percent of the total brain weight, but because the microscopic granule cells of the cerebellar cortex are tightly packed, the cerebellum has more neurons than the entire remainder of the brain. It weighs roughly 150 g and is the largest portion of the hind brain. It is preserved in the posterior cerebral fossa, behind the pons and medulla oblongata, beneath the tentorium cerebelli. Three enormous bundles of fiber known as cerebellar peduncles attach the cerebellum to the brainstem. The inferior peduncle connects to the medulla oblongata, the middle peduncle to the pons, and the superior peduncle to the cerebellum (Singh et al.) (Figs 13.1 and 13.2).

Parts

The cerebellum is made up of two sizable, opposite lobes known as cerebellar hemispheres. The median vermis, which resembles a worm, connects these two lobes. Superior and inferior vermis are the names for the superior and inferior aspects of the vermis, respectively.

Although the inferior vermis is divided from the hemispheres by a deep furrow, the superior vermis is continuous with the hemispheres (Fig. 13.3).

Surfaces

Two cerebellar hemispheres are contiguous with one another on the superior surface of the cerebellum, which is convex in shape. The inferior surface has a thick ridge called a vallecula that divides the two cerebellar hemispheres. The inferior vermis covers the vallecular floor.

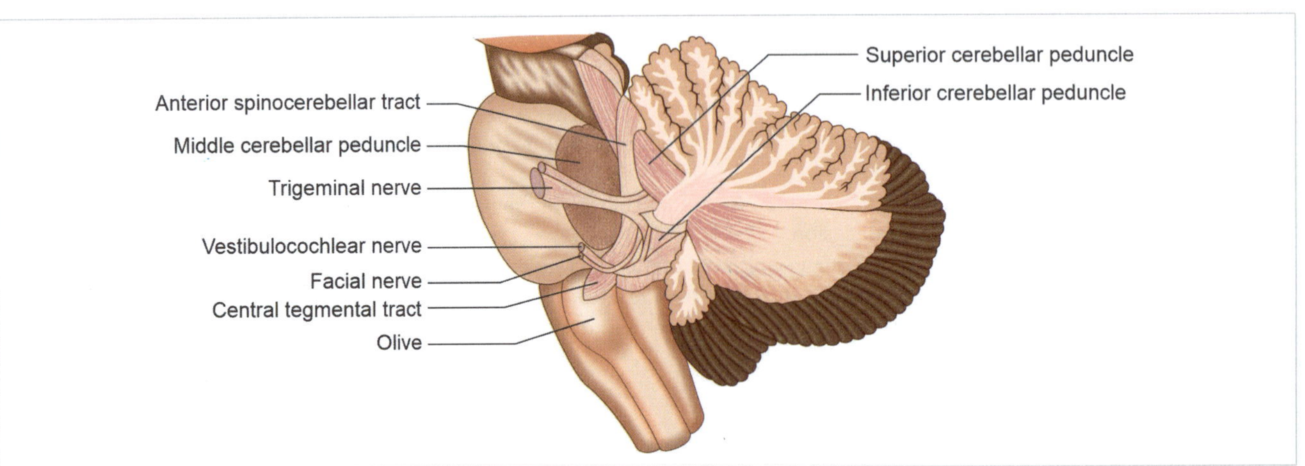

Fig. 13.2: Cerebellar peduncles

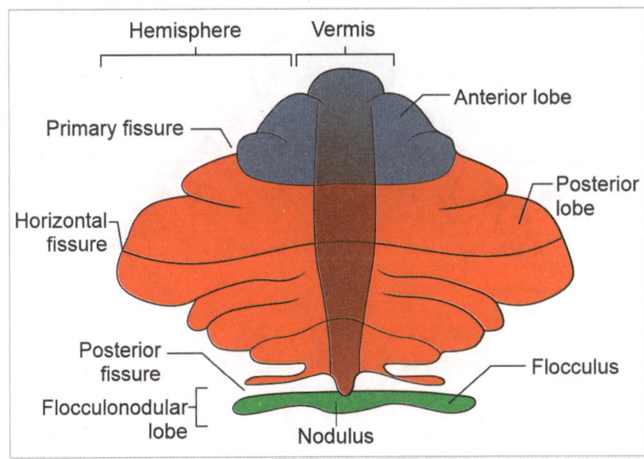

Fig. 13.3: Cerebellar fissures

Notches

On the anterior aspect of the cerebellum, there is a large, shallow gap known as the anterior cerebellar notch. The pons and medulla are lodged in the front cerebellar notch. Similar to this, the falx cerebelli are lodged in the posterior cerebellar notch.

Fissures

Cerebellum-related fissures include the horizontal, postero-lateral, and primary fissures (Fig. 13.3).

The most noticeable fissure is horizontal, and it runs along the lateral and posterior edges of the cerebellum. It divides the superior and inferior surfaces of the cerebellum. The flocculonodular lobe is divided from the rest of the cerebellum, also known as the corpus cerebelli, by the posterolateral fissure, which is found on the inferior surface of the brain. Primary fissure is situated on the superior surface and divides the corpus cerebelli into anterior and posterior (middle) lobes.

The flocculonodular lobe lies on the inferior surface, in front of the posterolateral fissure, consisting of the nodule of the inferior vermis and a pair of flocculi connected by peduncles.

The anterior lobe is situated on the superior surface, anterior to the primary fissure, composed of the lingual, central lobule, and culmen.

The posterior lobe is between the primary fissure on the superior surface and the posterolateral fissure on the inferior surface, encompassing both surfaces. Its superior surface comprises the declive and folium, while the inferior surface contains the tuber, pyramid, and uvula.

Cytoarchitecture

The cerebellar cortex, the inner layer of white matter, and the outer layer of gray matter make up the cerebellum. The white matter contains enormous amounts of intracerebellar nuclei and gray matter. The cerebellar cortex is folded into folia, which are little bands that resemble leaves. Each folium has a white matter core in the center that is encircled by a thin coating of gray matter. The arbour vitae cerebelli or central core of white matter is structured like a branching tree (Fig. 13.4).

Gray Tissue

The cerebellar cortex and intracerebellar nuclei are the two primary components of gray matter.

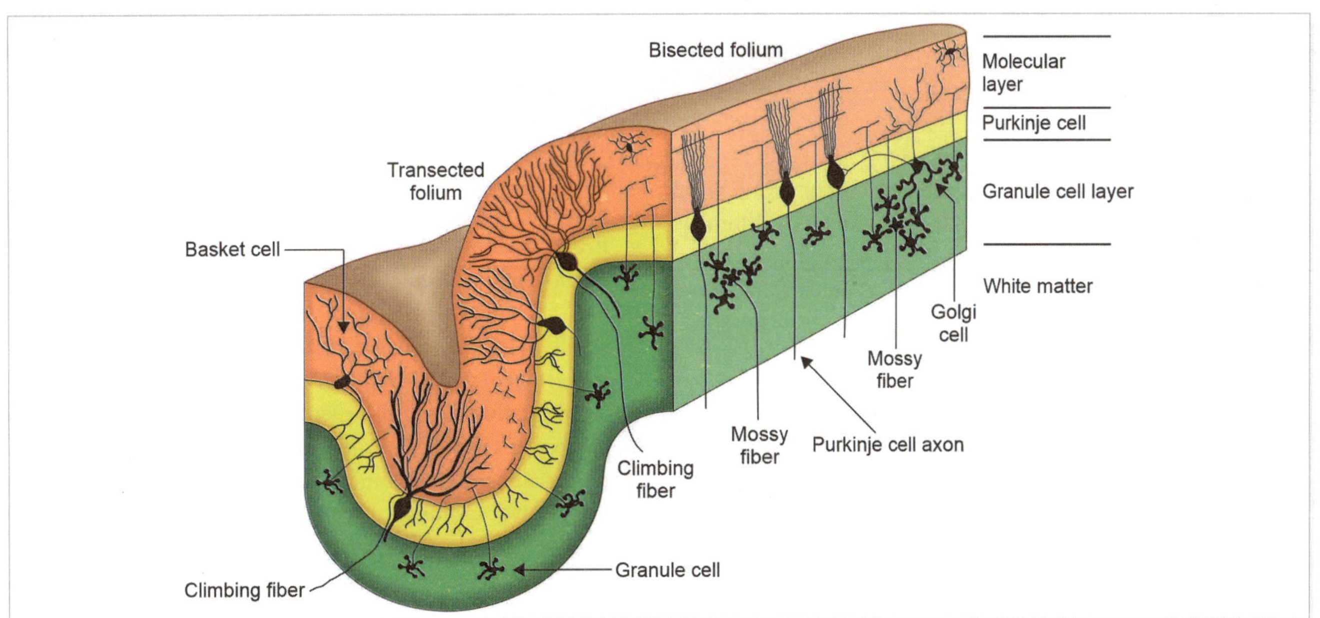

Fig. 13.4: Cerebellar cytoarchitecture

1. Cerebellar Cortex

The outside molecular layer, the intermediate Purkinje cell layer, and the inner granular layer are the three different layers that make up the cerebellar cortex.

1. **Molecular layer:** Unmyelinated nerve fibers from granule, stellate, and basket cells, as well as dendrites from Purkinje and Golgi cells, make up this layer. Additionally, it has basket and stellate cells. Stellate cells are dispersed at the surface. Axons of these cells make connections with Purkinje cell dendrites. Basket cells have numerous processes but little cytoplasm. Axons of these cells travel transversely along the cortical surface and connect with dendrites.
2. **Purkinje cell layer:** One layer of flask-shaped Purkinje cells makes up the Purkinje cell layer. Dendrites of these cells ascend into the molecular layer, where they undergo extensive branching. The axons of granule cells, ascending fibers, and collaterals of basket cells form synapses with the dendrites of Purkinje cells. Through the granular layer and into the white matter, axons of Purkinje cell connect with intracerebellar nuclei to form synaptic connections and have an inhibitory effect on these nuclei.
3. **Granular layer:** Numerous granule cells and a small number of Golgi cells make up the inner granular layer. Four to five dendrites on each granule cell synapse with mossy fibers. Axons of these cells travel into the molecular layer where they bifurcate and form branches that travel parallel to the long axis of the cerebellar folium. These strands connect to the dendrites of Purkinje cells and are referred to as parallel fibers.

2. Cerebellar Nuclei

The four paired deep gray matter nuclei in the fourth ventricle of the cerebellum make up the cerebellar nuclei. From lateral to medial, they are ordered as follows (Fig. 13.5):

1. The largest and most lateral nuclei are dentate nuclei.
2. Emboliform nuclei
3. Bulbous nuclei
4. Fastigial (most medial) nuclei

They are the sole output of the cerebellar cortex, along with the lateral vestibular nuclei (which also receives inhibitory input from Purkinje cells), and they receive inhibitory input from the axon of the Purkinje cells in the cerebellar cortex (Singh et al., Sillitoe et al.).

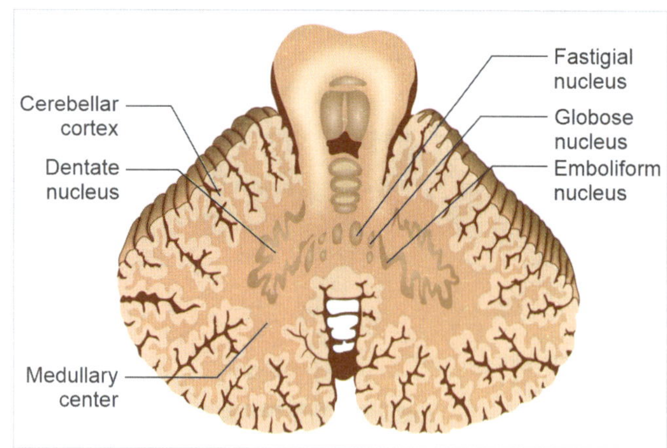

Fig. 13.5: Cerebellar nuclei

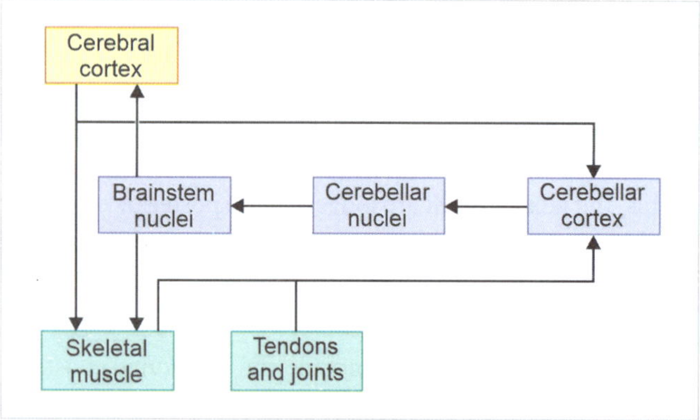

Fig. 13.6: Role of cerebellum as comparator

Functions of Cerebellum

The cerebellum regulates balance and posture by integrating sensory information from the vestibular system, proprioceptors in muscles and joints, and visual inputs.

- **Coordinating voluntary movements:** Cerebellum fine-tunes and coordinates voluntary motions by varying the timing, force, and direction of muscle contractions. This results in smooth, accurate movements.
- **Motor learning:** The cerebellum is critical to motor learning and skill gain. It allows for movement adaptation based on feedback and error correction, which contributes to the long-term development of motor abilities (Fig. 13.6).
- **Motor planning and execution:** It helps with the planning and execution of motor sequences, allowing for the initiation, timing, and synchronization of complicated movements.
- **Predictive control:** The cerebellum creates internal models of movement, predicting the effects of motor commands and altering them in real time to obtain the intended results. This predictive control helps to maintain accuracy and efficiency of movement.
- Coordination of cognitive work and physical actions by the cerebellum implies a unified mechanism underlying both motor and cognitive functions.
- Integrative models of cerebellar function propose its involvement in orchestrating complex behaviors by linking cognitive processes with motor output.

Recent Updates

Traditionally linked with motor control, recent findings suggest the involvement of cerebellum in cognitive functions including attention, language, and spatial processing.

- Emerging research highlights the role of cerebellum in coordinating both cognitive tasks and physical actions, suggesting an integrated role in brain function.
- The cerebellum appears to contribute to higher-order cognitive processes, indicating a broader influence on brain function.
- The involvement of cerebellum in attentional processes suggests a role in regulating focus and concentration.
- Studies implicating the cerebellum in language processing suggest its contribution to linguistic abilities such as comprehension and production.
- Spatial processing, crucial for navigation and perception, may also benefit from cerebellar involvement, enhancing our understanding of its cognitive role.

PHYSIOLOGY OF NEUROMUSCULAR COORDINATION

Movement coordination requires complex interactions between multiple brain areas and neuronal networks. Here is a brief overview of the physiology and systems involved in coordination:

- **The primary motor cortex:** It is located in the precentral gyrus of the frontal lobe and governs voluntary motions. It gets input from the somatosensory cortex, cerebellum, and basal ganglia. Brodmann area 4 primary motor cortex (PMC) is responsible for motor commands, whereas area 6 (premotor cortex) aids in movement planning and synchronization.
- **Descending motor pathways:** The corticospinal tract begins in the motor cortex and regulates skilled, fine motor actions, particularly in the distal limbs. Other descending routes, including the corticobulbar, tectospinal, reticulospinal, vestibulospinal, and rubrospinal tracts, play important roles in motor control and postural regulation.
- **The cerebellum:** It regulates movement, controls posture, and modulates muscle tone. It functions as a comparator and error-correcting system, comparing intended motor orders to actual performance and adjusting to ensure movement precision (Fig. 13.6).
- **The basal ganglia:** It includes the caudate nucleus, putamen, and globus pallidus, which are responsible for starting and controlling gross purposeful motions, planning complicated motor responses, and maintaining proper muscle tone. They have an indirect influence on motor function *via* connecting routes to cortical motor regions.
- **Dorsal column to medial lemniscal pathway:** This sensory route transmits discriminative sensations like touch, pressure, proprioception, and vibration, all of which are necessary for movement coordination. It sends sensory data from peripheral receptors to the somatic sensory brain *via* the thalamus.
- Neuromuscular coordination involves the integration of sensory input, central processing, and motor output to produce smooth and precise movements. Sensory receptors in the skin, muscles, and joints send information to the central nervous system (CNS), where structures like the cerebellum, basal ganglia, and motor cortex process and refine motor commands. These signals travel through motor pathways to stimulate muscles via neuromuscular junctions, where neurotransmitters like acetylcholine trigger contractions. Continuous feedback from proprioceptive sensors ensures adjustments for balance and accuracy, enabling coordinated and functional movement essential for daily activities and rehabilitation.

INCOORDINATION

Incoordination is defined as the inability to perform smooth, accurate, and controlled movements. This disorder is distinguished by awkward, unnecessary, uneven, or imprecise movements. It can be caused by abnormalities in somatosensory, visual or vestibular signals, as well as problems with the neuromuscular system, which runs from the motor brain to the spinal cord. Incoordination is characterized by difficulties with optimal muscle activation sequencing, timing, and grading, resulting in impairments in speed, distance, direction, timing, and muscular tension during movements.

Causes of Neuromuscular Incoordination

A variety of movement diseases that arise from a breakdown in the connection between the neurological system and the muscles are together referred to as neuromuscular incoordination. Neuromuscular incoordination can be brought on by a wide range of causes, including:

- **Neuromuscular conditions:** Disorders that impact the muscles, and the nerves that regulate them are known as neuromuscular problems, and they frequently result in muscle weakness, poor motor function, and loss of coordination. These disorders may be brought on by autoimmune, genetic, or degenerative factors that impact the muscles and neurological system.
- **Cerebellar lesions:** The cerebellum controls balance and motor coordination. Incoordination, tremors, and balance issues may develop from a lesion or other injury to the cerebellum. Specific causes of cerebellar dysfunction is shown in Table 13.1.
- **Lesions of the basal ganglia:** Basal ganglia dysfunction disrupts the brain's ability to regulate movement, coordination, and muscle control, leading to incoordination and abnormal motor patterns.
- **Lesions in the upper motor neurons:** The upper motor neurons regulate voluntary movements. Muscle weakness, stiffness, and issues with fine motor control can occur as a result of a lesion or injury to the higher motor neurons.

TABLE 13.1: Causes of cerebellar dysfunction

Etiology	Bilateral cerebellar dysfunction	Unilateral cerebellar dysfunction	Spastic paraparesis with cerebellar signs
Multiple sclerosis (demyelination)	✓	✓	✓
Posterior circulation stroke	✓	✓ (Part of lateral medullary syndrome)	
Bilateral cerebellar pontine angle lesions or space-occupying lesions (e.g., neurofibromatosis and schwannoma)	✓		
Paraneoplastic syndromes	✓		
Multiple system atrophy	✓	✓	
Toxins and drugs (e.g., alcohol, phenytoin, lithium, and carbamazepine)	✓		
Metabolic conditions (e.g., thyroid disorders, B_{12} deficiency, Wilson disease, and celiac disease)	✓		
Infectious etiologies including enteroviruses, HIV, neurosyphilis, toxoplasmosis, borreliosis, and Creutzfeldt–Jakob disease	✓		
Inflammatory conditions (e.g., Guillain-Barré syndrome)	✓		
Hereditary conditions including ataxia telangiectasia, Friedreich ataxia, Von Hippel-Lindau syndrome, spinocerebellar ataxias	✓		✓ (Friedreich ataxia, spinocerebellar ataxia)
Unilateral posterior circulation ischemic or hemorrhagic stroke		✓	
Space-occupying lesions in the posterior cranial fossa, including abscesses (e.g., tuberculosis and staphylococcal infection) and tumor		✓ (Abscesses, tumor)	
Unilateral cerebellar pontine angle lesions or space-occupying lesions (e.g., neurofibromatosis and schwannoma)		✓	
Syringomyelia, syringobulbia		✓	

- **Lesions in lower motor neurons:** Lower motor neurons carry messages from the spinal cord to the muscles. Muscle wasting, atrophy, and fasciculations (twitching) can occur as a result of a lesion or damage to the lower motor neurons.
- **Loss of kinesthesia:** Proprioceptive receptors, specialized sensory receptors found in the muscles, tendons, and joints of the body, are what this sense depends on. These receptors pick up changes in joint angle, muscle tension, and other variables and communicate with the brain to produce kinesthetic sensations. The kinesthetic sense can be lost or diminished under a variety of circumstances.
- **Aging:** Reduced strength, reduced reaction time, and restricted range of motion are all examples of age-related alterations that impede coordinated movement. Furthermore, as we age, postural alterations and decreased balance become more common, contributing to difficulties with motor function and stability.
- **Alcohol or drug use:** Alcohol and some substances can impair the neurological system, causing tremors, balance issues, and lack of coordination.

Neuromuscular Conditions

Like Parkinson's, multiple sclerosis, and cerebral palsy impair coordination through issues with motor control, muscle weakness, and nerve damage. Huntington's disease, muscular dystrophy, and other disorders also cause movement difficulties and balance problems.

Cerebellar Lesions

The cerebellum is a crucial component of the brain that controls balance and movement coordination. Incoordination, or the inability to carry out smooth, coordinated motions, can be brought on by a lesion or injury to the cerebellum.

Clinical Manifestations of Cerebellar Dysfunction

Clinical manifestations of cerebellar dysfunction are as follows (Aashrai et al.):

Common mnemonic to remember some of the cerebellar signs is DANISH.

- Dysdiadochokinesia/dysmetria
- Ataxia
- Nystagmus
- Intention tremor
- Speech-slurred or scanning
- Hypotonia

Complete description of various deficits seen in cerebellar dysfunction are shown in Table 13.2.

TABLE 13.2: Common deficits in cerebellar dysfunction

Deficit	Description
Motor deficits	
Dysmetria	Inaccurate judgment of distance and movement range is known dysmetria. A cerebellar lesion can impair a person's ability to coordinate and control limb movements, leading to inaccurate motions, over-or undershooting of targets, and difficulties carrying out tasks that call for precise movements.
Ataxia	Ataxia refers to a general lack of coordination or the inability to move smoothly and deliberately. Different types of ataxia, such as limb ataxia, truncal ataxia, and gait ataxia, can result from cerebellar lesions, causing symptoms like stumbling and swaying.
Dysdiadocho-kinesia	Dysdiadochokinesia is the inability to carry out quick, alternating movements, such as tapping fingers quickly on the hand's palm. Cerebellar damage disrupts the coordination of timing and direction of movements, leading to jerky and irregular movements.
Nystagmus	A cerebellar lesion can result in nystagmus, an uncontrollable eye movement. This occurs due to critical function of the cerebellum in regulating eye movements, which may be impaired by damage, resulting in nystagmus.
Hypotonia	Damage to the cerebellum may cause hypotonia, a decrease in muscular tone. The cerebellum provides feedback to motor neurons in the spinal cord to regulate muscle tone, but when damaged, this feedback is disrupted, resulting in decreased muscle tone.
Dyssynergia	Dyssynergia, also known as movement decomposition, refers to performing an action in distinct phases rather than seamlessly. For example, touching the nose may require separate, consecutive motions of the elbow, wrist, and shoulder.
Asynergia	Asynergia is the inability of muscles to coordinate complicated movements, leading to a loss of smooth, coordinated muscular action.
Rebound phenomenon	The rebound phenomenon is the inability to stop a forceful movement when resistance is suddenly removed. In cerebellar dysfunction, opposing muscles normally preventing sudden limb motion are ineffective, causing uncontrollable limb movement. This can lead to striking oneself or nearby objects.
Intention tremor	Intention tremor is an involuntary oscillatory movement that occurs during intentional motion of a limb, worsening as the limb approaches its destination or accelerates. It is caused by alternating contractions of opposing muscle groups and is reduced or absent during resting. Another type of tremor linked with cerebellar lesions is postural tremor, consisting of back-and-forth oscillatory movements of the body when standing or up-and-down motions of a limb against gravity.
Titubation	Titubation is defined as rhythmic oscillations of the head, whether side-to-side, forward-and-backward, or rotating. It can also affect the trunk, but this is less common. These motions are commonly observed in people with cerebellar abnormalities.
Extraocular movements	• **Nystagmus:** Central nystagmus is bidirectional and gaze-evoked, contrasting with peripheral nystagmus which is unidirectional and worsens with gaze toward the healthy ear (Alexander law). • **Impaired smooth pursuits:** Inability to track objects smoothly; catch-up saccades are common.
Adiadochokinesia	Difficulty performing rapid alternating movements, leading to jerky and irregular movements.
Ambulation	• **Stance and posture:** Broad-based stance with swaying (titubation). • **Gait:** Ataxic gait resembling acute alcohol intoxication. • **Tandem walk:** Inability to walk in a straight line with heel-to-toe pattern. • **Absence of Romberg's sign:** Disproportionate swaying or falling with eyes closed.
Language disorders	
Cerebellar mutism	Occurs post-central cerebellar injury, resulting in mutism ranging from days to indefinitely.
Dysarthria	It is a motor speech impairment that can be brought on by cerebellar damage. The speech muscles can be affected by dysarthria, leading to slurred or inaccurate speech as well as problems with articulation and pronunciation. A particular form of dysarthria known as ataxic dysarthria is linked to cerebellar injury. It may result in inconsistent speech patterns, labored speech, and trouble coordinating the motions required for speech.
Agrammatism	It is a lack of grammar and syntax that can result in word order issues, unfinished or disorganized phrases, and other grammatical mistakes.
Anomia	This is a problem with word retrieval or word search. People who have cerebellar lesions could have trouble finding the proper term or might use a word that sounds close instead.
Impaired verbal fluency	This refers to difficulty generating words or producing language spontaneously. It can result in a reduced ability to engage in conversations or to produce complex language.

Did You Know?

Neuromuscular incoordination can result from various neurological conditions beyond cerebellar dysfunction, emphasizing the complexity of motor control and the interconnectedness of different brain regions and neural pathways involved in movement regulation.

Incoordination Due to Basal Ganglia Pathology

Patients with basal ganglia lesions commonly exhibit distinct motor deficits, including (Table 13.3):

1. Reduced movement and slowness
2. Involuntary, extraneous movements
3. Changes in posture and muscle tone.

Lower Motor Neuron Lesions

Neuromuscular incoordination may develop from a lesion or other injury to the lower motor neurons, which are the nerves that emerge from the spinal cord and connect to the muscles. This is due to the fact that lower motor neurons are in charge of relaying information from the brain to the muscles, and any injury to these neurons can prevent this communication from occurring.

TABLE 13.3: Motor deficits in neuromuscular incoordination due to pathology of basal ganglia

Motor deficit	Description
Akinesia	Inability to initiate movement, seen in advanced Parkinson's disease, often accompanied by freezing episodes.
Athetosis	Slow, writhing movements, occasionally combined with spasticity or chorea.
Bradykinesia	Reduced amplitude and speed of voluntary movements, leading to decreased arm swing and shuffling gait.
Chorea	Involuntary, rapid, irregular movements involving multiple joints, difficult to control voluntarily.
Choreoathetosis	A blend of chorea and athetosis, manifesting as a movement disorder.
Dystonia	Sustained, involuntary muscle contractions causing abnormal postures or twisting movements.
Hemiballismus	Sudden, violent flailing of one side of the body due to contralateral subthalamic nucleus lesions.
Rigidity	Increased muscle tone resulting in resistance to passive movement, categorized as lead pipe or cogwheel rigidity.
Tremor	Involuntary, rhythmic movements, typically observed at rest and diminishing with purposeful movement.

The following processes can be used to explain the physiology of neuromuscular incoordination caused by lower motor neuron lesions:

- **Loss of muscle tone:** The degree of tension in a muscle at rest is determined by muscle tone, which is maintained by lower motor neurons. Flaccidity, commonly known as a decrease of muscular tone, can occur when these neurons are damaged. This can impair the ability of muscles to produce movement as well as make them appear limp and weak.
- **Muscle atrophy:** Muscle atrophy, or the fading away of muscle tissue as a result of inactivity, can result from injury to the lower motor neurons. This happens because the muscles are not getting the signals they need to contract and carry out their tasks, which can lead to a loss of muscular mass and strength.
- **Fibrillation and fasciculation:** Abnormal muscle contractions known as fibrillation and fasciculation can also occur as a result of injury to the lower motor neurons. Electromyography (EMG) can identify fibrillation, which is the spontaneous contraction of a single muscle fiber but is invisible to the unaided eye. Fasciculation is the term used to describe the observable twitching of muscle fiber clusters on the surface of skin.
- **Weakness and paralysis:** Loss of lower motor neurons can cause the affected muscles to become weak and paralyzed. The reason for this is because the muscles are no longer getting the signals they require to contract and carry out their tasks. The level of damage to lower motor neurons determines the degree of weakness and paralysis.

Upper Motor Neuron Lesions

Neuromuscular incoordination may develop from a lesion or other injury to the upper motor neurons, which are the nerves that leave the brain and travel to the spinal cord. The following processes can be used to explain the physiology of neuromuscular incoordination brought on by upper motor neuron lesions:

- **Spasticity:** One of the main consequences of damage to upper motor neurons is spasticity, which describes the potential for heightened muscular tone and rigidity in the affected muscles. This happens since the lower motor neurons are often inhibited by the upper motor neurons, which serves to control muscle tone. This inhibiting effect is lost when the higher motor neurons are destroyed, leading to uncontrolled muscular contraction and increased muscle tone.
- **Hyperreflexia:** Hyperreflexia is the term for an excessive reflex reaction when a muscle is stretched, and upper motor neuron injuries can also cause it. This happens because the reflex arc, the neurological system that regulates reflex

reactions, typically helps to moderate the sensitivity of the upper motor neurons. When the upper motor neurons are damaged, this regulation is disrupted, resulting in an exaggerated reflex response.

- **Clonus:** In the presence of upper motor neuron injuries, clonus is a sort of aberrant muscular contraction. The afflicted muscle undergoes a fast, alternating pattern of contractions and relaxations. The timing and strength of muscle contractions are ordinarily controlled by the higher motor neurons; however, when they are injured, this regulation is thrown off, which leads to clonus.
- **Weakness and paralysis:** Upper motor neuron lesions can also cause weakness and paralysis of the afflicted muscles, similar to lower motor neuron lesions. The method is different, though. With upper motor neuron lesions, the signals from the brain to the lower motor neurons are disrupted, resulting in weakness and paralysis. This alters the regular patterning of the muscles.

Loss of Kinesthesia

Kinesthesia is the sense that enables us to perceive the position and motion of our body parts in space. Proprioceptive receptors, specialized sensory receptors found in the muscles, tendons, and joints of the body, are what this sense depends on. These receptors pick up changes in joint angle, muscle tension, and other variables and communicate with the brain to produce kinesthetic sensations.

The kinesthetic sense can be lost or diminished under a variety of circumstances. Some of these consist of:

- **Peripheral neuropathy:** In peripheral neuropathy, the nerves that transmit information from the body to the brain suffer damage. Loss of sensation, especially kinesthetic sensation, may follow from this. Peripheral neuropathy patients may have trouble coordinating their movements and feel numbness or tingling in their limbs.
- **Cerebellar lesions:** Presence of a lesion in the cerebellum, is crucial for coordinating movements and preserving balance. Instability and a loss of kinesthetic sensation can occur as a result of cerebellar injury or lesions.
- **Proprioceptive disorders:** Several illnesses that damage the body's proprioceptive receptors might cause a loss of kinesthetic awareness. For instance, joint hypermobility syndrome causes joints to be more flexible than they should be, which can harm proprioceptive receptors and cause a loss of kinesthetic sensation.
- **Spinal cord injuries:** These injuries may cause a loss of sensation, particularly kinesthetic sensation, below the level of the injury. The degree of the sensory loss will depend on how severe the injury was.
- **Stroke:** Strokes can damage the brain regions in charge of processing kinesthetic information, which can lead to the loss of kinesthetic feeling. People may find it challenging to keep their balance and coordinate their motions as a result.
- **Tabes dorsalis:** Damage to the dorsal columns of the spinal cord, which carries sensory data, including kinesthetic sensations, to the brain, results in tabes dorsalis, a neurological disorder. Untreated or inadequately treated syphilis, which over time can harm the dorsal columns, is the most frequent cause of tabes dorsalis. As a result, patients with tabes dorsalis may experience other sensory problems in addition to a lack of kinesthetic sensation.
- **Syringomyelia:** It is a disorder in which the spinal cord develops a fluid-filled cyst (syrinx), harming the nerve fibers that carry sensory data, particularly kinesthetic sensations, to the brain. Disruption to the flow of cerebrospinal fluid (CSF) brought on by the cyst may result in extra harm to the spinal cord. As a result, people with syringomyelia may experience a loss of kinesthetic sensation, as well as other sensory and motor disturbances.
- **Leprosy:** Mainly affecting the skin and peripheral nerves, leprosy is a bacterial condition brought on by Mycobacterium leprae. The infection may damage the nerves that carry sensory data to the brain, including kinesthetic feelings. As a result, leprosy patients may suffer from various sensory and motor abnormalities in addition to a loss of kinesthetic sensation.

COORDINATION TESTS

We can assess coordination and balance using a variety of equilibrium and nonequilibrium tests. The following tests are some of the most popular ones:

Nonequilibrium Tests

The nonequilibrium coordination tests (Table 13.4) measure alternative motion, movement composition, accuracy, and limb holding without emphasizing balance maintenance. The complexity of the test increases from unilateral to multi-limb tasks, and there may be balance issues while transitioning from sitting to standing.

Equilibrium Tests

According to studies, equilibrium coordination tests are evaluation of a person's capacity to hold their balance and coordinate motions that support the stabilization of their body in various weight-bearing postures. The responses to different tasks, such as standing or moving on uneven terrain, are frequently measured by these tests. Few examples are mentioned below:

- Standing with normal base of support
- Standing with feet together (narrow base of support)
- Standing on one foot with or without altering arm position

TABLE 13.4: Nonequilibrium coordination tests

Coordination test	Description
Finger-to-nose	Extend the elbow and bring the index finger to the nose.
Finger-to-therapist's finger	Touch the index finger to the therapist's finger.
Finger-to-finger	Bring both hands together, touching index fingers (Refer to Fig. 6.8).
Alternate nose-to-finger	Alternate touching the nose and therapist's finger with the index finger (Refer to Figs 6.7A and B).
Finger opposition	Touch thumb to each finger in sequence.
Mass grasp	Alternate between fist opening and closing.
Pronation/supination	Alternately turn palms up and down (Refer to Figs 6.9A and B).
Rebound test	Apply resistance then release suddenly, observing the response of the opposite muscle.
Tapping (hand)	Tap the hand on the knee.
Tapping (foot)	Tap the ball of one foot on the floor without raising the knee, while the heel maintains contact with the ground (Refer to Figs 6.13A and B).
Pointing and past pointing	Flex shoulder and return to starting position, observing for "past pointing".
Alternate heel-to-knee; Heel-to-toe	Touch knee and big toe alternately with the heel.
Toe to examiner's finger	Touch the great toe to examiner's finger (Refer to Figs 6.10A and B).
Heel on shin	Slide heel up and down the shin of opposite lower extremity (Refer to Figs 6.11A and B).
Drawing a circle	Draw imaginary circle or figure-eight with extremity.
Fixation or position holding	Hold arms horizontally (upper extremity) or knee extended (lower extremity) while seated or standing.

- Functional reach while standing (forward trunk flexion with upper limb reach).
- Flex trunk to each side while standing
- Standing with eyes open to eyes closed (failure to maintain an upright posture when eyes closed is positive Romberg sign)
- March in place
- Tandem walking (heel to toe)
- Walk sideways, backward or cross stepping
- Alter speed of ambulatory activities (observe patient walking at normal speed, as fast as possible and as slow as possible)
- Start and stop walking abruptly on command
- Walk and pivot (turn 90°, 180° or 360°) on command
- Walk in circle, heels or on toes
- Walk on vertical or horizontal head turns on command
- Step over or around obstacles
- Stair climbing with or without using handrail
- Jumping jacks
- Sitting on a therapy ball while alternatively flexing and extending the knees

Clinical Correlation

When assessing incoordination, clinician needs to pay attention to the quality and characteristics of the patient's movements during the tests. Observing specific patterns of incoordination can provide valuable diagnostic insights.

COORDINATION EXERCISES

Coordination exercises are exercises designed to improve neuromuscular control and coordination of movements. They typically involve performing a series of movements that challenge the body's ability to control and coordinate muscle contractions, joint movements, and balance.

Commonly Used Exercises

Some examples of coordination exercises include:

- **Balancing exercises:** These exercises involve standing on one foot or on an unstable surface, such as a balance board or foam pad. They challenge the body's ability to maintain balance and stability (Fig. 13.7).
- **Agility drills:** These involve performing quick, movements in different directions, such as shuffling, jumping, or pivoting. These exercises challenge the body's ability to coordinate movements and change direction quickly.
- **Rhythm exercises:** Performing actions in time to music or a metronome is a rhythm workout. They put the body's capacity to match motions to an outside beat or rhythm to the test.
- **Proprioceptive exercises:** Such exercises involve moving while having the eyes closed or while standing on unsteady objects like a wobble board or balance disc. These exercises

Fig. 13.7: Coordination exercises

test the body's capacity for balance and control without the aid of external cues.

- **Eye hand coordination exercises:** Tasks requiring precise hand motions based on visual signals, such as catching a ball or striking a target, are included in eye-hand coordination exercises.
- **Cross body exercises:** Exercises that cross the body's midline, like touching the left hand to the right foot, are referred to as "cross-body" exercises. The body's capacity to coordinate motions across various body parts is put to the test by these workouts.
- **Frenkel's** exercises are explained ahead separately in this chapter.

Principles

The following are a few of the fundamentals of coordination exercises:

- **Specificity:** The goals and motions that an individual wants to develop should be taken into consideration while designing coordination exercises. A dancer may concentrate on balance and footwork, whereas an athlete may concentrate on agility and quickness.
- **Variability:** To push the body in various ways, coordination exercises should be varied and diversified, combining various sorts of movements, surfaces, and situations.
- **Feedback:** It is crucial to give feedback during coordination exercises so that people can become aware of their actions and make necessary corrections. This can be in the form of verbal, visual, or tactile feedback.
- **Repetition:** To encourage motor learning and to build muscle memory, coordination exercises should be performed repeatedly.
- **Rest and recovery:** The body needs sufficient rest and recovery in order to adjust to the demands of coordination exercises and improve.
- **Individualization:** Taking into account the person's age, fitness level, injury history, and goals, coordination exercises should be customized to the person's needs and abilities.
- **Progressive overload:** To guarantee that the body is continuously pushed and adapts to new movement

patterns, coordination exercises should be gradually increased in complexity.

Benefits

- **Enhances motor skills:** Coordination exercises put the body's capacity to coordinate movements to the test, which can enhance motor abilities including balance, agility, and proprioception.
- **Improves athletic performance:** Athletes who work on their coordination will perform better in sports that call for quick movements, accurate movements, and direction changes.
- **Improves body awareness:** Coordination exercises can assist a person become more aware of their body, which can aid in injury prevention and recovery.
- **Improves cognitive performance and brain function:** Coordination exercises activate neural networks, which can promote brain function (Kwot et al.).
- **Boosts self-assurance:** Better coordination can make someone feel more confident about their capacity to engage in physical activity, which can enhance their general quality of life.
- **Improves quality of life:** People who have better coordination may carry out daily tasks more efficiently and with less effort, which can enhance their general quality of life.

Clinical Correlation

Ensure that the environment is safe for performing exercises, especially for patients with balance impairments or fall risk. Use appropriate assistive devices, such as handrails or gait belts, and provide supervision and support as needed to prevent injuries.

FRENKEL'S EXERCISES

Physical therapy techniques like Frenkel's exercises are used to enhance balance and mobility. They were created by Dr Heinrich Frenkel in the 1920s as a means of treating patients with neurological disorders like Parkinson's disease or multiple sclerosis. The exercises usually involve a series of rhythmic and repetitive motions that are meant to test the patient's balance and motor coordination. They may include the use of weights or other equipment to increase resistance, and can be carried out while standing or sitting. The principles of neuroplasticity, or the brain's capacity to modify and reorganize itself in response to new experiences, serve as the foundation for Frenkel's exercises. Patients can improve the connections between their brain and body by doing specific motions regularly which can improve their coordination and overall physical function. Heel-to-toe walking, marching in place, and arm movements in various directions are a few typical Frenkel's exercises. Depending on their particular needs and skills, patients may also undergo activities that require balance on one leg or walking over obstacles. Numerous patients with conditions, including those with neurological diseases, orthopedic injuries, or balance issues, can benefit from Frenkel's exercises. They may be administered as a part of a more extensive physical therapy regimen, and they are frequently combined with other forms of treatment, like medicine or surgery.

Indications

Numerous physical and neurological problems can be treated using Frenkel's exercises, including:

- **Neurological diseases:** Frenkel's exercises are frequently used to help individuals with neurological illnesses including multiple sclerosis, Parkinson's disease, or stroke restore their motor function. These exercises can enhance brain-muscle communication, enhance coordination, and lessen tremors and spasticity. Results of study by Ghasemi et al showed that doing Frenkel's exercises continuously leads to improvement in ataxia, balance, activities of daily living (ADLs) and depression in patients with multiple sclerosis (MS).
- **Orthopedic problems:** Sprains, strains, and fractures can also be treated with Frenkel's exercises to aid in the patient's recovery. These exercises can be a useful component of a rehabilitation program since they can aid in the recovery of strength, flexibility, and balance.
- **Balance issues:** Frenkel's exercises can help patients who struggle with balance, especially those who have a risk of falling. By putting the patient's motor abilities and proprioception (the awareness of one's body's location in space) to the test, these exercises can aid in the improvement of balance and stability (Shivaprasad et al.). Frenkel's exercise improves sensory and balance recovery among subacute ischemic stroke patients with impaired proprioception and minimal lower limb motor weakness (Ko et al.).
- **Postsurgical recovery:** Frenkel's exercises can aid patients in getting better after surgery, especially if the treatment affected their balance or mobility. These exercises can speed up the healing process and help to regain the strength, flexibility, and coordination.

Principles

- **Begin with easy movements:** Frenkel's exercises frequently begin with easy motions to assist the patient become accustomed to the exercise. Easy motions can be as easy as tapping the feet, waving the arms, or swiveling the head.
- **Gradually raise the level of difficulty:** As the patient feels more at ease with the exercises, the level of difficulty is raised. This could entail adding more difficult movements or speeding up the movements.

- **Pay attention to precision and control:** Frenkel's exercises are meant to enhance proprioception and coordination. Therefore, it is crucial to pay attention to accuracy and control. Patients should move cautiously and slowly while being mindful of their posture and motions.
- **Employ visual cues:** When practicing Frenkel's exercises, visual cues might be useful. Patients might be instructed, for instance, to use their eyes to follow a line on the ground.

Technique

A series of therapeutic exercises known as the Frenkel's exercises are intended to enhance proprioception, balance, and coordination. The exercises are usually done slowly and deliberately, emphasizing accuracy and control. Here is a detailed description of how to carry out Frenkel's exercises:

- Suitable positioning and clothing of the patient is necessary. The limbs should be visible throughout exercise.
- Explanation and demonstration of exercise to patients.
- Patient must be attentive while performing the exercises.
- Speed should be rhythmic, as dictated by the therapist.
- Range of motion should be indicated by marking a spot on which foot or hand is to be placed.
- Concentration, precision and repetition of the movements should be emphasized.

Examples

- In lying position with head raised, patient will perform abduction and adduction of hip.
- In lying position, patient will perform one hip and knee flexion and extension (Fig. 13.8).
- In half lying position, patient will first abduct the hip to bring it to the side of the plinth, then bend the knee to put foot on the floor and then reverse the movement.
- In sitting, patient will extend legs to put the feet on the floor. This can be progressed to walking on the markings on the floor (Fig. 13.9).

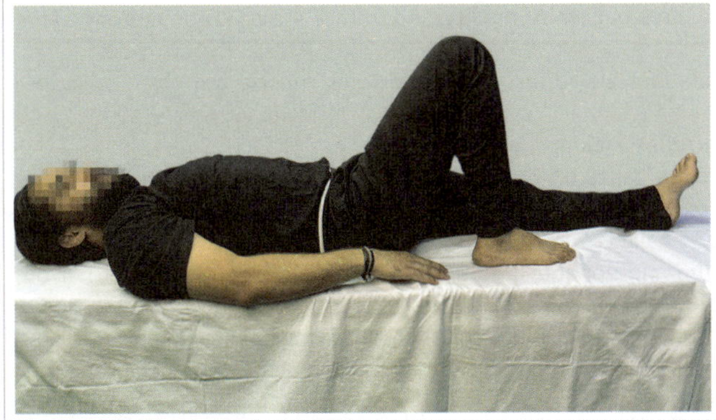

Fig. 13.8: Hip and knee flexion and extension in lying position

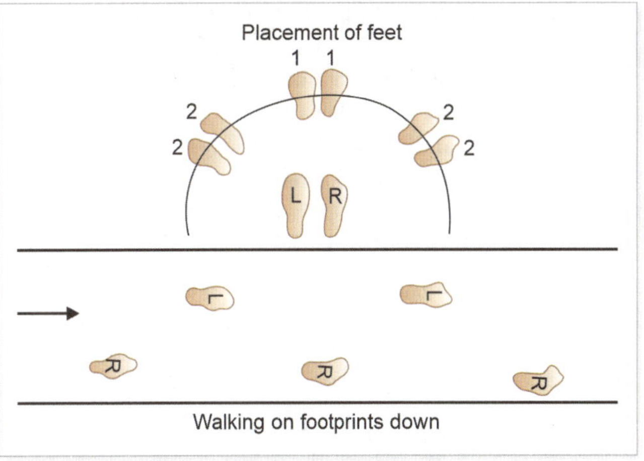

Fig. 13.9: Markings for Frenkel's exercise

- In sitting, patient will perform alternate leg raises and putting the foot back on floor (Fig. 13.10A).
- In sitting position, patient will perform shoulder flexion and extension and alternate leg movement can be added with upper limb movement (Fig. 13.10B).
- In sitting position, patient will stretch one arm to thread it through small loop or ring.
- In sitting position, patient will pick up objects and place them on specified marks.
- In sitting position, patient will perform one hip flexion and lean forward to touch the shoulder to the knee (Fig. 13.10C).
- In sitting position, the patient will lean forward as if putting weight on both feet and then sit back again (Fig. 13.10D).

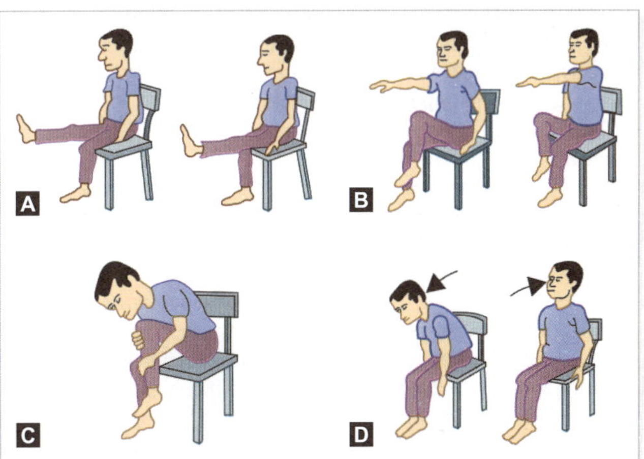

Figs 13.10A to D: **A:** Alternate leg stretching and lifting to put the foot; **B.** Alternate shoulder flexion and extension with alternate leg movement; **C.** Single hip flexion; **D.** Leaning forward to put weight on both feet and then sit back again

- In standing or walking position, patient will bounce and catch the ball.
  - In stride standing position, patient will transfer weight from one foot to another.
  - In stride standing position, patient will walk sideways and place foot on marks made on the floor.

Clinical Correlation

In conditions like Parkinson's disease, multiple sclerosis, or stroke, impaired mobility and balance can increase the risk of falls and related injuries. Frenkel's exercises help strengthen muscles, improve proprioception, and enhance postural stability, thereby reducing the risk of falls and preventing secondary complications such as fractures or immobility-related issues.

Progression

- Adjusting the speed, scope, and complexity of the exercises.
- Changing the pace of successive movements, incorporating pauses that entail starting and stopping on command.
- Beginning with broad and basic movements that engage major joints, then transitioning to movements involving smaller joints, a narrower range, and more frequent changes in direction.
- Basic movements are organized into sequences to develop specific actions that require coordination of multiple joints and limbs, such as walking.
- Depending on the individual's level of impairment, rehabilitation exercises begin with lying down with support for the head and limbs, progressing to seated exercises, and eventually to standing exercises.

SUMMARY

- Voluntary movements are coordinated through complicated interactions between multiple brain areas and neuronal networks. The primary motor cortex (PMC) in the frontal lobe initiates motor orders, whereas descending motor pathways, such as the corticospinal tract, govern fine motor movements and postural control.
- The cerebellum serves as a comparator and error corrector, assuring movement accuracy by comparing intended commands to actual execution.
  - Meanwhile, cortical motor areas provide information to the basal ganglia, which initiates and controls purposeful motions.
  - Sensory pathways, such as the dorsal column-medial lemniscal route, carry proprioceptive and tactile information required for movement coordination.
- The cerebellum, which accounts for approximately 10% of the total brain weight, is located in the posterior cerebral fossa and is made up of two massive hemispheres joined by the median vermis. Its complicated structure comprises surfaces, grooves, and fissures that define different anatomical regions.
  - Cytoarchitecturally, it consists of gray matter in the cortex and deep nuclei, as well as foliated folds called folia.
  - Functionally, the cerebellum maintains balance, organizes voluntary movements, and helps in motor learning and predictive control.
  - Recent study has also suggested its participation in cognitive activities such as attention, language processing, and spatial perception, emphasizing its importance in brain function.
- Symptoms of cerebellar dysfunction, such as ataxia, dysarthria, and cognitive impairment, can disturb the smooth execution of coordinated movements. Incoordination, characterized by awkward and inaccurate motions, is caused by anomalies in sensory inputs or neuromuscular dysfunction, altering movement timing, direction, and muscle activation.
- Neuromuscular incoordination arises from various conditions affecting the connection between the nervous system and muscles.
  - Cerebellar and basal ganglia lesions, upper and lower motor neuron injuries, loss of kinesthesia, aging, neurological conditions like Parkinson's disease, multiple sclerosis, cerebral palsy, Huntington's disease, muscular dystrophy, and substance use can contribute to this condition. Such dysfunctions often result in symptoms like tremors, ataxia, dysarthria, and impaired motor control.
  - Additionally, cerebellar lesions can lead to deficits such as dysmetria, ataxia, dysdiadochokinesia, nystagmus, hypotonia, dyssynergia, asynergia, rebound phenomenon, intention tremor, and titubation, along with language impairments like dysarthria, agrammatism, anomia, and impaired verbal fluency.
- Patients with basal ganglia lesions often display specific motor impairments, such as reduced movement initiation, involuntary movements, and alterations in posture and muscle tone.
  - Akinesia, athetosis, bradykinesia, chorea, choreoathetosis, dystonia, hemiballismus, rigidity, and tremor are common motor deficits associated with basal ganglia pathology.
  - Lower motor neuron lesions can lead to neuromuscular incoordination due to loss of muscle tone, muscle atrophy, fibrillation, fasciculation, weakness, and paralysis.
  - Conversely, upper motor neuron lesions may cause spasticity, hyperreflexia, clonus, weakness, and paralysis.
- Kinesthesia, the sense of body position and motion without visual cues, relies on proprioceptive receptors; its loss can occur due to conditions like peripheral neuropathy, cerebellar lesions, proprioceptive disorders, spinal cord injuries, stroke, tabes dorsalis, syringomyelia, and leprosy, resulting in difficulties with coordination and balance.
- Coordination and balance can be assessed through various equilibrium and nonequilibrium tests like, standing with normal/narrow base of support, marching, walking sideways, walking in circles, tandem walk, etc.
  - Nonequilibrium tests focus on motor tasks that require timing, control, and precision without emphasizing balance maintenance.
  - These tests include the finger-to-nose test, heel-to-shin test, alternating hand movements, rapidly switching movements, pronation-supination test, finger-to-finger test, and finger-to-thumb test, among others.
  - Equilibrium coordination tests are evaluation of a person's capacity to hold their balance and coordinate motions that support the stabilization of their body in various weight-bearing postures.

FURTHER READINGS

- De Smet HJ, Paquier P, Verhoeven J, Mariën P. The cerebellum: Its role in language and related cognitive and affective functions. Brain and language. 2013 Dec 1;127(3):334–42.
- Gardiner Dena M. Principles of Exercise Therapy. CBS Publishers and Distributors Pvt Ltd; 4th ed., 2023.
- Ghasemi E, Shaygannejad V, Ashtari F, Fazilati E, Fani M. The investigation of Frenkel's exercises effection on ataxia, balance, activity of daily living and depression in patients with multiple sclerosis. Journal of Research in Rehabilitation Sciences. 2008 Jun 1;4(1).
- Ko EJ, Chun MH, Kim DY, Kang Y, Lee SJ, Yi JH, Chang MC, Lee SY. Frenkel's exercise on lower limb sensation and balance in subacute ischemic stroke patients with impaired proprioception. Neurology Asia. 2018 Sep 1;23(3).Leiner HC, Leiner AL, Dow RS. Cognitive and language functions of the human cerebellum. Trends in neurosciences. 1993 Nov 1;16(11):444–7.
- Kwok TC, Lam KC, Wong PS, Chau WW, Yuen KS, Ting KT, Chung EW, Li JC, Ho FK. Effectiveness of coordination exercise in improving cognitive function in older adults: a prospective study. Clinical interventions in aging. 2011 Sep 29:261–7.
- O'Sullivan SB, Schmitz TJ, Fulk G. Physical rehabilitation. FA Davis; 2019 Jan 25.
- Shivaprasad MK, Rayjade A, Warude T, Jagtap V. Effectiveness of Chair Aerobics and Frenkel's Exercise In Geriatric Population On Balance and Coordination–Randomized Control Trial. International Neurourology Journal. 2023 Nov 1;27(4):265–70.
- Singh R. Cerebellum: Its anatomy, functions and diseases. In Neurodegenerative Diseases—Molecular Mechanisms and Current Therapeutic Approaches 2020 Jun 24. IntechOpen.
- Sillitoe RV, Fu Y, Watson C. Cerebellum. In the mouse nervous system 2012 Jan 1 (p. 360–397). Academic Press.

STUDENT ASSIGNMENT

LONG ANSWER QUESTIONS

1. Describe the functions of cerebellum, and enlist causes of dysfunction along with signs and symptoms.
2. Describe causes of incoordination in detail.
3. Enumerate and explain tests for coordination: Equilibrium test and nonequilibrium test.
4. Explain the principles of coordination exercises.
5. Describe the uses and technique of Frenkel's exercises.

SHORT ANSWER QUESTIONS

1. Write briefly about the structure of cerebellum with its pathways.
2. Define coordination and incoordination.
3. What is the mechanism of neuromuscular coordination?
4. Write in brief about Frenkel's exercises.
5. Write briefly about incoordination due to UMN and LMN lesions.

MULTIPLE CHOICE QUESTIONS

1. **Which brain structure is primarily responsible for maintaining balance, organizing voluntary movements, and aiding in motor learning?**
   a. Cerebral cortex b. Basal ganglia
   c. Cerebellum d. Thalamus
2. **What function does the primary motor cortex serve in voluntary motion coordination?**
   a. Initiates motor orders
   b. Corrects movement errors
   c. Maintains balance
   d. Processes sensory information
3. **Incoordination characterized by awkward and inaccurate movements is primarily caused by anomalies in:**
   a. Sensory inputs
   b. Neuromuscular junctions
   c. Basal ganglia function
   d. Cerebral cortex activity
4. **Which condition is NOT associated with cerebellar dysfunction?**
   a. Ataxia b. Dysarthria
   c. Muscle rigidity d. Cognitive impairment
5. **Patients with basal ganglia lesions commonly exhibit motor impairments such as:**
   a. Hyperreflexia b. Tremors
   c. Hypotonia d. Dysmetria
6. **Which test evaluates the ability to maintain balance while walking in a straight line from heel to toe?**
   a. Romberg test
   b. Tandem walking test
   c. Finger-to-nose test
   d. Sharpened Romberg test
7. **Neuromuscular incoordination due to lower motor neuron lesions can result in:**
   a. Increased muscle tone
   b. Fasciculations
   c. Spasticity
   d. Clonus
8. **Loss of kinesthesia, the sense of body position and motion without visual cues, can occur due to:**
   a. Cerebellar lesions
   b. Basal ganglia dysfunction
   c. Spinal cord injuries
   d. Peripheral neuropathy

9. Which motor deficit is characterized by sudden, violent flailing of one side of the body due to contralateral subthalamic nucleus lesions?

a. Akinesia
b. Hemiballismus
c. Chorea
d. Rigidity

10. Nonequilibrium tests for coordination primarily focus on:

a. Maintaining balance
b. Timing, control, and precision
c. Sensory inputs
d. Initiating motor orders

ANSWER KEY

1. c **2.** a **3.** a **4.** c **5.** b **6.** b **7.** b **8.** d **9.** b **10.** b

14 Group and Individual Exercises

Puneeta Ajmera, Sheetal Yadav, Sheetal Kalra

LEARNING OBJECTIVES

After the completion of the chapter, the readers will be able to:
- Define group and individual exercises and their types.
- Understand the structure and organization of group exercises.
- Discuss the effects and benefits of group and individual exercises.
- Explain home exercise program.
- Explain advantages and disadvantages of home-based exercises.

CHAPTER OUTLINE

- Introduction
- Group Exercises
- Individual Exercises
- Home Exercise Prescription

KEY TERMS

Cool down: Cooling down, sometimes referred to as limbering down or warming down, is the process of gradually reducing the intensity of the physical activity to a more normal level.

High-intensity interval training (HIIT): A warm-up, high-intensity exercise repetitions spaced out by medium-intensity workouts for recovery, and a cool-down phase often comprise an HIIT workout.

Pilates: Pilates is a low-impact workout that improves flexibility, mobility, core strength, balance, and even mood with its regulated movements.

Warm-up: Warming up is stretching and getting ready for physical activity or a performance by doing some light exercise or practicing beforehand; this is typically done before a practice or performance.

INTRODUCTION

Physical inactivity is a significant concern, as it contributes to the development of various illnesses and escalates healthcare costs. This underscores the necessity of promoting tested exercise programs that not only enhance health but also mitigate expenses. Healthcare providers often advocate for increased physical activity due to its myriad benefits on both physical and psychological health-related quality of life (HRQoL). However, there are still unanswered concerns regarding the best setting, level of supervision, delivery method (individual, group or both), level of intensity, length of time, and kind of exercise needed to reap these advantages. These issues are brought up by the variety of prescribed exercises that are performed in research, each with a different location and level of monitoring. According to reports, the recommended exercise consists of: Supervised group sessions at a center; individual home-based or center-based programs; home exercise programs with little to no supervision; and a mix of home exercise programs and supervised center-based sessions.

In recent years, group fitness classes have evolved to accommodate a diverse range of participants, including those with chronic conditions or injuries. Modalities like step aerobics, Pilates, and strength training offer distinct components of fitness and exert unique effects on HRQoL. These group exercises have proven to be particularly beneficial for individuals with various health conditions, such as multiple sclerosis (MS), cancer, fibromyalgia, and mild cognitive impairment.

GROUP EXERCISES

Group exercise therapy is the term used to describe systematic small-group exercise programs that incorporate a variety of rehabilitative modalities, including Pilates, functional movement, and strength and conditioning, etc. These classes are intended to help people with various disorders both therapeutically and in terms of improving their physical fitness. Every session is customized to each participant's unique needs, guaranteeing individualized attention in a welcoming group environment. Group exercise settings provide a structured and supportive environment wherein participants engage in physical activity while receiving social support. This supportive atmosphere fosters camaraderie, motivation, and accountability, which are crucial factors in maintaining long-term adherence to exercise regimens. Moreover, the camaraderie experienced in group settings often translates into improved mood and overall psychological well-being.

Studies have consistently demonstrated the positive impact of group exercises on both physical and psychological health outcomes. For instance, individuals with MS participating in group exercise classes have reported improvements in mobility, balance, and overall well-being. Similarly, women undergoing treatment for breast cancer have experienced functional references and psychological benefits from supervised group exercise programs. Furthermore, individuals with fibromyalgia and mild cognitive impairment have shown improvements in physical functioning and mood after engaging in group exercise interventions.

By combining the inclusive nature of group fitness classes with the tailored approach needed for individuals with specific health conditions, healthcare providers can offer holistic support for improving HRQoL. Group exercises not only facilitate physical activity but also provide a sense of community, support, and empowerment, contributing to overall well-being and quality of life.

Did You Know?

Research has shown that regular participation in group fitness activities, particularly those involving coordination and cognitive engagement like dance or interactive fitness games, can enhance memory, attention, and overall cognitive abilities.

Structure

Group exercises are typically organized into structured sessions led by a qualified instructor. These sessions can take place in various settings such as gyms, community centers, parks, or online platforms. The structure includes:

- **Warm-up:** Sessions generally begin with a warm-up to prepare the body for more intense activity. This includes light aerobic exercises and dynamic stretches to increase heart rate and loosen up muscles.
- **Main workout:** The core part of the session focuses on various types of exercises depending on the class type. It can range from HIIT, strength training, aerobic exercises, yoga, Pilates, dance fitness or circuit training.
- **Cool-down:** Concluding with a cool-down phase helps lower the heart rate gradually and includes static stretching to enhance flexibility and aid recovery.

Types

Group exercise (Fig. 14.1) classes can cater to different fitness levels and interests, providing a variety of formats such as:

1. **Aerobics:** High-energy classes that include rhythmic movements set to music, designed to improve cardiovascular fitness.
2. **Strength training:** Classes that use weights, resistance bands, or body-weight exercises to build muscle strength and endurance.
3. **Yoga and pilates:** These focus on flexibility, core strength, balance, and mental well-being through structured poses and movements.

Fig. 14.1: Group exercise

4. **Dance fitness:** It incorporates dance moves with fitness routines, often set to upbeat music, to make exercise fun and engaging.
5. **High-intensity interval training (HIIT):** Short bursts of intense activity followed by brief rest periods, aimed at maximizing cardiovascular efficiency and calorie burning. is called HIIT.
6. **Circuit training:** Rotating through a series of exercise stations that target different muscle groups for a comprehensive workout is circuit training.

Effective Group Exercise Program

To ensure a successful and beneficial group exercise program, the following elements are essential:

- **Qualified instructors:** Instructors should be certified, experienced, and capable of leading diverse groups with varying fitness levels.
- **Safety measures:** Proper attention to safety, including warm-ups, cool-downs, and the use of appropriate equipment, is crucial to prevent injuries.
- **Clear communication:** Instructors should clearly explain and demonstrate exercises, and be approachable for questions and assistance.
- **Inclusive environment:** Creating a welcoming and inclusive atmosphere encourages participation from people of all backgrounds and fitness levels.
- **Feedback and improvement:** Regular feedback from participants can help instructors refine the program to better meet the needs and preferences of the group.

Benefits

- **Improvement in mobility and balance:** Group exercise classes have been associated with improvements in mobility and balance, particularly in ambulant individuals with MS. These classes provide targeted exercises that help individuals enhance their motor skills, coordination, and overall physical function. Improved mobility and balance contribute to increased independence and confidence in daily activities.
- **Improved physical function and reduced falls:** Growing body of evidence supports the effectiveness of group-based exercise in promoting physical functioning and reducing falls among frail older individuals. The findings highlight the importance of tailored interventions for high-risk groups and emphasize the feasibility and potential benefits of implementing group exercise programs in community settings.
- **Enhanced physical well-being:** Participating in group exercise programs leads to improvements in physical dimensions such as fatigue and well-being. Individuals with MS, cancer, and fibromyalgia report feeling more energized, less fatigued, and generally healthier after engaging in group exercise sessions. These improvements in physical well-being contribute to a better quality of life and overall satisfaction with daily activities.
- **Psychological benefits:** Group exercise classes offer psychological benefits such as reduced stress, improved mood, and enhanced self-esteem. The supportive atmosphere and camaraderie among participants create a sense of belonging and social connection, which positively impacts mental health. Individuals with MS, cancer, and fibromyalgia often report feeling more motivated and emotionally uplifted after attending group exercise sessions.
- **Supportive environment:** Group exercise classes provide a supportive environment where individuals can share their experiences, challenges, and successes with others facing similar health conditions. This sense of camaraderie fosters encouragement, empathy, and understanding among participants, reducing feelings of isolation and loneliness often associated with chronic illnesses.
- **Improved adherence to exercise regimens:** The group setting encourages adherence to exercise regimens by providing accountability, motivation, and social support. Individuals are more likely to maintain regular attendance and engagement in group exercise classes compared to exercising alone. The sense of belonging and camaraderie fosters a commitment to achieving fitness goals and maintaining a healthy lifestyle.
- **Benefit from supervision:** Exercising alone often leads to skipping exercises that are less enjoyable or unsure in execution. In a small group class, the instructor designs the program to include the most beneficial exercises for the group and provides continuous supervision to ensure proper form and technique. This ensures participants have a safe, enjoyable, and effective workout.

- **Accountability in group workouts:** Many understand the benefits of their prescribed physiotherapy exercises but struggle with time management and motivation. Joining a group exercise class increases motivation and accountability to the group, the instructor, and oneself. Regular support from peers and a structured environment enhance consistency and commitment, promoting a more active role in rehabilitation.
- **Holistic approach to health:** Group exercise programs offer a holistic approach to health by addressing both physical and psychological aspects of well-being. By integrating exercise, social support, and education, these programs empower individuals to take control of their health and improve their overall quality of life. The combination of physical activity, social interaction, and emotional support contributes to comprehensive health promotion and disease management.

INDIVIDUAL EXERCISES

Individualized exercises (Figs 14.2A and B) are a cornerstone of physiotherapy, tailored specifically to meet the unique needs and goals of each patient. This personalized approach ensures that the exercise regimen is effective, safe, and aligned with the patient's health status, physical capabilities, and personal objectives.

Here is a detailed look at the components and benefits of individualized exercises:

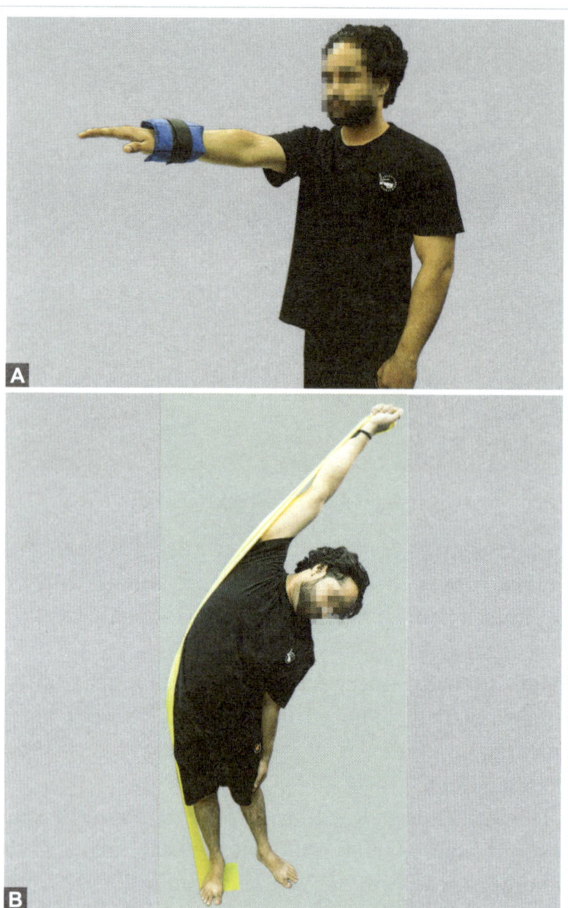

Figs 14.2A and B: Individual exercises

Initial Evaluation and Goal Setting

The process begins with a comprehensive initial evaluation conducted by a physiotherapist. This assessment includes:

- **Medical history review:** Understanding past injuries, surgeries, chronic conditions, and overall health status.
- **Physical examination:** Assessing strength, flexibility, range of motion, posture, and coordination.
- **Functional assessment:** Evaluating how well the patient performs daily activities and identifying any limitations or pain points.
- **Patient goals:** Discussing and setting realistic, specific, and measurable goals with the patient, such as reducing pain, improving mobility or enhancing athletic performance.

Designing the Customized Exercise Regimen

Based on the initial evaluation, the physiotherapist designs a tailored exercise program. This program typically includes:

- **Strength training:** Exercises that target specific muscle groups to improve overall strength and support joint stability. These may include weight lifting, resistance band exercises, or body-weight exercises.
- **Range of motion exercises:** Activities aimed at improving the movement of joints through their full range. This can include stretching exercises and joint mobilizations.
- **Cardiovascular fitness:** Aerobic exercises that improve heart and lung function, such as walking, cycling, swimming, or using a treadmill or elliptical machine.
- **Flexibility exercises:** Stretching routines that enhance muscle length and joint flexibility, reducing stiffness and preventing injuries.
- **Balance and coordination:** Exercises that improve proprioception and coordination, essential for reducing fall risk and enhancing overall functional ability. These might include balance board exercises, stability ball exercises, and agility drills.

Training and Assistance

The physiotherapist provides hands-on training and assistance to ensure that exercises are performed correctly and safely. This involves:

- **Demonstration and supervision:** Showing the patient how to perform each exercise correctly and supervising them during practice sessions.

- **Adjustments and modifications:** Modifying exercises based on the patient's response and progress, ensuring that the activities remain challenging yet achievable.
- **Education:** Teaching the patient about the importance of each exercise, how it contributes to achieve his goals, and encouraging adherence to the program.

Monitoring and Progress Tracking

Continuous monitoring and progress tracking are critical components of individualized exercises:

- **Regular assessments:** Periodic re-evaluations to measure progress against the initial goals and make necessary adjustments to the exercise regimen.
- **Feedback and encouragement:** Providing constructive feedback and motivation to help the patient stay committed to their exercise program.
- **Adapting to changes:** Modifying the exercise plan in response to any new developments, such as changes in the patient's condition, new injuries, or achieving set goals.

Benefits

- **Enhanced physical and mental health:** It was observed in a study by Arntzen et al., that participants experienced significant emotional and physical improvements when exercises were tailored to their specific needs, leading to a stronger sense of ownership and control over their movements.
- **Reduced fatigue levels:** Study by Burke et al., showed that individualized stretching exercises significantly reduced fatigue levels in patients with MS, highlighting the importance of customizing interventions based on each patient's condition.
- **Improved exercise adherence:** A study by Smith et al., concluded that individualized exercise prescriptions informed by reversal theory effectively enhance exercise consistency, motivation, and overall weight loss among participants.
- **Enhanced engagement:** In a study by Geraedts et al., it was found that individually tailored interventions effectively enhance adherence and engagement in physical activity programs for frail older adults.
- **Enhanced effectiveness of the treatment program:** Patients with chronic obstructive pulmonary disease (COPD) can benefit from home-based individualized pulmonary rehabilitation programs, which include breathing exercises to improve pulmonary function, respiratory muscle strength, exercise capacity, dyspnea, and health-related quality of life (Lu et al.).
- **Safety:** Personalized programs consider the patient's health status and physical capabilities, reducing the risk of injury. They are suitable for patients with various conditions who need to ensure that their exercise routines do not exacerbate their health issues.
- **Motivation and engagement:** Green et al., in a research study concluded that patients are more likely to stay motivated and engaged with a program that is designed specifically for them and aligns with their personal goals. Home-based individualized exercises are effective for patients with hip osteoarthritis, improving subjective and objective measures.

Limitations

- **Limited emphasis on education and exercise:** Individual interventions often include passive therapies with limited evidence of benefit, which can detract from the focus on effective education and exercise, potentially reducing their success.
- **Lack of comprehensive approach:** While individual exercises are tailored to specific needs, they may miss out on the thorough and regular application of fundamental exercise principles that group programs provide.
- **Neglect of biopsychosocial elements:** Musculoskeletal disorders involve a mix of physical, cognitive, psychological, and social factors. Individual exercise programs may not address this broad spectrum as effectively as group programs, which can inadvertently cater to these elements through social contact and support.
- **Inconsistent customization:** Despite the potential for more personalized care, individual physical therapy does not always achieve this in practice. Effective individual rehabilitation should integrate a biopsychosocial approach tailored to the patient's specific needs, but this does not always lead to better outcomes compared to group therapy.
- **Lack of social support:** Individual settings lack the inherent social interaction and support found in group environments, which can provide significant psychological and motivational benefits, enhancing overall outcomes.

HOME EXERCISE PRESCRIPTION

Home exercise programs are often recommended for various populations, including older adults and individuals recovering from injuries or surgeries (Green et al.). These programs offer a convenient way to engage in physical activity without the need to travel to a gym or rehabilitation center. However, the effectiveness of home exercise programs can vary based on several factors, including the nature of the exercises, the level of supervision, and the specific needs of the participants.

Benefits

- **Convenience and accessibility:**
  - Home exercise programs allow individuals to work out in the comfort of their own homes, which can be particularly beneficial for those with mobility issues, transportation challenges, busy schedules or during situations like quarantine.
  - Further home-based exercises eliminate the need to travel to a gym or rehabilitation center, saving time and reducing barriers to regular exercise.
  - Engaging in regular home-based aerobic and resistance exercises is essential for maintaining physical health and preventing musculoskeletal disorders during quarantine.
- **Cost-effectiveness:**
  - Home exercise programs can be less expensive than gym memberships or supervised rehabilitation sessions, making them a more affordable option for many individuals.
  - Minimal equipment is often required, and exercises can be performed using household items or body weight.
  - Study by Mccarthy et al., suggests that supplementing a home-based exercise program with an 8-week class-based program in patients with Osteoarthritis of knee is likely to be cost-effective, with slightly lower costs and higher quality-adjusted life-year gains.
- **Personalized and flexible:**
  - Exercises can be tailored to meet the specific needs and goals of the individual, allowing for a customized approach that can be adjusted as progress is made.
  - Flexibility in scheduling allows individuals to integrate exercise into their daily routines more easily.
- **Increased adherence and consistency:**
  - Research study by Bachmann et al., suggests that home-based exercise programs can have higher adherence rates compared to group programs, as individuals can exercise at their own pace and convenience.
  - Regular phone counseling and check-ins can enhance adherence by providing ongoing support and motivation.
- **Safety and comfort:**
  - Exercising at home can provide a safer environment for those who feel uncomfortable or self-conscious working out in public or group settings. Study by Ransdel et al., concludes that home-based physical activity programs are a safe and comfortable alternative for mothers and daughters, effectively increasing physical activity and improving health-related fitness without compromising participation or outcomes.
  - Reduced risk of exposure to contagious illnesses, which is especially important for older adults or individuals with compromised immune systems.

Disadvantages

- **Lack of supervision:**
  - Without professional supervision, there is an increased risk of performing exercises incorrectly, which can lead to injuries.
  - Limited feedback on exercise technique and progress can hinder improvements and lead to suboptimal outcomes.
- **Motivation and accountability:**
  - Exercising alone can require a higher level of self-motivation, and some individuals may struggle to stay consistent without the social support and accountability that group classes or supervised sessions provide.
  - Lack of regular interaction with a trainer or therapist can reduce motivation and adherence over time.
- **Limited resources and equipment:**
  - Home exercise programs may be constrained by the lack of specialized equipment or space, limiting the variety and intensity of exercises that can be performed.
  - Some individuals may not have access to necessary equipment, which can impact the effectiveness of the exercise program.
- **Potential for reduced effectiveness:**
  - Studies have shown that while home exercise programs can improve certain outcomes, they may not be as effective as supervised group-based programs, particularly in improving muscle strength, physical performance, and overall quality of life.
  - The lack of direct supervision and immediate feedback can result in less significant improvements in mobility and function.
- **Adherence challenges:**
  - Despite the potential for higher adherence, research study by Simek et al., showed a report significant dropout rates in home exercise programs, highlighting the challenge of maintaining long-term commitment.
  - Adherence to home exercise programs (HEPs) is crucial for physiotherapy success, yet rates are often poor (50–70% nonadherence in some conditions), risking recurrent injury and questioning treatment effectiveness.
  - Efforts are needed to identify barriers to adherence and implement effective strategies to improve patient compliance.
  - Factors such as home environment distractions, lack of social interaction, difficulty in self-monitoring progress, perceived obstacles like forgetting to work out or not having enough time, self-efficacy, threat beliefs about the condition, baseline physical activity, psychological symptoms like depression, and social support from family, friends, and therapists effect adherence to home exercise adherence.

SUMMARY

- Physical inactivity is a significant concern, leading to various illnesses and increased healthcare costs. Promoting effective exercise programs is crucial for enhancing health and reducing expenses.
- While healthcare providers advocate for increased physical activity to improve physical and psychological HRQoL, questions remain about the best exercise settings, supervision levels, delivery methods (individual, group or both), intensity, duration, and types of exercises.
- Current exercise recommendations include supervised group sessions at centers, individual home-based or center-based programs, minimally supervised home exercise programs, and combinations of home and supervised center-based sessions.
- Group fitness classes have diversified in recent years to include participants with chronic conditions or injuries, offering modalities like step aerobics, Pilates, and strength training that benefit HRQoL.
- Group exercise therapy involves small-group sessions with various rehabilitative modalities such as functional movement and strength conditioning, tailored to individual needs within a supportive group environment. This setting fosters camaraderie, motivation, and accountability, significantly improving physical and psychological health outcomes for individuals with conditions like MS, cancer, fibromyalgia, and mild cognitive impairment. Led by qualified instructors, group exercise sessions are structured into warm-up, main workout, and cool-down phases and can take place in various settings such as gyms, community centers, parks, or online platforms.
- Effective programs ensure safety, clear communication, inclusivity, and incorporate participant feedback. Benefits of group exercises include improved mobility and balance, reduced falls, enhanced physical well-being, psychological benefits like reduced stress and improved mood, a supportive environment, better adherence to exercise regimens, supervised execution ensuring proper form and technique, increased motivation and accountability, and a holistic approach addressing both physical and psychological health.
- By combining the inclusive nature of group fitness with individualized attention, healthcare providers can support holistic improvements in HRQoL through group exercise programs.
- Individualized exercises are a cornerstone of physiotherapy, tailored to meet the unique needs and goals of each patient, ensuring the regimen is effective, safe, and aligned with the patient's health status and capabilities.
- The process begins with a comprehensive evaluation, including medical history, physical and functional assessments, and goal setting. Based on this evaluation, a physiotherapist designs a customized exercise program encompassing strength training, range of motion exercises, cardiovascular fitness, flexibility exercises, and balance and coordination activities.
- The physiotherapist provides hands-on training, supervision, and education, continuously monitoring progress and making necessary adjustments. Individualized exercise programs offer several benefits, including enhanced physical and mental health, reduced fatigue, improved adherence and engagement, and increased effectiveness for specific conditions like COPD and hip osteoarthritis.
- They are tailored to individual needs, ensuring safety and promoting motivation. However, limitations include a lack of emphasis on education and comprehensive approaches, potential neglect of biopsychosocial factors, inconsistent customization in practice, and a deficiency in social support, which may diminish the overall effectiveness compared to group-based interventions.
- Home exercise programs offer numerous benefits, including convenience, cost-effectiveness, personalization, and increased adherence, making them particularly suitable for various populations, such as older adults and those recovering from injuries.
- They allow individuals to engage in physical activity in a comfortable setting, reducing barriers like transportation and costs associated with gym memberships. However, challenges include a lack of supervision, which may lead to incorrect exercise execution and increased injury risk, as well as potential declines in motivation and accountability without social support.
- Additionally, limited resources and equipment at home can restrict the variety of exercises, and while home programs can improve certain outcomes, they may not match the effectiveness of supervised group-based programs.
- Addressing adherence issues, such as distractions and perceived obstacles, is essential for maximizing the benefits of home exercise programs.

FURTHER READINGS

- Arntzen E C, Øberg G K, Gallagher S, Normann B. Group-based, individualized exercises can provide perceived bodily changes and strengthen aspects of self in individuals with MS: a qualitative interview study. Physiotherapy theory and practice. 2021 Oct 3;37(10):1080–95.
- Arthur H M, Smith K M, Kodis J, McKelvie R. A controlled trial of hospital versus home-based exercise in cardiac patients. Medicine & Science in Sports & Exercise. 2002 Oct 1;34(10):1544–50.
- Bachmann C, Oesch P, Bachmann S. Recommendations for improving adherence to home-based exercise: A systematic review. Physikalische Medizin, Rehabilitationsmedizin, Kurortmedizin. 2018 Jan;28(01):20–31.
- Bulat T, Hart-Hughes S, Ahmed S, Quigley P, Palacios P, Werner D C, Foulis P. Effect of a group-based exercise program on balance in elderly. Clinical interventions in aging. 2007 Jan 1;2(4):655–60.
- Burke S M, Carron A V, Eys M A, Ntoumanis N, Estabrooks P A. Group versus individual approach? A meta-analysis of the effectiveness of interventions to promote physical activity. Sport and Exercise Psychology Review. 2006 Jan 1;2(1):19–35.
- Caglar A T, Gurses H N, Mutluay F K, Kiziltan G Ü. Effects of home exercises on motor performance in patients with Parkinson's disease. Clinical rehabilitation. 2005 Dec;19(8):870–7.
- Cunha B, Ferreira R, Sousa A S. Home-based rehabilitation of the shoulder using auxiliary systems and artificial intelligence: An overview. Sensors. 2023 Aug 11;23(16):7100.
- Geraedts H A, Zijlstra W, Zhang W, Bulstra S, Stevens M. Adherence to and effectiveness of an individually tailored home-based exercise program for frail older adults, driven by mobility monitoring: Design of a prospective cohort study. BMC public health. 2014 Dec;14:1–7.
- Green J, McKenna F, Redfern E J, Chamberlain M A. Home exercises are as effective as outpatient hydrotherapy for osteoarthritis of the hip. Rheumatology. 1993 Sep 1;32(9):812–5.
- Iunes D H, Rocha C B, Borges N C, Marcon C O, Pereira V M, Carvalho L C. Self-care associated with home exercises in patients with type 2 diabetes mellitus. PloS one. 2014 Dec 5;9(12):e114151.
- Karbandi S, Gorji M A, Mazloum S R, Norian A, Aghaei N. Effectiveness of group versus individual yoga exercises on fatigue of patients with multiple sclerosis. North American journal of medical sciences. 2015 Jun;7(6):266.
- Keele-Smith R, Leon T. Evaluation of individually tailored interventions on exercise adherence. Western Journal of Nursing Research. 2003 Oct;25(6):623–40.
- Khalsa S S. Group exercises for enhancing social skills and self-esteem. Sarasota, F L: Professional Resource Press; 1996 Dec.
- Kruger C, McNeely M L, Bailey R J, Yavari M, Abraldes J G, Carbonneau M, Newnham K, DenHeyer V, Ma M, Thompson R, Paterson I. Home exercise training improves exercise capacity in cirrhosis patients: Role of exercise adherence. Scientific reports. 2018 Jan 8;8(1):99.
- Learmonth Y C, Paul L, Miller L, Mattison P, McFadyen A K. The effects of a 12–week leisure centre-based, group exercise intervention for people moderately affected with multiple sclerosis: A randomized controlled pilot study. Clinical rehabilitation. 2012 Jul;26(7):579–93.
- Lu Y, Li P, Li N, Wang Z, Li J, Liu X, Wu W. Effects of home-based breathing exercises in subjects with COPD. Respiratory care. 2020 Mar 1;65(3):377–87.
- McCarthy C J, Mills P M, Pullen R, Richardson G, Hawkins N, Roberts C R, Silman A J, Oldham J A. Supplementation of a home-based exercise programme with a class-based programme for people with osteoarthritis of the knees: A randomised controlled trial and health economic analysis. Health Technology Assessment (Winchester, England). 2004;8(46):iii-61.
- Mutrie N, Campbell A M, Whyte F, McConnachie A, Emslie C, Lee L, Kearney N, Walker A, Ritchie D. Benefits of supervised group exercise programme for women being treated for early stage breast cancer: Pragmatic randomised controlled trial. Bmj. 2007 Mar 8;334(7592):517.
- O'Reilly S C, Muir K R, Doherty M. Effectiveness of home exercise on pain and disability from osteoarthritis of the knee: A randomised controlled trial. Annals of the rheumatic diseases. 1999 Jan 1;58(1):15–9.
- O'Keeffe M, Hayes A, McCreesh K, Purtill H, O'Sullivan K. Are group-based and individual physiotherapy exercise programmes equally effective for musculoskeletal conditions? A systematic review and meta-analysis. Br J Sports Med. 2017 Jan;51(2):126–132. doi: 10.1136/bjsports-2015–095410. Epub 2016 Jun 24. PMID: 27343238.
- Ransdell L B, Taylor A, Oakland D, Schmidt J, Moyer-Mileur L, Shultz B. Daughters and mothers exercising together: Effects of home-and community-based programs. Medicine and Science in Sports and Exercise. 2003 Feb 1;35(2):286–96.
- Rooks D S, Gautam S, Romeling M, Cross M L, Stratigakis D, Evans B, Goldenberg D L, Iversen M D, Katz J N. Group exercise, education, and combination self-management in women with fibromyalgia: A randomized trial. Archives of internal medicine. 2007 Nov 12;167(20):2192–200.
- Rubenstein L Z, Josephson K R, Trueblood P R, Loy S, Harker J O, Pietruszka F M, Robbins A S. Effects of a group exercise program on strength, mobility, and falls among fall-prone elderly men. The Journals of Gerontology Series A: Biological Sciences and Medical Sciences. 2000 Jun 1;55(6):M317–21.
- Shariat A, Cleland J A, Hakakzadeh A. Home-based exercises during the COVID-19 quarantine situation for office workers: A commentary. Work. 2020 Jan 1;66(2):381–2.
- Simek E M, McPhate L, Haines T P. Adherence to and efficacy of home exercise programs to prevent falls: A systematic review and meta-analysis of the impact of exercise program characteristics. Preventive medicine. 2012 Oct 1;55(4):262–75.
- Tarakci E, Yeldan I, Huseyinsinoglu B E, Zenginler Y, Eraksoy M. Group exercise training for balance, functional status, spasticity, fatigue and quality of life in multiple sclerosis: A randomized controlled trial. Clinical rehabilitation. 2013 Sep;27(9):813–22.
- Williams P, Lord S R. Effects of group exercise on cognitive functioning and mood in older women. Australian and New Zealand journal of public health. 1997 Feb 1;21(1):45–52.

STUDENT ASSIGNMENT

LONG ANSWER QUESTIONS

1. Describe the effects and benefits of group and individual exercises.
2. Write about home-based exercise program in detail.
3. Discuss individual exercises in detail.
4. Explain the structure, types, and requirements to create an effective group exercise program.

SHORT ANSWER QUESTIONS

1. Define individual and group exercises.
2. What are the limitations of individual exercises?
3. Write a short note on group exercises.

MULTIPLE CHOICE QUESTIONS

1. **What is a significant concern associated with physical inactivity?**
   a. Improved healthcare costs
   b. Increased risk of various illnesses
   c. Enhanced psychological health
   d. Decreased exercise options
2. **Which type of exercise program is emphasized for participants with chronic conditions or injuries?**
   a. Individualized home exercise only
   b. Group exercise therapy
   c. High-intensity interval training
   d. Solo outdoor running
3. **What are some benefits of group exercise programs?**
   a. Increased isolation during workouts
   b. Enhanced physical well-being and reduced stress
   c. Higher costs due to additional resources
   d. Decreased motivation and accountability
4. **What is a primary advantage of individualized exercise programs?**
   a. Lack of need for evaluation
   b. Tailoring to meet unique patient needs
   c. Standardized approach for all patients
   d. Limited interaction with healthcare providers
5. **What is a common challenge faced by home exercise programs?**
   a. Excessive supervision
   b. High costs and barriers to access
   c. Lack of supervision leading to injury risk
   d. Increased social support
6. **Why is addressing adherence issues important in home exercise programs?**
   a. To increase the cost of the program
   b. To ensure participation in group settings
   c. To maximize the benefits of the exercise routine
   d. To discourage individual exercise efforts

ANSWER KEY

1. b **2.** b **3.** b **4.** b **5.** c **6.** c

Section IV

Therapeutic Techniques

SECTION OUTLINE

Proprioceptive Neuromuscular Facilitation

Paridhi Sharma

LEARNING OBJECTIVES

After the completion of the chapter, the readers will be able to:
- Describe the neurophysiological principles of proprioceptive neuromuscular facilitation (PNF).
- List various proprioceptive neuromuscular facilitation patterns.
- Understand the components of PNF.
- Explain the facilitation techniques of PNF.
- Describe the patterns of movements in detail.

CHAPTER OUTLINE

- Introduction
- Basic Principles
- Proprioceptive Neuromuscular Facilitation Patterns
- Head and Neck Patterns of Movement
- Trunk Patterns
- Proprioceptive Neuromuscular Facilitation Techniques

KEY TERMS

Agonist reversal: The agonist reversal approach (combination of isotonic) facilitates functional movement during the performance of a pattern or task. Contractions of the agonist muscle occur both eccentrically and concentrically. The main goal of this approach is to increase functional stability.

Contract relax: The contract-relax stretch is another popular PNF method. The only difference between it and hold-relax is that the muscle is contracted while moving, as opposed to being contracted while stationary. This is known as isotonic stretching at times.

Dynamic reversal: Using the isotonic technique known as Dynamic Reversal of Antagonist, the patient goes first in one direction and then the other without pausing.

Hold relax: The hold-relax method is the most often used kind of PNF stretch. This entails holding the target muscle for a short period of time while passively extending it. The same muscle will then be contracted (tightened) while the subject remains still. After 10–15 seconds of holding the contraction and stretch, subject releases the muscle.

Irradiation: A neurophysiologic process known as irradiation is characterized by an increase in muscle activity in connected muscles in response to resistance from external stimuli.

Proprioceptive neuromuscular facilitation (PNF): PNF is a stretching method used to increase muscle flexibility and has been demonstrated to improve both active and passive ranges of motion.

Resisted progression (RP): To promote pelvic motion and progression during locomotion, monitoring resistance, stretch, and approximation are physically applied during resisted progression. In order to prevent hindering the patient's motion, coordination, or velocity, the resistance is low. Applying RP *via* resistance using elastic bands is another method.

Rhythmic initiation: The technique aims to improve mobility that is impeded by deficiencies in the initiation, coordination, or relaxation of movements.

Rhythmic rotation: The use of passive movement in a rotational pattern defines rhythmic rotation.

Rhythmic stabilization: To establish rhythmic stability, an isometric contraction of the agonist is followed by an isometric contraction of the antagonist.

Slow reversal hold: An isometric contraction of the agonist occurs right after an isotonic contraction during a slow reversal-hold.

INTRODUCTION

One of the most widely utilized therapeutic approaches in neurological rehabilitation is proprioceptive neuromuscular facilitation. Proprioceptive neuromuscular facilitation (PNF) is a stretching method used to increase muscle flexibility and has been demonstrated to improve both active and passive ranges of motion. PNF was developed between 1945 and 1954 at the Kabat Kaiser Institute in California by Dr Kabat and Miss Knott. It employs a movement pattern based on natural movement, and it facilitates these activities by stimulating proprioceptors. The mechanism also takes advantage of higher righting and equilibrium reflex activation. PNF can be utilized to enhance functional task performance by enhancing strength, flexibility, and range of motion. The patient is helped by the integration of these advancement to:

- Develop head and trunk control.
- Begin and maintain motion.
- Control center-of-gravity shifts.
- Keep the extremities moving while keeping the pelvis and trunk in the middle.

BASIC PRINCIPLES

These techniques aim to promote the development of increasingly higher levels of competence and functional independence in bed mobility, transitional movements, sitting, standing, and walking. They are designed using the developmental sequence as a guide. These concepts are frequently referred to be the foundational ideas of PNF.

Clinical Implication

The PNF techniques can be adapted to various populations and conditions beyond neurological rehabilitation. Athletes, for example, often use PNF stretching to improve flexibility and enhance performance. Additionally, PNF principles have been incorporated into fitness training programs and even dance therapy to promote better movement patterns and coordination. This versatility highlights the broad applicability and effectiveness of PNF techniques across different domains of physical therapy and wellness.

Manual Contact

Placing the hands on the skin triggers pressure receptors and lets the patient know which way the motion is intended to go. Ideally, manual touches should be made in the direction of the planned action and on the skin covering the target muscle groups. For example, to assist in shoulder flexion, the physiotherapist may place one or both of their hands on the anterior and superior surface of the upper extremity. To control movement and offer the best resistance, particularly with regard to rotation, without applying too much pressure or causing discomfort, a lumbrical grip (Fig. 15.1) is ideal.

Body Positioning and Mechanics

Effective facilitation requires the physician to move dynamically in a manner that corresponds to the patient's motion. The clinician's hands, arms, shoulders, and pelvis should all be aligned with the motion. If this is not practicable, the clinician's hands and arms should be in line with the motion. The clinician uses their body weight to generate resistance, keeping their hands and arms mostly at ease.

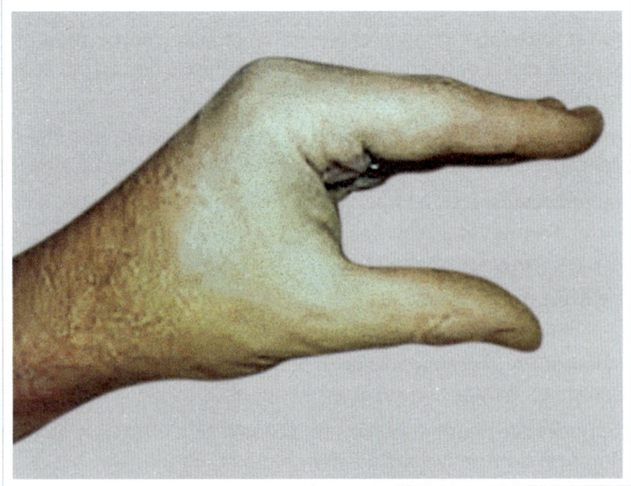

Fig. 15.1: Lumbrical grip

Stretch

Kabat proposed using the stretch reflex to encourage the activation of muscles. He proposed that when a muscle was in an elongated state, a slight movement farther into the elongated range may be used to trigger a stretch reflex. An extended muscle, the synergistic muscles at the same joint, and other associated muscles are all made easier by a stretch. Since a prolonged stretch may potentially reduce muscle activity, a rapid stretch tends to improve motor reactivity, hence it is important to carefully assess the patient's response. The use of facilitatory stretch is contraindicated in cases of joint hypermobility, pain, or fracture. Stretching, particularly quick stretching, needs to be done carefully when there is spasticity present since individual reactions might vary and may produce unwanted motor activity.

Manual Resistance

While some PNF techniques or exercises focus on increasing motor unit recruitment by providing external resistance, others focus on lowering internal resistance by altering the patterns of neuronal firing. Resistance can therefore be viewed in the context of PNF as a tool for promoting muscle strengthening or a method of neuromuscular facilitation. Through complex interactions between neural and contractile components, resistance may have an impact on motor learning, endurance, muscle hypertrophy, postural stability, and the onset of movement. The right amount of resistance enables the maximal motor response, which permits the correct accomplishment of the specified task. If the purpose of the intervention is mobility, the highest resistance that permits the patient to move within the permitted range of motion with ease and pain is deemed appropriate resistance. The applied force's magnitude and direction must alter to account for potential variations in patient capability and muscle function throughout the range. If the objective of the intervention is stability, the highest level of resistance that allows the patient to retain the specified position is considered acceptable resistance.

Irradiation

A neurophysiologic process known as irradiation is characterized by an increase in muscle activity in connected muscles in response to resistance from external stimuli. This phrase is often used to refer to both reinforcement and overflow. PNF uses the irradiation process to suppress the antagonist muscle groups that oppose it or increase the activity of the agonist muscle groups. Since everyone responds to resistance differently, overflow patterns vary amongst individuals. By watching the patient's response, the physiotherapist can choose the manual contacts and resistance level that will maximize the patient's ability to produce the desired movement. The following are some examples of activities and common patterns of reaction:

- Overflow into the dorsi-flexors of the ankle and hip results from resisted trunk flexion.
- Overflow into the hip and knee extensors occurs with resisted trunk extension.
- When upper extremity extension and adduction are resisted, the trunk flexors overflow.
- This causes overflow into the dorsi-flexors when hip flexion, adduction, and external rotation are resistively applied.

Joint Facilitation (Traction and Approximation)

Traction and approximation stimulate receptors in the joint and periarticular tissues. Traction causes a body segment to lengthen, which can be utilized to increase range of motion and reduce pain. Approximation produces compression of body structures and can be used to promote co-contraction of muscles and weight bearing. People respond to traction and approximation in different ways. These forces can be placed on body positions or applied during the performance of extremity patterns.

Timing of Movement

Smooth sequencing and gradation of muscle activation are necessary for normal movement. The majority of functional actions are timed from distal to proximal, like when taking up a pencil. The elbow and shoulder are used to place the pencil for use after being held in the hand. Another relevant point is that postural control develops from proximal to distal and from cephalad to caudal. When evaluating, facilitating, and instructing mobility techniques to neurologically challenged individuals, these concerns must be taken into account. Performing a specific functional task requires sufficient muscle strength and joint range of motion, but improper sequencing of movement patterns can still lead to inefficiency. Moreover, before mastering tasks that need precise motions of the distal joints, sufficient control of the trunk and proximal extremity joints must be accomplished.

Patterns of Movement

The PNF is distinguished by its distinctive diagonal movement patterns. It was recognized by Kabat and Knott that a group of muscles work synergistically in practical settings. They created PNF patterns by combining these connected movements. PNF simulates demands made during functional motions more accurately.

Visual Cues

An individual can control and make the right movements and body positions with the aid of visual cues. Head and body positioning are affected by eye movement. Stronger muscle contractions and appropriate alignment of body parts, such the head and trunk, can both be facilitated by postural reactions that are a result of feedback from the visual system.

Verbal Input

A directional indication should be included in the vocal instruction, which should be brief. Preparation, action, and correction are the three main phases of a verbal command. The patient is prepared for action during the preparatory phase. The action phase informs the patient about the desired activity and commands them to start moving. In the correction phase, the patient is instructed on how to adjust their action as appropriate. PNF uses its knowledge of how tone and loudness of voice influence behavior to promote the desired response, like more effort or relaxation.

PROPRIOCEPTIVE NEUROMUSCULAR FACILITATION PATTERNS

Analyzing typical movement patterns was a part of the early development of PNF approaches. The outcomes of these observations are incorporated into certain sets of joint movements known as patterns. Patterns concentrate on either the extremities or the trunk, though they are frequently blended in therapeutic practice. A combination of actions in three planes make up every PNF pattern. Particular attention should be paid to the rotation component, which should be recruited in the pattern's initial range. While eliciting more involvement from the trunk musculature, early rotation supports appropriate distal to proximal timing of extremities motions.

Clinical Correlation

By tailoring PNF interventions to target individualized goals and incorporating these patterns into therapy sessions, clinicians can optimize treatment outcomes and facilitate the transition to independent functional activities in daily life. Additionally, PNF patterns can be adapted and progressed as patients advance in their rehabilitation journey, providing a comprehensive and dynamic approach to promoting recovery and maximizing functional potential.

Extremity Patterns

Diagonal 1 (D_1) and diagonal 2 (D_2) are the names of the two extremity diagonal patterns. The movement that follows from performing an extremity pattern is designated by the direction of the proximal joint's movement, which is how extremity patterns get their names. Each diagonal is further divided into flexion and extension directions. For instance, the shoulder moves into flexion in D_1 flexion of the upper extremity (UE) and into extension in D_1 extension. One can flex or extend the middle or intermediate joint. The proximal component of the pattern is emphasized, and more trunk activity is recruited, by using straight arm and leg patterns. More attention can be put on the intermediate and distal components when the intermediate joints are flexed.

Upper Extremity Patterns

All of the movements are described using the right limbs and while supine. Any starting position, including forward lying, side lying, or sitting, is acceptable, but the patient must be positioned in these positions so that a full range of motion is possible.

UE D_1 Flexion (Shoulder Flexion/Adduction/External Rotation) (Fig. 15.2)

- **Starting position:** Beginning with the arm outstretched and slightly to the side, about a fist's width from the hip. The wrist is ulnarly deviated, the forearm is pronated, and the shoulder is extended, abducted, and internally rotated (Fig. 15.3A).
- **Grip of the physiotherapist:** The physiotherapist places their hand in the patient's palm. The right hand, which goes underneath the arm, is either positioned on the front of the arm over the biceps or the front of the forearm.

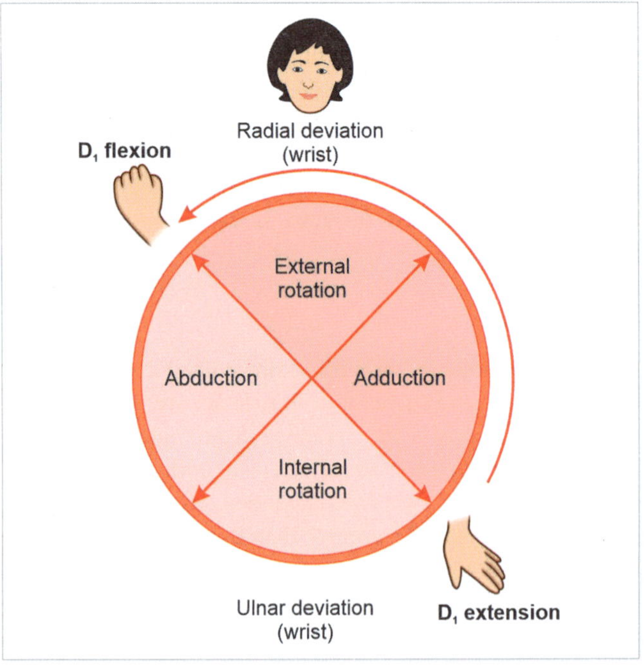

Fig. 15.2: Upper limb D_1 flexion

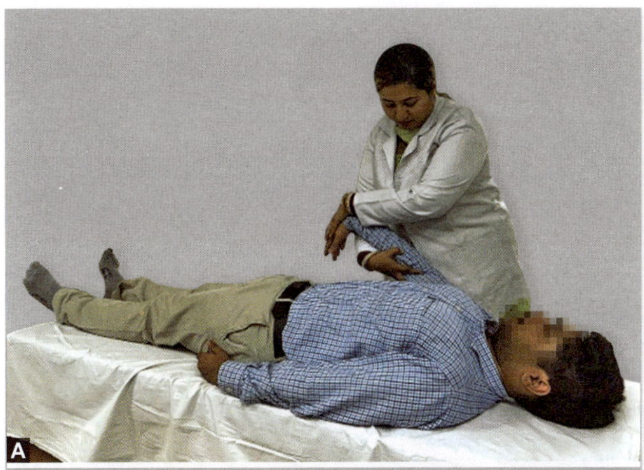

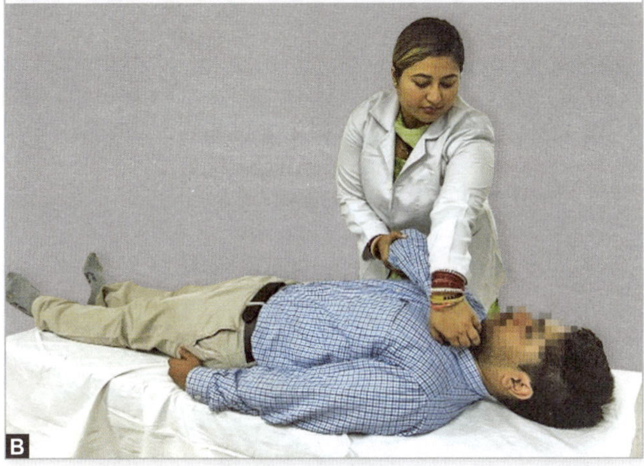

Figs 15.3A and B: Upper extremity D_1 flexion: **A.** Starting position; **B.** Final position

- Verbal command of the physiotherapist can be "squeeze my hand and pull up and cross over as if to bring a scarf over the opposite shoulder".
- **Direction of movement:** The arm is raised and brought inward across the face (Fig. 15.3B).

This pattern is functionally used for feeding.

Summary of UE D_1 flexion pattern is given in Table 15.1.

TABLE 15.1: Upper extremity D_1 flexion pattern

Joint	Starting position	Final position
Scapula	Posterior depression	Anterior elevation
Shoulder	Extension/abduction/ internal rotation	Flexion/adduction/ external rotation
Elbow	Flexion or extension	Flexion or extension
Forearm	Pronation	Supination
Wrist	External/ulnar deviation	Flexion/radial deviation
Fingers	Extension	Flexion

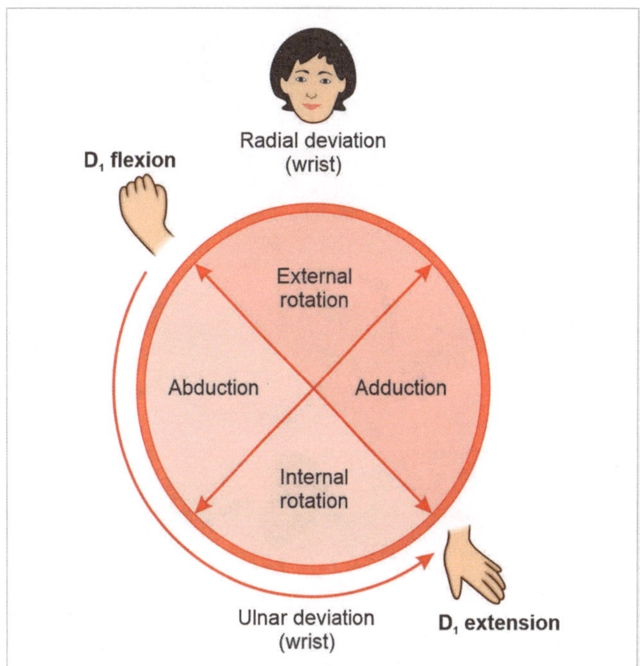

Fig. 15.4: Upper limb D_1 extension

UE D_1 Extension (Shoulder Extension/Abduction/External Rotation) (Fig. 15.4)

- **Starting position:** Beginning position for the patient involves an arm that is either flexed or extended, with the elbow roughly level with the patient's nose across the midline of the body. Supination of the forearm is seen, along with flexed fingers and wrists, and with wrist that is deviated radially (Fig. 15.5A).
- **Grip of the physiotherapist:** The physiotherapist's right hand is positioned on the ulnar side of the patient's hand dorsum and medial three fingers. The left hand, which passes beneath the patient's arm, is positioned on the extensor side of the forearm's ulnar surface.
- Verbal command of the physiotherapist can be "open your hand and push down and out".
- **Direction of movement:** Arm is extended downward and outward, as how one would extend their arm to protect themselves from falling (Fig. 15.5B).

This pattern is functional for performing a protective reaction when in a sitting position.

Summary of UE D_1 extension pattern is given in Table 15.2.

UE D_2 Flexion (Shoulder Flexion/Abduction/External Rotation) (Fig. 15.6)

- **Starting position:** Beginning with the arm extended across the body with elbow crossing the midline, the forearm is pronated, the wrist and fingers are flexed, and the wrist is deviated ulnarly (Fig. 15.7A).

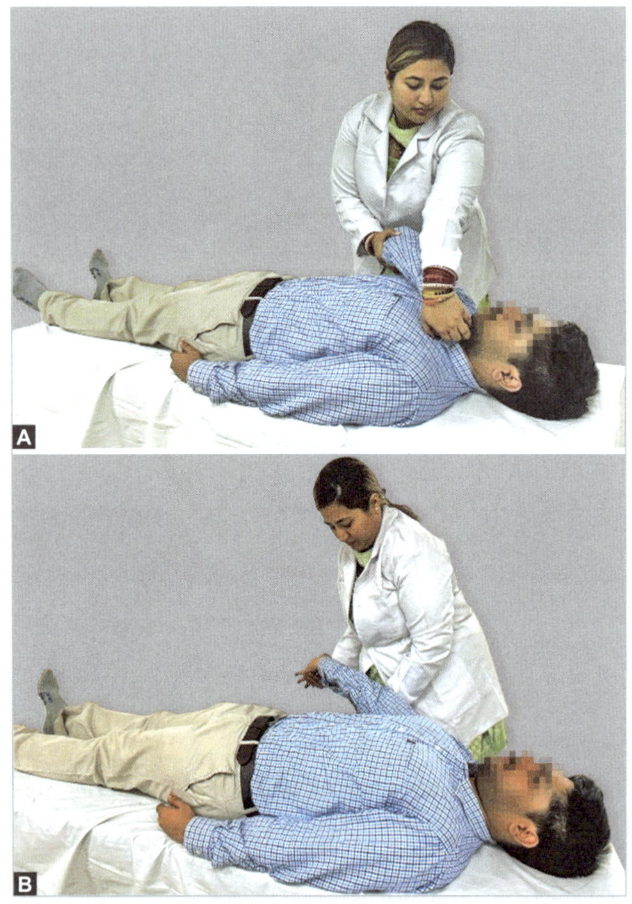

Figs 15.5A and B: Upper extremity D_1 extension: **A.** Starting position; **B.** Final position

TABLE 15.2: Upper extremity D_1 extension pattern

Joint	Starting position	Final position
Scapula	Anterior elevation	Posterior depression
Shoulder	Flexion/adduction/ external rotation	Extension/abduction/ internal rotation
Elbow	Flexion or extension	Flexion or extension
Forearm	Supination	Pronation
Wrist	Flexion/radial deviation	External/ulnar deviation
Fingers	Flexion	Extension

- **Grip of the physiotherapist:** The physiotherapist places her left hand on the radial side of the patient's hand dorsum and the radial three fingers. The right hand, that passes beneath the forearm, is positioned either on the arm's extensor surface or the forearm's extensor surface.
- Verbal command of the physiotherapist can be "lift your wrist and arm up".
- **Direction of movement:** The arm moves upwards and outwards (Fig. 15.7B).

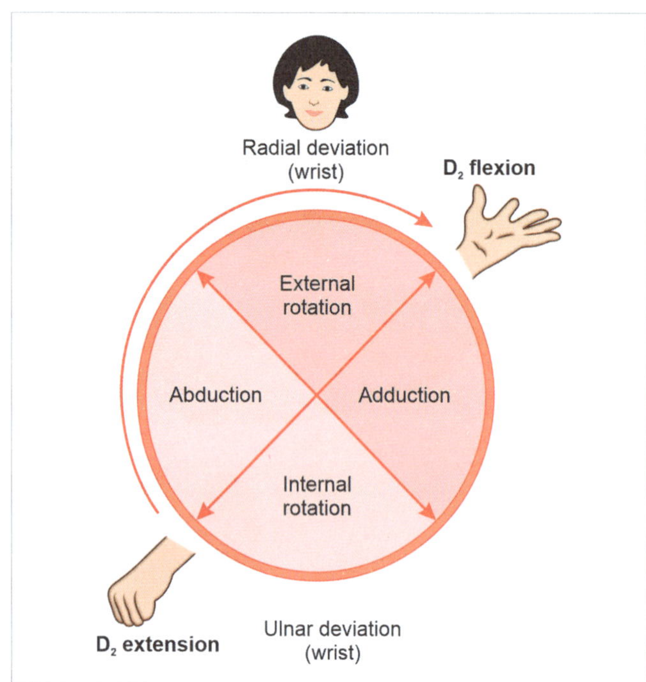

Fig. 15.6: Upper limb D_2 flexion

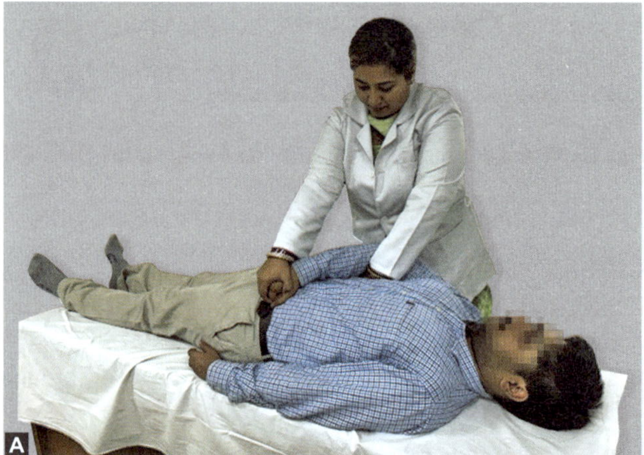

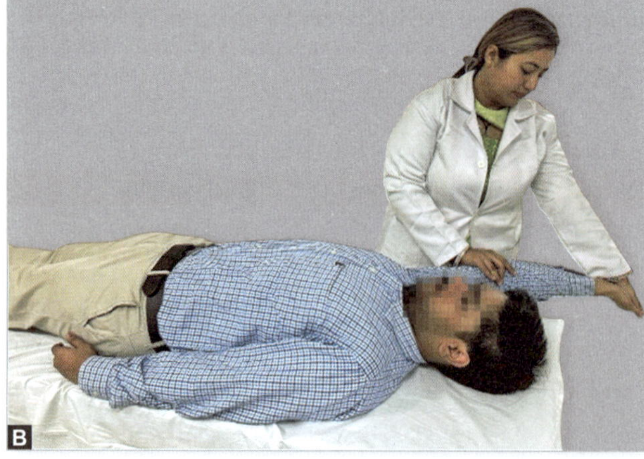

Figs 15.7A and B: Upper extremity D_2 flexion: **A.** Starting position; **B.** Final position

TABLE 15.3: Upper extremity D_2 flexion pattern

Joint	Starting position	Final position
Scapula	Anterior depression	Posterior elevation
Shoulder	Extension/adduction/ internal rotation	Flexion/abduction/ external rotation
Elbow	Flexion or extension	Flexion or extension
Forearm	Pronation	Supination
Wrist	Flexion/ulnar deviation	External/radial deviation
Fingers	Flexion	Extension

This pattern is functionally remembered as throwing a wedding bouquet over the same shoulder.

Summary of UE D_2 flexion pattern is given in Table 15.3.

UE D_2 Extension (Shoulder Extension/Adduction/ Internal Rotation) (Fig. 15.8)

- **Starting position:** At a fist's width lateral to the ipsilateral ear, the arm flexes initially. With the forearm supinated, the fingers and wrist extended, and the wrist deviated radially, the shoulder is externally rotated (Fig. 15.9A).
- **Grip of the physiotherapist:** The physiotherapist places her right hand in the patient's palm. The physiotherapist's left hand, passes underneath the patient's arm, grabs the forearm's or arm's flexor surface.
- Verbal command of the physiotherapist can be "squeeze my hand and pull down and across".
- **Direction of movement:** The arm descends and travels inside across the torso to the opposing hip as the hand grips (Fig. 15.9B).

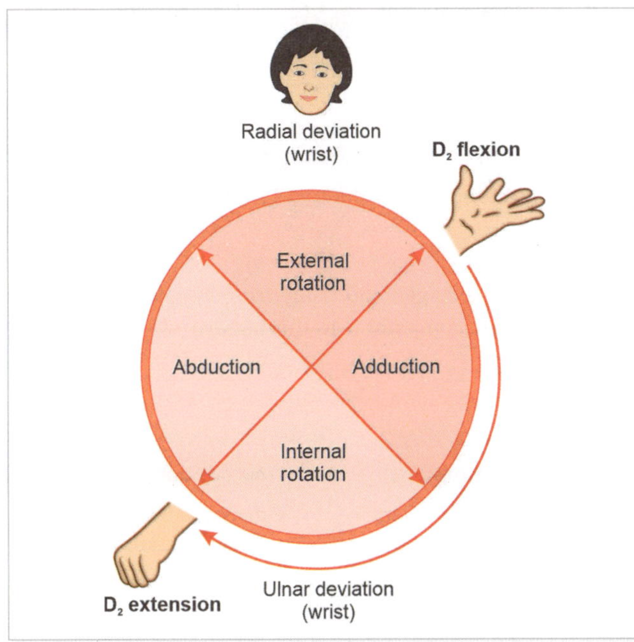

Fig. 15.8: Upper limb D_2 extension

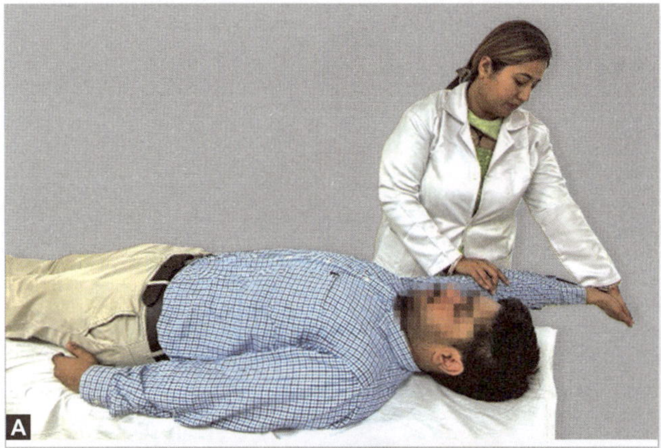

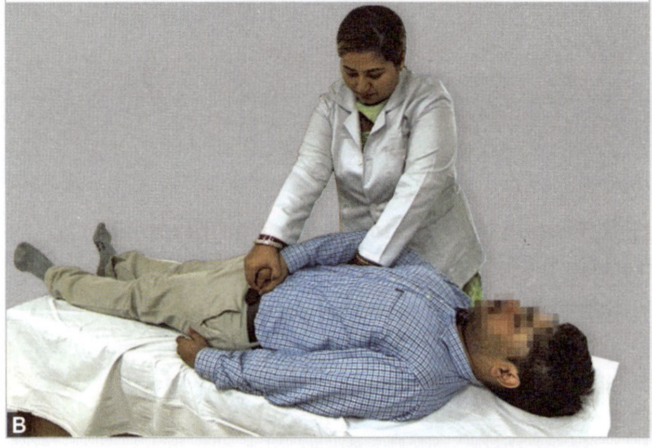

Figs 15.9A and B: Upper extremity D_2 extension: **A.** Starting position; **B.** Final position

TABLE 15.4: Upper extremity D_2 extension pattern

Joint	Starting position	Final position
Scapula	Posterior elevation	Anterior depression
Shoulder	Flexion/abduction/ external rotation	Extension/adduction/ internal rotation
Elbow	Flexion or extension	Flexion or extension
Forearm	Supination	Pronation
Wrist	External/radial deviation	Flexion/ulnar deviation
Fingers	Extension	Flexion

This pattern is functionally remembered as placing a sword in its sheath.

Summary of UE D_2 extension pattern is given in Table 15.4.

Lower Extremity Patterns

Despite being related to functional motions in sitting and standing, the lower extremity (LE) patterns are shown and described in the supine position. Four lower extremity designs along two diagonals will be described, each of which is analogous to the upper extremity.

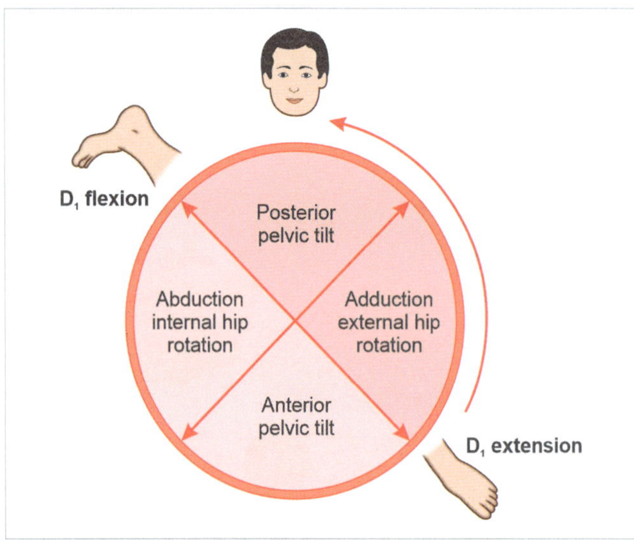

Fig. 15.10: Lower limb D_1 flexion

LE D_1 Flexion (Hip Flexion/Adduction/External Rotation) (Fig. 15.10)

- **Starting position:** With the heel aligned with the ipsilateral shoulder, the leg is supported by a surface. The hip is internally rotated and abducted. The foot is everted and plantar-flexed (Fig. 15.11A).
- **Grip of the physiotherapist:** The physiotherapist's right hand is placed on the medial side of the dorsum of the foot, directly over the insertion of the tibialis anterior. The left hand is positioned on the medial side of the thigh just above the knee.
- Verbal command of the physiotherapist is "pull your leg over while bringing your foot up and in".
- **Direction of movement:** To reach the opposing shoulder, the leg ascends and turns inward (Fig. 15.11B).

This motion is used to raise the foot up to the opposite hand to remove a shoe or to cross one leg over the other when sitting.

Summary of LE D_1 flexion pattern is given in Table 15.5.

LE D_1 Extension (Hip Extension/Abduction/Internal Rotation) (Fig. 15.12)

- **Starting position:** External hip rotation, knee flexion, and hip flexion. The foot is inverted and dorsi-flexed (Fig. 15.13A).
- **Grip of the physiotherapist:** The physiotherapist's right hand is positioned on the plantar surface at the lateral side of the ball of the foot. The left hand can be placed on either the lateral side of the heel or on the extensor surface of the leg over the insertion of the hamstrings.
- Verbal command of the physiotherapist can be "push your foot downwards and outwards".
- **Direction of movement:** The leg is pushed forward and downward (Fig. 15.13B).

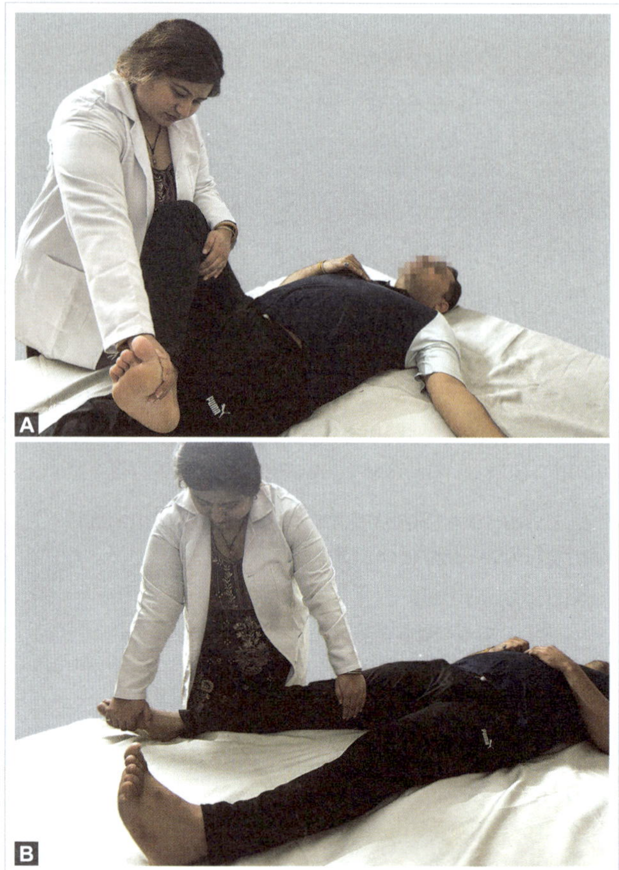

Figs 15.11A and B: Lower extremity D_1 flexion: **A.** Starting position; **B.** Final position

TABLE 15.5: Lower extremity D_1 flexion pattern

Joint	Starting position	Final position
Hip	Extension/abduction/ internal rotation	Flexion/adduction/ external rotation
Knee	Flexion or extension	Flexion or extension
Ankle	Plantar flexion/eversion	Dorsiflexion/inversion
Toes	Flexion	Extension

The stance phase of gait and rising from a seated position are motions that are comparable to this pattern.

Summary of LE D_1 extension pattern is given in Table 15.6.

TABLE 15.6: Lower extremity D_1 extension pattern

Joint	Starting position	Final position
Hip	Flexion/adduction/ external rotation	Extension/abduction/ internal rotation
Knee	Flexion or extension	Flexion or extension
Ankle	Dorsiflexion/inversion	Plantar flexion/eversion
Toes	Extension	Flexion

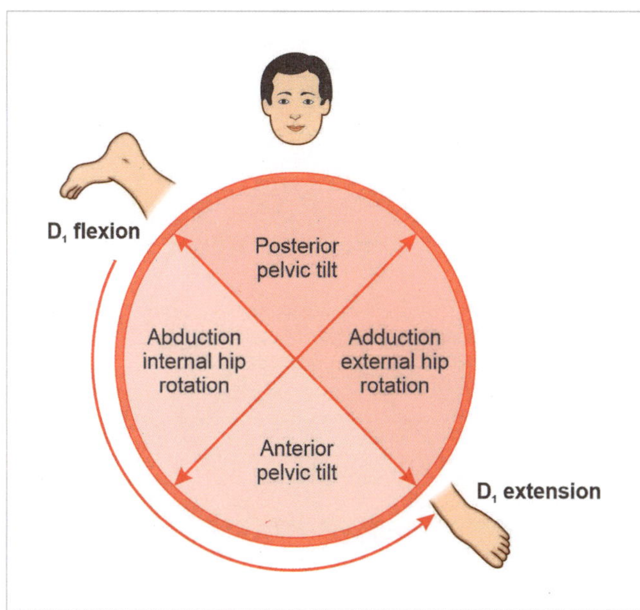

Fig. 15.12: Lower limb D_1 extension

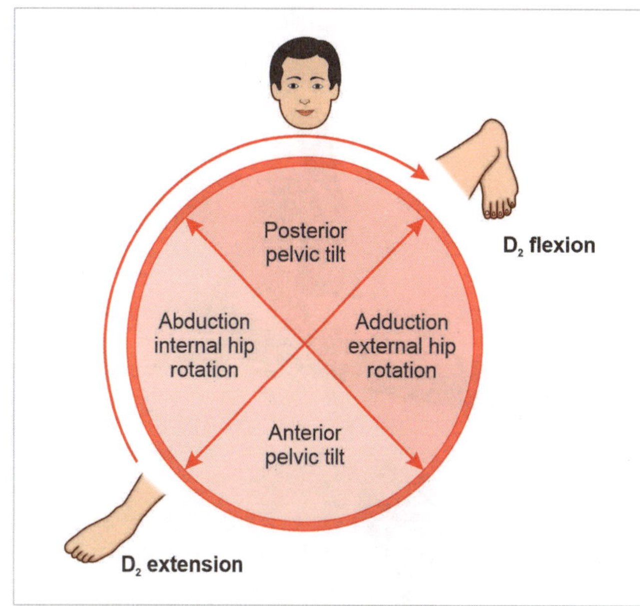

Fig. 15.14: Lower limb D_2 flexion

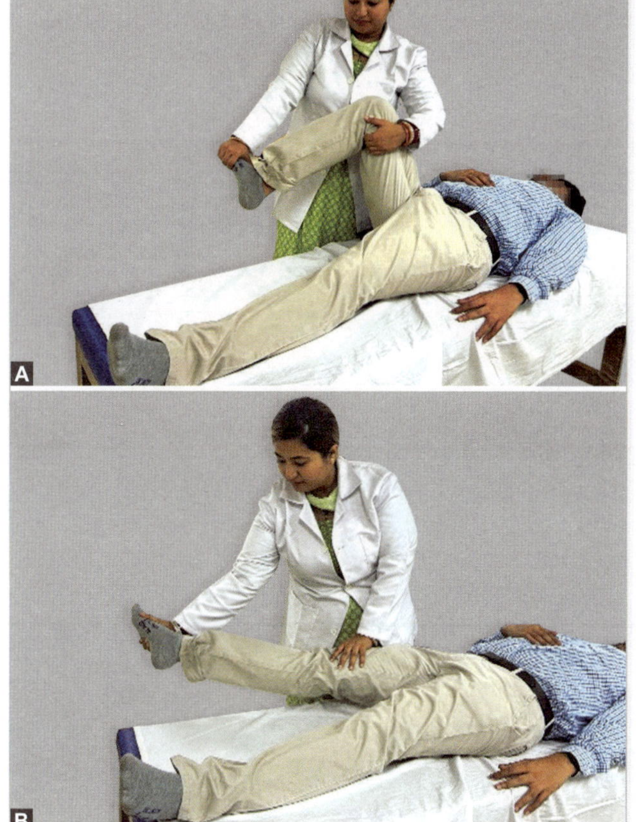

Figs 15.13A and B: Lower extremity D_1 extension: **A.** Starting position; **B.** Final position

LE D_2 Flexion (Hip Flexion/Abduction/Internal Rotation) (Fig. 15.14)

- **Starting position:** Leg starts with the hip externally rotated and the hip and knee extended. The leg not participating in the pattern is abducted in order to position the knee past the body's midline. The foot is inverted and plantar-flexed (Fig. 15.15A).
- **Grip of the physiotherapist:** The physiotherapist's right hand is positioned on the lateral side of the foot's dorsum. The left hand is either positioned on the lateral side of the heel or on the anterolateral side of the thigh.
- Verbal command of the physiotherapist can be "raise your foot upwards and outwards".
- **Direction of movement:** Away from the opposing ankle, the leg is raised upward and outward (Fig. 15.15B).
  Because the final position of this pattern mimics the gesture an animal makes to relieve itself, it has been euphemistically dubbed the "fire-hydrant" pattern.

Summary of LE D_2 flexion pattern is given in Table 15.7.

LE D_2 Extension (Hip Extension/Adduction/External Rotation) (Fig. 15.16)

- **Starting position:** With the hip abducted, the hip and knees are flexed. Internal rotation of the hip is done with caution to prevent valgus stress (Fig. 15.17A).
- **Grip of the physiotherapist:** The physiotherapist places her right hand on the medial side of the foot's plantar surface. The left hand, which passes under the leg, is either positioned on the medial aspect of the heel or placed on the extensor surface of the thigh, with the arm passing under the thigh from the lateral aspect.

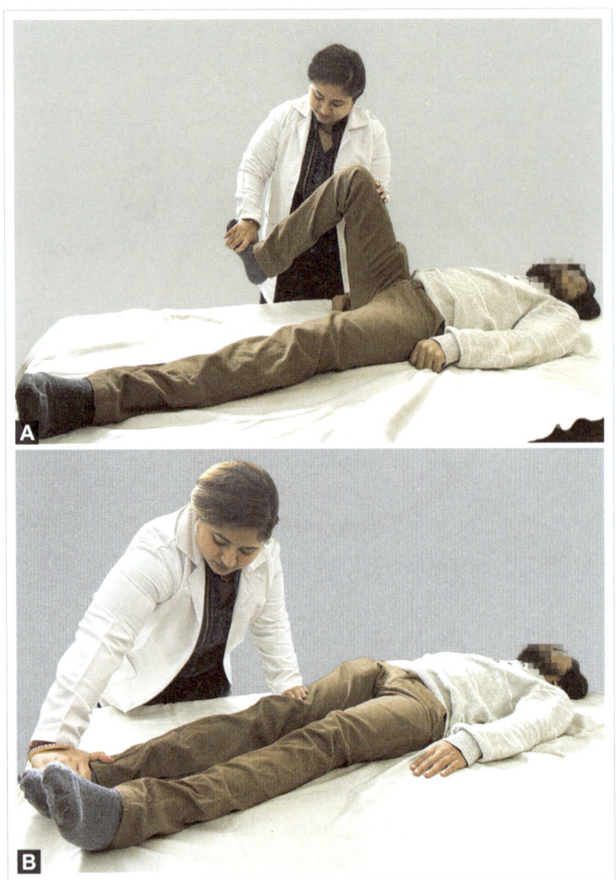

Figs 15.15A and B: Lower extremity D_2 flexion: **A.** Starting position; **B.** Final position

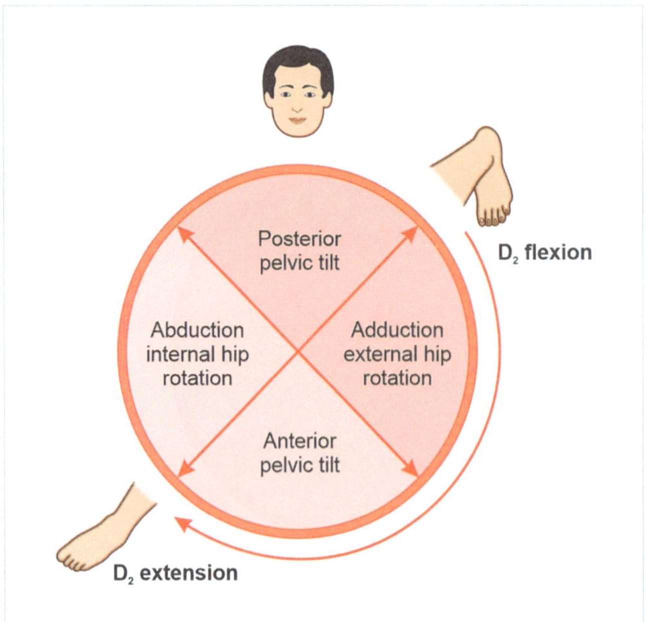

Fig 15.16: Lower limb D_2 extension

TABLE 15.7: Lower extremity D_2 flexion pattern

Joint	Starting position	Final position
Hip	Extension/adduction/ external rotation	Flexion/abduction/internal rotation
Knee	Flexion or extension	Flexion or extension
Ankle	Plantar flexion/inversion	Dorsiflexion/eversion
Toes	Flexion	Extension

- Verbal command of the physiotherapist can be "your foot should be pushed in and down".
- **Direction of movement:** Toward the opposing foot, the leg descends and turns inward. This pattern when in standing, mimics a soccer kick (Fig. 15.17B).

Summary of LE D_2 extension pattern is given in Table 15.8.

HEAD AND NECK PATTERNS OF MOVEMENT

In a diagonal plane, the head and neck also move. Flexion and extension are the two movement patterns, and both of these involve rotatory motion.

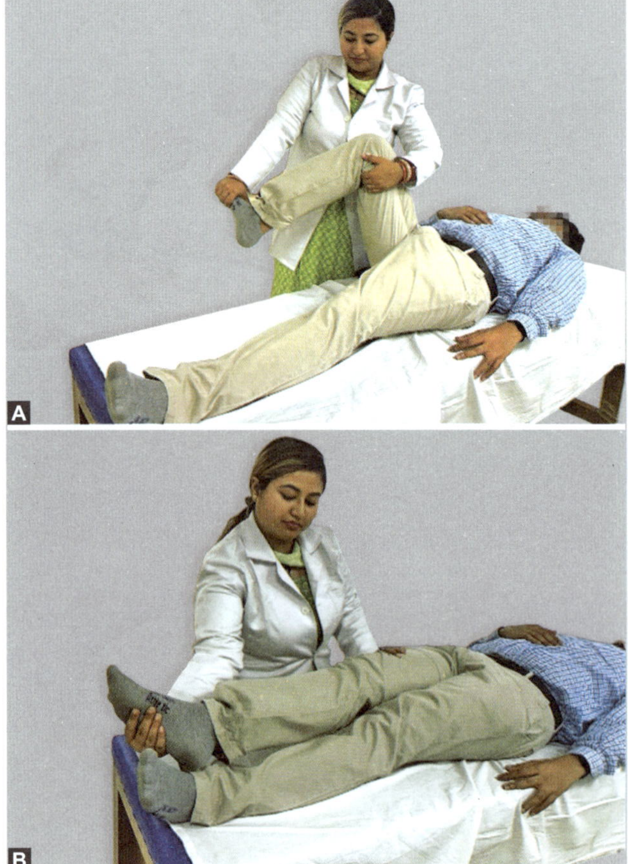

Figs 15.17A and B: Lower extremity D_2 extension: **A.** Starting position; **B.** Final position

TABLE 15.8: Lower extremity D_2 extension pattern

Joint	Starting position	Final position
Hip	Flexion/abduction/internal rotation	Extension/adduction/ external rotation
Knee	Flexion or extension	Flexion or extension
Ankle	Dorsiflexion/eversion	Plantar flexion/inversion
Toes	Extension	Flexion

Flexion of Head and Neck (Toward Right)

- **Starting position:** Extension of the head and neck to the left. The patient is lying on the back with their head extended over the plinth's end and the head and neck is turned to look toward the left shoulder (Fig. 15.18A).
- **Grip of the physiotherapist:** The physical physiotherapist adopts a walk-standing position behind the patient, parallel to the directions of movement. The one hand is positioned over the right mandible, under the chin. The occiput is grasped with the other hand.
- Verbal command of the physiotherapist can be "rotate, pull the head up and across, and pull the chin in".

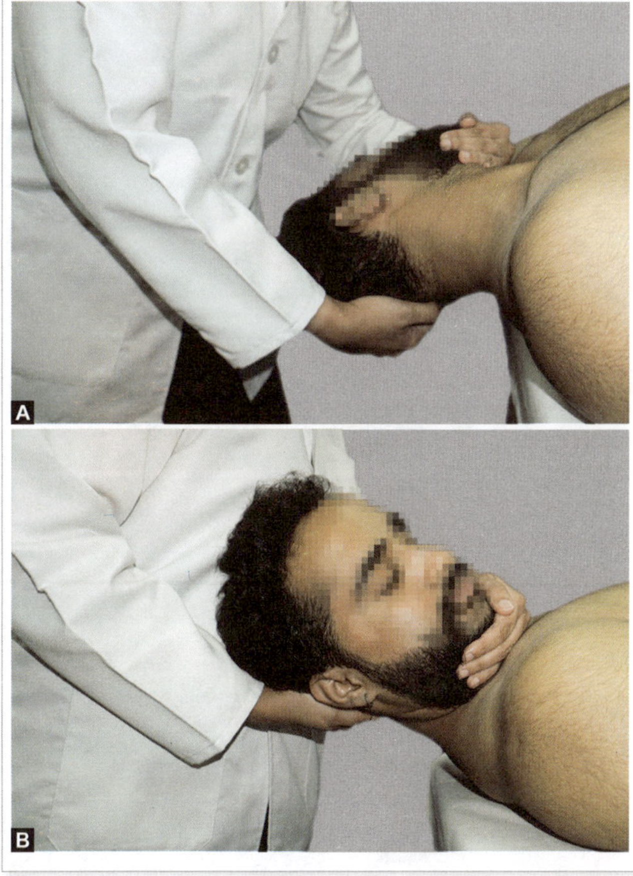

Figs 15.18A and B: Head and neck flexion pattern: **A.** Starting position; **B.** Final position

- **Direction of movement:** The head is elevated to cast a downward and rightward glance (Fig. 15.18B).

Extension of the Head and Neck (Toward Left)

- **Starting position:** Rightward flexion of the head and neck. The head casts a downward and diagonal glance at the right hip (Fig. 15.19A).
- **Grip of the physiotherapist:** The physiotherapist adopts a walk stance behind the patient, facing the direction of motion. The one hand of the physiotherapist is positioned under the occiput, whereas the other hand over the chin and the right mandible.
- Verbal command of the physiotherapist can be "bring up the chin. Rotate and push the head back, so as to glance over the left shoulder".
- **Direction of the movement:** To gaze over the opposing shoulder, the head is turned toward the left and pushed backward (Fig. 15.19B).

TRUNK PATTERNS

Either flexion or extension, together with rotation and side flexion, are the patterns of movement of the trunk. Upper and

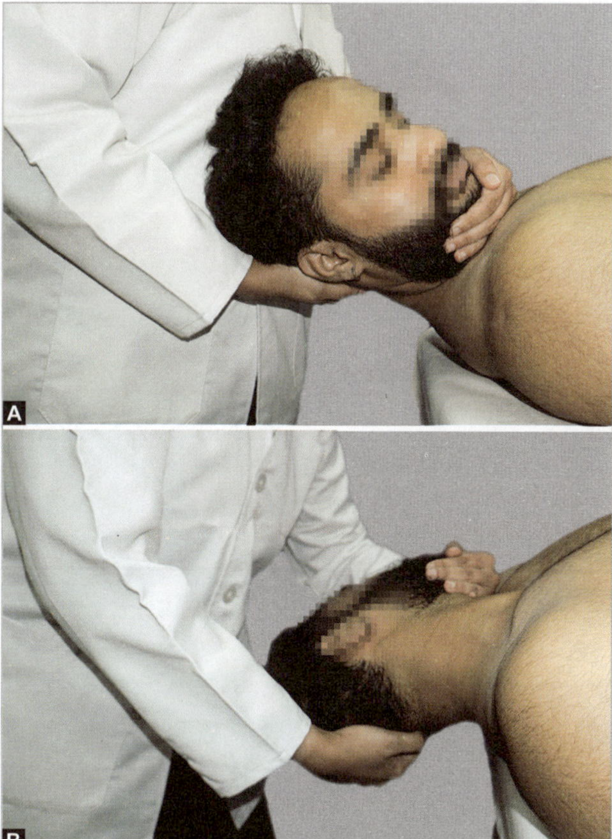

Figs 15.19A and B: Head and neck extension pattern: **A.** Starting position; **B.** Final position

lower trunk flexion patterns make up the flexion patterns, while upper and lower trunk extension patterns make up the extension patterns of movement. From a line drawn from the shoulder to the opposite hip, the movements occur in a diagonal plane. Any deviation from this path will result in excessive side flexion, which is an undesirable action.

Upper Trunk Flexion (Toward Right)

- **Starting position:** Extension of the upper trunk to the left. The patient's trunk is extended and rotated to the left while sitting, long-sitting, or supine lying with the head directed over the left shoulder (Fig. 15.20A).
- **Grip of the physiotherapist:** In keeping with the movement's orientation, the left hand is placed diagonally in front of the left shoulder, while the right hand is positioned on the forehead over the right eyebrow.
- Verbal command of the physiotherapist can be "turn the head, pull up and across" (if in the supine position).
- **Direction of movement:** The head and trunk are flexed and turned in the direction of the opposing hip (Fig. 15.20B).

Upper Trunk Extension (Toward Left)

- **Starting position:** Rightward upper trunk flexion. For this exercise, prone lying, side- lying, or sitting on a stool are the best positions but it can also be performed in supine position. The head is turned to face the opposing hip as the head and trunk are flexed and turned to the right (Fig. 15.21A).
- **Grip of the physiotherapist:** In the supine position, the physiotherapist places one hand on the patient's right forearm to guide the direction and provides additional support at the shoulder to stabilize and control the movement toward the left. In the prone and side-lying postures, the left hand is placed on the back of the patient's shoulder diagonally and in line with the movement while the right hand is placed on the back of the patient's head. The flexion-abduction pattern for the left arm may be used when the patient is sitting, in which case the physiotherapist's left hand is positioned on the dorsum of the patient's left hand and the right hand is placed on the back of the head.
- Verbal command of the physiotherapist can be "push up and back and look over the left shoulder".
- **Direction of movement:** The head is gazing over the left shoulder while the trunk and head are extended and rotated to the left (Fig. 15.21B).

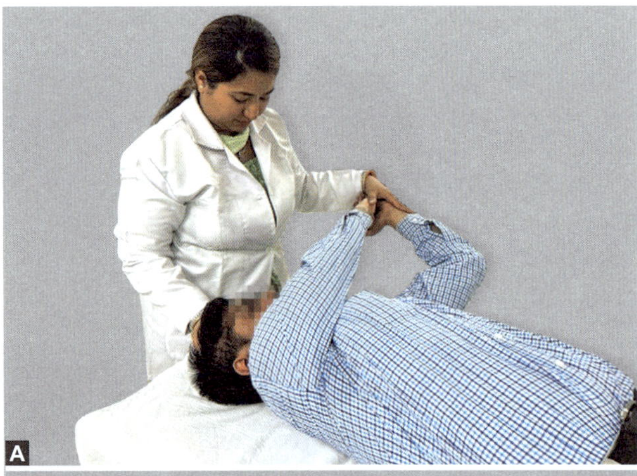

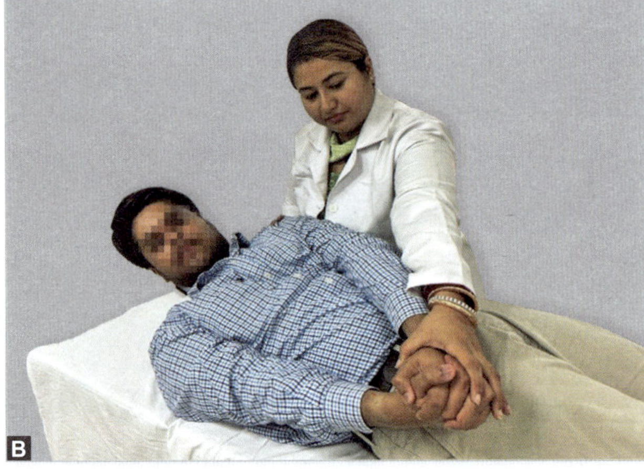

Figs 15.20A and B: Upper trunk flexion: **A.** Starting position; **B.** Final position

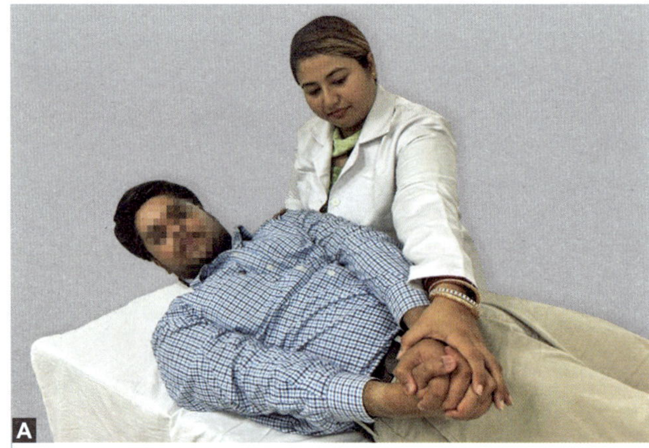

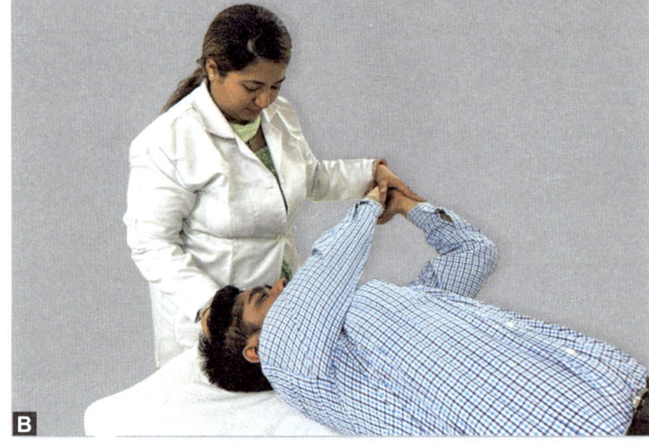

Figs 15.21A and B: Upper trunk extension pattern: **A.** Starting position; **B.** Final position

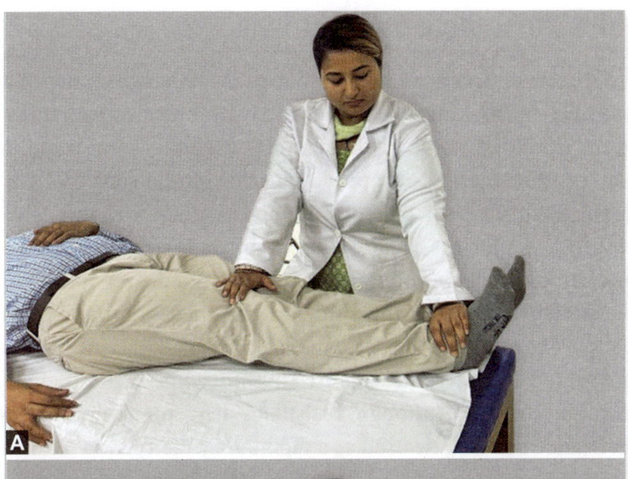
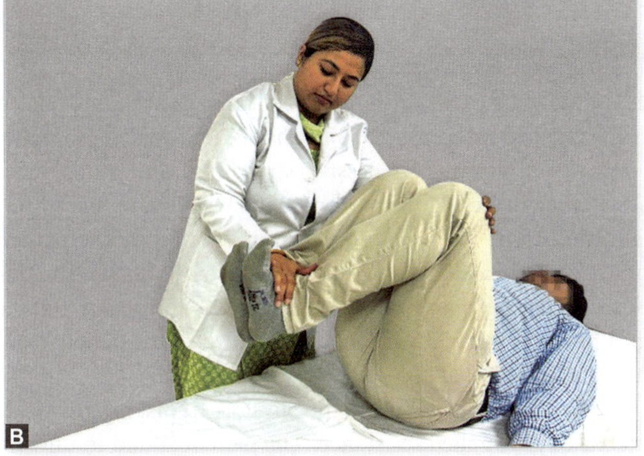

Figs 15.22A and B: Lower trunk flexion: **A.** Starting position; **B.** Final position

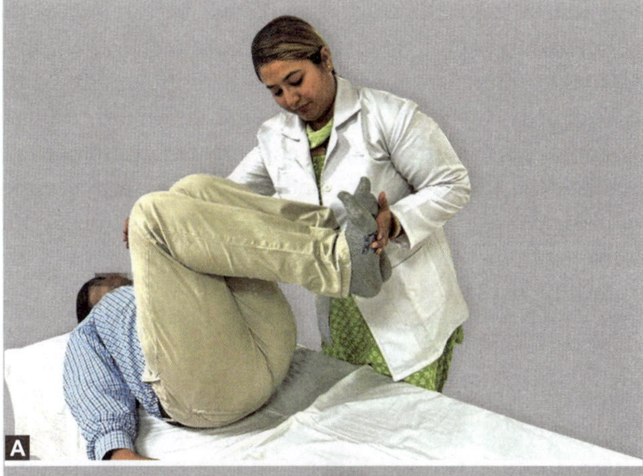
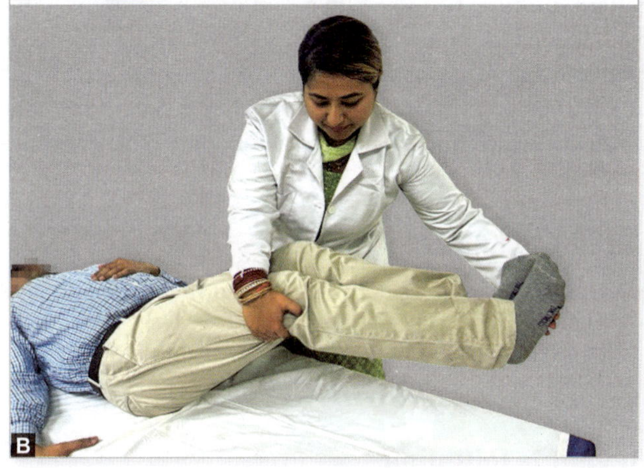

Figs 15.23A and B: Lower trunk extension: **A.** Starting position; **B.** Final position

Lower Trunk Flexion (Toward Right)

- **Starting position:** The lower trunk extension to the left side. In a supine position, the patient's legs are extended downwards and toward left as far as they can reach. The knees may be flexed or allowed to stay extended (Fig. 15.22A).
- **Grip of the physiotherapist:** The physiotherapist stands at the end of the plinth, facing the patient if the patient's legs are extended. The physiotherapist's left hand is placed on the medial side of the left heel, and the right hand is placed on the lateral side of the right heel. If the patient's knees are flexed during the exercise, the left hand is positioned over the anterior surface of both femurs, and the right hand is placed over the dorsum of both feet.
- Verbal command of the physiotherapist can be "pull up and over".
- **Direction of movement:** Both limbs are raised and shifted to the right (Fig. 15.22B).

Lower Trunk Extension (Toward Left)

- **Starting position:** Rightward flexion of the lower trunk. Patient is in a supine position, with both legs flexed as far to the right as possible. During the exercise, the knees are extended and can either stay that way or can be flexed (Fig. 15.23A).
- **Grip of the physiotherapist:** The physiotherapist adopts a stride standing stance on the patient's left side, aligned with the direction of the movement. The physiotherapist's right hand is placed under the thighs, left hand under the heels.
- Verbal command of the physiotherapist can be "push down and over to the opposite side".
- **Direction of movement:** Both extremities move downwards and over to the left (Fig. 15.23B).

PROPRIOCEPTIVE NEUROMUSCULAR FACILITATION TECHNIQUES

Proprioceptive pathways must accurately record muscle spindle tension and body location in space using the labyrinth in order for the motor centers of the central nervous system to produce a perfectly coordinated movement. The movement becomes uncomfortable, uncoordinated, and worthless if the afferent pathways are blocked. The motor cells are excited more when the afferent stimulation is increased, and over time, functional pathways that make up for the loss of the natural pathways brought on by any disease can be established within the central nervous system. The anterior horn cell must be maximally excited in order to elicit a maximum response in a muscle. Stretching a muscle and applying resistance to it stimulate the proprioceptors, and impulses from the upper motor centers also contribute to this excitement. A maximum discharge from the anterior horn cell will result from the stimulation of these variables.

The PNF Techniques aim to facilitate, inhibit, strengthen, or relax the muscular groups in order to enhance functional movement. These methods aim to promote or enhance a certain type of muscle activity associated with a desired pattern, posture, or task. While some methods use isotonic contractions to promote mobility within a functional range, others focus on isometric contractions to enhance stability in a particular position. While some methods address tissue shortness, which limits joint range of motion, others enhance movement initiation.

Did You Know?

One interesting aspect of proprioceptive neuromuscular facilitation (PNF) techniques is its ability to leverage the body's proprioceptive pathways to improve movement and functional outcomes.

Rhythmic Initiation

A technique called "rhythmic initiation" aims to improve mobility that is impeded by deficiencies in the initiation, coordination, or relaxation of movements. This method involves applying movements gradually—passively at first, then with assistance, and ultimately with active or mild resistance.

With any pattern or activity, this technique can be effectively applied. Patients who struggle to initiate functional movements due to hypertonicity are a prime choice for this technique. In men with proximal humerus gunshot injuries, positive changes in the active amplitude of movement occurs under the influence of PNF, which helps to break the cycle of pain and enhance upper extremity functionality.

Rhythmic Rotation

The use of passive movement in a rotational pattern defines rhythmic rotation. To encourage entire body relaxation or tone reduction, the exercise is slow and rhythmic. Lessening spasticity is intended to promote increased active or passive joint movement. The physiotherapist rotates the part slowly around its longitudinal axis. Both the hypertonicities that develop during efforts at active movement and the muscle tone at rest can be influenced by this technique. For example, lower trunk rotation in a hook lying.

Hold Relax Active Movement

Hold relax active movement (replication) enhances functional mobility by making it easier to activate muscle contraction in an extended range of the agonist. It emphasizes only one way of movement. To improve muscle spindle sensitivity, an agonist pattern isometric contraction is applied with resistance within the shortest range. The patient is instructed to relax once an optimal contraction has been attained. Depending on the patient's reaction, the clinician will then gradually move the part toward the lengthened position. The patient may be instructed to enter the agonist rhythm while also receiving a quick stretch. Although resistance isn't mandatory, light resistance can also be used as a facilitatory factor.

Hold Relax

The hold relax method is used to lessen movement-related pain and increase passive joint mobility. Resisted isometric contraction, active and passive stretching, and verbal cues are key elements of this technique. The joint or body part is moved to its full range of pain-free motion by the patient or the healthcare professional. The patient remains in this position as the physiotherapist fights the antagonist muscle group's isometric contraction, which prevents the patient from moving as intended.

"Hold" is given verbally as the physiotherapist progressively increases the applied resistance. The patient is told to unwind bit by bit. Whenever possible, a greater range of motion is granted to the joint or body segment. Though passive movement by the physiotherapist is acceptable, aggressive, controlled movement is recommended. Until the range of motion stops increasing, the entire procedure is repeated.

An option to the conventional method is to initiate an isometric contraction in the agonist muscle group instead of the antagonist, and then move either actively or passively into a greater range of motion. This method works well for extending the knee and improving hip flexion, as in a straight leg lift. According to a study by Gabriel in 2022, while unilateral lumbar mobilization techniques and hold-relax PNF both improved range of motion (ROM), hold-relax PNF was more effective at improving hamstring flexibility.

Contract Relax

Another approach to extend soft tissue length and passive joint range is the contract relax technique. It is most appropriate and successful when treating decreasing length in two-joint muscles and when discomfort is not a significant component. The main elements of the approach are verbal signals, an active or passive stretch, and resisted isometric and isotonic contractions of the small muscles.

The patient or a medical professional moves the joint or body part to the limit of its range of motion. The instruction to "turn and push or pull" is given verbally. Except for rotation, all motion is overcome by the resistance. The outcome is an isometric contraction of the remaining muscles and a resisted concentric contraction of the rotatory component. For a minimum of five seconds, a strong muscle contraction is generated and maintained. The joint or body segment is moved actively or passively to the new limit of passive range of motion after the patient relaxes after the contraction.

Similar to hold relax, the sequence is repeated until no more improvements are made. Changes in muscle tension using this approach are relatively abrupt, even though the muscles employed during hold relax gradually. According to a research by Rafal (2018), a single dose of hip adductor and abductor contract-relax proprioceptive neuromuscular facilitation stretching improved mediolateral dynamic balance.

Alternating Isometrics/Alternating Holds

Alternating isometrics, sometimes referred to as isotonic stabilizing reversals or alternating holds, is a technique that helps preserve strength, stability, and endurance in certain muscle groups or while maintaining a specific posture. Both agonist and antagonist muscle groups are made to contract isometrically alternately. The main facilitatory components are manual contacts and verbal commands. This technique is frequently used in developmental postures because trunk stability or proximal extremity joint is a prevalent emphasis, but it can also be used with bilateral or unilateral extremity patterns.

To promote the isometric contraction of agonist muscles, manual resistance is applied. The clinician then positions one hand over the antagonist muscles in a different way and gradually increases resistance in the desired direction until the desired reaction is achieved. The second hand can be withdrawn from the surface or shifted to the new place until the next change in resistance direction starts. Manual contacts are gently adjusted to encourage a progressive alternating contraction of the agonist and antagonist muscle groups. Alternating isometric exercises can help to improve trunk stability when seated unaided.

Rhythmic Stabilization

Through the contraction of muscles encircling the target joint(s), rhythmic stabilization (isometric stabilizing reversals) increases stability. Applying resistance encourages isometric contractions. Enhancing the patient's capacity to hold a particular developmental posture is frequently the aim. A rotating force is encouraged in order to facilitate concomitant contraction of the major stabilizers surrounding the affected joints. All that is required of the patient is maintaining the posture. In order to match the patient's effort, force is progressively raised while highlighting the rotatory aspect of the action. Once the patient develops muscle force in one direction, the physician again emphasizes rotation by moving one hand and gradually delivering force in a different direction. Depending on the requirements of the therapeutic scenario, rhythmic stabilization can be utilized to enhance strength and range of motion, reduce pain during movement, and improve stability and balance. Rhythmic stabilization is another technique that can be utilized to keep the trunk stable when sitting unsupported. The physiotherapist prevents trunk rotation by placing one hand on the posterior trunk and the other on the anterior trunk. It must be possible for the patient to sustain an erect trunk position isometrically. The phrase "Hold, don't let me move you" is used as an order. The right and left hands' relative postures are successively altered to produce opposing rotating forces. The patient does not intend for the movement to occur.

The patient retains the position dynamically by matching the resistance offered by the physiotherapist. It was found in a study by Lee, Young-Hun (2021) that the proprioceptive neuromuscular promotion technique of stable reversal and rhythmic stabilization, used in trunk stability exercises, can be an effective intervention for the respiratory function of stroke patients.

Dynamic Reversal

Slow reversal, sometimes referred to as dynamic reversals or reversal of antagonists, is a useful treatment method for a variety of patient problems, such as poor coordination, joint stiffness, and muscle weakness. Muscle contraction in an agonist pattern is facilitated by manual touches and verbal instructions. Changes in one or both hands' manual contact can aid in the antagonist pattern contracting concentrically at the intended end of range. Resistance is applied in both directions, with varying degrees of power based on the goals and capabilities of the patient. Resistance must adapt to changes in patient's effort because a patient's force output may vary throughout a pattern. Priority is given to fluid transitions between movement pattern directions, such as switching from upper limb D_2 flexion to D_2 extension. The mobility, controlled mobility, and skill stages of motor control can all be addressed with this approach. Smoothly switching from one way to another is of utmost importance during the skill stage. By rhythmically alternating between antagonist and agonist muscle groups, fatigue is reduced.

Slow Reversal Hold

A variation on the slow reversal technique is the slow reversal hold, which entails holding an isometric contraction against resistance near the end of the range in the chosen pattern or exercise. Depending on the patient's state or the intended outcome, movement may be limited to a smaller excursion or it may be continued through the available joint range. Movement happens as outlined for the slow reversal, but all involved muscles contract isometrically in resistance at the desired end points in each direction. By encouraging improved strength, balance, and endurance, this method helps with the transition of motor control from the mobility to stability phases. The slow reversal hold can be suitable for functional movements as well as single extremity or trunk patterns. A clinical example of the use of the slow reversal hold method is the upper extremity D_2 flexion as agonist pattern while kneeling. All of the muscles employed in the D_2 flexion (agonist) pattern are contracted concentrically throughout the target range. The patient is directed to use all the muscles involved in the flexion pattern without altering manual contacts, while maintaining the chosen end position. The proximal and distal hand locations are meticulously rebuilt to facilitate a smooth transition into the D_2 extension pattern. Progressive resistance is used throughout the D_2 extension design. The D_2 extension pattern is isometrically contracted while being held at the intended position.

Agonist Reversal

When performing a pattern or task, functional movement is facilitated by the agonist reversal technique (combination of isotonics). The agonist muscle is contracted eccentrically and concentrically. The primary objective of the approach is to promote functional stability in a seamless, controlled manner. Enhancing muscular strength and endurance, coordination, and eccentric control training are additional goals. During a certain pattern or activity, the approach is administered by opposing the agonist muscle group(s)' concentric contraction in a specific direction and range. At the targeted action endpoint, the patient maintains an isometric grip against the resistance. The physiotherapist then prevents the patient from slowly and deliberately returning to the beginning of the movement pattern in order to promote an eccentric contraction. The patient holds again at the end of the eccentric phase to encourage stability in this range. In summary, the technique consists of resistance to a concentric contraction at the beginning, stabilization, resistance to an eccentric contraction in the middle, and then another stabilizing hold at the end. The agonist muscle groups are the focus of this procedure. To superimpose the agonistic reversal technique, bridging is frequently an appropriate exercise.

Resisted Progression

The skill level task of locomotion is the main emphasis of the resisted progression technique. Resistance can be used to improve timing, build muscle strength and endurance, or enhance motor learning. This method can be used while walking, creeping, or crawling. Depending on the intended emphasis, different manual contacts are made to the upper or lower trunk, extremities, pelvis, and scapula. When a quadruped is moving backward (creeping), resisted progression can be used to successfully encourage proper recruitment of the hip extensors and pelvic rotators. Moving the opposing upper and lower extremities simultaneously or each extremity individually can cause a backward progression. This decision is based on the patient's motor skills, coordination, trunk control, strength, and cognitive state. Common manual contact points are the posterior thigh, posterior humerus, ischial tuberosity, and inferior angle of the scapula. Any configuration of contacts can be utilized, depending on the intended emphasis. For example, the physiotherapist's hands could rest on the bilateral ischial tuberosities, the right posterior humerus and the left posterior thigh, or the left scapula and the right ischial tuberosity. The physiotherapist kneels beside or behind the patient, facing their head.

SUMMARY

- Proprioceptive neuromuscular facilitation (PNF) is a therapeutic approach in neurological rehabilitation that increases muscle flexibility and improves active and passive ranges of motion. Developed between 1945 and 1954 at the Kabat Kaiser Institute, it uses natural movement patterns to stimulate proprioceptors and enhances functional task performance. PNF helps patients develop head and trunk control, control center-of-gravity shifts, and maintain balance. Skilled use of ten key elements improves motor learning.
- PNF is a method of promoting muscle strength and flexibility through manual contact, body positioning, and stretching. It involves placing hands on the skin to activate pressure receptors and inform the patient of the intended motion's direction. The clinicians should align their hands, arms, shoulders, and pelvis with the motion, using their body weight to generate resistance. Stretch can be used to promote muscle activation, but it should be done carefully to avoid causing discomfort or affecting motor responsiveness. Manual resistance can be used to improve motor unit recruitment or reduce internal resistance by changing neuronal firing patterns. Irradiation, a neurophysiologic phenomenon, can either boost muscular activity in linked muscles or suppress opposing antagonist muscle groups.
- Joint facilitation (traction and approximation) stimulates receptors in the joint and periarticular structures, allowing for weight bearing and muscle co-contraction. Smooth sequencing and gradation of muscle activation are necessary for normal movement, with most functional actions timed from distal to proximal. Postural control develops from proximal to distal and from cephalad to caudal.
- PNF is distinguished by its distinctive diagonal movement patterns, which simulate demands made during functional motions more accurately. Visual cues help control and make the right movements and body positions, with head and body positioning affected by eye movement. Verbal instruction should include directional indications, with the patient prepared for action during the preparatory phase, followed by action and correction.
- PNF patterns were initially developed by analyzing typical movement patterns, which are incorporated into certain sets of joint movements known as patterns. Each pattern focuses on either the extremities or the trunk, with rotation being a key component. Early rotation supports appropriate distal to proximal timing of extremity motions.
- Extremity patterns are defined by diagonal 1 (D_1) and diagonal 2 (D_2), which are the movements that follow from performing an extremity pattern. These patterns are divided into flexion and extension directions, with the proximal component being emphasized and more trunk activity recruited by using straight arm and leg patterns.
- Upper extremity patterns are described using the right limb and while supine. Any starting position, including forward lying, side lying, or sitting, is acceptable, but the patient must be positioned in these positions to ensure full range of motion.
- Upper limb D_1 Flexion involves an arm outstretched and slightly to the side, with the wrist deviated, the forearm pronated, and the shoulder extended, abducted, and internally rotated. The physiotherapist's grip is placed on the ulnar side of the hand's dorsum and medial three fingers, and the left hand that passes beneath the arm is positioned on the extensor side of the forearm's ulnar surface.
- Upper limb D_2 Flexion begins with the arm extended across the body with elbow crossing the midline, the forearm is pronated, the wrist and fingers are flexed, and the wrist is deviated toward ulnar. The physiotherapist's grip is placed on the radial side of the hand's dorsum and the radial three fingers.
- Lower extremity patterns are shown and described in the supine position. Four lower extremity designs along two diagonals are described, each of which is analogous to the upper extremity.
- LE D_1 flexion involves the hip being internally rotated and abducted, the foot everted and plantar-flexed, and the physiotherapist's grip is placed on the medial side of the dorsum of the foot. The direction of movement is to reach the opposing shoulder, the leg ascends and turns inward.
- The head and neck move in a diagonal plane, with flexion and extension being the two main patterns. Flexion involves extending the head and neck to the left, while extension involves extending the head and neck to the right. The physiotherapist's grip is positioned over the right mandible, under the chin, and the occiput is grasped with the left hand.
- Trunk movements occur in a diagonal plane from the shoulder to the opposite hip. Upper and lower trunk flexion patterns make up the flexion patterns, while upper and lower trunk extension patterns make up the extension patterns. The patient's trunk is extended and rotated to the left while sitting, long-sitting, or lying down, with the head directed over the left shoulder.
- Upper trunk flexion starts with the patient's trunk extended and rotated to the left while sitting, long-sitting, or lying down. The physiotherapist's grip is placed diagonally in front of the left shoulder, while the right hand is placed on the forehead over the right eyebrow. The head and trunk are flexed and turned in the direction of the opposing hip.
- Lower trunk flexion starts with the patient's legs extended downwards and toward the left as far as they can reach. The physiotherapist stands at the plinth's end facing the patient, placing their left hand on the medial side of the left heel and right hand on the lateral side of the right heel.
- Lower trunk extension begins with the patient in a supine position, with both legs flexed as far to the right as possible. The physiotherapist's grip is aligned with the direction of movement.
- Proprioceptive pathways play a crucial role in coordinating movement within the central nervous system. Blocking these pathways can lead to discomfort and incoordination. To enhance functional movement, PNF techniques facilitate, inhibit, strengthen, or relax muscular groups. Some techniques focus on isometric contractions to improve stability or mobility within a functional range. Rhythmic initiation is a technique that focuses on increasing mobility restricted by impairments in movement initiation, coordination, or relaxation. It involves progressive application of passive, active assisted, or slightly resisted movements.

- Rhythmic rotation uses passive movement in a rotational pattern to encourage entire body relaxation or tone reduction. It involves slow and rhythmic movement, reducing spasticity to promote increased active or passive joint movement. Hold relax active movement (replication) improves functional mobility by recruiting muscular contraction in a lengthened range of the agonist. This technique emphasizes one way of movement and can be used with light resistance or resistance as a facilitatory factor.
- The hold relax method is a technique used to reduce movement-related pain and increase passive joint mobility. It involves resisted isometric contraction, active and passive stretching, and verbal cues. The patient or practitioner moves the joint to its maximum range of pain-free motion, while the physiotherapist resists the isometric contraction of the antagonist muscle group. The patient is instructed to gradually relax, and the joint or body segment is given a wider range of motion whenever possible. This technique can be used to improve hip flexion and knee extension.
- The contract relax technique is another approach to extend soft tissue length and passive joint range. It involves resisted isotonic and isometric contractions of short muscles, verbal cues, and an active or passive stretch. The patient is moved to the limit of its range of motion, and the contraction is maintained for at least five seconds. The sequence is repeated until no more gains are achieved.
- Alternating isometrics, also known as alternating holds or isotonic stabilizing reversals, helps maintain strength, stability, and endurance in specific muscle groups or while holding a particular posture. It is often used in developmental postures and bilateral or unilateral extremity patterns.
- The techniques of rotary force, rhythmic stabilization, dynamic reversal, slow reversal hold, agonist reversal, and resisted progression are used to promote joint contraction, muscle weakness, coordination, and motor control. Stable reversal and rhythmic stabilization are effective interventions for respiratory function in stroke patients. Dynamic reversal is an effective technique for treating joint stiffness, muscle weakness, and poor coordination. Slow reversal hold involves holding an isometric contraction against resistance at the end of the range in the selected pattern or activity. This method helps with the transition of motor control from mobility to stability phases.
- Agonist reversal is a combination of isotonics that encourages functional stability in a smooth, controlled way. It involves resisting the concentric contraction of the agonist muscle group(s) in a particular direction and range during a particular pattern or task. The patient holds isometrically against resistance at the intended endpoint of the action, then resists the patient from making a slow, controlled return to the start of the movement pattern. At the end of the eccentric phase, the patient holds again to promote stability in this range.
- Resisted progression is the main emphasis of locomotion, using different manual contacts to the upper or lower trunk, extremities, pelvis, and scapula. This technique can be used while walking, creeping, or crawling, and can be used to encourage proper recruitment of hip extensors and pelvic rotators. The physiotherapist's hands can be positioned on the left scapula and right ischial tuberosity, the right posterior humerus and the left posterior thigh, or both ischial tuberosity bilaterally.

FURTHER READINGS

- Funk DC, Swank AM, Mikla BM, Fagen TA, Farr BK. Impact of Prior Exercise on Hamstring Flexibility: A Comparison of Proprioceptive Neuromuscular Facilitation and Static Stretching. Natl Str Cond Assoc J. 2003;17(3):489–492.
- Knott, Margaret, and Voss, D. E. (1968). Proprioceptive Neuromuscular Facilitation: Patterns and Techniques (2nd ed.). Bailliere Tindall.
- Oleksandra Kalinkina, Olena Lazareiva et al. Influence of physiotherapy on the active range of motion in proximal humerus gunshot patients. Sport Mont, 2021; 19(S2), 177–181.
- Sbardelotto, G. A. E. B., Weisshahn, N. K., Benincá, I. L., de Estéfani, D., E Lima, K.
- M. M., and Haupenthal, A. (2022). Hold-relax PNF is more effective than unilateral lumbar mobilization on increasing hamstring flexibility: A randomized clinical trial. Journal of Bodywork and Movement Therapies, 32, 36–42.
- Szafraniec, R., Chromik, K., Poborska, A., and Kawczyński, A. (2018). Acute effects of contract-relax proprioceptive neuromuscular facilitation stretching of hip abductors and adductors on dynamic balance. PeerJ, 6(e6108), e6108.
- Lee, Y.-H., and Cho, Y.-H. (2021). The effects of trunk stability exercise using stabilizing reversal and rhythmic stabilization techniques of PNF on trunk strength and respiratory ability in the elderly after stroke. PNF and movement, 19(1), 105–113.

STUDENT ASSIGNMENT

LONG ANSWER QUESTIONS

1. Describe PNF with its components in detail.
2. Explain the techniques of PNF.
3. Discuss in detail about the diagonal patterns of PNF in the upper limb.
4. Give the detailed description of the diagonal patterns of PNF in the lower limb.
5. Explain the patterns of PNF for the head, neck, and trunk.

SHORT ANSWER QUESTIONS

1. Write the PNF pattern of the upper limb which is commonly used for feeding.
2. What is rhythmic stabilization?
3. What is dynamic reversal?
4. Differentiate between hold relax and contract relax.
5. Write briefly about D_1 flexion and extension pattern for the lower limb.

MULTIPLE CHOICE QUESTIONS

1. **Which PNF technique would be the most appropriate to improve mobility and reduce the effects of rigidity in a patient with PD?**
   a. Alternating isometrics b. Rhythmic initiation
   c. Rhythmic stabilization d. Agonist reversal
2. **Which of the following combined motions make up a D_2 flexion UE PNF pattern?**
   a. Shoulder extension, abduction, ER
   b. Shoulder flexion, adduction, IR
   c. Shoulder extension, adduction, IR
   d. Shoulder flexion, abduction, ER
3. **What type of muscle contraction is utilized when performing rhythmic stabilization?**
   a. Isometric b. Isotonic
   c. Isokinetic d. Eccentric
4. **Which PNF technique is characterized by a patient with hypertonia first being passively moved through the ROM and then moving actively?**
   a. Dynamic reversals
   b. Rhythmic stabilization
   c. Rhythmic initiation
   d. Combination of isotonics
5. **Which PNF technique is performed while the patient performs isometric contractions?**
   a. Dynamic reversals b. Stabilizing reversals
   c. Rhythmic stabilization d. Approximation
6. **Which of the following is a PNF stretching technique that is characterized by passively stretching the target muscle followed by active contraction of the muscle opposite the joint?**
   a. Hold-relax
   b. Contract-relax
   c. Agonist-contraction
   d. Hold-relax with agonist contraction
7. **A PNF technique that uses slow and resisted concentric contractions of agonists and antagonists around a joint w/o rest between reversals is termed:**
   a. Alternating isometrics b. Rhythmic initiation
   c. Rhythmic stabilization d. Slow reversal
8. **Which UE PNF pattern moves the arm up and across the body?**
   a. D_1 flexion b. D_1 extension
   c. D_2 flexion d. D_2 extension

ANSWER KEY

1. b **2.** d **3.** a **4.** c **5.** c **6.** c **7.** d **8.** a

16 Manual Therapy and Peripheral Joint Mobilization

Kopal Pajnee

LEARNING OBJECTIVES

After the completion of the chapter, the readers will be able to:

- Understand and explain about the schools of manual therapy and their techniques:
  - Kaltenborn-Olaf
  - Maitland
  - Mulligan
  - McKenzie
  - Cyriax
  - Neural mobilization
- Explain and understand biomechanical basis for mobilization.
- Understand the effects of joint mobilization.
- Explain the grades of mobilization.
- Explain techniques of mobilization for upper limb and lower limb.

CHAPTER OUTLINE

- Introduction
- History of Manual Therapy
- Different Approaches of Manual Therapy
- Joint Mobilization

Joint Mobilization Techniques

- Shoulder Complex
- Elbow Complex
- Radioulnar Joints
- Wrist and Hand Complex
- Hip Joint
- Knee Complex
- Ankle and Foot Complex

KEY TERMS

Arthrokinematics: It is defined as the motion that occurs between two joint surfaces. It is classified into three: Rolling, spinning and sliding.

Centralization phenomenon: Any progressive decrease or resolution of pain from distal to proximal direction to the center of spine in response to treatment for spinal pain is called centralization phenomenon.

Derangement syndrome: The Derangement syndrome is characterized by a mechanical hindrance to joint motion. This syndrome is increasingly prevalent and well-known and its main feature is change and inconsistency.

Joint play: Joint play refers to joint action that an individual is unable to do. These motions, which coincide with a joint's natural movements, include roll, spin, and slide.

Manual therapy: Manual therapy is a medical strategy that relies on expert "hands on" treatment to reduce pain and increase nerve, soft tissue, and joint mobility.

Neural mobilization: According to Shacklock the term neurodynamics refers to the interaction of the nervous system with its various components and the surrounding structures.

Osteokinematics: It can be defined as movement between two bones at a joint. For example, flexion/extension, abduction/adduction.

Sustained natural apophyseal glide (SNAG): SNAG are a combination of application of sustained glide with movement with glide applied on the facet joint or the spinous process.

INTRODUCTION

Manual therapy is a clinical approach that uses specific, skilled hands-on techniques, such as manipulation and mobilization, to diagnose and treat soft tissues and joint structures. The goals of manual therapy are to: modulate pain; increase range of motion (ROM); decrease or eliminate soft tissue inflammation; induce relaxation; improve contractile and non-contractile tissue repair, extensibility, and/or stability; facilitate movement; and improve function.

Peripheral joint mobilization, a key subset of manual therapy, focuses on mobilizing joints in the limbs—shoulders, elbows, wrists, hips, knees, and ankles. By applying graded oscillatory or sustained forces to the joint, therapists can improve range of motion, decrease stiffness, and facilitate natural healing processes.

This technique is grounded in biomechanical principles and is highly adaptable, allowing therapists to tailor interventions to individual patient needs. Peripheral joint mobilization is especially effective in treating conditions such as arthritis, post-surgical stiffness, sports injuries, and movement disorders, serving as a cornerstone of comprehensive musculoskeletal care.

HISTORY OF MANUAL THERAPY

The history of manual therapy is tabulated as follows:

Sl. no.	Scientists	Contributions
1.	**Hippocrates** (460–355 BC)	First one to use joint manipulation and traction techniques. In one of his journals *On Setting Joints by Leverage*. He described a combination of extension (traction) and pressure (manipulation) on patients.
2.	**Claudius Galen** (131–202 AD)	Extensively researched on the work of Hippocrates and illustrated many of his techniques.
3.	**John Hunter** (1728–1793)	Advocated the use of movement and stretching to break adhesions which formed as result of inflammation after injury.
4.	**H Marsh and R Fox** (1882)	Proposed the use of the term "manipulation" instead of bone setting.
5.	**Andrew Taylor Still** (1828–1917)	Founded osteopathy in the year 1874. Still postulated when the mechanical locking of a joint is normalized, it resulted in improvement of patient's condition in quite a few medical conditions.
6.	**William Garner Sutherland** (1873–1954)	Gave the concept of cranial osteopathy based on assessment and treatment of cranial dysfunction.
7.	**Daniel David Palmer** (1845–1913)	Founded chiropractor techniques.
8.	**T Walmsley** (1927)	Gave the term "arthrokinematics".
9.	**Arthur Steindler** (1955)	Wrote a book titled *Kinesiology of the Human Body Under Normal and Pathological Condition* by combining previous research and arthrokinematic knowledge regarding both joint function and dysfunction
10.	**Cyriax** (1957)	• Published *Textbook of Orthopaedic Medicine* in two volumes focusing on of assessment soft tissues to diagnose joint dysfunction. • He was the first one to use the term end feel to diagnose soft tissue lesions.
11.	**S V Paris** (1963)	• Published an article titled *"The Theory and Technique of Specific Spinal Manipulation"*. • He gave the term dysfunction in which he explained that a loss of movement will result in degeneration and hypermobility of the segments above and below.
12.	**Geoffrey Maitland** (1964)	Wrote a book titled *"Vertebral Manipulation"*, in which he advocated the use of oscillatory joint mobilization techniques.
13.	**R Melzack and P D Wall** (1966)	Proposed the pain gate theory.
14.	**B. McCaleb** (1969)	• Published an article titled *"An Introduction to Spinal Manipulation"*, in which he stated that manipulation was helpful for joint dysfunction. • He described joint dysfunction as a partial absence or total absence of joint movement, called a joint lock.
15.	**Robin McKenzie** (late 1970's)	• Gave the McKenzie's concept for patients with low back pain. • He emphasized on the centralization phenomenon and the use of spinal extension exercises for treatment of patients with low back pain.
16.	**Brian Mulligan** (1990's)	Introduced the concept of NAGs, SAGs and mobilization with movement (MWM)
17.	**David Butler** (1991)	Wrote a book titled *"Mobilization of the nervous system"*. In his book, Butler described the examination and treatment of pathologies of the central and peripheral nervous system.

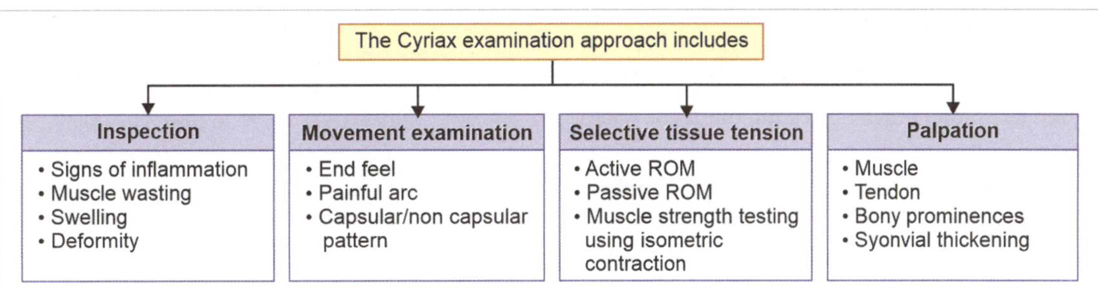

Fig. 16.1: Cyriax examination approach

Did You Know?

Manual therapy has evolved significantly over the centuries, drawing from various medical traditions and pioneering individuals. From Hippocrates' early techniques to modern innovations like Brian Mulligan's mobilization with movement, manual therapy continues to adapt and expand its understanding of the human body's mechanics and healing processes.

DIFFERENT APPROACHES OF MANUAL THERAPY

- Kaltenborn-Olaf Approach
- Cyriax Approach
- Maitland's Approach
- McKenzie's Approach
- Mulligan's Approach
- Neural Mobilization

Kaltenborn-Olaf Approach

Freddy Kaltenborn was a Norwegian Physical Therapist. In 1945, he started his career as a physical educator in Germany. In 1949, he completed his education in Physical Therapy in Germany. Soon he realized the limitations of the techniques used to treat patients and started to explore new concepts. In 1950, he travelled to London to learn the principles of treatment being taught by ***Dr James Mennell and Dr James Cyriax***. In 1955 he formed the Norwegian manipulation group. From 1958–1960, he collaborated with Cyriax to apply various biomechanical principles to assessment and treatment. In 1973, he worked with ***Olaf Evjenth*** and added methods of assessment and examination process and introduced symptom relief testing as a method for the localization of lesions. In 1974, he published a book titled "*Manual Mobilization of Joints*" based on his work.

The Kaltenborn-Olaf concept focuses on the use of arthrokinematics for assessment of joint lesion and the concept of treatment plane. Their treatment mainly focused on the use of glides, traction and compression techniques and the direction in which they should be applied to restore normal joint play. For the determination of the direction of glide, they proposed the concave convex rule which has become the foundation of joint mobilization techniques and focuses on the movement of bone and the joint glide. The grades and their applications will be further discussed in the section of joint mobilization.

Cyriax Approach

This approach is based on the work of ***Dr James Cyriax*** (Fig. 16.1). He developed a process of diagnosis by selective tension, which uses passive movements to test the inert structures and resisted movements to test the contractile structures (Table 16.1). The treatment techniques used in this approach fall into following categories:

- Transverse frictions (soft tissue or joint).
- Joint mobilization/manipulation.
- Traction and injection techniques.

According to this approach, every pain has a source. It is the responsibility of the therapist to follow the assessment principles, identify the nature and source of pain, localize the pathological structure, make a diagnosis and apply appropriate treatment technique for the resolution of patient's symptoms.

TABLE 16.1: Selective tissue tension testing

Evaluation	Aim	Structures affected
Passive ROM	Changes in symptoms, end feel, ROM	Non contractile structures such as joint capsule, ligaments, bursa, dura mater, dural nerve root sleeve
Active ROM	Change in symptoms, ROM, painful arc, willingness, capsular pattern	Both contractile and noncontractile structures
Resisted test	Pain, power	Contractile lesion (partial/complete tear), fracture, serious pathology

Techniques of Cyriax Concept

Transverse Frictions

It is a specialized technique which is used for joints, contractile and noncontractile structures.

The following points should be kept in mind while applying transverse friction:

- To ensure therapeutic efficacy, it must be applied precisely at the lesion site.
- To prevent skin damage, the patient's skin and the therapist's fingers must move in unison.
- Always apply to the afflicted tissue perpendicularly.
- Needs to be applied with enough sweep to treat the entire lesion.
- Should be used only to the extent that the patient can tolerate pain. Pain tolerance that has improved is regarded as encouraging.
- The patient has to be positioned correctly to reveal the area that has to be treated.
- The muscle is positioned so that it can be readily taken slack if the lesion is in the belly of the muscle. This facilitates the massage by keeping the muscle fibers apart
- To target as much of the tissues as possible, the tendon sheath must be positioned in a stretched position.
- The patient should use the procedure until the therapist notices a decrease in discomfort. Should the therapist fail to see this result, there may be a problem with the technique being applied, an incorrect spot or insufficient pressure being applied to the tissue. Refer to Chapter 17 for further details of this technique.

Graded Mobilization Techniques

The mobilization techniques of periphery and spine were graded into Grade A, Grade B or Grade C. Each technique has different indication and is applied to obtain specific therapeutic response. For further details, the reader can refer to Orthopedic medicine textbook written by Dr James Cyriax.

Injection Techniques

These techniques are usually used as an adjunct to other techniques. They are basically used to provide immediate relief and facilitate early joint movement. Usually a corticosteroid injection may be used alone or in combination with a local anesthetic.

Maitland's Approach

Geoffrey Maitland (1924–2010) was an Australian physiotherapist and is considered one pioneer in the field of manual therapy. In 1949, he completed his education in physical therapy. He was very much influenced by the works Mennell, Stoddard and Cyriax. He developed his own unique system of examination and treatment and started teaching manual therapy at University of South Australia in 1954. In 1962, he presented his research which advocated the used gentle passive mobilization techniques for the treatment of neuromusculoskeletal dysfunction. In 1964, he authored first edition of his book titled "*Vertebral Manipulation*" and in 1970 he wrote another book titled "*Peripheral Manipulation*". He was one of the cofounders of IFOMPT in 1974.

The key concepts of this approach include history taking, patient evaluation and clinical reasoning. Maitland advocated the use of re-evaluation of the most important signs and symptoms of the patient. He stressed on the fact that effective listening and communication with the patient in an open and nonjudgmental way is an important aspect to identify patient's problem. The therapist should always believe the patient and think that the information provided by the patient is relevant and valuable. He should always communicate in patient's own language or terminology. This helps the patient to communicate their condition in much better way. The therapist then assesses the patient and compartmentalizes all the information.

One of the most important aspects of this concept is the grading of the joint glides. Based on the therapist's assessment of joint arthrokinematics, appropriate treatment glide applied. The glides are passive accessory joint mobilization glides of different amplitudes which can be applied at different points in the available range of motion. These grades are further discussed in the joint mobilization section, in this chapter.

McKenzie's Approach

This concept was given in 1960's by ***Robin McKenzie*** from New Zealand. Initially it was developed for lumbar spine but in due course of this time it is scope expanded to cervical and thoracic spine and the periphery. The main features of this technique are as follows:

- Centralization phenomenon.
- Classification of spinal pathologies into postural, dysfunction and derangement syndromes.

Centralization Phenomenon

A number of patients with spinal pain may experience pain radiating to the extremities. Any progressive decrease or resolution of pain from distal to proximal direction to the center of spine in response to treatment for spinal pain is called centralization phenomenon. This reduction can be instant or may take some time to progressively resolve. It is also considered centralization if the pain felt in back localizes to the center of spine. This principle is applicable to derangement syndrome and points toward reduction in mechanical deformation. It can be achieved either by making postural correction or by active or passive exercises.

Classification of Spinal Pathologies into Postural, Dysfunction and Derangement Syndromes

The classification of spinal pathologies by McKenzie are tabulates in Table 16.2.

TABLE 16.2: Classification of spinal pathologies by McKenzie

Postural syndrome	• The patient experiences pain after sitting in a particular posture for long period of time and disappears when the patient changes the posture • There is no pain on any other activity • It occurs due to prolonged loading of normal tissues at end range
Dysfunction syndrome	• The mechanical deformation of damaged tissues, such as adhering or adaptively shortened tissue causes pain • The symptoms often last for eight to twelve weeks, during which the tissue can deteriorate and cause sporadic pain felt at the limit of a restricted range of motion. • Repetitive motions in the direction of the malfunction or the direction that causes the discomfort are part of the treatment. The goal is to use workouts to restructure that tissue, which restricts movement, until it eventually becomes pain-free
Derangement syndrome	• The symptoms can change side and be local, referral, radicular, or a combination of these • The severity of the symptoms can change throughout the day and can either start suddenly or gradually • Postures and routine everyday activities may have an impact on the symptoms • Directional preference: Recurring in a particular way or staying in a certain posture improves symptoms • Specific movements are used in treatments to reduce, centralize, and/or eliminate pain

McKenzie's Protocol for Lumbar Spine

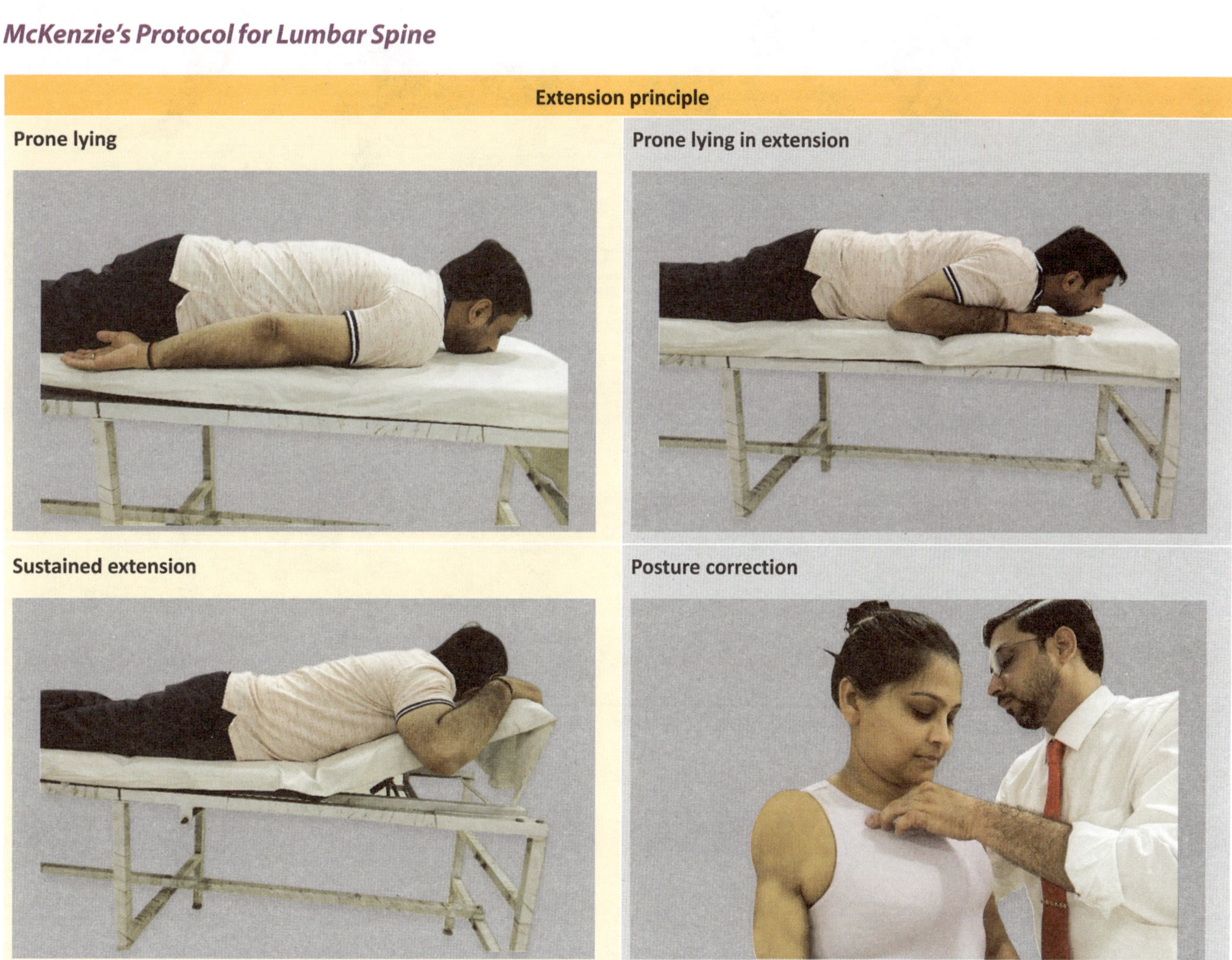

Contd...

Extension principle

Extension in lying

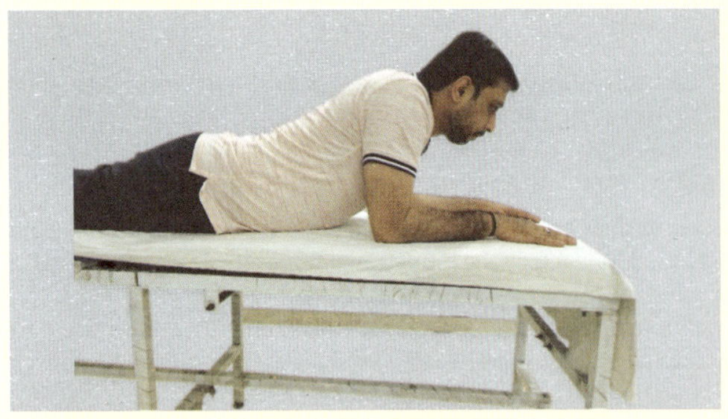

Extension in lying with patient overpressure

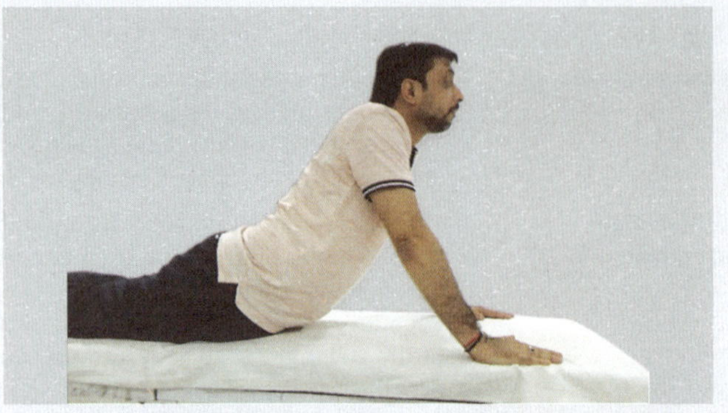

Extension in lying with clinician overpressure

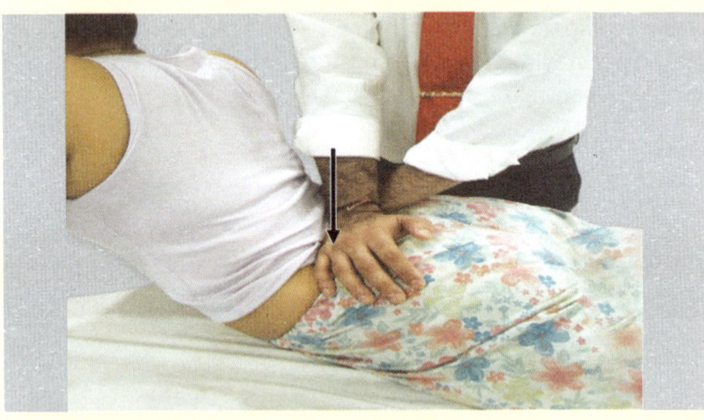

Extension in lying with belt fixation

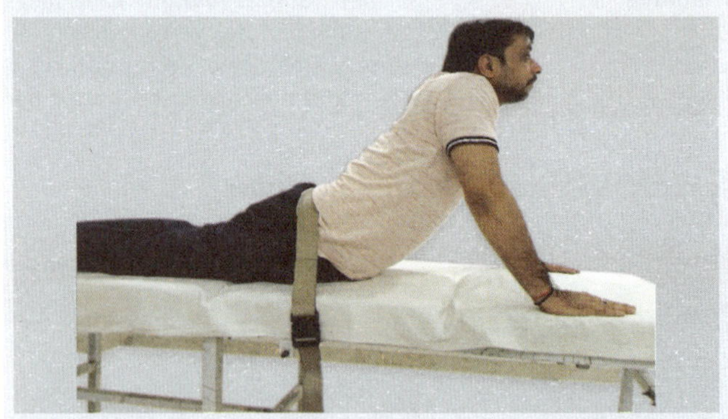

Extension mobilization in neutral or in extension

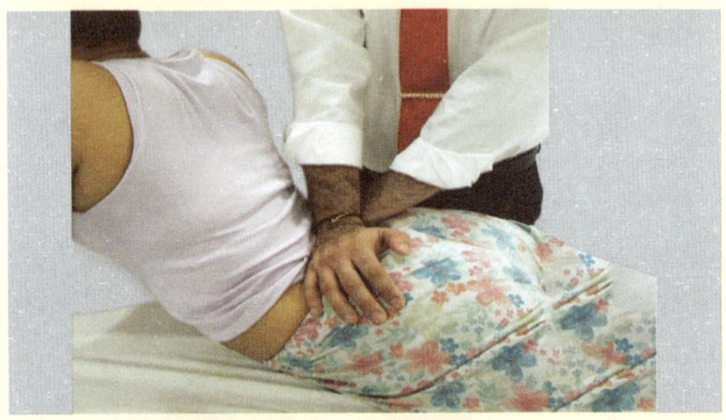

Extension manipulation

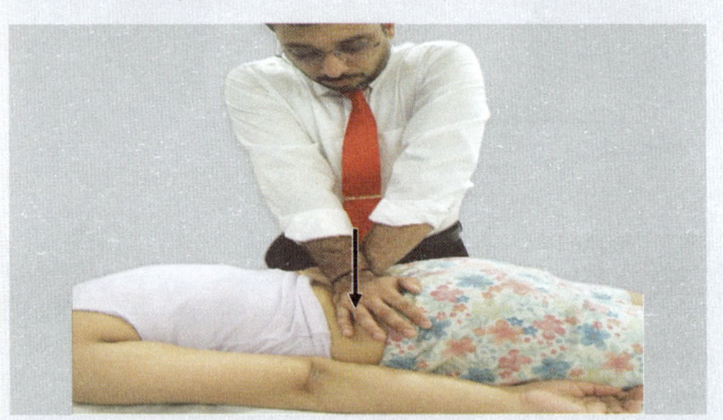

Extension in standing

Contd...

Flexion principle

Flexion in lying	Flexion in lying with clinician overpressure
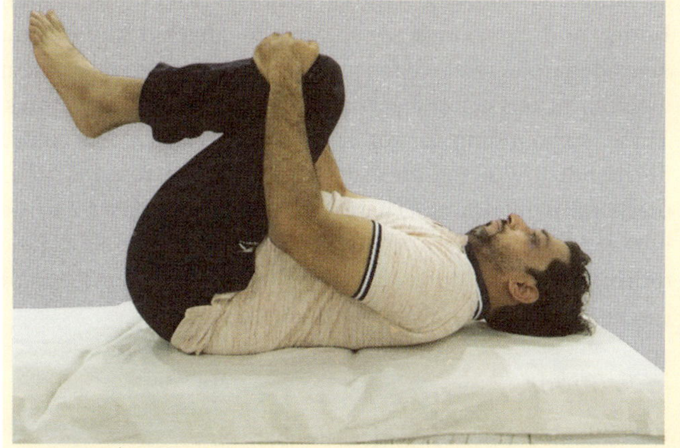	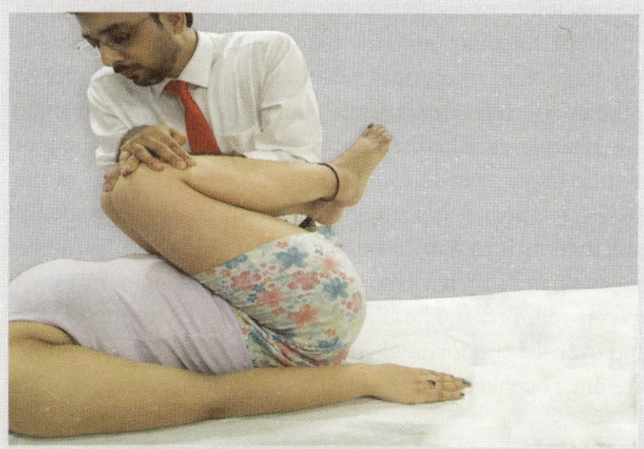
Flexion in sitting	**Flexion in standing**

Mulligan's Approach

The Mulligan concept of Manual Therapy was discovered and developed by Brian Mulligan. Brian Mulligan trained as a physiotherapist at the NZ School of Physiotherapy and qualified in 1954. His concept of manual therapy is based on the following:

- Correction of Positional Faults
- Mobilizations with Movement (MWM)
- Natural Apophyseal Glides (NAGs)
- Sustained Natural Apophyseal Glides (SNAGs)

Correction of Positional Faults

According to Mulligan, positional faults are minor biomechanical changes that take place in a joint due to any injury which causes restriction of normal physiological joint movement. This might result in pain during movement, loss of movement or pain in specific activities. A pain-free, sustained passive accessory joint glide is applied to the joint by the therapist and the patient is asked to perform the painful movement or activity within pain-free range. The therapist takes note of improvement in symptoms of the patient. Care must be taken that there should be no reproduction of pain and if there is any, then the technique should be terminated.

Mobilization with Movement

This innovative method involves the patient actively moving the afflicted limb to its maximum range while the therapist applies a prolonged passive joint glide. Finally, a passive overpressure is provided. If using mobilization belt, then the therapist can apply the pressure himself or the patient can be given proper guidance to apply it. The method should not cause the patient any pain.

Natural Apophyseal Glides

These are the mid to end range glides which are applied to the facet joints or spinous process from C2 to C7 according to patient's tolerance and are mainly indicated when there is gross restriction of cervical movements. The therapist stands on the unaffected side while the patient sits in a chair to apply these treatments. The therapist places the middle phalanx of the little finger over the spinous process or facet joint of the afflicted segment (supporting hand) and cradles the patient's head, holding it against his upper belly. The lateral edge of the opposite hand's thenar eminence (the mobilizing hand) then serves to reinforce it. The glide is then applied by the mobilizing hand over the supporting hand in anterior-superior direction. Care must be taken that the head remains still (Fig. 16.2A).

Sustained Natural Apophyseal Glides (SNAG)

The SNAGs are a combination of application of sustained glide with movement with glide applied on the facet joint or the spinous process. These glides can be applied from occiput to sacrum. For example, when using the technique to the cervical spine, the thumb of the mobilizing hand is used to reinforce the inner border distal phalanx of the stabilizing thumb, which is positioned over the afflicted section. All the other fingers rest over nape of neck on both sides. A sustained glide is then applied in anterior-superior direction and the patient is asked to perform the restricted movement and apply passive overpressure at the end. Care must be taken that the glide and patient's movement should be pain-free (Fig. 16.2B).

Clinical Correlation

Mulligan's approach to manual therapy offers innovative techniques for correcting positional faults and enhancing joint mobility without causing pain to the patient during treatment. By combining passive joint glides with active movements, Mulligan's mobilization with movement (MWM) and NAGs aim to restore normal physiological joint function and alleviate pain.

Neural Mobilization

The concept of neural mobilization is based on the works of eminent physiotherapists, namely **Elvey, Maitland, Butler and Shacklock**. According to Shacklock the term neurodynamics refers to the interaction of the nervous system with its various components and the surrounding structures. These surrounding anatomic structures which can move independently of the nervous system can be referred to as mechanical interfaces. These mechanical interfaces can be muscle, tendon, fascia, bone, disk or blood vessels. These techniques can be done actively by the patient themselves or can be applied passively by the therapist.

Indications

- Radiculopathies of cervical/lumbar spine
- Carpal tunnel syndrome
- Tennis elbow
- Cervicogenic headache
- Vertigo
- Piriformis syndrome
- Meralgia paraesthetica

Neurodynamic testing has become an integral part of physical therapist assessment (Table 16.3). The main aim is to put stress and strain over the nervous tissue to check its physiology and ability to bear mechanical stress. The testing involves moving spine or peripheral joint in a sequential pattern to provoke symptomatic or clinical responses. These tests are specific for each and every nerve and are termed as neural tissue tension test.

Precautions

Precautions to be taken while doing new neurodynamics testing:

- The patient should be relaxed and in a comfortable position. The procedure should be properly explained to the patient.

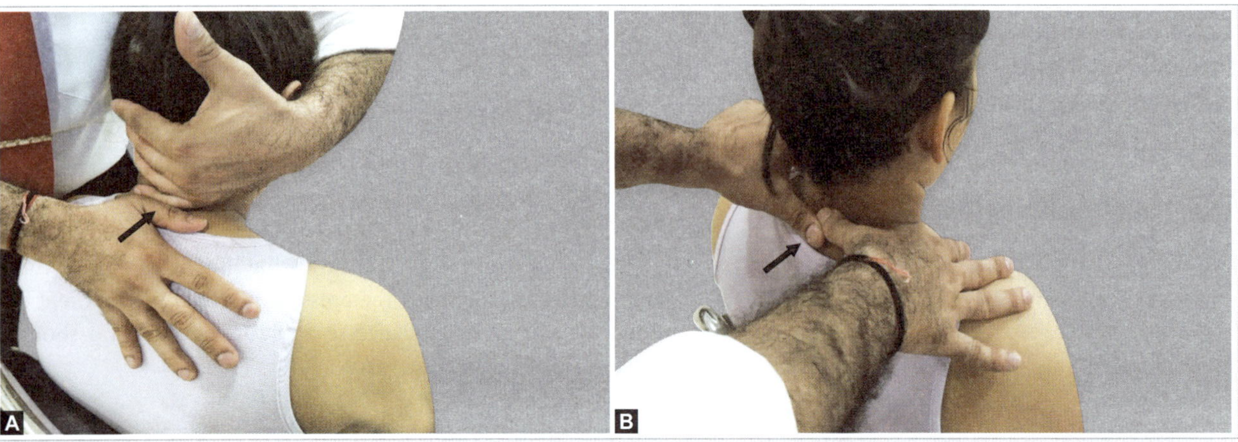

Figs 16.2A and B: **A.** Application of natural apophyseal glides; **B.** Application of sustained natural apophyseal glides

TABLE 16.3: Neurodynamic tests

Spine	Upper limb	Lower limb
Passive neck flexion (for cervical spine)	Median nerve test 1 (elbow flexion biased)	Straight leg raise (sciatic nerve)
Slump test (whole nervous system)	Median nerve test 2 (shoulder abduction biased)	Tibial nerve
Prone knee flexion	Radial nerve test	Sural nerve
	Ulnar nerve test	Common peroneal nerve

- Test the unaffected side first.
- There is no set pattern for sequence of movements. Usually the movements at the affected joints are performed first in the testing pattern. For example, in patients with carpal tunnel syndrome, the movements at the wrist joint are performed first before adding movements of other joint.
- The therapist should be able to maintain the position of the joint of the testing pattern. If the hold is loose then, stress over the nervous system is decreased and the test becomes inaccurate.
- The technique should be as gentle as possible.
- Reproduction of symptoms is considered positive test.

Clinical Correlation

Incorporating neural tissue tension tests into physical therapy assessments allows for the identification of neural dysfunction and the development of individualized treatment plans to address patients' unique needs. Additionally, the use of neural mobilization techniques can help reduce pain, improve nerve function, and enhance overall patient outcomes.

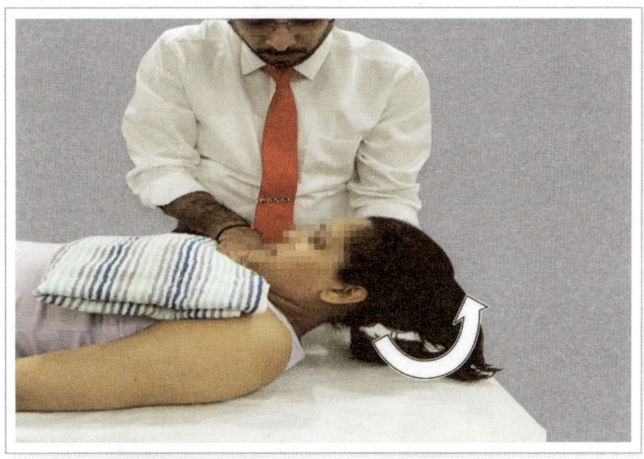

Fig. 16.3: Passive neck flexion test

Neurodynamics Testing-Spine

Passive Neck Flexion Test

The patient is in supine position. The therapist places one hand on the occiput and the other hand on the patient's sternum. The test is performed by moving the neck passively into cervical flexion (Fig. 16.3).

Slump Test

This test is used to sensitize the neural structures from the cervical spine to the big toe. This test places mechanical stress over CNS and PNS (Figs 16.4A to D). For performance of this test the patient is in high sitting position with both the hands behind the back and the therapist stands beside the patient.

- Procedure
  - Ask the patient to slouch → Add neck flexion → Add knee extension → Add ankle → Dorsiflexion
  - If the test provokes patient's symptoms, then the therapist can release the pressure on the nervous system by either bringing the neck to neutral position or the ankle in neutral position.

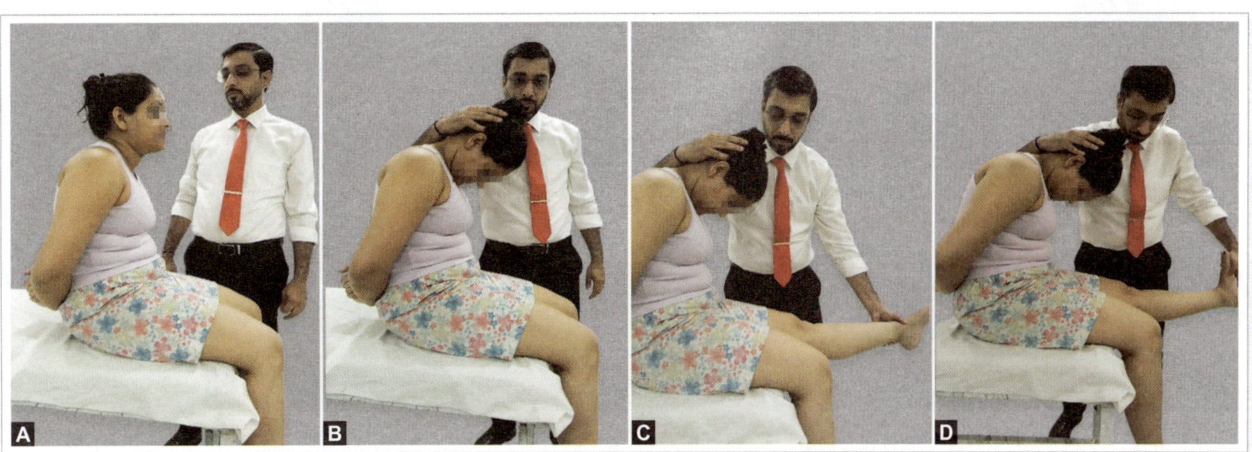

Figs 16.4A to D: Slump test: **A.** Ask the patient to slouch; **B.** Add neck flexion; **C.** Add knee flexion; **D.** Add ankle dorsiflexion

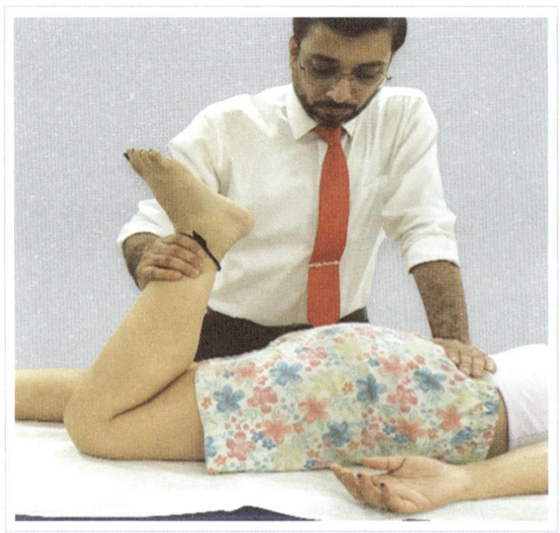

Fig. 16.5: Prone knee bend test

Prone Knee Bend (PKB) Test

This test is used to assess upper and mid lumbar spine and femoral nerve (Fig. 16.5). The patient is in prone lying and the therapist stands near the patient's pelvis. The proximal hand of the therapist stabilizes the pelvis of the patient while the distal hand grasps the anterior aspect of distal leg. The test is performed as passive knee flexion movement.

Neurodynamics Testing—Upper Limb

Median Nerve Test 1

- The patient is in supine lying. The therapist stands beside the patient facing her. The following movements are performed (Figs 16.6A to E):
  - Shoulder depression → Shoulder abduction → Shoulder external rotation → Elbow extension → Forearm supination → Wrist and finger extension.

Median Nerve Test 2

- The patient is in supine lying with the shoulder off the couch. The therapist stands near the shoulder level and supports the patient's shoulder on his thigh. The following sequence of movements is performed (Figs 16.7A to E):
  - Shoulder depression → Shoulder external rotation → Forearm supination → Wrist and finger extension → Shoulder abduction.

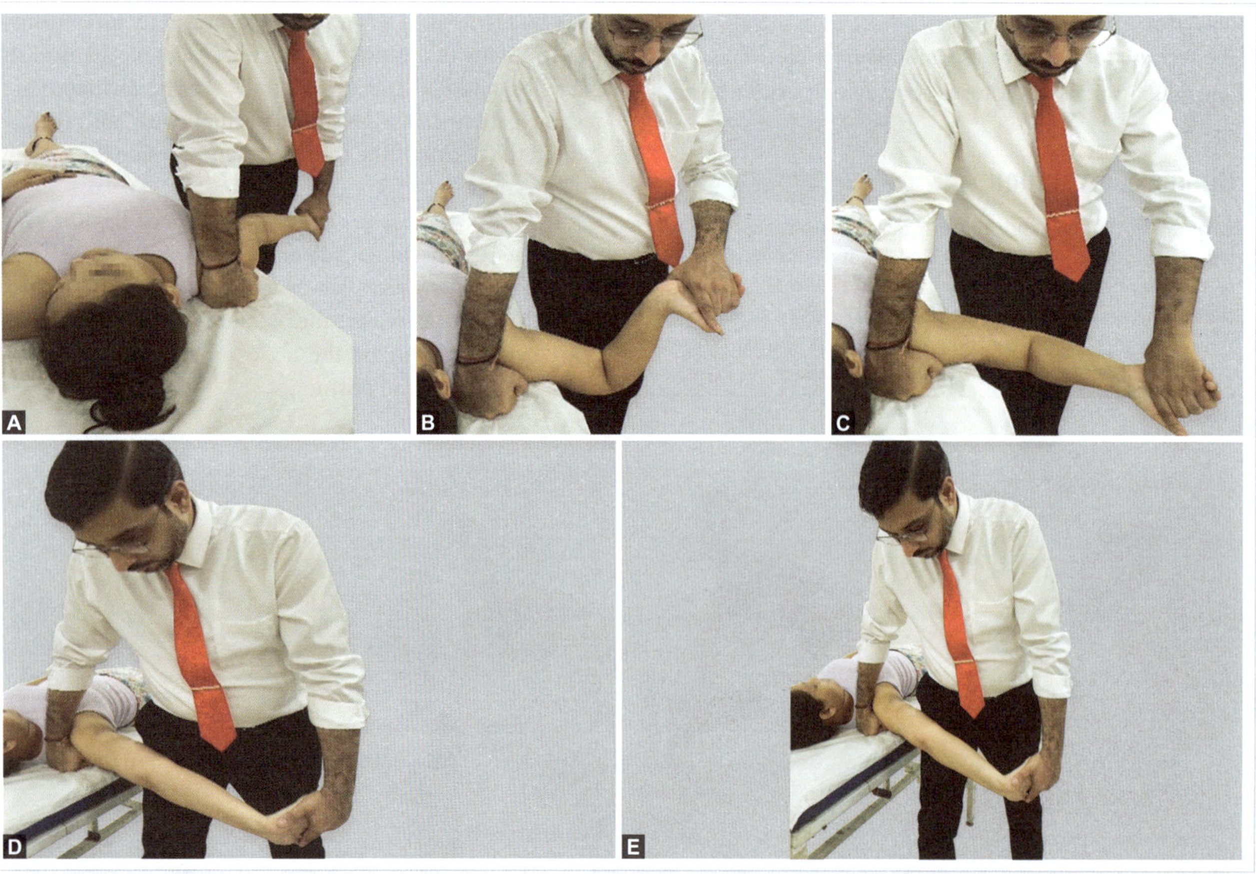

Figs 16.6A to E: Median nerve test 1: **A.** Shoulder depression; **B.** Shoulder abduction; **C.** Shoulder external rotation; **D.** Forearm supination; **E.** Wrist and finger flexion

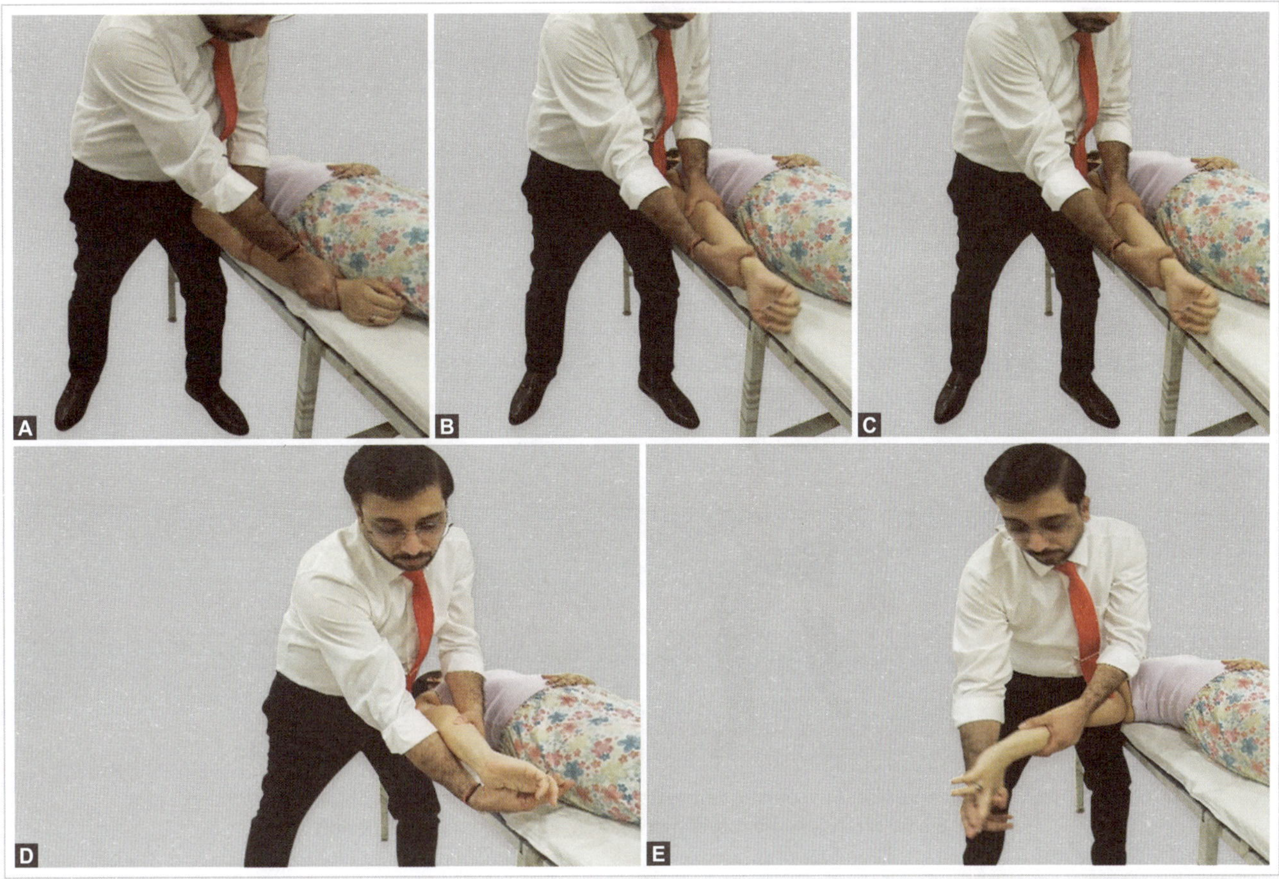

Figs 16.7A to E: Median nerve test 2: **A.** Shoulder depression; **B.** Shoulder external rotation; **C.** Forearm supination; **D.** Wrist and finger flexion; **E.** Shoulder abduction

Radial Nerve Test

- Starting position is same as median nerve test 2. The following sequence of movements is performed (Figs 16.8A to E):
  - Shoulder depression → Shoulder abduction → Shoulder internal rotation → Elbow extension and forearm pronation → Wrist flexion and finger flexion.

Ulnar Nerve Test

- The patient is in supine lying. The therapist stands beside the patient facing her. The following movements are performed (Figs 16.9A to E):
  - Shoulder depression → Shoulder abduction → Shoulder external rotation → Elbow flexion → Forearm pronation → Wrist extension → fourth and fifth fingers extension.

Neurodynamics Testing—Lower Limb

Straight Leg Raise (Sciatic Nerve Testing)

Patient is in supine position. In a stride standing stance, the therapist faces the patient. Immediately above the patella, the proximal hand is placed over the anterior thigh, just above the patella. Position the distal hand on the back of the distal leg. The test is completed by performing a passive hip flexion movement. If the test reproduces the patient's symptoms within the range of 30°–70° of hip flexion, it is deemed positive (Fig. 16.10).

Tibial Nerve Testing

Patient is in supine position. The therapist stands facing the patient's foot. The foot is then moved passively into dorsiflexion and eversion and the test is completed by adding passive hip flexion movement (Figs 16.11A and B).

Sural Nerve Testing

Patient is in supine position. The therapist stands facing the patient's foot. The foot is then moved passively into dorsiflexion and inversion and the test is completed by adding passive hip flexion movement (Figs 16.12A and B).

Common Peroneal Nerve Testing

Patient is in supine position. The therapist stands facing the patient's foot. The foot is then moved passively into

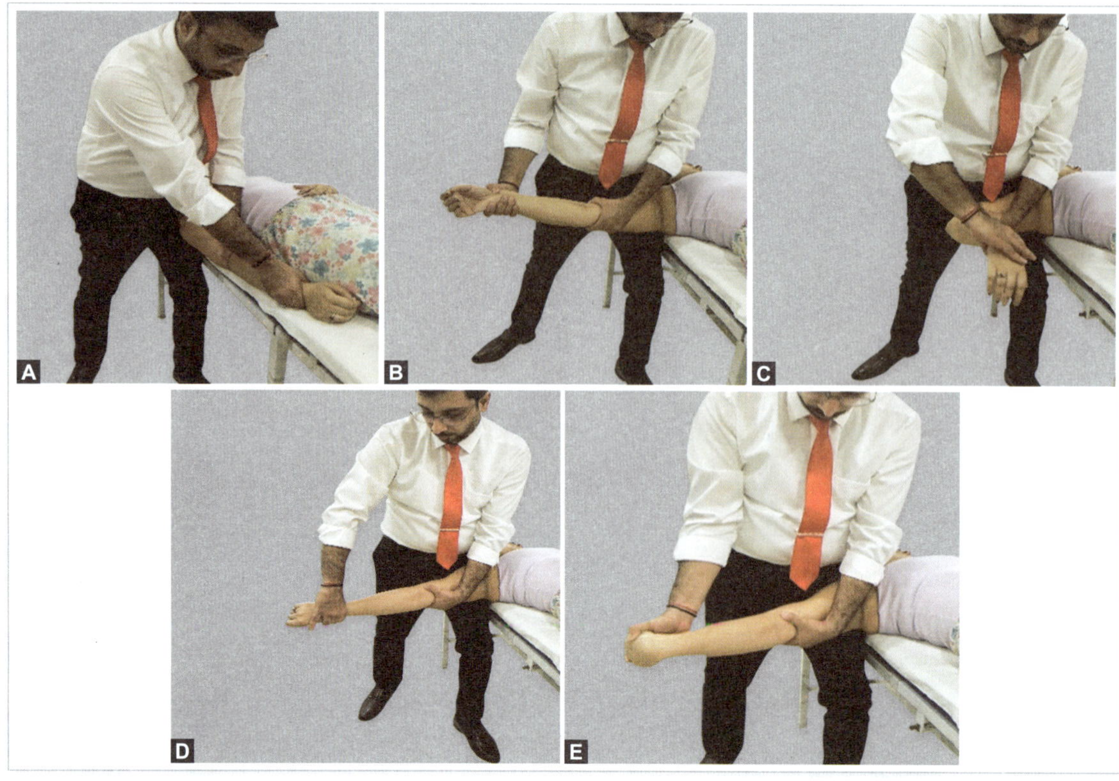

Figs 16.8A to E: Radial nerve test: **A.** Shoulder depression; **B.** Shoulder abduction; **C.** Shoulder internal rotation; **D.** Elbow extension and forearm pronation; **E.** Wrist and finger flexion

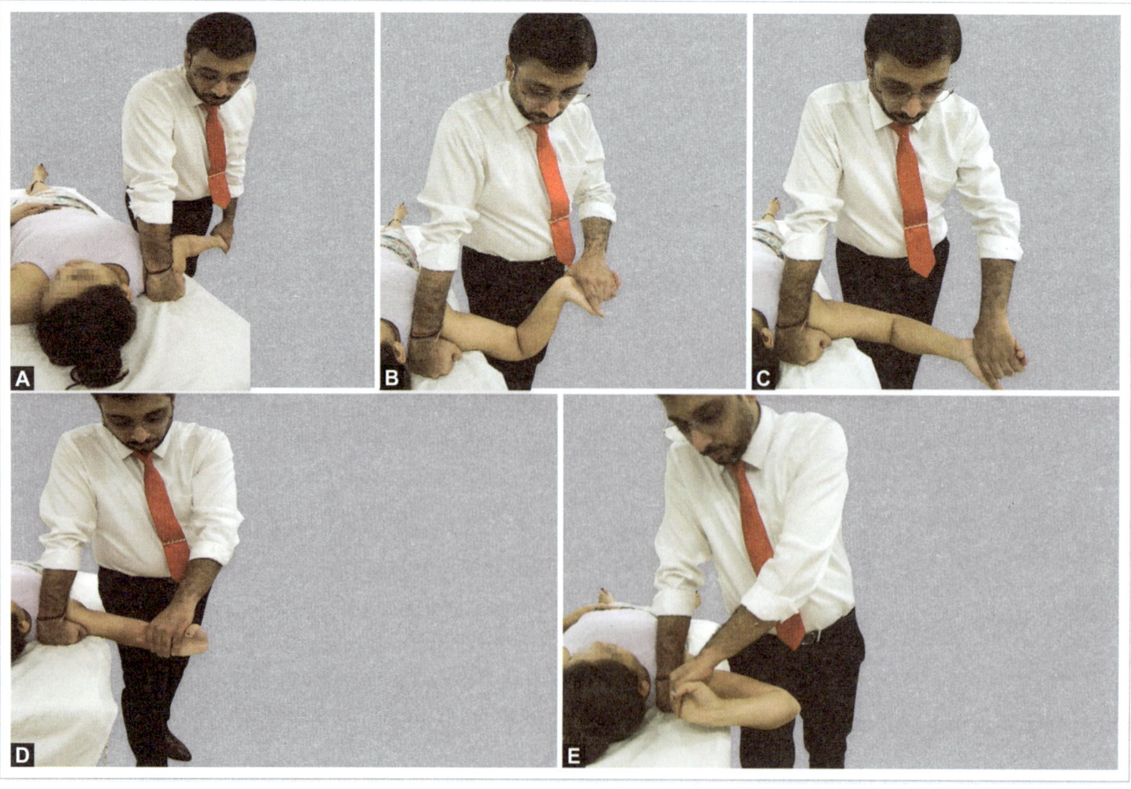

Figs 16.9A to E: Ulnar nerve test: **A.** Shoulder depression; **B.** Shoulder abduction; **C.** Shoulder external rotation; **D.** Forearm pronation and elbow flexion; **E.** Wrist and finger extension

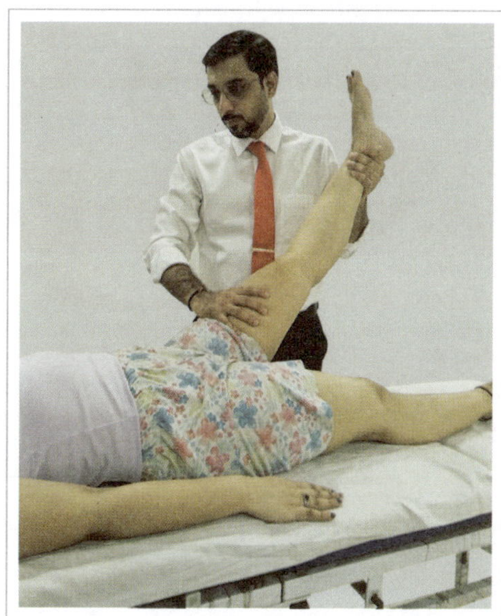

Fig. 16.10: Sciatic nerve testing

plantarflexion and inversion and the test is completed by adding passive hip flexion movement (Figs 16.13A and B).

JOINT MOBILIZATION

Joint mobilization is a treatment technique which is most commonly and widely used by therapist to treat musculoskeletal dysfunction. According to APTA Guide to Physical Therapist Practice, it is defined as *"a manual therapy technique comprised of a continuum of skilled passive movements that are applied at varying speeds and amplitudes, including a small amplitude/ high velocity therapeutic movement"*. International Federation of Orthopedic Manipulative Physical Therapists (IFOMPT) defined it as *"a manual therapy technique comprising a continuum of skilled passive movements that are applied at varying speeds and amplitudes to joints, muscles or nerves with the intent to restore optimal motion, function, and/or to reduce pain"*. Initially the term mobilization and manipulation were

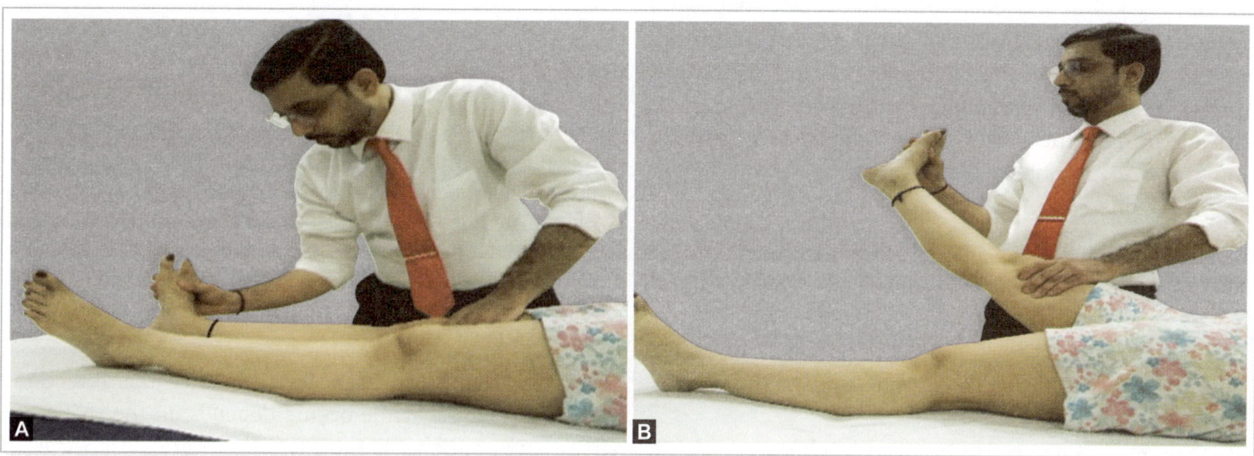

Figs 16.11A and B: Tibial nerve testing: **A.** Dorsiflexion and eversion; **B.** Hip flexion + dorsiflexion and eversion of foot

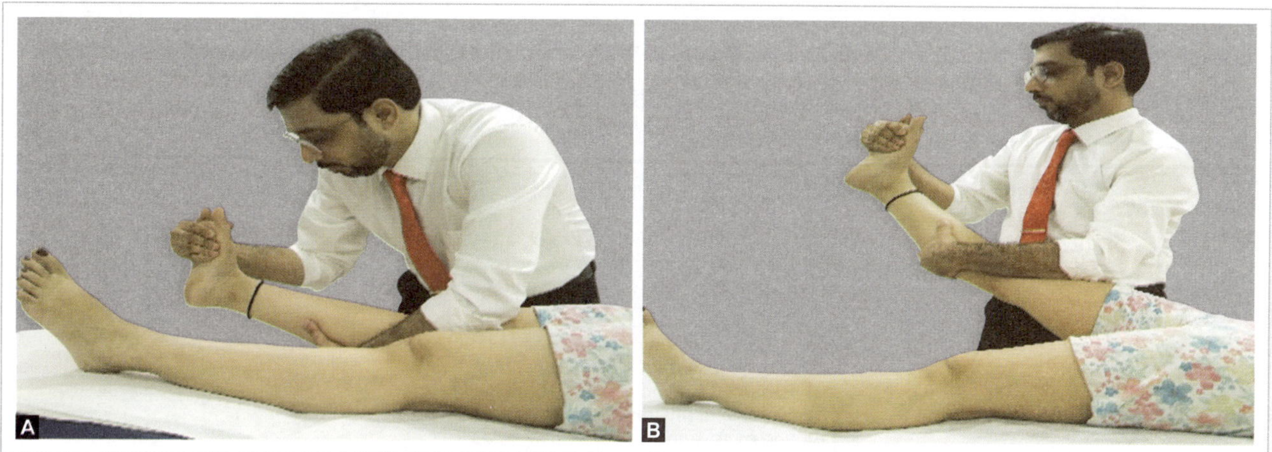

Figs 16.12A and B: Sural nerve testing: **A.** Dorsiflexion and inversion; **B.** Hip flexion + dorsiflexion and inversion of foot

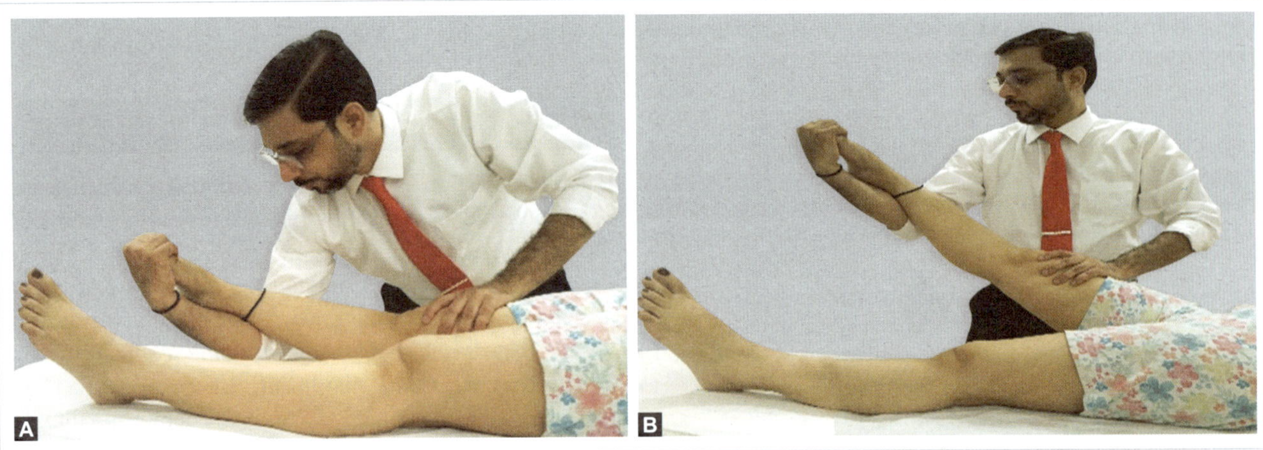

Figs 16.13A and B: Common peroneal nerve testing: A. Plantarflexion and inversion; B. Hip flexion + plantarflexion and inversion

used separately by lots of practitioners but in 2004 the Guide to Physical Therapist Practice released by APTA included both the terms under the same heading and are now used interchangeably. But still many people advocate that using these terms interchangeably might create a lot of confusion. A practitioner must have thorough knowledge of anatomical and biomechanical assessment and the pathological knowledge of the patient's condition to be able to clinically reason the use of mobilization techniques. Even though being one of the safest techniques of treatment, incorrect application can lead to potential damage to the joint and surrounding structures.

Basic Concepts

Osteokinematics

It can be defined as movement between two bones at a joint. For example, flexion/extension, abduction/adduction.

Arthrokinematics

It is defined as the motion that occurs between two joint surfaces. It is classified into three types: (1) Rolling, (2) Spinning and (3) Sliding (Fig. 16.14).

Did You Know?

There are only few joints where isolated movement of roll, spin or sliding takes place. Since majority of the joints of human body are incongruent, majority of joint movements are a combination of any of the two arthrokinematics motion which is quite important for joint mechanics.

Concavo-Convex Rule

Described by Kaltenborn it states that the direction in which the slide or glide occurs depends on the shape of the articulating surface. If the moving surface is concave, then the direction of glide is in the same direction of movement.

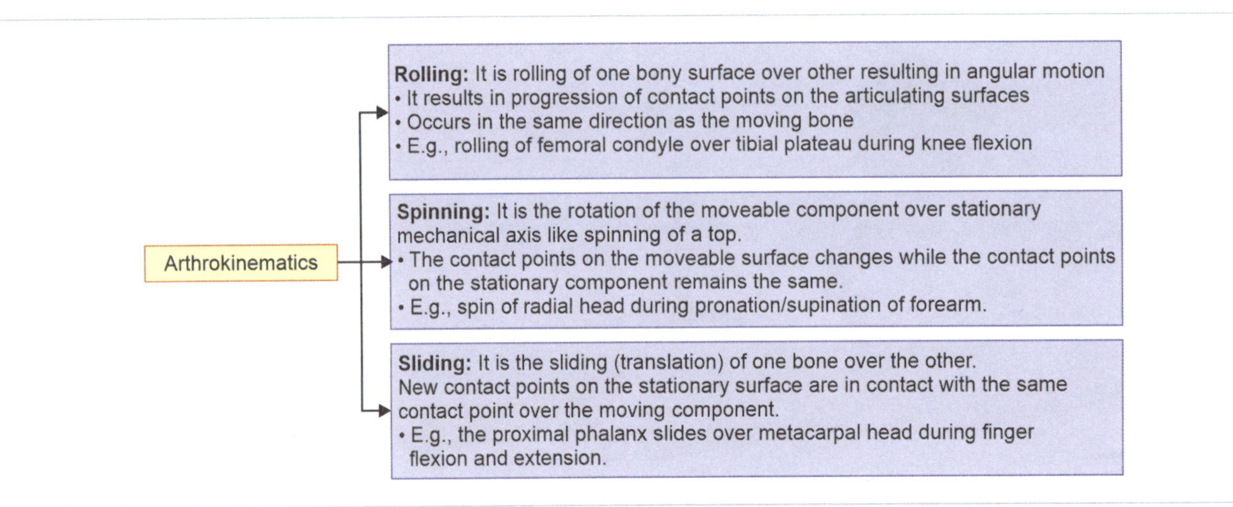

Fig. 16.14: Types of arthrokinematics

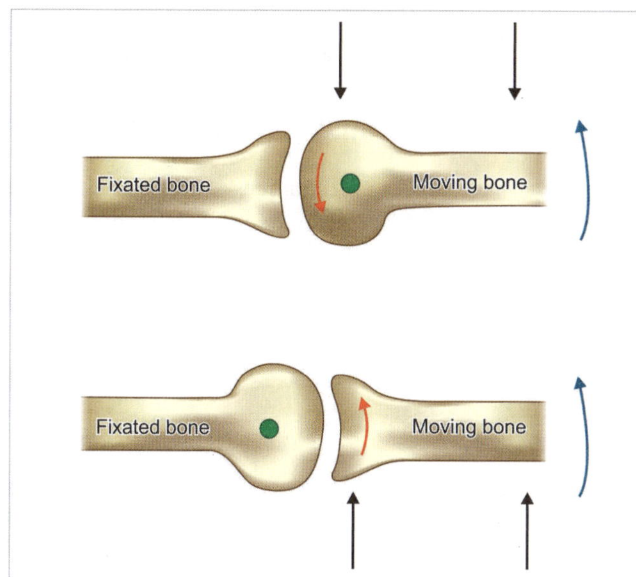

Fig. 16.15: Concavo-convex rule

If the moving surface is convex, then the glide takes place in opposite direction (Fig. 16.15).

Joint Play

Full and pain-free joint movements cannot be achieved unless certain precise, well-defined, small movements of joint play are present. These movements are independent of the action of voluntary muscles. It is the combination of movements of joint play and movements in the active range that complete movements.

Joint Shapes

Joint shapes may be sellar or ovoid type (Fig. 16.16)

Traction/Distraction

- **Traction:** It is applied at long axis of the bone resulting in glide at the joint.

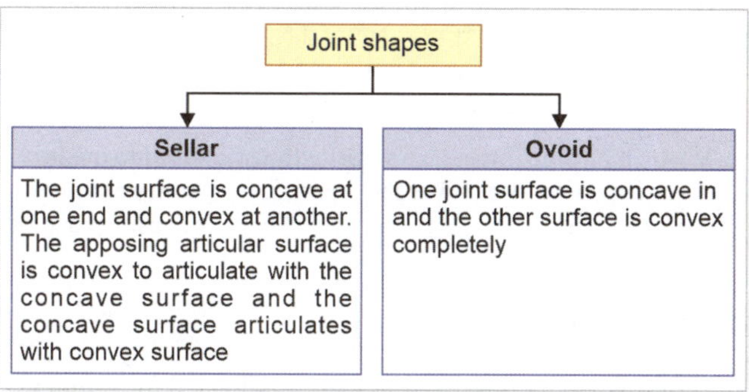

Fig. 16.16: Shapes of joints

- **Distraction:** It is a force which is applied at right angle to the articular surface resulting in the separation of joint surfaces.

Classification According to Kaltenborn

The technique uses a passive sustained stretch to improve joint mobility without compressing the articular surfaces (Fig. 16.17). The applied forces are categorized into grades I–III to progressively enhance joint movement which are shown in Table 16.4.

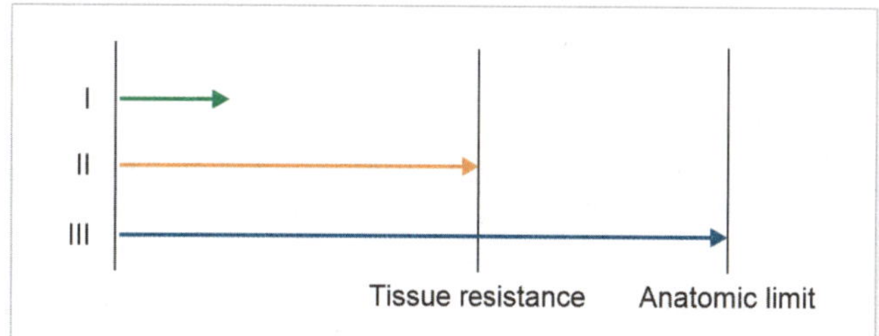

Fig. 16.17: Kaltenborn grading of joint mobilization

TABLE 16.4: Kaltenborn grading of joint mobilization

Grade	Application	Indications
Grade I (loosen)	Small-amplitude distraction is applied when no stress is placed on the capsule	Relieves pain with vibratory and oscillatory movements.
Grade II (tightening)	Takes up the slack in the tissues surrounding the joint and then tightens the tissues	Tests joint play traction and glide movements, relieves pain, increases or maintains movement
Grade III (stretching)	Movement is applied after the slack has been taken up and all tissues become taut	Tests joint play end-feel, increases mobility and joint play by stretching shortened tissues.

Effects

The effects of joint mobilization are given in Figure 16.18.

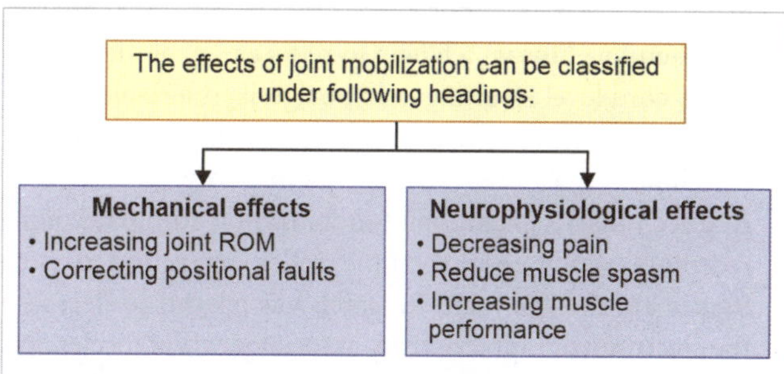

Fig. 16.18: Effects of joint mobilization

Mechanical

- **Increased joint ROM:** Due to pain or immobilization, adhesions develop in the articular or periarticular structures because of formation of new cross linkages in the collagen structure. This changes the biomechanics of the joint movement and decreases its functional ability. The synovial capsule also develops stiffness affecting the nutrition of the joint. There is loss of water content which leads to stiffness of articular cartilage, collagen structures and joint capsule. As a result, the new collagen fibers formed are less elastic and further develop adhesion with the existing collage fibers. All these factors contribute to change in arthokinematics of the joint resulting in less mobility.
  - Joint mobilization helps to restore joint function and range of motion. When applied with adequate force, it leads to reorganization of collagen fibers and helps to break adhesions.
  - It helps to improve joint arthrokinematics resulting in increased joint ROM. The contracted soft tissues are stretched helping in circulation of synovial fluid ultimately reducing the capsular stiffness and improving joint nutrition. As a result, the new collagen tissue formed is more elastic and aligned in the direction of force of joint.
- **Correction of positional faults:** Positional faults are considered minimal displacement or change in position of joint surfaces in relation to one another. Joint mobilization helps to realign and correct the joint surfaces. This can be achieved by mobilizing the joint in the direction of restriction of movement.

Neurophysiological

- **Decrease pain:** Joint mobilization may produce neurophysiological effects by stimulating the descending inhibitory pathways. The periaqueductal gray (PAG) area present in midbrain is mainly responsible for pain analgesia. When mobilizing force is applied to a joint, it results in firing of articular mechanoreceptors and proprioceptors. The stimulus through these receptors stimulate the areas of PAG to release endogenous opioids such enkephalins and β-endorphins which help to reduce pain at the level of spinal cord level resulting in hypoalgesia. It also affects the activity of alpha moto neurons decreasing pain pressure threshold.
- **Reduce muscle spasm:** Joint mobilization stimulates joint receptors which results in the reflex relaxation of soft tissues around the joint. Research has proven to decrease the electromyographic activity associated with muscles and joint impairment.
- **Increase muscle performance:** Joint effusion is one of the most common causes of decreased muscle performance. Joint mobilization helps to reduce joint effusion thereby improving joint mechanics and muscle performance.

Classification According to Maitland

A passive oscillatory technique, categorized into Grades I–IV based on intensity, is used to address pain and stiffness. (Fig. 16.19). These also, are called nonthrust oscillatory mobilization. According to Maitland, there are five grades of mobilization (Table 16.5):

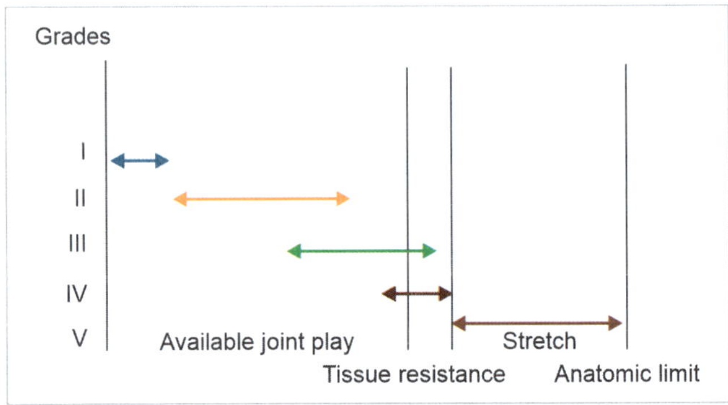

Fig. 16.19: Maitland grading of joint mobilization

TABLE 16.5: Maitland's grading of joint mobilization

Grade	Description	Use
I	Small-amplitude rhythmic oscillations applied at the beginning of the range	Used to decrease pain
II	Large-amplitude rhythmic oscillations applied within the range, not reaching the limit. Usually performed at 2–3 per second for 1–2 minutes.	
III	Large-amplitude rhythmic oscillations performed up to the limit of the available motion and are stressed into the tissue resistance. Usually performed at 2–3 per second for 1–2 minutes.	Used to increase joint ROM
IV	Small-amplitude rhythmic oscillations are performed at the limit of the available motion and stressed into the tissue resistance zone	
V	Small amplitude, high velocity thrust applied at the end of the range	Used to break adhesions or reposition a joint

Indications

- Improves a loss of accessory or physiological movement.
- Enhances motor function through reduction of pain and restoring articular relationships.

- Improves nutrition to intra-articular structures reducing capsular stiffness.
- Reduces muscle guarding.
- Reduces progressive loss of mobility associated with disease or injury.
- Encourages early mobility following injury.

Contraindications

Contraindications can be divided into absolute and relative as follows:

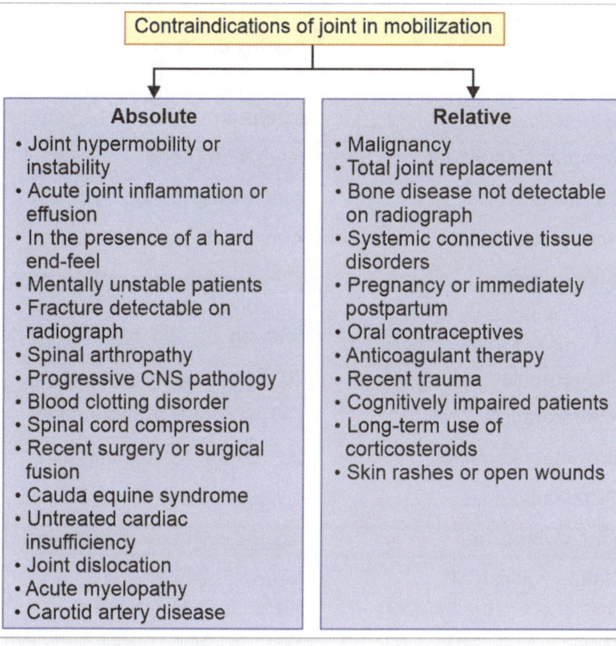

Patient Assessment

The evaluation of patient is done for:

- Joint excursion
- Pain
- End feel
- Capsular pattern

Joint Excursion

- It helps to identify mobility of the joint. It is assessed by applying a glide or a traction up to the level of tissue resistance.
- The therapist then applies the same glide on the unaffected side to comparison.
- The therapist can then categorize joint into one of the three categories, i.e., *hypomobile, normal or hypermobile.*

Pain

- Pain while applying accessory motion is an important part of evaluation as well as treatment.
- It helps to identify the recovery stage or progress the dosage of joint mobilization.
- Following clinical scenarios can be observed while evaluating pain:
  - Normal mobility without pain.
  - Normal mobility with pain.
  - Hypomobility without pain.
  - Hypomobility with pain.
  - Hypermobility.

End Feel

- It is defined as the sensation or feeling which the therapist detects when the joint is at the end of its available passive range of motion.
- It is determined while evaluating the accessory motion.
- They are graded into normal or abnormal end feel (Table 16.6).

TABLE 16.6: Normal or abnormal end feel

Type	End feel	Examples
Normal end feel		
Bone to bone	Hard feeling that is painless	Elbow extension
Soft tissue approximation	Painless restriction felt when tissue meets tissue	Elbow flexion Knee flexion
Tissue stretch	• Soft: Soft end feel without definite stopping point • Hard (capsular): Hard end feel with definite stopping point	• Wrist flexion • Knee extension
Abnormal end feel according to Cyriax		
Muscle spasm end feel	Painful, sudden restriction of movement	Upper trapezius muscle spasm
Bony end feel	Similar to bone to bone but the restriction comes before completion of normal range	Due to osteophyte formation
Empty end feel	The movement is stopped because of severe pain although no resistance is felt	Acute joint inflammation
Springy end feel	A rebound effect is felt before reaching end of ROM	Acute meniscus injury

Contd...

Type	End feel	Examples
Leathery end feel (Capsular stretch end feel)	The resistance starts at early ROM and increases as the range is increased. The movement stops abruptly	Synovitis, soft tissue edema
Soft end feel	Mushy with soft end feel	Mostly in acute conditions

Capsular Pattern

According to Cyriax, certain pathological conditions can affect the joint capsule and cause restriction of movement to passive movements in a pattern which are specific to a joint (Table 16.7).

TABLE 16.7: Common capsular patterns according to Cyriax

Joint	Capsular patterns
Temporomandibular	Mouth opening
Occipitoatlanto	Extension and side flexion equally limited
Cervical spine	Side flexion and rotations equally limited, extension
Glenohumeral	Lateral rotation, abduction, medial rotation
Sternoclavicular	Pain at extreme range of movement
Acromioclavicular	Pain at extreme range of movement
Humeroulnar	Flexion, extension
Radiohumeral	Flexion, extension, supination, pronation
Proximal radioulnar	Supination, pronation
Distal radioulnar	Pain at extremes of rotation
Wrist	Flexion and extension equally limited
Trapeziometacarpal	Abduction, extension
MCP and IP	Flexion, extension
Thoracic spine	Side flexion and rotation equally limited, extension
Lumbar spine	Side flexion and rotation equally limited, extension
SI, symphysis pubis, and sacrococcygeal	Pain when joints stressed
Hip	Flexion, Abduction, Medial rotation (order varies)
Knee	Flexion, extension
Tibiofibular	Pain when joint stressed
Talocrural	Plantar flexion, dorsiflexion
Subtalar (Talocalcaneal)	Limitation of varus range of movement
Midtarsal	Dorsiflexion, plantar flexion, adduction, medial rotation
First MTP	Extension, flexion
Second to fifth MTP	Variable
IP	Flexion, extension

Guidelines

- **Joint position:** Joint mobilization is usually done with joint in a resting position. In this position, the forces acting on a joint are minimal and the mobilization techniques are well tolerated. In this position, maximum joint play is available and the joint capsule and ligaments are most relaxed. Therefore, it is very easy to assess and treat the joints in this position. The mid range position is the most relaxed position for most of the joints (Table 16.8).

TABLE 16.8: Resting position of different joints of the body

Joint	Resting position
Vertebral	Mid range between flexion and extension
Temporomandibular	Jaw slightly open
Sternoclavicular	Arm resting by side
Acromioclavicular	Arm resting by side
Glenohumeral	55°–70° abduction; 30° horizontal adduction, neutral rotation
Humeroulnar	70° flexion; 10° supination
Humeroradial	Full extension and supination
Proximal radioulnar	70° flexion; 35° supination
Distal radioulnar	10° supination
Radio/ulnocarpal	Neutral, slight ulnar deviation
Hand – midcarpal	Neutral, slight flexion, slight ulnar deviation
Hand – carpometacarpal	Midway between flexion and extension
Hand – trapeziometacarpal	Midway between flexion and extension Midway between abduction and adduction
Hand – metacarpophalangeal	1st MCP: Slight flexion 2nd - 5th MCP: Slight flexion; ulnar deviation
Hand – interphalangeal	PIP: 10° flexion DIP: 30° flexion
Hip	30° flexion; 30° abduction; slight external rotation
Knee – tibiofemoral	25° flexion
Talocrural	Mid inversion/eversion; 10° plantar flexion
Subtalar	Midway; 10° plantar flexion
Midtarsal	Midway; 10° plantar flexion
Tarsometatarsal	Midway between supination/ pronation
Toes – metatarsophalangeal	Neutral
Toes – interphalangeal	Slight flexion

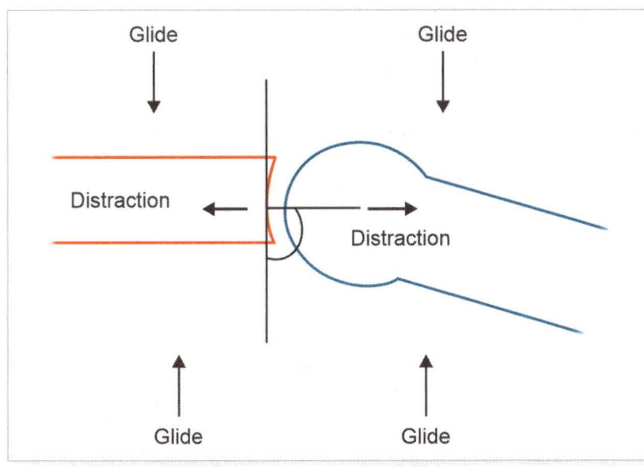

Fig. 16.20: Direction and force of movement of joint

- **Grades or dosage of movement:** Refer to Maitland's or Kaltenborn's grades of joint mobilization. The dosage of joint mobilization is usually 60–120 Hz/min, performed for 30 seconds to 1 minute, for 1–5 sets.
- **Direction and force of movement:** The direction of movement of glide is either parallel or perpendicular to the treatment plane. Treatment plane was defined by Kaltenborn as a plane perpendicular to a line running from the axis of rotation to the middle of the concave articular surface. Distraction is always applied in perpendicular direction to the treatment plane whereas glides are always applied in a direction parallel to the joint surface (Fig. 16.20).

Precautions

- The patient should be as relaxed as possible.
- When testing the joint, it should be in the resting position.
- The therapist should stabilize the proximal bone with their hands, a mobilization belt, a cloth or pillows.
- The mobilizing force over the distal bone can be applied by the therapist using the body weight wherever possible.
- The hand placement of the therapist should be as close to the joint surface.
- The patient's breathing should be timed with the procedure. During the mobilization/manipulation process, the patient exhales.
- While using the mobilization approach, the therapist is to keep an eye on the patient's level of pain. The therapist should reevaluate the patient, adjust the treatment method, or alter the treatment plan if the pain worsens.
- The therapist should follow the rule of one joint one movement at a time.
- The therapist should evaluate the patient before and after applying the technique. If there is positive response after the application of technique, then the patient should be evaluated again and if required the treatment plan can be modified.

JOINT MOBILIZATION TECHNIQUES

SHOULDER COMPLEX

The shoulder complex consists of the following articulations (Fig. 16.21):

1. Glenohumeral joint
2. Sternoclavicular joint
3. Scapulothoracic joint
4. Acromioclavicular joint

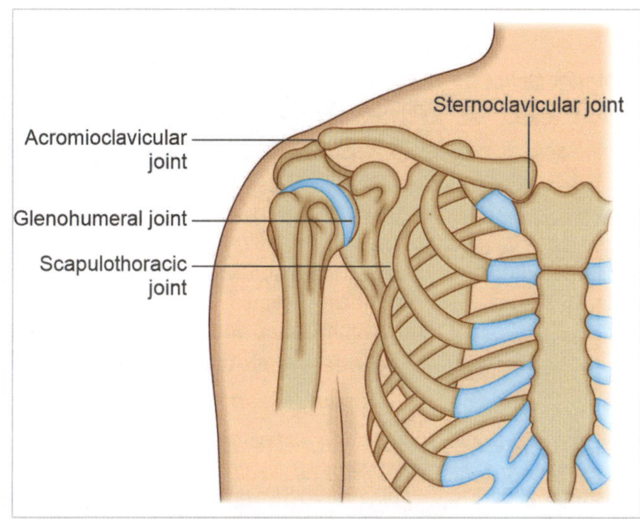

Fig. 16.21: Shoulder complex

Glenohumeral Joint

The mobilization techniques for glenohumeral joint are applied in following situations:

Technique	Indications
Distraction	To examine joint impairment, decrease pain and improve periarticular stiffness
Inferior glide	To improve shoulder abduction ROM
Anterior glide	To improve shoulder extension, external rotation and horizontal abduction ROM
Posterior glide	To improve shoulder flexion, internal rotation and horizontal adduction ROM

Mobilization Techniques of Glenohumeral (GH) Joint

The mobilization techniques applied for glenohumeral joint are as following:

Techniques	Patient's position	Therapist's position	Hand placement	Mobilizing force
Distraction (Fig. 16.22A)	Supine lying, stabilize the patient's scapula using mobilization belt	Places the patient's hand and forearm between the clinician's upper arm and trunk while standing with the affected side toward the GH joint.	Proximal humerus in close proximity to the axilla. The humerus is grasped by both hands from both the medial and lateral sides.	With the hand placed on medial side, the humerus is moved laterally in lateral, inferior or anterior direction perpendicular to the glenoid fossa.
Caudal glide/ Inferior glide (Fig. 16.22B)	Supine lying, stabilize the patient's scapula using mobilization belt	Stands on the affected side facing the GH joint, place the patient's hand and forearm between the clinician's upper arm and trunk to provide support.	The web space is positioned proximally over the humerus by the mobilizing hand. The stabilizing hand is positioned over the medial arm, directly below the axilla.	The mobilizing hand applies a caudal or inferior force. Note: Before applying inferior glide, a distraction force can be provided, allowing the shoulder to be abducted to varied angles.
Anterior glide (Fig. 16.22C)	Prone lying, stabilize the patient's scapula using mobilization belt	A towel is placed below the acromion and the arm of the patient is supported on therapist's thigh.	The mobilizing hand is placed just below the acromion over posterior humerus and stabilizing hand is placed over mid humerus.	The mobilizing hand applies force in anterior and medial direction.
Posterior glide 1 (Fig. 16.22D)	Supine lying, stabilize the patient's scapula using mobilization belt	Stand between patients arm and trunk with back facing the patient. The patient's arm is supported by therapist's trunk.	The stabilizing hand is placed on the distal humerus, directly above the epicondyles, and the mobilizing hand is placed on the anterior surface of the proximal humerus.	The mobilizing hand applies force is posterior direction.
Posterior glide 2 (Fig. 16.22E)	Supine, with the arm flexed to 90° and internally rotated and with the elbow flexed	Stands on the affected side facing the GH joint	The patient's elbow is used to mobilize the hand, and the anterior surface of the proximal humerus is used to stabilize the hand.	A mobilizing force is applied in posterior direction.

Sternoclavicular Joint

The mobilization techniques for sternoclavicular joint are applied in following situations:

Techniques	Indications
Superior glide	To increase depression of scapula
Inferior glide	To increase elevation of scapula
Anterior glide	To increase protraction of scapula
Posterior glide	To increase retraction of scapula

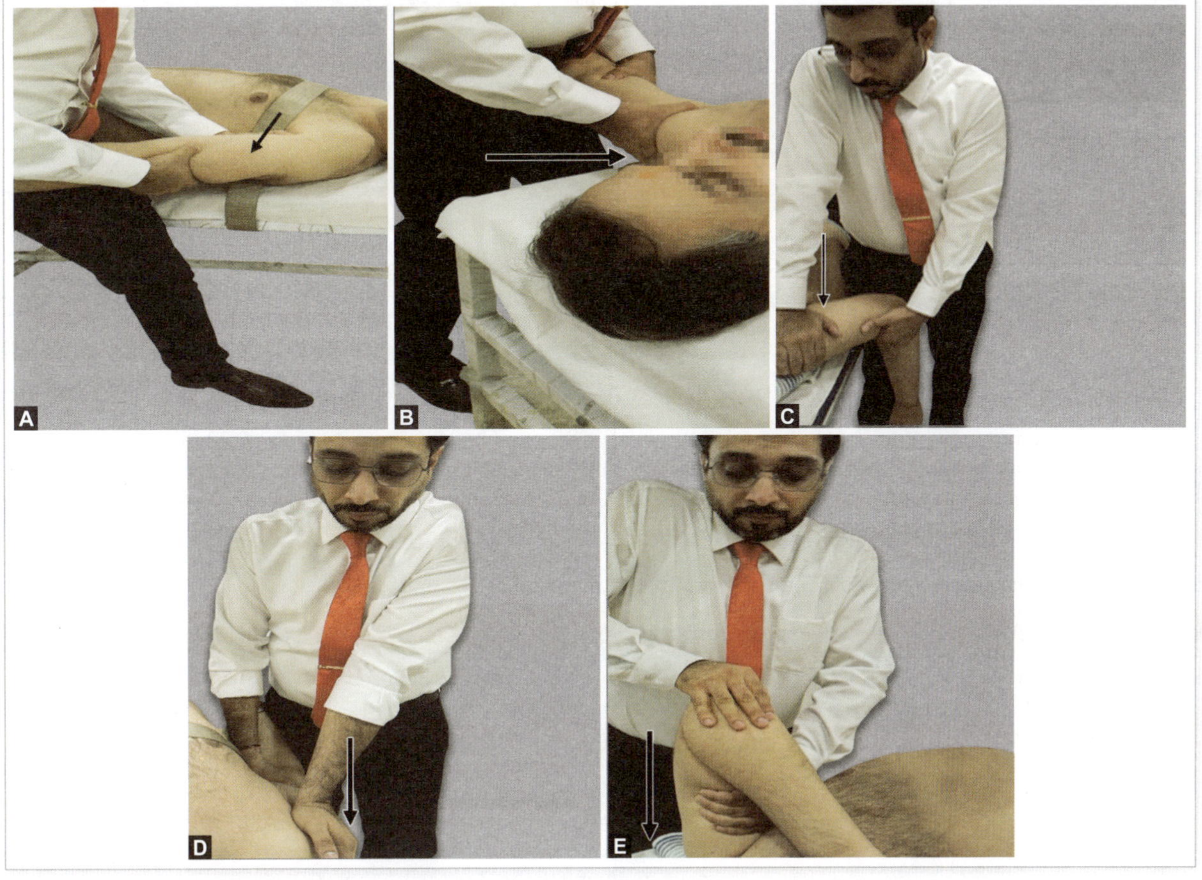

Figs 16.22A to E: Glenohumeral joint mobilization techniques: **A.** Distraction; **B.** Caudal glide/Inferior glide; **C.** Anterior glide; **D.** Posterior glide 1; **E.** Posterior glide 2

Mobilization Techniques of Sternoclavicular Joint

The mobilization techniques of sternoclavicular joint are as follows:

Techniques	Patient's position	Therapist's position	Hand placement	Mobilizing force
Anterior glide (Fig. 16.23A)	Supine lying with shoulder in neutral position	Stands near the head of the patient	The stabilizing hand grasps the clavicle while the mobilizing had is placed over the sternum.	The therapist applies a posterior glide over the sternum facilitating the anterior movement of clavicle
Posterior glide (Fig. 16.23B)	Supine lying with shoulder in neutral position	Stands beside facing the patient	The thumb of stabilizing hand is placed over medial end of the clavicle and the thumb of the mobilizing hand reinforces the thumb of the stabilizing hand	A glide in posterior direction is applied by the thumb of mobilizing hand
Superior glide (Fig. 16.23C)	Supine lying with shoulder in neutral position	Stands beside facing the patient	The stabilizing hand is positioned with the thumb over the inferior surface of the medial clavicle and the thumb of the mobilizing hand reinforces the thumb of the stabilizing hand	The mobilizing hand applies force in superior direction.
Inferior glide (Fig. 16.23D)	Supine lying with shoulder in neutral position	Stands beside facing the patient	The stabilizing hand is positioned with the thumb over the superior surface of the medial clavicle and the thumb of the mobilizing hand reinforces the thumb of the stabilizing hand	The mobilizing hand applies force in inferior direction.

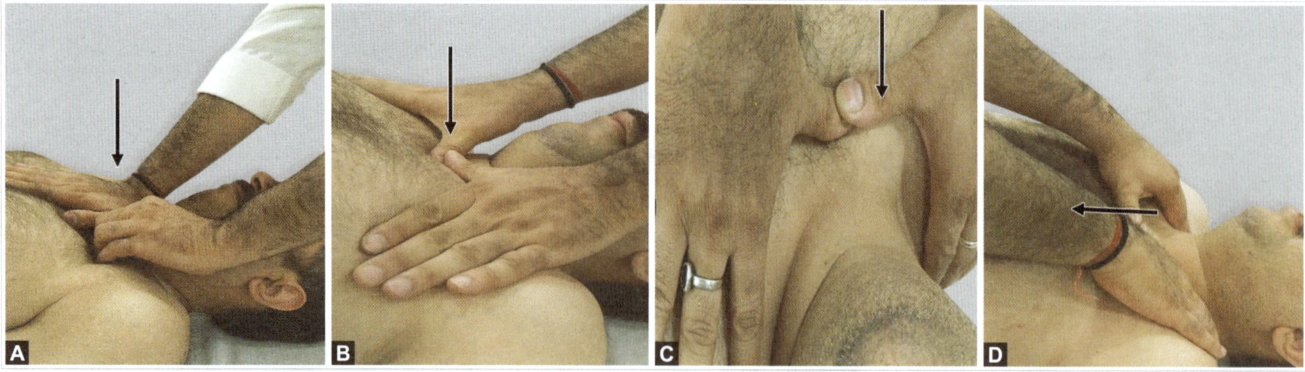

Figs 16.23A to D: Sternoclavicular joint mobilization techniques: **A.** Anterior glide; **B.** Posterior glide; **C.** Superior glide; **D.** Inferior glide

Acromioclavicular Joint

Mobilization Techniques of Acromioclavicular Joint

The mobilization techniques of acromioclavicular joint are as follows:

Techniques	Patient's position	Therapist's position	Hand placement	Mobilizing force
Anterior glide (Fig. 16.24A)	Sitting with shoulder in neutral position	Stands behind the patient	The stabilizing hand is placed over anterior acromion and over the anterior surface of the proximal humerus with the thumb placed over posterior clavicle. The thumb of the mobilizing hand reinforces the thumb of the stabilizing hand placed over the clavicle.	Anterior force is applied by the thumb of mobilizing hand.
Posterior glide (Fig. 16.24B)	Sitting with shoulder in neutral position	Stands in front of the patient facing him	The stabilizing hand is positioned over the posterior surface of the scapula. The mobilizing hand is positioned with the thumb over the thumb of the stabilizing hand.	Posterior force is applied by the thumb of mobilizing hand.

Scapulothoracic Joint

The mobilization techniques for scapulothoracic joint are applied in following situation:

Techniques	Indications
Superior glide	To increase elevation and lateral rotation of scapula
Inferior glide	To increase depression and medial rotation of scapula
Lateral glide	To increase protraction, elevation and lateral rotation of scapula
Medial glide	To increase retraction, depression and medial rotation of scapula
Distraction	For stretching of scapulo-thoracic soft tissues

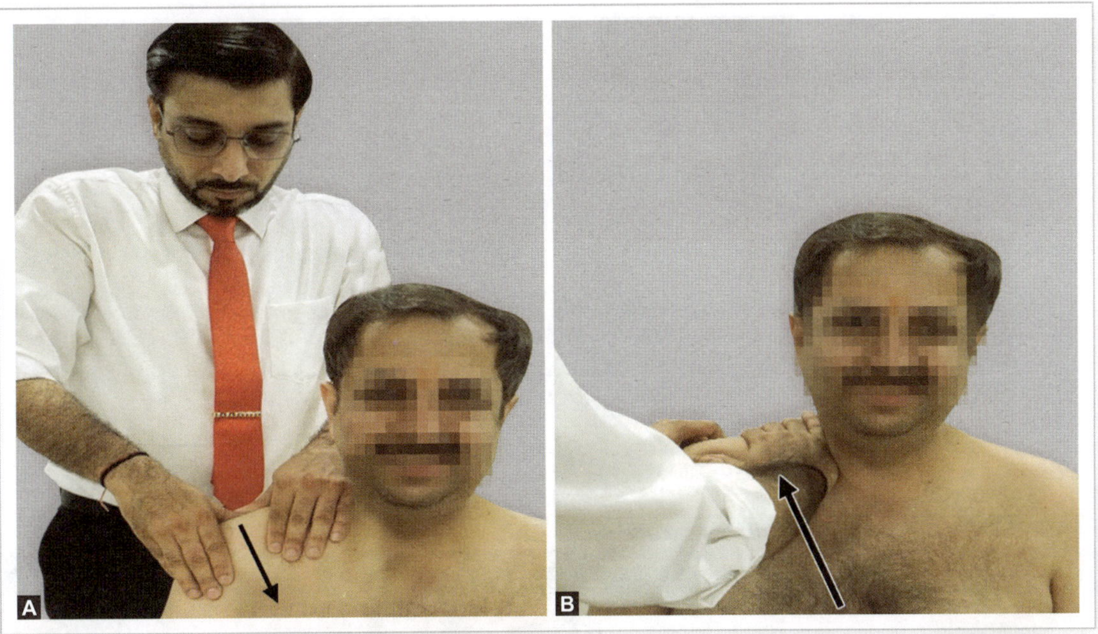

Figs 16.24A and B: Acromioclavicular joint mobilization techniques: **A.** Anterior glide; **B.** Posterior glide

Mobilization Techniques of Scapulothoracic Joint

The mobilization techniques of scapulothoracic joint are as follows:

Techniques	Patient's position	Therapist's position	Hand placement	Mobilizing force
Superior glide (Fig. 16.25A)	Side lying with arm in neutral position	In front of the patient facing the affected shoulder	The mobilizing hand is placed on the inferior angle of scapula and the stabilizing hand is placed over the acromion process of scapula	The mobilizing hand applies a force in superior direction
Inferior glide (Fig. 16.25B)	Side lying with arm in neutral position	In front of the patient facing the affected shoulder	The mobilizing hand is placed on the acromion and the stabilizing hand is placed over inferior angle of scapula	The mobilizing hand applies a force in inferior direction
Lateral glide (Fig. 16.25C)	Side lying with arm in neutral position	In front of the patient facing the affected shoulder	Both the hands are placed over the medial border of scapula	Both the hands apply force simultaneously to glide the scapula in lateral direction
Medial glide (Fig. 16.25D)	Side lying with arm in neutral position	In front of the patient facing the affected shoulder	One hand is placed over the acromion and the other is placed over the axillary border of scapula	Both the hands apply force simultaneously to glide the scapula in medial direction
Scapular distraction (Fig. 16.25E)	Side lying with affected side upwards and the arm in hand to back position	In front of the patient facing the affected shoulder	The mobilizing hand is placed on the acromion and the stabilizing hand is placed over inferior angle of scapula	The mobilizing hand applies a force in medial and inferior direction while the stabilizing hand lifts the scapula from the thorax

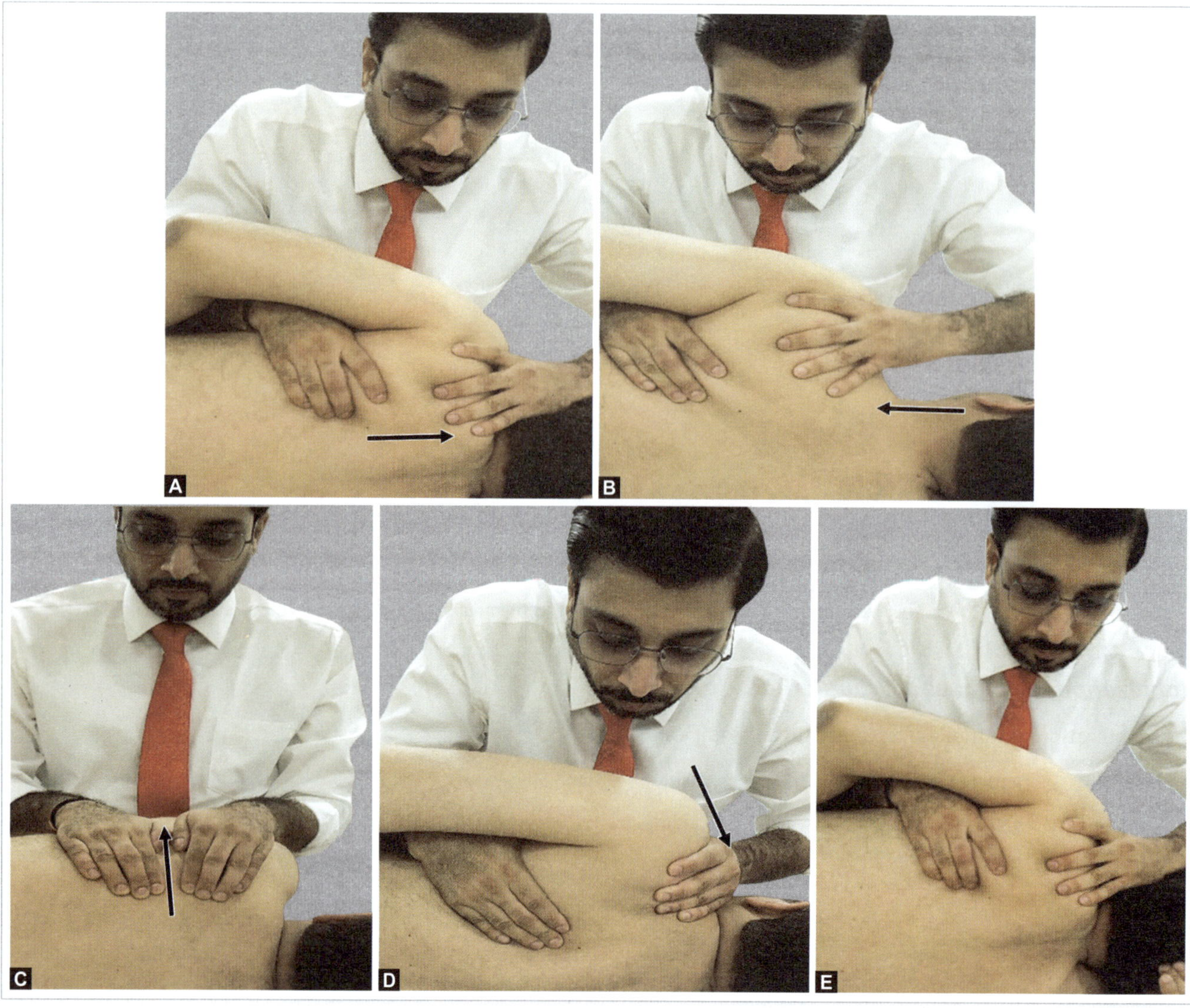

Figs 16.25A to E: Scapulothoracic joint mobilization techniques: **A.** Superior glide; **B.** Inferior glide; **C.** Lateral glide; **D.** Medial glide; **E.** Scapular distraction

ELBOW COMPLEX

The elbow complex consists of the following articulations (Fig. 16.26):

1. The humeroulnar joint
2. The humeroradial joint

Humeroulnar Joint

The mobilization techniques for humeroulnar joint are applied in following situation:

Techniques	Indications
Distraction	To improve ROM and decrease periarticular stiffness
Radial glide/lateral glide	To increase accessory varus movement helping in improving elbow flexion
Ulnar glide/medial glide	To increase accessory valgus movement helping in improving elbow extension
Distal glide	To increase flexion

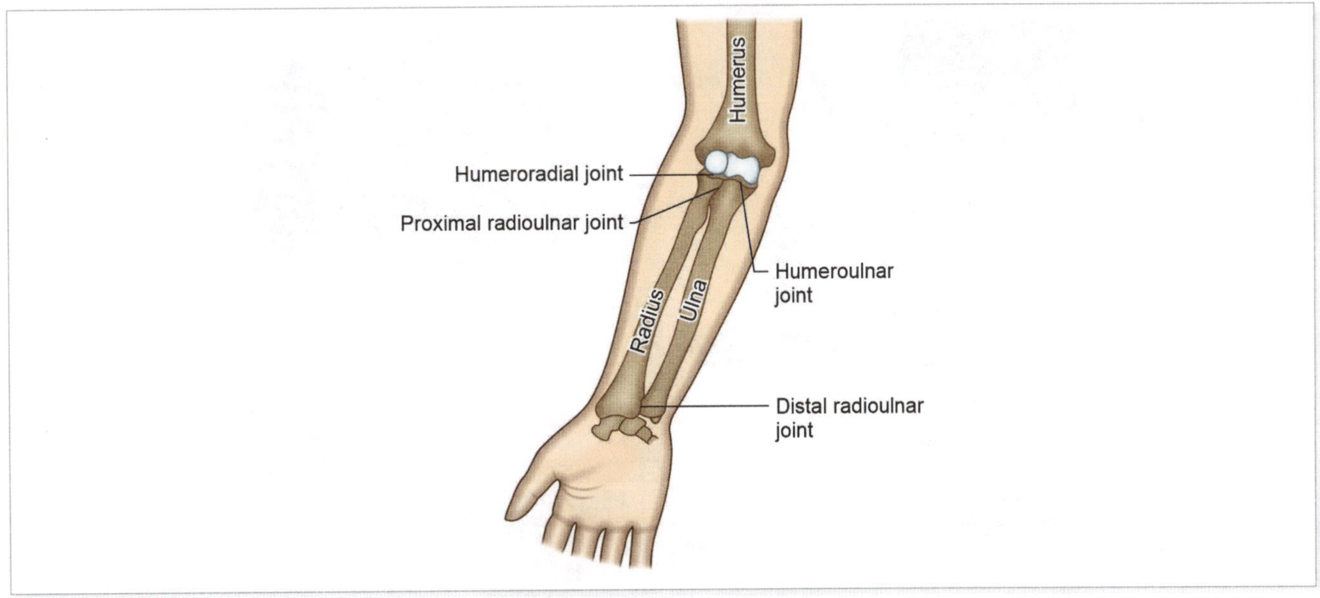

Fig. 16.26: Elbow complex

Mobilization Techniques for the Humeroulnar Joint

The mobilization techniques for the humeroulnar joint are as follows:

Techniques	Patient's position	Therapist's position	Hand placement	Mobilizing force
Distraction (Fig. 16.27A)	Supine lying with the patient's elbow off the couch. Stabilize it with a mobilization belt or place a towel below the elbow for support. Place the distal forearm and wrist over therapist's shoulder	Stands near the patient's hip facing the elbow joint.	The therapist places the medial hand over proximal ulna on the anterior aspect of elbow. The other hand is placed over the medial hand to reinforce it. Care must be take that there should be no contact with the radius	The therapists applies a distraction force at an angle of 45° to the shaft of ulna.
Radial glide/ lateral glide (Fig. 16.27B)	Supine lying	Stands between patient's arm and the trunk facing the patient. Patient's forearm is stabilized between therapist's arm and trunk.	The stabilizing hands holds the distal humerus from the lateral side and the mobilizing hand holds the proximal ulna from medial side.	The therapist applies a force in radial direction over the proximal ulna.
Ulnar glide/ medial glide (Fig. 16.27C)	Supine lying	Stands between patient's arm and the trunk facing the patient. Patient's forearm is stabilized between therapist's arm and trunk.	The stabilizing hands holds the distal humerus from the medial side and the mobilizing hand holds the proximal ulna from lateral side.	The therapist applies a force in medial direction over the proximal ulna.
Distal glide (Fig. 16.27D)	Supine lying with the patient's elbow off the couch. Stabilize it with a mobilization belt or place a towel below the elbow for support. Place the distal forearm and wrist over therapist's shoulder	Stands near the patient's hip, facing the elbow joint.	The therapist places the medial hand over proximal ulna on the anterior aspect of elbow. The other hand is placed over the medial hand to reinforce it. It must be ensured that there should be no contact with the radius	The therapists applies a distraction force in inferior direction and while maintaining it, then applies a force at an angle of 45° to the shaft of ulna.

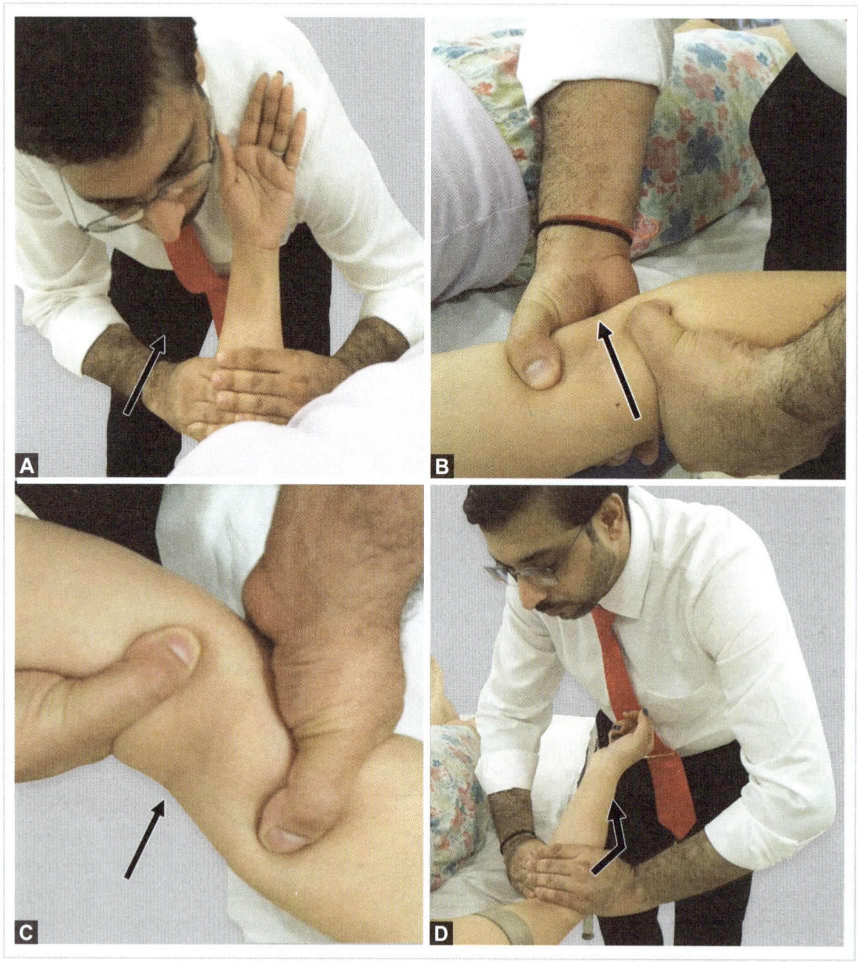

Figs 16.27A to D: Humeroulnar joint mobilization techniques: **A.** Distraction; **B.** Radial glide/lateral glide; **C.** Ulnar glide/medial glide; **D.** Distal glide

Humeroradial Joint

The mobilization of humeroradial joint is indicated in following situations:

Techniques	Indications
Distraction	To improve ROM and decrease periarticular stiffness
Dorsal glide	To improve elbow extension
Volar glide	To improve elbow flexion
Compression	To correct positional faults To correct pulled elbow subluxation.

Mobilization Techniques for the Humeroradial Joint

The mobilization techniques for the humeroradial joint are as follows:

Techniques	Patient's position	Therapist's position	Hand placement	Mobilizing force
Distraction (Fig. 16.28A)	Supine lying	Stands between patient's trunk and arm	The stabilizing hand is placed over distal humerus anteriorly. The mobilizing hand grips the distal radius using thenar eminence and fingers.	A distal mobilizing force along the long axis of the bone is applied.

Contd...

Techniques	Patient's position	Therapist's position	Hand placement	Mobilizing force
Dorsal glide (Fig. 16.28B)	Supine lying with elbow in neutral position	Stands near the patient' hip, facing the elbow joint	The stabilizing hand is placed over medial elbow. The palmar aspect mobilizing hand is placed over head of the radius anteriorly and the fingers cover the radial head posteriorly.	Using the palmar aspect of the mobilizing hand, a dorsally directed force is applied.
Volar glide (Fig. 16.28C)	Supine lying with elbow in neutral position	Stands near the patient' hip, facing the elbow joint	The stabilizing hand is placed over medial elbow. The palmar aspect mobilizing hand is placed over head of the radius anteriorly and the fingers cover the radial head posteriorly.	Using the fingers of the mobilizing hand, a dorsally volar directed force is applied.
Compression (Fig. 16.28D)	Supine lying, elbow flexed to 90°, forearm supinated	Stands beside the patient, facing the joint	The stabilizing hand is placed below the patient's elbow and the palmar aspect of the mobilizing hand grips the patient's hand.	A combination of mobilizing force is applied taking wrist into extension, downward force along shaft of radius while supinating the forearm simultaneously

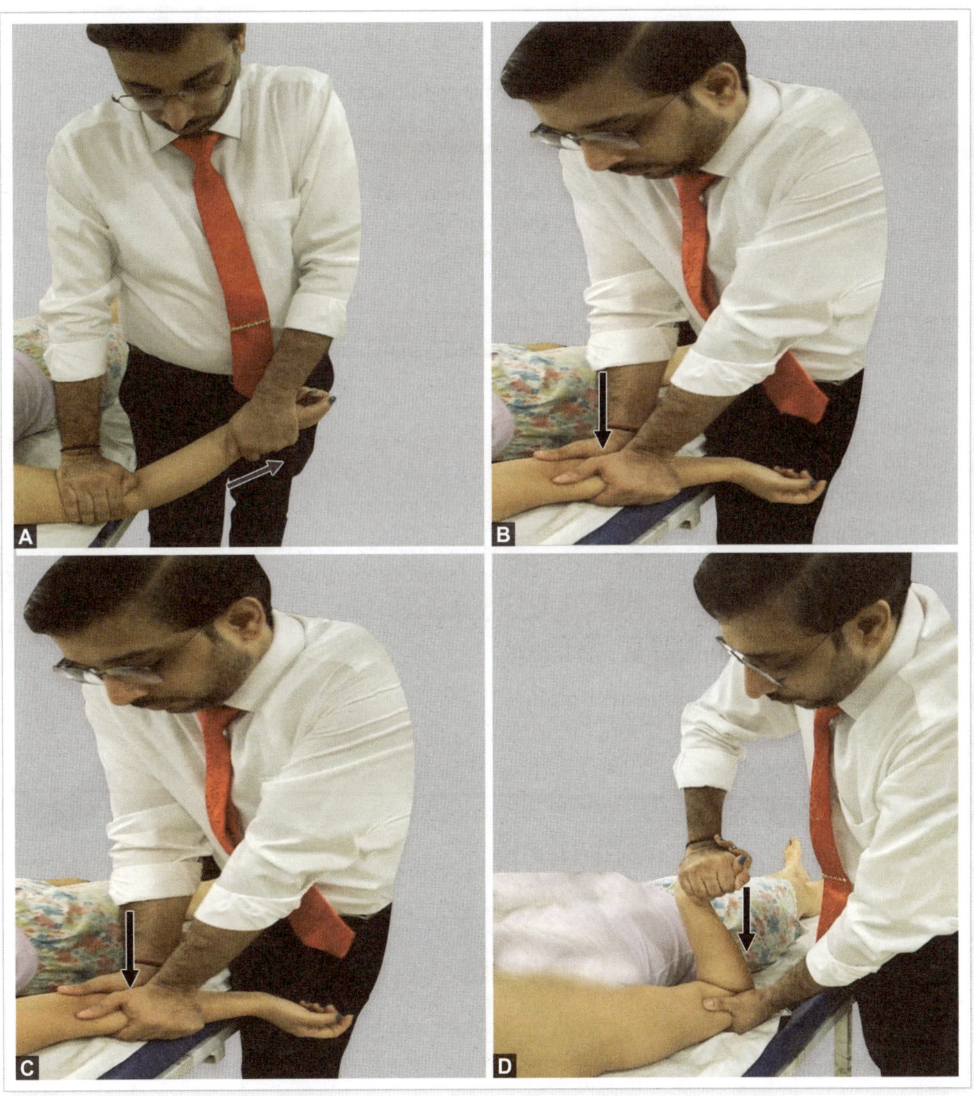

Figs 16.28A to D: Humeroradial joint mobilization techniques: **A.** Distraction; **B.** Dorsal glide; **C.** Volar glide; **D.** Compression

RADIOULNAR JOINTS

The types of radioulnar joint mobilization techniques can be classified as proximal and distal on broad basis (Fig. 16.29).

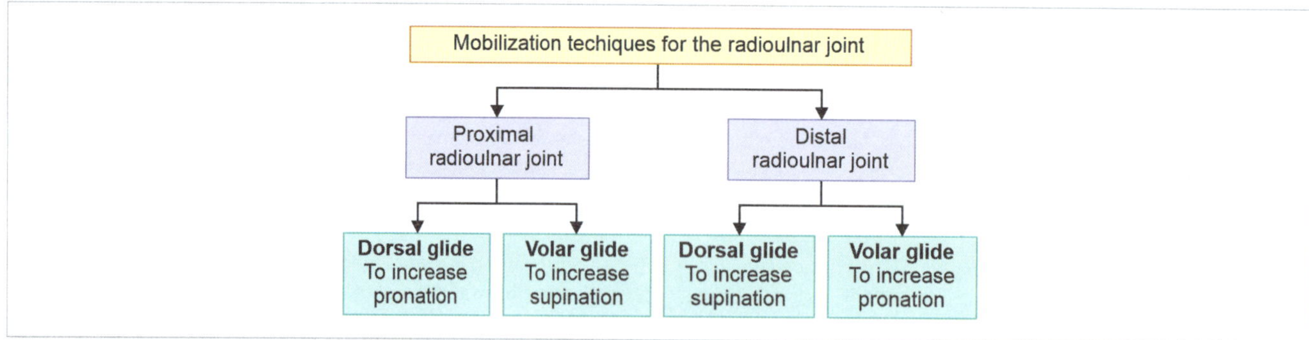

Fig. 16.29: Types of radioulnar joint mobilization

Proximal Radioulnar Joint

Mobilization Techniques for the Proximal Radioulnar Joint

The mobilization techniques for the proximal radioulnar joint are as follows:

Techniques	Patient's position	Therapist's position	Hand placement	Mobilizing force
Dorsal glide (Fig. 16.30A)	Supine lying or sitting, elbow is placed in 50° flexion and forearm in mid pronation	Stands beside the patient facing the proximal radioulnar joint	The stabilizing hand is placed over the proximal ulna and the mobilizing hand grasps the head of the radius.	A force is applied in the dorsal direction using the fingers
Volar glide (Fig. 16.30B)	Supine lying or sitting, elbow is placed in 50° flexion and forearm in mid pronation	Stands beside the patient facing the proximal radioulnar joint	The stabilizing hand is placed over the proximal ulna and the mobilizing hand grasps the head of the radius.	A force is applied in the volar direction using the palmar surface of the mobilizing hand

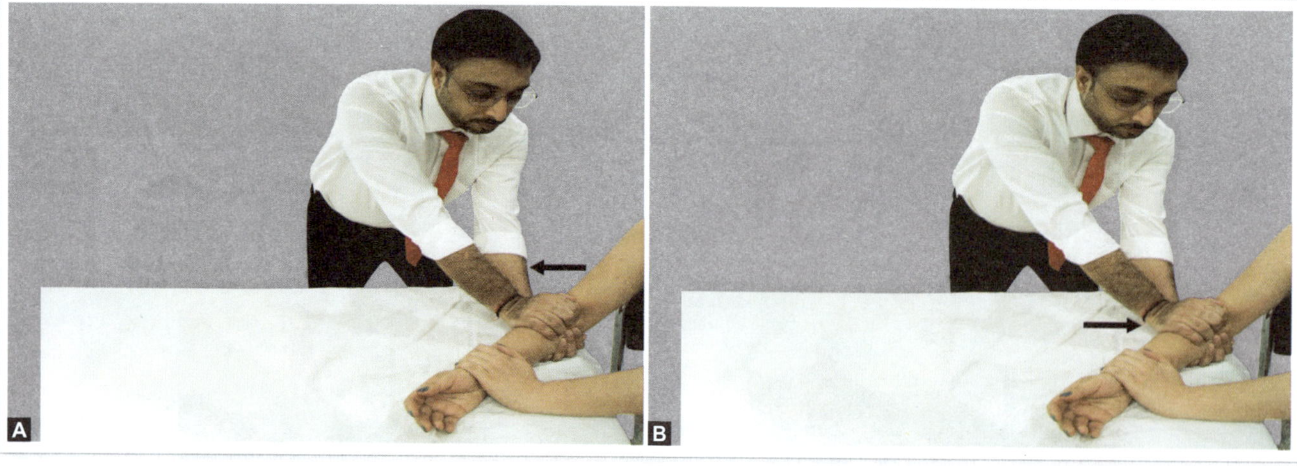

Figs 16.30A and B: Proximal radioulnar joint mobilization techniques: **A.** Dorsal glide; **B.** Volar glide

Distal Radioulnar Joint

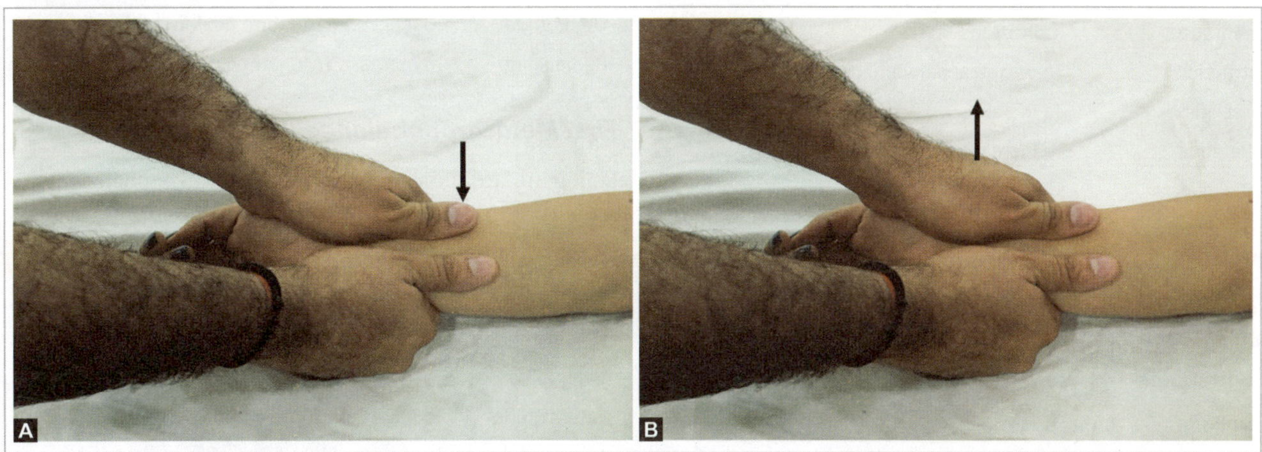

Figs 16.31A and B: Distal radioulnar joint mobilization techniques: **A.** Dorsal glide; **B.** Volar glide

Mobilization Techniques for the Distal Radioulnar Joint

The mobilization techniques for the distal radioulnar joint are as follows:

Techniques	Patient's position	Therapist's position	Hand placement	Mobilizing force
Dorsal glide (Fig. 16.31A)	Sitting with forearm in 10° supination	Standing or sitting near the patient's wrist	The stabilizing hand grasps distal ulna and the mobilizing hand grasps distal radius	Force is applied in dorsal direction
Volar glide (Fig. 16.31B)	Sitting with forearm in 10° supination	Standing or sitting near the patient's wrist	The stabilizing hand grasps distal ulna and the mobilizing hand grasps distal radius	Force is applied in volar direction

WRIST AND HAND COMPLEX

Constituent Joints

The wrist and hand are composed of several intricate joints (Fig. 16.32):

1. Radiocarpal joint
2. Ulnocarpal joint
3. Midcarpal joint
4. Trapeziometacarpal joint
5. First metacarpophalangeal joint
6. Metacarpophalengeal joint (II-V)
7. Interphalangeal joint

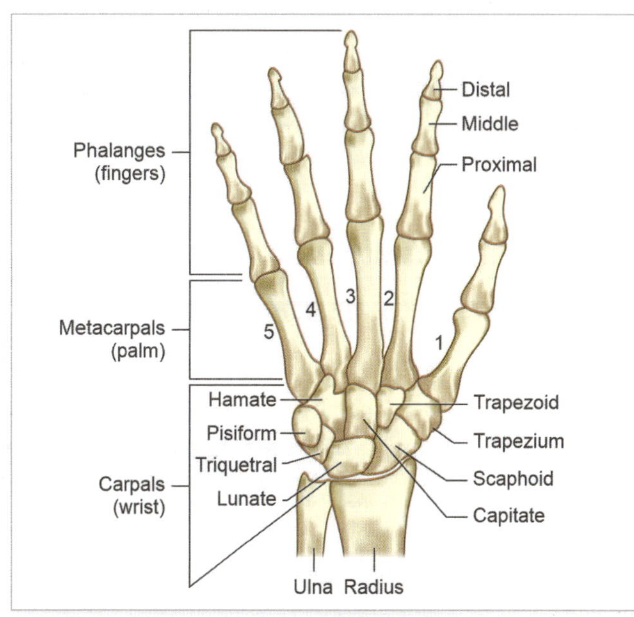

Fig. 16.32: Bones of hand and wrist

Radiocarpal Joint

Following are indications of radiocarpal joint mobilization:

Techniques	Indications
Distraction	To decrease pain and increase joint play
Dorsal glide	To increase wrist flexion
Volar glide	To increase wrist extension
Radial glide	To increase ulnar deviation
Ulnar glide	To increase radial deviation

Radiocarpal and Ulnocarpal Joint

Following are indications of radiocarpal and ulnocarpal joints mobilization:

Techniques	Indications
Anterior glide	To increase wrist extension
Posterior glide	To increase wrist flexion

Midcarpal Joint

Following are indications of midcarpal joint mobilization:

Techniques	Indications
Distraction	To decrease pain and increase joint play
Posterior glide	To increase posterior gliding of distal carpal row
Anterior glide	To increase anterior gliding of distal carpal row
Lateral glide	To increase wrist joint radial deviation
Medial glide	To increase wrist joint radial deviation

Trapeziometacarpal Joint

Following are indications of trapeziometacarpal joint mobilization:

Techniques	Indications
Distraction	To decrease pain and increase joint play
Posterior glide	To increase TMC joint abduction
Anterior glide	To increase TMC joint adduction
Ulnar glide	To increase TMC joint flexion
Radial glide	To increase TMC joint extension

First Metacarpophalangeal Joint

Following are indications of first metacarpophalangeal joint mobilization:

Techniques	Indications
Distraction	To decrease pain and increase joint play
Dorsal glide	To increase thumb adduction
Volar glide	To increase thumb abduction
Lateral glide	To increase thumb extension
Medial glide	To increase thumb flexion

Metacarpophalangeal Joint (II-V)

Following are indications of metacarpophalangeal joint mobilization:

Techniques	Indications
Distraction	To decrease pain and increase joint play
Dorsal glide	To increase MCP joint extension
Volar glide	To increase MCP joint flexion
Lateral glide	To increase finger radial deviation
Medial glide	To increase finger ulnar deviation

Interphalangeal Joint

Following are indications of interphalangeal joint mobilization:

Techniques	Indications
Distraction	To decrease pain and increase joint play
Anterior glide	To increase finger flexion
Posterior glide	To increase finger extension

Mobilization Techniques for Constituent Joints of Wrist and Hand Complex

Mobilization of Radiocarpal Joint

The mobilization techniques for the radiocarpal joints of wrist and hand complex are as follows:

Techniques	Patient's position	Therapist's position	Hand placement	Mobilizing force
Distraction (Fig. 16.33A)	Sitting on a chair with the forearm in pronation and hand at the edge of the table with a towel placed below the wrist.	The therapist stands facing the patient's wrist.	The stabilizing hand is placed over distal radius and ulna. The mobilizing hand is placed over distal carpal row.	The therapist applies a distraction force over distal carpal row in distal direction.
Dorsal glide (Fig. 16.33B)	Sitting on a chair with the forearm in pronation and hand at the edge of the table with a towel placed below the wrist.	The therapist stands facing the patient's wrist.	The stabilizing hand is placed over distal radius and ulna. The mobilizing hand is placed over distal carpal row.	The therapist applies force over distal carpal row in dorsal direction.

Contd...

Techniques	Patient's position	Therapist's position	Hand placement	Mobilizing force
Volar glide (Fig. 16.33C)	Sitting on a chair with the forearm in pronation and hand at the edge of the table with a towel placed below the wrist.	The therapist stands facing the patient's wrist.	The stabilizing hand is placed over distal radius and ulna. The mobilizing hand is placed over distal carpal row.	The therapist applies force over distal carpal row in volar direction.
Radial glide (Fig. 16.33D)	Sitting on a chair with the forearm in mid prone position and hand at the edge of the table with a towel placed below the wrist.	The therapist stands facing the patient's wrist.	The stabilizing hand is placed over distal radius and ulna. The mobilizing hand is placed over distal carpal row.	The therapist applies force over distal carpal row in radial direction.
Ulnar glide (Fig. 16.33E)	Sitting on a chair with the forearm in mid prone position and hand at the edge of the table with a towel placed below the wrist.	The therapist stands facing the patient's wrist.	The stabilizing hand is placed over distal radius and ulna. The mobilizing hand is placed over distal carpal row.	The therapist applies force over distal carpal row in ulnar direction.

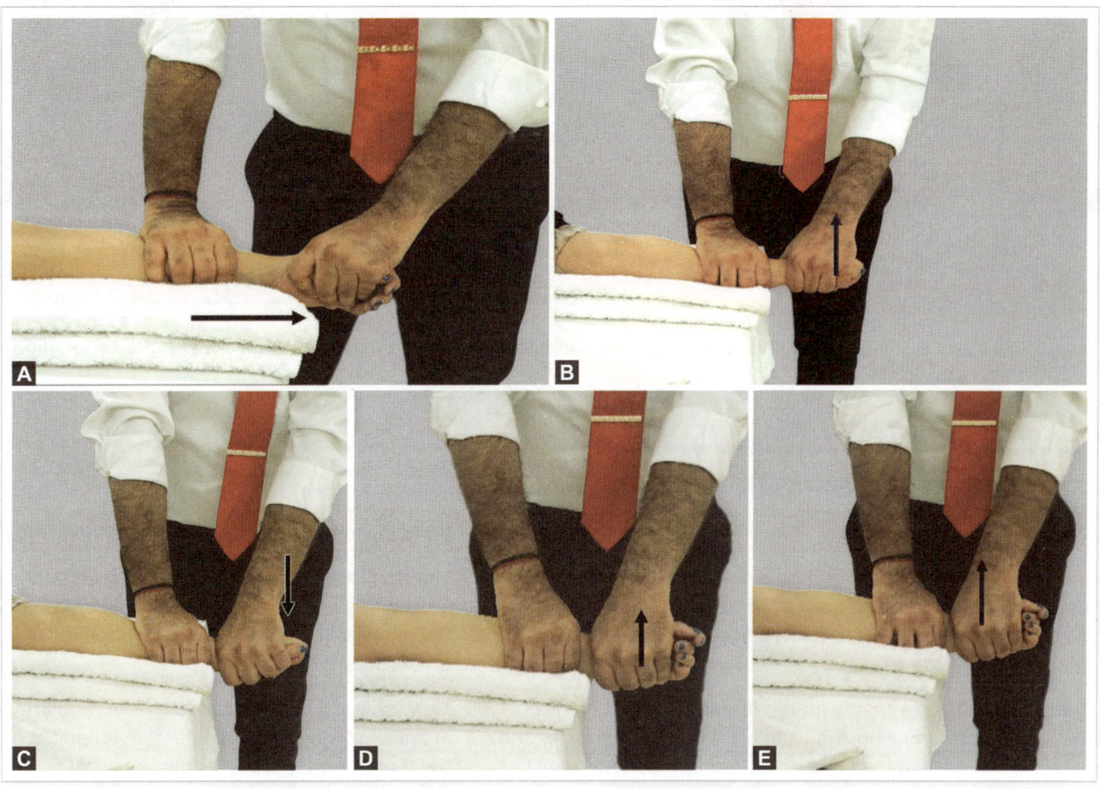

Figs 16.33A to E: Mobilization techniques for radiocarpal joints: **A.** Distraction; **B.** Dorsal glide; **C.** Volar glide; **D.** Radial glide; **E.** Ulnar glide

Mobilization of Radiocarpal and Ulnocarpal Joint

The mobilization techniques for radiocarpal and ulnocarpal joints are as follows:

Techniques	Patient's position	Therapist's position	Hand placement	Mobilizing force
Anterior glide (Fig. 16.34A)	Sitting on a chair with the forearm in pronation and hand at the edge of the table with a towel placed below the wrist.	The therapist stands facing the patient's wrist	The stabilizing hand holds the distal radius or ulna with the thumb on the dorsal surface and the index finger on the volar surface. The mobilizing hand grips the proximal row carpal bone with the thumb on the posterior surface and the index finger on the anterior surface.	The mobilizing force is applied on scaphoid in anterior direction on the radius, the lunate in anterior direction on the radius, and the triquetrum in an anterior direction on the ulna

Contd...

Techniques	Patient's position	Therapist's position	Hand placement	Mobilizing force
Posterior glide (Fig. 16.34B)	Sitting on a chair with the forearm in pronation and hand at the edge of the table with a towel placed below the wrist.	The therapist stands facing the patient's wrist	The stabilizing hand holds the distal radius or ulna with the thumb on the dorsal surface and the index finger on the volar surface. The mobilizing hand grips the proximal row carpal bone with the thumb on the posterior surface and the index finger on the anterior surface.	The mobilizing force is applied on scaphoid in a posterior direction on the radius, the lunate in posterior direction on the radius, and the triquetrum in a posterior direction on the ulna

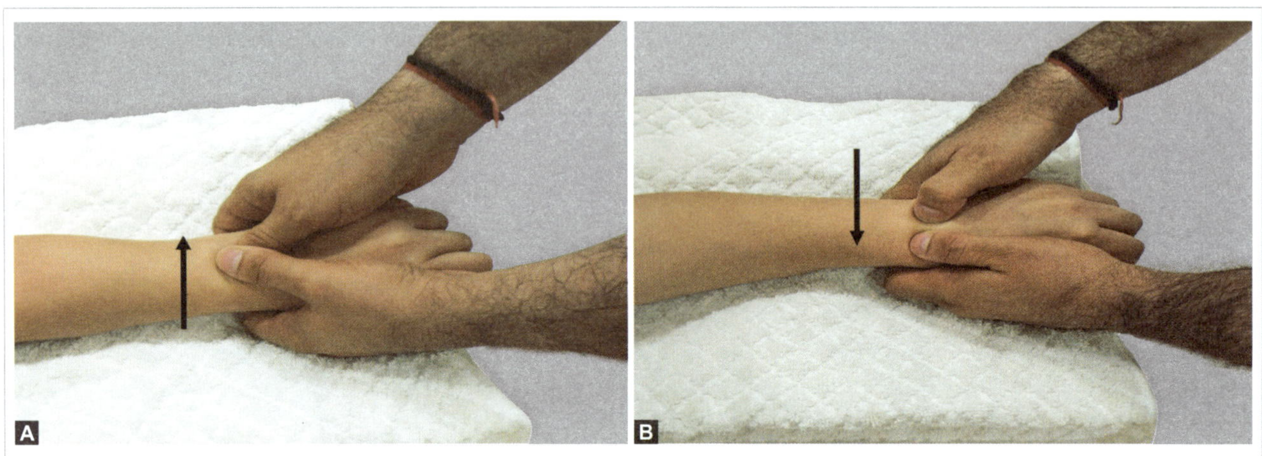

Figs 16.34A and B: Radiocarpal and ulnocarpal joints mobilization techniques: **A.** Anterior glide; **B.** Posterior glide

Mobilization of Midcarpal Joint

The mobilization techniques for midcarpal joints are as follows:

Techniques	Patient's position	Therapist's position	Hand placement	Mobilizing force
Distraction (Fig. 16.35A)	Sitting on a chair with the forearm in pronation and hand at the edge of the table with a towel placed below the wrist.	The therapist stands facing the patient's wrist	The stabilizing hand grasps the proximal row of carpals while the mobilizing hand grasps the distal row.	A perpendicular force in distal direction is applied.
Posterior glide (Fig. 16.35B)	Sitting on a chair with the forearm in mid prone position and hand at the edge of the table with a towel placed below the wrist.	The therapist stands facing the patient's wrist	The stabilizing hand grasps the proximal row of carpals while the mobilizing hand grasps the distal row.	A mobilizing force is applied over distal carpal row in posterior direction.
Anterior glide (Fig. 16.35C)	Sitting on a chair with the forearm in pronation and hand at the edge of the table with a towel placed below the wrist.	The therapist stands facing the patient's wrist	The stabilizing hand grasps the proximal row of carpals while the mobilizing hand grasps the distal row.	A mobilizing force is applied over distal carpal row in anterior direction.
Lateral glide (Fig. 16.35D)	Sitting on a chair with the forearm in mid pronation and hand at the edge of the table with a towel placed below the wrist.	The therapist stands facing the patient's wrist	The stabilizing hand grasps the proximal row of carpals while the mobilizing hand grasps the distal row.	A mobilizing force is applied over distal carpal row in radial direction.
Medial glide (Fig. 16.35E)	Sitting on a chair with the forearm in mid prone position and hand at the edge of the table with a towel placed below the wrist.	The therapist stands facing the patient's wrist	The stabilizing hand grasps the proximal row of carpals while the mobilizing hand grasps the distal row.	A mobilizing force is applied over distal carpal row in ulnar direction.

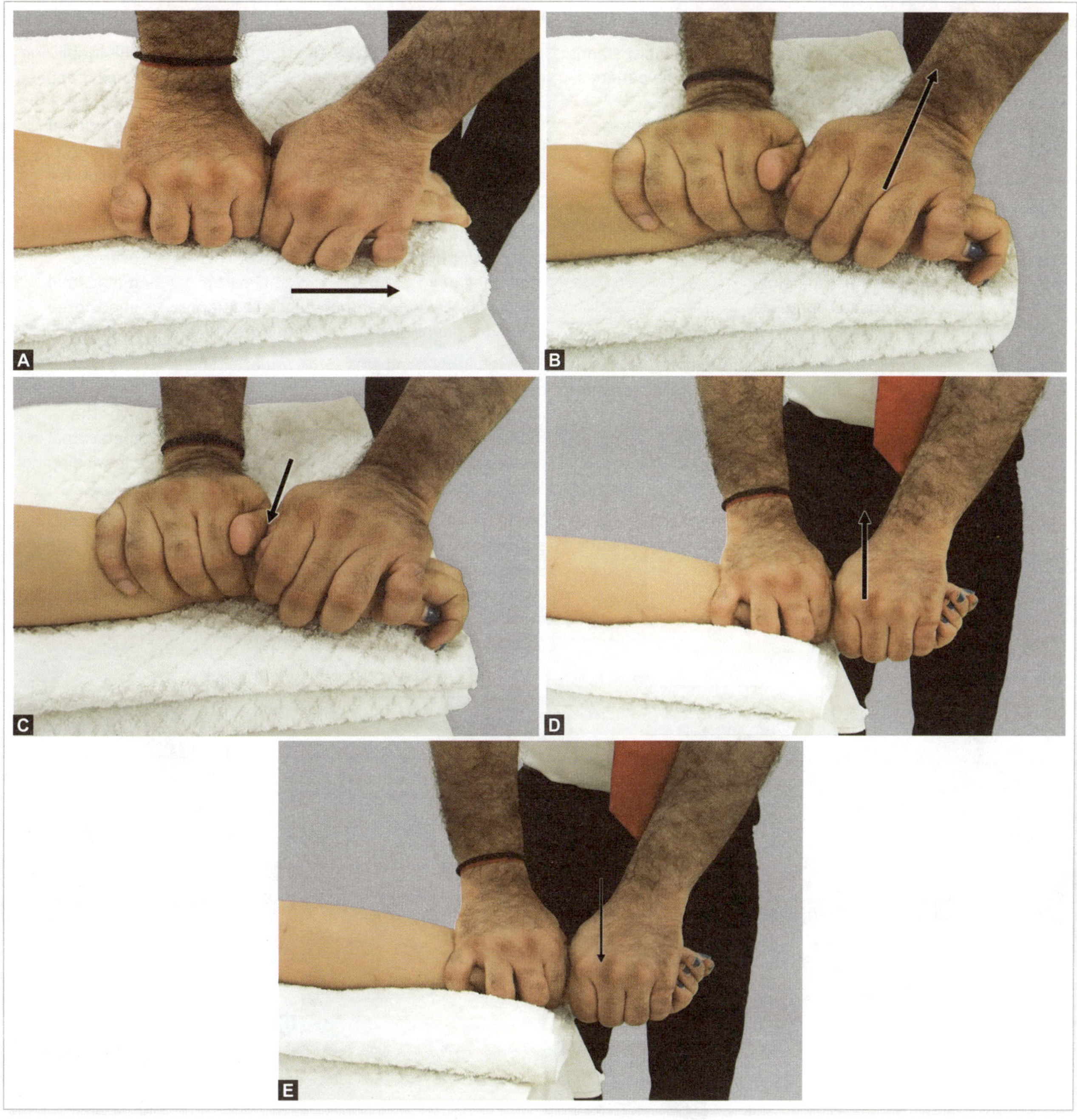

Figs 16.35A to E: Midcarpal joints mobilization techniques:
A. Distraction; **B.** Posterior glide; **C.** Anterior glide; **D.** Lateral glide; **E.** Medial glide

Mobilization of Trapeziometacarpal Joint

The mobilization techniques for trapeziometacarpal joints are as follows:

Techniques	Patient's position	Therapist's position	Hand placement	Mobilizing force
Distraction (Fig. 16.36A)	Sitting on a chair with the forearm in mid prone position and a towel is placed below the wrist for support.	The therapist stands facing the patient's wrist	The stabilizing hand grasps the trapezium between thumb (placed on dorsal aspect) and index finger (placed on dorsal aspect). The mobilizing hand grasps the proximal part of metacarpal between thumb (placed posteriorly) and index finger (placed anteriorly).	The mobilizing force is applied over the metacarpal by pulling it in a distal direction.

Contd...

Techniques	Patient's position	Therapist's position	Hand placement	Mobilizing force
Anterior glide (Fig. 16.36B)	Sitting on a chair with the forearm in mid prone position and a towel is placed below the wrist for support.	The therapist stands facing the patient's wrist	The stabilizing hand grasps the trapezium between thumb (placed on dorsal aspect) and index finger (placed on dorsal aspect). The mobilizing hand grasps the proximal part of metacarpal between thumb (placed posteriorly) and index finger (placed anteriorly).	The mobilizing force is applied over the metacarpal in anterior direction.
Posterior glide (Fig. 16.36C)	Sitting on a chair with the forearm in mid prone position and a towel is placed below the wrist for support.	The therapist stands facing the patient's wrist	The stabilizing hand grasps the trapezium between thumb (placed on dorsal aspect) and index finger (placed on dorsal aspect). The mobilizing hand grasps the proximal part of metacarpal between thumb (placed posteriorly) and index finger (placed anteriorly).	The mobilizing force is applied over the metacarpal in posterior direction
Radial glide (Fig. 16.36D)	Sitting on a chair with the forearm in mid prone position and a towel is placed below the wrist for support.	The therapist stands facing the patient's wrist	The stabilizing hand grasps the trapezium between thumb (placed on dorsal aspect) and index finger (placed on dorsal aspect). The mobilizing hand grasps the proximal part of metacarpal between thumb (placed posteriorly) and index finger (placed anteriorly).	The mobilizing force is applied over the metacarpal in lateral direction
Ulnar glide (Fig. 16.36E)	Sitting on a chair with the forearm in mid prone position and a towel is placed below the wrist for support.	The therapist stands facing the patient's wrist	The stabilizing hand grasps the trapezium between thumb (placed on dorsal aspect) and index finger (placed on dorsal aspect). The mobilizing hand grasps the proximal part of metacarpal between thumb (placed posteriorly) and index finger (placed anteriorly).	The mobilizing force is applied over the metacarpal in medial direction

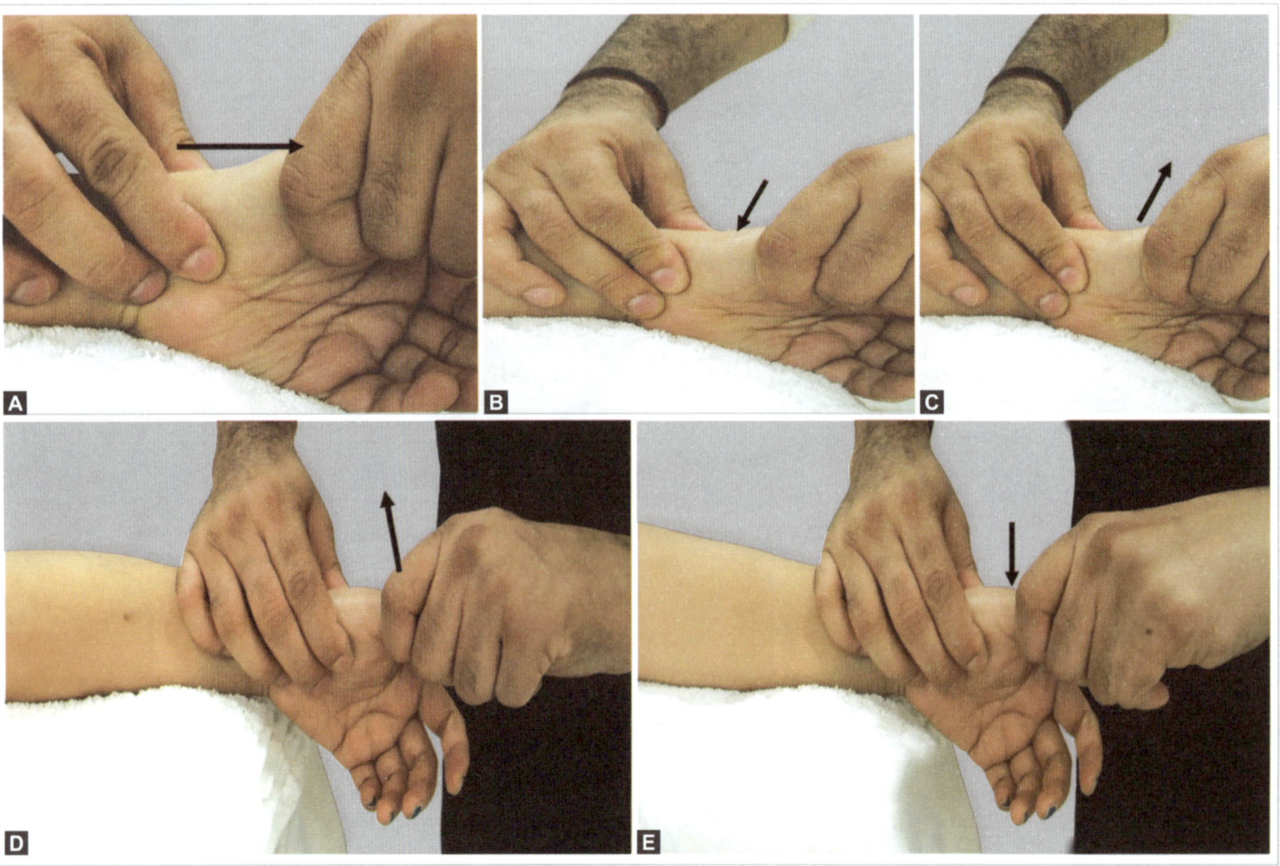

Figs 16.36A to E: Trapeziometacarpal joints mobilization techniques: **A.** Distraction; **B.** Anterior glide; **C.** Posterior glide; **D.** Radial glide; **E.** Ulnar glide

Mobilization of First Metacarpophalangeal Joint

The mobilization techniques of first metacarpophalangeal joints are as follows:

Techniques	Patient's position	Therapist's position	Hand placement	Mobilizing force
Distraction (Fig. 16.37A)	Sitting on a chair with the forearm in mid prone position and a towel is placed below the wrist for support.	The therapist stands facing the patient's wrist	The stabilizing hand grasps the head of the metacarpal while mobilizing hand grasps the proximal part of proximal phalanx	The therapist applies distraction by pulling the proximal phalanx perpendicularly in distal direction.
Dorsal glide (Fig. 16.37B)	Sitting on a chair with the forearm in supination and a towel is placed below the wrist for support.	The therapist stands facing the patient's wrist	The stabilizing hand grasps the distal carpal row while the thenar eminence of the mobilizing hand is placed over base of first metacarpal on the patient's hand	The mobilizing force is applied via thenar eminence over base of first meatacarpal in dorsal direction.
Volar glide (Fig. 16.37C)	Sitting on a chair with the forearm in mid prone position and a towel is placed below the wrist for support.	The therapist stands facing the patient's wrist	The stabilizing hand is placed below the patient's hand on the ulnar border and the mobilizing hand grasps the proximal part of proximal phalanx between thumb and index finger.	A mobilizing force is applied in volar direction.
Lateral glide (Fig. 16.37D)	Sitting on a chair with the forearm in supination and a towel is placed below the wrist for support.	The therapist stands facing the patient's wrist	The stabilizing hand grasps the distal carpal row while the thenar eminence of the mobilizing hand is placed over base of first metacarpal on the patient's hand	The mobilizing force is applied via thenar eminence over base of first meatacarpal in lateral/radial direction.
Medial glide (Fig. 16.37E)	Sitting on a chair with the forearm in mid prone position and a towel is placed below the wrist for support.	The therapist stands behind and faces the patient's wrist	The stabilizing hand grasps the distal carpal row while the thenar eminence of the mobilizing hand is placed posteriorly and grasps the base of first metacarpal and the fingers wrap around the thenar eminence of patient's hand.	The mobilizing force is applied via thenar eminence over base of first meatacarpal in medial/ulnar direction

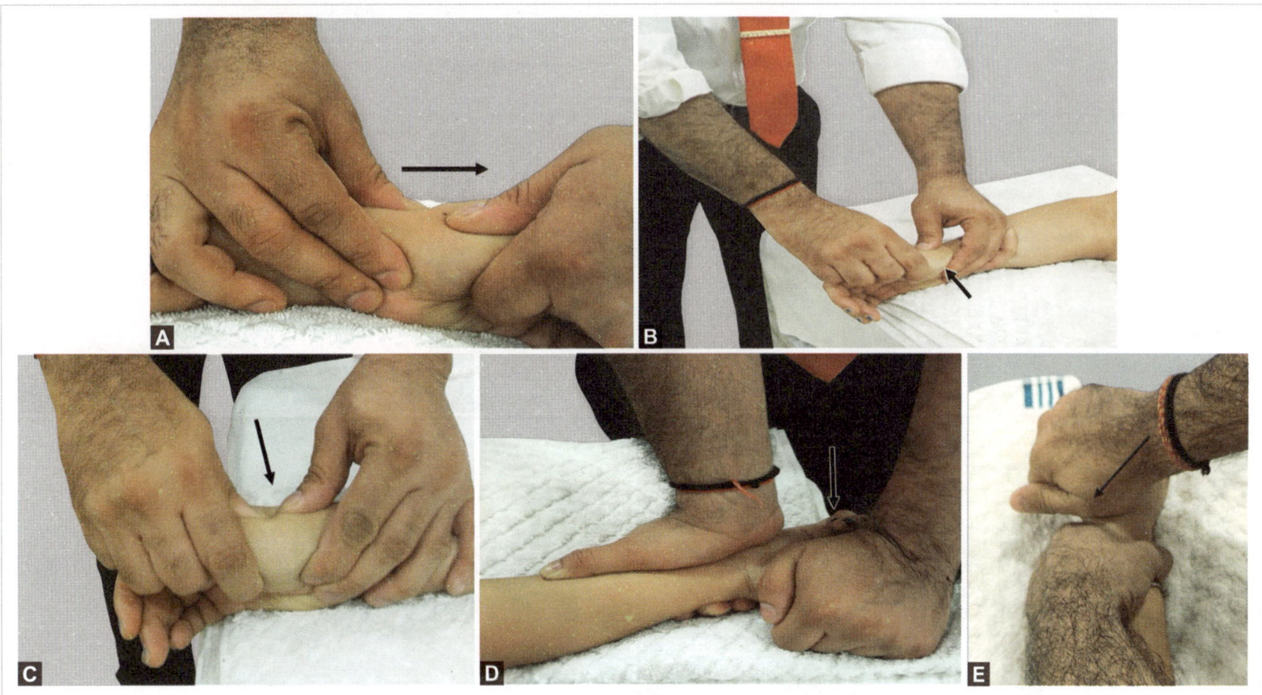

Figs 16.37A to E: First metacarpophalangeal joint mobilization techniques: **A.** Distraction; **B.** Dorsal glide; **C.** Volar glide; **D.** Lateral glide; **E.** Medial glide

Mobilization of Metacarpophalangeal Joint (II-V)

The mobilization techniques of metacarpophalangeal joints (II-V) are as follows:

Techniques	Patient's position	Therapist's position	Hand placement	Mobilizing force
Distraction (Fig. 16.38A)	Sitting with forearm in pronation with a towel placed below the wrist for support.	The therapist stands facing the patient's wrist	The stabilizing hand grasps the metacarpal hand while the mobilizing hand grasps the proximal part of proximal phalanx from medial and lateral surface.	A distally directed perpendicular force is applied over the proximal phalanx.
Dorsal glide (Fig. 16.38B)	Sitting with forearm in pronation with a towel placed below the wrist for support.	The therapist stands facing the patient's wrist	The stabilizing hand grasps the metacarpal hand while the mobilizing hand grasps the proximal part of proximal phalanx from anterior and posterior surface.	A dorsal/posteriorly directed force is applied over the proximal phalanx.
Volar glide (Fig. 16.38C)	Sitting with forearm in pronation with a towel placed below the wrist for support.	The therapist stands facing the patient's wrist	The stabilizing hand grasps the metacarpal hand while the mobilizing hand grasps the proximal part of proximal phalanx from anterior and posterior surface.	A volar/anteriorly directed force is applied over the proximal phalanx.
Lateral glide (Fig. 16.38D)	Sitting with forearm in pronation with a towel placed below the wrist for support.	The therapist stands facing the patient's wrist	The stabilizing hand grasps the metacarpal hand while the mobilizing hand grasps the proximal part of proximal phalanx from medial and lateral surface.	A radial/laterally directed force is applied over the proximal phalanx
Medial glide (Fig. 16.38E)	Sitting with forearm in pronation with a towel placed below the wrist for support.	The therapist stands facing the patient's wrist	The stabilizing hand grasps the metacarpal hand while the mobilizing hand grasps the proximal part of proximal phalanx from medial and lateral surface.	A ulnar/medially directed force is applied over the proximal phalanx

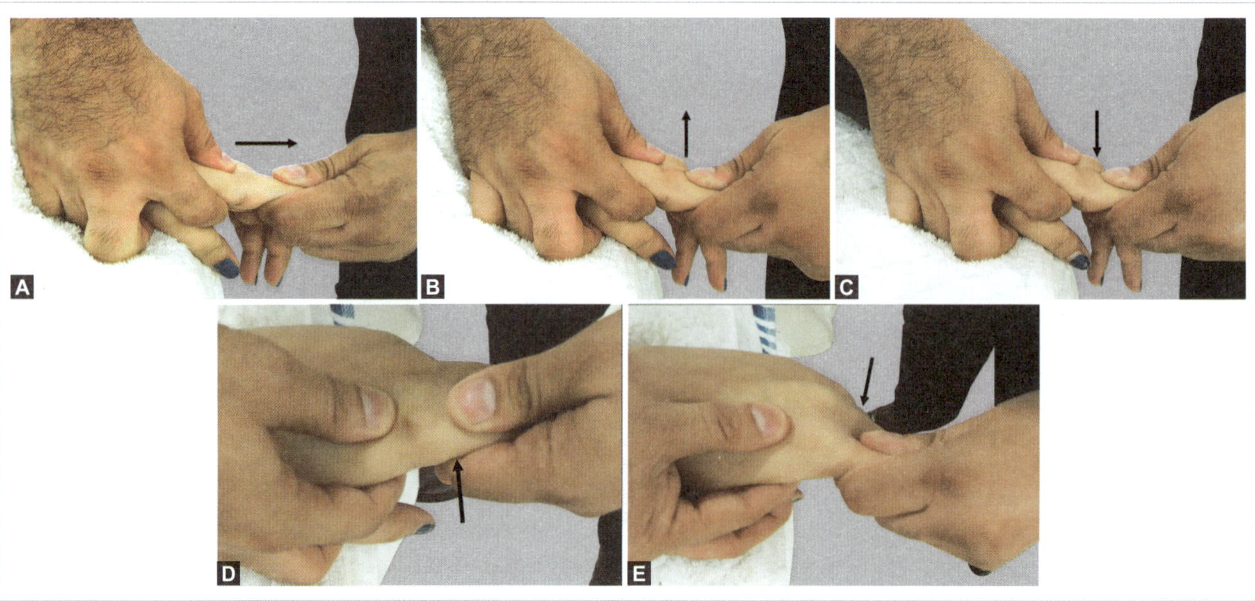

Figs 16.38A to E: Metacarpophalangeal joints mobilization techniques: **A.** Distraction; **B.** Dorsal glide; **C.** Volar glide; **D.** Lateral glide; **E.** Medial glide

Mobilization of Interphalangeal Joint

The mobilization techniques for interphalangeal joints are as follows:

Techniques	Patient's position	Therapist's position	Hand placement	Mobilizing force
Distraction (Fig. 16.39A)	Sitting with forearm in pronation with a towel placed below the wrist for support.	Stands facing the patient's hand.	The stabilizing hand grasps the distal part of the proximal phalanx while the mobilizing hand grasps the proximal part of the distal phalanx	Force is applied in distal direction along long axis
Dorsal glide (Fig. 16.39B)	Sitting with forearm in pronation with a towel placed below the wrist for support.	Stands facing the patient's hand	The stabilizing hand grasps the distal part of the proximal phalanx while the mobilizing hand grasps the proximal part of the distal phalanx	Force is applied in dorsal direction
Volar glide (Fig. 16.39C)	Sitting with forearm in pronation with a towel placed below the wrist for support.	Stands facing the patient's hand	The stabilizing hand grasps the distal part of the proximal phalanx while the mobilizing hand grasps the proximal part of the distal phalanx	Force is applied in volar direction

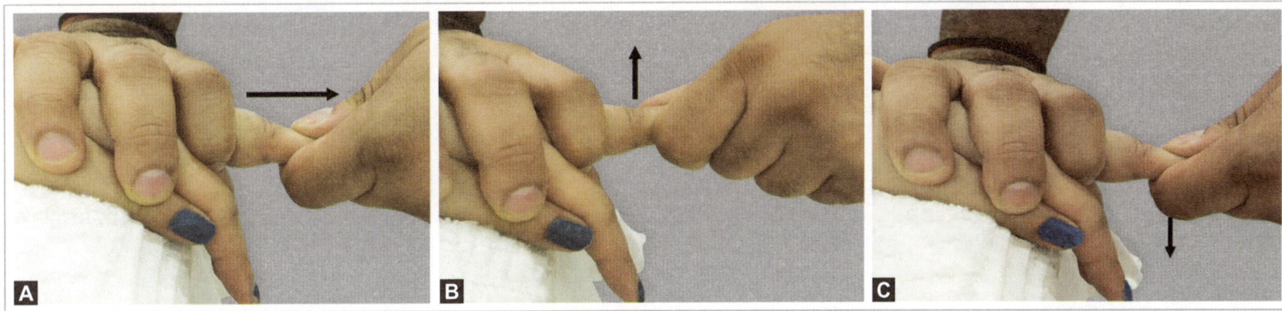

Figs 16.39A to C: Interphalangeal joints mobilization techniques: **A.** Distraction; **B.** Dorsal glide; **C.** Volar glide

HIP JOINT

The hip joint, also known as the acetabulofemoral joint, is a ball-and-socket synovial joint that connects the pelvis to the femur. The components of hip joint are given in Figure 16.40.

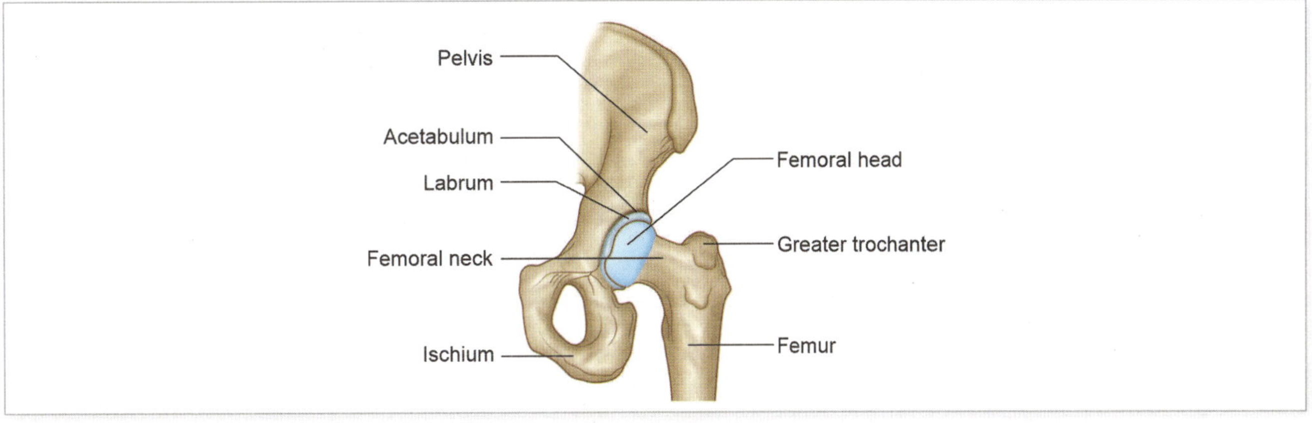

Fig. 16.40: Components of hip joint

Acetabulofemoral Joint

The mobilization techniques for hip joints are applied in the following situations:

Techniques	Indications
Distraction	To decrease pain and increase ROM
Caudal glide	To increase abduction; can be used as long axis traction
Anterior glide	Increase extension and external rotation
Posterior glide	Increase flexion and internal rotation

Mobilization Techniques of Hip Joint

The mobilization techniques for hip joint are as follows:

Techniques	Patient's position	Therapist's position	Hand placement	Mobilizing force
Distraction (Fig. 16.41A)	Supine lying with the patient's posterior thigh placed on therapist's shoulder. The pelvis is stabilized using a mobilization belt.	Stride standing facing the patient	The therapist wraps both the hands around medial thigh	The therapist applies a caudal force by leaning backwards
Caudal glide with belt (Fig. 16.41B)	Supine lying. The pelvis is stabilized using a mobilization belt.	At the end of the treatment table; places a belt around trunk, then crosses the belt over the patient's foot and around the ankle.	The hands are placed proximal to the malleoli, under the belt.	The therapist applies a caudal force by leaning backwards
Without belt (Fig. 16.41C)	Supine lying. The pelvis is stabilized using a mobilization belt.	Stands facing the patient near the distal thigh.	The therapist wraps both the hands around distal thigh	The therapist applies a caudal force by leaning backwards
Anterior glide with belt (Fig. 16.41D)	Prone with the pelvis on the treatment table and the legs off the end of the table.	At the foot of the treatment table facing the patient's hip. The therapist's wraps a belt around the patient's thigh. Patient's leg is supported between the clinician's arm and trunk.	The mobilizing hand is placed over the posterior thigh just below the buttocks and the stabilizing hand is placed over distal thigh anteriorly.	With the elbow extended, the therapist applies force in anterior direction.
Without belt (Fig. 16.41E)	Side lying with a pillow placed between the thighs	Stands behind the patient facing the hip joint.	The stabilizing hand is placed over ASIS and the mobilizing hand is placed over the greater trochanter posteriorly.	With the elbow extended, the therapist applies force in anterior direction.
Posterior glide with belt (Fig. 16.41F)	Supine, with hips at the end of the table. The patient flexes the opposite hip and holds the thigh against the chest	Stands on the medial side of the patient's thigh and places a belt around shoulder and under the patient's thigh Therapist places distal hand under the belt and distal thigh. He/she places proximal hand on the anterior surface of the proximal thigh.	The stabilizing hand is placed under the belt and distal thigh and the mobilizing hand is placed proximally on the anterior thigh	With the elbows extended, force is applied posteriorly.

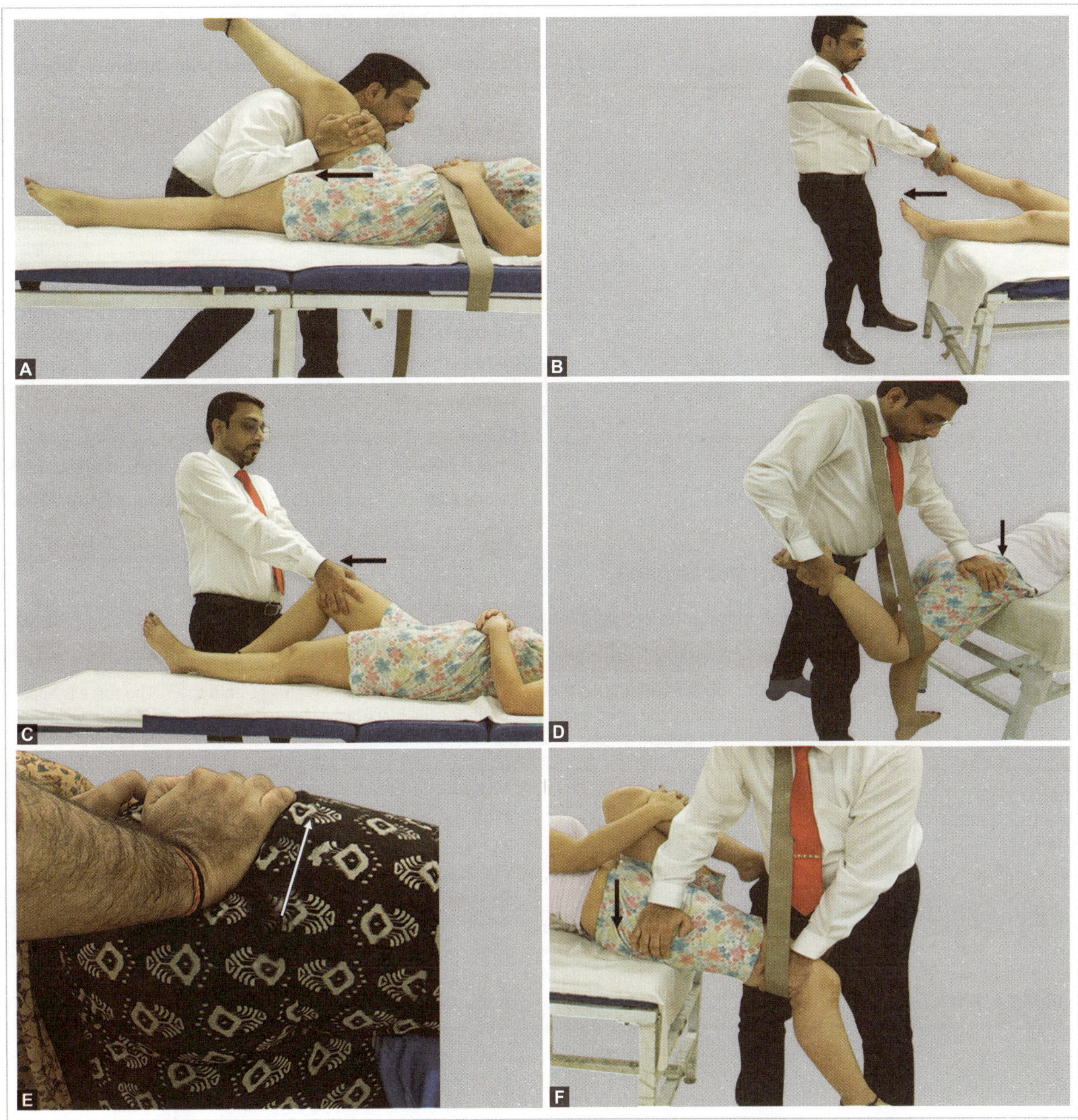

Figs 16.41A to F: Hip joints mobilization techniques: **A.** Distraction; **B.** Caudal glide with belt; **C.** Caudal glide without belt; **D.** Anterior glide with belt; **E.** Anterior glide without belt; **F.** Posterior glide with belt

KNEE COMPLEX

The knee complex consists of the following articulations (Fig. 16.42):

- Tibiofemoral joint
- Patellofemoral joint

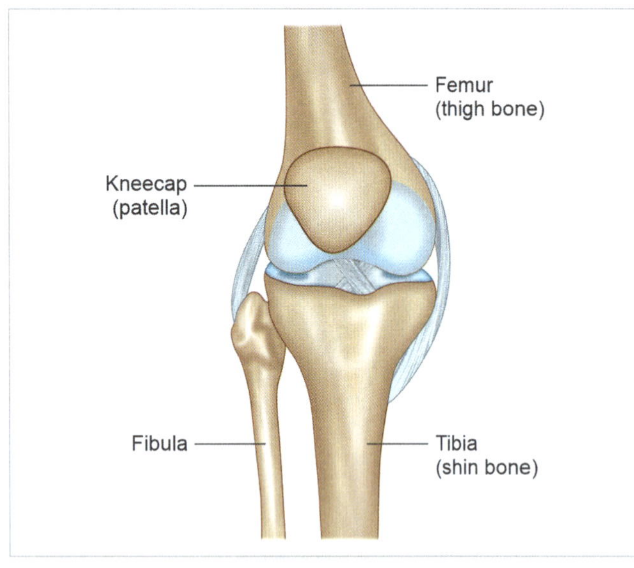

Fig. 16.42: Knee complex

Tibiofemoral Joint Mobilization

The indications for tibiofemoral joint mobilization are as follows:

Techniques	Indications
Distraction	To decrease pain and improve ROM
Anterior glide	To increase flexion
Posterior glide	To increase extension

Patellofemoral Joint Mobilization

The indications for mobilization of patellofemoral joint are as follows:

Techniques	Indications
Inferior	To increase knee flexion
Medial glide	To increase accessory motion of medial glide
Lateral glide	To increase accessory motion of lateral glide

Techniques for Tibiofemoral Joint Mobilization

The mobilization techniques for tibiofemoral joint are as follows:

Techniques	Patient's position	Therapist's position	Hand placement	Mobilizing force
Distraction (Fig. 16.43A)	Lying position with the leg hanging from the edge of the couch	Stride standing on the side of the knee facing the patient's knee	The therapist bends forwards to hold the patient's leg just above the malleolus.	The force is applied in downward direction.
Anterior glide Glide 1: Glide tibia on femur (Fig. 16.43B)	Supine lying with knee flexed.	On patient's foot facing the patient's knee.	Both the hands grasp the proximal part of tibia. The palmar surface of both the hands are placed on the anterior tibial plateau and the thumbs are placed on distal tibia. The finger grasps the posterior tibial plateau	Mobilizing force is applied in anterior direction.
Glide 2: Glide femur on tibia (Fig. 16.43C)	Supine lying	The side of the patient's leg facing the patient's knee. The therapist holds the patient's leg between trunk and medial aspect of upper arm	The stabilizing hand is placed over the posterior tibia and the mobilizing hand is placed over distal femur.	A posterior glide is applied over distal femur by the mobilizing hand.
Glide 3: Prone lying (Fig. 16.43D)	In prone lying. A small towel is placed below distal femur to prevent patellar compression	Stands beside facing the patient's knee	The stabilizing hand is placed over the anterior aspect of distal femur. The mobilizing hand is placed on the posterior aspect of proximal tibia	Mobilizing force is applied in an anterior direction.
Posterior glide Glide 1 (Fig. 16.43E)	Supine lying	On patient's foot facing the patient's knee	Both the hands grasp the proximal part of tibia. The palmar surface of both the hands are placed on the anterior tibial plateau and the thumbs are placed on distal tibia. The finger grasps the posterior tibial plateau.	Mobilizing force is applied in posterior direction.

Contd...

Techniques	Patient's position	Therapist's position	Hand placement	Mobilizing force
Glide 2 (Fig. 16.43F)	High sitting.	Stands facing the patient. If the knee flexion nears 90°, then the therapist can sit on the stool. Stabilizes the patient's distal leg between his knees.	Both the hands grasp the proximal part of tibia. The palmar surface of both the hands are placed on the anterior tibial plateau and the thumbs are placed on distal tibia. The finger grasps the posterior tibial plateau.	Mobilizing force is applied in posterior direction.

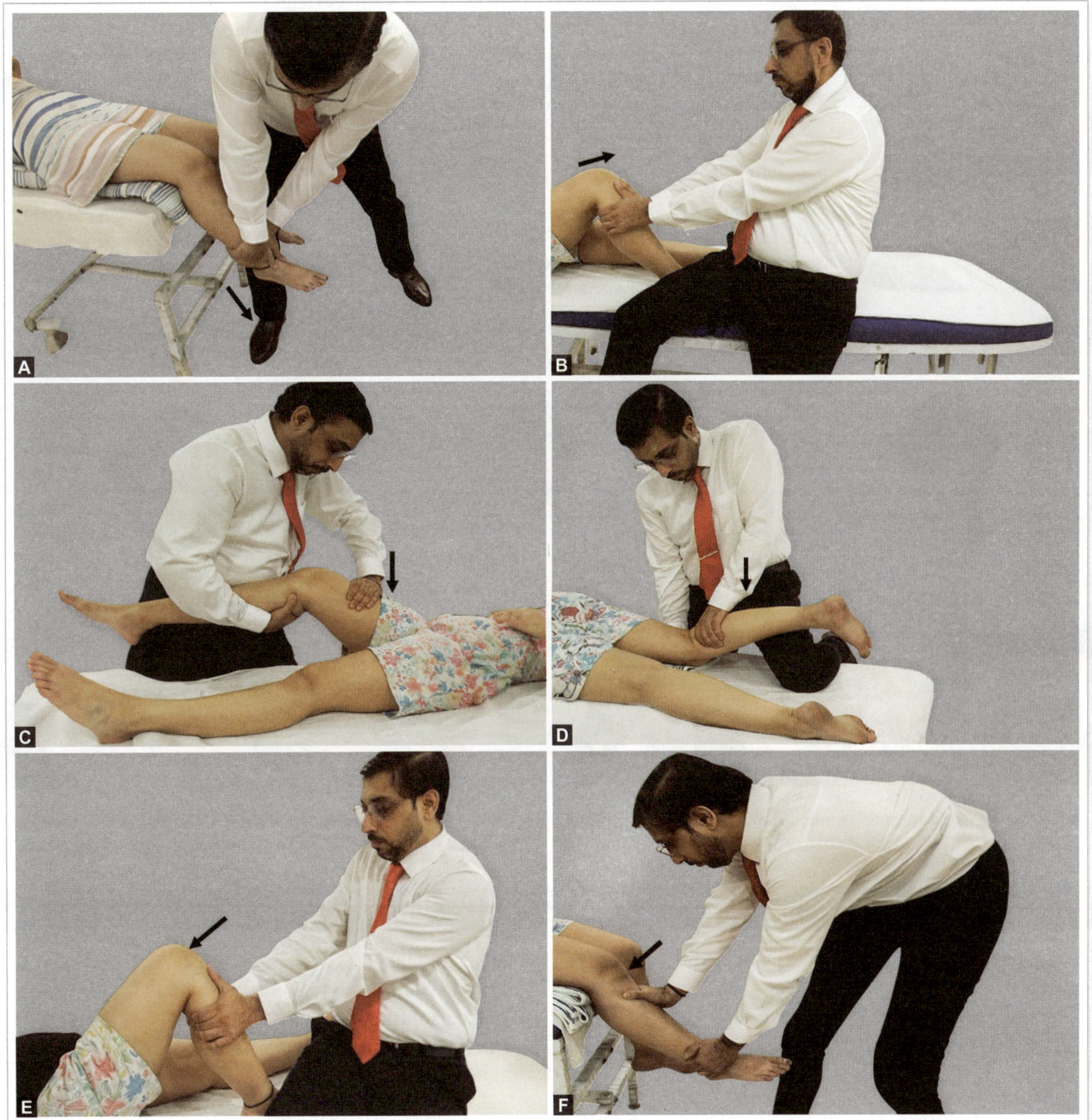

Figs 16.43A to F: Tibiofemoral joint mobilization techniques: **A.** Distraction; **B.** Anterior glide: glide 1; **C.** Anterior glide: glide 2; **D.** Anterior glide: glide 3; **E.** Posterior glide: glide 1; **F.** Posterior glide: glide 2

Techniques for Patellofemoral Joint Mobilization

The mobilization techniques for patellofemoral joint are as follows:

Techniques	Patient's position	Therapist's position	Hand placement	Mobilizing force
Inferior glide (Fig. 16.44A)	Supine lying. A towel Is placed below the patient's knee	Near patient's hip facing the patellofemoral joint	The web space of the mobilizing hand is placed over superior border of patella. The stabilizing hand is used for reinforcement and grasps the patient's wrist.	Mobilizing hand glides the patella in an inferior direction
Medial glide (Fig. 16.44B)	Supine lying. A towel Is placed below the patient's knee	Stands beside the patient facing the patellofemoral joint.	The stabilizing hand is placed in popliteal fossa and the heel of mobilizing hand is placed over lateral border of patella	Mobilizing hand glides the patella in an medial direction
Lateral glide (Fig. 16.44C)	Supine lying. A towel Is placed below the patient's knee	Stands on the unaffected side facing the patient's patellofemoral joint.	The stabilizing hand is placed in popliteal fossa and the heel of mobilizing hand is placed over medial border of patella	Mobilizing hand glides the patella in an lateral direction

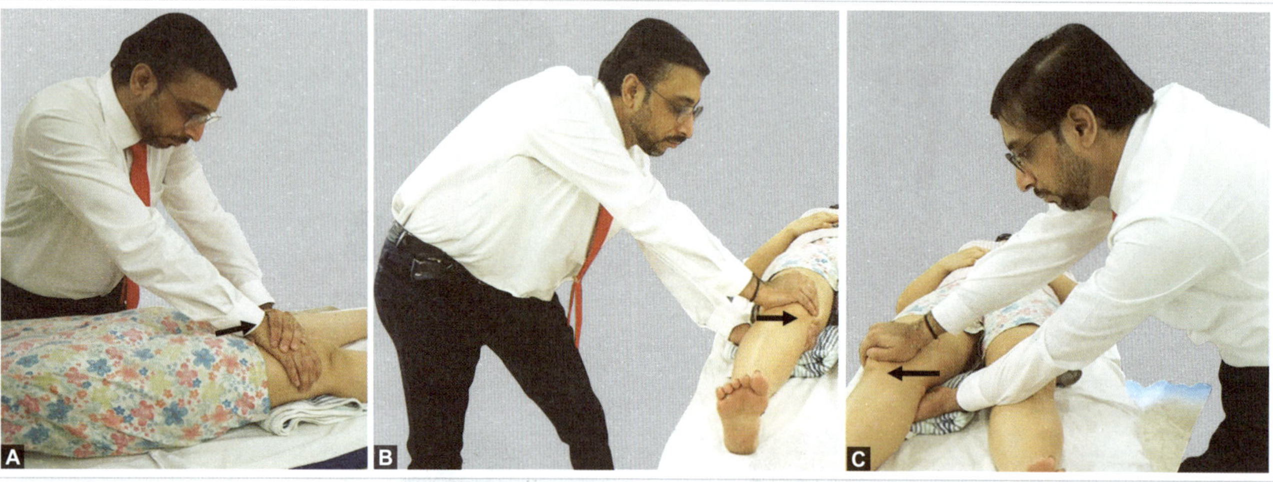

Figs 16.44A to C: Patellofemoral joint mobilization techniques: **A.** Inferior glide; **B.** Medial glide; **C.** Lateral glide

ANKLE AND FOOT COMPLEX

The ankle and foot complex (Figs 16.45A and B) consists of following articulations:

- Tibiofibular joint
- Talocrural joint
- Subtalar joint
- Intermetatarsal and tarsometatarsal joint
- Metatarsophalangeal joint
- Interphalangeal joint

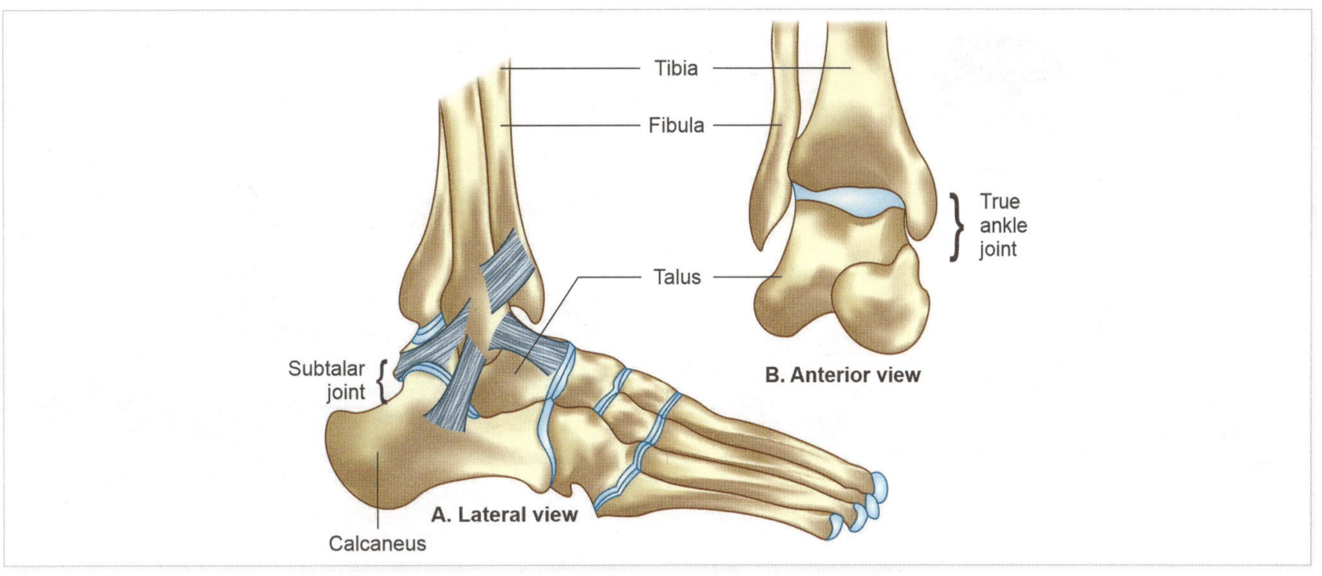

Figs 16.45A and B: Joints of ankle and foot: **A.** Lateral view; **B.** Anterior view

Mobilization Techniques of Tibiofibular Joint

Mobilization techniques of tibiofibular joints are proximal and distal (Fig. 16.46)

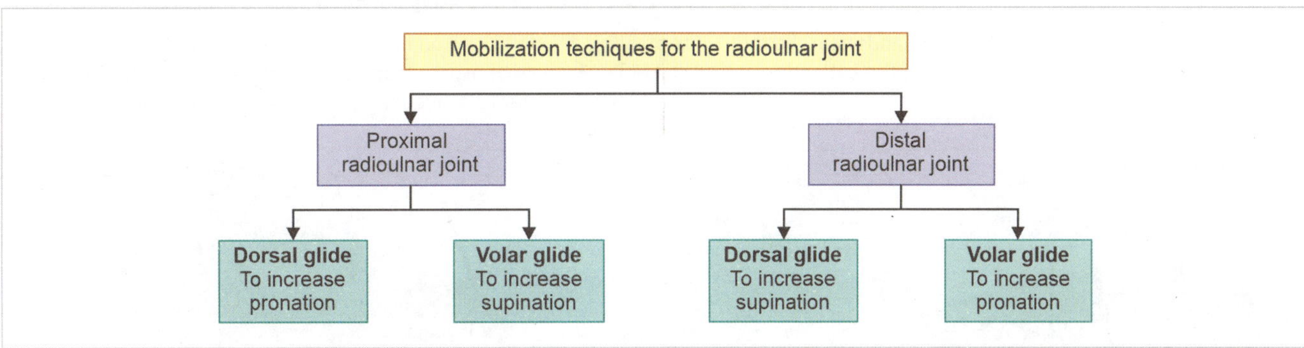

Fig. 16.46: Mobilization techniques of tibiofibular joint

Mobilization Techniques of Proximal Tibiofibular Joint

The mobilization techniques for proximal tibiofibular joints are as follows:

Techniques	Patient's position	Therapist's position	Hand placement	Mobilizing force
Anterior glide (Fig. 16.47A)	Side lying position with hip and knee flexed to 40–50° and a towel placed between the couch and medial knee	The therapist stands behind the patient is stride standing	The stabilizing hand grasps medial side of proximal tibia. The mobilizing hand is placed posteriorly on the leg over head of fibula	Using the palmar aspect of the mobilizing hand, the therapist applies a force in anterolateral direction.
Posterior glide (Fig. 16.47B)	Supine with the knee supported by a small pillow.	At patient's leg facing the knee	The stabilizing hand grasps medial and posterior side of proximal tibia. The mobilizing hand is placed anteriorly over head of fibula.	Using the palmar aspect of the mobilizing hand, the therapist applies a force in posterior direction.

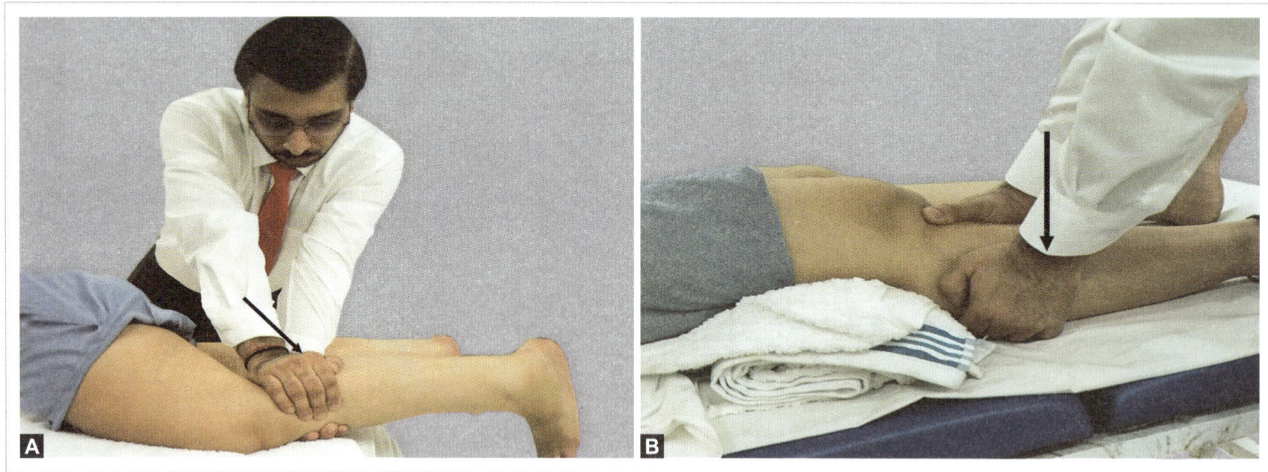

Figs 16.47A and B: Proximal tibiofibular joint mobilization techniques: **A.** Anterior glide; **B.** Posterior glide

Mobilization Techniques of Distal Tibiofibular Joint

The mobilization techniques for distal tibiofibular joint are as follows:

Techniques	Patient's position	Therapist's position	Hand placement	Mobilizing force
Anterior glide (Fig. 16.48A)	Prone lying with the foot off the couch	Stands near the foot of the patient	The stabilizing hand placed on medial aspect of distal tibia. The palmar aspect mobilizing hand is placed over the posterior aspect of distal fibula.	Force is applied over tibia in anterior direction
Posterior glide (Fig. 16.48B)	Supine lying	Stands near the foot of the patient	The stabilizing hand placed on medial aspect of distal tibia. The palmar aspect mobilizing hand is placed over the anterior aspect of distal fibula.	Force is applied over tibia in posterior direction.

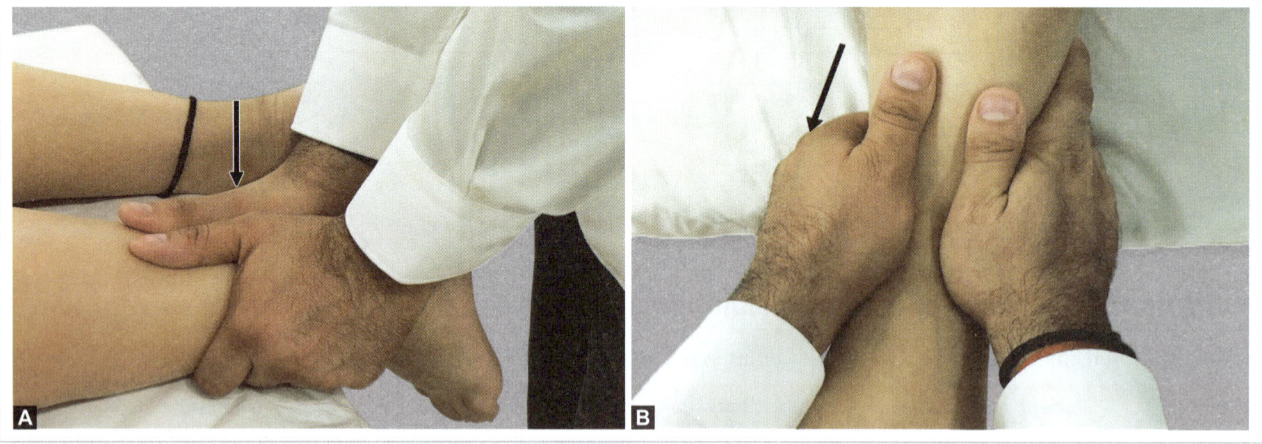

Figs 16.48A and B: Distal tibiofibular joint mobilization techniques: **A.** Anterior glide; **B.** Posterior glide

Mobilization Techniques of Talocrural Joint

The indications for mobilization of talocrural joint are as follows:

Techniques	Indications
Distraction	For pain and mobility
Anterior glide	To increase plantarflexion ROM
Posterior glide	To increase dorsiflexion ROM

The mobilization techniques for talocrural joint are as follows:

Techniques	Patient's position	Therapist's position	Hand placement	Mobilizing force
Distraction (Fig. 16.49A)	Supine lying, the patient's lower leg is immobilized using belt	The therapist stands at the edge of the table facing patient's foot	The patients grasp dorsal surface of foot using fingers of both the hands. The thumbs are placed over plantar surface.	The therapist applies long axis distraction force by leaning backwards.
Anterior glide (Fig. 16.49B)	Prone lying with patient's foot off the couch. Stabilize the lower leg using mobilizing belt.	The therapist stands near the foot of the patient	The stabilizing hand is placed at distal tibia just above the ankle mortice. The mobilizing hand is placed over posterior calcaneum.	The mobilizing force is applied by pushing the calcaneum anteriorly.
Posterior glide (Fig. 16.49C)	Supine lying with patient's foot off the couch. Stabilize the lower leg using mobilizing belt	The therapist stands near the side of foot of the patient	The web space of the stabilizing hand is placed just above the mortice and the web space of the mobilizing hand is placed over the mortice.	A mobilizing force is applied over the talus in posterior direction.

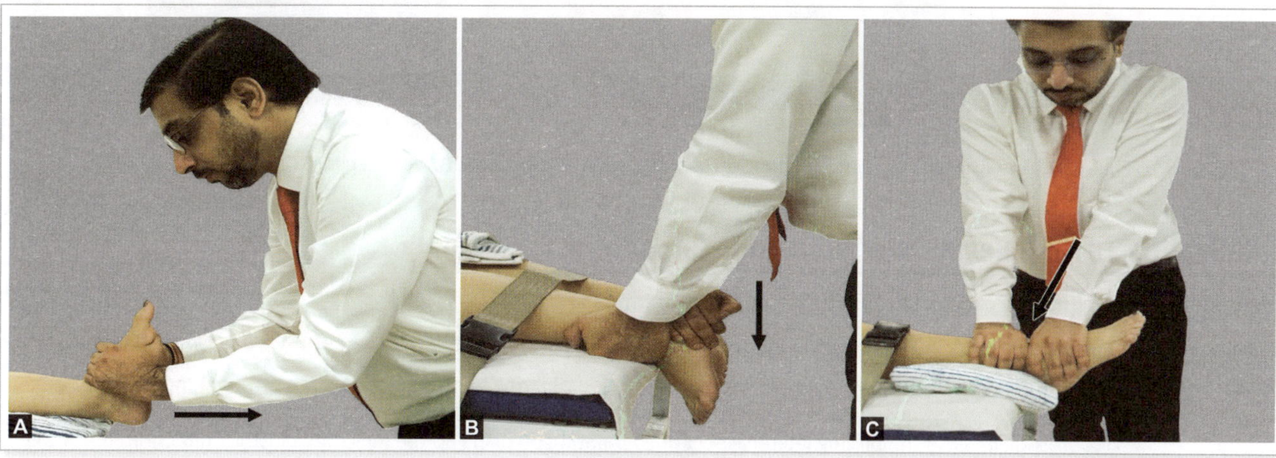

Figs 16.49A to C: Talocrural joint mobilization techniques: **A.** Distraction; **B.** Anterior glide; **C.** Posterior glide

Mobilization Techniques of Subtalar Joint

The indications of subtalar joint mobilization techniques are as follows:

Techniques	Indications
Distraction	For pain and mobility
Medial glide	To increase inversion
Lateral glide	To increase eversion

The mobilization techniques for subtalar joint are as follows:

Techniques	Patient's position	Therapist's position	Hand placement	Mobilizing force
Distraction (Fig. 16.50A)	Spine lying. The hip is externally rotated and the foot is stabilized in dorsiflexion by the therapist's thigh	At the end of the couch facing the patient's foot	The stabilizing hand is placed at distal tibia just above the ankle mortice. The mobilizing hand grasps the posterior calcaneum	The force is applied by pulling the calcaneum in the direction of the therapist.

Contd...

Techniques	Patient's position	Therapist's position	Hand placement	Mobilizing force
Medial glide (Fig. 16.50B)	Prone lying with the patient's foot off the couch.	Stands at the lateral side of the foot in stride standing	The stabilizing arm is placed anteriorly over the ankle mortice. The mobilizing hand grasps posterior calcaneum.	Keeping the elbow extended a mobilizing force is applied in medial direction.
Lateral glide (Fig. 16.50C)	Prone lying with the patient's foot off the couch.	Stands at the medial side of the foot in stride standing	The stabilizing arm is placed anteriorly over the ankle mortice. The mobilizing hand grasps posterior calcaneum	Keeping the elbow extended a mobilizing force is applied in lateral direction.

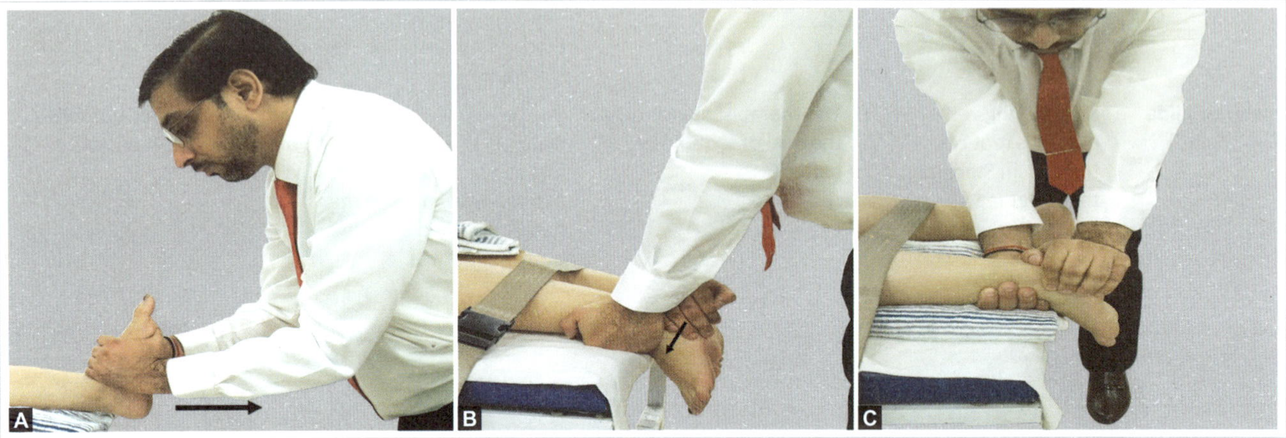

Figs 16.50A to C: Subtalar joint mobilization techniques: **A.** Distraction; **B.** Medial glide; **C.** Lateral glide

Mobilization Techniques of Intertarsal and Tarsometatarsal Joint

The mobilization techniques of intertarsal and tarsometatarsal joints are as follows:

Techniques	Patient's position	Therapist's position	Hand placement	Mobilizing force
Plantar glide (Fig. 16.51A)	High sitting with the patient's foot resting in therapist's lap	Sitting on a chair. Sit on medial side of foot if mobilizing lateral aspect of foot or sit on lateral side of foot if mobilizing medial aspect of foot.	The stabilizing hand is placed on the dorsum of the foot with the fingers pointing medially, so the index finger can be wrapped around and placed under the bone to be stabilized. The thenar eminence of the mobilizing hand is placed over the dorsal surface of the bone to be moved and the fingers are wrapped around the plantar surface.	A mobilizing force is applied in plantar direction.
Dorsal glide (Fig. 16.51B)	Prone, with knee flexed	• For lateral tarsal bones, the therapist stands on the inner side of patient's leg. • For medial tarsal bones, the therapist stands on the outer side of the leg	• For lateral tarsal bones mobilization, the stabilizing hand wraps around the lateral side of the foot • For medial tarsal bones mobilization, the stabilizing hand wraps around the medial side of the foot • The mobilizing hand is placed over the bone to be mobilized.	Mobilization force is applied in dorsal direction.

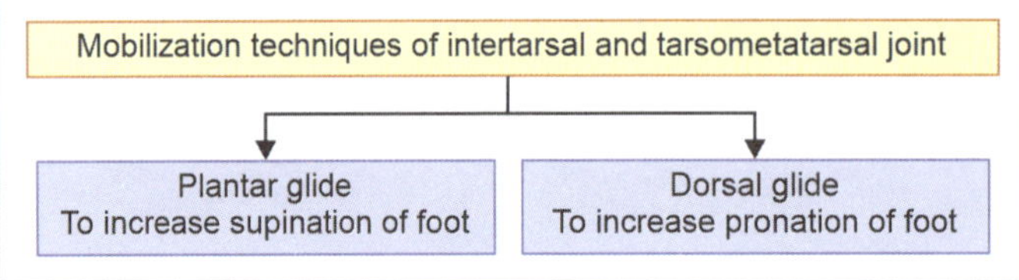

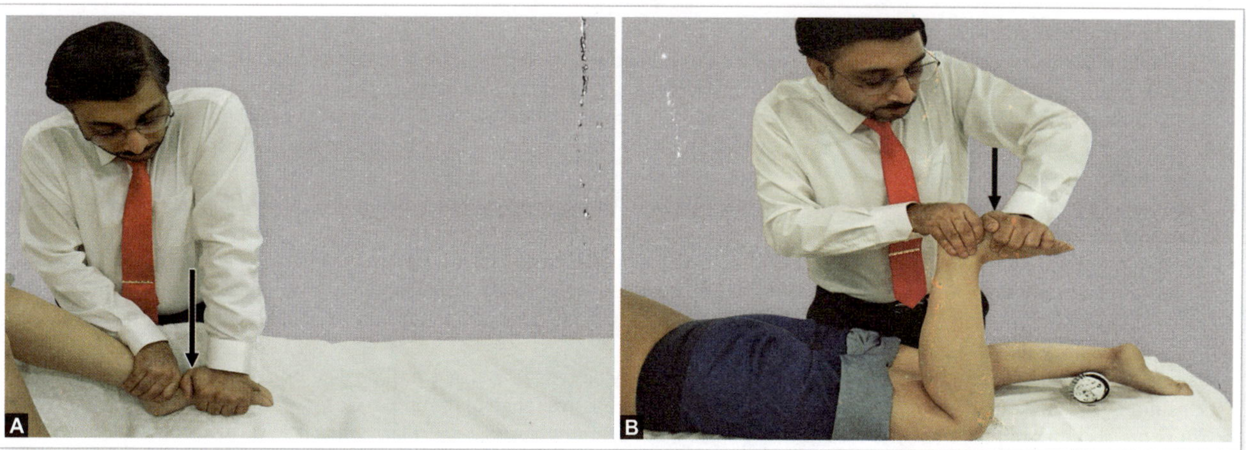

Figs 16.51A and B: Mobilization techniques of intertarsal and tarsometatarsal joint: **A.** Plantar glide; **B.** Dorsal glide

SUMMARY

- Manual therapy is a clinical approach that uses skilled, specific hands-on techniques to diagnose and treat soft tissues and joint structures. It has a history dating back to Hippocrates, who used joint manipulation and traction techniques.
  - Other notable figures include Galen, Hunter, Marsh and Fox, Still, Sutherland, Palmer, Walmsley, Steindler, Cyriax, Paris, Maitland, Melzack, Wall, and Butler.
  - The Kaltenborn-Olaf approach, developed by Freddy Kaltenborn, focuses on the use of arthrokinematics for joint lesion assessment and treatment plane.
  - The approach focuses on glides, traction, and compression techniques and the direction in which they should be applied to restore normal joint play.
  - The concave convex rule is used to determine the direction of glide, which has become the foundation of joint mobilization techniques.
- The Cyriax Approach, is based on the work of Dr James Cyriax. It involves a process of diagnosis by selective tension, using passive movements to test inert structures and resisted movements to test contractile structures.
  - Treatment techniques used in this approach include transverse frictions, joint mobilization/manipulation, traction, and injection techniques.
- Geoffrey Maitland, an Australian physiotherapist, was a pioneer in manual therapy. He developed his unique system of examination and treatment, teaching it at the University of South Australia in 1954.
- Maitland advocated for gentle passive mobilization techniques for treating neuro-musculoskeletal dysfunction. He co-founded IFOMT in 1974 and developed the Mulligan Concept of Manual Therapy, which includes correction of positional faults, mobilizations with movement (MWMs), natural apophyseal glides (NAGs), and sustained natural apophyseal glides (SNAGs).
- Mulligan's approach involves a combination of sustained passive joint glides and active movement of the affected limb, with no pain felt during the application. Natural apophyseal glides are mid to end range glides applied to the facet joints or spinous process, mainly indicated when there is gross restriction of cervical movements.
  - SNAGs are a combination of sustained glide with movement applied on the facet joint or spinous process.
- McKenzie's Approach, developed in the 1960s, focused on the centralization phenomenon and classification of spinal pathologies into postural, dysfunction, and derangement syndromes.
  - Centralization occurs when pain radiates to the extremities, and this reduction can be instant or take time to resolve. This principle is applicable to derangement syndrome and points toward reduction in mechanical deformation.
- Neural mobilization is a technique developed by physiotherapists like Elvey, Maitland, Butler, and Shacklock, which involves the interaction of the nervous system with surrounding structures, such as muscle, tendon, fascia, bone, disk, or blood vessels.
  - Neurodynamic testing involves moving spine or peripheral joints sequentially to provoke symptomatic or clinical responses.
- Joint mobilization is a widely used treatment technique for musculoskeletal dysfunction, involving a continuum of skilled passive movements applied at varying speeds and amplitudes to restore optimal motion, function, and reduce pain. It utilises osteokinematics (movement between two bones at a joint) and arthrokinematics (movement between two joint surfaces).
  - The concave-convex rule states that the direction of glide or glide depends on the shape of the articulating surface. Joint play is essential for full and pain-free joint movements, and joint shapes are determined by the force applied at the right angle to the articular surface.

- Joint mobilization can be classified:
    1. Maitland Mobilization: Utilizes oscillatory techniques graded I–IV to address pain and stiffness. Lower grades (I and II) focus on pain relief by stimulating joint receptors, while higher grades (III and IV) provide stretching to reduce joint stiffness.
    2. Kaltenborn Mobilization: Employs sustained stretch techniques graded I–III to improve joint mobility. Lower grades (I and II) relieve pain and stretch periarticular tissues, while Grade III applies sufficient force to stretch the joint capsule and enhance range of motion.
- The effects of joint mobilization can include increasing joint ROM, correcting positional faults, decreasing pain, reducing muscle spasm, and increasing muscle performance.
- Mechanical effects include increased joint ROM due to adhesions in the articular or periarticular structures, correction of positional faults, and neurophysiological effects such as decreased pain, reduced muscle spasm, and improved muscle performance.
- Joint excursion is a method used to assess joint mobility by applying glides or tractions up to tissue resistance. It helps categorize joints into hypo mobile, normal, or hypermobile.
- Pain is an important part of evaluation and treatment, identifying recovery stages and dosage of joint mobilization. End feel is a sensation or feeling detected when the joint is at the end of its passive range of motion.
- There are six types of abnormal end feels: Muscle spasm, bone, empty, springy, capsular, and soft.
- Capsular pattern refers to the pathological conditions that affect the joint capsule and cause restriction of movement to passive movements in a specific pattern. Joint mobilization is usually done with the joint in a resting position, with the mid-rage position being the most relaxed for most joints.
- The therapist should apply glides in a direction parallel or perpendicular to the treatment plane, with distraction applied in a perpendicular direction.

• The procedure for applying joint mobilization techniques involves the patient being relaxed, testing the joint in a resting position, stabilizing the proximal bone, applying mobilizing force over the distal bone, and timed with the patient's breathing.
    - The therapist should monitor pain during the mobilization technique, reassess the patient, modify the treatment technique, and follow the rule of one joint at a time.

FURTHER READINGS

- Butler DS, Jones MA, Gore R. Mobilisation of the Nervous System. Melbourne: Churchill Livingstone; 1991 Jan.
- Cyriax J., Cyriax P. Cyriax's illustrated manual of orthopaedic medicine. 2nd edition. Oxford; Boston: Butterworth-Heinemann, 1993.
- Edmond SL. Joint Mobilization/Manipulation. Extremity and Spinal Techniques- III edition. Mosby Inc. June 2016
- Hing W, Hall T, Mulligan B. The mulligan concept of manual therapy: Textbook of techniques. Elsevier Health Sciences; 2014 Jul 13.
- Kaltenborn F, Vollowitz E. Manual Mobilization of the Joints - Vol. 1: The Extremities, 8th Edition, 2014.
- Kaltenborn F. Manual Mobilization of the Joints, Volume II: The Spine; 7th edition, March 2018.
- Kisner C, Colby LA, Borstad J. Therapeutic Exercise: Foundations and Techniques. FA Davis; 2022 Oct 2017.
- Maitland G, Hengeveld E, Banks K. Maitland's Vertebral Manipulation: Management of Neuromusculoskeletal Disorders—Volume 1. Churchill Livingstone; 8th edition, 2013
- McKenzie R, May S. The lumbar spine: Mechanical Diagnosis and Therapy. Vol. 2. Spinal Publications New Zealand; 2003; 2nd edition, 2003.
- Shacklock M. Clinical Neurodynamics. A new system of neuromusculoskeletal treatment. Elsevier Health Sciences; 2005 May 6
- Ward T. Maitland's Peripheral Manipulation-Management of Neuromusculoskeletal Disorders-Volume Two. Churchill Livingstone; 5th edition, 2013

STUDENT ASSIGNMENT

LONG ANSWER QUESTIONS

1. Explain the principles, grades, indications and contraindications of joint mobilization.
2. Discuss the technique of mobilization of shoulder joint.
3. Describe the technique of mobilization of hip joint.
4. Describe the technique of mobilization of ankle joint.
5. Explain the technique of mobilization of knee joint.
6. Explain the principles of neurodynamic testing. Describe the various upper limb tension tests.

SHORT ANSWER QUESTIONS

1. Mention the concept of Maitland technique.
2. Write about the concept of Mulligan technique.
3. What do you know about the concept of Cyriax technique?
4. Briefly write about McKenzie's approach.
5. Write about the principles of joint mobilization.

MULTIPLE CHOICE QUESTIONS

1. The purpose of joint mobilization technique is to:
   a. Limit joint stability
   b. Improve passive range of motion
   c. Grading muscle strength
   d. Suppress active range of motion

2. Condition which is contraindicated of joint mobilization is:
   a. Atrophy b. Joint stiffness
   c. Hypermobility d. Adhesion capsular

3. Joint mobilization technique improves joint range of motion by increased flexibility of:
   a. Tendon b. Muscle
   c. Nerve supply d. Capsular

4. Joint mobilization with large amplitude at the inner range is referred to as grade:
   a. I b. II
   c. III d. IV

ANSWER KEY

1. b **2.** c **3.** d **4.** c

17 Soft Tissue Manipulation

Sheetal Kalra, Paridhi Sharma

LEARNING OBJECTIVES

After the completion of the chapter, the readers will be able to:

- Define massage and soft tissue manipulation.
- Understand the principles and techniques of soft tissue manipulation.
- Explain physiological and therapeutic uses of various manipulations.
- Enumerate the indications and contraindications.
- Classify the types of soft tissue manipulation.
- Define and describe various manipulation techniques used in massage.

CHAPTER OUTLINE

- Introduction
- History
- Classification
- Procedure
- Techniques
- Physiological Effects
- Indications
- Contraindications

KEY TERMS

Contact heel percussion: A variation of clapping. This technique involves applying rhythmic percussion with the heel of the hand fixed onto a specific area of soft tissue.

Effleurage: Effleurage is a massage method that uses smooth, continuous, gliding strokes with gentle pressure on the palms, fingers or thumbs.

Petrissage: It is a massage method that involves kneading, rolling, and squeezing the muscles and soft tissues.

Soft tissue manipulation: Refers to manual techniques used to treat soft tissues that are stiff, inflexible or have poor circulation due to inactivity or strain.

Tapotement: Technique that uses rhythmic tapping, pounding or striking movements on the body. This technique makes use of a sequence of fast and repetitive movements, which are typically executed with the edges of the hands, fists or fingertips.

Tenting: This method is a variant of clapping. The middle finger, which is slightly raised and put over the index and ring fingers, creates a concavity.

INTRODUCTION

All mechanical treatments for disease, as defined by Hoffa, are referred to as massage. Graham defined the term "massage" to a variety of techniques that are typically performed with the hands on the body's outer tissues, either for therapeutic, palliative, or hygienic purposes. Gertrude Beard, a leading physical therapist in the 1920s used the word 'massage' to describe specific soft tissue manipulations. The palmar aspect of the hand is the most efficient way to perform these manipulations, which aim to affect the neurological and muscular systems, as well as, the local and general circulation of blood and lymph.

HISTORY

The term "massage" has a rich history that spans across different cultures and civilizations. Its roots can be traced back to ancient Chinese, Indian, Greek, and Roman societies, where it was recognized for its therapeutic benefits.

In ancient China, around 1000 BC, the "Nei-Ching" documented the use of massage in treating paralysis and circulation issues. However, its popularity declined after the Sung Dynasty. Meanwhile, in India, massage was extensively used and mentioned in Ayurveda, dating back to the second millennium BC. The practice continues to be a vital component of traditional Indian medical systems.

Greek and Roman populations were avid supporters of physical therapy and massage. Hippocrates, who is considered the father of modern medicine, acknowledged the therapeutic benefits of massage and outlined its advantages and contraindications. Greek physician Asclepius even found that gentle stroking could induce sleep.

The Dark Ages saw a decline in the practice of massage, but it resurfaced in the fifteenth century. Surgeon Ambroise Pare in the sixteenth century pioneered the use of massage on surgical patients, and the term "kneading" was introduced by Fabricius-Ab-Acquapendente. The English Hippocrates, Thomas Sydenhams, supported physical therapy, and Nicholas Andry discussed the benefits of massage on circulation and skin color in 1741.

In the nineteenth century, Per Henrik Ling, a physical education instructor, made significant contributions by introducing massage as a branch of medical gymnastics. Swedish massage, with terms like effleurage and tapotement, gained global recognition by the end of the century.

The late 19th century saw more literature on gymnastics and massage, with significant research initiatives focusing on massage's impact on lymphatic flow, circulatory effects of vibration, histological impact on tissue injury, and physiological effects. The oldest organization of masseurs was established in Holland in 1889.

The 20th century brought new methods and systems, such as sports massage, external cardiac massage, reflex massage, acupressure, and connective tissue massage. Rosenthal, Cyriax, Graham, and Mennell were prominent figures in massage therapy during this period.

However, the first half of the 20th century witnessed a decline in the use of massage globally. Despite the growth in techniques, massage faced a love-hate relationship with the medical establishment. In the late 1960s and early 1970s, some physiotherapists began using the term "soft tissue manipulation" instead of "massage".

In the late 20th century, scientific interest in massage rekindled, leading to rigorous studies evaluating its effects using objective criteria. Techniques like plethysmography, radioactive isotope clearance rate, Doppler ultrasound, and electromyographic techniques were employed to assess how massage affects blood flow and the neuromuscular system.

The development of the Gate Control Theory of Pain in 1965 by Melzack and Wall provided fresh support for the role of massage in pain management. Researchers also sought to standardize pressure used during various massage maneuvers, utilizing high-tech pressure monitoring equipment to re-establish massage's scientific value.

From Massage to Soft Tissue Manipulation

Massage therapy terminology underwent a significant transition in the late 1960s and early 1970s. Physiotherapists and other healthcare practitioners started replacing the term "massage" with "soft tissue manipulation". This shift in nomenclature was motivated by a wish to separate the technique from negative connotations and attitudes that had developed over time. By using the phrase "soft tissue manipulation", practitioners want to emphasize the therapeutic and rehabilitative parts of their work, bringing it more in line with the developing scientific understanding of the effects of manual techniques on the body's soft tissues. This modest linguistic modification represented an intentional endeavor to strengthen the legitimacy and acceptance of the approach within the medical community, encouraging a more nuanced and clinically oriented understanding of that was once simply known as "massage".

Overview of Soft Tissue Manipulation

Musculoskeletal problems are widespread and economically costly, affecting diverse tissue types throughout life. Clinicians often use manual treatment techniques, with most of the physical therapists use manual therapy into their everyday practices. Soft tissue manipulation/mobilization (STM), often known as therapeutic massage, is a type of manual therapy that is done by hand or using rigid equipment. Instrument-assisted soft tissue mobilization (IASTM), a variant of STM, uses rigid instruments to impart precise stresses to tissues. Despite STM's ancient roots and shown benefits, much remains unknown about its underlying mechanobiology. In essence, STM is a type of mechanotherapy that uses mechanical stimuli to cause biological changes through mechanotransduction processes, ultimately improving function. Mechanotransduction pathways translate mechanical stimuli into cellular, molecular, and tissue responses. Almost all cells are mechanosensitive, meaning they respond to physical forces such as tension, compression, shear, hydrostatic pressure, fluid shear, and vibration, which affect tissue organization, growth, remodeling, maturation, and overall organism function. Most physical therapies, such as exercise, pulsed ultrasound, and vibration, use mechanical forces to effect desired changes. STM, as explained here, integrates multiple mechanical stimuli, which directly impact molecular pathways, cellular responses, tissue shape and function, as well as the processes of healing, repair, and regeneration. Terms massage and soft tissue manipulation will be used interchangeably in the entire chapter.

Did You Know?

Practitioners should recognize that the choice of terminology can influence patient perceptions, professional identity, and acceptance of manual therapy techniques. By using terminology such as "soft tissue manipulation", healthcare providers can convey a more nuanced understanding of the therapeutic benefits of manual therapy thereby enhancing patient trust and engagement in treatment.

CLASSIFICATION

The maneuvers of STM consist of the application of touch and pressure in different ways. The type of tissue that is approached during a specific technique determines the effects that are created, and the nature of the technique controls this. Any given manipulation method can be compared with other massage techniques in terms of: magnitude of the force used, force's direction, force's duration and force deployment methods. Therefore, one of the basis for categorization could be the character of techniques. Another way to categorize STM is according to the depth of tissue that is reached during a method. STM has also been categorized according to the bodily part being massaged. The hands of the massage therapist or different mechanical instruments can be used to perform massage maneuvers. This may serve as yet another classification scheme for STM (Fig. 17.1).

Based on the Character of Technique

The STM methods are divided into the following four basic groups based on their nature of character, and each group has multiple subgroups.

1. **Stroking manipulations:** Superficial stroking, effleurage (deep stroking).
2. **Pressure manipulations:** Kneading, friction, petrissage.
3. Tapotement/percussion manipulations.
4. Vibratory manipulations.

Based on the Depth of Tissue Approached

Massage can be categorized as follows depending on the profundity of tissue reached during manipulations:

- **Light massage techniques:** Since little power is used during the maneuver, only the superficial tissues are affected by the massage. For example, stroking and tapping, etc.
- **Deep massage techniques:** To ensure that the effects of massage reach the deeper tissues, like muscles, moderate to

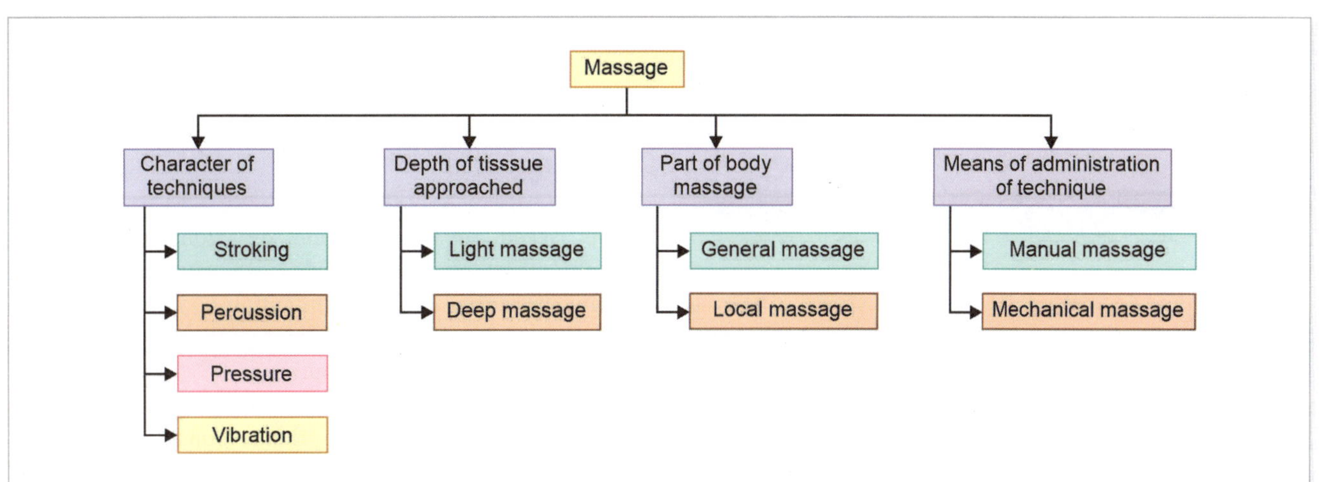

Fig. 17.1: Classification of massage techniques

deep pressures are used during the massage. For example, friction, kneading, etc.

Based on the Region

Massage can be categorized as follows according to the region to which it is given:

- **General massage:** The word "general massage" typically refers to massage that is performed on the entire body. However, massage that is given to major areas of the body, such as the back, a lower leg, etc., can fall under this heading. It is typically given to debilitated people after extended recumbency and to athletes after strenuous physical activity to make them feel better and provide comfort.
- **Local massage:** Local massage refers to the application of massage to a specific region of the body. This is used for managing the pathological conditions that are present locally. Local massage can include, for instance, friction to the lateral ligament of the foot after a sprain or massage of the wrist in tenosynovitis.

Based on the Means of Administration of Technique

Massage falls into one of the following two categories according to how it is administered:

1. **Manual massage:** The term "manual" describes the act of placing one's palm over another's body. Manual massage is a type of massage performed by using the therapist's hands or another part of the body. For instance, acupressure massage, connective tissue massage, trigger point massage, and traditional massage techniques.
2. **Mechanical massage:** Mechanical massage is done when the body of the patient is treated with mechanical energy delivered through mechanical devices based on the massage principle in order to manipulate soft tissue. As an example, consider compression devices, vibrator, pneumatic massage, etc.

PROCEDURE

Guidelines

- **Understanding underlying pathology:** In-depth knowledge of the underlying pathology is essential for informed and effective soft tissue manipulation. This includes awareness of specific conditions, injuries, or issues affecting the soft tissues.
- **Basic massage principles:** Practitioners must possess manual dexterity, coordination, and concentration. Additionally, patience and courteousness are crucial attributes for a positive therapeutic experience.
- **Joint care:** Avoiding constant hyperextension or hyperflexion of joints is vital to prevent hypermobility, ensuring the safety and comfort of the patient.
- **Hand hygiene and warmth:** Hands should be clean, warm, dry, and soft. Maintaining short and smooth nails is important for client comfort. Warm hands enhance the overall experience for the recipient.
- **Correct positioning and posture:** Proper positioning is necessary for relaxation, preventing fatigue, and allowing free movement of arms, hands, and body. Maintaining good posture is key to preventing practitioner fatigue and backache.
- **Optimal body position:** Ensuring a good position is crucial for the correct application of pressure and rhythmic strokes during the procedure. This contributes to the effectiveness and comfort of the soft tissue manipulation.
- **Duration guidelines:** The duration of the massage should be tailored to various factors, including pathology, the size of the area, speed of motion, age, size, and condition of the patient or athlete. Pain should be avoided during the procedure.
- **Direction of forces:** Forces should be applied in the direction of the muscle fibers to optimize the therapeutic impact of the soft tissue manipulation.
- **Session initiation and closure:** Each session should begin and end with care. Ensuring the patient is warm, comfortable, and properly draped sets the stage for a positive experience.
- **Sequential approach:** The massage session should commence with superficial stroking, gradually progressing to deeper techniques based on the patient's response and tolerance.
- **Joint and bony prominence considerations:** Bony prominences and painful joints should be avoided whenever possible to prevent discomfort and ensure a safe and effective soft tissue manipulation experience.

Variables

Depth, speed, rhythm, duration, direction, and frequency are all important variables when doing massage strokes. For inexperienced therapists, intentionally combining these elements is critical, but with practice, they become second nature, evolving into a fluid approach.

- Depth refers to the force exerted to the tissue, which varies depending on the desired result. Consider the client's tolerance when increasing pressure gradually and carefully. Signs of discomfort, such as facial strain or breath-holding, should be observed.

- Speed regulates the stroke's tempo, which influences the response—slow for relaxation or brisk for invigoration.
- Rhythm, like pace, can be slow or quick, affecting the overall tone of the massage. Avoiding jerky motions is essential.
- Duration includes the length of each stroke as well as the amount of time spent on a certain body area. Slower and longer strokes are great for relaxation, while faster strokes are energetic.
- Direction refers to the course of the stroke, which is normally centripetal toward the heart for maximum blood flow.
- Frequency refers to how many times a stroke is done. The rule of three applies, with each stroke performed three times before transferring to the next, ensuring a thorough and effective massage session. Overall, mastering these concepts helps to ensure that massage strokes are applied smoothly and therapeutically.

TECHNIQUES

In diverse ways, different strategies have been explained by various authors. The basic massage techniques covered in this chapter are stroking, pressure manipulation, tapotement, and vibration (Fig. 17.2). The following is a list of the fundamental characteristics of massage techniques:

- Technique should exert mechanical pressure on the body's soft tissues.
- The joint's position must not change as a result of these forces.
- A physiological and/or psychological response must be elicited by the approach in order to achieve the therapeutic, restorative, or preventive purpose.

Given these characteristics, massage can be defined as any manual or mechanical technique that applies mechanical energy to the body's soft tissues through the skin without changing the position of any joints in order to elicit a particular physiological or psychological effect that can be used for therapeutic, preventive or restorative purposes on both sick and healthy individuals.

Stroking Manipulations

- Stroking massage, sometimes called effleurage, is a gentle and rhythmic massage method that uses long, sweeping strokes on the body generally with the palms or fingers. This method is frequently employed at the start and conclusion of a massage session to encourage relaxation, stimulate blood circulation, and prepare the muscles for more advanced massage movements. Stroking massage creates a soothing and calming impact, making it an essential component in many massage methods.
- In this technique, one continuous linear movement of the hand or a portion of it is used to cover a single aspect of an entire segment. "Stroke" is the term used for describing the continuous linear movement of a hand throughout a segment's whole length. Equal rhythm, even pressure, and persistent hand-to-body contact between the therapist and patient are the key characteristics this category of techniques. This group comprises two fundamental stroke types, namely *superficial stroking* and *effleurage/deep*

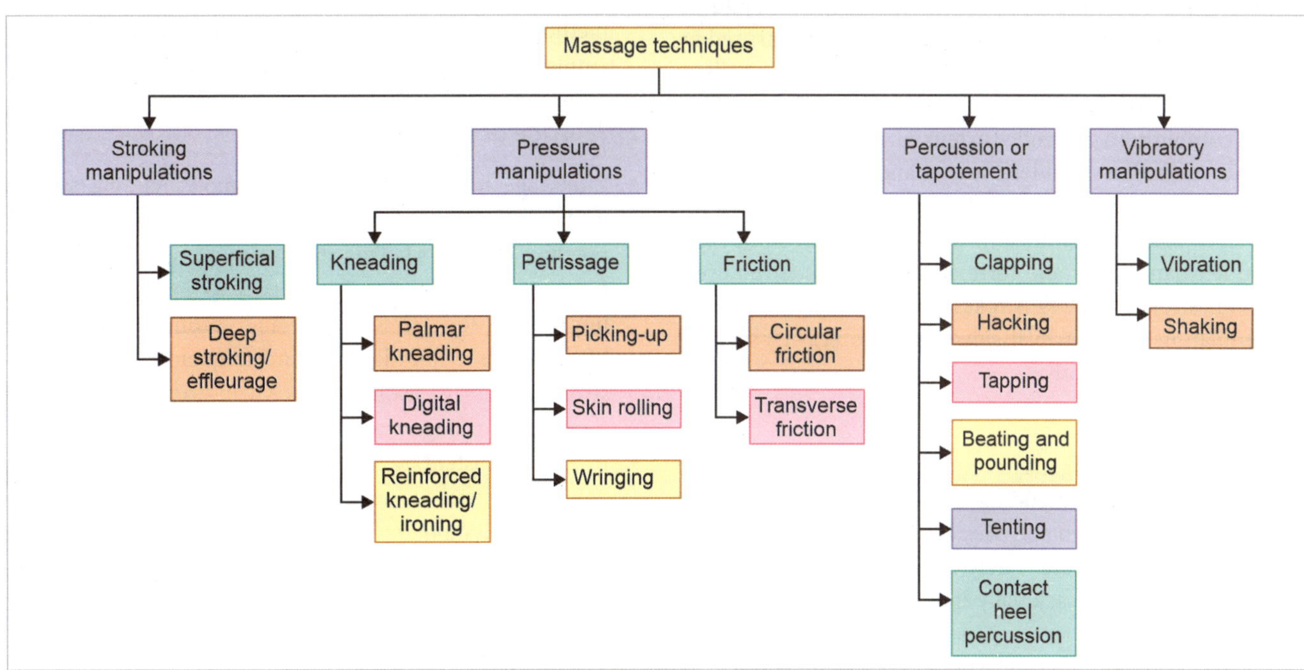

Fig. 17.2: Various massage techniques

stroking, based on the amount of pressure used and the direction of hand movement.

Superficial Stroking

- It is the rhythmic application of light pressure to the skin using the hand or parts of the hand over the skin in order to stimulate the senses. Either the proximal to distal portions or the opposite way around could be used for the strokes.
- Although the speed, or the number of strokes per minute, might be fast or slow, the rhythm must be maintained. The desired outcome determines how this approach changes in direction and speed.

Technique

1. A comfortable position for the therapist is recommended. The walk-out and fall-out standing techniques are advised since they usually cover the entire segment in a single move. The therapist needs to keep their hand completely relaxed. To be able to bend the fingers to the shape of the body, the fingers should be flexible and kept slightly apart.
2. The most proximal end of the area that has to be massaged should be where the therapist places their right hand pulp to start the process. The therapist flexes their right shoulder and extends their right elbow to do this. To hold the left hand in the air and position it toward the most distal point of the segment, the left elbow should be bent and the left shoulder should be extended (Fig. 17.3A).
3. To ensure that the hand softly lowers down and makes contact with the skin, the stroke is started with a controlled extension of the right shoulder. The contact should be light and practically pressure-free. Throughout the entire stroke, this "just touch, no pressure" contact should be kept. The right hand's massage over the skin is ensured by the controlled and coordinated flexion of the right elbow and extension of right shoulder. By carefully controlling and combining a controlled extension of the left elbow and flexion of the left shoulder, the left hand should travel in the proximal direction as the right hand advances toward the distal area. This rhythmically steady movement of the right and left hands back and forth across the skin is required. Faster strokes (30–40 strokes/min.) are preferable for stimulating effects whilst slower strokes (12–15 strokes/min.) are more sedative.

Types

The superficial stroking may be modified by using:

1. The entire palm, including the large, heavily muscled regions like the thigh and back (Fig. 17.3B).
2. Though knuckles are rarely used in general massage therapy, they are occasionally applied in sports massage for specific purposes.
3. The ball of the thumb can be used over a small region.
4. The thousand-hand technique, which does not span the entire length or section in a single motion. Instead, the method uses a number of brief, overlapping strokes. It is employed for relaxation purpose.

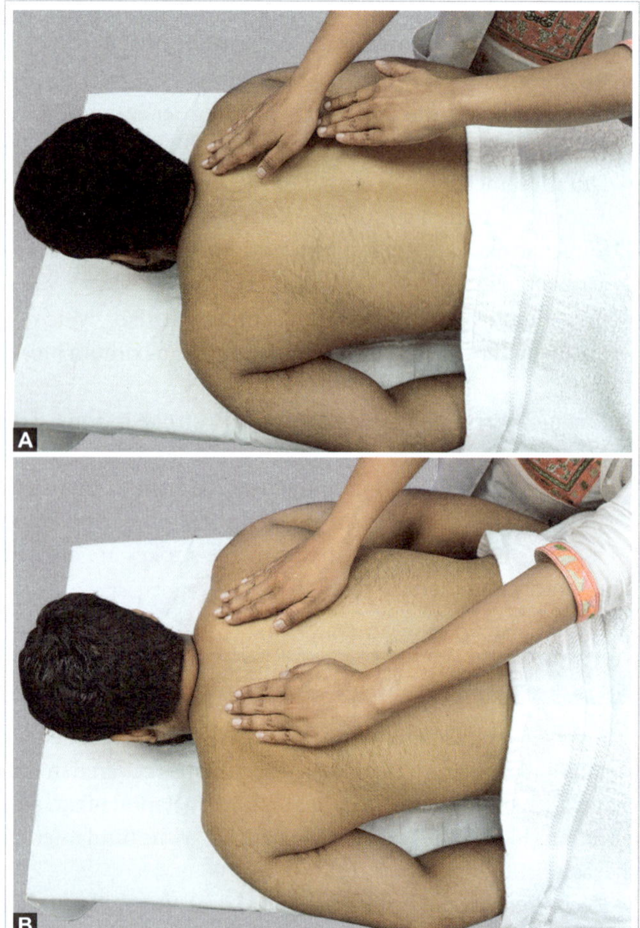

Figs 17.3A and B: Superficial stroking

Effleurage or Deep Stroking

- Effleurage involves smooth, continuous, gliding strokes using the palms, fingers, or thumbs, applying gentle pressure (Fig. 17.4). This rhythmic and relaxing action is frequently used in massage therapy to warm up muscles, improve blood circulation, and promote relaxation. Effleurage is a foundational and frequently introductory stroke in massage sessions, which helps to prepare the body for deeper massage maneuvers.
- It is the persistent, rather forceful movement of the hand's palmar side over the exterior of the body in the direction of venous and lymphatic drainage. It is one of the fundamental massage techniques.

Technique

1. The therapist can use one or both hands to apply the effleurage, depending on the area being massaged. Typically, the thumb, fingers, or palmar side of the hand

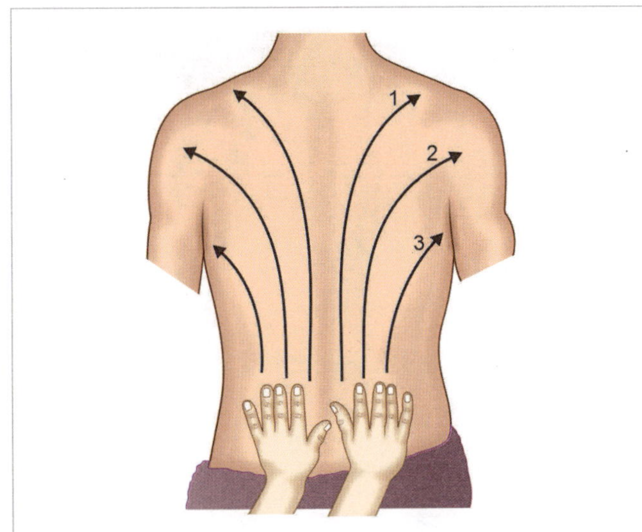

Fig. 17.4: Effleurage or deep stroking

are used. This method covers the entire segment in a single motion, much like superficial stroking. So, it is preferred that the therapist attains walk standing position.

2. The therapist's hand(s) should be laid back over the distal-most portion of the segment to be massaged.
3. The hand must be tailored to fit the contour of the component in order to provide uniform skin contact across the hand. The regulated, coordinated bending of the shoulder and extension of the elbow initiates the stroke that advances the hand over the skin from the distal to the proximal region.
4. Pressure is applied by the therapist applying their body weight through their upper extremity to the subject's skin.
5. As the hands approach, the body weight should be transferred from the back foot to the front foot to ensure proper pressure distribution on the skin. Weight should be totally transmitted from the rear foot to the front foot toward the finish of the stroke.
6. A small amount of overpressure applied close to the lymph node halts a stroke. A returning stroke is then used to return the hand to its initial position. This returning stroke is carried out by carefully extending the shoulder and flexing the elbow.
7. This returning stroke is essentially superficial in nature because it does not apply any pressure to the skin. Lessening the effects of effleurage, pressure during the returning stroke will impede the lymphatic and venous refilling.
8. Every effleurage stroke should be performed with as much pressure as is possible without the patient feeling uncomfortable. The stroke rate should essentially be slow (10–20 strokes/min) to provide the lymphatic and venous channels adequate time to fill up.

Types

Effleurage can be adjusted in the following ways, depending on the area that requires treatment:

1. One can finish the stroke fast by following one hand with the other hand, using both hands on the segment's opposing aspects, such as the medial and lateral side of the thigh (Fig. 17.5).
2. It is possible to utilize just one hand. The massaged portion is supported or moved around by the other hand. This variant is appropriate for use with the upper limb (Fig. 17.6).
3. Other techniques involve effleurage using ulnar border of hand, forearm, thumb and fist (Figs 17.7A to D).
4. To make the hand resemble the letter "C", limbs can be held in the initial web space between the thumb and other fingers (Fig. 17.8).
5. **Crosshand effleurage or knee effleurage:** Special hand positioning is utilized to drain the region around the knee joint.

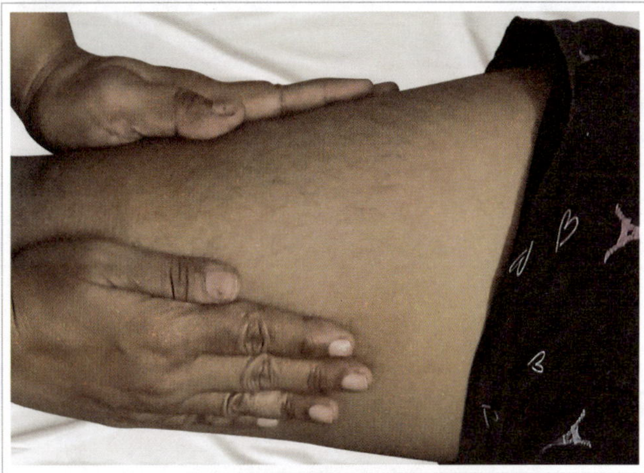

Fig. 17.5: Effleurage medial and lateral side of the thigh

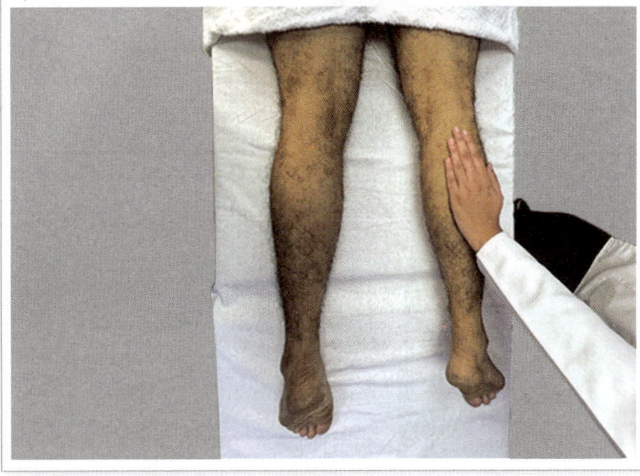

Fig. 17.6: Effleurage using one hand

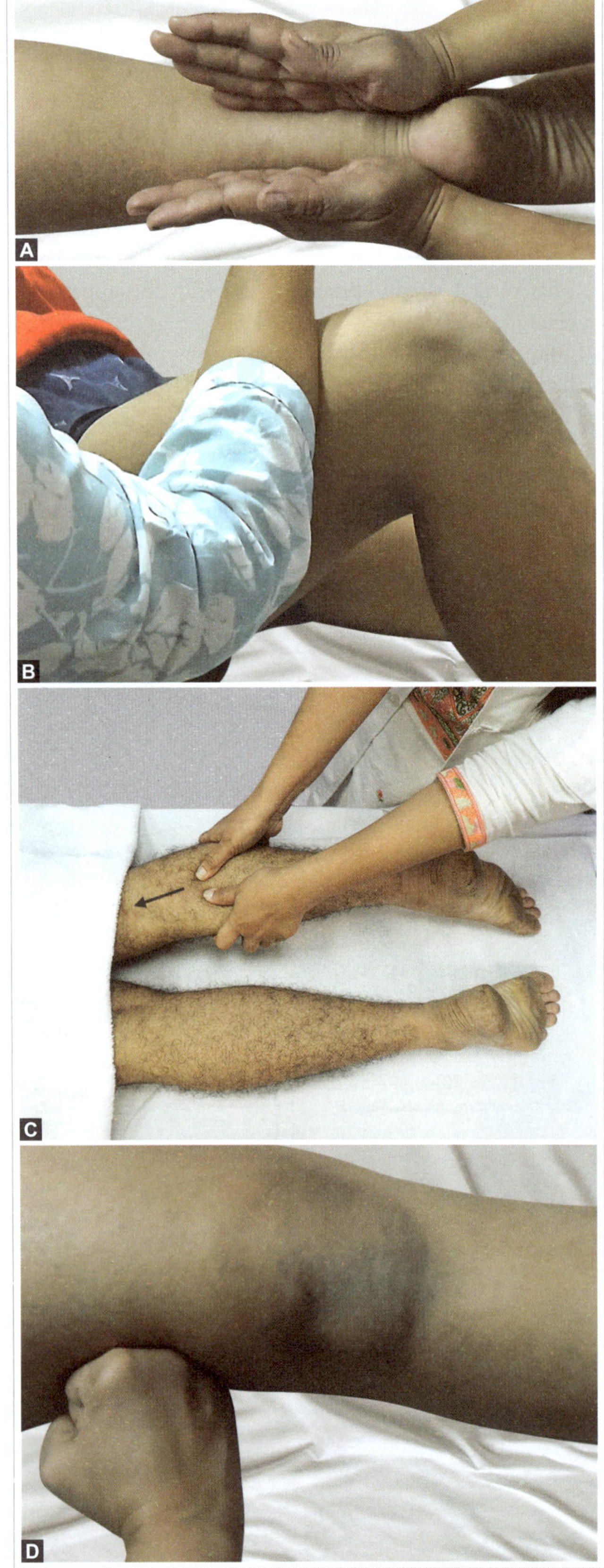

Figs 17.7A to D: Effleurage adjusted techniques: A. Using ulnar border; B. Using forearm; C. Using thumb; D. Using fist

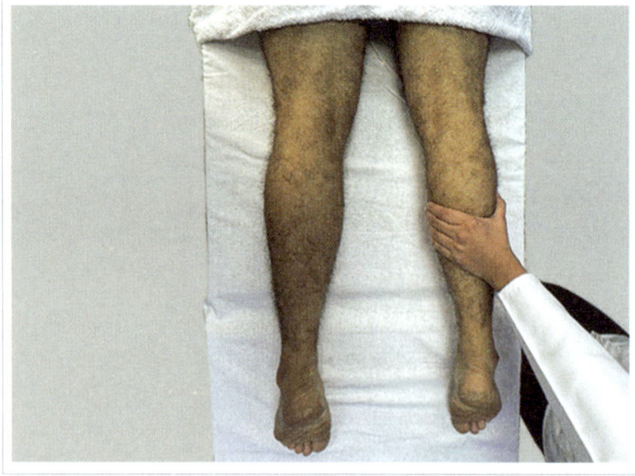

Fig. 17.8: Effleurage with C shaped hand

i. Just above the patella, both hands are positioned so that they cross one another without touching (Fig. 17.9A).
ii. During effleurage, both hands are drawn backward, one on each side, until the heels of both hands meet slightly below the patella (Fig. 17.9B).
iii. Following that, the stroke terminated in the popliteal fossa by moving the fingers posteriorly and forward (Fig. 17.9C).
iv. Effleurage should not be applied to wounds that are open or have recently healed scar tissue, as these situations increase the danger of the skin peeling off.

Pressure Manipulations

The application of deep compression to the body with persistent touch is the key component of this series of techniques. These methods focus in particular on the muscle tissue. By using deep localized pressure, these approaches enable the greatest mechanical movement possible between various fibers. Techniques from this group can be classified into three main subgroups based on the type and direction of applied pressure namely:
1. Kneading
2. Petrissage
3. Friction

Pressure is applied intermittently during kneading and petrissage, while pressure is applied continuously during friction.

Kneading

By applying pressure in a circular motion along the long axis of the underlying bone, this approach creates a vertical compression as tissues press on the underlying structures. The pressure is gradually increased and decreased.

Techniques

1. Hands that are at ease are put on the skin. The hands/fingers should be firmly placed on the skin to guarantee

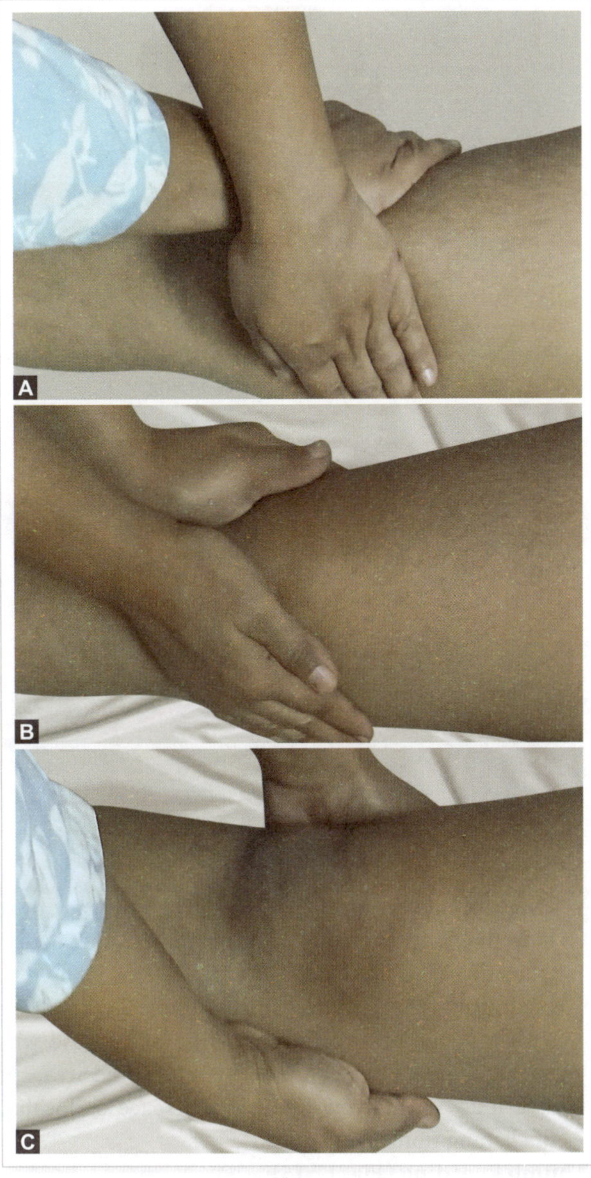

Figs 17.9A to C: Crosshand effleurage or knee effleurage

that the movement happens only in the deeper structure and not over the skin.

2. The entire maneuver is executed in a sequence of small, concentric circles that are parallel to the body surface, each of which overlaps the one before it.
3. For the first half of the circle, the pressure increases gradually until it reaches its maximum at the top. As in the first half, the pressure decreases gradually in the second half until it reaches its lowest point at the bottom of the circle.
4. The compression phase and the relaxation phase are the two phases that each circle in this system goes through.
5. The hand gently moves to the next adjacent site during the phase of relaxation when the pressure is at its lowest. The pressure is applied using body weight.

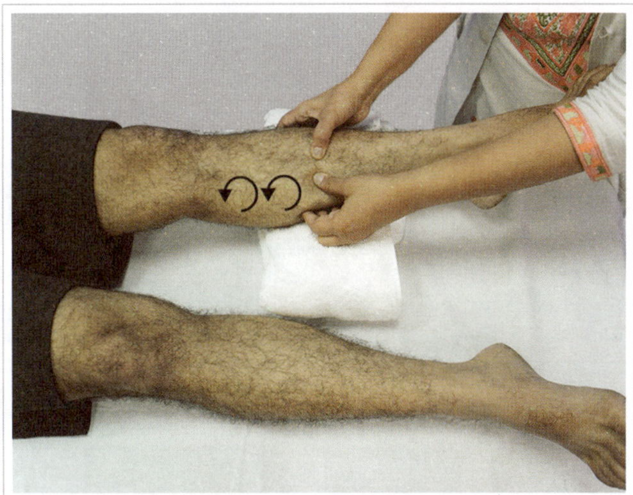

Fig. 17.10: Kneading

6. The therapist typically adopts a walking stance while standing, applying pressure by alternately shifting weight to both legs.
7. A therapist may start kneading at the segment's proximal or distal end (Fig. 17.10).
8. To knead, perform a quick, circular motion with one or both hands using fingers, thumbs, or palms. When making a circle with both hands, they can do so alternately or simultaneously. While the other hand circles, one hand may support the tissue.

Types

The following three kneading techniques can be categorized according to the part of the hand utilized during the maneuver:

1. Palmar kneading
2. Digital kneading
3. Reinforced kneading

Numerous modifications can be made to each technique:

Palmar Kneading

Either the whole palm or the heel of the hand is used to execute it. It can be done with either hand or both hands, and it is typically applied to large areas, like the thigh, calf, arm, etc., (Fig. 17.11).

Digital Kneading

There are two subcategories of this since the kneading movements are performed either with the thumb or the fingers:

1. **Finger kneading:** This method applies pressure by making contact with the palmar portion of a finger, either in its entirety or in part. To enhance the contact area, two or three fingers might be used in combination with one finger. Little finger is typically not used due to their low length and inability to exert enough pressure.

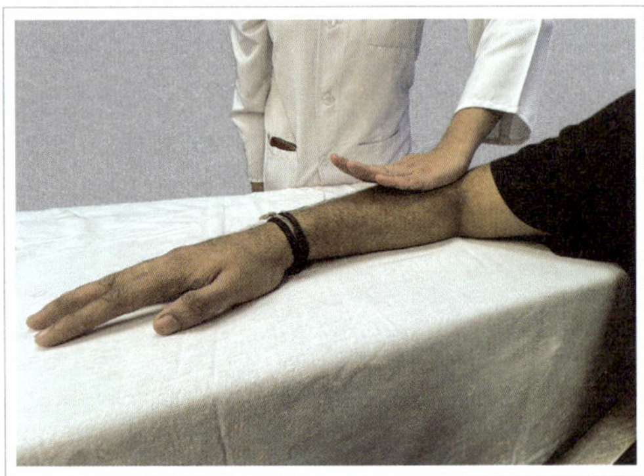

Fig. 17.11: Palmar kneading at forearm

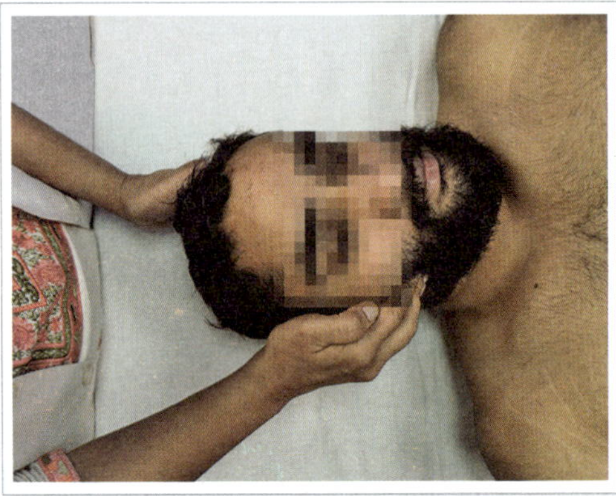

Fig. 17.12 Whole finger kneading

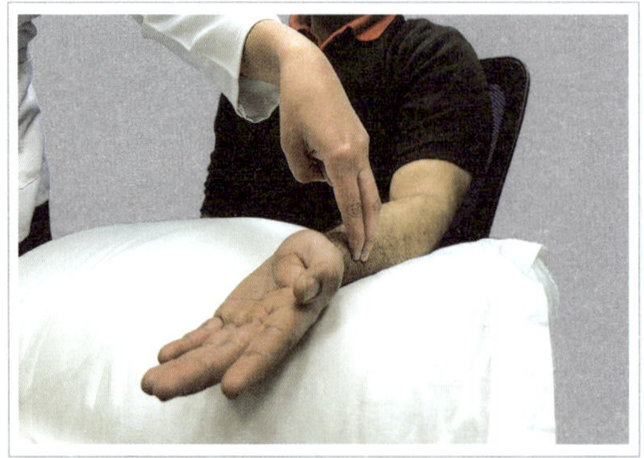

Fig. 17.13: Finger pad kneading

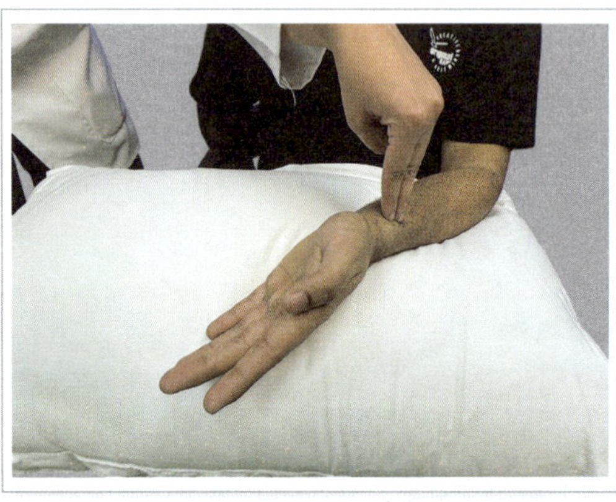

Fig. 17.14: Fingertip kneading

This has the following subgroups depending on which part of the finger stays in contact with the skin while kneading.

- **Whole finger kneading (Fig. 17.12):**
  - The palmar surfaces of the second, third, and fourth fingers are frequently used in this method, which is sometimes referred to as flat finger kneading, as they are held together to provide a large contact area.
  - The palm and thumb avoid making skin-to-skin contact.

  Usability: This is typically applied to the mandibles since they are less muscular or have less padding.
- **Finger pad kneading (Fig. 17.13):**
  - This method makes use of the contact of distal phalanx's palmar surface, commonly recognized as the finger pad or finger pulp.
  - A small area can be massaged with one finger, while a larger area can be massaged with two or three fingers held together.
  - The distal interphalangeal and proximal interphalangeal joints of the kneading finger should be kept slightly flexed during this technique to provide proper contact between the pulp and skin and to lessen strain on the interphalangeal joint ligaments.

  Usability: It is applied to the area around joint lines, deep scars, and along the ligament lines.
- **Fingertip kneading (Fig. 17.14):** The only part of the pulp that is in contact with the skin with this method is the tip.

  Usability: This method, which is typically conducted with just one or two fingers, is the one to use when kneading needs to be done across a localized area, such as a long and narrow interosseous space or localized thickening fibrositis nodules.

2. **Thumb kneading (Fig. 17.15):**
   - Depending on the extent of the area to be treated, the therapist can knead using the pad or the tips of one or

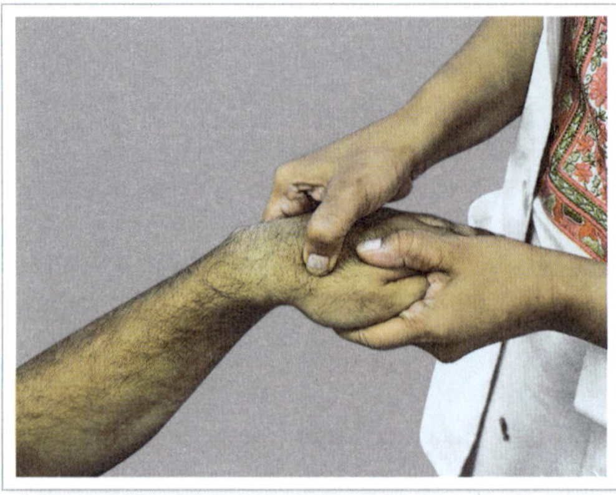

Fig. 17.15: Thumb kneading

both thumbs. The remaining fingers should be placed slightly away from the area that has to be massaged, preferably on the opposite side of the extremities, or they should refrain from making direct contact with the skin.

- Alternatively, to strengthen the thumb in contact, one thumb can be positioned over the other.

  The two variants of this method are as follows:

  i. **Thumb pad kneading:** The thumb pulp, consisting of the palmar surfaces of the proximal and distal phalanx, remains in touch with the skin. It is used on the thenar and hypothenar eminences, which are smaller, more muscular areas (Fig. 17.16A).

  ii. **Thumb tip kneading:** The technique is similar to fingertip kneading over long, thin places such as interosseous gaps, and is usually administered with the side of the thumb tip (Fig. 17.16B).

  Throughout these procedures, the therapist should keep their IP joint slightly flexed. By doing so, the palmar ligaments are protected, which could be damaged by lengthy finger-extended kneading.

Usability: These digital kneading techniques are most frequently applied to the face, the paravertebral region of the spine, localized areas like palm, and joints.

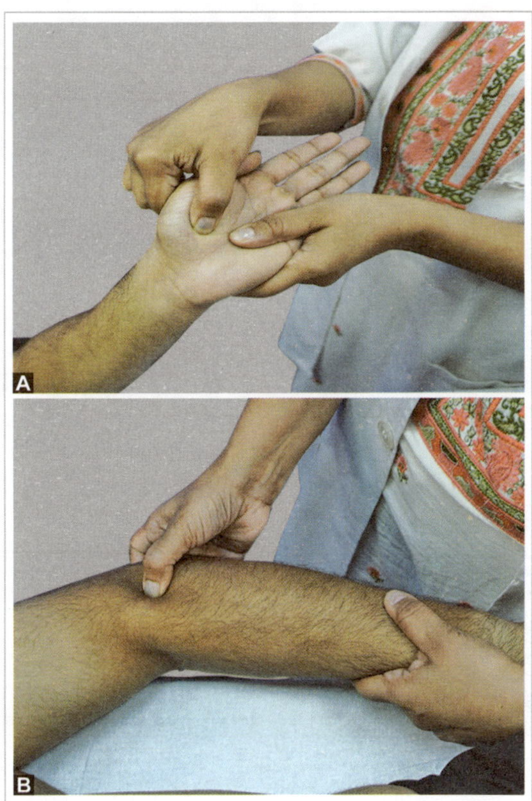

Figs 17.16A and B: **A.** Thumb pad kneading over the thenar eminence; **B.** Thumb tip kneading

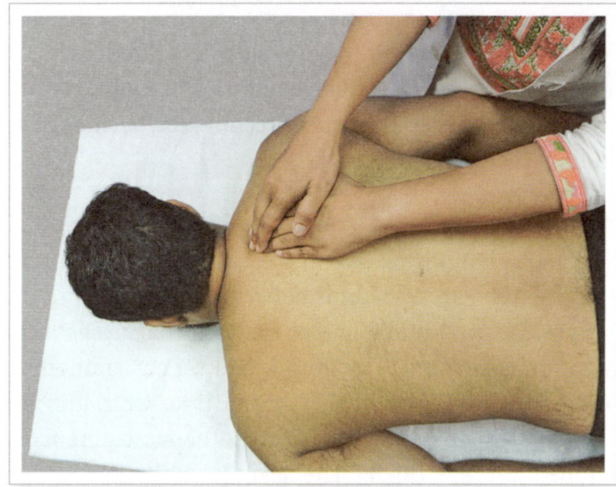

Fig. 17.17: Reinforced kneading

When a localized region needs to be mobilized, digital kneading maneuvers are the best option.

Reinforced Kneading

This technique consists of kneading with both hands and the technique is also known as ironing. When a deeper depth is needed, it is used to convey pressure to very deep palmar region and joints. This maneuvers is usually carried out over the back (Fig. 17.17).

Petrissage

The French term "Petrir", which means "to knead", is the source of the English word "petrissage". It is a massage method that involves kneading, rolling, and squeezing the muscles and soft tissues.

- Petrissage involves lifting and manipulating the underlying tissues using a tight hold and rhythmic motions of the therapist's hands, fingers or thumbs.
- This technique seeks to stimulate blood circulation, improve flexibility, and relieve muscle stress. Petrissage is widely used in massage therapy to relieve muscle tension and enhance total relaxation.
- Petrissage and kneading are only distinguished by the direction in which pressure is applied.

According to Mennels, petrissage requires picking up soft tissues while applying lateral compression, whereas kneading involves vertical compression. Rolling, lifting, and wringing the skin will move the tissues away from the tissues which are compressed against the underlying bone by kneading. Thus, these three techniques differ from kneading maneuvers by the direction in which pressure is applied, which is something they all have in common. So, it makes sense to classify these methods as petrissage.

As a result, in the petrissage set of techniques, the tissues are elevated from the underlying structures and pressure is periodically applied at a right angle to the long axis of the bone. This group uses a variety of techniques, such as picking up, rolling the skin, and wringing.

Picking-up

The therapist hand is placed on the afflicted area of the body so that the web gap that forms between the thumb and index finger crosses the middle line of the area that needs to have skin and muscle lifted. The thumb and thenar eminence are positioned on one side of the central line, while the index and middle fingers, which have hypothenar eminence, are positioned on the other. The arm is maintained slightly abducted and the elbow is maintained in a semi-flexed position. Using this technique, the tissue is raised at a straight angle to the bone underneath it, squeezed, and then released (Figs 17.18A to D).

Usability: It is among the most difficult massage techniques to learn. This technique is useful in the upper back, calf and lower back regions. One both hands can be used in this technique.

Technique

1. The wrist need to stay slightly extended.
2. The procedure is started by applying compression to the skin by transferring body weight through an upper extremity. At the same moment, the wrist extends slightly, and the grasp of the hand tightens. This causes the tissue to be simultaneously lifted and squeezed.
3. The compression is then once more released when the hold is removed and body weight is transferred.
4. The identical maneuver is repeated when the compression is removed, with the relaxed hand moving to the next close spot without losing touch or skin conformation. This is how lift, squeeze, and release are done.

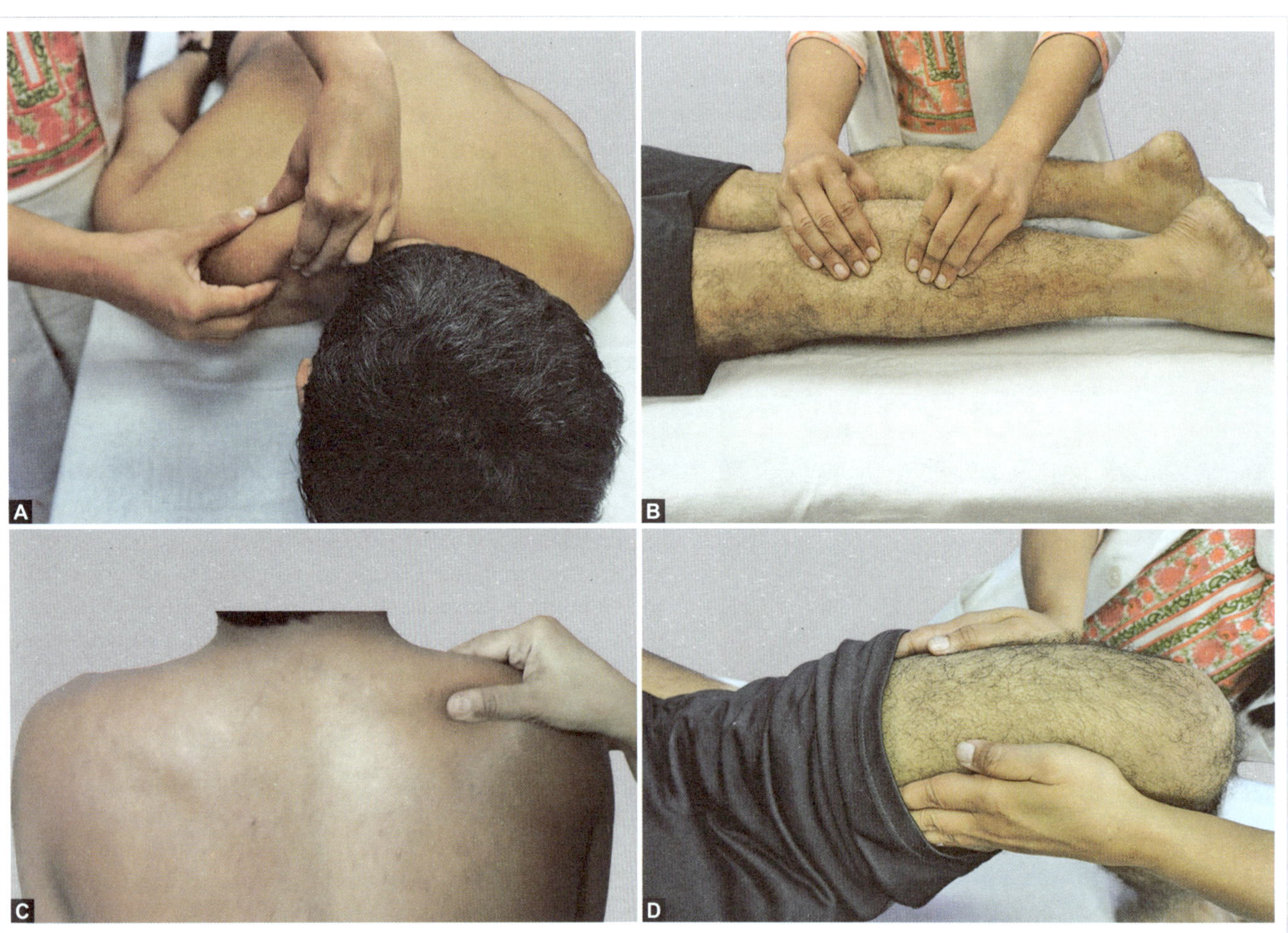

Figs 17.18A to D: Picking-up techniques

5. When taking up, always make sure the web area between the thumb and index finger is in contact with the skin.
   - The therapist should not make contact with the anterolateral aspect of the second metacarpal head or the anteromedial aspect of the thumb's proximal phalanx.
   - A localized force is transmitted to skin and causes discomfort when the bony prominences come into contact.
   - The pressure applied to the tissue being picked up should be uniformly distributed throughout.
   - Either the pressure applied or the amount of tissue being gripped should be reduced if pain is felt during the maneuver.
6. On the larger areas, like thigh, the technique can be modified. In order to provide a larger grasp, the hands may be placed on either side of the thigh. This results in a larger hold, and tissues are lifted and squeezed using the palm of both hands.

Skin Rolling

The skin between the thumb and the fingers is lifted, stretched, and moved over the subcutaneous tissue in this procedure. With one or both hands, the therapist raises and moves the skin and superficial fascia while maintaining a continuous roll of raised skin in front of the moving thumb (Figs 17.19A and B).

Technique

1. A full palmar contact is maintained with the skin by placing both hands on it with the thumbs abducted.
2. The method is started by drawing the fingers backwards firmly enough to lift the skin. If done correctly, the skin will be lifted.
3. Across the underlying skin, the thumb is simultaneously flexed and adducted while applying downward pressure. Lifting a roll of skin away from the underlying structures requires the coordination of finger and thumb movements.
4. The raised skin is then rolled against the fingers by moving the thumbs forward. Stretching the epidermal and subcutaneous tissue and inducing relaxation are the key effects of this therapy.

Usability: Most frequently, it is performed over the back. It can be applied in a modified form on the thickened and shortened scar tissue as well as the area around the superficial joints. One hand can also roll the skin across a small area, such the foot or wrist.

- Using a similar grasp, one can slightly lift the entire muscle over the arm, thigh, and calf rather than just the skin.
- This can be done by gently lifting the muscle between the tip of the fingers and the thumb.

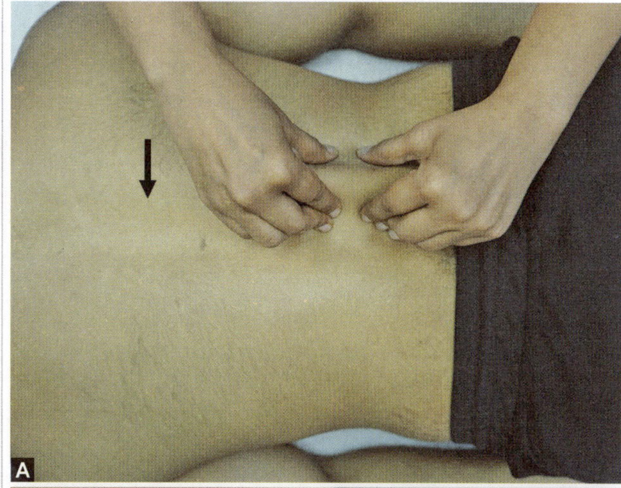

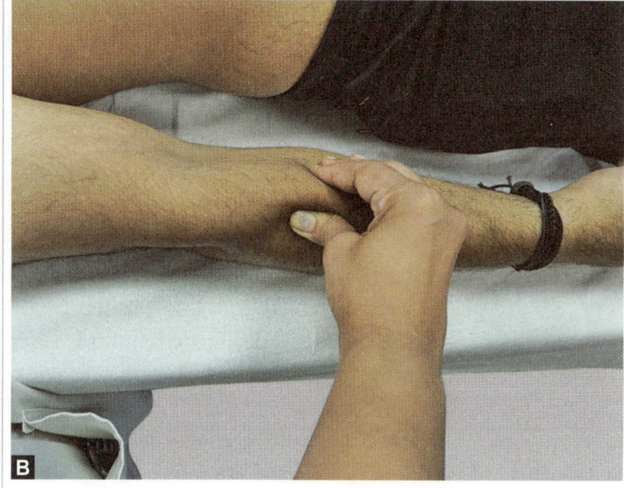

Figs 17.19A and B: Skin rolling: **A.** With both hands; **B.** Using single hand

- Then using the thumb and fingers to alternately apply and release pressure, the therapist can roll the muscle fibers from side to side.
- Muscle rolling is the term for this version.

Wringing

This method has some similarities to the act of twisting and compressing a damp towel. With each hand on the opposite side of the limb, move it in the opposite direction, moving one forward and the other backward. The tissues are lifted and twisted as a result of this (Fig. 17.20).

Technique

- With the exception of the fact that both hands are employed in this technique, the grasp and hand placement are identical to those used in picking up. The hands are placed so that the fingers and thumbs of both hands face one another but are slightly apart on the skin with the

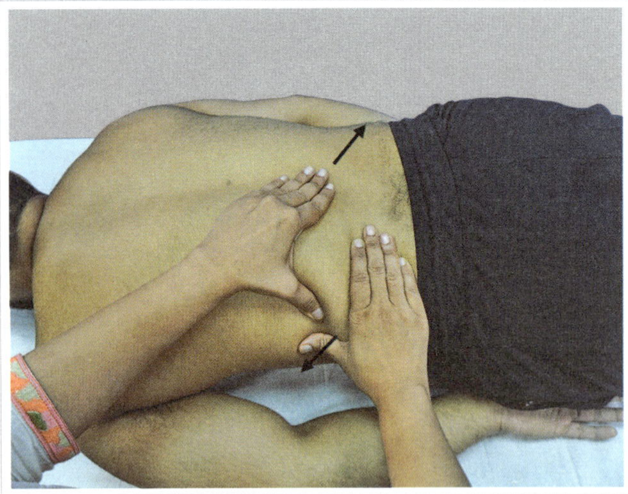

Fig. 17.20: Wringing

thumbs abducted. The therapist adopts walk standing stance while applying compression with body weight. The technique starts with the right hand's fingers being pulled, lifting the skin off the underlying tissues. Concurrently, the left thumb presses the skin in the opposite direction. As a result, lifted tissue experiences stretch. After that, the tissues are pulled by the right thumb and pushed by the left hand's finger. The process is performed multiple times before the hands move on to the next location. It works incredibly well to remove adhering skin.

- Overly stretching may have the reverse effect of what is intended in case of flaccid muscles, therefore deep pressure movements should be judiciously used over areas afflicted by flaccid paralysis.

Friction

Friction massage is a massage therapy technique that uses deep pressure and friction to target specific regions of the body. This technique is characterized by firm circular or transverse oscillatory movements made generally with the fingers, thumbs, or palm. By releasing adhesions in the muscles and connective tissue, friction massage improves blood flow, increases range of motion, and relieves tension in the muscles.

There are two kinds of friction which are impacted by the direction of movement:

1. Circular friction
2. Transverse friction

Circular Friction

Wood advocated this type of friction massage in 1974. This is similar to digital kneading, with the exception that there is no relaxation phase and the tissue is continuously subjected to deep pressure throughout the whole performance (Fig. 17.21A).

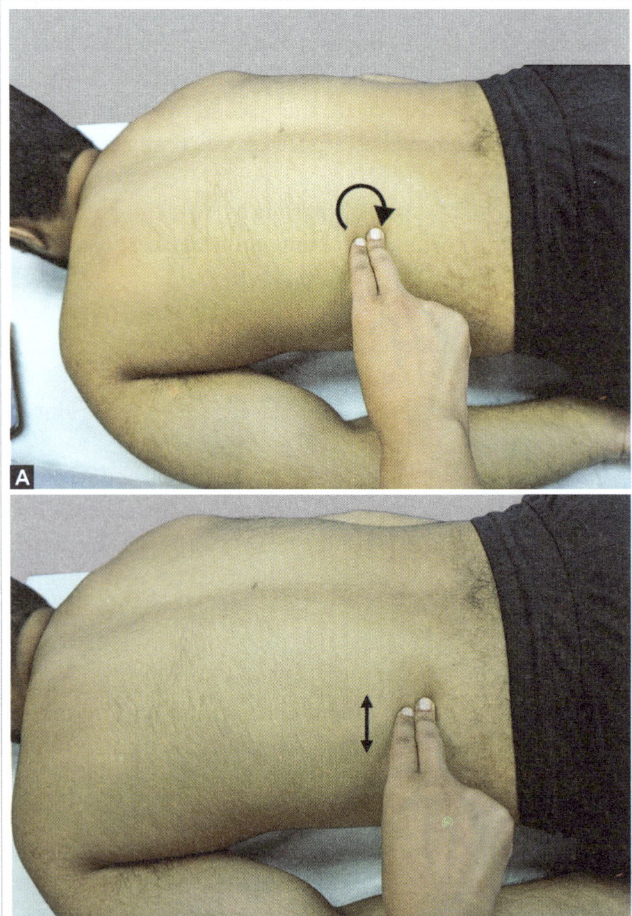

Figs 17.21A and B: Friction massage: **A.** Circular; **B.** Transverse

- Fingers are put over a certain spot, and a tiny amount of downward pressure is applied, causing the skin and fingers to move as one in a circular manner.
- The pressure is steadily increased as the surface structure relaxes.

  Usability:
  - This method is used on a specific targeted location, such as joints, muscle attachments, fibrositic nodules and their surroundings, etc.
  - It is utilized to produce a localized effect on muscles that have been under strain for a long time, such as paravertebral muscles.

Transverse Friction

Dr Cyriax of England promoted and popularized this method in the first half of the 20th century. Here, the movement that needs to be corrected is transverse, or across the long axis of the structure. It can be carried out using:

- The thumb's tip (Fig. 17.22A)
- The index or middle finger, which can be reinforced by overlapping them (Fig. 17.22B)

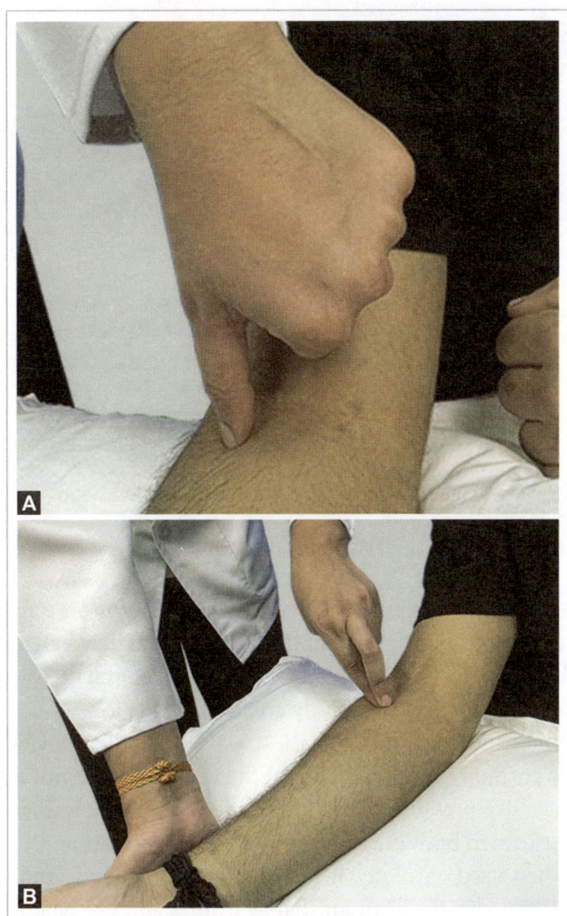

Figs 17.22A and B: **A.** Transverse friction using thumb's tip; **B.** Transverse friction using reinforced fingers

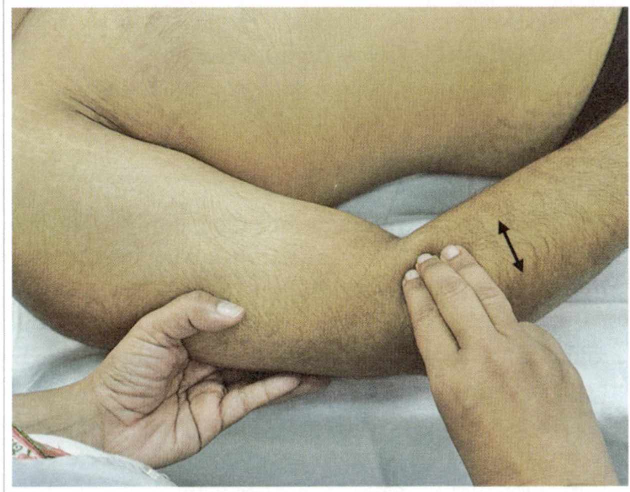

Fig. 17.23: Transverse friction technique using three fingers over wrist extensor muscles

- Two or three fingers (Fig. 17.21B and Fig. 17.23)
- The opposing thumb and finger

The skin should not be in contact with the rest of the hand. Across the length of the structure, the fingers are positioned. Movement is carried out in small ranges with pressure applied either forward or backward, causing the therapist's finger and the patient's skin to move together.

- The only parts that should move are the supporting elements and the impacted structures (ligaments, tendons, and muscles). It never ought to happen between the finger and the skin.
- The movement's direction is the single most crucial element that determines how effective this maneuver will be.
- The anatomical structures that must be moved are identified prior to initiating the friction, and the fingers are positioned transversely along the structure's long axis in accordance with the arrangement of the fibers within that structure.
- To guarantee flawless movement, considerable consideration should be given to the patient's positioning as well as the application of friction to the soft tissue.

In contrast to the contractile tissues, such as muscles, which should be totally relaxed, the non-contractile structures, such as ligaments and tendons, are held in a fully tense condition. Ligaments and tendons can be moved efficiently against an immovable foundation when they are stretched to their utmost extent. The patient should be made aware of the procedure's discomfort before the maneuver is applied. Although it is crucial that the target muscle is relaxed, failing to do so will prevent the forces of friction from penetrating deeply and prevent the expansion of individual fibers.

Usability: Subacute and chronic lesions of the muscles, tendons, ligaments, capsules, nodules, and adhesions are the main conditions it is used to treat. It is quite helpful for pain that is localized (trigger points).

Technique

The hand, forearm, and fingers should generally create a straight line while remaining parallel to the action applied.

1. The distal interphalangeal joint is only slightly flexed.
2. The finger or thumb can be made to flex and extend alternately, or the wrist, elbow, shoulder, or trunk can be moved while the finger or thumb is firmly pressed against the structure that has to be treated in order to create force.
3. Nevertheless, prolonged use of these motions is not recommended due to the excessive strain they place on the hand muscle. Cyriax favored using the latter because it produces greater power while requiring lesser effort.
4. **The following disorders should never be treated with this technique:** elbow traumatic arthritis, soft tissue calcification or ossification, over the joint capsule in case of rheumatoid arthritis, bursitis, pressure on the nerves, and inflammation brought on by bacterial infection.
5. **Applying friction requires caution:** The production of blisters may result from intense friction. It generally

happens on moist skin. So, by using spirit or powder, the area should be made dry before applying friction.

Percussion or Tapotement

These are French terms that signify "the striking of two objects against each other" when translated into English. It is sometimes known as tapotement, massage technique that uses rhythmic tapping, pounding or striking movements on the body. This technique makes use of a sequence of fast and repetitive movements, which are typically executed with the edges of the hands, fists or fingertips. Tapotement, which includes techniques like cupping, hacking, thumping or pummeling, is often used to stimulate the muscles, promote blood circulation, and rejuvenate the body. The application of intermittent contact and pressure to the body's surface is a distinguishing aspect of this category of techniques. Each method in this group uses a controlled wrist and forearm movement to rhythmically strike the patient's body surface. Mild blows are delivered in a variety of ways and with varying forces. The main idea behind these methods is to strike the body with various portions of the hand. According to which part of the hand was used to strike the surface, various techniques are given different names. This group uses a variety of techniques, including:

- Clapping
- Hacking
- Tapping
- Beating and pounding
- Tenting
- Contact heel percussion.

Clapping

The widely used and well-researched tapotement technique is important in the treatment of chronic respiratory disorders, which often lead to sputum retention. This method involves striking the chest wall repeatedly at a set tempo with slightly cupped hands (Fig. 17.24).

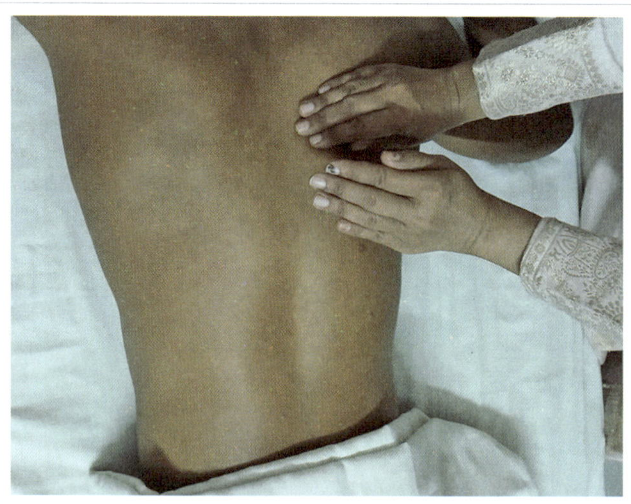

Fig. 17.24: Clapping

Technique

1. Stride standing stance is used by the therapist.
2. The elbow should be flexed at 90°, with the arm slightly abducted for optimal positioning.
3. The hands are cupped to allow air to be trapped in the hollow, as it touches the skin.
4. Only the edges, fingers, and heel of a hand make touch with the surface of the body when it is resting over it. In order to do this, the index, middle, and ring fingers' MCP joints are kept slightly flexed, and fingers and thumbs are kept adducted.
5. The procedure is carried out by quick, controlled wrist flexion and extension. Elbow movement should never be permitted.
6. A free fall of extended hand is necessary for an efficient correction. This is accomplished by deliberately extending the wrist and then letting it fall as a result of gravity without making an effort to actively create wrist flexion.
7. When the hand hits the chest wall, it should provide an air cushion between itself and the wall. It should not apply excessive pressure to the soft tissue of the chest wall and is performed both during inspiration and expiration.
8. When used correctly, it produces a distinctive sound that is immediately distinguishable from slapping, where the entire palmar surface of the hand makes contact with the skin. Having the ability to produce the appropriate sound is essential because it indicates that the required suctioning, or negative pressure, has been reached, which has the mechanical effect of releasing the secretion.
9. To avoid painful and sharp skin stimulation, it should be done across the chest wall, ideally over a blanket or towel.

Hacking

- For this percussion technique, the ulnar border is the only area of the medial three fingers (little, ring, and middle) that are used to strike the skin.
- During each stroke, the ulnar edges of the little, ring, and middle fingers should make slight contact with the skin by being held freely apart.
- A characteristic percussive sound results from this technique. The hacking is caused by alternate forearm supination and pronation, together with ulnar and radial wrist deviations, respectively.
- Elbow movement is completely prohibited. The shoulder motion causes the hands to move either forward, backward, or laterally as the hacking progresses.

Usability: This technique is applied to larger areas like the back (Fig. 17.25), thigh (Fig. 17.26), etc., and is highly beneficial from a relaxation perspective.

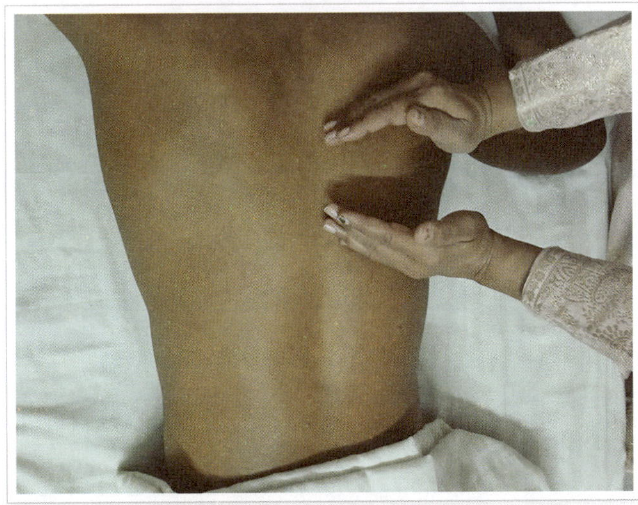
Fig. 17.25: Hacking on the back

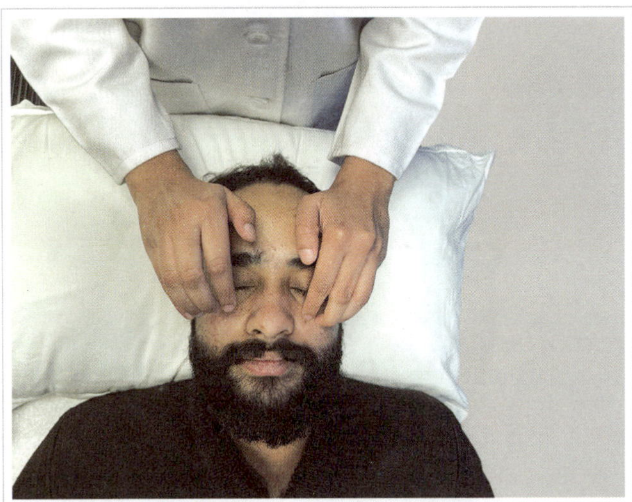
Fig. 17.27: Tapping technique

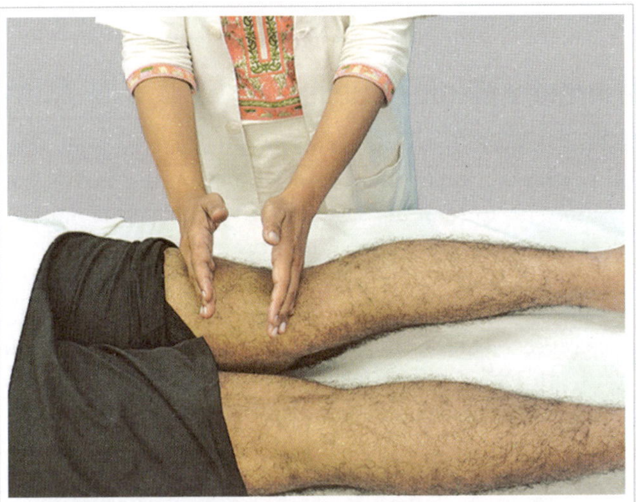
Fig. 17.26: Hacking on the posterior thigh

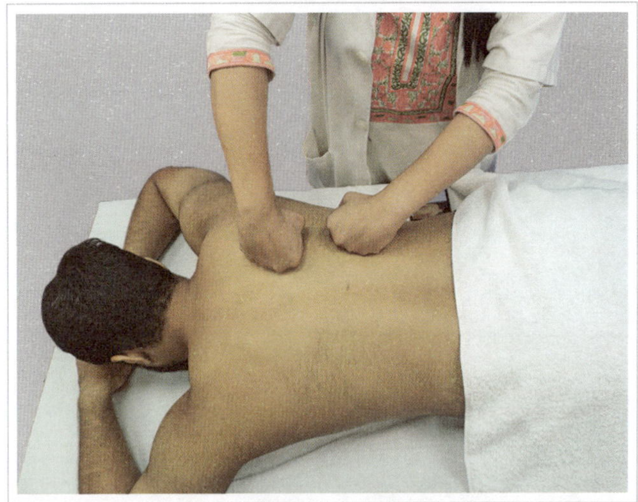
Fig. 17.28: Beating technique

Tapping

This is a useful technique for providing pressure and touch intermittently to a small region.

- This technique only impacts the bodily portion with the pulp of the fingers. It is possible to utilize one or both hands.
- The fingers are relaxed and held loosely. The tapping is caused by the metacarpophalangeal joint flexing and extending alternately. There should be no movement allowed in the joints of the wrist and fingers. It is frequently applied to the face, neck, and other smaller places. Practically, utilized for children (Fig. 17.27).

Beating and Pounding

These techniques involve striking the body part with a loosely clenched fist. It is possible to utilize one or both hands. Two techniques have different names depending on which area of the body the clenched fist strikes:

1. **Beating:** In this technique, the part is struck with the front of the fist. By alternately flexing and extending the wrist, like while clapping, beating is produced. Elbow movement is prohibited because it could impart too much force (Fig. 17.28).
2. **Pounding:** This technique involves striking the part with the side of the fist (Fig. 17.29). The therapist forms a fist with the thumb resting over the index finger. The pounding is caused by the forearm's supination and pronation, as well as the wrist's ulnar and radial deviation, respectively. The technique is the same as one used in hacking.

Usability: These two techniques are commonly used to relax the thighs, back, and other large, fleshy parts of the body.

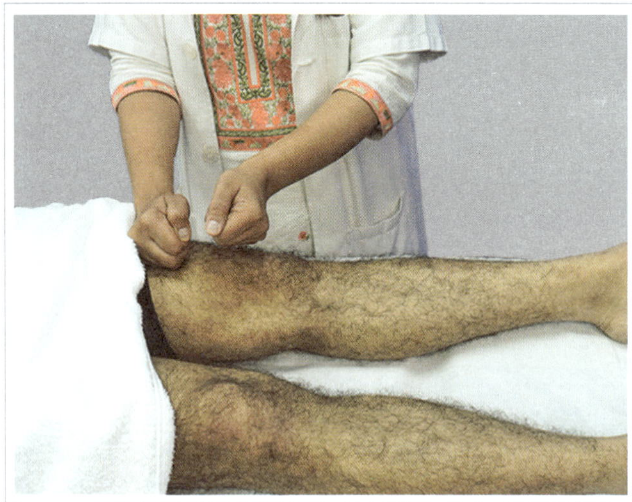

Fig. 17.29: Pounding technique

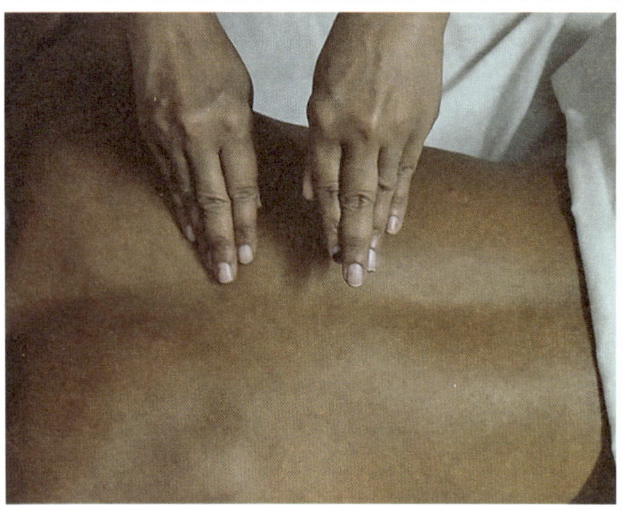

Fig. 17.30: Tenting

Tenting

This approach is a clapping variation. In this case, the concavity is produced by the middle finger, which is slightly elevated and placed over the index and ring fingers (Fig. 17.30).

Usability: This technique works especially well to release the visceral fluids from a preterm infant or newborn across its narrow chest.

Contact Heel Percussion

This is yet another variation of clapping. Here, a concavity formed between the thenar and hypothenar eminences strikes the chest wall (Fig. 17.31).

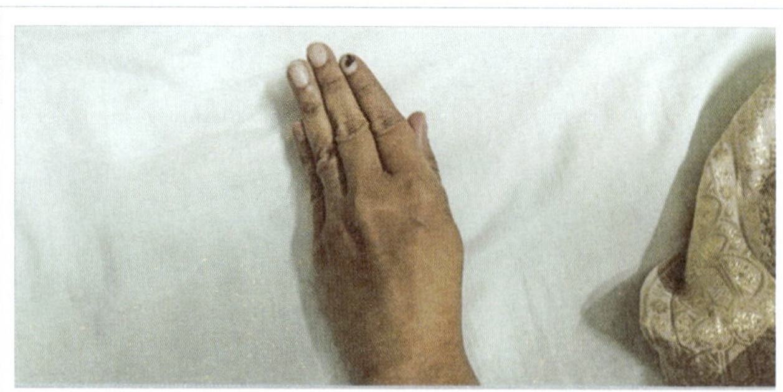

Fig. 17.31: Contact heel percussion

Clinical Correlation

Percussion should be avoided in cases of osteoporosis, rib fracture, hemoptysis, metastatic deposits in the ribs and spine, over surgical wounds, acute pulmonary tuberculosis, and pleuritic pain.

Vibratory Manipulations

Vibratory massage is a technique that involves the application of rhythmic vibrations or oscillations to the body's soft tissues. This form of massage aims to stimulate the muscles, increase blood circulation, and promote relaxation. The vibrating motion can be gentle or more intense, depending on the desired therapeutic outcome and the individual's preferences. The distal part of the upper limb vibrates to transfer mechanical energy to the body in this type of surgery. The widespread co-contraction of the upper limb muscles, which are in constant contact with the patient's skin, is used to produce vibration in the hands and fingers. However, what sets apart its two ways is how the forearm is held. When vibrating, the forearm is maintained in full pronation while the hands move upward and downward. In contrast, the forearm is retained in a mid-prone position when shaking.

Usability: This method focuses primarily on the lung and other hollow cavities.

Vibration

This technique involves applying quick intermittent pressure while keeping the therapist's hand or a portion of it in continual contact with the patient's skin.

Technique

1. For chest wall vibration, the therapist with their elbow completely extended and their shoulder locked in minimum flexion.
2. With both hands still in contact with the patient's chest wall, one is placed on top of the other (Fig. 17.32).

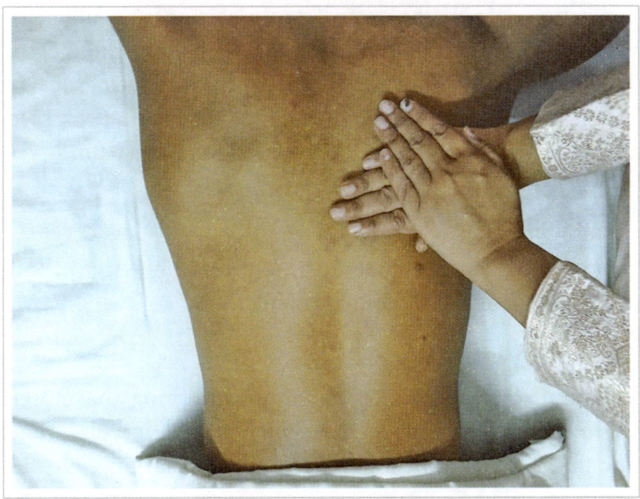

Fig. 17.32: Vibration technique

3. The tissues are alternately compressed and released by co-contracting the therapist's entire arm, causing an effect similar to trembling. This causes their hand to waver up and down and transmits mechanical energy to the patient's chest. Vibration is always produced during the expiratory stage of respiration.

Shaking

- Similar to vibration, this technique also applies oscillatory mechanical energy to the chest wall. However, there are two key characteristics that set this method apart from vibration.
- When compared to oscillation produced by vibration, shaking produces a coarser oscillation.
- Instead of the hand moving up and down, this approach oscillates in a sideways direction, which might be caused by the wrist deviating in the radial and ulnar directions.

Technique

The therapist either adopts walks standing stance or fall out standing stance.

1. It is done by placing both hands on the chest wall over the affected lobe. Elbow is kept slightly flexed and shoulder is adducted.
2. One of the following methods may be used to place the hand:
   - **Patient in a supine lying position:** Place both hands on the anterior chest wall on either side, or place one hand on the anterior and the other on the posterior wall of the same side.
   - **Patient in a lateral position:** Position both hands on the upper lateral chest wall if the patient is on their side, or place one hand on the anterior and the other on the posterior chest walls of the upper side.

Clinical Correlation

Only during the expiratory phase, this technique is performed. In cases of severe hemoptysis, osteoporosis, acute pleuritic pain, rib fracture, or active pulmonary tuberculosis, shaking should not be employed.

PHYSIOLOGICAL EFFECTS

The physiological effects of soft tissue manipulation can be discussed under the following headings:

- Effects on the circulatory system
- Effects on blood
- Effects on the exchange of nutritive elements
- Effects on metabolism
- Effects on the nervous system
- Effects on the mobility of soft tissue
- Effects on muscle strength
- Effects on the respiratory system
- Effects on the skin
- Effects on the adipose tissue
- Psychological effects
- Effects on immune system

Circulatory System

On Venous and Lymphatic Flow

The mechanical draining of the lymphatics and veins is aided by massage. It lessens the possibility of venous blood and lymph stagnating in tissue space by facilitating their forward motion. The action of the smooth muscles found in the vessel walls primarily controls the flow of lymphatic and venous channels from the extremities. The vessels' valves and the contraction of these muscles work together to provide a powerful pumping mechanism that keeps the tissue space free of excess fluid. The blood vessels are compressed when the skeletal muscles contract, applying pressure to the fluid that is inside. This rise in intravascular pressure drives smooth muscle contraction, which raises intravascular pressure even more. The valves open, and the fluid enters the following segment when the pressure exceeds the threshold. The flow of fluid *via* the valves is unidirectional, thus it cannot return to the empty segment. The segment is refilled with fluid from distal segments when the muscle has relaxed. In this manner, lymphatic and venous fluids can only flow in one direction.

On Arterial Flow

The blood flow in the area being massaged is improved. After massage, there is usually a noticeable vasodilation and increase in peripheral blood flow. The following things that

occur during massage may be responsible for this slight but steady increase in arterial flow:
- Activation of axon reflex.
- Release of vasodilators.
- Decrease of venous congestion.

Blood

Both healthy people and people with anemia experienced an increase in RBC count after massage. The neutrophil count increased after 30 minutes of massage given two hours after strenuous exercise, according to Smith et al. However, after 20 minutes of massage, no difference in the neutrophil count was found by Hilbert et al.

Metabolism

Massage encourages speedy flushing of waste materials and nutrient replenishment. Following a massage, there is an increase in arterial blood flow that carries more oxygen and nutrients as well as speeds up blood oxygenation. The lymphatic and venous flow are accelerated during massage, which encourages the quick elimination of metabolic waste. These modifications improve the efficiency of nutrient exchange at the cellular level. It may hasten the body's numerous metabolic processes by improving arterial blood flow and venous lymphatic drainage.

Nervous System

There are sensory, motor, and autonomic components to the nervous system. Different massage techniques have an impact on each of these elements.

On Sensory System

The various massage techniques stimulate the skin's and soft tissue's touch and pressure receptors, which are responsible for a variety of sensory experiences. Large diameter A-beta fibers, which are vital in inhibiting the sense of pain carried by A-delta and C fibers, carry these sensations.

Presynaptic inhibition at the level of the substantia gelatinosa of the spinal cord limits the pathway of pain perception when low threshold mechanoreceptors are simulated. This may be how light pressure massage techniques like stroking, hacking, tapping, etc. work to relieve pain.

Some massage techniques cause mild to moderate discomfort by stimulating the body's painful regions. According to some theories, it aids in the release of certain anti pain substances like enkephaline and beta endorphin, into the periaqueductal grey matter (PAG) at the level of the midbrain. These compounds then travel down to the spinal cord's dorsal horn, where they block the release of substance P, a pain-related neurotransmitter. This prevents the higher pain perception area of the brain from receiving pain impulses. Counter irritant effect is another name for this phenomenon. This might be one of the mechanisms at work when heavy pressure techniques like kneading, friction, petrissage, connective tissue massage, etc. are used to relieve pain.

On Motor System

Neuromuscular excitability can be affected by massage in both a facilitating and an inhibitory manner.

Facilitatory Effects

By stimulating the skin receptor or stretching the muscle spindle, massage can reflexively enhance muscular tone. For this aim, superficial strokes, tapping, hacking, etc. are frequently used.

Here are a few facilitation methods that make use of massage maneuvers:
- Skin stimulation *via* touch, when placed over the working muscle and/or the surface against which the movement must occur, functions as a guide for the movement and promotes motoneuron pool activities.
- Applying pressure to the muscular belly has the effect of activating the muscle spindle, most likely by distorting the shape of the muscle fibers and stimulating a stretch response.
- In order to raise the excitability of the motoneurons that supply the inhibited muscle, cutaneous stimulation using quick, light brushing, stroking, clapping, and pressure over the relevant dermatome is utilized as a preliminary facilitation.
- The excitatory action is local and primarily limited to superficial muscles in skin supplied by anterior primary rami. Deep muscles are excited when the skin supplied by the posterior primary rami is stimulated.

Inhibitory Effects

Muscle tone can also be decreased by massage treatments. It has been asserted that petrissage, or massage that involves kneading the muscles, can have an inhibitory effect on motoneuron. A proposed mechanism for these effects is the activation of the tension-dependent Golgi tendon organ, which inhibits the stretch reflex mechanism.

On Autonomic System

According to a number of authors, massage can affect the functioning of visceral organs by altering the autonomic nervous system through peripheral sensory input. It is generally acknowledged that massage raises skin temperature, stimulates sweat glands, and increases skin conductance (also known as the electrodermal response). By harmonizing the sympathetic and parasympathetic components of the autonomic nervous system, connective tissue massage enhances the blood flow to the target organ.

Soft Tissue

The elasticity, plasticity, and mobility of soft tissues are only a few of the characteristics that massage significantly affects. Muscles, sheath, ligaments, tendons, aponeurosis, joint capsules, and superficial as well as deep fascia are among the tissues that can be impacted by massage. The collagen fibers that make up these tissues are stretched in various directions by different massage techniques. Maximum mobility between fibers and nearby structures is ensured as the adhesions between the fibers are broken. A variety of massage techniques, especially transverse friction, mechanically separate the bonded muscle fibers and restore mobility.

Muscle Strength

The vast majority of experts on the subject agree that massage does not increase muscle strength. Only through active muscle contraction can a muscle be strengthened. At best, massage can get the muscle ready for contraction by boosting circulation and making it easier to get rid of metabolic waste.

Respiratory System

The clearance of secretion from the bigger airways is aided by massage treatments that use vibration and percussion. The loosening and central advancement of secretions from the airways caused by percussion and vibration can lead to increased secretion clearance following chest physical therapy in both the adult and pediatric groups.

Skin

In general, massage enhances the skin's nutritional condition. After a massage, skin temperature increases. The mobility of the skin over the subcutaneous structures is facilitated by massage. Skin gets softer, more supple, and finer as a result. Furthermore, the skin gets tougher, more elastic, and flexible after prolonged massage. By stimulating the sweat glands, massage increases perspiration, which increases heat dissipation. Additionally, it helps the skin's exocrine glands produce sebum, which enhances the lubrication and aesthetic appeal of the skin.

Psychological

Massage can reduce somatic and psychoemotional arousal, including tension and anxiety. It facilitates relaxation. Massage therapy has been shown in studies to effectively relieve anxiety, stress, and depressive symptoms. It elevates mood by raising serotonin and dopamine levels, reduces cortisol (the stress hormone), and improves sleep quality. Tactile stimulation promotes relaxation, releases éndorphins, and may improve symptoms in mental health conditions like PTSD.

Immune System

It is presently believed that massage therapy may have some influence over how the immune system functions. In an effort to support its use in treating various late-stage problems related to AIDS and cancer, recent research has focused on the exploration of the immunological effects of massage therapy. According to research, massage therapy may benefit the immune system in a variety of ways. It has been associated to enhanced natural killer cell activity, a rise in white blood cell count, and immunological marker modification such as cytokines and lymphocytes. Massage has also been shown to reduce inflammation, which helps to improve immunological function overall.

INDICATIONS

- Chronic pain syndromes.
- Stress and anxiety reduction.
- Reduced state anxiety, blood pressure, and heart rate.
- Massage in infancy improves growth and post massage sleep.
- Massage is widely used by the athletic population for a variety of purposes such as injury prevention, recovery from fatigue, relaxation, and to increase performance.
- Massage is believed to offer various benefits for athletes, including improved blood flow, reduced muscle pressure, and increased well-being through biomechanical, neurological, physiological, and psychological mechanisms.
- Massage induces musculoskeletal relaxation, improves quality of life, and sleep in children with cerebral palsy.
- Facilitates recovery from repetitive muscular contractions and sternous exercise.
- Scar management.
- Low back pain.
- Intensity of delayed onset muscle soreness reduced after application of massage.
- Reduction of anxiety, back pain, urinary stress hormonal levels, improved sleep, mood and less postnatal complications in pregnant women.
- Management of migraine.
- Massage therapy is associated with enhancement of the immune system's cytotoxic capacity.
- Improves balance, pain and function in patients with osteoarthritis of knee.

CONTRAINDICATIONS

General

In these situations, applying massage to any portion of the body carries some danger. As a result, even before to physical examination and patient positioning, one should make sure the patient does not have any of the following general contraindications to massage:

- Recent surgery
- Thrombosis/phlebitis
- Severe spasticity
- Osteoporosis
- Deep X-ray therapy
- High fever
- Severe renal or cardiac diseases
- Very hairy skin

Local

Massage should not be performed on the concerned body part if any of these conditions are present. The approaches can, however, be applied to other body sections, if necessary. For instance, in the event of an acute sprain, the athlete's thigh and knee can be massaged to aid in relaxation and healing rather than the ankle, which should not be treated right away. During the patient's physical examination prior to therapy, certain contraindications should be ruled out. Some of these are listed here:

- Open wound
- Malignancy
- Poisonous foci
- Thrombosis
- Myositis ossificans
- Atherosclerosis
- Acute inflammation
- Recent fractures
- Skin diseases
- Severe varicose veins

SUMMARY

- In the late 1960s and early 1970s, massage therapy nomenclature shifted, with practitioners using "soft tissue manipulation" to emphasize therapeutic features and disassociate themselves from negative connotations.
  - Soft tissue manipulation (STM) is a manual therapy that uses mechanical stimulation to cause biological changes.
  - It combines a variety of approaches to influence molecular pathways, cellular responses, and tissue function.
  - STM is classified depending on its nature (stroking, pressure, tapotement, vibratory), depth (superficial and deep), region (general and local), and administration (manual or mechanical).
- In massage, characteristics like as depth, speed, rhythm, duration, direction, and frequency are all important. Inexperienced therapists must purposefully mix these parts, eventually mastering them, to develop a fluid and effective approach that meets the client's goals.
- Stroking maneuvers in massage include superficial stroking and effleurage, which emphasize rhythmic and continuous strokes with variable pressure.
  - Superficial stroking stimulates the senses with light and quick strokes, whereas effleurage uses steady and moderately firm movements to warm up the muscles and enhance blood circulation.
  - Both procedures move the entire length of the section in one action. The therapist's body position and hand coordination are critical for maintaining even skin contact and producing therapeutic benefits.
- Pressure manipulations in soft tissue manipulation are characterized by deep compression and sustained touch, with a particular emphasis on muscular tissue.
  - The techniques, which include kneading, petrissage, and friction, are classified according to the type and direction of the applied pressure.
  - Kneading is a circular compression along the long axis of the underlying bone that progressively increases and decreases pressure.
  - Palmar kneading, digital kneading (finger and thumb), and reinforced kneading are all subcategories, each with a specific application.
- Petrissage, derived from the French word "Petrir", meaning "to knead", is a massage technique that involves kneading, rolling, and compressing muscles and soft tissues.
  - Petrissage, distinguished by lateral compression, is the lifting and manipulation of tissues to improve blood circulation, increase flexibility, and reduce muscle stress.
  - Petrissage procedures include pick-up, skin rolling, and wringing.
  - Picking up elevates tissue at a straight angle, while skin rolling comprises lifting, stretching, and moving the skin, and wringing simulates twisting and compressing a moist towel. These treatments are intended to increase overall relaxation and alleviate specific muscular tightness.
- Friction massage, which uses deep pressure and friction with fingers, thumbs or palms, tries to break down adhesions in muscles and connective tissue, enabling better blood flow, flexibility, and reduced muscle tension.
  - Circular friction provides constant deep pressure without a relaxation phase, which is useful for certain locations such as joints and fibrositic nodules.
  - Dr Cyriax popularized transverse friction, which provides pressure throughout the structure's long axis and is useful for treating subacute and chronic lesions, including muscles, tendons, ligaments, capsules, nodules, and adhesions.

- Tapotement, sometimes known as percussion, is a massage method that involves rhythmic tapping, pounding or striking movements with the edges of the palms, fists or fingers.
  - Clapping involves smacking the chest wall with slightly cupped hands, which is good for chronic respiratory disorders.
  - Hacking strikes larger areas, such as the back, with the ulnar border of the medial three fingers.
  - Tapping uses intermittent touch and pressure with the pulp of the fingers, which is appropriate for smaller regions and children.
  - Beating involves striking with the front of the hand, whereas pounding employs the side of the fist to relax larger body parts.
  - Tenting, a type of clapping, and touch heel percussion are specialized techniques with specific uses. Percussion should be avoided in certain cases, such as osteoporosis and rib fractures.
- Vibratory massage is a technique in which rhythmic vibrations or oscillations are applied to the body's soft tissues to stimulate muscular contraction, enhance blood circulation, and promote relaxation.
  - Vibration is a technique that uses quick intermittent pressure with the therapist's hand in constant touch with the patient's skin to produce a vibrating sensation.
  - It is frequently used during the expiratory stage.
  - In contrast to vibration, which moves up and down, shaking produces coarse oscillations in a sideways orientation.
  - During the expiratory phase, the therapist places hands on the chest wall while standing or walking, with caution advised for certain situations such as severe hemoptysis or rib fractures.
- Massage therapy has various physiological effects on the circulatory, nervous, soft tissue, muscular, respiratory, skin, adipose tissue, and psychological systems:
  - **Circulatory system:**
    - Venous and lymphatic flow: Massage aids in draining lymphatics and veins, preventing stagnation of blood and lymph in tissues.
    - Arterial flow: Massage improves arterial blood flow, induces vasodilation, and enhances peripheral blood flow.
  - **Blood:** Massage may influence red and white blood cell counts, but the literature lacks a clear consensus on the mechanisms behind these changes.
  - **Metabolism:** Massage speeds up various metabolic processes by improving arterial blood flow and venous lymphatic drainage. It accelerates nutrient exchange at the cellular level by increasing arterial blood flow and promoting the elimination of metabolic waste.
  - **Nervous system:**
    - Sensory system: Massage stimulates touch and pressure receptors, inhibiting pain perception and promoting the release of anti-pain substances.
    - Motor system: Massage can both facilitate and inhibit neuromuscular excitability, affecting muscle tone through mechanisms like skin stimulation and Golgi tendon organ activation.
    - Autonomic system: Massage can impact visceral organs by altering the autonomic nervous system, balancing sympathetic and parasympathetic components.
  - **Soft tissue:** Massage affects the elasticity, plasticity, and mobility of various soft tissues, including muscles, ligaments, tendons, and fascia, breaking adhesions and restoring mobility.
  - **Muscle strength:** Massage does not directly increase muscle strength but may prepare muscles for contraction by improving circulation and aiding in the removal of metabolic waste.
  - **Respiratory system:** Massage treatments using vibration and percussion assist in clearing secretions from larger airways, enhancing secretion clearance.
  - **Skin:** Massage improves skin nutrition, increases skin temperature, facilitates skin mobility, and promotes exocrine gland activity.
  - **Adipose tissue:** Massage may enhance muscle function and reduce fatigue, contributing to more intense workouts, potentially aiding in weight loss.
  - **Psychological effects:** Massage reduces tension, anxiety, and stress, promoting relaxation, elevating mood, and positively affecting sleep quality. It releases endorphins and improves symptoms in mental health conditions.
  - **Immunological effects:** Recent research suggests that massage therapy may influence the immune system, enhancing natural killer cell activity, white blood cell count, and modifying immunological markers. It has been associated with reduced inflammation.
- Soft tissue manipulation (STM) provides a wide range of advantages, from relieving chronic pain syndromes and stress to promoting newborn growth and assisting athletes in a variety of ways.
  - It is also excellent at treating scars, low back pain, and delayed onset muscle soreness.
  - Massage helps pregnant women feel less anxious, have less back discomfort, and sleep better.
  - Furthermore, STM has been linked to increasing the immune system's cytotoxic capacity and balance in osteoarthritis patients. However, caution is advised, taking into account contraindications such as recent surgery, thrombosis, open wounds, and other specific health conditions, to ensure safe and effective use.

FURTHER READINGS

- Agarwal KN, Gupta A, Pushkarna R, Bhargava SK, Faridi MM, Prabhu MK. Effects of massage & use of oil on growth, blood flow & Sleep. Indian J Med Res. 2000 Dec;112:212–7.
- Beard G: A history of Massage technique, Phys Ther, 1952; 32: 613–24.
- Best TM, Hunter R, Wilcox A, Haq F. Effectiveness of sports massage for recovery of skeletal muscle from strenuous exercise. Clinical journal of sport medicine. 2008 Sep 1;18(5):446–60.
- Corbin L. Safety and efficacy of massage therapy for patients with cancer. Cancer control. 2005 Jul;12(3):158–64.
- Cyriax J: Textbook of Orthopedic Medicine treatment by manipulation massage and injection. London: Balliere Tindall, 1998; 2.
- Field T, Hemandez-Reif M, Hart S, Theakston H, Schanberg S, Kuhn C. Pregnant women benefit from massage therapy. Journal of Psychosomatic Obstetrics & Gynecology. 1999 Jan 1;20(1):31–8.
- Furlan AD, Giraldo M, Baskwill A, Irvin E, Imamura M. Massage for low-back pain. Cochrane database of systematic reviews. 2015(9).
- Gasibat Q, Suwehli W. Determining the benefits of massage mechanisms: A review of literature. Rehabilitation Sciences. 2017 May 27;3(2):58–67.
- Glew GM, Fan MY, Hagland S, Bjornson K, Beider S, McLaughlin JF. Survey of the use of massage for children with cerebral palsy. International journal of therapeutic massage & bodywork. 2010;3(4):10.
- Graham D: Massage Manual Treatment and Remedial Movements. Philadelphia: JB Lippincott, 1913.
- Hemmings BJ. Physiological, psychological and performance effects of massage therapy in sport: A review of the literature. Physical therapy in sport. 2001 Nov 1;2(4):165–70.
- Hilbert J, Sforzo G, Swensen T. The effects of massage on delayed onset muscle soreness. British journal of sports medicine. 2003 Feb;37(1):72.
- Hilbert JE, Sforzo GA, Swensen T. The effect of massage on delayed onset muscle soreness. Br J Sports Med, 2003; 37(1):72–75.
- Hoffa AJ: Technik der Massage, atuttgart, Germany: Ferdinand enke. 1909; 14.
- Hollis M: Massage for therapists. Oxford England: Blackwell Scientific Publications, 1987.
- Ironson G, Field T, Scafidi F, Hashimoto M, Kumar M, Kumar A, Price A, Goncalves A, Burman I, Tetenman CY, Patarca R. Massage therapy is associated with enhancement of the immune system's cytotoxic capacity. International journal of neuroscience. 1996 Jan 1;84(1–4): 205–17.
- Lawler SP, Cameron LD. A randomized, controlled trial of massage therapy as a treatment for migraine. Annals of Behavioral Medicine. 2006 Aug 1;32(1):50–9.
- Lehn C, Prentice WE: Massage. In: Prentice WE (Ed), Therapeutic Modalities in sports medicine, St. Louis: Mosby- Year Book Inc, 1994; 1:335–63.
- Loghmani MT, Whitted M. Soft tissue manipulation: A powerful form of mechanotherapy. Physiother Rehabil. 2016;1(4):1000122.
- Melzack R and Wall PD: Pain Mechanisms: A new theory. Science, 1965; 150:971–79.
- Mennell JB: Physical Treatment. Philadelphia: Blakiston Co, 1945;5.
- Moyer CA, Rounds J, Hannum JW. A meta-analysis of massage therapy research. Psychological bulletin. 2004 Jan;130(1):3.
- Perlman AI, Sabina A, Williams AL, Njike VY, Katz DL. Massage therapy for osteoarthritis of the knee: A randomized controlled trial. Archives of internal medicine. 2006 Dec 11;166(22):2533–8.
- Shin TM, Bordeaux JS. The role of massage in scar management: a literature review. Dermatologic Surgery. 2012 Mar;38(3):414–23.
- Smith LL, Keating MN, Holbert D, Spratt DJ, Mc Cammon R, Smith SS, Israel RG: The effects of athletic massage on delayed onset muscle soreness, creatine kinase, and neutrophil count: A preliminary report. J Orthop Sports Phys Ther, 1994; 19(2):93–99.
- Walach H, Güthlin C, König M. Efficacy of massage therapy in chronic pain: A pragmatic randomized trial. The Journal of Alternative & Complementary Medicine. 2003 Dec 1;9(6):837–46.
- Wood EC, Becker PD: Beard's massage. Philadelpia: WB Saunders, 1981;3.

STUDENT ASSIGNMENT

LONG ANSWER QUESTIONS

1. Explain pressure manipulations in detail.
2. Discuss superficial and deep stroking techniques in detail.
3. What is percussion? Enumerate its types and explain their techniques.
4. Explain friction techniques in detail, including their technique, effects and contraindications.
5. Explain in detail about the physiological effects of massage.
6. Classify soft tissue manipulation techniques with a flowchart.

SHORT ANSWER QUESTIONS

1. Briefly write about the following techniques:
   a. Effleurage
   b. Kneading
   c. Wringing
   d. Friction massage
2. How does massage help in reducing pain?
3. Enumerate the general and local contraindications of massage.
4. How does massage positively affect the circulatory system?
5. Write about stroking with its technique.

MULTIPLE CHOICE QUESTIONS

1. **Which term emerged in the late 1960s and early 1970s to emphasize therapeutic aspects of massage therapy?**
   a. Deep tissue massage
   b. Soft tissue manipulation
   c. Vibratory massage
   d. Petrissage therapy
2. **Which massage technique involves rhythmic tapping, pounding or striking movements with the edges of the palms, fists or fingers?**
   a. Vibration
   b. Effleurage
   c. Tapotement
   d. Petrissage
3. **What does petrissage focus on in soft tissue manipulation?**
   a. Quick intermittent pressure
   b. Circular compression
   c. Lateral compression
   d. Deep pressure and friction
4. **What physiological effect does massage have on the circulatory system by enhancing nutrient exchange at the cellular level?**
   a. Improved arterial blood flow
   b. Accelerated lymphatic drainage
   c. Increased venous congestion
   d. Reduced blood circulation
5. **How does massage affect the nervous system's sensory components?**
   a. By inhibiting touch and pressure receptors
   b. By decreasing neuromuscular excitability
   c. By promoting pain perception
   d. By stimulating touch and pressure receptors
6. **What is a characteristic of the tapping technique in massage?**
   a. Circular compression
   b. Steady and moderately firm movements
   c. Rhythmic tapping or pounding
   d. Deep compression and sustained touch

7. **In soft tissue massage, what does superficial stroking primarily involve?**
    a. Deep compression
    b. Circular compression
    c. Lateral compression
    d. Rhythmic and continuous strokes
8. **What is the main focus of friction massage in soft tissue manipulation?**
    a. Lifting and manipulation of tissues
    b. Breaking down adhesions
    c. Circular compression along the long axis
    d. Quick intermittent pressure
9. **What physiological system does massage influence to enhance the immune system's cytotoxic capacity?**
    a. Respiratory system
    b. Nervous system
    c. Immune system
    d. Adipose tissue
10. **What should therapists be cautious about when applying tapotement techniques in massage?**
    a. Osteoporosis and rib fractures
    b. Recent surgery
    c. Severe spasticity
    d. Deep X-ray therapy

ANSWER KEY

1. b **2.** c **3.** c **4.** a **5.** d **6.** c **7.** d **8.** b **9.** c **10.** a

18 Relaxation

Sheetal Kalra

LEARNING OBJECTIVES

After the completion of the chapter, the readers will be able to:

- Define muscle tone, postural tone, voluntary movement and degrees of relaxation.
- Explain the causes of pathological tension in muscle.
- Understand stress mechanics, types of stresses and effects of stress on body mechanisms.
- Define fatigue and spasm, along with the general causes, signs and symptoms of fatigue.
- Understand indications of relaxation and the rationale of relaxation techniques.
- Explain and demonstrate methods and techniques of relaxation.
- Explain principles and uses of the general local, Jacobson's, Mitchell's, and autogenic relaxation methods.

CHAPTER OUTLINE

- Introduction
- Muscle Tone
- Stress and its Effects on Various Systems of the Body
- Fatigue
- Relaxation Exercises
- Techniques of Relaxation

KEY TERMS

Dystonia: Muscle contractions brought on by dystonia are an involuntary occurrence.

Guided imagery: The process of using visualization while working with a person or a recording to finish a task is called guided imagery.

Hypertonia: It is an abnormal increase in muscle tone, making passive movement difficult. It is independent of movement velocity and may occur with or without spasticity.

Hypotonia: It is defined as reduced muscle tone, where there is a diminished resistance to passive stretching of the muscles.

Muscle tone: The traditional definition of muscle tone is "tension in the relaxed muscle" or "resistance, felt by the examiner during passive stretching of a joint when the muscles are at rest".

Myotonia: It is a condition characterized by delayed relaxation of muscles after voluntary contraction, leading to prolonged muscle stiffness or difficulty relaxing muscles after movement.

Phasic tone: The alpha motor neuron system alone is responsible for mediating phasic tone, which is a short contraction brought on by a high-intensity stretch. The clinical examination involves the elicitation of muscular stretch reflexes.

Postural tone: In axial muscles, where gravity is the primary excitation component, postural tone is observed. It appears as prolonged muscle contraction and is caused by a continuous stretch on the tendons and muscles.

INTRODUCTION

"The tension in the relaxed muscle" or "the resistance, felt by the examiner during passive stretching of a joint when the muscles are at rest" are the standard definitions of muscle tone.

Muscle tone can be understood mathematically as the change in force or resistance per unit change in length (Δ force/Δ displacement of the tissue). However this explanation is based on the fact that muscle is completely relaxed which is practically impossible.

Bernstein emphasized that, contrary to popular belief, muscle tone may actually indicate a person's level of readiness for a movement. As a result, it might not be able to determine muscle tone when a person is instructed to remain still and stationary.

Bernstein postulated that muscle tone is an adaptive function of the neuromotor apparatus that responds to commands coming from upper levels of movement construction by fine-tuning the excitability of the sensory and motor cells for the tasks of active postural or movement control. This definition better defines muscle tone as an active contributor to movements and different postural tasks.

While assessing muscle tone following factors should be considered:

- Posture assessment
- The feel of the muscle
- Resistance that muscle offers to passive movements
- Reflex testing and available range of motion (ROM).

MUSCLE TONE

Muscle tone can be divided into "phasic" and "postural" categories.

Postural tone: The tone of the muscles involved in maintaining posture is referred to as postural tone. Axial muscles exhibit postural tone, with gravity serving as the primary initiating stimulus. In accordance with the body's constantly shifting attitudes and postures, postural tone is the regular contraction of the muscles necessary to sustain particular skeletal components.

The stretch reflex mechanism is responsible to keep the body in it is place. When muscles are stretched by gravity, sensory receptors within the muscles are stimulated. This results in the release of motor impulses into the same muscles, which causes a sufficient number of motor units to be contracted and an increase in muscle tension. Muscle tension rises as stretching brought on by an outside stimulus.

Phasic tone: Phasic tone, on the other hand, is typically evaluated clinically in the extremities as a quick and short duration response. It results from the rapid stretching of a tendon and attached muscle and more precisely, the muscle spindle.

Did You Know?

Muscle tone is not solely a passive state but rather an active and dynamic condition influenced by various factors such as neural input, motor control systems, and task demands. It's not just about the tension in relaxed muscles but also about the readiness for movement.

Clinical Correlation

Educating patients and caregivers about the nature of muscle tone and its role in movement control can help improve compliance with treatment plans and facilitate self-management strategies.

Voluntary Movement

These are the movements that the person is in control of. Movements are started and produced by muscles contraction. After one group of muscles contracts, another group of muscles relaxes. This factor is crucial to take into account when using muscle relaxation treatments.

Causes of Pathological Tension in Muscle

- **Hypertonia:** Lesions of the upper motor neuron, such as CVA, can cause hypertonia. The two main types of hypertonia are spasticity and rigidity.
- **Hypotonia:** Lower motor neuron lesions, such as spinal or peripheral nerve damage, can cause hypotonia.
- **Myotonia:** It is the delayed relaxation of muscles, and brought on by hypothyroidism, protracted exposure to cold, intense physical activity, and medicines like propranolol.
- **Dystonia:** A person with dystonia has repetitive motions or aberrant postures due to uncontrolled muscles contraction and involuntary twisting.

STRESS AND ITS EFFECTS ON VARIOUS SYSTEMS OF THE BODY

In response-based models, Cox initially presented stress as a dependent variable, where it is seen in terms of a person's reaction to upsetting or unpleasant settings, and then as an independent variable, where it is seen in terms of the stimulus properties of those environments (stimulus-based models). The third method of researching stress considers stress to be a reflection of an environment/person mismatch (interactional models).

Phases

The response-based models of stress was developed by Selye, who saw stress as a general (physiological) reaction to environmental demands. The General adaptation syndrome, as he called it, is a response that goes through three stages.

- **Alarm phase:** The body first displays the alterations typical of initial exposure to the stressor during the alarm phase, along with a decrease in the level of stressor resistance. If the stressor is strong enough, resistance may completely crumble.
- **Resistance phase:** As a result of continued exposure to the stressor, adaptation takes place in the second phase (the resistance phase), and the degree of resistance increases.
- **Exhaustion phase:** When the energy required for resistance to ongoing exposure to the stressor is depleted and final collapse takes place, the adaptation reaches its last stage (exhaustion).

According to Selye, these responses might overwhelm the body's defence mechanisms or cause an overactive reaction, both of which can lead to the so-called illnesses of adaptation.

Clinical Correlation

Interdisciplinary collaboration between healthcare providers, including psychologists, psychiatrists, and primary care physicians, is essential in addressing the complex interplay between stress, mental health, and physical well-being.

Physiological Effects

Even though Selye's beliefs on the connection between stress and physical disease have been established, they serve as the foundation for the current belief that stress from everyday life can cause disease. The parallel from engineering serves as the foundation for stimulus-based models of stress. Physical systems have an "elastic limit", beyond which permanent deformation will take place but within which the system will recover to its initial state if the stressing agent is removed. A "stress limit" also exists in the human mechanism. Stress can therefore be tolerated up to a certain point, but when it is excessive and/or protracted, both physiological and psychological harm may result.

When the apparent demand and the person's impression of his capacity to handle the demand are out of balance, stress may also result. Fundamentally, a person will experience high levels of stress if they believe their ability to manage is poor and their environmental demands are high. On the other hand, if the persons believe that their coping mechanisms are strong, they will be able to accept the same demands with less difficulty and experience less stress. This model includes the idea that stress may also be brought on by a person's sense of insufficient (as opposed to excessive) demand. Exercise-related neuroendocrine and physiological changes, as well as a strong emotional component, all contribute to the response.

The autonomic nervous system and the endocrine system regulates the physiological changes brought on by stress.

Cardiovascular System

- Blood is redistributed from the viscera to the voluntary muscles by the sympathetic branch of the autonomic nervous system, which also makes the heart beat faster. Heart rate, blood pressure, blood clotting time, and breathing rate all rise as a result of this.
- Moreover, the blood's glucose level rises, and the voluntary muscles' blood flow increases.

Sympathetic and Parasympathetic System

- Increased sympathetic activity may also be accompanied by observable symptoms including hyperventilation, cold sweats from increased sweat gland activity, and palpitations.
- The autonomic nervous system's parasympathetic branch becomes more active, which balances off sympathetic activity and returns the body to its resting condition.

Muscles

Psychological stress can lead to increased muscle tension because when the body is stressed, it responds by tensing up muscles in preparation for a fight or flight reactions. The tension occurs even in the absence of physical activity making it more likely that the individuals may experience muscle tightness and pain. Additionally, negative emotions can weaken muscle strength further impacting physical functions and recovery.

Endocrinal Changes

- The adrenal glands' activity is increased as the endocrine system reacts to stress. Noradrenaline, which causes changes linked to alertness, aggression, and fighting behavior as well as a pleasurable feeling of arousal (eustress, or good stress), and adrenaline, which is linked to anxiety and flight behavior as well as increased blood flow to the legs, feelings of threat, and diminished mental abilities (distress, or bad stress), are both released by the adrenal medulla.

  The balance of noradrenaline and adrenaline secretion appears to depend on whether an individual perceives the circumstance as being delightfully challenging (in which case noradrenaline predominates) or dangerous (in which case adrenaline predominates).
- Cortisol, a kind of glucocorticoid released by the adrenal cortex, works to keep the muscles' supply of glucose up to par, increasing the activity of the catecholamines (epinepherine and norepinepherine). These hormones support the fight-or-flight response by speeding up metabolic rate, diverting blood to muscles, and preserving coronary and cerebral blood flow.

- Anxiety, a sense of foreboding and anticipation of bodily damage, dread, rage and violence, hopelessness and helplessness are some of the emotions connected to the fight-flight response. Similar to physiological reactions, psychological reactions are intended to spur on adaptive behaviors in the face of threat, enabling the person to temporarily manage the situation. However, if the stressful event continues or happens frequently, the result could be pathogenic and lead to bodily and/or mental illness.
- The opposite of the emergency "fight/flight" or stress reaction, the relaxation response has been called a hypothalamic trophotropic response. The relaxation response is associated with reductions in all of these measurements, whereas the stress response is characterized by physiological manifestations of increased catecholamine production, oxygen consumption, blood pressure, heart rate, respiratory rate, arterial blood lactate, and skeletal muscle blood flow and tension. As a result, these physiological measurements are frequently utilized in studies to demonstrate the efficacy of relaxation treatments.

FATIGUE

There are numerous different approaches to classifying fatigue. Acute fatigue and chronic fatigue are two different types fatigue that vary in duration. Rest and lifestyle changes can quickly relieve acute fatigue, but chronic fatigue is a prolonged condition that lasts for at least six months and isn't improved by rest. It involves persistent tiredness that doesn't go away easily, even with adequate rest. Classification of fatigue is as follows:

Types

- **Muscle fatigue:** A reduction in the maximum force or power produced by a muscle in response to contractile activity is referred to as muscle fatigue. It can come from several points along the motor pathway and is typically split into central and peripheral parts. This definition of muscle exhaustion fatigue covers two distinct situations in which muscle strength may be diminished. Firstly, muscles may be less able to signal or be instructed to fully contract by bodies that are fatigued from exercise. Second, a muscle's actual contractions may get weaker and weaker for any particular "contraction strength" that the body demands.
- **Peripheral fatigue:** Changes at or distal to the neuromuscular junction cause peripheral fatigue.
- **Central fatigue:** The central nervous system (CNS), where central fatigue originates, reduces the neuronal drive to the muscle. Athletic performance and other strenuous or extended exercise are sometimes limited by the phenomenon of muscle fatigue. Under numerous pathological situations, such as neurological, muscular, and cardiovascular illnesses, as well as old age and frailty, it also worsens and hinders everyday living.
- **Mental fatigue:** The cognitive or perceptual parts of exhaustion are referred to as mental fatigue, and the functioning of the motor system is referred to as physical fatigue.

Causes

- **Lifestyle choices:** A sedentary lifestyle, poor diet, excessive drinking and drug use, high levels of stress, and poor dietary choices can all cause fatigue.
- **Medical conditions:** A wide range of diseases, ailments, and deficits that affect various body parts can all manifest fatigue as a symptom.
- **Illness and infection:** Multiple sclerosis, kidney disease, and cancer are just a few illnesses that might make one tired.
- Moreover, illnesses like mononucleosis, HIV, and the flu can manifest as fatigue.
- **Mental health issues:** The exhaustion brought on by anxiety or sadness can make doing daily tasks challenging or impossible.
- **Autoimmune disorders:** Many autoimmune diseases, such as diabetes, lupus, and rheumatoid arthritis, have fatigue as a symptom.
- **Hormone imbalances:** Issues with endocrine system (the glands that produce hormones in the body) might make one feel exhausted. Hypothyroidism is a common cause of fatigue.
- **Chronic conditions:** Fibromyalgia and Chronic Fatigue Syndrome, often known as CFS or Myalgic Encephalomyelitis
- **Heart and lung issues:** Heart disease, postural orthostatic tachycardia syndrome (POTS), chronic obstructive pulmonary disease (COPD), emphysema, and congestive heart failure are all cardiovascular disorders that can cause fatigue.
- **Deficiencies:** Fatigue is frequently brought on by anemia and other vitamin shortages (such as vitamin D or vitamin B_{12}). Fatigue can result from dehydration since the body needs a lot of fluids to function.
- Fatigue and other symptoms can result from eating disorders such as anorexia, bulimia, obesity, or underweight.

Causes of Fatigue at Cellular Level (Muscle Tissue)

- **Not enough oxygen:** For some types of activity, inadequate oxygen availability, either externally or internally, might hasten fatigue. A chemical called glucose is broken down by cells with the aid of oxygen, giving muscles the energy they require to "fire" and contract.

- **Insufficient energy:** Muscle fatigue can worsen if the amount of glucose and other energy sources in muscle cells is too low.
- **Impaired calcium release:** The release of calcium molecules during exercise is a key factor in signaling skeletal muscle fibers to "fire". Impaired release can cause muscle fatigue.
- **Metabolite and inorganic phosphate buildup:** When the body breaks down glucose and other energy sources, metabolites are the molecules that are left over in the cells. The buildup of these metabolites appears to contribute to increased muscular fatigue during exercise. For instance, the body frequently breaks down a molecule called creatine phosphate when it requires more energy, leaving behind an accumulation of inorganic phosphate as a metabolite. Because it disrupts the calcium signaling system, this phosphate accumulation appears to be a significant contributor to muscle tiredness.
- Inadequate amounts of vitamins, minerals, or electrolytes might hinder the ability of red blood cells to carry oxygen to muscles. Important deficiencies in substances like folate, iron, and vitamin B_{12} can have this effect. Charged electrolytes such as calcium, magnesium, potassium, and sodium help the body signal muscles to contract.
- **Not drinking enough water:** During exercise, dehydration can delay and restrict blood flow to the muscles.
- It is believed that lactic acid buildup in muscles during exercise is a significant contributor to muscular tiredness. At very high concentrations, it can lead to lactic acidosis, which can result in burning sensations and cramping in the muscles.

Symptoms of Muscle Fatigue

- Muscle exhaustion signs frequently progress over time.
- Reduction of maximum muscle power to total muscle exhaustion in a little period of time.
- Respiratory muscles will work hard if people are running quickly, and as a result, they are likely to experience muscle weakness as they get more and more out of breath.
- If people are doing heavy weight training, they will feel the muscular strength of the arms and legs depleting. It gets harder and harder to finish the lifting sets.

RELAXATION EXERCISES

Relaxation exercises are therapeutic activities created to help people reduce stress and anxiety, both physically and mentally. Relaxation technique has become a staple of psychotherapy and relaxation techniques can be used in other healthcare settings as complementary therapies to help patients who are suffering from a variety of distresses, including but not limited to anxiety, depression, pain, and tension.

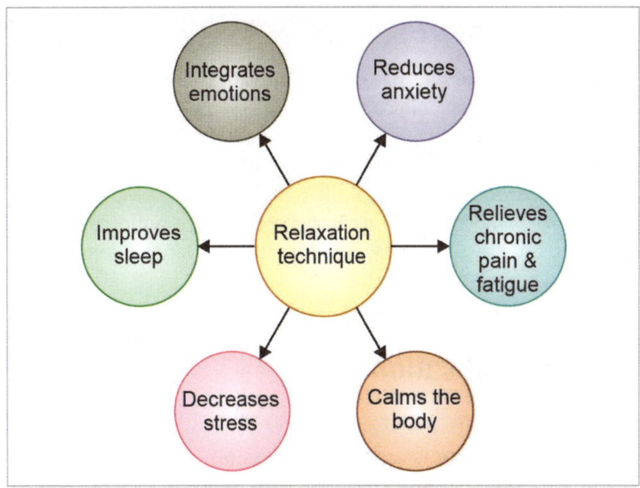

Fig. 18.1: Benefits of relaxation techniques

There are many different methods for relaxing that can help one feel more at ease and less stressed. In addition to the subjective emotional experience, physiological reactions to stress might include elevated heart rate, shortness of breath, and tight muscles. Relaxation techniques can help to lessen these symptoms (Fig. 18.1).

Benefits

The following are the added benefits of relaxation techniques, if practiced on regular basis:

- Decreasing stress hormone activity
- Increasing blood flow to major muscles
- Reducing muscle tension and chronic pain
- Increasing concentration and mood
- Improving sleep quality
- Lowering fatigue
- Lowering anger and frustration increasing confidence to handle issues
- Improves digestion

Indications

- Stress management
- Anxiety attacks
- Muscle fatigue
- Chronic myofacial pain syndrome
- Depression
- General well-being
- Headache
- High blood pressure
- Preparation for hypnosis
- Immune system support
- Insomnia
- Pain management

> **Did You Know?**
>
> **Relaxation techniques are useful adjuvant therapy for:**
> Chronic myofascial pain caused by:
> - Strains
> - Lack of coordination/poor exercise technique.
> - Poor posture
> - Change in muscle tension of psychological or emotional causes such as stress, anxiety, anger and annoyance, depressive mood.

Clinical Applications

- Most patients experience distress which is routinely present with pain, anxiety, depression, sleep disorders or inflammation. Relaxation therapy reduces the impact of stress-related conditions and encourages physiological and psychological equilibrium.
- Regular relaxation can delay the onset and progression of disease, reduce duration of illness and hasten a return to better health.
- Relaxation therapy complements and works well with other medical and pharmaceutical interventions in various medical, mental and musculoskeletal conditions.
- Patients who regularly use easy relaxation techniques report less pain, irritability, depression, insomnia, and inflammation. Research in psychoneuroimmunology (PNI) illustrates how relaxation therapy lowers sympathetic arousal and boosts parasympathetic responses to lessen chronic illnesses. PNI research also shows causal linkages between stress, inflammation, and disease. A crucial benefit of learning how to relax is that it gives patients an effective coping mechanism and more control over their own health and wellness.

TECHNIQUES OF RELAXATION

Techniques of relaxation are broadly divided into:

- General relaxation techniques
- Local relaxation techniques

Figure 18.2 illustrates the various relaxation techniques.

General Relaxation Techniques

Full support of the body in comfortable and restful atmosphere promotes general relaxation. Supine lying, half lying, prone lying, side lying, total body suspension promotes general body relaxation.

For general body relaxation firm surface should be used and sagging bed and soft mattress should be avoided. Proper position of pillows during relaxation, removal of constrictive clothing, belts etc., warm room, proper ventilation, restful atmosphere in treatment room, attitude and behavior of the therapist, easy explanation and instructions by the therapist will help in achieving general body relaxation.

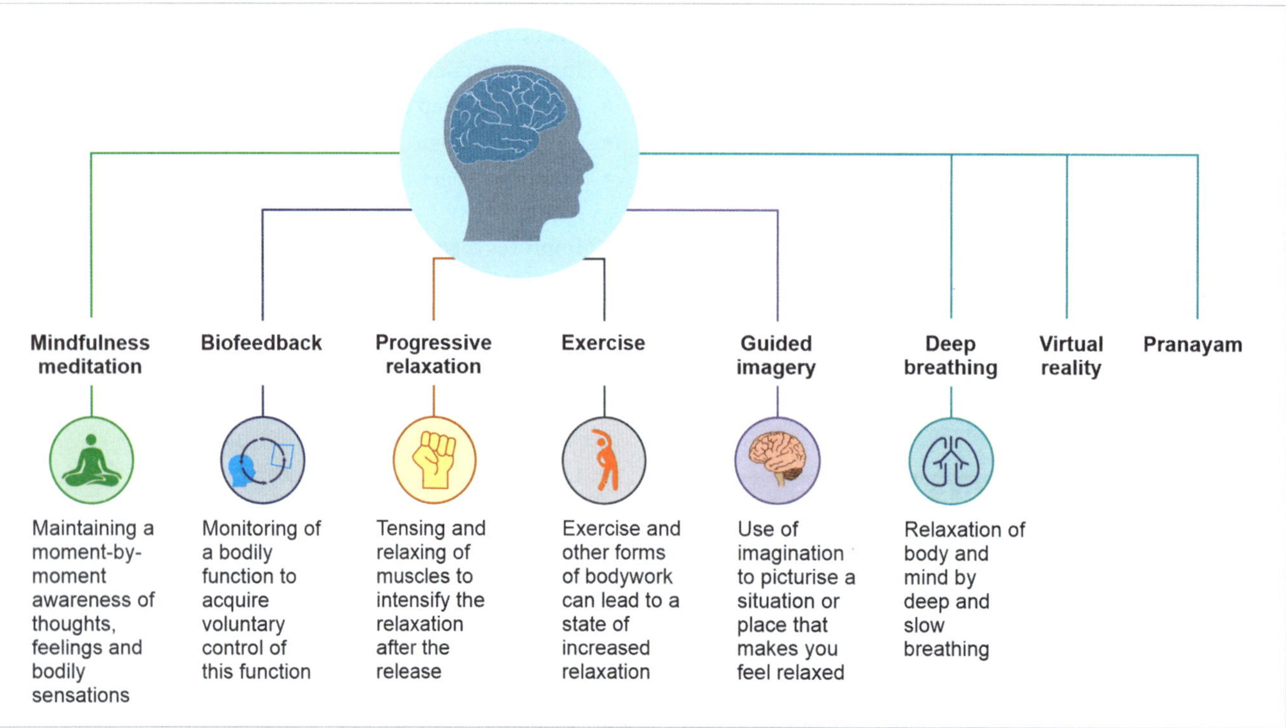

Fig. 18.2: Relaxation techniques

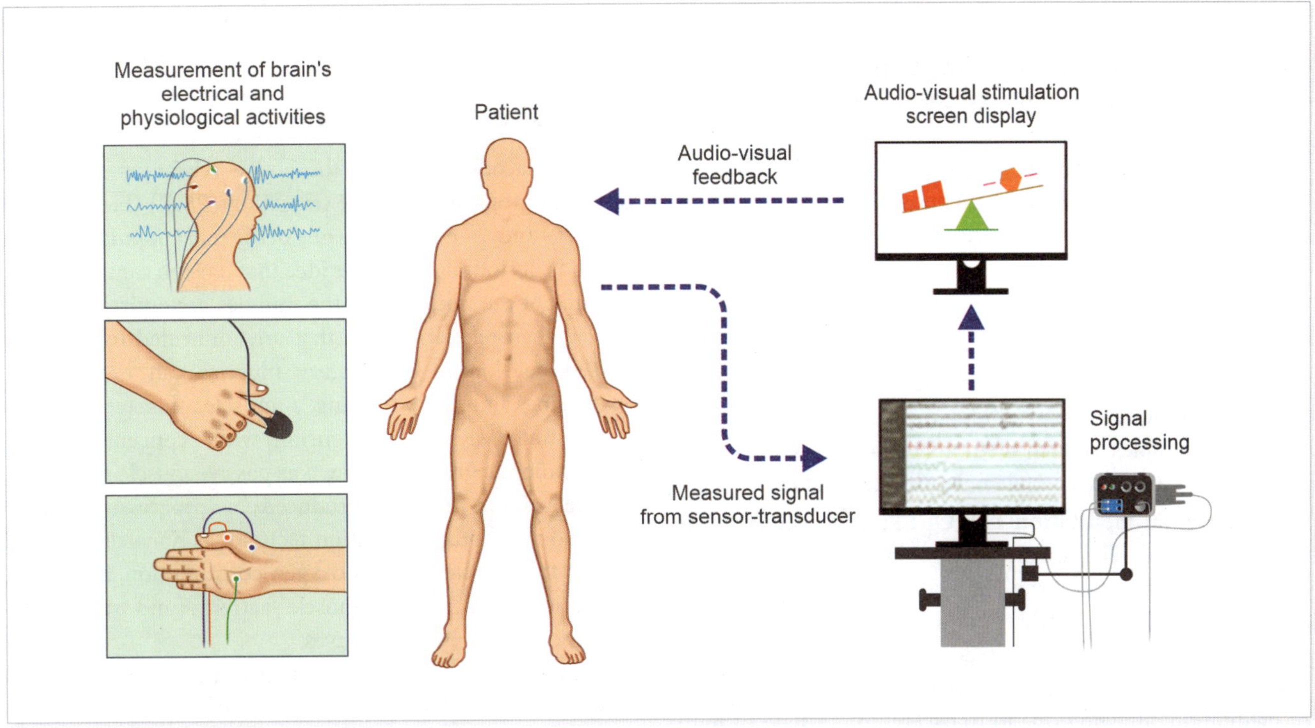

Fig. 18.3: Biofeedback

Biofeedback

Helps to learn, notice and control how the body responds through input, which is typically delivered by an electrical device. One can use the electronic device to monitor changes in the heart rate, blood pressure, or muscle tension in reaction to stress or relaxation (Fig. 18.3).

Deep Breathing

The idea behind deep breathing, sometimes referred to as diaphragmatic breathing, is that integrating the mind and body leads to calm. Participants in the technique must contract their diaphragms while inhaling and exhaling gently. In addition to massaging the internal organs in or around the belly and increasing blood oxygen levels, deep breathing may also activate the vagus nerve. Deep breathing has been demonstrated to have a beneficial effect on a number of issues, including stress, anxiety.

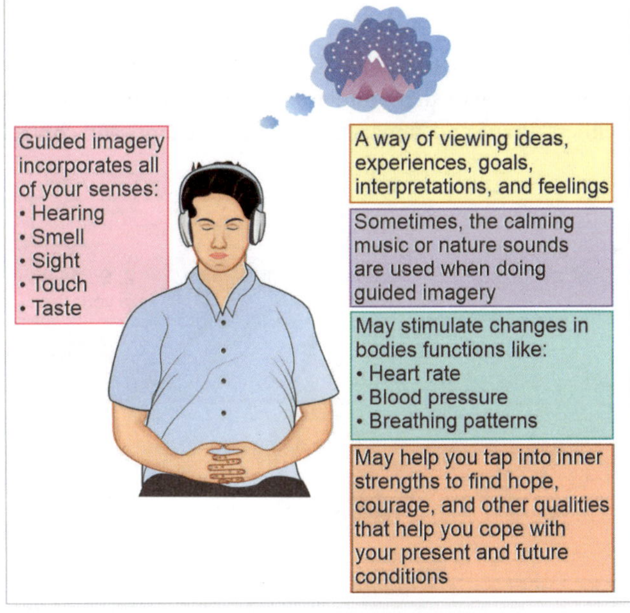

Fig. 18.4: Guided imagery

Guided Imagery

The internal experience of memories, dreams, fantasies, and visions is referred to as imagery (Fig. 18.4). It may involve one, several or all of the senses and act as a link between the physical body, mental processes, and spiritual realms. Visualization and imagery are similar concepts that can be used interchangeably. Guided imagery is a term used to describe the process of using visualization to enhance focus, reduce stress and improve performance by imagining successful outcomes related to a task at hand. Current advances in psychoneuroimmunology provide the theoretical underpinnings for the use of placebos, acupuncture, and visualization to reduce pain and affect physiological changes. When combined with behavioral and cognitive approaches to treatment, visualization can successfully reduce the arousal response to certain stresses or triggers and bring about the required behavioral adjustments for behaviors to be firmly substituted.

Hypnosis

This technique involves concentrated attention. A person is considered to have improved focus, concentration, and responsiveness to suggestions while under hypnosis. A series of instructions and suggestions known as a hypnotic induction normally precedes the onset of hypnosis. "Hypnotherapy" is the term for the therapeutic application of hypnosis.

Meditation

Meditation is a practise where someone utilizes a method to train their attention and awareness to reach a state of emotional tranquility, stable thought process and clear state of mind. Examples of such methods include mindfulness or focusing the mind on a certain object, topic or activity.

Transcendental meditation TM: One kind of meditation is the TM method. This involves silent repetition of a mantra in one's head while using the TM technique. The TM method focuses on calming body into a condition of peaceful alertness. Person is fully alert despite totally relaxed body and calm mind. The busy mind becomes still during this sort of meditation, allowing one to transcend into a state of pure consciousness. To transcend is to exceed.

Pranayama

Breath control is a technique of pranayama. It is a crucial part of yoga, an activity that promotes both physical and emotional wellness. Prana and Yama are terms for life energy and control respectively in Sanskrit. The practice of pranayama involves breathing exercises and patterns.

Progressive Muscle Relaxation

Edmund Jacobson created the active relaxation technique known as progressive muscle relaxation (PMR) (Fig. 18.5). In the 1920s and even today, it is one of the most popular relaxing methods. Its foundation is the idea that tension cannot exist in any area of the body where the muscles are entirely relaxed. Also, if the skeletal muscles that are connected to them are relaxed, tension in involuntary muscles can be lessened. By tensing and then relaxing a muscle, this technique is demonstrated. Release of tension while concentrating on specific muscle groups is known as passive muscle relaxation. The relaxation response produced by progressive muscle relaxation and other relaxation techniques normalizes blood flow to the muscles, lowers oxygen consumption, heart rate, respiration, and skeletal muscle activity, and raises skin resistance and alpha brain waves.

Participants in PMR actively tense their muscles to build tension, which they then gradually release. The exercise is done until the participants are totally relaxed. In order to produce results, this method applies the "top-down" and "bottom-up" processing principles of neurons. Participants in "top-down" processing use the cerebral cortex and cerebellum, higher-order parts of the nervous system, to contract muscles and gradually release tension. Holding and releasing physical tension results in proprioceptive input from peripheral

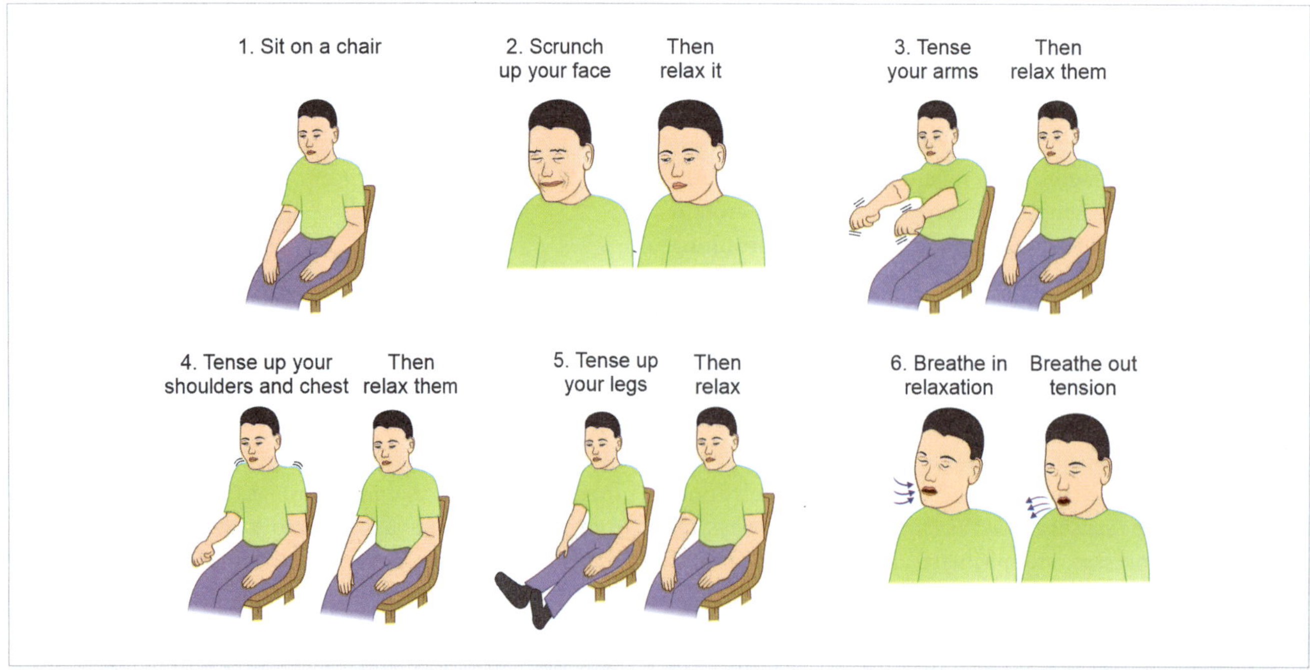

Fig. 18.5: Edmund Jacobson's relaxation routine

muscles, which ascends to the brain *via* the spinal cord and brainstem in "bottom-up" processing. PMR gives participants swift and immediate comfort when both stimulatory pathways are active.

The effects of this technique have been widely demonstrated in numerous studies. For instance, PMR can be useful in reducing stress and anxiety and depression.

Did You Know?

Progressive muscle relaxation (PMR) has been found to be effective not only for reducing stress, anxiety, and depression but also for improving sleep quality. Research suggests that regular practice of PMR can help individuals fall asleep faster, experience deeper sleep, and wake up feeling more refreshed and rejuvenated.

Mitchell Method of Relaxation

The physiological relaxation method developed by Mitchell (Fig. 18.6) is based on reciprocal inhibition and consists of a sequence of ordered isotonic contractions, as well as, diaphragmatic breathing. According to Mitchell (1977), the technique realigns posture by reversing the stress-related posture known as "the punching position". This posture, which typically involves a forward head, clenched jaw, rounded shoulders, bent elbows, and crossed legs, is used when working at a desk. The neurological and endocrine systems are hypothesized to be affected by the punching position's increased muscle strain. This causes the release of the hormones epinephrine and norepinephrine, which, if they are sustained, may cause the adrenal and lymphatic glands to grow and finally result in illness and diseases. Mitchell's relaxation technique is aimed at correcting nervous system imbalance by initiating a phenomenon known as the 'relaxation response'. This response is caused by the release of hormones, which have a widespread effect on the cardiorespiratory system. The relaxation response can be assessed by measuring diastolic and systolic blood pressure, oxygen consumption, heart rate or respiration rate; all of which have the potential to change.

Did You Know?

In a research study by Ganesh et al. in 2017, it was observed that the physiological concept of reciprocal inhibition forms the foundation of Mitchell's diaphragmatic breathing technique, wherein the activation of one muscle group acting on a joint corresponds with the relaxation of the opposing muscle group.

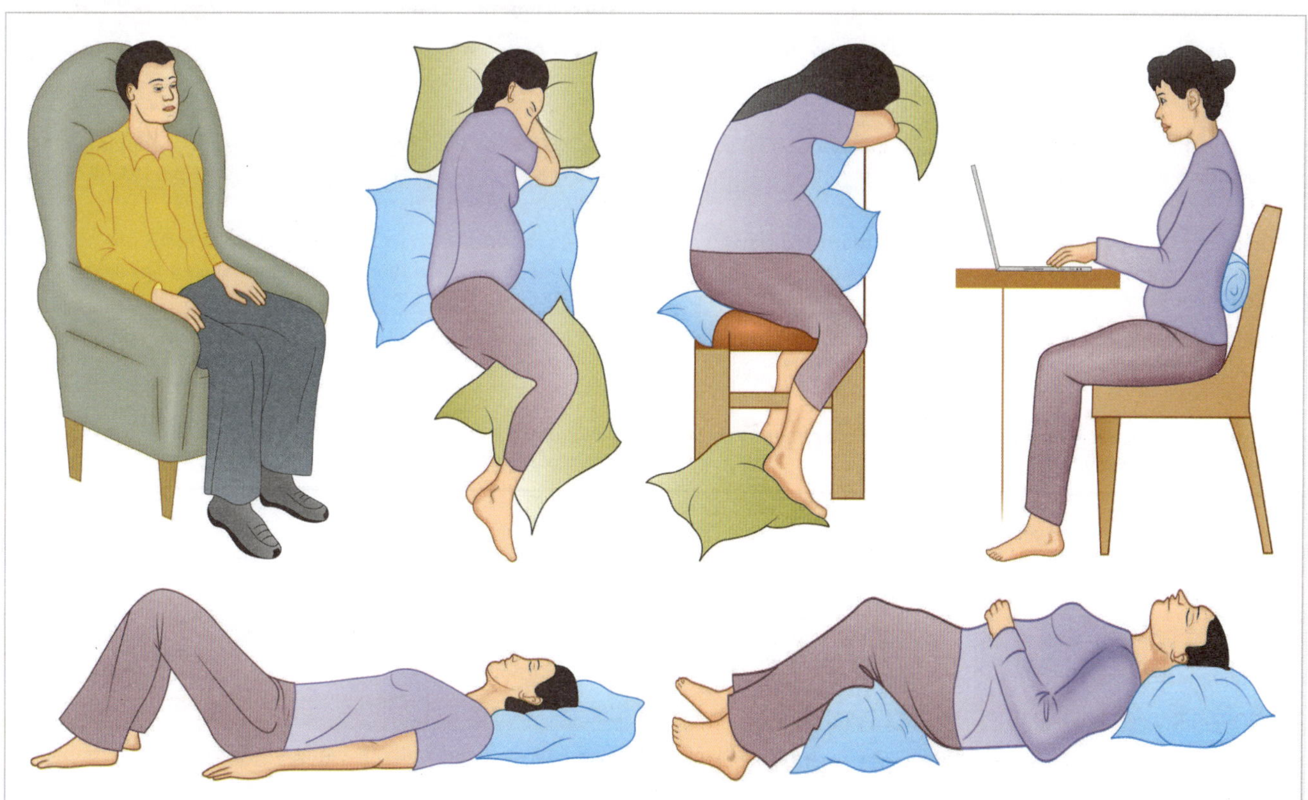

Fig. 18.6: Mitchell's method of relaxation

Technique following in Mitchell method is described as follows (Table 18.1):

TABLE 18.1: Technique of Mitchell method of relaxation

Instructions for upper limb

Instructions given to the patient	Verbal cues for the action performed
"Pull your shoulders towards your feet, making the neck longer." "STOP"	"Feel that there is a wider space between your shoulders and your ears."
"Elbows out and open your elbow joint by keeping your arms supported and slightly away from the sides." "STOP"	"Through the sensation of your skin, feel the position of your elbow and the pressure of your arms supported on them."
"Fingers and thumb long and supported extend your fingers and thumb, then rest your wrist on them." "STOP"	"Without allowing your hands to come into contact, feel your thumb and fingers resting on their support. While the hands are still, feel the pads of the fingers touching them. The conscious brain, which is where experience is recognized, is mostly occupied by the nerves in the hands. Use this to feel at peace and focus on the joy of feeling your hands while they rest."

Instructions for lower limb

Instructions given to the patient	Verbal cues for the action performed
"Turn your hips outward." "STOP"	"Feel your thighs and legs rolled outwards."
"Move your knees gently until they are at ease." "STOP"	"Sense how comfortable your knees are"
"Push your feet away from your face." "STOP"	"Your lower leg muscles are all relaxed, so feel the ankle joints on your feet go softer."

Instructions for body, head and face

Instructions given to the patient	Verbal cues for the action performed
"Use the floor, the chair's back, or the bed to press your body into the support, but avoid using the seat." STOP	"Feel your body's weight against the support."
"Push your head into the pillow or a chair, and notice how your neck moves when you do this." "STOP"	"In the hollow you have created, feel the weight of your head. As soon as the brain recognizes that the head is supported, all of the neck muscles become relaxed."
"Drag your jaw down, loosen your clenched teeth and slowly lower your jaw without opening your mouth." "STOP"	"Feel the space between your upper and lower teeth, the smooth skin on your cheeks, and the little pressure of your lips as they meet."
"Bring your tongue down and let it lie in the middle of your mouth" "STOP"	"Feel the tongue touching your lower teeth."
"Close your eyes. Let your eyelids close down over your eyes, do not shut them tightly." "STOP"	"Be aware of the darkness with your eyes are at rest."
"Smoothen the skin over your forehead from your eyebrows into the hairs, continue the movement over the top of your head and down the back of your neck." "STOP"	"As the major muscle in your skull relaxes and slackens, feel the skin on your forehead smoothing and the hair falling back."

Breathing Instructions

Instructions given to the patient	Action performed
"Take a deep breath feel your abdomen move outwards, slowly breath out" "STOP"	"Ribs will move in and out". "Breathe slowly".

> **Did You Know?**
>
> Mitchell's relaxation technique, by promoting relaxation and reducing stress, may also have positive effects on cognitive function and mental clarity. Research suggests that relaxation techniques can enhance cognitive performance, including improving focus, memory, and decision-making abilities. Therefore, Mitchell's method may offer benefits beyond physiological relaxation, extending to cognitive well-being as well.

Autogenic Relaxation

Autogenic therapy (AT) is a stress management technique developed by Schultz in 1932, focusing on facilitating autonomic self-regulation through a series of specific formulas aimed at inducing relaxation and promoting well-being. It encompasses various methods such as autogenic training, autogenic neutralization, meditative exercises, and graduated active hypnosis, all aimed at addressing functional disorders like stress or trauma.

Mechanism

The AT operates on the premise of homeostatic self-regulatory brain mechanisms, supporting natural self-healing processes. It integrates elements of psychotherapy, hypnosis, yoga, and Zen meditation to empower individuals to control their internal organs and improve self-efficacy. By promoting concentration, reducing tension, and promoting internal calm, AT aids in managing difficult situations and negative emotions. It aligns with Cannon's concept of autonomic homeostasis by balancing sympathetic and parasympathetic control to address psychosomatic dysfunctions.

Physiology

The exercises in AT induce physiological changes through specific formulas targeting different bodily functions. These include muscular relaxation, vascular dilation, heart function stabilization, breathing regulation, visceral organ regulation, and blood flow regulation in the head. By repetitively engaging in these formulas, individuals elicit sensations of heaviness, warmth, calmness, and regulated bodily functions, promoting relaxation and overall well-being.

Procedure

The AT procedure involves practicing a series of exercises twice daily, each focusing on a specific bodily sensation and function.

- The first AT exercise focuses on inducing muscular relaxation as a sensation of heaviness, starting with the dominant arm. The formula "The right (left) arm is heavy" is repeated six times, followed by "I am very quiet" once, for six cycles, practiced twice daily.
- The second exercise adds warmth through vascular dilation with the formula "The right (left) arm is pleasantly warm". Each step is practiced for about a minute, ensuring a total session length of 10–15 minutes, until both heaviness and warmth are consistently felt.
- The third AT exercise regulates heart activity with the formula "The heart is beating calmly and regularly".
- The fourth exercise integrates breathing with "It breathes me," ensuring autonomous breathing.
- The fifth exercise uses "Sunrays are streaming and warm" to regulate visceral organs.
- Sixth exercise induces vasoconstriction with "The forehead is cool". Each step is practiced progressively after mastering the previous ones.

Effects

The effects of AT include inducing relaxation, reducing stress, enhancing well-being, and improving overall mental and physical health. By engaging in the prescribed exercises, individuals experience physiological changes such as muscular relaxation, increased blood flow, and regulated heart activity. These changes promote a sense of calmness, improved self-efficacy, and better coping mechanisms for managing life's challenges and negative emotions. AT empowers individuals to achieve significant physiological changes through self-hypnosis, promoting autonomy and resilience in facing stressors.

Indications

- **Behavioral and emotional problems in children and adolescents:** AT can be effective in managing internalizing symptoms and certain externalizing symptoms like aggression, impulsivity, or attention deficits.
- **Multiple sclerosis (MS):** AT can improve Health-Related Quality of Life (HRQOL) and well-being in individuals with MS, including increased energy, vigor, and reduced limitations due to physical and emotional problems.
- **Insomnia:** AT has been found effective in treating insomnia, offering a potential non-pharmacological intervention for sleep disorders.
- **Anxiety in adolescents:** AT may help reduce anxiety levels among adolescents in school settings, providing a valuable tool for managing stress and promoting relaxation.
- **Irritable bowel syndrome (IBS):** AT may enhance self-control and contribute to general improvement in patients with IBS, potentially offering relief from symptoms associated with this condition.
- **Myofascial trigger point (TrP) pain:** AT may be effective in managing TrP pain, possibly mediated by the sympathetic nervous system and intrafusal muscle fibers. (Banks et al.).

- **Chronic pain:** AT has shown promise in reducing chronic pain, indicating its potential as a complementary approach in pain management strategies.
- **Improving mood states:** AT can positively impact mood states, as evidenced by its effects on young soccer players, suggesting its utility in enhancing emotional well-being.

Virtual Reality

Virtual reality (VR) technology could enhance a variety of relaxation techniques because they are simple to use, engaging, and practical. In a research by Mazgelyte et al., VR-based relaxing strategies raised galvanic skin response values while decreasing salivary steroid hormone levels (including cortisol, cortisone, and total glucocorticoid). There was marked decline in the degree of subjectively felt psychological strain. Recently developed VR-based relaxation techniques are potentially useful stress-reduction aids, and they may be especially well suited for people who are unable to adhere to a rigid and time-consuming stress management intervention regimen.

Local Relaxation

Sometimes it's impossible or undesirable to relax the entire body. The best local relaxation techniques depend on the tension's source and location.

- Relaxation can be achieved locally by massaging the affected area and doing passive motions nearby.
- Limited range, small-range and rhythmically supported pendular movements can alleviate the pain that comes with movement.
- Using appropriate positions that apply traction to tight muscles and reciprocal relaxation of the antagonist muscle will aid in stretching tight muscles without inducing the stretch reflex to treat adaptive shortening of a muscle group.
- **Passive movements:** Gentle, rhythmic passive movements performed in the available range produces relaxation of the part moved and the surrounding structures.

SUMMARY

- Muscle tone is a dynamic and complex condition controlled by various input and output systems, including the brainstem reticular system, spinal cord, muscle spindle, and cortex.
  - It is an adaptive function of the neuromotor apparatus that responds adequately to commands from upper levels of movement construction by fine-tuning the excitability of sensory and motor cells for active postural or movement control tasks.
  - Muscle tone can be divided into "phasic" and "postural" categories. Postural tone involves the regular contraction of muscles necessary to sustain specific skeletal components, while phasic tone is a quick and fleeting response resulting from rapid stretching of a tendon and attached muscle.
- Voluntary movement refers to movements that a person is in control of, starting and producing by muscles contracting. Causes of pathological tension in muscle include hypertonia, hypotonia, myotonia, and dystonia. Stress, on the other hand, is a general physiological reaction to environmental demands and can lead to illnesses of adaptation.
- Stress can be influenced by a person's perception of their ability to handle environmental demands and their belief in their capacity to handle them.
  - The autonomic nerve system and endocrine system regulate physiological changes brought on by stress, such as blood redistribution from the viscera to voluntary muscles, increased heart rate, blood pressure, blood clotting time, and breathing rate, increased blood glucose levels, and increased voluntary muscle blood flow.
- The autonomic nervous system's parasympathetic branch becomes more active, which balances off sympathetic activity and returns the body to its resting condition. Understanding the relationship between muscle tone and stress is crucial for effective muscle relaxation treatments.
- The adrenal glands release noradrenaline and adrenaline, which are involved in the fight-or-flight response. The balance of these hormones depends on the perceived challenge or danger of the situation.
  - Cortisol, a glucocorticoid, helps maintain glucose supply and increase catecholamine activity, supporting the fight-or-flight response.
  - Anxiety, fear, and hopelessness are associated with the fight-flight response, which can lead to physical and mental illness if the event persists. The relaxation response, a hypothalamic trophotropic response, reduces these physiological responses, while the stress response increases catecholamine production and oxygen consumption.
  - Fatigue is a common side effect of these responses.
  - Weariness can be classified into acute and chronic types, with acute fatigue lasting for four months and not alleviated by rest.
  - Muscle fatigue is a reduction in the maximum force or power produced by a muscle in response to contractile activity, often split into central and peripheral parts.
- Causes of fatigue include lifestyle choices, medical conditions, illnesses, infections, mental health issues, autoimmune disorders, hormone imbalances, chronic conditions, heart and lung issues, deficiencies, and eating disorders.
  - Causes at the cellular level include inadequate oxygen availability, insufficient energy, impaired calcium release, metabolite and inorganic phosphate buildup, inadequate amounts of vitamins, minerals, or electrolytes, not drinking enough water during exercise, and lactic acid buildup.
  - These factors can lead to muscle tiredness, which can progress over time and result in reduced muscle power, respiratory muscle debility, and difficulty in lifting sets.
- Symptoms of muscle exhaustion include a reduction in maximum muscle power, difficulty breathing, and difficulty in lifting sets due to heavy weight training.
  - To alleviate fatigue, individuals should make lifestyle adjustments and maintain a healthy diet.
  - Chronic fatigue, on the other hand, is characterized by persistent tiredness that lasts for four months and is not eased by rest.
- Relaxation exercises are therapeutic activities designed to help people reduce stress and anxiety, both physically and mentally. They can be used in healthcare settings as complementary therapies to help patients suffering from various distresses, including anxiety, depression, pain, and tension.
  - There are numerous relaxation techniques that can be used, and they can be learned through self-help or by health professionals.
  - Practicing relaxation techniques can have many benefits, such as improving digestion, maintaining normal blood sugar levels, decreasing stress hormone activity, increasing blood flow to major muscles, reducing muscle tension and chronic pain, increasing concentration and mood, improving sleep quality, lowering fatigue, lowering anger and frustration, and increasing confidence to handle issues.
  - Relaxation therapy is important in clinical practice because it reduces the impact of stress-related conditions and encourages physiological and psychological equilibrium.
  - Regular relaxation can delay the onset and progression of disease, reduce the duration of illness, and hasten a return to better health. It complements and works well with other medical and pharmaceutical interventions in various medical, mental, and musculoskeletal conditions.
  - Patients who regularly use easy relaxation techniques report less pain, irritability, depression, insomnia, and inflammation.
  - Research in psychoneuroimmunology (PNI) shows that relaxation therapy lowers sympathetic arousal and boosts parasympathetic responses to lessen chronic illnesses.
  - Learning how to relax gives patients an effective coping mechanism and more control over their own health and wellness.
- Edmund Jacobson developed progressive muscle relaxation (PMR) in the 1920s, which involves actively tensing muscles to build tension and gradually release it. This technique uses both top-down and bottom-up processing principles of neurons, reducing stress, anxiety, and depression.

- The Mitchell's method of relaxation is based on reciprocal inhibition and involves isotonic contractions and diaphragmatic breathing. It realigns posture by reversing the stress-related posture known as the punching position, which can cause the release of hormones that can cause physical sickness and death.
  - Mitchell's relaxation technique aims to correct nervous system imbalance by initiating the relaxation response.
- Virtual reality (VR) technology can enhance relaxation techniques by raising galvanic skin response values and decreasing salivary steroid hormone levels. These techniques are potentially useful stress-reduction aids and may be particularly beneficial for people who cannot adhere to rigid stress management interventions.
- Local relaxation techniques can be achieved by massaging the affected area and performing passive motions nearby. Limited-range rhythmically supported pendular movements can alleviate pain associated with movement, while appropriate positions can treat adaptive shortening of a muscle group.

FURTHER READINGS

- Atkins T, Hayes B. Evaluating the impact of an autogenic training relaxation intervention on levels of anxiety amongst adolescents in school. Educational and Child Psychology. 2019 Sep 1;36(3):33-51.
- Banks SL, Jacobs DW, Gevirtz R, Hubbard DR. Effects of autogenic relaxation training on electromyographic activity in active myofascial trigger points. Journal of Musculoskeletal pain. 1998 Jan 1;6(4):23-32. the efficacy of AT in individuals suffering from chronic pain (Kohlert et al.)
- Bell JA, Saltikov JB. Mitchell's relaxation technique: Is it effective? Physiotherapy. 2000 Sep 1;86(9):473-8.
- Camp bell WW, Dejong RN; Dejong's the neurologic examination. Lippincott Williams and Willkins; 2005.
- Cox T. The nature and measurement of stress. Ergonomics. 1985 Aug 1;28(8):1155-63.
- Ganesh B, Chodankar A, Parvatkar B. Comparative study of Laura Mitchell's Physiological Relaxation Technique versus Jacobson's Progressive Relaxation Technique on severity of pain and quality of life in primary dysmenorrhea: randomized clinical trial. J Med Sci Clin Res. 2017;5(7):25379-87.
- Ganguly et al. Ganguly J, Kulshreshtha D, Almotiri M, Jog M. Muscle tone physiology and abnormalities. Toxins. 2021 Apr 16;13(4):282.
- Gardiner MD. The principles of exercise therapy. Bell; 1957.
- Goldbeck L, Schmid K. Effectiveness of autogenic relaxation training on children and adolescents with behavioral and emotional problems. Journal of the American Academy of Child & Adolescent Psychiatry. 2003 Sep 1;42(9):1046-54.
- Hashim HA, Hanafi H, Yusof A. The effects of progressive muscle relaxation and autogenic relaxation on young soccer players' mood states. Asian journal of sports medicine. 2011 Jun;2(2):99.
- Kerr, Kate (2017). Critical Review on Relaxation Techniques. Critical Reviews in Physical and Rehabilitation Medicine, 29(1-4), 390–428. doi:10.1615/critrevphysrehabilmed.v29.i1-4.140
- KIM MK Choe YW, KIM SG, Choi EH. Relationship among stress, anxiety depression, muscle tone and hand grip strength in patients with chronic stroke: partial correction. Journal of the Korean Society of Physical Medicine. 2018; H3(4): 27–33.
- Kohlert A, Wick K, Rosendahl J. Autogenic training for reducing chronic pain: a systematic review and meta-analysis of randomized controlled trials. International journal of behavioral medicine. 2022 Oct 1:1-2.
- Linden W. Autogenic training: a narrative and quantitative review of clinical outcome. Biofeedback and Self-Regulation. 1994 Sep;19:227-64.
- Mazgelytė E, Rekienė V, Dereškevičiūtė E, Petrėnas T, Songailienė J, Utkus A, Chomentauskas G, Karčiauskaitė D. Effects of virtual reality-based relaxation techniques on psychological, physiological, and biochemical stress indicators. InHealthcare 2021 Dec 14 (Vol. 9, No. 12, p. 1729). MDPI.
- Nicassio P, Bootzin R. A comparison of progressive relaxation and autogenic training as treatments for insomnia. Journal of Abnormal Psychology. 1974 Jun;83(3):253.
- Selye H. What is stress. Metabolism. 1956 Sep;5(5):525-30.
- Shinozaki M, Kanazawa M, Kano M, Endo Y, Nakaya N, Hongo M, Fukudo S. Effect of autogenic training on general improvement in patients with irritable bowel syndrome: a randomized controlled trial. Applied psychophysiology and biofeedback. 2010 Sep;35:189-98.
- Simons DG, Mense S. Understanding and measurement of muscle tone as related to clinical muscle pain. Pain 1998 Mar I; 75(1): 1–7.
- Singh A, Singh T, Singh H. Autogenic training and progressive muscle relaxation interventions: effects on mental skills of females. European Journal of physical education and sport science. 2018 Nov 28.
- Sutherland G, Andersen MB, Morris T. Relaxation and health-related quality of life in multiple sclerosis: the example of autogenic training. Journal of behavioral medicine. 2005 Jun;28:249-56.
- Swaiman KF, Philleps J.5 – muscular tone and gait disturbances Swaiman's Pediatric Neurology' E book: Principles and practice. 2024 Sep 23:27.
- Toussaint, L., Nguyen, Q.A., Roettger, C., Dixon, K., Offenbächer, M., Kohls, N., Hirsch, J. and Sirois, F., 2021. Effectiveness of progressive muscle relaxation, deep breathing, and guided imagery in promoting psychological and physiological states of relaxation. Evidence-Based Complementary and Alternative Medicine, 2021.

STUDENT ASSIGNMENT

LONG ANSWER QUESTIONS

1. Explain cause of pathological tension in muscle.
2. Describe stress mechanics, types of stresses and the effects of stress on the body.
3. Describe fatigue, general causes, signs and symptoms of fatigue.
4. Explain the indications of relaxation techniques.
5. Explain the principles and uses of Jacobson and Mitchell's relaxation methods.

SHORT ANSWER QUESTIONS

1. Define the following:
   a. Muscle tone
   b. Muscle fatigue
   c. Postural tone
   d. Relaxation
2. Write about stress and its types.
3. Define autogenic relaxation.

MULTIPLE CHOICE QUESTIONS

1. **Which of the following is NOT a goal of relaxation therapy in exercise therapy?**
   a. Decreasing muscle tension
   b. Increasing heart rate
   c. Reducing stress and anxiety
   d. Improving sleep quality
2. **Which relaxation technique involves tensing and then relaxing specific muscle groups sequentially?**
   a. Progressive muscle relaxation
   b. Deep breathing
   c. Guided imagery
   d. Yoga
3. **Which of the following is a psychological benefit of relaxation therapy?**
   a. Increased muscle strength
   b. Reduced blood pressure
   c. Enhanced concentration
   d. Improved joint flexibility
4. **Which relaxation technique involves slow, deep breathing to promote relaxation?**
   a. Biofeedback
   b. Autogenic training
   c. Diaphragmatic breathing
   d. Meditation
5. **Which of the following is NOT a physiological effect of relaxation therapy?**
   a. Lowered blood pressure
   b. Reduced cortisol levels
   c. Increased sympathetic nervous system activity
   d. Slowed heart rate
6. **Which relaxation technique involves imagining oneself in a peaceful, relaxing environment?**
   a. Progressive muscle relaxation
   b. Guided imagery
   c. Autogenic training
   d. Mindfulness meditation

7. **Which relaxation technique focuses on achieving a state of deep relaxation through self-suggestion of warmth and heaviness in different parts of the body?**
   a. Progressive muscle relaxation
   b. Biofeedback
   c. Autogenic training
   d. Tai chi
8. **Which of the following is a potential application of relaxation therapy in physiotherapy?**
   a. Increasing joint stiffness
   b. Reducing postoperative pain
   c. Inducing muscle spasm
   d. Enhancing muscle fatigue
9. **Which relaxation technique emphasizes the mind-body connection and focuses on being present in the moment?**
   a. Progressive muscle relaxation
   b. Biofeedback
   c. Mindfulness meditation
   d. Autogenic training
10. **Which relaxation technique uses electronic monitoring to provide feedback on physiological processes such as heart rate and muscle tension?**
    a. Biofeedback
    b. Yoga
    c. Tai chi
    d. Pilates

ANSWER KEY

1. b	**2.** a	**3.** c	**4.** c	**5.** c	**6.** b	**7.** c	**8.** b	**9.** c	**10.** a

19 Suspension Therapy

Sheetal Kalra

LEARNING OBJECTIVES

After the completion of the chapter, the readers will be able to:

- Define and explain principles of suspension therapy.
- Describe the equipment and accessories used in suspension therapy.
- Discuss the indications, contraindications, and benefits of suspension therapy.
- Identify the different types of suspension therapy.
- Explain the indications and technique for each type of suspension.
- Demonstrate axial and vertical fixation for mobilizing, strengthening and re-education of various muscles and joints.
- Explain the techniques of suspension therapy for upper limb and lower limb.

CHAPTER OUTLINE

- Introduction
- Principles
- Equipment and Accessories
- Indications
- Contraindications
- Types of Suspension
- Advantages
- Disadvantages
- Techniques for Lower Limb
- Techniques for Upper Limb

KEY TERMS

Axial suspension: In axial suspension, the point of suspension is situated vertically above the joint's axis of movement, and the rope is secured from a point that is directly above the joint that needs to be moved.

Inflammation: The process by which white blood cells and the substances they produce protect the body against infection from external organisms like bacteria and viruses.

Mechanical advantage: In physics, the term "mechanical advantage" refers to a mechanism's ability to increase the force applied to it. Suspension therapy provides a mechanical advantage by reducing joint tension through the use of a harness or sling that allows the patients to swing their arms in different directions.

Pendular suspension: Pendular motion is utilized to strengthen, increase range of motion, and preserve muscular structure by minimizing the effects of gravity.

Suspension therapy: It is a technique in which ropes, pulleys, and slings are used to suspend a body part or the entire body to allow friction-free movement, which improves range of motion, muscle strength, and body support.

Vertical suspension: Vertical suspension is mostly utilized to support the body part because it restricts the movement to a narrow range of pendular motion on either side of the central resting position.

INTRODUCTION

Suspension therapy is a unique form of therapeutic exercise that involves suspending a person's body in the air, as they perform a variety of mobility exercises. The therapy can be administered while sitting or lying down, and the suspension devices could be ropes or slings. To help increase joint mobility, flexibility, and strength, physical therapists, chiropractors, and other medical practitioners frequently employ the technique. Additionally, it can be utilized to lessen inflammation, boost circulation, and relieve pain.

Traction, which involves gently pulling on the spine or other portions of the body, is the foundation of suspension therapy. According to the idea underlying this therapy, by suspending the person in midair, the force of gravity is diminished, which may aid in decompressing the joints and relieving strain on the painful joints or areas of the body.

Suspension frees the body from the friction of material upon which the body components may be resting and it permits free movement without resistance. The suspension apparatus was invented by Mrs F. Guthrie Smith in 1918. She developed this device to aid in the rehabilitation of injured soldiers during World War I.

Did You Know?

Suspension therapy also engages the proprioceptive system, which involves sensory receptors in the muscles and joints that provide feedback to the brain about body position and movement. This engagement helps improve proprioception, which can enhance coordination, balance, and overall body awareness, contributing to better movement patterns and injury prevention.

PRINCIPLES

It works under the principles of:

- Friction
- Pendulum
- Movement in the gravity eliminated plane
- Traction
- Suspension
- Resistance
- Control

Let's discuss these principles one by one in detail:

Friction

Frictional force appears when a particular surface moves over another. In suspension, friction is reduced, therefore causing smooth and easy movement.

The term "mechanical advantage" in physics describes a device's capacity to increase the force that is applied to it. In suspension therapy, the harness or sling that is used to suspend the patient's body, offers a mechanical benefit by lessening the stress on the joint, facilitating limb movement in various directions.

This mechanical advantage is created in largely due to friction. Gravity pulls the patient's body downward and toward the ground when they are suspended in a harness or sling. The upward force or tension force, which is produced by the tension in the harness or sling, supports the body and offloads some or all of its weight. The tension force produced is influenced by the degree of friction between the patient's body and the harness or sling. More tension force will be required if the friction is high in order to resist the gravitational pull downward. A lower tensile force will be required if the friction is low.

Pendulum

A heavy object suspended by a weightless string is called a pendulum. The pendulum moves to and fro when force is applied to it (Fig. 19.1A). Pendular motion in the human body mostly occurs in the shoulder and hip joints, as well as in leg movements and arm swings while walking.

According to this principle, a suspended object will swing back and forth in a predictable manner.

In suspension therapy, the patient is suspended in a sling or harness while having his limbs free to hang. Due to the pull of gravity, the body part will inevitably swing back and forth like a pendulum (Fig. 19.1B). This motion can encourage increased range of motion and reduce joint stiffness and pain by stimulating the affected joint and surrounding muscles.

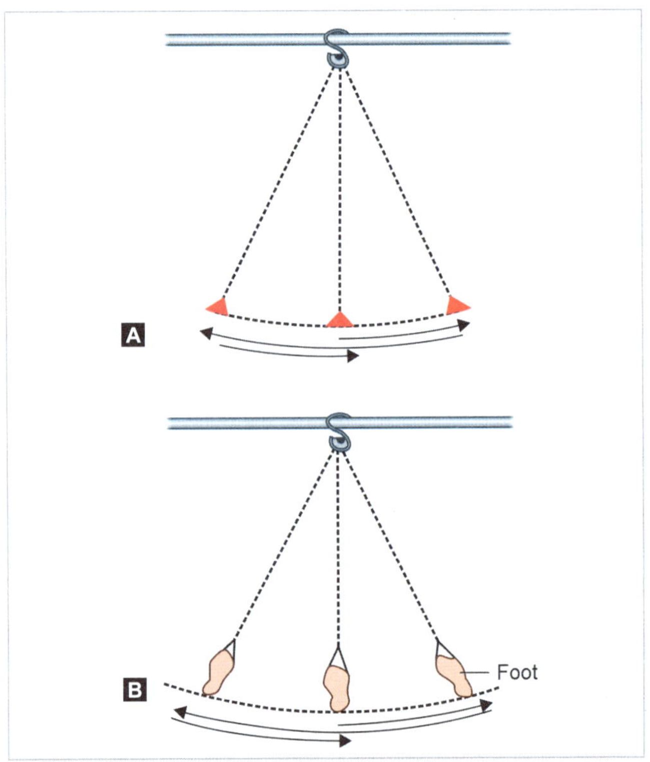

Figs 19.1A and B: Pendulum and a foot in suspension therapy swinging to and fro

In order to promote controlled mobility, the therapist may also direct the body part in particular directions. The pendulum effect can be utilized to maintain mobility in the body part and encourage relaxation of the muscles surrounding joints, especially the shoulder joint by lightly beginning the movement. This is really beneficial in patients with conditions such as frozen shoulder/painful hip where movement of the joint is limited.

The pendulum concept can also be used in conjunction with other methods to assist increase the flexibility and strength of the shoulder joint, such as gentle stretches and exercises. Overall, the pendulum concept is useful in suspension therapy because it offers a gentle and efficient technique to stimulate movement and healing in the affected joint.

Other uses are to maintain the muscle property, increase the range of movement, with strengthen muscles.

Movement in the Gravity Eliminated Plane

One of the fundamental ideas guiding suspension therapy is the elimination of gravity. According to this principle, the force of gravity acting on the patient's body can be reduced or completely eliminated by suspending them. This lessens the strain on the painful or weak joint and facilitates movement in various directions.

The therapist can reduce the weight that the joint must support by suspending the patient's body in a harness or sling, which can ease discomfort along with stiffness and increase range of motion. This is especially helpful for individuals who have diseases like frozen shoulder or rotator cuff injuries, where joint movement may be restricted, as a result of pain or inflammation.

The elimination of gravity also enables the therapist to regulate the level of resistance the patient encounters while performing exercises. The therapist can change the amount of weight the patient must lift by modifying the tension in the harness or sling, giving the muscles surrounding the joint a progressive and regulated resistance. Over time, this may contribute to increased muscle endurance and strength.

This approach can aid in the promotion of healing, the reduction of discomfort, and the improvement of function in patients by lowering the load on the joint and managing resistance.

Traction

Traction, similar to a gentle pulling force applied to the spine or other regions of the body, is a component of suspension therapy. Traction can help to decompress the joints and relieve pressure on the spinal discs by decreasing the force of gravity, which may help to lessen discomfort and increase mobility.

Suspension

The treatment involves suspending the patient's body in midair with the help of ropes, slings, or elastic bands. By lessening the effects of gravity on the body, this can assist people complete stretches and exercises that would be uncomfortable or difficult to do when standing or lying down under the influence of gravity.

Resistance

Suspension treatment can be used to add resistance to workouts, which can aid in strength development. Depending on the demands and capabilities of the individual, the suspension systems can be modified to offer various levels of resistance.

Control

Because the body is suspended in midair, suspension therapy calls for a high level of control and stability. In order to prevent damage, the exercises and regulated movements used in therapy must be carried out with the right form and technique.

Clinical Correlation

Suspension therapy can also be adapted for postural correction and rehabilitation. By suspending the body in a controlled manner, therapists can manipulate the positioning of the body to promote proper alignment and muscle activation. This can be especially helpful for individuals with postural imbalances or conditions such as scoliosis, where targeted exercises and positioning can help alleviate pain and improve overall posture and spinal alignment.

EQUIPMENT AND ACCESSORIES

Slings

During suspension therapy, the body is supported by slings and harnesses. They can usually be adjusted to accommodate the person's body size and form and are composed of sturdy, long-lasting materials. Harnesses and slings can be used on their own or in combination with other tools, such as suspension trainers.

Single Sling

A single sling has a D-shaped ring at each end and is constructed of canvas tied with soft webbing. It typically measures 17 cm in width and 68 cm in length. It is used for the knee and elbow. It can be folded to support the ankle and wrist (Fig. 19.2A).

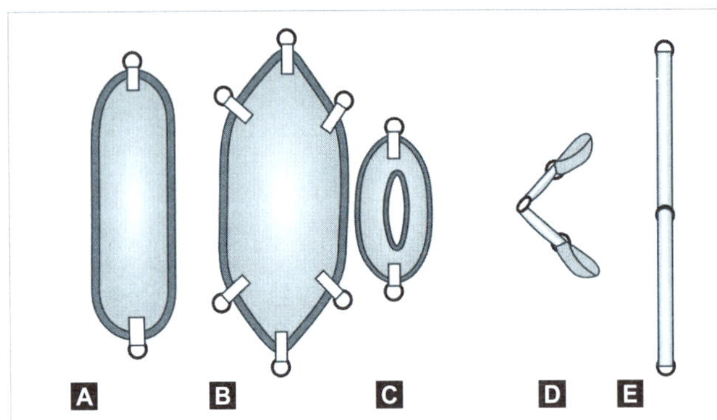

Figs 19.2A to E: Slings: **A.** Single sling; **B.** Double sling; **C.** Head sling; **D and E.** Three-ring sling

Double Sling

The double sling is larger than the single sling. The sling is broad and has D rings on both ends and used to support the thighs, pelvis, knee, and thorax, which are substantial body parts. It is 29 cm wide and 68 cm long (Fig. 19.2B).

Head Sling

This short split sling has a central slit created by the angle at which its two parts are stitched together. It is a wooden cleat-It is made up of wood and is used for altering of rope. It has 2–3 holes from where the rope passes. The rope itself holds the cleat from friction resistance. It is utilized for head support and has a slit in the center to accommodate the lower ear while side-laying or the occipital region when sleeping on one's back (Fig. 19.2C).

Three-ring Sling

Ring sling 75 cm long and 3–4 cm wide three-ring sling. It consists of three D rings, with two of them stationary at each end and one movable in the middle (Figs 19.2D and E).

Supporting Ropes

Ropes are essential components in suspension therapy, providing support and enabling controlled movement. They are usually made of 3-ply hemp to prevent slipping and are arranged in three main types: single rope, pulley rope, and double rope. These rope configurations are designed to ensure stability, adjustability, and support during various therapeutic exercises.

Single Rope

A single rope has a ring at one end for hanging, while the other end passes through a wooden cleat and a dog clip before being secured with a half-hitch knot. The cleat is used to adjust the rope's length and holds it in place through friction. The total length of the rope is 1.5 m, and further adjustments can be made without tying permanent knots. The dog clip is mounted on a pivot, allowing easy adjustments and reducing discomfort when slings are attached.

Pulley Rope

The pulley rope includes a dog clip at one end and passes through a pulley system before being attached to a cleat and another dog clip. This setup is commonly used for movements like abduction, adduction, flexion, and rotation, enabling more complex, three-dimensional limb exercises.

Double Rope

The double rope uses two pulleys to create a mechanical advantage, making it suitable for supporting heavier body parts like the pelvis or thorax. This arrangement allows for smoother suspension and easier load management by using gravity effectively.

Dog Clip

Dog clip is used to attach the supporting rope with mesh and to attach the sling with supporting rope (Figs 19.3A and B).

Wooden Cleat

Made of wood, it is used to change ropes. From where the rope passes, it has two to three holes. The rope itself prevents the cleat from rubbing against moving on its own (Figs 19.4A and B).

S hook

S hook can be used at either end and based on the fixed point's size (Fig. 19.5).

Storage Trolley

Storage of slings and ropes on wall frame is done with S-shaped hook (Fig. 19.6).

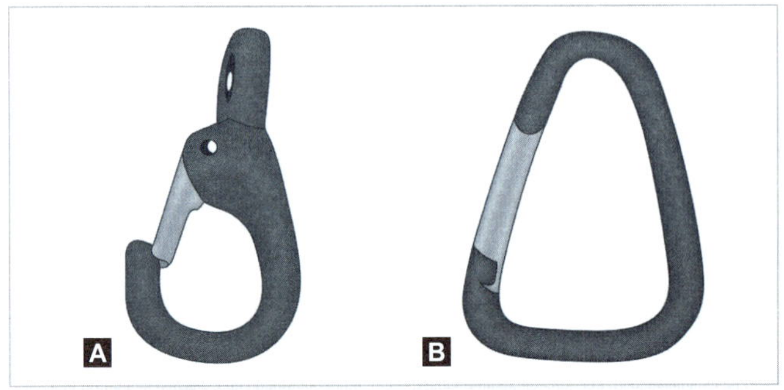

Figs 19.3A and B: **A.** Dog clips; **B.** Carabiner clip

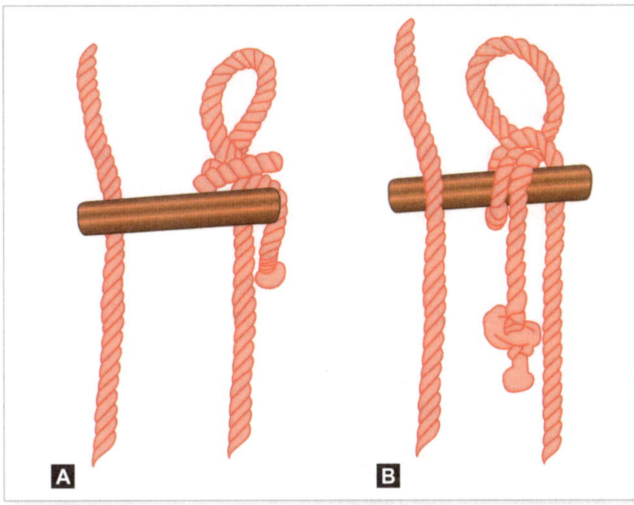

Figs 19.4A and B: Wooden cleat: A. Without knotting; B. With knotting

Fig. 19.5: "S" hook

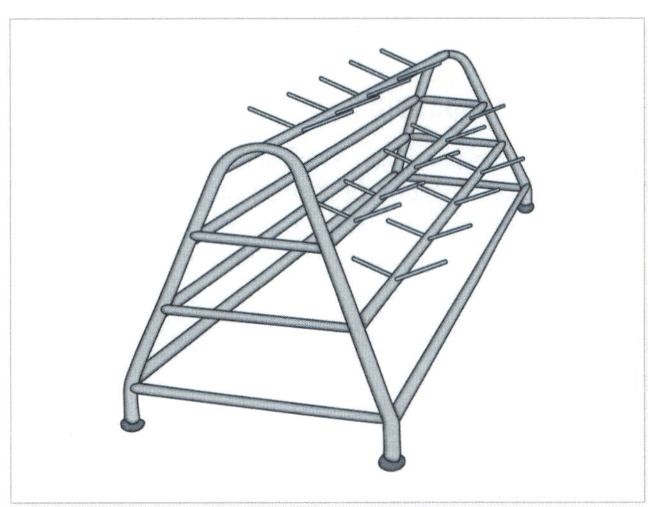

Fig. 19.6: Storage trolley

INDICATIONS

- **In musculoskeletal disorders:** Suspension therapy can be used to treat back pain brought on by conditions that put stress on the spinal discs, such as spinal compression. Suspension therapy may assist in decompressing the spine and easing pain by lowering the force of gravity and providing traction.
- **To relieve pain and improve mobility:** Suspension therapy can also be used to treat joint pain brought on by inflammatory diseases such as arthritis. Suspension therapy may assist to increase joint mobility and lessen inflammation by applying traction.
- **For muscle imbalances:** By giving specific exercises that concentrate on particular muscle groups, suspension therapy can be utilized to rectify muscle imbalances. This can assist in enhancing posture and avoiding injuries brought on by muscular imbalances.
- **For postoperative rehabilitation:** To aid people in recovering from surgery, an injury, or other ailments, suspension therapy can be incorporated into a rehabilitation program. The treatment can provide a low-impact workout while simultaneously enhancing mobility, strength, and flexibility.
- **In spinal cord injuries:** Suspension therapy can be a helpful technique for patients with spinal cord injuries (SCI) because it allows for safe and controlled movement while reducing the load on the affected spinal column.
- **For muscle strength and endurance:** Patients may develop muscle weakness following surgery as a result of injury or lack of use to the afflicted area. Through controlled, progressive resistance exercises, suspension treatment can assist to address this deficit and build muscle strength and endurance.

CONTRAINDICATIONS

- **Spinal instability:** Patients who have unstable spinal fractures or other spinal injuries that could be made worse by the traction forces used during therapy are not suitable for suspension therapy.
- **Severe osteoporosis:** Due to the stresses placed on the body during suspension therapy, people who have severe osteoporosis, a condition that results in weakened bones, may be at risk of fractures.
- **High blood pressure:** The traction forces applied to the body during suspension therapy can raise blood pressure, which may be troublesome for people who already have high blood pressure.
- **Pregnancy:** Due to the risk of falls, which could harm the growing fetus, suspension therapy is generally not advised for pregnant women.

- **Cardiovascular disease:** Due to the increased strain on the heart during suspension therapy, those with cardiovascular illness, such as heart failure or coronary artery disease, may be at risk of consequences.

Did You Know?

Benefits of suspension therapy
- It is simple to lift the limb, which lessens the load on the therapist.
- Active movement is simple and frictionless to accomplish.
- Slings and pulleys in the required position make it simpler to maintain the position of the limbs.

TYPES OF SUSPENSION

The three types of suspension used are:
1. Axial suspension
2. Vertical suspension
3. Pendular suspension

Axial Suspension

It can be defined as type of suspension in which the point of suspension is taken to be the joint axis. The slings are used to support the limb above the joint's axis and the limb is moved on both sides, if the movement is initiated (Fig. 19.7A). The limb moves parallel to the ground.

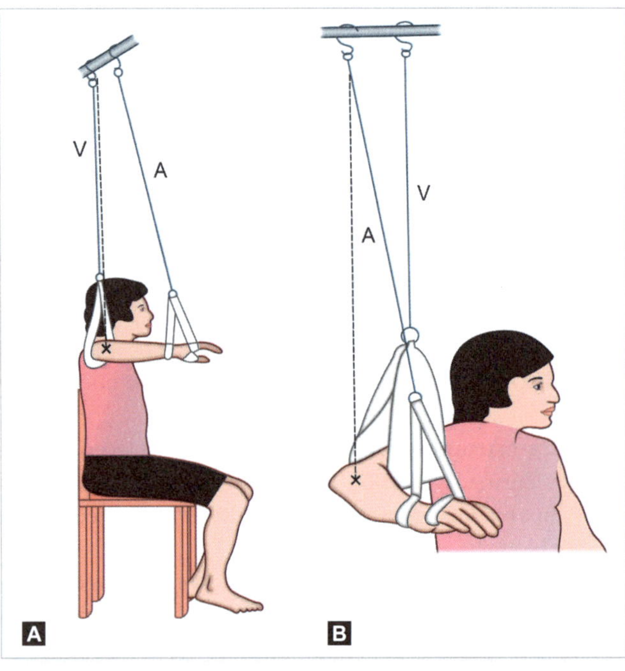

Figs 19.7A and B: A. Axial suspension; **B.** Vertical suspension
Abbreviations: A, axial suspension; V, vertical suspension

If lower limb is to be moved at the hip joint, two ropes—one to the foot and one at the area of the knee—will be used and fixed at a point immediately over the axis of the hip joint. In this method, all of the ropes supporting a part are attached to one "S" hook that is fixed to a point immediately above the center of the joint, that is to be moved. The "S" hook is moved away to a place from which it can provide resistance to the muscles that are contracting, if resistance is needed. Adduction must be resisted by moving the fixed point in the direction of the abductors, which causes the limb to adduct.

Uses

- Relaxation
- Increase muscle strength
- Maintain muscular property
- To increase the blood circulation
- Increase the venous drainage
- Increase the lymphatic drainage

Vertical Suspension

Rope is fixed in a manner such that it hangs vertically above the center of gravity (COG) of the body part which is suspended. COG of the body part or the body is taken as a point of suspension (Fig. 19.7B).

It is used to provide support to the body part of the patients. Vertical fixation is used primarily to support, e.g., the abducted upper limb when the elbow is to be moved is supported from above the center of gravity of the arm and axial fixation is used over the elbow for forearm movement.

Uses

- It limits the movement to small range pendular movement on each side of suspension.
- To support the body part
- To reduce the pressure sores.

Pendular Suspension

In this type of suspension, movement normally occurs against gravity, hence the point of suspension should be moved farther from the joint axis. If the axis is altered in the opposite direction of the movement, the muscle will experience resistance while moving.

Uses

- To strengthen the muscle.
- To maintain joint range of motion (ROM).
- To increase endurance.

ADVANTAGES

- Suspension therapy is one type of assisted exercise, and as patients must actively participate, they learn to employ the right movement for their intended movements.
- Following instructions, patients can work independently of the therapist and can easily perform active movements with minimal friction.
- Support with rhythmic motion of the body part promote relaxation.
- Less movement is required from the stabilized muscles because the body part is supported.
- Slings and pulleys in the required position make it simpler to maintain the position of the limbs.

Clinical Correlation

Suspension therapy can potentially be used as a complementary treatment for musculoskeletal conditions such as osteoarthritis, back pain, or sports injuries. By reducing the impact of gravity on the joints and providing gentle traction, suspension therapy may help alleviate pain, improve joint mobility, and promote tissue healing. Additionally, it can aid in restoring proper movement patterns and muscle balance, which are crucial for long-term management of musculoskeletal issues.

DISADVANTAGES

- Suspension therapy necessitates specialized equipment, such as a suspension harness or sling, which might not be available in all medical facilities. This may make this technique less available to some patients.
- **Injury risk:** If not done correctly or under the supervision of a qualified physiotherapist, suspension therapy involves the same risk of harm as any other type of exercise or therapy.
- Suspension therapy might not be suitable for all patients, including those with specific illnesses or injuries that make carrying out weight-bearing exercises risky or harmful. In addition, patients with severe muscle weakness or restricted range of motion might not benefit from suspension therapy.

TECHNIQUES FOR LOWER LIMB

Hip Abduction and Adduction

- **Patient setup:** The patient is positioned supine, in a suspension harness or sling that provides support for the trunk and pelvis. The harness should be securely attached to a suspension system.
- **Starting position:** The position to start the movement is supine lying.

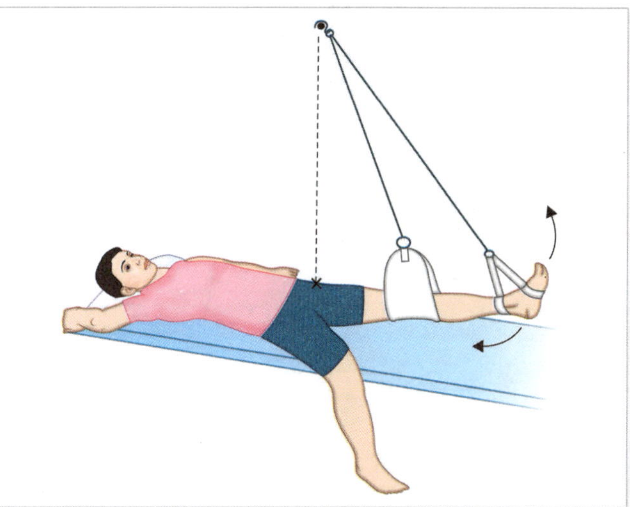

Fig. 19.8: Hip abduction-adduction suspension

- **Fixation point:** Axial fixation immediately above the joint axis.
- **Slings:** Two slings are used. One under the thigh and another three-ring sling under the foot and ankle.
- The limb is lifted just clear of the plinth.
- **Exercises:**
  - The patient moves the affected leg out to the side, away from the midline of the body, to do hip abduction exercise. During the exercise, the patient can use their hands to help stabilize the pelvis and maintain appropriate alignment.
  - The patient brings the injured leg back to the midline of the body and crosses it over the other leg to do hip adduction exercise.

Figure 19.8 shows the hip abduction-adduction suspension.

Hip Flexion and Extension

- **Patient setup:** The patient is positioned in side lying in a suspension harness or sling that provides support for the hip joint. The harness should be securely attached to a suspension system.
- **Starting position:** The position to start the movement is side lying with the underneath hip and knee flexed as much as possible.
- **Fixation point:** Axial fixation immediately above the hip joint axis.
- **Slings:** Two slings are used, One under the thigh and a three-ring sling under the foot and ankle.
- The limb is lifted just clear of the plinth.
- **Exercises:**
  - The patient brings the injured leg up toward the chest and bends at the hip joint to do hip flexion exercise. During the exercise, the patient can use their hands

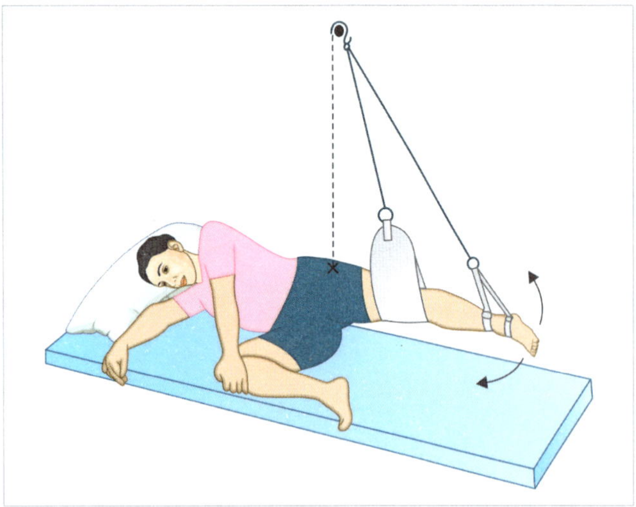

Fig. 19.9: Hip flexion-extension suspension

Fig. 19.10: Suspension therapy for hip medial and lateral rotation

to help stabilize the pelvis and maintain appropriate alignment.

- The patient extends the hip joint by moving the injured leg behind. To stabilize the pelvis and maintain appropriate alignment throughout the activity, the patient can use their hands.
- In order to overcome the passive insufficiency of the hamstrings during flexion, the knee and hip must be flexed slightly. The knee should be extended when performing extension to overcome active hamstring insufficiency.

Figure 19.9 shows hip flexion-extension suspension.

Hip Medial and Lateral Rotation

- **Patient setup:** The patient is positioned in a suspension harness or sling that provides support to the leg. The harness should be securely attached to a suspension system, ensuring proper alignment and stability.
- **Starting position:** The patient lies in a supine position with the hip and knee flexed to about 90°. Ensure the leg is suspended comfortably and the foot is supported.
- **Fixation point:** The suspension system is adjusted to align the axis of rotation with the hip joint.
- **Slings:** One or two slings are used to support the foot and ankle, preventing compensatory movements and maintaining alignment of the thigh.
- The limb is in 90°–90° position and is kept clear of the surface of the plinth.
- **Exercises**
  - The therapist or the patient moves the foot inward towards the midline, causing the hip to rotate laterally. This movement is performed smoothly within the pain-free range of motion.
  - The foot is moved outward, away from resulting in medial rotation of the hip. Ensure controlled and smooth execution during the movement.
  - During both movements, the patient can actively or passively engage in the exercises depending on their condition. The suspension system minimizes gravitational forces, allowing the patient to observe the movement arc and enhance coordination. This position is effective for mobilizing the hip joint or strengthening the internal and external rotators. It also prepares the patient for functional movements such as walking and turning.

Figure 19.10 shows suspension therapy for hip medial and lateral rotation.

Knee Flexion and Extension

- **Patient setup:** The patient is positioned in side lying in a suspension harness or sling that provides support to the knee joint. The harness should be securely attached to a suspension system.
- **Starting position:** The position to start the movement is side lying. One or two pillows can be kept between the distal parts of thigh.
- **Fixation point:** Immediately above the knee joint axis.
- **Slings:** One three-ring sling is placed to the foot and ankle.
- The limb is lifted just clear of the plinth.
- **Exercises:**
  - The patient bends the knee of the injured leg, bringing the heel toward the buttocks.
  - The patient straightens the injured leg at the knee joint and extends the leg in front.
  - During both movements, by keeping the hip slightly flexed on the trunk the foot can be seen each time the

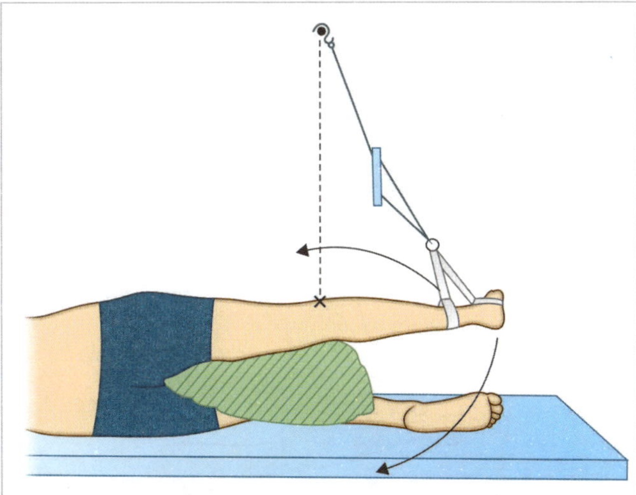

Fig. 19.11: Knee flexion-extension suspension

knee is extended and part of the arc of movement is thus observed by the patient (Fig. 19.11). This position may be used to mobilize the knee joint or to work the flexors or extensors of the knee.

Clinical Correlation

By simulating weight-bearing activities in a controlled environment, suspension therapy helps in restoring functional abilities essential for activities of daily living. This includes tasks such as standing, walking, ascending/descending stairs, and transitioning between different positions.

TECHNIQUES FOR UPPER LIMB

Shoulder Abduction and Adduction

- **Patient setup:** The patient is positioned in a suspension harness or sling that provides support for the shoulder joint. The harness should be securely attached to a suspension system.
- **Starting position:** The position of the patient is generally supine lying with, quarter turned toward the side which is to be moved. Alternatively the patient can be placed in prone lying position, quarter turned toward side lying with pillow under the trunk on the side of arm to be moved.
- **Fixation point:** Immediately above the shoulder joint axis.
- **Slings:** It takes two single ropes, one tied to a single sling worn under the elbow and the other to a three-ring sling worn around the wrist and hand.
- The limb is lifted just clear of the plinth.
- **Exercises:**
  - The patient elevates the arm away from the body and out to the side and back to do exercises for the shoulder abduction and adduction.
  - The muscles that move the shoulder girdle can all be exercised to retrain the glenohumeral rhythm.
  - This position may be used to mobilize the shoulder joint or to work the abductors of the shoulder joint (Fig. 19.12A).

Shoulder Flexion and Extension

- **Patient setup:** The patient is positioned in a suspension harness or sling that provides support for the shoulder joint. The harness should be securely attached to a suspension system.
- **Starting position:** Side-lying on pillows with one fourth of patient's back turned is the starting posture.
- **Fixation point:** Immediately above the shoulder joint axis.
- **Slings:** It takes two single ropes, one tied to a single sling worn under the elbow and the other to a three-ring sling worn around the wrist and hand.
- The limb is lifted just clear of the plinth.
- **Exercises:**
  - The affected arm is raised above the head as the patient performs shoulder flexion exercise.
  - The patient lowers the injured arm below the head to execute extension exercise (Fig. 19.12B). This position may be used to mobilize the shoulder joint or to work the flexors and extensors of the shoulder joint.

Shoulder Medial and Lateral Rotation

- **Patient setup:** The patient is positioned in a suspension harness or sling that provides support for the shoulder joint. The harness should be securely attached to a suspension system.
- **Starting position:** It depends on the movement performed.
- **Fixation point:** It is a axial fixation above the shoulder joint axis.
- **Slings:** A single pulley rope should be tied to the fixed point above the shoulder, and only one sling should be utilized at elbow level. Sling ends are fastened to each end of the pulley circuit.
- The limb is lifted just clear of the plinth.
- **Exercises:**
  - With the patient facing 15° to the right, the movements of extension/adduction/medial rotation and flexion/ abduction/lateral rotation can be accomplished using

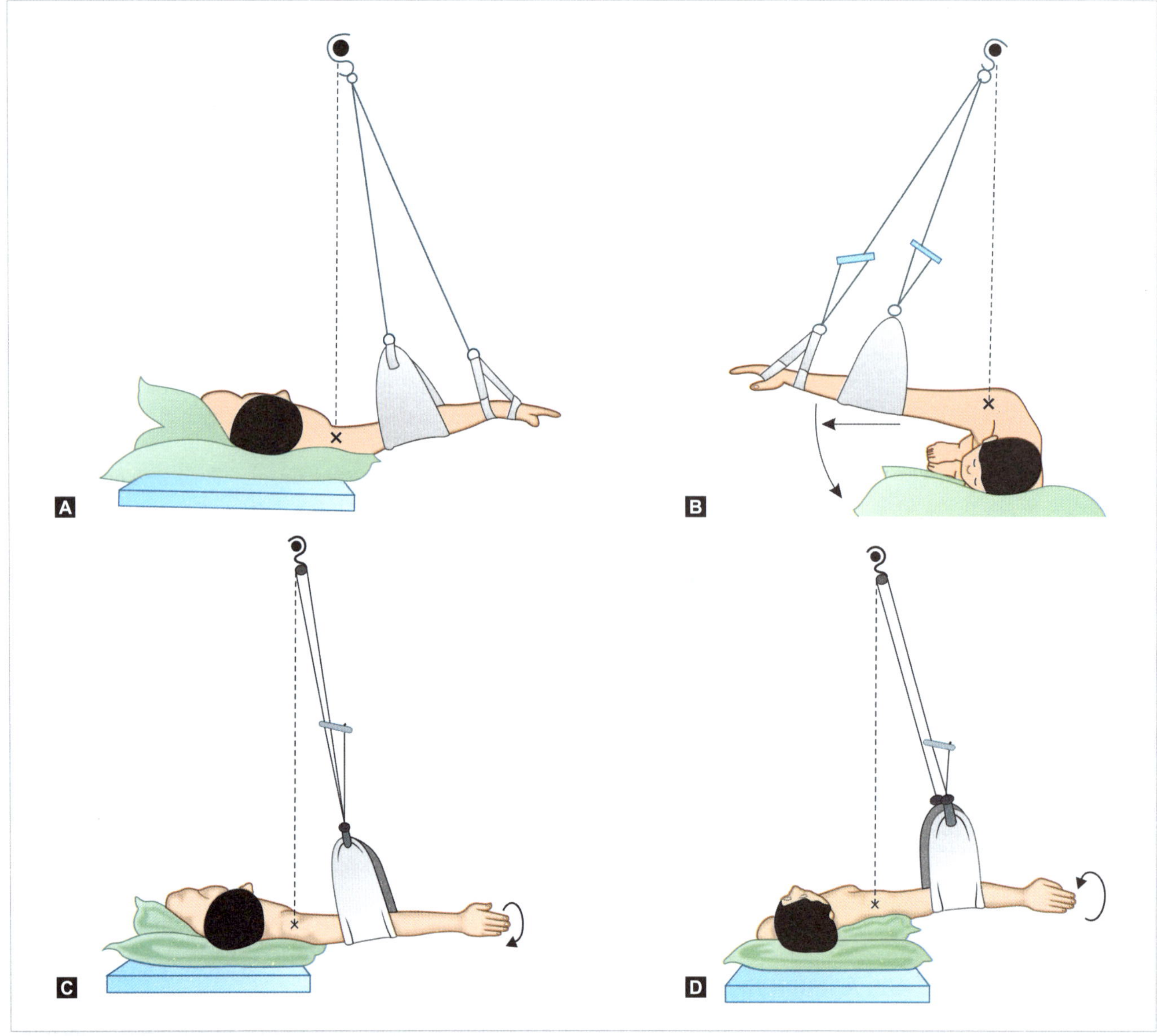

Figs 19.12A to D: Shoulder suspension: **A.** Abduction and adduction; **B.** Flexion and extension; **C.** Lateral rotation; **D.** Medial rotation

an axial fixation over the right glenohumeral joint and a single pulley rope (Fig. 19.12C).

- With the patient turned 15° to the left, the movements of extension/abduction/medial rotation and flexion/adduction/lateral rotation can be carried out using axial fixation over the right glenohumeral joint and a single pulley rope (Fig. 19.12D).
- This position may be used to mobilize the shoulder joint or to work the rotators of the shoulder joint.

Elbow Flexion and Extension

- **Patient setup:** The patient is positioned in a suspension harness or sling that provides support for the elbow joint. The harness should be securely attached to a suspension system.
- **Starting position:** The sitting position is opted on a low backed chair with the arm suspended on abduction.
- **Fixation point:** Vertical fixation by rope for the arm; axial fixation by rope for the forearm.
- **Slings:** A three-ring sling and a single rope are secured to a point above the elbow joint, and a single sling and rope support the arm in a vertical fixation.
- **Exercises:**
  - The patient bends his elbows while performing exercise for elbow flexion, bringing his forearm toward his upper arm.

- The patient straightens their elbow during this exercise, separating his forearm from the upper arm.
- This position may be used to mobilize the elbow joint or to work the flexors and extensors of the elbow joint.

The elbow just flexion and extension suspension is shown in Figures 19.13A and B.

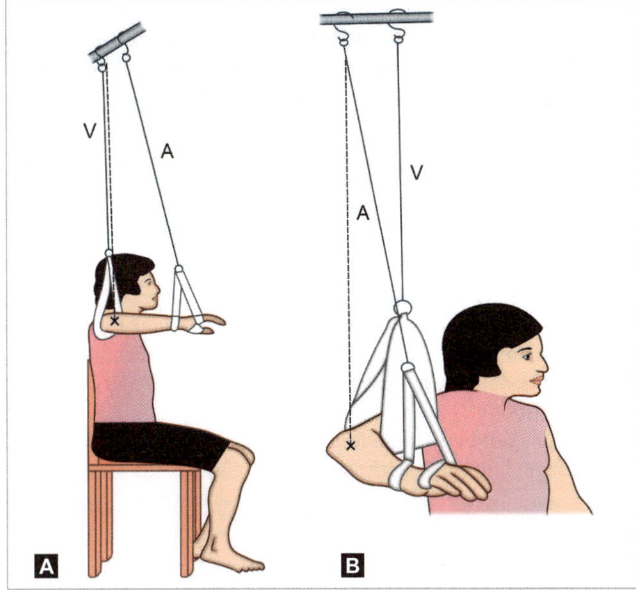

Figs 19.13A and B: Elbow flexion-extension suspension

Clinical Correlation

Suspension therapy allows individuals to practice functional tasks involving the upper limbs, such as reaching, lifting, carrying, and manipulating objects. By simulating real-life activities in a controlled environment, patients can work on improving coordination, motor skills, and task-specific movements.

SUMMARY

- Suspension therapy is a therapeutic exercise that involves suspending a person's body in the air while performing various movements and exercises.
  - This technique can be administered while sitting or lying down, and the suspension devices can be ropes or slings. It is often used by physical therapists, chiropractors, and other medical practitioners to increase joint mobility, flexibility, and strength.
  - The theory behind suspension therapy is that by suspending the person in midair, the force of gravity is diminished, which can help decompress joints and relieve strain on painful joints.
- The principles of suspension therapy include friction, pendulum, movement in the gravity eliminated plane, tension, resistance, and control.
- Friction reduces the stress on the joint, allowing limb movement in various directions.
- Pendular motion, which occurs in the shoulder and hip joints, encourages increased range of motion and lessens stiffness and pain by stimulating affected muscles.
  - The pendulum concept can also be used in conjunction with other methods to increase flexibility and strength of the shoulder joint.
- Movement in the gravity eliminated plane is another principle guiding suspension therapy.
  - By suspending the patient's body, the force of gravity can be reduced or completely eliminated, reducing the strain on the painful or weak joint and facilitating movement in various directions.
  - This is particularly beneficial for patients with conditions like frozen shoulder or rotator cuff injuries where joint movement may be restricted due to pain or inflammation.
  - The elimination of gravitational forces also allows the therapist to regulate resistance levels during exercises, contributing to increased muscle endurance and strength.
- Suspension therapy is a method of reducing the force of gravity on the body, which can help to decompress joints and relieve pressure on the spinal discs.
  - It involves suspending the patient's body in midair using ropes, slings, or elastic bands to reduce discomfort and increase mobility.
  - Suspension therapy can also add resistance to workouts, aiding in strength development.
- Equipment and accessories used in suspension therapy include slings and harnesses, which can be adjusted to accommodate the person's size and form.
  - Slings can be single, double, three-ring, head, wooden, dog clip, wooden cleat, S hook, and storage trolley.
- Suspension therapy can treat back discomfort, joint discomfort, muscle imbalances, sports injuries, rehabilitation, and postoperative muscle weakness.
  - It can be beneficial for patients with spinal cord injuries (SCI) as it allows for safe and controlled movement while reducing the load on the affected spinal column.
  - It can also help address postoperative muscle weakness by providing a safe setting for activity and recovery.
- However, there are some limitations to suspension therapy, such as spinal instability, severe osteoporosis, high blood pressure, pregnancy, and cardiovascular disease.

- Suspension therapy is a type of assisted exercise that involves lifting a limb and performing active movements with minimal friction. It has several types, including axial, vertical, and pendular suspensions.
  - Axial suspension involves the joint axis, with slings supporting the limb above it.
  - It is used for relaxation, muscle strength, muscle maintenance, blood circulation, venous and lymphatic drainage, and muscle strength.
  - Vertical suspension uses a rope fixed above the body part's center of gravity, limiting movement to small range pendular movement and reducing pressure sores.
  - Pendular suspension, on the other hand, occurs against gravity and requires the point of suspension to be moved farther from the joint axis.
- Advantages of suspension therapy include the ability to work independently, promote relaxation, and require less movement from stabilized muscles.
  - Slings and pulleys in the required position make it easier to maintain limb position. However, it requires specialized equipment, which may not be available in all medical facilities.
  - Additionally, suspension therapy involves an injury risk, especially if not done correctly or under the supervision of a qualified practitioner.
  - It may not be suitable for all patients, including those with specific illnesses or injuries that make weight-bearing exercises risky or harmful, or those with severe muscle weakness or restricted range of motion.

FURTHER READINGS

- El-Meniawy GH, Kamal HM, Elshemy SA. Role of treadmill training versus suspension therapy on balance in children with Down syndrome. Egyptian Journal of Medical Human Genetics. 2012;13(1):37-43.
- Gao, B., Rong, X., Liang, D. and Li, L. (2008) The Effect of Sling Exercise Therapy on Low Back Pain Caused by Exercises Training. Chinese Journal of Rehabilitation Medicine, 23, 1095-1097.
- Gardiner MD. The Principles of Exercise Therapy. 4th edition, 2023.
- Hollis M, Fletcher-Cook P. Practical Exercise Therapy. Blackwell Science; 2018 Jul 17.
- Nasb M, Li Z. Sling suspension therapy utilization in musculoskeletal rehabilitation. Open Journal of Therapy and Rehabilitation. 2016;4(03):99.
- Olama KA, Thabit NS. Effect of vibration versus suspension therapy on balance in children with hemiparetic cerebral palsy. Egyptian Journal of Medical Human Genetics. 2012;13(2):219-26.

STUDENT ASSIGNMENT

LONG ANSWER QUESTIONS

1. Explain the techniques of suspension therapy for the upper limb and lower limb.
2. Explain indications, contraindications and benefits of suspension therapy.
3. Discuss in detail the principles of suspension therapy.

SHORT ANSWER QUESTIONS

1. Define suspension therapy.
2. Enlist the equipment and accessories used in suspension therapy.
3. Define the types of suspension.
4. What are the benefits of suspension therapy?

MULTIPLE CHOICE QUESTIONS

1. Which is an indication of suspension therapy?
   a. Relieve pain
   b. Improve mobility
   c. Enhance muscle strength
   d. All of the above

2. Purpose of suspension therapy is:
   a. To decrease ROM
   b. To decrease muscle power
   c. Remove body support
   d. All of the above

3. Which type of suspension therapy is incorrect:
   a. Vertical
   b. Horizontal
   c. Axial
   d. Pendular

4. Regarding axial suspension which statement is incorrect?
   a. It is the most common type.
   b. Joint is taken as a point of suspension.
   c. Gravity is not eliminated.
   d. Limb is supported by the slings above the joint.

5. Which statement is incorrect regarding pendular suspension therapy?
   a. Point of suspension should be shifted away from the joint axis.
   b. Movement usually takes place against gravity.
   c. Muscles will be getting resistance while moving if the axis is shifted opposite to that movement.
   d. Point of suspension should be fixed with the joint.

6. Following are the supporting ropes except:
   a. Single rope
   b. Double rope
   c. Pulley rope
   d. Triple rope

ANSWER KEY

1. d **2.** d **3.** b **4.** c **5.** d **6.** d

Note

20 Hydrotherapy

Sheetal Kalra

LEARNING OBJECTIVES

After the completion of the chapter, the readers will be able to:

- Understand history of hydrotherapy.
- Define hydrotherapy and describe its techniques.
- Explain mechanism of action of hydrotherapy.
- Explain physical properties of water and its physiological effects.
- Mention indications, effects and uses of hydrotherapy.
- Mention disadvantages, contraindications and precautions of hydrotherapy.
- Explain the application of hydrotherapy in upper limb, lower limb and spinal injuries.
- Explain types of hydrotherapy.
- Understand the designing and safety features of hydrotherapy pool.

CHAPTER OUTLINE

- History
- Physical Properties of Water
- Techniques
- Mechanisms of Action
- Physiological Effects
- Disadvantages
- Contraindications and Precautions
- Adverse Effects
- Clinical Uses
- Procedure
- Types of Baths/Tanks
- Pool Design Considerations

KEY TERMS

Ai Chi: A specialized approach to water therapy. Essentially, Ai Chi strengthens and relaxes the body with progressive resistance training in water and breathing exercises.

Aqua jogging: Running in the deep end of the pool is known as "aqua jogging", and it has several advantages because it closely resembles the running motion.

Bad Ragaz Ring Method (BRRM): A concept for active one-on-one water physical therapy is the Bad Ragaz Ring Method. The patient receives the resistive fix points from the therapist. Therapists using this technique must possess a high level of competence and accuracy.

Balneotherapy: The use of baths and bathing, particularly in naturally occurring mineral waters, to treat illnesses, wounds, and other bodily conditions.

Buoyancy: Upward force that a fluid exerts against an object submerged wholly or partially, counteracting the object's weight.

Contrast bath: Targeted body portions are immersed in cold water for a short while, followed by a simultaneous immersion in warm water for a few minutes, as part of contrast bath therapy.

Hydrotherapy: Any activity done in water to aid in rehabilitation and recuperation from, rigorous training or a serious injury, is called hydrotherapy, often known as aqua therapy.

Naturopathic treatment: A comprehensive approach to wellbeing is naturopathy. The fundamental tenets of naturopathy include the significance of sunlight, exercise, stress reduction, a clean, fresh diet, and stress management.

Sitz bath: A sitz bath is a shallow, heated bath that one can sit in to ease discomfort in the perineal area.

Watsu: Watsu is a type of aquatic massage that is mostly applied in therapeutic and relaxing contexts. This method entails one-on-one sessions in which a trained professional or therapist gently supports, moves, lengthens, and massages a patient in warm, chest-deep water.

HISTORY

Private hot baths were a thriving element of daily life in ancient Greece. A warm or hot spring bath was thought to improve one's physical and mental health, purify the soul, and reenergize the spirit. The ancient Greeks placed a high value on water and built public baths. These public bathing areas were revered as hallowed ground and dedicated to specific deities.

Alcmaeon of Croton, a renowned Greek medical theorist and philosopher, was the first to assert in the fifth century BC that a person's health may be influenced by the quality of his water, while Hippocrates (460–377 BC) was the first to address the issue of water, its use, its effects on the human body, and its relationship to disease.

The majority of the time, a whole therapeutic cycle required more than a purifying bath, and salt water rich in minerals had also been employed. Hippocrates understood the significance of the climate and other natural components for the sustenance of life. Hippocrates advised using hot or cold water to cure skin infections including herpes and empyema as well as pneumonia, pleurisy, and hepatitis.

Rome was the "Mother of Spa Therapies" in antiquity. Asclepiades of Bithynia (124–40 BC) brought water as a medical treatment in Rome during the first century AD. The Etruscans had been utilizing it to maintain health and treat illnesses. Celsus, a Roman encyclopaedist who lived from 25 BC to around 50 AD, emphasised the value of bathing for personal hygiene and suggested drinking water to go along with a therapeutic bath to promote diuresis and get rid of "bad humours".

Years later, in the field of medicine, Paracelsus (1493–1541), a prominent renaissance physician, recommended hot mineral baths rather than herbal remedies.

A new era in hydrotherapy began in the 18th century thanks to the work of English doctors John Floyer (1649–1734) and James Currie (1756–1805). In his essay on cold bathing, Floyer highlighted the advantages of hydrotherapy. Floyer also constructed a water treatment facility in Litchfield, UK. Two rooms were used for the treatment; one was used for dry packs and hot baths, and the other was used for cold baths.

Athens Medical School's first female graduate, Angeliki Panagiotatou (1878–1954), wrote a historical analysis on hydrotherapy in ancient Greece at the turn of the 20th century in Greece. The French Academy of Sciences honoured her for her masterpiece, "Hygiene in Ancient Greece", which was published in 1924. Panagiotatou popularized the water therapy in Greece by highlighting its healing properties. Another physician, Emmanuel Mandalakis, made a substantial contribution to the growth of thermal medicine tourism and hydrotherapy at the same time period. Mandalakis supplied crucial epidemiological data on thermal springs and disorders treated, disclosed the curative virtues and physicochemical and biological features of Greek thermal springs, and offered many treatment strategies.

Currently, hydrotherapy is used to treat a variety of ailments, including poor circulation, skin disorders, chronic fatigue, moderate depression, and chronic fatigue syndrome. It is widely used in orthopedic and sports medicine as a kind of rehabilitation.

PHYSICAL PROPERTIES OF WATER

Buoyancy

- Buoyancy is the upward force that water exerts on the body, which makes one feel lighter. According to Archimedes' Principle, when one is in water, some water is pushed out of the way, and the upward force felt is equal to the weight of the water displaced (Fig. 20.1). This reduces the pressure on the joints:
  - In water up to the neck, one will experience about a **90% reduction** in joint compression.
  - In water up to the waist, it's about a **50% reduction**.
  - Being submerged opens up joint capsules, allowing for greater flexibility and movement.

Clinical Correlation

Buoyancy helps to reduce pressure on joints and muscles, making it easier for people with weak muscles or injuries to exercise without strain.

Archimedes' Principle

- This principle explains that whether an object floats or sinks depends on its density compared to the water's density.
- If density of body is less than density of fluid, it will displace a smaller volume of fluid *(Floats)*.
- Body will float higher if density is further reduced (by adding salt to water or air-filled items at rest).

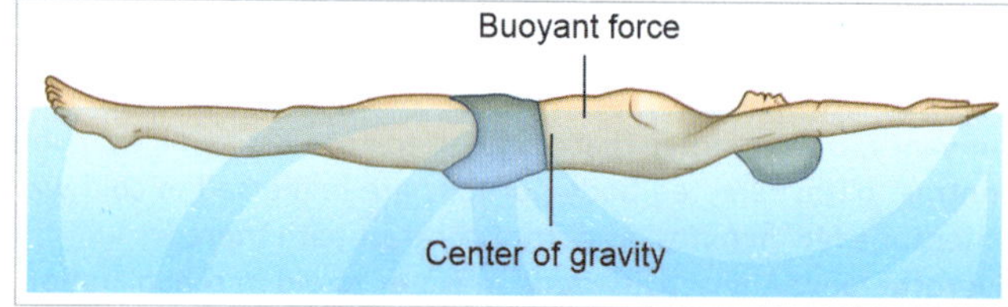

Fig. 20.1: Buoyancy

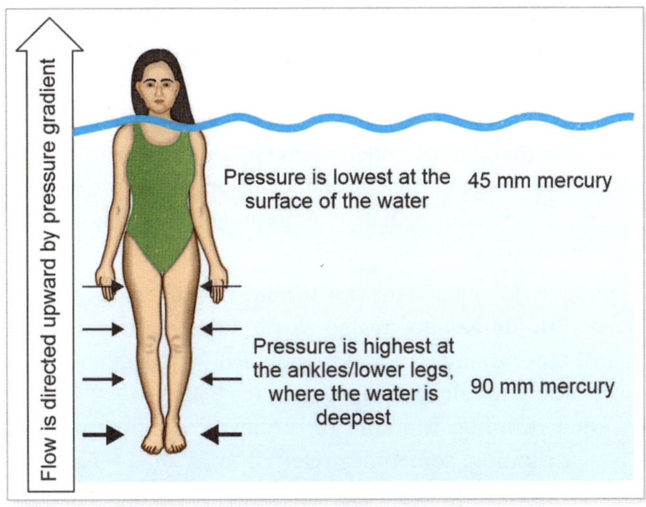

Fig. 20.2: Hydrostatic pressure—under water

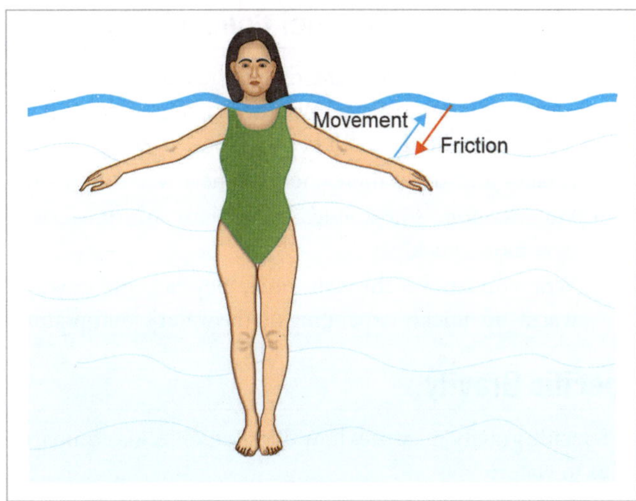

Fig. 20.3: Resistance felt in water as per law of inertia

Hydrostatic Force

- When one is submerged, water exerts pressure on their body. According to Pascal's Law, this pressure increases with depth (Fig. 20.2):
    - The deeper they go, the more pressure they feel.
    - This pressure can improve blood circulation and help reduce swelling. For example:
        - Blood flow can increase from **1.8 mL/min to 4.1 mL/min/100 g of tissue**.
        - There is about **10% increase** in blood flow to the kidneys, helping remove waste from the body.
        - Hydrostatic pressure can be similar to using bandages or compression wraps to reduce swelling.
    - The effects are strongest when they are in an upright position. Laying down or having limbs hanging down may reduce these benefits.

Hydrodynamics

Law of Inertia

- **What it means:** An object will stay still or keep moving in a straight line unless something pushes or pulls it.
- **In water:** When one moves through water, it's harder to get started, stop, or change direction. This is because water pushes back (Fig. 20.3).
- **Energy use:** Since the water resists the movement, one has to use more energy to swim or move through it.

Law of Acceleration

- **What it means:** How fast something speeds up depends on two things: The amount of force one applies and the weight of the object.
    1. If one pushes harder (apply more force), they can go faster.
    2. If the object is heavier (more mass), it's harder to make it move quickly.
- **In water:** When swimming, instead of trying to go faster, it's often better to focus on using more strength (force) to move.
- **Therapeutic setting:** If in a therapeutic situation (like physical therapy), increasing the force is usually better than trying to swim faster, as it can help avoid injury.

Viscosity

- Viscosity refers to how thick a liquid is and how it resists movement. When one moves through water, it creates resistance, which can slow them down:
    - The more surface area they have in contact with the water, the greater the resistance.
    - Resistance increases based on how fast they move and how much of their body is touching the water.
    - One of the important factors of viscosity is drag.

Drag

The force that resists an object's motion as it moves through a fluid.

Types of Drag

- **Cohesive force:** This is the force created by water molecules sticking together.
- **Form drag:** Resistance based on the shape and size of the body.
- **Wave drag:** Resistance caused by turbulence in the water.
- **Frictional drag:** Resistance due to surface tension in the water.

Understanding these forces is important for rehabilitation, as they can affect joint stress and movement.

Specific Heat and Thermal Conductivity

- Specific heat is the amount of heat needed to raise the temperature of a substance. Water has a high specific heat, meaning it can hold a lot of heat:
    - It takes four times more energy to heat water than air.
    - Water conducts heat much better than air, transferring heat more quickly.
    - When one moves through water, they can cool down or warm-up quickly depending on the water's temperature.

Specific Gravity

- Specific gravity measures how dense an object is compared as to water:
    - Water has a specific gravity of 1. Objects will float if their specific gravity is <1 and sink if it is >1.
    - The human body has a specific gravity of about 0.95–0.97, which is why most people can float.
- **Factors affecting specific gravity include:**
    - The ratio of bone to muscle weight.
    - The amount and distribution of fat in the body.
    - The air in the lungs can lower the specific gravity, helping the head and chest float higher in water.

Clinical Correlation

For patients dealing with mental health disorders such as anxiety, depression or PTSD, hydrotherapy can serve as a complementary therapy to traditional mental health interventions. The calming effect of water combined with therapeutic exercises can help alleviate symptoms, improve mood, and promote a sense of relaxation and well-being.

TECHNIQUES

The application of water in any form—water, ice, or steam—either internally or externally with the purpose of promoting health or treating a range of conditions is known as hydrotherapy. This popular naturopathic treatment technique was used in ancient China, Egypt, India, and other countries.

The purpose of hydrotherapy is to improve blood circulation. This is significant because blood nourishes tissues and organs and removes waste. Degeneration of the tissues and organs results from inadequate or sluggish circulation, which prevents the delivery of therapeutic nutrients and the removal of toxins. More nutrients are made available to cells for usage and toxins are handled more effectively by improving blood quality. Swimming, running, walking, cycling, stretching and underwater treadmill or parallel bars are some of the aquatic exercises used in rehabilitation of patients and athletic injuries.

Some aquatic therapies include Halliwick, the Bad Ragaz Ring Method, Watsu, Aqua Running, and Ai Chi. Aquatic therapies are treatments and exercises that are done in water for the purposes of relaxation, fitness, physical rehabilitation, and other therapeutic benefits. Pressure ulcers, neurological conditions, musculoskeletal pain, postoperative orthopedic rehabilitation, and pediatric impairments are among the therapeutic uses.

Techniques for aquatic therapy include the following:

- **Ai Chi:** Jun Konno created Ai Chi in 1993. It uses active resistance training in water combined with diaphragmatic breathing to calm and strengthen the body.
- **Aqua running:** It a kind of cardiovascular fitness called aqua running, sometimes referred to as aqua jogging or deep water running. Both injured sportsmen and people seeking a low-impact cardio exercise can benefit from it. Aqua running involves swimming in deep water while donning a floating apparatus (a belt or vest) to keep the head above the surface (Fig. 20.4).
- **The Bad Ragaz Ring Method (BRRM):** The method focuses on the restoration of neuromuscular function through patterns of therapist-assisted exercise performed while the patient lies horizontally in water with support from rings or floats around the neck, arms, pelvis, and knees (Fig. 20.5).

Fig. 20.4: Aqua running

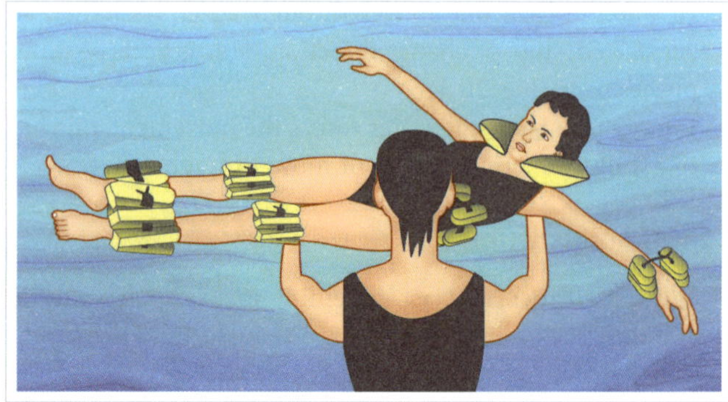

Fig. 20.5: Bad Ragaz Ring Method

- **Watsu:** Watsu is an aquatic bodywork technique that was developed in the early 1980s. It involves an aquatic therapist guiding and supporting the patient through a series of flowing movements and stretches that have therapeutic effect and it deepens relaxation.

MECHANISMS OF ACTION

- Temperature, resistance, buoyancy, and immersion all have significant effects. The pain gate theory states that the pressure and warmth of the water on the skin could be the cause of the pain reduction.
- Immersion in water raises plasma levels of methionine-encephalin while suppressing levels of prolactin, corticotropin, and β endorphin.
- Reduced joint edema and relaxed muscles.
- Significant improvements in mood and tension may contribute to the results.
- By reducing the impact of gravity, hydrotherapy increases joint flexibility and range of motion (ROM). A supported body component will float, simplifying and relieving discomfort during movement. Additionally, the warm water helps to relax stiff or sore muscles.
- Weaker muscles can be strengthened by the 600–700 times greater resistance of water compared to air. Since water resists rapid movement, equipment with a bigger surface area is used to build strength in the pool.
- Warm water helps relieve discomfort by promoting relaxation and increasing blood flow to sore muscles. Reducing weight on body areas that are weaker also improves comfort. These systems aid in the lessening of pain.
- Because of the constant buoyancy and water pressure supporting the body, one can have more time to react without fear of falling or getting injured.

Clinical Correlation

Hydrotherapy also helps in improving circulation and cardiovascular function. The pressure of water can assist in blood flow, which can aid in reducing swelling and promoting healing.

PHYSIOLOGICAL EFFECTS

Hydrotherapy produces a range of physiological effects through properties such as buoyancy, hydrostatic pressure, and temperature. These effects influence various body systems, making hydrotherapy a valuable modality in rehabilitation and wellness.

Effects on Musculoskeletal System

- Decreases weight-bearing (WB) on the painful joints. Reduction of stress and pain is helpful in patients with arthritis.
- Up to 75% immersion reduces WB by 75%.
- There is Increased blood flow to muscles that causes reduction in fatigue.
- The properties of buoyancy and viscosity helps in Strengthening of the muscles, decreased bone density loss.
- Compared to other forms of exercise, hydrotherapy is associated with a lower rate of fat reduction.
- Excellent for obese people due to nonweight-bearing activities.

Effects on Cardiovascular System

- **Cold exposure:** When one exposes a small area of the body to cold, deeper blood vessels dilate (widen). This helps increase blood flow to the tissues below the cold area, keeping them warm.
- **Warm water bathing and saunas:** Bathing in warm water or using a sauna can improve heart function, especially in people with chronic heart failure (CHF). Studies show that this can lead to:
  - Better heart performance (measured by the left ventricular ejection fraction).
  - Increased ability to walk for 6 minutes without getting tired.
  - Improved blood vessel function.
  - Lower levels of stress hormones like norepinephrine.
  - Healthier cholesterol levels, which can help prevent heart disease.
- **Sauna therapy:** Regular sauna use can increase the activity of a protein called endothelial nitric oxide synthase (eNOS), which helps improve blood flow. This is particularly beneficial for:
  - People with heart failure.
  - Patients recovering from heart attacks, as it helps repair the heart and improve blood flow in the heart muscle.
- **Risks of rapid cooling:** Alternating between hot and cold (like going from a sauna to a cold bath) can cause rapid changes in blood vessel behavior, which might be risky for people with heart issues.
- **Cold water immersion:** Jumping into cold water can speed up the heart rate and raise blood pressure temporarily. It also affects blood flow to the brain and can lead to changes in metabolism.
- **Warm water immersion:** Being in warm water can help improve heart function and has minimal effects on blood clotting. However, it is important to monitor for any potential complications.

- **Contrast baths:** Alternating between hot and cold baths can help increase blood flow, but the effects depend on how long one spends in each temperature.
- **Carbon dioxide (CO_2) enriched water:** Bathing in water enriched with CO_2 can reduce free radicals (harmful molecules) in the blood and improve blood flow in small blood vessels.

Effects on Respiratory System

- Increased work of breathing has been noticed due to hydrostatic pressure on lungs (up to 60%) (Fig. 20.6).
- Therapist may need to be very careful with respiratory and/or cardiac patients.
- Research reports that there is decreased exercise-induced asthma in children treated with hydrotherapy
- Martin-Valero et al. concluded that breathing exercises performed in water can improve both pulmonary functions and cardiac functions in patients with chronic obstructive pulmonary disease (COPD).
- The study by Kurabayashi et al., demonstrated significant increases in ejection fraction and improvements in lung capacity after a 2-month aquatic exercise program.

Effects on Nervous System

- Cold treatments, like ice packs and cold water immersion, can lower skin temperature and slow down nerve signals. This might help with pain relief, as cold can block pain signals effectively.
- Hydrotherapy can also stimulate certain areas in the brain that help activate muscles and improve movement.
- Specific aquatic exercises, like the Ai Chi program, can help people with conditions like multiple sclerosis by reducing pain and fatigue and improving their overall mood and independence.
- For people with Parkinson's disease, aquatic therapy has shown better results in improving balance and stability compared to land-based exercises.
- Cold exposure can boost certain brain chemicals that help with mood and pain relief without causing side effects or addiction.
- Kesiktas et al., in a study concluded that hydrotherapy significantly reduces spasticity and improves functional independence in spinal cord injury patients, while also decreasing the need for oral baclofen.

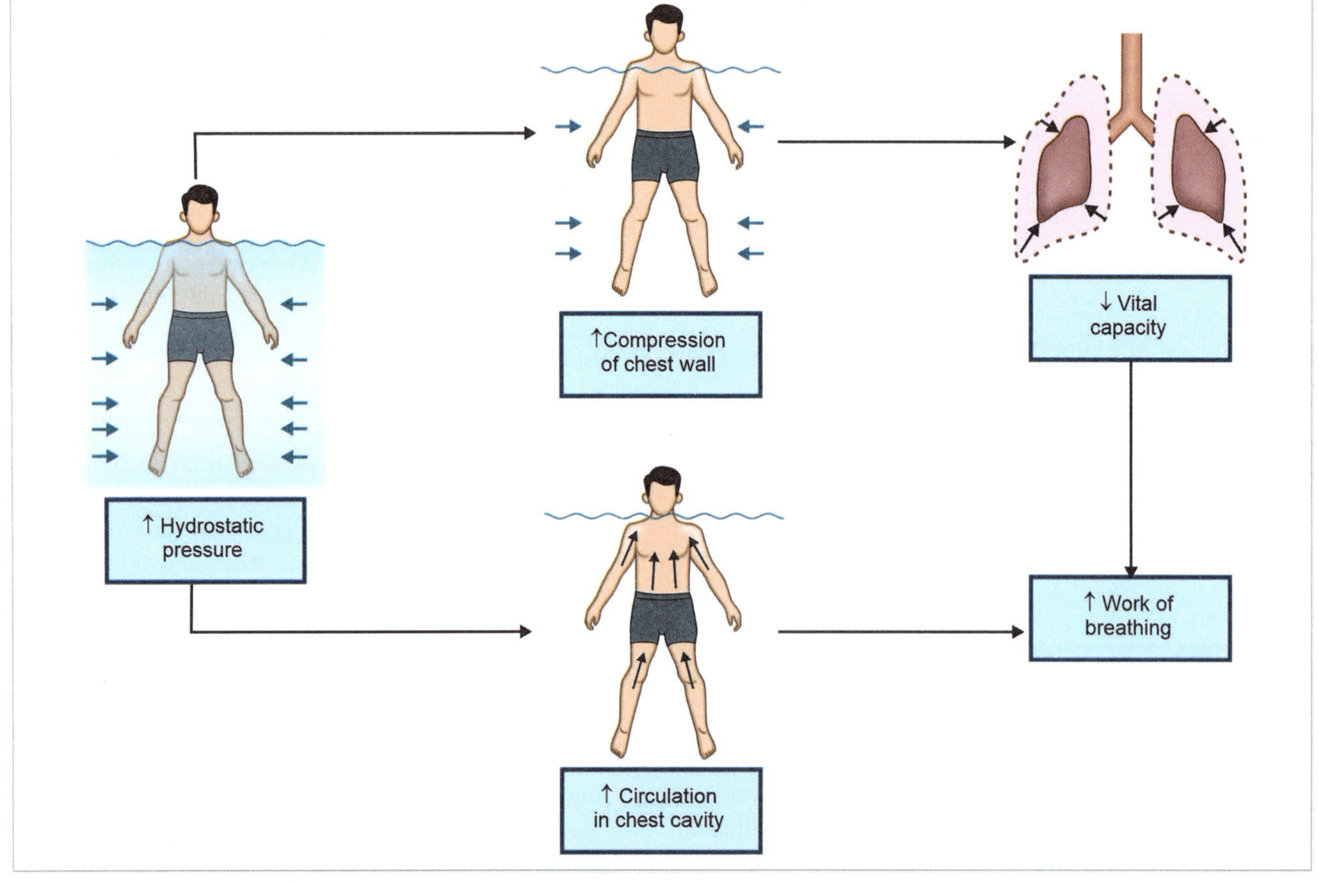

Fig. 20.6: Effects of hydrotherapy on respiratory system

- Another study by Marinho-Buzelli et al., mentioned that aquatic therapy shows "fair" evidence of improving dynamic balance and gait speed in adults with neurological conditions, making it a valuable rehabilitation intervention.

Effects on Reproductive System

- Immersion in water during childbirth can shorten labor and reduce pain, leading to less need for pain relief methods like epidurals.
- Women giving birth in water typically experience more effective contractions without harming the baby or mother.
- Ergin et al., concluded that hydrotherapy during the first stage of labor is safe and effective, reducing pain perception and increasing comfort while promoting vaginal births without serious complications.
- Gautham et al. found that hydrotherapy enhances maternal comfort, lowers pain perception, and increases the rate of vaginal births, without negatively affecting maternal or neonatal health outcomes.

Effects on Hematology/Immunology

- There have been reports of subsequent cold exposure-induced increases in leukocytes, granulocytes, interleukin (IL)-6 levels in the blood, natural killer (NK) cells, and their activity.
- Cold exposure can increase the number of certain white blood cells that help fight infections. It can also enhance the immune response when combined with exercise.
- However, very cold water can cause temporary issues like swelling in the lungs, which might worsen certain infections.

Effects on Endocrine/Hormonal System

- Cold exposure can raise levels of norepinephrine (a stress hormone) and may help improve the function of the body's stress response system over time.
- This cold stress can lead to temporary increases in cortisol and other hormones that help manage stress and pain.
- Interestingly, while cold stress may lower serotonin levels (a mood-regulating hormone), it can also boost natural pain-relieving chemicals in the body.

Effects on Urinary System

- Boosts the output of urine, increases excretion of sodium and potassium.
- Can be beneficial for patients with peripheral edema and hypertension (Fig. 20.7).

Effects on Psychology

- **Relaxing:** Warm water can help one feel calm and relaxed.
- **Energizing:** Cold water can make one feel more awake and energized.

Did You Know?

Hydrotherapy can be a valuable adjunctive therapy for individuals dealing with lymphatic disorders or those who have undergone lymph node removal surgery.

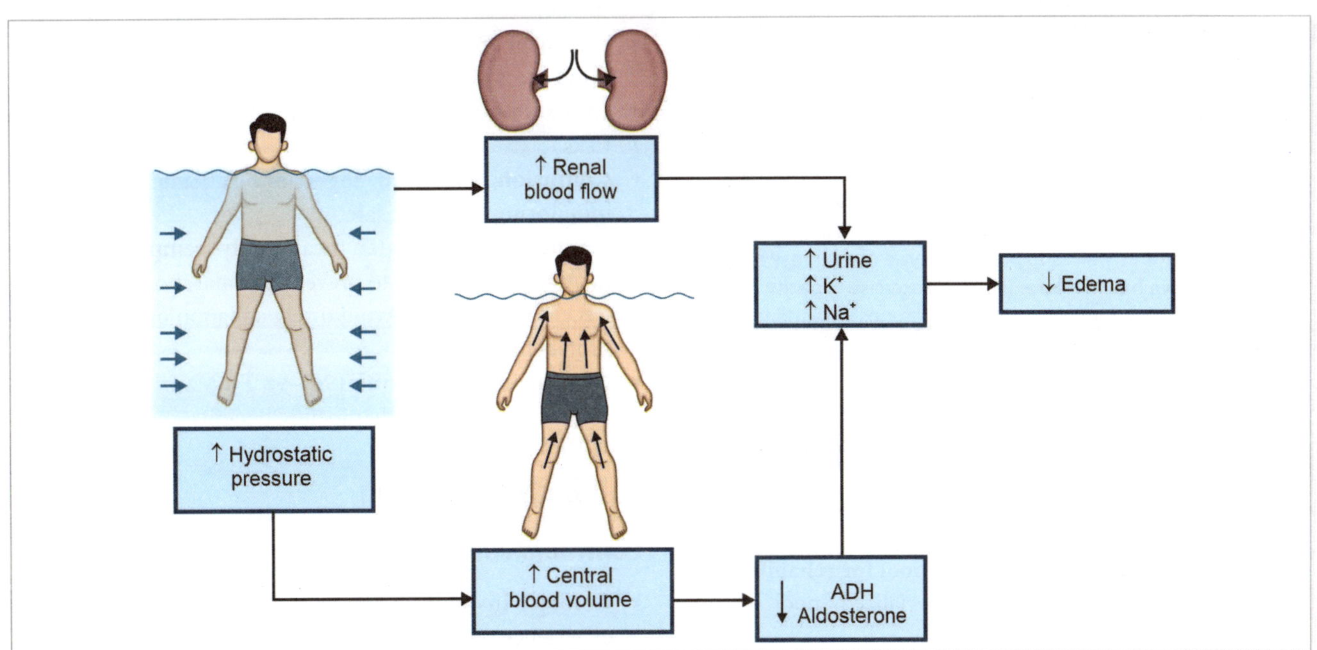

Fig. 20.7: Effects of hydrotherapy on urinary system

Recent Updates

Research Study Conclusions on Hydrotherapy

- **Arthritic pain:** Those who are suffering from painful, arthritic joints, access to hydrotherapy can provide large and sustained improvements in physical functions, for many older, sedentary individuals with chronic hip or knee OA (Fransen et al.).
- **Prevention of falls among elderly:** Hydrotherapy sessions led to improved balance that gave rise to an increase in balance and a possible reduction in the risk of falls among these aged men.
- Reduction in localized edema (Vaile et al.).
- **Multiple sclerosis:** According to these findings, an Ai-Chi aquatic exercise program improves pain, spasms, disability, fatigue, depression, and autonomy in MS patients (Castro-Sánchez et al.).
- **Chronic pain:** Access to either hydrotherapy or can provide large and sustained improvements in physical function for many elders, sedentary individuals with chronic hip or knee osteoarthritis (OA) (Fransen et al.).
- **Spinal cord injuries:** Hydrotherapy improves persons with spinal cord injuries underwater gait-kinematics, cardiorespiratory and thermoregulatory responses and reduces spasticity (Hammill et al.).
- **Back and leg pain:** The findings offer qualified support to anecdotal evidence that group hydrotherapy can benefit subjects with CLBP or back and leg pain (McIlveen et al.).
- **Total hip arthroplasty:** In an RCT pain, stiffness, function improved in patients with total hip arthroplasty compared to conventional treatment (Giaquinto et al.).
- **Cerebral palsy:** In patients with cerebral palsy-both as an individual treatment and in combination with a standard land-based exercise program, can significantly contribute to the improvement of gross mobility, the improvement of exercise endurance and the improvement of the quality of life of people with CP (Bairaktaridou et al.).
- **Pressure ulcer:** Healing of pressure ulcer was faster in group treated with conventional treatment and whirlpool bath. Also useful in wound bed preparation, burn wound bacterial contamination (Burke et al.).
- **Stress related disorders:** Hydrotherapy is used to treat stress-related disorders, which can lead to a variety of psychological and physical issues such stomach complaints, high blood pressure, panic attacks, melancholy, migraines, and sleeplessness as well as to break the cycle of everyday stress.
- **Incongestive heart failure:** Improves exercise capacity as well as muscle function in small muscle groups in patients with congestive heart failure (Cider et al.).

DISADVANTAGES

- **Cost:** Hydrotherapy can be expensive due to facility fees and equipment maintenance.
- **Constructing and maintaining a pool for rehabilitation:** Building and upkeep of a specialized pool require significant investment and resources.
- **Area and personnel:** Requires adequate space and trained staff for effective treatment.
- **Thermoregulation:** Maintaining optimal water temperature is crucial for patient safety and comfort.
- **Effect on the body's core temperature:** Water immersion can influence core temperature, necessitating careful monitoring.
- **May affect one's capacity to participate in heat:** Some patients may struggle with heat sensitivity during therapy.
- **Injury-related contraindications and possible open wounds:** Patients with injuries or open wounds may not be suitable candidates for hydrotherapy.

CONTRAINDICATIONS AND PRECAUTIONS

Local Immersion

Contraindications

- **Maceration:** Excessive moisture can worsen skin conditions and increase wound size.
- **Increased maceration, increased size of wound:** Moisture can lead to further skin breakdown.
- **Bleeding:** Immersion is not safe near areas with active bleeding.

Precautions

- **Impaired thermal sensation:** Patients may not feel water temperature, risking burns or discomfort.
- **Check temperature of water with thermometer first:** Always verify water temperature to prevent injury.
- **Infection:** Maintain strict hygiene to prevent infection spread.
- **Universal precautions:** Follow standard safety measures to avoid cross-contamination.
- **Clean water:** Ensure water is clean to minimize infection risks.
- **Confusion, impaired cognition:** Patients require close monitoring to ensure safety.
- **Temperature of water near body temperature:** Use neutral temperatures to prevent thermal shock.
- **Recent skin grafts:** Avoid direct agitation on graft sites to protect healing skin.
- **Neutral or mild warmth (96°–98°F):** Maintain a comfortable temperature for therapeutic benefits.

Full Body Immersion

Contraindications

- **Cardiac instability:** Patients with heart issues may experience dangerous complications.
- **Uncontrolled hypertension or heart failure:** Risk of exacerbating cardiovascular conditions.

- **Infectious conditions that spread with water:** Conditions like UTIs pose a risk of transmission in pools.
- **Bowel or bladder incontinence:** Immersion is not advisable for patients unable to control bowel or bladder.
- **Severe epilepsy:** High risk of drowning due to seizure potential.
- **Suicidal patients:** Risk of self-harm in a water setting.

Precautions

- **Confusion, disorientation:** Maintain head above water and ensure constant supervision.
- **Low temperature:** Keep water cool to avoid adverse effects on health.
- **After ingestion of alcohol:** Alcohol can impair judgment and increase drowning risk.
- **Hypotensive effects:** Monitor patients as immersion can lower blood pressure.
- **Patients with limited strength, endurance, balance, ROM (Hands-on approach required):** Requires close assistance to ensure safety.
- **Patients on medications (Physician clearance sometimes required):** Certain medications may affect immersion safety.
- **Patients with fear of water:** Anxiety may lead to panic, necessitating extra support.
- **Patients with respiratory problems:** Monitor for distress, particularly in warm water settings.
- **Patients that are pregnant:** Special care needed, especially in the first trimester, to prevent overheating.
- **Temperature >88°F may cause increased fatigue, weakness:** High temperatures can lead to overheating and decreased endurance.
- **Patients with poor thermal regulation:** Elderly or infants may struggle to maintain a safe body temperature.

ADVERSE EFFECTS

- **Fainting:** Patients may experience fainting due to sudden changes in body temperature or blood pressure while immersed in water.
- **Hypotension:**
  - **Immersion in warm/hot water:** Can cause blood vessels to dilate, leading to decreased blood pressure.
  - **Hypertension medications:** Patients on these medications may experience exacerbated hypotension in warm water.
- **Hyponatremia (burns):** Prolonged immersion can lead to dilution of sodium levels in the body, particularly in patients with burns.
- **Increased edema:**
  - **Use of hot water:** Heat can cause swelling in tissues due to increased blood flow and fluid accumulation.
  - **Dependent positioning:** Patients in dependent positions (e.g., sitting) may experience increased swelling in lower extremities.

CLINICAL USES

- Enhanced environment for exercise
- Health maintenance/disease prevention
- Movement skill and strength can be enhanced
- Lessening of joint compression
- Reduces muscle guarding
- By supplying nutrition and oxygen to all of the cells and tissues, a hydrobath has a unique ability to increase circulation.
- Strengthens immunity through traditional cleansing makes it easier to fight off viruses, germs, and illnesses. Additionally, it aids in the recovery from minor infections, including everything from the common cold to fatal diseases.
- A conducive setting that gives the user more time to manage his movements and improves proprioception.
- Provides a gradual transition from non-WB to full WB.
- Acts as a tactile sensory stimulant and a destabilizer.
- Psychologically, the improved function made possible by water may lead to a rise in confidence.
- Increase ROM
- Initiate resistance training.
- Make weight-bearing tasks easier
- Improve manual technique delivery
- Access patients in three dimensions and encourage cardiovascular fitness
- Initiate functional activity
- Reduce risk of injury and re-injury during rehabilitation
- Increase patient comfort
- Decrease weight-bearing on joints
- Relaxation/psychological well-being
- Improve or maintain function
- Useful for enhancing fitness and movement
  - Aerobic capacity/endurance conditioning
  - Balance, coordination and agility body mechanics and postural stabilization
  - Flexibility
  - Gait and locomotion
  - Muscle strength, power, and endurance

Wound Care

- **Rehydration is facilitated by cleaning properties:** Hydrotherapy helps maintain moisture in the wound, promoting healing.
- **Cleaning up wound debris:** The flow of water aids in removing dirt and dead tissue from the wound site.

- **Softening and debridement of necrotic tissue:** Warm water helps soften dead tissue, making it easier to remove (Fig. 20.8).
- **Heat and hydrostatic pressure promote circulation:** Increased blood flow enhances nutrient delivery and waste removal at the wound site (Fig. 20.9).
- **Ensures a moist environment for healing:** Keeping wounds moist helps accelerate the healing process and reduces scarring.

Immersion Technique (Whirlpool Therapy)

Whirlpool therapy is a traditional medical practice used to treat burn wounds and other wounds. The water in the whirlpool is agitated to help remove the debris and clean the wound. Antibacterial agents can also be added to the water to help with cleansing.

- **Extensive thick exudate:** Water pressure is effective for managing large amounts of drainage.
- **Slough or necrotic tissue:** Helps in softening and removing dead tissue.
- **Gross purulence:** Ideal for cleansing infected or highly contaminated wounds.
- **Dry eschar:** Assists in softening and debriding dry, hard tissue.

Discontinue all forms of hydrotherapy once the wound is clean, as many antimicrobial products can be cytotoxic to healthy tissue unless diluted. It's crucial to thoroughly clean and disinfect the tank and turbine to prevent infections.

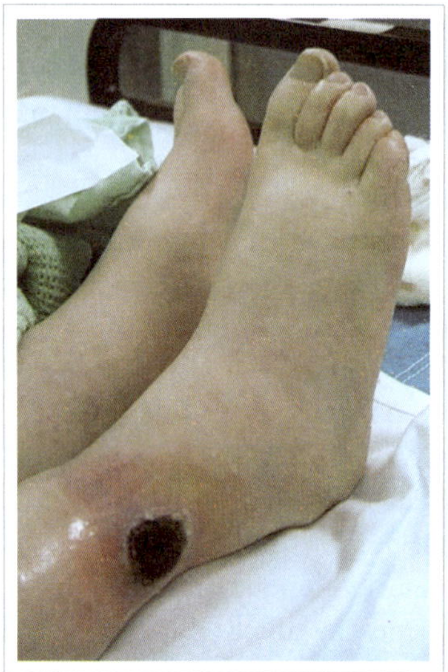

Fig. 20.8: Ischemic arterial ulcer

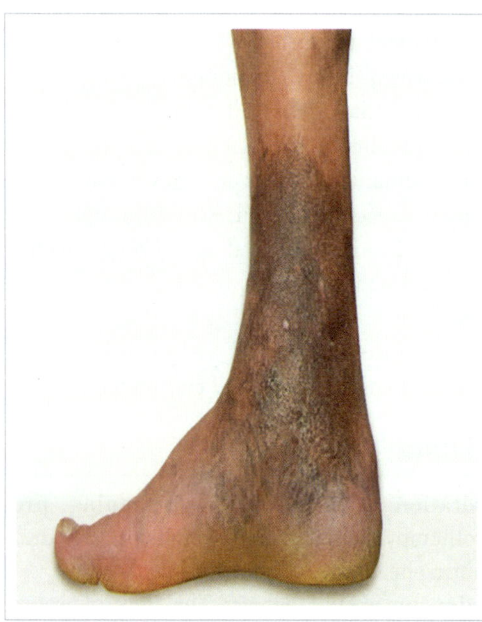

Fig. 20.9: Venous insufficiency ulcer

Nonimmersion Techniques

Nonimmersion hydrotherapy techniques include methods such as jet lavage, where a high-pressure water stream cleans wounds; hydrogel dressings that keep wounds moist to promote healing and reduce pain; hydrotherapy baths, where warm or cold water is applied to specific areas using a sponge or cloth for pain relief and inflammation reduction; and Negative Pressure Wound Therapy (NPWT), which uses a vacuum device to promote healing by applying negative pressure to the wound area. These techniques are ideal for patients who cannot tolerate full immersion due to pain, mobility issues, or other medical conditions.

- **Fluid delivered at pressure of 4–15 psi:** This pressure range effectively removes bacteria without causing wound trauma.
- **Below this, bacteria not removed:** Insufficient pressure may not adequately cleanse the wound.
- **Above this, wound trauma may occur, or bacteria driven into wound:** Excess pressure can cause further injury or introduce bacteria into the wound.
- **Saline squeeze bottle, Water Pik:** Recommended for removing necrotic or nonviable tissue and debris.

Continue using nonimmersion techniques until all debris is removed and a complete granulation bed is present. For both types of treatments, thoroughly drying the intact skin in the surrounding area is essential to prevent maceration.

Clinical Correlation

Factors favouring nonimmersion therapy for wound care:

- **Nonimmersion therapy becoming more popular:** This method is gaining attraction due to concerns about the risks associated with full immersion.
- **Concern for increased pressure on regenerating tissues by water and turbine:** Full immersion can place excessive pressure on healing tissues, potentially delaying recovery.
- **Potential for infection in contaminated tank:** Immersion in water can pose an infection risk if the tank is not properly maintained.

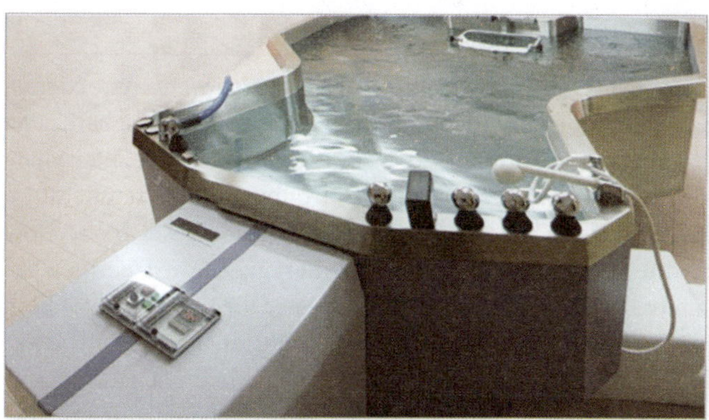

Fig. 20.10: Hubbard tank

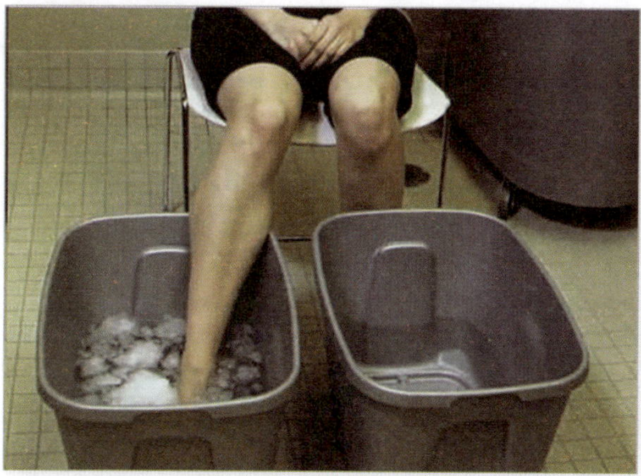

Fig. 20.11: Contrast bath

Burns

- **Pain management:** Burns can be particularly painful during debridement, requiring close monitoring of the patient, often necessitating high-dose analgesics.
- **Wound characteristics:** Burns may be less deep than other wounds, with intact sensory nerves that can increase sensitivity.
- **Hubbard tank:** This is useful for treating large body surface areas but carries risks of contamination and sodium loss (salt should be added to the water) (Fig. 20.10).
- **Shower (Nonimmersion):** Recommended in the early stages to clean and manage burns without immersion risks.
- **Post-re-epithelialization:** Once the skin has healed, water exercises can help with rehabilitation and mobility.

Pain Control

- **Increased sensory stimulation:** Hydrotherapy provides sensory input to peripheral mechanoreceptors, helping to mask pain.
- **Pain gate theory:** Hydrotherapy may activate the gate control mechanism, reducing pain perception by stimulating competing sensory pathways.
- **Cold water effects:** Cold water immersion helps decrease inflammation and provides a numbing effect that reduces pain.
- **Decreased weight-bearing:** Water supports body weight, increasing ease of movement and reducing stress on painful areas.

Edema Control

- **Hydrostatic pressure:** Water creates pressure that can help reduce swelling in affected areas.
- **Cold water effects:**
  - **Vasoconstriction:** Cold water causes blood vessels to constrict, reducing blood flow and decreasing edema.
  - **Decreased vascular permeability:** Cold water minimizes leakage of fluids into surrounding tissues.
- **Hot water effects:** Hot water can increase edema by promoting vasodilation and enhancing blood flow to the area.
- **Contrast baths:** Alternating between hot and cold water trains smooth muscle in blood vessels, promoting a cycle of vasoconstriction and vasodilation to help manage swelling (Fig. 20.11).

PROCEDURE

- Assess problem and set goals of treatment
- Determine the most appropriate treatment
- Make sure no contraindications
- Select appropriate form of hydrotherapy
  - Whirlpool
  - Hubbard Tank
  - Contrast bath
  - Nonimmersion device
  - Pool
- Explain the procedure, purpose, check sensations
- Apply appropriate form of hydrotherapy
- Assess outcome
- Document

Selection of Hydrotherapy

The decision of selecting the appropriate form of hydrotherapy is based on:

- Desired effects
- Size of area to be treated
- Allowance for safety, control of infection
- Cost-effectiveness

TYPES OF BATHS/TANKS

- Whirlpool bath
- Hubbard Tank
- Contrast bath
- Sitz bath
- Balneotherapy

Whirlpool Bath and Hubbard Tank

- **Design and functionality:** Both systems are equipped with a turbine, agitator, and adjustable height and direction for targeted treatment. They should also be properly grounded for safety.
- **Portability:** Smaller, portable whirlpool baths are used to treat single extremities, while the Hubbard tank accommodates a larger volume of water.
- **Temperature range:** Whirlpool baths can operate at temperatures between **43°C and 46°C (109.4°F to 114.8°F)** for specific conditions, while the Hubbard tank is limited to a maximum temperature of **39°C (102.2°F)** for full-body immersion.
- **Agitation effects:** The agitation of water in these baths produces a massaging effect that is beneficial for various therapeutic applications, including the treatment of arthritic conditions, regional pain syndrome, delayed onset muscle soreness, wound healing, postoperative pain, and myofascial pain syndrome.
- **Irrigation and debridement:** The Hubbard tank is equipped with multiple handheld shower heads and water jets, making it suitable for local treatments such as irrigation and deep wound debridement.

Table 20.1 shows the features of hubbard tank and whirlpool bath.

TABLE 20.1: Features of hubbard tank and whirlpool bath

Feature	Hubbard tank	Whirlpool bath
Design and structure	Large, rectangular tank for full-body immersion.	Smaller, portable unit for localized treatment
Treatment focus	Full-body treatments; ideal for extensive conditions.	Localized treatments for specific body parts.
Water temperature	Maximum of 39°C; gradual temperature changes allowed.	Can operate up to 46°C; higher temperatures for intense treatment.
Agitation	Gentle agitation; uniform therapeutic effect.	Stronger agitation from concentrated jets for massage effect.
Patient positioning	Patients can sit or lie down for comfort.	Patients typically sit in a tub-like structure or immerse limbs.
Applications	Comprehensive rehabilitation, wound care, burn treatment.	Pain relief, stiffness, muscle tension, and minor injuries.
Movement	More freedom of movement, beneficial for extensive rehabilitation.	Limited range of movement; focuses on specific areas.
Portability	Generally fixed in place and larger in size.	Often portable and can be used in various settings.

Contrast Bath

This kind of hydrotherapy used to relieve pain, edema, inflammation, and to rebuild strength is the contrast bath. The extremities are immersed in hot and cold water alternately during a contrast bath. Controlling body temperature contributes to the pumping action that increases blood flow to the extremities. During the therapy, the extremities are alternated between hot and cold water for five cycles totaling 20 minutes. Acute limb or joint injuries, upper and lower extremity fractures, rheumatoid arthritis, diabetic neuropathy, chronic pain syndromes, etc., can all benefit from this treatment.

Technique: The temperature range for hot water is 42°–45°C, whereas the temperature range for cold water is 8.5°–12.5°C. It is generally a 20–30 minutes session. Should start with immersion of extremity in warm water for 10 minutes followed by alternate 1–4 minutes of immersion in cold water and 4–6 minutes of immersion in warm water. The session should end in a cool soaking.

Sitz Bath

This treatment is used to treat post partum perineal pain, perineal injury, uterus cramps, hemorrhoids, constipation, inflammation, prostate issues, and other gastrointestinal issues by soaking simply the hips in cold or warm plain water or saline solution. Although it requires more work than other hydrotherapy procedures, this has a significant impact on the pelvic and abdominal organs. Involves sitting in water in a temperature range between 40°–50°C.

Balneotherapy

Balneotherapy is a term that comes from the Latin word *balneum*, which means "bath". Historically, it referred specifically to bathing in thermal or mineral waters, but today it generally means any kind of water-based treatment.

In balneotherapy, the water is usually heated to around **34°C (93.2° F)**. This can be thermal water from natural springs, seawater, or even regular tap water. The idea is that when one is in the water, it reduces the weight on their joints. This effect, known as hydrostatic force (or Archimedes' principle), helps to lessen pain by lowering the pressure of gravity on painful areas.

For many years, taking baths, also known as balneotherapy or spa therapy, has been a common practice in medicine to treat various illnesses. The main goals of balneotherapy are to:

- Reduce pain
- Increase mobility
- Strengthen muscles
- Relieve muscle spasms
- Improve the range of motion in joints.

POOL DESIGN CONSIDERATIONS

Location

The pool should be conveniently located near hospitals and rehabilitation centres with easy access to public transportation and adequate parking area for patients and caregivers.

Access

Important factors to take into account include ramp placement, proper signage, and the width of doors for wheelchair users. Being at ground level for the pool has benefits. Pools on the upper levels and basements will require access through stairs and elevators. Plans for emergency evacuation must be considered. As a health and safety precaution to prevent inadvertent immersion in deeper water, ingress into the pool hall should ideally be at the shallow end of pools with varied depths.

Internal Building Design

The area inside the building should include not just the pool hall but also the necessary reception/rest area, locker rooms, storage. The space must accommodate the pool's planned uses, the probable number of users, their level of supervision, and any special needs they may have. The layout of the facility is crucial, too, with cleanliness and infection control being the key priorities.

Circulation of Water in Hydrotherapy Pool

Pools will either have a freeboard or deck level. The pool is at deck level when the water is level with the surround. Water spills over and enters a transfer canal. Pools with a freeboard have a perimeter that is higher than the water. The best circulation mechanisms and thus the best elimination of water pollution are said to be found in deck level pools. They are the most secure for emergency exits from hydrotherapy pools.

Dimensions of Pool

It will vary depending upon requirement. A pool of 5.5 m × 8.5 m or larger is required to allow space for various exercises and activities. Stretchers, wheelchairs, and the hoist's turning circle should all be able to pass through the pool concourse, which needs to be at least 2 m wide on two sides. Allowing for emergency access on the other two sides by a minimum of 1.5 m.

The depth of the pool may vary depending upon the requirement. Depth of at least 4 feet is generally recommended to allow for full body immersion and exercises.

Steps can be made of metal marine grade stainless steel. Tiles must be nonslippery and must withstand prolonged high temperatures. Tiles are clearly marked at the edge with contrasting colors.

Exit and Entry to Pool

- Steps should lead to the shallowest part of the pool;
- Stainless steel handrails on both sides; and
- Plastic must be able to withstand continuous higher temperature and maximum weight expected of mobile patients. Wood is not recommended because it does not last long and can become slippery if pool is not maintained properly.
- Since ramp access takes up a lot of room, it is uncommon to see one in hydrotherapy pools.
- For autonomous wheelchair access, they can be acceptable in specialized environments like spinal cord injury centers. Ramps should deliver to the shallowest area of the pool and have a maximum gradient of 1:14.

Fixtures of Pool

Hydrotherapy pools must have handrails in order to facilitate therapeutic treatment methods that call for the patient to lean on them for support. They ought to extend continuously from the handrail for the step entry and along at least three sides of the pool.

Changing Room

Individual or group sessions at a hydrotherapy pool run on an appointment basis, and users frequently need more time to adjust. As a result, more room is needed than at a recreation center. Based on the most likely number of people using the changing rooms at once, the amount of space needed for the changing rooms should be determined. There can be individual disability or open space changing rooms.

Pool Equipment

Pool equipment can be broadly divided into safety equipment, exercise equipment, and resistive devices (water shoes, webbed gloves).

- Neck collars, floats, noodles, fins, hand bats, aqua jogging belts, hanging plinths, submersible steps, beach balls, and toys are examples of therapeutic pool accessories.
- Laminar flow devices, underwater treadmills, rowing machines, and water turbulence machines can all be helpful for high-level or specialized functional rehabilitation and sports injuries.

Insulation

Since the water in a hydrotherapy pool is warmer than that in a recreational swimming pool, the heat loss and humidity are higher. One method to reduce the utility expense is to use solar panels to generate the electricity needed to heat the pool.

Safety Measures

- An experienced therapist should be present for the duration of the treatment session.
- A maximum relative humidity of 55% should be maintained while the department is in operation, and the temperature of the pool water should be thermostatically controlled and monitored to ensure that it does not exceed 35.5°C. For all patients, the treatment time should be kept to a maximum of 30 minutes to 1 hour.
- The pool surrounding should be kept as dry as possible, personnel and patients should wear shoes with rubber grooves, and the floor tile is made to minimize the possibility of slip and fall. One of the staff members who is on duty should accompany patients to and from the pool area.
- The physiotherapist should evaluate the patient's capability to use the steps to enter the pool in cooperation with the patient. To facilitate entry and exit, a sitting hoist or stretcher should be supplied.
- Controlling the pH of the pool will reduce eye and skin irritation and avoid skin rashes. To guarantee efficient disinfection and to prevent corrosive damage, the pH should be kept between 7.2 and 7.8, and the water is continuously filtered.

Cleaning of Hydrotherapy Pool

- Thorough cleaning of the poolside and wearing individualized footwear around the pool minimizes the risk of foot infections like athletes' foot, and anyone with verrucas should be asked to wear foot protection in the pool in order to limit susceptibility to infections.
- Make sure the water used for hydrotherapy is clean and safe for the patients' health and well-being.
- After each patient uses hydrotherapy equipment (such as Hubbard tanks, tubs, whirlpools, whirlpool spas, or birthing tanks), drain and clean it, and then the surfaces should be disinfected.
- If a registered water treatment product is not available sodium hypochorite can be used as an alternative for disinfection treatment product.
    - Maintain a residual chlorine level of 15 ppm in the water, Hubbard tanks, and small hydrotherapy tanks.
    - Keep the chlorine residual in the water of whirlpools and whirlpool spas between 2 and 5 ppm.

Hygiene of Pool and Water

The following actions will increase the facility's structural longevity and cut down on operating expenses while reducing health concerns to users, such as infection transmission and pain.

- Keep the temperature in the hydrotherapy room between 35.5°C and 36°C at all times, never going above 38°C. Temperature measurement is essential because patients with certain medical diseases, such as neuropathy, have lost the ability to sense temperature changes.
- The pH of the water has a big impact on how well chlorine and other disinfectants work. At increasing pH levels, chlorine loses its potential for disinfection. The pH of the water must be kept within a specific range to allow the disinfectants to work as effectively as possible. pH should be checked at the start of each day, then every 2 hours, and finally at night. It should lie between 7.2 and 7.8.
- For ensuring disinfection control chlorine is the most commonly used chemical. At least three times per day, free chlorine levels should be checked, and they should be in the range of 1.5 and 5.0 mg/L. While a facility is in use, total chlorine levels must not exceed 10 mg/L and should be tested at the same time as free chlorine. It is advised that facilities operate with combined chlorine levels no >30% of the free chlorine (total chlorine minus free chlorine).
- The alkalinity is a gauge of the water's resistance to pH fluctuations.
- The alkalinity should be kept between 60 and 200 mg/L. The calcium hardness must to be kept between 50 and 400 mg/L. Water turns corrosive if the calcium hardness is too low.
- Microbiological contamination should be monitored. Hydrotherapy pools should be examined twice a week since users and staff may be more susceptible to infection due to extended immersion times.
- Daily cleaning of the poolside area with pool water and weekly cleaning with a solution containing 200 mg/L of free chlorine should be done with the proper dilution of chlorine-releasing tablets. A regular inspection and maintenance procedure should be carried out in conjunction with pool emptying.

SUMMARY

- Hydrotherapy has a long history dating back to ancient Greece, where private hot baths were believed to improve physical and mental health.
  - The Hippocratic Corpus states that water can cure everything, and it was also used to treat skin infections.
  - In the 18th century, English doctors John Floyer and James Currie highlighted the advantages of cold bathing and constructed a water treatment facility in Litchfield, UK. Angeliki Panagiotatou, Athens Medical School's first female graduate, popularized hydrotherapy in Greece by highlighting its healing properties.
- Currently, hydrotherapy is used to treat various ailments, including poor circulation, skin disorders, chronic fatigue, moderate depression, and chronic fatigue syndrome.
  - Aquatic therapies, such as Halliwick, the Bad Ragaz Ring Method, Watsu, Aqua Running, and Ai Chi, are used for relaxation, fitness, physical rehabilitation, and other therapeutic benefits.
  - Techniques include Ai Chi, which uses active resistance training in water combined with diaphragmatic breathing, Aqua Running, and Bad Ragaz Ring Method (BRRM), which focuses on neuromuscular function rehabilitation.
- Hydrodynamics is a field of study that involves the laws of inertia, acceleration, buoyancy, and hydrostatic force.
  - Cold water immersion (CWI) has various physiological and biochemical effects on the body, including reducing weight-bearing on painful joints, improving cardiovascular health, and enhancing respiratory function.
  - It can also help patients with multiple sclerosis, Parkinson's disease, and other painful conditions.
- Hydrotherapy can improve physical function, prevent falls, and reduce the risk of injury in various conditions, particularly for individuals with painful, arthritic joints, chronic hip or knee OA, overweight or inactive individuals, those recovering from illness or surgery, and those with multiple sclerosis.
  - It can also enhance fitness and movement, aerobic capacity, endurance conditioning, balance, coordination, agility, body mechanics, postural stabilization, flexibility, and muscle strength, power, and endurance.
- Hydrotherapy is a therapeutic intervention that can be used to build strength in patients, improve bilateral coordination, and treat various injuries and conditions.
  - The program should include warm-up, strength training, mobility exercises, endurance/cardiovascular, and cool down/stretch.
  - The chosen form of hydrotherapy depends on desired effects, size of area to be treated, safety, control of infection, and cost-effectiveness.
- Various types of tanks are available, including whirlpool baths, Hubbard tanks, contrast baths, sitz baths, and balneotherapy.
- The hydrotherapy pool building design should consider the neighborhood's accessibility needs, ramp placement, proper signage, and the width of doors for wheelchair users.
  - Pools can have freeboard or deck level types, with deck level pools having better circulation mechanisms and water pollution elimination.
- Safety features include an experienced physiotherapist present during treatment sessions, maintaining a maximum relative humidity of 55%, and limiting treatment time to 30 minutes.
  - The pool surround should be kept dry, personnel and patients wear rubber grooved shoes, and staff members should accompany patients to and from the pool area.
- Maintaining a temperature between 35.5°C and 36°C, maintaining a pH of 7.2–7.8, and monitoring microbial contamination are essential. Regular inspections and maintenance are also recommended.

FURTHER READINGS

- Aquatic Therapy Association for Chartered Physiotherapists (ATACP) publication; Guidance on Good Practice in Aquatic Physiotherapy 2015
- Bairaktaridou A, Lytras D, Kottaras A, Iakovidis P, Chatziprodromidou IP, Moutaftsis K. The effect of hydrotherapy on the functioning and quality of life of children and young adults with cerebral palsy. Int J Adv ResMed. 2021;3:21-4.
- Burke DT, Ho CH, Saucier MA, Stewart G. Effects of hydrotherapy on pressure ulcer healing. American journal of physical medicine and rehabilitation. 1998 Sep 1;77(5):394–8.
- Cider Å, Schaufelberger M, Sunnerhagen KS, Andersson B. Hydrotherapy—a new approach to improve function in the older patient with chronic heart failure. European Journal of Heart Failure. 2003 Aug;5(4):527–35.
- Fransen M, Nairn L, Winstanley J, Lam P, Edmonds J. Physical activity for osteoarthritis management: a randomized controlled clinical trial evaluating hydrotherapy or Tai Chi classes. Arthritis Care and Research. 2007 Apr 15;57(3):407–14.
- Gautham K, Devi S. A case series on maternal and neonatal outcomes of hydrotherapy during labor and childbirth. Indian Journal of Obstetrics and Gynecology Research. 2020;7(2):257-62.
- Giaquinto S, Ciotola E, Dall'Armi V, Margutti F. Hydrotherapy after total hip arthroplasty: a follow-up study. Archives of gerontology and geriatrics. 2010 Jan 1;50(1):92-5.
- Hammill HV, Ellapen TJ, Strydom GL, Swanepoel M. The benefits of hydrotherapy to patients with spinal cord injuries. African journal of disability. 2018 Jan 1;7(1):1–8.
- Kesiktas N, Paker N, Erdogan N, Gülsen G, Biçki D, Yilmaz H. The use of hydrotherapy for the management of spasticity. Neurorehabilitation and neural repair. 2004 Dec;18(4):268-73.
- Kurabayashi H, Machida I, Kubota K. Improvement in ejection fraction by hydrotherapy as rehabilitation in patients with chronic pulmonary emphysema. Physiotherapy Research International. 1998 Nov;3(4):284-91.
- Langhorst J, Musial F, Klose P, Häuser W. Efficacy of hydrotherapy in fibromyalgia syndrome—a meta-analysis of randomized controlled clinical trials. Rheumatology. 2009 Sep 1;48(9):1155–9.
- Linhares GM, Machado AV, Malachias MV. Hydrotherapy reduces arterial stiffness in pregnant women with chronic hypertension. Arquivos brasileiros de cardiologia. 2020 Mar 13;114:647-54.
- Marinho-Buzelli AR, Bonnyman AM, Verrier MC. The effects of aquatic therapy on mobility of individuals with neurological diseases: a systematic review. Clinical rehabilitation. 2015 Aug;29(8):741-51.
- Martin-Valero R, Cuesta-Vargas AI, Labajos-Manzanares MT. Evidence-based review of hydrotherapy studies on chronic obstructive pulmonary disease patients. International Journal of Aquatic Research and Education. 2012;6(3):8.
- McIlveen B, Robertson VJ. A randomised controlled study of the outcome of hydrotherapy for subjects with low back or back and leg pain. Physiotherapy. 1998 Jan 1;84(1):17–26.
- Michalsen A, Lüdtke R, Bühring M, Spahn G, Langhorst J, Dobos GJ. Thermal hydrotherapy improves quality of life and hemodynamic function in patients with chronic heart failure. American heart journal. 2003 Oct 1;146(4):728-33.
- Mooventhan A, Nivethitha L. Scientific evidence-based effects of hydrotherapy on various systems of the body. North American journal of medical sciences. 2014 May;6(5):199.
- MunicinÃ A, Nicolino A, Milanese M, Gronda E, Andreuzzi B, Oliva F, Chiarella F. Hydrotherapy in advanced heart failure: the cardio-HKT pilot study. Monaldi Archives for Chest Disease. 2006;66(4).
- Pinto C, Salazar AP, Marchese RR, Stein C, Pagnussat AS. The effects of hydrotherapy on balance, functional mobility, motor status, and quality of life in patients with Parkinson disease: a systematic review and meta-analysis. PM&R. 2019 Mar;11(3):278-91.
- Pinto C, Salazar AP, Marchese RR, Stein C, Pagnussat AS. The effects of hydrotherapy on balance, functional mobility, motor status, and quality of life in patients with Parkinson disease: a systematic review and meta-analysis. PM&R. 2019 Mar;11(3):278-91.
- Sánchez AM, Matarán-Peñarrocha GA, Lara-Palomo I, Saavedra-Hernández M, Arroyo-Morales M, Moreno-Lorenzo C. Hydrotherapy for the treatment of pain in people with multiple sclerosis: a randomized controlled trial. Evidence-based complementary and alternative medicine. 2012 Oct; 2012.
- Tom EK, Joseph V. Contrast Bath. International Journal of Nursing Education and Research. 2019;7(3):415-7.
- Vaile J, Halson S, Gill N, Dawson B. Effect of hydrotherapy on recovery from fatigue. International journal of sports medicine. 2008 Jul;29(07):539–44.
- Vaile J, Halson S, Gill N, Dawson B. Effect of hydrotherapy on the signs and symptoms of delayed onset muscle soreness. European journal of applied physiology. 2008 Mar;102(4):447–55.
- Verhagen AP, Bierma-Zeinstra SM, Boers M, Cardoso JR, Lambeck J, de Bie R, de Vet HC. Balneotherapy for osteoarthritis. Cochrane database of systematic reviews. 2007(4)
- Zamunér AR, Andrade CP, Forti M, Marchi A, Milan J, Avila MA, Catai AM, Porta A, Silva E. Effects of a hydrotherapy programme on symbolic and complexity dynamics of heart rate variability and aerobic capacity in fibromyalgia patients. Clin Exp Rheumatol. 2015 Feb;33(1 Suppl 88):S73-81.

STUDENT ASSIGNMENT

LONG ANSWER QUESTIONS

1. Define hydrotherapy. Write its indications, effects and uses.
2. Explain in detail the physiological effects of hydrotherapy.
3. Discuss various types of hydrotherapy.
4. Describe the history of hydrotherapy.

SHORT ANSWER QUESTIONS

1. Write effects and uses of hydrotherapy.
2. Write about precautions and contraindications related to hydrotherapy.
3. Mention the general rules for application of hydrotherapy.
4. Write about mechanism of action of hydrotherapy.
5. What are the physical properties of water?
6. Write notes on:
   a. Safety features of hydrotherapy pool.
   b. Maintenance of hydrotherapy pool.

MULTIPLE CHOICE QUESTIONS

1. **Hydrotherapy can assist in the following:**
   a. Boosting immune system
   b. Easing of tight muscles
   c. Controlling the production of stress hormone
   d. All of the above
2. **In the 1800's, hydrotherapy was used to treat a variety of minor physical illnesses by:**
   a. Increasing circulation to the area
   b. Removing toxins
   c. Preventing infections
   d. All of the above
3. **Temperature often plays a role in hydrotherapy as does________________.**
   a. Water pressure
   b. Elevation
   c. Time
   d. Training
4. **Which variables are most relevant when administering hydrotherapy?**
   a. Elevation and time
   b. Temperature and pressure
   c. Temperature and elevation
   d. Pressure and time
5. **Which statement regarding hydrotherapy safety precautions is not true?**
   a. Hydrotherapy units require a ground fault circuit interrupter
   b. Hubbard tank temperatures should not exceed 100°F
   c. Patients with open wounds are not candidates for hydrotherapy
   d. The hydrotherapy tank must be drained and cleaned after each treatment

ANSWER KEY

1. d **2.** d **3.** a **4.** b **5.** d

Section V

Movement and Alignment

SECTION OUTLINE

21

Posture

Sheetal Kalra, Deepak Raghav

LEARNING OBJECTIVES

After the completion of the chapter, the readers will be able to:

- Define active and inactive postures, postural mechanism, patterns of posture.
- Explain principles of re-education: Corrective methods and techniques, patient education.
- Describe posture (static and dynamic).
- Define good posture and understand which muscles are responsible for good posture.
- Discuss abnormal posture.
- Demonstrate methods of assessment of abnormal posture.
- Describe and demonstrate postural correction by: Strengthening of muscles, mobilization of trunk, relaxation. Active correction of the deformities, passive correction (traction), postural awareness, abdominals and back extensors.
- Outline principles of bracing for the trunk and surgical correction.

CHAPTER OUTLINE

- Introduction
- Types of Posture
- Postural Control Mechanism
- Good Posture
- Poor Posture
- Posture Assessment
- Treatment for Poor Posture

KEY TERMS

Arcometer: The arcometer is a recognized instrument for lordosis and kyphosis measurement. A ruled bar with three perpendicular arms—the first fixed at one end, the second movable on two axes, and the third movable on just one axis—makes up the manual arcometer.

Debrunner kyphometer: The Debrunner Kyphometer is designed to measure an external Cobb angle by positioning its feet over the T2/3 and T11/12 interspaces.

Exercise prescription: It is an exercise regimen that specifies the kinds of exercises to be performed and their frequency, duration, and intensity.

Flexicurve index: The flexicurve is used by physiotherapists as portable tools for measuring thoracic kyphosis. A flexible ruler is placed on the back to fit the lumbar and thoracic spinal curves. The ruler's shape is then sketched on paper to get the kyphosis index.

Forward head posture: When the cervical spine is hyperextended and the head protrudes forward, it is known as forward head posture, which is an improper neck posture.

Global postural re-education (GPR): The therapeutic postures used while lying, sitting, or standing—and which should be sustained for 15–20 minutes each are part of the GPR approach. During sessions, postures can be blended in different ways. A number of factors, including the patient's age, load capability, degree of pain, and the muscles that need to be stretched, are taken into consideration when selecting a pose.

Kyphosis: Exaggerated forward rounding of the upper back is called kyphosis.

Lordosis: It is an excessive inward bend of the spine. People of all ages are susceptible to lordosis.

Posture: Posture is the position that the body assumes when it is supported during periods of muscle inactivity or when numerous muscles function in unison to preserve stability.

Scoliosis: A sideways bend in the spine is called scoliosis. The curve's angle could be tiny, huge or in the middle. Also, scoliosis is defined as any X-ray measurement >10°.

Sway back posture: It is characterized by an exaggerated lumbar curve and a forward pelvic tilt. This posture typically results in the hips being pushed forward and the chest leaning back, forming an accentuated "S" shape. It often leads to muscle imbalances, discomfort, and inefficient movement.

INTRODUCTION

The attitude of the body is referred to as posture. It can be supported by the activity of muscles, achieved through coordinated muscle action or it might create an important foundation that is continuously adaptable to movement overlaid on it. Good posture is essential for maintaining overall health and well-being as it supports spinal health, ensures musculoskeletal efficiency, enhances respiratory and circulatory functions, aids digestion, and prevents nerve compression. Proper alignment promotes balanced muscle engagement, reduces strain, and helps prevent overuse injuries and chronic conditions. Additionally, good posture positively influences psychological well-being by boosting self-esteem and confidence. Therefore, awareness and practice of proper posture are crucial in both clinical settings and everyday life.

Did You Know?

Research suggests that sitting or standing with proper alignment can enhance confidence, reduce stress, and increase productivity. For future physiotherapists, understanding the holistic benefits of good posture can empower them to educate and motivate their patients toward healthier habits.

TYPES OF POSTURE

Inactive Posture

Describes the positions taken when sleeping or resting. These are better suited for this use when there is minimal necessity of muscular activity to sustain life.

Characteristic Features

- Posture adopted for rest or sleep
- Postures make minimal demand upon muscles
- Responsible for essential body functions such as respiration and circulation postures.

Active Postures

Active postures are the postures maintained by the integrated actions of groups of many muscles.

Types

1. **Static postures:** These are the postures adopted during static positions of the body. Standing, sitting, lying, and kneeling are a few examples.
2. **Dynamic postures:** These are the postures adopted by the body while performing dynamic movement. For example walking, running, jumping, throwing, and lifting involve the body or its parts moving. These postures are necessary for a stable foundation for movement.

POSTURAL CONTROL MECHANISM

To maintain the center of gravity within the base of support, corrective motions are required. The coordination of the sensory, skeletal muscular, and central neurological systems is required to accomplish this goal. Table 21.1 lists the components of the postural control system. Figure 21.1 displays the posture control mechanism.

TABLE 21.1: Components of posture control system

Sensory system	Skeletal muscle system	CNS
Vestibular system located in the inner ear (semicircular canals, otoliths, macules)	Muscles of the upper and lower extremities	Stretch reflex
Vision (retina)	Trunk muscles	Long-loop reflexes
Proprioceptive system (muscle spindle-type I and II, Golgi tendon organ, joint receptors)	Neck muscles	Preprogramed reactions (Learned skills)
Cutaneous receptors		Synergistic action

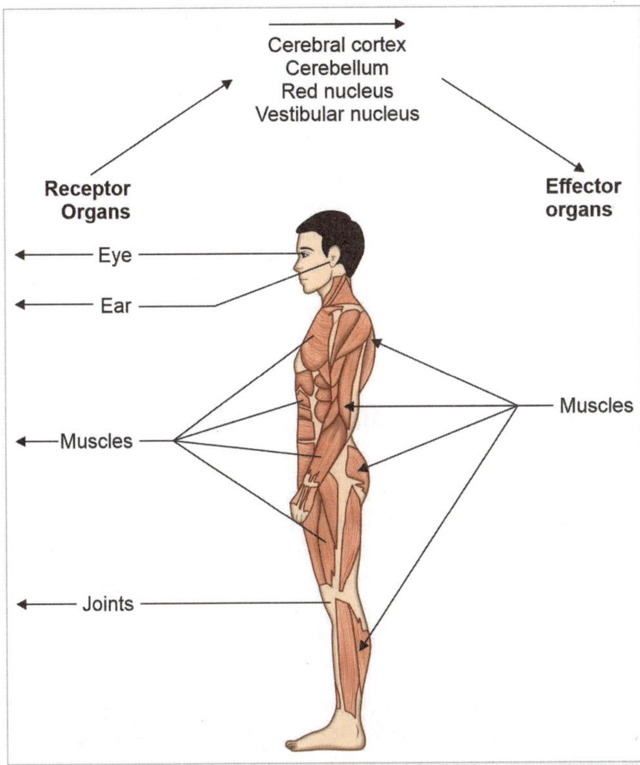

Fig. 21.1: Posture control mechanism

Sensory Systems

The fundamental purpose of sensory systems is to tell the system about its own status as well as that of its environment. Afferent routes carry the information from the sensory receptors to the central nervous system. Light, pressure, temperature, and sound are just a few examples of the energy that sensory receptors can transform.

Visual System

- The focal system, which is thought to be specialized for object recognition, and the ambient system, which is thought to be specialized for movement control, are the two regions in the brain where visual information is conveyed from the retina.
- Research has also indicated that the latter has a significant impact on balance and stability. Vision appears to affect balance *via* responding to motion as a relative picture shift on the retina, as well as by activating the muscles necessary for postural adjustments.
- Visual acuity, visual contrast, object distances, and room illumination all affect how well one can control his posture.

Vestibular System

- The semicircular canals are active during the beginning and end of movement because they are sensitive to velocity changes; in contrast, the otoliths work at low frequencies of <5 Hz and offer information on linear acceleration, such as gravity.
- The vestibular nuclei in the brainstem get information from additional sensory sources in addition to the information from the otoliths and semicircular canals. According to a research in 1993 by Baloh et al. it was observed that the vestibulo-ocular reflex stabilizes vision by causing the eyes to move in the opposite direction when the head is turned, while the primary function of the vestibulospinal reflex is to stabilize the head and body.

Proprioceptive and Exteroceptive Systems

Proprioceptors and exteroceptive receptors in the somatosensory system convey information about the position of the body. The proprioceptive receptors, which are found in muscles, tendons, and joints, provide information about the location of the body and limbs as well as the degree of muscle distension.

- Muscle spindles (types Ia and II), Golgi tendon organs (type Ib), and joint receptors are all examples of proprioceptors. Several pressoreceptors present on the sole of the foot provide exteroceptive data. Skin and subcutaneous tissues include exteroceptive receptors. Meissner corpuscles and Merkel discs, which are found closest to the skin's surface, and Ruffini ending and Pacinian corpuscles, which are found farther below the skin's surface, are the two main types of cutaneous receptors.
- Information about the movements and positions of the bodily parts in relation to one another is provided by the receptors in joint capsules, The muscle spindles can be engaged by passively extending the entire muscle, and they provide information regarding changes in muscle length and tension (dynamic stretch). The motor neurons provide efferent input to the intrafusal fibers in the muscle spindles in addition to an afferent pathway. The bodily sway is picked up by the pressoreceptors.

Clinical Correlation

Recognizing how environmental factors, such as lighting and surface textures, impact postural control can inform recommendations for modifying patients' surroundings to optimize safety and stability. For example, ensuring adequate lighting and minimizing clutter can enhance visual input and reduce fall risk.

Musculoskeletal System

The primary mechanism of the postural control system receives a signal from the proprioceptive receptors in the muscle and tendon whenever the muscles are stretched.

For example, in order to produce sufficient muscular contractions, postural control requires coordinated muscle movement. The ankle, knee, and hip joints, in particular, play a crucial role in maintaining the body's balance when the muscles act on the joints. Although the calf muscles are stimulated first to provide postural control during body movements, other "primary postural muscles" such as the muscle of neck, hamstring, soleus, and supraspinatus muscles also become active, simultaneously.

Influence of Muscles on Stability

- Depending on the posture, different muscle groups must work harder or differently to support both static and dynamic postures.
- The muscle groups known as the anti-gravity muscles are needed to maintain the upright position.
- Without the dynamic stabilizing activity from the trunk muscles, the spine would collapse in the upright position. The muscles of the neck and trunk operate as prime movers or as antagonists to movement produced by gravity during dynamic exercise.
  - To maintain the upright posture, both superficial (global) and deep (core) muscles are involved.

Muscles controlling the spine

Superficial and deep muscle play critical role in providing stability and maintaining the upright posture.

- **Lumbar region**
  - **Superficial muscles**
    - Rectus abdominis
    - External and internal obliques
    - Quadratus lumborum (lateral portion)
    - Erector spinae
    - Iliopsoas
  - **Deep muscles**
    - Transversus abdominis
    - Multifidus
    - Quadratus lumborum (deep portion)
    - Deep rotators
- **Cervical region**
  - **Superficial muscles**
    - Sternocleidomastoid
    - Scalene
    - Levator scapulae
    - Upper trapezius
    - Erector spinae
  - **Deep muscles**
    - Rectus capitis anterior and lateralis
    - Longus colli

Neurological Control

Neuromuscular coordination allows for the maintenance or adaptation of postures, with the appropriate muscle being activated by way of an extremely intricate reflex process.

Postural Reflexes

Reflexes are efferent responses to afferent stimuli by definition.

The antigravity muscle is the main effector organ in this case, and the efferent reaction is a motor one. The body's afferent stimuli come from a multitude of sources. The postural reflex is made up of a complicated series of efferent responses that are coordinated by the central nervous system, which includes the cerebellum, red nucleus, vestibular nucleus, and ex-cerebral cortex.

Clinical Correlation

Motor control exercises focusing on activating deep stabilizing muscles, such as the transversus abdominis and multifidus, can help restore spinal stability and reduce pain. Similarly, individuals with cervical spine disorders may benefit from exercises that target the coordination of deep cervical flexors to improve neck posture and reduce strain on the cervical spine.

GOOD POSTURE

The position in which minimum stress is applied to each joint is considered a good posture.

Did You Know?

Good Posture Requirements

A good posture requires:

- Good muscle flexibility
- Normal motion in the joints
- Strong postural muscles
- A balance of muscles on both sides of the spine
- Fulfilling the purpose for which it is attained
- Minimal muscular effort
- Aesthetically pleasing pose
- Body parts aligned with respect to each other

Example of good sitting posture: Shoulders in line over the hips, feet flat on the floor, low back support provided, and chin aligned over the chest.

Standing

- Weight should be evenly distributed through feet, knee's face forward.
- Arms on the sides with palms facing forward.
- Chest slightly lifted with shoulders back.
- Chin tucked in (imagine a string is tied to the top of the head and is pulling it upwards).
- The head and spine seem straight when viewed from the rear, with no left or right curvature (Fig. 21.2).

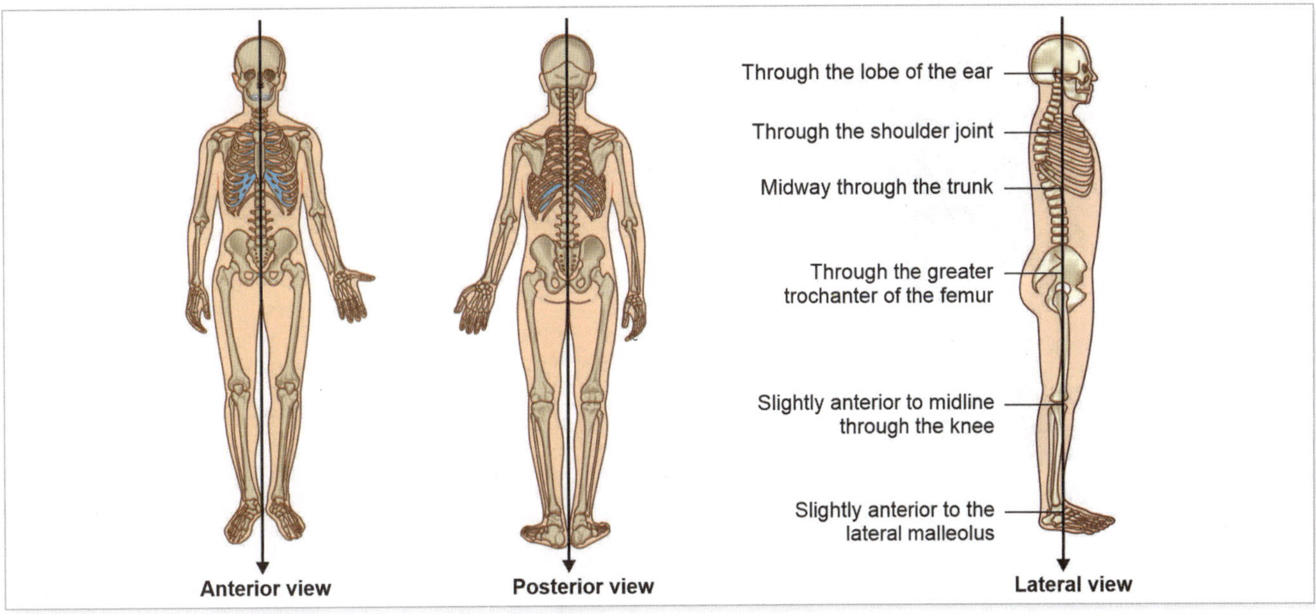

Fig. 21.2: Good standing posture

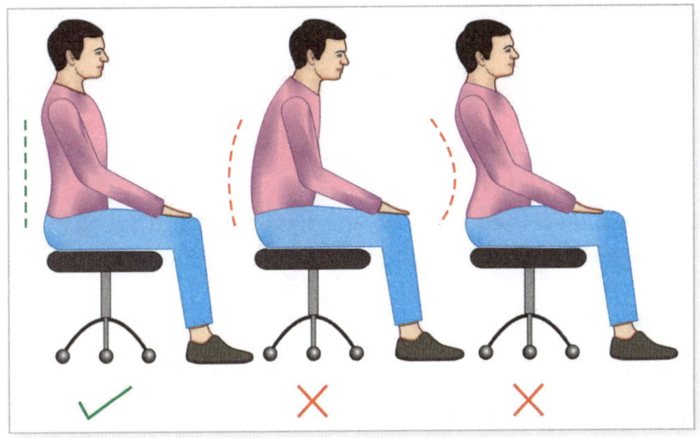

Fig. 21.3: Good sitting posture

Sitting

- Shoulders held back, chin tucked in and head held high.
- Abdominal muscles gently pulled inward and up toward the spine.
- Lower back's arch to be maintained in place. It may be easy to keep this posture by using a chair or lumbar roll (Fig. 21.3).
- Hips and knees are bent at right angle with feet supported on the ground.

Good Sitting Posture at Desk

Shoulders aligning over hips with supported low back posture is the ideal position. Knees and elbows flexed to 90°, and feet level on the ground or supported by a stool form a good posture while sitting at a desk. The computer screen should be positioned such that the ears and shoulders are in line. The elbow should stay bent if the computer mouse is placed close enough. Lastly, phones and pens should be kept at a distance of 14–16 inches (Fig. 21.4).

Fig. 21.4: Good sitting posture at desk

Other Common Activities

Sit to Stand

Knees are aligned with the ankles. Toes are in line with the feet. Upper body is brought anteriorly by bending forward and bringing the nose over and beyond the plane of the toes. A careful push-up from the chair is required. When necessary, the hands may be used.

Tying Shoes

To prevent leaning too far forward, the ankle can be placed on the opposite knee (Fig. 21.5). This posture is helpful in condition like acute lower back pain (LBP) where bending forward is not advised.

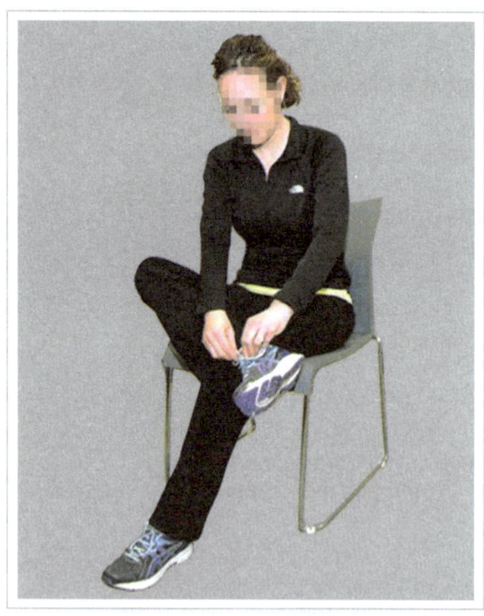

Fig. 21.5: Good posture while tying shoes

Clinical Correlation

Proper bending and reaching technique (Golfer's lift)

Care should be taken in patients with lumbar fusion or kyphoplasty surgeries as lifting and bending are not advised in these cases. This method is meant for those infrequent times when one have to pick up something from the ground using this posture with a flat back, one can use the leg as a lever arm to raise the body back to standing posture.

Golfer's lift

Proper Sleeping Posture

It is recommended to use a pillow support beneath the knees when lying on the back. A pillow can also be used to support the head.

Proper Side Lying Posture

In this one pillow supports the lumbar region, another is placed between the knees, and a third supports the head and neck (Fig. 21.6).

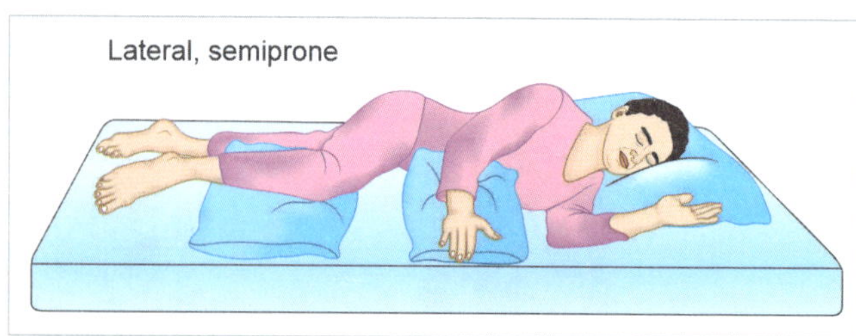

Fig. 21.6: Proper side lying posture

Log Roll

- With starting position of lying on the back, the knees are bent.
- Shoulders and hips are kept together as a single unit as one rolls onto the side.
- The top hand is placed on the bed and used for push up while lowering the legs to the floor.
- The body is raised slowly while lowering the legs to assume an upright position.

Getting into the Car

- Gently lowering to the seat and the vehicle is entered, one leg at a time.
- As one gets into the car, one moves back into the seat until it can be felt behind the legs.
- With the other hand on the dashboard for balance, the back of the seat is reached by moving backwards.

Getting out of the Car

- For maximum leg clearance, fully recline the seat.
- To prevent twisting at the spine, each leg one at a time is brought out while rotating the hips and shoulders as a whole.
- One hand is placed on the dashboard or door frame and the other on the back of the seat.
- Standing posture is ascended to, without dragging the body.

POOR POSTURE

When the spine is positioned in an unnatural way, with the curves highlighted, the joints, muscles, and vertebrae are put under stress, which is known as "postural dysfunction" or "poor posture". Pressure builds up on these tissues as a result of the prolonged inadequate posture. Aches and pains in the body, such as pain in the arms, neck, shoulders, or lower back; pain in the lower limbs, such as the leg, hip, knee, or ankle muscle exhaustion are likely to be observed. One of the symptoms of poor posture is headache brought up by a build-up of tension in the shoulders, neck, and upper back.

> **Did You Know?**
>
> **Factors which Predispose to Poor Posture**
> - Mental attitudes of the patient
> - Illness leading to weakness
> - Factors such as pain, muscular weakness
> - Occupational stresses
> - Muscle imbalances

Symptoms

- Muscle imbalance, decreased blood flow, decreased mobility in the spine, and increased pressure and degeneration in the spinal disks (mostly in the neck and lower back) result from poor posture.
- Symptoms may also include:
  - Pain across the shoulders or neck
  - Tension headaches
  - Nerve pain, heavy feeling or numbness in the arms
  - Muscle spasm
  - Constant ache in the lower back or neck
  - General fatigue
  - Repetitive strain injuries in the wrist or hand

Types of Postural Deviations

Figure 21.7 shows the types of postural deviations.

Lumbar Lordosis

The term "lordosis" describes the spine's normal inward (posterior concavity) curvature. Exaggerated versions of this curve are commonly referred to as hyperlordosis. Typically, the pelvis is tipped anteriorly.

Sway Back

Sway back posture is characterized by a forward head, hyperextended cervical and thoracic spines, flexion of the thoracic spine, extended lumbar spine, posterior tilt of the pelvis, hyperextended hips and knees, and slightly plantarflexed ankles.

Flat Back

Also known as hypolordosis, this posture involves a loss of the lumbar spine's natural curvature, making it appear straight. This can cause upper back pain by putting more pressure on the thoracolumbar region.

Forward Head Posture

Forward head posture is described by the forward tilt of the head with the chin protruding. It is brought on by increased extension of the upper cervical spine and the occiput on C_1, as well as increased flexion of the upper thoracic and lower cervical spines.

Scoliosis

A divergence from the spine's usual vertical line, resulting from the vertebrae's rotation and lateral curvature leads to scoliosis. When there is spinal angulation on the posterior-anterior radiograph accompanied with vertebral rotation of at least 10°, it is called scoliosis. This is the sideways curve of the spine, curved like a C or S in three dimensions.

Kyphosis

The thoracic or sacral areas of the spine exhibit an enhanced convex curve which is known as kyphosis. A lateral curvature of the spine and rib rotation are nearly invariably linked leading to kyphoscoliosis.

> **Did You Know?**
>
> Forward head posture often occurs due to prolonged smartphone and computer use and can lead to neck, shoulder, and back pain.

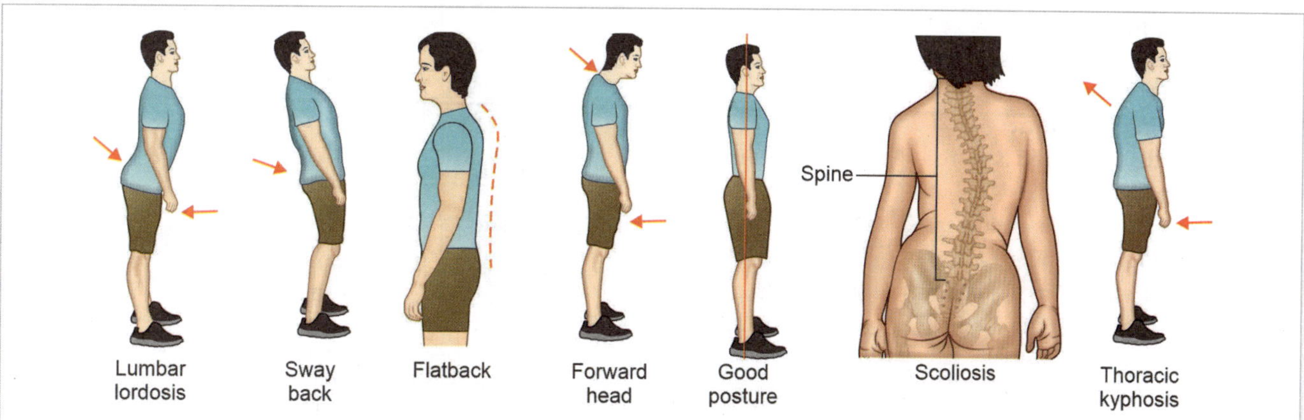

Fig. 21.7: Types of postural deviations

POSTURE ASSESSMENT

The basic foundation of physiotherapy assessment is posture assessment. It aids in locating anatomical flaws that cause a range of musculoskeletal issues. Postural assessment is a valuable tool for evaluating the causes of various postural deviations.

Conventional Methods

Visual Observation Method

It is the most common method which is used to assess posture in clinical practice. The one and only advantage of this method is that it does not require any equipment. With this method, quantitative data cannot be obtained. Thus, minor postural alterations cannot be detected. Also, it has been reported to have a poor interrater agreement. All these limitations discourage the use of this method for scientific research purposes.

In this method, subject is made to wear minimum clothes and observations are made from the front, back and side views which are analyzed according to predetermined guide. For example, in the ideal sagittal alignment, the gravitational line passes through the external acoustic meatus, the bodies of the cervical vertebrae, the tip of the shoulder, the mid-point of the thorax, slightly behind the hip joint, slightly in front of the knee joint and immediately preceding the lateral malleolus.

Following observations should be made in the visual observation method.

- **Shoulder:**
  - **Symmetry:** Are the shoulders level from anterior and posterior views?
  - **Roundedness:** Assessed anteriorly and from the side.
  - Are the shoulders in internal rotation?
  - Is there anterior translation of the humeral head?
- **Cervical spine:**
  - **Side view:** Are there any abnormalities in the spine's curvature, such as cervical kyphosis or lordosis?
  - **Posterior view:** Is there any observable muscle wasting?
- **Thoracic spine:**
  - **Side view:** Is an increased or decreased thoracic hyper-kyphosis present?
- **General spinal curvature:**
  - **Assessed posteriorly:** Is there a spinal scoliosis present?
- **Lumbar spine:**
  - **Side view:** Is there an increased lumbar lordosis or a flattened lumbar spine?
  - **Posterior view:** Any visible muscle spasm? Hinging at the thoracolumbar junction?
- **Pelvis:**
  - Assessed from the front, rear and side.
  - Assess levels of anterior superior iliac spine (ASIS) and posterior superior iliac spine (PSIS).
  - Assess levels of iliac crests.
  - Is pelvis in anterior or posterior tilt?
- **Hips:**
  - **Symmetry:** Are hips at level?
  - Are hips in internal or external rotation?
  - Is there a visible gluteal bulk?
  - **Side view:** Are the hips in extension or flexion?
- **Knees:**
  - **Side view:** Are the knees in hyperextension?

Plumb Line Method

It is again a common method used for postural assessment. A plumb line is used for posture evaluation. The physiotherapist looks for deviation from the ideal alignment in the side and back views, with a plumb line being placed along the midline of the body. The disadvantage associated with this method is again it cannot provide quantitative data of deviation from normal (Fig. 21.8). A variety of other postural assessment methods and applications have been described in literatures.

Movement Assessment Tool (MAT)

Using a tool requires manual assistance to identify or to analyze movement patterns and postures.

Arcometer

An approved tool for kyphosis and lordosis measurement is the arcometer (Fig. 21.9). The manual arcometer consists of a ruled bar with three perpendicular arms, the first of which is fixed at one end, the second of which is mobile on two axes, and the third of which is mobile on just one axis.

The ends of the three arms designate three sites from which a single circumference can be calculated. Therapist can

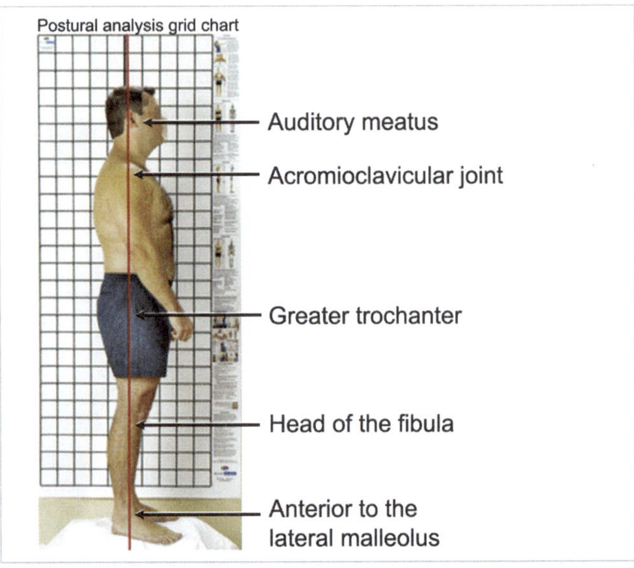

Fig. 21.8: Plumb line method

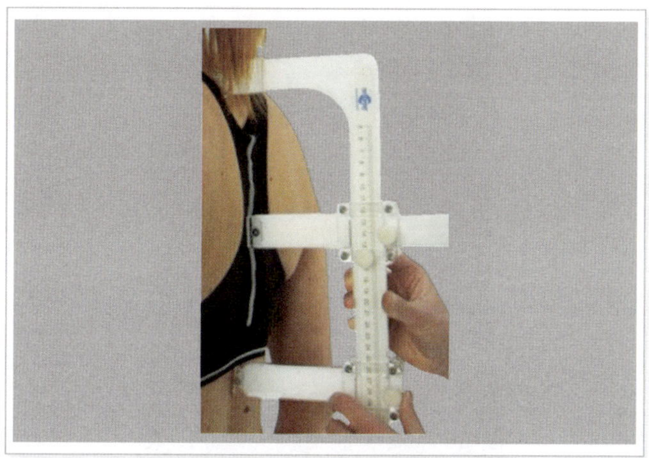

Fig. 21.9: Arcometer

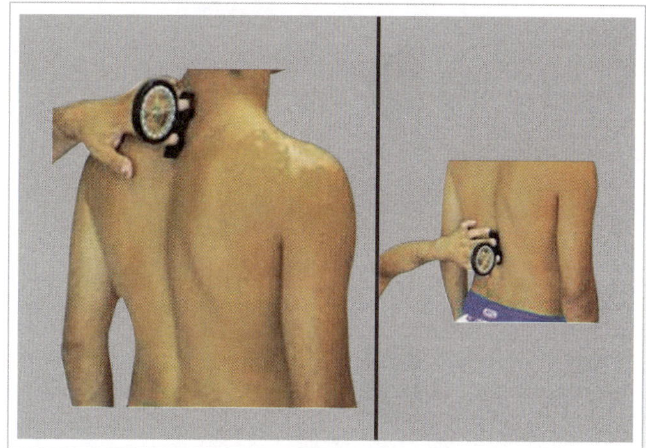

Fig. 21.11: Manual inclinometer

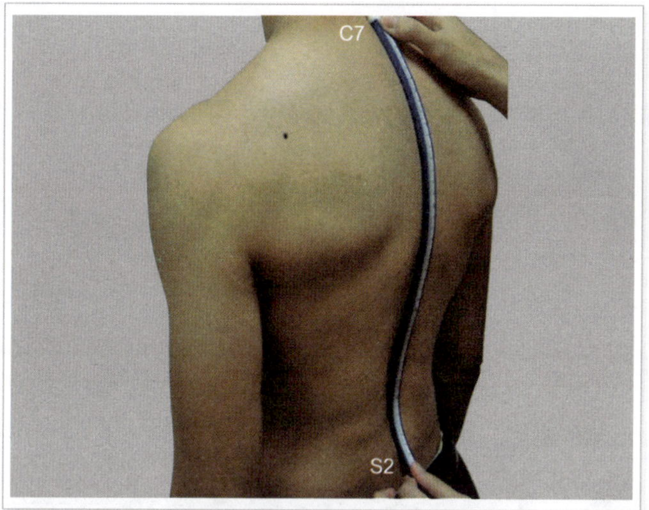

Fig. 21.10: Flexicurve index

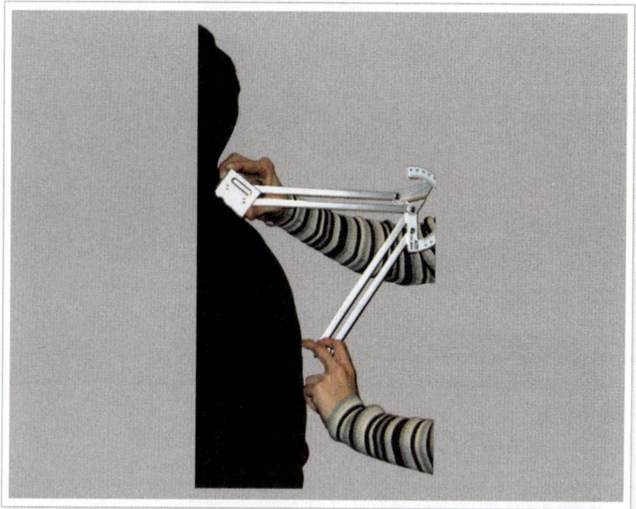

Fig. 21.12: Debrunner kyphometer

calculate the chord and rise using the arcometer and can determine the radius of the curve and the Cobb angle using these two measurements.

Flexicurve Index

Physiotherapists frequently utilize the flexicurve and manual inclinometer as handheld instruments to measure thoracic kyphosis. Flexible ruler is placed on back to adopt contours of thoracic and lumbar spine. On paper shape of the ruler is then drawn to calculate kyphosis index (Fig. 21.10).

Flexicurve Angle

Method similar to flexicurve index is used but kyphosis angle is calculated using geometric formulas.

Manual Inclinometer

An instrument called an inclinometer, is used to measure the angles of slope, elevation, or depression of an object with regard to the direction of gravity (Fig. 21.11).

Debrunner Kyphometer

It consists of two arm protractor positioned at specified bone markers to predict kyphosis angle (Fig. 21.12). The chord is the straight line connecting two points on a curve, while the rise is the vertical distance from the midpoint of the chord to the highest point of the curve.

Goniometers

These are used in physiotherapy practice, not only to measure joint ROM but also for the assessment of posture. Measurement of postural angles, such as neck inclination angle (craniovertebral angle) and cranial rotation angle (sagittal head tilt) by using manual goniometry has been reported in the literature.

Spinal Wheel

It is a plastic device with a reflective marker at the center that is guided along the midline from S-1 to occiput (Fig. 21.13).

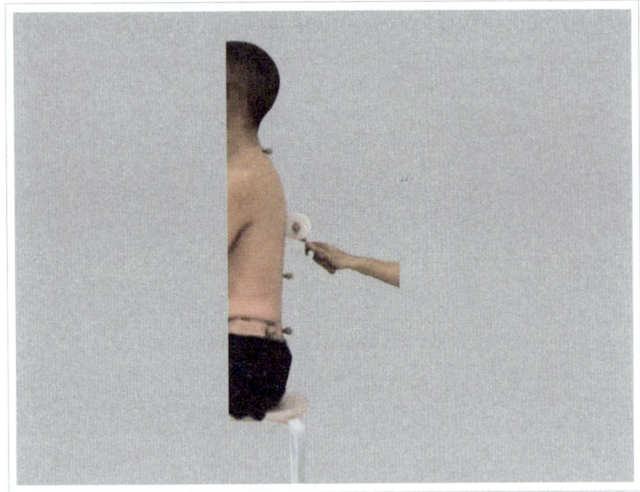

Fig. 21.13: Spinal wheel

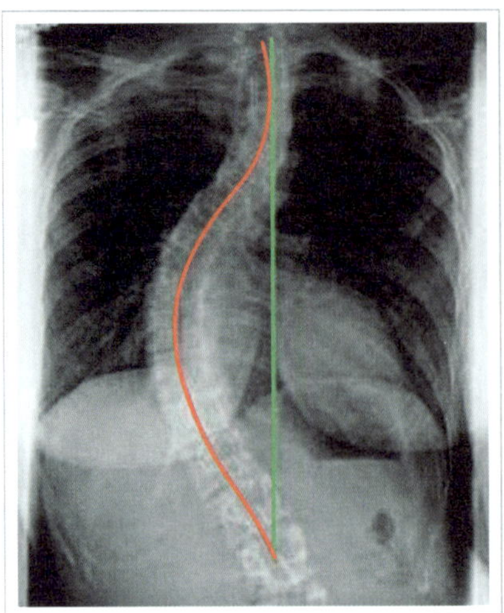

Fig. 21.15: X-ray showing scoliosis

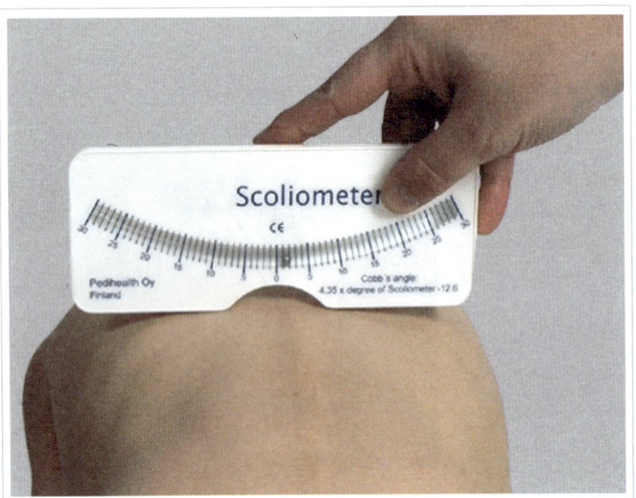

Fig. 21.14: Scoliometer

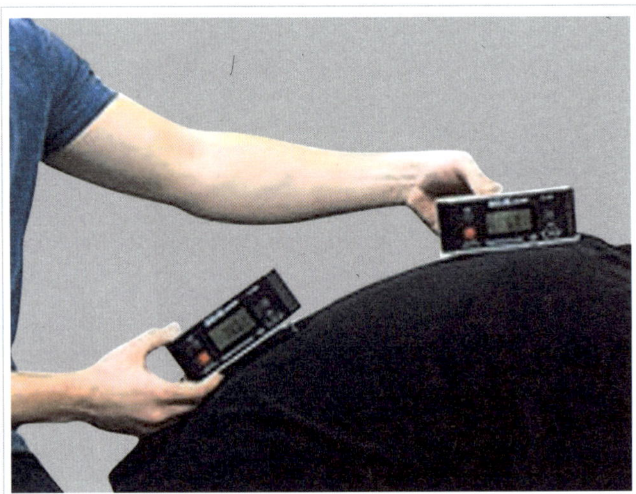

Fig. 21.16: DAT instrument

Scoliometer

The scoliometer is a type of protractor used to measure vertebral rotation and rib hump that is seen in scoliosis with forward bending test (Fig. 21.14).

Kypholordometer

It is made up of aluminium and 39 horizontally shaped and deformable rods. These rods are pointed at the back and contours of the curvature of the spine which are then drawn on paper which is attached to the back of instrument.

Radiograph Method

X-ray examinations (XR) are routinely used to measure curvatures of the vertebral column and to analyze vertebral conditions. X-rays have been considered the golden standard regarding the observation of posture deviations, despite being an invasive examination in which the individual is exposed to radiation. Spine angles can be quantified from calculations performed from the vertebra visible through X-ray using Cobbs angle (Fig. 21.15).

Modern Methods

DAT Instruments

Digital assessment tools (DATs) calculate body inclinations and deviations to present them on a digital display. This category does not include software (Fig. 21.16).

Digital Inclinometer

A digital inclinometer has sensors for capturing inclination of body.

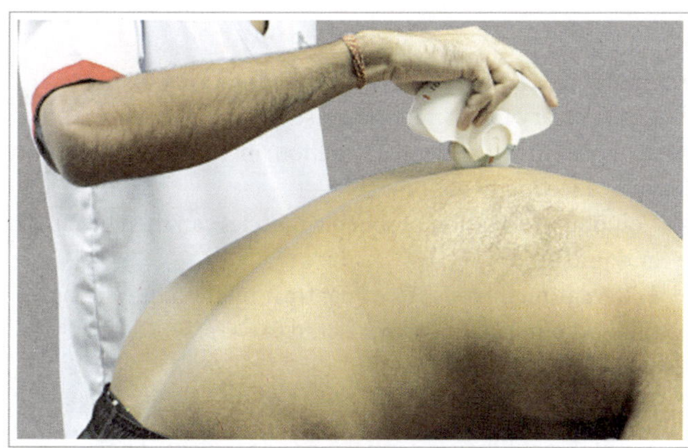

Fig. 21.17: Spinal mouse

Spinal Mouse

A device called a Spinal Mouse (SM) (Fig. 21.17) is moved manually along the skin of the spine. The spine is rebuilt in the sagittal and frontal planes in both neutral and extreme postures. Real measurements taken from the patient are used to create the photos.

- **Sagittal plane:**
  - Standing straight
  - Complete flexion
  - Complete extension.
- **Frontal level:**
  - Standing straight
  - Leaning to the left
  - Leaning to the right

Electrogoniometer

An electrical tool used to evaluate a joint's flexibility and mobility is called an electrogoniometer. Using topical attachments situated at numerous joints throughout the body, it tracks angular positions through distinct planes of motion in order to monitor transduction, the conversion of voltage signals in response to dynamic movements.

SAT Proposals

The Software-aided Assessment Tools (SATs) have been classified as those methods or instruments which are supported by a computing system.

3D Ultrasound

It is a method of identifying a column of reference points by means of a point marker. The setup includes a computer with specialized software that has normative data on vertebrate distances, a directional microphone, a reference marker, and the ability to calculate angles and distances of measured 3D data that are turned into degrees of range of motion.

Rasterstereography

It is possible to perform a 3D examination of spinal curvatures. This system coordinates the data of the back surface points and the line of symmetry between them in order to identify anatomical reference points, conspicuous vertebrae, and two superior iliac spines.

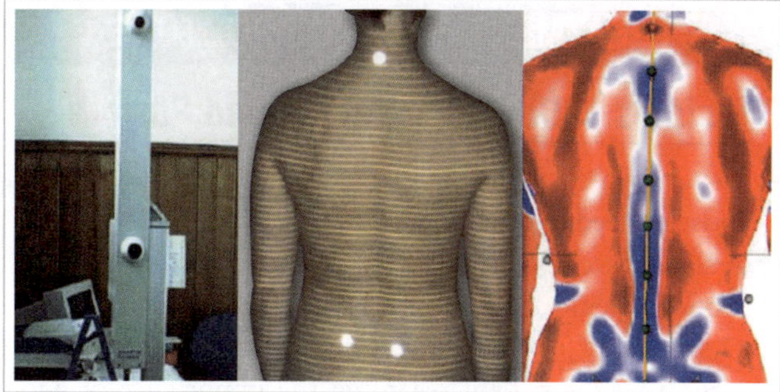

Fig. 21.18: Stereovideography

Stereovideography

Stereovideography is a technique for creating stereoscopic videos and films using a variety of equipment (Fig. 21.18). Most stereoscopic methods present a pair of two-dimensional images to the viewer. When viewed, the human brain perceives the images as a single 3D view, giving the viewer the perception of 3D depth.

Photogrammetry

In this method, photographs of the subjects are taken in frontal or sagittal plane with a camera which is mounted on a levelled tripod stand, which is placed at some distance from the subjects. This distance varies amongst various researches. The photographs thus obtained are transferred to a computer system. They are used to calculate postural angles with the help of some software which has been installed in the computer system (Fig. 21.19).

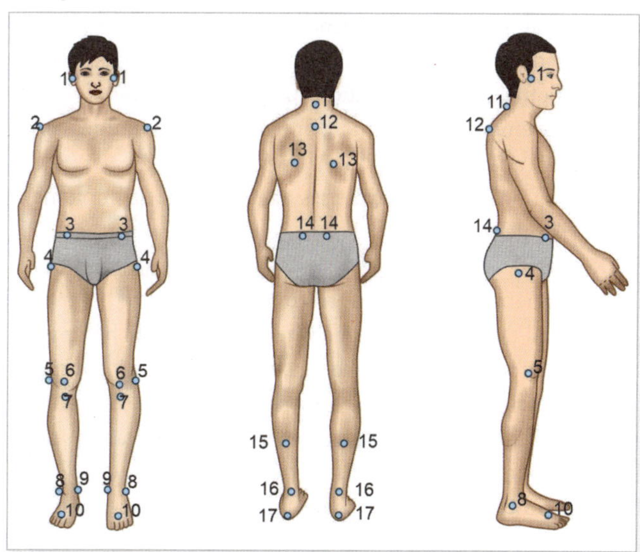

Fig. 21.19: Photogrammetry

TREATMENT FOR POOR POSTURE

Principles of Development of Good Posture

- Factors which predispose to health and development of muscles need to be considered.
- A stable psychological background
- Good hygienic conditions (nutrition, sleep)
- Opportunity for plenty of free movements
- Educating the patient regarding the importance of good posture. A mirror, posture recorder, photographs, video tapes can be used for this purpose.
- Assessment should include examination of weak trunk muscles, joint hypermobility, limb-length discrepancy, spine and pelvic alignment.
- Based on the needs, symptoms, and posture examination, a customized treatment plan for correction of posture should be created.

Strategies to Improve Posture

- Postural re-education and training
- Relaxation and mobility exercises
- Corrective exercises to improve strength and flexibility
- Other methods:
  - Bracing or strapping to assist maintaining a good posture
  - Workstation assessments, ergonomics, moving and handling techniques
  - Activity modifications, to prevent reoccurrences and promote healthy lifestyle.

Postural Re-education

- Postural training largely depends on the cause.
- Alteration in the habitual mental attitude plays a crucial role in postural re-education and training.
- Improvement of hygienic conditions has as influence on posture and its correction.
- Relaxation, mobility exercises are important complementary strategies.
- Repeated presentation of satisfactory posture is a fundamental principle for re-education.
- General debility and fatigue when cause should be treated.
- Alleviation of pain, muscle weakness, balance by appropriate exercises must also be undertaken.
- Occupational strains can be relieved by analysis of movement and substitution of new pattern.
- Physiotherapist plays an important role in reeducation. For this, the cooperation of the patient must be gained.
- Individual instructions are important.
- Group activities can be helpful.

Relaxation

- Ability to relax is important for reeducation.
- Unnecessary tension in the muscles can be relieved.
- To begin with, general relaxation in horizontal position gives a feeling of alignment.
- Voluntary relaxation of specific muscles can then be taught and practised.
- Different methods such as contrast method, physiological method of relaxation can be used.

Mobility

- Normal mobility is essential for wide variety of postures.
- General free exercises, full range of motion (ROM) exercises should be practiced.
- Full extension should be emphasized.
- Joint stiffness should be treated.

Corrective Exercises to Improve Strength and Flexibility

In people with poor posture, some muscles become weak and longer whereas others may be overused and become tight due to poor postures. To correct poor or painful posture, an exercise prescription is necessary. Correcting imbalances, strengthening muscles, improving joint mobility, and improving postural awareness can all be achieved by an exercise program.

Strengthening Exercises

The goal of strengthening exercises is to make specific muscles or groups of muscles stronger. Strength increases as a result of a muscle's growth being encouraged by force and overload. Weak muscles might make it more likely that nearby joints and soft tissues will sustain damage.

There are many different benefits of strengthening exercises. Some of the benefits of a strengthening exercise program are:

- Improved muscle strength
- Decreased energy expenditure as muscles are more efficient
- Reduced risk of injury
- Improved posture and function
- Improved quality of movement

Did You Know?

Exercise programs aimed at strengthening particular muscles or groups comprise a range of activities. As the strength increases, these exercises get easier. Some of the exercises may involve the following:

- Exercising against gravity
- Exercising against the resistance of water
- Exercising against a resistance band
- Exercising with weight
- Exercising using own body weight as the load.

Stretching Exercises

Depending on the desired outcome, stretching exercises can be customized to target specific problem areas.

It is crucial to hold each stretch for 10–15 seconds in order to release the initial tension in the muscle tissue and modify the tissue's flexibility over the time. The main rehabilitation benefits of stretching programs include:

- Increased range of movement (ROM)
- Decreased muscle tension/tone
- Increased proprioception, awareness of body position and postures
- Improved neuromuscular coordination
- Reduced the risk of muscular and soft tissue injury
- Improved sweep of synovial fluid around joint capsule, supplying nutrients

Global Postural Reeducation (GPR)

Developed empirically by Philippe Suchard in 1981, this approach is currently in use in Portugal, Spain, Brazil, and France.

Basic Principles

Three basic principles underpin the GPR philosophy:

1. Individuality, which views every person as unique.
2. Causality, which asserts that a musculoskeletal condition may have distant causative sites.
3. Totality, which dictates that a body should be treated as a whole.

Additionally, GPR takes into account the existence of many muscle chains, which are made up of linked muscles that form a continuum along the body and have distinct functional responsibilities. The anterior diaphragmatic chain and the posterior static chain are the two primary muscle chains. These ideas lead to the assumption that retractions in the muscle chains could cause pathological diseases. As a result, unique static postures are applied to each patient in order to lengthen shortened muscle chains and improve the antagonists' co-contraction. The ultimate objective of this strategy is to promote postural symmetry, which is thought to mediate the reduction of pain and impairment, by lengthening the shortened muscles and strengthening the contraction of the antagonists. Research has been done on the clinical application of GPR for TMJ disorders, low back pain, neck discomfort, and ankylosing spondylitis.

Bracing

The purpose of bracing in correction of postural deformities is to keep the body upright and prevent progression of the curve while the patient is growing and awaiting possible need for operative intervention. This is achieved through stabilization and maintenance of spinal alignment; and prevention correction of spinal deformities.

Principles

- **Three-point pressure control system:**
  1. **Support around the circumference:** When the orthosis encloses the trunk, it creates a semirigid cylinder that encloses the vertebral column and bridges the gap between the pelvic brim and the lower rib cage. Moreover, the abdominal contents are compressed.
  2. **Irritant:** The orthotic device is designed to compel the user to adopt the preferred posture in order to alleviate pain (kinesthetic feedback) or to be reminded to voluntarily limit motion.
  3. **Skeletal fixation:** Orthotic devices have been shown to reduce spinal segment mobility.
- All orthotic devices must have a three-point pressure control system, at the very least. Three points of contact with evenly balanced opposing forces in a certain plane are necessary for this.
- A three-point pressure system generates a corrective moment (force).
- The "Law of Equilibrium" states that the forces on each side of a structure must be equal or balanced. There are tissues that can withstand pressure and tissues that cannot, such as those covering nerves or bony prominences.
- Pressure tolerance can be altered by adjusting the size of the contact pad or the lever arm's length.
- In essence, orthotic devices are lever arms that generate corrective angular forces.
- The total force divided by the area of application of force yields the pressure over a given area.

SUMMARY

- Posture is the body's attitude, supported by muscle activity or an adaptable foundation.
- It can be inactive or active, and control mechanisms involve the coordination of sensory, skeletal, muscular, and central neurological systems.
- The vestibular system provides information on linear acceleration, while the somatosensory system conveys information about the body's position through muscle spindles, Golgi tendon organs, joint receptors, skin and subcutaneous tissue.
- The postural control system is a mechanism that uses coordinated muscle movement to maintain balance and stability. It involves the ankle, knee, hip joints, calf muscles, and spine support. Neurological control, coordinated by the central nervous system, allows posture adaptation.
- Posture is crucial for health and well-being, involving muscle flexibility, joint motion, and alignment of body parts. Standing, sitting, sleeping, and car entry require specific postures and techniques.
- Poor posture, caused by unnatural spine positions, can cause stress on joints, muscles, and vertebrae, leading to pain, fatigue, and muscle imbalances. Factors include mental attitudes, illness, and occupational stresses. Treatment involves addressing health and muscle development factors, maintaining a stable psychological background, promoting good posture, and using tools like mirrors and posture recorders.
- Postural assessment is essential for pain management and promoting a healthy lifestyle. It involves evaluating muscles, joint stiffness, and imbalances.
- Postural assessment is crucial in physiotherapy, identifying musculoskeletal defects and preventing injuries using various methods.
- Postural re-education, training, relaxation, mobility exercises, corrective exercises, manual therapy, soft tissue massage, joint mobilization, ergonomic work states, and stretching are recommended. Treatment involves physiotherapists, patient cooperation, individual instruction, and group activities.
- Strengthening exercises target specific muscles, while stretching exercises improve function and movement quality.
- Global Postural Re-education (GPR) is a method developed by Philippe Suchard in 1981, used in countries like Brazil, Spain, France, and Portugal. It focuses on individuality, causality, and totality, aiming to improve postural symmetry and reduce pain and disability.

FURTHER READINGS

- Ferreira GE, Barreto RG, Robinson CC, Plentz RD, Silva MF. Global Postural Reeducation for patients with musculoskeletal conditions: A systematic review of randomized controlled trials. Braz J Phys Ther. 2016 Apr 1;20(3):194–205. doi: 10.1590/bjpt-rbf.2014.0153. PMID: 27437710; PMCID: PMC4946835.
- Gardiner Dena M. Principles of Exercise Therapy. CBS Publishers and Distributors Pvt Ltd; 4th ed., 2023.
- Moreira R, Teles A, Fialho R, Baluz R, Santos TC, Goulart-Filho R, Rocha L, Silva FJ, Gupta N, Bastos VH, Teixeira S. Mobile applications for assessing human posture: A systematic literature review. Electronics. 2020 Jul 25;9(8):1196.Goniometry

STUDENT ASSIGNMENT

LONG ANSWER QUESTIONS

1. Explain posture and its types. Describe the postural control mechanism.
2. Define abnormal posture and mention its types.
3. Discuss the methods of assessment of abnormal posture.
4. Describe various strategies for posture correction.
5. Explain the purpose and principles of bracing of the trunk for correction of Cobb's angle.

SHORT ANSWER QUESTIONS

1. Define active and inactive postures.
2. What are the components of postural control system?
3. Define good and bad posture.
4. Write about the principles of posture-education.
5. Write about static and dynamic posture.
6. What is global postural re-education technique?

MULTIPLE CHOICE QUESTIONS

1. **Bad posture can affect the functioning of:**
   a. Musculoskeletal system
   b. Respiratory system
   c. Neurological system
   d. All of the above
2. **An awkward prolonged posture that may cause headache due to trigger points in the muscles involved is:**
   a. Forward head posture
   b. Rounded shoulders
   c. Shrugged shoulders
   d. All of the above
3. **Postural analysis can be done by means of:**
   a. Observation
   b. Photography
   c. Videography
   d. All of these
4. **Postural analysis can be done by:**
   a. Kyphometer
   b. Pelvic inclinometer
   c. Plumb line
   d. All of these
5. **Posture with hyper extended hips is known as:**
   a. Sway back posture
   b. Flat back posture
   c. Lordotic posture
   d. Kyphotic posture

ANSWER KEY

1. d **2.** d **3.** d **4.** d **5.** a

22 Gait

Sheetal Kalra

LEARNING OBJECTIVES

After the completion of the chapter, the readers will be able to:

- Define gait and the center of gravity of the human body.
- Describe the muscles responsible for normal gait and the six determinants of gait.
- Describe the gait cycle, including the stance phase and the swing phase.
- Describe various pathological gaits.
- Identify pathological gait and provide proper gait training.

CHAPTER OUTLINE

- Introduction
- Gait Cycle
- Determinants of Gait
- Pathological Gaits

KEY TERMS

Antalgic gait: It is a maladaptive walking pattern resulting from discomfort that eventually results in a limp, with a reduced stance phase compared to a longer swing phase.

Gait cycle: The cyclical pattern of movement that takes place during walking is referred to as the gait cycle. When one foot's heel meets the ground for the first time and then again, that foot's gait cycle comes to an end.

Parkinsonian gait: It is characterized by small, shuffling steps and generalized slowness of movement (hypokinesia) or in the most extreme circumstances, a total loss of movement (akinesia).

Pathological gait: It can have musculoskeletal and neuromuscular etiologies, is caused by reduced strength, range of motion, proprioception, discomfort or balance along with mechanical compensations.

Stance: The duration that the foot is in touch with the ground is known as the stance phase. Heel-strike (first foot-floor contact), foot-flat, mid-stance, push-off, and toe-off are the five stages that make up the stance phase.

Steppage gait: It is the result of weak muscles that produce the ankle joint's dorsiflexion, which prevents the foot from being raised during walking.

Swing: The second phase of gait during which, the foot can travel forward freely is called the swing phase. It is defined as the time between the heel strike and toe off.

Trendelenburg gait: Unilateral weakening in the hip abductors, which mainly affects the gluteal musculature, results in the Trendelenburg gait. A lesion in the fifth lumbar spine or injury to the superior gluteal nerve could be the cause of this disability.

Vaulting: It is a compensatory walking pattern in which an individual elevates himself on the toes of the unaffected leg to allow the other leg to swing through without touching the ground.

INTRODUCTION

Gait or the manner of walking can be characterized as a form of locomotion in which the two legs are used alternately for support and propulsion. In order to move the body forward while simultaneously preserving stance stability, formal walking involves a repetitive sequence of limb action.

GAIT CYCLE

One limb serves as a source of support while the other limb moves to a new support location as the body moves forward. Afterwards the limbs switch directions. Each limb reciprocally timed repeats this sequence of actions till the person reaches his target. *A gait cycle (GC) can be defined as cyclic patterns of moves that takes place during walking.*

Stance and swing, commonly known as gait phases, are the two segments that make up each gait cycle (Fig. 22.1). The entire time that the foot is on the ground is referred to as the stance. The period of time the foot is in the air during limb progress is referred to as the swing. Swing starts as soon as the foot is lifted from the floor (toe-off).

Each foot makes one ground contact (stance phase) throughout a gait cycle and remains on the ground for roughly 60–62% of the whole gait cycle. As a result, the portion of the gait cycle during which the foot is lifted off the ground (the swing phase) makes up around 38–40%. Walking involves two moments of double support, during which both feet are in contact with the ground. These occur during the initial and terminal phases of the stance period, collectively accounting for approximately 20–25% of the gait cycle at normal walking speeds. In contrast, running eliminates the double support phase and introduces a flight phase, where neither foot is in contact with the ground.

Phases

Two widely used terminologies for describing these phases are Traditional and Rancho Los Amigos (RLA). While Traditional terminology focuses on specific events, RLA terminology emphasizes functional tasks and dynamic processes.

In the stance phase, traditional terms like heel strike, foot flat, midstance, heel off, and toe off correspond to RLA terms: initial contact, loading response, midstance, terminal stance, and pre-swing. Similarly, in the swing phase, traditional terms (acceleration, midswing, and deceleration) align with RLA terms: Initial swing, midswing, and terminal swing. The RLA terminology is more descriptive of the body's motion, making it widely used in clinical and rehabilitation contexts. These terms are described below:

Initial Contact

The moment the foot makes initial contact with the ground, the gait cycle begins. Phases of gait cycle are shown in Figure 22.1. The heel strikes the ground and absorbs some of the impact power.

Loading Response

Following initial contact, the body weight shifts to the stance limb. The foot pronates slightly, which aids in shock absorption and adaption to uneven surfaces.

Midstance

This phase happens when the body's weight is directly over the stance leg and the foot is flat on the ground.

Terminal Stance

During this phase, the body moves forward over the stance limb. The heel lifts off the ground, and the forefoot and toes sustain the majority of the body weight.

Pre-Swing

This phase is the transition from single-limb support to swing. As the stance limb begins to push off the ground, the body's weight shifts to the opposing side.

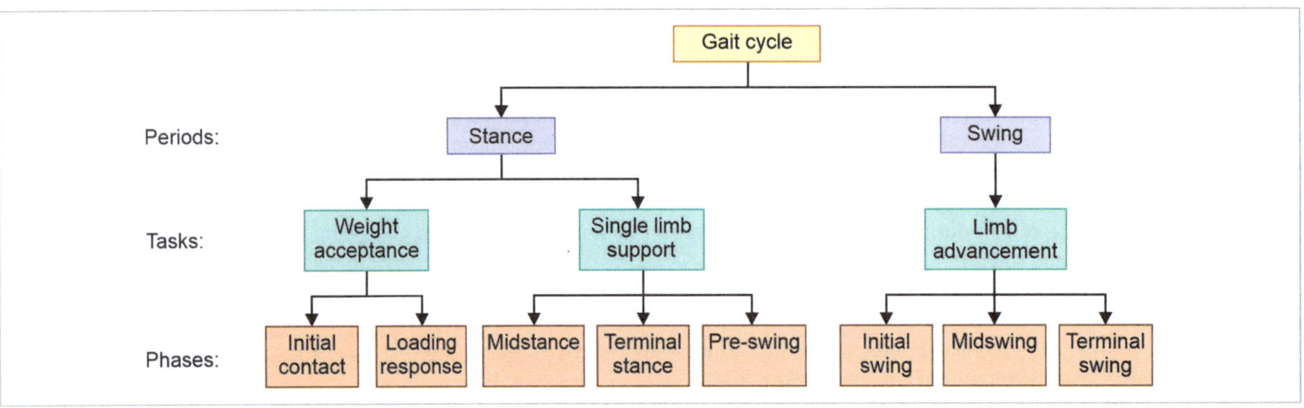

Fig. 22.1: Phases of gait cycle

Initial Swing

As the foot leaves the ground, the limb swings forward in preparation for the following stride. This phase requires bending at the hip, knee, and ankle joints to clear the ground.

Midswing

The limb continues to swing forward, achieving maximum flexion. This phase is distinguished by the limb passing beneath the body with minimum impediment.

Terminal Swing

As the limb prepares for initial contact, it extends, preparing to make contact with the ground again.

Kinematics and Kinetics

In gait analysis, the kinematic system records body segment position and orientation, joint angles, linear and angular velocity, and acceleration.

Stance Phase

Heel Strike or Initial Contact (0–2% of GC)

- At heel strike, the hip is slightly flexed (about 20–30°).
- The knee is extended with slight flexion (0–5°).
- The ankle is dorsiflexed.

Foot Flat or Loading Response (2–12% of GC)

- The hip moves slowly into extension as a result of a contraction of the adductor magnus and gluteus maximus muscles. The angle of hip flexion decreases to 15°.
    - Knee flexion increases to 15–20°.
    - Ankle plantarflexion rises to 5°.
    - The body pronates to absorb the impact of the foot.

Midstance (12–31% of GC)

- The gluteus medius muscle contracts to cause the hip to move from a 15° flexion to an extension position (0°).
- Knee has 5° of flexion before starting to extend.
- The body is supported by one single leg during this phase, and the ankle becomes supinate and dorsiflexed (5°).
- The body now switches from absorbing force at impact to being propelled forward by force.

Heel off or Terminal Stance (31–50% of GC)

- Starts when the heel lifts off the ground.
- The metatarsal heads distribute the weight of the body.
- Hip hyperextension of 10–20° followed by flexion.
- Knee flexes (0–5°)
- Plantar flexion and supination of the ankle.

Toe off or Pre-Swing (50–60% of GC)

- Hip extension slightly decreases up to 10°
- Knee flexes up to 30°
- Plantar flexion at ankle increases up to 20°.

Swing Phase

Acceleration or Initial Swing (60–75% of GC)

- The hip flexes by 20° with lateral rotation
- Knee flexion ranges from 40° to 60°.
- The ankle moves from 20° of plantar toward dorsiflexion (10° plantar flexion) before returning to neutral.

Midswing (75–85% of GC)

- During midswing, the hip is flexed at approximately 20–30°.
- The knee flexion reduces to 30°.
- The ankle is in a neutral position.

Terminal Swing or Deceleration (85–100% of GC)

- Hip flexion of 25–30°
- Locked extension of the knee (0°)
- Neutral position of the ankle

Percentage of gait cycle for both Traditional and RLA classification is depicted in (Figs 22.2A to C). The various joint positions during stance and swing phase are shown in (Fig. 22.3). Swing phases percentages in the GC is depicted in (Fig. 22.4).

Muscle Activity

Overview of muscle activity during different phases of gait cycle is shown in (Fig. 22.5).

At Hip Joint

During the swing phase of gait, the hip joint flexes, and during the stance phase, it extends. The hip flexor muscles cause hip flexion, which in turn causes the lower extremities to swing forward. The iliopsoas muscle is primarily responsible for this motion. The hamstring muscles' eccentric contraction facilitates this motion (biceps femoris, semimembranosus, semitendinosus). Gluteus maximus and hamstring muscles, which are hip extensors, work to control the body's forward momentum when the weight is transferred forward over the stance leg and to stabilize the pelvis during shock absorption. The gluteus medius and gluteus minimus, which make up the majority of the hip abductor muscles, help to maintain the pelvis and prevent pelvic lowering to the side of the free leg.

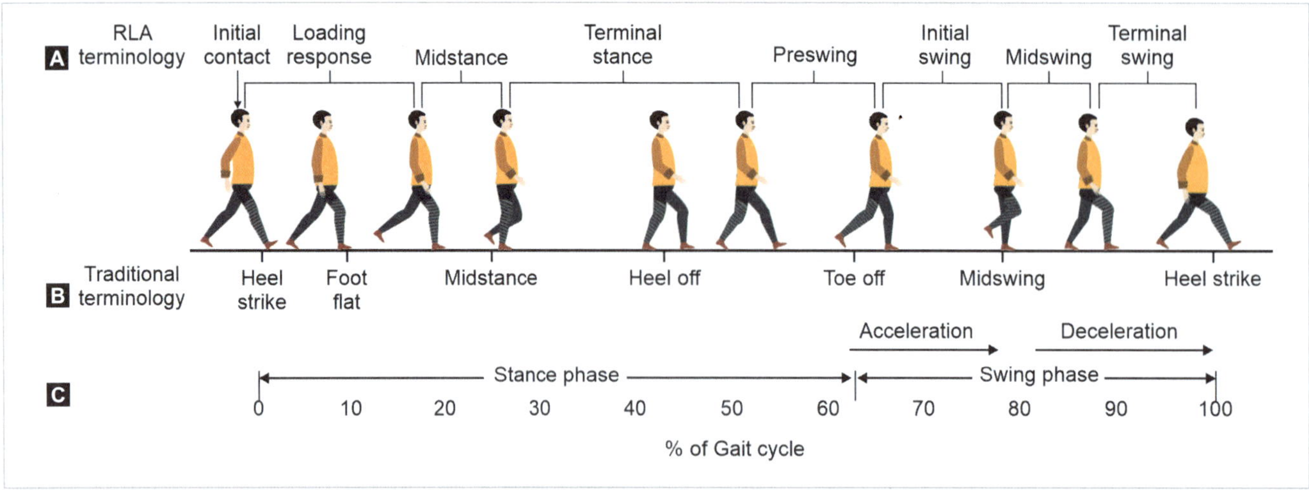

Figs 22.2A to C: Phases with percentage of gait cycle

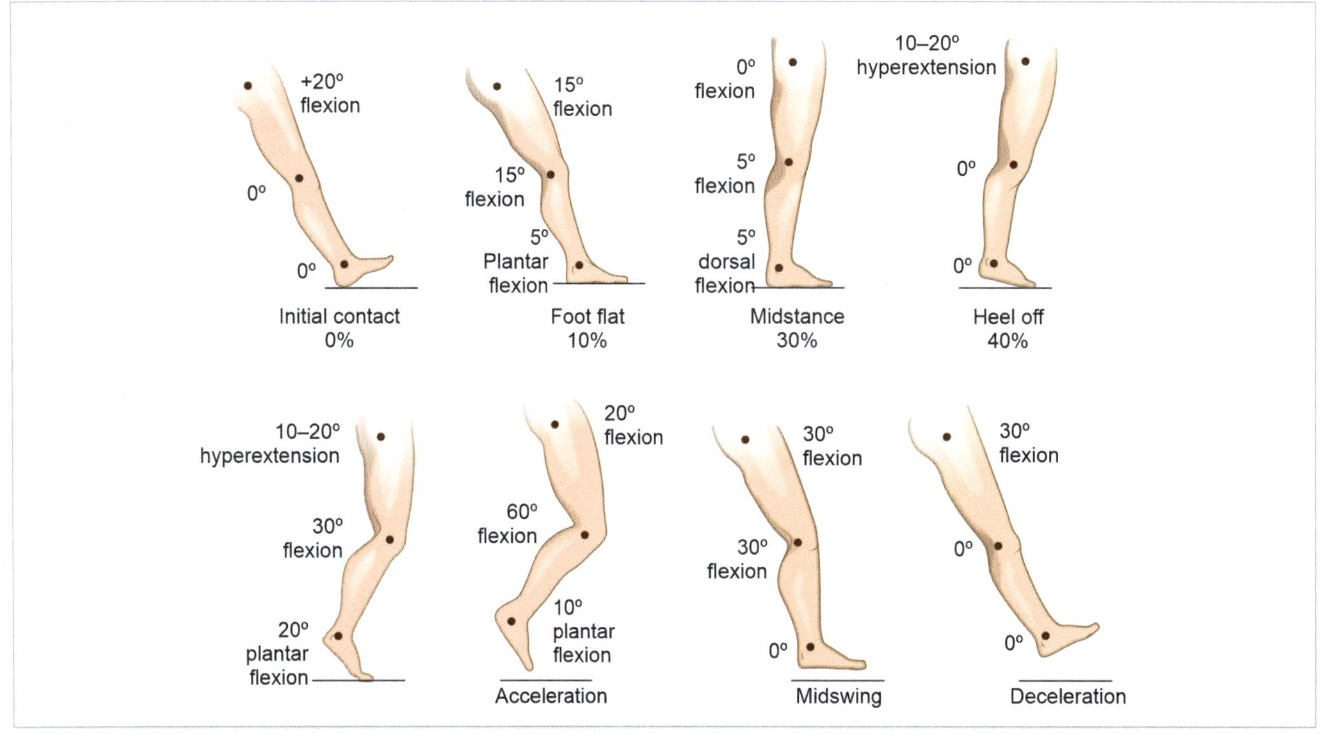

Fig. 22.3: Joint motions during stance and swing phases

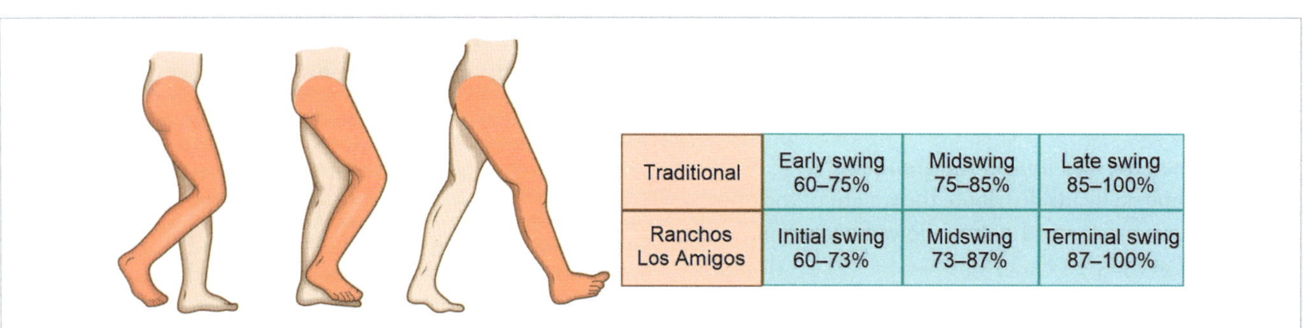

Traditional	Early swing 60–75%	Midswing 75–85%	Late swing 85–100%
Ranchos Los Amigos	Initial swing 60–73%	Midswing 73–87%	Terminal swing 87–100%

Fig. 22.4: Swing phase and percentages of gait cycle

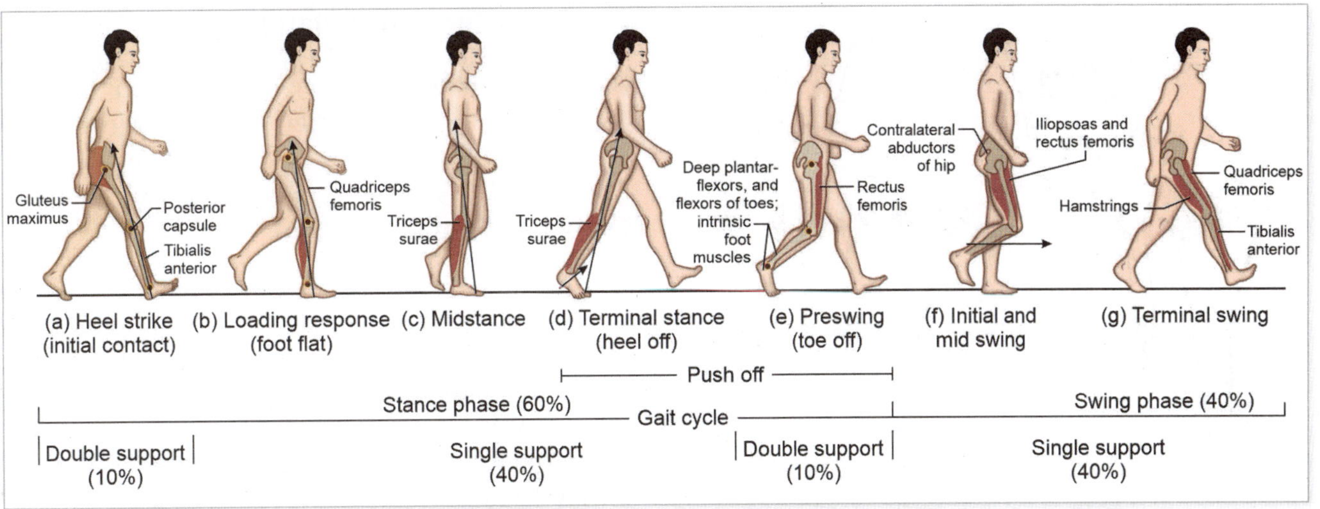

Fig. 22.5: Muscle activity during different phases of gait cycle

The hip adductor muscles regulate the balance in the weight-bearing leg.

At Knee Joint

Concentric contraction of the quadriceps and eccentric contraction of the hamstring muscles hold the knee in a stable extended position as the foot makes contact with the ground during heel strike. To enable steady weight-bearing by the lower limb throughout the stance phase, the knee stays (predominantly) extended the entire time. The knee, however, exhibits a slight "flexion wave" movement as the lower leg absorbs shock as it comes into touch with the ground. The knee transitions from extension to flexion and back to extension throughout the swing phase. To ensure controlled leg acceleration and deceleration, the knee flexor and extensor muscles work in coordination to produce this smooth motion.

At Foot and Ankle

As the foot swings through to establish heel contact with the ground, the ankle dorsiflexes. The foot then rolls forward into the flat position under control of an eccentric contraction of the ankle dorsiflexor muscles (tibialis anterior, extensor digitorum longus, extensor hallucis longus, and fibularis tertius). The gastrocnemius and soleus muscles subsequently exert forceful plantarflexion, producing propulsion throughout the heel-off and toe-off periods.

DETERMINANTS OF GAIT

The determinants of gait refer to a series of biomechanical adjustments made during walking that help minimize vertical and lateral movement of the body's center of gravity (COG). The key determinants are mentioned below:

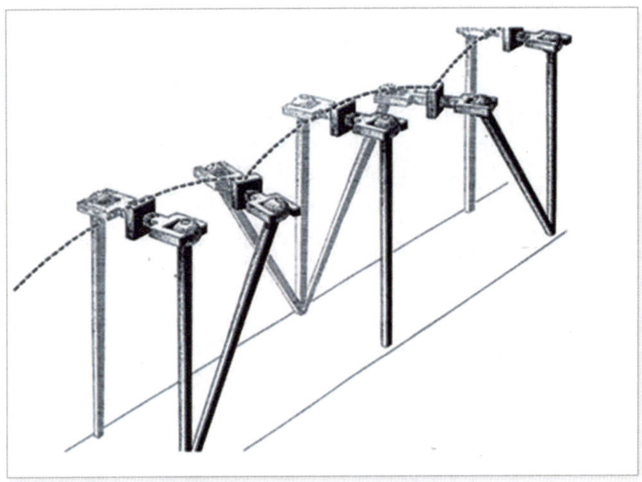

Fig. 22.6: Pelvic rotation

Pelvic Rotation

The pelvis rotates alternately to the right and left, averaging about 8° total. This rotation helps flatten the trajectory of the COG and minimizes vertical displacement by approximately 3/8 inches (Fig. 22.6).

Pelvic Tilt (Obliquity)

During normal walking, the hip of the swing leg drops lower than the hip of the stance leg, with a maximum tilt of about 5° at midswing. This adjustment lowers the COG by 1/8 inch, conserving energy through reduced vertical displacement (Fig. 22.7A).

Lateral Displacement of the Pelvis

The greatest lateral movement occurs on the side of the weight-bearing leg, typically measuring around 5 cm.

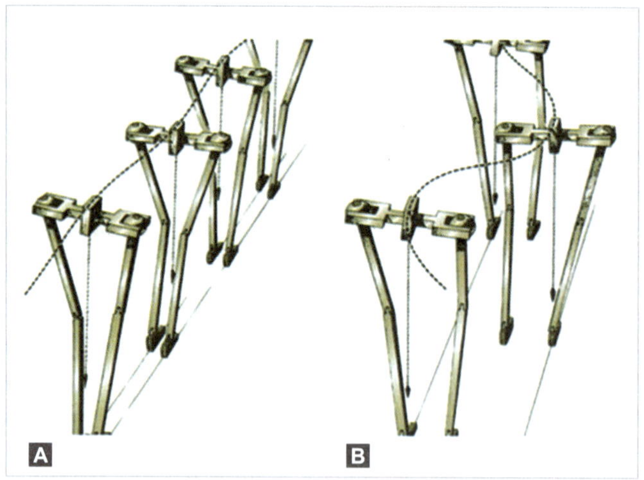

Figs 22.7A and B: **A.** Pelvic tilting and **B.** Lateral displacement of pelvis

Maintaining balance during this lateral shift is crucial to prevent falls(Fig. 22.7B).

Knee Flexion in Stance Phase

The knee flexes at heel strike and again during mid-stance, lowering the hip joint's height and aiding in shock absorption. This interaction among the foot, ankle, and knee minimizes COG displacement and energy expenditure (Figs 22.8A and B).

Trunk and Shoulder Rotation

The upper body also contributes to gait efficiency through coordinated rotations that counterbalance the movements of the lower body.

Ankle and Foot Mechanism

During the early stance phase, the knee is nearly fully extended with the foot dorsiflexed, maximizing the length of the lower limb and lowering the COG. As the stance progresses, the knee begins to flex and the foot plantar-flexes, helping maintain the COG's position.

Overall, the determinants operate independently and simultaneously to produce smooth vertical, horizontal and lateral sinusoidal path, which is the overall displacement of the path of the COG.

Clinical Correlation

Gait analysis serves as a fundamental tool for biomechanical assessment in individuals with lower limb biomechanical abnormalities such as overpronation (excessive inward rolling of the foot), supination (excessive outward rolling of the foot), leg length discrepancies, or abnormal joint angles.

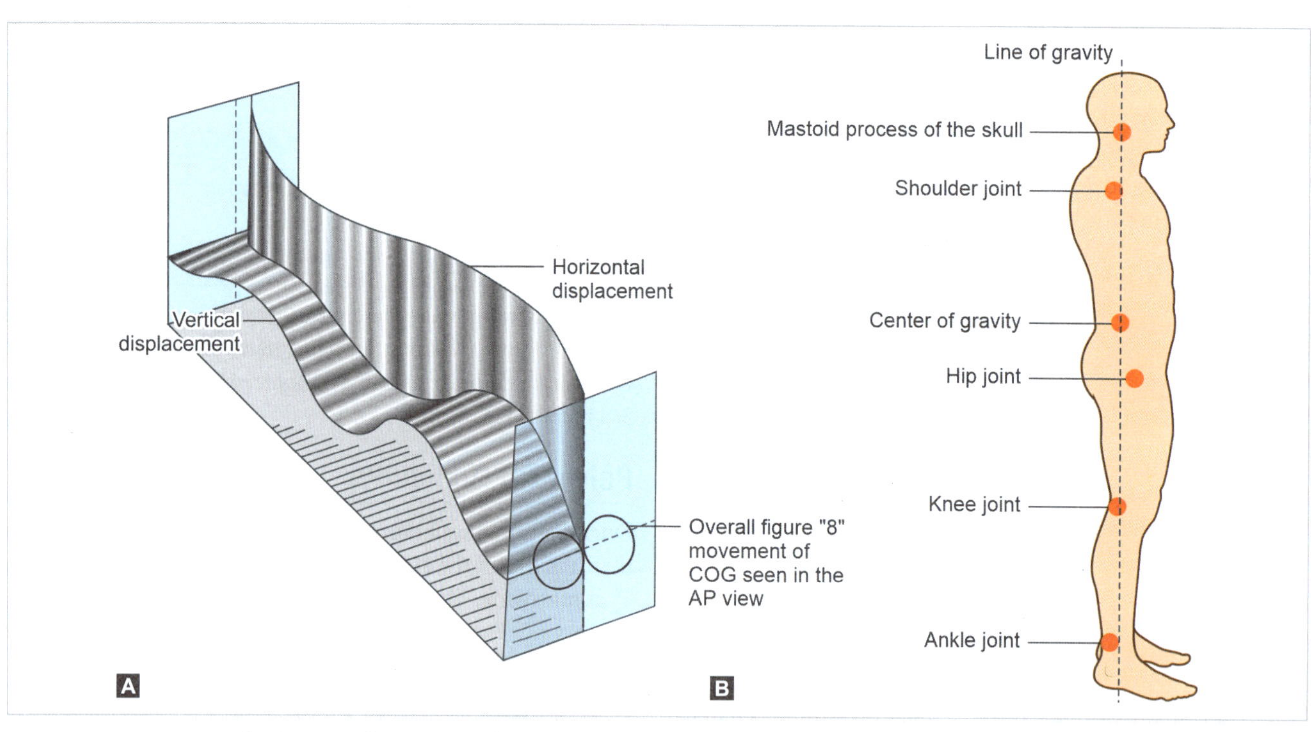

Figs 22.8A and B: **A.** Overall displacement of center of gravity; **B.** Path of center of gravity

PATHOLOGICAL GAITS

Pathological gait is characterized as an abnormal walking pattern brought on by physical weaknesses, deformities or other disabilities. The following table lists some of the causes of gait deviations as per different etiologies (Tekin et al, Fujisawa et al, Cavallieri et al, Gupta et al, Soh et al, Holroyd et al) (Table 22.1).

TABLE 22.1: Gait deviations as per different etiologies

Conditions	Descriptions
Musculoskeletal etiologies	
Injuries to the legs or feet	Trauma leading to pain, swelling, and altered movement patterns.
Arthritis	Joint inflammation causing pain, stiffness, and reduced range of motion.
Muscle contracture or tightness	Abnormal shortening of muscles, affecting movement.
Broken bones in feet and legs	Fractures causing pain and instability.
Tendonitis	Inflammation of tendons leading to pain and limited movement.
Shin splints	Pain along the shin bone due to overuse.
Soft tissue imbalance, joint alignment or bony abnormalities	Discrepancies affecting gait mechanics.
Deformities of foot	Structural abnormalities causing altered gait.
Pain in hip, knee or ankle	Discomfort leading to compensatory gait patterns.
Impaired strength, balance, range of motion or proprioception	Deficits leading to instability and altered movement patterns.
Congenital and developmental etiologies	
Birth defects	Congenital abnormalities affecting lower limb structure and function.
Length discrepancy	Differences in limb length due to asymmetrical pelvic, tibia, or femur length or other reasons such as scoliosis or contractures, resulting in pelvic dip and compensatory movements.
Neurological etiologies	
Cerebral palsy	Neuromuscular disorder affecting muscle tone and coordination.
Stroke	Neurological impairment leading to muscle weakness and spasticity.
Lesions of cerebellum or basal ganglia	Damage to brain areas controlling coordination and movement.
Conversion disorder or other psychological disorders	Mental health conditions manifesting as physical gait abnormalities.
Parkinson's disease	Shuffling gait with small steps, reduced arm swing, and postural instability.
Huntington's disease	Neurodegenerative disorder causing gait disturbances.
Normal pressure hydrocephalus	Condition affecting neurocognitive functions and gait.
Muscular dystrophy	Progressive muscle weakness affecting gait.
Charcot Marie Tooth disease	Peripheral nerve disorder causing gait abnormalities.
Ataxia-telangiectasia	Genetic disorder affecting coordination and gait.
Spinal muscular atrophy	Genetic disorder leading to muscle weakness and gait issues.
Peroneal neuropathy	Nerve damage leading to muscle weakness and altered gait.
Microvascular white-matter disease	Small vessel disease in the brain affecting gait.
Infectious etiologies	
Infections in the soft tissue of the legs	Inflammation and pain affecting gait.
Infections in the inner ear	Affecting balance and coordination, leading to altered gait.
Electrolyte imbalances	
Hyponatremia	Causes severe neurological symptoms affecting gait.
Hypokalemia	Electrolyte disorder affecting gait.
Hypomagnesemia	Electrolyte imbalance impacting gait.

Contd...

Conditions	Descriptions
Vitamin deficiencies	
Folate deficiency	Causes neurological dysfunctions affecting gait.
Vitamin B_{12} deficiency	Causes subacute combined degeneration of the spinal cord, leading to gait disturbances.
Vitamin E deficiency	Results in neurological issues affecting gait.
Copper deficiency	Leads to neurological dysfunction impacting gait.

Clinical Correlation

By analyzing the characteristics of pathological gaits, clinicians can localize lesions within the central nervous system (CNS) or peripheral nervous system (PNS). For instance, a spastic gait with stiff, jerky movements and increased muscle tone typically indicates upper motor neuron lesions, whereas a flaccid gait with weak, floppy movements suggests lower motor neuron lesions.

Trendelenburg Gait

Trendelenburg gait is characterized by the lateral bending of the trunk toward the supporting limb during the stance phase of walking. It usually suggests weakness or dysfunction in the hip abductor muscles, including the gluteus medius and minimus or problems with the hip joint itself.

Pathophysiology

The hip joint and its abductor mechanism act as a class 1 lever, with the effort (muscle force) and load (body weight) on opposite sides of the fulcrum (hip joint). Any disorder affecting the fulcrum, lever or effort produces a positive Trendelenburg sign and gait. Typically, during the swing phase of gait, the body is unsupported on one side, causing the pelvis to sink. To prevent this, the hip abductors contract and support the pelvis and trunk. Damage or weakening in this process causes a pelvic drop on the opposite side, which is known as a positive Trendelenburg sign. The Trendelenburg gait results from repeated pelvic dropping during the stance phase (Fig. 22.9). To compensate the patient laterally bends the trunk on the opposite side (lateral trunk bending gait).

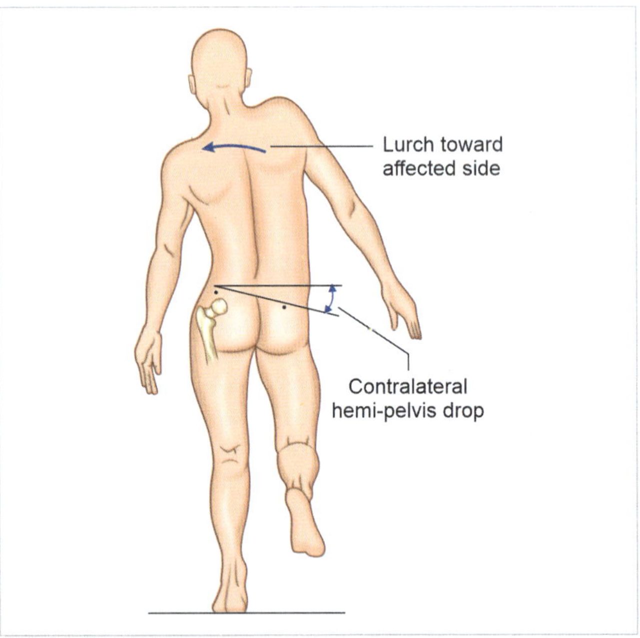

Fig. 22.9: Trendelenburg gait

Causes

The causes of trendelenburg gait are presented in Table 22.2.

Clinical Features

During mid stance, there is an apparent lateral hip movement toward the stance leg and a pelvic tilt toward the swing leg. This is most prevalent with mild to moderate isolated hip abductor weakness. Bilateral hip issues result in a waddling gait due to side-to-side trunk movement.

Treatment

Addressing Underlying Causes

Treating the primary pathology: Correcting the underlying condition causing the Trendelenburg gait is essential for long-term improvement and includes surgical and medical interventions as needed.

Physical Therapy

- Strengthening exercises for hip abductors:
  - Non-weight-bearing standing abduction
  - Weight-bearing standing abduction
  - Side-lying abduction
  - Resisted side-stepping exercises
  - Lateral side stepping and balance exercises
  - Using therabands for resistance training
- Weight-bearing exercises have shown better functional recovery compared to non-weight-bearing exercises.
- Assistive devices: Use of assistive devices in the contralateral hand helps develop a torque opposing the pelvic drop, stabilizing the gait.

TABLE 22.2: Causes of Trendelenburg gait

Causes	Specific conditions
Failure of the fulcrum	Osteonecrosis of the hip
	Legg-Calvé-Perthes disease
	Developmental dysplasia of the hip
	Chronically dislocated hips secondary to trauma
	Chronically dislocated hips secondary to infections (e.g., tuberculosis of the hip)
Failure of the lever mechanics	Greater trochanteric avulsion
	Non-union of the neck of the femur
	Coxa vara
Failure of effort (weakness of abductor muscles)	Poliomyelitis
	L5 radiculopathy
	Superior gluteal nerve damage
	Gluteus medius and minimus tendinitis
	Gluteus medius and minimus abscess
	Post-total hip arthroplasty
Additional causes	Aching hip due to osteoarthritis or rheumatoid arthritis
	Congenital hip dislocation, coxa vara, slipped femoral epiphysis
	Leg length discrepancy
	Proximal muscle weakness from conditions like plexopathies, myopathies, anterior horn cell disease
	Prolonged bed rest
	Upper motor neuron (UMN) gait disorders

Quadriceps Gait

Knee extensor weakness, which mostly affects the quadriceps muscles, causes gait abnormalities such as knee buckling (uncompensated) and genu recurvatum (compensated). This syndrome causes trouble maintaining knee stability during the stance phase of walking, typically necessitating compensatory actions to avoid knee collapse (Fig 22.10).

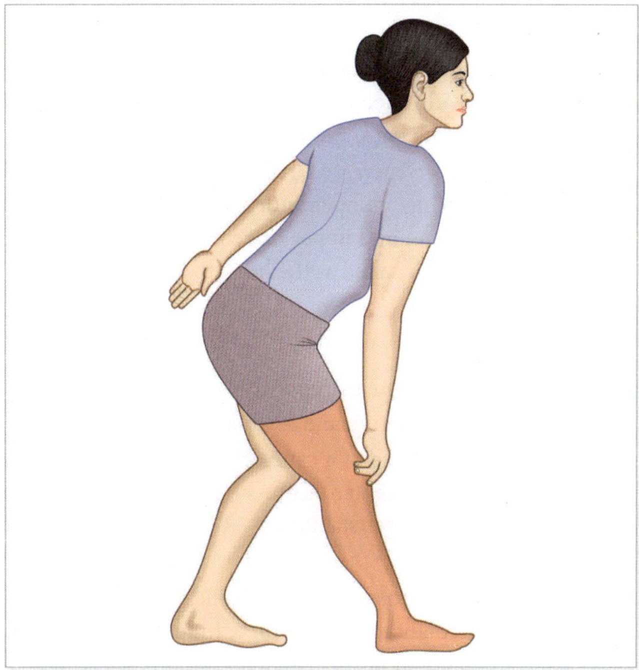

Fig. 22.10: Quadriceps gait

Pathophysiology

During normal walking, the quadriceps eccentrically contract upon heel strike to regulate limb loads and prevent excessive knee flexion. When the quadriceps are weak, the knee may remain completely extended during the stance phase to prevent buckling. In early stance, an anterior trunk lean can shift the line of gravity closer to the axis of rotation of the knee, allowing the knee to remain extended without the need of the knee extensors. This compensatory mechanism may cause excessive stretching of the posterior knee capsule, potentially causing further complications. With mild to severe quadriceps weakness, knee flexion is reduced or abolished, and the knee is extended at or before heel strike. Vigorous extension may result in the knee snapping back into hyperextension. Normal hip extensor and plantar flexor strength can help preserve knee extension using closed kinetic chain dynamics. Severe quadriceps weakness raises the risk of knee instability and collapse, necessitating extra compensatory techniques, such as upper extremity support and forward trunk lean, to maintain knee posture and ensure the ground reaction force (GRF) line passes anterior to the knee.

Causes

The causes of quadriceps gait due to knee extensor weakness are given in Table 22.3.

TABLE 22.3: Causes of quadriceps gait due to knee extensor weakness

Causes	Specific conditions
Traumatic	Femoral neuropathy
	Lumbar plexopathies
Neurological	L3/4 radiculopathies
	Diabetic amyotrophy/mononeuropathy
	Anterior horn cell (AHC) diseases

Clinical Features

- **Mild to moderate weakness:** Reduced or eliminated knee flexion with knee extension at or before heel strike, potential snapping back into full extension.
- **Severe weakness:** Increased risk of knee instability and collapse, requiring compensatory mechanisms such as forward trunk lean and upper extremity support for knee control (this is also called as the hand to knee gait).

Treatment

- **Physical therapy:** Quadriceps strengthening exercises to improve muscle function and knee stability.
- **Assistive devices:** Use of a cane or other upper extremity aids to transfer the center of mass (COM) anterior to the knee, providing greater stability.
- **Orthotic devices:**
  - **Ankle foot orthosis (AFO):** A solid AFO set at a few degrees of plantar flexion or a hinged AFO with a neutral dorsiflexion stop can help stabilize the knee.
  - **Bracing:** For cases of recurvatum or total paralysis, bracing is often necessary to maintain knee stability.

Gluteus Maximus Gait

Gluteus maximus gait or posterior trunk bending gait, is a compensatory gait pattern where the individual leans the trunk backward during the stance phase of walking (Fig. 22.11). This compensatory movement occurs primarily due to weakness or paralysis of the hip extensors, particularly the gluteus maximus muscle.

Pathophysiology

- **Compensation for hip extensor weakness:** During the early stance phase, normally the ground reaction force (GRF) passes in front of the hip joint, creating a flexor moment opposed by the hip extensors. If the hip extensors are weak or paralyzed, the trunk moves backward, shifting the GRF line behind the hip joint axis. This adjustment reduces the external flexor moment, thus decreasing the need for hip extensor muscle activity to maintain joint stability.
- **Swing phase adaptation:** In cases of hip flexor weakness or hip extensor spasticity, the trunk may be thrown backward at the start of the swing phase to help propel the leg forward. This maneuver compensates for the difficulty in accelerating the femur.
- **Hip ankylosis:** When the hip is fused, backward trunk movement helps in advancing the thigh forward during walking.
- **Loading response:** The backward lean during loading response inclines the GRF vector posteriorly, shortening its moment arm and reducing the hip flexor moment.

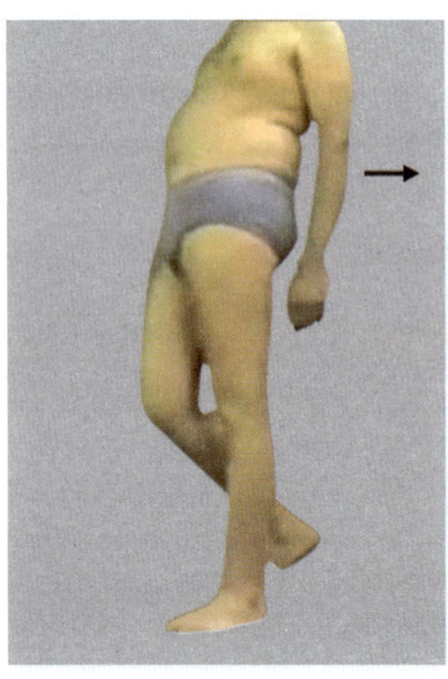

Fig. 22.11: Gluteus maximus gait

Causes

- **Injury to gluteal nerves:** Damage to the nerves supplying the gluteus maximus.
- **Proximal weakness:**
  - Plexopathies
  - Myopathies
  - Anterior horn cell (AHC) diseases
  - Myelodysplasia

Clinical Features

- **Early stance phase:** Backward trunk lean compensates for inadequate hip extensor activity, reducing the forward flexor moment at the hip.
- **Swing phase:** Trunk thrown backward to assist in leg propulsion when there is hip flexor weakness or knee extension limitation.
- **Overall walking speed:** Decreased walking speed to minimize forward momentum, especially when weakness is bilateral.
- **Lumbar extension:** Increased lumbar extension to relocate the center of mass posteriorly, aiding in hip stability.
- **Backward thrust:** Noticeable backward thrust of the trunk at heel strike, particularly when hip extension is limited.

Treatment

- Physical therapy includes strengthening exercises which are focused on hip extensors and overall proximal muscle groups to improve muscle function.
- **Assistive devices:** Use of upper extremity assistive devices like canes or walkers to provide trunk stability and enhance walking safety.

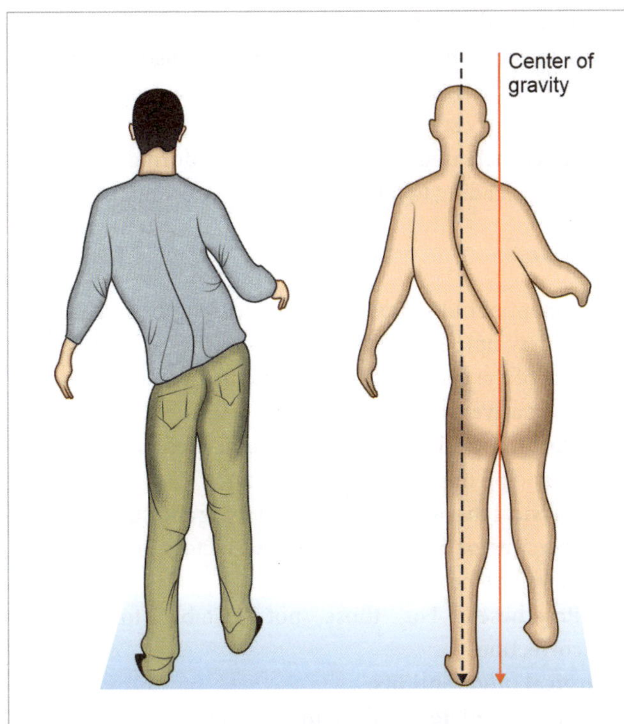

Fig. 22.12: Antalgic gait

- **Gait training:** Techniques to optimize compensatory strategies and reduce the risk of falls.

Antalgic Gait

Antalgic gait is a walking pattern characterized by a pain-induced limp (Fig. 22.12). This gait adaptation features a shorter stance phase compared to the swing phase in order to reduce pain and prevent putting undue strain on problematic areas. It is one of the most common types of altered gait seen in emergency rooms and general care clinics.

Pathophysiology

Antalgic gait is a compensatory technique that reduces discomfort by limiting the weight and movement on the affected limb. This gait change is caused by pain, which can range from minor discomfort to severe reactions such as flexor withdrawal. The patient automatically changes his walking pattern to reduce joint compressive pressures and limit muscle activation, which exacerbates pain.

Causes

- **Traumatic:** Sprains, fractures, soft tissue damage, bone injuries
- **Infectious:** Joint infections, osteomyelitis
- **Inflammatory:** Osteoarthritis (OA), degenerative joint disease (DJD), rheumatoid arthritis
- **Vascular:** Peripheral arterial disease, deep vein thrombosis
- **Neoplastic:** Bone tumors, metastatic disease
- **Other:** Heel spurs, sciatica, lumbar radiculopathy

Clinical Features

- **Shortened stance phase:** Reduced duration of weight-bearing on the affected limb.
- **Reduced walking speed:** Slower overall gait to minimize pain.
- **Asymmetry:** Noticeable difference in the gait cycle between the affected and unaffected limbs.
- **Lack of weight shift:** Minimal lateral weight shift over the stance limb to avoid pain.
- **Stiffened limb:** Tendency to keep the affected limb stiff to reduce joint excursion.
- **Decreased push-off:** Less forceful foot contact and push-off from the affected limb.

Treatment

Management of antalgic gait focuses on identifying and treating the underlying cause of pain. Treatment options according to management, may include:

- **Traumatic injuries:** Immobilization with splints or casts, use of crutches or walking boots, ice, elevation, and non-steroidal anti-inflammatory drugs (NSAIDs).
- **Inflammatory conditions:** NSAIDs, physical therapy, rheumatology follow-up, orthotics for deformities, weight loss to reduce joint stress.
- **Vascular conditions:** Specialist referral, blood-thinning medications.
- **Infections:** Prompt diagnosis, orthopedic consultation, joint aspiration, intravenous antibiotics, possible surgery.
- **Neoplastic causes:** Oncology and neurosurgical consultation, appropriate oncologic treatments.

Gait Deviation Due to Functional Leg Length Discrepancy

Leg length discrepancy (LLD) refers to a condition where there is a notable difference in the lengths of an individual's lower extremities. This discrepancy can be either structural, due to actual differences in bone lengths, or functional, arising from factors such as joint contractures or asymmetries in muscle strength.

Pathophysiology

The presence of LLD leads to biomechanical compensations in the body to maintain balance and functionality. Individuals develop strategies involving the feet, ankles, knees, hips, and pelvis to minimize energy expenditure:

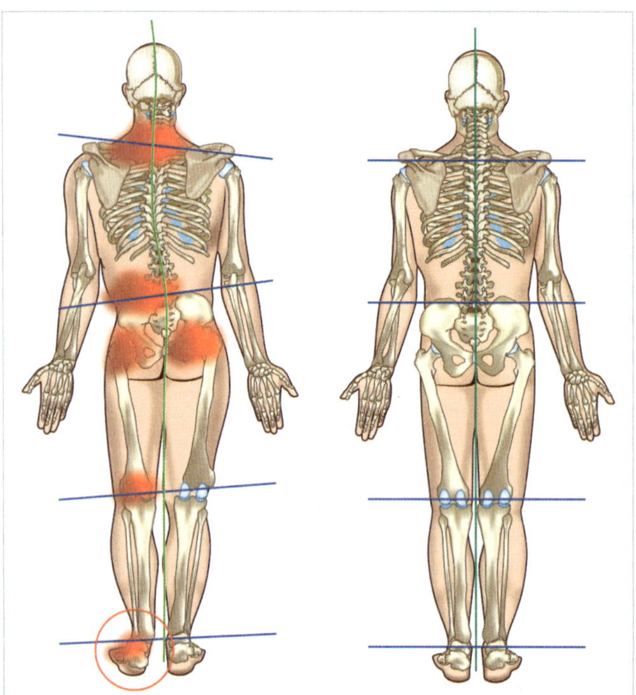

Fig. 22.13: Deviation due to functional leg length discrepancy

- **Biomechanical changes:** A discrepancy of 3 cm results in a significant shift in the center of pressure during standing, and changes in gait such as increased step length and width of the longer leg, and altered step time.
- **Compensatory mechanisms:** The body may adjust through pelvic tilt, spinal curvature, and altered gait patterns to balance the leg lengths. Over time, these adjustments can lead to secondary complications such as osteoarthritis, scoliosis, intervertebral disc degeneration, and increased stress on the joints and soft tissues (Fig. 22.13).

Causes

The LLD can be attributed to a variety of congenital and acquired factors:

- **Congenital causes:** These include conditions present at birth such as hemihypertrophy, congenital hip dislocation, and limb deficiency syndromes.
- **Acquired causes:** These develop later in life due to reasons such as trauma (fractures, growth plate injuries), infections (osteomyelitis), tumors, or surgical interventions (total hip arthroplasty, growth plate surgeries).

Clinical Features

Patients with LLD may present with a variety of symptoms and signs, including:

- **Gait abnormalities:** Noticeable limping, increased step length and width on the longer leg, and altered cadence.
- **Pain and discomfort:** Chronic pain in the lower back, hips, knees, or feet due to uneven weight distribution.
- **Postural changes:** Pelvic tilt, functional scoliosis, and compensatory spinal curvature.
- **Fatigue:** Increased energy expenditure during walking or standing, leading to fatigue.

Treatment

Management of LLD aims to equalize leg lengths and alleviate associated symptoms. Treatment approaches vary based on the extent of the discrepancy and patient-specific factors:

- **Conservative treatments:**
  - **Shoe lifts and inserts:** Useful for discrepancies between 20 and 50 mm to balance leg length.
  - **Physical therapy:** Includes stretching and strengthening exercises for the hip abductors, adductors, and extensors, as well as for the back.
  - **Prostheses:** For those not suitable for surgical correction.
- **Surgical interventions:**
  - **Growth plate modulation:** Epiphysiodesis to slow growth in the longer leg.
  - **Leg lengthening procedures:** Osteotomy followed by distraction osteogenesis using either internal or external fixators for larger discrepancies.

Hip Hiking and Circumduction

Hip hiking and circumduction are compensatory gait patterns used to achieve appropriate foot clearance during the swing phase of walking in the presence of limitations.

Pathophysiology

- Hip hiking is defined as maintaining an unaffected coronal hip and/or pelvic angle while the affected leg is in midswing. The lateral abdominal wall and spinal muscles contract, raising the pelvis on the side of the swinging leg (Fig. 22.14A).
- Circumduction is defined as a higher-than-normal coronal thigh angle during the affected limb's midswing. Patients with a circumduction gait abduct their thigh and swing their leg in a semicircular motion to clear the foot, which is commonly caused by insufficient hip or knee flexion or ankle dorsiflexion (Fig. 22.14B).

Causes

The primary causes of hip hiking and circumduction gait patterns include:

- **Hip flexion difficulty:** Due to weak hip flexors, possibly from L2–L3 nerve compression or upper motor neuron lesions.

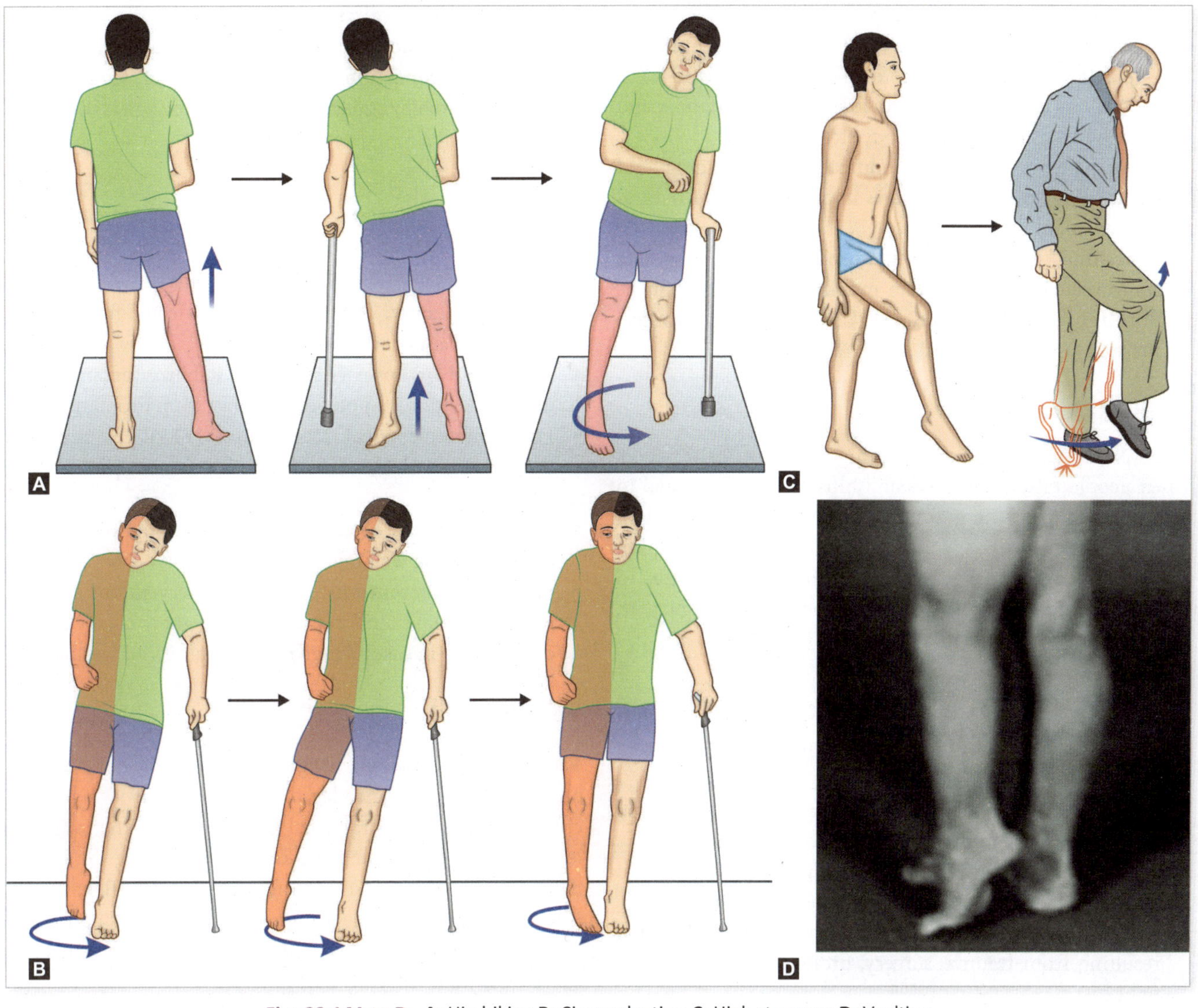

Figs 22.14A to D: **A:** Hip hiking **B:** Circumduction **C:** High steppage **D:** Vaulting

- **Knee flexion difficulty:** Impaired knee flexion can prevent the foot from clearing the ground during the swing phase.
- **Ankle dorsiflexion difficulty:** Issues such as foot drop can impede the ability to lift the foot, necessitating compensatory movements.
- **Neurological conditions:** Conditions like cerebral palsy, multiple sclerosis, or hemiplegia post-stroke often result in these gait patterns.
- **Leg length discrepancy (LLD):** LLD can cause uneven limb movement, leading to circumduction to compensate for the shorter leg.

Clinical Features

- **Hip hiking gait:** The pelvis on the affected side visibly elevates during the swing phase, creating an exaggerated upward movement. This is often accompanied by reduced knee flexion or ankle dorsiflexion, making the gait appear asymmetrical.
- **Circumduction gait:** The affected leg moves in a wide, circular arc during the swing phase. This gait pattern is often observed with stiff knees, weak hip flexors, or foot drop, leading to lateral trunk movement for balance.

Treatment

Treatment aims to improve gait quality and address the underlying causes:

- **Physical therapy:**
  - **Muscle strengthening and range of motion exercises:** Targeting the hip, knee, and ankle to improve overall function.
  - **Gait training:** Focused exercises to improve walking patterns.
  - **Robot-assisted training:** To enhance functional movement.

- **Orthosis and assistive devices:**
  - **Ankle-foot orthotics (AFOs):** To prevent excessive plantar flexion and promote better foot contact.
  - **Walkers and canes:** To provide additional support and stability, utilizing upper body strength.
  - As patients improve, modifications to bracing and other devices may be made to facilitate more natural movement patterns.
- **Functional electrical stimulation (FES):** To stimulate appropriate muscle activation and improve gait mechanics.
- **Botulinum toxin injections:** Used to reduce muscle spasticity in cases where it is a significant factor.

High-Steppage Gait

Neuropathic gait, also known as high-steppage gait, occurs when people raise their knees and hips higher than usual to keep their foot off the ground when walking (Fig. 22.14C). This gait pattern is largely used to compensate for foot drop, a condition in which the ankle cannot be dorsiflexed due to muscle weakness or paralysis, leading the foot to dangle with the toes pointed down.

Causes

Steppage gait can result from various underlying conditions and factors, including:

- **Foot drop:** The primary cause of steppage gait, characterized by weakness or paralysis of the anterior tibialis muscle.
- **Neuropathy:** Peripheral neuropathies, often due to diabetes, leading to unilateral or bilateral foot drop.
- **Nerve injuries:** Common peroneal nerve injury, often resulting from trauma, surgery, prolonged bed rest, or a tight cast.
- **Radiculopathy:** L5 radiculopathy can cause foot drop with preserved ankle inversion.
- **Neurological disorders:** Conditions such as cerebral palsy, multiple sclerosis, and upper motor neuron lesions, which may also present with hemiplegia and aphasia.
- **Collagen vascular diseases:** Leading to nerve ischemia and subsequent foot drop.

Clinical Features

Patients with steppage gait typically present with:

- **High-stepping motion:** Exaggerated knee and hip flexion during walking to prevent the toes from dragging.
- **Foot drop:** Inability to dorsiflex the ankle, leading to a hanging foot.
- **Slapping foot:** Audible sound as the foot lands due to lack of control.
- **Prolonged stance time:** Compared to individuals without gait abnormalities.
- **Weakness on examination:** Inability to lift the foot during physical tests, more evident during heel walking.
- **Associated conditions:** History of trauma, surgery, prolonged immobilization, diabetes, or neurological disorders.

Treatment

The treatment of steppage gait focuses on addressing the underlying causes such as, diabetes, neuropathy, or other contributing factors and improving mobility:

- **Physical therapy:** To strengthen the affected muscles and improve overall gait mechanics.
- **Ankle-foot orthoses (AFOs):** To stabilize the ankle in a neutral position and prevent the foot from dropping.
- **Functional electrical stimulation (FES):** To stimulate the anterior tibialis muscle and promote proper foot movement.
- **Assistive devices:** Walkers and canes provide stability and support during walking.
- **Surgical interventions:** Surgical intervention such as posterior tibialis tendon transfer is sought to achieve active dorsiflexion in cases where nerve repair is not feasible.

Vaulting Gait

Vaulting gait is a walking irregularity in which a person raises on the toes of the stance leg to ensure that the swinging leg clears the floor. Under typical circumstances, the swinging leg "shortens" by bending the knee and elevating the foot. If the knee cannot bend and the foot cannot dorsiflex, the leg is too lengthy to clear the ground. Vaulting makes up for this by lifting the body, but it is an inefficient gait pattern that can cause early tiredness (Fig. 22.14D).

Causes

Vaulting gait can result from various underlying issues that prevent the swinging leg from shortening properly:

- **Spasticity in the foot:** Involuntary downward pointing of the foot, making the leg functionally longer.
- **Extensor synergy pattern:** Abnormal movement pattern causing the hip and knee to remain straight.
- **Foot drop:** Weakness or paralysis of the muscles responsible for dorsiflexion, resulting in the foot pointing downward.
- **Limb length discrepancy:** A difference in leg lengths, requiring the shorter leg to tiptoe during stance to allow the longer leg to swing through.
- **Prosthetic issues:** In amputees, problems such as a too-long prosthesis, poor suspension, or insufficient knee flexion can lead to vaulting.
- **Other factors:** Habit, fear of toe catching, and decreased confidence in knee flexion can also contribute to vaulting.

Did You Know?

Vaulting gait can also occur as a result of limb length discrepancy. When one leg is shorter than the other, individuals may exhibit a vaulting gait to compensate for the height difference.

Treatment

Treatment for vaulting gait focuses on addressing the underlying causes and improving gait efficiency:

- **Physical therapy:** Strengthening and flexibility exercises to improve knee and ankle function.
- **Orthotic devices:** Use of shoe lifts for limb length discrepancies or ankle-foot orthoses (AFOs) for foot drop.
- **Spasticity management:** Medications or interventions such as Botox injections to reduce muscle spasticity.
- **Prosthetic adjustments:** Ensuring proper fit and function of prosthetic limbs, including appropriate length and knee mechanics.
- **Surgical interventions:** In cases of significant limb length discrepancy or other structural issues, surgery may be considered to correct the underlying problem.
- **Gait training:** Comprehensive rehabilitation to improve overall walking mechanics and reduce compensatory patterns.

Gait Deviations Due to Abnormal Foot Contact

The foot may be unusually burdened if it only bears weight in one of its four quadrants. The following gait deviations are seen in case of abnormal foot contact:

Calcaneal Gait

Calcaneal gait, also known as heel-walking or ankle plantar flexor weakness gait, is a walking abnormality characterized by prolonged heel contact with the ground during terminal stance and a lack of push-off (Fig. 22.15). This gait pattern results from weakened or absent ankle plantar flexors, leading to increased reliance on the quadriceps to prevent knee buckling and an overall inefficient walking pattern.

Causes

Calcaneal gait can arise from various conditions that weaken the plantar flexors or alter their functions:

- **Achilles tendon issues:** Lengthening or rupture of the Achilles tendon, often due to injury, over-corrective surgery (common in cerebral palsy), or calcaneal fractures.
- **Neuromuscular diseases:** Conditions such as low-level spina bifida that affect the posterior muscle group.
- **Trauma:** Displaced calcaneal avulsion fractures causing slackness of the Achilles tendon.
- **Cerebral palsy:** Over 30% of patients with CP may develop calcaneal gait even without prior surgical intervention.

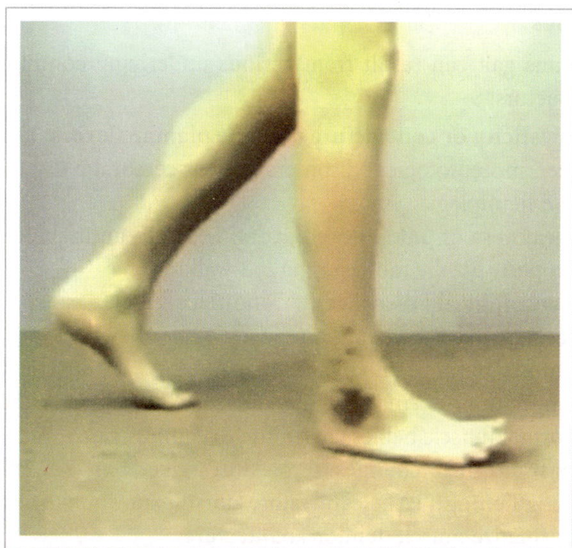

Fig. 22.15: Calcaneal gait

- **Weight gain and overstretching:** Significant weight gain and tendencies toward excessive passive dorsiflexion with the knee flexed can contribute to the development of calcaneal gait.

Treatment

Treatment for calcaneal gait focuses on managing symptoms and addressing the underlying causes:

- **Physical therapy:** To maintain or improve plantar flexor strength and prevent overstretching. This may include exercises specifically targeting the strengthening of plantar flexors.
- **Orthotic devices:** Using a hinged ankle-foot orthosis (AFO) with a dorsiflexion stop or a solid AFO set at a few degrees of plantar flexion to shift the GRF anterior to the knee, preventing buckling.
- **Serial casting:** For nonsurgical management of triceps surae contractures to avoid hastening the development of calcaneal gait.
- **Surgical interventions:** Addressing structural issues such as calcaneal fractures or correcting previous over-corrective surgeries, if necessary.
- **Pain management:** Addressing chronic pain and managing ulcerations to improve quality of life.

Equinus Gait

Equinus gait, often known as toe-walking, is a walking anomaly marked by sustained plantar flexion of the ankle during the stance phase, accompanied by extended hips and knees. This gait pattern can be caused by plantar flexor spasticity or contracture, as well as dorsiflexor weakness. It is also possible to have compensatory genu recurvatum (knee hyperextension).

Causes

Equinus gait can result from various underlying conditions and factors:

- **Spasticity or contracture of ankle plantar flexors:** Often seen in neurological disorders such as cerebral palsy (CP) and hemiplegia.
- **Weakness of ankle dorsiflexors:** Leads to an inability to properly lift the foot, causing toe-walking.
- **Neurological disorders:** Conditions such as spastic cerebral palsy are common causes.
- **Muscle-tendon imbalance:** In children with spastic CP, longer-than-normal Achilles tendons and shorter-than-normal muscle bellies contribute to equinus gait.
- **Post-surgical changes:** Procedures like Tendo-Achilles Lengthening (TAL) can restore dorsiflexion but may not normalize muscle-tendon architecture.

Treatment

Management of equinus gait aims to address the underlying causes and improve gait mechanics:

- **Physical therapy:** Targeted exercises to strengthen ankle dorsiflexors and improve overall lower limb function.
- **Orthotic devices:** Use of a hinged ankle-foot orthosis (AFO) with dorsiflexion assist and/or plantar flexion stop to support proper foot positioning.
- **Chemoneurolysis:** Injections such as botulinum toxin to reduce spasticity in the gastroc-soleus muscle.
- **Surgical interventions:** Procedures like Tendo-Achilles Lengthening (TAL) to increase dorsiflexion and correct the equinus deformity.
- **Management of muscle-tendon imbalance:** Specific therapies to maintain or improve muscle-tendon architecture, especially in children with cerebral palsy.
- **Gait training:** Rehabilitation programs to improve walking mechanics and reduce compensatory patterns.

Excessive Medial or Lateral Foot Contact Gait

Excessive medial or lateral foot contact is a gait abnormality where the foot excessively contacts the ground on either the medial or lateral side (Fig. 22.16). This can result from muscular imbalances such as spasticity or weakness in the muscles responsible for foot inversion and eversion.

Causes

- **Spasticity of evertors:** Increased muscle tone in the evertors (peroneal muscles) can cause excessive medial foot contact by forcing the foot into eversion.
- **Weakness of invertors:** Weakness in the invertors (tibialis anterior and posterior) can also result in excessive medial foot contact due to the inability to maintain proper foot alignment.

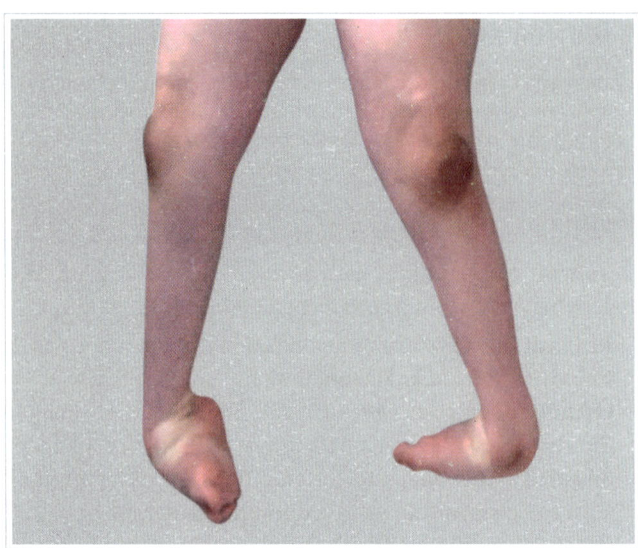

Fig. 22.16: Excessive medial or lateral foot contact gait

- **Spasticity of invertors:** Increased muscle tone in the invertors can cause excessive lateral foot contact by forcing the foot into inversion.
- **Weakness of evertors:** Weakness in the evertors can lead to excessive lateral foot contact due to inadequate eversion control.

Clinical Features

- **Gait deviations:** Patients exhibit an abnormal gait with either excessive medial or lateral foot contact.
- **Foot deformities:** Conditions like Talipes equinovarus (clubfoot) may be present, where the foot is excessively turned inward.
- **Instability:** Difficulty in maintaining balance due to improper foot contact with the ground.
- **Pain and discomfort:** Chronic pain and discomfort in the foot, ankle, and possibly higher up the kinetic chain (knee, hip, back) due to altered gait mechanics.

Treatment

- **Physical therapy:**
  - **Strengthening exercises:** Focused on strengthening weak invertors or evertors to restore muscle balance.
  - **Stretching exercises:** To reduce spasticity in overactive muscles.
- **Orthotic interventions:**
  - **Custom orthotics:** Provide support to correct foot alignment and distribute pressure evenly across the foot.
  - **Ankle-foot orthosis (AFO):** To stabilize the ankle and foot, preventing excessive inversion or eversion.
- **Pharmacological treatments:** Use of botulinum toxin injections to reduce spasticity in the involved muscles.

- **Surgical interventions:**
  - **Tendon transfers:** To balance muscle forces around the foot and ankle.
  - **Release surgeries:** For severely spastic muscles to improve range of motion and function.
- **Gait training:**
  - **Rehabilitation programs:** Designed to improve overall gait mechanics, focusing on correcting foot placement during walking.

Parkinson's Gait

Parkinson's gait is a characteristic walking abnormality observed in individuals with Parkinson's disease (PD). It is often marked by shuffling steps, reduced arm swing, and a stooped posture. Advanced stages may present with complex gait disturbances such as freezing of gait (FoG) and festination (Fig. 22.17).

Pathophysiology

Parkinson's gait, also known as shuffling gait, results from the degeneration of dopaminergic neurons in the substantia nigra, leading to reduced dopamine levels in the basal ganglia. This disrupts motor control, causing bradykinesia (slowness of movement), rigidity, and impaired postural reflexes. The characteristic features include a stooped posture, short shuffling steps, difficulty initiating movement (freezing), and reduced arm swing.

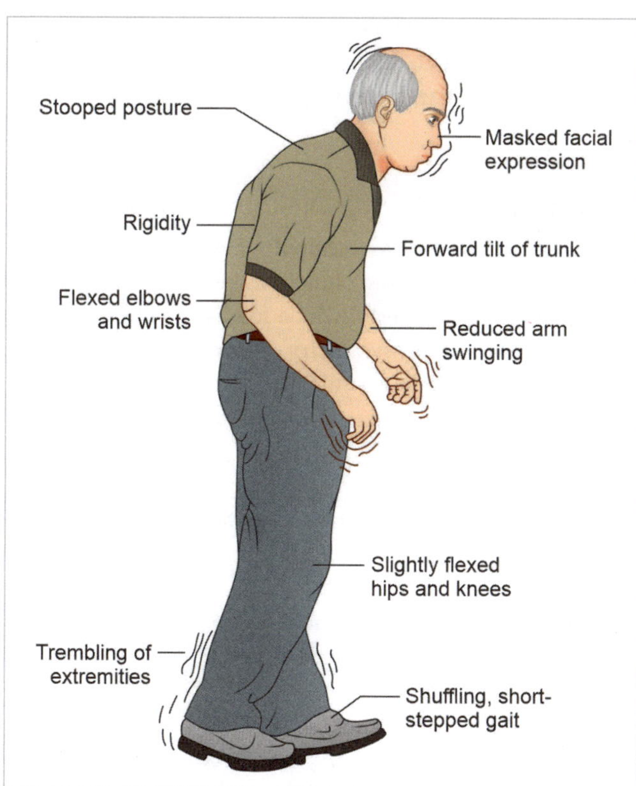

Fig. 22.17: Typical features of Parkinson's gait

Causes

- **Dopaminergic deficiency:** Primary cause due to the loss of dopamine-producing neurons in the substantia nigra.
- **Medication response:** Gait issues can be exacerbated during "off" medication periods when the effects of dopaminergic drugs wane.
- **Non-dopaminergic systems:** Involvement of other neurotransmitter systems such as noradrenaline, serotonin, and acetylcholine.
- **Neuromuscular complications:** Secondary effects of muscle rigidity and reduced coordination.

Clinical Features

- **Hypokinetic rigid gait:** Characterized by a decrease in gait speed and step amplitude with an unchanged or slightly increased cadence. It is primarily due to a deficit in internal generation of step length, rather than an inability to increase cadence.
- **Festination:** Involves rapid, small steps as the patient attempts to keep their center of gravity within their base of support. It is often associated with a forward-leaning trunk and may lead to falls.
- **Freezing of gait (FoG):** Episodic, transient inability to move the feet forward despite the intention to walk. It is more common in advanced stages and can occur in both "on" and "off" medication states.
- **Postural instability:** Related to hypertonia and a flexed, stiffened trunk posture throughout the gait cycle.

Did You Know?

Portable cueing devices, such as smartphone apps or wearable devices have been developed to provide visual cues on-demand to individuals with Parkinson's disease.

Treatment

Physical Therapy

- **Rehabilitation:** Includes compensatory strategies with sensory cues to improve gait hypokinesia and FoG.
- **Aerobic and resistance training:** Provides complementary benefits but requires long-term adherence for sustained impact.
- **Novel therapies:** Approaches like tai chi, high-amplitude movement exercises (LSVT-BIG), and virtual reality-based therapies show promise.

Pharmacological Approaches

- **Dopaminergic drugs:** Optimization of dopamine replacement therapies like Levodopa remains central, though its efficacy diminishes in advanced stages.
- **Methylphenidate:** Inhibits presynaptic dopamine and noradrenaline transporters but shows limited and inconsistent benefits.
- **Memantine:** Decreases glutamatergic hyperactivity, yet shows minimal improvement in axial symptoms.
- **Cholinesterase inhibitors:** Drugs like donepezil can reduce fall frequency in a subset of patients.

Deep Brain Stimulation (DBS)

- **Effectiveness:** Shown to improve motor symptoms but may have mixed effects on gait disturbances and FoG.
- **Programming:** Electrode placement and stimulation frequency significantly impact outcomes.

Other Interventions

- **Deprescribing:** Reducing use of benzodiazepines and anticholinergic drugs to mitigate postural instability and gait disturbance (PIGD) features.
- **Droxidopa:** Can reduce falls, particularly in cases of symptomatic neurogenic hypotension.
- **Noninvasive neurostimulation:** Techniques like transcranial direct current stimulation and vagal nerve stimulation are being explored.
- **Closed loop cueing:** Emerging technologies to assist with FoG, focusing on reducing dependency on attentional cues.

Hemiplegic Gait

Hemiplegic gait is a type of walking abnormality often seen in patients who have experienced a stroke, leading to one-sided weakness or paralysis (hemiplegia). This gait is characterized by asymmetry in spatial and temporal walking parameters, poor motor control, and disrupted balance.

Pathophysiology

- **Normal gait symmetry:** Typically, normal gait is symmetrical with <6% interlimb differences in vertical force and temporal parameters.
- **Hemiplegic gait asymmetry:** Characterized by marked asymmetry with poor motor control and compensatory movements. The paretic limb often exhibits prolonged swing time and reduced stance time, while the non-paretic limb shows prolonged stance time.
  - **Intralimb and interlimb coordination:** In hemiplegic gait, coordinated limb movements are replaced by mass movement patterns (synergies), necessitating compensatory pelvic and non-paretic side adjustments.

Causes

- **Stroke:** The primary cause, leading to hemiparesis or hemiplegia due to brain damage.
- **Muscle weakness and spasticity:** Resulting from neurological deficits post-stroke.
- **Joint range of motion and trunk control limitations:** Complications such as limited joint range of motion and poor trunk control significantly influence gait asymmetry and overall dysfunction.

Clinical Features

Temporal Features

- **Reduced walking velocity:** Hemiplegic patients generally exhibit a reduced walking velocity due to impaired motor recovery and balance.
- **Altered stance and swing times:** The paretic limb spends less time in stance and more in swing, whereas the non-paretic limb exhibits the opposite pattern.
- **Stride time and cadence:** Increased stride time and reduced cadence are characteristic temporal features.

Spatial Features

- **Step length asymmetry:** Differences in paretic and non-paretic step lengths are common, with paretic step length often being longer due to compensatory mechanisms.
- **Propulsive force generation:** Patients with less paretic propulsion tend to have longer paretic steps, indicating reliance on the non-paretic leg for propulsion.

Treatment

- **Physical therapy:**
  - **Comprehensive assessment:** Essential to evaluate deficits and remaining functions before therapy.
  - **Various techniques:** Include force exercises, spasticity reduction, gait symmetry training, utilization of equilibrium reflexes, stepping automation, endurance training, and rhythmic movement repetition, etc.
- **Orthotics:**
  - Ankle-foot orthoses (AFOs) are used to improve walking patterns in patients with mild to moderate spasticity by stabilizing the foot and ankle.
- **Pharmacotherapy:**
  - **Locomotor pharmacotherapy:** Includes medications aimed at enhancing motor functions and reducing spasticity.
- **Advanced therapies:**
  - **Treadmill training with partial body weight support:** Effective in restoring gait patterns through enforced stepping movements.

- **Botulinum toxin injections:** Used to selectively reduce spasticity in leg muscles, thereby improving gait functions.
- **Musical biofeedback:** Auditory rhythmic cues significantly improve weight-bearing stance time on the paretic side and enhance stride symmetry.
- **Multichannel functional electrical stimulation (MFES):** Shows potential benefits in gait rehabilitation for patients with severe hemiplegia.

Diplegic or Spastic Gait

Diplegic or spastic gait is defined by bilateral lower extremity involvement, with the lower limbs often having more spasticity than the upper limbs. It is frequently associated with a narrow base, dragging of both legs, and scraping of the toes while walking. In addition, individuals may exhibit a variety of atypical gait patterns, such as equinus gait, jump gait, apparent equinus, and crouch gait.

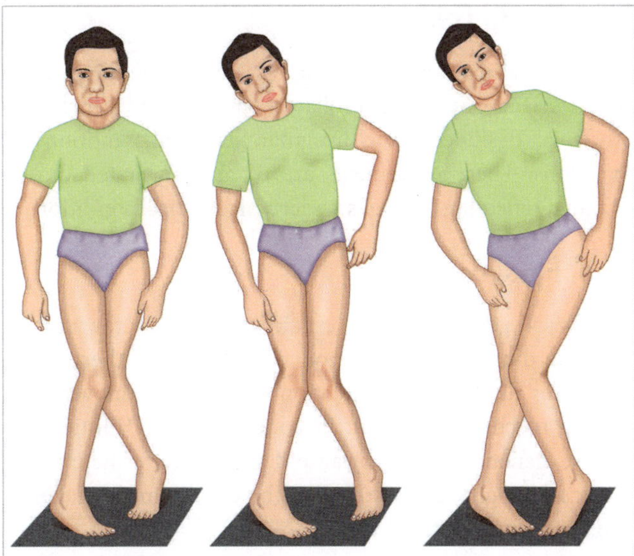

Fig. 22.18: Scissors gait

Causes

- Cerebral palsy (CP)
- Brain damage
- Perinatal factors
- Prenatal infections
- Genetic factors
- Traumatic brain injury
- Hypoxic-ischemic encephalopathy (HIE)
- Intracranial hemorrhage
- Maternal health factors

Clinical Features

- **Bilateral lower extremity involvement:** Patients often exhibit worse spasticity in the lower extremities than the upper extremities.
- **Abnormal gait patterns:**
  - **Narrow base:** Patients walk with an abnormally narrow base and may drag both legs while scraping the toes.
  - **Scissors gait:** Extreme tightness of hip adductors can cause legs to cross the midline, resulting in a scissors-like gait (Fig. 22.18).
  - **Equinus gait patterns:**
    - **True equinus:** Ankle remains in plantarflexion throughout stance phase, with extended hips and knees.
    - **Jump gait:** Characterized by ankle equinus, knee and hip flexion, anterior tilt, and increased lumbar lordosis.
    - **Apparent equinus:** Normal ankle dorsiflexion range, but excessive hip and knee flexion during stance phase.
    - **Crouch gait:** Excessive dorsiflexion at the ankle combined with knee and hip flexion.
- **Asymmetric and mild gait groups:** Some patients may exhibit asymmetry between limbs or have gait patterns within normal limits.

Treatment

- **Physical therapy:** Focuses on maintaining range of motion, preserving or restoring strength, and improving balance and coordination.
- **Orthoses:** Orthoses are the primary conservative treatment option for control of dynamic equinus in spastic cerebral palsy
- **Spasticity management:**
  - **Botulinum neurotoxin-A (BoNT-A) injections:** Used to treat local spasticity, with caution advised for long-term implications.
  - **Selective dorsal rhizotomy:** Reserved for selected patients with diplegic CP, providing permanent global impact on lower limb spasticity.
  - **Intrathecal baclofen:** Treats severe spasticity with a reversible effect.
- **Surgical interventions:**
  - **Addressing bony deformities and muscle contractures:** Surgery aims to preserve or restore lever arms to efficiently use muscle forces.
  - **Single event multi-level surgery (SEMLS):** Manages musculoskeletal deformities comprehensively in both lower limbs in a single surgical procedure, preventing "birthday syndrome" of correcting one deviation after another with growth.

Choreiform Gait

Choreiform gait also known as hyperkinetic gait, is characterized by uncontrollable movements, including oro-facial dyskinesia and irregular, dance-like motions of the limbs. These movements are often exaggerated during walking, reflecting a serious underlying medical condition.

Pathophysiology

The basal ganglia play a crucial role in motor control and coordination. Dysfunction or damage to these brain structures disrupts the regulation of voluntary movements, leading to choreiform gait. In conditions like Huntington's disease, genetic mutations cause degeneration of neurons in the basal ganglia, contributing to the development of chorea and choreiform gait.

Causes

Choreiform gait is associated with various basal ganglia disorders, such as Sydenham's chorea, Huntington's disease, and other forms of chorea, athetosis, or dystonia. These conditions result in abnormal muscle contractions, leading to involuntary writhing movements.

Did You Know?

Choreiform gait can sometimes be a side effect of certain medications, particularly neuroleptic drugs used to treat psychiatric disorders such as schizophrenia or bipolar disorder.

Treatment

- **Physical therapy:**
  - Correction of abnormal posture through positioning or splinting to prevent deformity.
  - Stretching of contracted muscles to reduce spasticity.
  - Soft tissue and joint mobilization to maintain mobility.
  - Proximal stability training, balance exercises, and functional movement training to improve body awareness and stability.
- **Medications:** Muscle relaxers and anti-spasticity medications can help reduce muscle overactivity and involuntary movements.
- **Rest and fatigue management:**
  - Adequate rest is essential to prevent fatigue-induced falls.
  - Therapeutic interventions such as aqua therapy and therapeutic horseback riding can provide relief and improve mobility.
- **Other approaches:**
  - Address abnormal posture and gait patterns with appropriate exercises and assistive devices.
  - **Range of motion exercises:** Maintain joint function and reduce pain associated with muscular imbalance.
  - **Safety measures:** Ensure safety during daily activities, especially walking, with the use of walking aids and assistance.

Cerebellar Ataxic Gait

Cerebellar ataxia is characterized by a combination of dysmetria, dyssynergia, dysdiadochokinesia, dysrhythmia, and intention tremor. Dysmetria involves inadequate force rate and amplitude misscaling, while dyssynergia leads to abnormal multijoint movement paths. Dysrhythmia refers to abnormal timing and coupling of movements, often visualized in angle-angle plots.

Pathophysiology

Cerebellar ataxic gait is characterized by impaired motor coordination due to dysfunction or degeneration of the cerebellum, which plays a critical role in regulating movement and balance. This pathology results in a broad-based, unsteady gait with irregular, staggering steps. Affected individuals often demonstrate balance difficulties, lateral veering, and dysmetria, indicative of disrupted integration of sensory inputs and motor outputs.

Causes

- Brain damage
- Perinatal factors
- Prenatal infections
- Genetic factors
- Traumatic brain injury
- Hypoxic-ischemic encephalopathy (HIE)
- Intracranial hemorrhage
- Maternal health factors
- In some cases it could be an indication of an underlying disease like hyperthyroidism, alcoholism, stroke, or multiple sclerosis (MS). Cerebellar ataxia can also be hereditary, resulting from inherited mutations. However, many cases are idiopathic, which means that cerebellar degeneration develops without any apparent explanation.

Clinical Features

Cerebellar ataxia is characterized by poor balance, dysmetria, dyssynergia, and dysrhythmia. It involves widened base, increased foot rotation angles, and variability in gait parameters, particularly evident during tandem gait. Trunk instability leads to increased gait variability, necessitating compensatory adjustments in step width, duration of foot contact, and cadence (Fig. 22.19).

Treatment

- **Physical therapy:** Coordination and balance exercises, and stability and motor coordination training improve motor performance and reduce ataxia symptoms.
- **Pharmacologic interventions:** Rehabilitation, including high-intensity motor coordination training, improves function, mobility, ataxia, and balance.
- **Surgical or other interventional therapies:** Transcranial stimulation possibly improves cerebellar motor signs, but rigorous clinical trials are needed due to the challenges of studying ataxia patients.

Fig. 22.19: Cerebellar ataxic gait

SUMMARY

- The gait cycle, comprised of stance and swing phases, represents the repetitive sequence of limb actions during walking.
  - Stance refers to the period when the foot is on the ground, while swing denotes the time when the foot is in the air.
  - Typically, each foot undergoes one stance phase and one swing phase per gait cycle, with stance accounting for approximately 60%–62% of the cycle and swing constituting about 38%–40%.
- The gait cycle consists of several distinct phases. Initial Contact marks the beginning of the cycle, occurring when the foot makes contact with the ground and the heel absorbs impact force.
  - Loading response follows, with the body weight shifting to the stance limb and the foot pronating for shock absorption.
  - Mid stance ensues when the body's weight is directly over the stance leg and stability is maintained.
  - Terminal stance sees the body moving forward over the stance limb, with the heel lifting off the ground.
- Pre-swing signifies the transition from single-limb support to swing, as the stance limb begins to push off the ground.
  - Initial swing involves the limb swinging forward in preparation for the next stride, requiring bending at the hip, knee, and ankle joints.
  - Midswing continues the forward swing with maximum flexion, while terminal swing prepares the limb for initial contact with the ground again.
  - Each phase of the gait cycle contributes to the smooth and coordinated movement of walking, ensuring both propulsion and stability throughout the process.
- The gait cycle consists of stance and swing phases, each marked by specific kinematic and kinetic characteristics.
- During the stance phase:
  - Heel strike (0–2% of GC) initiates the cycle, with the hip slightly flexed, knee extended with slight flexion, and ankle dorsiflexed.
  - Foot flat (2–12% of GC) sees hip extension and knee and ankle flexion to absorb impact and adapt to the ground.
  - Midstance (12–31% of GC) involves hip extension, knee reaching maximum flexion before starting to extend, and ankle becoming supinated and dorsiflexed.
  - Heel off (31–50% of GC) starts as the heel lifts off the ground, and weight distribution shifts to the metatarsal heads.
  - Toe off/pre-swing (50–60% of GC) involves hip extension, knee flexion, and ankle plantar flexion as the toes lift off the ground.
- In the swing phase:
  - Early swing (60–75% of GC) sees hip flexion, knee flexion, and ankle moving toward dorsiflexion.
  - Midswing (75–85% of GC) involves hip flexion, knee flexion reducing, and ankle in a neutral position.
  - Late swing/deceleration (85–100% of GC) features hip flexion, locked knee extension, and ankle in a neutral position.
- Muscle activity varies throughout the gait cycle:
  - Hip flexors, extensors, abductors, and adductors control hip movement and stability.
  - Quadriceps and hamstrings stabilize the knee during weight-bearing and smooth transitions.
  - Ankle dorsiflexors and plantarflexors control foot positioning and propulsion.

- The determinants of gait are factors that work together to minimize vertical and lateral displacement of the body's center of gravity (COG) during walking, thus contributing to energy efficiency and smoothness of movement.
  - These determinants include pelvic rotation, pelvic tilting, lateral displacement of the pelvis, knee flexion in the stance phase, and foot and knee mechanisms.
  - Pelvic rotation involves alternating right and left rotations of the pelvis on a horizontal plane, aiding in acceleration and deceleration during gait.
  - Pelvic tilting refers to the hip of the swing phase leg being lower than that of the stance phase leg, reducing vertical COG displacement.
  - Lateral displacement of the pelvis occurs at mid-distance, helping maintain balance and prevent falls.
  - Knee flexion in the stance phase involves flexing the knee at heel strike and midstance, providing shock absorption and minimizing COG displacement.
  - Foot and knee mechanisms ensure stability and smooth progression throughout the stance phase, with the knee extending early in stance and flexing later to maintain COG position.
- Trendelenburg gait, characterized by lateral trunk bending toward the supporting limb during walking, often indicates weakness or dysfunction in the hip abductor muscles or hip joint issues.
  - This gait pattern can be attributed to various underlying conditions affecting the fulcrum, lever, or effort in the hip joint's class 1 lever system.
  - Treatment typically involves physical therapy to strengthen hip abductors, assistive devices for stability, and addressing the underlying cause through medical or surgical interventions.
- Knee extensor weakness, primarily affecting the quadriceps muscles, leads to gait abnormalities such as knee buckling or genu recurvatum.
  - This weakness interferes with maintaining knee stability during the stance phase, requiring compensatory actions to prevent knee collapse.
  - Treatment approaches include physical therapy to strengthen the quadriceps, the use of assistive devices like canes, orthotic devices such as ankle-foot orthosis or bracing, and addressing underlying neurological or traumatic causes.
- Gluteus maximus gait, characterized by posterior trunk bending during the stance phase of walking, is a compensatory pattern resulting from weakness or paralysis of the hip extensor muscles, particularly the gluteus maximus.
  - This gait alteration is caused by various factors such as injury to gluteal nerves, proximal weakness due to plexopathies or myopathies, and hip ankylosis.
  - Clinical presentation includes a backward trunk lean during the stance phase and reduced walking speed.
  - Treatment involves strengthening exercises, assistive devices for stability, physical therapy, and gait training to improve muscle function and mechanics.
- Antalgic gait, on the other hand, is a walking pattern induced by pain, featuring a shortened stance phase to minimize discomfort.
  - Causes range from traumatic injuries to neoplastic conditions, with clinical features including reduced walking speed, asymmetry in the gait cycle, and a stiffened limb.
  - Pathophysiology involves compensatory techniques to reduce joint compressive pressures and muscle activation, such as shortening the stance phase and stiffening the limb.
  - Treatment focuses on addressing the underlying cause of pain, with management options including immobilization, anti-inflammatory medications, physical therapy, and specialized referrals as needed.
- Functional leg length discrepancy (LLD) results from differences in leg lengths during specific phases of the gait cycle due to factors such as joint contractures or muscle strength asymmetries, rather than actual structural differences in bone lengths.
  - This condition leads to biomechanical compensations to maintain balance and functionality, involving various body parts like the feet, ankles, knees, hips, and pelvis.
  - Patients with LLD often present with noticeable gait abnormalities, including limping, increased step length and width on the longer leg, altered cadence, and chronic pain in the lower back, hips, knees or feet.
  - Management strategies range from conservative treatments like shoe lifts and physical therapy to surgical interventions such as growth plate modulation or leg lengthening procedures, depending on the extent of the discrepancy and individual patient factors.
- Hip hiking and circumduction gait patterns are compensatory mechanisms used to achieve adequate foot clearance during the swing phase of walking, typically seen in individuals with difficulties in hip or knee flexion or ankle dorsiflexion.
  - These gait deviations can arise from various causes, including weakness or neurological conditions, and involve biomechanical adjustments such as pelvic tilt, spinal curvature, and altered gait patterns.
  - Treatment focuses on addressing the underlying causes through bracing, assistive devices, physical therapy, and other interventions to improve gait quality and functionality.
- Steppage gait, characterized by heightened knee and hip flexion to lift the foot higher than usual, is commonly observed in individuals with foot drop, neurological disorders or nerve injuries.
  - This compensatory pattern aims to prevent tripping by avoiding dragging the toes during the swing phase, but can lead to clinical features such as slapping foot sounds and increased stance time.

  - Treatment strategies include ankle-foot orthoses, physical therapy, functional electrical stimulation, and surgical interventions to address the underlying conditions and improve mobility.
- Vaulting gait, where individuals raise on the toes of the stance leg to ensure foot clearance of the swinging leg, is typically seen in cases of ineffective knee flexion, ankle plantar flexion, limb length discrepancies or prosthetic issues.
  - This compensatory mechanism aims to shorten the swinging leg to prevent dragging but can lead to inefficiencies and early fatigue.
  - Treatment involves addressing underlying causes through physical therapy, orthotic devices, spasticity management, prosthetic adjustments, surgical interventions, and comprehensive gait training to improve walking mechanics and reduce compensatory patterns.
- Abnormal foot contact during the gait cycle can lead to various gait deviations, each with distinct characteristics and underlying causes.
  - Calcaneal gait, characterized by prolonged heel contact with the ground and lack of push-off due to weakened ankle plantar flexors, often results from conditions such as Achilles tendon issues, neuromuscular diseases, trauma or cerebral palsy.
  - Treatment involves orthotic devices, physical therapy, serial casting, surgical interventions, and pain management to address symptoms and underlying causes.
- Equinus gait, marked by sustained plantar flexion of the ankle during the stance phase, can arise from spasticity or contracture of ankle plantar flexors or weakness of ankle dorsiflexors.
  - It is commonly seen in neurological disorders like cerebral palsy, leading to compensatory gait patterns such as genu recurvatum.
  - Treatment includes orthotic devices, chemoneurolysis, surgical interventions, physical therapy, and gait training to improve walking mechanics.
- Excessive medial or lateral foot contact gait abnormalities result from muscular imbalances such as spasticity or weakness in the muscles responsible for foot inversion and eversion.
  - These imbalances can lead to conditions like talipes equinus or talipes equinovarus.
  - Treatment involves orthotic interventions, physical therapy, pharmacological treatments, surgical interventions, and gait training to correct foot alignment and restore muscle balance.
- Parkinson's gait, observed in Parkinson's disease (PD) patients, features shuffling steps, reduced arm swing, and a stooped posture, often progressing to freezing of gait (FoG) and festination.
  - The primary cause is dopaminergic deficiency due to the loss of dopamine-producing neurons.
  - Treatment involves dopaminergic drugs, deep brain stimulation, and exercise therapies.
- Hemiplegic gait, common after stroke, exhibits asymmetry in spatial and temporal parameters due to weakness or paralysis on one side.
  - Orthotics, physiotherapy, and advanced therapies like botulinum toxin injections are used for management.
- Diplegic or spastic gait, often seen in conditions like cerebral palsy, involves bilateral lower extremity involvement with more severe spasticity in the lower limbs.
  - Abnormal gait patterns include a narrow base, dragging of both legs, and scraping of the toes while walking.
  - Treatment approaches include physiotherapy, orthoses, spasticity management through interventions like botulinum toxin injections and selective dorsal rhizotomy, and surgical interventions addressing bony deformities and muscle contractures.
- Choreiform gait, also known as hyperkinetic gait, is characterized by uncontrollable movements, including orofacial dyskinesia and irregular, dance-like motions of the limbs.
  - It is associated with various basal ganglia disorders such as Sydenham's chorea, Huntington's disease, and other forms of chorea, athetosis, or dystonia.
  - Treatment involves physical therapy, medications to reduce muscle overactivity, rest and fatigue management, propulsive gait correction, range of motion exercises, and safety measures.
- Ataxic gait, associated with cerebellar ataxia, involves poor balance, dysmetria, dyssynergia, and dysrhythmia.
  - It can be caused by brain damage, perinatal factors, genetic factors, and other underlying diseases like hyperthyroidism, alcoholism, stroke or multiple sclerosis.
  - Treatment includes physical therapy, rehabilitation interventions, pharmacologic interventions, and possibly surgical or other interventional therapies like transcranial stimulation.

FURTHER READINGS

- Auerbach N, Tadi P. Antalgic Gait in Adults. In: StatPearls. StatPearls Publishing, Treasure Island (FL); 2023. PMID: 32644669.
- Balaban B, Tok F. Gait disturbances in patients with stroke. Pmandr. 2014 Jul 1;6(7):635-42.
- Bogataj U, Gros N, Kljajić M, Aćimović R, Malezič M. The rehabilitation of gait in patients with hemiplegia: A comparison between conventional therapy and multichannel functional electrical stimulation therapy. Physical therapy. 1995 Jun 1;75(6):490–502.
- Buckley E, Mazzà C, McNeill A. A systematic review of the gait characteristics associated with Cerebellar Ataxia. Gait and posture. 2018 Feb 1;60:154–63.
- Cavallieri F, Fini N, Contardi S, Fiorini M, Corradini E, Valzania F. Subacute copper deficiency myelopathy in a patient with occult celiac disease. J Spinal Cord Med. 2017 Jul;40(4):489–491.
- Gardiner Dena M. Principles of exercise therapy. CBS Publishers and Distributors Pvt Ltd; 4th ed., 2023.
- Ebersbach G, Moreau C, Gandor F, Defebvre L, Devos D. Clinical syndromes: Parkinsonian gait. Movement Disorders. 2013 Sep 15;28(11):1552–9.
- Fujisawa H, Sugimura Y, Takagi H, Mizoguchi H, Takeuchi H, Izumida H, Nakashima K, Ochiai H, Takeuchi S, Kiyota A, Fukumoto K, Iwama S, Takagishi Y, Hayashi Y, Arima H, Komatsu Y, Murata Y, Oiso Y. Chronic Hyponatremia Causes Neurologic and Psychologic Impairments. J Am Soc Nephrol. 2016 Mar;27(3):766–80.
- Gandbhir VN, Lam JC, Lui F, Rayi A. Trendelenburg gait. InStatPearls [Internet] 2024 Feb 29. StatPearls Publishing.
- Grabli D, Karachi C, Welter ML, Lau B, Hirsch EC, Vidailhet M, François C. Normal and pathological gait: what we learn from Parkinson's disease. Journal of Neurology, Neurosurgery and Psychiatry. 2012 Oct 1;83(10):979–85.
- Gupta N, Carmichael MF. Zinc-Induced Copper Deficiency as a Rare Cause of Neurological Deficit and Anemia. Cureus. 2023 Aug;15(8):e43856.
- Holroyd KB, Berkowitz AL. Metabolic and Toxic Myelopathies. Continuum (Minneap Minn). 2024 Feb 01;30(1):199–223.
- Jamshidi N, Rostami M, Najarian S, Menhaj MB, Saadatnia M, Salami F. Assessment of ground reaction forces of steppage gait in comparison with normal gait. Journal of Musculoskeletal Research. 2009 Mar;12(01):45–52.
- Kerrigan DC, Frates EP, Rogan S, Riley PO. Hip hiking and circumduction: Quantitative definitions. American journal of physical medicine and rehabilitation. 2000 May 1;79(3):247–52.
- Levangie PK, Norkin CC. Joint structure and function: a comprehensive analysis. FA Davis; 2011 Mar 9.
- Lovaglio A, Di Masi G, Socolovsky M, Bonilla G, Bataglia D, Barillaro K. Posterior tibialis tendon transfer for steppage gait: Functional results and indications. Neurology, Psychiatry and Brain Research. 2018 Jun 1;28:13–8.
- Manca M, Ferraresi G, Cosma M, Cavazzuti L, Morelli M, Benedetti MG. Gait patterns in hemiplegic patients with equinus foot deformity. BioMed research international. 2014 Jan 1;2014.
- Mauritz KH. Gait training in hemiplegia. European journal of Neurology. 2002 May;9:23–9.
- Mirek E, Filip M, Chwała W, Banaszkiewicz K, Rudzinska-Bar M, Szymura J, Pasiut S, Szczudlik A. Three-Dimensional Trunk and Lower Limbs Characteristics during Gait in Patients with Huntington's Disease. Front Neurosci. 2017 Oct 12;11:566. doi: 10.3389/fnins.2017.00566. PMID: 29075175; PMCID: PMC5643481.
- Müller ML, Marusic U, van Emde Boas M, Weiss D, Bohnen NI. Treatment options for postural instability and gait difficulties in Parkinson's disease. Expert Review of Neurotherapeutics. 2019 Dec 2;19(12):1229–51.
- Nori SL. Steppage gait.
- Sobel EL, Glockenberg A. Calcaneal gait. Etiology and clinical presentation. Journal of the American Podiatric Medical Association. 1999 Jan 1;89(1):39–49.
- Soh Y, Won CW. Association between frailty and vitamin B_{12} in the older Korean population. Medicine (Baltimore). 2020 Oct 23;99(43):e22327.
- Stolze H, Klebe S, Petersen G, Raethjen J, Wenzelburger R, Witt K, Deuschl G. Typical features of cerebellar ataxic gait. Journal of Neurology, Neurosurgery and Psychiatry. 2002 Sep 1;73(3):310–2.
- Tekin HG, Edem P, Özyılmaz B. Spinal muscular atrophy with predominant lower extremity (SMA-LED) with no signs other than pure motor symptoms at the intersection of multiple overlap syndrome. Brain Dev. 2022 Apr;44(4):294–298.
- Wang RY. Effect of proprioceptive neuromuscular facilitation on the gait of patients with hemiplegia of long and short duration. Physical Therapy. 1994 Dec 1;74(12):1108–15.
- Wren TA, Cheatwood AP, Rethlefsen SA, Hara R, Perez FJ, Kay RM. Achilles tendon length and medial gastrocnemius architecture in children with cerebral palsy and equinus gait. Journal of Pediatric Orthopaedics. 2010 Jul 1;30(5):479–84.

STUDENT ASSIGNMENT

LONG ANSWER QUESTIONS

1. Describe various phases of gait cycle in detail.
2. Explain kinematics and kinetics of the gait cycle.
3. Discuss the determinants of gait in detail.
4. Describe the various pathological gaits in detail.
5. Discuss muscle activity during different phases of gait cycle.

SHORT ANSWER QUESTIONS

1. Define gait cycle.
2. What are the causes of pathological gait?
3. Write about the Trendelenburg gait.
4. What is hemiplegic gait?
5. Write about anterior trunk bending gait.
6. What is Parkinson's gait?
7. Define ataxic gait.

MULTIPLE CHOICE QUESTIONS

1. **What percentage of the gait cycle is typically allocated to the stance phase?**
   a. 20–25% b. 40–45%
   c. 60–62% d. 80–85%
2. **Which phase of the gait cycle involves the body weight shifting to the stance limb and the foot pronating for shock absorption?**
   a. Midstance b. Loading response
   c. Heel strike d. Terminal stance
3. **Trendelenburg gait is often indicative of weakness or dysfunction in which muscle group?**
   a. Hip extensors
   b. Hip abductors
   c. Quadriceps
   d. Hamstrings
4. **Which gait pattern is characterized by posterior trunk bending during the stance phase?**
   a. Steppage gait
   b. Equinus gait
   c. Gluteus Maximus gait
   d. Antalgic gait
5. **Antalgic gait is primarily induced by:**
   a. Weakness in the hip abductors
   b. Pain
   c. Weakness in the quadriceps
   d. Dysfunction in the hip flexors
6. **Functional leg length discrepancy results from:**
   a. Structural differences in bone lengths
   b. Joint contractures
   c. Muscle strength asymmetries
   d. Neurological disorders
7. **Which gait deviation involves raising on the toes of the stance leg to ensure foot clearance of the swinging leg?**
   a. Vaulting gait b. Steppage gait
   c. Trendelenburg gait d. Gluteus maximus gait
8. **Calcaneal gait is characterized by:**
   a. Sustained plantar flexion of the ankle during the stance phase
   b. Prolonged heel contact with the ground
   c. Excessive medial or lateral foot contact
   d. Heightened knee and hip flexion

9. **Equinus gait is marked by:**
   a. Lateral trunk bending
   b. Shortened stance phase
   c. Sustained plantar flexion of the ankle
   d. Excessive medial or lateral foot contact

10. **Which gait pattern involves lateral trunk bending toward the supporting limb during walking?**
    a. Steppage gait
    b. Trendelenburg gait
    c. Equinus gait
    d. Antalgic gait

11. **Which muscle group weakness primarily leads to genu recurvatum during walking?**
    a. Hip abductors
    b. Quadriceps
    c. Hip extensors
    d. Hip flexors

12. **What percentage of the gait cycle is typically allocated to the swing phase?**
    a. 20–25%
    b. 40–45%
    c. 60–62%
    d. 80–85%

ANSWER KEY

1. c **2.** b **3.** b **4.** c **5.** b **6.** c **7.** a **8.** b **9.** c **10.** b
11. b **12.** b

23

Gait Training and Walking Aids

Sheetal Kalra

LEARNING OBJECTIVES

After the completion of the chapter, the readers will be able to:

- Define and explain gait training, its purpose and different methods of gait training.
- Explain different types of walking aids, indications, principles and training with walking aids, measurement, gait training methods.
- Describe components, classification, characters of good crutch.
- Explain the preparation of a patient for crutch walking, crutch walking muscles, measurement of crutches, crutch stance, crutch palsy, types of crutch walking (ground, stairs, ramp).

CHAPTER OUTLINE

- Introduction
- Gait Training
- Walking Aids or Assistive Devices
- Strength Training
- Balance Drills

KEY TERMS

Body-weight support systems: These are also known as unweighting devices which allow a user to reduce the amount of weight borne by their lower extremities during walking or other weight-bearing activities.

Crutch palsy: It is characterized by motor and sensory involvement at the axilla level, which can result in loss of triceps function, weakness in supination, and other deficits.

Gait training: A treatment technique aimed at enhancing a person's walking skill.

Walking aids: A variety tools the patients may be given to enhance their gait, balance or safety when moving around on their own.

INTRODUCTION

The proper functioning of the complete central nervous system, which comprises the brain and spinal cord, is necessary for the complex process of synchronization of all bodily activities during a typical human stride. Any disease process that affects the brain, spinal cord, or the peripheral nerves that branch out of them and supply the muscles can cause irregularities in gait. Physical medicine and rehabilitation (PM and R) consultants, physiotherapists, occupational therapists, and other allied specialists typically assist patients in relearning how to walk.

GAIT TRAINING

Gait training can be defined as a treatment technique aimed at enhancing a person's walking skill. To improve the biomechanics, coordination, strength, and endurance needed for safe and effective walking, this could involve a variety of methods and activities. Programs for gait training are frequently customized to meet the unique requirements and skills of each person, taking into consideration factors like medical issues, the intensity of injuries, and functional objectives. This field of study investigates many methods of gait training, such as using technology or walking aids to improve gait mechanics and functional results. Different outcome measures, including walking speed, endurance, balance, and quality of movement, are used to assess the efficacy of gait training therapies and provide guidance.

Indications

Gait training may be recommended when ability to walk is impaired or lost. For this ICF classification is helpful to design and guide gait training strategies. It also helps to select appropriate assessments.

Walking aids have been demonstrated to facilitate numerous domains of activity and engagement when used in daily life. The clinical features of the subject's residual motor capacity, cognitive function, vision, vestibular function, muscle force (trunk and limbs), degenerative status of lower and upper limb joints, overall physical condition of the patient, as well as additional characteristics of the environment in which the patient lives and interacts, must be carefully analyzed and interpreted in order to select the most suitable model of assistive device. Significant flaws in one or more of these features can jeopardize the device's ability to be used safely and raise the risk of falls or energy consumption.

Based on the Severity of Movement

Patients can be divided into two major functional groups based on the severity of their movement restrictions:

1. Those who have lost the ability to move around
2. Those who have lost some of their movements

The first group consists of people who still have some residual motor function that can be enhanced by an assistive device. In other words, by utilizing the user's remaining motor abilities, the use of augmentative devices tries to strengthen the user's natural ways of mobility. The second category of rehabilitation aids includes external aids like canes, crutches, and walkers as well as wearable orthoses and prostheses. Research on intelligent assistive devices for gait training has grown over the past ten years, with an emphasis on the use of cutting-edge robotic technologies for the benefit of individuals with disabilities.

Did You Know?

Gait training, which focuses on enhancing walking skills, can be tailored to meet the unique needs of each individual. Factors such as medical conditions, severity of injuries, and functional goals are taken into account when designing gait training programs. These programs often incorporate various methods and activities, including the use of technology, assistive devices, and specific exercises to improve biomechanics, coordination, strength, and endurance for safe and effective walking.

Based on International Classification of Functioning, Disability and Health (ICF)

As per ICF classification indications for gait training are as follows:

Impairments

Muscular Impairments

The goal of gait training for individuals with muscle weakness is to enhance the endurance, coordination, and strength of the muscles needed for walking. Through the implementation of targeted exercises and procedures, gait training assists persons who suffer from muscle weakness in regaining or improving their capacity to walk safely and effectively.

Joint Stiffness

Gait training is crucial for treating limited range of motion and mobility impairments in people with joint stiffness. Gait training aims to increase flexibility and decrease stiffness by combining joint mobilization techniques and stretching exercises, enabling more comfortable and fluid walking motion.

Impaired Balance

Gait training is essential to alleviate instability and lower the risk of falls in patients with impaired balance. Gait training makes it possible for people with poor balance to walk more securely and independently by improving stability and coordination.

Activity Limitations

Difficulty in Walking

Gait training is designed to enhance walking mechanics and restore functional mobility by addressing specific problems that people have as a result of injuries, neurological issues, or musculoskeletal illnesses. The use of assistive devices, task-specific workouts, and gait retraining are some methods to improve walking ability and independence in daily tasks.

Reduced Endurance

Gait training addresses reduced endurance by incorporating exercises and progressively increasing walking duration and intensity, thereby enhancing cardiovascular fitness and stamina for prolonged walking activities. Through structured training programs, individuals build endurance, enabling them to walk longer distances with less fatigue and improved overall functional capacity.

Participation Restrictions

Limited Community Participation

Enhancing walking ability, self-assurance, and safety through gait training allows people to participate in social activities and move around the community more independently and comfortably. This promotes community involvement. Gait training gives people the ability to actively engage in social, professional, and recreational activities in their communities by resolving mobility constraints.

Safety Concerns

In order to maintain independence and prevent injuries, gait training promotes confidence in mobility and reduces the chance of falls. It also improves walking mechanics and addresses balance impairments.

Environmental Factors

Accessibility Barriers

To assist people in navigating the environmental barriers and hurdles they confront in their daily lives, gait training may be recommended.

Assistive Device Training

In order to maximize the use of assistive devices, increase mobility, and prevent compensatory behaviors, gait training is crucial for persons who use them. This ensures safe and effective walking while gradually reducing dependency on the device.

Personal Factors

Motivation and Goals

Persons undergoing rehabilitation have different needs, motivations, and goals, and that they should be considered in gait training programs. By taking these things into account, the program may be tailored to the person's requirements, improving their involvement and commitment to the recovery process and increasing its effectiveness.

Clinical Correlation

Some common conditions where gait training is given:

- Amputation and after prosthetic fitment
- Osteoarthritis
- Muscular dystrophy
- Cerebral palsy
- Stroke
- Polio
- Spinal cord injury
- Parkinson's disease
- Multiple sclerosis
- Brain and spinal cord injuries
- After surgery
- Sports injury
- Elderly people to prevent falls

WALKING AIDS OR ASSISTIVE DEVICES

Patients who struggle to maintain a regular gait cycle or balance because of an injury to one or both of their legs are given walking aids or assistive devices (ADs). Among other signs, loss of leg perception, leg weakness, walking pain, and a history of falling are some conditions where assistive devices are prescribed. An AD not only offers additional support, but it can also protect the hurt leg and stop it from getting worse because of the need to bear weight.

Purposes

- Increase area of support or base of support
- Maintain center of gravity over base of support
- Redistribute weight-bearing area by decreasing force on injured or inflamed part or limb
- Compensate for weak muscles
- Decrease pain
- Improve balance

Selection

A patient's weight-bearing status, or how much of their body weight can be supported on his legs, coordination, and strength, determine the type of walking aid that will be given to him/her. The doctor decides the weight-bearing status for each patient. A patient's ability to bear weight typically changes as his treatment progresses. To assess the weight bearing level, two scales are used, one under each foot, and adjusting the weight on each foot until the affected foot is supported. The appropriate amount of weight is frequently the simplest way to keep track of a patient's ability to bear weight.

There are different weight-bearing statuses.

- **Nonweight bearing (NWB):** The patient in NWB is not permitted to put any weight on the affected leg (Figs 23.1A and B).
- **Touch-down weight bearing (TDWB) or toe-touch weight bearing (TTWB):** With TTWB, the weight on

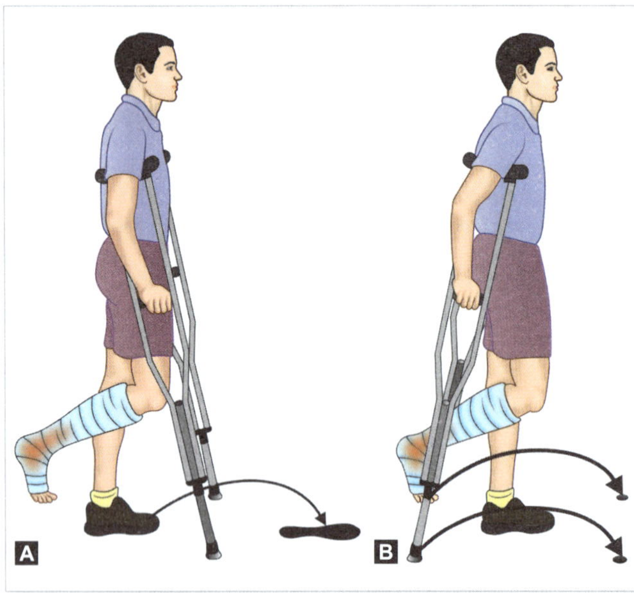

Figs 23.1A and B: Nonweight bearing ambulation—swing through gait

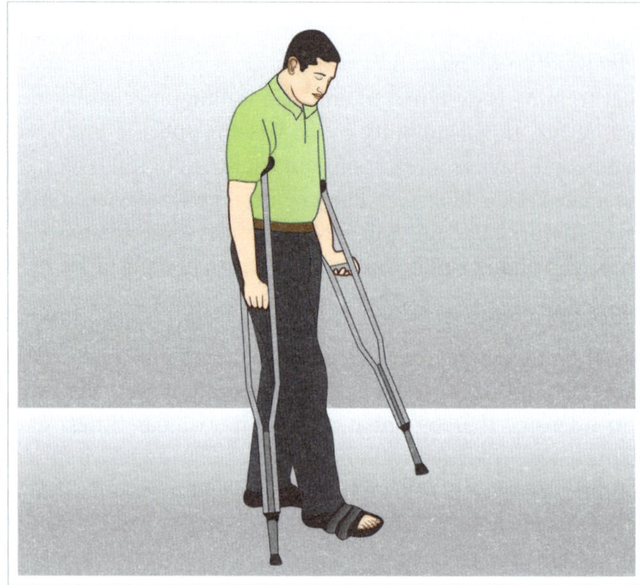

Fig. 23.3: Partial weight bearing

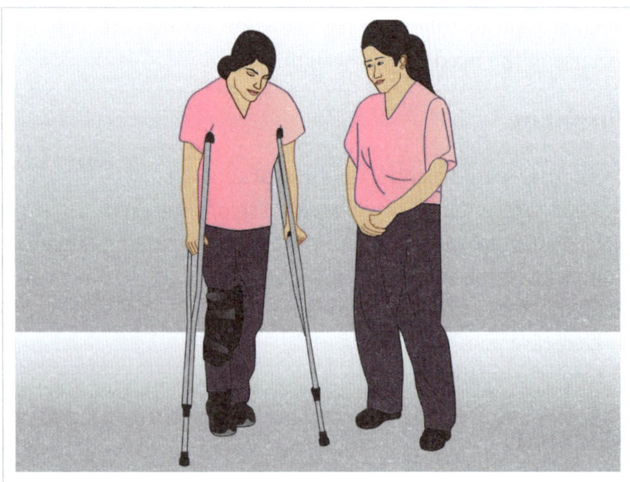

Fig. 23.2: Toe-touch weight bearing

the leg can be described as a weight around 20% of body weight (Fig. 23.2).

- **Partial weight bearing (PWB):** In PWB, the amount of weight that may be supported by the affected leg is typically expressed as a percentage, such as 25% or 50% (Fig. 23.3).
- **Weight bearing as tolerated (WBAT):** If the pain level does not increase during the WBAT, the patient is permitted to carry as much weight as the discomfort permits.

Body Weight Support

Extensive research has been conducted on body-weight support (BWS) systems, commonly referred to as unweighting devices, which are beginning to gain popularity.

- It is feasible to use BWS devices before the patient develops sufficient strength or motor control to bear their full weight.
- A trunk harness with movable straps is used to secure the patient to an overhead suspension system (Fig. 23.4).
- A portion of the patient's weight is supported by the harness and its attachments. The ability of gait training methods involving a BWS system to enhance and possibly recover walking capability has been proven in patients with partial spinal cord injuries.
- A BWS system can be used outside or on a treadmill for gait training.
- Body-weight-supported treadmill training (BWSTT) allows people who have motor deficits that limit their

Fig. 23.4: Body weight support—unweighting device for gait training

ability to sustain their own body weight, to practise and experience locomotion at physiological speeds.

Types

Devices like canes, crutches, and walkers increase balance, help walking, reduce weight on the lower limbs, communicate sensory cues, and allow for mobility in areas inaccessible to wheelchairs.

Canes

These are the assistive devices that assist in ambulation by enhancing balance and base of support. As a mobility aid, assistive canes or walking sticks, transfer weight away from weakened or injured lower limbs. The basic cane consists of a handle for a secure grasp, a collar to join the handle to the shaft, a shaft to carry the weight burden, and a ferrule (tip) to operate as a tactile interface that adapts to the ground surface.

Canes or walking sticks are designed to:

- Reduce pain
- Provide stability and balance in standing and walking
- Take pressure off one or both legs
- Improve sensory feedback for safety and security when walking

Types

Standard Cane

1. Straight canes, sometimes referred to as standard canes, are typically constructed of wood or aluminum. They are affordable and lightweight (Fig. 23.5).
2. To ensure that wooden standard canes are of proper length for each user, they must be specially fitted. There is no need for custom fitting because aluminum standard canes contain pins that allow for length adjustment.
3. These common canes are helpful for patients who just require an extra point of contact with the ground to maintain balance, broadening the base of support. Little to no weight bearing through the cane is necessary.
4. **Uses and indications:** Patients with modest sensory or coordination issues, such as those caused by visual, auditory, vestibular, peripheral proprioceptive, or central cerebellar illness, can utilize this cane.

Fig. 23.5: Standard cane

Quad Canes

- Quad canes have four feet at the bottom of the stick, making them heavier and more stable. Because of this, they are excellent walking aids for people who have more severe balance or stability problems.
- Compared to the more popular single point walking stick, they offer a wider base of support (Fig. 23.6) and is often constructed of aluminum.
- **Uses and indications:** This cane enables the patient to bear more weight, broadens the base of support, and promotes stability. The patient can utilize his hands by allowing the device to stand on its own. Quad canes can be prescribed for hemiplegic patients or patients with moderate to severe antalgic gait from osteoarthritis.
- **Drawback:** A drawback is that the cane's four legs must make contact with the ground for support during gait, which slows down quick movements. They could also produce potential tripping hazard due to their increased width of the base.

Clinical Correlation

In patients with mild disability brought on by hemiparesis during standing, a quad cane appears to be more useful than a normal cane in reducing postural sway. When the assistive device is contralateral to the foot that is positioned forward, postural sway is reduced the most.

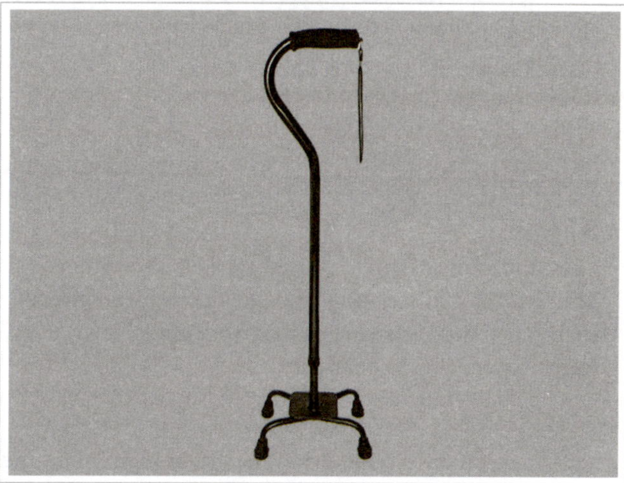

Fig. 23.6: Quad cane

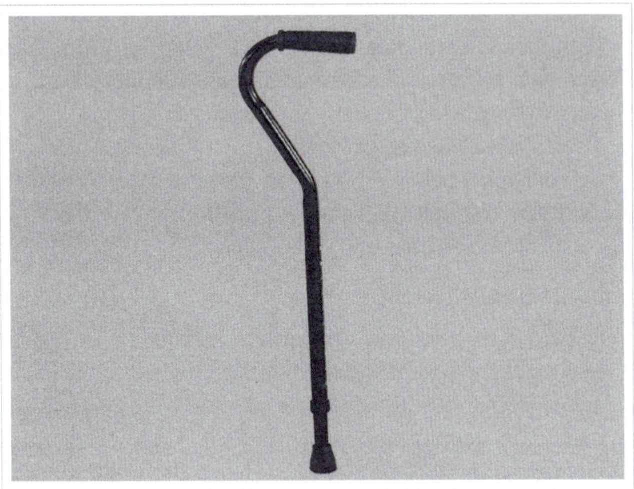

Fig. 23.7: Forearm or offset cane

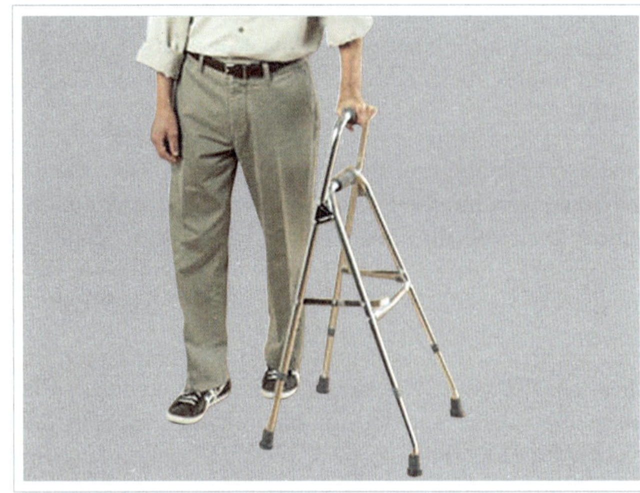

Fig. 23.8: Hemi cane or stroll cane

Forearm Canes

- A special kind of cane called a forearm cane supports the forearms more by transferring some of the weight from the hands and wrists to the upper arm.
- Also known as offset cane (Fig. 23.7), these are constructed of aluminum.
- They may be adjusted in length, thus bespoke fittings are not necessary.
- **Uses and indications:** These canes allow for the patient's weight to be displaced over the shaft of the cane. More stability is provided by this cane, which can also be used sporadically for weight bearing. Those with painful gait abnormalities, such as mild to moderate antalgic gait linked to hip or knee osteoarthritis, benefit from an offset cane.

Hemi Cane

- Also referred to as stroll cane, it is made of aluminum and has two parts. The vertical part has a handle and two-legs, and the horizontal part has two angled legs away from the patient (Fig. 23.8).
- Comparatively, this one offers a wider base of support.
- There are several different handle designs available, including t-shaped, offset, crook and swan neck. Ergonomic (or arthritic) handles are made to give extra support under the palm.
- **Uses and indications:** It is used by patients who need to bear weight continuously through one upper extremity, such as hemiparetic stroke patients with moderate to severe lower extremity (LE) paralysis.

Rolling Cane

- The wheeled base cane, made mostly of aluminum and aluminum tube, represents a significant leap in mobility aid technology (Fig. 23.9).

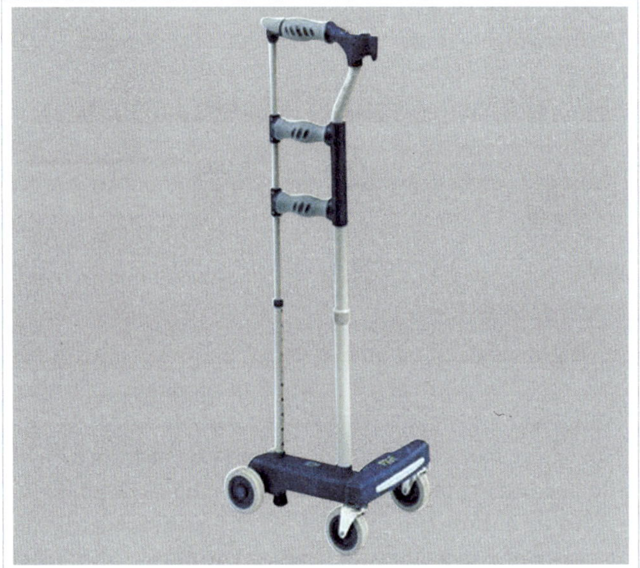

Fig. 23.9: Wheeled rolling cane

- **Uses and indications:** Its wide base allows for continuous forward movement, avoiding the need to constantly lift and place the cane. This function not only improves stability but also promotes smoother and faster movement, making navigation more effective for users.
  - The cane provides tailored support with a curved handgrip for comfort and adjustable height to accommodate diverse user preferences.
  - Furthermore, incorporating a pressure-sensitive brake into the handle provides an additional degree of safety and control.
  - Many users regard the wheeled base cane as a significant improvement in mobility support, combining stability, efficiency, and comfort. Gait improvements are seen in patients with hemiparesis using rolling canes.

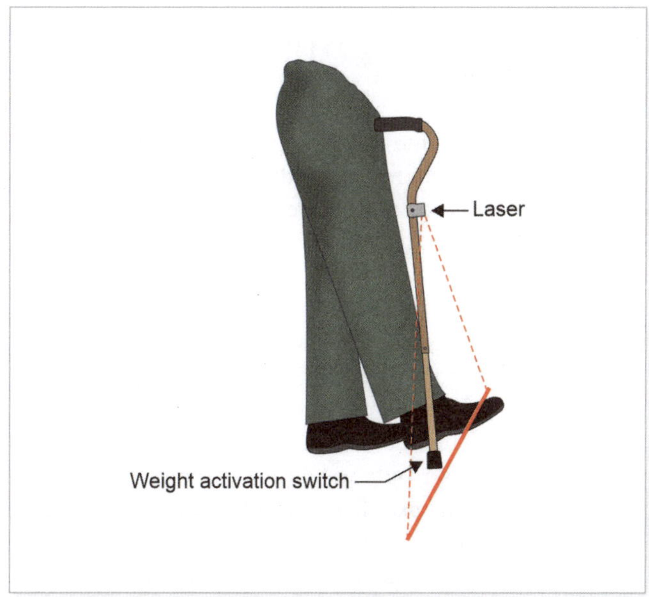

Fig. 23.10: Laser cane

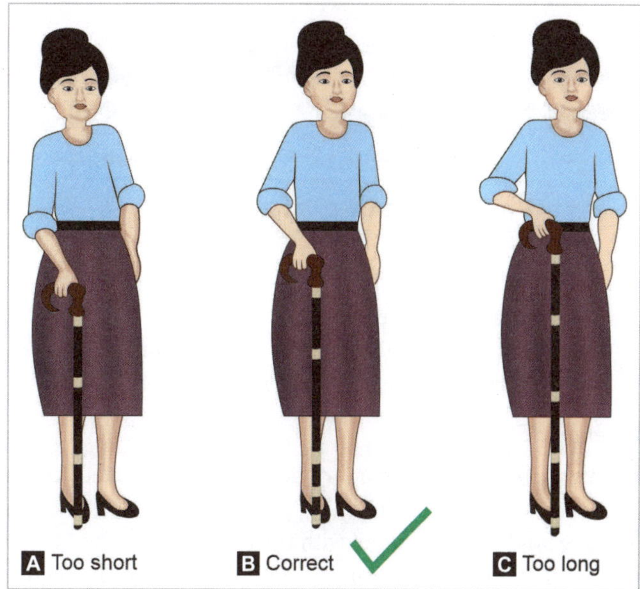

Figs 23.11A to C: Measurement of cane

- **Drawbacks:**
  - Despite its benefits, the cane has a higher cost than typical quadruped canes and requires sufficient upper extremity power to engage the brake mechanism.
  - Furthermore, its design may not be appropriate for those with certain gait patterns, such as those exhibited in Parkinson's disease.

Laser Cane

- A laser cane is a mobility-aid device used to help people with visual impairments or balance concerns when walking (Fig. 23.10).
- It usually consists of a normal cane with a laser pointer on the tip.
- **Uses and indications:**
  - The laser shines a small, visible beam onto the ground in front of the user, delivering a visual signal and improving spatial awareness. This can help users notice impediments or changes in terrain, allowing them to move more safely.
  - Laser canes are especially useful for people with impaired vision or who have difficulties keeping balance while walking.
  - They boost independence and confidence by combining traditional tactile input with visual hints.

Measurement

The patient should be standing with their feet comfortably apart and shoulders relaxed in order to determine the right cane height. The cane's tip or center is positioned 15 cm (6 inches) laterally from the toes. The elbow should be flexed to roughly 20–30°, and the top of the cane should be levelled with the greater trochanter of the hip. The most significant factor affecting cane height is elbow flexion because of variations in trunk and limb length (Figs 23.11A to C). The arm length can alter throughout the various walking phases when the elbow is flexed 20–30°, and this also absorbs floor forces.

Positioning while Walking

- Maintain good posture with the head up and look forward to avoid tripping on objects on the floor.
- Be sure the cane is out of the way but within reach when sitting down or rising from a chair.
- The top of cane should touch the wrist crease while standing upright. Holding the cane properly requires a small bend in the elbow.
- The cane is held in the hand that is not on the side that requires assistance. For example, the cane is held in left hand, for instance, if right leg is hurt.

Gait Patterns

- The majority of patients propel the cane forward while simultaneously moving the affected lower extremity first, followed by the unaffected lower extremity.
- For patients with bilateral weakness, the use of two canes may be desired. A two-point or four-point gait can be used in this situation.
  - **Two point gait:** A two-point gait involves simultaneously moving one cane and the more affected lower extremity forward, then the second cane and the less affected lower extremity (Figs 23.12A to C).
  - **Four point gait:** In a four-point gait pattern, the cane is moved forward first, then the lower extremity that is most affected, the other cane, and finally the lower

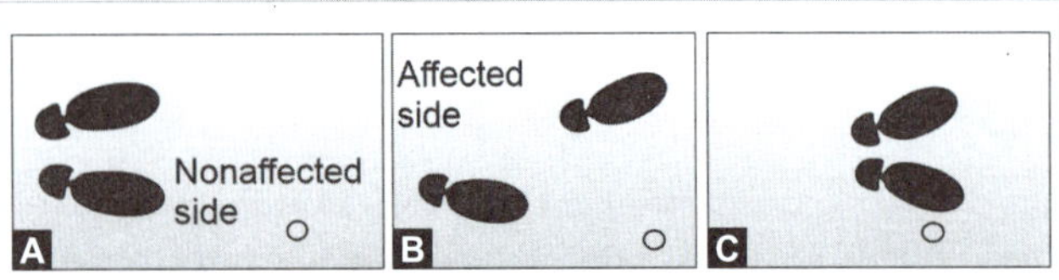

Figs 23.12A to C: Two point gait

extremity that is less affected (e.g., if the right side is affected, then the sequence is right cane, left lower extremity, left cane, right lower extremity).

Ascending Stairs

- Put the cane in the hand opposite to the affected leg to ascend stairs (Fig. 23.13A).
- Take hold of the handrail with the free hand.
- Rise first on the healthy limb (Fig. 23.13B), then rise on the injured leg (Fig. 23.13C).

Descending Stairs

- Place the cane on the step first when descending stairs (Fig. 23.13D).
- Next, place the hurt leg on the step (Fig. 23.13E).
- Lastly, place the healthy leg—the one that supports body weight—on the step (Fig. 23.13F).

Walker

- An aid that has four points of contact with the ground is a walker. It is more frequently used to stabilize people with poor balance and movement or lower extremity impairment since it offers a wider base of support than a walking stick.
- Walkers have a metal frame with four legs, commonly known as Zimmer frames. On the two front legs, they may contain wheels to facilitate forward motion.
- Walkers have far wider contact with the ground than canes and crutches, offer more stability and balance.
- Typically, the user puts weight on the frame's two sides, benefiting from excellent stability while retaining a respectable level of independence.
- Even though they are bigger than canes and crutches, walkers are typically nonetheless lightweight and collapsible.

Maintenance

- The brakes are checked regularly to make sure they are functioning properly.

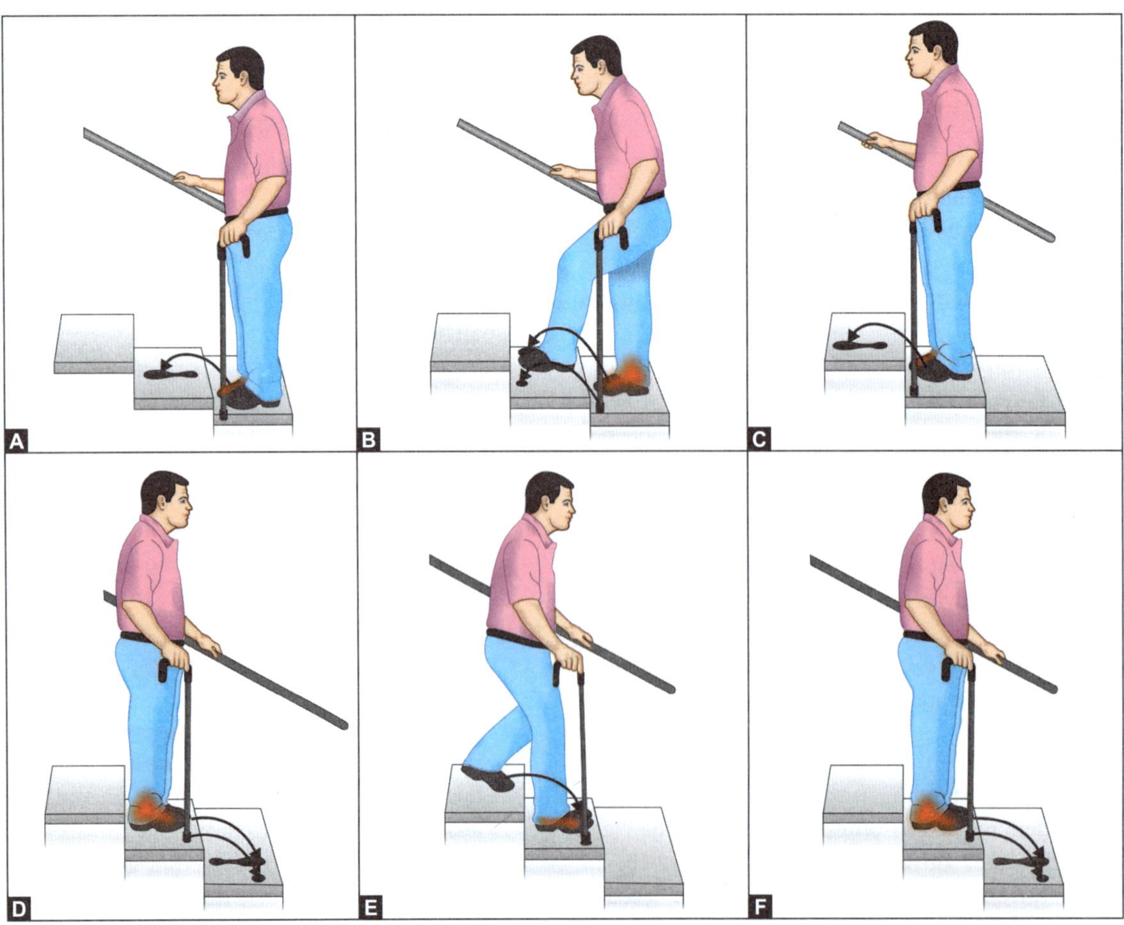

Figs 23.13A to F: Gait training on stairs using cane: **A to C.** Ascending; **D to F.** Descending (**Note:** The affected side is marked with red.)

- Wheels should be examined frequently, and whenever user notices tracking problems, such as a frame that pulls to one side. Wheels should also be inspected since occasionally, forceful braking that occurs repeatedly might cause the wheel to lose its roundness.
- Ferrules should be frequently inspected and replaced as needed. If the walker is used outside, wear and tear could happen more quickly.
- Periodically the frame's height is checked to make sure it is at the right height for the user's posture. If the user is receiving rehabilitation and changes in the balance, strength, or posture as anticipated or shown during assessment, it is extremely crucial to get rid of this issue at the earliest.

Types

Standard Walker

1. Four rubber-tipped, nonskid legs on this walker offer stability.
2. To move, it must be picked up.
3. A standard walker is identical to two wheeled walker, but without wheels on the back legs.

Wheel Walker

- This walker is helpful when patient requires some, but not continuous, weight-bearing assistance.
- It includes wheels on the two front legs two back legs which are rubber-tipped, nonskid, the frame of the walker being created from aluminum (Fig. 23.14).
- Normally it can be folded flat for easy storage and transportation.
- There are no brakes.

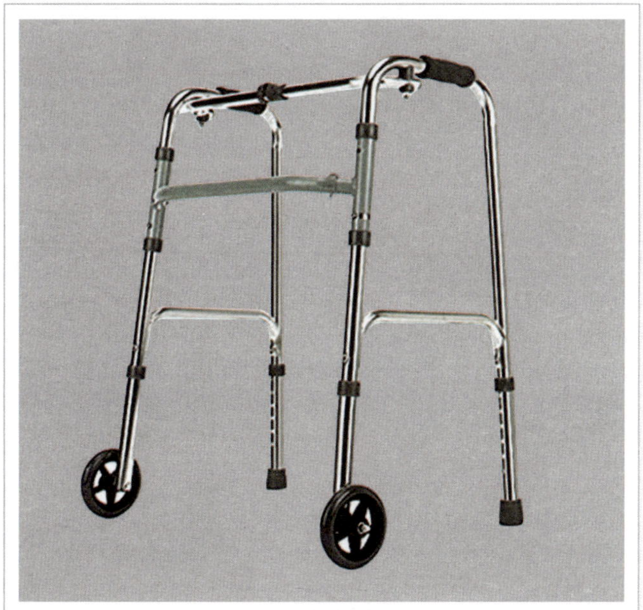

Fig. 23.14: Wheel walker

Fig. 23.15: Three-wheeled walker

Three-Wheel Walker

The walker has three-wheels attached at bottom. It supports balance but is lighter and easier to move (Fig. 23.15).

Rollator

- These are four-wheeled walkers designed for people who don't need to rely on them for balance. Big wheels are for outdoor use and small wheels for indoor use (Fig. 23.16).
- It can be folded.
- A rollator is constructed of lightweight materials, such as aluminum or carbon fiber, which makes transporting the frame simpler.
- Seats, baskets, and trays can be connected.
- Other adjustments, such as a light that displays a red line on the ground to direct step length in Parkinson's patients may be present.

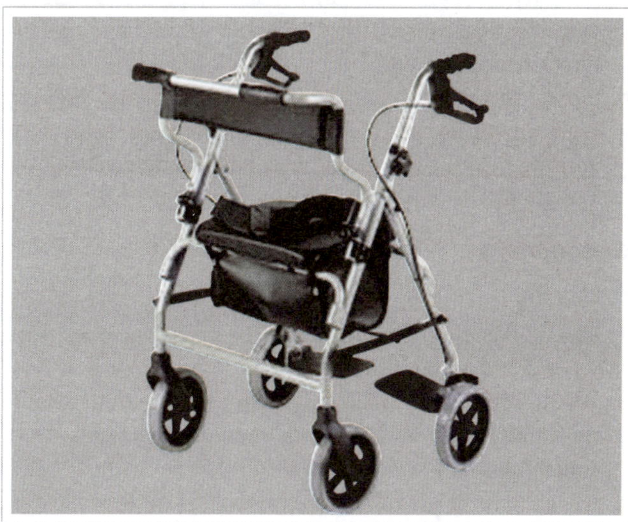

Fig. 23.16: Rollator

Fig. 23.17: Knee scooter

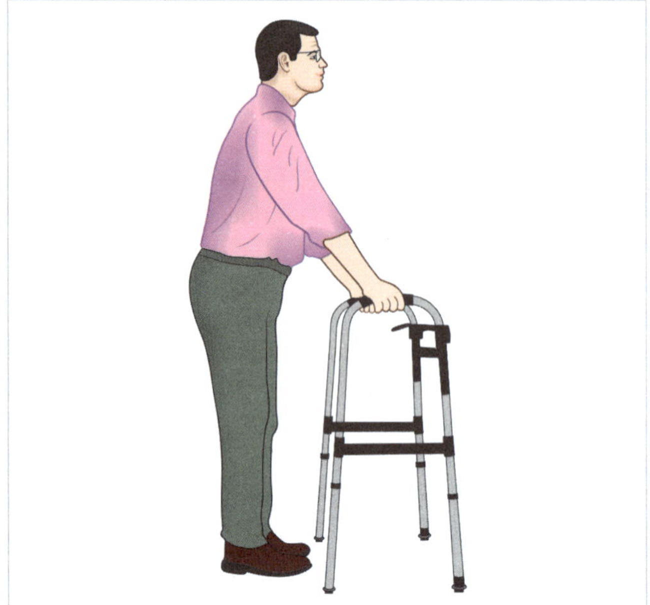

Fig. 23.18: Starting position with standard walker

- Given the number of wheels, rollators typically feature hand brakes for added safety.
- Seniors who require more walking support than average but do not wish to contemplate a wheelchair for the time being find rollators very helpful. They offer the user excellent safety and stability, but similar to walkers, they can be challenging to use on a daily basis because of their size, weight, and mass.

Knee Scooter

- Although similar to a rollator, a knee scooter/walker allows the user to rest one knee on a cushion and maneuver the walker with one leg (Fig. 23.17).
- They are not the greatest choice for people with general weakness or restricted movement.
- Mobility scooters are larger, battery-powered vehicles with steering controls that resemble wheelchairs and are typically used outside.
- Similar to wheelchairs, mobility scooters are "walking replacement devices" rather than "walking help devices". Yet, they can benefit those who have trouble getting out and about.

Gait Training

Positioning

- The walker's top should touch the crease in one's wrist when standing up straight.
- When holding the walker's handgrips, the elbows should be slightly bent, and the back should be straight. Avoid slouching over the walker.
- Check to see if the rubber tips on the legs of the walker are in good condition.

Walking

- Place the walker in a forward position, about one step make sure that the walker's four legs are on levelled ground affected (Fig. 23.18).
- Place the affected leg in the middle of the walker and support by holding onto the top of the walker with both hands. Do not take a whole step forward (Figs 23.19A and B).
- As the healthy leg is raised to align with the affected leg, push firmly down on the walker's handholds. Always move carefully and in short increments when turning.

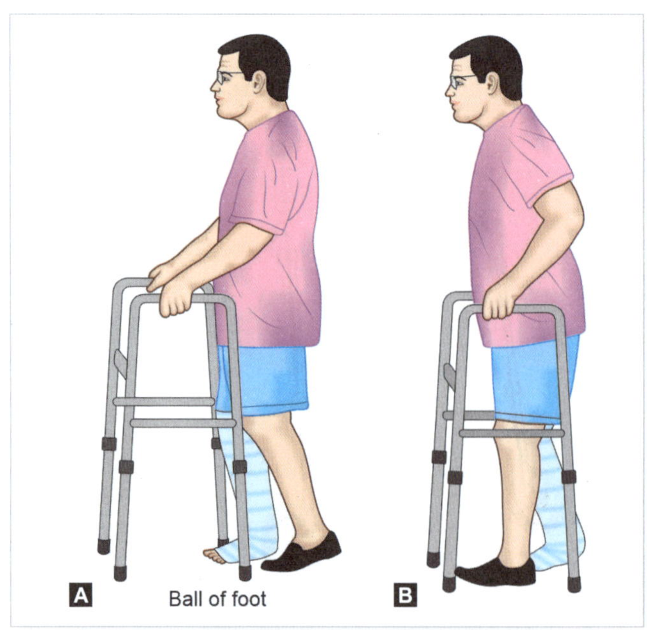

Figs 23.19A and B: Gait training with walker

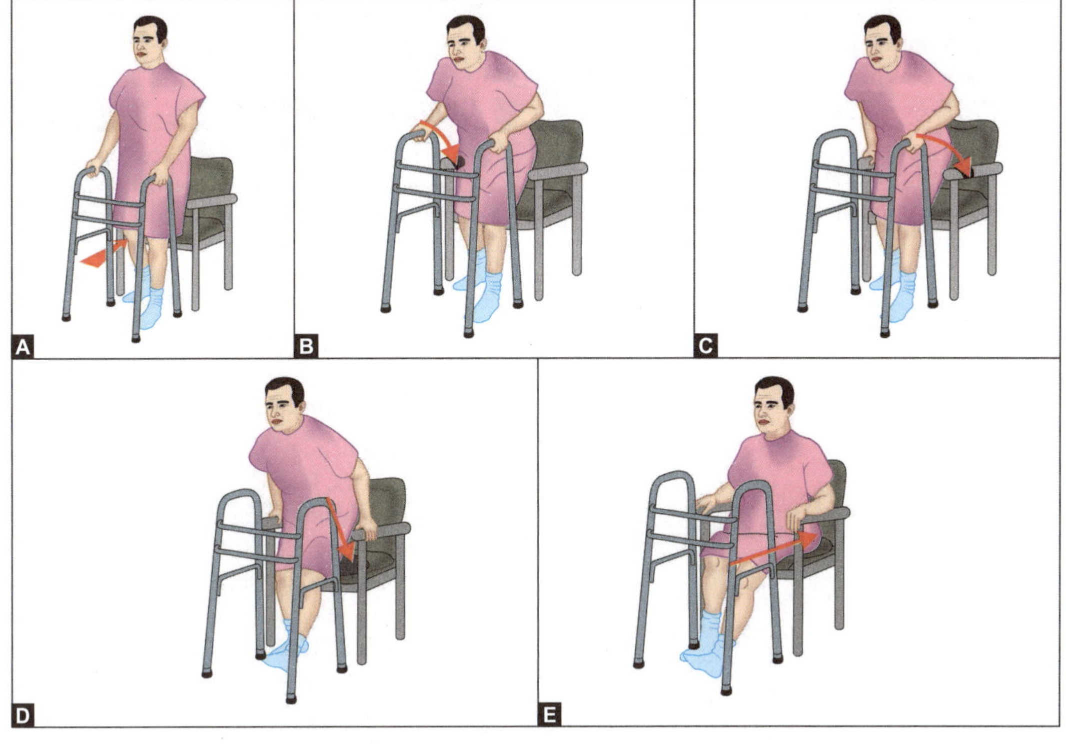

Figs 23.20A to E: Gait training—sitting with the walker

Sitting

- Lean back until the legs are in contact with the chair.
- To locate the chair's seat behind, the hands can be used.
- Gradually recline in the chair (Figs 23.20A to E).
- Use the arm strength to propel oneself up and grasp onto the walker's handgrips in order to stand.
- Refrain from tilting or tugging the walker to help stand up.

Crutches

Crutches are walking aids made of wood or metal that stretch from the floor to the area around the elbow or axilla. These devices are intended to lessen weight bearing on one or both legs and to provide assistance in situations where balance and strength are compromised.

Similar to canes, crutches shift part of the weight from the legs to the upper body. They are frequently used following a leg injury since they work best as a temporary walking aid.

Prerequisites

- Good strength of upper limb muscles is required.
- Range of motion of upper limb should be good.
- Muscle groups which should be strong are as follows:
  - Shoulder flexors, extensors and depressors
  - Shoulder adductors
  - Elbow and wrist extensors
  - Finger flexors

Types

Axillary or Underarm Crutches

1. They have a rubber ferrule, an axillary pad, and are composed of wood or metal (Fig. 23.21).
2. The length and handle height of the crutch are easily adjustable. An axillary piece on top connects two upright shafts.

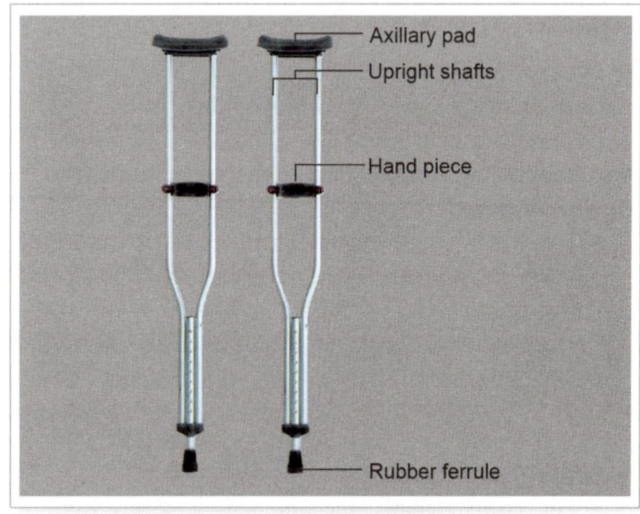

Fig. 23.21: Axillary crutches/underarm crutches

3. To allow full contact with the floor, an extension piece is affixed to a big suction tip (rubber ferrule).
4. While using axillary crutches, the axillary pad should be placed behind the apex of the axilla when no weight is being supported.
5. The elbow extends and weight is transferred down the arm to the hand piece when weight is passed through the axillary pad.

Advantages

Axillary crutches are convenient for short-term injuries, provide a lot of support for the lower body, are inexpensive, and let the user go through a wider range of gait patterns and move more quickly.

Disadvantages

- Limited movement in the upper body.
- Patients using axillary crutches must have strong standing balance.
- Improper use of the crutch can injure the axillary region and generate tension in the arms and upper torso, which can result in paralysis. The brachial plexus and radial nerve under the arms are compressed, and there is also a chance of losing one's equilibrium.
- The upper body power required to utilize axillary crutches may be absent in elderly patients, or they may feel anxious.

Precautions

- Until the patient gets acclimated to using the crutches, he must have an assistant nearby.
- Regular checks to make sure all the pads are securely in place are required.
- Crutch tips must be made free of stones and debris by cleaning them out.
- Any small, loose carpets from the walkways are removed.
- An eye out for icy, waxed or wet flooring must be kept.
- Crowded areas must be steered clear off.
- A backpack should be used instead holding anything in the hands.

Measurement of crutch length:

- **In lying with shoes off:** Measurement is taken using a measuring tape from the axilla's apex to the medial malleolus' lower margin (this is the most accurate way).
- **With shoes on:**
  - For crutch length, measurement is taken from 5 cm below the axilla's peak to 20 cm laterally to the heel of the shoes.
  - For hand piece, a spot is marked 5 cm below the axilla's apex and measurement taken until the ulnar styloid process to determine the distance from the axillary pad to the hand grasp keeping the elbow flexed to 15°.
- **While in standing position:**
  - For crutch length, measurement is taken from 6″ anterior to the foot to 2″ lateral to the axilla.
  - The hand piece should be set to enable 20°–30° of elbow flexion with the shoulder relaxed.

Clinical Correlation

Users of axillary crutches who place their weight on the shoulder rest can develop **Crutch Palsy**. The radial and ulnar nerves may become paralyzed as a result of strain on the brachial plexus. Crutch palsy can be avoided by adding more padding to the shoulder rest. Patients should be educated about the dangers of placing their body weight on the shoulder rest. This damage may be treatable with care.

Elbow or Forearm or Lofstrand Crutches

- They have a handgrip on an aluminum tubular shaft and a metal or plastic forearm band.
- The forearm component is stretched to 2 inches below the elbow and bent backward.
- Given that the handgrip and forearm component of the crutches may both be extended using a press clip or metal button, these devices are suitable for patients with strong arms and good balance and coordination. The same method of weight transmission as with axillary crutches is used (Fig. 23.22).
- The crutch may be adjusted both proximally to change the forearm cuff's position and distally to change the crutch's height. A push button mechanism is an option.
- Forearm crutches let the hands be used without taking the crutch off.
- They enable for practical stair climbing activity and are simply adjustable.

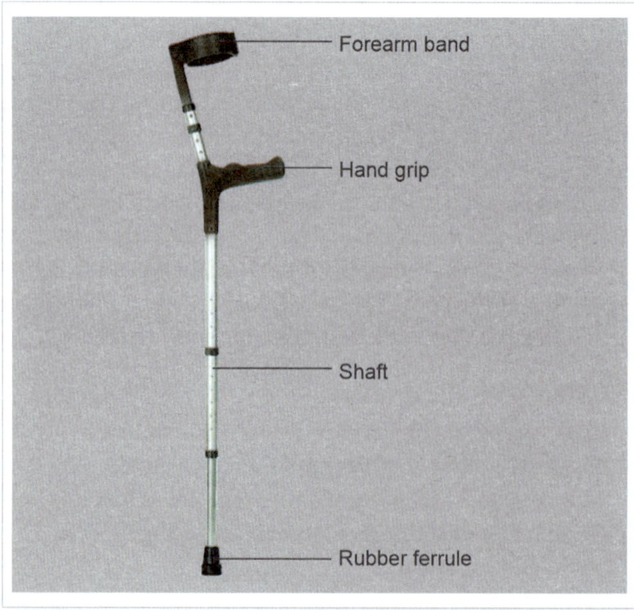

Fig. 23.22: Elbow crutch

Advantages

- Compared to axillary crutches, using forearm crutches does not require more energy, more oxygen, or a higher heart rate.
- Being portable and easy to store, forearm crutches are easily adjustable, lightweight, and suitable for manual tasks.
- There is no chance of injury to the neurovascular structures in the axillary region when using this type of crutch.

Disadvantages

- Forearm crutches are less stable and require strong upper body muscles and good standing balance.
- There are times when older patients feel uncomfortable using these crutches. It is plausible that they do not possess the necessary upper-body strength to utilize forearm crutches.

Measurement of crutch length:

- **In lying position:** The measurement is made from the ulnar styloid process to a position 20 cm laterally to the heel of the shoe when the subject is lying down with shoes on.
- **While standing:**
  - Measurement is taken from 6″ anterior and 2″ lateral to the foot.
  - Height should be set to allow for 20°–30 of elbow flexion while shoulders are relaxed.

Proper position for the arm cuff:

- The mechanical advantage increases with the position of arm cuff on the forearm. As one bends down to pick something up off the ground, the cuff will bite into upper arm if it is placed too high up.
- The forearm's proximal third, or about 1–1.5 inches below the elbow, should be measured.

Forearm Support or Gutter or Platform Crutches

- They have a padded forearm support platform, a velcro strap, an adjustable hand piece, and a rubber ferrule. They are composed of metal.
- Patients with painful wrist and hand conditions, elbow contractures, or weak hand grips can utilize these.
- With the elbow flexed 90°, the hand rests on a grip which can be oriented suitably, depending on the user's disability.
- These are prescribed mostly for supporting people with rheumatoid arthritis, cerebral palsy, or other diseases as due to pain or deformity, the patient cannot bear weight on his hands, wrists, or elbows under these circumstances (Fig. 23.23).

Measurement of crutch length:

In lying position: Measurement is taken from the fixed elbow position to 20 cm laterally to the heel while wearing shoes.

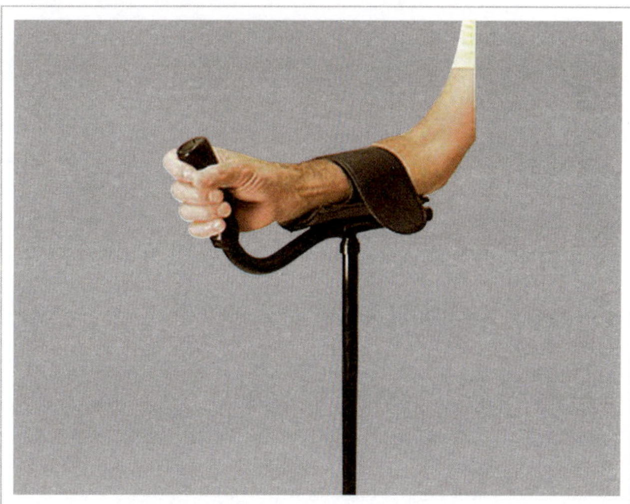

Fig. 23.23: Forearm support crutch

Advantages

- They are easily adjusted, similar to elbow crutches.
- Better looking than other crutches.

Disadvantages

- Provide less lateral support due to absence of axillary pads
- Cuffs are difficult to remove
- These might be costly.

Triceps Crutches

- Triceps crutches are meant to give the wearer more assistance while walking by acting like the triceps muscles. These are similar to the axillary crutch, but with a proximal end at mid-arm level.
- Originally created for polio sufferers, triceps crutches (also known as Warm Springs crutches) are now primarily used by people with other disabilities.
- The crutch is intended for use by people with paralyzed shoulder muscles.
- Each crutch has medial and lateral uprights that are connected by a pair of posterior bands at the proximal end. These bands maintain the elbow extended by simulating the action of the triceps muscle. A single shaft with a rubber tip at the base makes up the distal portion of the crutch.

Gait Training

Progression is made to crutch walking only when the technique between the bars is good.

The change from walking in bars to crutch walking is considerable, and all patients are initially unstable and fearful. A high degree of strength and balance skill is essential and this is only achieved with perseverance and much practice.

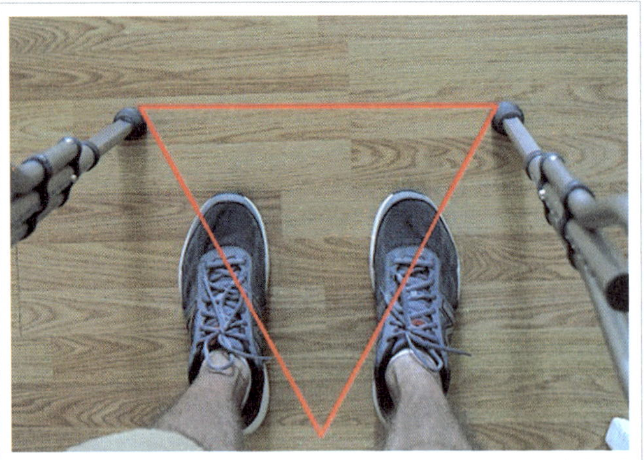

Fig. 23.24: Standing with tripod stance

Standing

When one stands still, the tripod position gives the body the most support and takes the weight off the injured leg (Fig. 23.24).

Place the crutches in front of the body at a 45° angle to the good foot, keeping it firmly planted on the ground.

One does not get the assistance they need if they move the crutches too near to the body or too far away from it.

Walking

There are different types of gaits possible when using crutches. The "Four-Point Gait", "Partial Weight-Bearing Three-Point Gait", and "Three Point Swing-through and Swing-to Gait" are the three primary gaits (Fig. 23.25).

Single Crutch Gait

This walking type simply makes use of one crutch. The crutch is placed besides the unaffected lower extremity. The affected leg is carried forward while using the crutch to support the weight, following which the unaffected leg continues forward.

Two-point Gait

Alternate crutch and leg are used simultaneously to move forward. That is, right crutch and left leg advance first, then the left crutch and right leg.

Four-point Gait

The four-point gait is considered the most stable walking pattern. This is most often used to help with walking in cases where one or both legs are weak. Crutch on the affected side (considering right) is moved forward followed by the unaffected leg (left). Then the crutch on the unaffected side (left) is moved forward followed by the affected leg (right). Left crutch-right foot and right crutch-left foot are the patterns.

Partial Weight-bearing Three-point Gait

This is used when the injured foot can bear some weight, but still need the assistance of crutches. To do this, both crutches are placed in front at a 45° angle, just like the tripod stance. Affected foot is stepped forward up to the crutches and then the good foot steps past the crutches.

Three-point Swing-through and Swing-to Gait

This is used to keep all pressure off the injured lower extremity (LE). These gaits are commonly employed for bilateral LE involvement, such as in spinal cord injury (SCI). The swing-to gait involves moving both crutches forward, shifting weight to the hands, and 'swinging to' the crutches. The swing-through gait involves moving the crutches forward together, shifting weight to the hands, and swinging the lower extremities beyond the crutches.

Use of Crutches on Stairs

When on stairs and no railing is available or at a curb, the following instructions are used.

- **Ascending/Up:**
  1. With the weight on the hand grips, take a powerful step up. Next is the weak or operated leg (Fig. 23.26A).
  2. Do not bear any weight on the affected or weak leg.
  3. Finally, the crutches are brought up (Fig. 23.26B). Make sure the toes pass the steps if in a cast before moving on to the next one (Fig. 23.26C).
- **Descending/Down:** Make sure the toes are in close proximity to the stair edge in order to reverse the process.
  1. Position both crutches on the lower stairway (Fig. 23.27A).
  2. Use the weak or affected leg to descend.
  3. Use the stronger leg to step down and apply pressure on the hand grips (Fig. 23.27B).
  4. Set down the crutches once more and move forward (Fig. 23.27C).

If a railing is available, the following steps should be followed:

- **Ascending/up:**
  1. Put the crutches on the side that is not facing the railing (Fig. 23.28A).
  2. Step up with the stronger leg while holding to the railing and crutches (Fig. 23.28B).
  3. Next, take the crutches and the weak or injured leg up to the same step (Fig. 23.28C).
- **Descending/down:**
  1. Put the crutches in the opposite direction of the railing.
  2. Lower the crutches first (Fig. 23.29A).
  3. Use the leg that is weaker or operated on to get down.
  4. Shift the weight onto the crutches and railing.
  5. Take the next step down using the strong leg (Figs 23.29B and C).

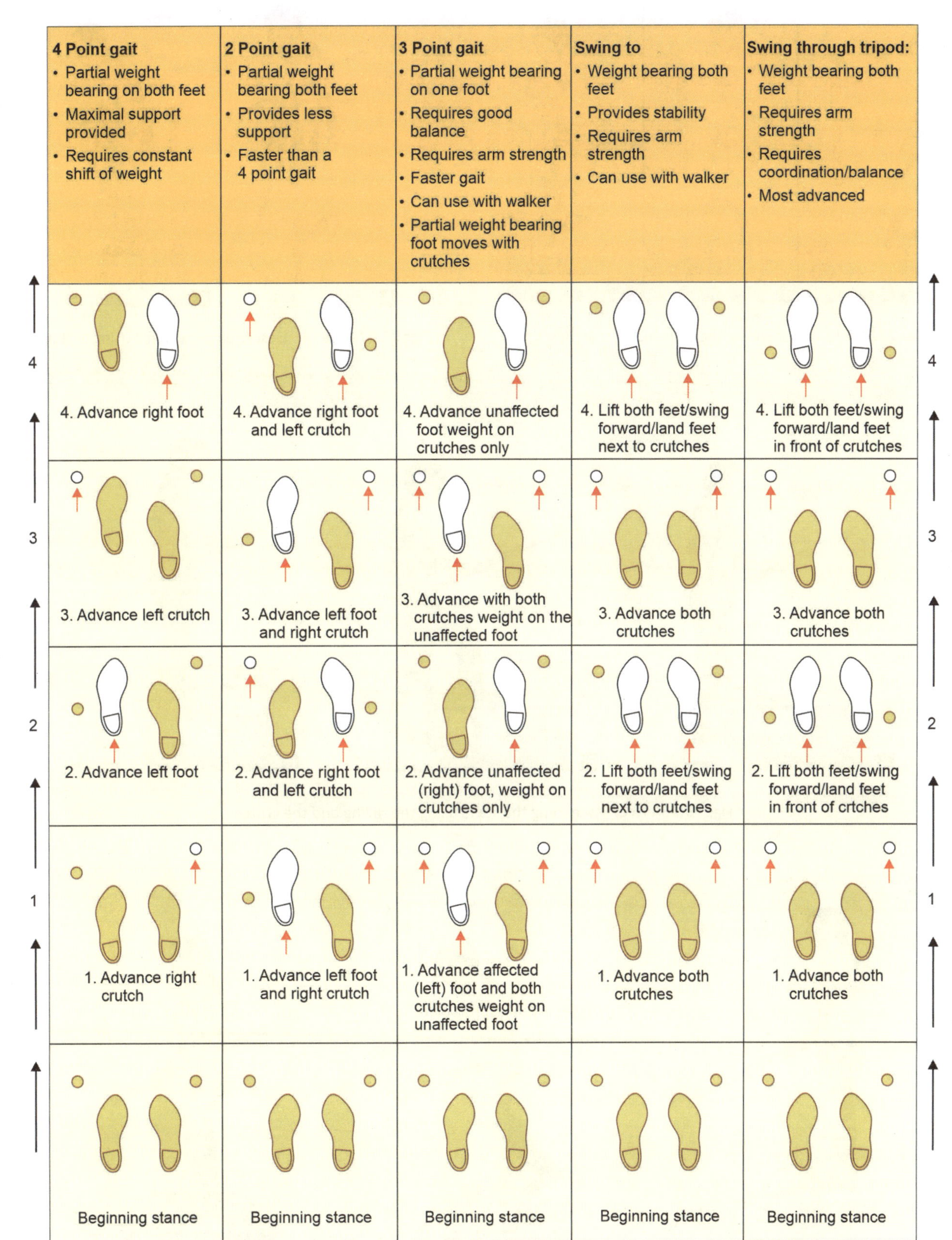

Fig. 23.25: Gait pattern when walking with crutches

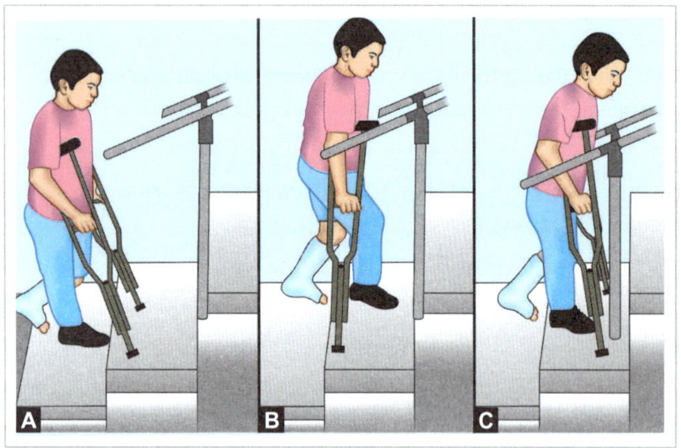

Figs 23.26A to C: Ascending the stairs using crutches

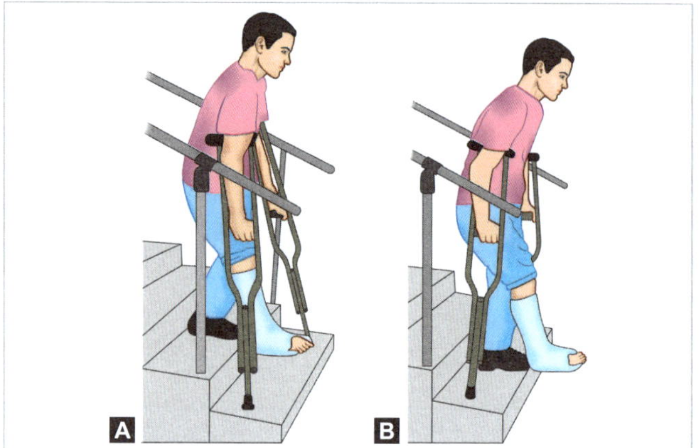

Figs 23.27A and B: Descending the stairs using crutches

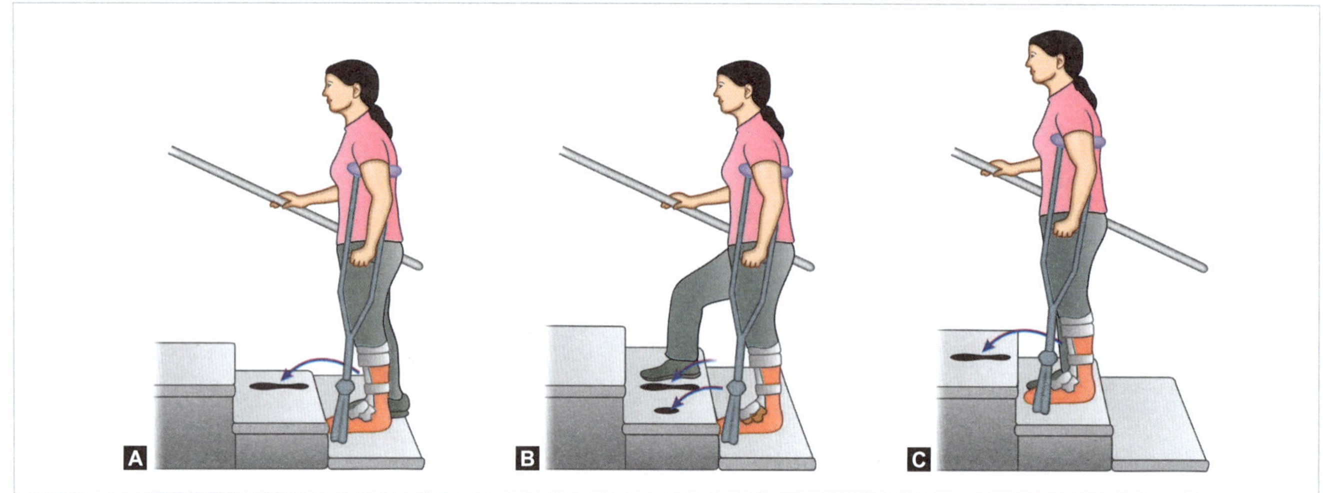

Figs 23.28A to C: Ascending the stairs using the railing and the crutch

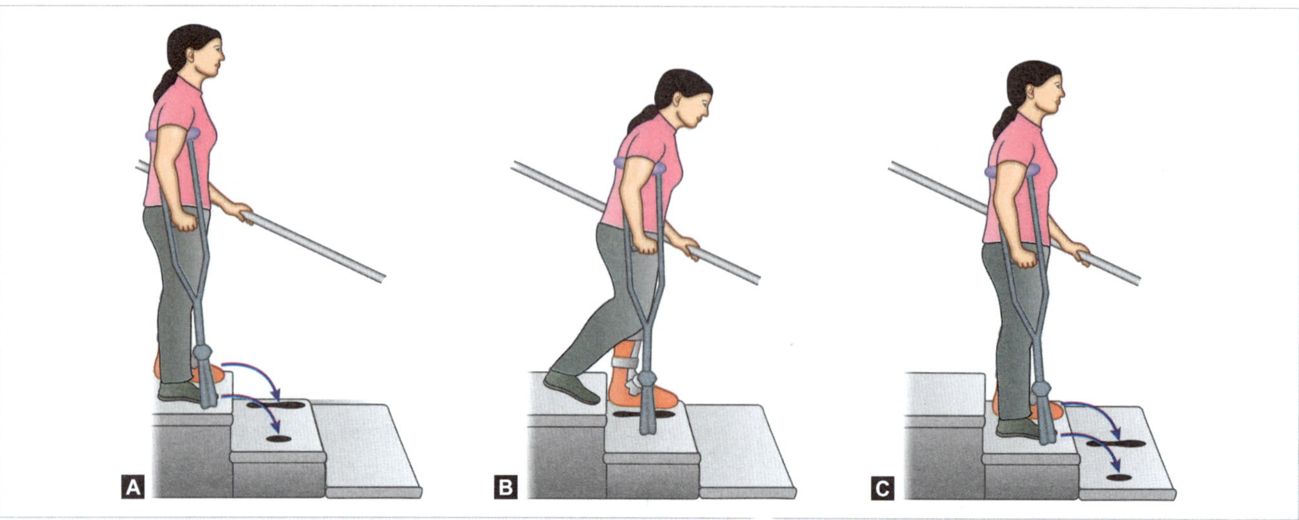

Figs 23.29A to C: Descending the stairs using the railing and the crutch

STRENGTH TRAINING

Exercises that prepare the patient to walk while using crutches are performed. These exercises are designed to strengthen the muscles in the upper body that support the patient's weight when using crutches to walk. The following muscle groups are essential for using crutches:

Shoulder Depressors

Muscles in the shoulders that depress the shoulders to stabilize the upper extremities and stop shoulder elevation (Figs 23.30A and B).

Shoulder Adductors

Muscles in the shoulders that act as crutch holders on the chest wall (Figs 23.31A and B).

Shoulder Flexor, Extensor, Abductors

Crutches can be moved forward, backward, and sideways using the arm flexor, extensor, and abductor muscles (Figs 23.32A to D).

Wrist Extensors

Weight bearing on hand pieces is made possible by the wrist extensor muscles (Figs 23.33A and B)

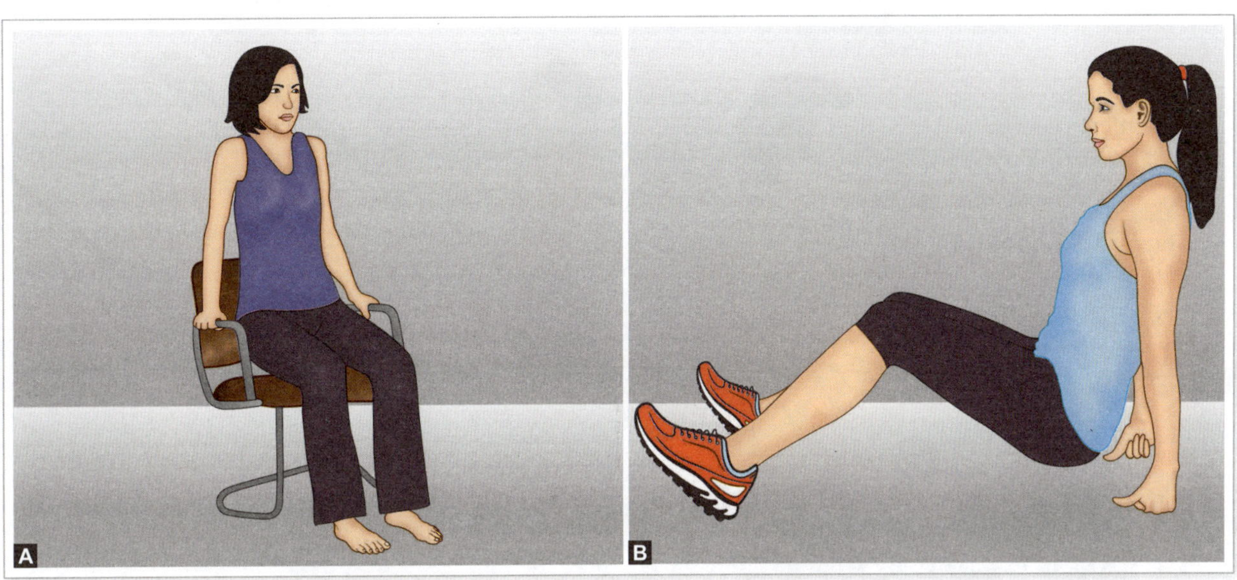

Figs 23.30A and B: Strengthening of shoulder depressors

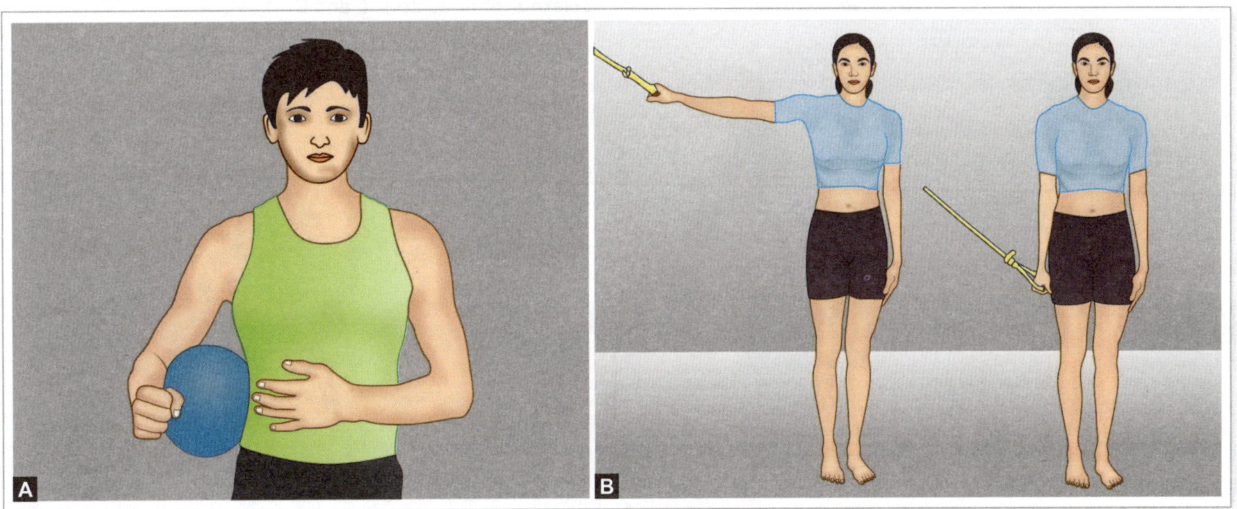

Figs 23.31A and B: Shoulder adductors strength training

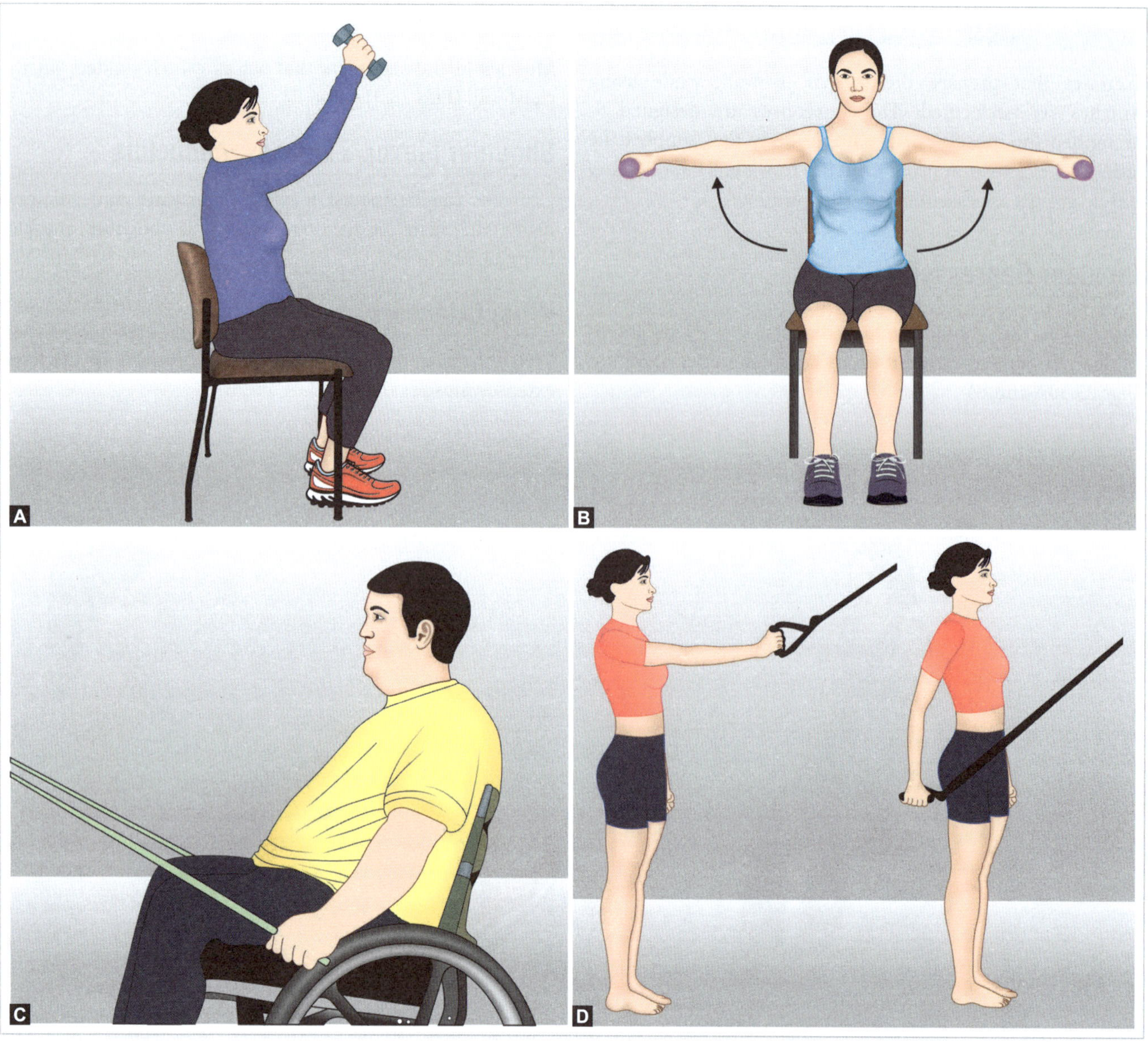

Figs 23.32A to D: Strengthening of shoulder: **A.** Flexors; **B.** Abductors; **C and D.** Extensor

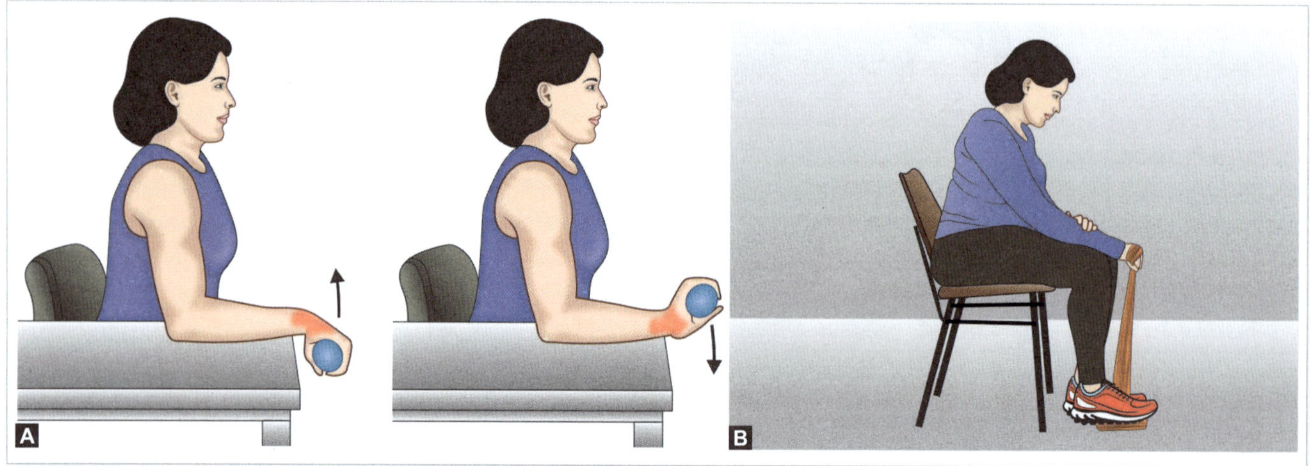

Figs 23.33A and B: Strengthening of wrist extensors

Finger and Thumb Flexors

The hand piece is grasped by the flexor muscles of the finger and thumb (Fig. 23.34).

Fig. 23.34: Strengthening of finger and thumb flexors

BALANCE DRILLS

To help the patient regain sufficient control over the trunk and pelvis, balance exercises are administered (Figs 23.35A to F).

Clinical Correlation

- Elbow triceps function loss (loss of elbow extension) together with weakness during supination, loss of thumb extension, radial abduction, and digit MP extension are the main symptoms of crutch palsy (axilla level, motor and sensory involvement). While elbow extension for functional tasks can be aided by gravity, the therapist should keep an eye out for the development of elbow flexion contractures, which are caused by the unopposed tightening and adaptive shortening of elbow flexors.
- The utilization of axillary crutches has been correlated with axillobrachial arterial complications, Conversely, the use of forearm crutches has been predominantly linked to compressive neuropathies stemming from pressure exerted by the forearm cuff.
- To reduce injury risk, doctors should prescribe crutches that are appropriately fitted, engage in appropriate gait training, be aware of prevalent crutch-related injuries, and examine potential patient-specific injury risk factors.

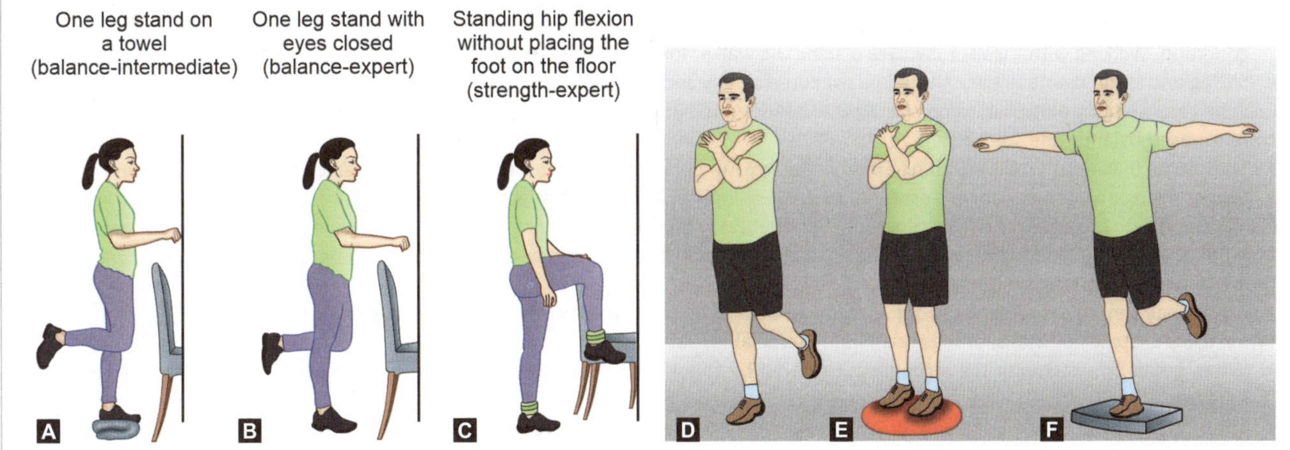

Figs 23.35A to F: Balance drills

SUMMARY

- Gait training is a specific rehabilitation therapy that improves walking ability by addressing biomechanics, coordination, strength, and endurance. It is tailored to each individual's needs, taking into account medical issues and functional goals.
- Gait training uses a variety of strategies, including technology, assistive equipment, and exercises, to increase mobility and lower the risk of falling. It is recommended for symptoms such as muscle weakness, joint stiffness, poor balance, and limited community engagement.
- Walking aids, also known as assistive devices (ADs), are provided to patients who have difficulties maintaining a regular gait cycle or balance as a result of lower limb injuries.
  - They offer additional support, protect injured legs, and aid in weight distribution. Walking aids serve several functions, including expanding the base of support, maintaining the center of gravity, transferring weight bearing, compensating for weak muscles, reducing pain, and enhancing balance.
  - The kind of gait instructed is determined by the patient's weight-bearing state, which can range from non-weight bearing (NWB) to weight bearing as tolerated.
  - Body-weight support (BWS) systems, such as unweighting devices, are also used to help patients walk before regaining full motor control and strength. Gait training with BWS devices have shown potential in improving walking ability, particularly in patients with partial spinal cord injuries. Common types of walking aids include canes, walkers/frames, and crutches.
- Canes, a sort of assistive device, help with walking by improving balance and support; they are commonly used for leg injuries. They alleviate discomfort, promote stability, and distribute weight.
  - Standard canes, quad canes, forearm canes, hemi canes, rolling canes, and laser canes all have their own set of characteristics and benefits.
  - Proper measurement promotes a good fit, which usually aligns with the wrist crease and the greater trochanter of the hip.
  - When walking with a cane, it is critical to maintain proper posture and alignment.
  - Gait patterns consist of thrusting the cane forward while moving the affected leg, with adjustments for bilateral weakness. For stairs, position the cane opposite the injured leg, with the free hand clutching the handrail.
- A walker, an assistive device with four points of contact with the ground, aids individuals with poor balance or lower extremity impairment by offering a wider base of support. It enhances stability and independence while being lightweight and collapsible.
  - Maintenance involves regular checks on brakes, wheels, and ferrules.
  - Types include standard walkers, wheel walkers, three-wheel walkers, rollators, and knee scooters, each designed for specific needs and preferences.
  - Gait training involves proper positioning, walking techniques, and sitting and standing maneuvers.
  - Walker tops should align with wrist creases, elbows should be slightly bent when gripping, and rubber tips should be in good condition. Walking with a walker involves placing the walker forward, stepping carefully, and maintaining stability during movement.
  - Sitting and standing maneuvers should be executed slowly and with arm support.
- Crutches are mobility aids used to reduce weight-bearing on injured or weak legs, providing support and stability for individuals with compromised balance and strength.
  - They come in various types, including axillary, forearm, and platform crutches, each offering different levels of support and comfort.
  - Proper fitting and adjustment are essential to prevent injuries and ensure effectiveness.
- Gait training with crutches involves strengthening upper body muscles, improving balance, and learning different gait patterns such as four-point, two-point, and swing to and swing through gait. Patients must practice standing and walking with crutches while maintaining proper posture and stability.
  - Additionally, exercises targeting shoulder, arm, wrist, and hand muscles are crucial for effective use of crutches.
- When using crutches on stairs, proper technique is crucial for safety.
  - When going up, start with the strong leg, followed by the weak or operated leg, while placing weight on the hand grips of the crutches.
  - Then, bring the crutches up. Going down, place both crutches on the lower step, step down with the weak leg, followed by the strong leg while using the hand grips for support.
- If a railing is available, utilize it for added stability.
  - When going up, place the crutches on the opposite side of the railing, hold onto the railing, step up with the strong leg, then bring up the weak leg and crutches.
  - When going down, place the crutches opposite the railing, bring them down first, step down with the weak leg, and then with the strong leg while using the railing and crutches for support.
- Crutch palsy, characterized by motor and sensory involvement at the axilla level, can result in loss of triceps function, weakness in supination, and other deficits.
  - Proper monitoring and adaptation are essential to prevent complications.

FURTHER READINGS

- Bertrand K, Raymond MH, Miller WC, Ginis KA, Demers L. Walking aids for enabling activity and participation: a systematic review. American Journal of Physical Medicine and Rehabilitation. 2017 Dec 1;96(12):894-903.
- Deltombe T, Leeuwerck M, Jamart J, Frederick A, Dellicour G. Gait improvement in adults with hemiparesis using a rolling cane: A cross-over trial. Journal of rehabilitation medicine. 2020 Jul 29;52(7):1–5.
- Edelstein J. Canes, crutches, and walkers. InAtlas of orthoses and assistive devices. Elsevier. 2019 Jan 1 (p. 377–382).
- Edelstein JE. Assistive devices for mobility: canes, crutches, walkers and wheelchairs. Physical rehabilitation: evidence-based examination, evaluation, and intervention. 1st ed: Saunders. 2007:877–96.
- Gardiner Dena M. Principles of exercise therapy. CBS Publishers and Distributors Pvt Ltd; 4th ed., 2023.
- Joyce BM, Kirby RL. Canes, crutches and walkers. American family physician. 1991 Feb 1;43(2):535–42.
- Kisner C and Colby LA. Therapeutic Exercises: Foundations and Techniques. 7th Edition, 2019.
- Laufer Y. The effect of walking aids on balance and weight-bearing patterns of patients with hemiparesis in various stance positions. Physical therapy. 2003 Feb 1;83(2):112–22.
- Manocha RH, MacGillivray MK, Eshraghi M, Sawatzky BJ. Injuries associated with crutch use: a narrative review. PM and R. 2021 Oct;13(10):1176–92.
- Neto AF, Elias A, Cifuentes C, Rodriguez C, Bastos T, Carelli R. Smart Walkers: Advanced Robotic Human Walking-Aid Systems. Intelligent Assistive Robots: Recent Advances in Assistive Robotics for Everyday Activities. 2015:103–31.
- O'Sullivan SB, Schmitz, Thomas J, George D, Physical Rehabilitation, 6th edition (2014). Faculty Bookshelf. 85.
- Rasouli F, Reed KB. Walking assistance using crutches: A stae-of-the-art review. Journal of biomechanics. 2020 Jan 2;98:109489.
- Warees WM, Clayton L, Slane M. Crutches. InStatPearls [Internet] 2022 Apr 30. StatPearls Publishing.

STUDENT ASSIGNMENT

LONG ANSWER QUESTIONS

1. Explain types of crutches with their respective measurements, indications and gait training on ground and stairs.
2. Discuss types of canes, indications, measurements and gait training methods.
3. Describe walking frames and their types.

SHORT ANSWER QUESTIONS

1. Define gait training and enumerate the indications.
2. Write about the different types of walking aids.
3. Enlist the precautions to be taken while using axillary crutches.
4. Write notes on:
   a. Axillary crutches
   b. Types of cane
   c. Gait training with crutches
   d. Crutch walking muscles

MULTIPLE CHOICE QUESTIONS

1. Which type of walking aid provides the widest base of support?

a. Canes
b. Walkers
c. Crutches
d. Knee scooters

2. What type of gait pattern is commonly used with crutches when both legs are weakened?

a. Two-point gait
b. Four-point gait
c. Swing-through gait
d. Partial weight-bearing gait

3. What is the primary purpose of using forearm crutches over axillary crutches?

a. Greater stability
b. Easier maneuverability
c. Reduced upper body strain
d. Enhanced weight-bearing capacity

4. Which condition may occur due to improper usage of axillary crutches?

a. Wrist sprain
b. Elbow contracture
c. Shoulder dislocation
d. Hip fracture

5. How should crutches be positioned when ascending stairs without a railing?

a. Placed on the same side as the stronger leg
b. Placed on the opposite side of the weaker leg
c. Held in one hand while using the railing with the other
d. Used to push off the stairs while both legs step up simultaneously

6. Which type of crutch is particularly beneficial for individuals with painful wrist conditions?

a. Axillary crutches
b. Elbow crutches
c. Forearm support crutches
d. Triceps crutches

7. What should be checked regularly to ensure the safety while using a walker?

a. Wheel alignment
b. Hand grip padding
c. Frame height
d. Ferrule condition

8. **Which gait pattern involves moving both crutches forward together while swinging the lower extremities beyond them?**
   a. Two-point gait
   b. Four-point gait
   c. Swing-through gait
   d. Partial weight-bearing gait
9. **What is the primary purpose of using a knee scooter?**
   a. Enhancing balance
   b. Reducing upper body strain
   c. Resting one knee while moving
   d. Supporting partial weight-bearing
10. **Which nerve is at risk of compression and paralysis due to improper axillary crutch usage?**
   a. Median nerve
   b. Radial nerve
   c. Ulnar nerve
   d. Brachial plexus
11. **How should crutches be positioned when descending stairs without a railing?**
   a. Placed on the same side as the stronger leg
   b. Placed on the opposite side of the weaker leg
   c. Held in one hand while using the railing with the other
   d. Used to push off the stairs while both legs step down simultaneously
12. **Which type of crutch is commonly used by individuals with severe lower extremity paralysis?**
   a. Axillary crutches
   b. Elbow crutches
   c. Forearm support crutches
   d. Triceps crutches
13. **What is a potential consequence of crutch palsy?**
   a. Loss of elbow extension
   b. Loss of shoulder flexion
   c. Loss of knee flexion
   d. Loss of ankle dorsiflexion
14. **Which gait pattern involves moving both crutches forward simultaneously with each step?**
   a. Two-point gait
   b. Four-point gait
   c. Swing-through gait
   d. Partial weight-bearing gait
15. **What is the recommended position of the hand grip when using forearm crutches?**
   a. Directly under the elbow
   b. At the level of the wrist
   c. At the level of the ulnar styloid process
   d. Proximal to the elbow joint
16. **What is the primary function of a walker?**
   a. Reducing upper body strain
   b. Providing enhanced maneuverability
   c. Offering a wider base of support
   d. Facilitating rapid movement
17. **Which type of cane is particularly beneficial for individuals with mild to moderate antalgic gait linked to hip or knee osteoarthritis?**
   a. Standard cane
   b. Quad cane
   c. Forearm cane
   d. Hemi cane
18. **What should be checked regularly to ensure the safety of using a cane?**
   a. Ferrule condition
   b. Hand grip padding
   c. Collar tightness
   d. Handle flexibility
19. **Which type of walking aid is typically made of wood or metal and extends from the floor to the area around the elbow or axilla?**
   a. Canes
   b. Walkers
   c. Crutches
   d. Knee scooters
20. **What is the recommended positioning of the cane's handle when standing upright?**
   a. Directly aligned with the shoulder
   b. Level with the greater trochanter of the hip
   c. Touching the wrist crease
   d. Parallel to the ground

ANSWER KEY

1. b	**2.** b	**3.** c	**4.** b	**5.** b	**6.** c	**7.** d	**8.** c	**9.** c	**10.** d
11. a	**12.** d	**13.** a	**14.** a	**15.** c	**16.** c	**17.** b	**18.** a	**19.** c	**20.** c

24 Balance

Sheetal Kalra, Paridhi Sharma

LEARNING OBJECTIVES

After the completion of the chapter, the readers will be able to:

- Define Balance.
- Understand the physiology of balance.
- Explain various components of balance.
- Know about the causes for impaired balance.
- Understand the examination and evaluation of impaired balance.
- Discuss the activities for treating impaired balance.
- Describe balance retraining.

CHAPTER OUTLINE

- Introduction
- Types of Balance
- Mechanism
- Impaired Balance
- Interventions for Improving Balance
- Hip, Ankle and Stepping Strategies

KEY TERMS

Dynamic balance: Dynamic balance is the ability to maintain postural stability and alignment with the center of mass above the base of support while the body's parts are moving.

Kinesthesia: Kinesthesia, sometimes referred to as kinesthetic sense, is a basic sensory perception ability that allows us to detect and be cognizant of the position, acceleration, and movement of our body parts.

Static balance: Static balance refers to the ability to maintain postural stability and alignment with the center of mass over the base of support and the body at rest.

INTRODUCTION

Balance in biomechanics, is the ability to keep the body's line of gravity (an imaginary line perpendicular to the horizontal, that is, in the direction of gravity, and passing through the center of mass) within a support base with minimal postural fluctuations.

In other terms, balance is the ability of an individual to maintain a position, which means even distribution of weight on each side of the vertical axis, between opposing, interacting or contrasting elements, which results in equilibrium.

TYPES OF BALANCE

Static Balance

Static balance can be defined as the ability to hold the body in a fixed position. It is the ability of the body to maintain its state of equilibrium whenever at rest that means maintaining the center of gravity over the base of support resulting in minimal postural sway while at rest.

Dynamic Balance

The ability to transfer the line of gravity around the supporting base is termed as dynamic balance. It is the ability of a body to maintain the center of gravity over a base of support while the body is in motion resulting in orientation and postural stability during motion.

A well-functioning balance system allows people to observe their surroundings, recognize their orientation with respect to gravity, determine the direction of movement, and maintain postural stability under a variety of daily life situations and activities.

MECHANISM

Postural control is based on synchronization of multiple systems: sensory (vestibular, visual, somatosensory), cognitive (central nervous system), and musculoskeletal (Figs 24.1A and B). A system of sensorimotor controls, which integrates sensory information from proprioception, vision, and vestibular system (equilibrium, motion, and spatial orientation) and provides motor output to muscles of the body and eye, is responsible for maintaining and achieving balance.

For maintaining a proper balanced posture, the sensory system, the cognitive system and the musculoskeletal system work synchronously. Three peripheral sources of information—muscles and joints, eyes, and vestibular organs—send nerve impulses to the brainstem, which processes and combines it with previously learned information from the cerebellum, the center for coordination, and the cerebral cortex, the memory and thought center. An individual can sustain and have clear vision while moving because the brainstem transmits impulses to muscles that govern the movements of the head and neck, eyes, legs, and trunk as soon as sensory integration occurs.

Did You Know?

The three main components for maintaining balance of an individual are:

1. Vision
2. Vestibular function
3. Proprioception

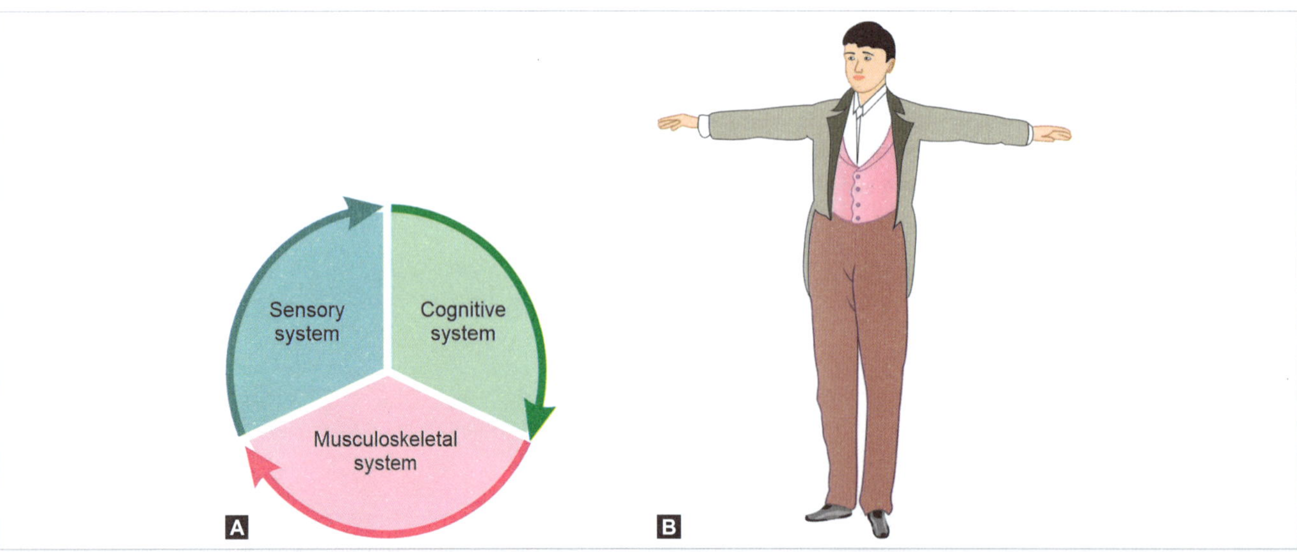

Figs 24.1A and B: Synchronization of sensory and cognitive and muscular skeletal system to maintain balanced posture of an individual

Components for Maintaining Balance

The main components that synchronize (Fig. 24.2) to maintain balance are as follows.

Vision

One of the main components of the system maintaining balance is vision, particularly static balance. Conditions hampering the perception of visual stimuli such as cataract, strabismus, etc. may disturb static balance. People with various degrees of dysfunction and in different conditions can compensate for the inability to receive visual stimuli while maintaining balance. Rods and cones are the sensory receptors present in retina and their main function is to have better vision in low light situations, say in nighttime and for identifying color respectively. The brain, after getting impulses from these receptors as visual cues, identifies how a person is oriented with respect to other objects. For example, on observing any obstacle across the path, an individual changes his/her direction while walking.

Vestibular System

The vestibular system (sensory organ of balance) is in the inner ear of all vertebrates and provides an organism with accurate information on its body movements. This information is transmitted to the brain for participation in muscular reflex arcs that function to stabilize the trunk, head and visual field during postural and locomotor activities. The major roles of peripheral vestibular system are:

- To stabilize visual images on retina's fovea during head's movements to allow clear vision
- To maintain postural stability, especially during movement of the head
- To provide information used for spatial orientation.

Proprioception

Proprioception involves the position of sense and awareness of joint at rest, along with movement's awareness which is also termed as kinesthesia; both are responsible for maintaining the static as well as the dynamic balance. The proprioceptive information which is received from spinocerebellar pathway, is processed unconsciously in the cerebellum and is responsible for controlling postural balance.

Clinical Correlation

In addition to vision, the vestibular system, and proprioception, another crucial component of balance is muscle strength. Muscles throughout the body, especially those in the legs and core, play a vital role in maintaining stability and controlling movement. Weakness in these muscles can significantly impact balance and increase the risk of falls, particularly in older adults.

Additional Factors

A few factors can be applied to maximize the body's stability, enhance equilibrium and thereby achieving balance biomechanically, factors like base of support (BOS), center of gravity (COG), body mass and friction between the body and the surface in contact affect the stability of that body.

- **Base of support (BOS):** BOS is directly proportional to stability, that means larger the base of support, greater will be the stability of that body.

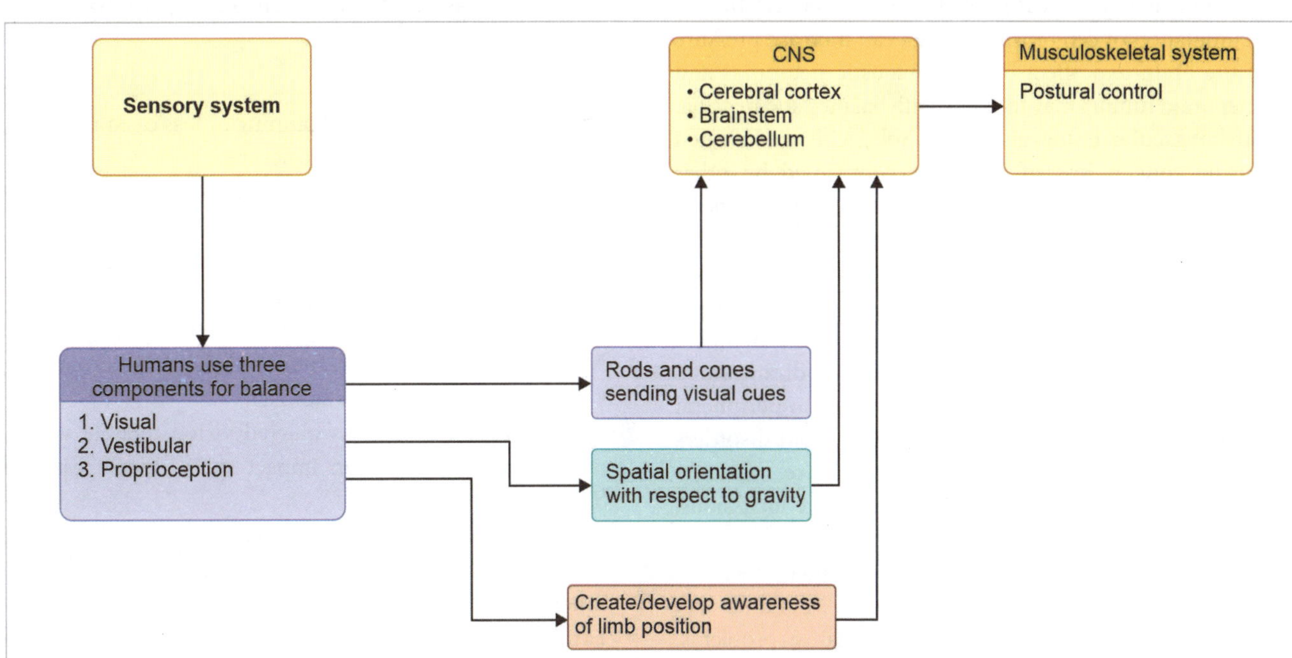

Fig. 24.2: Mechanism of postural control

- **Center of gravity (COG):** COG is inversely proportional to stability that means lower the center of gravity with respect to the BOS, more stable or balanced the body will be.
- **Line of gravity (LOG):** It is an imaginary line drawn from the center of mass in the direction that gravity acts, and must be inside the base of support for an object to be stable.

IMPAIRED BALANCE

Causes

Various balance disorders can arise from a variety of causes:

- **Inner ear disorders** include conditions such as labyrinthitis, which is the inflammation of the labyrinth that affects both hearing and balance. Vestibular neuritis refers to inflammation of the vestibulocochlear nerve in the inner ear, while benign paroxysmal positional vertigo (BPPV) is characterized by vertigo triggered by changes in head position. Ménière's disease leads to sudden episodes of vertigo, and an acoustic neuroma is a noncancerous tumor that presses on inner ear nerves, impacting balance and hearing.
- **Neurological conditions** that can lead to balance disorders include Alzheimer's disease, which impairs cognitive functions and can affect balance; Parkinson's disease, a neurological disorder that often results in balance issues; and multiple sclerosis, which affects the central nervous system and can lead to coordination problems. Traumatic brain injury can disrupt balance due to damage to the brain, while cerebellar diseases specifically affect the cerebellum, the part of the brain responsible for balance and coordination. Hydrocephalus, characterized by the accumulation of cerebrospinal fluid, can increase pressure on the brain and affect balance. Acoustic neuromas and other brain tumors may interfere with balance and hearing.
- **Cardiovascular issues** also play a role, with stroke often causing sudden loss of balance accompanied by other symptoms like weakness and confusion. Abnormal heart rhythms can lead to dizziness or faintness, and congestive heart failure may cause dizziness due to ineffective blood circulation.
- **Blood disorders** such as anemia, characterized by a low red blood cell count, can result in fatigue and dizziness.
- **Endocrine disorders** like diabetes can lead to peripheral neuropathy, which affects balance, and thyroid disorders may also present symptoms that impact balance.
- **Trauma**, particularly head injuries such as concussions, can disrupt normal balance and coordination.
- **Infections**, whether viral or bacterial, can affect the inner ear or brain, leading to balance issues. Migraines can also cause dizziness associated with balance problems, occurring with or without headaches.
- **Musculoskeletal conditions** such as osteoarthritis can result in joint pain and stiffness, leading to decreased range of motion and instability. Rheumatoid arthritis, characterized by joint inflammation, can also impair mobility and balance. Spondylosis involves degeneration of the spine, which may compress nerves and affect balance and coordination. Muscle weakness in the legs or core can compromise stability, increasing the risk of falls, while foot deformities, such as bunions or flat feet, can alter walking mechanics and affect balance.
- **Other medical conditions** include peripheral neuropathy, which involves damage to nerves outside the brain and spinal cord, impairing balance. Polypharmacy refers to medication interactions that may lead to dizziness or balance issues.
- **Environmental factors** such as alcohol consumption can impair reaction time, judgment, and coordination, affecting balance. Stressful events may also lead to temporary balance issues or lightheadedness.
- **Specific syndromes** like mal de débarquement syndrome (MdDS) involve a sensation of movement or swaying after being on a moving surface. A perilymph fistula is a fluid leak between the inner and middle ear that can cause dizziness and unsteadiness. Balance dysfunction can persist and worsen with age in individuals with cerebral palsy (CP).
- **Functional limitations** such as tibial neuropathy (tarsal tunnel syndrome) may cause impaired balance or a perception of instability due to diminished sensation or pain in the foot.
- **Age-related factors** can contribute to balance issues, which may increase as a part of the normal aging process.

Symptoms

- **Dizziness** is a sensation of spinning or loss of equilibrium, which can range from mild to severe.
- **Light-headedness** refers to a feeling of faintness or near-fainting.
- **Unsteadiness** involves difficulty maintaining stability while standing or walking, which can result in swaying or staggering.
- **Nausea** can accompany dizziness or balance issues, causing a feeling of sickness in the stomach.
- **Vision problems** such as blurred vision, double vision, or difficulty focusing can impact spatial awareness and balance.
- **Tinnitus**, which is ringing, buzzing, or other noises in the ears, may occur alongside balance disorders related to the inner ear.
- **Fatigue**, or general tiredness and weakness, can contribute to balance problems, particularly in older adults.

- **Sensory disturbances** include altered sensations in the feet or legs, such as numbness or tingling, which can affect the ability to sense the ground. Increased sensitivity to motion is another symptom, often seen in conditions like Mal de Débarquement Syndrome (MdDS), where individuals have heightened sensitivity to movement or changes in position.
- **Gait changes** can manifest as altered walking patterns, such as shuffling or a wider stance, which may affect stability and balance.
- Anxiety **or fear** related to the risk of falling.

Did You Know?

Certain medications, particularly those that affect the central nervous system, can cause impaired balance as a side effect. Examples include sedatives, anticonvulsants, antihistamines, and some antidepressants. These medications can affect neurotransmitters or interfere with signals between the brain and muscles, leading to instability and difficulty maintaining balance.

Assessment

- **Patient history**
  - **Medical history:** Review of past medical conditions, medications (including polypharmacy), surgeries, and any history of falls.
  - **Symptom assessment:** Document symptoms (dizziness, lightheadedness, unsteadiness) and their frequency, duration, and triggers.
  - **Functional limitations:** Identify specific activities that cause difficulties or fears.
- **Physical examination**
  - **Neurological examination:** Assess cognitive function, cranial nerves, motor strength, and coordination.
  - **Musculoskeletal assessment:** Check for joint stability, muscle strength, range of motion, and any deformities (e.g., bunions, flat feet).
  - **Cardiovascular assessment:** Measure blood pressure in different positions (lying, sitting, standing) to check for postural hypotension.
- **Balance assessment**
  - **Static balance tests:** Evaluate unipedal stance (standing on one leg) for a certain duration; use tandem stance (heel-to-toe standing) as an additional test.
  - **Dynamic balance tests:** Perform functional tests like the Timed Up and Go (TUG) test (described below).
  - **Balance reaction tests:** Assess responses to perturbations or challenges (e.g., unexpected pushes).
- **Gait analysis**
  - Observe the patient's walking pattern for abnormalities (e.g., shuffling gait, wide stance).
  - Assess for any assistive devices used (e.g., cane, walker) and their effectiveness.
- **Vestibular testing**
  - **Dix-Hallpike maneuver:** Used to diagnose Benign Paroxysmal Positional Vertigo (BPPV).
  - **Caloric testing:** Assesses vestibular function by irrigating the ear canal with warm and cold water.
- **Functional assessment**
  - Evaluate the impact of balance issues on daily living activities using tools like the Activities-specific Balance Confidence (ABC) scale or the Berg Balance Scale.
- **Referral for imaging**
  - Consider imaging studies (e.g., MRI or CT scan) if structural abnormalities (e.g., tumors, lesions) are suspected based on neurological exam findings.
- **Multidisciplinary approach**
  - Collaborate with specialists (e.g., neurologists, physiotherapists, audiologists) for comprehensive evaluation and management.

The goal of a functional test of balance is to ensure that both static and dynamic equilibriums are maintained, regardless of the type of perturbation or shift in center of mass, as well as during silent posture. Physical therapists use standardized tests of balance to evaluate a patient's postural control. Some functional balance tests that are available include the following.

Romberg Test

This is employed in the analysis of the proprioceptive role in upholding an upright equilibrium. Subject keeps his eyes open (Fig. 24.3A) and stays upright as a normal response; followed by eyes closed (Fig. 24.3B). A more sensitive version, the Sharpened Romberg's test can also be used (Figs 24.3C and D). This causes the subject's center of mass to rise and their base

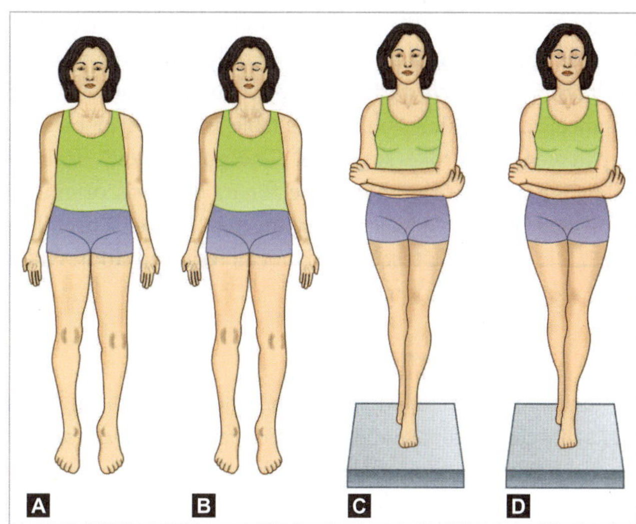

Figs 24.3A to D: Romberg test: **A.** Romberg (eyes open); **B.** Romberg (eyes closed); **C.** Sharpened Romberg (eyes open); **D.** Sharpened Romberg (eyes closed)

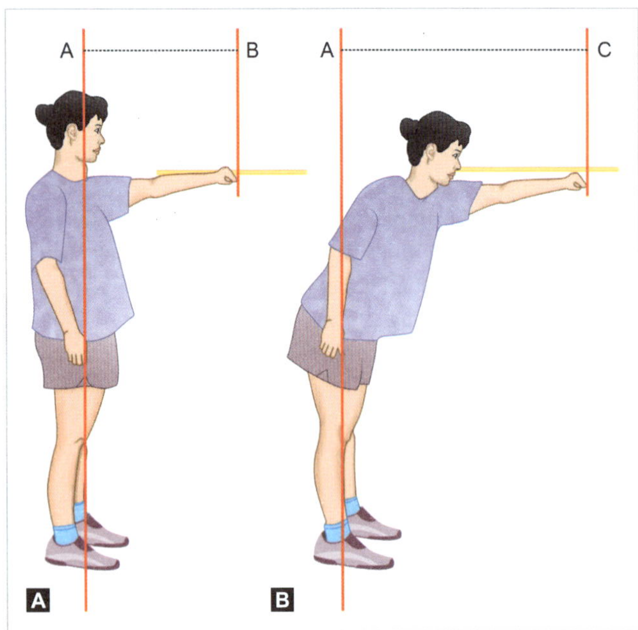

Figs 24.4A and B: Functional reach test

Fig. 24.5: Star excursion test

of support to decrease, preventing them from utilizing their arms to aid with balance.

Functional Reach Test

While maintaining a standing position with the feet firmly planted and an outstretched hand, the furthest distance that can be stretched beyond arm's length is measured (Figs 24.4A and B).

Berg Balance Scale

Berg Balance Scale (BBS) is a 14-item performance measure which is used with specific functional tasks that are typically completed in daily life to assess both static and dynamic balance abilities. According to one study, the most popular evaluation tool used during stroke rehabilitation is the Berg Balance Scale, which is also a reliable indicator of balance impairment in stroke survivors. The golden test, or Berg balancing scale, is another name for it. BBS was originally published in 1989, and it is amazing that it is still useful in 2022. It is a true gold test because not all tests and studies have been around for this long.

Performance-Oriented Mobility Assessment

Performance-Oriented Mobility Assessment (POMA) includes tasks that assess balance and gait which can be used to measure both static and dynamic balance.

Timed Up and Go (TUG) Test

TUG test measures dynamic balance and mobility. TUG test measures how long does a person take to get up from a chair, walk to a marker 3 m away, turn, walk back and sit down on the chair again.

Star Excursion Test

A dynamic balance test that quantifies the greatest range in many directions of a single posture (Fig. 24.5). Star Excursion Test (SET) has been found to have good reliability and validity for assessing dynamic balance and lower extremity function. Studies have shown that it can effectively detect deficits in balance and stability, making it a valuable tool for identifying injury risk and guiding rehabilitation programs in various populations, including athletes, individuals with lower limb injuries, and those undergoing physical therapy interventions.

Balance Evaluation Systems Test

Balance Evaluation Systems Test (BesTest) encompasses assessments for six distinct balance control techniques to identify particular balance impairments and provide a personalized rehabilitation plan.

Mini-Balance Evaluation Systems Test (Mini-BESTest)

The Mini-BESTest is a shortened version of the Balance Evaluation Systems Test, which is extensively employed in research and clinical settings. Exam items for dynamic balance consist of 14 items and are broken down into four subcomponents: anticipatory postural adjustments, reactive postural control, sensory orientation, and dynamic gait. The exam is designed to evaluate impairments in balance. Numerous disorders, primarily neurological ones, have been tested for with Mini-BESTest, though not exclusively. According to an analysis of the test's psychometric qualities, which validates its validity, responsiveness, and reliability, it can be regarded as a typical balancing measure.

Balance Error Scoring System

The Balance Error Scoring System or BESS, is a widely used method for evaluating balance. In sporting contexts, the BESS is used to evaluate how a mild to moderate head injury affects a person's postural stability. For a total of six tests, the BESS looks at three distinct stances (single leg, double leg, and tandem) while on two different surfaces (hard surface and medium density foam). Every exam lasts for 20 seconds, and the assessment takes about 5–7 minutes in total. The double leg stance is the first posture. The participant is told to stand on a solid surface and place his hands and feet side by side with eyes closed. The one-leg stance is the second posture. The individual is told to adopt this stance by placing his hands on his hips, closing his eyes, and standing on his recessive foot on a stable surface. The tandem stance is the third position. With hands on hips and closed eyes, the participant takes a heel-to-toe stance on a solid surface. The first, second, and third stances are repeated in order in the fourth, fifth, and sixth stances that are executed on a medium-density foam surface. An examiner who searches for departures from the correct postures scores the BESS. When a participant exhibits any of the following during testing, a deviation is recorded: Opening the eyes, taking hands off the hips, abduction or flexion of the hips greater than thirty degrees, tripping or falling, lifting the forefoot or heel off the test surface, or being unable to maintain the correct position for longer than five seconds.

Treatment

Exercise interventions aimed at enhancing balance are often those in which participants move and stand in positions that are more challenging than others in order to test their bodies' capacity to anticipate and react to demands of various tasks or surroundings. In order to keep the body's center of mass within manageable bounds of the base of support or in motion to a new base of support, participants must exercise their muscles (and neuromuscular responses) against an external force, as a result of voluntary movement, or in response to an unexpected perturbation or stimulus. Walking, cycling, functional static and dynamic standing balance training, strengthening exercises, computerized balance training, dance, Tai Chi, yoga, and whole-body exercises are some examples of exercise therapies.

Objectives

The objectives of balance training programs are to:

- Maintain balance while enhancing control during regular tasks such as increased walking speed
- Boosts fall-related self-efficacy, and lessens fear of falling.
- Improve physical activity.
- Improve quality of life.

Clinical Correlation

Patient education is crucial in promoting self-management and preventing falls. This may involve teaching strategies to enhance awareness of balance, proper body mechanics, safe mobility techniques, and home exercise programs. Education on fall prevention measures, such as proper footwear, home safety modifications, and strategies for getting up safely after a fall is also important.

Posture

- Core stability training to improve center of gravity control and sense of trunk posture.
- Over sessions, use a variety of arm positions, unstable surfaces and single leg stance.

Movement

- Adding movement patterns to acquired stable static postures increase balance challenges.
- Add anterior-posterior sway to increase stability limits.
- Trunk rotations and alterations in head positions alter vestibular inputs.
- Stepping back and forward assists in re-stabilization exercises.

Progression

- Begin with weight shifts on a stable surface.
- Gradually increase sway.
- Increase surface challenges.

Common Exercises

Examples of balance exercises (Fig. 24.6) include:

- While standing, shift weight to the other leg and lift one leg to the side or back.
- Adopt a tandem stance, where one heel is directly in front of toes of the other foot.
- Getting up from a chair and sitting down without using hands.
- Alternating raises of the knees with each step while walking.
- Doing yoga or tai chi.

Did You Know?

Tai Chi is a traditional Chinese martial technique that incorporates slow, rhythmic motions in the lower limbs, such as core rotation, weight shifting, coordination, and a progressive increase in stance tightness. It is now widely acknowledged as an excellent form of exercise for senior citizens. More so than other workouts, Tai Chi increases postural stability, according to studies. It also offers several advantages for the heart and musculoskeletal system.

Fig. 24.6: Balance exercises

- It can be difficult to maintain balance while using equipment like the Bosu or balancing board with an inflatable dome on top of the circular platform.
- Repetitive postural disturbance interventions aimed at improving control of rapid balance responses is a type of perturbation-based balance training. It has been demonstrated that this exercise enhances reactive balance control in poststroke patients during the subacute phase (Fig. 24.7).

Progression

Over the time, the exercises described above can be progressed by:

- Holding the position for a longer amount of time
- Walking tandem stance with support then without support.
- Closing eyes
- Letting go of chair or other support.

INTERVENTIONS FOR IMPROVING BALANCE

- **Frenkel exercises:** Frenkel exercises improve static and dynamic balance in the elderly in different settings. For details refer chapter 13 (Vafaeenasab et al.).
- **Aerobic exercises** are effective in improving balance in patients with diabetic neuropathy (Kiani et al.).
- **Pilates training:** It has been found to be effective in improving balance, muscle strength and mobility in patients with multiple sclerosis (Guclu-Gunduz et al.); and improved balance and fall risk in community dwelling older adults (Josephs et al.).
- **Core strengthening:** Strength and conditioning of the regional and global muscles that cooperate to maintain the spine are prioritized during core training. A study showed

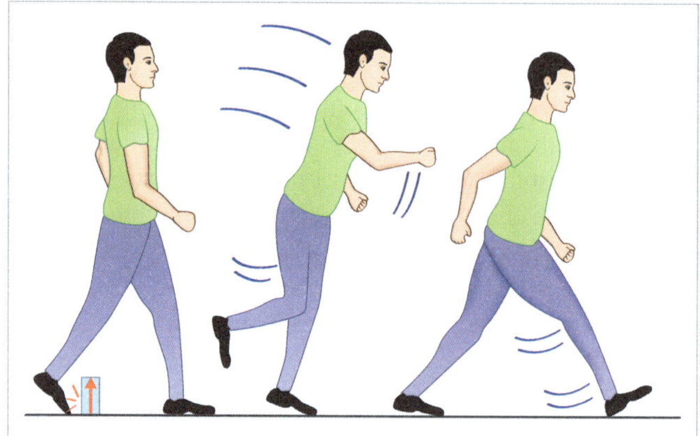

Fig. 24.7: Perturbation-based balance training

improvement in balance and strength training following Swiss ball core strength training in sedentary women (Sekendiz et al.).
- **Video game-based exercises:** Such exercises have been found to be effective in improving balance in a patient with balance dysfunction (Betker et al.).
- **Vestibular rehabilitation:** Improvements in gaze stabilization, eye/head coordination, retraining of the use of sensory information and movement strategies for balance, sensory replacement, and adaptation of methods are the objectives of vestibular rehabilitation techniques. Exercises for vestibular adaptation have also evolved into a crucial part of vestibular therapy. The movement of an image across the retina causes adaptation, which is the gradual enhancement of the vestibular system's capacity to adjust to head movement. Exercises that use both head and visual information appear to yield the highest results. After vestibular therapy, significant improvements in balance and vertigo have been recorded. The central nervous system's habituation or adaptation, sensory substitution, or reweighting of the sensory systems are thought to be the neurophysiologic bases of these alterations. The size of the response to recurrent sensory stimuli decreases with habituation. Numerous physiological processes most likely play a role in this process. In habituation exercises, the provoking position or stimuli is repeated at regular intervals until the person no longer responds to the stimuli (Wrisley et al.).
- **Aquatic therapy:** People with balance issues can benefit from aquatic physical therapy because the water's buoyancy produces equal resistance across all working muscle groups and increases stability (Fig. 24.8). Children's balance improves in the water environment, much like it does for adults with coordination and/or balance issues. Due to the concentration of excitement in the relevant central nervous system zones during water therapy, the labyrinthine functions are enhanced. The capacity to perform daily tasks is improved when the vestibular-ocular

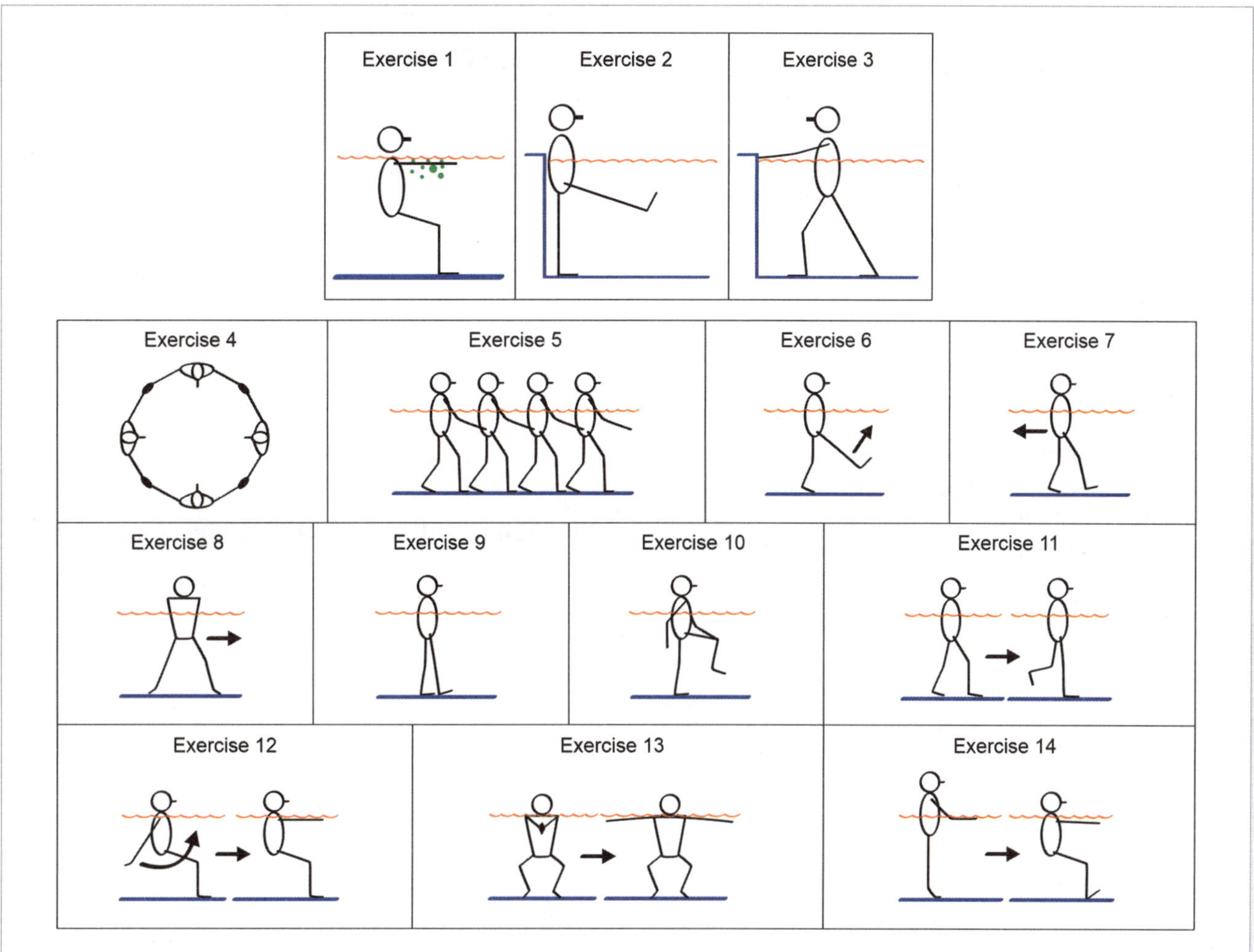

Fig. 24.8: Aquatic balance exercises

and vestibular-spinal reflexes are recovered utilizing an adaptive aquatic program that includes substitutive and even habitual drills and different motor structures on a fluctuant support (Vasile et al.).

- A research showed after receiving aquatic therapy based on the Halliwick and Ai Chi approaches, stroke survivors' postural balance and knee flexor strength improved.
- **Virtual reality**-based treatments have been found to be effective in improving balance in patients with chronic stroke (Morel et al.).

Precautions

- Pain should not occur during exercises.
- Do not initiate exercises at maximum level of difficulty.
- Discontinue exercises if patient starts to feel pain, dizziness or unusual shortness of breath.
- Remove unsafe instruments from the training area.

HIP, ANKLE AND STEPPING STRATEGIES

The dynamic process of postural control involves proactive, reactive, and steady-state balance, all of which interact with task and environmental restrictions. Stability is maintained by steady-state balance under foreseeable circumstances, whereas reactive balance uses techniques like the ankle, hip, and stepping strategy to regain stability following unforeseen disturbances. The stepping strategy, which involves taking a step to re-establish balance, is used when the perturbation is too big to recover with fixed support. Balance that is proactive foresees possible destabilizing movements. By choosing the best reactions depending on the nature of the disturbance, the central nervous system maintains functional equilibrium in a variety of settings and activities.

The **ankle**, **hip** and **stepping** strategies are two key postural control mechanisms that humans use to maintain balance, particularly during perturbations to their upright stance.

Ankle Strategy

- **Description:** This strategy involves the activation of muscles in a distal-to-proximal order, starting with the ankle muscles, followed by the calf, thigh, and trunk muscles. It's primarily used for **slow, low-amplitude perturbations** or when standing on a **wide, stable surface**.
- **Mechanics:** The body acts like an inverted pendulum, rotating around the ankle joint, with minimal motion at the knee, hip, and trunk. It relies on ankle joint torque to control posture.
- **Limitations:** The ankle strategy is constrained by the ability of the feet to exert torque against the ground. If the surface is narrow, friction is insufficient, or the perturbation is too large or fast, this strategy becomes less effective.
- **Activation pattern:** In response to backward platform movements, the ankle strategy causes activation of **ankle plantar flexors**, followed by **knee flexors** and **hip extensors**.

Hip Strategy

- **Description:** This strategy involves a more proximal-to-distal activation order, with the trunk and hip muscles activated before the lower leg muscles. It is generally employed for **fast, large-amplitude perturbations** or when standing on a narrow or unstable surface.
- **Mechanics:** The body rotates at the hip joint, with the trunk moving forward and downward, creating a backward rotation at the lower body. This strategy allows for greater control of the center of mass (COM) by reducing the body's moment of inertia at the ankle, thus enabling a greater angular acceleration from a given torque.
- **Limitations:** It is constrained by surface friction and the ability to generate horizontal forces at the support surface. However, it can tolerate **greater perturbations** and provide a **faster response** than the ankle strategy.
- **Activation pattern:** In response to a backward platform movement, the hip strategy involves activation of **knee extensors** and **hip flexors** to control balance.

Overlap and Mixed Strategies

In practice, **mixed strategies** often occur, especially in slower or less intense perturbations. For example, a **hip strategy** may involve some **ankle movement**, and vice versa. This overlap reflects the **coordination** between both strategies, as compensatory **hip torques** or **ankle torques** are required to resist or assist the rotational movements at the other joint.

Central Nervous System Control and Task Dependence

- The central nervous system (CNS) chooses between the two strategies based on the **nature of the perturbation**, the **task**, and the **available sensory information**. The **hip strategy** is typically more **efficient** in terms of muscle activation and can tolerate higher perturbations and time latencies, which is why it is used in more dynamic or perturbed conditions. The **ankle strategy**, on the other hand, is often preferred in **unperturbed, stable standing**, likely due to its **efficiency** and **minimal muscle activation**.
- Furthermore, the interaction between balance control and other tasks (e.g., visual or cognitive demands) can influence which strategy is selected.

Emergent Control and Efficiency

The choice of strategy is also influenced by **efficiency** and the **cost of muscle activation**. The hip strategy, while more efficient for larger perturbations, may not always be the best choice during **unperturbed standing**, where the goal may not always be minimizing muscle activation, but balancing various tasks and maintaining stability.

Stepping Strategy

When the ankle or hip methods are not enough to correct a perturbation, the stepping strategy is an essential postural control mechanism that is employed to restore balance. It involves moving towards the destabilizing force in order to expand the base of support and restore balance. Frequently employed in reaction to lateral or forward-backward shifts in the center of mass, this tactic is especially crucial in situations involving significant or quick disturbances. Factors such as step size, step velocity, and step latency affect how effective the stepping approach is. A higher risk of falls can result from delayed or inadequate stepping reactions in people with balance issues, such as those with Parkinson's disease.

Mille et al., 2005 in a research concluded that lateral stability is frequently impaired in older persons, which makes it challenging to implement efficient lateral stepping techniques in reaction to disturbances. The stepping approach is essential for regaining balance, but delayed or ineffective steps might be caused by aging-related deficiencies in hip torque and lateral trunk control. Increasing the elderly's capacity for lateral stepping may lower their risk of falling and improve their general postural stability.

SUMMARY

- Balance involves maintaining equilibrium by maintaining an even distribution of weight on each side of the vertical axis. Balance can be static or dynamic, and is synchronized by sensory, cognitive, and musculoskeletal systems.
  - Sensorimotor control systems, including vision, vestibular function, and proprioception, help maintain static balance, stabilize visual images, and control postural stability.
- Balance disorders, caused by various factors, can cause symptoms like dizziness, falling, and fear. Functional tests assess impaired balance, focusing on static and dynamic balance.
- Balance training, including perturbation-based exercises, is effective in post-stroke individuals, elderly, diabetic neuropathy, multiple sclerosis, and balance dysfunction patients. It also includes Frenkel exercises, aerobic exercises, pilates, and virtual reality treatments.

FURTHER READINGS

- Alamer, Abayneh; Getie, Kefale; Melese, Haimanot; Mazea, Habtamu (2020-08-17). "Effectiveness of Body Awareness Therapy in Stroke Survivors: A Systematic Review of Randomized Controlled Trials". Open Access Journal of Clinical Trials. 12: 23–32. doi:10.2147/OAJCT.S260476. S2CID 225364826.
- Appeadu M, Bordoni B. Falls and Fall Prevention. StatPearls [Internet]. 2020 Jul 8. Available from:https://www.ncbi.nlm.nih.gov/books/NBK560761/ (last accessed 17.10.2020)
- Bannister R: Brain's Clinical Neurology, edition 7. New York, NY, Oxford University Press, Inc, 1993.
- Bell DR, Guskiewicz KM, Clark MA, Padua DA (May 2011). "Systematic review of the balance error scoring system". Sports Health. 3 (3): 287–295. Doi: 10.1177/1941738111403122. PMC 3445164. PMID 23016020.
- Betker AL, Szturm T, Moussavi ZK, Nett C. Video game–based exercises for balance rehabilitation: A single-subject design. Archives of physical medicine and rehabilitation. 2006 Aug 1;87(8):1141–9.
- Blenkinsop GM, Pain MT, Hiley MJ. Balance control strategies during perturbed and unperturbed balance in standing and handstand. Royal Society open science. 2017 Jul 26;4(7):161018.
- Blum L, Korner-Bitensky N (May 2008). "Usefulness of the Berg Balance Scale in stroke rehabilitation: A systematic review". Physical Therapy. 88 (5): 559-566. doi:10.2522/ptj.20070205. PMID 18292215.
- Cynthia Lions, Emmanuel Bui, Sylvette Wiener: Postural Control in Strabismic Children: Importance of Proprioceptive Information: Front Physio|2014;5:156.
- Di Carlo S, Bravini E, Vercelli S, Massazza G, Ferriero G (June 2016). "The Mini-BESTest: A Review of Psychometric Properties". International Journal of Rehabilitation Research. 39 (2): 97–105. doi:10.1097/MRR.0000000000000153. PMID 26795715. S2CID 9649113.
- Franz, JR, Francis, CA, Allen, MS, O'Connor, SM, and Thelen, DG (2015). Advanced age brings a greater reliance on visual feedback to maintain balance during walking. Hum. Mov. Sci. 40, 381–392. doi: 10.1016/j.humov.2015.01.012.
- GOLDIE PA, BACH TM, EVANS OM. Force platform measures for evaluating postural control: Reliability and validity. Arch Phys Med Rehabil. 1989; 70:510–517.
- Grzegorz Bednarczuk, Ida Wiszomirska, European Journal of Physical and Rehabilitation Medicine 2021 August; 57(4):593–9.
- Guclu-Gunduz A, Citaker S, Irkec C, Nazliel B, Batur-Caglayan HZ. The effects of pilates on balance, mobility and strength in patients with multiple sclerosis. Neurorehabilitation. 2014 Jan 1;34(2):337–42.
- Horak FB, Wrisley DM, Frank J (May 2009). "The Balance Evaluation Systems Test (BESTest) to differentiate balance deficits". Physical Therapy. 89 (5): 484–498. doi:10.2522/ptj.20080071. PMC 2676433. PMID 19329772.
- Hrysomallis C (March 2011). "Balance ability and athletic performance". Sports Medicine. 41 (3): 221–232. doi: 10.2165/11538560-000000000–00000. PMID 21395364. S2CID 24522106.
- Josephs S, Pratt ML, Meadows EC, Thurmond S, Wagner A. The effectiveness of Pilates on balance and falls in community dwelling older adults. Journal of bodywork and movement therapies. 2016 Oct 1;20(4):815–23.
- Kiani N, Marryam M, Malik AN, Amjad I. The effect of aerobic exercises on balance in diabetic neuropathy patients. Journal of Medical Sciences. 2018 Jun 17;26(2):141–5.
- King LA, Horak FB. Lateral stepping for postural correction in Parkinson's disease. Archives of physical medicine and rehabilitation. 2008 Mar 1;89(3):492–9.
- Mille ML, Johnson ME, Martinez KM, Rogers MW. Age-dependent differences in lateral balance recovery through protective stepping. Clinical Biomechanics. 2005 Jul 1;20(6):607–16.
- Morel M, Bideau B, Lardy J, Kulpa R. Advantages and limitations of virtual reality for balance assessment and rehabilitation. Neurophysiologie Clinique/Clinical Neurophysiology. 2015 Nov 1;45(4-5):315–26.
- O'Sullivan S, Schmitz T (2007). Physical Rehabilitation (Fifth edition). Philadelphia: FA Davis Company. p. 254–259.

- Rubenstein, LZ (2006). Falls in older people: Epidemiology, risk factors and strategies for prevention. Age Ageing 35(Suppl. 2), 37–41. doi: 10.1093/ageing/afl084.
- Schinkel-Ivy, A., Huntley, AH, Danells, CJ, Inness, EL and Mansfield, A., 2020. Improvements in balance reaction impairments following reactive balance training in individuals with sub-acute stroke: A prospective cohort study with historical control. Topics in stroke rehabilitation, 27(4), p. 262–271.
- Sekendiz B, Cug M, Korkusuz F. Effects of Swiss-ball core strength training on strength, endurance, flexibility, and balance in sedentary women. The Journal of Strength and Conditioning Research. 2010 Nov 1;24(11):3032–40.
- Shumway-Cook A, Anson D, Haller S (June 1988). "Postural sway biofeedback: Its effect on reestablishing stance stability in hemiplegic patients". Archives of Physical Medicine and Rehabilitation. 69 (6): 395–400. PMID 3377664.
- Sipla, JS, and Spoor, F. (2008). The physics and physiology of balance. In sensory evolution on the threshold adaptations in secondarily aquatic vertebrates (p. 226–232). University of California Press.
- Susan B O Sullivan, Leslie G Portney. Physical Rehabilitation: Sixth Edition. Philadelphia: FA Davis 2014.
- Tomomitsu, M., Alonso, A., Morimoto, E., Bobbio, T., and Greve, J. (2013). Static and dynamic postural control in low-vision and normal-vision adults. Clinics 68, 517–521. doi: 10.6061/clinics/2013(04)13.
- Vafaeenasab MR, Amiri A, Morowatisharifabad MA, Namayande SM, Abbaszade Tehrani H. Comparative study of balance exercises (frenkel) and aerobic exercises (walking) on improving balance in the elderly. Elderly Health Journal. 2018 Dec 10;4(2):43–8.
- Valovich TC, Perrin DH, Gansneder BM (March 2003). "Repeat Administration Elicits a Practice Effect with the Balance Error Scoring System but not with the Standardized Assessment of Concussion in High School Athletes". Journal of Athletic Training. 38 (1): 51–56. PMC 155511. PMID 12937472.
- Vasile L, Stănescu M. The aquatic therapy in balance coordination disorders. Procedia-Social and Behavioral Sciences. 2013 Oct 10;92: 997–1002.
- Wrisley, Diane M.; Pavlou, Marousa (2005). Physical Therapy for Balance Disorders. Neurologic Clinics, 23(3), 855–874. doi:10.1016/j.ncl.2005.01.005

STUDENT ASSIGNMENT

LONG ANSWER QUESTIONS

1. Define balance and describe its mechanism.
2. Discuss causes of impaired balance and its symptoms.
3. Describe various methods and tests for balance assessment.
4. Evaluate balance training methods in details.

SHORT ANSWER QUESTIONS

1. Define static and dynamic balance.
2. What do you understand by balance mechanism?
3. What are the causes of impaired balance?

MULTIPLE CHOICE QUESTIONS

1. **In progressive balancing exercises, which of the following exercises would come last?**
   a. Seated: Eyes open, then eyes closed
   b. Single-leg standing: Eyes open, then eyes closed
   c. Minitrampoline: Double-leg standing
   d. Balance board
2. **Which of the following are included in a comprehensive balance assessment and training?**
   a. Progressive tasks of incremental difficulty
   b. Static stance with varying bases of support
   c. Progressive tasks with unexpected perturbations
   d. All of the above
3. **What term is defined as the perception of knowing joint position, movement, and movement resistance?**
   a. Coordination b. Neuromuscular control
   c. Proprioception d. Balance
4. **For which training tasks would a wobble board be used?**
   a. Balance training for the lower extremities
   b. Coordination training for the upper extremities
   c. Proprioceptive training for the upper extremities
   d. Flexibility training for the lower extremities
5. **Which of the following factors does NOT contribute to balance dysfunction?**
   a. ROM b. Coordination
   c. Perception d. CKC exercises

ANSWER KEY

1. d **2.** d **3.** c **4.** a **5.** d

25 Introduction to Yoga

Sheetal Kalra

LEARNING OBJECTIVES

After the completion of the chapter, the readers will be able to:

- Understand concept and types of yoga.
- Explain physiology and therapeutic effects of yoga.
- Understand effects and physiological basis of Yogasanas, Pranayamas, and Transcendental Meditation.
- Explain application of yoga in fitness and flexibility.
- Describe common poses with name, position, specific effects and uses.
- Interpret yoga as holistic approach.

CHAPTER OUTLINE

- Introduction
- What is Yoga?
- Types of Yoga
- Psychophysiological Mechanisms of Action
- Principles of Therapeutic Approach of Yoga
- Therapeutic Effects
- Benefits of Yoga
- Yoga: A Holistic Approach

KEY TERMS

Pranayama: Pranayama is a fundamental part of yoga, a physical and mental health practice. "Yama" signifies control and "prana" is the Sanskrit word for life energy. Breathing exercises and patterns are part of pranayama practice.

Transcendental meditation (TM): Transcendental meditation is a method for preventing distracting thoughts and encouraging a calm state of awareness.

Yogasanas: Postures are called asanas. The body can adopt countless different positions. Of them, some positions are known as "yoga asanas" or yogasanas. A greater dimension or a higher understanding of life is what is meant to be reached through "yoga". Consequently, a "yogasana" is a position that raises one's potential.

INTRODUCTION

The word "yoga" derives from the Sanskrit root "yuj", which also means 'to join' or 'to unite'. Regular yoga practice cultivates qualities of friendliness, compassion, and more self-control while encouraging strength, endurance, and flexibility. It also improves a sense of peace and well-being. Sustained practice also produces significant results, such as shifts in perspective on life, increased self-awareness, and more vitality to live life to the fullest and with genuine happiness.

A condition of balance and oneness between the mind and body can be attained because yoga practice results in a physiological state that is the reverse of the flight-or-fight stress reaction.

WHAT IS YOGA?

Yoga is a type of mind-body exercise that combines physical exercises with a conscious interior emphasizing on awareness of the self, the breath, and energy (Fig. 25.1).

In the famous work known as the Yoga Sutras, which is usually regarded as the canonical work on yoga, Patanjali first provided a description of the philosophy and practice of yoga. Ashtanga, which literally translates to "eight limbs", (Fig. 25.2) is the eight-fold route Patanjali describes in the Yoga Sutras for achieving consciousness and enlightenment. The eight limbs are composed of moral guidelines for leading a fulfilling life; they act as a guide for moral behavior and self-discipline, and they emphasize physical well-being while also recognizing one's spiritual nature. Any one of the eight limbs may be practiced independently, but according to yoga philosophy, physical poses and breathing techniques help to prepare the body and mind for meditation and spiritual growth. There are numerous distinct yogic disciplines that have been created based on Patanjali's eight limbs. Each uses a different method for both disease prevention and treatment.

Fig. 25.1: Yoga

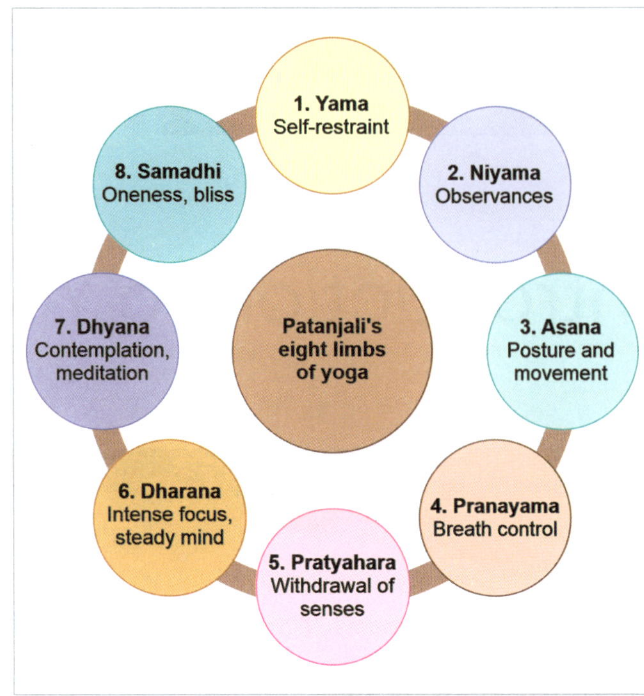

Fig. 25.2: Eight limbs of yoga

Did You Know?

Recognizing the interconnectedness of the mind, body, and spirit, healthcare providers can integrate yogic principles and practices into holistic treatment plans to address the multifaceted needs of patients.

TYPES OF YOGA

Ashtanga Yoga

- This type of yoga practice uses ancient yoga teachings. However, it became popular during the 1970s.
- Ashtanga applies the same poses and sequences that rapidly link every movement to breath.

Bikram Yoga

People practice Bikram yoga, also known as hot yoga, in artificially heated rooms at a temperature of nearly 105°F and 40% humidity. It consists of 26 poses and a sequence of two breathing exercises.

SECTION V Movement and Alignment

Hatha Yoga

The purposeful prolonging of inhalation, breath retention, and exhalation is the main focus of Hatha yoga breathing methods. Blockages in the body's energy channels are removed, and the energy system of the body becomes more balanced, by combining the physical body, breath, and focus while doing the postures and motions.

Did You Know?

Hatha yoga, often associated with gentle and slow-paced movements, actually encompasses various styles ranging from the more meditative and therapeutic to the vigorous and physically challenging. It is a versatile system that caters to different needs and preferences, making it accessible to practitioners of all levels and abilities.

Iyengar Yoga

This style of yoga emphasizes using a variety of props, including chairs, bolsters, blankets, blocks, and straps, to assist practitioners achieve the proper alignment in each posture.

Kripalu Yoga

- Kripalu yoga educates practitioners to understand, accept, and draw lessons from their bodies. By going inward, a Kripalu yoga student discovers his own level of practice.
- It is a gentler form of Hatha yoga.
- Breathing exercises and gentle stretches commonly start the classes. Individual poses are then practiced, culminating in a final relaxation.

Kundalini Yoga

- Kundalini yoga is a type of meditation that is intended to help release stored energy.
- Chanting usually opens a Kundalini yoga class, while singing closes it. It includes asana, pranayama, and meditation aimed at achieving a particular result.

Power Yoga

This vigorous and athletic style of yoga was created by practitioners in the late 1980s, drawing inspiration from the Ashtanga method.

Sivananda Yoga

- The core of this method is a five-point concept which emphasizes that a healthy yogic lifestyle may be achieved by appropriate breathing, relaxation, food, exercise, and positive thought.
- Practitioners of Sivananda Yoga perform 12 fundamental asanas, Sun salutations coming first and Savasana last.

Did You Know?

Sivananda yoga was founded by Swami Sivananda, who was not only a yogi but also a physician. He synchronized traditional yoga teachings with modern medical knowledge to create a holistic approach to health and well-being, emphasizing not just physical postures but also proper breathing, relaxation, diet, and positive thinking.

Viniyoga

Viniyoga emphasizes the art and science of sequencing form over function, breath and adaptation, repetition and holding.

Yin Yoga

Holding passive positions for extended periods of time is the main focus of yin yoga. Deep tissues, ligaments, joints, bones, and fascia are all targeted by this type of yoga.

Prenatal Yoga

Prenatal yoga practitioners design poses specifically with expectant mothers in mind. This type of yoga can promote health throughout pregnancy and assist individuals in getting back into shape after giving birth.

Restorative Yoga

This style of yoga is calming. During a restorative yoga class, an individual performs four or five basic poses, utilizing props like blankets and bolsters to help them sink into deep relaxation and hold the poses without having to work at all.

PSYCHOPHYSIOLOGICAL MECHANISMS OF ACTION

Yoga is likely to affect the neural system, cardiovascular system, psycho-neuro-endocrine axis, neurotransmitter levels, molecular and gene expression, among other things.

At Mental Level

- According to published research, yoga practice stimulates the vagal nerve, increasing parasympathetic function of the autonomic nervous system and brain Gamma-Aminobutyric acid (GABA) activity. Studies showing long-term yoga practitioners' gene expression profiles compared to controls demonstrate that yoga has a favorable impact on immune cell gene expression profiles.
- Regular yoga practice has also been proven to alter the brain, increasing the left prefrontal cortex's level of activation. Numerous studies have demonstrated that yoga-based techniques can lead to creating new brain pathways.

- Yoga activities have an impact on resonance circuits that thicken the pre-frontal medial cortex and insula, especially on the right side, which aids in the processing of empathy, apprehension, reasoning, and intuition. These activities frequently lower stress, which improves mood and lessens emotional suffering.
- The autonomic processes, neurophysiologic mechanisms, neurochemical mechanisms, cognitive mechanisms, and spiritual factors have all been the focus of scientific studies on yoga.
- The effects of yoga in conditions including depression, schizophrenia, attention-deficit/hyperactivity disorder (ADHD), and other psychological illnesses seem to be best explained by neurochemical pathways. According to studies, yoga has a biological basis for treating depression since it lowers cortisol levels and increases brain-derived neurotrophic factor (BDNF) serum levels (Telles et al., 2020).

At Physical Level

- Pancreatic cells are rejuvenated or regenerated as a result of abdominal stretching during yoga exercise, which could boost the enzymatic process that increases the use and metabolism of glucose in peripheral tissues, the liver, and adipose tissues.
- Muscle development, relaxation, and better blood flow to the muscles may boost insulin receptor expression in the muscles, resulting in higher glucose uptake by the muscles and lowering of blood sugar.
- The rise in hepatic lipase and lipoprotein lipase at the cellular level, which alter lipoprotein metabolism and consequently increase uptake of triglycerides by adipose tissues, may be the cause of the improvement in lipid levels after yoga.
- The cumulative effect of doing the postures can improve insulin sensitivity and the β-cells' sensitivity to the glucose signal in the pancreas, which in turn can improve glucose sensitivity.
- The ability of the pancreas to make insulin can be restored by the postures' direct stimulation of the organ. Yoga poses that stimulate the pancreatic meridian and workouts that increase blood circulation in the pancreas region may help some diabetes patients regenerate pancreatic β-cells.
- Pranayama exercises strain the lung tissue, which results in inhibitory signals from the activity of hyperpolarizing currents and slowly adapting receptors. These inhibitory signals from the vagi-involved cardiorespiratory region are thought to coordinate neuronal components in the brain, changing the autonomic nervous system and producing a state marked by decreased metabolism and a predominance of the parasympathetic nervous system.
- Higher melatonin levels could be one mechanism through which the claimed health promoting effects of meditation occur.

Did You Know?

Effects of Transcendental Meditation

Researchers identified considerably more gray matter in the meditators' right orbitofrontal cortex, right thalamus, and left inferior temporal gyrus in a study to determine the morphological correlates of long-term meditation.

An increase in alpha activity was observed in both the major depressive patient group and the healthy individuals in a comparative investigation of the effects of Sahaja Yoga meditation on Electroencephalogram (EEG) in both groups after two months of Sahaja yoga meditation practice.

Long-term practitioners of meditation have structural variations in brainstem regions related to cardiorespiratory regulation, according to a Danish study that found experienced meditators had higher gray matter density in lower brain stem regions than age-matched nonmeditators.

Another study carried out in Germany with eight Buddhist nuns and monks trained in meditation showed that regular meditation practice increased the speed at which attention may be distributed, increasing the depth of information processing and decreasing reaction latency.

Long-term Sahaja Yoga meditation practitioners reported higher quality of life (QoL) and functional health than the general population, according to a cross-sectional survey with 347 respondents (Balaji et al., 2012).

Research indicates that regular practice of TM is associated with changes in gene expression related to immune regulation, as well as increased antibody production.

PRINCIPLES OF THERAPEUTIC APPROACH OF YOGA

The therapeutic approach of yoga is based on four fundamental principles:

1. The first principle is that the human body is a holistic system made up of multiple interconnected dimensions that are inextricably linked to one another, and that the health or illness of any one dimension has an impact on all the others.
2. The second principle is that each person is different, and that each person has different needs. As a result, each person should be treated as an individual, and practices should be adjusted as necessary.
3. The third principle of yoga empowers the practitioners to be their own healer. Yoga involves the learner in the healing process; by actively participating in their quest for health, the healing occurs within rather than from an external source, and a greater sense of autonomy is developed.
4. The fourth principle is that healing depends greatly on a person's mental makeup and condition. Recovery proceeds more rapidly when the person is in a positive mental state; conversely, if the person is in a negative mental state, healing can take longer.

THERAPEUTIC EFFECTS

As per research studies, following are some of the therapeutic effects of yoga.

Mental Health

- Regular practice is effective in reducing symptoms of depression, anxiety, fatigue, stress, post-traumatic stress disorder.
- **Reduction in anxiety:** Participation in yoga leads to significant reduction in perceived levels of anxiety in women who suffer from anxiety disorders (Javnbakht et al., 2009).
- **Antidepressant effects:** Yoga alone produced substantial antidepressant effects correlated with the elevation of serum BDNF levels. The findings argue for a neuroplastic mechanism of antidepressant action for Yoga (Naveen et al., 2013).
- **Reduces stress:** Yoga is valuable in helping to achieve relaxation and diminish stress, it also helps cancer patients perform daily and routine activities, and increases the quality of life in cancer patients (Ulger et al., 2010).

Physical Fitness

- Positive effects are seen on body composition, muscle strength, endurance, flexibility and balance.
- **Improve health and fitness:** Yogic practices enhance muscular strength and body flexibility, promote and improve respiratory and cardiovascular function, expedite recovery from and treatment of addiction, reduce stress, anxiety, depression, and chronic pain, improve sleep patterns, and enhance overall well-being and quality of life (Rayal et al., 2021).
- **Yoga and cardiopulmonary conditions:** Regular practice of yoga improves cardiopulmonary parameters.
  - **Reduction in blood pressure:** Yoga has been found to decrease blood pressure as well as the levels of oxidative stress in patients with hypertension (Dhameja et al., 2013).
  - **Improvements in pulmonary functions:** A study reported significant increase in FVC, FEV-l and peak expiratory flow rate (PEFR) at the end of 12 weeks of yogic practice in young healthy females (Yadav et al., 2001).
- **Patients with neurological disorders:** Yoga might be considered an effective adjuvant for the patients with various neurological disorder (poststroke aphasia, epilepsy, Alzheimer's, dementia, myelopathy, Guillain-Barré syndrome, diabetic neuropathy, amyotrophic lateral sclerosis (ALS)) (Mooventhan et al., 2017).
- **Yoga and musculoskeletal conditions:**
  - Yoga improves musculoskeletal function and pain perception.
  - Regular yoga practice decreases frequency, intensity and degree of interference due to musculoskeletal discomfort, increases hand grip strength, coordination and flexibility (Telles et al., 2009).
- **Therapeutic effects for children (Galantino et al., 2008):**
  - Children with ADHD-show improved sensory motor skills, health and performance.
  - Reduction in asthmatic attacks.
- **Special conditions:**
  - **Improved QoL** and physiological functions in stages 1–2 of Parkinson's disease (Sharma et al., 2015).
  - **Reduction in symptoms of dysmenorrhea:** Yoga intervention was found to be associated with reductions in severity of dysmenorrhea and may be effective in lowering serum homocysteine levels after an intervention period of 8 weeks (Chien et al., 2013).
  - **Irritable bowel syndrome:** Yoga improves symptoms in patients with IBS (Schumann et al., 2016).
  - **Non-insulin-dependent diabetes mellitus (NIDDM):** Better glycemic control and pulmonary functions can be obtained in NIDDM cases with yoga asanas and pranayama. The exact mechanism as to how these postures and controlled breathing, interact with somato-neuro-endocrine mechanism affecting metabolic and pulmonary functions remains to be worked out (Sharma et al., 2002).

Clinical Correlation

Yoga can complement traditional rehabilitation programs by improving flexibility, strength, and balance. Understanding styles like Viniyoga, which prioritize adaptation and gradual progression, can be particularly useful in designing rehabilitation plans for patients recovering from injuries or surgeries.

BENEFITS OF YOGA

- **Physical advantages:**
  - A predisposition toward parasympathetic nervous system dominance rather than the typical stress-induced sympathetic nervous system dominance; stable autonomic nervous system homeostasis.
  - Blood pressure falls.
  - Pulse rate reduces.
  - Respiratory rate decreases.
  - Galvanic skin response (GSR) rises, EEG alpha waves rise (theta, delta, and beta waves also rise throughout the meditation process), electromyography (EMG) activity falls, cardiovascular efficiency rises, respiratory efficiency rises (breathing amplitude and smoothness rises, vital capacity rises, breath-holding time rises), gastrointestinal function normalizes, endocrine function normalizes, excretory function normalizes.
  - Musculoskeletal flexibility and joint range of motion increase; posture is improved; strength and resilience rise; endurance soars; energy levels soar; weight returns to normal; sleep is enhanced; immunity soars; and pain levels decline.

- **Mental health benefits:**
  - A rise in somatic and kinesthetic awareness; an uptick in mood and subjective well-being
  - An increase in social adjustment, a decrease in anxiety and despair, a rise in self-acceptance and self-actualization, and a decline in hostility.
- **Psychomotor functions:** Grip strength, dexterity, fine motor skills, eye-hand coordination, reaction time, steadiness, depth perception, balance, and integrated body part functioning are just a few of the psychomotor improvements. The cognitive functions like depth perception, symbol coding, attention, concentration, memory, and learning effectiveness increase.
- **Effects on biochemistry:** The metabolic profile improves, showing a stress-relieving and antioxidant effect that is crucial for preventing degenerative illnesses, reduced levels of glucose, sodium, and total cholesterol, reduced triglycerides, increased high-density lipoprotein (HDL) cholesterol, decreased low-density lipoprotein (LDL) cholesterol, and decreased very low-density lipoprotein (VLDL) cholesterol are all positive effects. Cholinesterase levels rise, catecholamine levels fall, and ATPase levels rise, Lymphocyte count increases, hemoglobin levels rise, total white blood cell count falls, thyroxin levels rise, vitamin C levels rise, total serum protein levels rise, oxytocin levels rise, prolactin levels rise, and oxygen levels in the brain rise.

Yoga for Fitness

Fitness can be defined as an individual's capability to perform any activity with full vigor and alertness without fatigue. Yoga can be practiced by people of all ages to improve fitness, and the asanas can even be modified to accommodate those with particular needs or limitations. The positions increase muscle strength, coordination, flexibility, and agility. The National Institutes of Health claim that when people consciously try to lessen their stress levels through mental stillness, their bodies frequently start to heal. In this sense, yoga may be viewed as a method of self-healing as well as a way to get physically and mentally fit.

Asanas, which are aimed to promote strength, flexibility, balance, and the coordination of the mind, body, and breath through controlled breathing exercises (pranayama), as well as meditation, are common components of many types of yoga. The development of a strong, flexible body that is pain-free, the development of a balanced neurological system that allows all physiological systems to work effectively with a clear and quiet mind, and the development and integration of the body, mind, and breath are the general goals of yoga practice.

Research-Based Evidences

- 8 weeks of Hatha yoga practice improved health related fitness of subjects—study by Tran et al., in 2001. Isokinetic muscular strength, isometric muscular endurance, ankle, shoulder, trunk flexibility, absolute and relative maximal oxygen uptake increased.
- A 90-minutes short duration session of Bikram Yoga increased deadlift strength, lower back/hamstring flexibility, shoulder flexibility, and decreased body fat compared with control group—study by Tracy et al., in 2013.
- Regular yoga practice results in enhanced flexibility very rapidly as this process involves gentle stretching of muscles and connective tissues around bones and joints—study by Woodyard in 2011. Yoga also has profound effect on balance, muscular strength, endurance and coordination because of its highly structural activity and involvement.

Yoga for Flexibility

Flexibility is an important components of fitness which can be defined as the capability of a joint to move through its full ROM. Good flexibility is essential for optimal movement, range of motion and prevention of injury. Both competitive and recreational athletes stretch their muscles to increase flexibility. In order to complete daily duties, maintain proper posture, and relax the muscles, one needs a sufficient range of motion. This can also improve performance and lower the risk of injury.

One of the first and most noticeable advantages of yoga is increased flexibility. Yoga leads to gradual relaxation of the muscles and connective tissues surrounding the bones and joints that occurs with repeated practice.

Yoga positions for flexibility can improve health in a variety of ways. Stretching helps to enhance range of motion, increase mobility, and lower risk of injury. Any fitness program should include flexible exercises. Tension and stress can be relieved by stretching. It enhances both mental and physical health generally.

Yoga stretching is thought to assist discharge of lactic acid from the muscle cells into the bloodstream so that it does not interfere with muscular contraction, which may help the body become more flexible and increase the range of motion of the joints. Some yogasanas for flexibility are shown in Figures 25.3A to G.

Research-Based Evidences

- A study showed regular yoga training may improve the balance and flexibility of shooting athletes even within short period of time (6 weeks) and can also improve the athletic performances that demand high flexibility and balance—study by Iftekhar et al., in 2017.
- Yoga exercises increased spinal mobility and flexibility of the hamstring muscles regardless of age as was demonstrated in a study by Grabara et al., in 2015.
- Study by Tekur et al., showed seven (7) days of a residential intensive yoga-based lifestyle program reduced pain-related

Figs 25.3A to G: Yoga asanas for flexibility

disability and improved spinal flexibility in patients with chronic low back pain better than a physical exercise regimen.

YOGA: A HOLISTIC APPROACH

Yoga is a holistic science that embodies the interdependence of life; it is a physical, mental, and spiritual practice that has its roots in ancient India and links emotional stability and mental equilibrium to physical well-being. Yoga has been shown to help control chronic conditions like obesity, cancer, metabolic syndrome, cardiovascular, respiratory, and endocrine disorders as well as boost immunity. Yoga poses or asanas, such as Dhyana for meditation and Pranayama for breathing control enhance innate immunity, reduce inflammation, and delay the onset of chronic diseases. Yoga also improves joint flexibility and microcirculation, which reduces the symptoms of chronic arthritis. Neurotransmitters, neuropeptides, hormones, and cytokines that mediate interactions between the immune system and the central nervous system are regulated by yoga and meditation. The psychological and physical impacts of ongoing stress are lessened by these methods. It has been demonstrated that the direct release of serotonin, oxytocin, and melatonin during yoga practice helps people more effectively handle anxiety and panic, particularly during pandemics.

Physical activity, relaxation, breathing exercises, meditation, a healthy diet, and purifying detoxification methods are used to achieve this comprehensive approach. Health, calmness, tranquilly, positive thinking, mental stability, energy, and inner power are all improved through these techniques.

Clinical Correlation

Research suggests that yoga practice may help alleviate symptoms of various physical and mental health conditions, including stress, anxiety, depression, chronic pain, hypertension, and insomnia. Moreover, incorporating yoga into clinical interventions has shown promising results in improving overall quality of life and promoting holistic wellness.

SUMMARY

- Yoga, derived from the Sanskrit root "yuj", is a mind-body exercise that promotes friendliness, compassion, self-control, strength, endurance, flexibility, and a sense of peace.
- It can lead to significant results such as shifts in perspective, increased self-awareness, and more vitality.
- The Yoga Sutras, Patanjali's canonical work on yoga, describes the eight limbs of yoga philosophy.
- The therapeutic approach of yoga is based on four fundamental concepts: The human body is a holistic system with multiple interconnected dimensions; each person has different needs and should be treated as an individual; yoga empowers the practitioner to be his/her own healer; and healing depends greatly on a person's mental makeup and condition.
- Ashtanga yoga uses ancient teachings, while Bikram yoga involves 26 poses and two breathing exercises.
- Hatha yoga focuses on prolonging inhalation, breath retention, and exhalation to balance the body's energy system. Iyengar yoga focuses on finding correct alignment in each pose using props.
- Kripalu yoga teaches practitioners to know, accept, and learn from the body, while Kundalini yoga aims to release pent-up energy.
- Power yoga, developed in the late 1980s, is an active and athletic type of yoga based on the traditional Ashtanga system.
- Yoga has numerous benefits for mental health, including reducing symptoms of depression, anxiety, fatigue, stress, and post-traumatic stress disorder. It also has antidepressant effects, reduced stress, and improved physical fitness.
- Yoga can also help patients with cardiovascular conditions like hypertension and musculoskeletal conditions, such as ADHD.
- Therapeutic effects for children include improved sensory motor performance as well as reduced asthmatic attacks.
- Yoga has been shown to improve QoL and physiological functions in Parkinson's disease patients, reduce symptoms of dysmenorrhea, and improve symptoms in IBS patients. It can also improve glycemic control and pulmonary functions in NIDDM cases.
- Physical advantages of yoga include a predisposition toward parasympathetic nervous system dominance and stable autonomic nervous system homeostasis. Blood pressure falls, pulse rate reduces, respiratory rate decreases, and GSR increases ; normalization of EEG alpha waves, EMG activity, cardiovascular efficiency, respiratory efficiency, gastrointestinal function, endocrine function, excretory function, and pain levels.
- Psychomotor functions improve, with improvements in grip strength, dexterity, fine motor skills, eye-hand coordination, reaction time, steadiness, depth perception, balance, and integrated body part functioning. Cognitive functions also improve.
- Yoga has potential psycho-physiological mechanisms, such as stimulating the vagal nerve, increasing parasympathetic function of the autonomic nervous system and brain GABA activity.
- Studies have shown that yoga has a positive impact on immune cell gene expression profiles and alters the brain by increasing the left prefrontal cortex activation level.
- Yoga activities also aid in processing empathy, apprehension, reasoning, and intuition, lowering stress and improving mood.
- Yoga has numerous health benefits, including rejuvenating pancreatic cells through abdominal stretching, boosting insulin receptor expression, and improving insulin sensitivity.
- Transcendental meditation has been found to have health-promoting effects, with long-term practitioners reporting higher QoL and functional health than the general population.
- Yoga can be practiced for fitness and flexibility, with asanas aimed at promoting strength, flexibility, balance, and coordination of the mind, body, and breath through controlled breathing exercises (pranayama).
- Yoga is also beneficial for flexibility, as it leads to gradual relaxation of muscles and connective tissues surrounding bones and joints. Stretching helps enhance range of motion, increase mobility, and lower the risk of injury.
- Yoga poses and asanas, such as dhyana for meditation and pranayama for breathing control, enhance innate immunity, reduce inflammation, and delay the onset of chronic diseases.
- Yoga has been shown to help control chronic conditions like obesity, cancer, metabolic syndrome, cardiovascular, respiratory, and endocrine disorders, and boost immunity.
- Yoga and meditation regulate neurotransmitters, neuropeptides, hormones, and cytokines that mediate interactions between the immune system and the central nervous system, reducing the psychological and physical impacts of ongoing stress.

FURTHER READINGS

- Balaji P A, Varne S R, Ali S S. Physiological effects of yogic practices and transcendental meditation in health and disease. North American Journal of Medical Sciences. 2012 Oct;4(10):442.
- Chien L W, Chang H C, Liu C F. Effect of yoga on serum homocysteine and nitric oxide levels in adolescent women with and without dysmenorrhea. The Journal of Alternative and Complementary Medicine. 2013 Jan 1;19(1):20–3.
- Dhameja K, Singh S, Mustafa M D, Singh K P, Banerjee B D, Agarwal M, Ahmed R S. Therapeutic effect of yoga in patients with hypertension with reference to GST gene polymorphism. The Journal of Alternative and Complementary Medicine. 2013 Mar 1;19(3):243–9.
- Galantino M L, Galbavy R, Quinn L. Therapeutic effects of yoga for children: a systematic review of the literature. Pediatric Physical Therapy. 2008 Apr 1;20(1):66–80.
- Grabara M, Szopa J. Effects of hatha yoga exercises on spine flexibility in women over 50-year-old. Journal of physical therapy science. 2015;27(2):361–5.

- Iftekher S N, Bakhtiar M, Rahaman K S. Effects of yoga on flexibility and balance: a quasi-experimental study. Asian Journal of Medical and Biological Research. 2017 Aug 29;3(2):276–81.H
- Lamb T. Health benefits of yoga. Yoga world. 2001 Jan;16:6.
- Mooventhan A, Nivethitha L. Evidence-based effects of yoga in neurological disorders. Journal of Clinical Neuroscience. 2017 Sep 1;43:61–7.
- Naveen G H, Thirthalli J, Rao MG, Varambally S, Christopher R, Gangadhar BN. Positive therapeutic and neurotropic effects of yoga in depression: A comparative study. Indian Journal of Psychiatry. 2013 Jul 1;55(Suppl 3):S400–4.
- Rayal S P, Singh B, Jain P. Exploring the Therapeutic Effects of Yoga and its Ability to Increase Quality of Life. International Journal for Modern Trends in Science and Technology. 2021.
- Schumann D, Anheyer D, Lauche R, Dobos G, Langhorst J, Cramer H. Effect of yoga in the therapy of irritable bowel syndrome: A systematic review. Clinical Gastroenterology and Hepatology. 2016 Dec 1;14(12):1720–31.
- Sharma N K, Robbins K, Wagner K, Colgrove Y M. A randomized controlled pilot study of the therapeutic effects of yoga in people with Parkinson's disease. International Journal of Yoga. 2015 Jan;8(1):74.
- Sharma S B, Madhu S, Ta'ldon O P. Study of yoga asanas in assessment of pulmonary function in NIDDM patients. Indian J Physiol Pharmacol: lrlnacol. 2002;46(3).
- Tekur P, Singphow C, Nagendra HR, Raghuram N. Effect of short-term intensive yoga program on pain, functional disability and spinal flexibility in chronic low back pain: A randomized control study. The Journal of Alternative and Complementary Medicine. 2008 Jul 1;14(6):637–44
- Telles S, Dash M, Naveen K V. Effect of yoga on musculoskeletal discomfort and motor functions in professional computer users. Work. 2009 Jan 1;33(3):297–306.
- Telles S, Gupta R K, editors. Handbook of Research on Evidence-Based Perspectives on the Psychophysiology of Yoga and Its Applications. IGI Global; 2020 Aug 28.
- T Javnbakht M, Kenari R H, Ghasemi M. Effects of yoga on depression and anxiety of women. Complementary therapies in clinical practice. 2009 May 1;1.
- Tracy B L, Hart C E. Bikram yoga training and physical fitness in healthy young adults. The Journal of Strength and Conditioning Research. 2013 Mar 1;27(3):822–30.
- Tran M D, Holly R G, Lashbrook J, Amsterdam E A. Effects of Hatha yoga practice on the health-related aspects of physical fitness. Preventive cardiology. 2001 Oct;4(4):165–70.
- Ülger Ö, Yağlı N V. Effects of yoga on the quality of life in cancer patients. Complementary therapies in clinical practice. 2010 May 1;16(2):60–3.
- Woodyard C. Exploring the therapeutic effects of yoga and its ability to increase quality of life. International journal of yoga. 2011 Jul;4(2):49.
- Yadav R K, Das S. Effect of yogic practice on pulmonary functions in young females. Indian Journal of Physiology and Pharmacology. 2001 Oct 1;45(4):493–6.

STUDENT ASSIGNMENT

LONG ANSWER QUESTIONS

1. Explain the therapeutic effects of yoga.
2. Describe the psychophysiological mechanisms of action of yoga.
3. Explain how yoga improves flexibility.
4. What are types of yoga and its benefits? Elaborate

SHORT ANSWER QUESTIONS

1. Define yoga and principles.
2. Write note on yoga as a holistic approach.
3. What is the role of yoga in fitness enhancement?

MULTIPLE CHOICE QUESTIONS

1. **Yoga is derived from a Sanskrit word. What does that word mean?**
   a. Diffusion
   b. Breaking into pieces
   c. Surya Namaskar
   d. Union
2. **What would not be helpful when meditating?**
   a. Concentrate on breathing
   b. Picture a peaceful place
   c. Thinking of problems
   d. Concentrate on a color
3. **Pranayama is cutting down the speed of:**
   a. Mind
   b. Inhalation-Exhalation
   c. Anger
   d. Jealousy
4. **Which one comes under Antaranga Yoga?**
   a. Dharana b. Pratyahara
   c. Niyama d. Asana
5. **The yoga that emphasizes use of prop is:**
   a. Ashtanga yoga b. Hatha yoga
   c. Iyengar yoga d. Kundalini yoga

ANSWER KEY

1. d **2.** c **3.** b **4.** c **5.** d

Annexure

Bed Rest

Sheetal Kalra

BENEFITS OF REST

Muscles can adapt and change during rest to meet the needs of a certain exercise. Rest promotes the natural healing of weakened or injured tissue. Together with resting in bed, sleep is necessary for proper neurologic and immune function. The average human sleeps 6–9 hours a day and may take extra, shorter naps throughout the day. When people are ill, they may sleep and rest more than usual.

Since it has the potential to have a lot of beneficial effects, increased rest is a common kind of therapy for unwell patients.

Potential Beneficial Effects of Increased Body Rest in Humans

- Preserves metabolic resources for use in recovery and healing
- Muscle oxygen consumption is decreased, and oxygen delivery is directed toward injured tissues and organs that require more oxygen.
- Improves blood flow to the central nervous system.
- Reduces stress on the heart.
- Prevents ischemia and dysrhythmias.
- Reduces requirements for ventilation.
- Reduces risk of ventilator-induced lung injury.
- Reduces risk of oxygen toxicity.
- Reduces pain from and further injury to an injured body part.

INDICATIONS FOR PROLONGED BED REST

- Following a major fracture such as lumbar, spine, femoral neck, etc. Patients with exacerbations of inflammatory conditions like rheumatoid arthritis to reduce joint inflammation and pain and promote resolution of the acute flare.
- In primary care settings, the three most common causes of acute low back pain are lumbosacral disc disease, spondylolisthesis, and muscular strain, which may require prolonged bed rest.
- Patients with head injuries, spinal cord injuries, comatosed patients.
- In past, bed rest has been prescribed for clinical conditions like acute myocardial infarction. Patients were advised to limit their physical activity for a few weeks. It was believed that minimizing stress to the damaged myocardium would aid in healing, lower the chance of further acute ischemia episodes, and stop harmful cardiac dysrhythmias.

COMPLICATIONS OF BED REST

Physical activity has positive benefits on various facets of organs and organ systems performance throughout the day. As a result, increasing rest is typically linked to a reduction in the health benefits of exercise. Bed rest itself can result in major consequences, too.

Complications that can arise as a result of prolonged bed rest are depicted in Figure 1.

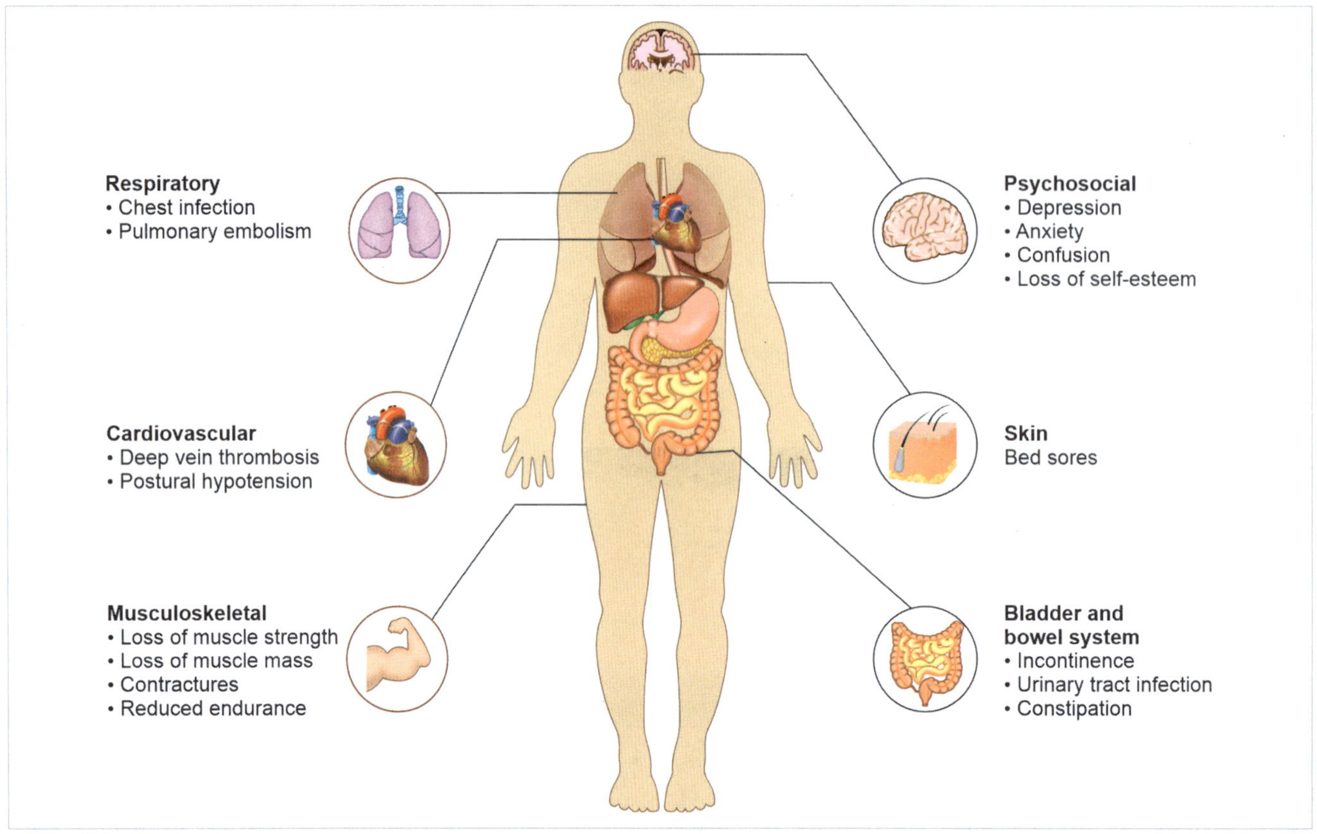

Fig. 1: Complications of prolonged bed rest

Musculoskeletal Complications

- **Skeletal muscle weakness:** It occurs during critical illness as a result of a number of causes, including sepsis-induced vascular and metabolic abnormalities, starvation, neuropathy, myopathy, pharmacologic dosages of corticosteroids, and protracted inactivity. Reduced muscle fiber size is the main cause of muscle mass loss following inactivity.
- **Joint contractures:** Joints lose their range of motion when they are not subjected to regular mobility and stress.

Vascular Complications

- **Microvascular dysfunction:** The side effects of vascular dysfunction include skin ulcers, lactic acidosis, gastrointestinal hemorrhage, intestinal ischemia, and numerous organ dysfunction. The breakdown of the skin, which typically takes place at pressure points between the skin and the bed, is known as a skin ulcer. In addition to unrelieved pressure, other factors that might cause skin ulcers include poor microcirculation, starvation, shear stress at the points of contact, and dampness.
- **Pressure sores:** Localized regions of cellular necrosis include pressure sores and decubitus ulcers. They are typically present on bony prominences that have experienced extended external pressure greater than capillary pressure. Individuals with spinal cord injuries and elderly patients are two populations who are regularly immobile and are more likely to develop pressure sores (Fig. 2).
- **Venous stasis:** Bed resting for an extended period of time tends to encourage venous stasis. Stasis may also be caused by sustained compression of veins, which happens when limbs are in close proximity to the bed for an extended period of time. This can likewise harm the vascular endothelium. Bed rest is a significant risk factor for thromboembolic disease as a result.

Respiratory Complications

- **Atelectasis:** Chest radiographs taken within 48 hours of recumbency frequently show partial or total atelectasis of the left lower lobe in critically ill patients. This might be brought on by the supine position's combination of the heart's gravitation-induced dorsal displacement and the cephalad shift of the diaphragm. Atelectasis increases pulmonary vascular resistance and may increase the risk of pneumonia. In patients with or at risk for acute lung injury and acute respiratory distress syndrome, loss of the

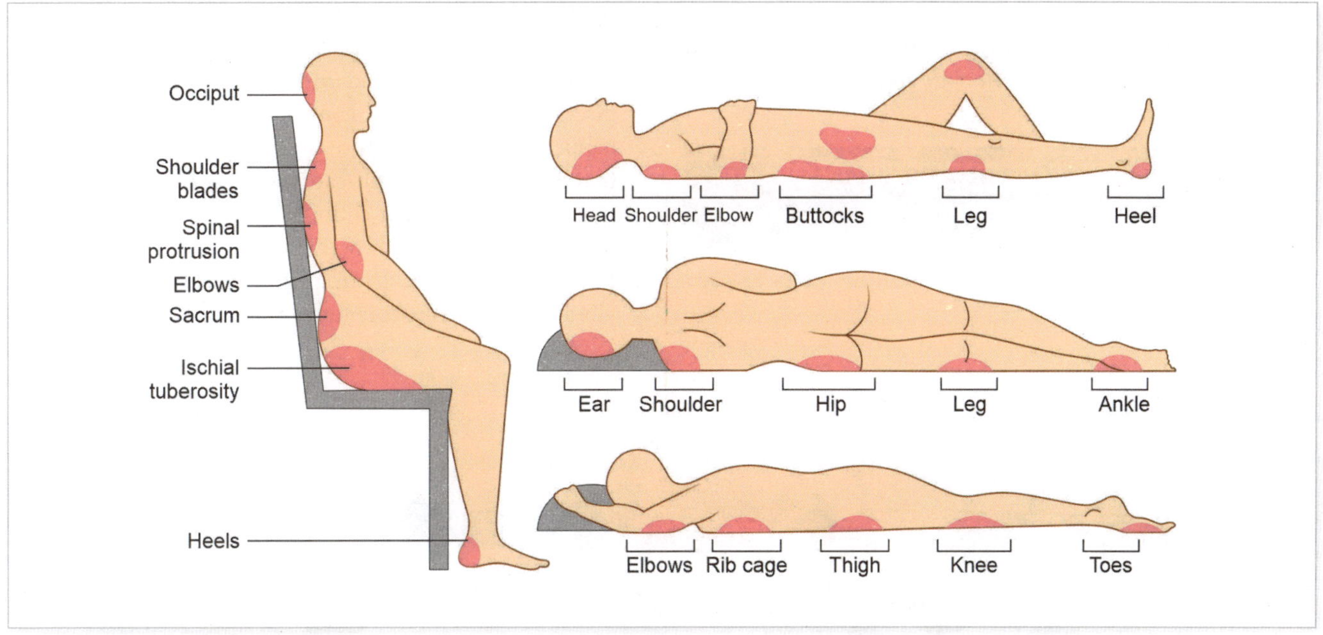

Fig. 2: Common locations of pressure sores

aerated lung due to lobar atelectasis may increase the risk of ventilator-induced lung injury from the overdistension of the aerated lung.

- **Decreased ventilation:** Patients who are immobile exhibit decreased minute ventilatory volume and tidal volume. Patients who are supine and inactive have trouble contracting their ventilatory muscles enough to complete a full inspiration. The respiratory muscles experience a loss of strength while immobile.
- **Pneumonia:** Immobilization can significantly decrease the body's ability to remove secretions (sometimes exacerbated by motor weakness). And then, secretions build up in the lower bronchial tree, obstructing airways and finally, leading to atelectasis and hypostatic pneumonia.

Endocrinal Complications

- **Fall in BMR:** The basal metabolic rate (BMR) falls throughout the immobilization period.
- **Natriuresis:** Prolonged bed rest can lead to fluid shifts and decreased kidney perfusion, causing increased sodium excretion (natriuresis). This is worsened by suppressed antidiuretic hormone, leading to excessive diuresis and potential electrolyte imbalances.
- **Nitrogen loss:** Inactivity causes the entire body to lose nitrogen. The typical nitrogen loss through urine may be 2 g/dL. An increase in protein catabolism and a corresponding decline in protein synthesis are the causes of this nitrogen loss.
- **Insulin resistance:** Type II diabetes mellitus, which is characterized by insulin resistance, is linked to sedentary lifestyles. Even in critically ill patients without a history of diabetes, insulin resistance can develop. Insulin levels can increase to twice normal levels, which suggests that the increased tissue resistance to endogenous insulin is the cause of the glucose intolerance. Muscles appear to contain fewer sites for insulin binding.
- **Renal complications:** Stone formation in the kidneys or bladder is frequently brought on by the trifecta of hypercalciuria, urine stasis, and a urinary tract infection.

Gastrointestinal Complications

- **Anorexia:** Reduced calorie requirements, hormone adjustments, anxiety, and despair all contribute to loss of appetite in anorexia. However, inactivity frequently leads to weight increase.
- **Constipation:** Decreased peristalsis and constricted sphincters cause constipation, which is prevalent in immobile individuals. Constipation is also brought on by low-fiber diets and a drop in fluid consumption. A high-fiber diet, stool softeners, laxatives, increasing fluid intake, and a regular bowel practise are the best ways to relieve constipation. Also, usually, oral lactulose works well, reducing the frequency of enemas.

Central Nervous System Complications

Complications of the central nervous system might result in sensory deprivation, mental decline, behavioral problems, and lack of sensation. Patients who are immobilized are more likely to develop consequences from sensory deprivation if they are elderly or have cognitive impairment (stroke, brain injury, dementia). Intellectual regression, sadness, a limited attention span, and low motivation are some of these issues.

EXERCISES FOR PREVENTION OF COMPLICATIONS

Exercises for prevention of complications from prolonged bed rest (Figs 3A to K).

- **ROM exercises:** As a means of avoiding musculoskeletal issues health professionals recommend range of motion exercises as part of standard treatment. Although proper placement cannot stop contractures, it can make them happen at a less unfavorable angle. Early ambulation, access to the bathroom, and self-care tasks that call for patients to use and maintain their entire range of motion are the best forms of prevention (Brower et al.).
- **Resistance exercises:** During prolonged bed rest, a resistive vibration exercise countermeasures decreased lumbar

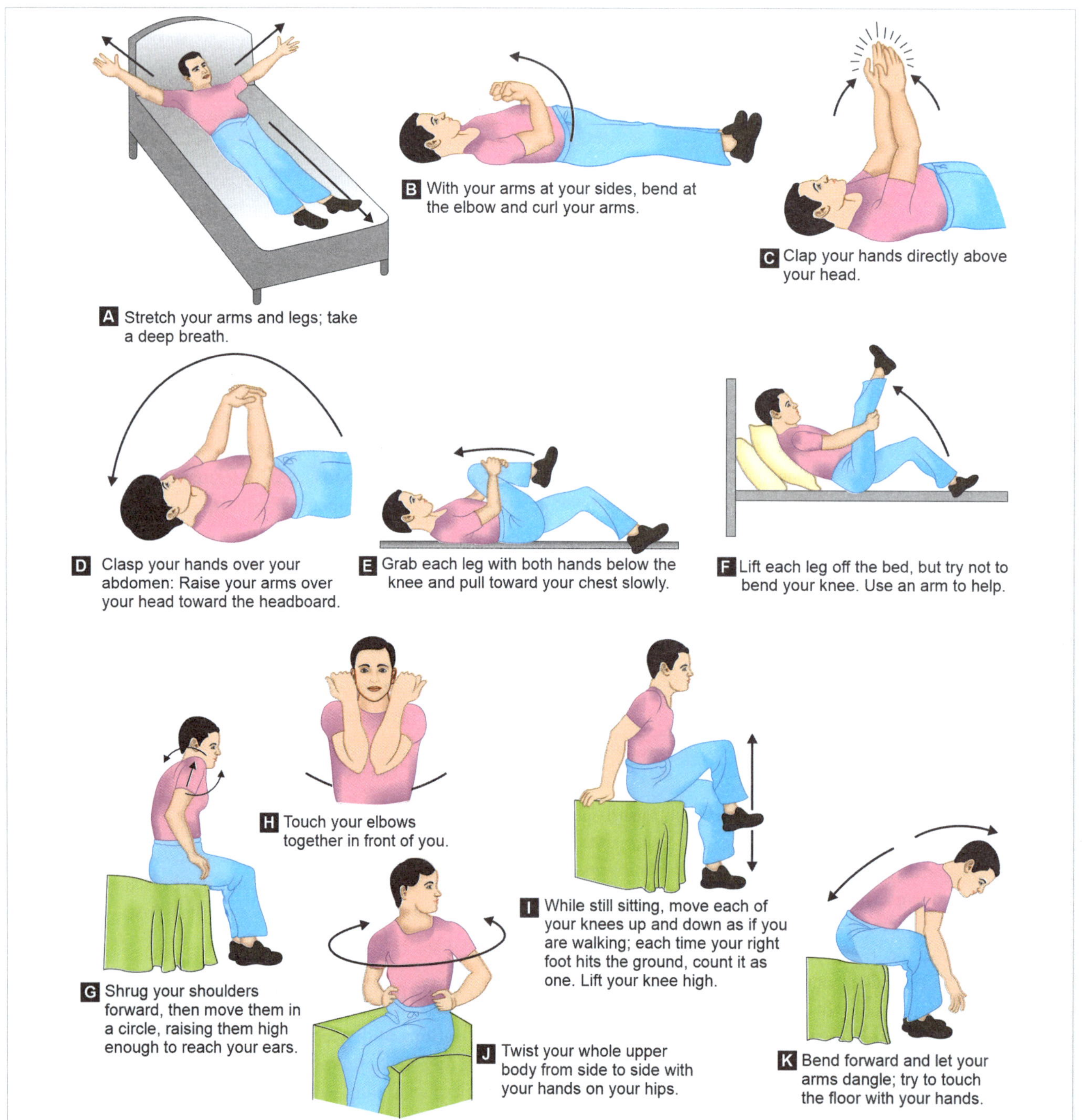

Figs 3A to K: Exercises to prevent complications from prolonged bed rest

multifidus muscle atrophy, and leads to increase in amount of spinal lengthening, and disc area (Belavý DL, Hides JA, Wilson SJ, Stanton W, Dimeo FC, Rittweger J, Felsenberg D, Richardson CA). Resistive simulated weight-bearing exercise with whole body vibration reduces lumbar spine deconditioning in bed-rest.

- **Rowing-based exercise:** Short bursts of high-intensity rowing exercise are adequate to counteract the circulatory deconditioning and cardiac atrophy brought on by extended (5 weeks) bed rest (Hasting et al.).
- **Whole body vibration exercise:** It appears to be a safe and promising training intervention for maintaining bone density, muscle strength and size in the lower half of the body in patients on prolonged bed rest (Blizzard et al.).
- **Supine lower body negative pressure (LBNP) treadmill exercise:** Similar to the integrated cardiovascular and skeletal stress produced by upright exercise, LBNP helps to maintain exercise capacity. The LBNP exercise can be used as a preventative measure for long-term deconditioning brought on by microgravity (Watenpaugh et al.).
- **Chest physiotherapy:** Such as deep breathing and coughing, vibration, postural drainage, and incentive spirometry are all forms of treatment for preventing respiratory issues.
- **Buerger exercises:** These are the exercises that are meant to stimulate collateral circulation in legs. The mechanism uses gravitational changes in positions that work on the smooth muscles of vessels and to the vasculature bed. Gravity helps to alternately empty and fill the blood columns which eventually increases blood transportation to them (Figs 4A to C).
  - Benefits of buerger exercises
    - Buerger exercises are beneficial for diabetes patients as preventive workouts in order to improve circulation and stop future issues such as with diabetic foot.
    - Reduce the peripheral neuropathy symptoms and signs in diabetic individuals.
    - Encourage collateral blood to circulate.
    - Increase the skin's perfusion pressure.
    - An increase in peripheral blood flow.
    - Shorten the time of necrosis and edema.
    - Improve ability to walk.
    - Enhance postoperative local circulation after gynecological and orthopedic procedures.

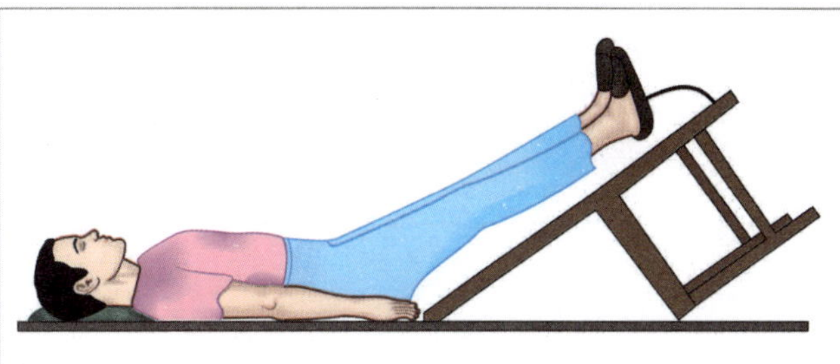

A Subject lies supine with legs elevated up to 45° angle on a chair or stool.

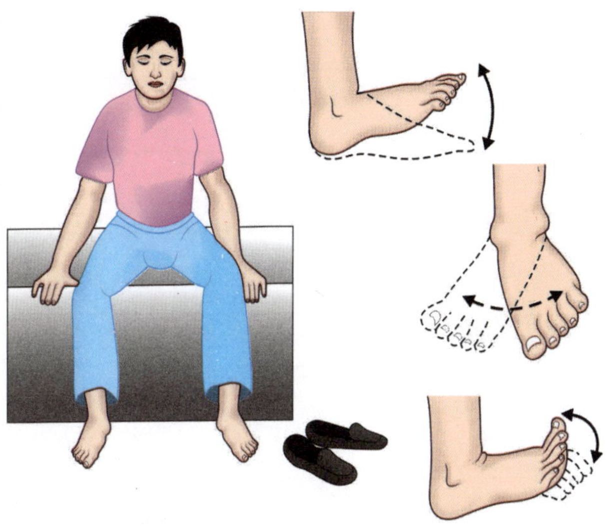

B The subject sits on the chair (up to 3 minutes) in a relaxed position followed by ankle and foot exercises (flexion, extension, pronation, supination).

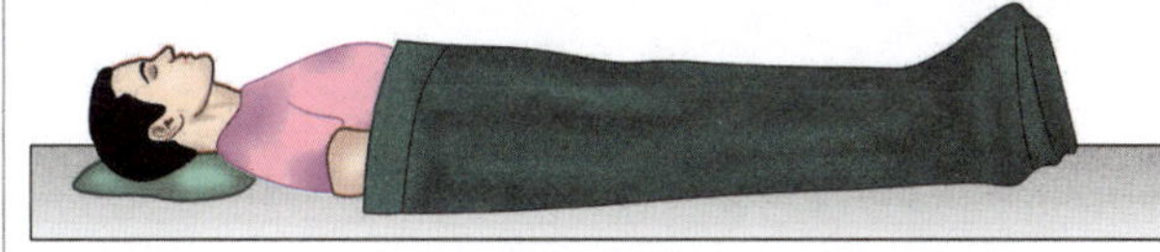

C Subject lies supine with legs covered in a blanket up to 5 minutes.

Figs 4A to C: Buerger exercises

Index

Refer 'f' for figure and 't' for table respectively.

A

D

E

H

I

M